JOHN D. ROBERTS
California Institute of Technology

MARJORIE C. CASERIO
University of California, Irvine

A Study Guide to BASIC PRINCIPLES
OF ORGANIC
CHEMISTRY

D1213580

W. A. BENJAMIN, INC.
Menlo Park, California

The authors and publisher are pleased to acknowledge the skills and assistance of Rosemary Kissel, who typed the text. We are also indebted to Charles Kissel, who drafted the many thousands of chemical structures, and to Margaret Kline and Jhong Kim for their assistance in proofreading.

PREFACE

This Study Guide attempts to help the student of organic chemistry in two major ways. First, to provide a concise summary of the most important topics covered in each chapter of <u>Basic Principles of Organic Chemistry</u>. Second, to provide reasonably detailed solutions to the many exercises in the parent text.

The chapter summaries are designed to highlight major concepts developed in the text and to display, in an abbreviated format, the chemical reactions discussed. They are included here as a rapid and convenient way to recall the material covered without the encumbrance of descriptive detail. Cross-references to appropriate Sections in the main text are given frequently to assist the reader in locating the more detailed coverage. Certain chapter summaries utilize tables to present summary information; and to avoid confusion with the tables in the parent text, we have appended the letter S to the table numbers in the summary (as in Table 14-1S).

A few comments are appropriate on the numerous exercises presented in the text. We have tried to select problems that are both straightforward and instructive and that illustrate the topics discussed. Some of these problems fall in the category of simple exercises to provide practice with and to reinforce the material learned. Other problems are more challenging. But regardless of the type of problem, answers are provided for all of them. However, we offer more than just answers. The solutions to most of the problems are given in sufficient detail to illustrate the approach and the logic used. We hope thereby that the solutions will be clear and helpful.

Even though there are over one thousand problems in the text, we offer <u>additional</u> problems at the end of several chapters of the Study Guide, at least through Chapter 13. These extra problems are specifically designed to be simple and straightforward exercises requiring short answers. They are included here to allow the student to get as much opportunity to learn by the problem-solving approach as is reasonably possible - given the limitations of space and time imposed on all of us.

Because chemical structure is best communicated by means of structural drawings, we have included an introductory chapter on the representation of organic structure. In the concluding chapter, we offer a brief description of the chemical literature and how to use it to retrieve information about the subject of organic chemistry.

We sincerely hope that the wide variety of problems, their solutions, the chapter summaries and ancillary chapters will make this Study Guide a useful adjunct to the parent text.

M. C. Caserio
J. D. Roberts
1977

TABLE OF CONTENTS

REPRESENTATION OF ORGANIC STRUCTURES

Structural formulas are the primary means of communication between organic chemists, and the importance of being able to draw neat clear structural formulas can hardly be overemphasized. A number of stencils are available for assistance in written work and are very useful in the preparation of manuscripts. However, stencils are too slow to be practical for direct communication using pencil and paper and are impossible at the chalkboard. Skill at freehand drawing of organic structures including ones that show three-dimensional perspective can be readily acquired with a little practice, and it is the purpose of the present discussion to offer a few suggestions and guidelines as to how this may be done.

Molecular Models

A set of molecular models is a necessary tool for the organic chemist. Structural drawings showing conformational perspective often cannot be made without reference to molecular models, and this is one of several reasons why the purchase of a suitable set of models is strongly recommended. In selecting a set of molecular models, ideally one requires that they be made to scale - that is, the relative lengths of the "bonds" in the model should correspond to the relative distances of the chemical bonds which they represent; the assembled model must be rigid enough to hold together when manipulated but flexible enough to simulate the motions in real molecules, particularly bond rotation; they must be able to withstand a reasonable amount of strain imposed by bending the "bonds" and they should not fatigue rapidly with use. There are several inexpensive types of models available at the present time, the best being the Maruzen models distributed by W. A. Benjamin Inc., and the framework molecular models distributed by Prentice-Hall.

Three-Dimensional Structures

The structures that are most commonly represented in three dimensions include methane, ethane and cyclohexane derivatives. If you can draw methane to appear tetrahedral, you can quickly learn how to represent the more complex structures of ethane and cyclohexane derivatives. Methane is shown here as it might appear if a model were placed on a flat surface. If this drawing were perfect, one bond should be vertically in the plane of the paper, two should appear to recede slightly behind the paper, and one should project in front of the paper - the bond being deflected slightly down and to the left as if you were looking at the model from above and to one side.

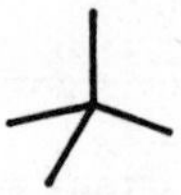

methane

The easiest way to make a reasonably good drawing of methane viewed in this way is to first draw three lines at angles of about 120° to each other with one of them pointing vertically up.

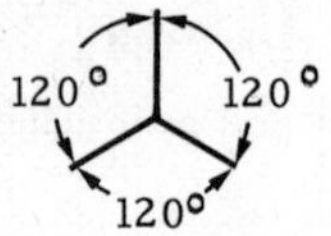

The fourth bond may now be included to project down and slightly to the left or to the right as you face it.

The illusion of three dimensions may be enhanced by slightly tapering the forward bond. Similarly, a dashed line can be drawn to imply a receding bond.

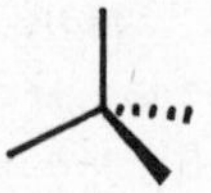

Sawhorse Structures

A sawhorse structure of ethane showing the staggered conformation may be drawn starting with the drawing of tetrahedral methane described above. Simply lengthen the bond projecting towards you and connect it to three lines at angles of 120° to each other - one of the three pointing vertically down.

staggered ethane

To draw an eclipsed conformation, the forward three bonds inclined at 120° are drawn so that one of them points vertically up. The forward bonds will now be parallel to the rear bonds.

eclipsed ethane

Note that there should be a gap in the left rear bond to designate clearly that the bonds to the forward carbon are in front of and do not intersect with the bonds to the rear carbon. Remember that line drawings for chemical structures infer the presence of a carbon atom at each intersection. It can therefore be a serious mistake to draw ✕ when you mean ✕ !

The conformations of linear hydrocarbon chains or branched chains can be drawn using the basic sawhorse unit for staggered ethane. The two staggered forms of butane may accordingly be represented as follows:

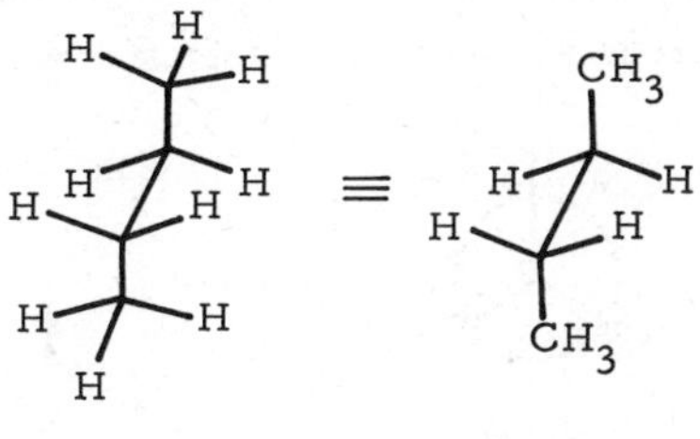

anti   or   trans   butane

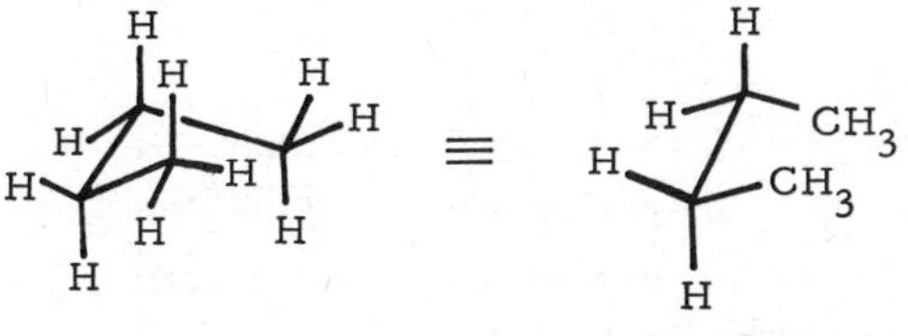

syn   or   gauche   butane

Newman Projection Formulas

A Newman projection of staggered ethane may be drawn by viewing the sawhorse structure directly down the C—C bond axis and projecting what you see on paper.

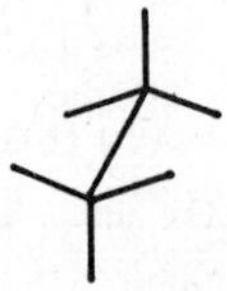

staggered sawhorse   projection of staggered sawhorse

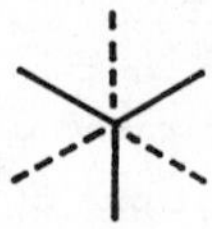

To designate that the three groups on the rear carbon are behind the plane of the paper, draw a circle (diameter ~ 1.3 cm) so that the rear bonds appear to be emerging from behind it, as shown below.

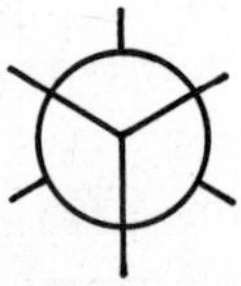

staggered ethane
in Newman projection

If the same procedure were followed for the representation of eclipsed ethane in Newman projection, the rear bonds would not be visible. To obviate this, the groups on the rear carbon are slightly offset from the eclipsed position.

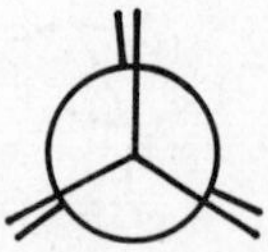

eclipsed ethane
in Newman projection

Chair Form of Cyclohexane

When it comes to interpreting conformational drawings for cyclohexane derivatives, many students understandably conclude they are looking at some many-legged insect or otherworld crustacean. With a little care and thought malformations of this kind can be avoided and proper communication established between student and instructor. To draw a respectable-looking chair form of

cyclohexane, we suggest that one start with the staggered sawhorse form of
ethane. The two vertical bonds will become two axial C—H bonds, and the two
rear-horizontal bonds on one side of the C—C bond axis will become two equatorial C—H bonds in cyclohexane.

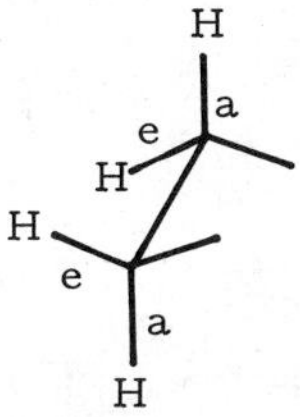

The two remaining equatorial bonds are used to complete the ring structure.
This is done by drawing each C—C bond <u>parallel</u> to the third C—C bond removed
from it. This is shown in a stepwise fashion below.

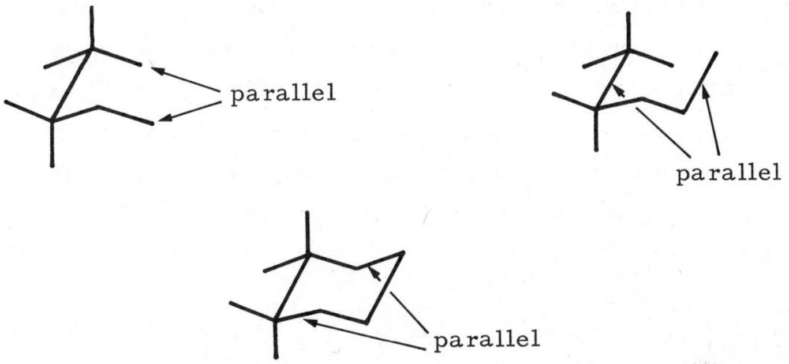

The remaining axial and equatorial bonds may now be drawn in. The axial bonds
will all be directed either vertically up or vertically down, and the equatorial
bonds will each be parallel to the C—C bond once removed from it. Perspective
can be introduced by making the forward ring bonds slightly bolder (heavier line)
than the rear bonds.

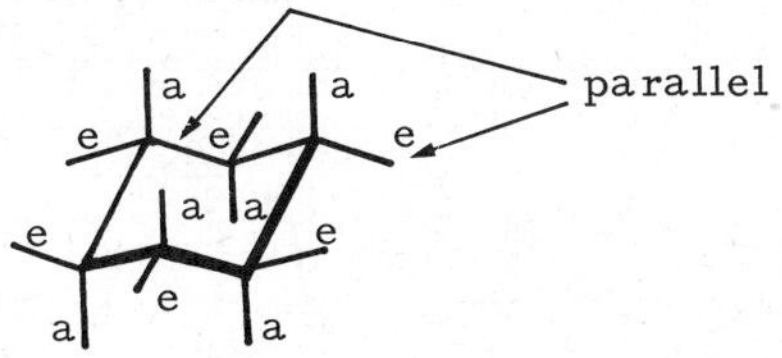

chair form of cyclohexane

More complex polycyclic ring systems can be easily drawn starting with a good basic drawing of chair cyclohexane. For example, <u>trans</u>-decalin can be drawn by appending a second chair form by way of two adjacent equatorial bonds; and <u>cis</u>-decalin can be drawn by adding a second chair by way of one equatorial bond and an adjacent axial bond.

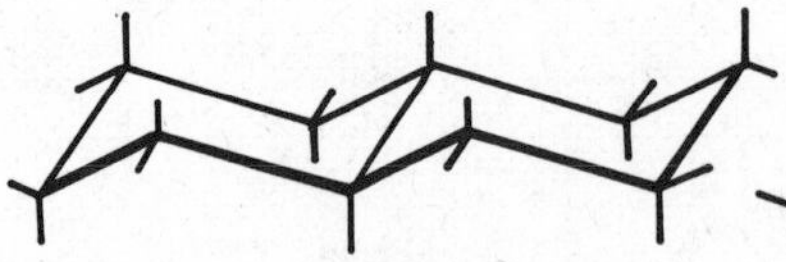

<u>trans</u> - decalin   <u>cis</u> - decalin

Cyclopentane

The cyclopentane ring system is not planar, although it is often represented as a regular pentagon. The necessity arises not infrequently to draw the five-membered ring in a nonplanar form showing a staggered relation-ship between substituents. This may be done as with cyclohexane by starting with the staggered form of ethane in a sawhorse representation. The ring may be completed by connecting two adjacent equatorial-type bonds on one side by two additional bonds.

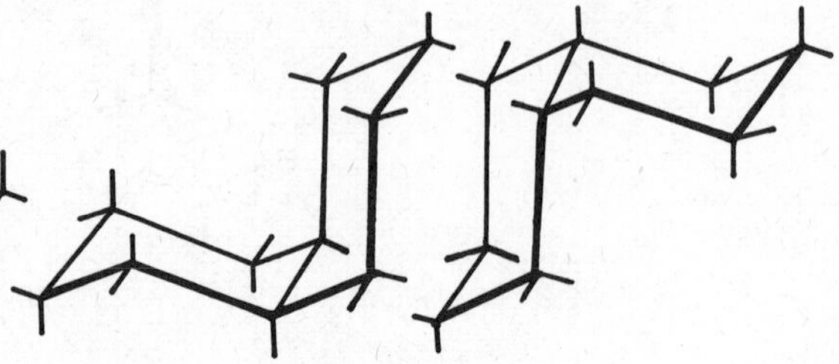

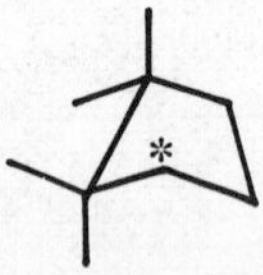

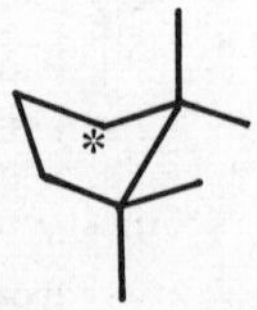

The representation corresponds to a conformation in which the ring carbon marked with an asterisk is out of the plane containing the other four carbons. Substituents on adjacent carbons may be represented as being pseudo-equatorial or pseudo-axial and this representation is more realistic than that in which they are shown as eclipsed.

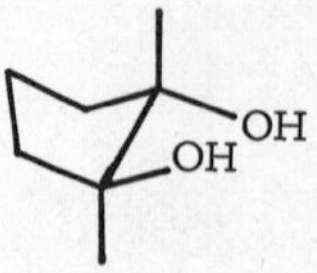

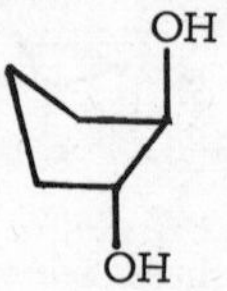

<u>trans</u>- 1, 2-cyclopentanediol

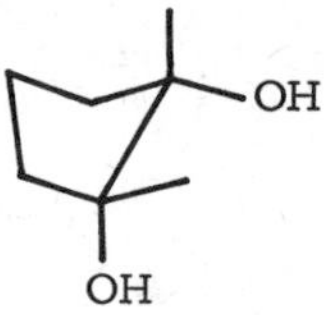

cis-1, 2-cyclopentanediol

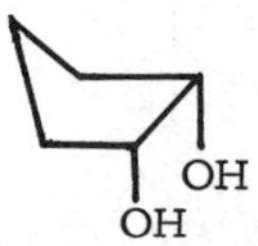

Two-Dimensional Drawings and Planar Structures

Planar molecules and planar functional groups present no special problem in representation other than how best to signify the bond angles. Since doubly-bonded carbon (i.e. sp^2 - hybridized carbon) forms three coplanar bonds at angles equal to or near 120°, compounds of this type are best represented to conform with this shape. For example:

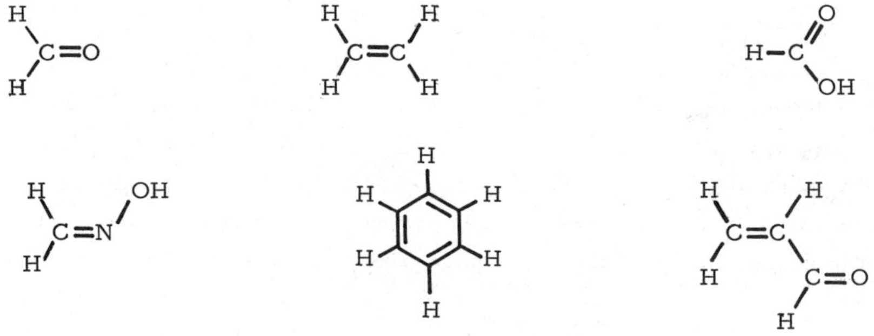

(all bond angles are 120°)

Space limitations often preclude writing structures in the manner shown above, and frequently it becomes expedient to write structures in a distorted linear manner for example:

$$CH_3\overset{O}{\overset{\|}{C}}CH_2CH_3 \qquad CH_2{=}\overset{CH_3}{\overset{|}{C}}CH{=}CH_2 \qquad H_2NCH_2\overset{O}{\overset{\|}{C}}NHCH_2\overset{O}{\overset{\|}{C}}OH$$

These are essentially abbreviated structures that are so commonly used that they have become an accepted form of structure representation. If possible, however, the windswept appearance of $CH_3\overset{O}{C}CH_2CH_3$ should be very definitely avoided.

Note that the bonds to an aromatic ring should be drawn to coincide with a symmetry axis. Even in saturated cyclic systems, the rings are best represented as a regular polygon with substituents drawn along a symmetry axis.

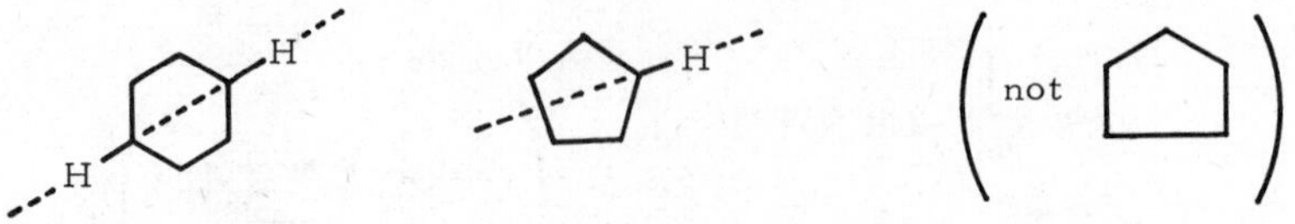

An edge view of a planar ring system is usually shown as a distorted polygon in which the forward edges are heavier and slightly tapered to provide the necessary perspective. The amount of tapering must not be too exaggerated and the lines should not be too heavy in order to avoid an unpleasant and unrealistic "wedged" appearance.

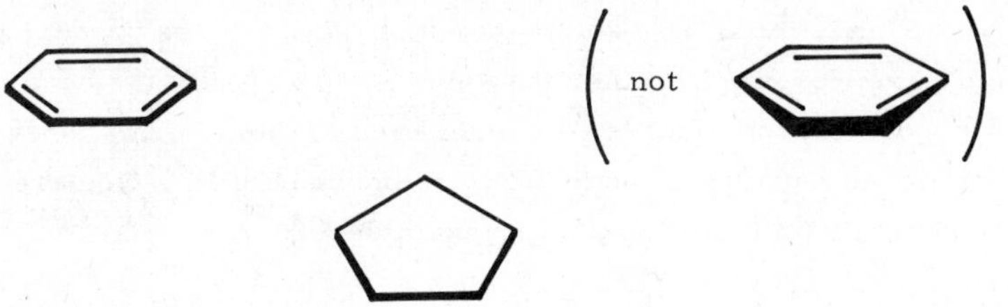

A common malpractice that dies hard is the misrepresentation of five-membered ring systems. A two-dimensional structure for a five-membered ring is best drawn as a regular pentagon but it is often depicted as a distorted or irregular pentagon. While, for example, imidazole is undoubtedly not a regular pentagon with equal bond lengths and bond angles, it is more nearly that than a child's impression of a two-story house with a chimney on top.

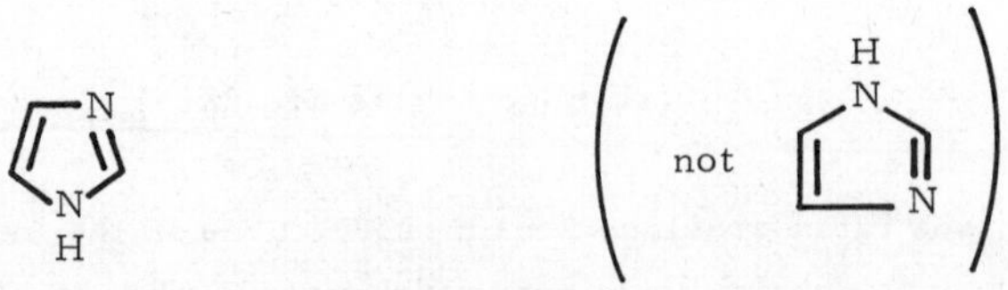

Likewise, three-membered rings are best drawn as equilateral triangles. Six-membered rings should be drawn hexagonal - not rectangular, as in the classic example of barbituric acid and uric acid which for years were presented in the published literature in the distorted forms shown on the left side below.

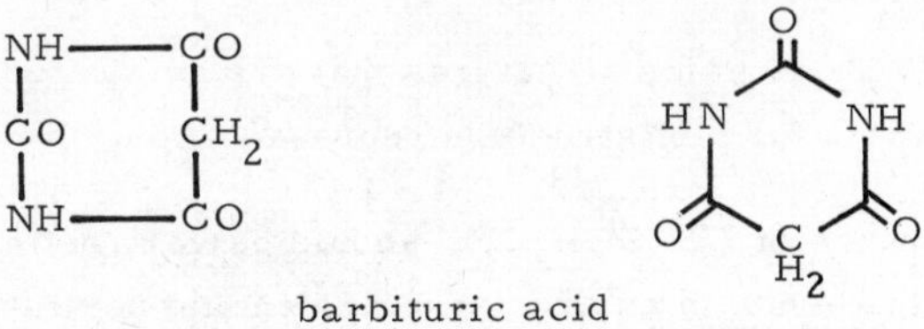

$$NH\text{———}CO$$
$$CO \qquad CH_2$$
$$NH\text{———}CO$$

barbituric acid

NH———CO
CO   C——NH
               CO
NH———C——NH

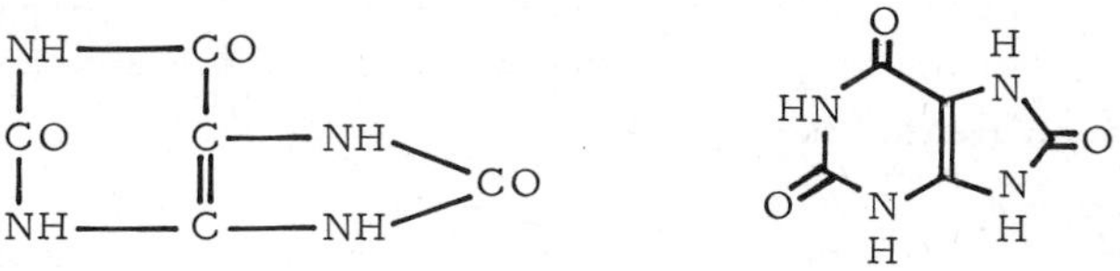

uric acid

The distortion is sufficiently serious in the case of uric and barbituric acid to disguise the fact that these substances are actually pyrimidine derivatives.

Fischer Projection Formulas

The use of Fischer projection formulas to designate configuration of asymmetric molecules is not as common as in years previous but their use continues to be sufficiently widespread to warrant referring to them in basic organic text books. It should be emphasized that, because of the special stereochemical meaning of Fischer projections, carelessness in representing them can lead to a misinterpretation of configuration. According to the rules for writing Fischer projections, the molecule must be viewed so that the asymmetric atom(s) is contained by the plane of projection and substituent groups along the vertical lie behind the plane while substituent groups along the horizontal lie in front of the plane. The projection is then written in the form of a horizontal and verticle cross. This is illustrated below for L-cysteine.

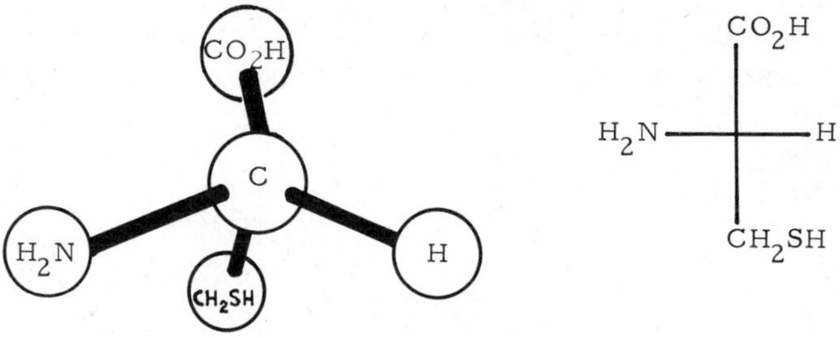

It is implicit in the given projection that the two horizontal groups project towards the viewer and the two vertical groups recede from the viewer. This is sometimes more forcefully designated by a broken line for the vertical or receding groups.

$$CO_2H$$

$$H_2N \cdots\!\!|\!\!\cdots H$$

$$CH_2SH$$

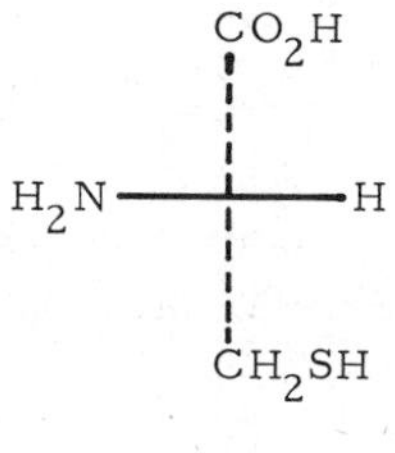

The ambiguity comes when a broken line is drawn along the horizontal since this implies that the structure corresponding to the Fischer projection of L-cysteine has been rotated through 90°.

$$\text{H}_2\text{N} \overset{\displaystyle \text{CO}_2\text{H}}{\underset{\displaystyle \text{CH}_2\text{SH}}{\vline}} \text{H} \qquad \xrightarrow{\quad 90° \quad} \qquad \text{HO}_2\text{C} \overset{\displaystyle \text{H}}{\underset{\displaystyle \text{NH}_2}{\vline}} \text{CH}_2\text{SH}$$

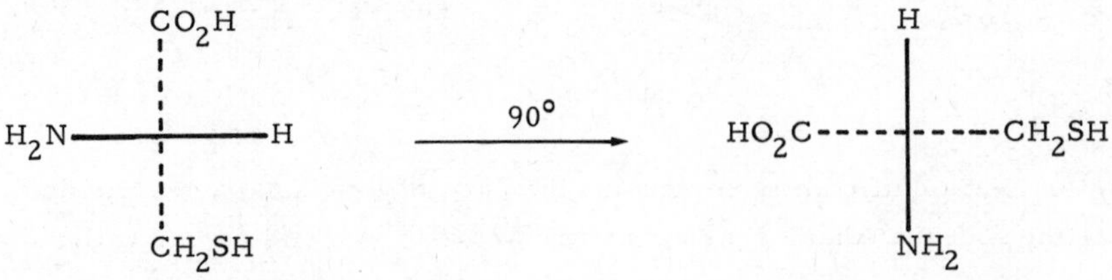

A genuine Fischer projection with full horizontal and vertical lines must <u>not</u> be rotated through 90° in the plane of the paper since this amounts to an inversion of configuration at the asymmetric atom. Interchange of any two groups will also lead to a projection of the inverted configuration.

Also, a projection formula that is written as a diagonal cross is meaningless as far as configuration is concerned.

$$\begin{array}{cc} \text{HO}_2\text{C} & \text{H} \\ & \times \\ \text{H}_2\text{N} & \text{CH}_2\text{SH} \end{array}$$

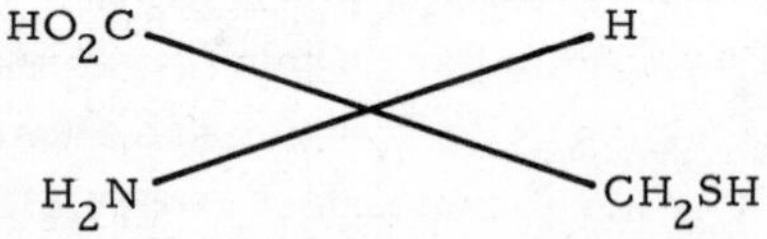

Miscellaneous Comments

A common mistake in structural drawings is to make multiple bonds, particularly carbon-carbon double bonds, too long or with right angles instead of 120° angles.

$$\text{C=O} \qquad (\text{not} \quad \text{C}=\!=\!\text{O}, \qquad \text{not} \quad \text{C}=\!=\!\text{O})$$

$$\text{C=C} \qquad (\text{not} \quad \text{C}=\!=\!\text{C} , \qquad \text{not} \quad \text{C}=\!=\!\text{C})$$

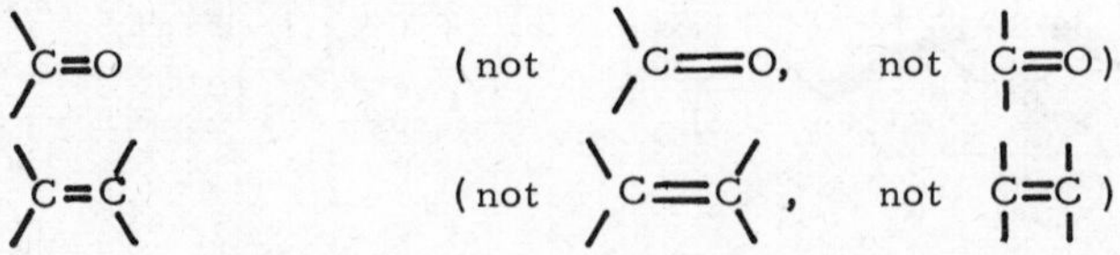

Care should also be taken to position a bond <u>between</u> the two bonded atoms rather than between nonbonded atoms.

$$\begin{array}{cc} \text{CH}_3\text{O} & \text{CN} \\ & \text{C=C} \\ \text{H}_3\text{C} & \text{C}_6\text{H}_5 \end{array} \qquad \text{not} \qquad \begin{array}{cc} \text{CH}_3\text{O} & \text{CN} \\ & \text{C=C} \\ \text{CH}_3 & \text{C}_6\text{H}_5 \end{array}$$

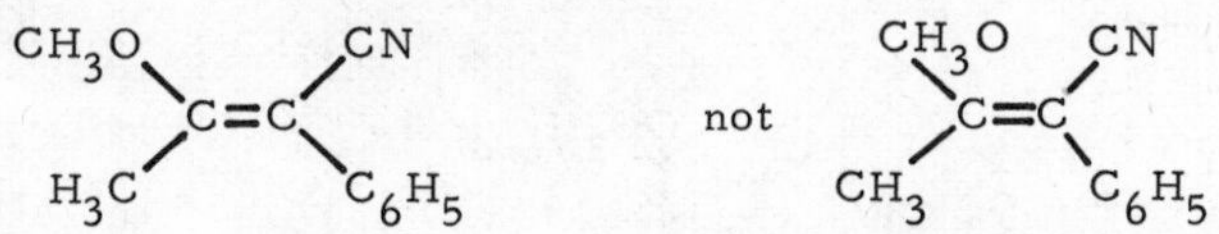

Finally, try to space your structures on the written page so that they do not crowd each other or the written text. Make large drawings if you can, and give them plenty of space.

CHAPTER 1

INTRODUCTION.   WHAT IS ORGANIC CHEMISTRY ALL ABOUT?

Organic chemistry is the chemistry of carbon compounds. The theory of organic chemistry is based on the idea that organic compounds are made up of molecules comprised of atoms held together by chemical bonds to form three-dimensional structures.

The first step in establishing the structure of an organic compound is to determine the percentages of the elements it contains. These percentages, in conjunction with the molecular weight, permit determination of the <u>molecular formula</u>. Valence theory and the assumption that the valences of carbon are directed to the corners of a regular tetrahedron may allow for only one, or for many more than one, possible structure corresponding to a given molecular formula. Different compounds with the same molecular formula are called <u>isomers</u>.

The early organic chemists were able to use molecular formulas, valence theory, the tetrahedral carbon, the number of possible isomers formed on substitution, the idea of <u>free rotation</u> around carbon-carbon single bonds, and the <u>principle of least structural change</u> to establish structures for almost all of the simple compounds they encountered.

The special character of organic chemistry is the ability of carbon to form bonds to itself. No other element has such a rich and diverse chemistry. The <u>polarity</u> of the bonds and <u>hydrogen bonding</u> play an important role in determining the differences in physical properties of organic compounds such as these.

Most of the facts of organic chemistry can be correlated with aid of three different kinds of structural influences - the <u>steric effect</u>, which depends on the physical bulk of atoms or groups of atoms, <u>electrical effects</u> which are based on electrostatic attraction or repulsion, and <u>quantum mechanical effects</u>. The last of these has little root in practical experience.

ANSWERS TO EXERCISES

1-1   There are <u>two</u> isomers of formula $C_2H_2Br_4$, A and B. Substitution of
one bromine with a different atom or group would give <u>two</u> products from B
and <u>one</u> product from A. When the atom is hydrogen, A gives C, B gives
C and D.

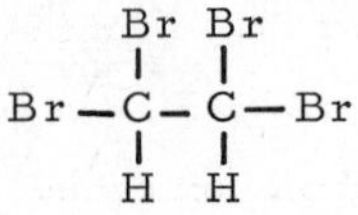

```
      Br Br                    Br Br
      |  |                     |  |
Br — C — C — Br          Br — C — C — H
      |  |                     |  |
      H  H                     Br H

        A                        B

              Br Br                    Br H
              |  |                     |  |
        Br — C — C — H          Br — C — C — H
              |  |                     |  |
              H  H                     Br H

                C                        D
```

1-2

```
      H  Br H                 Br Br H
      |  |  |                 |  |  |
H — C — C — C — H   →   H — C — C — C — H
      |  |  |                 |  |  |
      H  Br H                 H  Br H
```

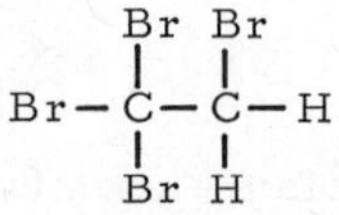

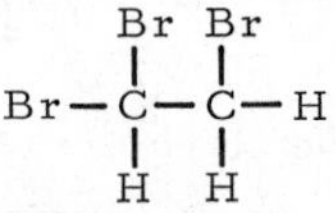

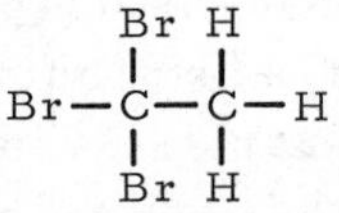

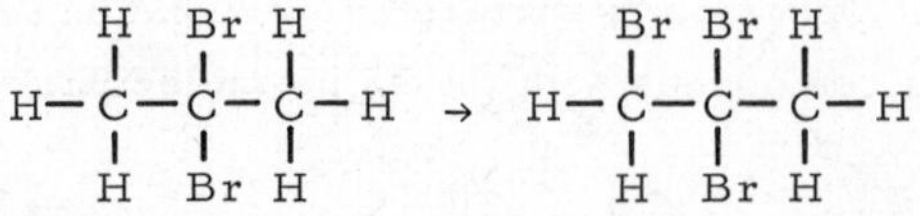

Notice that for $C_3H_6Br_2$ to give only one $C_3H_5Br_3$ isomer, all six hydrogens
must be equivalent.

1-3   All twelve hydrogens of C_5H_{12} must be equivalent.

```
         H                              H
         |                              |
      H — C — H                     H — C — H
         |                              |
  H      |      H              H        |        H
  |      |      |              |        |        |
H — C — C — C — H    →    Br — C — C — C — H
  |      |      |              |        |        |
  H      |      H              H        |        H
         |                              |
      H — C — H                     H — C — H
         |                              |
         H                              H
```

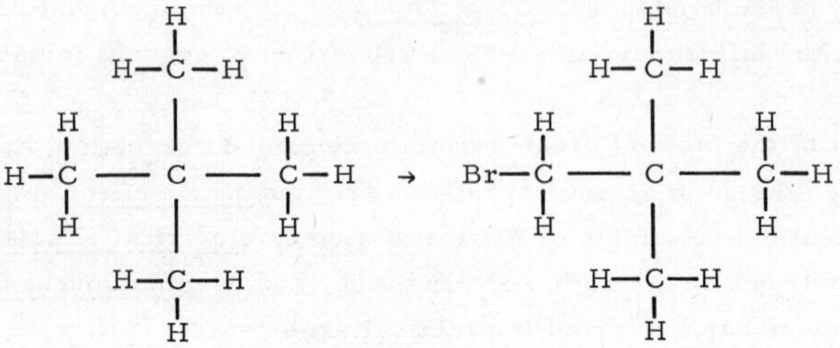

1-4

$$H_2C=CH_2 \xrightarrow{Br_2} H-\overset{\overset{\displaystyle Br}{|}}{\underset{\underset{\displaystyle H}{|}}{C}}-\overset{\overset{\displaystyle Br}{|}}{\underset{\underset{\displaystyle H}{|}}{C}}-H \rightarrow H-\overset{\overset{\displaystyle Br}{|}}{\underset{\underset{\displaystyle H}{|}}{C}}-\overset{\overset{\displaystyle Br}{|}}{\underset{\underset{\displaystyle H}{|}}{C}}-Br$$

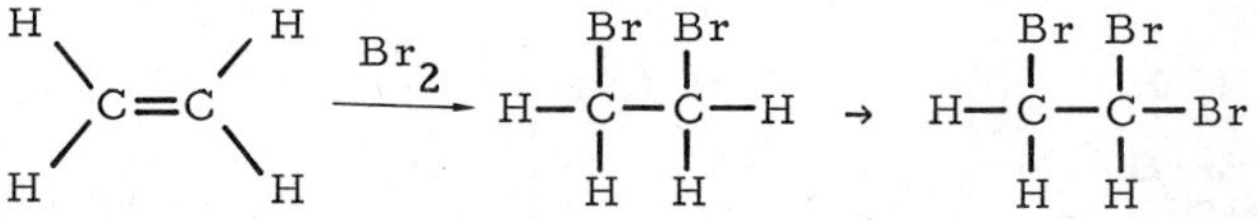

1-5   Compound 14 has <u>two</u> nonequivalent hydrogens and gives <u>two</u> isomers

of formula $C_6H_3Br_3$, A and B. Compound 15 has <u>three</u> nonequivalent hydro-

gens and gives <u>three</u> isomers, A, B and C. Compound 16 has one type of

hydrogen and gives one isomer A.

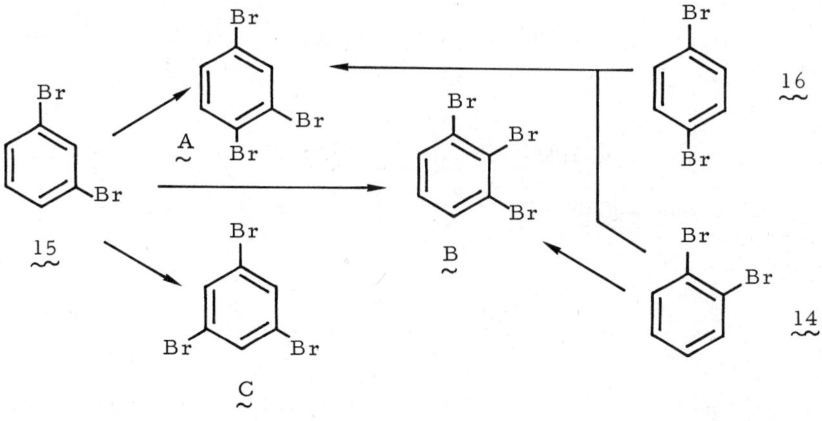

The only possible isomer from 16 has structure A ; structure of B follows

because 14 gives A and B (A is known); structure of C follows because

15 gives A, B and C (A and B are known).

1-6   Structure 17 would give <u>one</u> monobromine-substituted isomer, <u>two</u>

dibromine-substituted and <u>two</u> tribromine-substituted isomers. In contrast,

13 would give one, three, and three, mono-, di- and tri-bromine-substituted

isomers, respectively.

1-7   Structures A, B and C are isomers of formula C_3H_8O

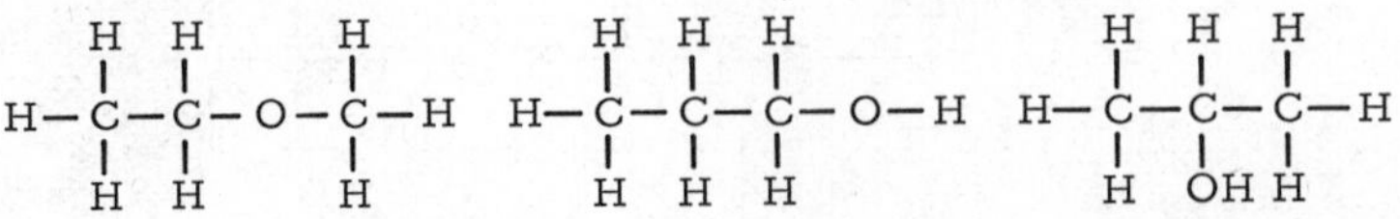

$$
\begin{array}{ccc}
\underset{\text{A}}{
\text{H}-\overset{\overset{\text{H}}{|}}{\underset{\underset{\text{H}}{|}}{\text{C}}}-\overset{\overset{\text{H}}{|}}{\underset{\underset{\text{H}}{|}}{\text{C}}}-\text{O}-\overset{\overset{\text{H}}{|}}{\underset{\underset{\text{H}}{|}}{\text{C}}}-\text{H}
} &
\underset{\text{B}}{
\text{H}-\overset{\overset{\text{H}}{|}}{\underset{\underset{\text{H}}{|}}{\text{C}}}-\overset{\overset{\text{H}}{|}}{\underset{\underset{\text{H}}{|}}{\text{C}}}-\overset{\overset{\text{H}}{|}}{\underset{\underset{\text{H}}{|}}{\text{C}}}-\text{O}-\text{H}
} &
\underset{\text{C}}{
\text{H}-\overset{\overset{\text{H}}{|}}{\underset{\underset{\text{H}}{|}}{\text{C}}}-\overset{\overset{\text{H}}{|}}{\underset{\underset{\text{OH}}{|}}{\text{C}}}-\overset{\overset{\text{H}}{|}}{\underset{\underset{\text{H}}{|}}{\text{C}}}-\text{H}
}
\end{array}
$$

Of these, only A could be formed from C_2H_5Br and CH_4O by the principle of least structural change.

$$
\text{H}-\overset{\overset{\text{H}}{|}}{\underset{\underset{\text{H}}{|}}{\text{C}}}-\overset{\overset{\text{H}}{|}}{\underset{\underset{\text{H}}{|}}{\text{C}}}-\text{Br} + \text{H}-\text{O}-\overset{\overset{\text{H}}{|}}{\underset{\underset{\text{H}}{|}}{\text{C}}}-\text{H} \rightleftharpoons \text{H}-\overset{\overset{\text{H}}{|}}{\underset{\underset{\text{H}}{|}}{\text{C}}}-\overset{\overset{\text{H}}{|}}{\underset{\underset{\text{H}}{|}}{\text{C}}}-\text{O}-\overset{\overset{\text{H}}{|}}{\underset{\underset{\text{H}}{|}}{\text{C}}}-\text{H} + \text{HBr}
$$

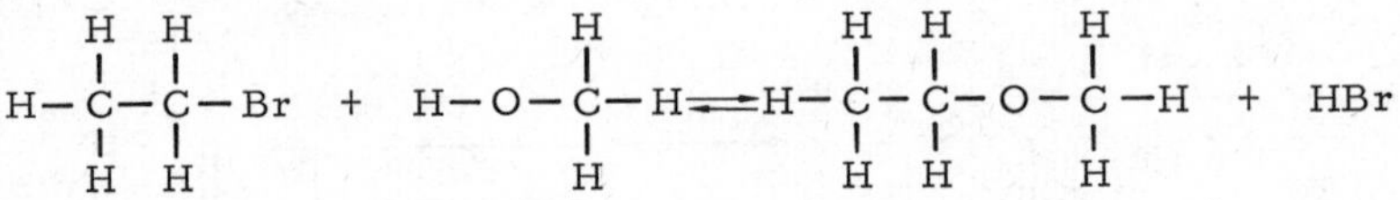

High concentrations of HBr with A would reverse the above reaction. B and C with HBr would give E and F, respectively. (The hydrogens are implied).

$$
\underset{\text{E}}{\text{C}-\text{C}-\text{C}-\text{Br}} \qquad\qquad \underset{\text{F}}{\text{C}-\overset{}{\underset{\underset{\text{Br}}{|}}{\text{C}}}-\text{C}}
$$

1-8   All the hydrogens of C_5H_{10} are equivalent in a structure of this formula that gives only one C_5H_9Br compound. A cyclic compound is indicated:

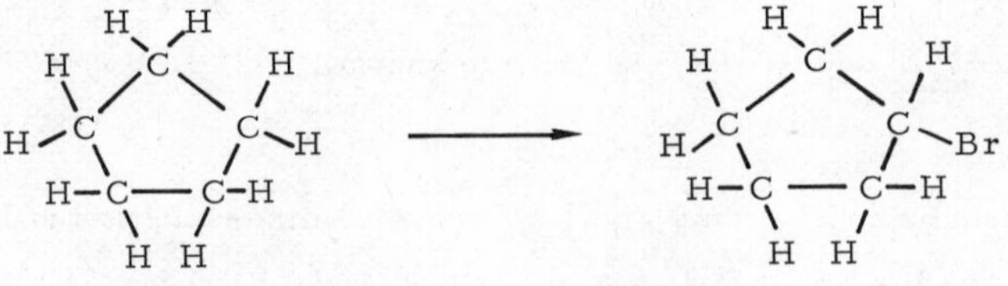

1-9   The atom with the lowest ionization energy (lowest electronegativity) will lose an electron most easily to become a positive ion. This atom is lithium-- not hydrogen. $Li \rightarrow Li^{\oplus} + e$. An electron is lost more easily from lithium than from hydrogen because the valence electron is further from the nucleus and is shielded from the nucleus by the inner-shell (1s) electrons.

1-10   The stronger acid will be the compound that donates protons more completely to the reference base (water). The more electronegative (electron-attracting) the atom is that carries the acidic hydrogen, the stronger will be the acid. Thus,

	electronegativity	acidity
a.	$F > Li$	$HF > LiH$
b.	$O > N$	$H_2O > NH_3$
c.	$O > H$	$H_2O_2 > H_2O$
d.	$F > H$	$CF_3H > CH_4$

1-11

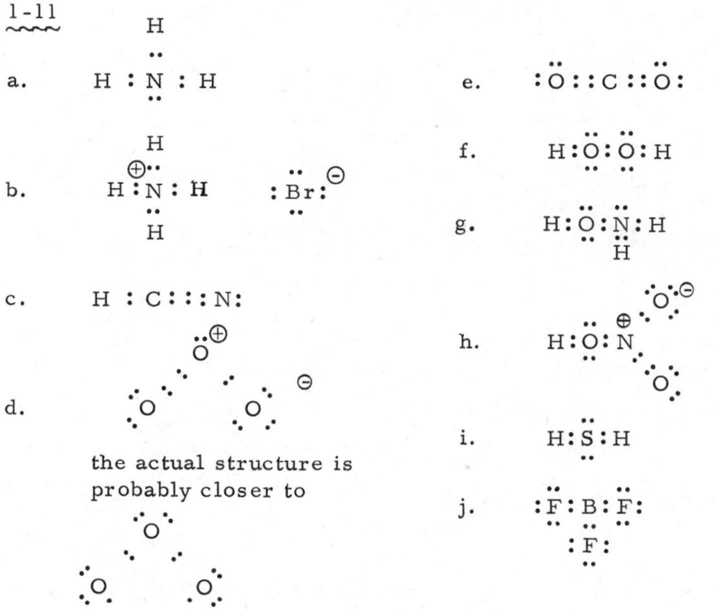

a. H : N̈ : H (with H above, lone pair on N)

b. H : N̈⊕ : H (with H above and below)   : B̈r :⊖

c. H : C ⋮⋮ N :

d.

the actual structure is
probably closer to

e. : Ö :: C :: Ö :

f. H : Ö : Ö : H

g. H : Ö : N̈ : H (with H below N)

h.

i. H : S̈ : H

j. : F̈ : B : F̈ :
    : F̈ :

1-12

a.   First and third are identical; second and fourth are identical.

b.   First, second and fourth are identical.

c.   Second, third and fourth are identical.

1-13

a.

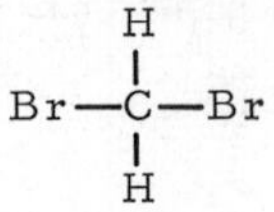

$$Br-\underset{\underset{H}{|}}{\overset{\overset{H}{|}}{C}}-Br$$

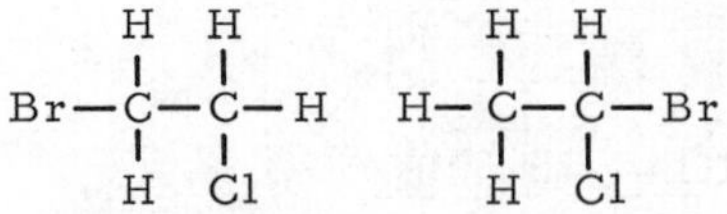

c.

$$CH_3-\underset{\underset{Br}{|}}{\overset{\overset{CH_3}{|}}{C}}-CH_3 \qquad CH_3-\underset{\underset{H}{|}}{\overset{\overset{H_3C}{|}}{C}}\underset{\underset{H}{|}}{\overset{\overset{H}{|}}{C}}-Br$$

b.

$$Br-\underset{\underset{H}{|}}{\overset{\overset{H}{|}}{C}}\underset{\underset{Cl}{|}}{\overset{\overset{H}{|}}{C}}-H \qquad H-\underset{\underset{H}{|}}{\overset{\overset{H}{|}}{C}}\underset{\underset{Cl}{|}}{\overset{\overset{H}{|}}{C}}-Br$$

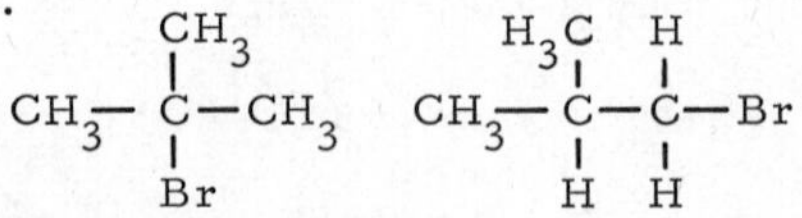

d. See Exercise 1-3.

e.

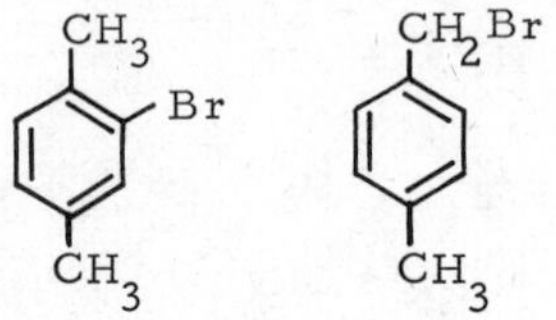

(remaining hydrogens are implied)

f.

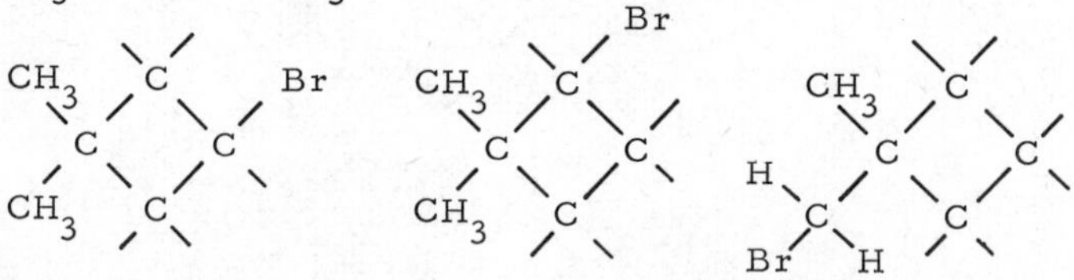

(remaining hydrogens are implied)

1-14

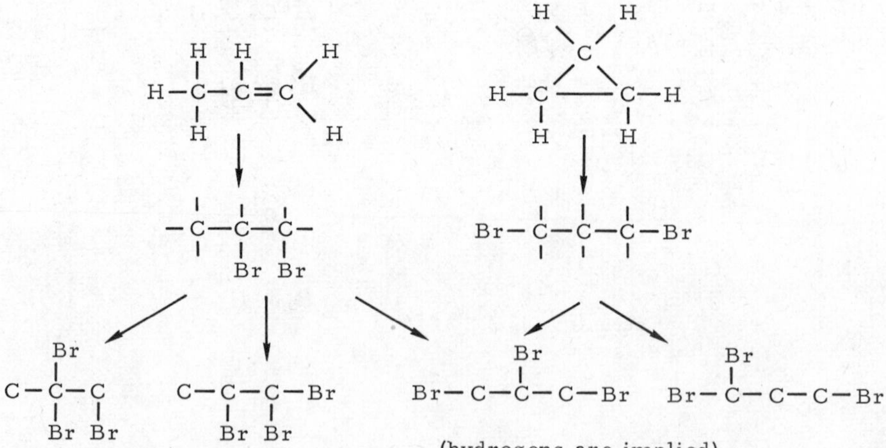

(hydrogens are implied)

1-15   The first step is to determine the empirical formula.

$$C_xH_yO_z + O_2 \rightarrow xCO_2 + \tfrac{y}{2} H_2O$$

$$\text{moles of } CO_2 \text{ in } 0.01222g = \frac{0.01222}{44} = x$$

$$\text{moles of } H_2O \text{ in } 0.00499g = \frac{0.00499}{18} = \frac{y}{2}$$

$$x = 0.000278, \quad y = 0.000554$$

percent carbon in sample = $\left(\dfrac{12x}{0.005372}\right) \cdot 100 = 62.04$

percent hydrogen $= \left(\dfrac{1.008y}{0.005372}\right) \cdot 100 = 10.4$

percent oxygen (by difference) = $\qquad = 27.56$

ratio of g atoms $C:H:O = x:y:z = \dfrac{62.04}{12} : \dfrac{10.4}{1.008} : \dfrac{27.56}{16}$

$$= 5.17 : 10.3 : 1.72$$

$$= 3 : 6 : 1$$

Therefore, the formula is $(C_3H_6O)_N$. The second step is to determine the molecular weight, $\underline{M}$, in order to find the value of $\underline{N}$, 1, 2, 3, or so on.

$PV = \underline{n}RT$ where $P = 728mm$, $V = 15.2ml$, $R = 62,400$, $T = 300°K$

Hence $\underline{n} = 5.91 \times 10^{-4}$

$$\underline{M} = \frac{m}{\underline{n}} = \frac{0.0343}{5.91 \times 10^{-4}} = 58$$

Hence correct formula is C_3H_6O .

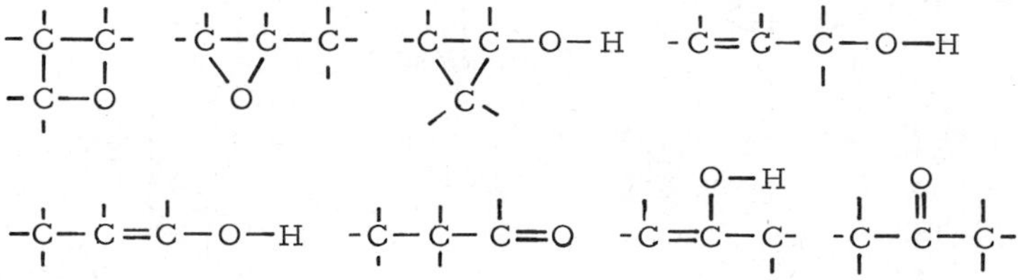

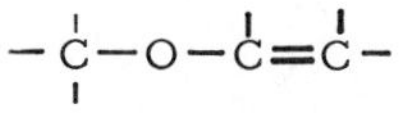

(hydrogens are implied)

<u>1-16</u>

g atoms of carbon in 100g $= \dfrac{45}{12.01} = 3.75$

g atoms of hydrogen in 100g $= \dfrac{7.50}{1.008} = 7.44$

g atoms of fluorine in 100g $= \dfrac{47.45}{19} = 2.50$

ratio of g atoms $C, H, F = 3.75 : 7.44 : 2.50$

$$= 1.5 : 3.0 : 1$$

$$= 3 : 6 : 2 \text{ or } C_3H_6F_2$$

$$\begin{array}{ccc} & \overset{\text{F}}{\underset{|}{}} & & \overset{\text{F}}{\underset{|}{}} \\ \text{F}-\text{C}-\text{C}-\text{C} & \quad \text{F}-\text{C}-\text{C}-\text{C} & \quad \text{F}-\text{C}-\text{C}-\text{C}-\text{F} \end{array}$$

(hydrogens are implied)

1-17   The O—H bonds in water are polar. Therefore, water molecules are associated in the liquid and solid phases through hydrogen bonding,

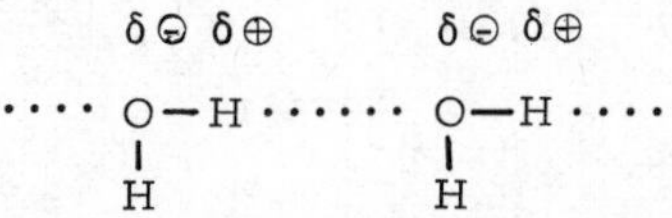

$$\cdots \overset{\delta\ominus}{\underset{\underset{\text{H}}{|}}{\text{O}}}\overset{\delta\oplus}{-\text{H}} \cdots \cdots \overset{\delta\ominus}{\underset{\underset{\text{H}}{|}}{\text{O}}}\overset{\delta\oplus}{-\text{H}} \cdots$$ (See p. 20 of text.) To vaporize water, energy

is required to break the hydrogen bonds. Hence, the boiling point is higher than that of methane, which is nonpolar and nonassociated.

1-18   The large difference in volatility suggests a substantial difference in the polarity of the bonds. The bonds in dimethylmercury (bp $91°$) must be nearly nonpolar. The high melting point of mercuric fluoride indicates ionic bonding, $\text{F}^\ominus \text{Hg}^{2\oplus} \text{F}^\ominus$, or $\text{F}^\ominus {}^\oplus\text{Hg}-\text{F}$.

1-19

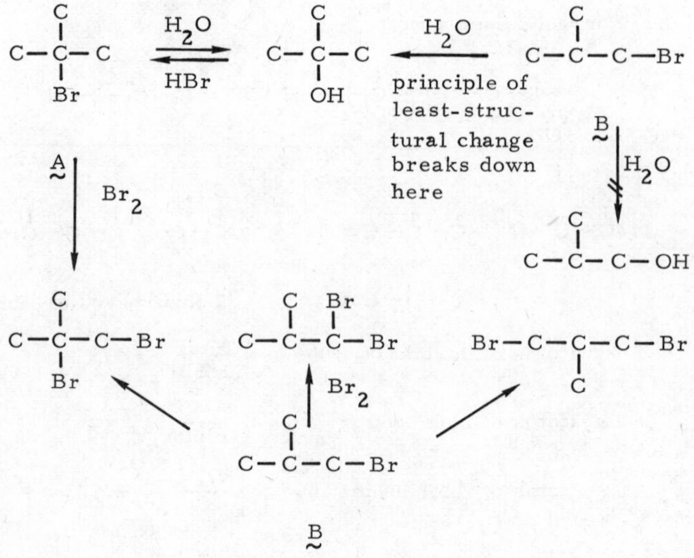

(hydrogens are assumed).

ADDITIONAL EXERCISES

1A-1   Deduce the molecular formula of a compound that has a molecular weight of 108 and by combustion analysis has 66.3% carbon, 3.5% hydrogen, and 30.2% oxygen.

1A-2   Write possible structures for the compound in 1A-1 given that it can add hydrogen (be reduced) to give 1,4-benzenediol (hydroquinone).

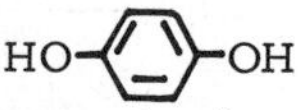

1A-3   Which of the following $C-X$ bonds would be most ionic: HO—⬡—OH
$C-Li$, $C-Na$, $C-K$? Would the carbon be electropositive or electronegative?

1A-4   Which of the following $C-Y$ bonds would be most ionic (polar): $C-I$, $C-Cl$, $C-Br$, $C-F$? Would the carbon be electronegative or electropositive?

1A-5   Write an equation to show the expected outcome of a reaction between:
a. CH_3Na and HCl,   b. CH_3Na and CH_3Cl,   c. CH_3Na and H_2O,   d. CH_3Na and CH_3OH.

1A-6   Which of the following compounds would you expect to associate through hydrogen bonding: benzene, CCl_4, CH_3OCH_3, CH_3CH_2OH, $(CH_3)_3N$, CH_3NH_2?

1A-7   Why is water, H_2O, less volatile (bp $100°$) than hydrogen sulfide, H_2S (bp $60.7°$)? Which would have the higher boiling point, methanol, CH_3OH or methanethiol, CH_3SH?

1A-8   Trimethylamine, $(CH_3)_3N$, is soluble in water whereas the hydrocarbon $(CH_3)_3CH$ is not. Explain.

1A-9   Which compound in each pair would have the greater water solubility?
a. CH_3OH, CH_3CH_3 b. CH_3OH, CH_3OCH_3 c. CH_3OH, CH_3SH d. CH_3OH, $C_{10}H_{21}OH$ e. CH_3SH, CH_3SNa

1A-10   Based on electronegativity arguments predict which compound in each group is the stronger acid.   a. BH_3, CH_4, NH_3 b. H_2O, $ClOH$ c. H_3CH, Cl_3CH.

1A-11   A compound Z of formula C_5H_{12} gave on bromination a mixture of four different monobromo compounds, $C_5H_{11}Br$. Elimination of HBr from the $C_5H_{11}Br$ mixture gave three compounds of formula C_5H_{10}. What is the structure of Z?

1A-12   Draw the structure of a compound having the formula C_9H_{20} and only two different kinds of hydrogens in the molecule in the ratio of 9:1. How many different kinds of carbon atoms are there in the molecule?

CHAPTER 2

STRUCTURAL ORGANIC CHEMISTRY.
THE SHAPES OF MOLECULES, FUNCTIONAL GROUPS

Single, double, and triple covalent bonds involving sharing of electron pairs can be represented by Lewis structures or by formulas with appropriate numbers of lines connecting the atom symbols. Representations with lines denoting bonds are called <u>structural formulas.</u> Care should be taken to remember the unshared electron pairs, which usually are not shown in the common structural formulas, because these often play an important role in the chemistry. It is often preferable to include the electron pairs as in $CH_3-\overset{..}{\underset{..}{O}}-H$ instead of simply writing CH_3-O-H.

Condensed formulas in which no bonds are shown explicitly are used very commonly to save space in chemical publications. Another form of abbreviation is simply to use a line to represent the unit $C-C$ in structural formulas and not write in any hydrogen that is present in the form of a $C-H$ bond. In the condensed and line formulas, normal valences are considered to be used for each atom unless explicit notation is made to the contrary.

For fluorocyclopropane, we now have several possible ways to write the structure (all of which mean precisely the same thing):

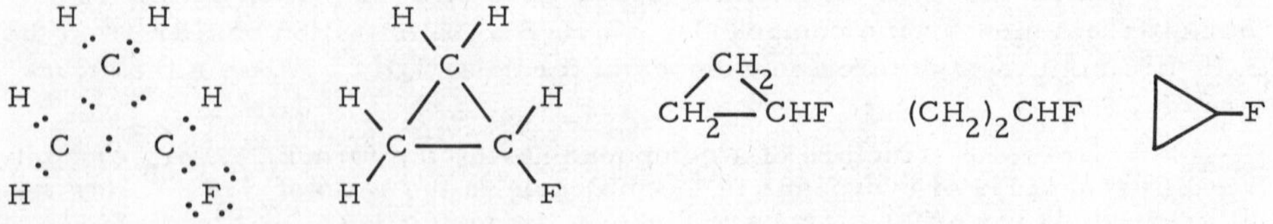

Which will be used in practice will depend considerably on what we are interested in. In this book, at one time or the other, we will use all of the styles of representation.

Structural formulas may be expanded by the use of molecular models to show the relative positions of the atoms in space. Ball-and-stick models are excellent for this purpose, but if it is desired to see the relative sizes of the atoms, the so-called space-filling models are used.

Position isomers are compounds with the same formula but different bonding arrangements between the atoms.   Three isomers are known of formula C_5H_{12}.

$$CH_3CH_2CH_2CH_2CH_3 \qquad CH_3CH_2\overset{\overset{\displaystyle CH_3}{|}}{C}HCH_3 \qquad CH_3-\overset{\overset{\displaystyle CH_3}{|}}{\underset{\underset{\displaystyle CH_3}{|}}{C}}-CH_3$$

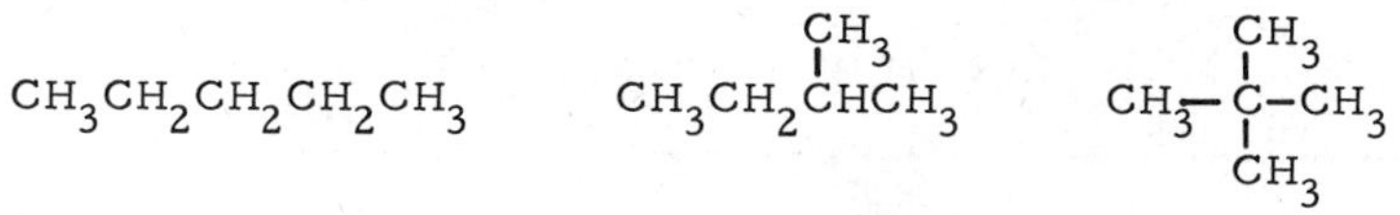

The number of possible isomers increases very rapidly with the number of carbons in the molecule.   An example is nonane, C_9H_{20}, which has 35 possible isomers, all of which have been synthesized.

A traditional and still very widely used classification scheme of organic compounds is by functional group.   A <u>functional group</u> is a distinct structural entity that confers particular and usually reasonably predictable properties to a molecule.   An example is the functional group $-\overset{\overset{\displaystyle H}{|}}{C}=O$.   When this group is present we call the molecule an <u>aldehyde.</u>   Thus methanal, $H_2C=O$, ethanal, $CH_3CH=O$, and decanal, $CH_3(CH_2)_8CH=O$ all are aldehydes and they have similar chemical properties - at least in those reactions that occur at the $-HC=O$ part of the molecule.   Examples of the most common functional groups are given in Table 2-2.

Most reactions of organic compounds can be classified as acid-base, substitution, addition, elimination, rearrangement, oxidation, or reduction processes.   Examples are given in Section 2-3B.

ANSWERS TO EXERCISES

2-1

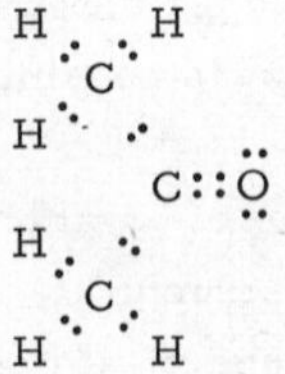

H H
 · ·
H · C ·
 · · ·
H · · C :: Ö   a. $H-\overset{H}{\underset{H}{C}}-\overset{H}{\underset{H}{C}}-C\overset{O}{\underset{H}{}}$ b. $H-\overset{H}{C}-O$, $H-\overset{H}{\underset{H}{C}}-\overset{H}{C}-O$
H · C · $H-\overset{}{\underset{H}{C}}-\overset{}{\underset{H}{C}}-H$ $\overset{}{\underset{H}{C}}$
 · · ·   H   H
H   H

c.

$\overset{H}{\underset{H}{}}C=C\overset{H}{\underset{C}{}}\overset{H}{\underset{H}{}}O-H$, $\overset{H}{\underset{H}{}}C=C\overset{O-H}{\underset{C}{}}\overset{H}{\underset{H}{}}H$, $\overset{H-O}{\underset{H}{}}C=C\overset{H}{\underset{C}{}}H$ or $\overset{H}{\underset{H-O}{}}C=C\overset{H}{\underset{C}{}}\overset{H}{\underset{H}{}}H$

d. $H-\overset{H}{\underset{}{C}}-\overset{H}{\underset{}{C}}-O-H$
          $\overset{}{\underset{H}{C}}$
          H   H

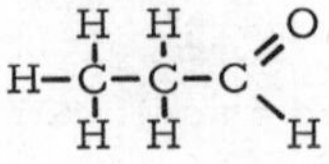

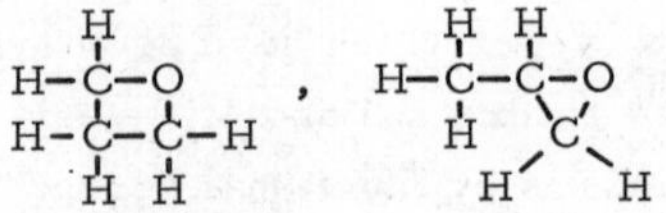

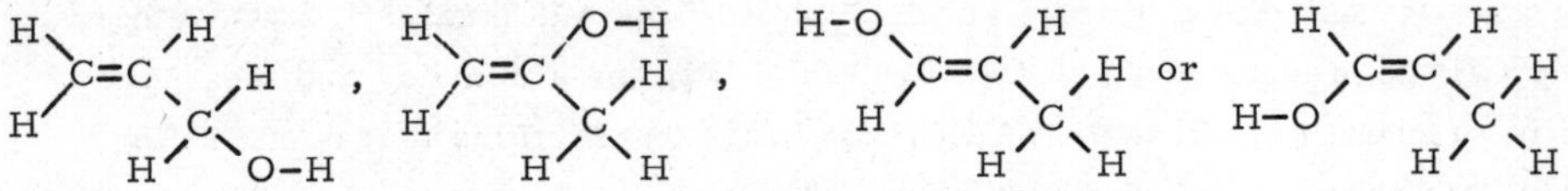

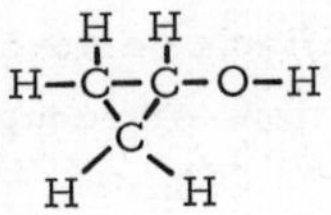

2-2

a.	acid-base (proton transfer)	f.	rearrangement
b.	substitution	g.	elimination
c.	addition, oxidation-reduction	h.	acid-base (proton transfer)
d.	substitution	i.	substitution
e.	rearrangement		

2-3

$$C_6H_6 + \frac{15}{2} O_2 \rightarrow 6CO_2 + 3H_2O \; ; \; C_6H_6 + \frac{9}{2} O_2 \rightarrow 6CO + 3H_2O$$

2-4

a.   114.3g   b.   246.2g

2-5

a. $Na + H-O-H \longrightarrow Na^{\oplus} + {}^{\ominus}OH + \frac{1}{2}H_2$

b. $Na + CH_3-O-H \longrightarrow Na^{\oplus} + {}^{\ominus}OCH_3 + \frac{1}{2}H_2$

c. $Na^{\oplus}H^{\ominus} + H-O-H \longrightarrow Na^{\oplus} + {}^{\ominus}OH + H_2$

d. $Na^{\oplus}H^{\ominus} + CH_3CH_2-O-H \longrightarrow Na^{\oplus} + {}^{\ominus}OCH_2CH_3 + H_2$

2-6

a. $\overset{..}{N}H_3 + HO-SO_2-OH \longrightarrow \overset{\oplus}{N}H_4 + {}^{\ominus}O-SO_2-OH$

or $2\overset{..}{N}H_3 + HO-SO_2-OH \longrightarrow 2\overset{\oplus}{N}H_4 + SO_4{}^{2\ominus}$

b. $CH_3CH_2\overset{..}{N}H_2 + HO-SO_2-OH \longrightarrow CH_3CH_2\overset{\oplus}{N}H_3 + {}^{\ominus}O-SO_2-OH$

or $2CH_3CH_2\overset{..}{N}H_2 + HO-SO_2-OH \longrightarrow 2CH_3CH_2\overset{\oplus}{N}H_3 + SO_4{}^{2\ominus}$

c. $Na^{\oplus}\,{}^{\ominus}OH + NH_4{}^{\oplus}\,{}^{\ominus}Cl \longrightarrow Na^{\oplus}\,{}^{\ominus}Cl + H_2O + \overset{..}{N}H_3$

or ${}^{\ominus}OH + NH_4{}^{\oplus} \longrightarrow H_2O + \overset{..}{N}H_3$

d. $Na^{\oplus}\,{}^{\ominus}OH + CH_3CH_2NH_3{}^{\oplus}\,{}^{\ominus}Cl \longrightarrow Na^{\oplus}\,{}^{\ominus}Cl + H_2O + CH_3CH_2\overset{..}{N}H_2$

or ${}^{\ominus}OH + CH_3CH_2NH_3{}^{\oplus} \longrightarrow H_2O + CH_3CH_2\overset{..}{N}H_2$

2-7

The hydrogens on carbon are inferred in the following structures:

a.   See structures, pp. 9 and 14 of text and answer to Exercise 1-3

b.
$$\overset{\displaystyle Br}{\underset{\displaystyle |}{}}\,\,\,\,\,\,\,\,\overset{\displaystyle Br}{\underset{\displaystyle |}{}}$$

c.

d.

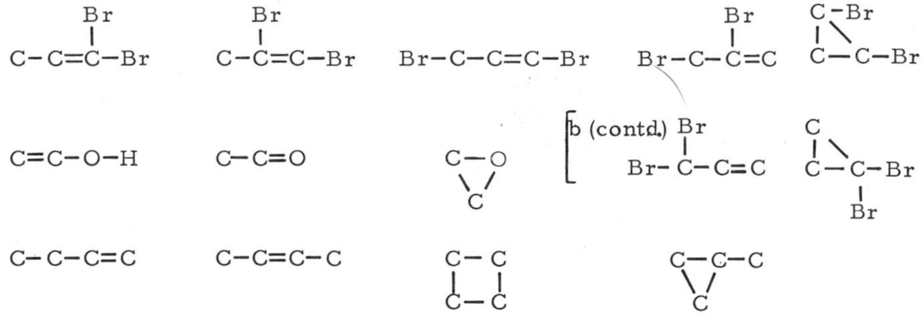

2-7 (cont.)

e. $C-C-C-NH_2$ $\underset{\underset{NH_2}{|}}{C-C-C}$ $\underset{\underset{H}{|}}{C-N-C-C}$ $C-N\underset{C}{\overset{C}{<}}$

f. $C-C\equiv C-Cl$ $C\equiv C-C-Cl$ $\underset{\underset{Cl}{|}}{C-C\equiv C}$ $C-C-Cl$ (with cyclopropane C below)

g. $C\equiv C-C-C-C\equiv C$ (four-membered ring with C atoms) etc.

h. $HO-C\equiv C-O-H$ $O=C-C-O-H$ $\underset{\underset{O-H}{}}{C-C\overset{\overset{O}{\|}}{}}$ $C-C-O-H$ (epoxide)

  $\begin{matrix}C-O\\ C-O\end{matrix}$ $\begin{matrix}C-O\\ O-C\end{matrix}$ etc.

i. $C\equiv C-C=O$ $C-C\equiv C=O$ $\begin{matrix}C-C\\ C-O\end{matrix}$ $\underset{O}{C=C-C}$ (epoxide) $\underset{C}{C-C=O}$ (epoxide) etc.

j. $C\equiv C-NH_2$ $C=C=NH$ $C-C\equiv N$ $\underset{C}{\overset{C}{\|}}NH$ etc.

2-8

a. $\overset{\overset{H\;H\;H}{}}{H:C:C:C:H}$ (with lone pairs)
   $\underset{H\;H\;H}{}$

d. $\overset{H}{\underset{H}{H:C:::C:C:H}}$

g. $\overset{H}{\underset{H}{H:C:O:C:H}}$ (with lone pairs) $\underset{H\;\;H}{}$

b. $\overset{H}{\underset{}{H:C:H}}$
   $H:C\;:\;\overset{H}{C:H}$
   $H:C$
   $\underset{H}{}$

e. (ring of C atoms with H)

h. $\overset{H}{\underset{H\;\;H}{H:C:C}}\overset{O}{}$

c. $\overset{H}{}\;C::C::C\;\overset{H}{}$
   $\underset{H}{}\qquad\underset{H}{}$

f. $\overset{F\;\;\;F}{C::C}\underset{F\;\;\;F}{}$

i. $\overset{H}{\underset{H}{H:C:C}}\overset{O}{\underset{O}{}}H$

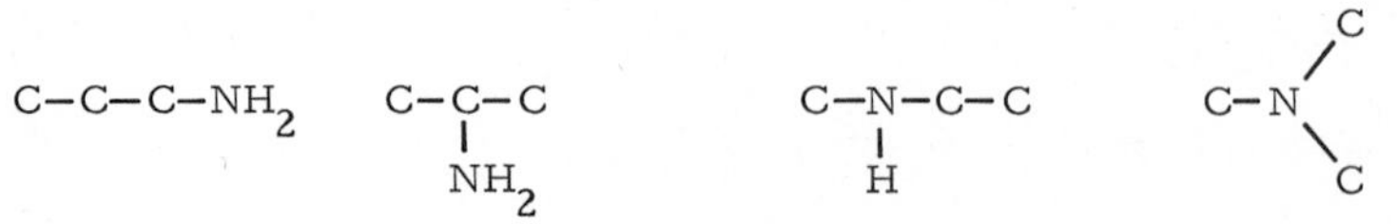

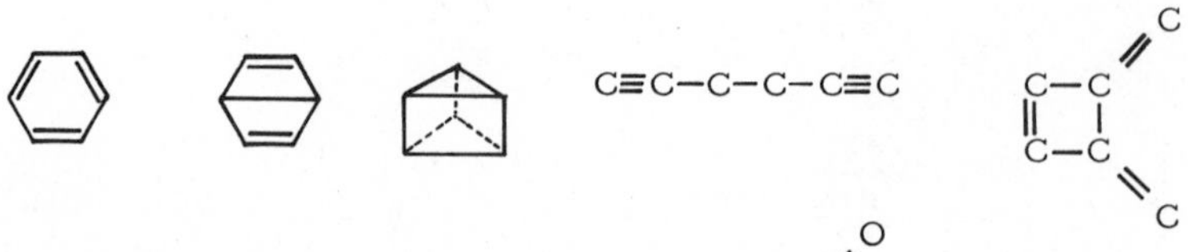

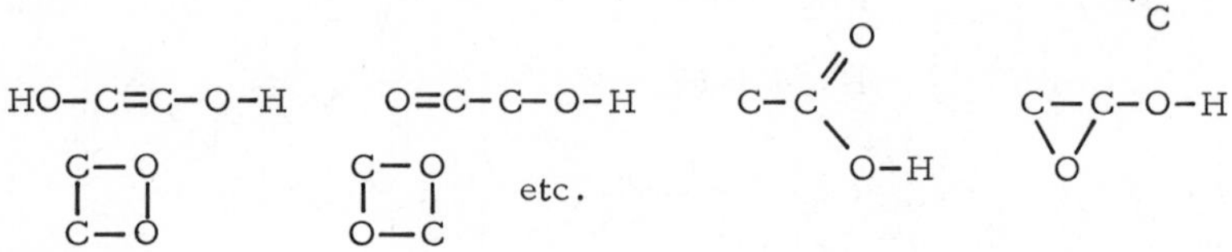

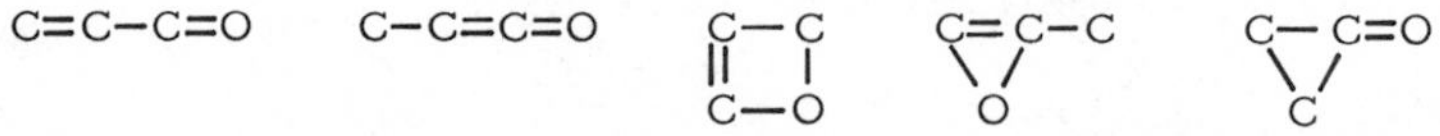

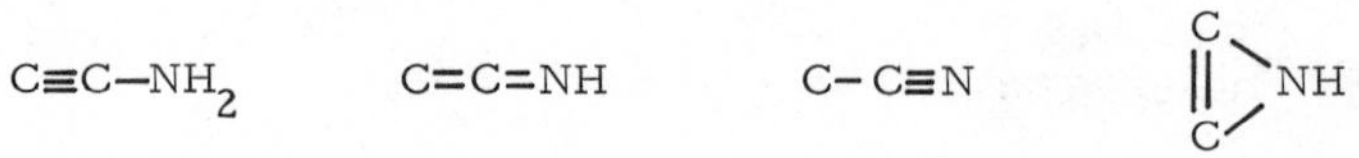

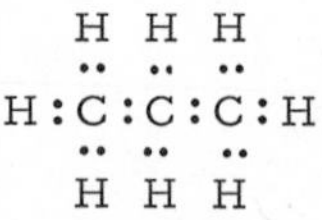

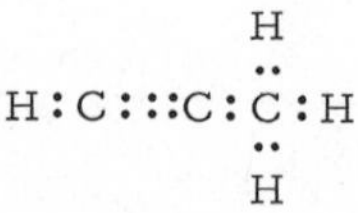

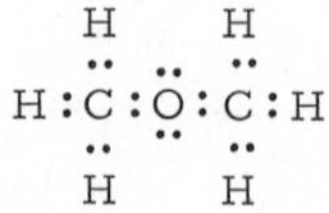

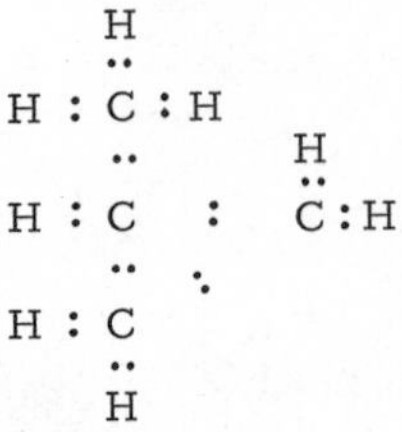

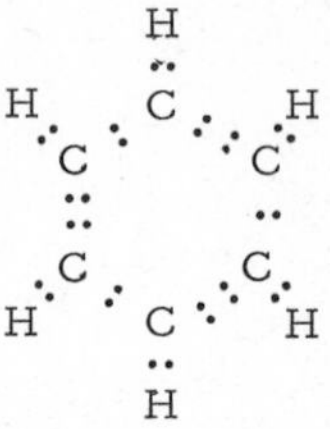

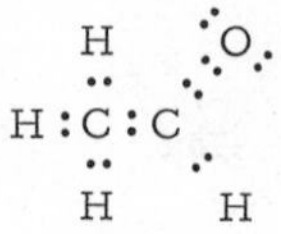

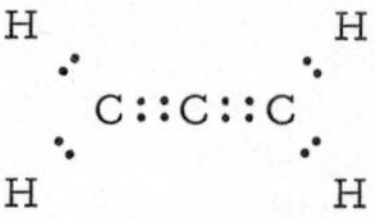

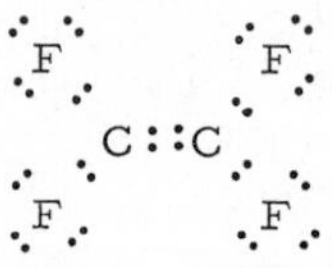

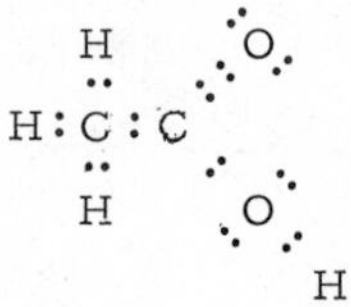

2-8 (cont.)

j.   As in (e) with one H replaced with $\overset{H}{\underset{\cdot\cdot}{\text{N}}}\text{:}H$

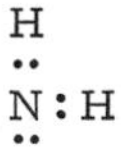

k. 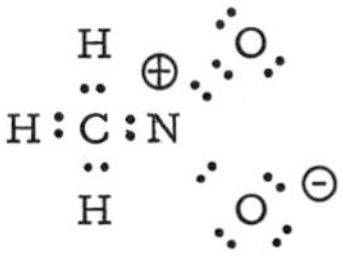 $\text{H}\overset{H}{\underset{H}{\text{:}\overset{\cdot\cdot}{\underset{\cdot\cdot}{\text{C}}}\text{:N}}}\overset{\oplus\;\;\overset{\cdot\cdot}{\underset{\cdot\cdot}{\text{O}}}\text{:}}{\underset{\overset{\cdot\cdot}{\underset{\cdot\cdot}{\text{O}}}\text{:}\;\ominus}{}}$

l.   As in (e) with one H replaced with $\text{C}\text{:::N}\text{:}$

2-9   The hydrogens on carbon are inferred in the following structures:

a. 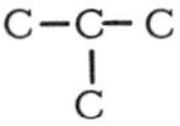 $\underset{\overset{|}{\text{C}}}{\text{C}-\text{C}-\text{C}}$

b. $\text{C}-\text{C}{\equiv}\text{C}-\text{C}$

c.

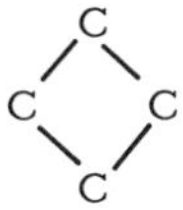

d.

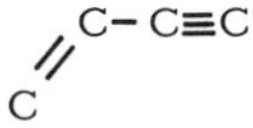

e.

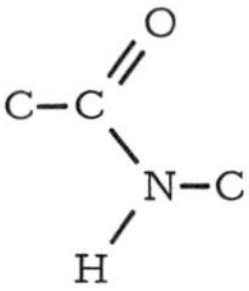

f.

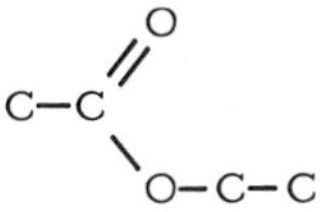

g.

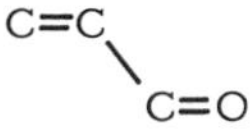

h.

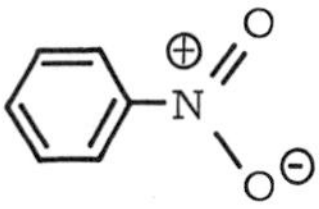

i. $\text{C}-\text{C}-\text{C}{\equiv}\text{N}$

j.

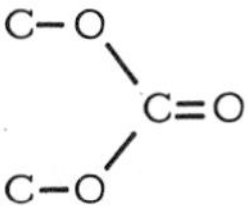

2-10

a.   identical

b.   identical

c.   first three are identical and different from the fourth

d.   identical

e.   first two are identical and different from the third

2-11

a. $\text{CH}_3-\text{CH}{=}\text{CH}_2$

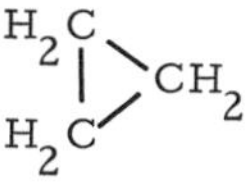

b. $\text{CH}_3-\text{C}{\equiv}\text{CH}$ $\text{CH}_2{=}\text{C}{=}\text{CH}_2$

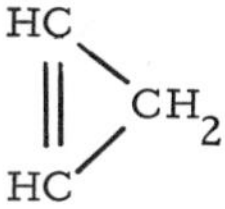

2-11 (cont.)

c. $CH_2{=}CH{-}OH$ $CH_3{-}CH{=}O$

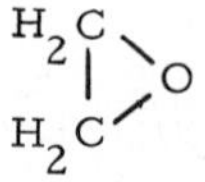

d. $CH_3{-}CH{-}F$ $F{-}CH_2{-}CH_2{-}Cl$
            |
            Cl

e. $CH_3{-}CH_2{-}CH_2{-}NH_2$ $CH_3{-}CH_2{-}NH{-}CH_3$ $CH_3{-}\overset{\overset{\displaystyle CH_3}{|}}{N}{-}CH_3$ $CH_3{-}\overset{\overset{\displaystyle CH_3}{|}}{C}H{-}NH_2$

f.

$CH_3{-}CH_2{-}CH_2{-}CH_2{-}Cl$ $CH_3{-}CH_2{-}\underset{\underset{\displaystyle Cl}{|}}{C}H{-}CH_3$ $CH_3{-}\underset{\underset{\displaystyle CH_3}{|}}{C}H{-}CH_2{-}Cl$ $CH_3{-}\overset{\overset{\displaystyle CH_3}{|}}{\underset{\underset{\displaystyle Cl}{|}}{C}}{-}CH_3$

2-12

a. $CH_3{-}C\overset{\displaystyle O}{\underset{\displaystyle OH}{\big\Vert}}$ $HO{-}CH_2{-}C\overset{\displaystyle O}{\underset{\displaystyle H}{\big\Vert}}$ $CH_3{-}O{-}C\overset{\displaystyle O}{\underset{\displaystyle H}{\big\Vert}}$

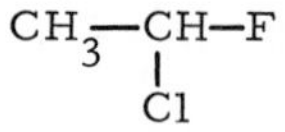

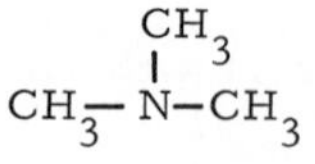

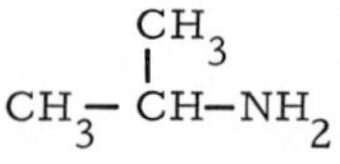

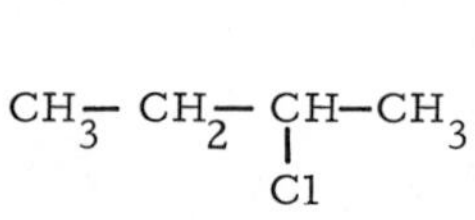

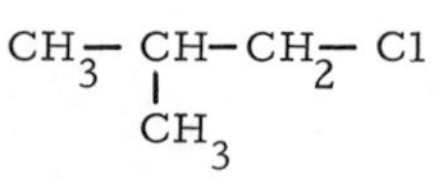

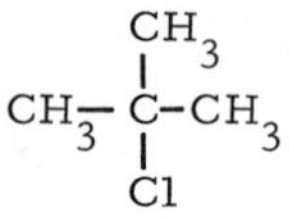

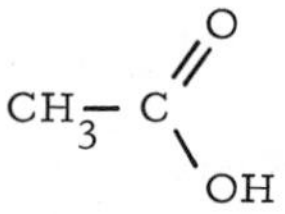

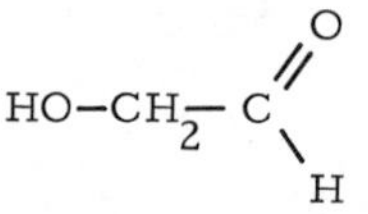

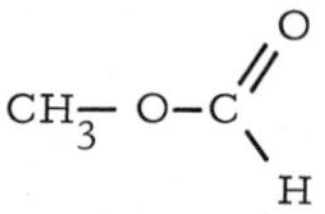

b.   As in Exercise 2-11f --substitute Cl for OH.

c.   See Exercise 1-3.

d. $CH_3{-}C\overset{\displaystyle O}{\underset{\displaystyle NHCH_3}{\big\Vert}}$

e. $CH_3{-}CH_2{-}CH_2{-}C\overset{\displaystyle O}{\underset{\displaystyle H}{\big\Vert}}$ $CH_3CH_2{-}C\overset{\displaystyle O}{\underset{\displaystyle CH_3}{\big\Vert}}$

f. $CH_3{-}CH_2{-}C\overset{\displaystyle O}{\underset{\displaystyle OH}{\big\Vert}}$ $CH_3{-}C\overset{\displaystyle O}{\underset{\displaystyle O{-}CH_3}{\big\Vert}}$

g. $CH_2{=}CH{-}C\overset{\displaystyle O}{\underset{\displaystyle H}{\big\Vert}}$ h.

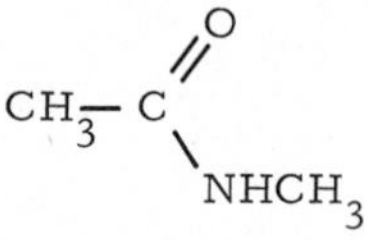

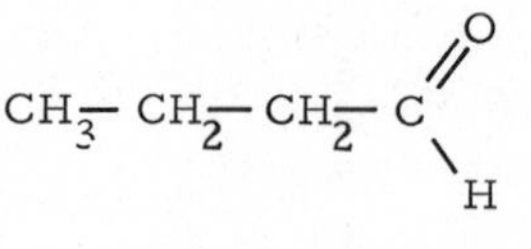

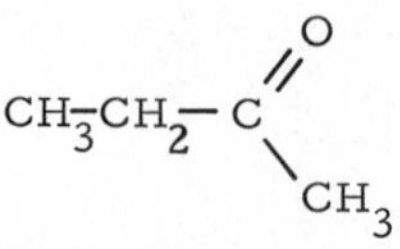

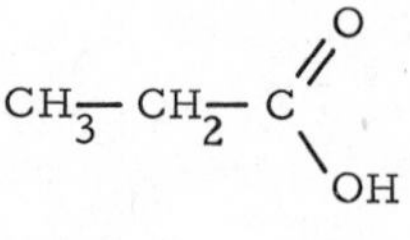

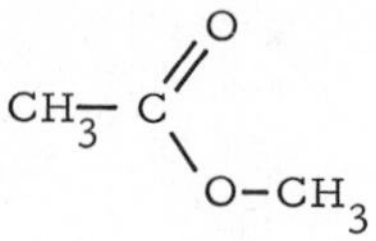

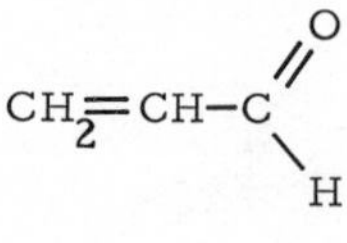

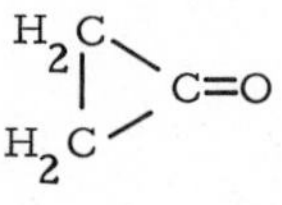

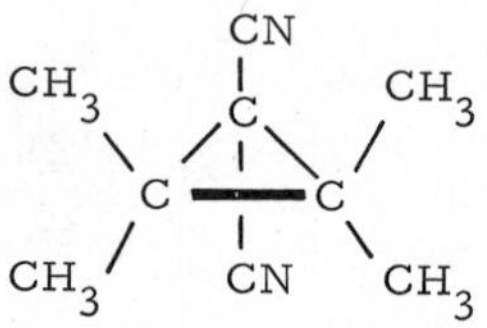

ADDITIONAL EXERCISES

2A-1   Many, if not most, naturally occurring compounds have complex structures with several functional groups. Examples are shown below. For each compound identify the main functional groups present using Table 2-2 as a guide.

a.

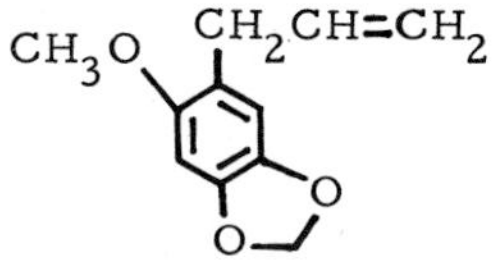

myristin
(from nutmeg)

b.

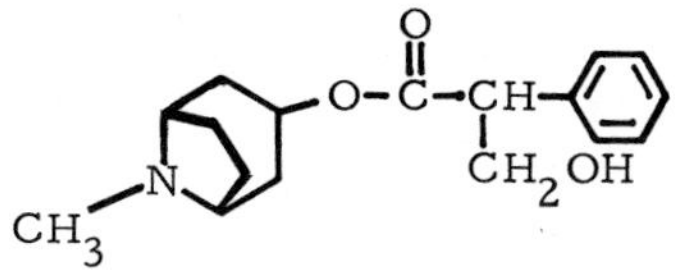

atropine
(from belladonna or deadly nightshade; a
potent poison; used to dilate pupils and
relieve spasms)

c.

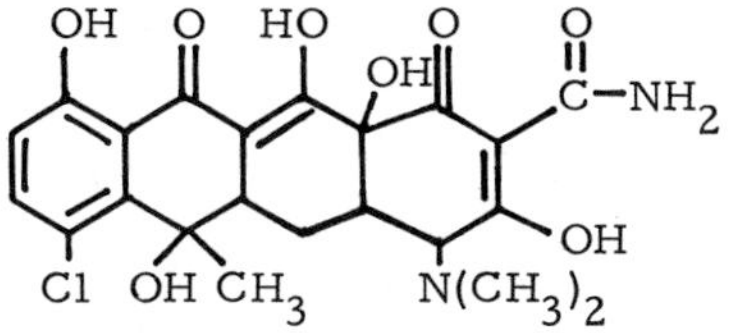

chlortetracycline
(Aureomycin; antibiotic)

d.

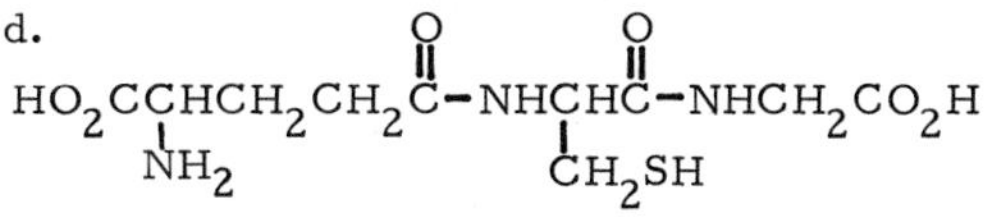

$HO_2CCHCH_2CH_2C-NHCHC-NHCH_2CO_2H$

glutathione
(peptide hormone)

2A-2   All nine hydrogens in a compound M of formula C_3H_9N are equivalent. Also, M reacts with hydrochloric acid to form an ionic compound N of formula $C_3H_{10}NCl$. Draw structures for M and N.

2A-3   How many monochloro compounds are possible by way of addition of one mole of HCl to one mole of $CH_2=CHCH_2CH=CHCH_3$?

2A-4   In the following sequence of reactions classify each step as either addition, elimination, substitution, or rearrangement. Also indicate the steps that may also be classified as reduction.

$CH_3CH(OH)CH_2CHO \xrightarrow{1} CH_3CH=CHCHO \xrightarrow{2} CH_3CH=CHCH_2OH \xrightarrow{3}$
$CH_3CH=CHCH_2Br \xrightarrow{4} CH_3CH=CHCH_2C\equiv N \xrightarrow{5} CH_3CH=CHCH_2CONH_2 \xrightarrow{6}$
$CH_3CH_2CH=CHCONH_2$

CHAPTER 3

ORGANIC NOMENCLATURE

The IUPAC system for naming alkanes (C_nH_{2n+2}), cycloalkenes, alkenes (C_nH_{2n}), alkynes (C_nH_{2n-2}), and some arenes is described.

The nomenclature rules are used to derive the structure from the name and to write the name for the structure such that no ambiguity exists. The system is best illustrated by example. In naming the compound with the following structure, the structure first should be inspected and the functional groups noted:

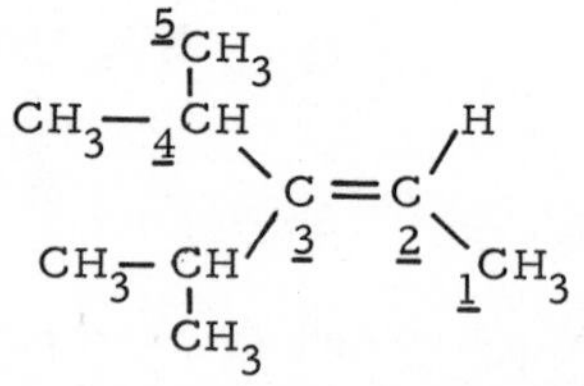

In this example there is one carbon-carbon double bond; therefore the compound is an <u>alkene</u>. The parent alkene is determined by the length of the longest chain passing through the double bond, which in this case is a five-carbon chain, or a <u>pentene</u>. The chain now must be numbered to give the double bond the <u>lowest</u> possible number. The compound is then a <u>2-pentene</u>. The name is completed by identifying the remaining substituents and their positions along the chain. The name for the alkyl group at C3 is 1-methylethyl, or <u>isopropyl</u>; that at C4 is <u>methyl</u>. Therefore, the complete name is, 3-(1-methylethyl)-4-methyl-2-pentene or, more commonly, 3-isopropyl-4-methyl-2-pentene.

In writing a structure to fit the name, consider the compound 5-ethyl-1-methylcyclohexene. The parent compound is identified at the <u>end</u> of the name as cyclohexene, which always is numbered such that the double bond is between C1 and C2:

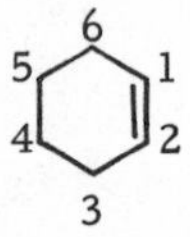

The rest of the name identifies the nature and location of the substituents in the ring. There is a methyl group at C1 and an ethyl group at C5. Hence the complete structure is

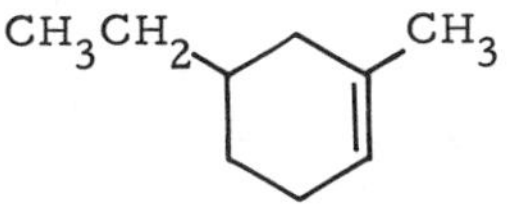

The IUPAC nomenclature system for other types of compounds is given in Chapter 7 and is based on the fundamental rules described here. Nonsystematic names in common use are included, where possible, parenthetically.

The terms primary, secondary, tertiary, and quaternary refer to carbons of an alkane chain that are bonded to one, two, three, and four other cabons, respectively.

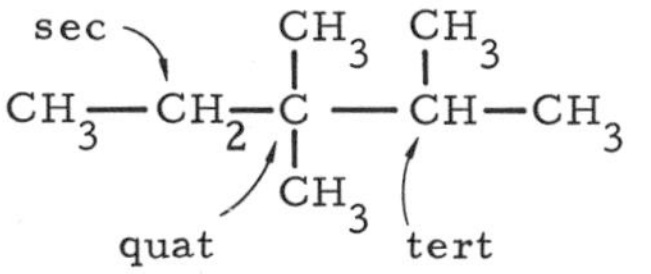

All other carbons are primary

2,3, 3-trimethylpentane

Alkyl groups are primary, secondary or tertiary according to the number of carbons at the point of attachment, one two or three, respectively.

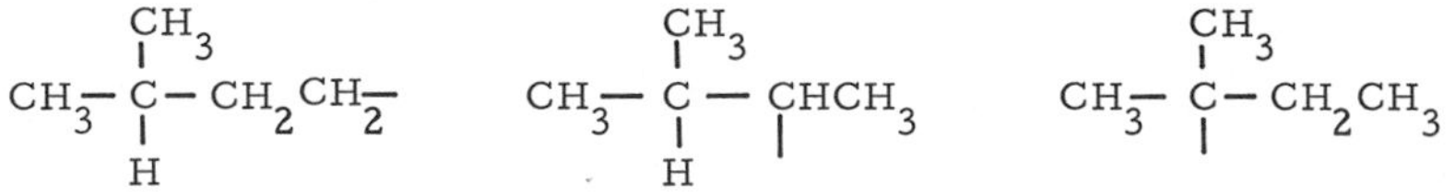

primary   secondary   tertiary

ANSWERS TO EXERCISES

3-1

a. 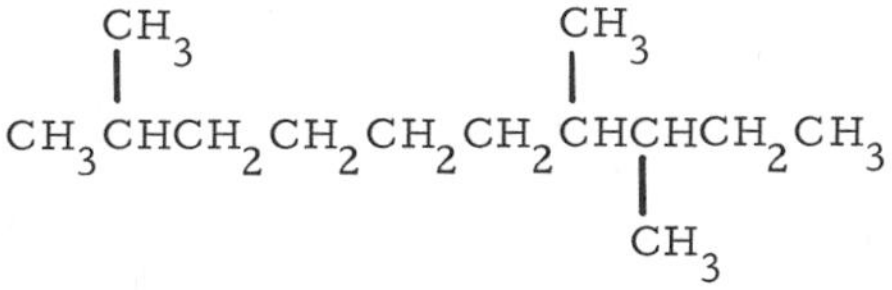 $CH_3CHCH_2CH_2CH_2CH_2CHCHCH_2CH_3$
     with CH_3 substituents and CH_3 branch

b. 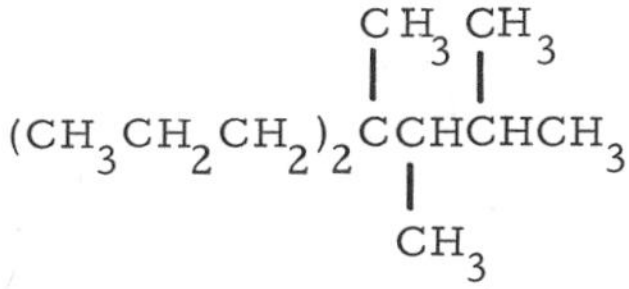 $(CH_3CH_2CH_2)_2CCHCHCH_3$
     with CH_3, CH_3, CH_3 substituents

c. 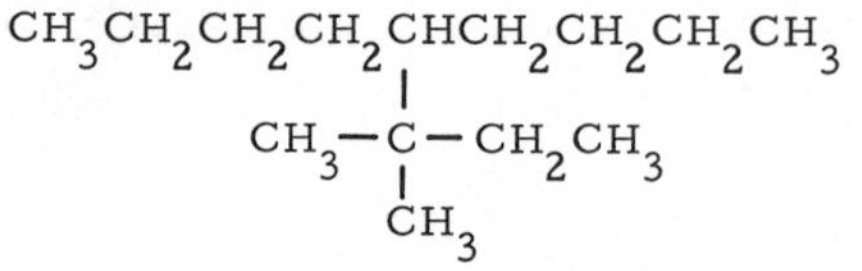
$$CH_3CH_2CH_2CH_2CHCH_2CH_2CH_2CH_3$$
$$CH_3-C-CH_2CH_3$$
$$CH_3$$

d. 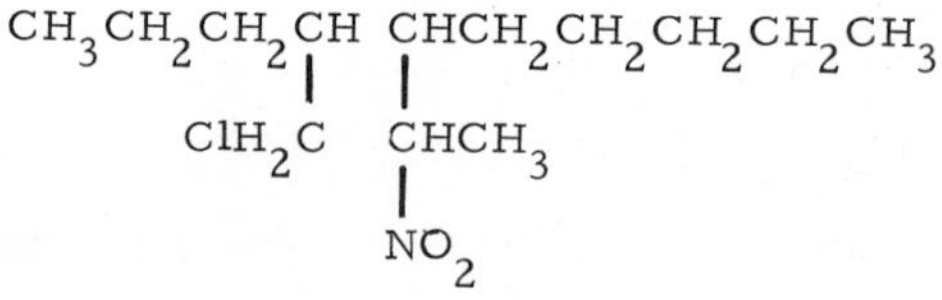
$$CH_3CH_2CH_2CH \quad CHCH_2CH_2CH_2CH_2CH_3$$
$$ClH_2C \quad CHCH_3$$
$$NO_2$$

3-2

a. 2, 3, 5-trimethylhexane

c. 4-ethyl-2-methylheptane

b. 4-ethyl-5-methyloctane

d. 2, 3, 4, 5-tetramethylhexane

3-3

a. Longest chain is C_6 (3-ethyl-2-methylhexane).

b. Substituents are to be cited alphabetically (3-chloro-3-methylpentane).

c. Preferred numbering is 2, 2, 6, 6, 7 because, if we compare the numbers term-by-term, 2, 2 is lower than 2, 3 (see Footnote 2, Chapter 3).

d. With chains of same length, preference goes to the one with the largest number of substituents (3-ethyl-2, 2-dimethylpentane).

3-4

a.

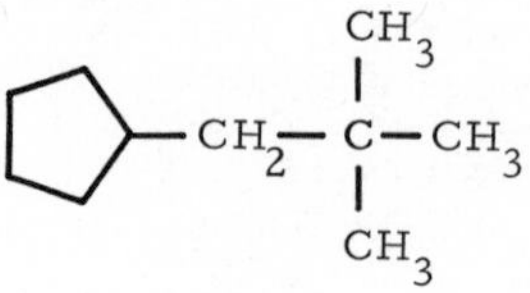

c.

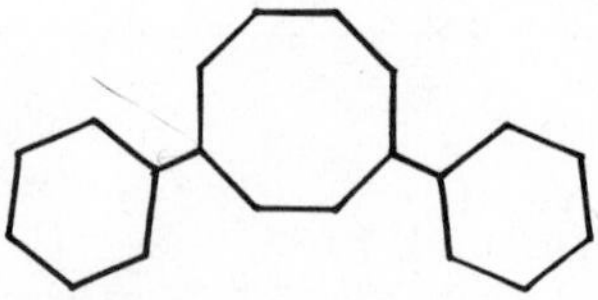

b.

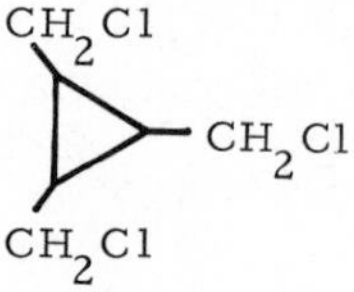

d.

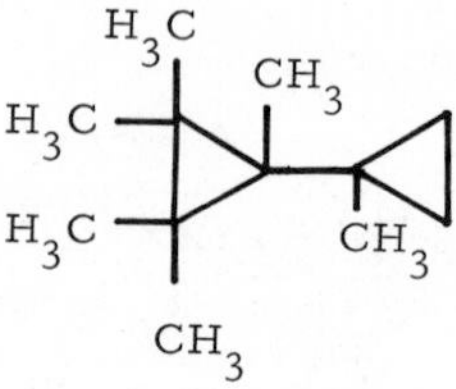

3-5

a.   1, 2, 5-trimethylcyclohexane

b.   4-(1-methylethyl)- or 4-isopropyl-1, 2, 5-trimethylcyclohexane

c.   1-(3-chlorocyclobutyl)-3-(2-methylpropyl) cyclohexane.

d.   cyclopentylcyclodecane

3-6

a.   (structure: cyclohexene with CH$_3$ groups)   b.   (structure: cyclopropene with Cl groups)

c. CH_2=C-CH$_2$-CH$_2$-C=CH$_2$ (with CH$_3$ groups)

d.   CH$_2$= (methylenecyclohexene structure)

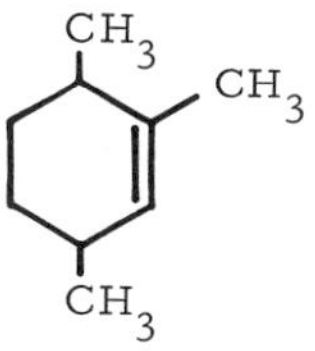

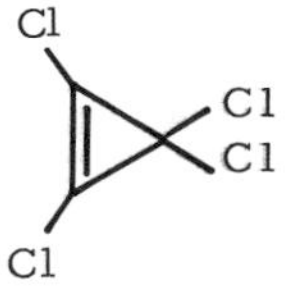

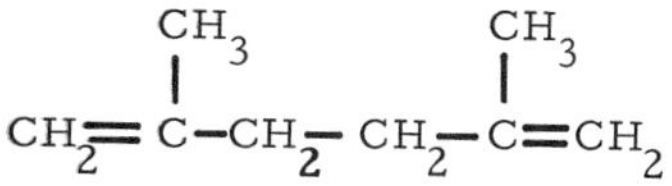

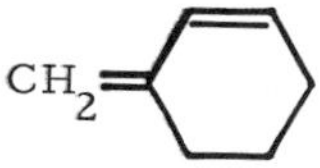

3-7

a.   5-ethyl-1-methylcyclopentene   b.   2-propyl-1, 4-pentadiene

c.   2, 7-dimethyl-4, 5-dimethylidene-2, 6-octadiene

d.   cyclopropylidenecyclopropane

3-8

a. CH_2= CHCH= CHC≡CH   c. CH_2= CHCH= CH–CH–CH=CH$_2$ with C≡CH

b.   (cyclic structure)   d.   (cyclooctyne structure with =CH$_2$)

3-9   C-C-C-C-C-C-C   C-C-C-C-C-C   C-C-C-C-C-C
                                 | |
                                 C   C

heptane   2-methylhexane   3-methylhexane

C-C-C-C-C (with C branches)   C-C-C-C-C (with C branches)   C-C-C-C-C (with C branches)

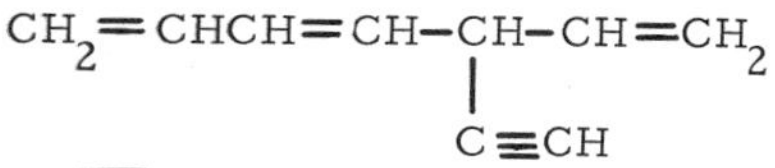

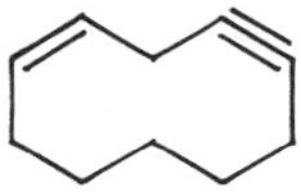

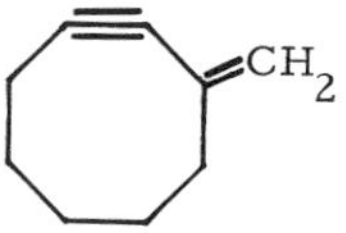

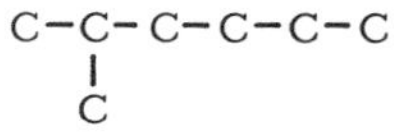

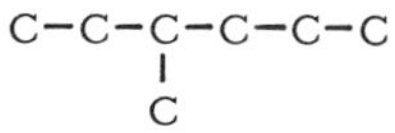

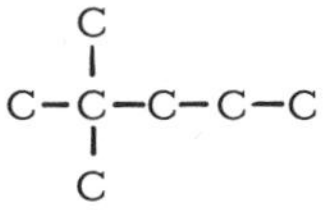

2, 2-dimethylpentane   3, 3-dimethylpentane   2, 4-dimethylpentane

C-C-C-C-C (with C branches)   C-C-C-C-C (with C-C branch)   C-C-C-C (with C branches)

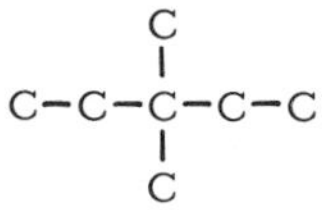

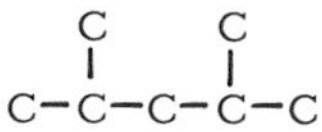

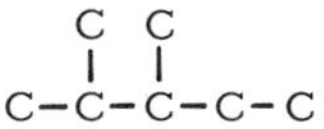

2, 3-dimethylpentane   3-ethylpentane   2, 2, 3-trimethylbutane
(Hydrogens are inferred)

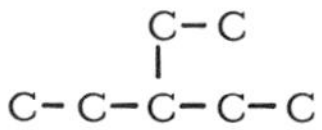

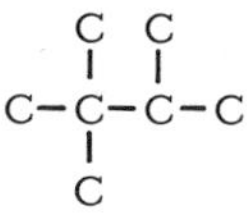

3-10

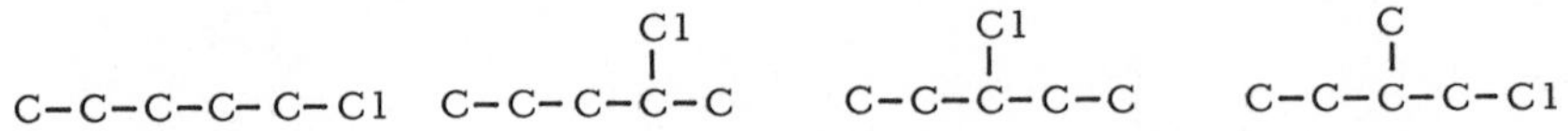

C—C—C—C—C—Cl

1-chloropentane

$$\begin{array}{c} \text{Cl} \\ | \\ \text{C—C—C—C—C} \end{array}$$

2-chloropentane

$$\begin{array}{c} \text{Cl} \\ | \\ \text{C—C—C—C—C} \end{array}$$

3-chloropentane

$$\begin{array}{c} \text{C} \\ | \\ \text{C—C—C—C—Cl} \end{array}$$

1-chloro-2-
methylbutane

$$\begin{array}{c} \text{C} \\ | \\ \text{C—C—C—C} \\ | \\ \text{Cl} \end{array}$$

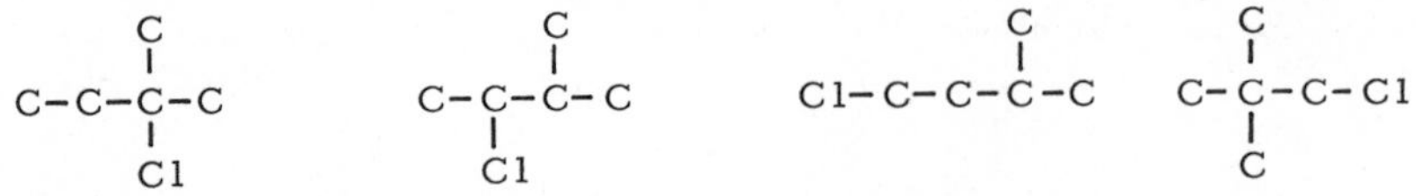

2-chloro-2-
methylbutane

$$\begin{array}{c} \text{C} \\ | \\ \text{C—C—C—C} \\ | \\ \text{Cl} \end{array}$$

2-chloro-3-
methylbutane

$$\begin{array}{c} \text{C} \\ | \\ \text{Cl—C—C—C—C} \end{array}$$

1-chloro-3-
methylbutane

$$\begin{array}{c} \text{C} \\ | \\ \text{C—C—C—Cl} \\ | \\ \text{C} \end{array}$$

1-chloro-2,2-
dimethyl-
propane

3-11

a.   2,5-dimethylhexane

b.   2,2,4-trimethylpentane

c.   2,4-dimethylhexane

d.   3-methylpentane

e.   4,4-dipropylheptane

f.   2-methyl-5-(2-methylpropyl)nonane

3-12

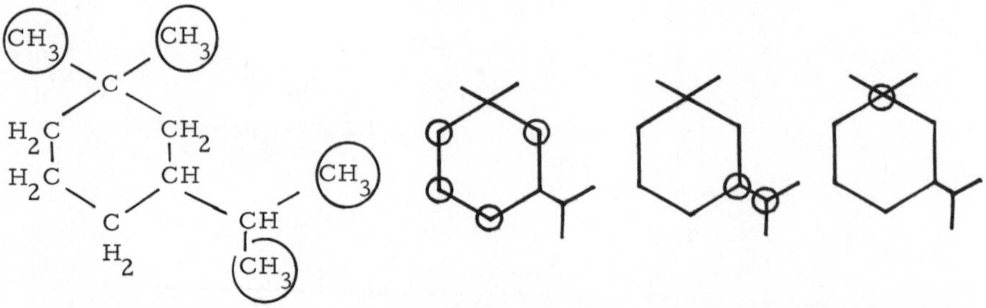

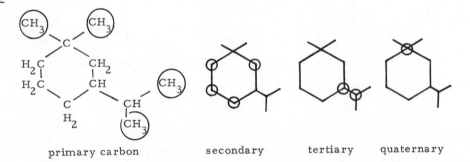

primary carbon   secondary   tertiary   quaternary

3-13

a.

$CH_3CH_2CH_2CH_2CH_2-$

$$\begin{array}{c} \text{CH}_3 \\ | \\ \text{CH}_3\text{CHCH}_2\text{CH}_2- \end{array}$$

$$\begin{array}{c} \text{CH}_3 \\ | \\ \text{CH}_3\text{CH}_2\text{CHCH}_2- \end{array}$$

$$\begin{array}{c} \text{CH}_3 \\ | \\ \text{CH}_3-\text{C}-\text{CH}_2- \\ | \\ \text{CH}_3 \end{array}$$

b.

$CH_3CH_2CH=CHCH_2-$

$CH_3CH=CHCH_2CH_2-$

$CH_2=CHCH_2CH_2CH_2-$

$$\begin{array}{c} \text{CH}_3\text{CH}_2-\text{C}-\text{CH}_2- \\ || \\ \text{CH}_2 \end{array}$$

$$\begin{array}{c} \text{CH}_3\text{CH}=\text{C}-\text{CH}_2- \\ | \\ \text{CH}_3 \end{array}$$

$$\begin{array}{c} \text{CH}_2=\text{CHCHCH}_2- \\ | \\ \text{CH}_3 \end{array}$$

$$\begin{array}{c} \text{CH}_3\text{C}=\text{CHCH}_2- \\ | \\ \text{CH}_3 \end{array}$$

$$\begin{array}{c} \text{CH}_2=\text{CCH}_2\text{CH}_2- \\ | \\ \text{CH}_3 \end{array}$$

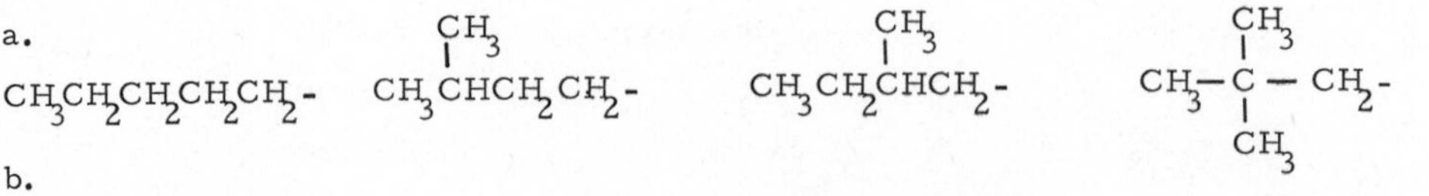

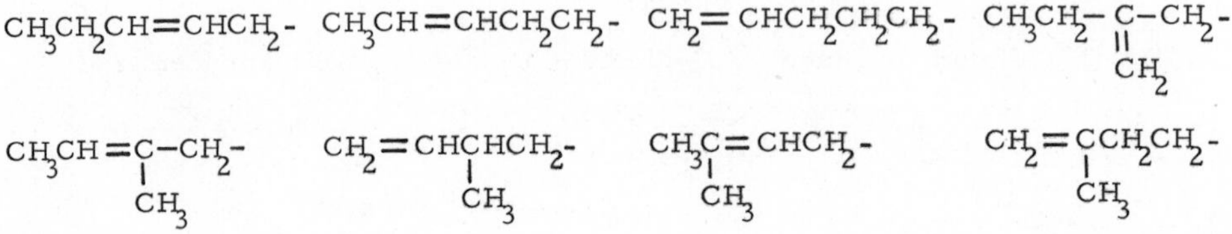

3-14

a. $ClCH_2-$

b. $CH_2=C\overset{Cl}{\diagdown}$

c. $(CH_3)_2CHCH_2CH_2-$

d. $(CH_3)_2CHCHCH_3$

e. $CH_2=CHCHCH_3$

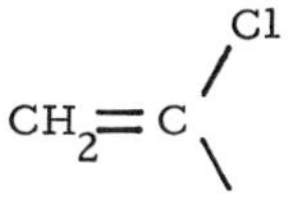

f. $(CH_3)_2C=CH-$

g. $HC\equiv CCH=CH-$

h.

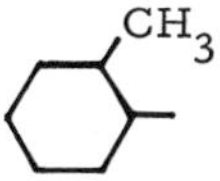

i.

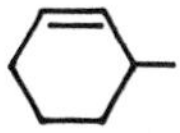

j. $-CH_2-$

k. O_2N-

l. $Cl-$ ... Cl

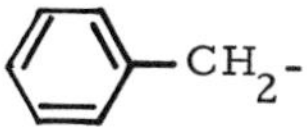

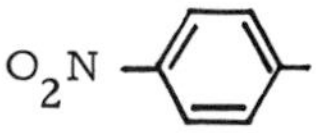

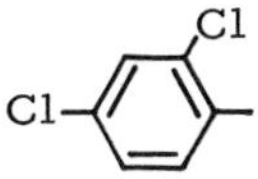

m. $CH_3CH_2CH=$

3-15

a. 3-bromopropyl (primary)

b. 1,1-diethylpropyl (tertiary)

c. 2-butynyl (primary)

d. 2-isopropylcyclopentyl or 2-(1-methylethyl)cyclopentyl (secondary)

e. 4-tert-butyl-2-cyclohexenyl or 4-(1,1-dimethylethyl)-2-cyclohexenyl (secondary)

f. 2,6-dimethylphenyl (aryl)

g. 4-(1-propenyl)phenyl (aryl)

3-16

a. $CH_3(CH_2)_5CH=CH_2$

b. $CH_3CH=CHCH_2CH=CH_2$

c.

d. $\underset{H}{\overset{}{C}}=C=\underset{H}{\overset{}{C}}$

e. $(CH_3)_2\underset{Cl}{\overset{}{C}}-C\equiv CH$

f.

g. $CH_2=\underset{Cl}{\overset{}{C}}-CH=CH_2$

h. $CH_3C\equiv CC=CHCH_3$ with CH_3

i. phenyl $-\underset{CH_3}{\overset{CH_3}{C}}-CH_3$

j. phenyl $-\underset{}{CHCH_2CH_3}$ with CH_3

k.

l. H_3C

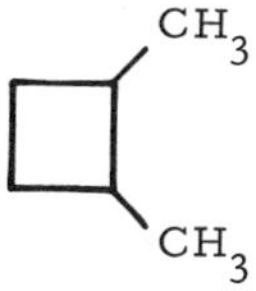

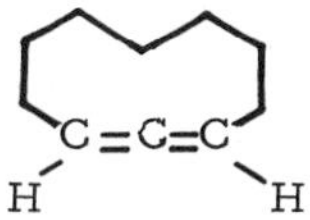

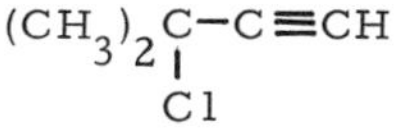

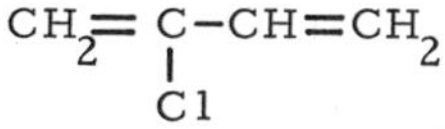

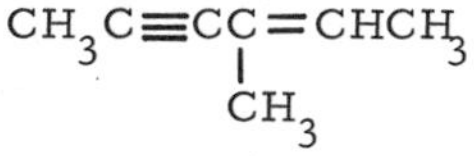

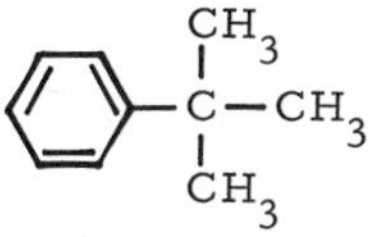

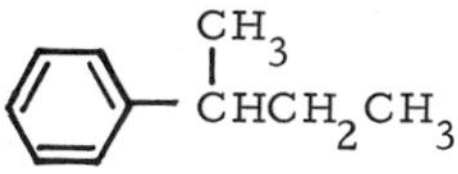

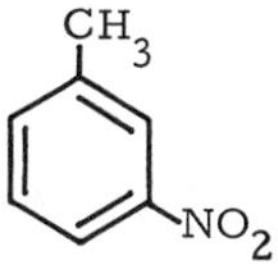

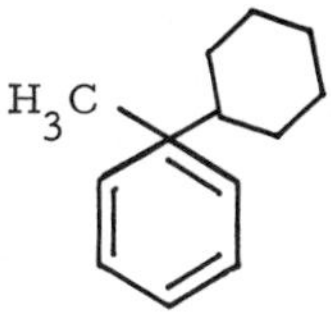

m.

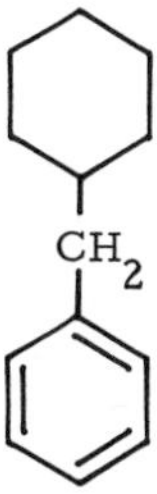

3-17

a. 2-methyl-2-butene

b. 1,1-dichloro-2-methyl-1-propene

c. 1-bromo-3-methyl-1,2-butadiene

d. 3,4-diethyl-1,3,5-hexatriene

e. 3,3-dimethyl-1-penten-4-yne

f. 1,3-cyclobutadiene

g. ethenylcyclopropane (not cyclopropylethene)

h. methylidenecyclohexane

i. 1,2-dichlorobenzene or ortho-dichlorobenzene

j. 2,4-dichloro-1-(1-propenyl)benzene

3-18

a. 2,3-dimethylpentane

b. 1,1,2,2-tetrachloroethane

c. 3-methylcyclohexene

d. 3-chloropropene

e. 3-methyl-3-penten-1-yne

f. propylbenzene

g. 1,4-dinitrobenzene or para-dinitrobenzene

h. 2-methylbutane

3-19

a. 1-methyl-4-(1-methylethenyl)cyclohexene

b. 2-methyl-1,3-butadiene (isoprene)

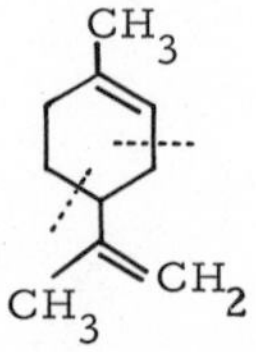

c.

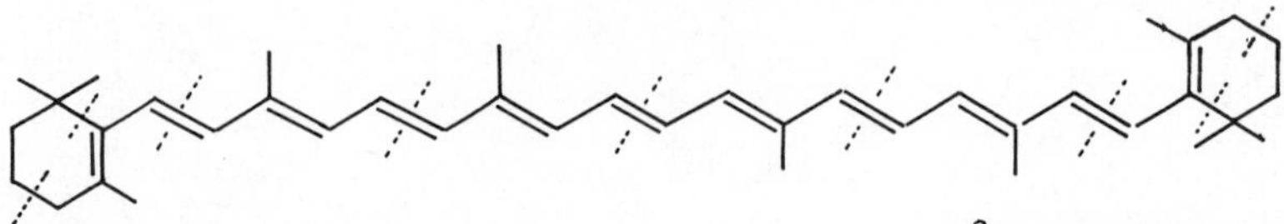

β-carotene

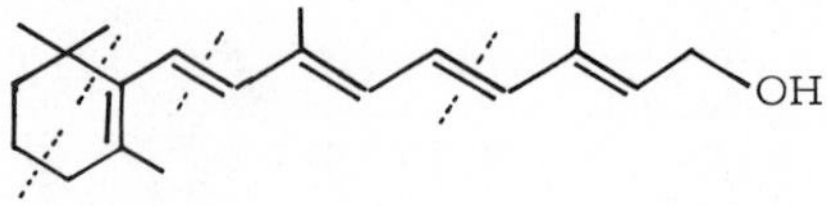

vitamin A

ADDITIONAL EXERCISES

3A-1   How many primary, secondary, tertiary, and quaternary carbons has 2, 3, 3, 5-tetramethylhexane?

3A-2   Draw the structure of a compound of formula $C_5H_{11}Br$ in which the bromine substituent is bonded to a tertiary carbon. Name the compound by the IUPAC system.

3A-3   Name the position isomers of formula C_5H_9Cl having a cyclopropane ring and a chlorine attached to a secondary carbon.

3A-4   Draw the structure of a trimethylbenzene isomer that gives two different monochloro compounds on substitution of hydrogen for chlorine, $C_9H_{12} \longrightarrow C_9H_{11}Cl$. Name the hydrocarbon and the two monochloro compounds.

3A-5   Draw the structure of each of the compounds named: a. 2-methyl-2-butene b. 2-chloro-1, 3-pentadiene c. 3-isopropylcyclopentene d. 1, 2-pentadien-4-yne e. 3-chloro-3-methyl-1-butyne f. 1-tert-butyl-4-ethylbenzene

3A-6   Draw structures for all the possible position isomers of methylpentadiene having conjugated double bonds. Name each by the IUPAC system. How many position isomers of methylpentadiene are possible having cumulated double bonds?

3A-7   The compounds shown below are examples of naturally occurring isoprenoid compounds (see also Exercise 3-19). Name each by the IUPAC system.

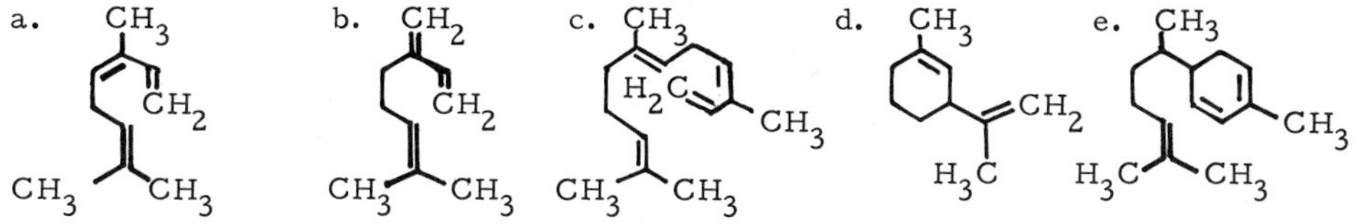

a. (ocimene)   b. (myrcene)   c. (α-farnesene)   d. (sylvestrene)   e. (zingiberene)

CHAPTER 4

ALKANES

The alkanes are hydrocarbons without multiple bonds or rings and have the general formula C_nH_{2n+2}. They are called variously paraffin, saturated open-chain, or acyclic hydrocarbons.

The continuous-chain alkanes, $CH_3(CH_2)_{n-2}CH_3$, with $\underline{n}$ carbons in the chain, are nonpolar substances with quite regular changes in physical properties with increasing number of CH_2 groups. Such a series of compounds is said to be a homologous series. The concept of homology is useful in correlating the properties of the individual members of the many series of organic substances that are known. A homologous series of compounds, such as $CH_3(CH_2)_{n-2}CO_2H$, containing both polar and nonpolar groups is likely to show dramatic decreases in polar properties, such as water solubility, as $\underline{n}$ increases and hence the contribution of the polar part of the molecule to the total molecular weight decreases. Polarity of uncharged molecules is associated with electrically dissymmetric bonds.

Increased chain branching in alkanes leads to isomers of lower boiling points and higher melting points. Octane has bp 125.7^0 and mp -57^0 whereas its isomer, 2,2,3,3-tetramethylbutane, has bp 106.5^0 and mp 101^0.

The so-called saturated character of alkanes arises from their resistance at ordinary temperatures to strong acids, such as sulfuric acid, H_2SO_4; oxidizing agents, such as bromine, oxygen, or potassium permanganate, $KMnO_4$; or reducing agents, such as hydrogen in the presence of platinum, palladium, or nickel. Alkanes are oxidized by oxygen at elevated temperatures, and this kind of reaction in the form of combustion to carbon dioxide and water is the major source of energy for the highly industrialized nations. Petroleum consists of a wide variety of alkanes and closely related compounds.

Heats of organic reactions at constant pressure, such as combustion of alkanes, can be calculated for vapor-phase reactants from bond-energy data. The heat change for a process at constant pressure is called the enthalpy change, or $\Delta \underline{H}$, and is positive if heat is absorbed (endothermic reaction) and negative if heat is evolved (exothermic reaction).

Bond energies represent either the heat evolved when the bond is formed or the heat _absorbed_ when the bond is broken (for the vapor state). Thermal dissociation of covalent bonds in the gas phase produces atoms or radicals, not ions. For molecules with more than one bond, the bond energies given in Table 4-3 are average values for complete dissociation to the gaseous atoms and should not be taken as accurate values for the dissociation of any one particular bond. To illustrate, the average C—H bond energy for methane is 98.7 kcal but ΔH^0 for the breaking of just one C—H bond per molecule, $CH_4 \longrightarrow CH_3 \cdot + H \cdot$, is 104 kcal. Bond energies for the dissociation of _one_ bond in a polyatomic molecule are given in Table 4-6.

The relationship between the energies of reactants, products, and the transition state by which the reactants and products are interconverted, are useful and can be visualized from diagrams in which energy is shown as a function of some suitable reaction coordinate. The reaction coordinate is chosen so as to provide a measure of how far the reaction has proceeded (Figure 4-4). The difference in energy between the average energy of the reactant molecules and the transition state is called the _activation energy_. In the absence of an external stimulus, only those reactant molecules will react that have the requisite excess energy above the average.

In an equilibrium reaction, products are favored over reactants at equilibrium when the equilibrium constant, $\underline{K}_{eq}$, is favorable (i.e., when $\underline{K}_{eq} \stackrel{\sim}{>} 1$). The position of equilibrium in a reaction is directly related to $\Delta \underline{G}^0$, the Gibbs standard free energy change, which is defined by

$$\Delta \underline{G}^0 = -2.303 \, \underline{RT} \log \underline{K}_{eq}$$

in which $\underline{R}$ is the gas constant (1.987 cal deg^{-1} mole^{-1}), $\underline{T}$ is the absolute temperature, and $\underline{G}^0$ is in cal mole^{-1}.

The enthalpy change, $\Delta \underline{H}^0$, and $\Delta \underline{G}^0$ differ by $\underline{T} \Delta \underline{S}^0$, in which ΔS^0 is the entropy change and has the units e.u. = cal deg^{-1}:

$$\Delta \underline{G}^0 = \Delta \underline{H}^0 - \underline{T} \Delta \underline{S}^0 .$$

Entropy is a measure of the disorder, or randomness, of a chemical system. The entropy change, $\Delta \underline{S}^0$, for a reaction can be understood as a measure of the difference in randomness, or disorder, between the reactants and the products. A long hydrocarbon chain can be disordered as the result of the variety of ways the chain can be oriented by rotation around the C—C bonds. If it is made into a ring by fastening the ends together, it has to become less disordered, thus ring formation will have a _negative_ $\Delta \underline{S}^0$ value. Large negative entropy changes are associated with conversion of a gas to a liquid or a gas to a solid. Smaller negative entropy changes result on conversion of a liquid to a solid.

If $T \Delta S^o$ is small, ΔH^0 will be a reasonable measure of ΔG^0 and K_{eq}. For simple reactions involving gases, if ΔH^0 (which can be calculated from bond energies) is more positive than 15 kcal, it is likely that $K_{eq} < 1$. In contrast, if ΔH^0 is more negative than -15 kcal, then it is likely that $K_{eq} > 1$.

Handbooks contain tables of ΔG^0 values of formation of many simple organic compounds from the elements (usually called ΔF_f^0), which often can be used to calculate precise equilibrium constants. If ΔG^0 (consequently K_{eq}) is unfavorable, a satisfactory conversion may be obtained by selectively removing one or more of the reaction products. Otherwise, alternative routes for going from reactants to products must be sought.

The details of the way in which reactants become transformed to products is the <u>reaction mechanism</u>. Generally, those reaction mechanisms involving simultaneous or concerted making and breaking of more than two bonds are uncommon. The mechanisms of most reactions, wherein the overall change results in several bonds being broken and made, usually involve several simpler steps. One of these steps is the slowest and is called the <u>rate-determining step</u>.

Irradiation of a gaseous mixture of methane and chlorine with violet or ultraviolet light can cause a rapid or even explosive reaction to give chloromethane and hydrogen chloride. The light causes dissociation of chlorine molecules to chlorine atoms:

$$Cl_2 \longrightarrow 2Cl\cdot$$

A radical-chain reaction then ensues, with the steps $CH_4 + Cl\cdot \longrightarrow CH_3\cdot + HCl$ and $CH_3\cdot + Cl_2 \longrightarrow CH_3Cl + Cl\cdot$; the sequence consumes $Cl\cdot$ in the first step but regenerates $Cl\cdot$ in the second step. This cycle continues until the reagents are consumed or else atom or radical combinations produce species incapable of carrying the chain. Chain-termination reactions destroy radicals or atoms and include, for methane chlorination,

$$CH_3\cdot + Cl\cdot \longrightarrow CH_3Cl \qquad 2CH_3\cdot \longrightarrow CH_3CH_3$$

and

$$Cl\cdot + Cl\cdot \longrightarrow Cl_2$$

Chain-termination reactions are exceedingly fast and usually occur whenever the atoms or radicals encounter one another. This is <u>diffusion control</u>, which means that the limit of the rate is how fast the atoms or radicals can reach each other through the molecules surrounding them. For diffusion-controlled processes the activation energy is effectively zero.

Reactions such as $CH_4 + Cl\cdot \longrightarrow CH_3\cdot + HCl$ have nonzero activation energies, and their activation energies generally will be smaller the weaker the bond that is being broken and the stronger the bond that is being made.

The general case of a hydrocarbon RH reacting with a reagent XY by a radical-chain mechanism may be expressed as follows.

Initiation: Light, heat, or peroxides generate X· or R· .
                 The explicit reactions vary from case to case.

Propagation: R:H + ·X → R· + H:X $\Delta \underline{H}^o_1$

                R· + Y:X → R:Y + ·X $\Delta \underline{H}^o_2$

Termination: 2R· → RR   2X· → X$_2$ R· + ·X → RX

Noteworthy features of a radical-chain process are: (1) the <u>net</u> reaction RH + XY → RY + HX is determined by the propagation steps and <u>not</u> by initiation or termination steps; thus $\Delta \underline{H}^o_{net} = \Delta \underline{H}^o_1 + \Delta \underline{H}^o_2$; (2) in any radical-chain reaction, the atom or radical consumed in one propagation step <u>must</u> be regenerated in another propagation step; (3) propagation steps always involve reactions between one of the reagents (RH or XY) and a very reactive species (X· or R·); (4) the reactive species (X· or R·) must be at very low concentrations to minimize termination and maximize propagation; (5) if any one of the propagation steps is energetically unfavorable ($\Delta \underline{H}^o_1$ or $\Delta \underline{H}^o_2$ positive), that step will be slow and an efficient chain reaction will not be sustained.

The types of halogen-containing reagents XY considered in this chapter are the halogens, F_2, Cl_2, Br_2, and I_2, and reagents with weak bonds to halogens such as SO_2Cl_2, $(CH_3)_3COCl$, $BrCCl_3$, and <u>N</u>-bromosuccinimide.

Practical halogenation of alkanes are carried out best with chlorine. Fluorine is very difficult to control, bromine reacts slowly, and iodine is unreactive. Polychlorination is easily achieved; in fact, if monosubstitution is desired, it is advantageous to use an excess of hydrocarbon. Alkanes such as 2-methylbutane give mixtures of isomers on chlorination, although the product distribution favors those isomers that correspond to Cl· attacking the <u>weakest</u> C—H bonds. The rates of attack of atoms and radicals at hydrogen in C—H bonds decrease in the order tertiary C—H > secondary C—H > primary C—H. Greater selectivity in substitution of alkanes is possible with bromine or with chlorine in solution in the presence of arenes. Sulfuryl chloride (SO_2Cl_2) is an effective laboratory reagent for use in the chlorination of alkanes by a radical-chain mechanism.

Alkanes react with nitric acid (HNO_3) in the vapor phase at high temperatures ($>400°$) by a radical-chain mechanism to form nitroalkanes, $R—NO_2$. The reaction is complex, and carbon-carbon bond breaking occurs along with simple substitution.

ANSWERS TO EXERCISES

4-1

Tetradecane 252°C; heptadecane 299°C; 2-methylhexane 93°C; 2,2-dimethyl-pentane 80°C.

4-2

Part	a	b	c	d	e
Lower bp	$CH_3(CH_2)_2CH_3$	CH_3OCH_3	$(CH_3)_3CQH$	$\overset{\overset{O}{\|\|}}{HCOCH_3}$	$\overset{\overset{O}{\|\|}}{C_7H_{15}COH}$
Higher water-solubility	$H_2N(CH_2)_2NH_2$	CH_3CH_2OH	$(CH_3)_3COH$	$\overset{\overset{O}{\|\|}}{CH_3COH}$	$\overset{\overset{O}{\|\|}}{C_7H_{15}COH}$

4-3

$$C_{10}H_{22}(g) + \frac{31}{2}O_2(g) \rightarrow 10CO_2(g) + 11H_2O(g) \qquad \Delta \underline{H}^\circ = 1516.3 \text{ kcal}$$

4-4   Combustion of 1000g methane(g) gives 12,000 kcal; combustion of 1000g of decane(l) gives 11,400 kcal.

4-5

a.   -487.5 kcal   b.   -126.6 kcal   c.   -46.5 kcal

4-6   +171.0 kcal

4-7   221.2 - 119.9 = 101.3 kcal

4-8

a. $CH_4 + Br\cdot \rightarrow CH_3\cdot + HBr$

     (x)   (-87.4) = (+17)

     x = 104.4 kcal

b. $CH_4 + 2O_2 \rightarrow CO_2 + 2H_2O$

     4x + 2(118.9) + 2(-192) + 4(-110.6) = -192

     x = 99 kcal

c.   Bond energies in parts (a) and (b) should not be the same in theory or practice. That in part (b) is an average value of C—H bond energies of four C—H bonds; that in part (a) is the bond dissociation energy of the first bond.

4-9   The reaction is virtually complete, and the concentrations of participants at equilibrium correspond to pressures of one atmosphere each for HCl and CH_3Cl and zero pressure for CH_4 and Cl_2.

4-10

a. $\Delta \underline{H}^\circ = -19.4$ kcal; $\underline{K} = 5.05 \times 10^9$; $\underline{K} = 1.7 \times 10^{14}$ if $\Delta \underline{S}^\circ = 0$.

b.   When $\Delta \underline{H}^\circ = -15$ kcal, then $\underline{K}$ becomes less than unity when $\Delta \underline{H}^\circ - T\Delta \underline{S}^\circ < 0$ or when $\Delta \underline{S}^\circ < \Delta \underline{H}^\circ / T < -50$ eu.

4-11   From bond energies $\Delta \underline{H}^\circ = +54$ kcal; from $\underline{K}$ and $\Delta \underline{H}^\circ$, the calculated value of $\Delta \underline{S}^\circ = -198$ eu for the process $C_9H_{20}(l) \rightarrow 9C(s) + 10H_2O(g)$. Spontaneous decomposition of nonane does not occur because energy $(\Delta \underline{H}^*)$ in excess of 54 kcal must be supplied to start the process going. There is no easy mechanism for decomposition to occur without dissociating individually strong bonds.

4-12 $CH_4 + Cl_2 \rightarrow CH_3{}^\bullet + HCl + Cl^\bullet$ $\Delta \underline{H}_1^\circ = +59$ kcal

        $CH_3{}^\bullet + Cl^\bullet \rightarrow CH_3Cl$ $\Delta \underline{H}_2^\circ = -84$ kcal

The suggested mechanism is <u>not</u> a chain reaction. The first step is unlikely to occur (the $\Delta \underline{H}^\circ$ is large and positive) even though the second step is very favorable. The corresponding reaction with fluorine is more likely because the first step is less endothermic.

        $CH_4 + F_2 \rightarrow CH_3{}^\bullet + HF + F^\bullet$ $\Delta \underline{H}^\circ = +5.6$ kcal

4-13 $CH_4 + Cl^\bullet \rightarrow CH_3Cl + H^\bullet$ $\Delta \underline{H}_1^\circ = +20$ kcal

        $Cl_2 + H^\bullet \rightarrow HCl + Cl^\bullet$ $\Delta \underline{H}_2^\circ = -45$ kcal

The first step is unfavorable and is unlikely to occur at a practical rate. Alternative propagation steps are $CH_4 + Cl^\bullet \rightarrow CH_3{}^\bullet + HCl$ $(\Delta \underline{H}^\circ = +1$ kcal) and $CH_3{}^\bullet + Cl_2 \rightarrow CH_3Cl + Cl^\bullet$ $(\Delta \underline{H}^\circ = -26$ kcal) are more probable because neither step is strongly endothermic.

4-14   The possible monochlorination products are 2-chloro-1,1-dimethylcyclo-propane and 1-chloromethyl-1-methylcyclopropane. Because the slow step in radical-chain chlorination is the reaction $-\overset{|}{\underset{|}{C}}-H + Cl\cdot \rightarrow -\overset{|}{\underset{|}{C}}\cdot + HCl$ the product composition is determined by which C—H bond is broken most rapidly (is the weakest). From Table 4-6, methyl C—H bonds (as in ethane) are weaker than cyclopropane C—H bonds. Hence the preferred product is expected to be:

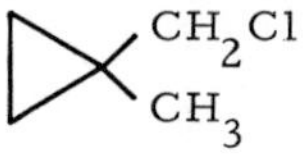

1-chloromethyl-1-methylcyclopropane

4-15

a.   The methyl C—H bond strengths are the same because the product ratio $4:5$ equals the ratio of the number of methyl C—H bonds (2 : 1).

b.   Relative rates of attack of Cl· on individual primary, secondary, and tertiary C—H bonds are:

$$\frac{\text{percent product}}{\text{number of C—H bonds giving product}} = \frac{45}{9} \text{ (prim)} : \frac{33}{2} \text{ (sec)} : \frac{22}{1} \text{ (tert)}$$

$$= 1 : 3.3 : 4.4$$

As a check, the percent 4 formed can be calculated as follows:

$$\text{Percent } 4 = \frac{\text{(Rate of attack at 6 equivalent methyl hydrogens)}}{\text{Overall rate of attack at all positions}} \; 100$$

$$= \frac{6 \times 1 \times 100}{(6\times1) + (3\times1) + (2\times3.3) + (1\times4.4)} = 30\%$$

c.   Ratio of products = Relative rates of formation

$$CH_3CH_2CH_3 \xrightarrow{\;Cl_2, h\nu\;} CH_3CH(Cl)CH_3 + CH_3CH_2CH_2Cl$$

$$= 2 \times 3.3 : 6 \times 1 = 1.1 : 1$$

$$(CH_3)_3CH \xrightarrow{\;Cl_2, h\nu\;} CH_3\overset{CH_3}{\underset{|}{C}}HCH_2Cl + (CH_3)_3C-Cl$$

$$2 \quad : \quad 1$$

4-15 (cont.)

$$(CH_3)_3CCH_2CH_3 \xrightarrow{Cl_2, h\nu} ClCH_2\underset{\underset{CH_3}{|}}{\overset{\overset{CH_3}{|}}{C}}-CH_2CH_3 + CH_3\underset{\underset{CH_3}{|}}{\overset{\overset{CH_3}{|}}{C}}-CH_2CH_2Cl + CH_3\underset{\underset{H_3C}{|}\ \underset{Cl}{|}}{\overset{\overset{CH_3}{|}}{C}}-CH-CH_3$$

$$3 \quad : \quad 1 \quad : \quad 2 \cdot 2$$

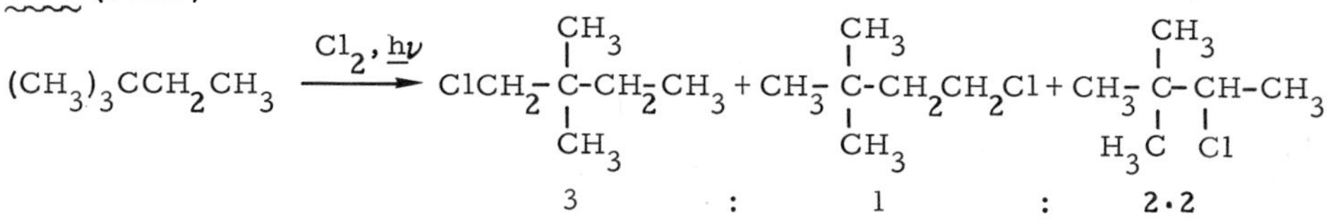

4-16

a. $Br_2 \xrightarrow{h\nu} 2Br\cdot$ initiation

$RH + Br\cdot \rightarrow R\cdot + HBr$ ⎫
$\qquad\qquad\qquad\qquad\qquad\qquad$ ⎬ propagation
$R\cdot + Br_2 \rightarrow RBr + Br\cdot$ ⎭

$2Br\cdot \rightarrow Br_2$ ⎫
$\qquad\qquad$ ⎪
$2R\cdot \rightarrow RR$ ⎬ termination
$\qquad\qquad$ ⎪
$R\cdot + Br\cdot \rightarrow RBr$ ⎭

Atoms and radicals are present at very low concentrations. Therefore, the rates of atom-atom or radical-radical chain-termination reactions are slow and the products of such reactions are formed in trace amounts only.

b.

$$(CH_3)_2CHCH_2CH_3 + Br\cdot$$

$$\xrightarrow[\underset{kcal}{\underline{\Delta H}^{\circ} = +4\cdot 6}]{-HBr} (CH_3)_2\dot{C}CH_2CH_3 \xrightarrow[\underset{kcal}{\underline{\Delta H}^{\circ} = -16\cdot 6}]{Br_2} (CH_3)_2\overset{\overset{Br}{|}}{C}CH_2CH_3$$

$$\xrightarrow[\underset{kcal}{\underline{\Delta H}^{\circ} = +7\cdot 6}]{-HBr} (CH_3)_2CH\dot{C}HCH_3 \xrightarrow[\underset{kcal}{\underline{\Delta H}^{\circ} = -21\cdot 6}]{Br_2} (CH_3)_2CH\overset{\overset{Br}{|}}{C}HCH_3$$

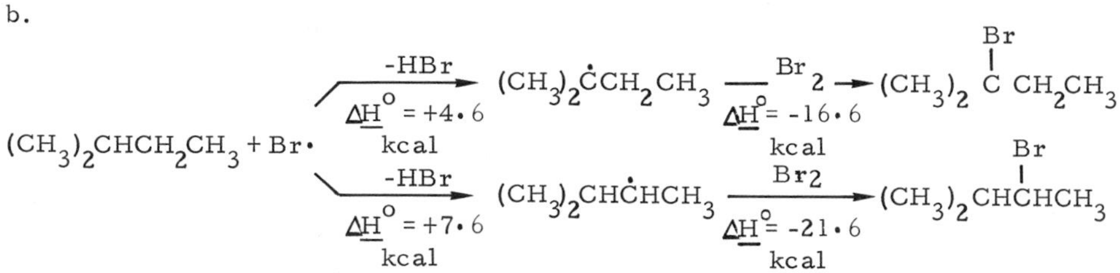

The slow step is the hydrogen abstraction step, $C-H + Br\cdot \rightarrow C\cdot + H-Br$.

c.   Rate ratio = Product ratio = $93 \cdot 5 : 6 \cdot 3 = x : 2y$ where $x : y$ is the relative rate of attack of $Br\cdot$ at tertiary _versus_ secondary $C-H$. Hence $x : y = 2 \times 93 \cdot 5 : 6 \cdot 3 = 30 : 1$.

The observed relative rate is qualitatively the same as predicted from the $\underline{\Delta H}^{\circ}$ data. The major product expected and found is 2-bromo-2-methylbutane.

4-17

$$\text{ROOR} \xrightarrow{\text{heat}} 2RO\cdot$$

$$ArCH_3 + RO\cdot \rightarrow ArCH_2\cdot + ROH$$

} initiation

$$ArCH_2\cdot + BrCCl_3 \rightarrow ArCH_2Br + \cdot CCl_3$$

$$ArCH_3 + \cdot CCl_3 \rightarrow ArCH_2\cdot + HCCl_3$$

} propagation

$$2Cl_3C\cdot \rightarrow Cl_3CCCl_3$$

$$ArCH_2\cdot + \cdot CCl_3 \rightarrow ArCH_2CCl_3$$

$$2ArCH_2\cdot \rightarrow ArCH_2CH_2Ar$$

} termination

The overall $\Delta \underline{H}^{\circ}$ is the sum of the $\Delta \underline{H}^{\circ}$'s for the two propagation steps. To obtain $\Delta \underline{H}^{\circ}$ we need to know the C—Br bond energy of $ArCH_2Br$. This value is not included in Table 4-6 but can be estimated by assuming that the difference in C—X bond energies for CH_3—X and $ArCH_2$—X is independent of X. Hence if $(CH_3Cl - ArCH_2Cl) = (84 - 69) = 15$ kcal, then $(CH_3Br - ArCH_2Br)$ also is 15 kcal - from which we derive that C—Br in $ArCH_2Br$ is 70-15 = 55 kcal. The overall $\Delta \underline{H}^{\circ}$ of bromination of methylbenzene is then -12 kcal mole^{-1}.

Bromotrichloromethane is a highly selective brominating agent. The propagation step for bromine abstraction is favorable only when the strength of the C—Br bond to be made is 54 kcal or more. The hydrogen abstraction step requires that the C—H bond to be broken is weaker than 96 kcal.

4-18

$$\text{ROOR} \rightarrow 2RO\cdot$$

$$ArCH_3 + RO\cdot \rightarrow ArCH_2\cdot + ROH$$

or

$$(CH_3)_3COCl \xrightarrow{h\nu} (CH_3)_3CO\cdot + Cl\cdot$$

} initiation

$$ArCH_3 + (CH_3)_3CO\cdot \rightarrow ArCH_2\cdot + (CH_3)_3COH \qquad \Delta \underline{H}^{\circ} = -17 \text{ kcal}$$

$$ArCH_2\cdot + (CH_3)_3COCl \rightarrow ArCH_2Cl + (CH_3)_3CO\cdot \qquad \Delta \underline{H}^{\circ} = -8 \text{ kcal}$$

} propagation

4-18 (cont.)

The net reaction is exothermic and therefore energetically feasible. Each pro-

pagation step is exothermic, and reaction should therefore be kinetically feasible.

4-19

a. $RH + Br \cdot \rightarrow R \cdot + HBr$

$\underset{/}{\overset{\backslash}{\,}} N{-}Br + HBr \rightarrow \underset{/}{\overset{\backslash}{\,}} N{-}H + Br_2$

$R \cdot + Br_2 \rightarrow RBr + Br \cdot$

b. $(CH_3)_2 CBrCH_2 CH_3$

4-20

a.   -98.5 kcal per mole of CH_4 d.   -7 kcal

b.   -340 kcal   e.   +98.8 kcal

c.   -10.6 kcal

4-21

a. ΔH^o for reaction 4-20e to give C(s) would be 98.8-171 = -72 kcal.

b.   The ΔH values for hydrogenation and bromination of ethane indicate that

the C—C bond in ethane is not completely saturated--at least in the sense that

K for these reactions is likely to be greater than unity.

4-22 ΔH^o C—F = 115 kcal, using C—H, F_2 and HF bond energies in

Table 4-3.

4-23

$C_6H_6(g) + 15/2\ O_2(g) \rightarrow 6CO_2(g) + 3H_2O(g) \qquad \Delta H^o = -757.6\ kcal$

From bond energies, $\Delta H^o = -798.4$ kcal. Thus combustion of benzene evolves

40 kcal less heat than calculated which means that benzene is more stable by

40 kcal than predicted from average bond energies.

4-24   For the ΔH^o values to be independent of which set of bond energies we use, the strengths of the C—Br bonds must differ with structure to the same extent as the C—H bonds. Since $=\overset{|}{C}-H$ and $-\overset{|}{\underset{|}{C}}-H$ differ by 104-98 = 6 kcal, the $=\overset{|}{\underset{|}{C}}-Br$ and $-\overset{|}{\underset{|}{C}}-Br$ bonds will differ similarly. If we take $-\overset{|}{\underset{|}{C}}-Br$ as 68 kcal (Table 4-3), then $=\overset{|}{C}-Br$ will be 74 kcal.

4-25   The entropy ΔS^o decreases progressively (becomes a large, negative number) from methane to nonane formation.

4-26   The formation of cyclohexane from 1-hexene is accompanied by a much larger decrease in entropy than is the formation of hydrogen chloride.

4-27   The large decrease in entropy in the formation of chloroethane from ethene and hydrogen chloride must result largely from constraining two molecules to react to give one molecule--thereby losing degrees of freedom.

4-28

$$Br_2 \longrightarrow 2Br\cdot \qquad \Delta H^o = 46.4 \text{ kcal} \qquad \text{(feasible to occur with light)}$$

$$CH_3{:}H + Br\cdot \rightarrow CH_3\cdot + HBr \qquad \Delta H^o = +16.6 \text{ kcal} \qquad \text{(much more positive than } Cl_2)$$

$$CH_3\cdot + Br_2 \rightarrow CH_3Br + Br\cdot \qquad \Delta H^o = -23.6 \text{ kcal} \qquad \text{(comparable to } Cl_2)$$

$$CH_3{:}H + Br_2 \rightarrow CH_3Br + HBr \qquad \Delta H^o = -7 \quad \text{kcal} \qquad (\underline{K} \text{ probably} > 1)$$

The alternate chain sequence has an even worse balance of ΔH^o values for each step.

$$CH_3{:}H + Br\cdot \rightarrow CH_3Br + H\cdot \qquad \Delta H^o = +34.7 \text{ kcal}$$

$$Br_2 + H\cdot \rightarrow HBr + Br\cdot \qquad \Delta H^o = -41.9 \text{ kcal}$$

For iodine, substitution is not favorable because here $\underline{K}$ is likely to be < 1 and ΔH^o for one step is large and positive.

4-28 (cont.)

$$I_2 \longrightarrow 2I\cdot \qquad \Delta \underline{H}^{o} = +36.5 \text{ kcal} \quad \text{(very easy with light)}$$

$$CH_3\colon H + I\cdot \rightarrow CH_3\cdot + HI \qquad \Delta \underline{H}^{o} = +32.6 \text{ kcal}$$

$$CH_3\cdot + I_2 \rightarrow CH_3I + I\cdot \qquad \Delta \underline{H}^{o} = -19.5 \text{ kcal}$$

$$CH_3\colon H + I_2 \rightarrow CH_3I + HI \qquad \Delta \underline{H}^{o} = +13.1 \text{ kcal}$$

4-29 Average C—C bond energy for cyclopropane is calculated to be 73 kcal assuming the C—H bonds are normal (98.7 kcal). Notice that this is signfi-cantly lower than the normal value of 82.6 kcal.

4-30

$$Cl_2 \xrightarrow{h\nu} 2Cl\cdot \; ; \qquad CH_4 + Cl\cdot \rightarrow CH_3\cdot + HCl;$$

$$CH_3\cdot + Cl_2 \rightarrow CH_3Cl + Cl\cdot \rightarrow CH_2Cl\cdot + HCl;$$

$$CH_2Cl\cdot + Cl_2 \rightarrow CH_2Cl_2 + Cl\cdot \rightarrow CHCl_2\cdot + HCl;$$

$$CHCl_2\cdot + Cl_2 \rightarrow CHCl_3 + Cl\cdot \rightarrow CCl_3\cdot + HCl;$$

$$2CCl_3\cdot \longrightarrow CCl_3 - CCl_3$$

4-31

a. $(CH_3)_3CBr$
b. ⌬—CH_2Br
c.

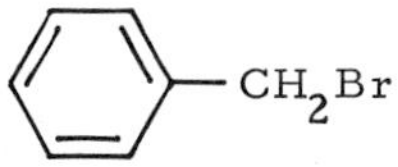

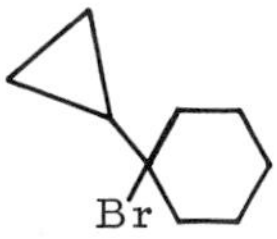

4-32

a. $CH_2 = CH-CH_2\cdot + HCl$
c. $CH_3S\cdot + CH_3CH_3$
b. $CCl_3\cdot + CH_3CH_2Br$
d. $2HO\cdot$

When the bond energy data is incomplete so that $\Delta \underline{H}^{o}$ of atom abstraction steps cannot be estimated, then the feasibility of a reaction cannot be predicted unam-biguously. In reaction c, an alternate pathway is $CH_3CH_2\cdot + CH_3SH \rightarrow$ $CH_3CH_3 + \cdot CH_2SH$, but we cannot evaluate it because the methyl C—H bond energy of CH_3SH is not given in Table 4-6. In fact, the S—H bond is weaker

4-32 (cont.)

than the methyl C—H. (For methanol, CH_3OH, the methyl C—H is weaker

than O—H, and hydrogen abstraction reactions occur at carbon--not oxygen).

4-33

a. $R^{\cdot} + O_2 \rightarrow ROO^{\cdot}$, $ROO^{\cdot} + RH \rightarrow ROOH + R^{\cdot}$

Antioxidants scavenge the chain-propagating radicals, $R^{\cdot}$ and $ROO^{\cdot}$, thus

terminating the chains.

b.

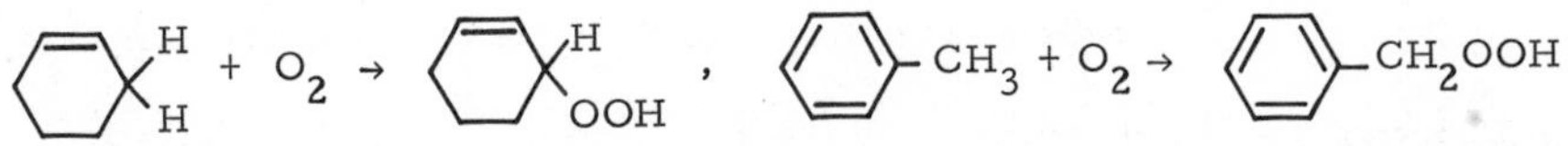

c.   Compounds with weak C—H bonds are susceptible to autoxidation pro-

moted by air and light. Table 4-6 shows that ethanal has an aldehyde C—H bond

of 86 kcal; abstraction of this hydrogen is therefore feasible, $R^{\cdot} + CH_3CH=O \rightarrow$

$RH + CH_3\overset{\cdot}{C}=O$, $\Delta H \sim$ 12 kcal. Once the $CH_3\overset{\cdot}{CO}$ radical is formed, it can add

oxygen in a radical-chain process, $CH_3\overset{\cdot}{CO} + O_2 \rightarrow CH_3\overset{O}{\overset{\|}{C}}-OO^{\cdot} \xrightarrow{\ CH_3CH=O\ }$

$CH_3-\overset{O}{\overset{\|}{C}}-OOH + CH_3\overset{\cdot}{CO}$. Accumulation of peroxidic products occurs in time,

creating a hazard. Ethers, like aldehydes, have weak C—H bonds on carbon

attached to oxygen.

4-34   The $C-SO_2^-Cl$ bonds can be formed by termination steps involving $R^{\cdot}$

and $^{\cdot}SO_2Cl$ radicals. This suggests a chain mechanism involving $^{\cdot}SO_2Cl$

radicals.

    Initiation $Cl-SO_2^-Cl \longrightarrow Cl-SO_2^{\cdot} + Cl^{\cdot}$

    Propagation $RH + Cl-SO_2^{\cdot} \longrightarrow R^{\cdot} + HCl + SO_2$

                  $R^{\cdot} + Cl-SO_2^-Cl \longrightarrow RCl + ^{\cdot}SO_2^-Cl$

    Termination $R^{\cdot} + ^{\cdot}SO_2^-Cl \longrightarrow R-SO_2^-Cl$

Another possibility is breakdown of the $^{\cdot}SO_2Cl$ radical to $SO_2 + ^{\cdot}Cl$ and a

subsequent set of propagation steps.

4-34 (cont.)

$$RH + Cl\cdot \longrightarrow R\cdot + HCl$$

$$R\cdot + SO_2 \longrightarrow RSO_2\cdot$$

$$RSO_2\cdot + ClSO_2Cl \rightarrow RSO_2Cl + \cdot SO_2Cl$$

4-35 The two possible hydrogen abstraction steps are:

$$RH + \cdot Cl \longrightarrow R\cdot + HCl$$

$$RH + Cl-SO_2\cdot \rightarrow R\cdot + HCl + SO_2$$

The two steps have different $\Delta \underline{H}^o$ values and occur at different rates. There-fore, when product mixtures are possible, attack by the two different radicals will not necessarily give the same product distributions.

4-36

Initiation $ROCl \xrightarrow{h\nu} RO\cdot + Cl\cdot$ ($R = \underline{tert}$ - butyl)

$+ RO\cdot \longrightarrow$ $\cdot + ROH$

Propagation $\cdot + BrCCl_3 \rightarrow$ $-Br + \cdot CCl_3$

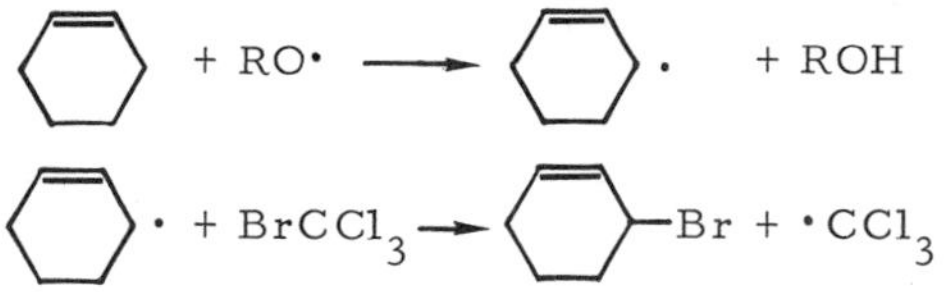

$$ROCl + \cdot CCl_3 \longrightarrow RO\cdot + CCl_4$$

If $BrCCl_3$ is not in excess, chlorination with $ROCl$ will compete successfully with bromination. $-\overset{|}{\underset{|}{\overset{\cdot}{C}}}-H + ROCl \longrightarrow \overset{H}{\underset{Cl}{C}} + RO\cdot$

4-37

a.   Initiation $\underset{Ar_3Sn-H}{ROOR \rightarrow 2RO\cdot} \longrightarrow Ar_3Sn\cdot + ROH$

$\xrightarrow{\underline{ArCH}_3} ArCH_2\cdot + ROH$

Propagation

(1) $Ar_3Sn\cdot + ArCH_2-Cl \rightarrow Ar_3Sn-Cl + ArCH_2\cdot$ $\Delta\underline{H}^o = -51$ kcal

(2) $ArCH_2\cdot + Ar_3Sn-H \longrightarrow ArCH_3 + Ar_3Sn\cdot$ $\Delta\underline{H}^o = -5$ kcal

net $ArCH_2-Cl + Ar_3Sn-H \longrightarrow ArCH_3 + Ar_3Sn-Cl$ $\Delta\underline{H}^o = -56$ kcal

<u>4-37</u> (cont.)

b.

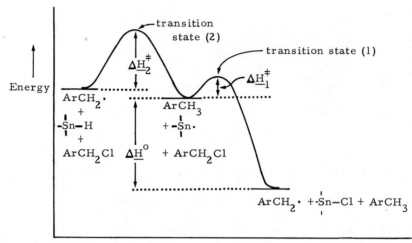

Of the two propagation steps, the most exothermic is likely to be the fastest step.

ADDITIONAL EXERCISES

<u>4A-1</u> If you had a mixture of one mole each of benzene, hexane, 3-methyl-pentane and bromine which you exposed to light, what monobromo compound would you find in the reaction mixture in largest amount? Name the product.

4A-2   Which product mixture, A, B, C or D would you expect to obtain on reaction of isopropylbenzene with $BrCCl_3$ in the presence of a peroxide, ROOR? Write initiation and propagation steps for this reaction.

A. $C_6H_5C(Cl)(CH_3)_2 + HCBrCl_2$ B. $C_6H_5CH(CH_3)CH_2Br + HCCl_3$

C. $C_6H_5C(Br)(CH_3)_2 + HCCl_3$ D. $C_6H_5C(CCl_3)(CH_3)_2 + HBr$

4A-3   Which compound A - F would you expect to react most rapidly with a chlorine atom? A. benzene, B. methane, C. chloromethane, D. dichloromethane, E. trichloromethane, F. trichloroethene.

4A-4   If the relative rate of chlorination at primary, secondary and tertiary C–H bonds is 1:3.3:4.4, calculate the percentage composition of monochloro compounds formed in the radical chlorination of 1,3-dimethylcyclopentane. Name each compound.

4A-5   Indicate which compound A - D would be most likely to function successfully as an inhibitor of a chain reaction by terminating the growing chain with an atom transfer reaction of the type $R\cdot + HX \longrightarrow RH + X\cdot$

A. HF,   B. H_2S,   C. H_2O,   D. NH_3

4A-6   A compound of formula C_6H_{10} adds <u>one</u> mole of bromine and subsequently undergoes a substitution reaction to give only one of three possible tribromo position isomers, $C_6H_9Br_3$. Draw the structure of the compound C_6H_{10} and write equations to show the formation of $C_6H_9Br_3$. Mechanism need not be shown.

4A-7   A compound Z of formula C_6H_{14} reacts with chlorine in sunlight to give <u>two</u> monochloro compounds, X and Y. The halogen is attached to a tertiary carbon in X and a primary carbon in Y. Draw structures for X, Y and Z, and name each by the IUPAC system.

4A-8   There has been concern in recent years that large fleet operations of supersonic aircraft (SST's) might contribute significant quantities of NO to the stratosphere. Nitric oxide in the stratosphere catalyzes the destruction of ozone by atomic oxygen, $O_3 + O \xrightarrow{NO} 2 O_2$. Ozone in the stratosphere protects us from harmful uv radiation, and the implications of depleting the ozone layer by whatever means are serious indeed.

    To understand what may be involved, first write Lewis structures for the odd-electron molecules, NO and NO_2. Second, write the propagation steps (two) of a radical chain process for the net reaction $O_3 + O \longrightarrow 2 O_2$ that is catalyzed by NO.

4A-9   Large quantities of chlorofluoromethanes are used world-wide as aerosol propellants and refrigerants. The two most widely used compounds are $CFCl_3$ and CF_2Cl_2. Both of these compounds are normally inert but there is much concern that, when they are expelled into the atmosphere, they will ultimately reach the stratosphere where, on exposure to strong ultraviolet radiation, they will dissociate as follows:

$$CFCl_3 \xrightarrow{h\nu} \dot{C}FCl_2 + Cl\cdot \quad , \quad CF_2Cl_2 \xrightarrow{h\nu} \dot{C}F_2Cl + Cl\cdot$$

The production of chlorine atoms could then catalyze the destruction of ozone, $O_3 + O \xrightarrow{Cl\cdot} 2 O_2$. Write the chain propagating steps for this reaction. (See previous exercise).

CHAPTER 5

STEREOISOMERISM OF ORGANIC MOLECULES

A summary of the kinds of isomers possible for organic compounds follows:

1. <u>Position isomers</u> have different arrangements of bonds. Examples are $CH_3CH_2CH_2CH_2Cl$ and $(CH_3)_2CHCH_2Cl$. Position isomers also are called <u>constitutional isomers</u>.

2. <u>Stereoisomers</u> have the same kind of bonds connecting the atoms, but differ in the way the bonds are oriented in space. That is, they have different <u>spatial</u> arrangements. Stereoisomers have widely different degrees of ability to maintain their particular spatial arrangements. Those having stable spatial arrangements under normal conditions are called <u>configurational isomers</u>.

A. <u>Geometric</u> or <u>cis-trans isomers</u> are configurational isomers of compounds with double bonds or rings. Examples are <u>cis</u>- and <u>trans</u>-1,2-dibromoethene and <u>cis</u>- and <u>trans</u>-1,2-dimethylcyclopropane.

| <u>cis</u> | <u>trans</u> | <u>cis</u> | <u>trans</u> |

1,2-dibromoethene   1,2-dimethylcyclopropane

B. <u>Chiral isomers.</u> A substance is <u>chiral</u> if its molecules are not identical with their mirror images; that is, they lack reflection symmetry. An achiral substance is one whose molecules are identical with their mirror images or, where conformational equilibria are involved, there is at least one conformation that is identical with its mirror image. Examples are:

<u>chiral</u>, $CHFClBr$, $CH_3CH_2CHClCH_3$

<u>achiral</u>, $CH_2{=}CH_2$, $CH_3CH_2CH_2CH_3$

Chiral molecules can be either <u>asymmetric</u> or <u>dissymmetric</u>. An asymmetric (NOT assymmetric!) molecule has no symmetry at all - it will

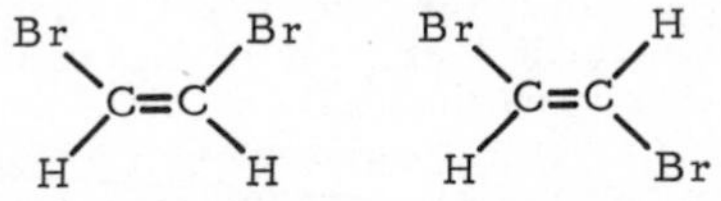

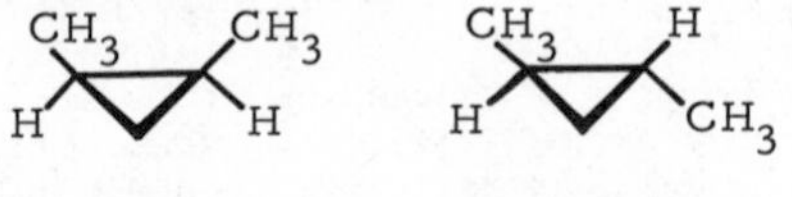

have a different appearance from each angle of view. A shoe, at least one that will fit only on a right (or left) foot is an asymmetric object. A dissymmetric molecule is chiral but will have the same appearance from more than one angle of view. A right- (or left-) handed helix (such as a coil spring or a headless screw) is an example of a dissymmetric object. It is not identical with its mirror image but looks the same when viewed from either end.

A chiral molecule and its mirror-image isomer constitute a pair of enantiomers. Samples corresponding to each member of a pair of enantiomers will, in principle, rotate the plane of polarized light in opposite directions; that is, the samples will be optically active. In practice, the optical rotations may be too small to measure. Whether or not the rotations are measurable, the enantiomers commonly are referred to as optical antipodes or as optical isomers. Although all enantiomers are optical isomers, not all optical isomers are enantiomers.

A mixture of equal amounts of each enantiomer will not be optically active because the rotation produced by one enantiomer will be cancelled by the opposite rotation produced by the other. Such a mixture is called a racemic mixture. A process that converts an enantiomer to a racemic mixture is called racemization and separation of one or both enantiomers from a racemic mixture is called resolution.

Optical isomers that rotate the plane of polarization of polarized light to the right are called d or (+) isomers, whereas those that rotate the plane to the left are called l or (-) isomers. The designations d and l should not be confused with D and L (see below), which are used to denote configurations, not rotations.

Projection formulas are used widely to represent particular configurations of chiral carbons. The horizontal bonds from a chiral carbon in a projection formula are understood to extend out of the plane of the paper toward you, while the vertical bonds extend behind the plane of the paper away from you. All of the following three formulas represent the same configuration:

$$\underset{CH_2CH_3}{\overset{CH_3}{H-\!\!\!\!\mid\!\!\!\!-OH}} \quad\equiv\quad \underset{CH_2CH_3}{\overset{CH_3}{H\!-\!C\!-\!OH}} \quad\equiv\quad \underset{H\quad OH}{\overset{CH_3}{C_{\textit{IIII}}C_2H_5}}$$

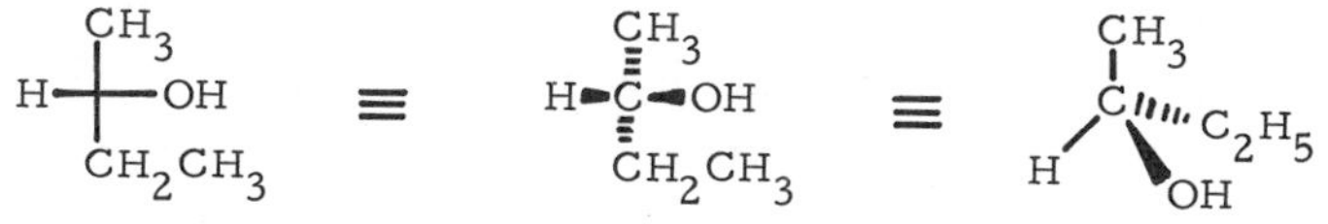

To establish the configuration of a chiral carbon by the D, L system, the molecule must be thought of as oriented with the bonds in the main chain vertical and the lowest numbered atom at the top. When properly oriented, the bonds between the chain atoms recede and those on the left and right of the chiral carbon project toward you, just as in the projection formulas. The L configuration is correct if the main substituent is on the left, whereas the $\underline{\underline{D}}$ configuration is correct with the main substituent on the right. The following examples are represented in Fischer projections:

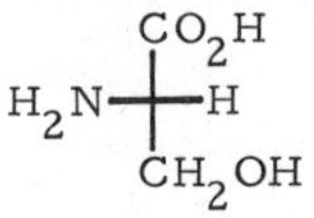

$$
\begin{array}{cc}
\text{CO}_2\text{H} & \text{CO}_2\text{H} \\
\text{H}_2\text{N}\!\!-\!\!\!\!\mid\!\!-\!\!\text{H} & \text{H}\!\!-\!\!\!\!\mid\!\!-\!\!\text{OH} \\
\text{CH}_2\text{OH} & \text{CH}_3
\end{array}
$$

$\underline{\underline{L}}$-serine $\underline{\underline{D}}$-lactic acid

Stereoisomers that differ only in configuration at chiral centers yet are not enantiomers (mirror images) are called <u>diastereomers</u>. Diastereomers have different physical and chemical properties. In general, we expect that there will be $2^{\underline{n}}$ stereoisomers for $\underline{n}$ chiral centers. Molecules having chiral centers, but which are actually achiral (identical with their mirror images) are called <u>meso</u> compounds. This situation arises when the chiral centers are similarly substituted, as in the tartaric acids. Molecules are achiral when their projection formulas indicate a plane of symmetry in the molecule, or if any rotational conformation of the molecule is identical with its mirror image (or has a plane or center of symmetry).

$$
\begin{array}{ccc}
\text{CO}_2\text{H} & \text{CO}_2\text{H} & \text{CO}_2\text{H} \\
\text{H}\!\!-\!\!\text{OH} & \text{HO}\!\!-\!\!\text{H} & \text{H}\!\!-\!\!\text{OH} \\
\text{HO}\!\!-\!\!\text{H} & \text{H}\!\!-\!\!\text{OH} & \text{H}\!\!-\!\!\text{OH} \\
\text{CO}_2\text{H} & \text{CO}_2\text{H} & \text{CO}_2\text{H}
\end{array}
$$

         enantiomers   meso (achiral)

              diastereomers

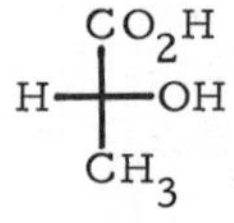

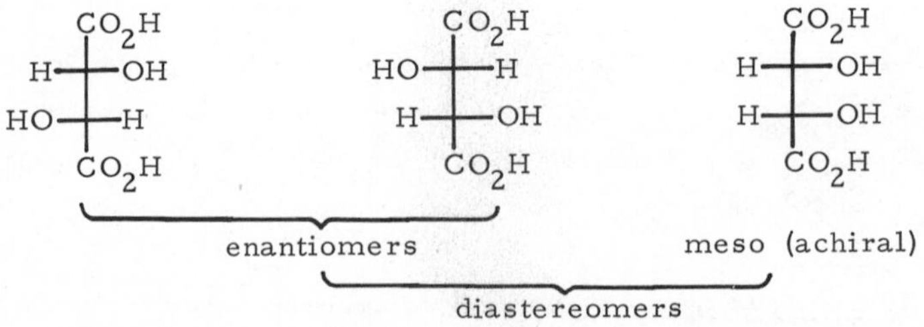

If <u>meso</u> forms are possible, then each <u>meso</u> form reduces the $2^{\underline{n}}$ possible stereoisomers by one. Thus, if there are four chiral carbons and two possible <u>meso</u> forms, the number of nonidentical stereoisomers will be $2^4 - 2 = 14$.

Stereoisomerism is of great importance in biology and, where there is a possibility of more than one configurational isomer, it generally is true that only one of the configurational isomers is biologically active.

C. <u>Conformational isomers</u> have different spatial arrangements of molecules resulting from rotation(s) around single bonds and normally are converted rapidly, one into the other, at room temperature. With few exceptions <u>staggered</u> conformations are more stable than <u>eclipsed</u> conformations. The eclipsed and staggered conformations of 1,2-dibromoethane obtained by

successive 60° changes in the torsional angle about the C—C bond can be used
as examples:

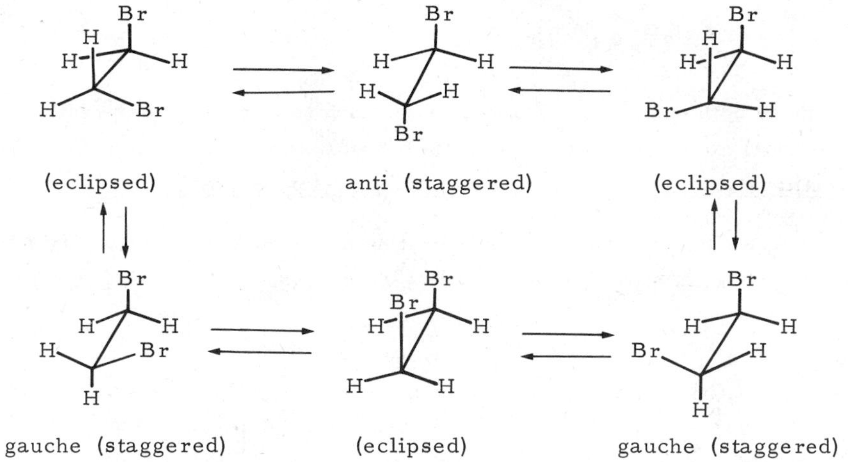

(eclipsed)   anti (staggered)   (eclipsed)

gauche (staggered)   (eclipsed)   gauche (staggered)

    The staggered gauche forms are dissymmetric; however, we do <u>not</u>
think of 1,2-dibromoethane as a chiral molecule because the gauche forms are
in rapid equilibrium with the achiral anti form (see above).

    It is important to recognize that the kinds of stereoisomers are not
mutually exclusive. Thus there will be conformational as well as geometric
isomers of 2-pentene, and conformational isomers as well as enantiomers of
2-butanol. In many situations, all three will be involved. Thus for 4-chloro-
2-pentene, there is a cis compound that exists as a pair of enantiomers, and a
trans compound that also exists as two enantiomers. Thus there are four
configurational isomers of 4-chloro-2-pentene and each of these can exist in
several possible conformations:

<u>cis</u> enantiomers

<u>trans</u> enantiomers

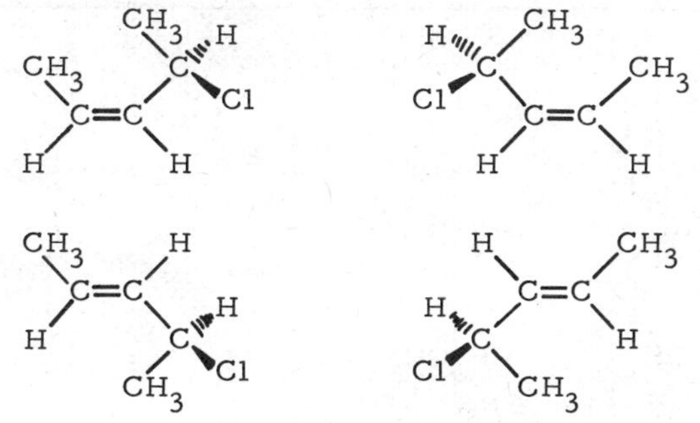

two of the possible
conformations of one
of the cis enantiomers

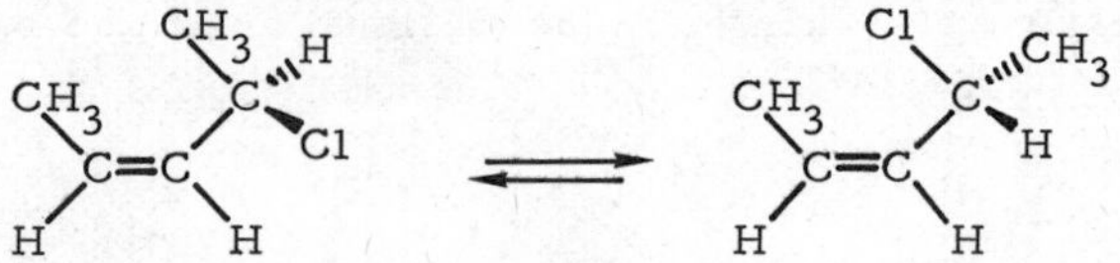

Where both configurational and conformational isomers are possible, the conformational isomers will be conformations of a particular configuration. Conformational changes do not alter the configuration of a molecule.

As an added example, the configuration of meso- 2,3-dibromobutane is shown below in three different staggered conformations.

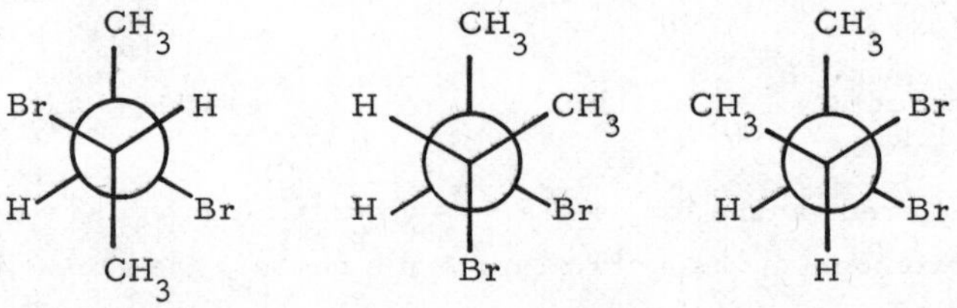

ANSWERS TO EXERCISES

5-1

a.

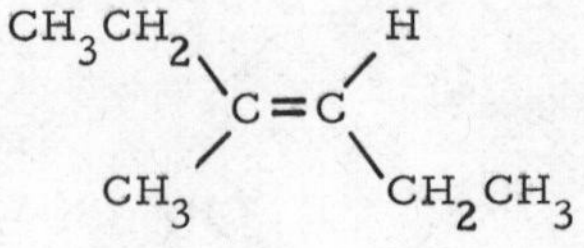

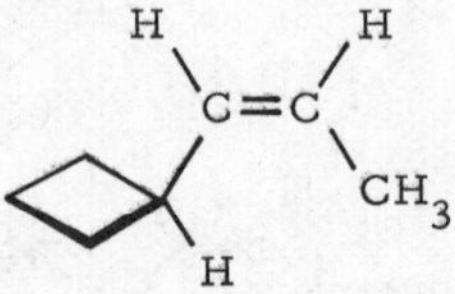

b.

c.

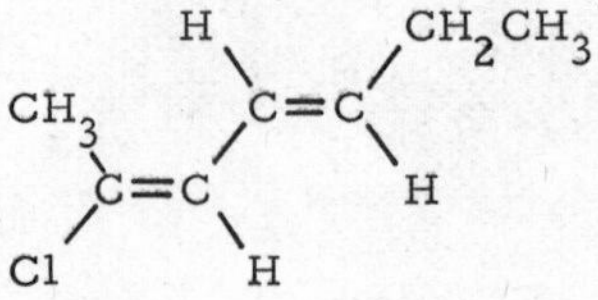

d.

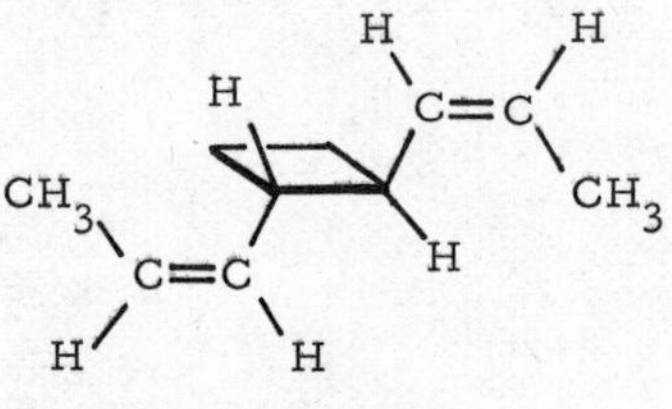

5-2   Structure C

5-3   CH$_3$$\overset{*}{C}$HBrCH = CH$_2$ d.

b.   achiral

c.   HOCH$_2$$\overset{*}{C}$H(NH$_2$)CO$_2$H   e.   achiral

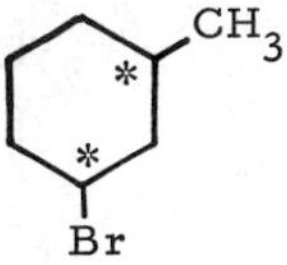

f.

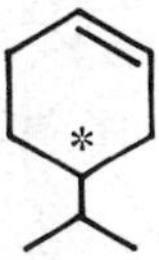

5-4   3, 8, 9, 10, 13, 14, 17, 20

5-5

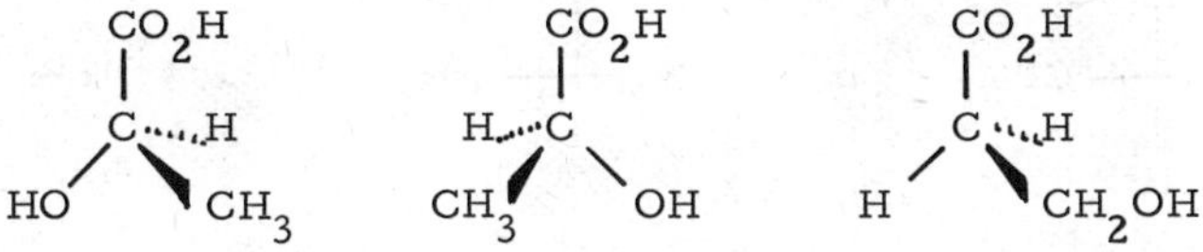

A(or B)   B(or A)   C

A and B are enantiomers; isomer C is a <u>position</u> isomer of A and B.

5-6   All trans; achiral.

5-7   a.

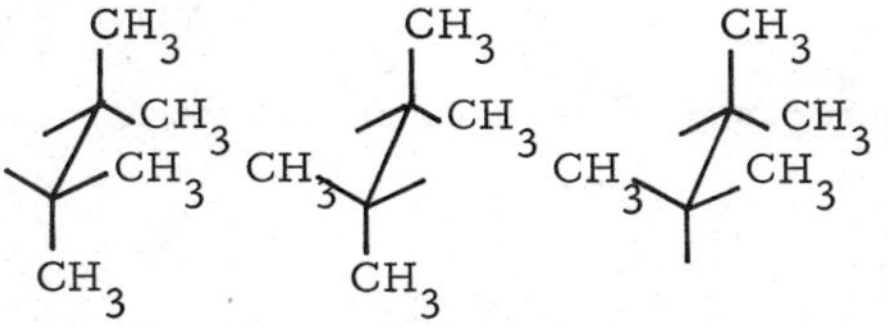

b.

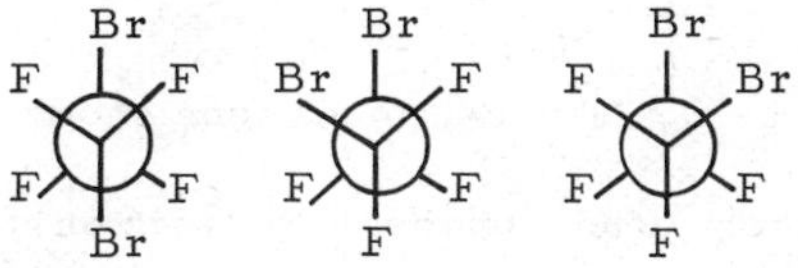

c.

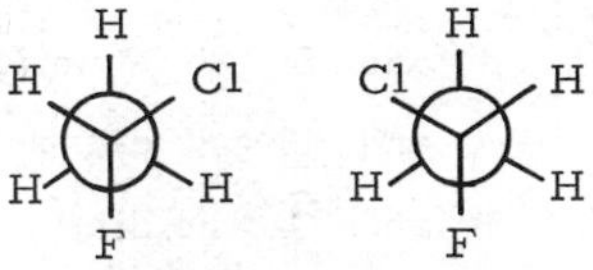

(There are three equivalent staggered conformations possible for each enantiomer.)

5-8

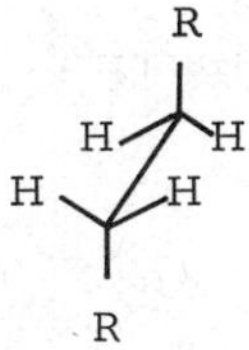

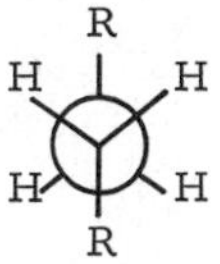

R = CH$_3$—$\overset{\underset{\displaystyle CH_3}{|}}{\overset{\displaystyle \overset{CH_3}{|}}{C}}$—

5-9

a.

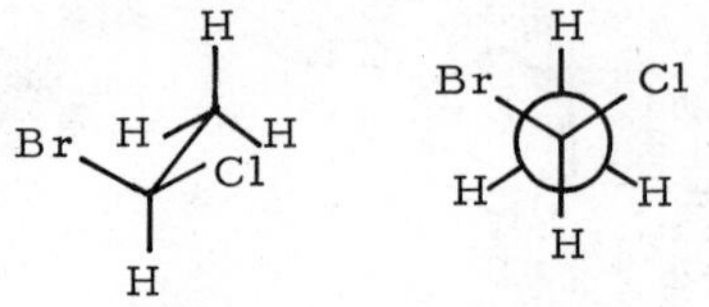

c.

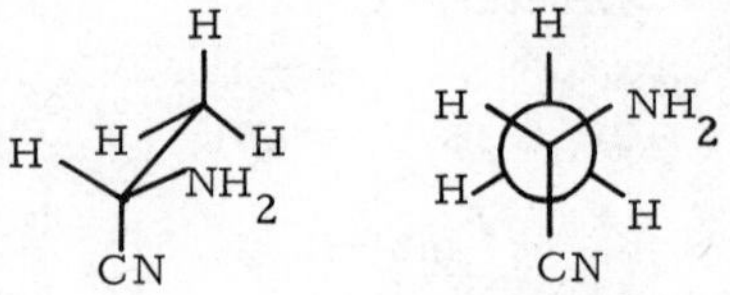

b.

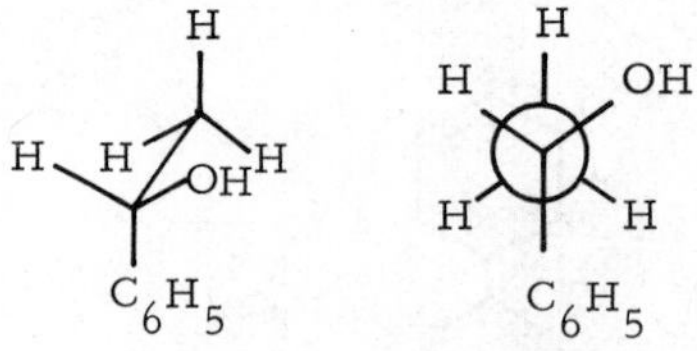

5-10

a.

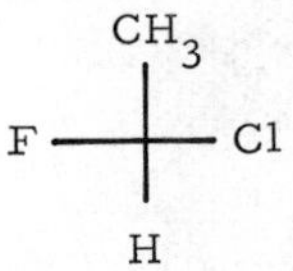

$$F \underset{H}{\overset{CH_3}{\underset{|}{\overline{}}}} Cl$$

b.

$$H \underset{CH_2CH_3}{\overset{Br}{\underset{|}{\overline{}}}} CH_3$$

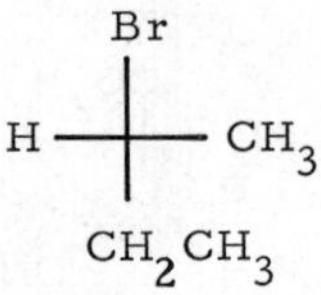

c.

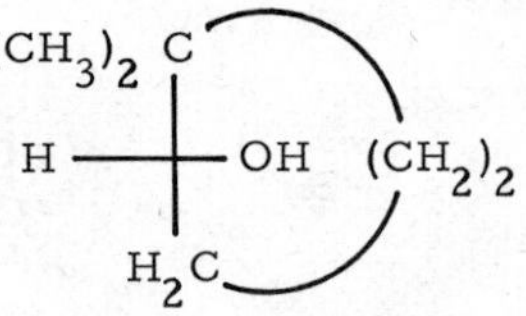

Part c. illustrates that projection formulas are not easily adapted to represent configurations of cyclic compounds without distorting the appearance of the ring.

5-11   Operations a, b, d and e change (invert) configuration; operation c. retains configuration.

5-12   The conformations shown are mirror images; they represent two conformations of meso-tartaric acid.

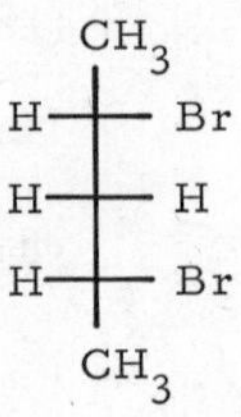

5-13

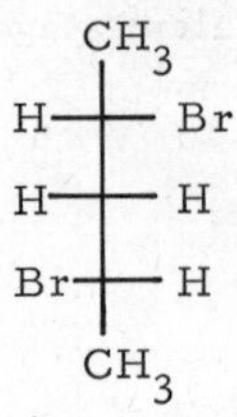

A and B are enantiomers;
A and C are diastereomers;
B and C are diastereomers;
equal amounts of A and B
would give a racemic mixture; C is achiral.

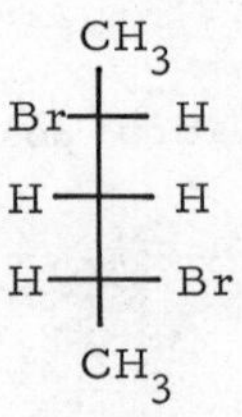

5-14

a. (structure: cyclopropane-like ring with H, H on top and Cl, Cl on bottom)

e.
```
      CH₃              CH₃
  H ──┼── Cl      H ──┼── Cl
  H ──┼── Cl      Cl──┼── H
  H ──┼── Cl      H ──┼── Cl
      CH₃              CH₃
```

f.
```
      CH₃              CH₃
  H ──┼── Cl      H ──┼── Cl
  H ──┼── Cl      Cl──┼── H
  H ──┼── Cl      Cl──┼── H
  H ──┼── Cl      H ──┼── Cl
      CH₃              CH₃
```

c. (ring structure with H, H, Cl, Cl)

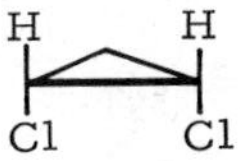

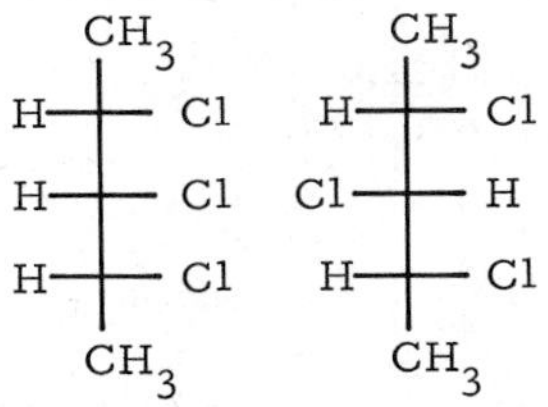

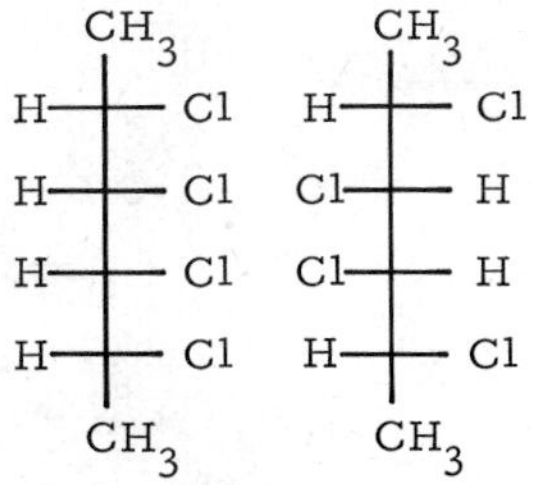

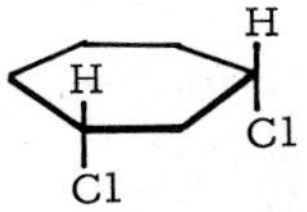

b and d do not have <u>meso</u> configurations.

5-15   a, c, d, e, and f have identical pairs; b has nonidentical <u>cis-trans</u>

isomers; g first and third structures are identical (<u>trans</u> configuration); middle

structure has <u>cis</u> configuration.

5-16   a.   positional   d.   configurational

b.   configurational   e.   conformational (notice they are

c.   conformational   interconverted by rotation about

the central C—C bond).

5-17   a and d only.

5-18

enantiomers	cis-trans isomers	cis enantiomers, trans-enantiomers
3-chloro-1-butyne	1-chloro-1,3-butadiene	4-chloro-2-pentene
	5-chloro-2-pentene	

5-19

a. (structure: C_6H_5, C_6H_5, H, H on C=C)

b. (structure: CH_3, H, Cl, CH_3 on C=C)

c. (structure: H, CH_3, H_5C_6, H on C=C)

d. (structure: CH_3 ... CH_3)

e. (structure: CH_3 ... CH_3)

f. (structure: CH_3 ... CH_3)

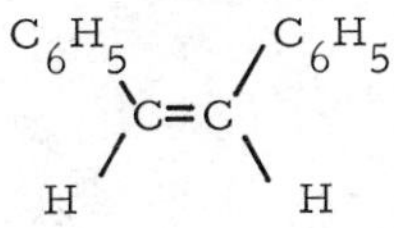

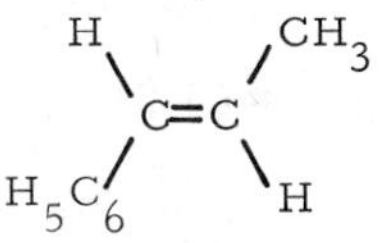

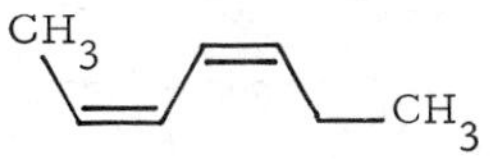

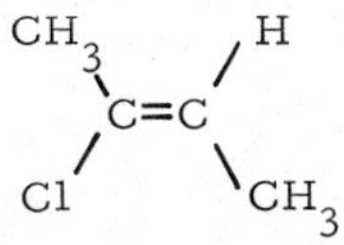

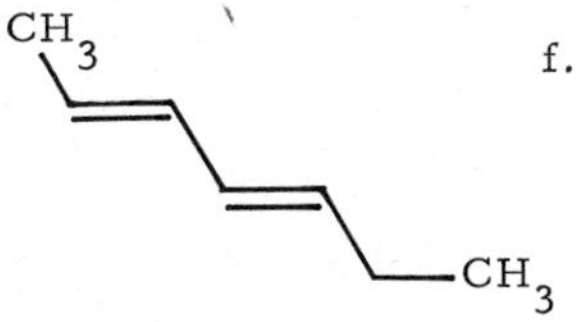

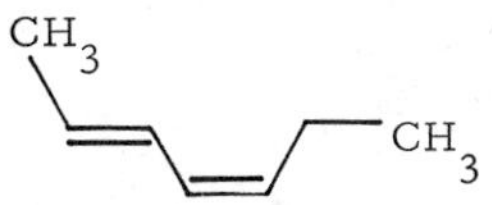

5-19 (cont'd)

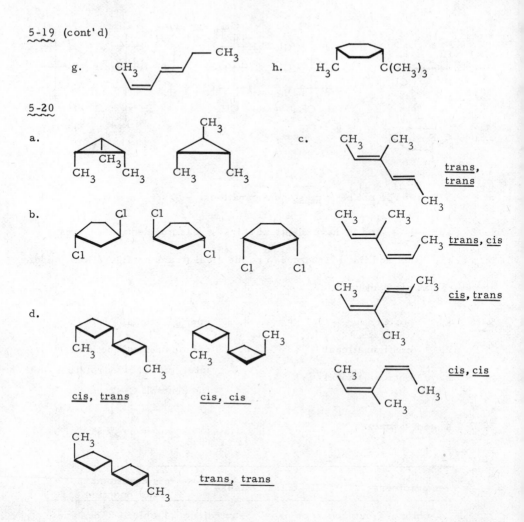

g.

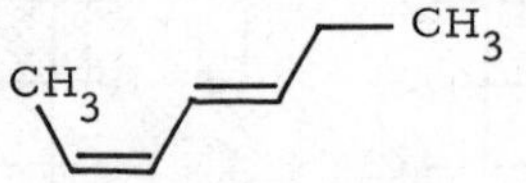

h.

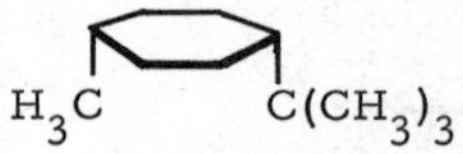

5-20

a.

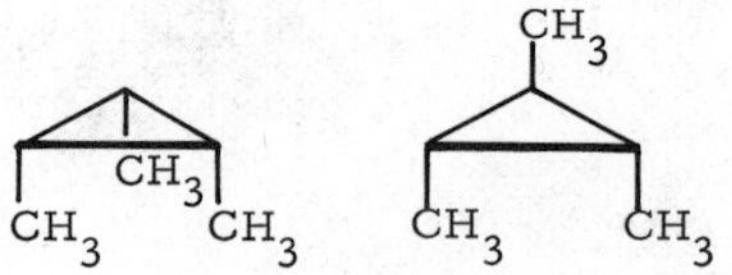

c.   trans, trans

b.   trans, cis

cis, trans

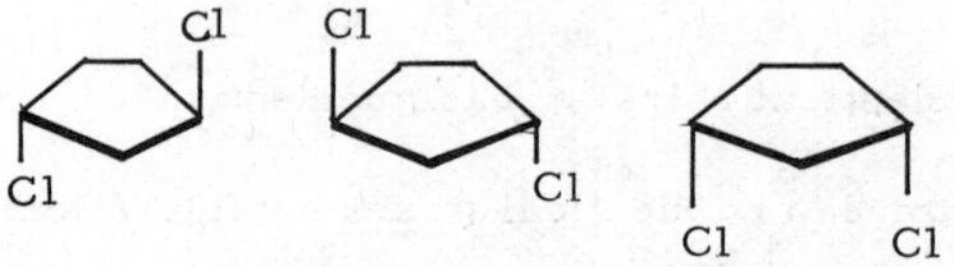

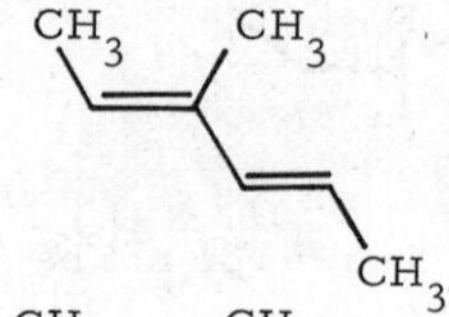

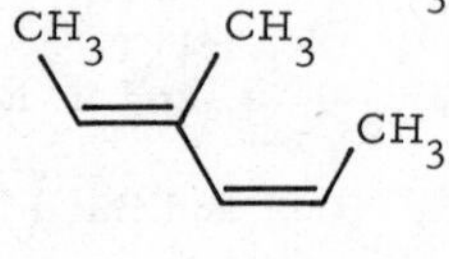

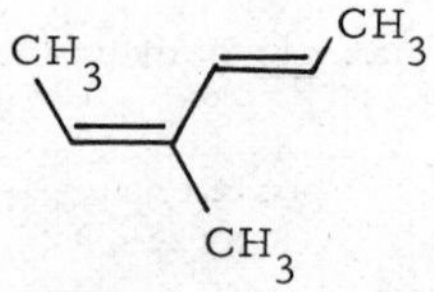

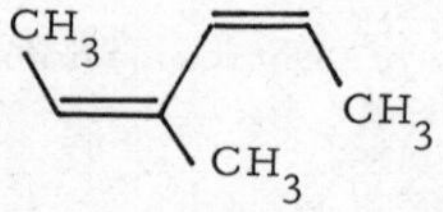

d.

cis, trans   cis, cis   cis, cis

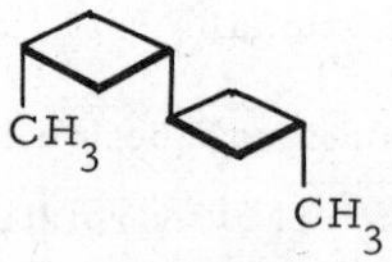

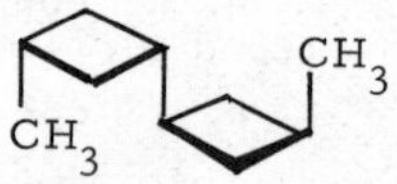

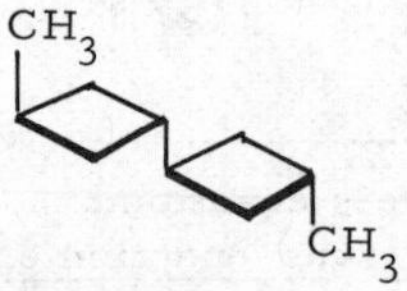

trans, trans

5-21   The trans isomer is more stable than the cis isomer because the methyl groups are further apart (less steric hindrance).

5-22

a. $CH_2 = CHCH_2Cl$ $ClCH = CHCH_3$ $CH_2 = C(Cl)CH_3$ $\begin{array}{c} CH_2 \\ | \\ CH_2 \end{array}\Big\rangle CHCl$

(cis and trans)

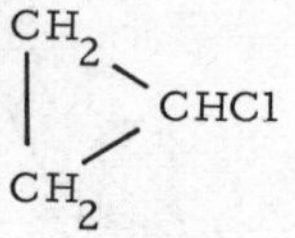

5-22 (cont'd)

b.

C-C-C-C≡C   C-C-C=C-C   (cis and trans)   C-C-C≡C

C-C≡C-C   C≡C-C-C

 -C-C

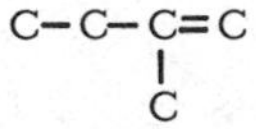

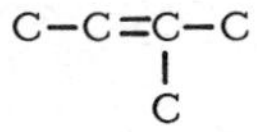

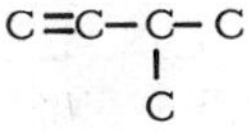

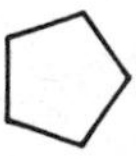

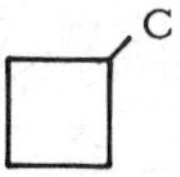

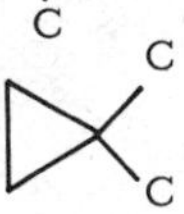

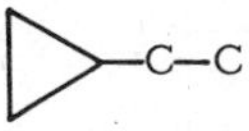

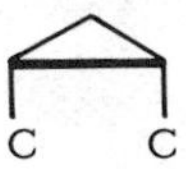

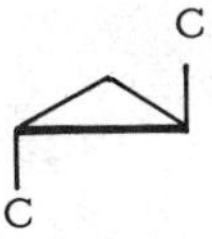

(pair of enantiomers)

c.

C-C-C≡C-Cl   C-C-C≡C   C-C-C≡C   Cl-C-C-C≡C

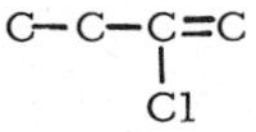

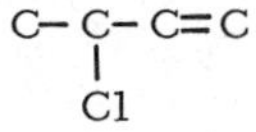

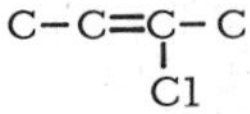

(cis and trans)

(enantiomers)

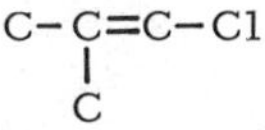

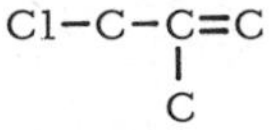

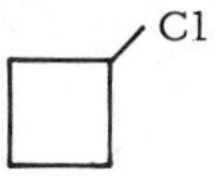

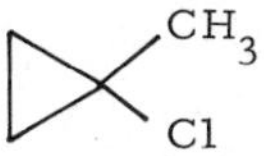

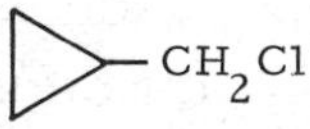

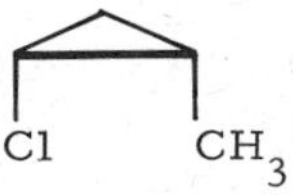

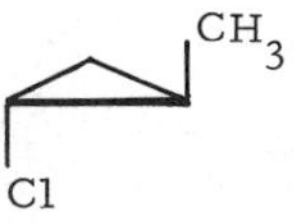

C-C≡C-C-Cl   C-C≡C-C   C-C=C-Cl   Cl-C-C≡C

(cis and trans)   (cis and trans)

(enantiomers) (enantiomers)

5-23

a.

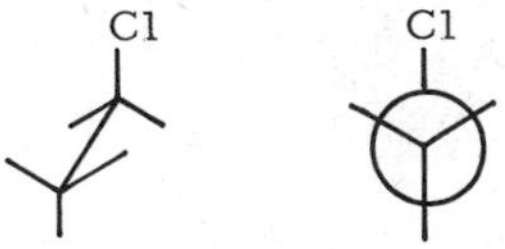

(hydrogens are assumed)

b.

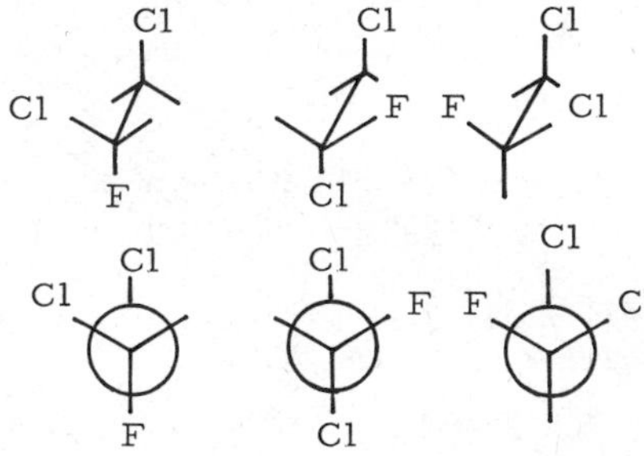

(only one enantiomeric
configuration is shown)

5-23 (cont'd)   c. and d.

c.

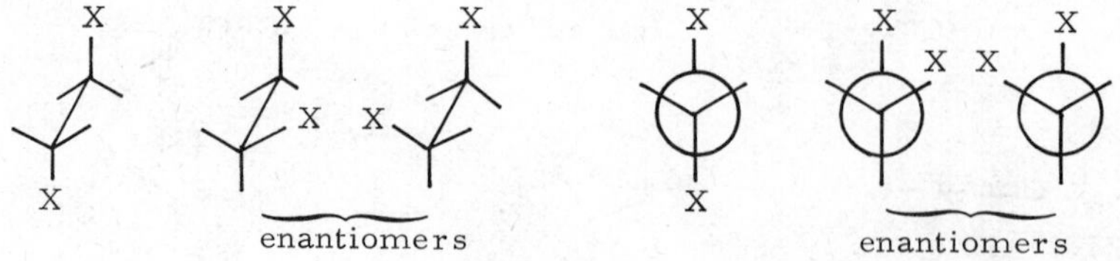

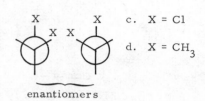

c. X = Cl

d. X = CH₃

enantiomers   enantiomers

5-24

a.
$$\underset{*}{C-\overset{\overset{\displaystyle C}{|}}{C}-\overset{\overset{\displaystyle C}{|}}{C}-C-C}$$

d. [cyclohexanone with Cl, * marked]

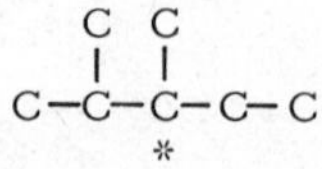

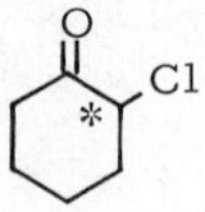

f.   achiral

b.   achiral

c.
$$C-\overset{\overset{\displaystyle Br}{|}}{\underset{*}{C}}-\overset{\overset{\displaystyle Cl}{|}}{\underset{*}{C}}-C$$

e. [cyclohexanone with Cl, * marked]

g. [cyclohexane ring with $\overset{*}{C}$(Cl)(CH₃)(H) substituent and CH₃, * marked]

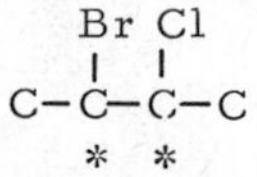

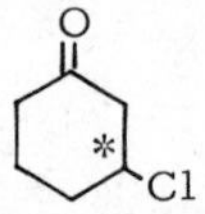

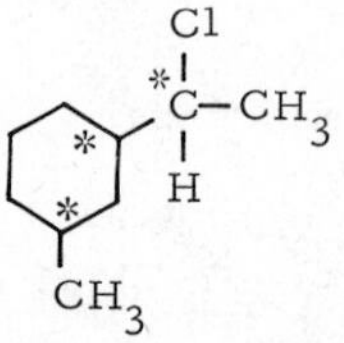

5-25   a.   cis-1,4-, trans-1,4-, cis-1,2-, or cis-1,3-dimethylcyclohexane

b. $CH_3CH_2CH_2\overset{*}{C}HCH_3$ $CH_3\overset{*}{C}H\overset{CH_3}{\overset{|}{C}}HCH_3$ $CH_3\overset{*}{C}HCH_2CH_3$
              $\underset{|}{OH}$ $\underset{|}{OH}$ $\underset{|}{O-CH_3}$

c. $\begin{array}{c} HC=\!\!=CH \\ |\!\!* \quad\quad | \\ H-C-CH_2 \\ | \\ Cl \end{array}$

d.   [Newman projection with CH(CH₃)₂, H, H, H, H, CH(CH₃)₂]

5-26   a.   enantiomers   c.   cis-trans isomers
             (see Section 13-5)
                                        d.   enantiomers (2 pairs) and
       b.   cis-trans isomers   two meso forms
             (see Section 13-5)

5-27   a.   cis   c.   trans   e.   3,4-dimethyl-cis-2,
                                              trans-4-hexadiene
       b.   cis   d.   cis

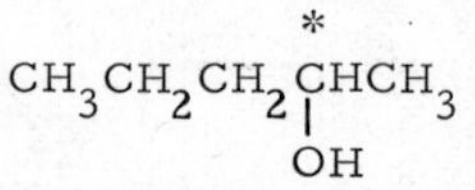

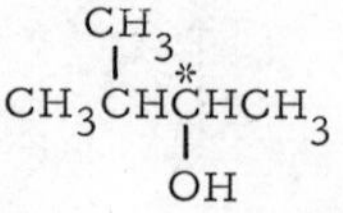

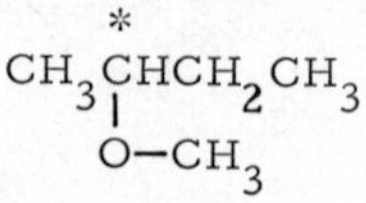

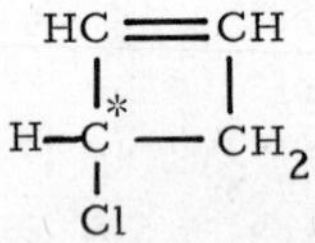

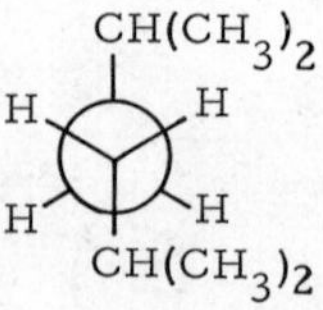

5-28

a.

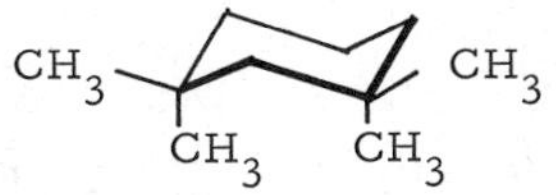

b.

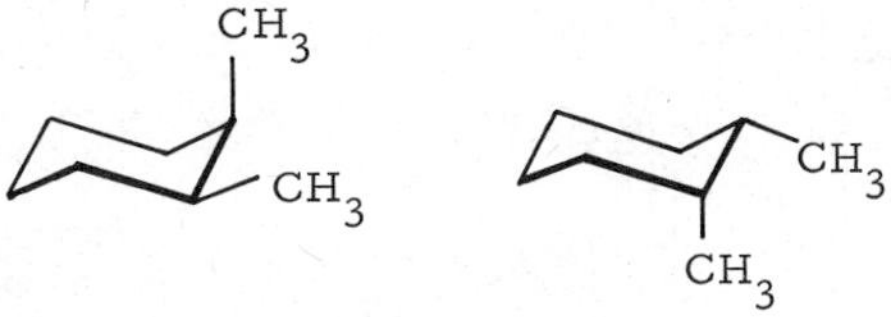

(These are enantiomeric conformations; see Section 12-3D)

c.

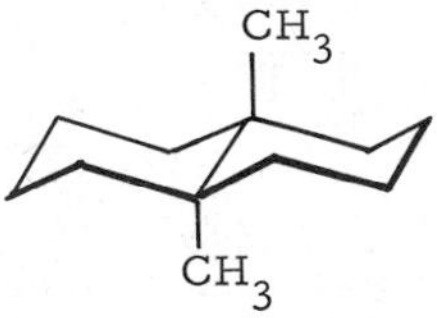

(Hydrogens are not shown)

5-29   a.   enantiomers   c.   enantiomers   e.   cis-trans isomers

       b.   diastereomers   d.   identical isomers

                                                           f.   enantiomers

5-30   a.   identical   c.   diastereomers   e.   diastereomers

       b.   enantiomers   d.   positional isomers   f.   conformational isomers

5-31

a.

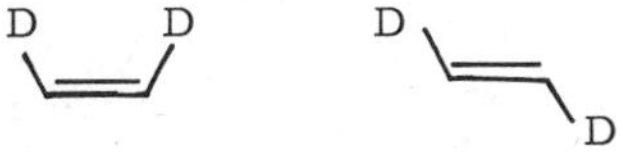

b.

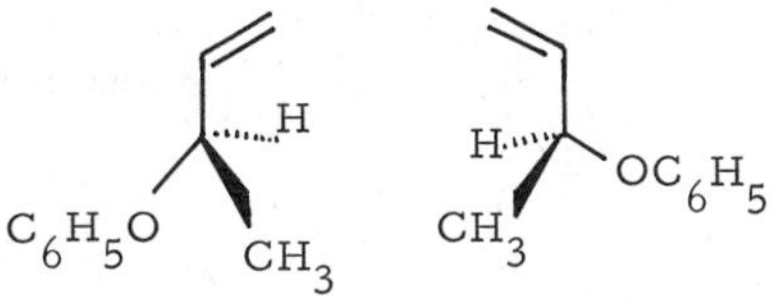

c.

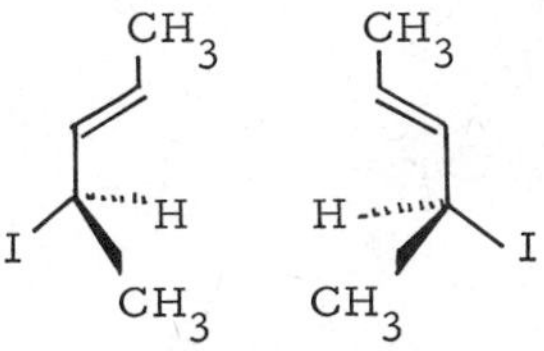

(trans) enantiomers 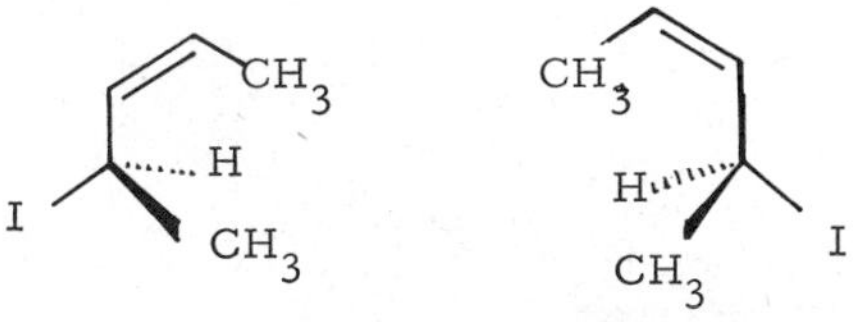   (cis) enantiomers

d.

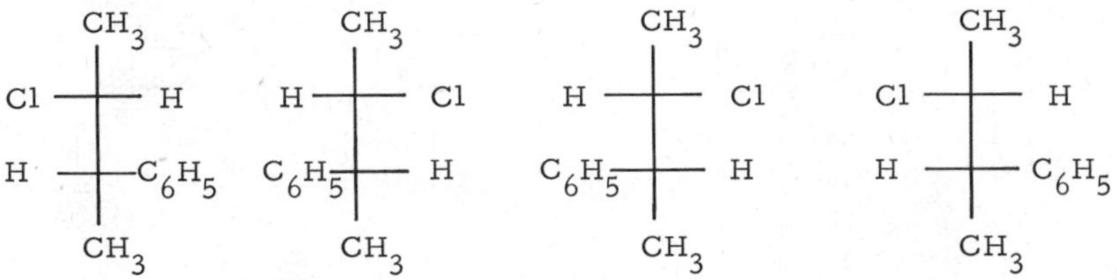

enantiomers   enantiomers

5-31 (cont'd)

e.

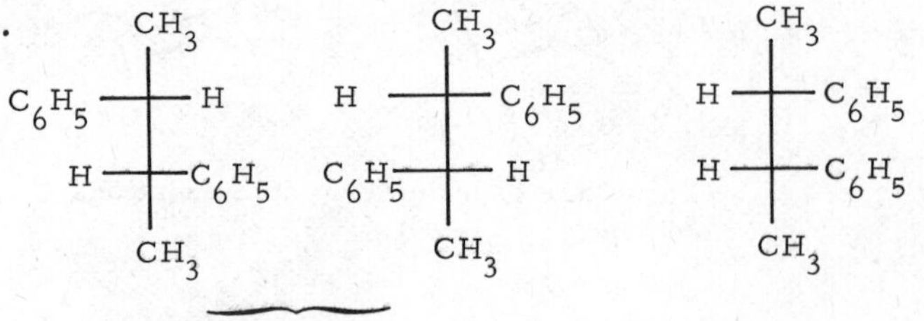

enantiomers

meso

f.

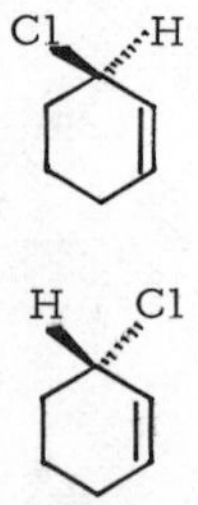

(Note that trans
enantiomers
are sterically
impossible)

g.   In contrast to part f, the eight-membered ring is large enough to fit

cis and trans double bond configurations.

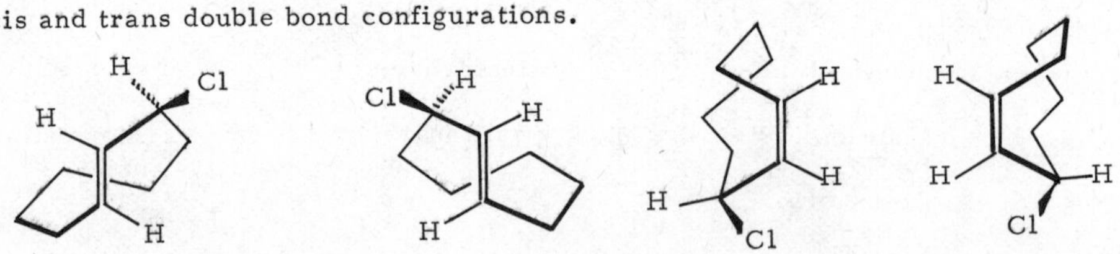

trans enantiomers   cis enantiomers

h.   i.

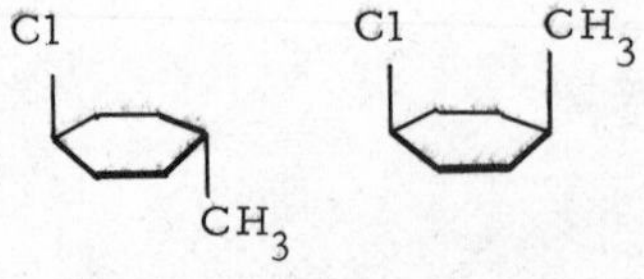

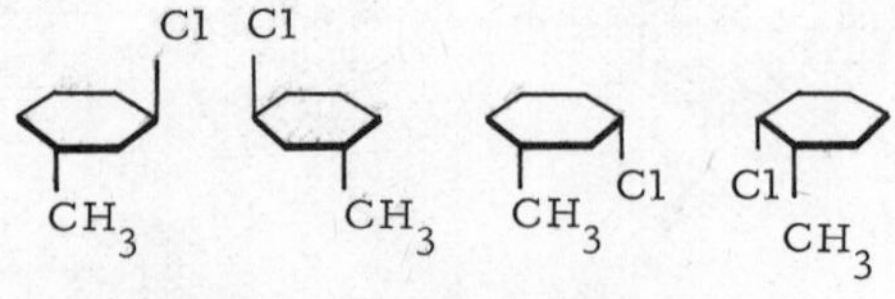

(trans) enantiomers   (cis)
                          enantiomers

j.

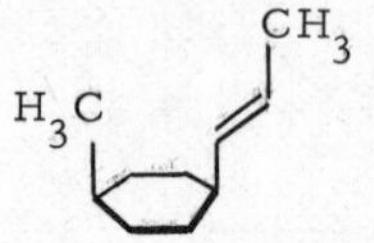

cis, trans   cis, cis   trans, cis

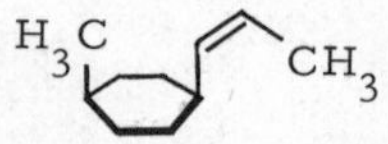

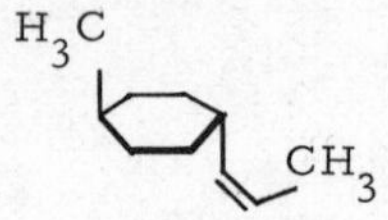

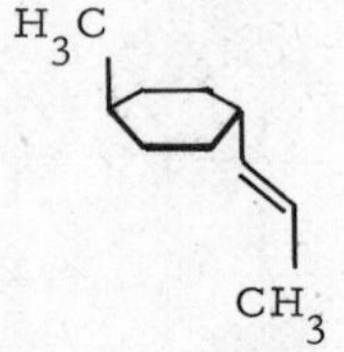

trans, trans

5-31 (cont'd)

k.   As in part j, there will be <u>cis</u> -<u>cis</u>, <u>cis</u> -<u>trans</u>, <u>trans</u>- <u>trans</u> , and

<u>trans</u>- <u>cis</u> configurations, but each of these can exist as a pair of enantiomers.

So there will be 8 stereoisomers in all.

5-32

a.   Nonidentical with mirror image;   d.   Identical (meso) with mirror
       remains chiral on free rotation   image

b.   Identical with mirror image   e. and f. Nonidentical with mirror
       (center of symmetry)   image; remains chiral in all

c.   Nonidentical with mirror image   conformations
       but becomes achiral on free
       rotation

5-33

a.   CH$_3$ c.   CO$_2$H   d.   NH$_2$

       H——Br   HO——H   H⟋\ulcorner$_{\shortmid}$C$_2$H$_5$
                                                                           CH$_3$
       C$_2$H$_5$ HO——H

b.   CO$_2$H   CH$_2$OH   e.   CO$_2$H

       H$_2$N——H   H⟋—OH
                                                        H——⟍—OH
       H——OH   CH$_2$OH

       CH$_3$

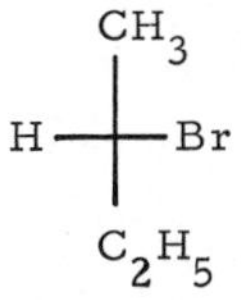

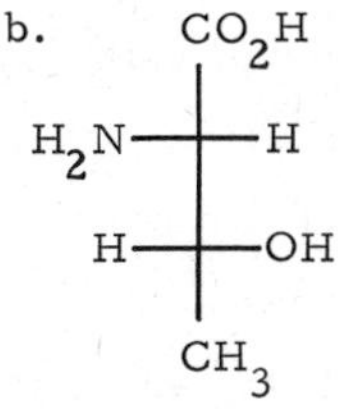

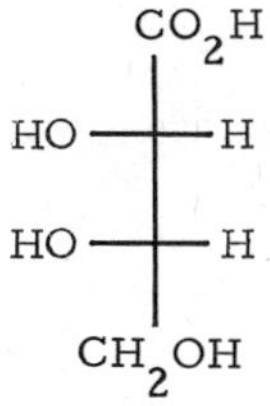

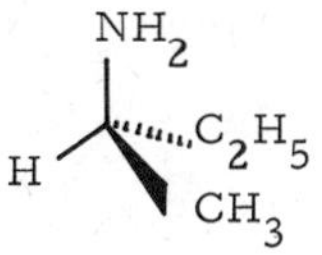

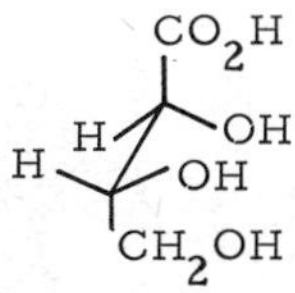

5-34

a.   <u>D</u>   b.   2<u>L</u>, 3<u>D</u>   c.   2<u>L</u>, 3<u>L</u>   d.   <u>L</u>   e.   2<u>L</u>, 3<u>D</u>

5-35

       a.   (1) chiral, asymmetric; (2) achiral, symmetric; (3) chiral,

dissymmetric

5-35 (cont'd)

   b.   cup, achiral, symmetric; shirt, chiral, asymmetric (provided

it has buttons and button-holes); bicycle, achiral, symmetric (provided the

pedals rotate to interconvert the chiral forms);tennis racket, achiral, sym-

metric; automobile, chiral, asymmetric; penny, chiral (taking account of the

asymmetry of the design); scissors, chiral, asymmetric (if you are uncertain

try using a pair of scissors in your right and then your left hand--do they cut

similarly?); flat spiral, achiral, symmetric; conical spring, chiral,

asymmetric.

5-36

camphor $2^2 = 4$ isomers; only 2 are known because a trans bridge across the six-membered ring is sterically impossible.

ribose $2^4 = 16$ isomers

codeine $2^5 = 32$ isomers; not all of these will exist because some are sterically impossible. Carbons marked ⊗ belong to one set that can exist only in the configuration shown or its mirror image. Carbon marked * can exist in all possible configurations. The result is $2^2 = 4$ isomers

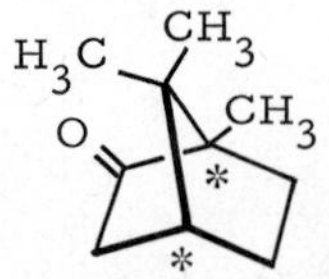

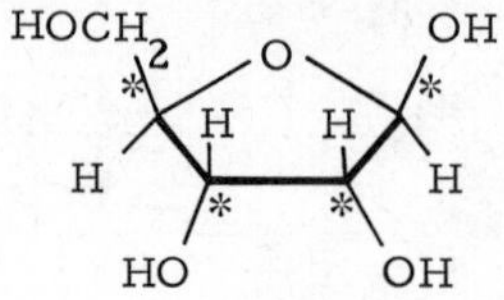

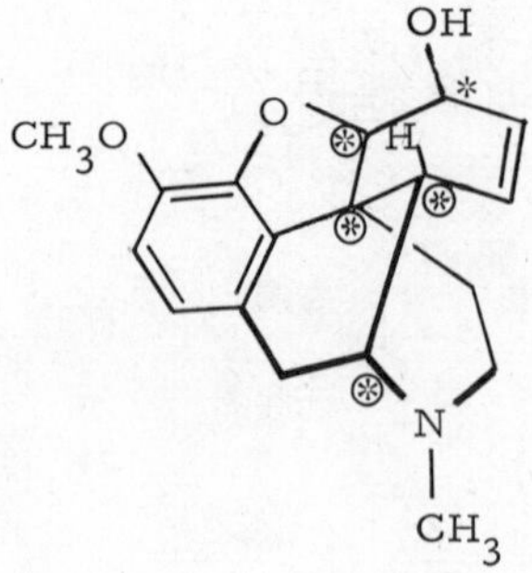

5-36

quinine

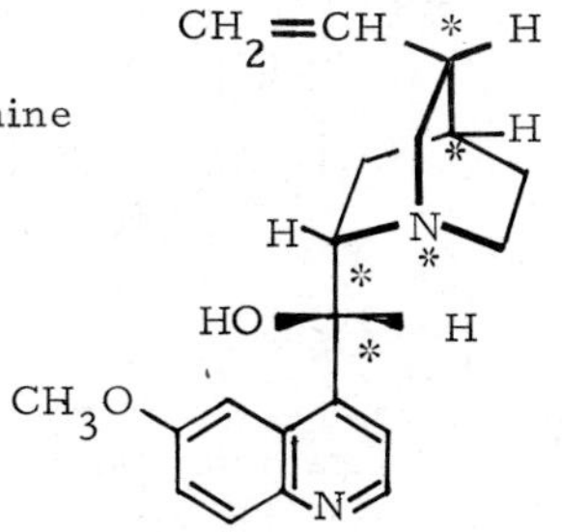

$2^5 = 32$ isomers; but because a trans bridge across the six-membered ring is not possible, there are only $2^4 = 16$ possible isomers.

prostaglandin F

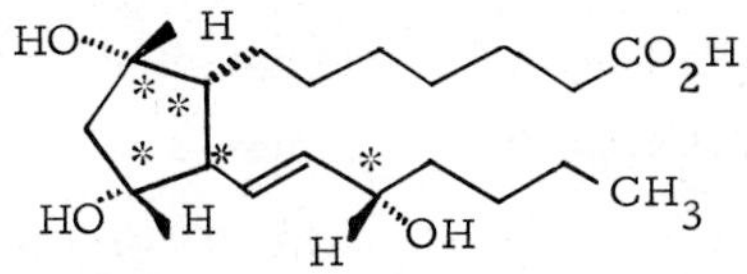

5 chiral centers and one double bond makes $2^6 = 64$ isomers.

ADDITIONAL EXERCISES

5A-1   The major product of radical chlorination of a hydrocarbon of formula C_5H_{12} is a chiral monochloro compound $C_5H_{11}Cl$. Given that the relative rates of chlorination at primary, secondary, and tertiary C—H bonds are 1: 3.3: 4.4, identify by name both the hydrocarbon and the monochloro product.

5A-2   It is possible to add molecular hydrogen to the double bond of an alkene in the presence of a metal catalyst. Furthermore, both hydrogen atoms come in on the same face of the double bond. In the case of trans-3, 4-dimethyl-3-hexene, would you expect to get meso, D or DL-3, 4-dimethylhexane?

5A-3   It is possible to transform a nitro group to an amino group ($NO_2 \longrightarrow NH_2$) by catalytic hydrogenation or by reduction with iron and hydrochloric acid. Would you expect to get DL, D, or L-1-amino-1-phenylethane by the reduction of L-1-nitro-1-phenylethane?

5A-4   Molecular bromine adds to 1-alkenes to give 1, 2-dibromoalkanes. The two bromine atoms add on opposite faces of the double bond. Would Br_2 add to cis-2-butene to give meso, D, L, or DL-2, 3-dibromobutane? Draw the structure of the product showing the configuration of the product in its most stable conformation.

5A-5   How many configurational isomers are conceivable from the addition of molecular bromine to DL-3-methylcyclohexene to give 1, 2-dibromo-3-methyl-cyclohexane? Draw their structures showing the configurations.

5A-6   If the hydrogen labeled as H_a in the structure of 2-bromobutane shown below were replaced by bromine, would the product be meso, DL, DD, LD, or LL-2, 3-dibromobutane?

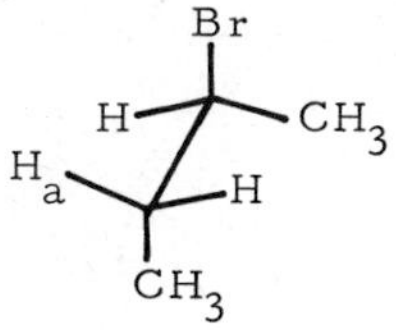

5A-7   The male sex pheromone of the boll weevil is a mixture of compounds having structures 1, 2 and 3. How many configurational isomers are possible for each of these structures? Which are chiral and which are achiral?

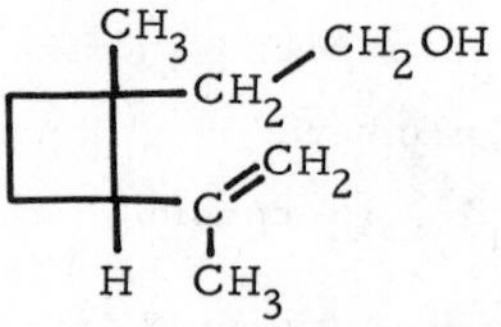

1   2   3

5A-8   Oleic, linoleic, and linolenic acids are unsaturated acids (fatty acids) that occur naturally as esters of glycerol in the seeds of certain plants. These esters are important constituents of vegetable oils (e.g., linseed oil). How many configurational isomers are possible for each of these fatty acids? (The natural acids occur in the all-cis configuration.)

$$CH_3(CH_2)_7CH=CH(CH_2)_7CO_2H$$ oleic acid

$$CH_3(CH_2)_4CH=CHCH_2CH=CH(CH_2)_7CO_2H$$ linoleic acid

$$CH_3CH_2CH=CHCH_2CH=CHCH_2CH=CH(CH_2)_7CO_2H$$ linolenic acid

5A-9   The antibiotic substance, streptomycin, is a metabolite of certain soil bacteria. Streptomycin is a carbohydrate that has an unusual sugar component called L-streptose. Whereas most sugars are derived from straight chain aldehydes or ketones, the chain in streptose is branched.

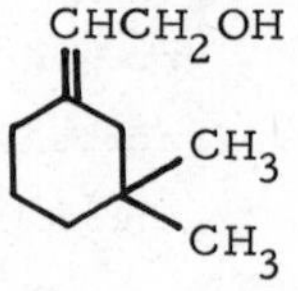

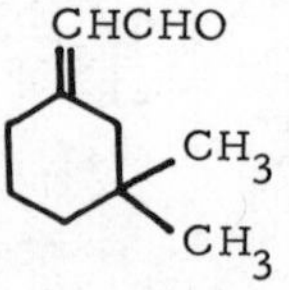

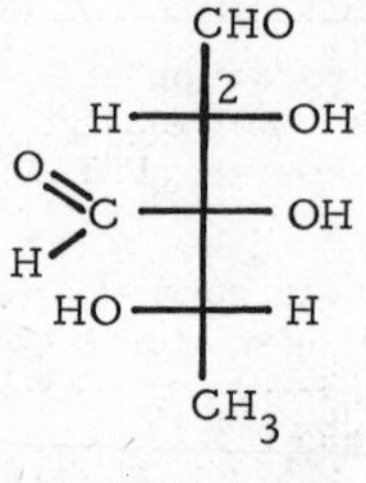

L-streptose

a. How many configurational isomers are possible for the streptose structure?

b. If the configuration at C2 were inverted, would the product be an enantiomer or a diastereomer of L-streptose? Would the product be chiral or achiral?

5A-10   The principle pigment in paprika is a compound (a carotenoid) called capsanthin. Inspect the structure of capsanthin and deduce the number of chiral carbons, and the configurations of the double bonds in the main chain.

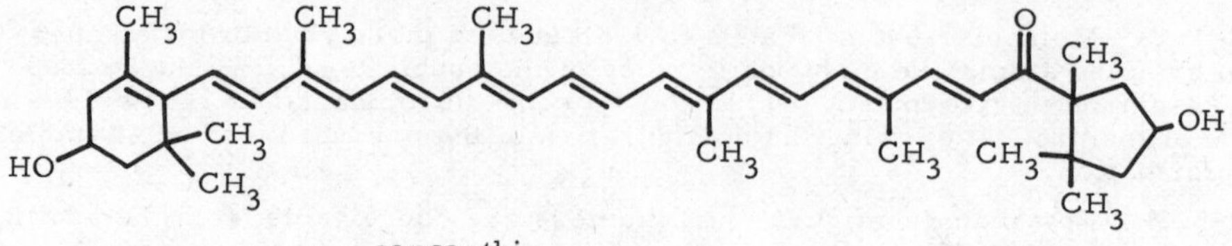

capsanthin

CHAPTER 6

BONDING IN ORGANIC MOLECULES. ATOMIC ORBITAL MODELS

The atomic orbitals of polyelectron atoms may be considered to resemble the s, p, d, and f atomic orbitals of the hydrogen atom. Populating the orbitals of lowest energy with electrons in accordance with the Pauli principle and Hund's rule gives the electronic configuration of the atoms of the periodic table. Bonding arises from the overlap of atomic orbitals centered on different atoms, and the directional character of bonding orbitals confers shape to molecules. The concept of <u>orbital hybridization</u> was developed because of the need to account for the numbers of bonds per atom and the general pattern of bond angles observed for the compounds of the first main row of elements in the periodic table. Atomic-orbital models for organic compounds utilize sp^3 or tetrahedral carbon orbitals in compounds in which carbon is bonded to four ligands; sp^2 or triangular orbitals for carbon bonded to three ligands; sp or linear orbitals for carbon bonded to two ligands. Carbon is considered to use pure p atomic orbitals to form π bonds.

Although hybridization of atomic orbitals goes a long way toward explaining the molecular geometries of organic and other types of compounds, it does not provide qualitative understanding of why bond angles deviate from the ideal values of $109.5°$, $120°$, $180°$, and $90°$ expected for sp^3, sp^2, sp, and p orbitals, respectively. An alternative, in fact complementary, approach to explaining molecular geometry uses arguments based on electrostatic repulsion between bonding and unshared electron pairs associated with the same atom. These arguments assume that electrons will seek to be as far apart as possible. Electron repulsion between unshared pairs is greater than between bonding pairs, and electron-attracting (electronegative) ligands reduce the repulsion effect of bonding pairs. The two electron pairs in a multiple bond exert more repulsion than an electron pair in a single bond.

Electron delocalization, or resonance, arises when atomic orbitals overlap sufficiently to permit electrons to become distributed over <u>more than two</u> atomic nuclei. Benzene may be formulated as having a planar σ framework of twelve localized electron-pair bonds (6 C—H and 6 C—C) and a π-bond system of six electrons associated equally with six carbon nuclei. Electron delocalization can be denoted by writing <u>resonance structures</u> or by writing hybrid structures with dashed bonds:

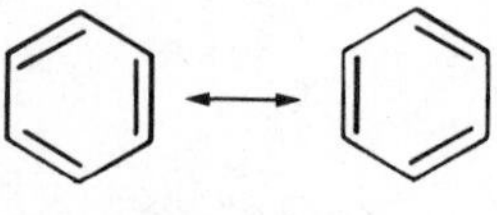

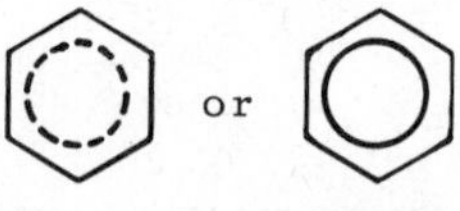

resonance structures   hybrid structures

<u>Worked Exercise.</u> Draw an atomic-orbital diagram of nitromethane, CH_3NO_2, clearly labeling all the bonding and nonbonding orbitals. Indicate the expected bond angles.

<u>Step 1.</u> Decide on the probable hybridization of each of the bonded atoms. Since the geometry is not specified, a reasonable guess must be made based on the number of bonded and nonbonded electron pairs. It is not necessary to be very detailed about the hybridization of the oxygen orbitals because there is no "bond angle" or geometry to consider. Nonetheless, we will consider the oxygen to form sp^2 σ bonds, and its unshared pairs will be designated as $(\underline{n})^2$. Inspection of a Lewis structure for nitromethane indicates sp^3 hybridization at carbon (4 bonded atoms) and sp^2 hybridization at nitrogen (3 bonded atoms). The nitrogen is written as $\oplus$ because it has a half share in eight electrons in the four covalent bonds to it; that is one less than is required for electrical neutrality. Similarly, one of the oxygens has an extra electron and will be written as $\ominus$:

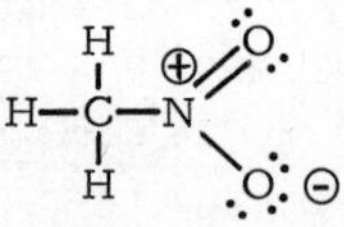

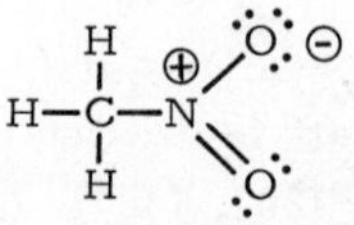

<u>Step 2.</u> To construct the orbital diagram, first draw the <u>orbitals</u> before populating them with the appropriate number of valence electrons. The sigma framework should be drawn first:

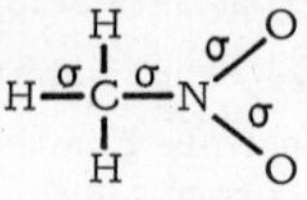

$$2\ \text{N—O}\ (2sp^2 - 2sp^2)\sigma$$
$$1\ \text{C—N}\ (2sp^3 - 2sp^2)\sigma$$
$$3\ \text{C—H}\ (2sp^3 - 1s)\sigma$$

H—C—H = 109°   H—C—N = 109°

C—N—O = 120°   O—N—O = 120°

Now draw the remaining orbitals. There will be one <u>p</u> orbital on nitrogen, and one <u>p</u> orbital and two $\underline{sp}^2$ orbitals on each oxygen. Notice that the model can be set up as to have three parallel, adjacent $\underline{p}$ orbitals that may overlap in a π manner to give O—N—O π bonding:

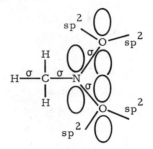

Step 3. Populate the orbitals with the valence electrons. There are 24 valence electrons to be accounted for - 4 from carbon, 3 from three hydrogens, 5 from nitrogen, and 12 from two oxygens. Populate the σ orbitals first, two per orbital. Next, populate <u>nonbonding</u> orbitals on oxygen, each oxygen having two $(\underline{n})^2$ unshared pairs. The remaining electrons are used to populate the π bonding orbitals.

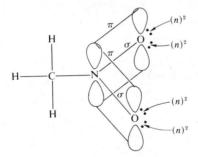

6 σ bonds accounts for	12 electrons
4 nonbonding orbitals accounts for	8 electrons
π bonding over three $\underline{p}$ orbitals	4 electrons
TOTAL	24 electrons

Notice that this formalism implies that there are four electrons associated with three $\underline{p}$ orbitals, or that π bonding extends over three nuclei. This accounts for the fact that the oxygens are <u>equivalent</u> in nitromethane. The nitrogen-oxygen bonds are neither double nor single, as implied by a Lewis structure. The resonance hybrid of nitromethane is represented best by two equivalent Lewis valence-bond structures:

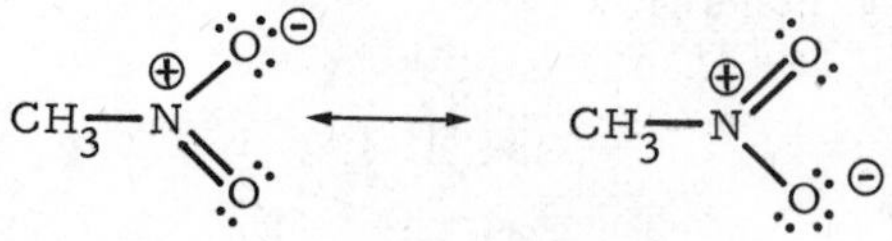

$$CH_3-\overset{\oplus}{N}\underset{\ddot{\ddot{O}}}{\overset{\ddot{O}:^{\ominus}}{<}} \longleftrightarrow CH_3-\overset{\oplus}{N}\underset{\ddot{\ddot{O}}:^{\ominus}}{\overset{\ddot{O}}{<}}$$

(You will notice that the terms Lewis structure, electron-pairing scheme, valence-bond structure are used more-or-less interchangeably.)

The general rules for writing an acceptable set of structures contributing to a hybrid structure are:

All structures must have the same relative locations of the atoms in space, the same number of paired or unpaired electrons, and they must have reasonably low energies. They differ only in the pairing schemes for the electrons.

We have not specified what structures are energetically unreasonable, but as a guiding principle, structures with unnecessary and unlikely charge separations, excessively long bonds, abnormal bond angles, or with electronegative atoms having less than an octet of electrons, may be considered as high-energy structures and unlikely to contribute to the hybrid of a stable molecule, radical, or ion. Some examples follow:

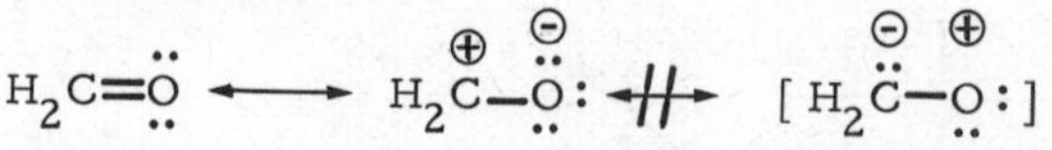

$$H_2C=\ddot{\ddot{O}} \longleftrightarrow H_2\overset{\oplus}{C}-\overset{\ominus}{\ddot{\ddot{O}}}: \; \text{\textbardbl\kern-0.4em/} \; [\; H_2\overset{\ominus}{\ddot{C}}-\overset{\oplus}{\underset{\cdot\cdot}{O}}: \;]$$

(unlikely charge separation and only six electrons on oxygen)

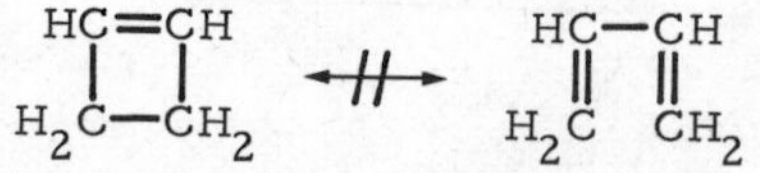

$$\underset{H_2C-CH_2}{HC=CH} \; \text{/\kern-0.5em/} \; \underset{H_2C \quad CH_2}{HC-CH}$$

(excessively distorted molecular geometry)

ANSWERS TO EXERCISES

6-1 $(1s)^1(2s)^1$ is $He\uparrow\uparrow$; the two unpaired electrons must be in different orbitals, and the two of lowest energy are 1s and 2s.

6-2

He_2 $He_2^{\oplus}$

1so*   (↑↓)   (↑)

1so   (↑↓)   (↑↓)

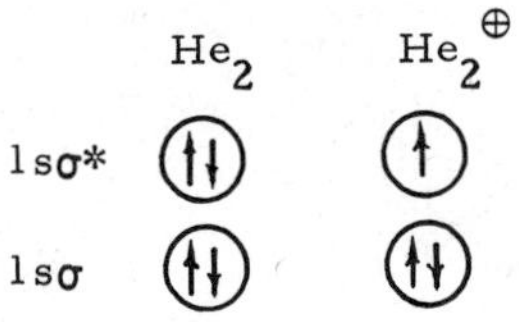

The nuclear charge in $He_2^{\oplus}$ is 4, the nuclear charge in $H_2^{\ominus}$ is 2. Conse-quently the interelectronic repulsion in $H_2^{\ominus}$ is much less compensated by attraction from the atomic nuclei than in $He_2^{\oplus}$.

6-3   1so*   (↑)   (↑)   (↑↓)

          1so   (↑)   (↓)   ( )

1so(↑)1so*(↑)   1so(↓)1so*(↑)   (so*)₂

———————————————→
decreasing stability

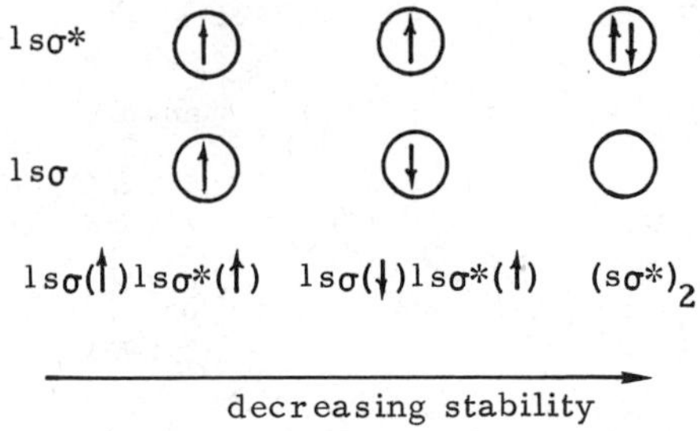

6-4   a.   sp   d. sp^3 g.   sp

       b.   sp   e. sp^3 h. sp^2

       c. sp^2 f. sp^2

6-5

a.   nonplanar, tetrahedral (109.5°)   e.   nonplanar, tetrahedral at oxygen and carbon

b.   planar, triangular (120°)

                                          f.   planar, triangular (120°)

c.   nonplanar, tetrahedral at C3, linear at C1 and C2

                                          g.   nonplanar, tetrahedral at carbon (see Section 12-3)

d.   nonplanar, triangular (120°) at C1 and C3, linear (180°) at C2 (see Section 13-5)

6-6   The H—H distance (2x) is 2.06A in PH_3 and 1.63A in NH_3.

Since the distance is greater in PH_3 than NH_3 the repulsion is less.

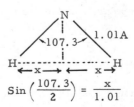

$$\sin\left(\frac{107.3}{2}\right) = \frac{x}{1.01}$$

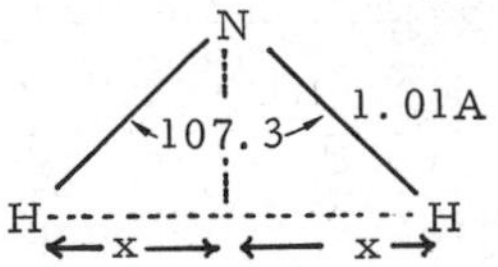

6-7

a. 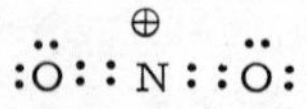 :Ö::N::Ö: (N with ⊕ above)

linear

b. 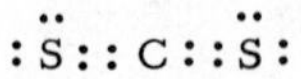 :S̈::C::S̈:

linear

c. 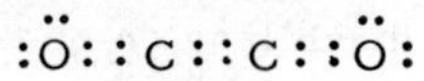 :Ö::C::C::Ö:

linear

d. 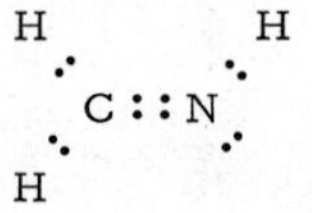
H   H
    C::N
H

triangular-planar

e. 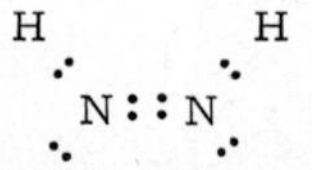
H   H
    N::N

angular

f. 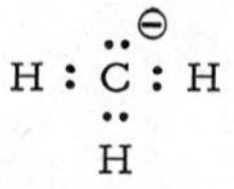
H : C̈ : H (⊖ above)
    H

pyramidal

g. 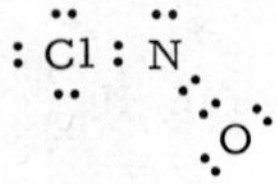
: C̈l : N
         Ö

angular

h. 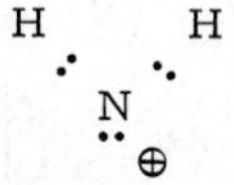
H   H
    N
    ⊕

angular

i. 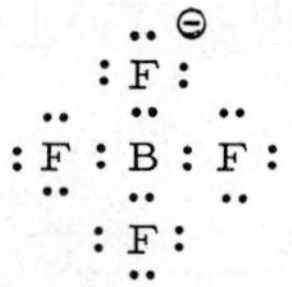
    : F̈ : (⊖ above)
:F̈ : B : F̈:
    : F̈ :

tetrahedral

6-8

a.   Because the unshared pair of electrons in :NH$_3$ exerts more repulsion on the N—H bonding pairs than does a bonding pair in H—NH$_3$ (⊕), the HNH angle is compressed in NH$_3$ relative to NH$_4$ (⊕).

b.   The nitrogen of NF$_3$ has more N$^{\oplus}$ character than in NH$_3$ because of the greater electronegativity of fluorine. There is correspondingly less repulsion between the bonding pairs in NF$_3$, and the bond angle is less.

c.   The same argument applies here as in part b.

d.   The polarity of the carbon-oxygen double bond in CH$_2$=O makes carbon relatively positive. Therefore, electron repulsion between the double-bond electrons and C—H electron pairs is reduced compared to that in ethene, and the HCH angle opens up slightly.

6-9

(net charge = -1)

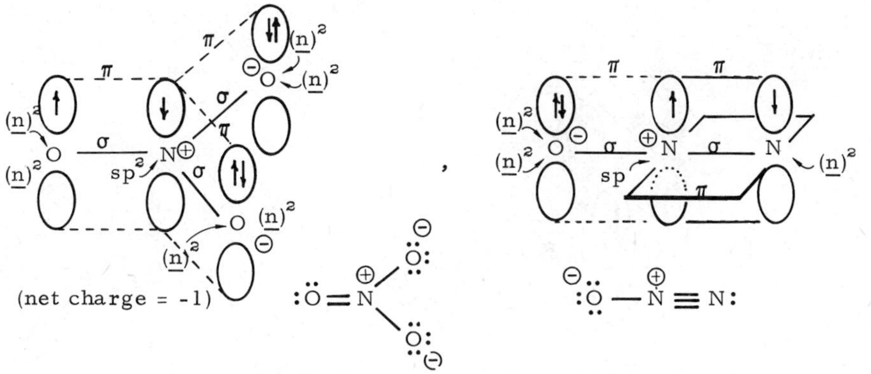

The orbital model of $CO_3^{2\ominus}$ is the same as for $NO_3^{\ominus}$ except that the nitrogen is replaced with carbon.

6-10

a.   See Figure 6-21 for the atomic-orbital model of the 2-propenyl radical; that for the cation is the same but with one less π electron.

b.

π electron delocalization is not significant

c.

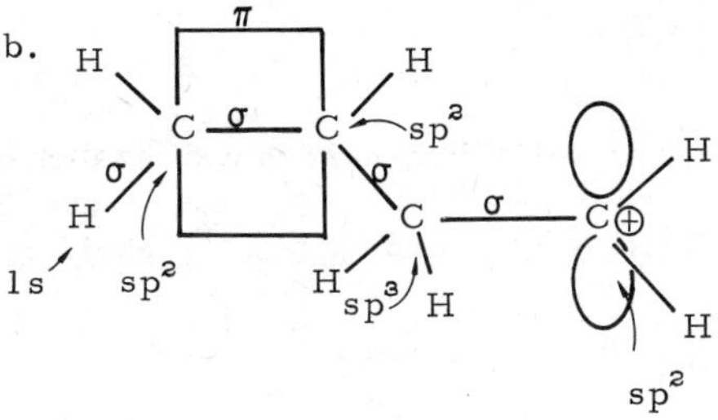

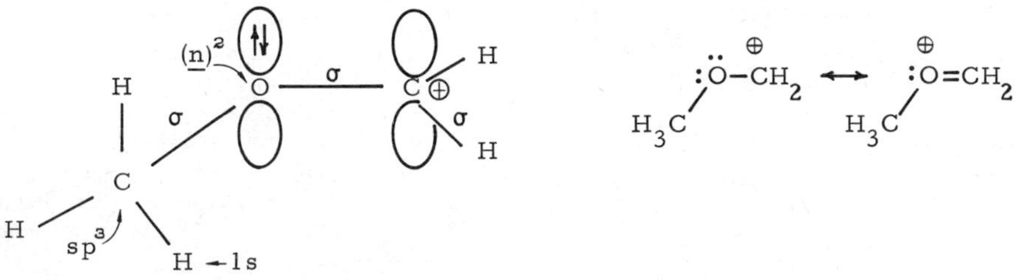

6-10 (cont'd)

d.

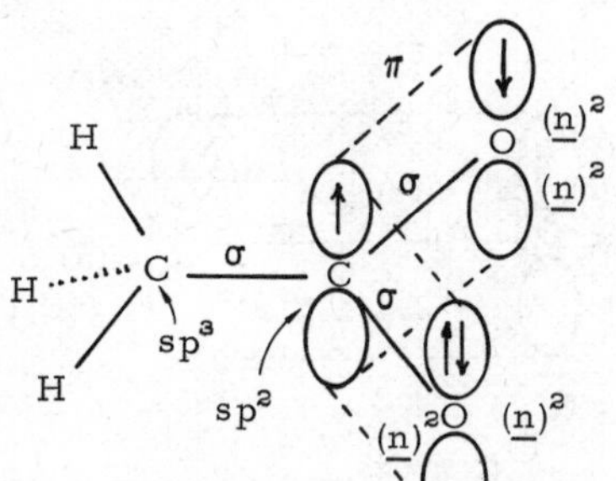

(net charge = -1)

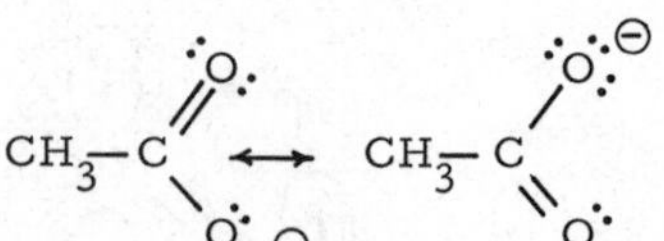

$$CH_3-C \overset{\cdots \overset{\cdots}{O}}{\underset{\overset{\cdots}{O}: \ominus}{}} \longleftrightarrow CH_3-C \overset{\overset{\cdots}{O}: \ominus}{\underset{\cdots \overset{\cdots}{O}}{}}$$

e.

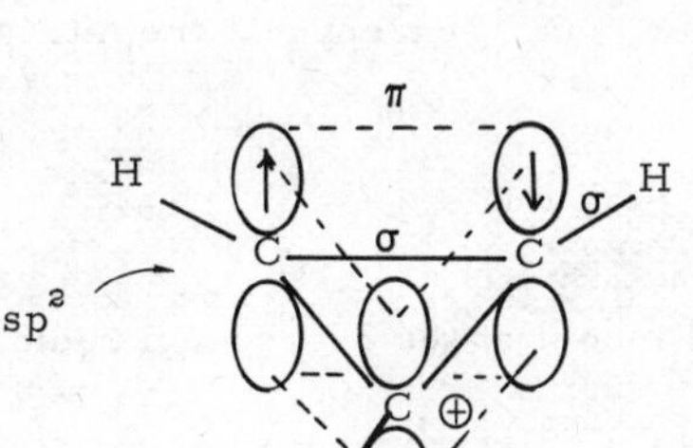

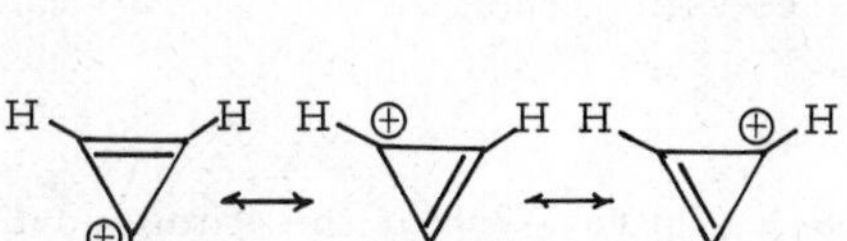

f.   See Figure 6-19 for an atomic-orbital model of benzene; that of azabenzene (pyridine) is similar except the ring C—H is replaced with sp^2-hybridized nitrogen N: The Kekulé structures are:

6-11

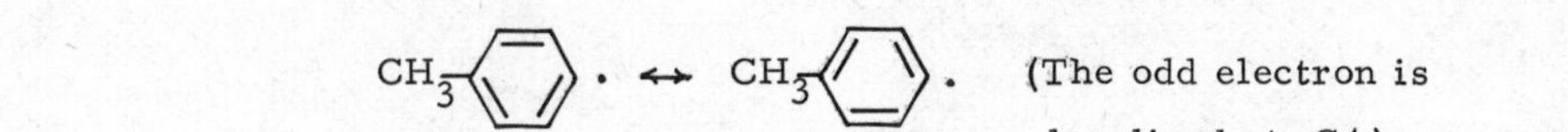

(The odd electron is
localized at C4)

6-11 (cont'd)

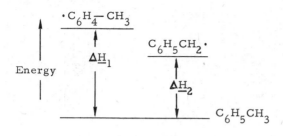

$\Delta\underline{H}_2 < \Delta\underline{H}_1$ ...because of the stabilization of $C_6H_5CH_2\cdot$ by electron-delocalization.

6-12   A molecule Be_2 would have the valence electronic configuration $(2s\sigma)^2(2s\sigma*)^2$ and no net bonding.

For Be to form $BeCl_2$ means that Be uses hybrid sp bonding orbitals. The configuration is $(2sp\sigma)^2(2sp\sigma)^2$. Helium cannot form $HeCl_2$ because there are no low-lying p orbitals in its valence shell with which to form sp hybrid orbitals. Unhybridized orbitals would have the configuration $(1s\sigma)^2$ $(1s\sigma*)^2$ and no net bonding.

6-13   a.   C — C — C
               sp^3 sp^3 sp^3

       b.   C — C ≡ C
               sp^3 sp^2 sp^2

       c.   C ≡ C — C ═ O
               sp   sp   sp^2

       d.   C — C ═ O
               sp^3 sp^2

       e.   C ═ C ═ C
               sp^2 sp sp^2

6-14

a.

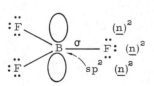

B—F $(sp^2 - p)\sigma$, planar, triangular $(120°)$

6-14 (cont'd)

b.

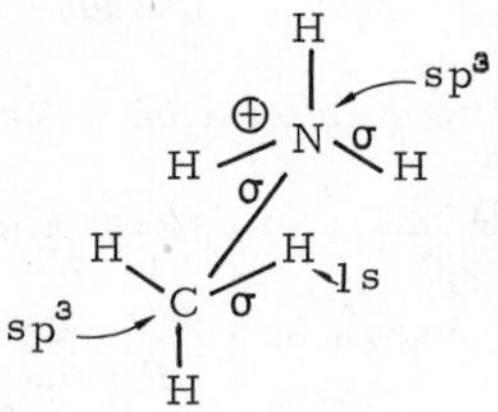

tetrahedral
nonplanar

d.

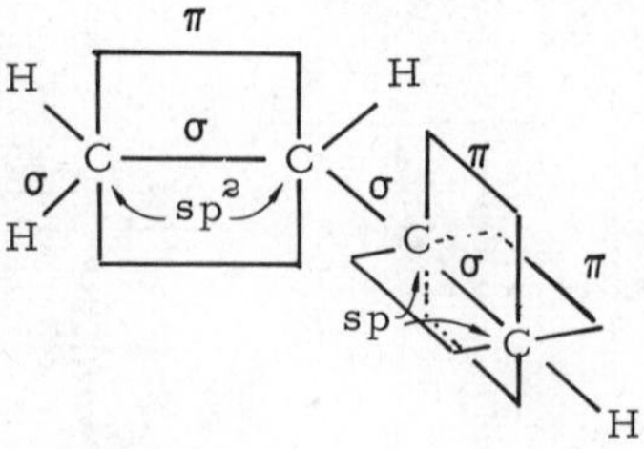

planar,
angles of 120°, 180°

c.

H   ↑   σ   ↓   H
 \ ( )   ( ) /
  C — N⊕
 /   sp²   \
H   ( )   ( )   H
         ⊖

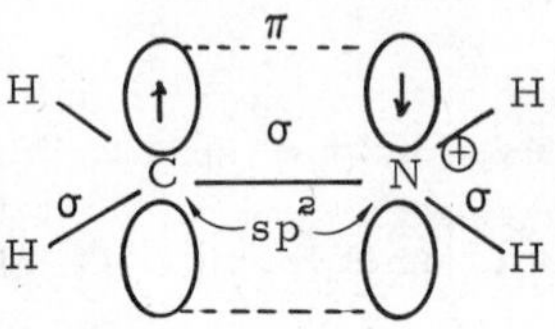

planar, triangular (120°)

e.

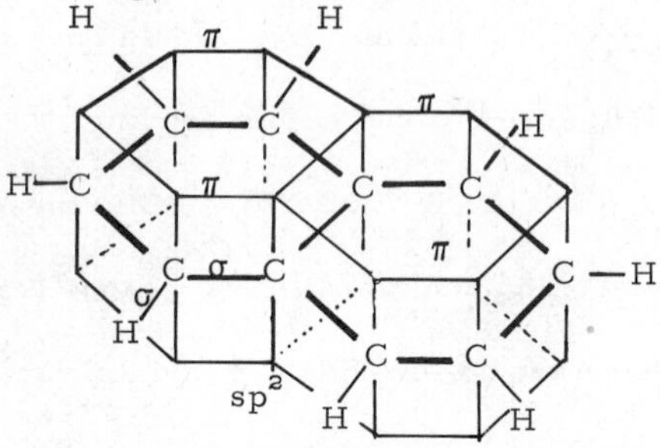

all angles 120°

6-15

a.   :O::C::O:

      linear

b.   :N:::C:O: ⊖

      linear

c.   H
       ·
        C::C::O:
       ·
      H

      triangular (planar)

d.   H·⊕·H
         C
         ··
         H

      triangular (planar)

e.   F   H
       · ·
        C::C
       · ·
      F   H

      triangular (planar)
      at each carbon

f.   H
      ··
      H:C:C:::N:
      ··
      H

      tetrahedral CH₃
      carbon ∠C-C-N 180°

g.   :F:
      ··
      :F:Si:F:
      ··
      :F:

      tetrahedral

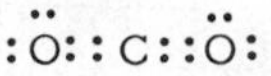

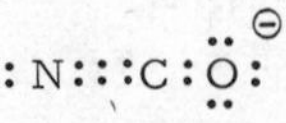

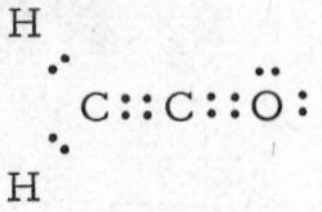

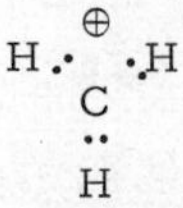

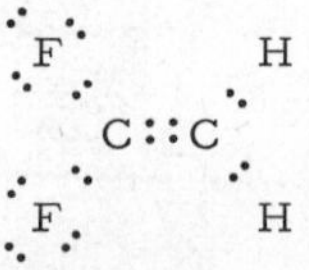

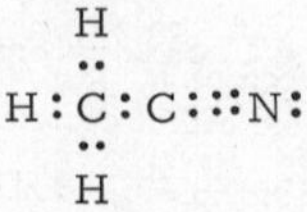

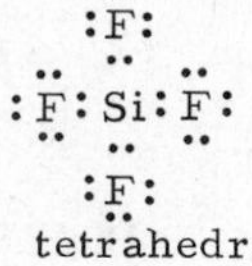

<u>6-15</u> (cont, d)

h.

angle C-O-H bent

triangular (planar) at carbon

j.

    H
    .. ..
H : C : S :
    .. ..
    H H

tetrahedral at carbon
angular at sulfur

i.   H : O : H
         H

pyramidal

k.

triangular (planar)

<u>6-16</u>

a.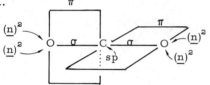

b.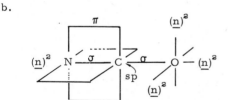

c.   As in part a., except replace
one oxygen   H
with the grouping   C — σ
                 sp²

d.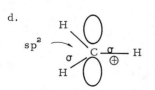

e.   Similar to ethene (Figure 6-14)

but with : F — σ in place of H — σ

at one terminus

f.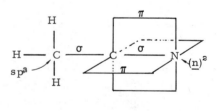

g.

Si — 3sp³
        (n)²
    σ   F : (n)²
        (n)²

Si—F are (3sp³ - 2p) σ

6-16 (cont'd)

h.

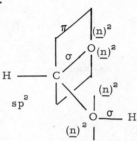

i.

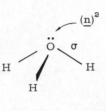

O—H are $(sp^3 - 1s)\sigma$

j.   As in Figure 6-13, except replace $:\ddot{O}$—H with $:\ddot{S}$—H and utilize sulfur $3 sp^3$ hybrid orbitals.

k.

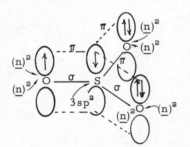

Note similarity to $NO_3^{\ominus}$ and $CO_3^{2\ominus}$, Ex. 6-9

6-17   See Figure 13-4; substitute C—H for C—Cl , one at each end. The structure is dissymmetric and therefore chiral (the two ends lie in perpendicular planes, Section 13-5). The isomers are enantiomers.

6-18

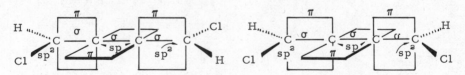

trans isomer   cis isomer

<u>6-19</u>   Electron repulsion between three separate sets of electrons around

oxygen (or any atom) is minimized at or near angles of 120°.

<u>6-20</u>

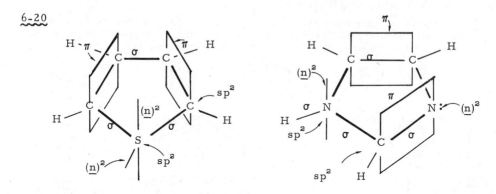

<u>6-21</u>   a.   Nonplanar; each boron is associated with four separate pairs of

bonding electrons, and repulsion between them is minimized in a nonplanar

arrangement (which would be tetrahedral if the four pairs were equivalent).

b.   Assume sp^3 orbitals for boron:

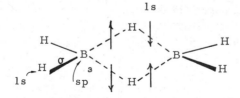

c.   The three-center-electron pair bonds offer less electron repul-

sion than the external two-center B—H bonding electrons. Therefore, the

terminal H—B—H opens up to greater than tetrahedral.

<u>6-22</u>   a. $:N{\equiv}N:$ b. $\underset{H}{\overset{H}{>}}C=\overset{..}{O}:$ c. $H_3C-\overset{..}{\underset{..}{O}}-H$ d. $NO_3^{\ominus}$

(See Ex. 6-9)

ADDITIONAL EXERCISES

(The exercises that follow are problems related to the theory of resonance as described in Section 6-5.)

6A-1   Draw valence bond structures for each of the following ions to show that the charge is distributed over more than one atom. Draw only those structures that are the major contributors to the resonance hybrid.

a. $CH_3 \overset{\oplus}{C}O$

b.

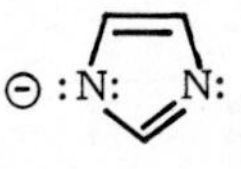

c.

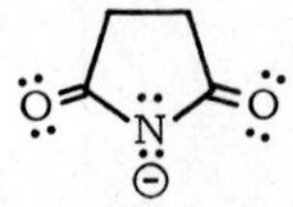

d.

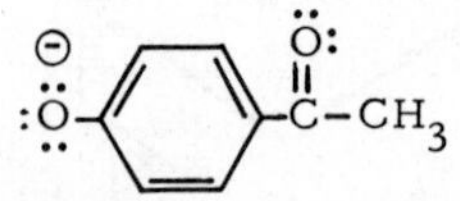

e.

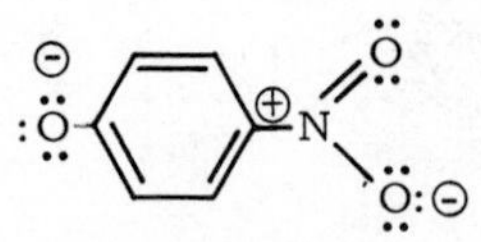

f.

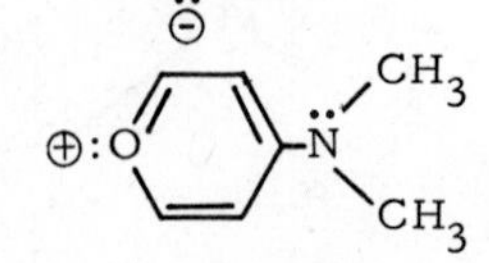

6A-2   Acid dissociation of HA is represented by the general equation, $HA \rightleftharpoons H^{\oplus} + A^{\ominus}$. The more stable the conjugate base $A^{\ominus}$ the stronger is the acid HA. Which compound in each of the following pairs would you expect to be the stronger acid? Draw the important valence bond structures that contribute to the resonance hybrid structure (and hence stability) of the conjugate base.

a. 1,3-cyclopentadiene, cyclopentane;   b. ethyne, hydrogen cyanide;
c. ethanoic (acetic) acid, CH_3CO_2H, and ethanamide (acetamide), CH_3CONH_2;
d. benzene, methylbenzene (toluene).

6A-3   Which is the most acidic hydrogen in each compound listed below? Base your judgement on the degree of resonance stabilization gained or lost on forming the conjugate base ($A^{\ominus}$) from the acid (HA). Draw valence bond structures for the conjugate base obtained on removal of the most acidic hydrogen.

a. $HOCH_2CO_2H$ b. 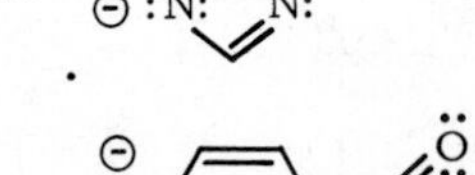 HO—⟨ ⟩—CH_2OH c. $CH_3COCH_2CO_2CH_3$

6A-4   Which is the weakest C—H bond in each compound listed below? Base your judgement on the degree of stabilization gained or lost on forming a radical by removal of a hydrogen atom. Draw valence bond structures for the radical.

a. $(CH_3)_2CHCN$ b. 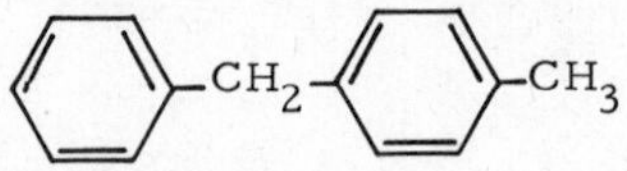   c.

6A-5   Which structure would contribute least to the resonance hybrid of each compound shown below?

a. $CH_2{=}CH{-}\overset{..}{N}H_2 \longleftrightarrow :CH_2^{\ominus}{-}CH{=}\overset{\oplus}{N}H_2 \longleftrightarrow \overset{\oplus}{C}H_2{-}CH{=}\overset{\ominus}{N}H_2$

b. 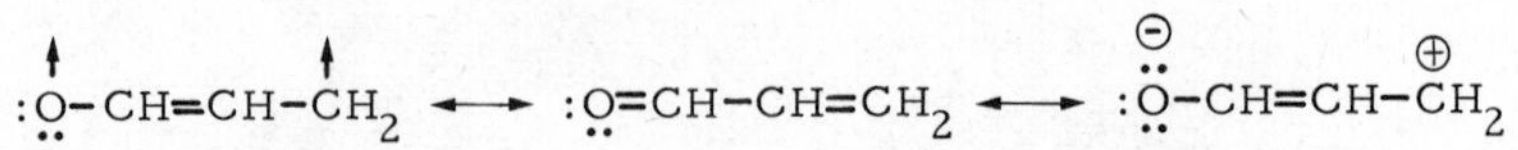 $:\overset{..}{O}{-}CH{=}CH{-}\overset{\uparrow}{C}H_2 \longleftrightarrow :\overset{..}{O}{=}CH{-}CH{=}CH_2 \longleftrightarrow :\overset{\ominus}{\underset{..}{O}}{-}CH{=}CH{-}\overset{\oplus}{C}H_2$

c.

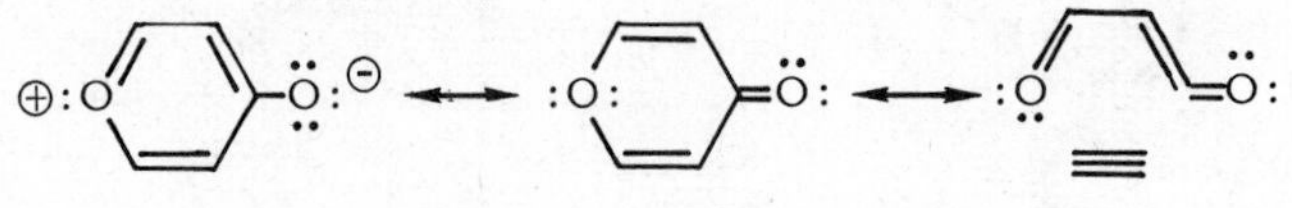

CHAPTER 7

MORE ON NOMENCLATURE. COMPOUNDS OTHER THAN HYDROCARBONS

A number of principles important for naming compounds of more complex structures than hydrocarbons or their simple derivatives are discussed. Probably the most important single idea is how to decide (for the purpose of having a specific name) whether a compound is classified as a hydrocarbon, alkene, alcohol, carboxylic acid, and so on. A precedence table is given for the important types of functional groups. A fairly general procedure for working from a name to structure is illustrated, as are the special features of the naming of alcohols, phenols, ethers, aldehydes, ketones, carboxylic acids, acyl derivatives (such as esters, anhydrides, halides, and amides), amines, and nitriles. The solution to the problem of how to decide when the name of an organic compound is one or two words is outlined in terms of specific examples.

ANSWERS TO EXERCISES

7-1 a. ester   c. nitrile   e. carboxylic   g. aldehyde
                                       acid
     b. alcohol   d. ketone   f. carboxylic   h. hydrocarbon
                                       acid   (benzene)

7-2

a. 1-propanol   d. 3-cyclohexylpropanal   g. 1-chloro-4-methoxy-
                                                             2-butanol
b. 4-penten-2-ol   e. 4-ethylcyclohexanol
                                                         h. 2-chloro-4-hydroxy-
c. 2,2-dimethyl-1-   f. 4-penten-2-yn-1-ol   butanal
    butanol

7-3 a.

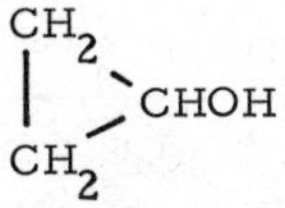

$$CH_2 - CH_2 \diagdown CHOH$$

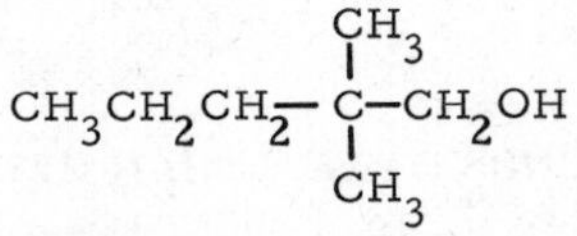

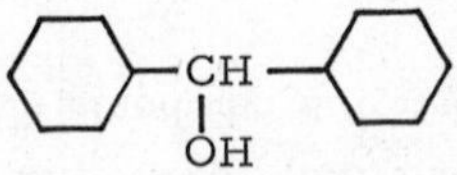

(cyclopropane ring with CHOH)

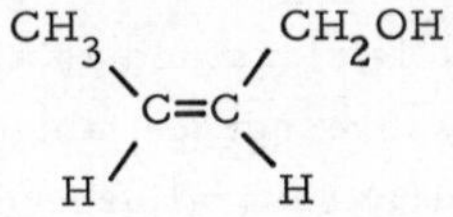

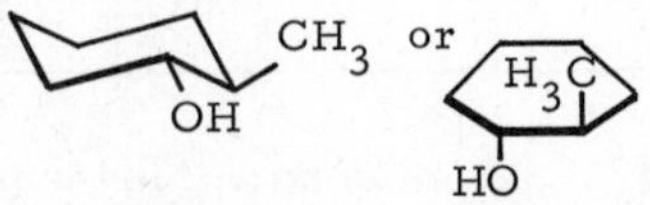

e. $(CH_3O)_4C$

b.

$$CH_3CH_2CH_2-\underset{\underset{CH_3}{|}}{\overset{\overset{CH_3}{|}}{C}}-CH_2OH$$

f.

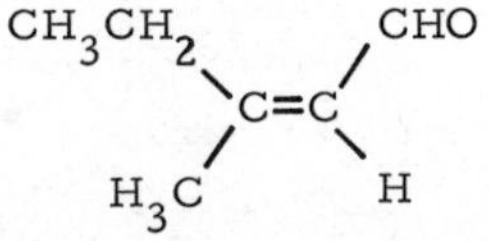

$$\underset{H_3C}{\overset{CH_3CH_2}{>}}C=C\underset{H}{\overset{CHO}{<}}$$

c.

(cyclohexyl)-CH(OH)-(cyclohexyl)

g. (cyclohexane with CH$_3$ and OH) or (cyclopentane with H$_3$C and HO)

d.

$$\underset{H}{\overset{CH_3}{>}}C=C\underset{H}{\overset{CH_2OH}{<}}$$

h.

(Fischer projection: CH$_3$, HO—H, H—OH, CH$_3$) or (Fischer projection: CH$_3$, H—OH, H—OH, CH$_3$)

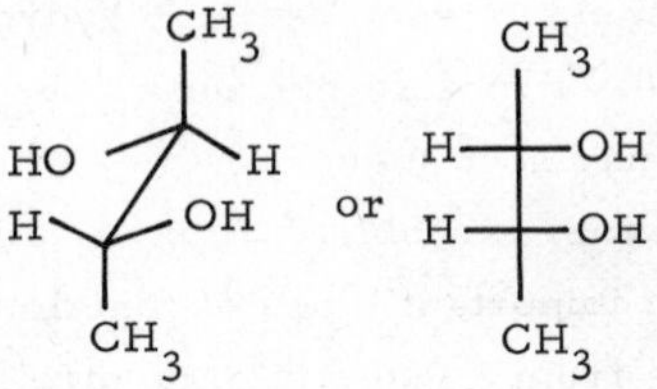

7-4

a. 2-hexanol

b. 4-tert-butylbenzenol
or 4-tert-butyl-1-hydroxybenzene
(4-tert-butylphenol)

c. 1,2,3-propanetriol (glycerol)

d. 2-cyclohexyl-1-cyclopropanol

e. tri-(4-methylphenyl)methanol

f. 4-hydroxyphenylmethanol

7-5 a. 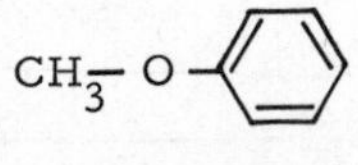 CH_3-O-(phenyl)

b. $CH_3OCH_2CH_2OH$

c. $ClCH_2OCH_2CH_2OH$

d. $(CH_3CF_2CH_2)_2O$

e. 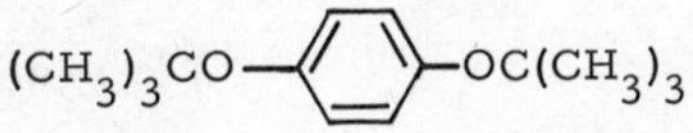 $(CH_3)_3CO-$(phenyl)$-OC(CH_3)_3$

f.

(phenyl)$-O-C\underset{\underset{CH_3}{|}}{\overset{H}{=}}C\overset{H}{<}$

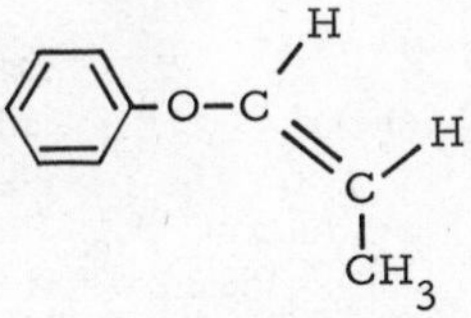

7-6

a. phenylethanal

b. propynal

c. 4-cyclohexene-cis-1,2-dicarbaldehyde

d. 2,3,4-trihydroxybutanal

e. 3-methanoylpentanedial (or 3-formylpentanedial)

f. 4-methyl-2-nitrobenzene-carbaldehyde

g. 2-D-hydroxybutanal

d. $CH_2 = CHOH$

7-7

a.

$$CH_3-CH-\overset{\overset{\displaystyle O}{\|}}{C}-CH_3$$
$$\underset{\displaystyle CH_3}{\mid}$$

e.

b.

$$CH_2=CH-\overset{\overset{\displaystyle O}{\|}}{C}-\triangle$$

c.

$$CH_3-\overset{\overset{\displaystyle O}{\|}}{C}-CHO$$

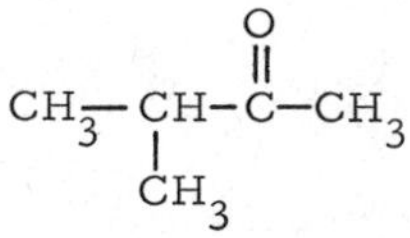

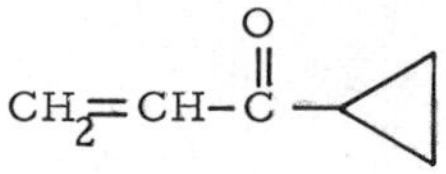

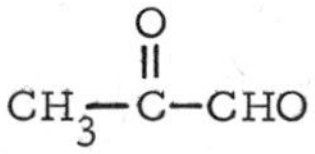

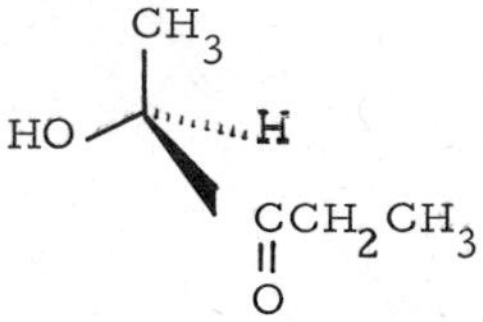

c.

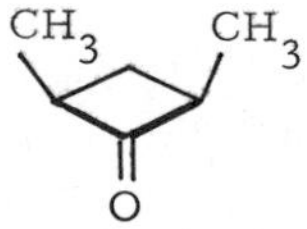

7-8

a.   3,3-dimethylbutanoic acid

b.   3-hydroxypentanedioic acid

c.   cis-butenedioic acid
      (maleic acid)

d.   4-methanoyl- or 4-formyl-
      benzenecarboxylic acid

e.   calcium propenoate

f.   2-carboxybutenedioic acid

7-9

a.   ethyl 2-methylbutanoate

b.   2-butenyl ethanoate

c.   methyl 4-methylbenzene-
      carboxylate

d.   methanoic anhydride

e.   benzenecarboxylic anhydride
      (benzoic anhydride)

f.   benzenecarboxylic methanoic
      anhydride

g.   propanoyl chloride

h.   4-methanoylbenzenecarbonyl
      chloride

i.   N-phenylmethanamide

j.   benzenecarboxamide

k.   N-ethyl-N-methylethanamide

l.   methyl 4-ethanoylbenzene-
      carboxylate

7-10

a. $$H-\overset{\overset{\displaystyle O}{\|}}{C}-NHCH_3$$

b. $$CH_3CH_2\overset{\overset{\displaystyle O}{\|}}{C}-O-\overset{\overset{\displaystyle O}{\|}}{C}CH_2CH_3$$

c. $$CH_3\overset{\overset{\displaystyle O}{\|}}{C}-O-CH_2$$
      $$H-\!\!\!\underset{\displaystyle C_2H_5}{\overset{\displaystyle \mid}{\rule{1cm}{0.4pt}}}\!\!\!-CH_3$$

d. $$CH_3\underset{\displaystyle CHO}{\overset{\displaystyle \mid}{C}}HC-OCH_3$$

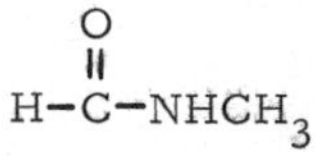

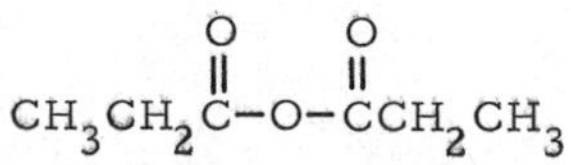

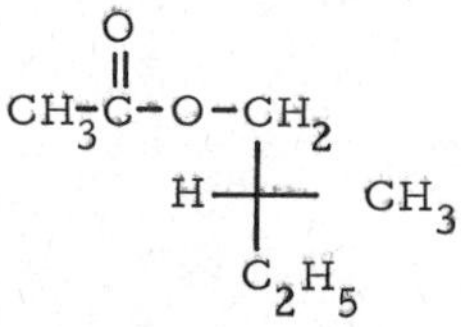

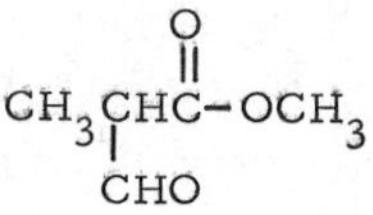

7-10 (cont'd)

e.

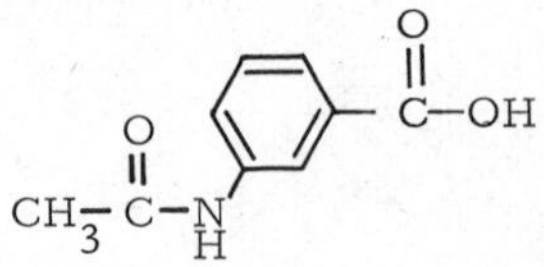

f.

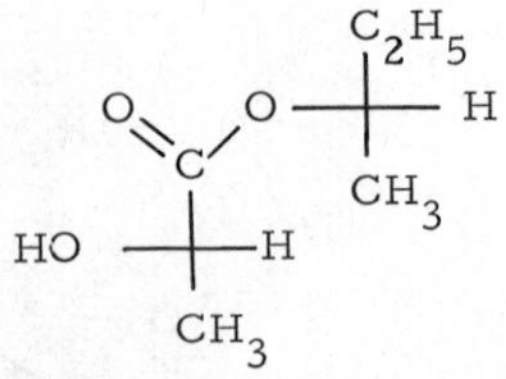

7-11

a.   3-butenamine

b.   N,N-dimethylpropanamine

c.   4-aminobenzenecarboxylic acid

d.   N-(2-aminoethyl)ethanamide

e.   2,5-dimethyl-1,4-benzenediamine

f.   trans-2-aminocyclohexanol

7-12

a.   3-methylbutanenitrile

b.   3-chloro-5-oxopentanenitrile

c.   hexanedinitrile

d.   2-methyl-1,4-benzenedicarbonitrile

e.   trans-1,2-cyclobutanedicarbonitrile

7-13

a.   3,3-dimethyl-1-butanol

b.   cyclohexylcyclopentyl-methanol

c.   methyl 3-bromobutanoate

d.   2-bromoethyl ethenyl ether or 2-bromoethoxyethene

e.   4-nitro-2-pentene

f.   N-methyl-N-nitroso-benzenamine

g.   1,4-cyclohexanedione

h.   2-methylcyclopentanecarbaldehyde

i.   3-(2-oxopropyl)benzenecarbaldehyde

j.   propenoyl chloride

k.   3-butynenitrile

l.   ethyl 4-cyano-2-oxopentanoate

m.   N-methyl-N-(4-methylphenyl) benzenamine

7-14

a. 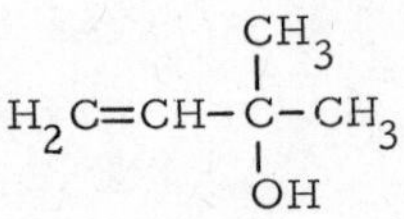 $H_2C=CH-\underset{\underset{OH}{|}}{\overset{\overset{CH_3}{|}}{C}}-CH_3$

b. $\underset{Br\ \ Br}{CH_2CH}-C\overset{O}{\underset{OH}{\big\langle}}$

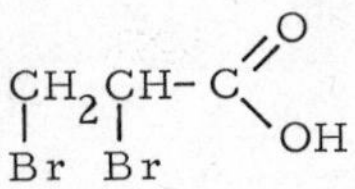

c.

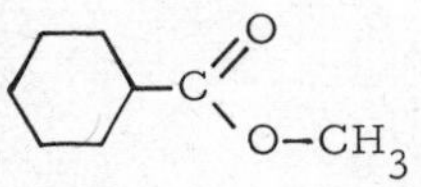

7-14 (cont'd)

d. $CH_3CH_2CH_2CH=CH-CHO$

l.

e.

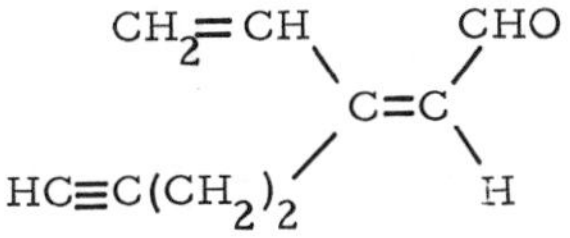

f.

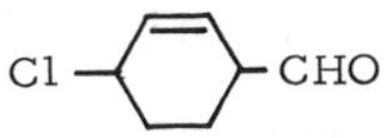

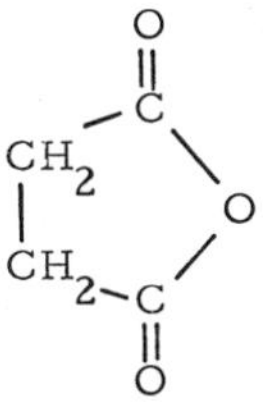

m. 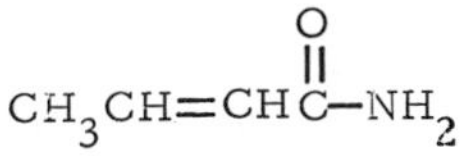 $CH_3CH=CHC-NH_2$ (with C=O)

g.

n. $CH_3(CH_2)_5CN$

h. $HC\equiv C-CH_2-C-CH_3$ (with C=O)

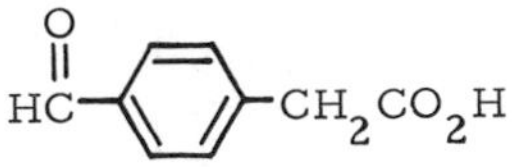

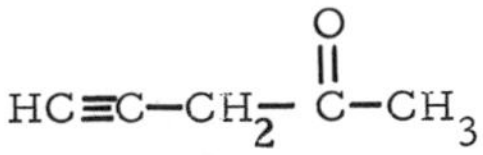

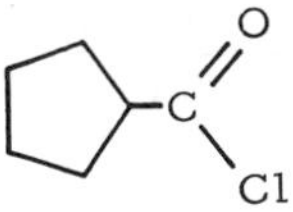

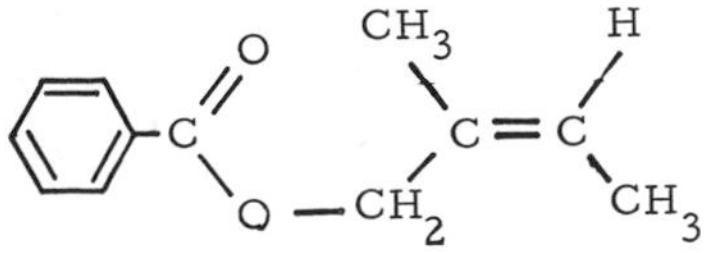

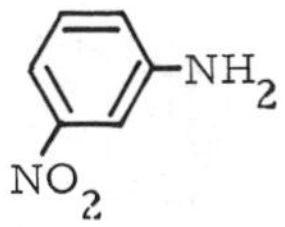

o. $CH_3CH_2CHC-C-OH$ (with two C=O and $HNCH_3$)

i.

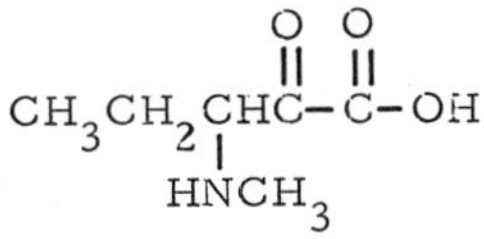

p.

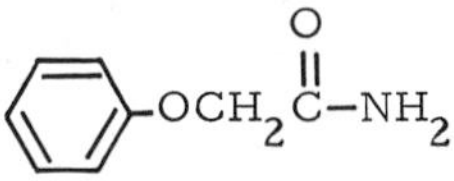

j.

q.

k.

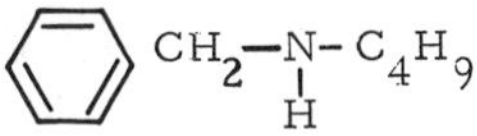

r.

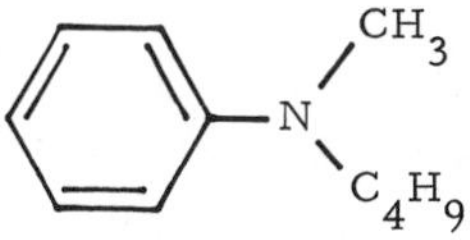

ADDITIONAL EXERCISES

7A-1   A number of compounds that are familiar, widely-used naturally occurring compounds or industrial chemicals are listed below. Their structures are shown together with their common names, their source and uses. For each compound, write a suitable name based on the systematic nomenclature rules discussed in Chapters 3 and 7.

a. diacetone alcohol $(CH_3)_2C(OH)CH_2COCH_3$ industrial chemical, solvent

b. mesityl oxide $(CH_3)_2C=CHCOCH_3$ industrial chemical, solvent

c. methyl malonate $CH_2(CO_2CH_3)_2$ synthetic intermediate

d. isovaleraldehyde $(CH_3)_2CHCH_2CHO$ occurs in orange, lemon, peppermint, eucalyptus and other oils.

e. citric acid $HO_2CCH_2\overset{\overset{\displaystyle OH}{|}}{\underset{\underset{\displaystyle CO_2H}{|}}{C}}CH_2CO_2H$ occurs widely in plants, fruits, especially citrus

f. sorbitol $HOCH_2(CHOH)_4CH_2OH$ occurs in fruit, seaweed, algae; used in candy manufacture, chemical and pharmaceutical industries

g. malic acid $HO_2CCH(OH)CH_2CO_2H$

h. oxalacetic acid $HO_2CCOCH_2CO_2H$ important intermediates in cellular metabolism

i. α-ketoglutaric acid $HO_2CCH_2CH_2COCO_2H$

j. valine $(CH_3)_2CHCH(NH_2)CO_2H$

k. serine $HOCH_2CH(NH_2)CO_2H$

l. cysteine $HSCH_2CH(NH_2)CO_2H$ natural α-amino acids from hydrolysis of proteins

m. lysine $H_2N(CH_2)_4CH(NH_2)CO_2H$

n. monosodium glutamate   from hydrolysis of proteins; imparts meat flavor to food.
$HO_2CCH(NH_2)CH_2CH_2CO_2^{\ominus}Na^{\oplus}$

o. acetylcholine chloride $(CH_3)_3\overset{\oplus}{N}CH_2CH_2O-\underset{\underset{\displaystyle O}{||}}{C}-CH_3$ neurotransmitter
$Cl^{\ominus}$

p. acetylenecarboxamide $H_2N\underset{\underset{\displaystyle O}{||}}{C}C\equiv C\underset{\underset{\displaystyle O}{||}}{C}NH_2$ antibiotic substance

q. acetylsalicyclic acid   aspirin

r. DDT   insecticide

s. methacrylonitrile $CH_2=C(CH_3)CN$ polymer intermediate

t. isopropenyl acetate $CH_2=C(CH_3)O\underset{\underset{\displaystyle O}{||}}{C}CH_3$ transfers CH_3CO group to enols

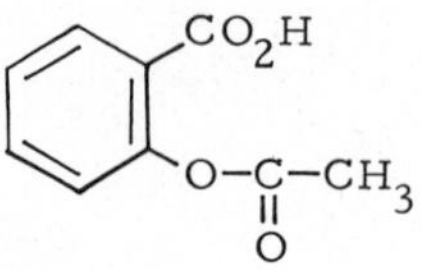

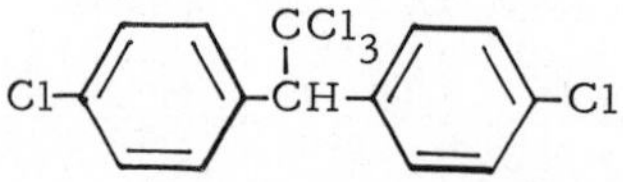

u.

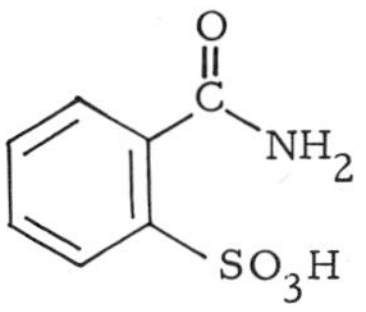

precursor to saccharin

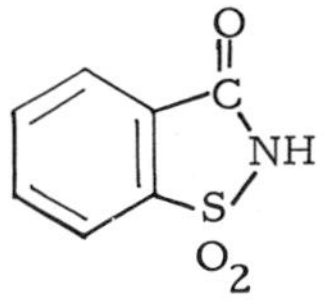

v.   shikimic acid

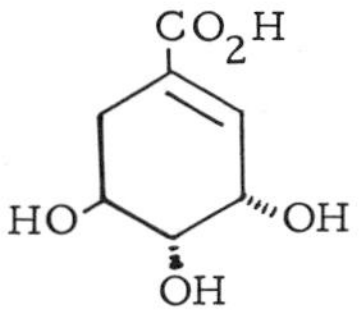

constituent of plants; an important biosynthetic precursor to aromatic compounds

w.   vanillin

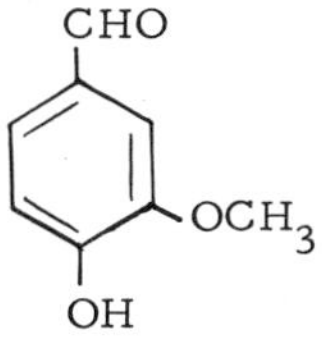

occurs in vanilla pods; flavoring agent

x.   picric acid

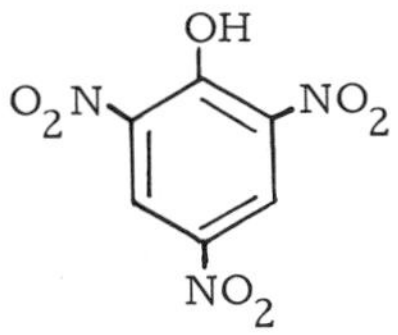

used as an explosive, in matches, in leather industry, as textile mordant.

y.   tiglic acid

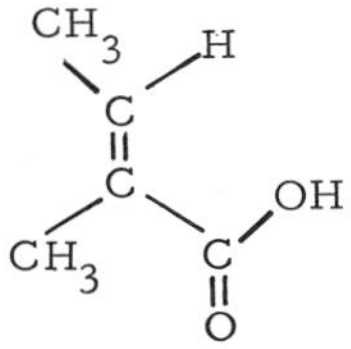

occurs in the form of esters in many plants

angelic acid

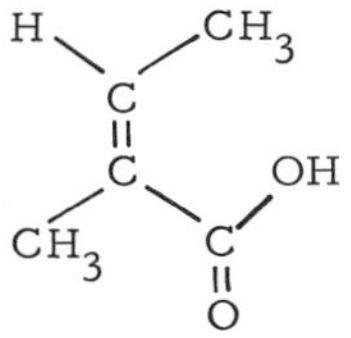

occurs in the form of esters in the root of _Angelica archangelica_

z.   geranyl tiglate (oil of geranium)

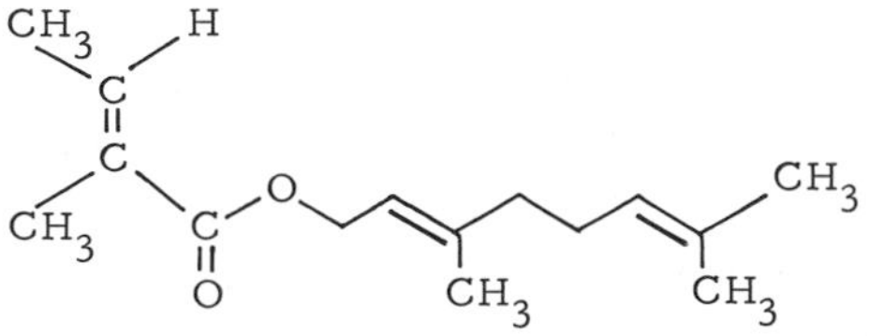

CHAPTER 8

NUCLEOPHILIC SUBSTITUTION AND ELIMINATION REACTIONS

Nucleophilic substitution S_N reactions are of the type

$$X-C{\cdots} + Y: \longrightarrow \overset{\oplus}{Y}-C{\cdots} + \overset{\ominus}{X}:$$

in which X and Y are more electronegative atoms than carbon. For simple
alkyl compounds there are two principal mechanisms of displacement, classified as S_N1 and S_N2. The S_N2 reaction is a one-step (concerted) reaction in
which a nucleophile Y: (or Y:$^\ominus$) forms a new bond to carbon as another breaks
with the departure of the attached group X along with the bonding electron pair.
Without exception, S_N2 reactions occur with <u>inversion</u> of configuration at the
reacting carbon because the entering nucleophile attacks carbon from the side
opposite to the leaving group.

<u>S_N2</u> (<u>inversion mechanism</u>):

$$Y: + {\cdots}C:X \longrightarrow \overset{\oplus}{Y}:C{\cdots} + :\overset{\ominus}{X}$$

The structural requirements for S_N2 reactions are well defined. The
reagent Y: must be nucleophilic; the group X must leave readily with an electron pair; and the reacting carbon should be free of steric encumbrances.
Hence, the reactivity order <u>primary</u> > <u>secondary</u> > <u>tertiary</u>. If Y: is an
anion, a mutual solvent for Y:$^\ominus$ and RX is needed and the choice of solvent
can be critical to attaining practical reaction rates. The best solvents for
promoting S_N2 reactions are those that solvate cations well and anions poorly.
An example is methylsulfinylmethane [dimethyl sulfoxide, $(CH_3)_2S{=}O$].

Nucleophilic substitution is of great practical value for the synthesis of
alkyl halides, alcohols, thiols, ethers, thioethers, nitriles, esters, alkynes,
amines, nitrates, ammonium and sulfonium compounds, and many other types
of compounds. Tabulations of S_N2 reactions are given in Tables 8-1S and 8-2S.
The reactions of Table 8-1S show the range of nucleophiles commonly used;
those in Table 8-2S show some of the range of the leaving groups X of RX
compounds.

Table 8-1S   Typical S_N Displacement Reactions of Alkyl Halides, RX

1. $$R \,\vdots\, X + Y : ^{\ominus} \longrightarrow R : Y + X : ^{\ominus}$$

Nucleophilic agent	Product	Product name, R=CH$_3$	Useful solvents
Cl$^{\ominus}$	RCl	methyl chloride	2-propanone (acetone), ethanol
Br$^{\ominus}$	RBr	methyl bromide	2-propanone, ethanol
I$^{\ominus}$	RI	methyl iodide	2-propanone, ethanol
$^{\ominus}$OH	ROH	methanol	water, 1,4-dioxacyclohexane (dioxane) water
$^{\ominus}$OCH$_3$	ROCH$_3$	dimethyl ether	methanol
$^{\ominus}$SCH$_3$	RSCH$_3$	dimethyl sulfide	ethanol
CH$_3$—C(=O)O$^{\ominus}$	RO—C(=O)CH$_3$	methyl ethanoate (methyl acetate)	ethanoic acid (acetic acid), ethanol
$^{\ominus}$: C≡N	RCN	ethanenitrile (acetonitrile)	2-propanone, methyl sulfinylmethane (dimethyl sulfoxide)
HC≡C: $^{\ominus}$	RC≡CH	propyne	liquid ammonia
$^{\ominus}$: CH(CO$_2$C$_2$H$_5$)$_2$	RCH(CO$_2$C$_2$H$_5$)$_2$	diethyl methyl-propanedioate (diethyl methylmalonate)	ethanol
$^{\ominus}$: N̈H$_2$	RNH$_2$	methanamine (methylamine)	liquid ammonia
: N$^{\ominus}$=N$^{\oplus}$=N$^{\ominus}$:	RN$_3$	azidomethane (methyl azide)	2-propanone
(phthalimide N$^{\ominus}$)	(RN-phthalimide)	N-methylphthalimide	N,N-dimethyl-methanamide [HCON(CH$_3$)$_2$]
NO$_2$$^{\ominus}$	RNO$_2$	nitromethane	N,N-dimethyl-methanamide

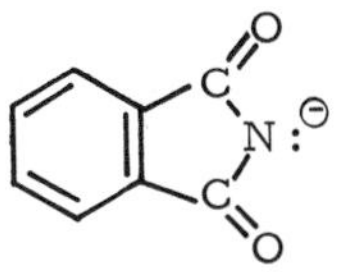

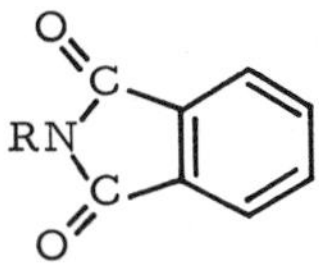

Table 8-1S (cont'd)

2.			$R \vdots X + Y : \longrightarrow \overset{\oplus}{R} : Y + \overset{\ominus}{X} :$	

Nucleophilic agent	Product	Product name, $R = CH_3$, $X = Cl$	Useful solvents
$(CH_3)_3N:$	$\overset{\oplus}{R}N(CH_3)_3\overset{\ominus}{X}$	tetramethylam- monium chloride	diethyl ether, benzene
$(C_6H_5)_3P:$	$\overset{\oplus}{R}P(C_6H_5)_3\overset{\ominus}{X}$	triphenylmethyl- phosphonium chloride	diethyl ether, benzene
$(CH_3)_2S:$	$\overset{\oplus}{R}S(CH_3)_2\overset{\ominus}{X}$	trimethylsulfonium chloride	diethyl ether, benzene

3.		$R \vdots X + H : Y: \longrightarrow R : \overset{\oplus}{Y} : H + X : \overset{\ominus}{} \longrightarrow R : Y : + H : X$	

Nucleophilic agent	Product	Product name, $R = CH_3$	Useful solvents
H_2O	ROH	methanol	water, 1,4-dioxa- cyclohexane (dioxane)-water
CH_3OH	$ROCH_3$	dimethyl ether	methanol
CH_3CO_2H	$RO\overset{\overset{O}{\|\|}}{C}CH_3$	methyl ethanoate	ethanoic acid
NH_3	RNH_2	methanamine (methylamine)	ammonia, methanol

Table 8-2S S_N Displacement Reactions of Various Types of Compounds, RX

Type of compound, RX	Illustrative reaction
alkyl chloride	$R-Cl + I^{\ominus} \rightleftharpoons RI + Cl^{\ominus}$
alkyl bromide	$R-Br + I^{\ominus} \rightleftharpoons RI + Br^{\ominus}$
alkyl iodide	$R-I + CH_3O^{\ominus} \rightleftharpoons ROCH_3 + I^{\ominus}$
dialkyl sulfate	$R-OSO-R + CH_3O^{\ominus} \longrightarrow ROCH_3 + {}^{\ominus}OSOR$ (with O above and below each S)
alkyl benzenesulfonate	$R-OS(C_6H_5) + H_2O \longrightarrow ROH + HOS(C_6H_5)$ (with O above and below each S)
alcohol	$R-OH + HBr \longrightarrow RBr + H_2O$
ether	$R-OR' + HBr \longrightarrow RBr + R'OH$
ammonium ion	$R-NR'_3{}^{\oplus} + HO^{\ominus} \longrightarrow ROH + NR'_3$
iodonium ion	$R-I-R'{}^{\oplus} + OH^{\ominus} \longrightarrow ROH + R'I$
diazonium ion	$R-N{\equiv}N^{\oplus} + H_2O \longrightarrow ROH + H^{\oplus} + N_2$

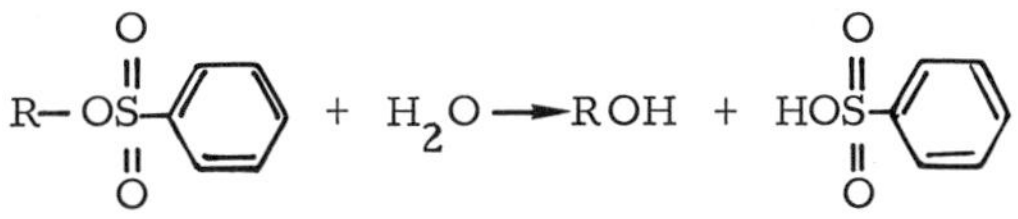

Substitution by the S_N2 mechanism often is accompanied by (and complicated by) a bimolecular elimination of HX to form an alkene.

E2 elimination mechanism:

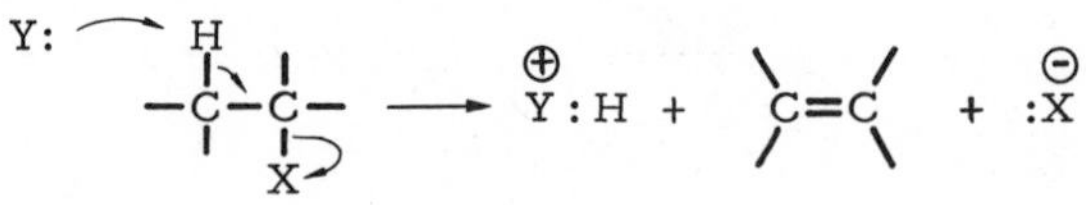

The E2 elimination is facilitated if Y: is a strong base and X is a good leaving group. A β hydrogen is required. The reactivity sequence for E2 is tertiary > secondary > primary. The preferred stereochemistry leads to antarafacial elimination by way of a transition state in which $H-C_\beta-C_\alpha-X$ are coplanar with the H and X trans to each other. However, suprafacial elimination also can occur, especially where the structure of the compound can prevent the transition state for antarafacial elimination from being formed. In general, E2 elimination is favored over S_N2 substitution when the attacking reagent is a strong base, a bulky base, or the reaction temperature is high (100° or higher).

Substitution by the S_N1 mechanism is a stepwise process. The critical first step is the ionic cleavage of the C—X bond. The carbocations so formed usually are highly reactive intermediates, which in subsequent steps react competitively with available nucleophiles in the reaction mixture.

S $_N$1 mechanism:

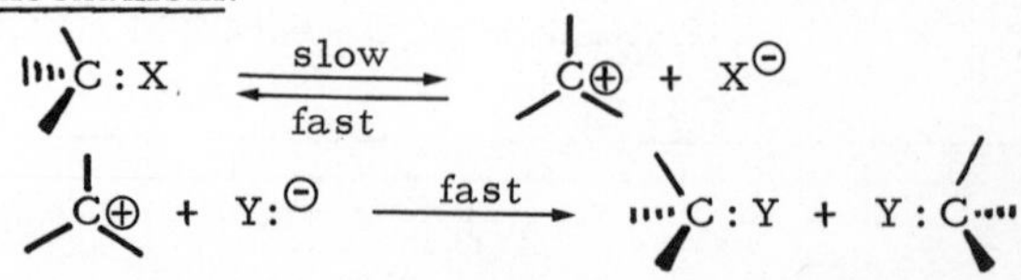

For alkyl compounds with a given leaving group, the ease of ionization depends on the stability of the carbocations that are formed. The usual reactivity order is tertiary > secondary > primary. In all cases, ionization requires a good leaving group such as halide, sulfonate ester, or onium ion, and a good ionizing solvent. The stereochemistry of substitution depends on the degree to which the solvent separates the ions, but generally there is some preference for the product of inverted configuration. In some cases, complete racemization occurs. Strongly nucleophilic or basic reagents may interfere with S_N1 substitution by promoting S_N2 or E2 reactions.

The hydrolysis of tert-butyl bromide in 2-propanone-water solution at 25° is a typical S_N1 reaction. The 2-propanone helps dissolve tert-butyl bromide while the water promotes the ionization. The reaction occurs at a reasonable rate because the _tert_-butyl carbocation is tertiary, bromide is a good leaving group, and the solvent is a good ionizing solvent. Lastly, no strong bases are present to cause E2 eliminations.

$$(CH_3)_3C-Br \; \xrightleftharpoons[]{H_2O \;(2\text{-propanone})} \; (CH_3)_3C^{\oplus} + Br^{\ominus}$$

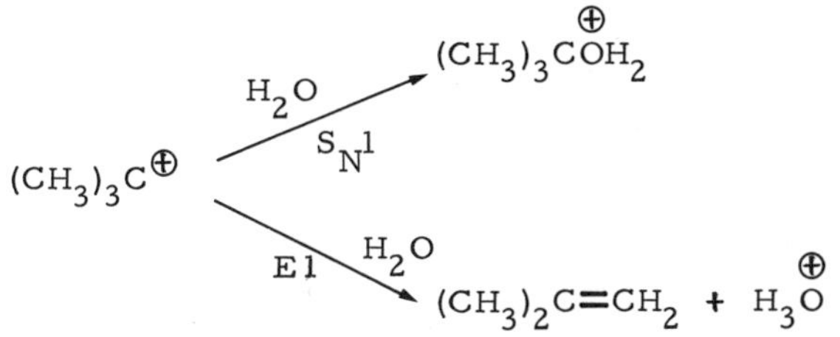

$$(CH_3)_3C^{\oplus} \underset{\underset{E1}{\searrow}\; H_2O}{\overset{\overset{H_2O}{\nearrow}\; S_N1}{}}$$

$(CH_3)_3C\overset{\oplus}{O}H_2$ substitution

$(CH_3)_2C{=}CH_2 + H_3\overset{\oplus}{O}$ elimination

Compounds of structure $\diagdown{\!\!C}{=}C{-}\overset{|}{\underset{|}{C}}{-}X$ and $-O-\overset{|}{\underset{|}{C}}{-}X$ have high S_N1 reactivity because they lead to especially stabilized cations:

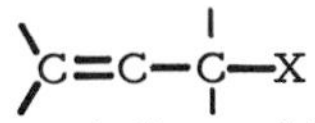

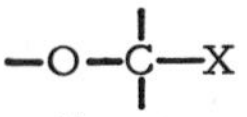

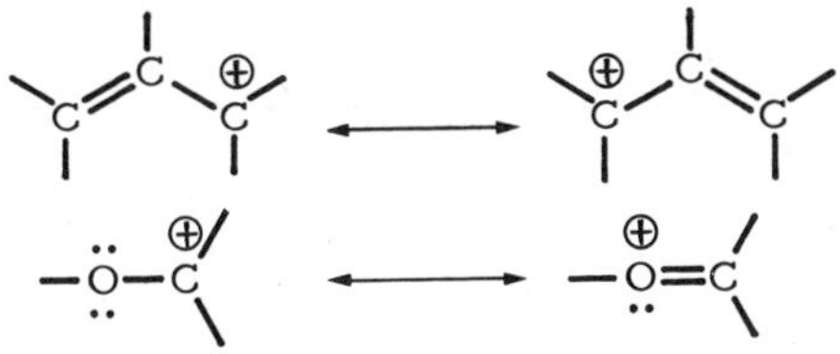

In any ionization reaction involving carbocations, the cation may react with a nucleophile (including the solvent) at carbon to give substitution products (S_N1), or it may react at a β hydrogen (if one is present) to give elimination products (E1), or it may rearrange to a more stable ion and give products of rearranged structure. Clearly, mixtures of substitution and elimination products as well as mixtures of products with different carbon skeletons can result. For this reason, S_N1 or E1 reactions do not have the synthetic importance of S_N2 or E2 reactions.

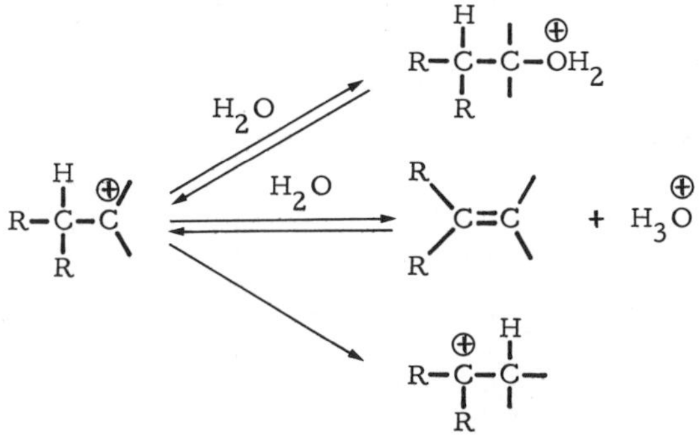

The structural and medium effects that so profoundly influence both substitution and elimination reactions are summarized for reference in Table 8-3S. This table will be useful for evaluating the outcome of reactions of alkyl compounds with nucleophiles and in devising suitable syntheses involving substitution and elimination.

Table 8-3S   Summary of Conditions for Substitution and Elimination Reactions

	S_N1	S_N2
Mechanism	Two step $$R{-}X \xrightarrow[\text{slow}]{(-X^{\ominus})} R^{\oplus} \xrightarrow[\text{fast}]{Nu:} R{-}Nu^{\oplus}$$	One step $$Nu: + R{-}X \rightarrow \overset{\oplus}{Nu}{-}R + :X^{\ominus}$$
Kinetics	First order (unimolecular) Rate = $\underline{k}$ [RX] Ionization is rate determining; second step does not affect rate as long as it is fast	A bimolecular reaction (usually second order) Rate = $\underline{k}$ [Nu:] [RX] If Nu: is the solvent, rate becomes first order but the mechanism remains S_N2
Stereochemistry	Racemization expected but inversion normally predominates	Inversion
Solvent Effects	Much favored by ionizing (polar) solvents such as H_2O, CH_3OH, CH_3CH_2OH, SO_2, HCO_2H	Polarity of solvent not so important. Nonpolar solvents favor S_N2 by suppressing S_N1. Aprotic (non-H-bonding) solvents are particularly good. Examples are 2-propanone, CH_3COCH_3, and methylsulfinylmethane, CH_3SOCH_3

Table 8-3S (cont.)

	E1	E2
Mechanism	Two-step process:	One-step process:
Kinetics	First order (unimolecular) Rate = $\underline{k}$ [RX] Ionization is rate determining; Second step does not affect overall rate as long as it is fast	Second order (bimolecular) Rate = $\underline{k}$ [B:] [RX]
Stereochemistry	When cis-trans isomers are possible, both usually are formed - low stereospecificity	HX eliminated most rapidly by way of a trans-coplanar transition state:
Solvent Effects	As in S_N1	Poor anion-solvating solvents can give very rapid E2 reactions

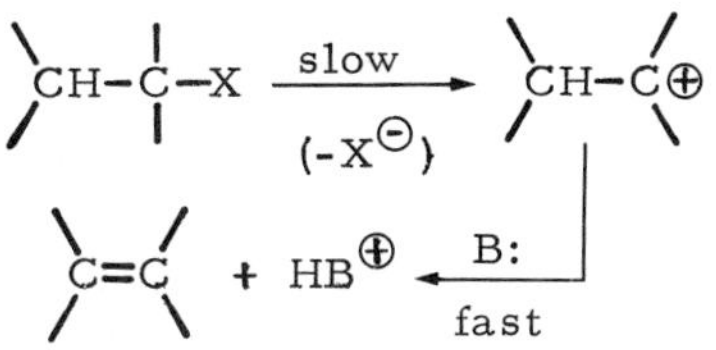

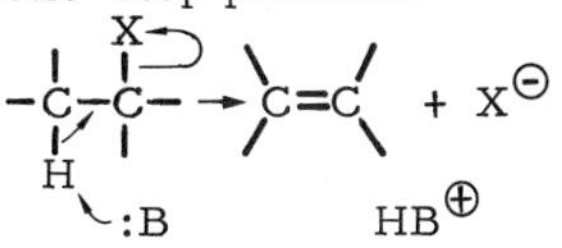

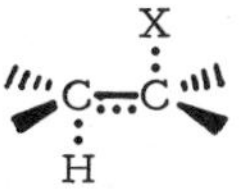

Table 8-3S (cont.)

	S_N1	S_N2
Effect of R	Resonance stabilization of intermediate ions favors reaction $\underline{tert}$-R > $\underline{sec}$-R >> $\underline{prim}$-R C_6H_5X almost never reacts	Steric hindrance largely controls the reaction rate CH_3 > $\underline{prim}$-R > $\underline{sec}$-R >> $\underline{tert}$-R C_6H_5 does not react unless activating groups are present
Competitive reactions	E1 or E2 elimination, rearrangement	E2 elimination by strongly basic nucleophiles
Catalysis	$H^\oplus$ with R—OH, R—OR $Ag^\oplus$ with R—X	$H^\oplus$ with R—OH, R—OR
Leaving Group	Good leaving groups usually are weak bases. Good leaving groups include: $-O-\overset{\overset{O}{\|}}{\underset{\underset{O}{\|}}{S}}-R$, —I, —Br, —Cl, $\overset{\oplus}{-}OH_2$, $\overset{\oplus}{-}NR_3$, $\overset{\oplus}{-}N_2$, $-O-\overset{\overset{O}{\|}}{C}-CF_3$ Trends: —I > —Br > —Cl >> —F Poor leaving groups: —OH, —OR, —H, —CR$_3$, —NR$_2$ (i.e., strong bases)	As for S_N1 Strong bases usually are good nucleophiles but the reverse need not be true. In fact, large (weakly basic, but polarizable) ions are particularly good.
Nucleophile (Nu:)	As for S_N2. However, the concentration of Nu: does not directly affect the S_N1 rate. The products may depend on the nature of Nu:; S_N1 reactions can be favored by using weak nucleophiles that are ineffective for S_N2.	Good nucleophiles: $I^\ominus$, $Br^\ominus$, $Cl^\ominus$, $RO^\ominus$, $HO^\ominus$, $RS^\ominus$, $RCO_2{}^\ominus$, $NO_2{}^\ominus$, $R_2N^\ominus$, $R\overset{..}{S}R$, $R_3\overset{..}{N}$, $R_3\overset{..}{P}$ Fair nucleophiles: $H\overset{..}{O}H$, $R\overset{..}{O}H$, $H-\overset{..}{O}-\overset{\overset{O}{\|}}{C}R$ Trends: $I^\ominus$ > $Br^\ominus$ > $Cl^\ominus$ >> $F^\ominus$, $RS^\ominus$ > $RO^\ominus$, R_3P: > R_3N: Solvation plays an important role in determining nucleophilicity

Table 8-3S (cont.)

	E1	E2
Effect of R	As for S_N1	tert-R > sec-R > prim-R (must have β hydrogen)
Competitive reactions	S_N1 (but E1 favored over S_N1 with increasing temperature) rearrangement is common	S_N2 Bulky bases, such as $^\ominus OC(CH_3)_3$, promote E2 by suppressing S_N2; higher reaction temperatures favor E2
Catalysis		
Leaving Group (X:)	As for S_N1	As for S_N1
Base (B:)	The base acts only to remove $H^\oplus$ from the carbocation. The proton-removal step usually is fast. However, if strong base is present the E2 mechanism tends to operate	Requires strong base: $HO^\ominus$, $RO^\ominus$, $H_2N^\ominus$, $H_3C^\ominus$ (as CH_3Li), $H^\ominus$ (as NaH)

ANSWERS TO EXERCISES

8-1   E stands for electrophilic, N for nucleophilic

a.

$$H:N:H \quad \xrightarrow{\qquad} \quad NH_4^{\oplus} \quad + H-\ddot{O}-H \text{ (N)}$$

with $H-\overset{H}{\underset{|}{\ddot{O}}}{}^{\oplus}H$

$$\xrightarrow{H-\ddot{O}:{}^{\ominus}} \quad H-\ddot{N}-H + H-\ddot{O}-H \text{ (E)}$$

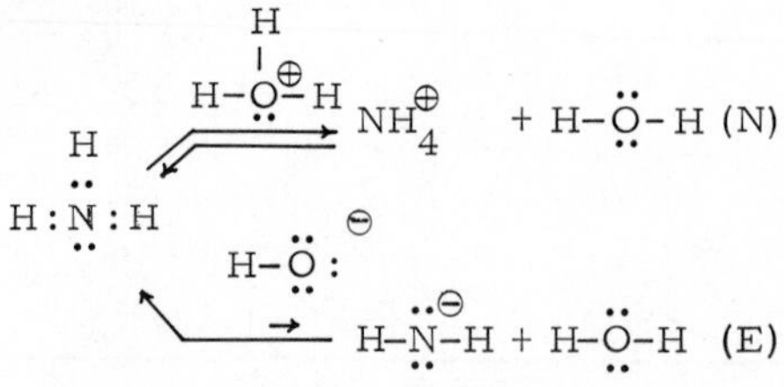

b. $H:\ddot{N}:H^{\ominus} + H_3O^{\oplus} \longrightarrow :\ddot{N}H_3 + H_2O$ (N)

c. $Na^{\oplus}$ neither (E) nor (N)

d. $:\ddot{\underset{..}{Cl}}:{}^{\ominus} + H_3O^{\oplus} \rightleftharpoons H-\ddot{\underset{..}{Cl}}: + H_2O$ (N)

e. $:\ddot{\underset{..}{Cl}}:\overset{|}{\underset{|}{\ddot{\underset{..}{Cl}}}}: + HO^{\ominus} \rightleftharpoons Cl^{\ominus} + Cl-OH$ (E)

f. Neither (E) nor (N), see Section 1-3

g. $:C:::N:{}^{\ominus} + H_3O^{\oplus} \rightleftharpoons H-C\equiv N + H_2O$ (N)

h.

$$H:C:\ddot{O}:H \xrightarrow{H_3O^{\oplus}} CH_3-\ddot{O}H_2^{\oplus} + H_2O \text{ (N)}$$

$$\xrightarrow{HO^{\ominus}} CH_3-\ddot{O}:{}^{\ominus} + H_2O \text{ (E)}$$

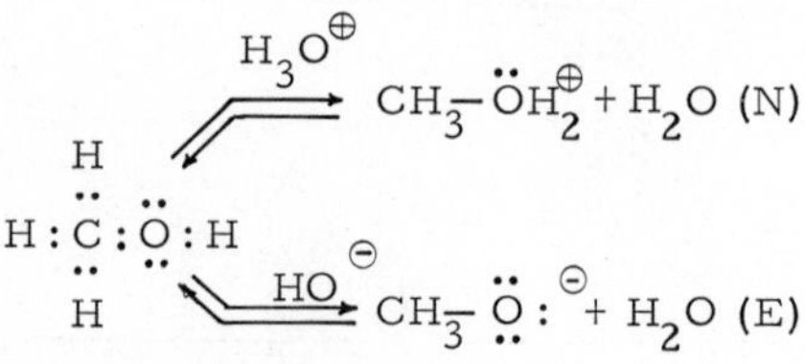

i. $CH_3-\overset{\oplus}{\ddot{O}}H_2 + HO^{\ominus} \longrightarrow CH_3\ddot{O}H + H_2O$ (E)

j. Neither (E) nor (N)   .

k. $H:\ddot{\underset{..}{Br}}: + HO^{\ominus} \longrightarrow :\ddot{\underset{..}{Br}}:{}^{\ominus} + H_2O$ (E)

l. $H:C:::C:{}^{\ominus} + H_3O^{\oplus} \longrightarrow$

$$H-C\equiv C-H + H_2O \quad \text{(N)}$$

8-1 (cont'd)

m. $H:\overset{..}{C}:H + HO^{\ominus} \longrightarrow H-\overset{\ominus}{\underset{\underset{OH}{|}}{C}}-H$ (E)

o. $:\overset{..}{\underset{..}{O}}\overset{\oplus}{\underset{..}{S}}:\overset{..}{\underset{..}{O}}:^{\ominus} + HO^{\ominus} \longrightarrow HO-\overset{O}{\underset{O}{\overset{\|}{\underset{\|}{S}}}}-\overset{..}{\underset{..}{O}}:^{\ominus}$ (E)

n. $F:\overset{:O:}{\underset{:O:}{\overset{..}{\underset{..}{S}}}}:\overset{..}{O}H + HO^{\ominus} \longrightarrow H_2O + F-\overset{:O:}{\underset{:O:}{\overset{\|}{\underset{\|}{S}}}}-\overset{..}{O}:^{\ominus}$ (E)

8-2

	E	N		E	N
a.	CH_3I	$CH_3O^{\ominus}$	c.	CH_3I	$CH_3\overset{..}{N}H_2$
b.	Br_2	$CH_2{=}CH_2$	d.	$CH_3\overset{\oplus}{O}H_2$	$Br^{\ominus}$

8-3 $CH_3Br(g) + H_2O(g) \longrightarrow CH_3OH(g) + HBr(g)$ $\quad\quad \underline{\Delta H} = + 5.7$ kcal

$H_2O(l) \longrightarrow H_2O(g)$ $\quad\quad \underline{\Delta H} = +10.8$ kcal

$HBr(g) \longrightarrow HBr(aq)$ $\quad\quad \underline{\Delta H} = -20 \quad$ kcal

$CH_3Br(aq) \longrightarrow CH_3Br(g)$ $\quad\quad \underline{\Delta H} = + 1 \quad$ kcal

$CH_3OH(g) \longrightarrow CH_3OH(aq)$ $\quad\quad \underline{\Delta H} = -10 \quad$ kcal

net $CH_3Br(aq) + H_2O(l) \longrightarrow CH_3OH(aq) + HBr(aq)$ $\quad\quad \underline{\Delta H} = -12.5$ kcal

8-4   a.   Rate $= \underline{k}[C_2H_5Cl][KI]$

$= \underline{k}[0.1][0.1] = 5.44 \times 10^{-7}$, or $\underline{k} = 5.44 \times 10^{-5}$

At 0.01 M, rate $= 5.44 \times 10^{-5} [0.01][0.01]$

$= 5.44 \times 10^{-9}$

b.   Rate $= \underline{k}[C_2H_5Cl][KI]^2$

$\underline{k} = \dfrac{5.44 \times 10^{-7}}{(0.1)^3} = 5.44 \times 10^{-4}$

At 0.01 M, rate $= 5.44 \times 10^{-4} \times (0.01)^3 = 5.44 \times 10^{-10}$

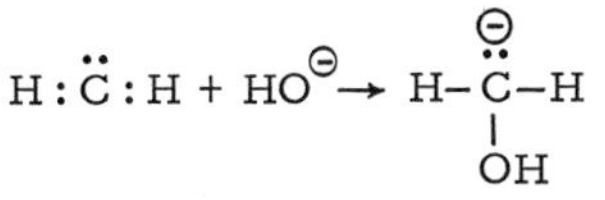

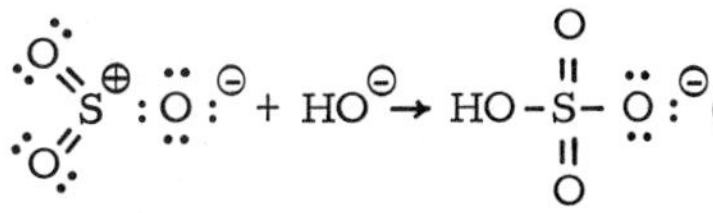

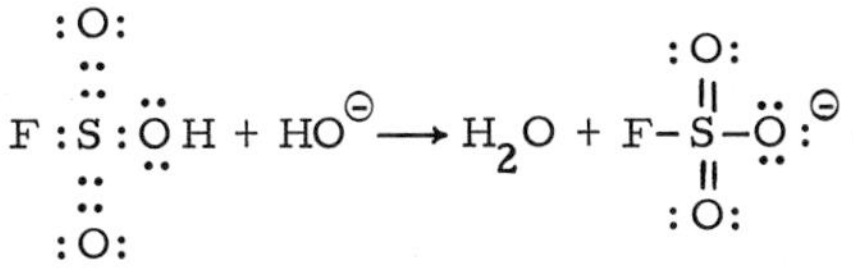

8-4 (cont'd)

c.

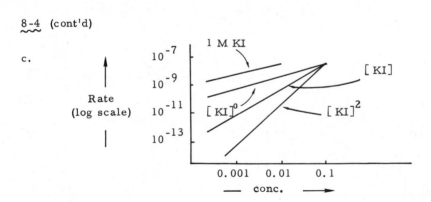

d.   If we were to change the concentration of just the potassium iodide, the

rate would remain <u>unchanged</u> if it were zeroth-order in potassium iodide; but it

would change markedly ($\propto$[KI]) if first-order in potassium iodide.

e.   Initially, the rate is the same as in Part c. for $[KI]^0$ and $[KI]$ but

decreases as if it were first order in ethyl chloride and zeroth order in KI.

The reason for the apparent first order rate overall is that the KI is in 100-

fold excess, and therefore its concentration does not vary significantly with time.

8-5   An S_N2 mechanism of solvolysis can be ruled unlikely because the rate

of disappearance of RCl by an S_N2 reaction should <u>increase</u> with added nucleo-

philes (azide).

$$RCl + H_2O \xrightarrow{k_1} ROH + HCl \qquad RCl + N_3^\ominus \xrightarrow{k_2} RN_3 + Cl^\ominus$$

$$\text{Rate} = k_1[RCl] + k_2[RCl][N_3^\ominus]$$

An S_N1 mechanism is implied because the slow step (ionization) precedes

the step involving the nucleophile.

$$RCl \xrightarrow[k]{slow} R^\oplus + Cl^\ominus \qquad R^\oplus \begin{array}{c} \overset{H_2O}{\nearrow} \; ROH + H^\oplus \\ \text{fast} \\ \underset{\text{fast}}{\searrow} \; R-N_3 \\ N_3^\ominus \end{array}$$

$$\text{Rate} = k[RCl]$$

8-6   An S_N1 mechanism is implied because added chloride can reverse the ionization of RCl. The more $R^{\oplus}$ that is trapped by $Cl^{\ominus}$ the less hydrolysis product is formed.

$$R-Cl \underset{}{\overset{slow}{\rightleftarrows}} R^{\oplus} + Cl^{\ominus}, \quad R^{\oplus} + H_2O \xrightarrow{fast} ROH + H^{\oplus}$$

This has the net effect of retarding the rate of hydrolysis of RCl.

8-7   Reactions 8-3 and 8-4 do not affect the chiral center and therefore occur with retention of configuration. Reaction 8-5 gives the same product as reaction 8-3 but with the opposite configuration, which means that inversion of configuration occurs in reaction 8-5.

8-8   A reaction must occur in which $Br^{\ominus}$ displaces the bromine from 2-bromobutane by an S_N2 mechanism. Substitution inverts the configuration, and because the reaction is reversible ($\underline{K} = 1$), at equilibrium there will be equal amounts of $\underline{D}$ and $\underline{L}$ enantiomers.

8-9   The experimental facts are that racemization of the chiral bromide takes place twice as fast as the rate of incorporation of radioactive $^*Br^{\ominus}$. These results mean that every act of substitution must invert configuration. Thus, if we start with 100 molecules of the $\underline{L}$ enantiomer and convert 5 of them to $\underline{D}$ enantiomers by the substitution reaction $RBr + {^*Br}^{\ominus} \longrightarrow RBr^* + Br^{\ominus}$, the measured optical activity will be reduced by 10 per cent (because the rotation of the 5 $\underline{D}$ molecules will cancel out the optical rotation of 5 $\underline{L}$ molecules). The actual rate of substitution is the rate of incorporation of $^*Br^{\ominus}$, and this rate can only

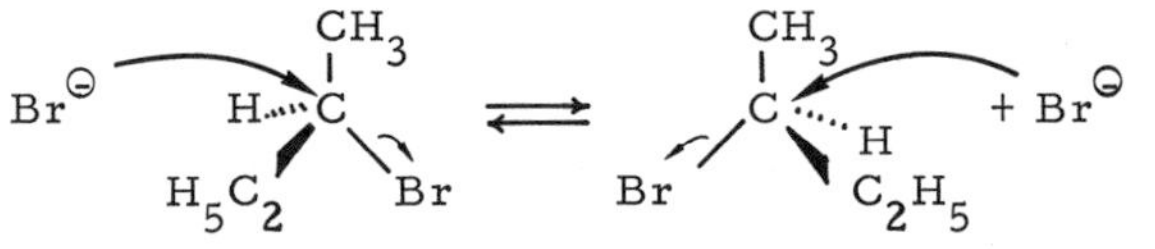

8-9 (cont'd)

be half the rate of racemization provided inversion of configuration occurs at

every pass.

8-10   Solvolysis results in <u>inversion</u> of configuration at the primary carbon,

consistent with an S_N2 reaction.

8-11   a. $\underline{\underline{L}}$ b. $\underline{\underline{D}}$ and $\underline{\underline{L}}$ with $\underline{\underline{D}} < \underline{\underline{L}}$

8-12

	Most S_N2 reactive	Reason	Product name
a.	$CH_3CH_2CH_2CH_2Cl$	least steric hindrance	1-iodobutane
b.	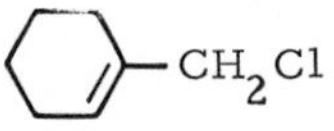 $-CH_2Cl$	primary halide and activated by adjacent double bond	1-iodomethylcyclo-hexene
c.	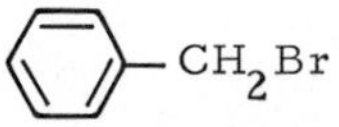 $-CH_2Br$	primary halide activated by adjacent phenyl group	iodomethylbenzene

8-13   a. 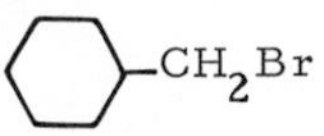 $-CH_2Br$ b.

8-14   a. 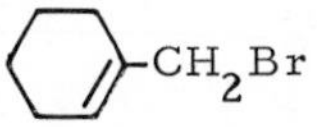 $-CH_2Br$ b.

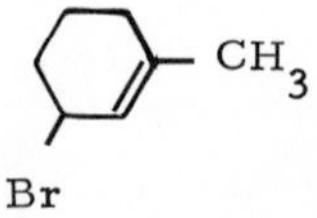

8-15   b and c only.

8-16

a.   The inertness of apocamphyl chloride on solvolysis suggests that it can-

not ionize to a stable tertiary carbocation. The reason for the instability is that

the bicyclic structure of the carbon framework prevents the positive carbon and

the three carbons attached to it from achieving a planar configuration. (Make a

model to see the constraints imposed.) Likewise, the rigid structure precludes

substitution with inversion of configuration (S_N2).

8-16 (cont'd)

b.   The high $S_N 1$ reactivity and retardation of solvolysis by lithium

chloride for $ROCH_2Cl$ indicates that compounds of this type form relatively

stable carbocations on ionization.   The ions are stabilized by electron delocali-

zation involving the electrons on oxygen.

$$R-\overset{..}{\underset{..}{O}}-\overset{\oplus}{C}H_2 \longleftrightarrow R-\overset{\oplus}{\underset{..}{O}}=CH_2$$

8-17

Reactive	Slow	Unreactive
a, b, e, f	d, g	c, h

(Based on pK_a data for H_2SO_4, $HONO_2$, HCN, HF, HI, $HOSO_2F$,

HO_2CCH_3, and H_2O of <0, <0, 9.3, 3.45, 0.77, <0, 4.75, and 7

respectively).

8-18

a.   A strong acid (HA) acts as an electrophilic catalyst by assisting the

cleavage of the C—F bond.

$$-\overset{|}{\underset{|}{C}}-\overset{..}{\underset{..}{F}}: + \text{ HA} \longrightarrow -\overset{|}{\underset{|}{C}}---\overset{..}{\underset{..}{:F:}}---H\overset{|}{:}A \longrightarrow -\overset{|}{C}\overset{\oplus}{} + :\overset{..}{\underset{..}{F}}:H + :A^\ominus$$

b.   Reversible ionization of the alkyl chloride by $HgCl_2$ (acting as an

electrophile) racemizes the alkyl chloride.

$$(+)-R-Cl + HgCl_2 \rightleftarrows R^\oplus ---- Cl-\overset{\ominus}{H}gCl_2 \rightleftarrows (-)-R-Cl + HgCl_2$$
$$\underset{}{\overset{}{\big\downarrow}} \text{H}_2\text{O} \quad (\pm)- ROH + HCl + HgCl_2$$

This process results in RCl racemizing <u>before</u> RCl hydrolyzes to ROH.

c.   Acid is required to activate the hydroxyl group for displacement by

bromide ion.

$$ROH \underset{}{\overset{\overset{\oplus}{H}}{\rightleftarrows}} RO\overset{\oplus}{H}_2 \overset{\overset{\ominus}{Br}}{\longrightarrow} RBr + H_2O .$$

8-18 (cont'd)

d. $C_6H_5O^\ominus$ is a better leaving group than $C_2H_5O^\ominus$ for the same reason

that C_6H_5OH is a stronger acid than C_2H_5OH; the $C_6H_5O^\ominus$ anion is stabil-

ized by electron delocalization involving the π electrons of the ring. Ethoxide

can not stabilize in the same way.

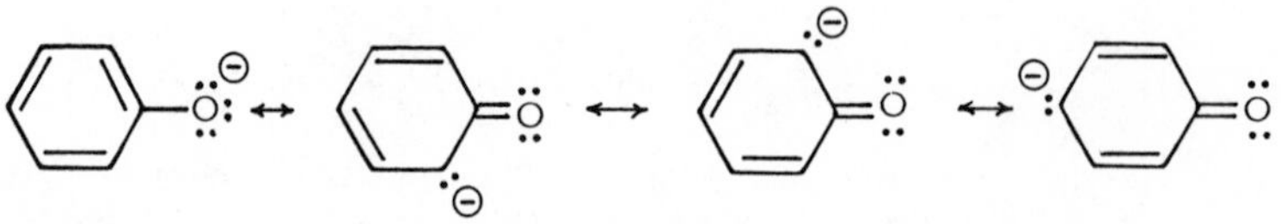

8-19   a. $C_6H_5O-CH_3 + HI \xrightarrow{S_N2} C_6H_5OH + CH_3I$

b. $(CH_3)_3C-OCH_3 + HCl \xrightarrow{S_N1} (CH_3)_3C-Cl + CH_3OH$

c. $CF_3O-CH_3 + HBr \xrightarrow{S_N2} CF_3OH + CH_3Br$

d. $CH_2=CHCH_2-\overset{\oplus}{\underset{CH_3}{S}}-C_3H_7 + \overset{\ominus}{I} \longrightarrow CH_3I + CH_2=CHCH_2SC_3H_7$

8-20   Silver ion coordinates with the iodine of CH_3I, which makes iodide a

better leaving group.

$$CH_3I + \overset{\oplus}{Ag} \rightleftharpoons CH_3-\overset{\oplus}{I}-Ag \xrightarrow[S_N2]{ROH} CH_3OR + \overset{\oplus}{H} + AgI .$$

8-21

a.   The result is evidence that the reaction is readily reversible. It pro-

ceeds to the right in benzene only because the product is insoluble in that sol-

vent. On solution in ethanol, the reaction reverses.

$$(CH_3)_2S:CH_2-Cl \underset{C_2H_5OH}{\overset{C_6H_6}{\rightleftharpoons}} (CH_3)_2\overset{\oplus}{S}-CH_2 \quad :\overset{\ominus}{Cl}:$$

(with R below both CH_2 groups)

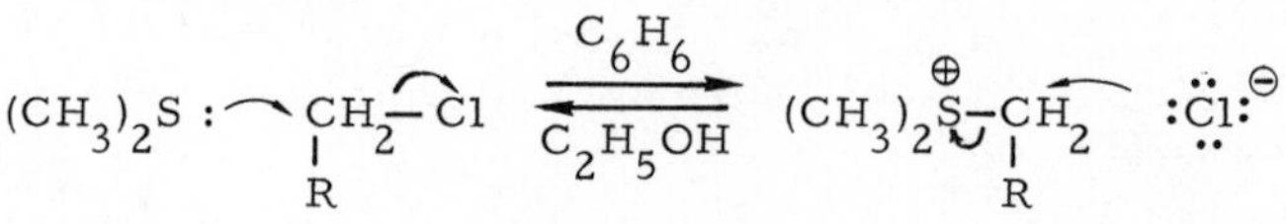

b.   Iodide catalysis involves <u>two</u> S_N2 displacements--one to convert RCl

to RI, the second to convert RI to ROH. Since inversion occurs at each step,

8-21 (cont'd)

the product alcohol by the catalytic sequence retains the configuration of the RCl.

$$\underline{D}\text{-RCl} \xrightarrow{\text{I}^{\ominus}} \underline{L}\text{-RI} \xrightarrow{\text{HO}^{\ominus}} \underline{D}\text{-ROH} \qquad \text{fast}$$

$$\xrightarrow{\text{HO}^{\ominus}} \underline{L}\text{-ROH} \qquad \text{slow}$$

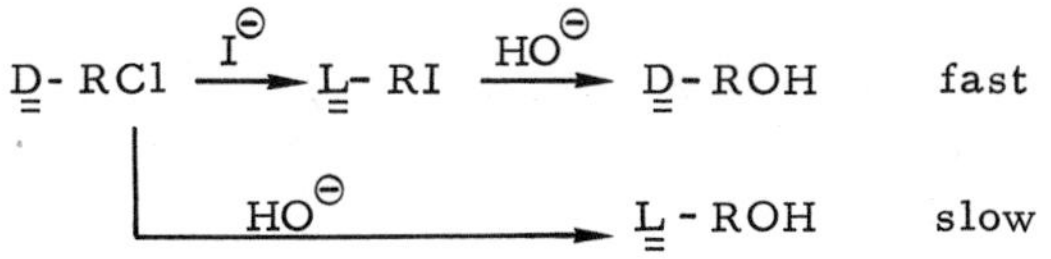

Direct hydrolysis inverts configuration. The net result is much less than 100%
inversion. (See Exercise 8-42).

c. The CF_3 groups are strongly electronegative (electron withdrawing) and
reduce the availability of the unshared pair of electrons on nitrogen for
bonding with electrophiles.

8-22

a. Fair to good for cations,

$$\overset{\delta\oplus}{\diagdown}\overset{\delta\ominus}{C}=\overset{}{O}\cdots M\overset{\oplus}{\cdots}\overset{\delta\ominus}{O}=\overset{\delta\oplus}{C}\diagup$$

b. Poor for both

c. Good for both

$$H:\overset{..}{\underset{..}{F}}:\cdots M^{\oplus}, \quad \overset{\delta\ominus}{F}—\overset{\delta\oplus}{H} \;\; :\overset{\ominus}{X}$$

d. Poor, but $CHCl_3$ does hydrogen
bond to anions, $\overset{\delta\ominus}{Cl_3}\overset{\delta\oplus}{C}—\overset{}{H}\cdots:\overset{\ominus}{X}$

e. Moderately good for cations,

$$R_3N:\cdots M^{\oplus}$$

f. Good for cations $R_3\overset{\oplus}{N}-\overset{\ominus}{O}:\cdots M^{\oplus}$

8-23   Faster. The solvent, $(CH_3)_2S=O$, solvates cations, but not anions,
efficiently. Therefore, $CN^{\ominus}$ is poorly solvated and more reactive in
$(CH_3)_2S=O$ compared to ethanol.

8-24

a.   Rate $= \underline{k}\left[:\overset{\ominus}{C}H_2CH_2Cl\right]$ and $\left[:\overset{\ominus}{C}H_2CH_2Cl\right] = \underline{K}\left[CH_3CH_2Cl\right]\left[HO^{\ominus}\right]$ hence,

rate $= \underline{k}\,\underline{K}\left[CH_3CH_2Cl\right]\left[HO^{\ominus}\right]$, which is overall second order.

b.   The alternative mechanism would produce deuterium exchange in the
starting halide.

$$CH_3CH_2Cl + \overset{\ominus}{O}D \rightleftarrows :\overset{\ominus}{C}H_2CH_2Cl + HOD \overset{D_2O}{\rightleftharpoons} D—CH_2CH_2Cl + \overset{\ominus}{O}D\,.$$

8-24 (Cont'd)

If no exchange is observed, the mechanism can be excluded.

c.   The test in Part b does not apply if the <u>slow</u> step is proton removal.
Under these circumstances, the rate would be second order and deuterium
exchange would not be observed.

8-25   Under basic conditions, the most prominent reactions for secondary
halide are E2 and S_N2 reactions. The nucleophile is $\overset{\ominus}{O}C_2H_5$ because of the
equilibrium $HO^{\ominus}$ + $HOC_2H_5 \rightleftharpoons HOH$ + $\overset{\ominus}{O}C_2H_5$.
    (excess)

$$CH_3CH_2CHClCH_3 + \overset{\ominus}{O}C_2H_5 \begin{cases} S_N2 \rightarrow CH_3CH_2CH(OC_2H_5)CH_3 \\ E2 \rightarrow CH_3CH=CHCH_3 \text{ and } CH_3CH_2CH=CH_2 \end{cases}$$

8-26

a.   Potassium <u>tert</u>-butoxide is a strong base; ethanamine is a weak base.
Strong bases promote E2 reactions. Also, the <u>tert</u>-butoxide ion is bulky and
therefore does not give substitution products as does ethanamine.

b. $(CH_3)_2S=O$ solvates cations efficiently but not anions. Therefore the
<u>tert</u>-butoxide ion in $(CH_3)_2S=O$ is effectively unsolvated and hence more
reactive. (See also Exercise 8-23.)

8-27 a. $(CH_3)_3CCl \longrightarrow (CH_3)_2C=CH_2$ 2-methylpropene

      b. $(CH_3)_2CHCH_2Cl \longrightarrow (CH_3)_2C=CH_2$ 2-methylpropene

      c.   chlorocyclohexane $\longrightarrow$ cyclohexene

8-28 a. $(CH_3)_3CCl \longrightarrow (CH_3)_2C=CH_2$ 2-methylpropene

      b.   bromocyclohexane $\longrightarrow$ cyclohexene

      c. $\longrightarrow$ cyclohexene + $N(CH_3)_3$

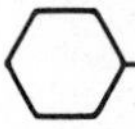

8-29

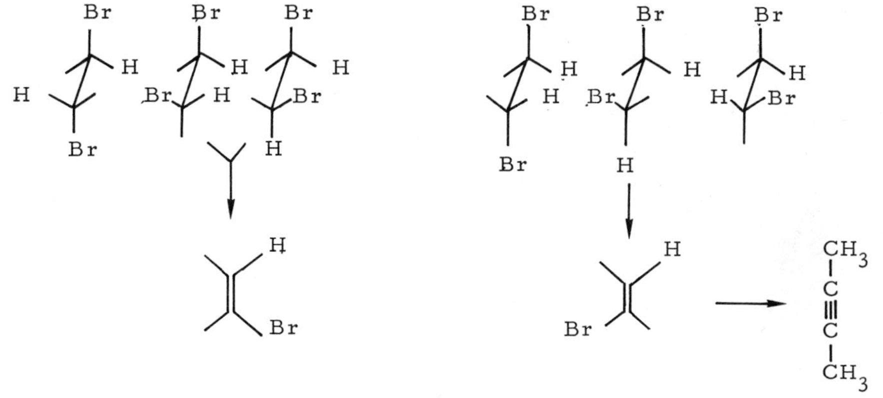

(CH$_3$ groups are not explicitly shown).

8-30

a.   Since the products are derived from a common intermediate, $(CH_3)_3C^{\oplus}$, which is formed __after__ heterolysis of the C—X bond, the product ratio should be independent of X, the leaving group.

b.   By an E2 mechanism, the product ratio will vary with X because the product-forming step is also the step involving cleavage of the C—X bond.

8-31   Mechanism changes from El-S$_N$l to E2; elimination products dominate.

8-32   The product of a direct El reaction would not contain deuterium, $(CH_3)_2C=CHCH_3$. The product of rearrangement and elimination would be a mixture of $(CH_3)_2C=CHCH_3$ and $(CH_3)_2C=CDCH_3$.

8-33

a. $(CH_3CH_2)_2\overset{\underset{\textstyle |}{CH_3}}{C}-OH$ $(CH_3CH_2)_2C=CH_2$ $CH_3CH=C(CH_3)CH_2CH_3$

b. $(CH_3)_3C\underset{\underset{\textstyle OH}{|}}{C}HCH_3$ $(CH_3)_2\underset{\underset{\textstyle OH}{|}}{C}CH(CH_3)_2$ $(CH_3)_3CCH=CH_2$

$(CH_3)_2C=C(CH_3)_2$ $CH_2=\underset{\underset{\textstyle CH_3}{|}}{C}CH(CH_3)_2$

8-33 (cont'd)

c.   ⌖–CH₂OH   ⌖⟨CH₃ / OH   ⌖=CH₂   ⌖–CH₃

(structures: cyclobutane rings with substituents)

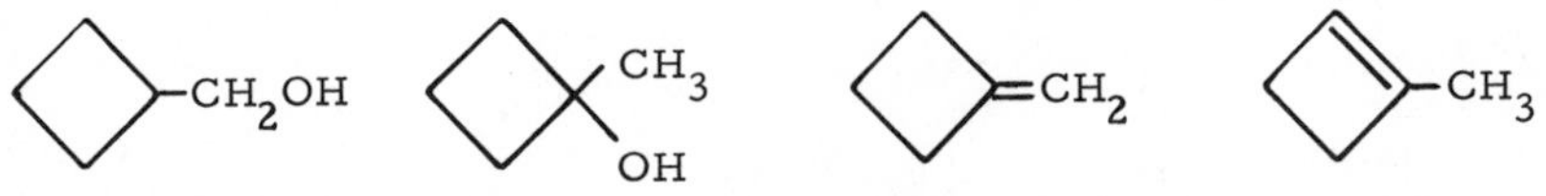

8-34   a.   ROH $\xrightarrow{\text{Na or NaH}}$ RO$^{\ominus}$ $\xrightarrow{\text{R}'\text{I}}$ ROR$'$

b.   ROH $\xrightarrow{\text{H}_2\text{SO}_4}$ R$\overset{\oplus}{\text{O}}H_2$ $\xrightarrow{\text{heat}}$ alkene + H$_3$O$^{\oplus}$

　　　$\llcorner$ $\xrightarrow{\text{HCl}}$ RCl $\xrightarrow{\ominus\text{OH}}$ alkene + H$_2$O + Cl$^{\ominus}$

c.   ROH $\xrightarrow{\text{HCl or SOCl}_2 \text{ or PCl}_5}$ RCl

d.   ROH $\longrightarrow$ RCl $\xrightarrow{\text{NaCN}}$ RCN + NaCl

e.   RCl $\xrightarrow{\text{R}_2'\text{S}}$ R$\overset{\oplus}{\text{S}}R_2'$ Cl$^{\ominus}$ $\xrightarrow{\ominus\text{OH}}$ alkene + HCl + R$_2'$S

8-35   The least-hindered carbon carrying a good leaving group is the methyl

carbon on sulfur. This carbon is attacked preferentially by nucleophiles in an

S$_N$2 reaction. For this reason, S-adenosylmethionine functions as a methylat-

ing agent.

$$R-\overset{\oplus}{\underset{R'}{S}}-CH_3 + H_2\overset{..}{N}CH_2CH_2OPO_3H_2 \longrightarrow RSR' + H_2\overset{\oplus}{N}CH_2CH_2OPO_3H_2 \ (CH_3)$$

8-36

a.   (CH₃)₂S=O ; this solvent dissolves both organic and ionic reagents yet

leaves the anion CN$^{\ominus}$ unsolvated and hence reactive.

b.   The most S$_N$2 reactive isomer is the primary allyl-type bromide,

⬡–CH₂Br .

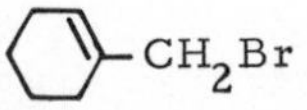

c.   The best leaving group X is that derived from the strongest conjugate

acid HX. X = $\cdot$O–$\overset{\text{O}}{\underset{\text{O}}{\overset{\|}{\underset{\|}{S}}}}$–CH₃, HX = HO–$\overset{\text{O}}{\underset{\text{O}}{\overset{\|}{\underset{\|}{S}}}}$–CH₃ p$\underline{K}_a$ ∼ 0.7. The other HX alter-

natives are weak or very weak acids.

8-37

a.   The acid is required to convert the $-OH$ group to $-\overset{\oplus}{O}H_2$, which is a

much better leaving group.

b.   Elimination competes with substitution if isopropyl iodide is used rather

than methyl iodide.

c.   The reaction is readily reversible and proceeds to the left only if the

nucleophilic bromide ion is removed (as $\underline{AgBr}$) as it is formed. (See also

Exercise 8-21a.)

d.

$$CH_3CH{=}CHCH_2Br \xrightarrow{H_2O,\,S_N1} \left[\, CH_3CH{=}CH{-}\overset{\oplus}{C}H_2 \leftrightarrow CH_3\overset{\oplus}{C}H{-}CH{=}CH_2 \,\right]$$

$$\Big\downarrow H_2O,-H^{\oplus}$$

$$CH_2{=}CH{-}CH{=}CH_2 + CH_3CH{=}CHCH_2OH + CH_3CH(OH)CH{=}CH_2$$

8-38

a. $CH_2{=}C(CH_3)C_6H_5 + C_6H_5{-}C(CH_3)_2OH$ (S_N1 conditions favors

                                                                                              tertiary bromide)

b. $C_6H_5CH_2CH_2I$ (S_N2 conditions favors primary bromide)

c. $CH_2{=}C(CH_3)C_6H_5$ (E2 conditions favors tertiary bromide)

d. $CH_2{=}CH_2 + CH_3OH + N(CH_3)_3$ (E2 favors strong base)

8-39

a. 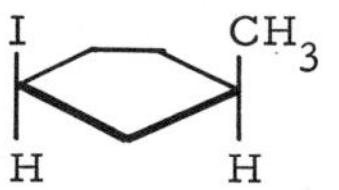   c.

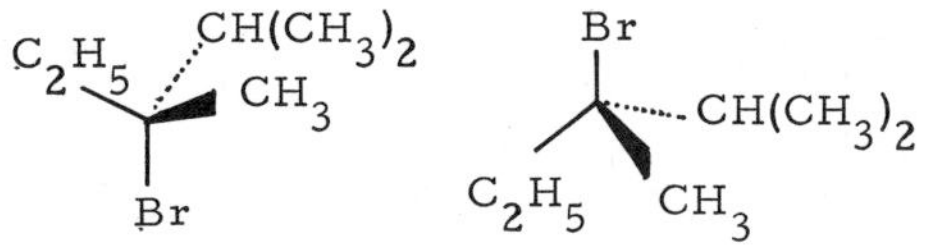

b.

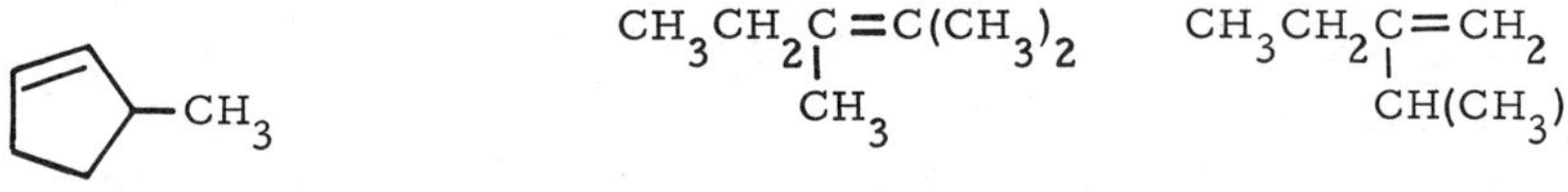

$$\underset{CH_3}{\underset{|}{CH_3CH_2C}}{=}C(CH_3)_2 \qquad \underset{CH(CH_3)_2}{\underset{|}{CH_3CH_2C}}{=}CH_2$$

8-39 (cont'd)

d.

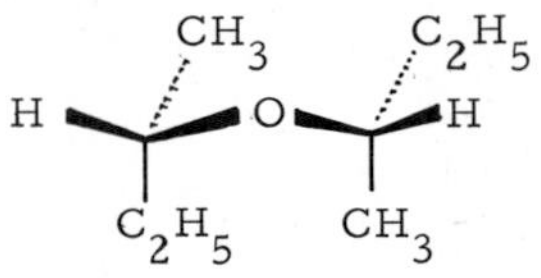

f.

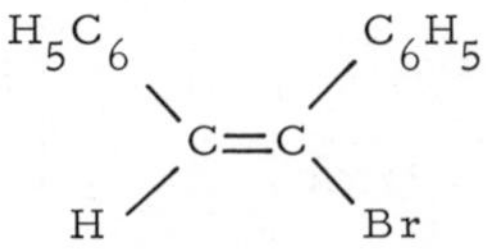

e. $CH_2\!\!=\!\!C\!-\!CH(CH_3)_2$
            $\overset{\displaystyle |}{CH_3}$

$(CH_3)_2C\!\!=\!\!C(CH_3)_2$

8-40 $RX \xrightarrow{\text{slow}} R^{\oplus} + X^{\ominus}$

$\underset{\displaystyle \xrightarrow{N_3^{\ominus}}}{\qquad} RN_3$

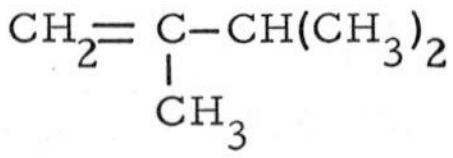

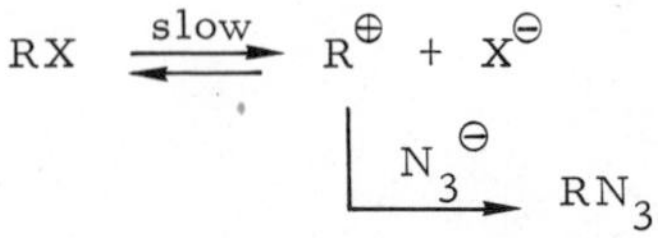

Added $X^{\ominus}$ retards reaction by increasing rate of reversal to RX.
However, if another nucleophile $(N_3^{\ominus})$ captures $R^{\oplus}$ more rapidly than does
$X^{\ominus}$, then retardation of the rate of disappearance of RX will be diminished.

8-41

a. $C_4H_9Cl + I^{\ominus} \rightleftharpoons C_4H_9I + Cl^{\ominus}$ (a)

$C_4H_9I + {}^{\ominus}OH \longrightarrow C_4H_9OH + I^{\ominus}$ (b)

$\overline{\phantom{C_4H_9Cl + {}^{\ominus}OH \longrightarrow C_4H_9OH + Cl^{\ominus}}}$

$C_4H_9Cl + {}^{\ominus}OH \longrightarrow C_4H_9OH + Cl^{\ominus}$ (c)

The catalytic effect of $I^{\ominus}$ in reaction (c) is the result of reactions (a)
and (b) which are faster than (c) when (c) is carried out in the absence of $I^{\ominus}$.

Uncatalyzed reaction (c) would lead to $CH_3CH_2CH_2\!\!-\!\!\overset{\displaystyle D}{\underset{\displaystyle H}{C}}\!\!-\!OH$ because
the reaction is a straightforward nucleophilic displacement reaction leading to
inversion of configuration. The catalyzed reaction would lead to
$CH_3CH_2CH_2\!\!-\!\!\overset{\displaystyle H}{\underset{\displaystyle D}{C}}\!\!\blacktriangleleft OH$ because two inverting displacement reactions, (a) and (b),
are involved, and the product would have the same configuration as the starting

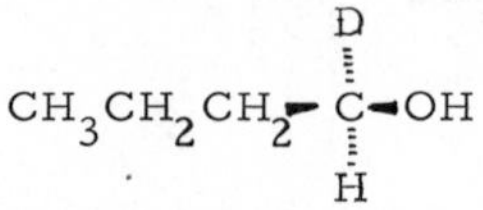

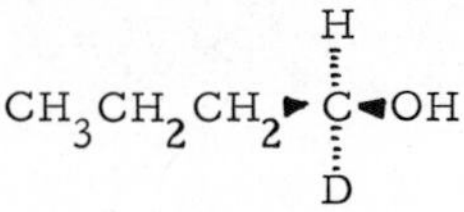

8-41 (cont'd)

material. Complications would result if $CH_3CH_2CH_2CHDI$ (left-handed) +

$\overset{\ominus}{I} \longrightarrow CH_3CH_2CH_2CDHI$ (right-handed) + $\overset{\ominus}{I}$ is fast compared to (b). If very

fast compared to (b), the alcohol would be racemic (see Exercise 8-8), but,

if very slow compared to (b), then the alcohol would be right-handed.

b.   The answer in part (a) clearly shows that S_N2-type reactions leading

to net retention of configuration do not automatically preclude operation of the

usual inversion mechanisms.

8-42

a.

$$HOAc \rightleftharpoons H^{\oplus} + OAc^{\ominus}, \; \underline{K}_1 = 1.8 \times 10^{-5}$$

$$H^{\oplus} + {}^{\ominus}OH \rightleftharpoons H_2O \qquad , \; [H^{\oplus}][O\overset{\ominus}{H}] = 10^{-14}$$

$$HOAc + {}^{\ominus}OH \rightleftharpoons H_2O + OAc^{\ominus}, \; \underline{K}$$

$$\underline{K} = \frac{[H_2O][OAc^{\ominus}]}{[HOAc][OH^{\ominus}]} = \frac{1.8 \times 10^{-5}[H_2O]}{10^{-14}} . \qquad \begin{array}{l}(\text{Abbreviation} \\ \text{OAc means} \\ \text{ethanoate, or} \\ \text{acetate})\end{array}$$

Let x be the amount of hydroxide ion present

$$\frac{1-x}{x^2} = 1.8 \times 10^9$$

$$x = 2.2 \times 10^{-5} \text{ moles per liter}$$

b.   Initial rate of = $\underline{k}[CH_3Br][NaOAc]$
         formation of
         methyl acetate = $1.0 \times 10^{-4} \times 10^{-2} \times 1.0$

                        = 1×10^{-6} moles per liter per sec.

         of methanol   = $\underline{k}'[CH_3Br][O\overset{\ominus}{H}]$

                        = $30 \times 10^{-4} \times 10^{-2} \times 2.2 \times 10^{-5}$

                        = 6.6×10^{-10} moles per liter per sec.

8-42 (cont'd)

c.   Assuming that the ratio of rates of reaction of CH_3Br with $OAc^{\ominus}$ and $OH^{\ominus}$ remains constant and equal to the initial rate (note the $OAc^{\ominus}$ is in large excess over CH_3I) then

$$\frac{[CH_3CO_2CH_3]}{[CH_3OH]} = \frac{1 \times 10^{-6}}{6.6 \times 10^{-10}} = 1.5 \times 10^3$$

d.   The following information would be required:

1.   position of the equilibrium $CH_3OH + NaOH \rightleftharpoons CH_3ONa + H_2O$

2.   rate constants for the reactions:

$$CH_3Br + CH_3OH \longrightarrow CH_3OCH_3 + HBr$$

$$CH_3Br + NaOH \longrightarrow CH_3OH + NaBr$$

$$CH_3Br + CH_3ONa \longrightarrow CH_3OCH_3 + NaBr$$

3.   initial concentration of sodium hydroxide

8-43

a. $RH \xrightarrow{Cl_2, \underline{h}\nu} RCl \xrightarrow{\overset{\oplus}{Na} \overset{\ominus}{O}\overset{O}{\overset{\|}{C}}CH_3} R-O-\overset{O}{\overset{\|}{C}}CH_3 + NaCl$

b. $CH_3CH_2OH \xrightarrow[-H_2]{NaH} CH_3CH_2O^{\ominus} \xrightarrow{CH_3I} CH_3CH_2OCH_3 + I^{\ominus}$

c. $(CH_3)_3COH \xrightarrow[-H_2O]{H_2SO_4} (CH_3)_3C^{\oplus} \xrightarrow[-H^{\oplus}]{CH_3OH} (CH_3)_3COCH_3$

d.   ⬡ $\xrightarrow{Cl_2, \underline{h}\nu}$ ⬡-Cl $\xrightarrow{KOH}$ ⬡

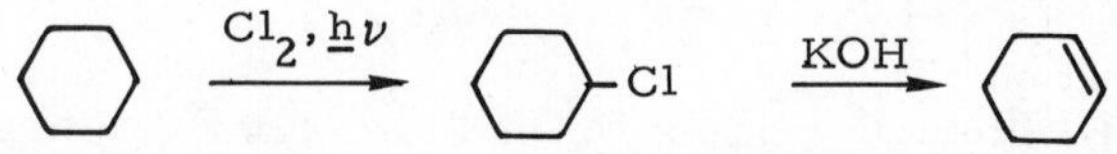

8-44 $(S_N2)KI, CH_3COCH_3$ $(E2)NaOH, C_2H_5OH$ $(S_N1)AgNO_3, H_2O, C_2H_5OH$

a. CH_3Cl $(CH_3)_2CHCl$ $(CH_3)_2CHCl$

b. CH_3Cl $(CH_3)_3CCl$ $(CH_3)_3CCl$

c.   — $(CH_3)_2CCH_2F$ with Cl below $(CH_3)_3CCl$

8-44 (cont'd)

d. $CH_3CH=CHCH_2Cl$ $CH_2=CHCH_2CH_2Cl$ $CH_3CH=CHCH_2Cl$

8-45

	Yield	Rate	Side Reactions
a.	Zero; S_N2 attack of a bulky nucleophile at tertiary carbon does not occur	zero	E2 elimination of HCl to yield 2-methylpropene
b.	Zero; unfavorable ΔH of -29 kcal; cleavage in the sense $C—\vdots I$ does not occur readily	zero	Unlikely to be any very rapid reaction
c.	Low; water is not a strong enough base to cause E2 elimination	slow	Hydrolysis by S_N1 process to give $(CH_3)_2C=CHCH_3$ and $(CH_3)_2C(OH)CH_2CH_3$
d.	Moderate to good; unlikely to go to completion unless NaCl precipitates	slow (secondary halide)	Inversion of configuration by exchange 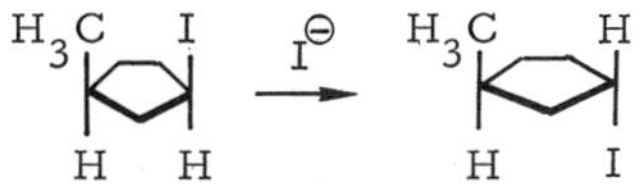
e.	Zero; S_N2 attack does not occur at a practical rate because of hindrance by flanking methyl groups	zero	none

8-45 (cont'd)

	Yield	Rate	Side Reactions
f.	Zero; diethyl ether is a very weak nucleophile and $CH_3CH_2^{\ominus}$ is an exceedingly poor leaving group	zero	none
g.	Poor; displacement of halogen at sp^2 carbon is difficult	poor	none
h.	Good	moderate	Some 3-methylcyclopentene is expected by <u>trans</u> elimination

8-46

	Bottle A	Bottle B
a.	No reaction with $AgNO_3$ aqueous ethanol even on warming. No reaction with KI in CH_3COCH_3	AgCl would precipitate on warming with $AgNO_3$. Reacts with KI in CH_3COCH_3. $$RCl + Ag^{\oplus} \xrightarrow{H_2O} ROH + \underline{AgCl} + H^{\oplus}$$ $$RCl + I^{\ominus} \longrightarrow RI + Cl^{\ominus}$$
b.	Immediate precipitation of <u>AgCl</u> with $AgNO_3$ in aqueous ethanol	Immediate precipitation of <u>AgBr</u> with $AgNO_3$ in aqueous ethanol
c.	Immediate precipitation of <u>AgCl</u> from $AgNO_3$. No reaction with KI in CH_3COCH_3	Slow to react with $AgNO_3$ or with KI in CH_3COCH_3

8-46   (cont.)

Bottle A	Bottle B

d. No reaction with $AgNO_3$ or   Immediate precipitation of AgCl.

with KI in CH_3COCH_3 Rapid reaction with KI in

CH_3COCH_3

e. No reaction under conditions   Elimination of HCl under E2

of E2 elimination   conditions

$$CH_3CH_2CH = CHCl \xrightarrow[\text{heat}]{\text{KOH}} CH_3CH_2C \equiv CH$$

f. As in Part d   AgCl would precipitate on warming

with $AgNO_3$ in aqueous ethanol.

Reacts with KI in CH_3COCH_3

8-47   The methylene hydrogens at C3 are flanked by five bulky methyl groups,

making it difficult for a basic reagent to remove a proton from the C3 position

to give 2,4,4-trimethyl-2-pentene. Similar hindrance is not felt at C1 and

therefore the major product is 2,4,4-trimethyl-1-pentene, which is also the
most stable product.

ADDITIONAL EXERCISES

8A-1   Draw the structure of a chiral compound of formula $C_7H_{15}Cl$ and D
configuration that reacts at 25° in aqueous acetone to give an alcohol $C_7H_{15}OH$
as a racemic mixture.

8A-2   Draw the structure of a compound of formula $C_9H_{11}Cl$ that can be
prepared by photochlorination of an alkylbenzene and which reacts in aqueous
ethanol and (more rapidly) in a solution of sodium ethoxide in ethanol to give a
compound of formula C_9H_{10}.

8A-3   Which isomer of formula C_4H_9Cl would react most rapidly with
potassium iodide in 2-propanone?

8A-4   If one mole of 1,2-propanediol were to react with one mole of triphenylmethyl chloride, what major organic product would you expect to obtain?

8A-5   Which compound in each group would be the most reactive in the
presence of the given reagents? Write an equation to show the product(s)
expected.

8A-5 (cont.)

a. 1-chloro-3-methylpentane, 2-chloro-3-methylpentane, 3-chloro-3-methyl-
pentane, in 50% aqueous 2-propanone.

b. The compounds in Part a in a solution of KI in 2-propanone.

c. The compounds in Part a in a solution of $NaOC_2H_5$ in ethanol at 60°.

d. 1-chloro-3-methylpentane, 3-chloromethylpentane, 1-chloro-2,2-dimethyl-
butane, in a solution of KI in 2-propanone.

e. $CH_3CH=CHCH_2Br$, $CH_3CH_2CH=CHBr$, $CH_3CH_2CH_2CH_2Br$ in a solution
of NaCN in 2-propanone.

f. $BrCH_2COCH_3$, $BrCH_2CO_2CH_3$, $BrCH_2C(CH_3)=CH_2$ in a solution of
silver nitrate in ethanol.

g.

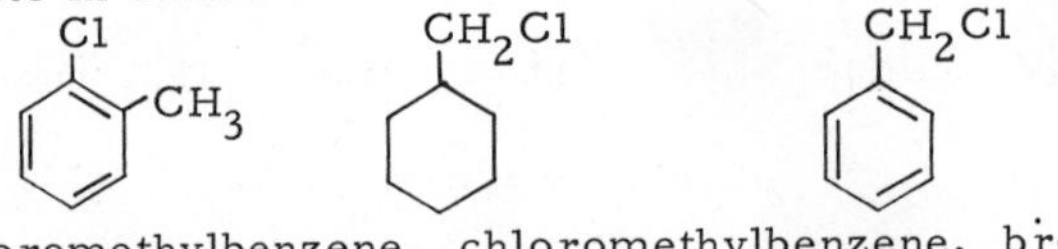

in a solution of silver
nitrate in ethanol

h. Fluoromethylbenzene, chloromethylbenzene, bromomethylbenzene, iodo-
methylbenzene, in $CH_3S^{\ominus}Na^{\oplus}$ in ethanol.

i. RCH_2OH, RCH_2OCH_3, $RCH_2OS(O_2)CH_3$, $RCH_2O\overset{\overset{O}{\|}}{C}CH_3$ in a solution of
NaCN in $CH_3S(O)CH_3$. (R is phenyl)

j. NaI, HI, LiI, CH_3I towards $C_6H_5OCH_3$

k.

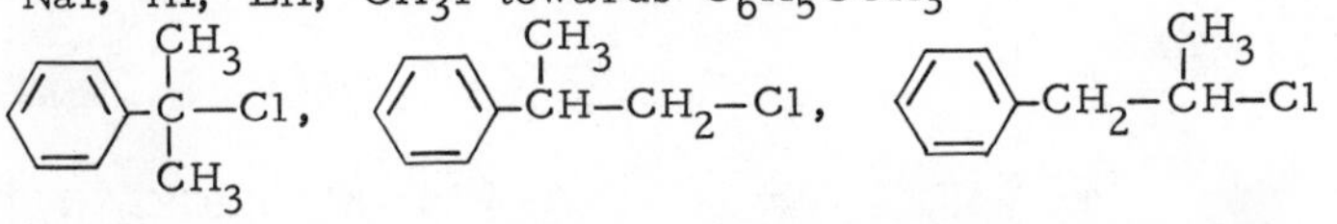

in a solution of
KOH in C_2H_5OH

l. HI, HOH, HNH_2, HONa towards $(CH_3)_3CCl$

m. $(CH_3)_2NH$, $(CH_3)_2N^{\ominus}Li^{\oplus}$, $(CH_3)_2NH_2^{\oplus}Cl^{\ominus}$ towards $(CH_3)_3CCl$

n. $(CH_3)_3CO^{\ominus}K^{\oplus}$, $(CH_3)_3COH$, $(CH_3)_3COH_2^{\oplus}$ towards CH_3CHCl in $(CH_3)_2SO$

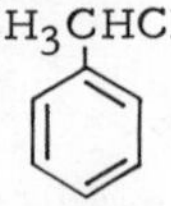

o. LiBr, LiF, LiCl towards 1-pentanol in concentrated H_2SO_4

p. $(CH_3)_3S^{\oplus}Cl^{\ominus}$, CH_3SH, $CH_3S^{\ominus}Na^{\oplus}$, CH_3SCH_3 towards 1-phenyl-2-
bromo-1-ethanone in benzene.

8A-6   In which solvent would you expect bromide ion to be least well solvated?

a. 2-propanone or isopropyl alcohol,   b. ethanol or diethyl ether,
c. dimethyl sulfide $(CH_3)_2S$ or dimethyl sulfoxide $(CH_3)_2SO$

8A-7   An optically active compound A of formula $C_5H_{10}O$ of the L-configuration reacts with 48% HBr to give a racemic bromide B of formula C_5H_9Br. If
compound A reacts with para-toluenesulfonyl chloride, $CH_3-C_6H_4-S(O_2)Cl$, a
sulfonate ester C is formed which reacts with LiBr in propanone to give
optically active B. Draw the structures and configurations of A, B and C.

CHAPTER 9

SEPARATION AND PURIFICATION. IDENTIFICATION
OF ORGANIC COMPOUNDS BY SPECTROSCOPIC TECHNIQUES

Chromatographic methods, especially gas-liquid and liquid-solid,
provide extraordinarily sensitive and efficient means to separate mixtures of
organic compounds.

Electron microscopy has developed to the point of being able to provide
structural information on relatively large, very stable, nonvolatile molecules,
or on substances that can be stained with heavy metal atoms (DNA, viruses,
etc.).

X-ray diffraction studies on organic crystals can give detailed structural determinations, including information on inter- as well as intramolecular
interactions. X-ray diffraction is becoming a valuable structural tool, but
elaborate facilities are required, which are not yet routinely available.

Spectroscopy is concerned with the excitation of matter by electromag-
netic radiation or energetic particles such as electrons.

Spectra are the results of searches for changes in energy over a range
of wavelengths (or frequencies). The energy change (ΔE) associated with a
transition between two different states is given by $\Delta E = h\nu$ or $\Delta E = hc/\lambda$,
in which h is Planck's constant, c is the velocity of light, ν is the frequency
in Hz (cycles per second), and λ is the wavelength of the absorbed radiation.
In more chemical terms, ΔE (kcal mole^{-1}) = 28,600$/\lambda$ (nm), in which λ is
in nanometers (1 nm = 10^{-9} meters) and ΔE corresponds to one einstein of
radiation (absorption of 6.02×10^{23} quanta of wavelength λ). The energy
changes involved may be of many different kinds - electronic, magnetic,
nuclear, and so on. For the purposes of several forms of spectroscopy, the
energy of organic molecules may be resolved into electronic energy, vibrational energy, and rotational energy. All of these kinds of energy are quan-
tized - only certain specific energy levels are possible.

Microwave spectroscopy involves changes in rotational energy levels.
The energy changes are relatively small, 10^{-4} - 10^{-2} kcal mole^{-1}, and useful
spectra can be obtained only with gaseous substances. Microwave spectra can
be used to obtain rotational moments of inertia, and it is possible to use these
spectra to obtain very accurate bond distances and bond angles. The units of

microwave spectroscopy are usually GHz (gigahertz, 10^9 Hz, 10^9 cycles per second).

Infrared spectroscopy is the usual work-horse spectroscopic method for the organic laboratory and is applicable to gases, liquids, and solids. The energy changes represented by infrared spectra are between vibrational energy levels and involve 1- 10 kcal mole^{-1}. The units of infrared spectra are wave numbers (cm^{-1}; that is, $1/\lambda$, where λ is in cm). The region of infrared spectra from 500 cm^{-1} to 1250 cm^{-1} is associated with the vibrational energies of particular bonds and is thus rather characteristic of the type of molecule involved. In contrast, the region between 1250 cm^{-1} and 650 cm^{-1} is more characteristic of special molecular vibrations and generally provides a rather good "fingerprint" of the particular molecule. Infrared spectroscopy is very useful for qualitative analysis of organic compounds. Characteristic absorption frequencies of types of chemical bonds are listed in Table 9-2.

Raman spectroscopy involves the same kind of energy changes as does infrared spectroscopy, and the units also are wave numbers. An important difference between infrared and Raman arises from the selection rules that govern the intensities of the transitions. For example, symmetrically substituted carbon-carbon double bonds, as in $Cl_2C{=}CCl_2$, do not show infrared absorption at the stretching frequency but do show very strong Raman bonds.

Electronic spectra involve excitation of molecules to higher electronic states. The units now most commonly used are nanometers ($1 nm = 10^{-9}$ m), although the literature is full of data in angstroms ($1 A = 10^{-10}$ m). The important transitions usually are $\underline{n} \to \pi *$ and $\pi \to \pi *$. Conjugated multiple bonds in organic molecules lead to longer wavelength absorption than for isolated multiple bonds through stabilization that results from electron delocalization in the excited electronic states. Electronic transitions usually involve 10 - 250 kcal mole^{-1}.

Nuclear magnetic resonance spectroscopy (nmr) measures the energy required to change the alignment of magnetic atomic nuclei in a magnetic field. The units are hertz or megahertz ($1 MHz = 10^6$ Hz, or 10^6 cycles per second). The energy differences between the states usually are very small, 10^{-6} kcal mole^{-1}, and we may be concerned with relative differences in energy as small as 10^{-12} kcal mole^{-1}.

For routine use with organic molecules, the nmr spectra of the protons of the bound hydrogen atoms, ^{1}H, are by far the most important. Other nuclei that are useful are ^{19}F and ^{13}C. The abundant isotopes of oxygen, ^{16}O, and carbon, ^{12}C, give no nmr signals.

Each nmr spectrum involves three principal elements - chemical shift, spin-spin splitting (or couplings), and kinetic processes.

Chemical shifts are differences in resonance line positions because of magnetic shielding of the nuclei by the electrons surrounding them or near them.   These shifts are directly proportional to the observing frequency and usually are expressed in frequency units relative to a standard, such as tetramethylsilane (TMS).   In order that results with spectrometers with different magnetic fields can be compared easily, it is customary to report chemical shifts divided by the observing frequency x 10^6, which gives a dimensionless quantity called δ, in units of parts per million (ppm).   Thus,

$$[\Delta \nu \text{ (relative to TMS)} / \nu \text{ (observing frequency)}] \times 10^6 = \delta \text{ (ppm)}.$$

On the usual charts, which have <u>increasing</u> magnetic field (or <u>decreasing</u> ν) from <u>left</u> to <u>right</u>, δ is <u>positive</u> if the peak comes to the left of TMS and <u>nega-</u><u>tive</u> if it comes to the <u>right</u> of TMS.

Chemical shifts of O—H protons are very sensitive to hydrogen bonding. For C—H protons, the shifts depend in a fairly regular way on the electronegativities of the groups attached to the carbon.   Chemical shifts are summarized in Table 9-4.

Spin-spin splittings arise from intramolecular magnetic interactions between different kinds of nuclei or between the same kind of nuclei with different chemical shifts.   The <u>first-order</u> splittings are independent of the chemical-shift differences between the interacting nuclei.   A proton that is coupled to <u>n</u> equivalent protons will give <u>n+1</u> equally spaced lines in its nmr spectrum. <u>Second-order</u> splittings, which give extra lines or complex line spacings, can be very important when the chemical-shift differences are small.   Secondorder splittings tend to disappear with increasing chemical-shift differences.

For protons (and other nuclei with spin $\underline{I} = \frac{1}{2}$), spin-spin splittings between sets of equivalent nuclei display patterns of lines in accord with the binomial coefficients.   Thus two nuclei tend to produce a 1:2:1 pattern, three nuclei a 1:3:3:1 pattern, four nuclei a 1:4:6:4:1 pattern, and so on.   The asymmetry of spin-spin splitting patterns, when the chemical shift is not large compared to the splittings, can give information as to which resonances are split by which nuclei.

Three-bond H—C—C—H couplings are very sensitive to the rotational angle about the C—C bond and can be used to assist conformational analysis. The magnitude of the coupling constant $\underline{J}$ depends on structure.   Some of the approximate values for the important couplings useful for structure determinations follow:

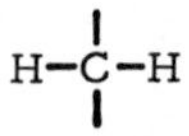

$$H-\overset{|}{\underset{|}{C}}-H$$

10 - 12 Hz

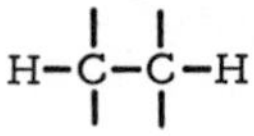

$$H-\overset{|}{\underset{|}{C}}-\overset{|}{\underset{|}{C}}-H$$

7 Hz

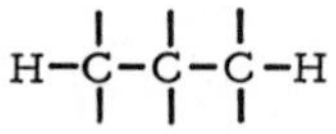

$$H-\overset{|}{\underset{|}{C}}-\overset{|}{\underset{|}{C}}-\overset{|}{\underset{|}{C}}-H$$

0 Hz

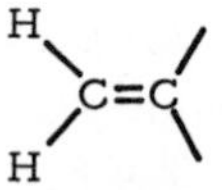

$$\underset{H}{\overset{H}{>}}C=C\underset{\diagdown}{\diagup}$$

0 - 3 Hz

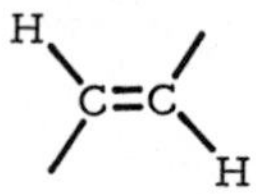

$$\underset{\diagup}{\overset{H}{\diagdown}}C=C\underset{H}{\diagup}$$

11 - 19 Hz

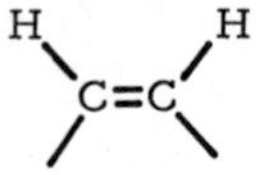

$$\underset{\diagup}{\overset{H}{\diagdown}}C=C\underset{\diagdown}{\overset{H}{\diagup}}$$

7 - 10 Hz

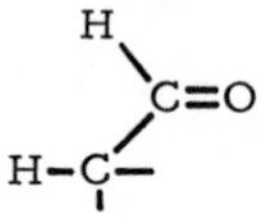

$$\underset{H-\overset{|}{C}-}{\overset{H}{\diagdown}}C=O$$

1 - 3 Hz

$$\underset{H-\overset{|}{C}-}{\overset{H}{\diagdown}}C=C\underset{\diagdown}{\diagup}$$

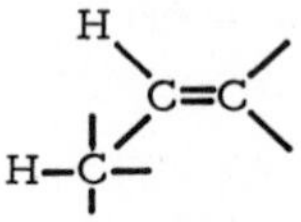

4 - 10 Hz

Mass spectroscopy involves determining the masses and abundances of the ions produced by electron impact. The mass of the $\underline{M}^{\oplus}$ peak that corresponds to the parent molecule is particularly helpful for molecular-weight determinations. Molecular formulas often can be obtained, and, to a considerable degree, molecular fragmentation produced by electron impact leads to ions by reasonable chemical processes.

ANSWERS TO EXERCISES

9-1 ν = 0.067 Hz, $\tilde{\nu}$ = 75 km^{-1}, λ = 13.3 m, $\underline{c}$ = 0.89 m sec^{-1}

9-2 λ = 481 nm, ν = 6.24 × 10^{14} Hz

9-3   48.5 kcal; the energy requirements of many chemical reactions are within this range--hence the energy absorbed by sodium vapor has the possibility of causing chemical reactions. (See Section 4-4D for an example.)

9-4   a. λ = 3 × 10^{11} nm, $\underline{E}$ = 10^{-7} kcal

       b. $\lambda \sim 1$ nm (check with Figure 9.7 to see that your estimates for Exercises 9-1 to 9-4 are in the right range).

9-5   There are two low-energy conformations about the single bond between the carbonyl and carbon-carbon double bonds. In both of these conformations the double bonds are coplanar so that π-orbital overlap and hence electron delocalization is maximized. In one conformation the double bonds are cis with respect to each other, in the other they are trans

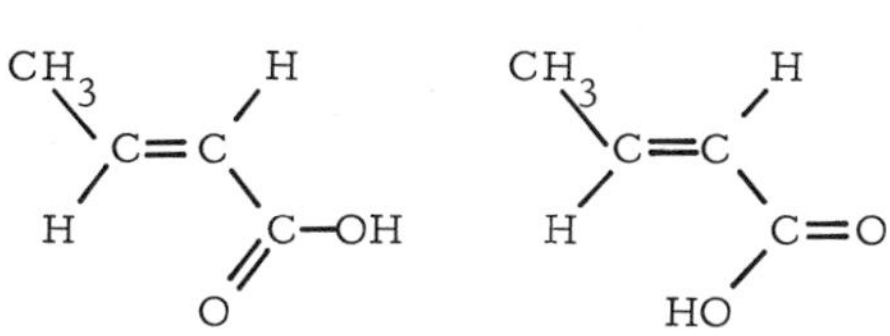

9-6

a.   From Equation 9-3, $\tilde{\nu}$ for a series of similar compounds RX increases as the mass of X decreases. Hence $\tilde{\nu}_{RF} > \tilde{\nu}_{RCl} > \tilde{\nu}_{RBr}$. The bond strengths are in the order $RF > RCl > RBr$ which means that $\underline{k}$ should change in a way to give the same order of $\tilde{\nu}$ as expected for the mass changes.

b.   Other things being nearly equal, $\tilde{\nu}$ increases with bond strength. Hence $\tilde{\nu}_{C\equiv N} > \tilde{\nu}_{C=NH} > \tilde{\nu}_{C-NH_2}$.

9-7   The compound with the most polar bonds (highest dipole moment) or with the least symmetry will very likely have the most intense ir absorption band.

a. $(CH_3)_2C=O$ c. $CH_3C\equiv CH$

b. CH_3OCH_3 d.   HCl

9-8   a.   3   c.   3

        b.   3

        stretching   stretching   bending

9-9 $\nu = 9 \times 10^{13}$ sec^{-1}, $\lambda = 3300$ nm,   33,000 A,   3.3 μ

        $\Delta \underline{E} = 8.7$ kcal

        Consider the process $\underline{M} \underset{}{\overset{h\nu}{\rightleftharpoons}} \underline{M}^*$, where $\nu = 3000$ cm^{-1}. Assuming $\Delta S = 0$, then $\Delta \underline{E} = \Delta \underline{H} = -2.303$ RT log $\underline{K}$ where $\underline{K} = [\underline{M}^*]/[\underline{M}]$

                $-2.303 \times 1.987 \times 298$ log $\underline{K} = 8700$ cal

                        $\underline{K} = 4.19 \times 10^{-7}$

9-10

a. $C=O$, 1650 - 1730 cm^{-1} (s) ; C—H, 2800 - 3100 cm^{-1} (s);

        C—D, ~ 2200 cm^{-1}; (νC—C are omitted here).

b. $C\equiv C$, 2050 - 2260 cm^{-1} (m) to (w); $\equiv$C—H, 3200 - 3350 cm^{-1}(s),

        C—H, 2800 - 3100 cm^{-1}(s).

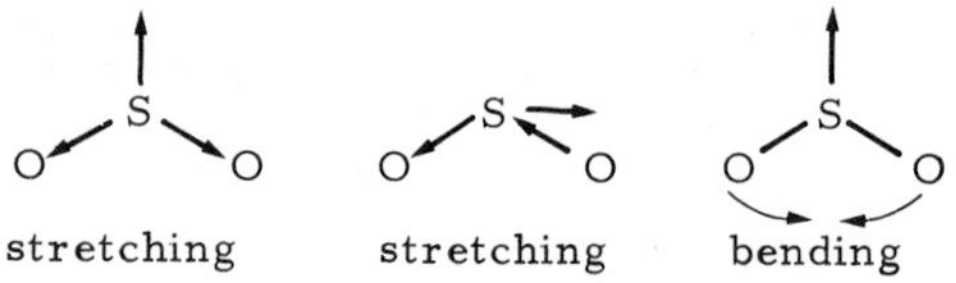

9-10 (cont'd)

c.   C=O, 1710 - 1780 cm^{-1}(s); C—O, 1035 - 1300 cm^{-1}(s), two bands;

    C—H, 2800 - 3100 cm^{-1}(s).

d.   C≡N, 2200 - 2400 cm^{-1}(m) to (w); C=C, 1600 - 1680 cm^{-1}(m) to (w);

    =C—H, 3000 - 3100 cm^{-1}(m).

e.   C=O, 1650 - 1730 cm^{-1}(s); C=O, 1710 - 1780 cm^{-1}(s);

    C—O, 1350 - 1400 cm^{-1}(m) to (w); O—H, 2500-3300 cm^{-1} (broad);

    C—H, 2800 - 3100 cm^{-1}(s).

f.   pure liquid: C—H, 2800 - 3100 cm^{-1}(s); O—H, 3200 - 3400 cm^{-1}(s)

    broad); O—H, 1000 - 1410 cm^{-1}(s); C—O, 980 - 1250 cm^{-1}(s).

    dilute solution in CCl$_4$: O–H 3400-3700 cm^{-1}(sharp, no H-bonding).

9-11

a.   The spectrum shows O—H (3400 cm^{-1}, strong, broad); C—H (2900 cm^{-1});

and C=C (1660 cm^{-1}). The data indicates the compound is an alkenol. In

support of this there are strong bands in the spectrum in the region 980-1250

cm^{-1} for O—H bond. Possible structures are:

    H$_2$C=CHCH$_2$OH   H$_3$CCH=CHOH   H$_3$CC=CH$_2$
                                                       |
                                                     OH

        A   B   C

    Structures B and C are known to be unstable and rapidly reaarange

to carbonyl compounds, CH$_3$CH$_2$CHO and CH$_3$COCH$_3$. The correct structure

is A.

b.   The spectrum shows C=O (1750 cm^{-1}, strong), C—O (1200 cm^{-1},

strong) and C—H (2800-2900 cm^{-1}). There are no bands for OH -- hence the

compound cannot be a carboxylic acid but rather a carboxylic ester. Two

structures are possible;

9-11 (cont'd)

$$CH_3-\overset{\overset{\displaystyle O}{\|}}{C}-OCH_3 \quad \text{and} \quad H-\overset{\overset{\displaystyle O}{\|}}{C}-OCH_2CH_3$$

9-12

		Infrared	Raman
a.	$HC\equiv CH$	no $C\equiv C$ stretch	strong $C\equiv C$
b.	ICl	strong $I-Cl$ stretch	strong $I-Cl$
c.	CO	strong $C=O$ stretch	strong $C=O$
d.	$CF_2=CH_2$	strong	strong
e.	$(CH_3)_2C=CH_2$	medium	strong
	$CH_3CH=CHCH_3$	none	strong

9-13

$$\overset{\leftarrow\quad\rightarrow}{O=C=O} \qquad \overset{\rightarrow\leftarrow\quad\rightarrow}{O=C=O} \qquad \overset{\uparrow\quad\uparrow}{\underset{\downarrow}{O=C=O}}$$

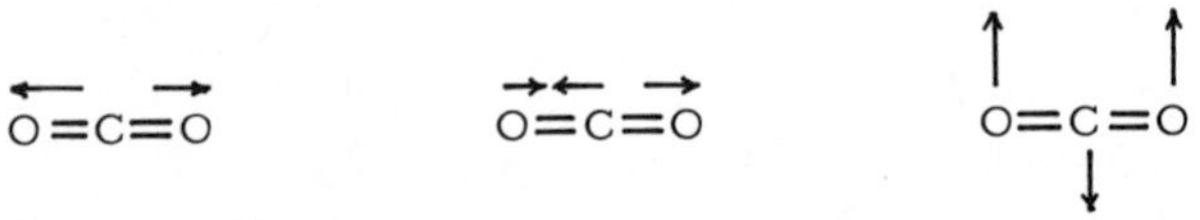

Raman active

infrared inactive   infrared active   infrared active

9-14 $\underline{n}\to\pi^*$, $\lambda = 280 + 43.5 \sim 324$ nm

$\pi\to\pi^*$, $\lambda = 190 + 43.5 \sim 234$ nm   energy

$\underline{n}\to\sigma^*$, $\lambda = 156 + 43.5 \sim 200$ nm

$\left.\begin{array}{l} \pi\to\sigma^* \\[4pt] \sigma\to\pi^* \\[4pt] \sigma\to\sigma^* \end{array}\right\}$ high energy, short wavelengths.

Wavelength of the transitions can be estimated crudely by assuming the same ordering of the $H_2C=NH_2$ transitions as for the $(CH_3)_2C=O$ transitions given in Table 9-3. A second assumption is that each will be at a longer wavelength than the corresponding $C=O$ transition by 43.5 nm which is the amount $n\to\sigma^*$ for $C-\overset{..}{N}$ differs from $n\to\sigma^*$ $C-\overset{..}{\overset{..}{O}}$.

9-15 $\text{Absorbance} = \log I_o/I = \epsilon \underline{c} \underline{1}$

λ_{280}, $\epsilon = 15$, $\underline{1} = 0.1$ cm, $\underline{c} = 0.01$ M

hence: percent transmission = I/I_o = 97%, percent absorption = 3%

λ_{190}, $\epsilon = 1,100$, $\underline{1} = 0.1$ cm, $\underline{c} = 0.01$ M

hence: percent transmission = 8%, percent absorption = 92%

9-16   Absorption at 227.3 nm for $(CH_3)_3\overset{..}{N}$ is an $\underline{n} \to \sigma^*$ transition, which is precluded in acid solution because the electron pair on nitrogen becomes involved in bond formation with a proton.

$$(CH_3)_3\overset{..}{N} + HA \rightleftharpoons (CH_3)_3\overset{\oplus}{N}H + \overset{\ominus}{A}$$

The analogous transition then would be a $\sigma \to \sigma^*$ transition.

9-17   The data indicate C_4H_6O is a conjugated aldehyde or ketone. Possible structures are:

$$CH_3-CH=CH-CH=O \qquad CH_2=\overset{\overset{\textstyle CH_3}{\textstyle |}}{C}-CH=O \qquad CH_2=CH-\overset{\overset{\textstyle O}{\textstyle ||}}{C}-CH_3$$

9-18

a.   The infrared bands at $3000\text{-}2700$ cm^{-1} and 1613 cm^{-1} are those expected for the O—H and conjugated carbonyl group of 4-hydroxy-3-penten-2-one, respectively. The uv band at 272 nm also is consistent for the grouping C=C-C=O. Therefore, in the liquid mixture and in cyclohexane the equilibrium lies far to the right.

b.   Because the 272 nm band decreases in aqueous solution, water must solvate the diketone form more than does cyclohexane. This is the result of hydrogen-bonding between the carbonyl oxygen and water. The hydroxylic form is comparatively stabilized in cyclohexane by internal hydrogen bonding:

9-18 (cont'd)

$$\overset{\delta\oplus\ \delta\ominus}{\underset{/}{\overset{\backslash}{C}}=O\text{-----}H\text{-}\overset{\delta\oplus\ \delta\ominus}{O}H}$$

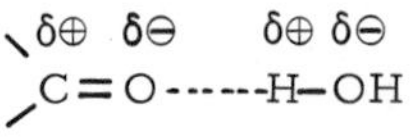

hydrogen-bonding with water

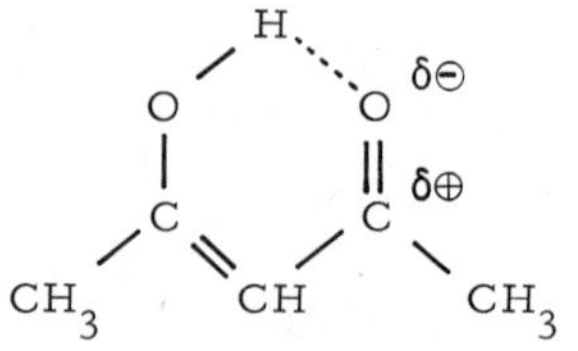

internal hydrogen-bonding

9-19   The fact that the anion absorbs at longer wavelengths implies that resonance is more important in the anion than in the neutral phenol. The resonance
form shown for the phenol has <u>an unfavorable charge separation</u> that is not
present in the anion. This form is not then an effective contributor to the
ground or excited state of the neutral phenol. In contrast, that shown for the
anion is a low-energy favorable form which contributes significantly to the
structure and stability of the anion.

9-20   Absorbance $= (\epsilon \cdot c \cdot l)_{NAD\oplus} + (\epsilon' \cdot c' \cdot l)_{NADH}$

$\lambda = 340$ nm, $0.311 = 0 + 6220 \cdot c' \cdot l$

$\lambda = 260$ nm, $1.2 = 18,000 \cdot c \cdot l + 15,000 \cdot c' \cdot l$

hence, concentration of $NAD^{\oplus}$ $(c) = 2.5 \times 10^{-5}$ l. $mole^{-1}$

concentration of NADH $(c') = 5 \times 10^{-5}$ l. $mole^{-1}$

9-21

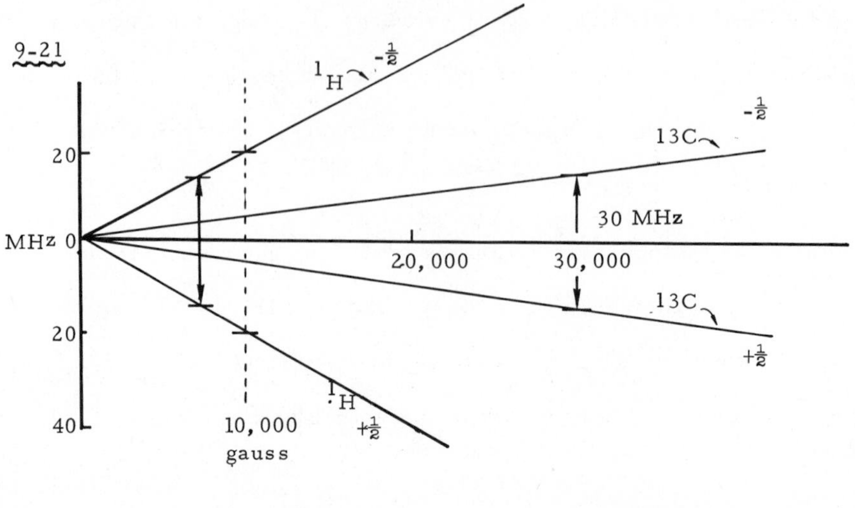

9-21   (cont'd)

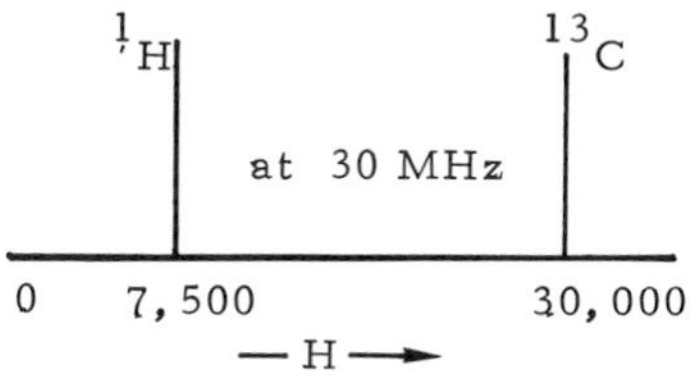

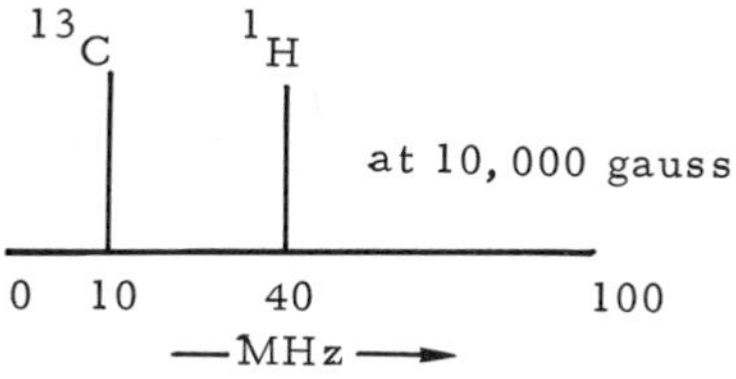

9-22 $\Delta E = h\nu = 6.62 \times 10^{-27}$ (erg-sec) $\times$ 1 Hz $= 6.62 \times 10^{27}$ erg

or $6.62 \times 10^{-27} \times \underbrace{6.023 \times 10^{23}}_{\text{Avogadro's No.}}$ erg mole^{-1}

and converting ergs to kcal:

$$\Delta E = 9.5 \times 10^{-14} \text{ kcal mole}^{-1}$$

or

$$\Delta E = \frac{28,600}{3 \times 10^{8} \times 10^{9}} = 9.5 \times 10^{-14} \text{ kcal mole}^{-1} \quad \text{(from Equation 9-2)}$$

9-23   Using the relationship $\Delta G = \Delta H - T\Delta S = -2303\ RT \log K$ and assuming

$\Delta S = 0$, we have (from Exercise 9-22):

$$\Delta E \equiv \Delta H = 180 \times 10^{6} \times 9.5 \times 10^{-11} \text{ cal mole}^{-1} = -2.303 \times 1.987 \times 183 \log K$$

$K = 0.999953$ for $H(+\tfrac{1}{2}) \rightleftarrows H(-\tfrac{1}{2})$

or

$K = 1.000047$ for $H(-\tfrac{1}{2}) \rightleftarrows H(+\tfrac{1}{2})$

9-24

a.   (i)

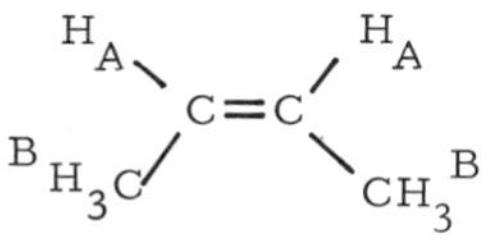

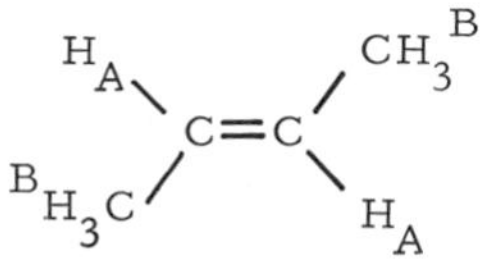

9-24 (cont'd)

(ii)

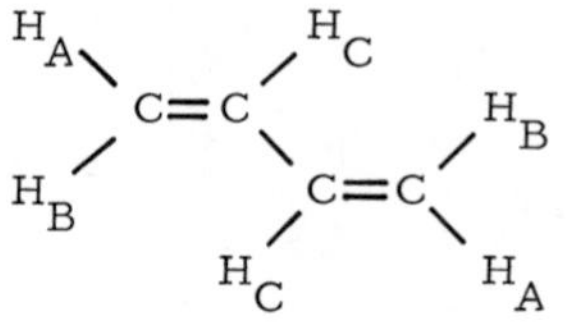

$$\underset{H_B}{\overset{H_A}{\diagdown}} C=C \underset{C=C}{\overset{H_C}{\diagup}} \underset{H_C}{\diagdown} \underset{H_A}{\overset{H_B}{\diagup}}$$

(iv)

$$CH_3 \overset{F}{\underset{H_E}{\overset{H_D}{\underset{|}{C}}}} \overset{H_A}{\underset{OH_B}{\overset{|}{C}}} CH_3 \quad C$$

(iii)

$$\overset{C}{CH_3} \overset{B}{-} CH_2 \overset{CH_3}{\underset{CH_3}{\overset{|}{\underset{|}{C}}}} \overset{A}{-} CH_2 \overset{}{-} Cl$$

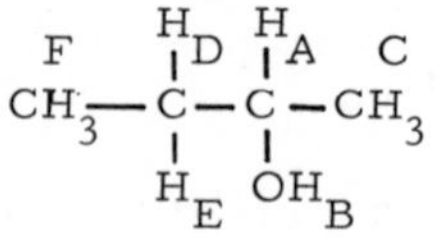

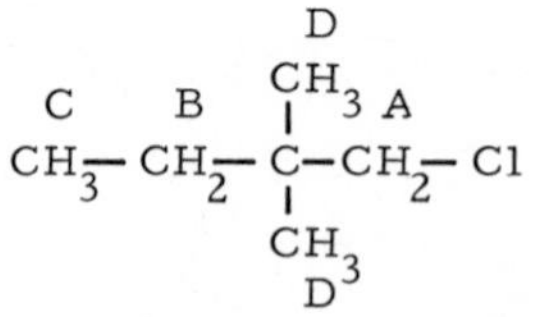

(v)

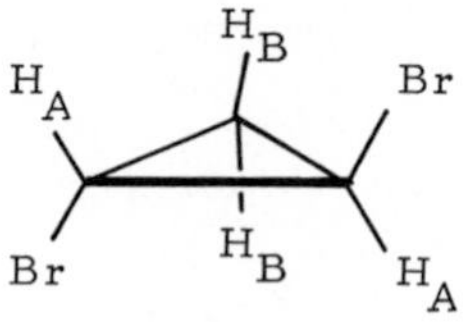

b.   The methyl groups at C3 are different from each other and from the
methyl group at Cl.   They are diastereotopic.

c.   In (iii), the methyl groups at C2, the methylene protons
at Cl and at C3 are enantiotopic.
In (iv), the protons at C3 are diastereotopic.
In (v), the methylene protons at C3 are diastereotopic.

9-25   a.   5.29 ppm   b.   5. 12 ppm   c.   3. 63 ppm

9-26 $\delta = \underline{n}_1 \delta_1 + \underline{n}_2 \delta_2$, $\underline{n}_1 = 2\underline{n}_2$ = fraction of the population at each site

δ_1 and δ_2 = chemical shifts

since $\underline{n}_1 + \underline{n}_2 = 1$, δ = 0. 66 X 1. 1 + 0. 33 X 3.2 = 1. 78 ppm

9-27

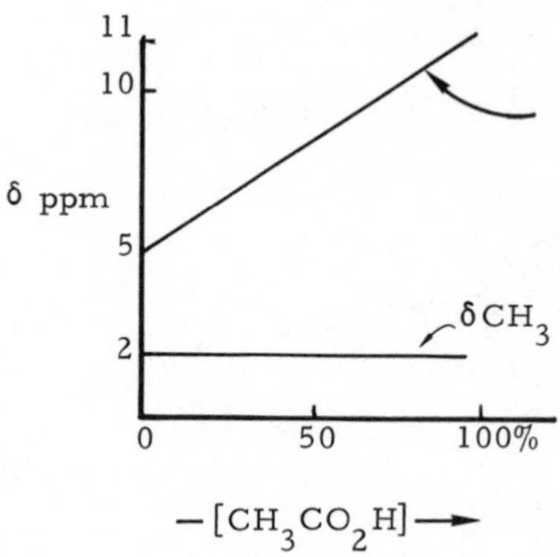

average chemical shift of H_2O and
$-CO_2H$ protons which are rapidly
exchanging. See Exercise 9-26 for
relationship.

9-28   Analysis of nmr spectrum gives:

δ 1.25 ppm   area 6   assignment $(CH_3)_2$

2.2   3 CH_3

2.6   2 CH_2

3.9 (broad)   1   OH

Infrared spectrum indicates OH (3300 cm^{-1}, broad) and C=O(1700 cm^{-1})

Structure is
$$CH_3-\underset{\underset{OH}{|}}{\overset{\overset{CH_3}{|}}{C}}-CH_2-\overset{\overset{O}{||}}{C}-CH_3,$$
4-hydroxy-4-methyl-2-pentanone

9-29

a.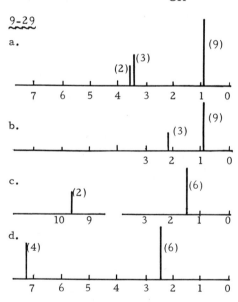

e.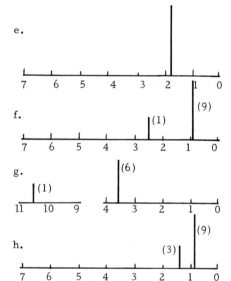

9-30

a. CH_3-O-CH_3

b.   cyclohexane

c.   2,2-dimethylpropane

d.
H$_2$C—O—CH$_2$ / H$_2$C—CH$_2$ (tetrahydrofuran ring)

e. $(CH_3)_2 CHCO_2 H$ (there are several other possibilities)

f. $Cl_3C-CCl=CCl-CCl_3$

9-31

a.

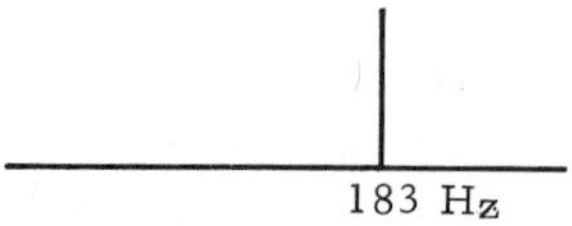

183 Hz

f.

318   294   Hz

b.

204   54

g.

270   126   54

c.

240   54

h.

230   135   54

d.

204   > 54

i.

204   192   > 75

e.

54

j.

204   > 90

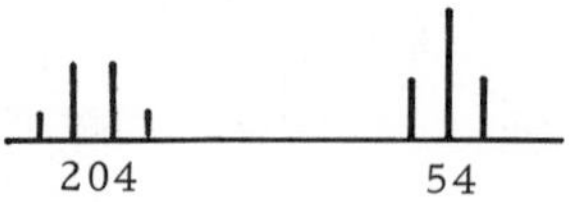

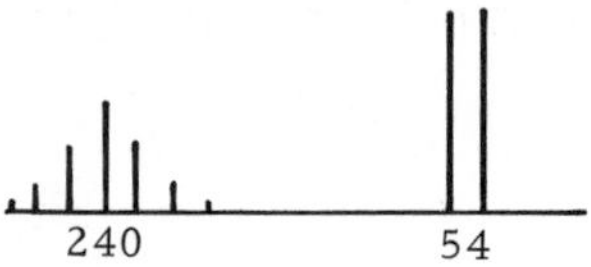

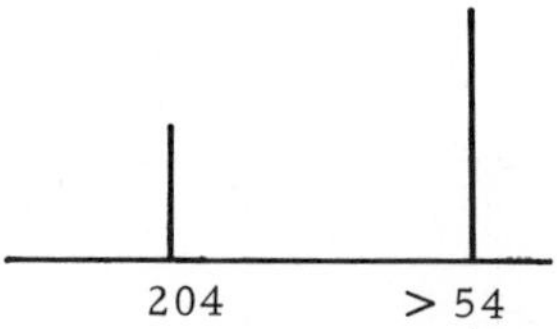

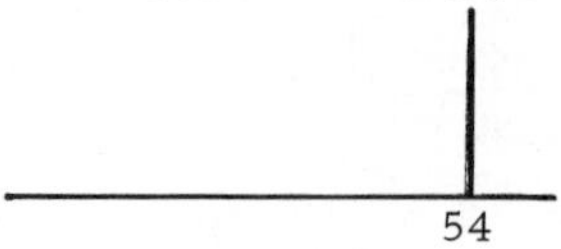

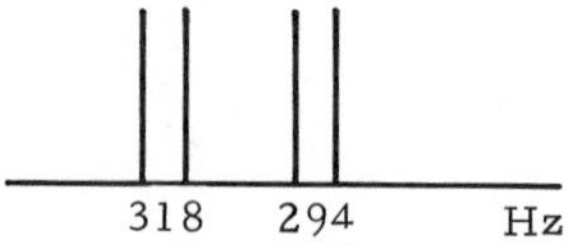

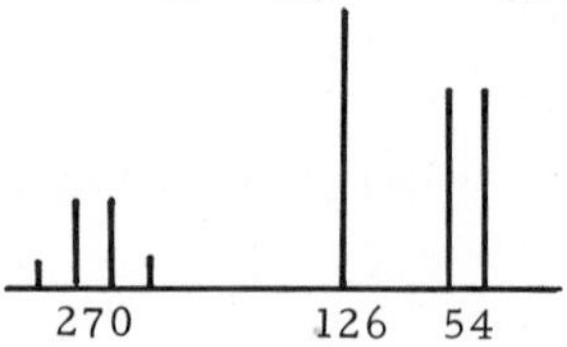

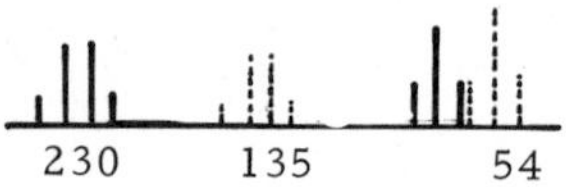

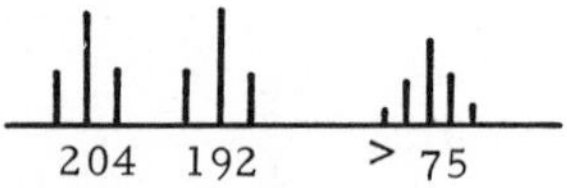

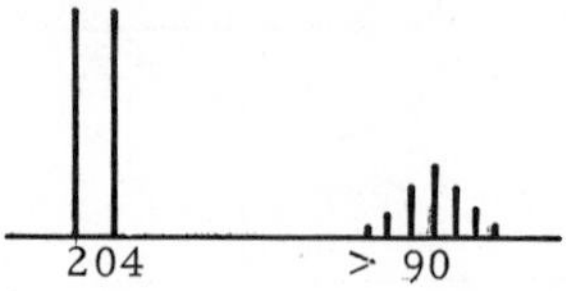

9-32   The observed J value of 3.1 Hz is the average of two gauche forms
(population = x) and one trans form (population = y) such that x + y = 1. Because
the observed J is <u>smaller</u> than either the common value of J(gauche) = 4.5 Hz
or J(trans) = 12 Hz, the predominant conformation must be gauche.

Cl
Cl   H
Cl   H
Cl

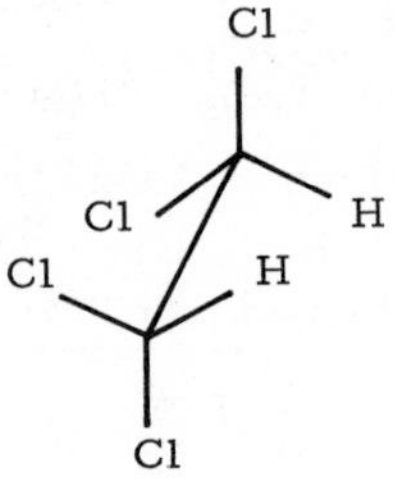

preferred conformation

9-33

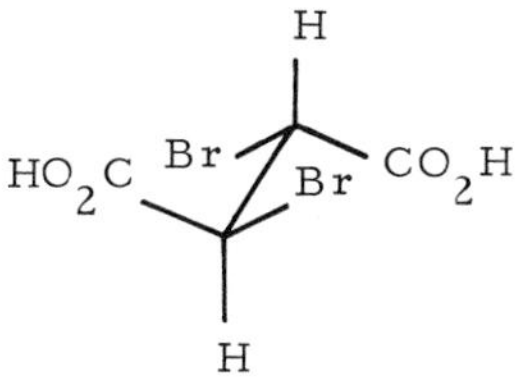

9-34

a.

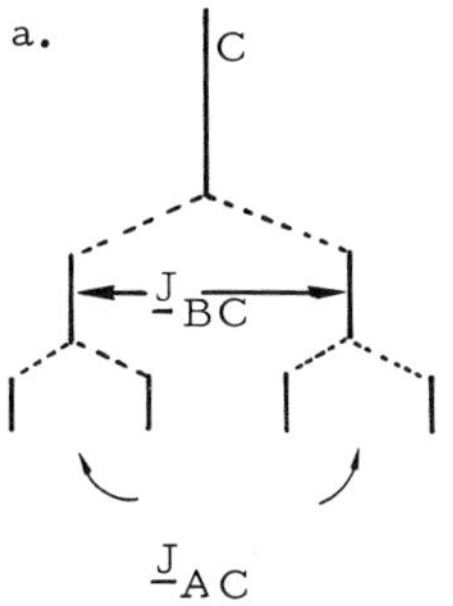

C

$\leftarrow \underline{J}_{BC} \rightarrow$

$\underline{J}_{AC}$

when $2\underline{J}_{AC} = \underline{J}_{BC}$

b.

A

←16 Hz→

4 Hz

B

←16 Hz→

C

4Hz

	Assignment	δ ppm	Area	$\underline{J}$ (multiplicity)
9-35				
a.	$CH_3C=C$	1.9	3	singlet
	$CH_3C=O$	2.1	3	singlet
	$CH_2=$	4.8	2	singlet
b.	$-CH_2-$	3.0	2	doublet (1:1) $-CH_2CH-$
	$(CH_3O)_2$	3.4	6	singlet
	$-CH-$	4.6	1	triplet (1:2:1) $-CH_2CH-$
	C_6H_5-	7.6	5	singlet

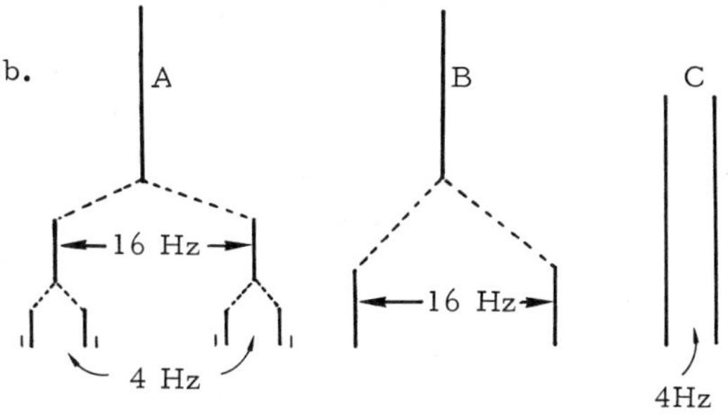

9-35 (cont'd)

	Assignment	δ ppm	Area	J (multiplicity)	
c.	$(CH_3)_2C$	1.2	6	doublet (1:1)	$(C\underline{H}_3)_2CH$
	$CH_3C\!=\!O$	2.1	3	singlet	
	$-\overset{\shortmid}{C}H-O-$	5.0	1	heptet	$(CH_3)_2C\underline{H}$
d.	$C\underline{H}_3CH_2$	1.0	3	triplet (1:2:1)	$C\underline{H}_3CH_2$
	$C\underline{H}_3CH$	1.6	} 5	doublet (1:1)	$C\underline{H}_3CH$
	$-CH\underline{}_2-$	~1.6		multiplet	$C\overset{\shortmid}{H}_3C\underline{H}_2CH$
	$-\overset{\shortmid}{C}HBr$	4.1	1	multiplet	$CH_3C\underline{H}CH_2$

9-36

	δ ppm	Area	Coupling	Assignment	Structure
a.	3.9	2	doublet	$C\underline{H}_2CH$	$Cl-C\underline{H}_2-CHCl_2$
	5.7	1	triplet	$CH_2C\underline{H}$	
b.	1.05	6	doublet	$(C\underline{H}_3)_2CH$	
	1.90	1	multiplet	$-\overset{\shortmid}{C}\underline{H}-$	$CH_3-\overset{\overset{\displaystyle CH_3}{\shortmid}}{\underset{\underset{\displaystyle H}{\shortmid}}{C}}-C\underline{H}_2-Br$
	3.3	2	doublet	$Br-C\underline{H}_2CH$	
c.	~1.9	2	quintet	$CH_2C\underline{H}_2CH_2$	$BrCH_2CH_2CH_2I$
	~3.6	6	two over-lapping triplets	$C\underline{H}_2CH_2C\underline{H}_2$	

9-37	1.9	3	closely-spaced multiplet	$CH_3C\!=$	
	3.7	3	singlet	CH_3O	$\overset{\displaystyle H}{\diagdown}C\!=\!C\overset{\diagup CH_3}{\underset{\diagdown C-OCH_3}{}}$
	5.5	1	closely-spaced multiplet	$CH\!=$	$\overset{\displaystyle H}{\diagup}\qquad\overset{\shortparallel}{O}$
	6.1	1	singlet	$CH\!=$	

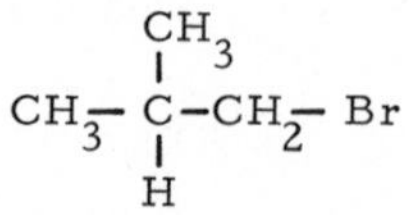

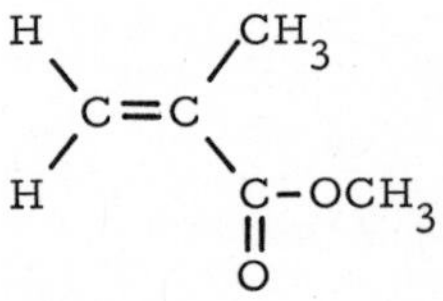

9-37   (cont'd)

(Notice that the alternate cis and trans isomers, $CH_3CH{=}CHCO_2CH_3$, can be eliminated because the couplings between either cis or trans protons - $CH{\equiv}CH$- are larger than the observed couplings, which are essentially zero and consistent with $\underset{H}{\overset{H}{\diagdown}}C{=}\quad$.)

9-38   A   hexachlorophene   D   1-phenylethanamine

           B   1,3-dimethyluracil   E   caffeine

           C   DDT   F   phenacetin

9-39   Multiplets at 5.8 and 4.1 ppm are interrelated by spin-spin coupling. This is evident from the slightly greater intensities of the high-field peak of triplet at 5.8 ppm and the low-field peak of doublet at 4.1 ppm than expected by the first order treatment. Similarly, peaks at 3.4 and 1.1 are interrelated.

$$C_2H_3Br_3 \equiv Br_2CHCH_2Br \qquad C_4H_{10}O \equiv (CH_3CH_2)_2O$$

9-40   There are two chiral centers in compounds of structure 16 or 17 ; isomer 16 would be the enantiomer of 17 only if the configuration at <u>both</u> chiral centers were opposite. Changing configuration at only one center gives a diastereomer (epimer), not a mirror image.

9-41   The weak proton signals in Figure 9-49 are due to $^{13}CHCl_3$ present in 1% natural abundance. The proton signal appears as a doublet because it is coupled to the ^{13}C nucleus with $\underline{J}$ = 210 Hz.

9-42   Assuming that a methyl carbon in an alkane has a chemical shift of 20 ppm downfield from TMS, then the following shifts can be estimated using the data in Figures 9-47 and 9-48 as well as the discussion.

9-42 (cont'd)

$(CH_3)_3C-CH_2OH$ $CCl_3-CH_2-\overset{\overset{\textstyle O}{\|}}{C}-O-CH_3$ 60 ppm (20+40)

~20 ~20 ~60   ~80   ~30   ~220

(20 + 40)   (20+3×20) (Figure 9-48,   (C=O in

C12)   warfarin)

9-43 ^{1}H spectrum ^{13}C spectrum

δ 1.8 ppm area 6, multiplet   δ 209 ppm area 1,   C=O

δ 2.3 ppm area 4, multiplet   δ 42 ppm area 2, $\underline{C}H_2CO\underline{C}H_2$

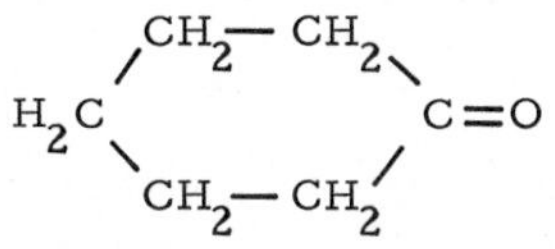

δ 27.5 ppm area 2, 2(CH$_2$)

assigned structure: δ 25.5 ppm area 1, CH$_2$

$$H_2C \overset{\displaystyle CH_2-CH_2}{\underset{\displaystyle CH_2-CH_2}{\diagdown \diagup}} C=O$$

9-44 The exact mass of $C_2H_5^{\oplus}$ (based on $^{12}C = 12.000000$) is 29.039125 ; that of $CHO^{\oplus}$ is 29.002740. The mass difference is 0.036385, and can be easily measured by instruments of resolving power of 1 part in 50,000.

9-45 a. $\dfrac{(M + 1)^{\oplus}}{M^{\oplus}} = \dfrac{3.75}{100}$ $\dfrac{(M + 2)^{\oplus}}{M^{\oplus}} = \dfrac{0.436}{100}$

b. C_8H_8O

c. $\dfrac{(M + 1)^{\oplus}}{M^{\oplus}} = 0.015 \times \dfrac{14}{100} = \dfrac{0.21}{100}$

9-46 The main peak at highest mass is taken to correspond to the molecular weight of the compound. In Figure 9-52, we have (a) 72, (b) 58, (c) 58.

9-46 (cont'd)

$$CH_3CH_2-\overset{\overset{\displaystyle :\overset{\oplus}{O}:}{\|}}{C}\!\cdot\!|\cdot H \xrightarrow{-H\cdot}$$

$$CH_3CH_2-\overset{\overset{\displaystyle :\overset{\oplus}{O}:}{\|}}{C}\!\cdot\!|\cdot CH_3 \xrightarrow{-CH_3\cdot}$$

$$CH_3CH_2\overset{\oplus}{C}=\overset{..}{\underset{..}{O}} \;\longleftrightarrow\; CH_3CH_2C\equiv\overset{\oplus}{\underset{..}{O}}$$

$$\underline{m/e}\ 57$$

$$CH_3CH_2\!\cdot\!|\cdot\overset{\overset{\displaystyle :\overset{\oplus}{O}:}{\|}}{C}-CH_3 \xrightarrow{-CH_3CH_2\cdot}$$

$$CH_3\!\cdot\!|\cdot\overset{\overset{\displaystyle :\overset{\oplus}{O}:}{\|}}{C}-CH_3 \xrightarrow{-CH_3\cdot}$$

$$CH_3\!-\!\overset{\oplus}{C}=\overset{..}{\underset{..}{O}} \;\longleftrightarrow\; CH_3\!-\!C\equiv\overset{\oplus}{\underset{..}{O}}$$

$$\underline{m/e}\ 43$$

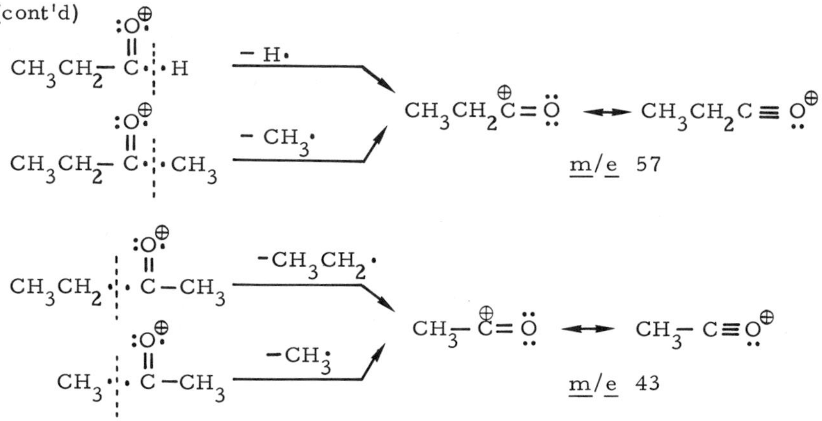

9-47   A peak at mass number 92 corresponds to a loss of 28 mass units from the molecular ion $(M^{\oplus} = 120)$. The neutral fragment 28 corresponds to C_2H_4 (ethene). A likely process for its function follows:

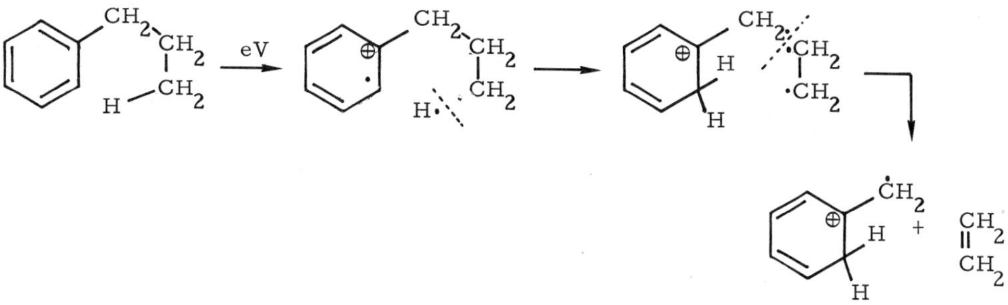

9-48   1,2-elimination from ethanol, and 1,3-elimination from 2-butanol.

$$CH_3-CD_2-\overset{..}{\underset{..}{O}}H \xrightarrow{eV} CH_3-CD_2-\overset{\cdot\oplus}{\underset{..}{O}}H \longrightarrow \overset{\cdot}{C}H_2-CD_2-\overset{..}{\underset{..}{O}}-H \xrightarrow{\overset{H\overset{\oplus}{}\ H_2\overset{..}{O}:}{}} \overset{\cdot}{C}H_2-\overset{\oplus}{C}D_2$$

$$M - 18$$

9-49

m/e 118

$$\underline{m/e}\ 90$$

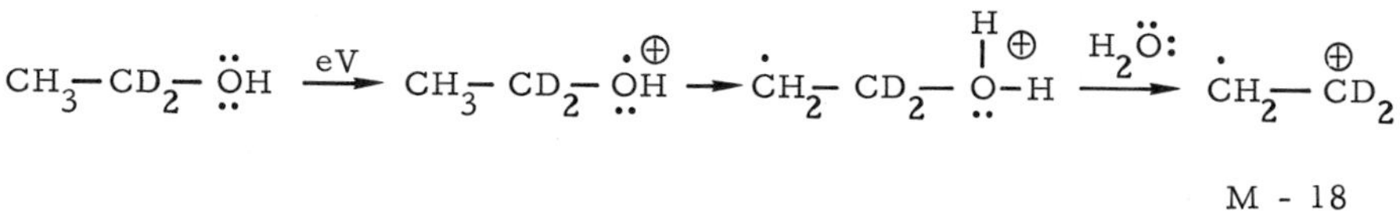

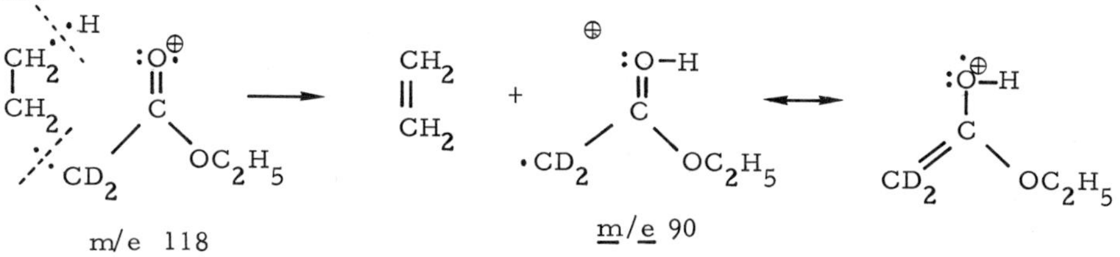

9-49 (cont'd)

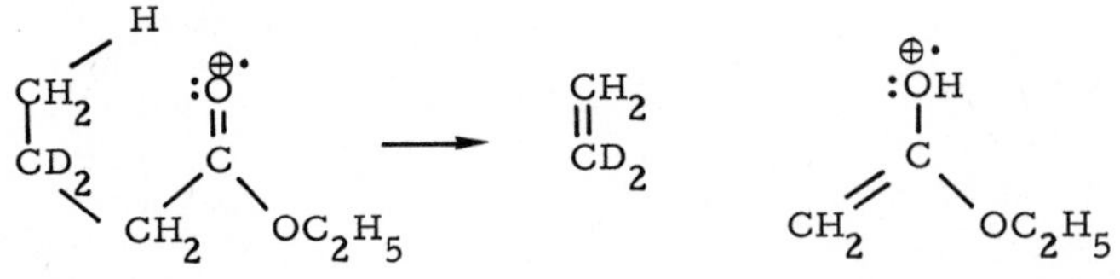

$\underline{m}/\underline{e}$ 118 $\underline{m}/\underline{e}$ 88

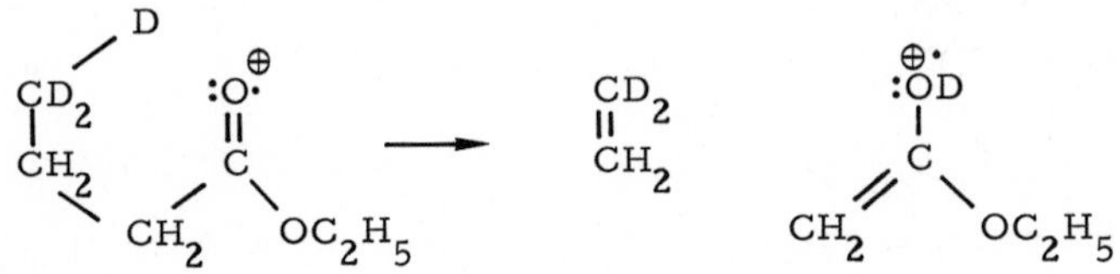

$\underline{m}/\underline{e}$ 119 $\underline{m}/\underline{e}$ 89

9-50

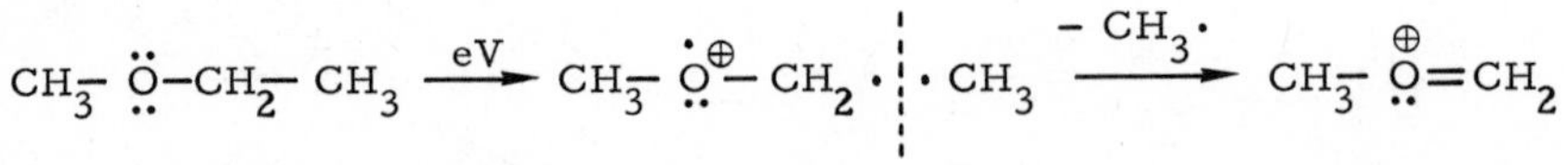

$\underline{m}/\underline{e}$ 45

9-51   Compound must contain one bromine atom to give two $M^{\oplus}$ peaks of equal intensities.

$$M^{\oplus} - Br = 136 - 79 = 138 - 81 = 57 \equiv C_4H_9^{\oplus}.$$

For C_4H_9Br to give only a single proton nmr resonance, all nine protons must be equivalent. The only possible structure is:

$$(CH_3)_3C-Br$$

9-52

a. $M^{\oplus} = 86$, $C_5H_{10}O$; $\dfrac{(M+1)^{\oplus}}{M^{\oplus}} = \dfrac{0.55 + 0.004}{10} \sim \dfrac{0.55}{10}$

$$\dfrac{(M+2)^{\oplus}}{M^{\oplus}} = \dfrac{0.012 + 0.02}{10} \sim \dfrac{0.03}{10}$$

$$CH_3CH_2CH_2 \overset{:\overset{\oplus}{O}}{\underset{\text{m/e } 43}{\vert}} \overset{\vert\vert}{\underset{\text{m/e } 43}{C}} - CH_3 \qquad\qquad CH_3CH_2CH_2 \overset{:\overset{\oplus}{O}}{\underset{\text{m/e } 71}{-}} \overset{\vert\vert}{C} \overset{}{\underset{\text{m/e } 15}{\vert}} CH_3$$

$\dfrac{(M+1)^{\oplus}}{M^{\oplus}} = \dfrac{2.2 + 0.04}{100} = \dfrac{2.24}{100}$ (exptl is $\dfrac{2.47}{100}$ for $C_2H_3O^{\oplus}$)

$\dfrac{(M+2)^{\oplus}}{M^{\oplus}} = \dfrac{0.012 + 0.2}{100} = \dfrac{0.212}{100}$ (exptl is $\dfrac{0.98}{100}$ for $C_2H_3O^{\oplus}$)

$\dfrac{(M+1)^{\oplus}}{M^{\oplus}} = \dfrac{3.3}{100} = \dfrac{3.3}{100}$ for $C_3H_7^{\oplus}$

$\dfrac{(M+2)^{\oplus}}{M^{\oplus}} = \dfrac{0.036}{100} = \dfrac{0.036}{100}$ for $C_3H_7^{\oplus}$

The agreement is better for $CH_3\overset{\oplus}{C}=\overset{..}{O}:$ $\underline{m}/\underline{e}$ 43

b.   Compound B : $M^{\oplus} = 112$, and 114 in ratio of 3:1

$$(M^{\oplus} - Cl) = 112 - 35 = 114 - 37 = 77$$

From Table 9-5 $\dfrac{(M+1)^{\oplus}}{M^{\oplus}}$ intensity indicates C_6 and therefore formula

of C_6H_5Cl.

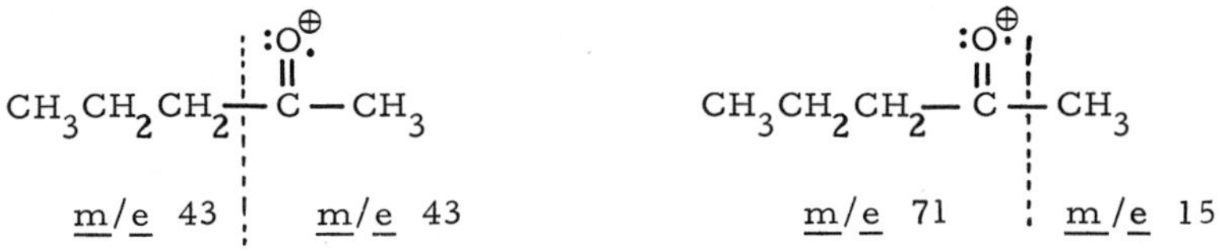

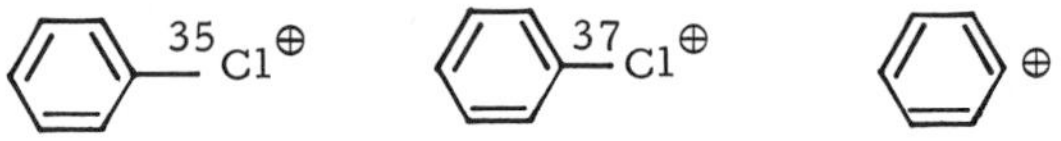

112   114   77

Compound C: $M^{\oplus} = 46$; $\dfrac{(M+1)^{\oplus}}{M^{\oplus}}$ intensity $\dfrac{0.43}{18.89} \times 100 = 2.3$

indicates C_2H_6O

$$CH_3 \overset{\oplus\bullet}{-O} - CH_2 \vert H \xrightarrow{-H\bullet} CH_3 \overset{\oplus}{-O} = CH_2 \quad \underline{m}/\underline{e} \ 45$$

$$CH_3 \overset{\oplus\bullet}{-O} \vert \xrightarrow{-CH_3\bullet} CH_3O^{\oplus} \quad \underline{m}/\underline{e} \ 31$$

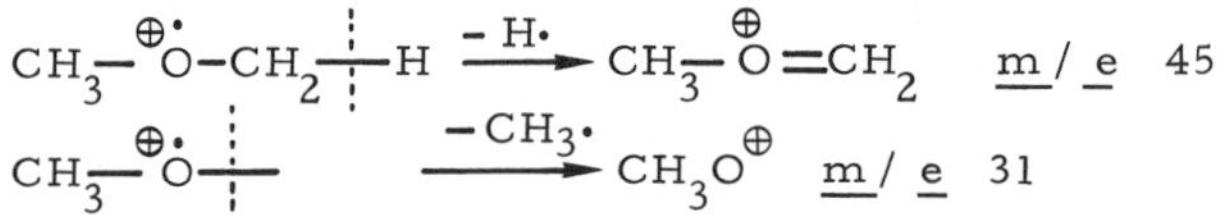

ADDITIONAL EXERCISES

9A-1   How many different proton chemical shifts are possible for each of the following compounds? Do not overlook stereochemical differences.

    a. $(CH_3)_2CHCH_2CH_3$ b. $(CH_3)_2C=CHCH_3$ c. $(CH_3)_2C(Br)CH(Br)CH_3$

    d.   4-methoxy-1-methylbenzene   e.   2-methoxy-1-methylbenzene

9A-2   Which physical or spectroscopic property A-I listed below would be most useful in distinguishing between each compound in the pairs of structures a-k? Base your judgement on the convenience, simplicity and definitiveness of the method.

A. boiling point

B. melting point

C. chemical shifts

D. nmr signal intensities (areas)

E. magnitude of spin-coupling ($\underline{J}$)

F. spin-coupling patterns (multiplicity)

G. infrared absorption

H. ultraviolet absorption

I. mass spectra

a. $CH_3SCH_2CH_3$, $CH_3OCH_2CH_3$

b. $CH_3(CH_2)_5CH_3$, $CH_3(CH_2)_7CH_3$

c. $\underline{cis}$-$C_6H_5CH=CHCO_2H$, and

      $\underline{trans}$-$C_6H_5CH=CHCO_2H$

d. CH_3COCD_3, a mixture of

     CH_3COCH_3 and CD_3COCD_3

e.

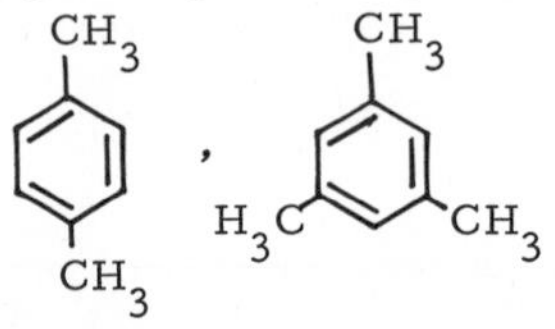

f.

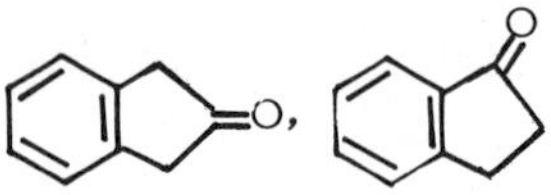

g.

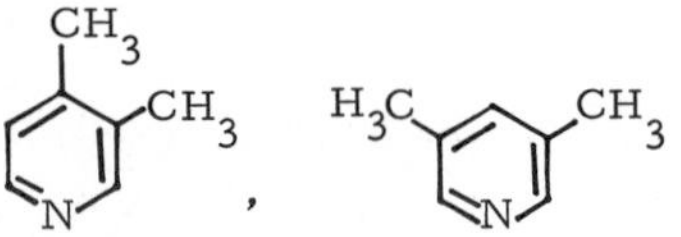

h.

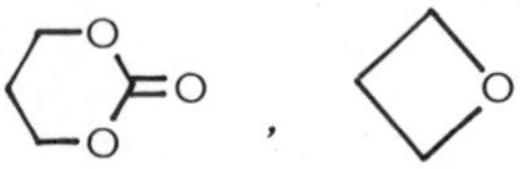

i. $(CH_3)_2CHOCH_3$, $CH_3CH_2CH_2OCH_3$ j. $Cl-\overset{O}{\overset{||}{C}}-\overset{O}{\overset{||}{C}}-Cl$, $Cl-\overset{O}{\overset{||}{C}}-Cl$

k.

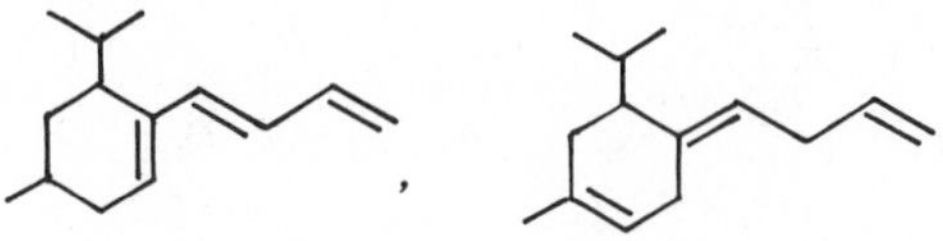

9A-3   The following spectral data refers to compounds of "unknown" structure. Deduce the structure of each compound from the spectroscopic information given.

    a.   Compound A gave a mass spectrum having $\underline{M}^+$ at $m/e = 76$ and a ratio of ion intensity $\underline{M} : \underline{M+1} : \underline{M+2} = 100 : 4.0 : 4.5$. The integrated nmr spectrum of the compound is sketched at the top of the following page.

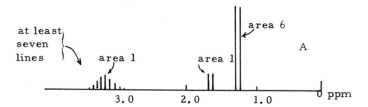

b. Compound B of formula $C_4H_8O_2$ shows strong infrared absorption at 1730 cm^{-1}, and 1300 cm^{-1}.   The nmr spectrum is sketched below.

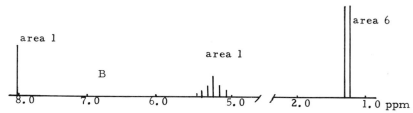

c. Compound C of formula $C_6H_{13}NO_2$ shows principal infrared absorption at 2900 cm^{-1}, 1750, 1250. 1430, and 1380 cm^{-1}. The strongest bands are at 1750 cm^{-1} and 1250 cm^{-1}. The nmr spectrum is sketched below.

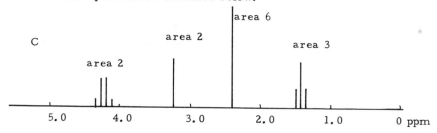

d. Compound D gives a mass spectrum having two mass peaks at 120 and 122 in the ratio of 3:1. The infrared spectrum of the compound shows major absorptions at 3000cm^{-1} (strong, broad), 1695 cm^{-1}, 1600 cm^{-1}, and 1220 cm^{-1} (strong). The nmr spectrum of the compound is sketched below. Is the structure completely specified with this information? If not, what are some alternatives?

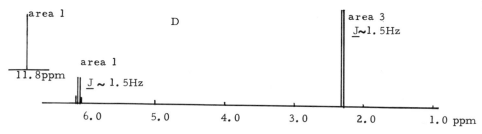

e.   Compound E gave a mass spectrum with peaks of highest mass and
     equal intensity at m/e of 168 and 170. The infrared spectrum showed
     no significant absorption in the region 1600 - 4000 cm^{-1} other than
     C - H absorption around 2900 cm^{-1}. The nmr spectrum of the comp-
     ound is sketched below.

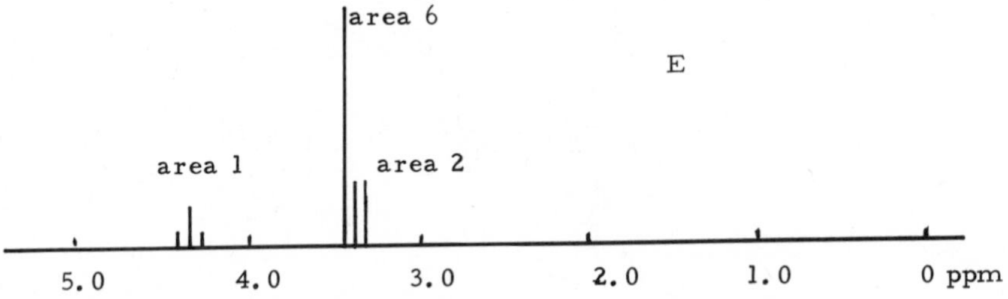

f.   Compound F of formula C_7H_{12} shows infrared absorption at 3100 cm^{-1},
     2900 cm^{-1}, 1645 cm^{-1}, and 1430 cm^{-1}. Note the absence of absorption
     at 1375 cm^{-1}. The nmr spectrum of the compound is sketched below.

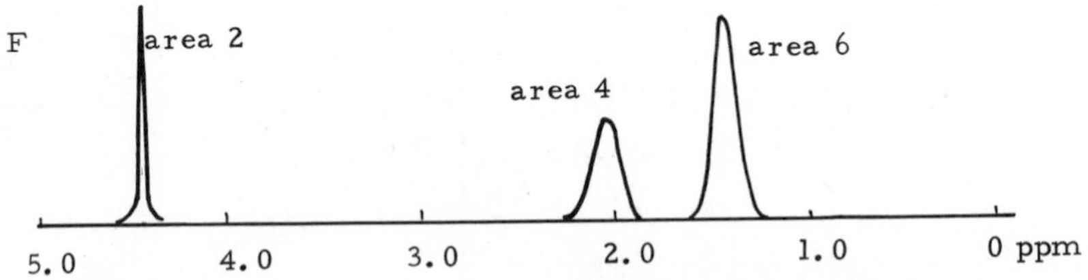

9A-4   Make a diagram similar to the diagrams shown in question 9A-3
of the nmr spectrum expected for each of the compounds listed below. Indicate
the relative peak intensities, chemical shifts, multiplicity, and approximate
J values for each type of hydrogen.

a.   4- isopropyl-1-methylbenzene   f.   methylidenepropanedioic acid
b.   (1-methylpropyl)benzene   g.   1-methoxy-1-buten-3-yne
c.   diethyl butanedioate   h.
d.   3-methoxypropanenitrile
e.   2-(dimethylamino)ethanol

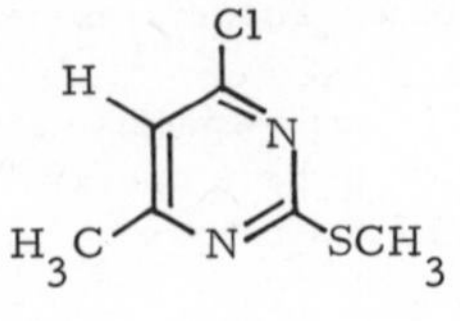

CHAPTER 10

ALKENES AND ALKYNES I.
IONIC AND RADICAL ADDITION REACTIONS

Carbon-carbon double bonds of alkenes are more reactive than single bonds. The most common reactions of alkenes are those of addition to the double bond by either ionic or radical mechanisms:

$$\text{\Large$\diagdown$}C=C\text{\Large$\diagup$} \quad + \quad XY \quad \longrightarrow \quad -\overset{\displaystyle X}{\underset{\displaystyle }{C}}-\overset{\displaystyle }{\underset{\displaystyle Y}{C}}-$$

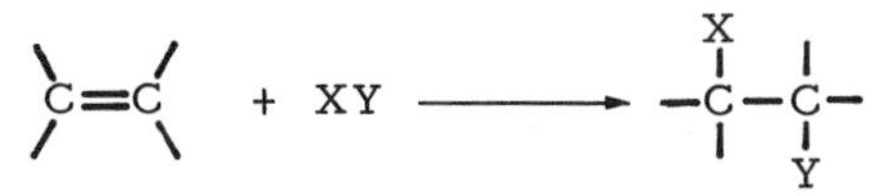

In the ionic reaction, the reagent XY generally is electron-seeking or <u>electrophilic</u>, while the double bond is electron-donating or <u>nucleophilic</u>. Addition usually involves at least two steps: The first step is electrophilic attack of $\overset{\oplus}{X}$ $\overset{\ominus}{Y}$ to give a carbocation intermediate; the second is nucleophilic attack at the positive carbon of the intermediate:

$$\text{\Large$\diagdown$}C=C\text{\Large$\diagup$} + X\!:\!Y \longrightarrow -\overset{X}{C}-\overset{}{\underset{\oplus}{C}}- + :Y^{\ominus} \longrightarrow -\overset{X}{C}-\overset{}{\underset{Y}{C}}-$$

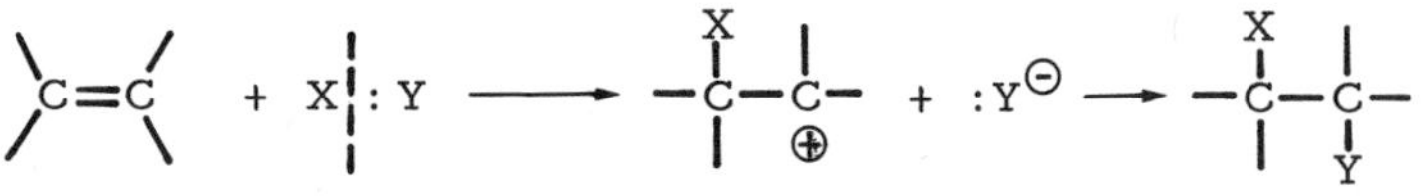

Antarafacial addition of XY commonly is observed.

Electrophilic reagents XY include many acids, HF, HCl, HBr, HI, $H_3O^{\oplus}$, H_2SO_4, in which the potential electrophile is $H^{\oplus}$; electrophilic halogen reagents include Cl_2, Br_2, I_2, HOCl, HOBr, HOI, ICl, and others (Table 10-2). The reaction has value in the preparation of many classes of compounds, such as, haloalkanes, alcohols, dihaloalkanes, and sulfate esters.

Electrophilic addition reactions usually are run under conditions of <u>kinetic control</u> rather than <u>thermodynamic control,</u> meaning that they are not reversible and do not lead to the equilibrium products. Where more than one addition product is possible, the major product is the one formed most rapidly and is not necessarily the one that is the most thermodynamically stable. Also, <u>the most rapidly formed product generally is derived from the most stable of the carbocation intermediates.</u> Thus the acid-catalyzed hydration of 2-methylpropene gives 2-methyl-2-propanol, not 2-methyl-1-propanol, because the tertiary cation is more stable (more rapidly formed) than the primary cation:

$$(CH_3)_2C{=}CH_2 + H^{\oplus} \begin{cases} \xrightarrow{\text{fast}} (CH_3)_2\overset{\oplus}{C}{-}CH_3 \xrightarrow{H_2O} (CH_3)_2\overset{OH}{\underset{|}{C}}{-}CH_3 + H^{\oplus} \quad \text{2-methyl-2-propanol} \\[2em] \xrightarrow[\text{slow}]{} (CH_3)_2CH{-}\overset{\oplus}{C}H_2 \xrightarrow{H_2O} (CH_3)_2CHCH_2OH + H^{\oplus} \quad \substack{\text{2-methyl-1-propanol}\\ \text{(not observed)}} \end{cases}$$

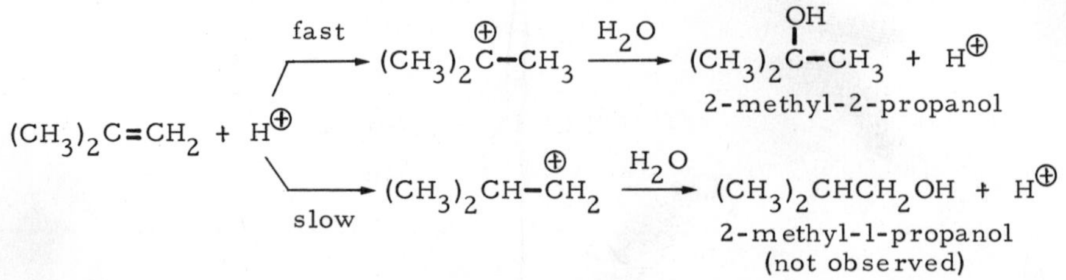

Stability of carbocations is enhanced by alkyl substitution and therefore decreases in the order tertiary ion > secondary ion > primary ion. Almost all substituents stabilize a carbocation relative to hydrogen. The few exceptions include $-CF_3$, $-NO_2$, $-\overset{\oplus}{N}R_3$, and $-\overset{\oplus}{S}R_2$.

An unsymmetrical reagent X—Y undergoes ionic cleavage such that the electrons of the bonding pair go with the more electronegative atom (Figure 10-11).

Alkynes are less reactive than alkenes in electrophilic addition. Hydration of a triple bond requires both an acid catalyst and a mercuric salt. The hydration product (alkenol) rapidly rearranges to a ketone:

$$C_6H_5C{\equiv}CH + H_2O \xrightarrow[H_2SO_4]{HgSO_4} \left[C_6H_5\overset{OH}{\underset{|}{C}}{=}CH_2 \right] \longrightarrow C_6H_5\overset{O}{\overset{||}{C}}CH_3$$

phenylethyne   alkenol   phenylethanone
                                                                            (acetophenone)

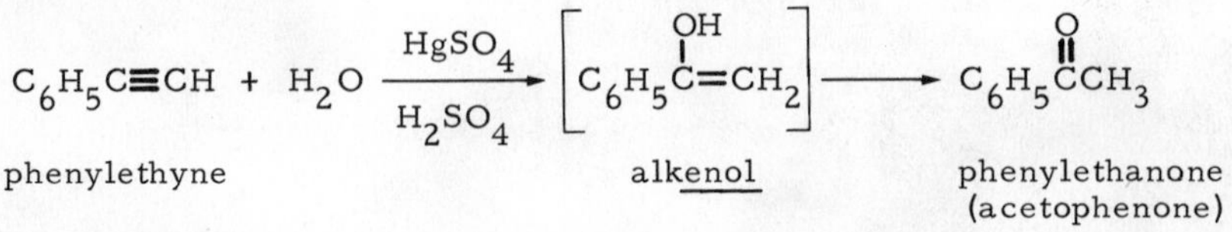

Other ionic additions, particularly those under basic conditions (high pH), are initiated by nucleophilic attack. The intermediate, a carbanion, is rapidly protonated:

$$\underset{/}{\overset{\backslash}{{}}}C{=}C\underset{\backslash}{\overset{/}{{}}} + X^{\ominus} \longrightarrow -\overset{X}{\underset{|}{C}}{-}\overset{\ominus}{\underset{|}{\overset{..}{C}}}{-} \xrightarrow{HX} -\overset{X}{\underset{|}{C}}{-}\overset{H}{\underset{|}{C}}{-} + X^{\ominus}$$

carbanion

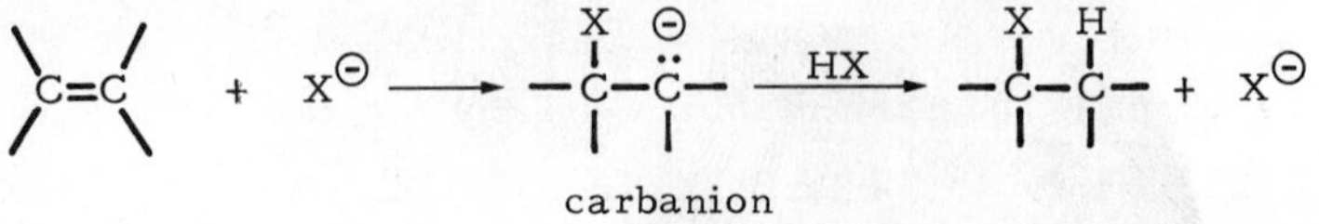

The HX reagent can be H_2O, ROH, RSH, RNH_2, or HCN. These reactions are observed only with alkenes that are substituted at the double bond with electron-withdrawing groups, such as CN, CO_2R, COR, and NO_2, which can stabilize carbanion intermediates. Alkynes are more reactive towards nucleophiles than are alkenes.

Radical-chain addition is common with reagents with bonds weak enough to form radicals by thermal, chemical, or light-induced processes. The mechanistic steps are

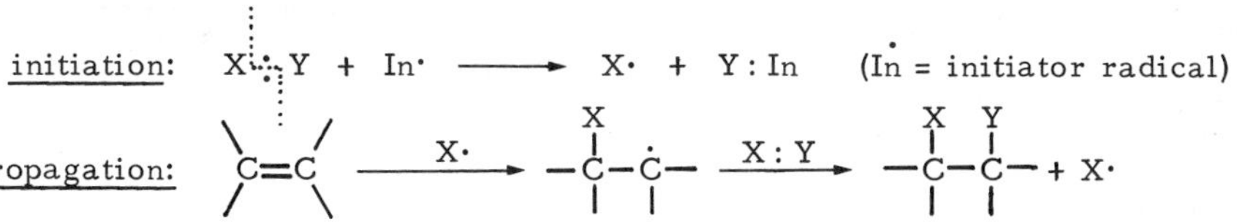

initiation: $X : Y + In\cdot \longrightarrow X\cdot + Y : In$ (In = initiator radical)

propagation: $\begin{array}{c}\diagdown \quad \diagup \\ C=C \\ \diagup \quad \diagdown\end{array} \xrightarrow{\quad X\cdot \quad} -\overset{X}{\underset{|}{C}}-\overset{\cdot}{\underset{|}{C}}- \xrightarrow{\quad X:Y \quad} -\overset{X}{\underset{|}{C}}-\overset{Y}{\underset{|}{C}}- + X\cdot$

Reagents XY include H_2S, RSH, ROH, HBr, Br_2, $BrCCl_3$, RCHO, RCH_2CO_2H, and others (Table 10-3). The orientation of addition to an unsymmetrical alkene or alkyne is determined by the stability of the intermediate carbon radical. Stability increases with increasing alkyl substitution. The order of stability is tertiary radical > secondary radical > primary radical. Most substituents, alkyl or otherwise, are stabilizing relative to hydrogen.

Addition polymerization of alkenes, but not alkynes, is commercially important (Table 10-4) and can occur by way of radical, cation, or anion intermediates. The polymerization of chloroethene (vinyl chloride) to polyvinyl chloride (PVC) is shown below.

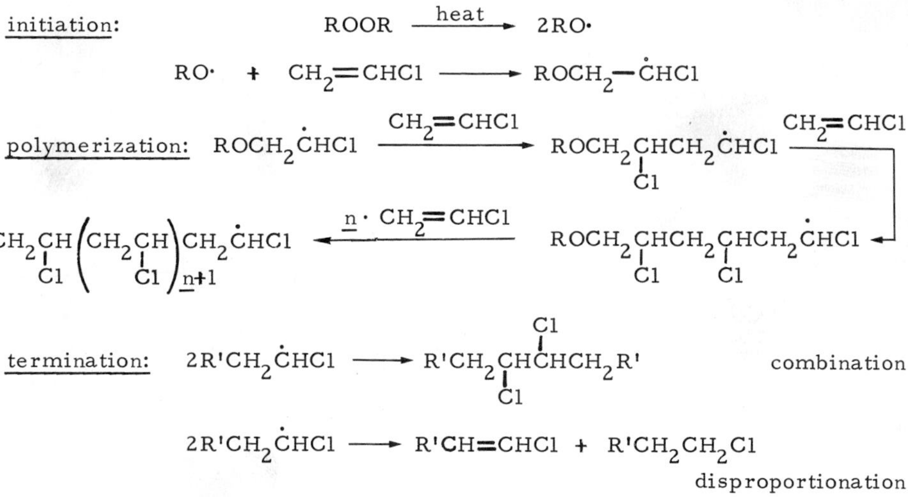

initiation: $ROOR \xrightarrow{\text{heat}} 2RO\cdot$

$RO\cdot + CH_2{=}CHCl \longrightarrow ROCH_2{-}\overset{\cdot}{C}HCl$

polymerization: $ROCH_2\overset{\cdot}{C}HCl \xrightarrow{CH_2{=}CHCl} ROCH_2CH\underset{Cl}{\overset{|}{C}}H_2\overset{\cdot}{C}HCl \xrightarrow{CH_2{=}CHCl}$

$ROCH_2\underset{Cl}{\overset{|}{C}}H\left(CH_2\underset{Cl}{\overset{|}{C}}H\right)_{\underline{n+1}}CH_2\overset{\cdot}{C}HCl \xleftarrow{n\cdot CH_2{=}CHCl} ROCH_2\underset{Cl}{\overset{|}{C}}HCH_2\underset{Cl}{\overset{|}{C}}HCH_2\overset{\cdot}{C}HCl$

termination: $2R'CH_2\overset{\cdot}{C}HCl \longrightarrow R'CH_2\underset{\overset{|}{Cl}}{\overset{\overset{\textstyle Cl}{|}}{C}}HCHCH_2R'$ combination

$2R'CH_2\overset{\cdot}{C}HCl \longrightarrow R'CH{=}CHCl + R'CH_2CH_2Cl$
                                                    disproportionation

Cationic polymerization is restricted to alkenes that form stable carbocations. A related process is the alkylation of alkenes,

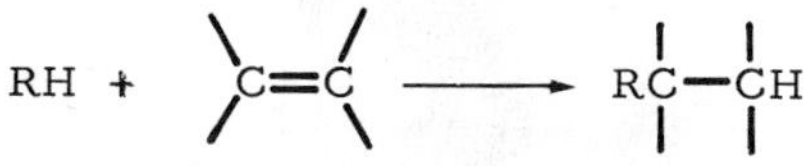

$RH + \begin{array}{c}\diagdown \quad \diagup \\ C=C \\ \diagup \quad \diagdown\end{array} \longrightarrow R\overset{|}{\underset{|}{C}}-\overset{|}{\underset{|}{C}}H$

ANSWERS TO EXERCISES

<u>10-1</u>   <u>Top</u>: Spectrum is very similar to that of 1-butene (Figure 10-1); absorptions at 3100, 1650, 1420, 1000 and 915 cm^{-1} indicate $-CH=CH_2$; there are two possible structures, $CH_3CH_2CH_2CH=CH_2$ and $(CH_3)_2CHCH=CH_2$; the band at 1375 cm^{-1} indicates CH_3 groups - hence, <u>3-methyl-1-butene</u> seems the most likely structure.

Middle: C_6H_{12} corresponds to an alkene or a cycloalkane. Because there are no double bond absorptions, the compound is most likely a cyclo-alkane. The absence of a prominent band at 1375 cm^{-1} suggests that there are <u>no</u> methyl groups. The most likely structure is <u>cyclohexane</u>.

Bottom: Formula C_5H_{10} and weak band at 1675 cm^{-1} indicates an alkene. The absence of bands at 3100, 1000 and 915 cm^{-1} means that the grouping $-CH=CH_2$ or $=CH_2$ is not present. The presence of a strong band at 1375 cm^{-1} implies the presence of CH_3 groups. Most likely structure is <u>2-methyl-2-butene</u>, $(CH_3)_2C=CHCH_3$.

10-2

C_5H_{10} (top)	δ	area	multiplicity	J (Hz)	assignment
	5.2	1	4	~ 6	$=C\underline{H}CH_3$
	~ 1.6	9	complex		$=C-C\underline{H}_3$ (3)

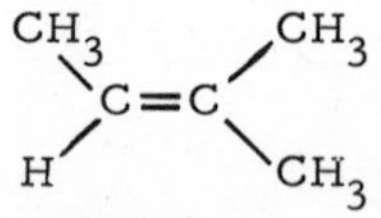

C_5H_{10} (middle)	δ	area	multiplicity	J (Hz)	assignment
	4.7	2	1	-	$C\underline{H}_2=$
	2.0	2	4	7	$C\underline{H}_2CH_3$
	1.7	3	1	-	$C\underline{H}_3-C=$
	1.0	3	3	7	$CH_2C\underline{H}_3$

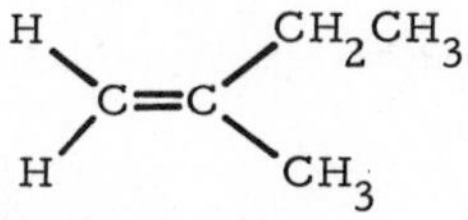

(The alkenic hydrogens are nonequivalent, but their chemical shifts are coincidentally the same.)

C_4H_8O	δ	area	multiplicity	$\underline{J}$ (Hz)	assignment
	6.5	1	4		$C\underline{H}=CH_2$
~	4.0	2	complex		$CH=C\underline{H}_2$
	3.6	2	4	7	$CH_3C\underline{H}_2-O-$
	1.1	3	3	7	$C\underline{H}_3CH_2-O-$

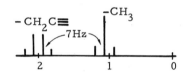

(In this case, the terminal alkenic hydrogens have different
chemical shifts.)

10-3   nmr spectra follow:

a.

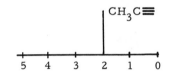

c.

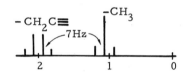

b.

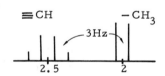

d.

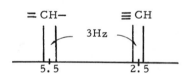

(cis or trans would be the same
although the shifts of the $-CH=CH-$
for the cis isomer are unlikely to be
the same as those of the trans isomer.)

Infrared bands include:

a.   C–H near 2900 cm^{-1}, 1375 cm^{-1} and 1420 cm^{-1}. No C≡C stretch
observable.

b.   ≡C–H at 3300 cm^{-1} (strong, sharp); C≡C near 2100 cm^{-1} (weak); C–H
near 2900 cm^{-1} for CH_3 group.

c.   same as in a.

d.   ≡C–H at 3300 cm^{-1} (strong, sharp); =CH, 3020 cm^{-1}, 1420 cm^{-1} and
1000 cm^{-1}; C≡C, below 2100 cm^{-1} because of conjugation; C=C,
below 1650 cm^{-1} because of conjugation. Unambiguous differentiation
between cis and trans isomers is not possible on the basis of infrared
evidence alone.

10-4   Infrared: bands at 3320 cm^{-1}, 2120 cm^{-1} indicate –C≡CH; bands
at 2950 cm^{-1}, 1460 cm^{-1}, 1375 cm^{-1} indicate CH_2 and CH_3.

Nmr: $CH_3-CH_2-CH_2-CH_2-C\equiv C-H$

δ   0.93   1.18 –   2.16   1.73 ppm
      (triplet) 1.65   (complex)   (triplet)
             (complex)

Mass spec: $\left[CH_3(CH_2)_2 \!\mid\! CH_2C\equiv CH \right]^{\oplus \cdot} \longrightarrow \overset{\oplus}{CH_2}-C\equiv CH$

$M^{\oplus} = 82$

$CH_2=C=CH$ (⊕)

$\underline{m/e}$ 39

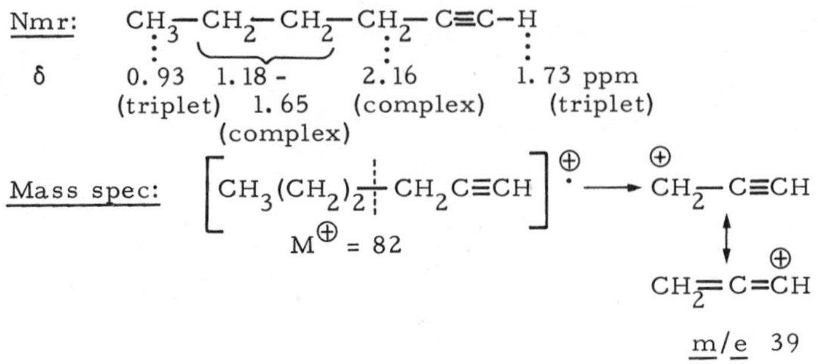

Conclusion: All spectral evidence is consistent with 1-hexyne as the structure.

10-5 Br_2, -26.4 kcal; Cl_2, -40.7 kcal; I_2, -2.3 kcal; HOCl, -51.3 kcal; HCl, -13.4 kcal; HBr, -16.1 kcal; HI, -15.1 kcal; H_2O, -10.4 kcal.

In aqueous solution, the hydrogen halides HX (X = Cl, Br, I) are dissociated into solvated ions, $HX + H_2O \longrightarrow H_3O^{\oplus} + X^{\ominus}$; the solvation energy is large and favorable. Addition to ethene is now unfavorable because of the loss in solvation energy that would be required for addition to occur.

$$CH_2=CH_2 + H_3O^{\oplus} + X^{\ominus} \not\longrightarrow CH_3CH_2X + H_2O$$

For example, the literature shows that $\Delta \underline{H}$ for dissolution of HCl (g) in water is -18 kcal/mole.

10-6

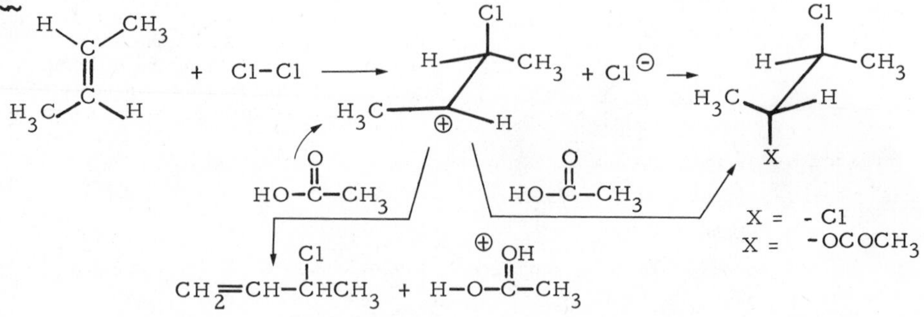

Addition of Cl_2 to cis-2-butene would give the same products as trans-2-butene, but with the configuration reversed at one of the two chiral centers. That is to say, the products from cis-2-butene would be the diastereomers of the products from trans-2-butene.

10-7   A Lewis acid or an electrophile removing $F^{\ominus}$; $SbF_6{}^{\ominus}$ is a very weak nucleophile because the Sb atom has no valence electrons to donate; the only nucleophilic behavior it could show would be to have $F^{\ominus}$ removed by an electrophile (e.g., reverse of ethenebromonium ion formation).

$$F_5\overset{\ominus}{Sb}\!\!-\!\!\ddot{\ddot{F}}\!: \,+\, E \longrightarrow SbF_5 \,+\, :\ddot{\ddot{F}}\!\!-\!\!E^{\ominus}$$

10-8

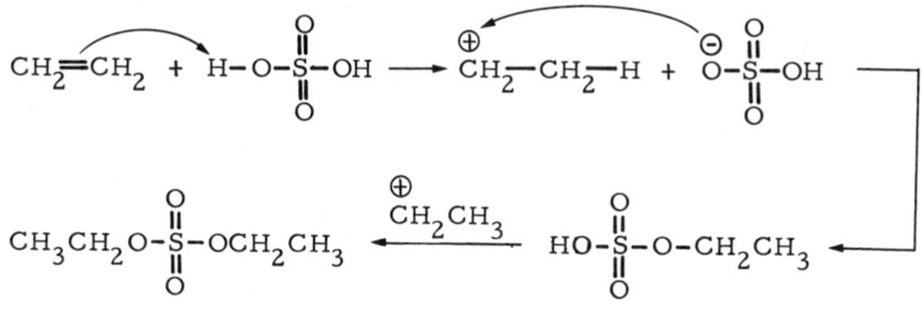

diethyl sulfate   ethyl hydrogen sulfate

10-9   First reaction would be:

$$H_2O \,+\, \text{excess } HSbF_6 \longrightarrow H_3O^{\oplus} \,+\, SbF_6^{\ominus}.$$

This would be followed by:

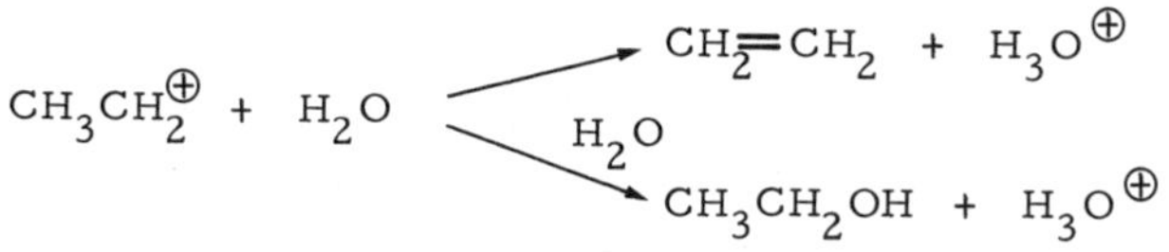

$$CH_3CH_2^{\oplus} \,+\, H_2O \,\Big\langle \begin{array}{l} CH_2\!\!=\!\!CH_2 \,+\, H_3O^{\oplus} \\[1mm] H_2O \\[1mm] CH_3CH_2OH \,+\, H_3O^{\oplus} \end{array}$$

10-10 $CH_3CH\!\!=\!\!CH_2 \,+\, \ddot{N}H_3 \longrightarrow CH_3\underset{:NH_2}{CHCH_3}$ $\Delta\underline{H}°$= -15 kcal

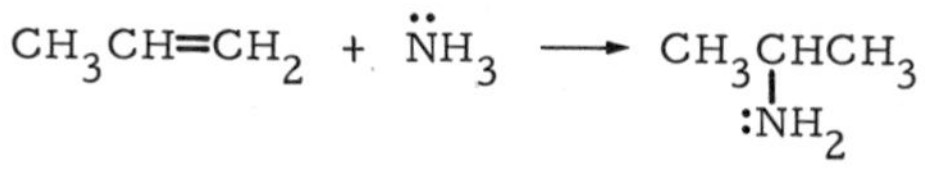

The reaction is energetically feasible but is incompatible with an acid catalyst because the ammonia is virtually wholly converted to ammonium $NH_4^{\oplus}$ in strong acid solution, and ammonium ion is neither a nucleophile nor a good electrophile.

10-11

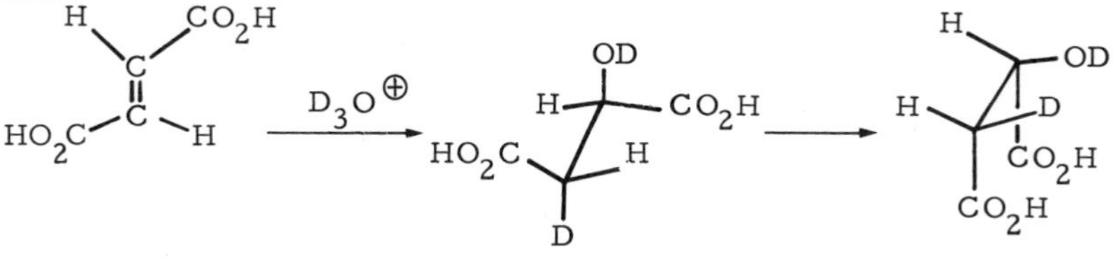

which is

equivalent to

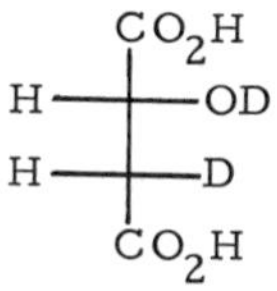

(and the $\underline{\underline{L}}$ enantiomer)

10-12   The configuration of the hydration product of antarafacial addition to fumaric acid is shown in Exercise 10-11. The enzyme-catalyzed reaction gives the same diastereomer (L enantiomer) from which it may be concluded that the biological hydration is also a stereospecific antarafacial addition.   The reverse reaction, dehydration, is an antarafacial elimination.

10-13   By Markownikoff's rule the H of HX adds to the carbon with the most hydrogens. The modern view is that H of HX adds most rapidly to that carbon which leads to the most stable carbocation intermediate. Clearly, by adding a proton to the carbon with the most hydrogens places the positive charge on the carbon with the fewest hydrogens, and the fewer hydrogens the more stable is the ion (tert > sec > prim).

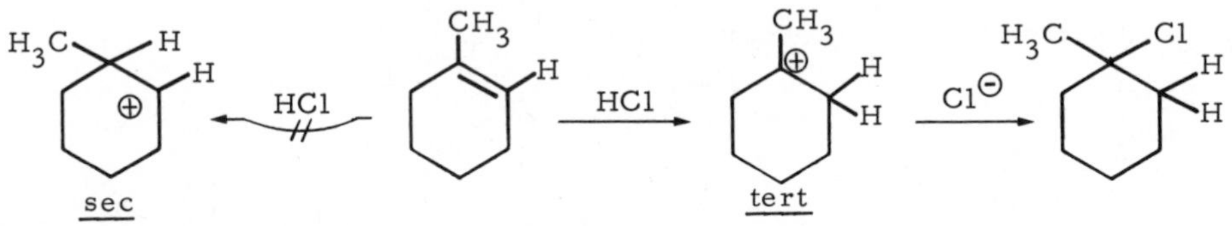

10-14   a. $CH_3CH_2CHCH_3$
                       OSO_2OH

          b. $(CH_3)_3COH$

          c. $(CH_3)_2C-CHCH_3$ + $(CH_3)_2C-CHCH_3$ no$\left[(CH_3)_2C-CHCH_3\right]$
                       Br Br CH_3O Br   Br OCH_3

10-15   The order of reactivity is expected to parallel the order of stability of the carbocations that can be formed. That is, tert > sec > prim and $(CH_3)_2C{=}CH_2$ > $CH_3CH{=}CH_2$ > $CH_2{=}CH_2$.

10-16   a. $(CH_3)_2C-CHCH_3$ $(CH_3)_2C-CHCH_3$
                       Cl I O_2N I

          b. $\overset{\delta\oplus}{C}={\overset{\delta\ominus}{N}}$ $\xrightarrow{\ H_2O\ }$ $-C-NH-$
                                                               OH

          c.   d. $(CH_3)_2C-CH_2OH$
                                                          F

          e. $(CH_3)_2C-CH_2Br$
                       F

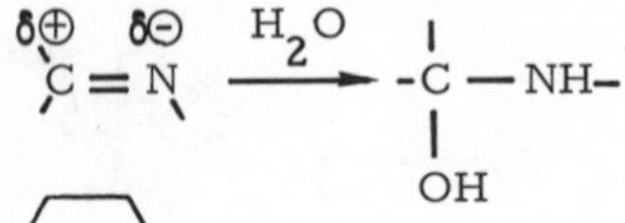

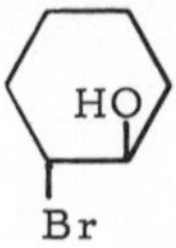

10-17

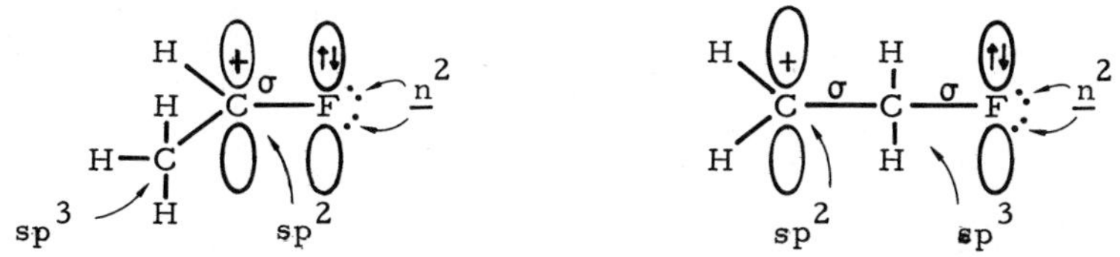

The 1-fluoroethyl carbocation is formed more rapidly because it is stabilized by π overlap with a filled fluorine p orbital. Similar overlap for the 2-fluoroethyl is not possible.

10-18   a. CH_3CCl_3 c. $CF_3CH_2CH(Cl)CH_3$

b. $(CH_3)_2\underset{\underset{Cl}{|}}{C}-CHCl_2$ d. $CH_3O\underset{\underset{Cl}{|}}{C}HCH_2F$

10-19 The $\Delta \underline{H}^{\circ}$ difference for these reactions can be estimated by calculating $\Delta \underline{H}$ values for the respective homolytic processes producing $CH_2{=}\overset{\cdot}{C}H$ and $Cl\cdot$ from each starting material. The $\Delta \underline{H}^{\circ}$ for electron-transfer to give ions will then be the same.

$$CH{\equiv}CH + HCl \longrightarrow CH_2{=}\overset{\cdot}{C}H + \overset{\cdot}{C}l \qquad \Delta\underline{H}^{\circ} = +58 \text{ kcal}$$

$$CH_2{=}CHCl \longrightarrow CH_2{=}\overset{\cdot}{C}H + \overset{\cdot}{C}l \qquad \Delta \underline{H}^{\circ} = +81 \text{ kcal}$$

$$CH_2{=}\overset{\cdot}{C}H + \overset{\cdot}{C}l \longrightarrow CH_2{=}\overset{\oplus}{C}H + \overset{\ominus}{C}l$$

Clearly, the addition route is 23 kcal more favorable than the solvolysis route.

10-20
$$CH_2{=}CH-\overset{..}{\underset{..}{O}}H + H\overset{..}{\underset{..}{O}}H \longrightarrow CH_2{=}CH-\overset{..}{\underset{..}{O}}{:}^{\ominus} + H_3\overset{\oplus}{O}{:}$$
proton acceptor

$$H\overset{..}{\underset{..}{O}}{:}^{\ominus} + H{:}CH_2{-}CH{=}\overset{..}{\underset{..}{O}}{:} \xleftarrow{\quad H\overset{..}{\underset{..}{O}}H \quad} {:}^{\ominus}CH_2{-}CH{=}\overset{..}{\underset{..}{O}}{:}$$
proton donor

10-21
a.

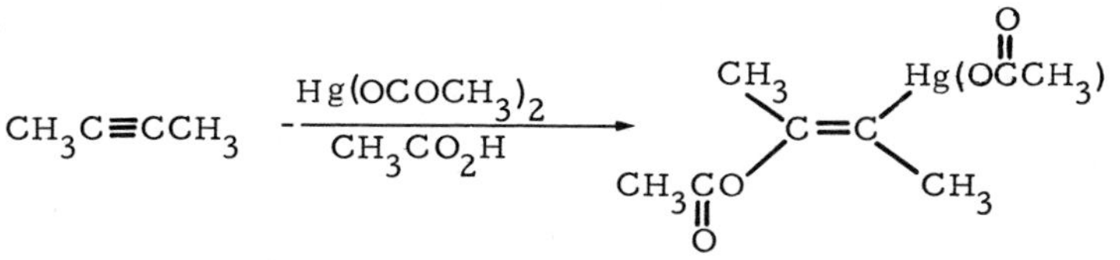

b.

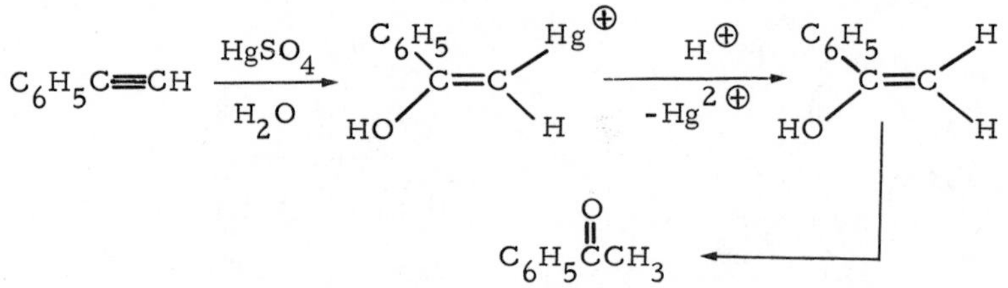

$$C_6H_5C{\equiv}CH \xrightarrow[\text{H}_2\text{O}]{\text{HgSO}_4} \underset{\text{HO}}{\overset{C_6H_5}{C}}{=}\underset{\text{H}}{\overset{Hg^{\oplus}}{C}} \xrightarrow[-Hg^{2\oplus}]{H^{\oplus}} \underset{\text{HO}}{\overset{C_6H_5}{C}}{=}\underset{\text{H}}{\overset{H}{C}}$$

$$C_6H_5\overset{\overset{\text{O}}{\|}}{C}CH_3 \longleftarrow$$

c.

$$HC{\equiv}CH \xrightarrow[\text{CH}_3\text{OH}]{\text{HgCl}_2} \underset{\text{CH}_3\text{O}}{\overset{H}{C}}{=}\underset{\text{H}}{\overset{HgCl}{C}}$$

10-22 $CH_3C{\equiv}C{-}CH{=}CHCH_3 \longrightarrow CH_3C{\overset{..}{=}}\overset{\ominus}{C}{-}CH{=}CHCH_3$ (resonance-

$CH_3O^{\ominus}$ $\underset{CH_3\overset{|}{O}}{} \quad H{-}OCH_3$ stabilized)

$$CH_3\underset{\overset{|}{CH_3O}}{C}{=}CH{-}CH{=}CHCH_3$$

10-23 Chloride is not a strong enough nucleophile to add to carbon and form $CH_3CCl{=}CHCH{=}CHCH_3$ by a mechanism analogous to that shown in Exercise 10-22.

10-24

Initiation: $ROOR \longrightarrow 2RO{\cdot}$, $RO{\cdot} + HCl \longrightarrow ROH + Cl{\cdot}$ $\Delta \underline{H}^{\circ} = -7.5$ kcal

          or $RO{\cdot} + HCl \longrightarrow ROCl + H{\cdot}$ $\Delta \underline{H}^{\circ} = +51$ kcal

Propagation: $RCH{=}CH_2 + Cl{\cdot} \xrightarrow{\Delta \underline{H}^{\circ} = -17.8 \text{ kcal}} R\overset{\cdot}{C}HCH_2Cl \xrightarrow[\Delta \underline{H}^{\circ} = +4.4 \text{ kcal}]{HCl}$

$$RCH_2CH_2Cl + Cl{\cdot}$$

or, $RCH{=}CH_2 + H{\cdot} \xrightarrow{\Delta \underline{H}^{\circ} = -35.5 \text{ kcal}} R\overset{\cdot}{C}HCH_3 \xrightarrow[\Delta \underline{H}^{\circ} = +22 \text{ kcal}]{HCl}$

$$\underset{\overset{|}{Cl}}{R}CHCH_3 + H{\cdot}$$

Neither mechanism has favorable energetics in __both__ steps. Propagation by $H{\cdot}$ is particularly unlikely.

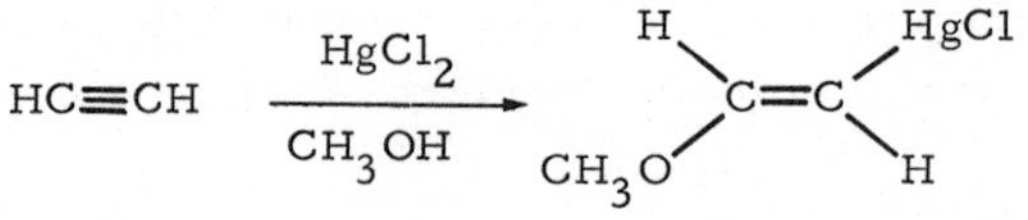

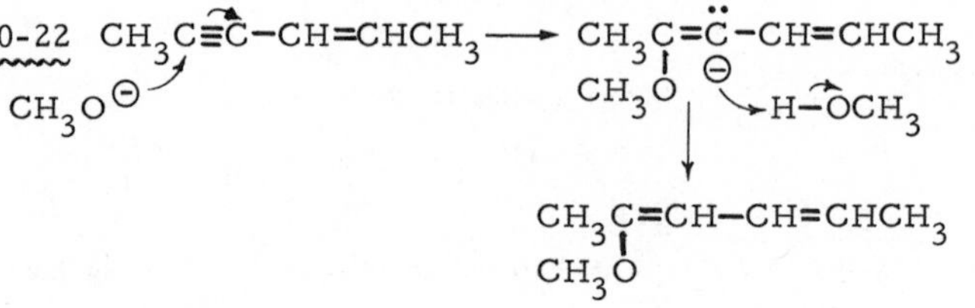

10-25

X	Initiation, $\Delta \underline{H}^{\circ}$ kcal	Propagation, $\Delta \underline{H}^{\circ}$ kcal		Comment
	$RO\cdot + HX \longrightarrow$ $ROH + X\cdot \longleftarrow$	$C{=}C + \overset{\cdot}{X} \longrightarrow$ $\cdot C{-}C{-}X \longleftarrow$	$\cdot C{-}C{-}X + HX \longrightarrow$ $HC{-}CX + X\cdot \longleftarrow$	
F	+25.3	-52.8	+37.2	bad
Cl	-7.5	-17.8	+4.4	fair
I	-39.2	+12.2	-27.3	poor

None of the hydrogen halides, except HBr, add readily to alkenes by a radical-chain mechanism.

10-26 a.   The observed stereochemistry is consistent with antarafacial addition of Br_2 to both esters by an electrophilic mechanism.

b. $Br_2 \xrightarrow{h\nu} 2Br\cdot$

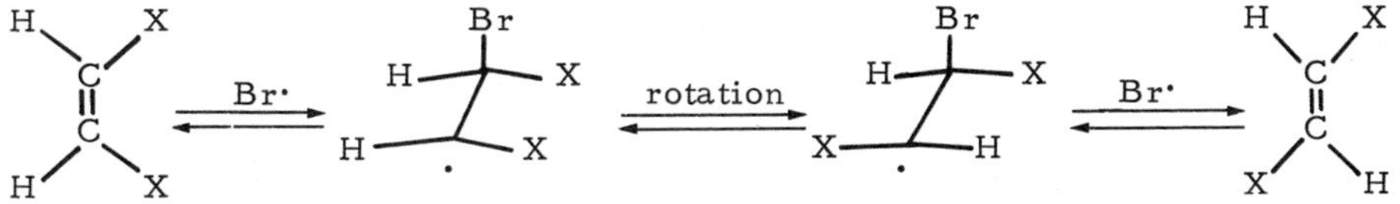

X = $CO_2C_2H_5$ A   B

The above mechanism accounts for the rearrangement of the <u>cis</u> to the <u>trans</u> ester on irradiation in the presence of <u>traces</u> of Br_2. Clearly, attack by Br· is reversible.

When the bromine and ester are equimolar, the intermediate radicals A and/or B are trapped rapidly and irreversibly to give dibromo adducts.

$$A + Br_2 \longrightarrow \underline{\underline{D}}, \underline{\underline{L}} \text{ adduct} + Br\cdot$$
$$B + Br_2 \longrightarrow \underline{meso} \text{ adduct} + Br\cdot$$

The fact that the trans ester gives only the <u>meso</u> adduct implies that B reacts with Br_2 stereospecifically and that B does <u>not</u> rotate to give A as rapidly as it reacts with Br_2 to give product. In contrast, the fact that the cis ester gives a mixture of diastereomers means that A rearranges to B at a rate competitive with its reaction with Br_2.

10-27 Electron-repulsion arguments (Sections 6-3 and 6-4F) suggest an angular (or bent) geometry.

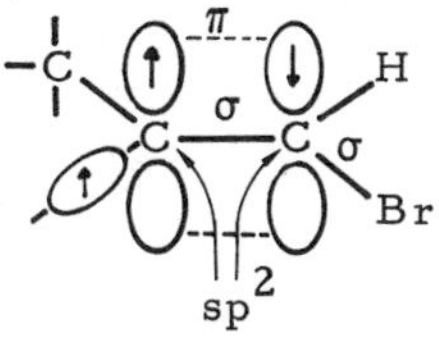

10-28 Initiation generates some radical species $(R \cdot)$ that reacts with $BrCCl_3$ as follows:

$$R \cdot + Br-CCl_3 \longrightarrow R-Br + \cdot CCl_3 \qquad\qquad \Delta \underline{H}^{\circ} = -14 \text{ kcal}$$

Propagation: $C_6H_{13}CH{=}CH_2 + \cdot CCl_3 \longrightarrow C_6H_{13}\overset{\cdot}{C}H{-}CH_2CCl_3$

$$\Delta \underline{H}^{\circ} = -19.4 \text{ kcal}$$

$$C_6H_{13}\overset{\cdot}{C}H{-}CH_2CCl_3 + BrCCl_3 \longrightarrow C_6H_{13}\underset{Br}{\overset{\mid}{C}H}{-}CH_2CCl_3 + \cdot CCl_3$$

$$\Delta \underline{H}^{\circ} = -14 \text{ kcal}$$

$$RC{\equiv}CH + BrCCl_3 \longrightarrow RCBr{=}CHCCl_3$$

10-29 Under the given conditions, $CN^{\ominus}$ adds to the double bond, and the resultant anion propagates addition polymerization.

$$CN^{\ominus} + CH_2{=}CHCN \longrightarrow NCCH_2\underset{CN}{\overset{\mid}{\overset{..}{C}}H}{}^{\ominus} \xrightarrow{\quad CH_2{=}CHCN \quad} NCCH_2\underset{CN}{\overset{\mid}{C}H}CH_2\underset{CN}{\overset{\mid}{\overset{..}{C}}H}{}^{\ominus}$$

$$NCCH_2\underset{CN}{\overset{\mid}{C}H}{-}{\Big(}CH_2\underset{CN}{\overset{\mid}{C}H}{\Big)}_{\underline{n}}CH_2{-}\underset{CN}{\overset{\mid}{\overset{..}{C}}H}{}^{\ominus} \xleftarrow{\quad \underline{n}\,CH_2{=}CHCN \quad}$$

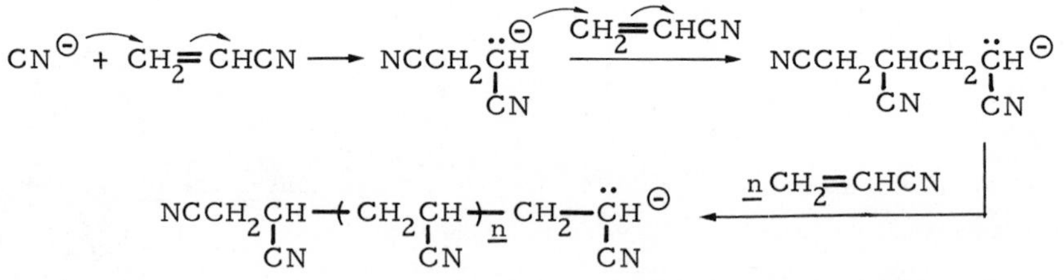

A polar solvent is necessary to solubilize the initiator $(NaCN)$. Propene does not polymerize similarly because $CN^{\ominus}$ does not add to propene; $CN^{\ominus}$ adds to propenenitrile because the resulting anion is resonance-stabilized.

$$NC{-}CH_2{-}\overset{\ominus}{\underset{}{\overset{..}{C}}}H{-}C{\equiv}N: \longleftrightarrow NC{-}CH_2{-}CH{=}C{=}\overset{..}{N}:{}^{\ominus}$$

10-30 $RCH_2\overset{\cdot}{C}H_2 + H{\mid}\underset{R}{\overset{\mid}{C}}H{-}\overset{\cdot}{C}H_2 \longrightarrow RCH_2CH_2{-}H + RCH{=}CH_2$

$$\Delta \underline{H}^{\circ} \text{ (disproportionation)} = -63.2 \text{ kcal}$$
$$\Delta \underline{H}^{\circ} \text{ (combination)} \qquad = -82.6 \text{ kcal} \quad \text{(more favorable)}$$

10-31 $\underset{H_3C}{\overset{H_3C}{>}}C{=}CH_2 + H_2SO_4 \rightleftharpoons H_3C{-}\underset{CH_3}{\overset{\oplus}{\underset{\mid}{C}}}{-}CH_3 + HSO_4^{\ominus}$

$$C-\overset{\oplus}{\underset{\underset{C}{|}}{C}}-C \;+\; C-\overset{\overset{H}{|}}{\underset{\underset{C}{|}}{C}}-C^{13} \;\;\rightleftharpoons\;\; C-\overset{\overset{H}{|}}{\underset{\underset{C}{|}}{C}}-C \;+\; C-\overset{\oplus}{\underset{\underset{C}{|}}{C}}-C^{13}$$

$$C-\overset{\oplus}{\underset{\underset{C}{|}}{C}}-C^{13} \;+\; C-\overset{\overset{D}{|}}{\underset{\underset{C}{|}}{C}}-C \;\;\rightleftharpoons\;\; C-\overset{\overset{D}{|}}{\underset{\underset{C}{|}}{C}}-C^{13} \;+\; C-\overset{\oplus}{\underset{\underset{C}{|}}{C}}-C$$

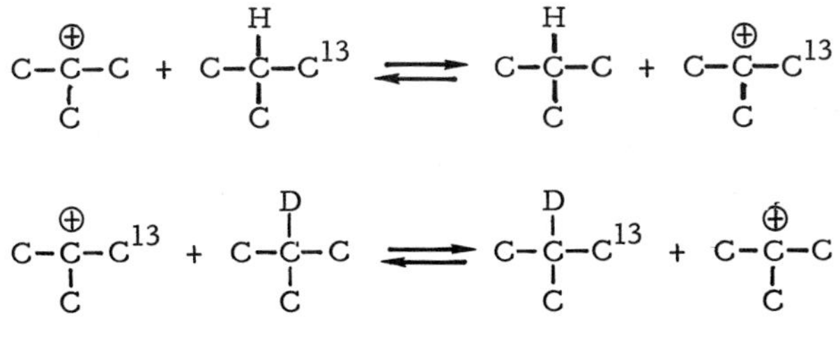

By the above mechanism, D_2SO_4 is not expected to form. Deuterium is transferred only from one tertiary carbon to another.

10-32   a. $H_3O^{\oplus}$ (E), $HO^{\ominus}$(N)   b. $H_2OI^{\oplus}$(E), $I^{\ominus}$ (N)

     c. $CH_3^{\oplus}C(OH)_2$ (E), $CH_3CO_2^{\ominus}$ (N)

     d. $CH_3NH_3^{\oplus}$ (E), $CH_3\overset{..}{N}H^{\ominus}$ (N)

     e. $\overset{\oplus}{N}O_2$ (E), $^{\ominus}ONO_2$ (N)   f. $HF + SbF_5$ (E), $F^{\ominus}$ (N)

     g. FSO_3H (E), $HOSO_3^{\ominus}$ (N).

10-33   a. $H_2C\overset{\oplus}{=}OH$ (E), H_2O (N)   b. Br_2 (E), $H_2C=CH_2$ (N)

     c. $:CH_2$ (E), $H_2C=CH_2$ (N)

     d. $CH_2=CHCN$ (E), $:\overset{..}{N}H_2^{\ominus}$ (N)

10-34   a.   HCl in a non-nucleophilic solvent

     b.   aqueous H_2SO_4 or H_3PO_4

     c. Cl_2 in H_2O

     d. CH_3CO_2H with an acid catalyst (H_2SO_4)

     e.   HF and <u>N</u>-bromosuccinimide in pyridine.   (See Table 10-2.)

10-35   a. $H_3C-\!|\,H$ b. $H_3Si\!|-H$ c. $CH_3^-\!|\,Li$

     d. $CH_3^-\!|\,MgCH_3$ e. $CH_3S\!|-Cl$ f. $H_2N-\!|SCH_3$

     g. $H_2N\!|-OH$ h. $H_2N-\!|Br$ i. $H_2P\!|-Cl$

     j. $(CH_3)_3Si\!|-Cl$

10-36

     a.   b. $CH_3\underset{\underset{F}{|}}{C}HCH_3$ c.

$$\underset{\underset{D}{|}}{\overset{Br}{\underset{\substack{H-\\ H-}}{\diagup}}}\overset{\diagdown C_6H_5}{\diagdown CH_3}$$

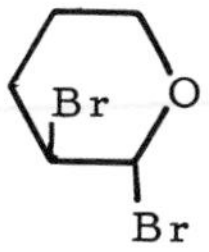

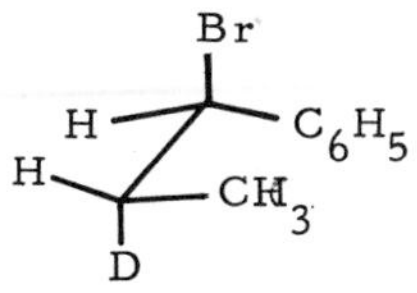

(and enantiomer)

d.

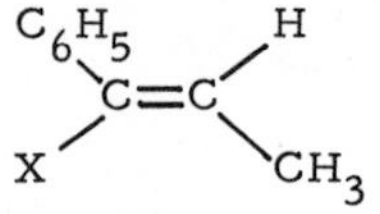

$$\underset{X}{\overset{C_6H_5}{\diagdown}}C=C\underset{CH_3}{\overset{H}{\diagup}}\qquad X = Cl,\quad O-\overset{\overset{O}{\|}}{C}-CH_3$$

e.

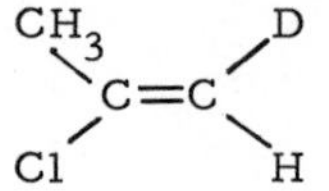

$$\underset{Cl}{\overset{CH_3}{\diagdown}}C=C\underset{H}{\overset{D}{\diagup}}\qquad f.\ \ C_6H_5CH(Cl)CH_2SCH_3 \qquad g.$$

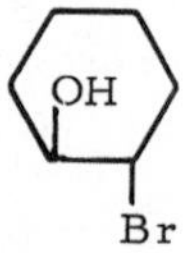

10-37 Addition is strongly exothermic ($\Delta \underline{H}^{\,o} = -131$ kcal) partly because the F—F bond is exceptionally weak and C—F bonds are exceptionally strong. The heat evolved is more than sufficient to dissociate F_2 to fluorine atoms – which then propagate chain reactions involving hydrogen abstraction,

$$-\overset{|}{\underset{|}{C}}H + F \longrightarrow -\overset{|}{\underset{|}{C}}\cdot + HF, \qquad \Delta \underline{H}^{\,o} = -36 \text{ kcal} .$$

Overall, product mixtures are obtained that reflect addition, substitution and C—C cleavage reactions.

10-38 The reaction can be explained as proceeding through carbocation intermediates A and B , or their equivalent C. Attack of a nucleophile (CH_3OH, $Cl^{\ominus}$) can occur at C2 or C3, leading to the observed product mixtures.

A. $C-C-\overset{\overset{\displaystyle Cl}{|}}{C}-\overset{}{\underset{\oplus}{C}}-C$
B. $C-C-\overset{\overset{\displaystyle Cl}{|}}{\underset{\underset{\displaystyle \oplus}{}}{C}}-C-C$
C. $C-C-\overset{\overset{\displaystyle \oplus}{\overset{\displaystyle Cl}{/\diagdown}}}{C}-C-C$

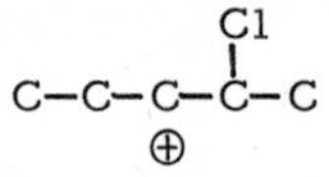

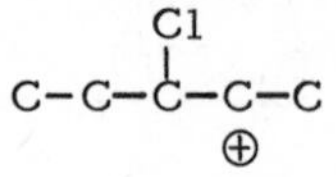

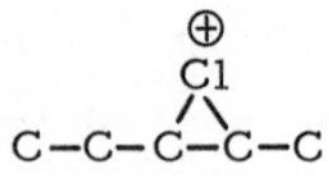

10-39

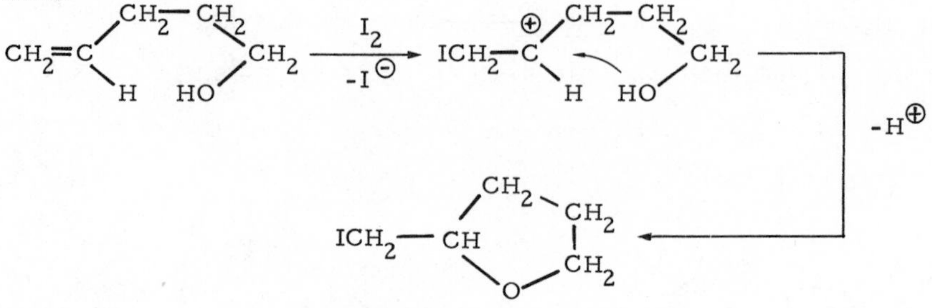

(internal nucleophilic attack)

10-40 Aluminum bromide functions to make the bromine molecule more electrophilic. This is necessary because the chlorines greatly reduce the nucleophilicity of the double bond.

$$Br-Br + AlBr_3 \rightleftharpoons Br^{\oplus} + AlBr_4^{\ominus}$$

$$Br^{\oplus} + Cl_2C=CCl_2 \longrightarrow Cl_2\overset{\oplus}{C}-CBrCl_2 \underset{-AlBr_3}{\overset{AlBr_4^{\ominus}}{\rightleftharpoons}} Cl_2CBr-CBrCl_2$$

10-41 Indicated product is unimportant because attack of $D^{\oplus}$ at tertiary carbon in preference to secondary carbon is very unlikely. Probable products would be formed by the sequence:

2- and 6- deuterium labelled cyclohexanol

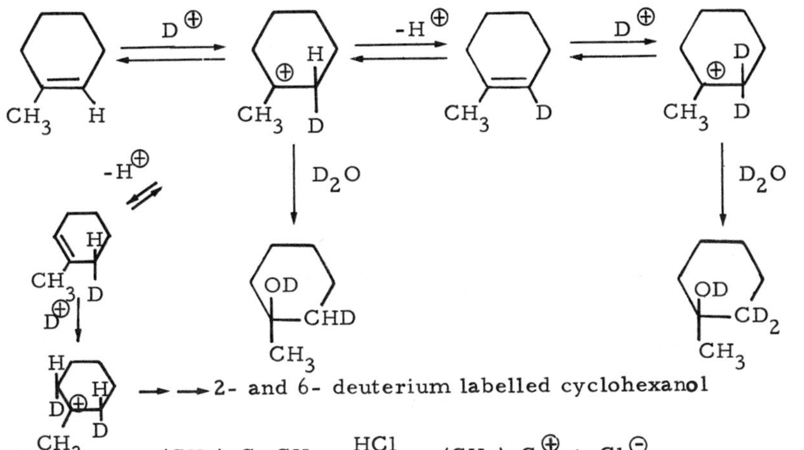

10-42
$$(CH_3)_2C=CH_2 \overset{HCl}{\rightleftharpoons} (CH_3)_3C^{\oplus} + Cl^{\ominus}$$

$$(CH_3)_3C^{\oplus} + CH_2=CH_2 \longrightarrow (CH_3)_3C\overset{\oplus}{C}H_2\overset{\oplus}{C}H_2 \overset{Cl^{\ominus}}{\longrightarrow} (CH_3)_3CCH_2CH_2Cl$$

10-43 a. $(CH_3)_2C=CH_2 \overset{D^{\oplus}}{\rightleftharpoons} (CH_3)_2\overset{\oplus}{C}CH_2D \overset{-H^{\oplus}}{\rightleftharpoons} (CH_3)_2C=CHD$

$(CH_3)_2\overset{\oplus}{C}CH_2D + (CH_3)_3CH \rightleftharpoons (CH_3)_2CHCH_2D + (CH_3)_3C^{\oplus}$

$(CH_3)_2C=CHD \overset{D^{\oplus}}{\rightleftharpoons} (CH_3)_2\overset{\oplus}{C}CHD_2$

$(CH_3)_2\overset{\oplus}{C}CHD_2 + (CH_3)_3CH \rightleftharpoons (CH_3)_2CHCHD_2$ etc.

Note that the hydrogen on the tertiary carbon undergoes intermolecular exchange and does not exchange with the deuterium of D_2SO_4.

    b. This can be so because exchange of the protons of the tert-butyl cation with D_2SO_4 is faster than the hydrogen-transfer reaction between

2-methylpropane and the tert-butyl cation.

$(CH_3)_2C=CH_2 \xrightarrow{D^\oplus} (CH_3)_2\overset{\oplus}{C}CH_2D \xrightarrow{H^\oplus} (CH_3)_2C=CHD \xrightarrow{D^\oplus} (CH_3)_2\overset{\oplus}{C}CHD_2$

$\xrightarrow{-H^\oplus} (CH_3)_2C=CD_2$ etc.... $\longrightarrow (CD_3)_3C^\oplus + (CH_3)_3CH$

$\rightleftharpoons (CD_3)_3CH + (CH_3)_3C^\oplus$

10-44   a, b.

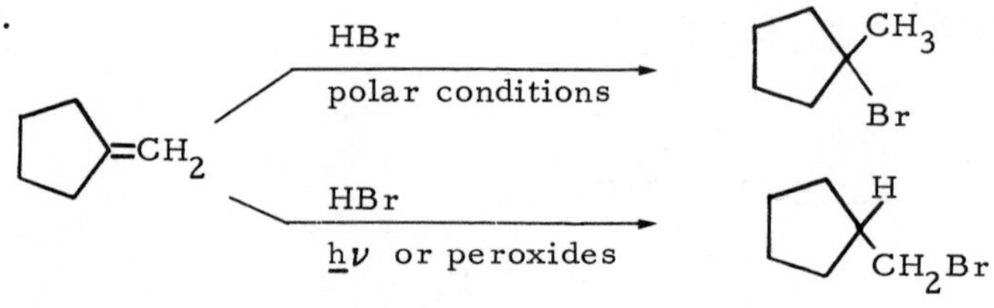

c.   1-pentene + $BrCCl_3 \xrightarrow{h\nu} CH_3CH_2CH_2CHBrCH_2CCl_3$

d. $C_6H_5C\equiv CCH_3 + H_2O \xrightarrow[HgSO_4]{H_2SO_4} C_6H_5COCH_2CH_3$

e. $C_6H_5CH=CH_2 + H_2O \xrightarrow{10\% \ H_2SO_4} C_6H_5CH(OH)CH_3$

f. $CH\equiv CH + 2Cl_2 \longrightarrow Cl_2CHCHCl_2$

g. $CH\equiv CH + Hg(OCOCH_3)_2 + CH_3CO_2H \xrightarrow{H_2SO_4} H_2C=CHOCOCH_3$

h.   The product from Part g. can be polymerized by radical addition polymerization initiated by peroxides.

10-45   a.   The $\Delta \overset{o}{H}$ is the difference between the strength of a single C—C bond (82.6 kcal) and the energy required to break one-half of the double bond (63.2 kcal) multiplied by n. That is -19.4 n kcal.

b.   -19.4 kcal

10-46   HOH + RO· $\longrightarrow$ ROH + HO· $\Delta \overset{o}{H} =$ 0   kcal

$CH_3CH=CH_2 + HO· \longrightarrow CH_3\overset{·}{C}HCH_2OH$ $\Delta \overset{o}{H} =$ -22.3 kcal

$CH_3\overset{·}{C}HCH_2OH + HOH \longrightarrow CH_3CH_2CH_2OH + HO·$ $\Delta \overset{o}{H} =$ +11.9 kcal

Energetic balance in two propagation steps in the addition of water to propene are such that the chains are not likely to be very long. A non-chain

mechanism is probably more likely since the individual chain addition steps
are strongly exothermic.

$$CH_3CH=CH_2$$

$$\xrightarrow{H\cdot} CH_3\overset{\cdot}{C}HCH_3 \xrightarrow{HO\cdot} (CH_3)_2CHOH$$

$$\Delta\underline{H}^\circ = -35.5 \qquad \Delta\underline{H}^\circ = -85.5$$

$$\xrightarrow{HO\cdot} CH_3\overset{\cdot}{C}HCH_2OH \xrightarrow{H\cdot} CH_3CH_2CH_2OH$$

$$\Delta\underline{H}^\circ = -22.3 \qquad \Delta\underline{H}^\circ = -98.7$$

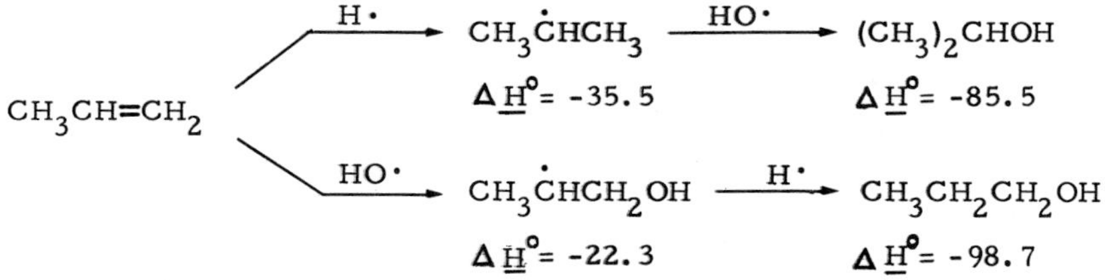

Prior addition of H· should give 2-propanol while of ·OH should give
1-propanol. In addition to 1-propanol and 2-propanol, propane and 1, 2-
propanediol might be expected.

10-47

$$NH_3 \xrightarrow{RO\cdot} \cdot NH_2 + ROH \qquad \Delta H^\circ = -17.2 \text{ kcal}$$

$$\overset{\diagdown}{\underset{\diagup}{C}}=\overset{\diagup}{\underset{\diagdown}{C}} + \cdot NH_2 \longrightarrow \overset{\diagdown}{\underset{\diagup}{\overset{\cdot\cdot}{C}}}-\overset{\mid}{\underset{\mid}{C}}-NH_2 \qquad \Delta\underline{H}^\circ = -9.6 \text{ kcal}$$

$$\overset{\diagdown}{\underset{\diagup}{\overset{\cdot\cdot}{C}}}-\overset{\mid}{\underset{\mid}{C}}-NH_2 + NH_3 \longrightarrow \overset{\diagdown}{\underset{\diagup}{C}}H-\overset{\mid}{\underset{\mid}{C}}-NH_2 + \overset{\cdot}{N}H_2 \qquad \Delta\underline{H}^\circ = -5.3 \text{ kcal}$$

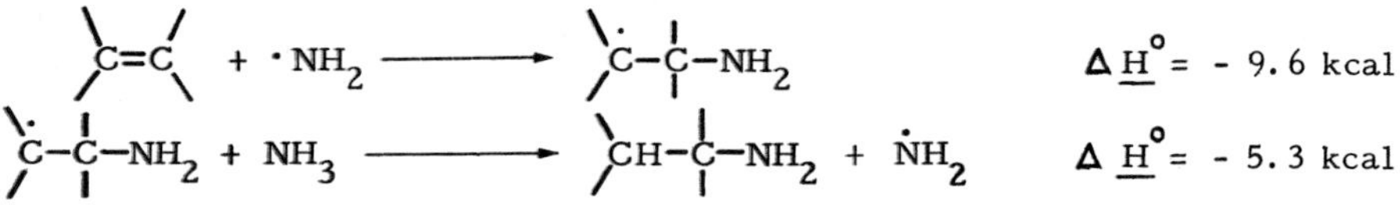

The reaction appears to be feasible (overall $\Delta H^\circ = -14.9$ kcal) with a
reasonable balance between the propagation steps.

$$CH_3CH=CH_2 + NH_3 \longrightarrow CH_3CH_2CH_2NH_2$$

10-48

transition state for
slow step

Energy

transition state
for
preequilibrium

$\Delta\underline{H}^{\ddagger}$

A-B-C+D

A+BC+D

$\Delta\underline{H}^\circ$

AB+CD

Progress of reaction

10-49 (See Table 10-3)

a. $(C_2H_5)_3SiCH_2CH_2CN$

b.

SCH_2CH_2OH

H

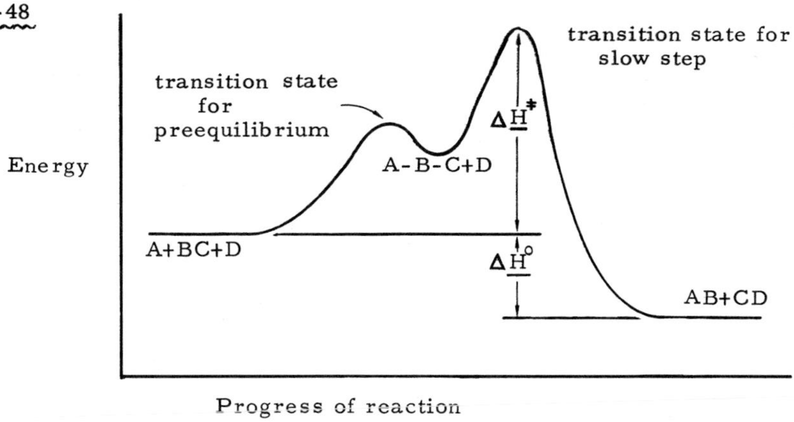

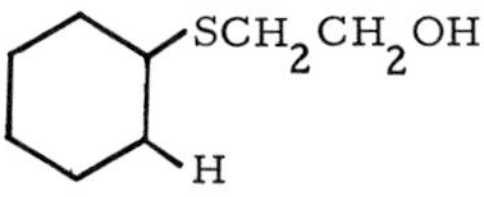

c.

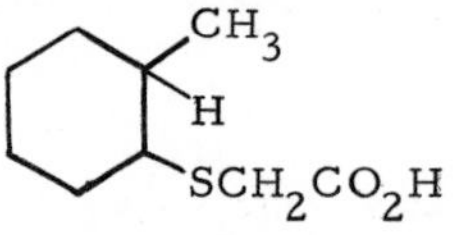

d. $C_6H_{13}-CH_2CH_2SiCl_3$

e. $(CH_3)_2\overset{\displaystyle |}{\underset{\displaystyle Cl}{C}}-CH_2CCl_3$

f. $Cl_3CCH=C(Br)CH_2O\overset{\displaystyle O}{\overset{\|}{C}}CH_3$

g. $CH_3CI=CHCF_3$

h.

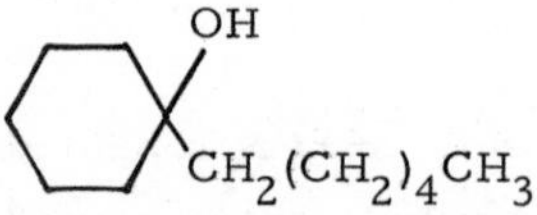

i.

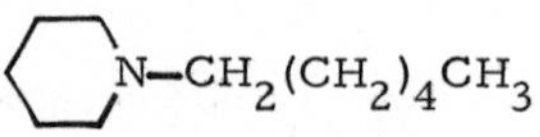

j. $HO_2CCH_2CH_2(CH_2)_4CH_3$

10-50   Formation of 2-bromo-1-trichloromethylcyclohexane is the product of radical-chain addition; 3-bromocyclohexene is the product of radical-chain substitution.

addition:

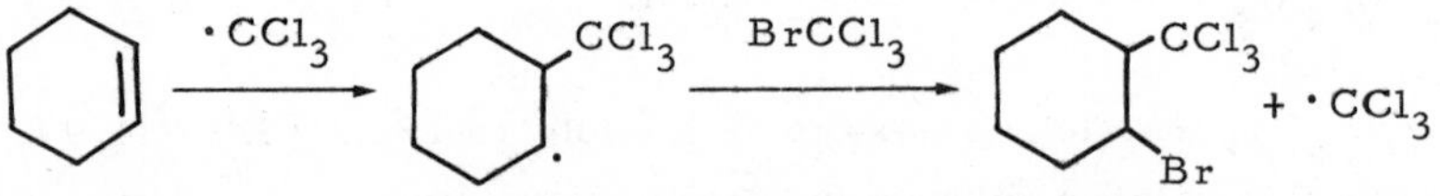

substitution:

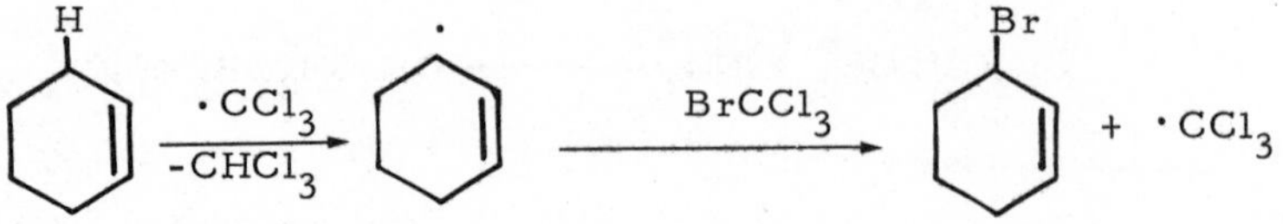

10-51   (See Table 10-3)

a. $HC\equiv CH$ $\xrightarrow[\text{ROOR}]{CH_3CO_2H}$ $H_2C=CHCH_2CO_2H$ $\xrightarrow[\text{ROOR}]{CH_3CO_2H}$ $HO_2C(CH_2)_4CO_2H$

b.

c. $C_6H_{13}CH=CH_2$ $\xrightarrow[\text{ROOR}]{CH_3CHO}$ $C_6H_{13}CH_2CH_2\overset{\displaystyle O}{\overset{\|}{C}}CH_3$

d. $RC\equiv CH$ $\xrightarrow[\text{HgSO}_4]{H_2O, H_2SO_4}$ $\left[R\overset{\displaystyle OH}{\overset{|}{C}}=CH_2 \right]\longrightarrow RCOCH_3$

10-52 polymer structure $\begin{array}{c} CO_2H \\ | \\ \!\!\!\!\left(CH_2\!\!-\!\!CH\right)_{\underline{n}} \end{array}$

The thiol RSH acts to reduce the molecular weight by terminating the growing radical chains by hydrogen atom transfer.

$$-CH_2-\overset{\cdot}{C}HCO_2H + RSH \longrightarrow -CH_2-CH_2CO_2H + RS\cdot$$

Thiols are efficient hydrogen-atom transfer agents because the S-H bond is weak (83 kcal) relative to C-H (98.7 kcal). Alcohols would not function similarly because of their high OH bond strength (110.6 kcal). Initiator decomposes best in basic solution.

$$^{\ominus}O_3S-O\!\!\mid\!\!O-SO_3^{\ominus} \longrightarrow 2\overset{\ominus}{O}_3S-O\cdot$$

(electrostatic repulsion between charged oxygen)

10-53   The mechanism is probably as follows:

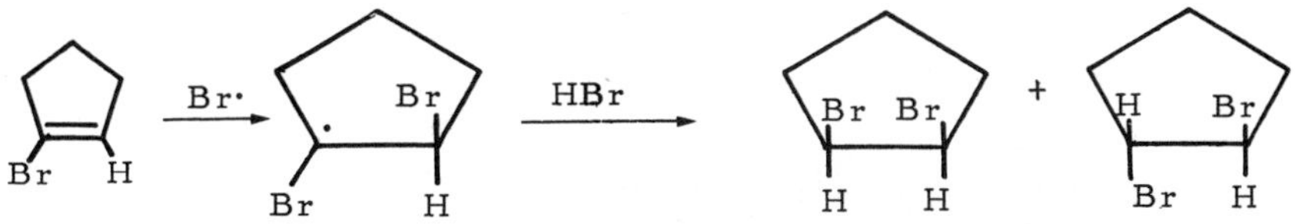

94%   6%

a.   The intermediate carbon radical may be a bridged radical, in which case hydrogen abstraction from HBr would give the cis product. Alternatively, the product reflects that steric hindrance to hydrogen abstraction is less severe on the side remote from the entering ·Br. The 1,1-dibromide is not formed because a Br-$\overset{\mid}{\underset{\mid}{C}}$· radical is more stable (formed more rapidly) than the H-$\overset{\mid}{C}$· radical.

b.   (i) $CH_3CHBrCHBrCH_3$ ($\underline{D}$, $\underline{L}$)   (ii)

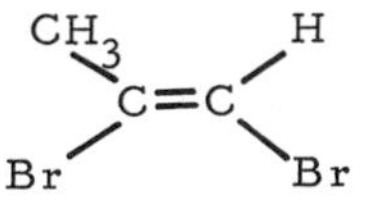

(iii)

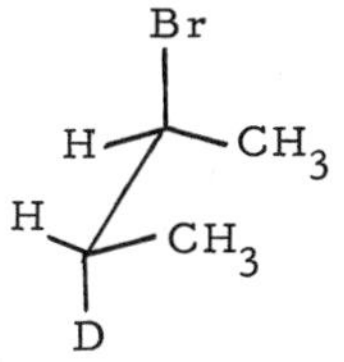

(and enantiomer)

(iv)

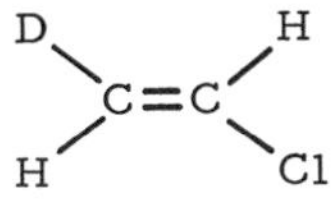

(For Additional Exercises, see p. 175)

CHAPTER 11

ALKENES AND ALKYNES II

OXIDATION AND REDUCTION REACTIONS

ACIDITY OF ALKYNES

An arbitrary scale of oxidation states of carbon is defined as follows: Each bond to a more electropositive atom (H, metals) contributes -1 to the oxidation state; each bond to a more electronegative atom (F, Cl, Br, I, O, N, S) contributes +1 to the oxidation state. Carbon is considered oxidized if a reaction increases its oxidation state. It is reduced if a reaction decreases its oxidation state.

The addition of hydrogen to a multiple bond is a reduction and can be achieved in various ways as summarized in Table 11-1S.

Table 11-1S

Some Methods of Hydrogenation of Carbon—Carbon Multiple Bonds

Reaction	Comment
Heterogeneous catalytic hydrogenation 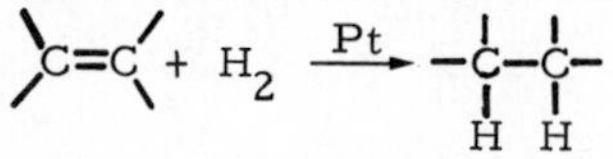 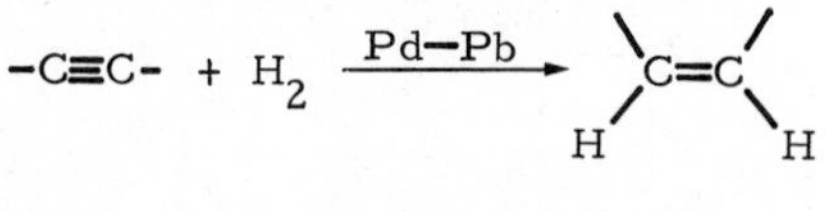 	Requires a transition-metal catalyst, Pt, Pd, Ni, etc. Addition is suprafacial from least hindered side. Rearrangements can occur. Alkynes are reduced to cis-alkenes over Lindlar catalyst, Pd—Pb (Section 11-2).
Homogeneous catalytic hydrogenation 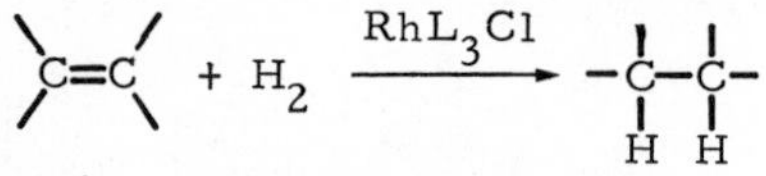 (L = triphenylphosphine)	Catalyst is a soluble complex salt of rhodium or ruthenium; suprafacial addition occurs to the least hindered double bond (Section 11-4).
Diimide reduction 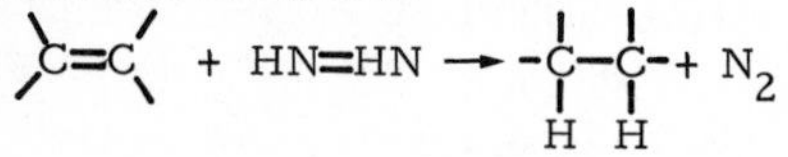 	Diimide is generated in situ by oxidation of H_2NNH_2. Suprafacial addition occurs (Section 11-5).

Table 11-1S continued

Reaction	Comment
Hydroboration-Protolysis	Hydroboration is suprafacial; protolysis occurs with retention (Section 11-6).

$$\text{C=C} + R_2BH \longrightarrow -\overset{|}{\underset{H}{C}}-\overset{|}{\underset{BR_2}{C}}- \xrightarrow{H^{\oplus}}$$

$$-\overset{|}{\underset{H}{C}}-\overset{|}{\underset{H}{C}}-$$

Addition of HB compounds to multiple bonds (hydroboration) leads to alkylboranes, R_3B, which are useful reagents in synthesis. They can be isomerized, converted to 1-alkenes, alkanes, alcohols, or amines. Other reactions are discussed in Chapter 16.

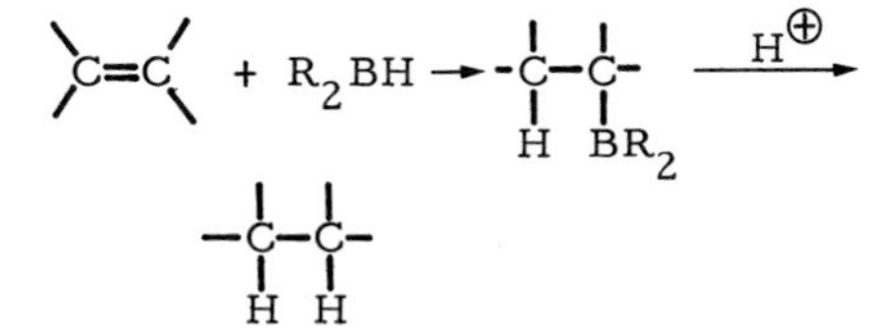

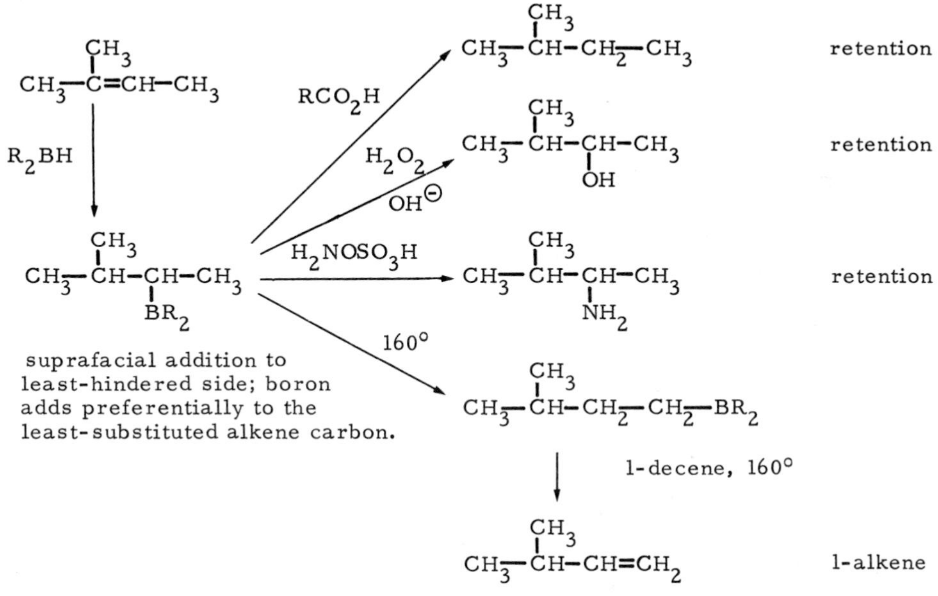

Commonly used HB compounds include the following:

B_2H_6 $H_3B:O$⟨⟩ $H_3B:S(CH_3)_2$

diborane   "borane-THF"   "borane-sulfide"

If the substituent groups on the alkene are bulky, addition of BH_3 may stop with formation of RBH_2 and R_2BH. These products are useful selective reagents

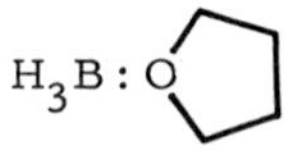

for addition to unhindered alkenes.

Hydroboration of alkynes yields alkenylboranes, which may be oxidized to aldehydes and ketones.

$$RC\equiv CR \xrightarrow{R_2BH} \underset{H}{\overset{R}{\diagdown}} C=C \underset{BR_2}{\overset{R}{\diagup}} \xrightarrow{H_2O_2,\ ^{\ominus}OH} \left[\underset{H}{\overset{R}{\diagdown}} C=C \underset{OH}{\overset{R}{\diagup}} \right] \longrightarrow RCH_2\overset{O}{\overset{\|}{C}}R$$

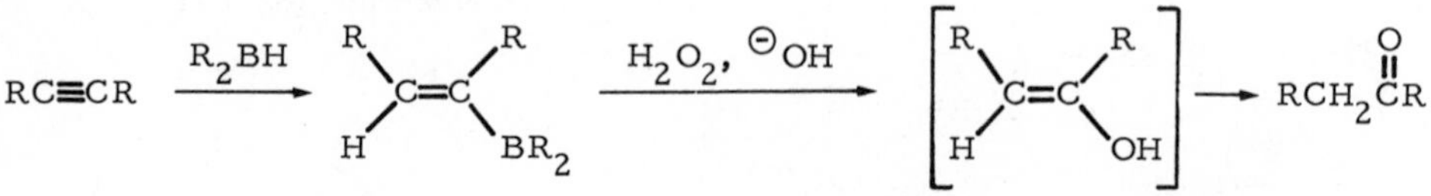

Oxidation reactions of alkenes and alkynes can lead to cleavage at the multiple bond. Ozone reacts with alkenes to give carbonyl compounds. Such reactions are useful for locating the position of a double bond in a hydrocarbon chain:

$$R_2C{=}CHR' \xrightarrow[\text{2. } H_2O,\ Zn]{\text{1. } O_3} R_2C{=}O + O{=}CHR'$$

Hydroxylation of alkenes with permanganate or osmium tetroxide gives diols by suprafacial addition:

$$\overset{}{\underset{}{\diagup}}C{=}C\overset{}{\underset{}{\diagdown}} \xrightarrow[\text{(or } OsO_4)]{M\overset{\ominus}{n}O_4,\ \overset{\ominus}{O}H,\ H_2O} \underset{OH\ \ OH}{-\overset{|}{C}-\overset{|}{C}-} \qquad \text{suprafacial addition}$$

Formation of diols by antarafacial hydroxylation can be achieved in a two-step oxidation with peroxyacids followed by hydrolysis:

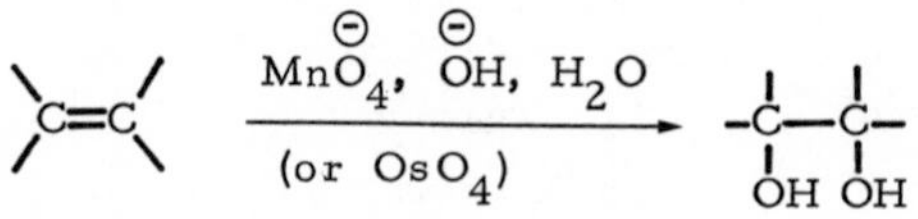

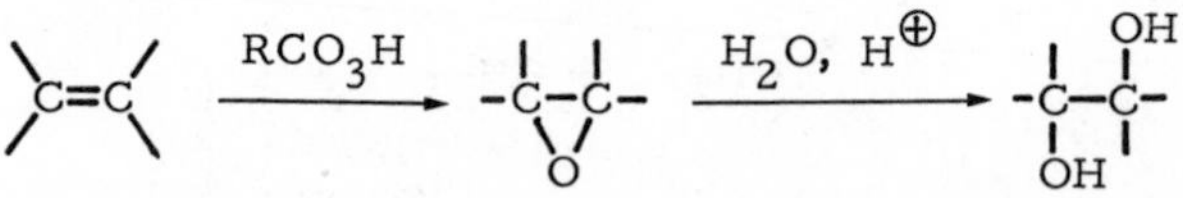

$$\overset{}{\underset{}{\diagup}}C{=}C\overset{}{\underset{}{\diagdown}} \xrightarrow{RCO_3H} \underset{O}{-\overset{|}{C}\underset{\diagdown\diagup}{}\overset{|}{C}-} \xrightarrow{H_2O,\ H^{\oplus}} \underset{OH}{-\overset{\overset{OH}{|}}{C}-\overset{|}{C}-} \qquad \begin{array}{l}\text{antarafacial}\\ \text{addition}\end{array}$$

1-Alkynes are very weak acids but will form salt-like metal derivatives with strong bases. The sodium, potassium, and magnesium alkynide salts are excellent nucleophiles for use in displacement reactions with compounds that undergo S_N2 reactions readily:

$$RC\equiv CH \xrightarrow[(-NH_3)]{KNH_2} RC\equiv C:^{\ominus}K^{\oplus} \xrightarrow[(-K^{\oplus}X^{\ominus})]{R'X} RC\equiv CR'$$

1-Alkynes form insoluble silver derivatives with silver-ammonia solution, a reaction that is the basis of separating 1-alkynes from other hydrocarbons.

Copper derivatives are useful in promoting coupling reactions of 1-alkynes:

$$RC{\equiv}CH \xrightarrow{\text{Cu}^{II}} RC{\equiv}C-C{\equiv}CR + (Cu^{I}) \quad \text{oxidative coupling}$$

$$\xrightarrow{\text{Cu}^{I}} RC{\equiv}C-CH{=}CHR$$

ANSWERS TO EXERCISES

11-1 a. $\overset{-4}{CH_4} \longrightarrow \overset{-2}{CH_3Cl}$ oxidation

  b. $CH_3\overset{-1}{CH}{=}\overset{-2}{CH_2} \longrightarrow CH_3\overset{0}{CH}Cl\overset{-3}{CH_3}$ no net change

  c. oxidation   d. no net change   e. no change   f. oxidation

11-2 a. $RCH{=}\overset{-1}{CH_2} \overset{-2}{} + 3H_2O \longrightarrow R\overset{+3}{C}O_2^{\ominus} + \overset{0}{CH_2}{=}O + 7H^{\oplus} + 6e^{\ominus}$

  $2K^{\oplus}MnO_4^{\ominus} + 8H^{\oplus} + 6e^{\ominus} \longrightarrow 2K^{\oplus} + 2MnO_2 + 4H_2O$

  $RCH{=}CH_2 + 2KMnO_4 + H^{\oplus} \longrightarrow$

  $\qquad\qquad RCO_2^{\ominus}K^{\oplus} + CH_2{=}O + 2MnO_2 + H_2O + K^{\oplus}$

  b. $C_6H_5CH_2CH_3 + 2CrO_3 + 6H^{\oplus} \longrightarrow C_6H_5CH_2CO_2H + 2Cr^{3\oplus} + 4H_2O$

11-3 a. $C_6H_5CH_2CH_3$ d.

  b.

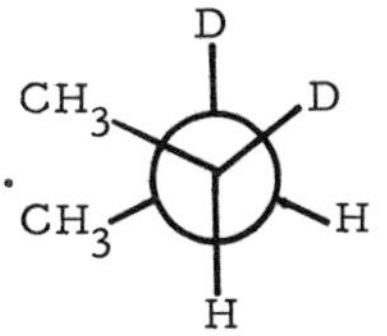

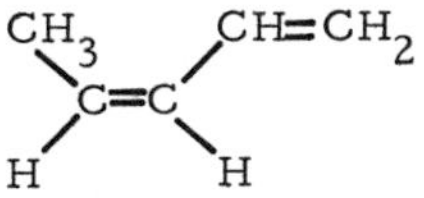

  e. cis-1,2-diphenylethene

  f. <u>cis</u>-1,3-dimethylcyclopentane
      (mostly)

  c. $CH_3CH_2COCH_3$

11-4

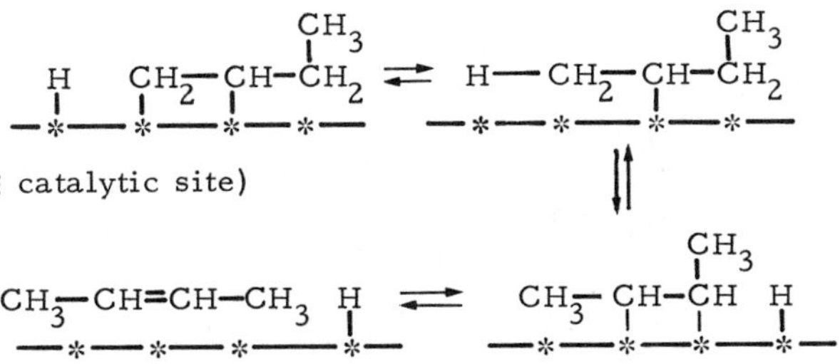

(* ≡ catalytic site)

If 1-butene and 2-butene could be completely equilibrated by the above mechanism in the presence of D_2, deuterium would appear at all four positions along the chain. If, as is more likely, complete statistical equilibration is not achieved, then the butane will be mostly $CDH_2CDHCH_2CH_3$ (from 1-butene) and $CDH_2CHDCHDCH_3$ (from 2-butene as $CDH_2CH=CHCH_3$).

11-5   a.   The energy required to break only one C—C bond in ethyne $(C{\equiv}C \rightarrow \overset{\cdot}{C}{=}\overset{\cdot}{C})$ is 199. 6 - 145. 8 = 53. 8 kcal; the comparable process for nitrogen is $N{\equiv}N \rightarrow \overset{\cdot}{N}{=}\overset{\cdot}{N}$, or 226. 8 - 100 = 126. 8 kcal, which is 2. 4 times that of ethyne. Therefore, addition to ethyne is much easier than to nitrogen.

b.   Less energy is required to break one C—C bond in ethyne (53. 8 kcal) than in ethene (63. 2 kcal). Thus, ethyne adds H_2 more easily than ethene.

c.   Less energy is required to break a C—C bond in ethene (63. 2 kcal) than a C—O bond in methanal (166 - 85. 5 = 80. 5 kcal).

d.   The bond energy of N—N is low (39 kcal) compared to that of C—C (82. 6 kcal), which makes the process $H_2N{-}NH_2 \xrightarrow{H_2} 2NH_3$ more favorable (by 33 kcal) than the process $H_3C{-}CH_3 \xrightarrow{H_2} 2CH_4$.

11-6   a.   The energy required to break one bond of $-C{\equiv}N$ is 212. 6 - 147 = 65. 6 kcal, which is 11. 8 kcal more than required to break $C{\equiv}C$ to $\overset{\cdot}{C}{=}\overset{\cdot}{C}$. Hence, $C{\equiv}N$ is less readily hydrogenated than $C{\equiv}C$.

b.   The bulky alkyl groups prevent the alkene from approaching the catalyst surface close enough to enable the double-bond carbons to bond to the metal.

11-7   a.   1-Butene, $\Delta \underline{S}^\circ$ = -57. 6 eu; cis-2-butene, $\Delta \underline{S}^\circ$ = -56. 5 eu; trans-2-butene, $\Delta \underline{S}^\circ$ = -55. 5 eu; the entropy change is large and negative because there is a significant loss in degrees of freedom when two molecules add to form one molecule.

b.

	$\Delta \underline{G}^\circ$ cal	$\Delta \underline{H}^\circ$ cal	ΔS° eu
$CH_2{=}CHCH_2CH_3 + H_2 \rightarrow CH_3CH_2CH_2CH_3$	-12,940	-30,120	-57. 6
$CH_3CH_2CH_2CH_3 \leftarrow H_2 + CH_3CH{=}^{t}CHCH_3$	+10,950	+27,480	+55. 5
net			
$CH_2{=}CHCH_2CH_3 \rightarrow CH_3CH{=}^{t}CHCH_3$	-1,990	-2,640	-2

Likewise, for the process cis-2-butene $\rightarrow$ trans-2-butene, $\Delta \underline{G}^\circ$ = -690 cal

$\Delta \underline{H}^{o}$ = -1000 cal, $\Delta \underline{S}^{o}$ = -1 eu.   The $\Delta \underline{S}$ values for rearrangement are small, as we would expect for a reaction with no change in the number of molecules, or in the number and type of bonds.   Therefore, even though $\Delta \underline{S}^{o}$ of hydrogenation is large, the value is almost the same for different alkenes.   The difference in alkene stabilities is actually given by the difference in ΔG^{o} values, but because the ΔS^{o} component cancels out, we are left with the result that differences in $\Delta \underline{H}^{o}$ values are a reliable estimate of relative alkene stabilities.

11-8

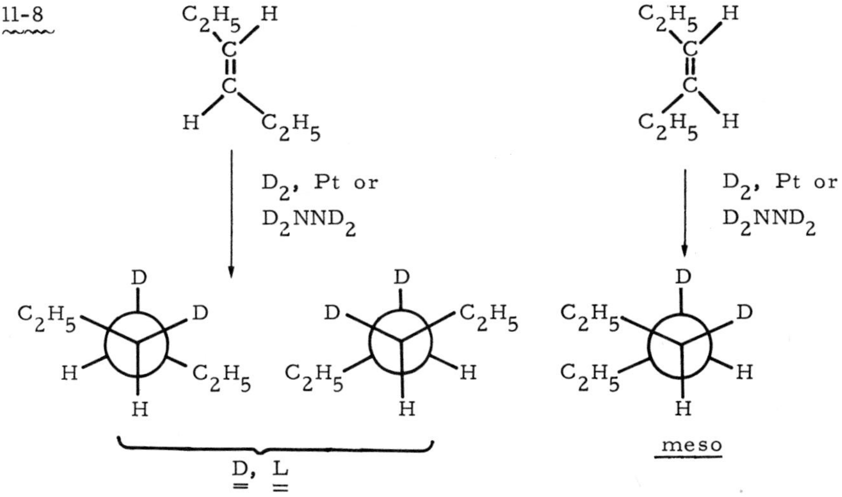

11-9   a.

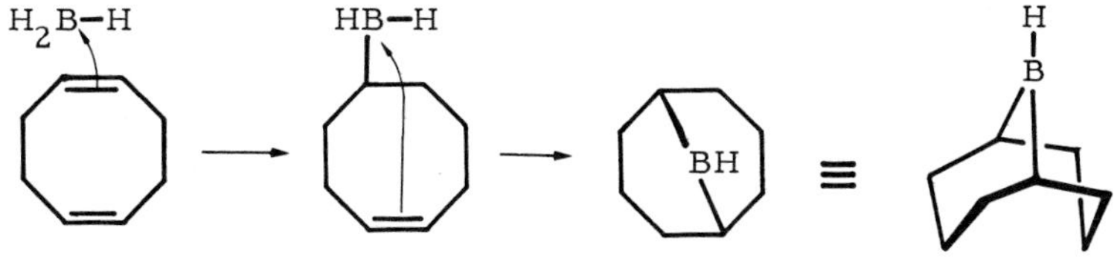

b.

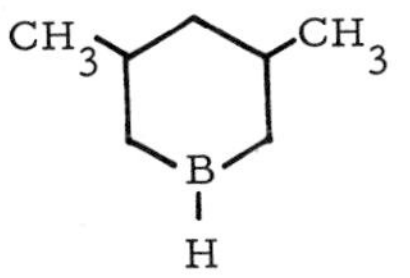

c.

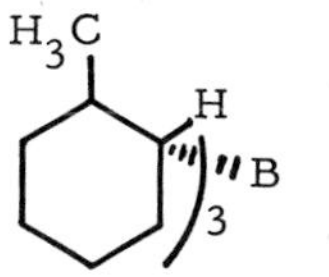

The trans isomer is formed because this is a suprafacial addition by what appears to be a concerted mechanism.

11-10   a. $CH_3CH_2CH_2CH_2BR_2$ b. $(CH_3)_3CCH_2CH(CH_3)CH_2CH_2BR_2$

c. $(CH_3)_3CCH(CH_3)CH_2CH_2BR_2$ d.

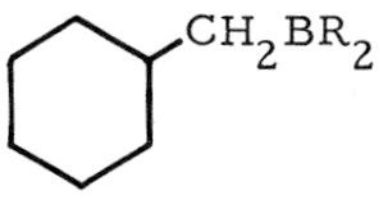

Note: in borane isomerization, the boron can pass a tertiary carbon - but not a quaternary carbon.

11-11   a.

b., c.
$$\left[(CH_3)_2CHCH\!-\!\underset{\underset{CH_3}{|}}{B}H \right]_2$$

$$\underset{H}{\overset{C_2H_5}{\diagup}}C=C\underset{BR_2}{\overset{C_2H_5}{\diagdown}} \longrightarrow C_2H_5CH_2COC_2H_5 + ROH$$

$$\underset{H}{\overset{H_5C_2}{\diagup}}C=C\underset{D}{\overset{C_2H_5}{\diagdown}} + RD$$

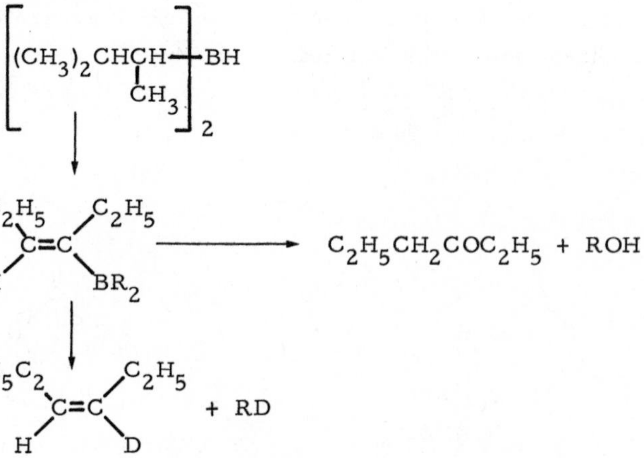

11-12

$$C_4H_9C\equiv C-Br \xrightarrow{(C_2H_5)_2BH} \underset{H}{\overset{C_4H_9}{\diagup}}C=C\underset{B(C_2H_5)_2}{\overset{Br}{\diagdown}} \xrightarrow{\ominus OCH_3}$$

$$\underset{H_5C_2}{\overset{C_4H_9}{\diagup}}C=C\underset{\underset{C_2H_5}{B^\ominus-OCH_3}}{\overset{Br}{\diagdown}}$$

$$\underset{H}{\overset{C_4H_9}{\diagup}}C=C\underset{C_2H_5}{\overset{B-}{\diagdown}} \xleftarrow{-Br^\ominus}$$

$$\underset{H}{\overset{C_4H_9}{\diagup}}C=C\underset{C_2H_5}{\overset{H}{\diagdown}}$$

with $H^\oplus$

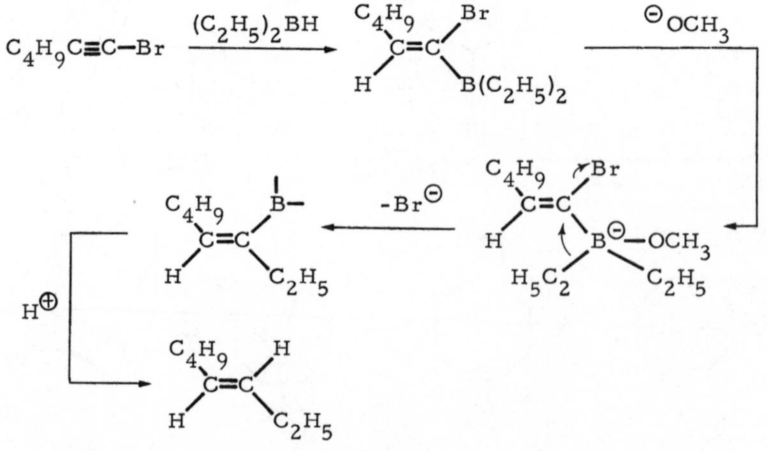

11-13   The positions of the carbonyl groups mark the positions of the carbon-carbon double bonds in the hydrocarbon.

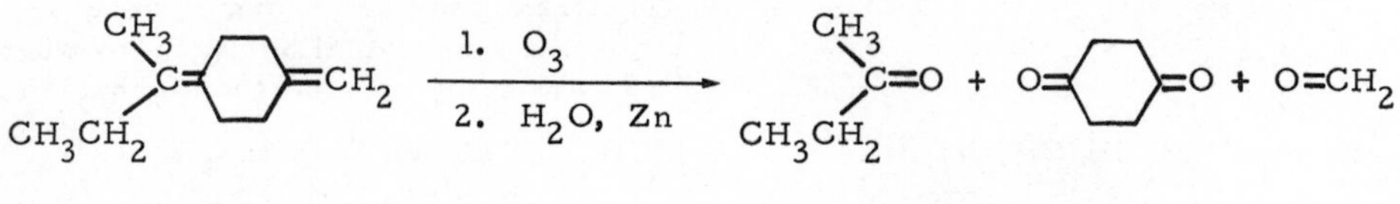

$$\underset{CH_3CH_2}{\overset{CH_3}{\diagdown}}C=\!\!\bigcirc\!\!=CH_2 \xrightarrow[\text{2. } H_2O,\ Zn]{\text{1. } O_3} \underset{CH_3CH_2}{\overset{CH_3}{\diagdown}}C=O + O=\!\!\bigcirc\!\!=O + O=CH_2$$

11-14

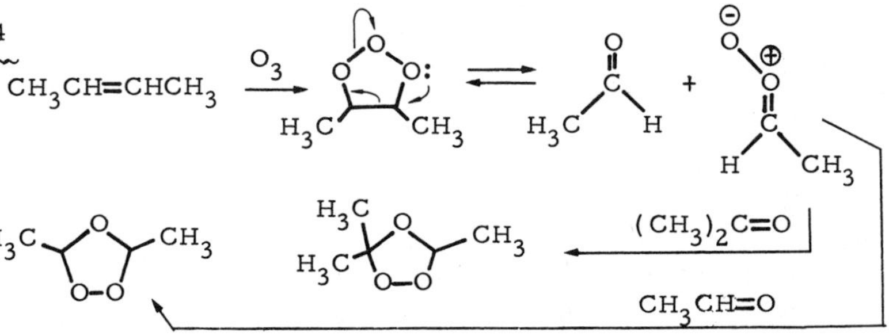

$$CH_3CH=CHCH_3 \xrightarrow{O_3}$$

11-15   a. $\underset{R-C\equiv C-R}{0\ \ 0} \longrightarrow \underset{\underset{O\ \ O}{\overset{+2\ +2}{R-C-C-R}}}{}$ (net change = +4)

b. $Mn^{+7} \longrightarrow Mn^{+4}$ (net change = -3); 4/3 mole $KMnO_4$ required
per mole of alkyne.

11-16   a.

b. $KMnO_4,\ \overset{\ominus}{OH},\ H_2O$

c. HCO_2H
     35% H_2O_2

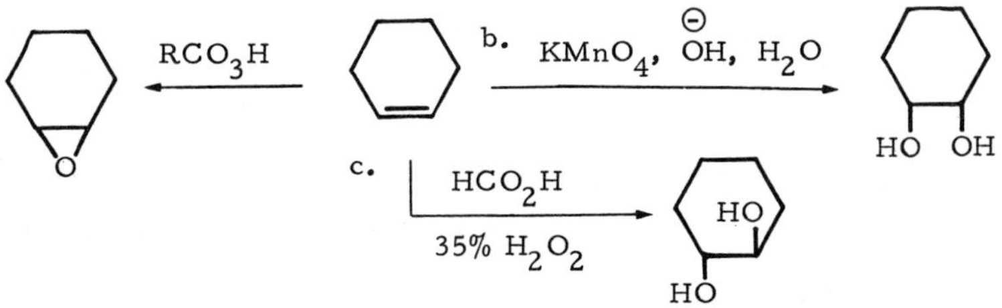

11-17   (i) Pentane could be easily identified by its failure to decolorize $KMnO_4$
solutions and Br_2 in CCl_4.   (ii) 1-Pentyne would give a precipitate of
silver salt with silver-ammonia solution.   (iii) 1-Pentene may be distinguished
from 2-pentyne by the amount of Br_2 taken up by known amounts of each
hydrocarbon.

11-18 $Na\ (\underline{s}) + 1/2\ Cl_2\ (\underline{g}) \longrightarrow Na^{\oplus}\ (aq) + Cl^{\ominus}\ (aq)$ $\Delta H^\circ = -97$ kcal

$e + Na^{\oplus}\ (\underline{g}) \longrightarrow Na\ (\underline{g})$ $\Delta H^\circ = -118$ kcal

$Cl^{\ominus}\ (\underline{g}) \longrightarrow Cl\cdot\ (\underline{g}) + e$ $\Delta H^\circ = +83$ kcal

$Cl\cdot\ (\underline{g}) \longrightarrow 1/2\ Cl_2\ (\underline{g})$ $\Delta H^\circ = -29$ kcal

$Na\ (\underline{g}) \longrightarrow Na\ (\underline{s})$ $\Delta H^\circ = -26$ kcal

net $Na^{\oplus}\ (\underline{g}) + Cl^{\ominus}\ (\underline{g}) \longrightarrow Na^{\oplus}\ (aq) + Cl^{\ominus}\ (aq)$ $\Delta H^\circ = -187$ kcal

11-19   Ethyne can add to itself to give butenyne; which in turn can add to
another mole of ethyne to give 1,5-hexadiene-3-yne.

$HC\equiv CH \xrightarrow{HC\equiv CH} H_2C=CH-C\equiv CH \xrightarrow{HC\equiv CH} H_2C=CH-C\equiv C-CH=CH_2$

11-20   The molecular formula of A and B can be estimated from the data presented as follows:

g-atoms of carbon in 100g = 85.63 / 12.01 = 7.13
g-atoms of hydrogen in 100g = 14.34 / 1.008 = 14.2
ratio C:H = 7.13 : 14.2 ~1:2

From the mp and bp and the data of Table 10-1, A and B are C_6 hydrocarbons; hence their formula is C_6H_{12}. The chemical properties indicate A and B are alkenes, and because they give the <u>same</u> products with O_3, they must be stereoisomers and not position isomers. Possible structures follow:

C—C—C—C≡C—C   (cis and trans)   C—C—C≡C—C—C   (cis and trans)

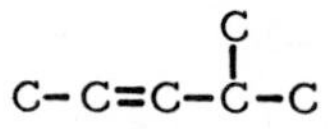

```
         C                                    C
         |                                    |
C—C≡C—C—C   (cis and trans)      C—C≡C—C—C   (cis and trans)
```

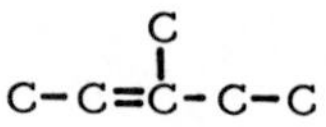

The proton nmr spectra of A and B would enable both structure and configuration to be determined from among the given possibilities.

11-21   a.   <u>cis</u>- and <u>trans</u>-cyclooctene
        b.   Different products would be obtained on hydroxylation with $KMnO_4$.

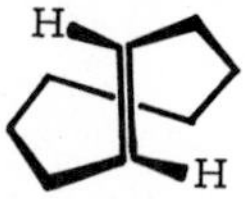

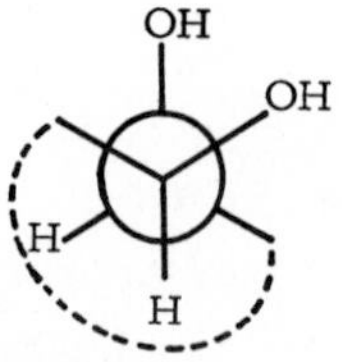

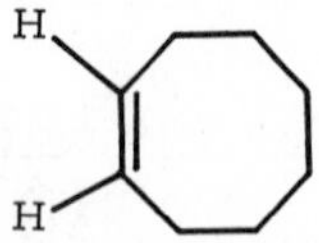

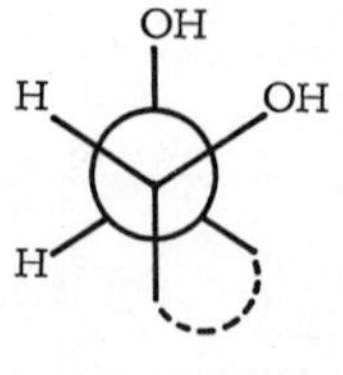

11-22

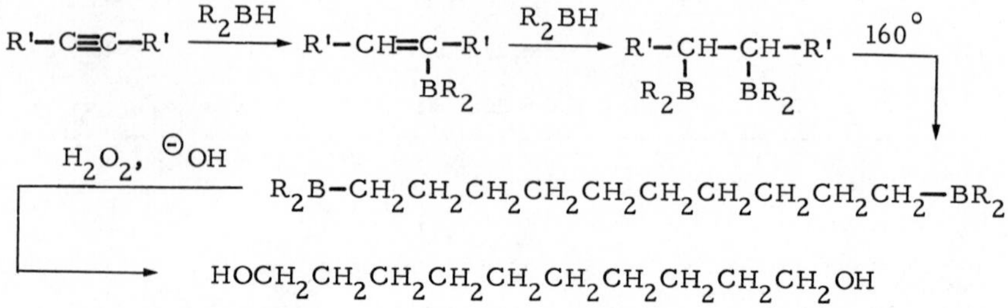

11-23

a.

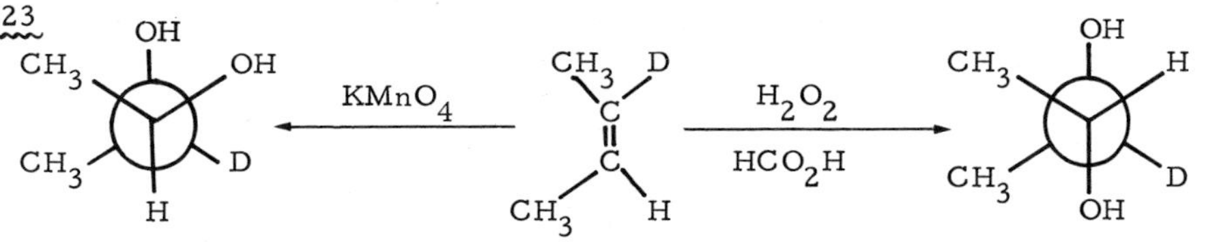

A (and enantiomer)   B (and enantiomer)

b.   The two products, A and B, are diastereomers; A is formed by supra-facial addition; B is formed by antarafacial addition.

11-24

a.   cis-2-butene $\longrightarrow$ by suprafacial addition

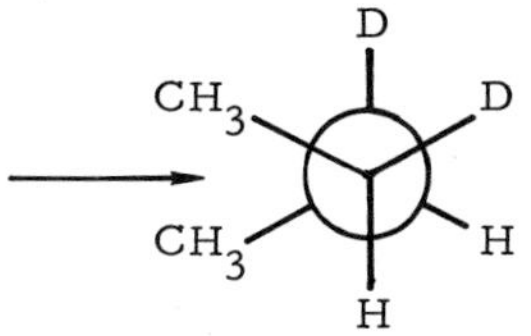

b.

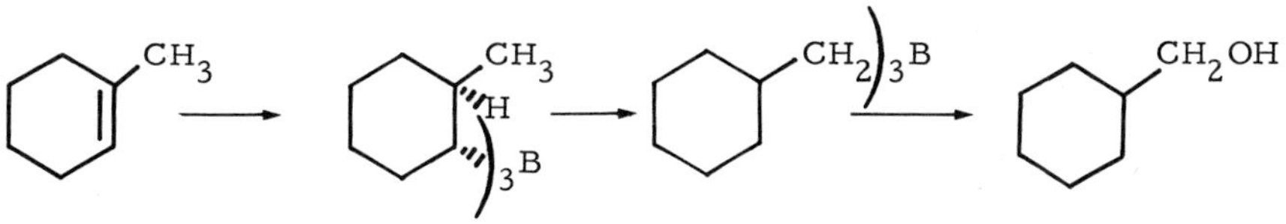

c.

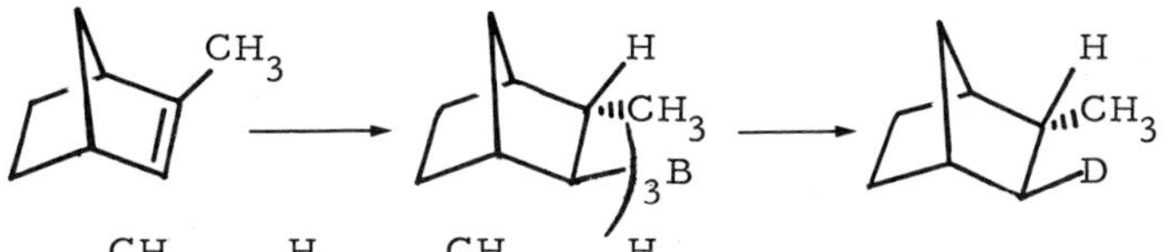

d. 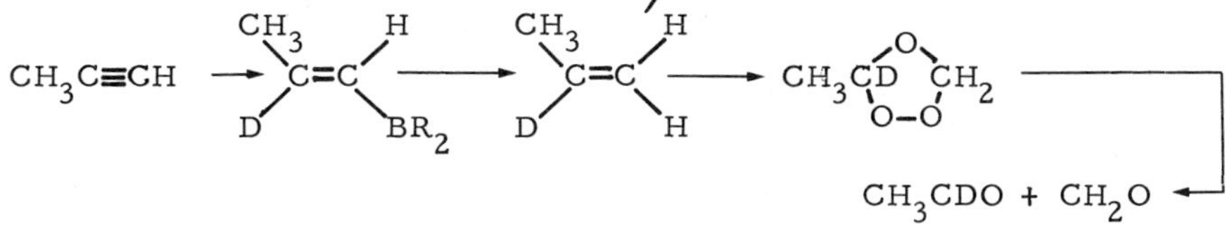 $CH_3C\equiv CH \longrightarrow$

$CH_3CDO + CH_2O$

e. 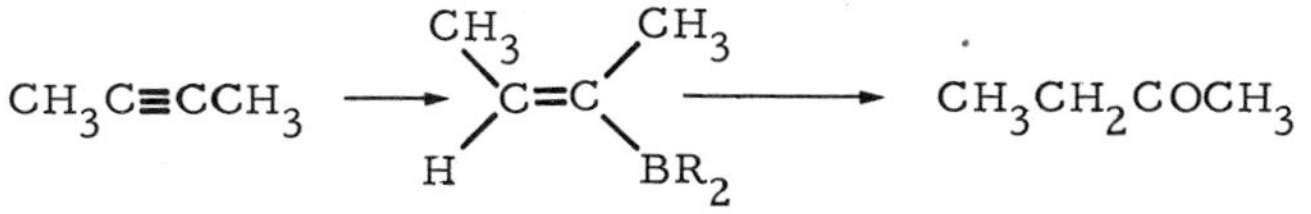 $CH_3C\equiv CCH_3 \longrightarrow$ $\longrightarrow CH_3CH_2COCH_3$

f. 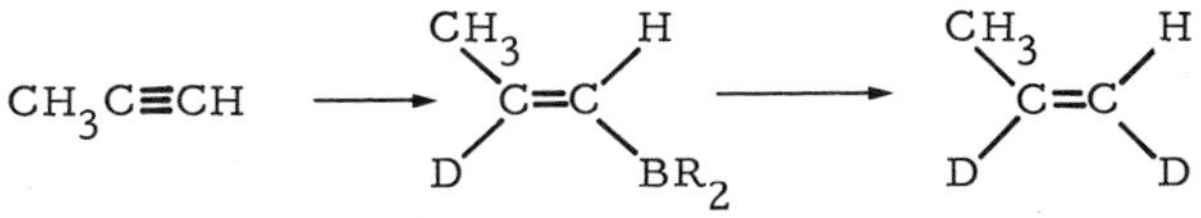 $CH_3C\equiv CH \longrightarrow$

g.

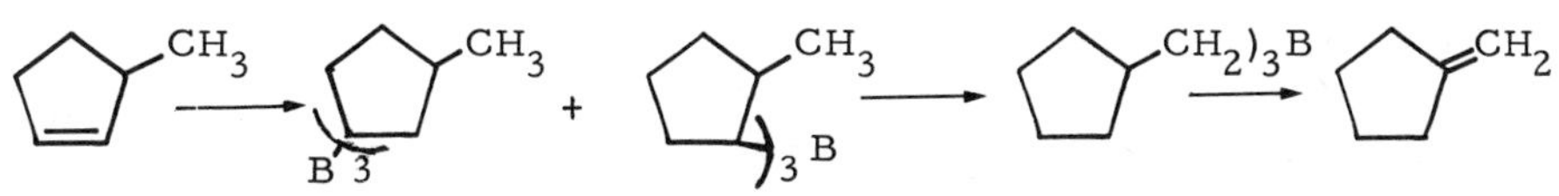

h. $CH_3CH_2CH_2CH{=}CH_2$ $\longrightarrow$ $CH_3CH_2CH_2CH_2CH_2Br$ $\longrightarrow$

$\longrightarrow$ $C_5H_{11}CHO + CH_3CHO$

i. $CH_3CH_2CH_2CH{=}CH_2$ $\longrightarrow$ $(CH_3CH_2CH_2CH_2CH_2)_3B$ $\longrightarrow$

$CH_3CH_2CH_2CH_2CH_2NH_2$ $\longleftarrow$

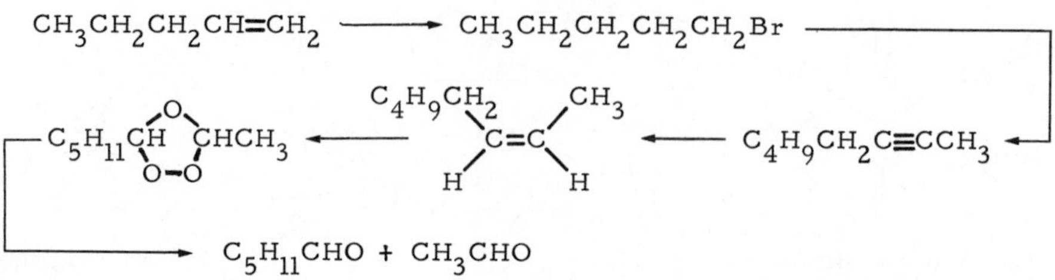

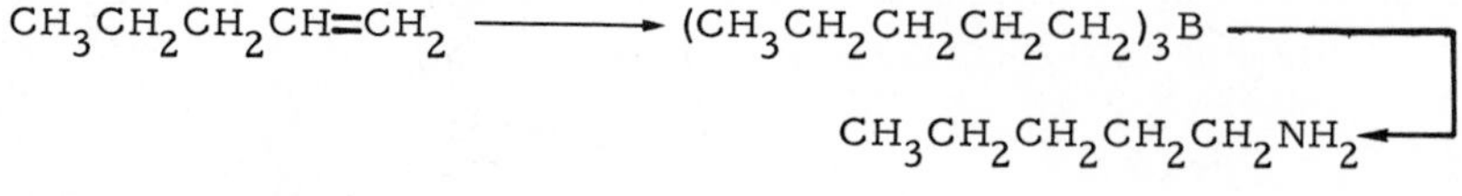

11-25   Of the possible structures A, B, C and D with two double bonds in a six-membered ring of formula C_6H_8, only A and B are stable. The conjugated isomer A would absorb ultraviolet radiation of longer wavelength than the nonconjugated isomer (Section 9-9B).

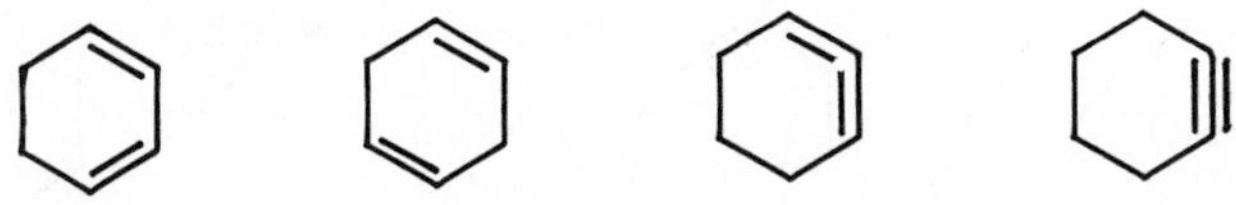

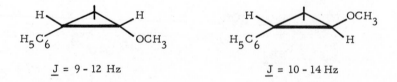

A   B   C   D

11-26   The possible structures could be distinguished from the magnitude of the spin-coupling constants of the ring protons (Section 9-10H).

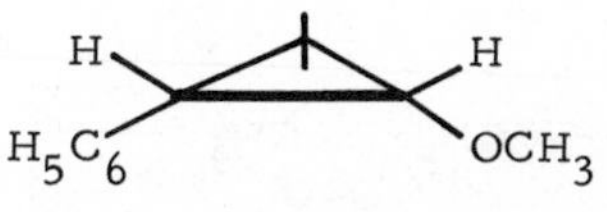

$\underline{J} = 9 - 12$ Hz $\underline{J} = 10 - 14$ Hz

11-27   For overall suprafacial addition to be observed, each step must be stereospecific in forward and reverse directions and the C—H bonds must be made at the expense of the C—metal bonds with <u>retention</u> of configuration at carbon.

11-28   Racemization occurs when an achiral intermediate is formed reversibly. The only achiral intermediate in the hydrogenation sequence of Figure 11-2 as applied to 3-methylhexane is an alkene. Deuterium exchange also occurs at the alkene stage, as shown below:

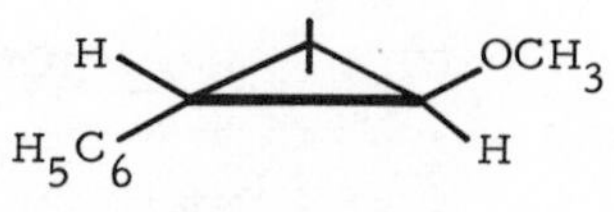

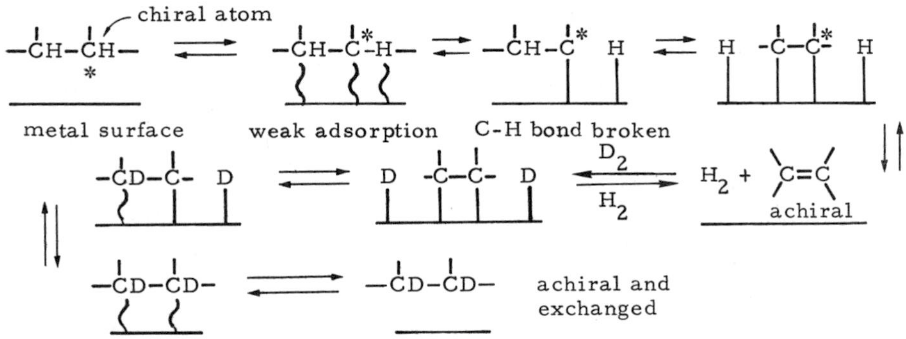

chiral atom

metal surface   weak adsorption   C-H bond broken

11-29 $\Delta \underline{H}^\circ$ = -69.4 kcal for hydrogenation of propyne and -60 kcal for 1,2-propadiene, as calculated from bond energies of Table 4-3. The calculated $\Delta \underline{H}^\circ$ of rearrangement $CH_3C\equiv CH \longrightarrow CH_2=C=CH_2$ is $[-69.4 - (-60)]$ = -9.4 kcal. The $\Delta \underline{H}^\circ$ value determined from measured heats of hydrogenation is $[-69.1 - (-71.3)]$ = +2.2 kcal. Thus, the calculated and experimental values for rearrangement do not agree very well. The reason is that the calculated heat of hydrogenation of 1,2-propadiene is lower than the measured value. It is lower because the bond energies used do not take into account the strain inherent in having two double bonds to one carbon.

11-30
a.

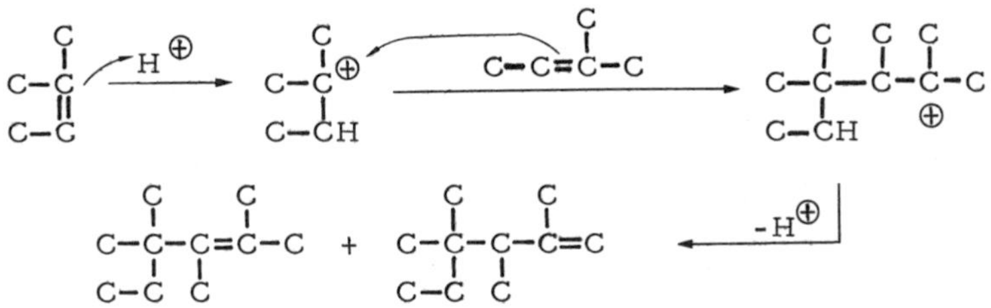

b.

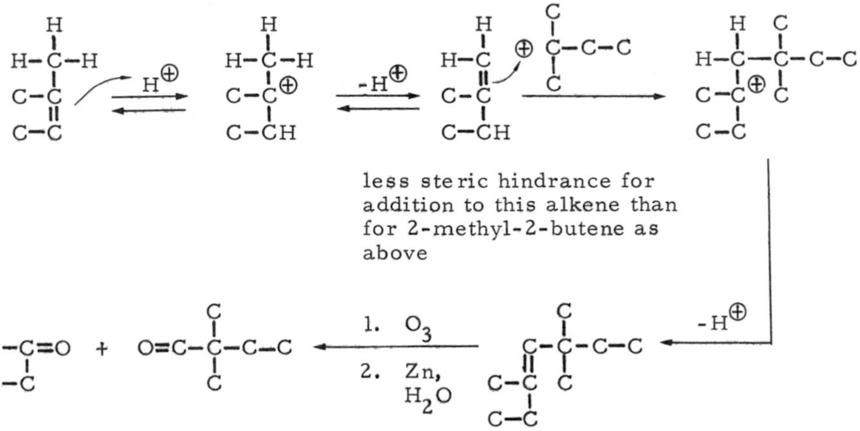

less steric hindrance for
addition to this alkene than
for 2-methyl-2-butene as
above

11-31   a.   1-Butyne would give a precipitate with silver-ammonia solution.

b.   as in Part a.

c.   1,2-Diphenylethene would decolorize $KMnO_4$ solutions and Br_2 in CCl_4.

11-32   a.   1-Alkynes are identified easily from their infrared spectra which show strong $\equiv C-H$ stretch at 3300 cm^{-1} and weak to medium $C\equiv C$ stretch in the region 2050 - 2260 cm^{-1}. The $\equiv CH$ proton of 1-butyne gives rise to a proton nmr signal at 2.5 ppm which appears as a triplet. In contrast, 2-butyne would have no $C\equiv C$ stretch in the infrared, and only a single resonance in the nmr around 2.0 ppm.

b.   As in Part a for 1-butyne.

c.   Nmr would show two distinct chemically shifted sets of protons for $C_6H_5CH_2CH_2C_6H_5$, and only one for $C_6H_5C\equiv CC_6H_5$.

ADDITIONAL EXERCISES

(The following problems pertain to the chemistry described in Chapters 10 and 11)

11A-1   Specify the reagents, catalysts, and appropriate reaction conditions necessary for each of the transformations of methylenecyclohexane indicated below. Note that more than one step may be required.

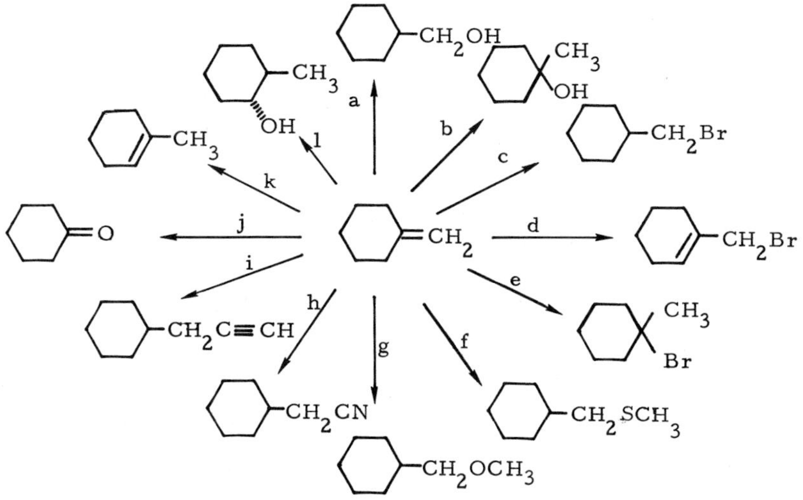

11A-2   Which compound in each pair would react most rapidly with an electrophilic reagent such as HCl or Br_2?

a. $CH_3CH_2CH_2CH_3$, $CH_3CH=CHCH_3$ e. $Cl_2C=CCl_2$, $H_2C=CH_2$

b. $CH_3CH=CHCH_3$, $CH_3CH_2C\equiv CH$ f. $CH_3C\equiv CH$, $CH_3C\equiv N$

c. $(CH_3)_2C=C(CH_3)_2$, $(CH_3)_2C=CH_2$ g. $H_2C=CHNO_2$, $H_2C=CHCH_2NO_2$

d. $(CH_3)_2C=CH_2$, $CH_3CH=CHCH_3$ h.

$$H_2C=C\begin{smallmatrix}OCCH_3\\ \\CH_3\end{smallmatrix},\ H_2C=C\begin{smallmatrix}COCH_3\\ \\CH_3\end{smallmatrix}$$

11A-3   A hydrocarbon of formula $C_{10}H_8$, when hydrogenated over a palladium-lead catalyst, gave a hydrocarbon of formula $C_{10}H_{10}$. The hydrocarbon of formula $C_{10}H_{10}$ reacted rapidly with two moles of ozone and, when the diozonide was decomposed by catalytic hydrogenation (H_2/Pt), two moles of methanal (CH_2O) and one mole of a dialdehyde of formula $C_8H_6O_2$ were formed. The dialdehyde showed two singlet resonances in its proton nmr spectrum. The singlets were in the ratio of 1:2 at 10 ppm and 7.8 ppm, respectively. Suggest suitable structures for the two hydrocarbons and the dialdehyde and write equations to show the reactions by which they are formed.

CHAPTER 12

CYCLOALKANES, CYCLOALKENES, AND CYCLOALKYNES

Cyclohexane would be considerably strained in a planar conformation, because the $C-C-C$ bond angles would be larger than their normal values, and because adjacent hydrogens attached to the carbons would be eclipsed. The stable conformation of cyclohexane is the chair form, with the twist-boat being next most stable (~5 kcal/mole^{-1} above the chair).

The chair conformation of cyclohexane has equatorial and axial substituent positions and, in general, the equatorial positions represent the most favorable steric location for a substituent, especially if it is a bulky substituent such as tert-butyl. The conformation with a substituent in the equatorial position normally is in equilibrium with the conformation in the axial position, even if this is substantially less stable. Ring inversion interconverts the substituent position between equatorial and axial, usually on the order of 100,000 times per second at room temperature.

The free-energy difference, ΔG, for a substituent in the axial and equatorial positions has been determined for many substituents. It appears that, along with repulsive forces due to steric effects, small contributions to conformational equilibrium differences can be made by van der Waals attractive forces. Chlorocyclohexane crystallizes at low temperatures as the pure equatorial conformation.

The cis-trans isomerism of cyclohexane derivatives is complicated by conformational isomerism. With trans-1,4-disubstituted cyclohexanes (unless the substituents are polar groups such as F, Cl, OH, etc.), the diequatorial conformation is favored. With cis-1,4-disubstituted cyclohexanes, normally the bulkiest group is equatorial and the less bulky group is axial. The tert-butyl group is used widely as a conformational "holding" group.

Cyclopentane and cyclobutane have nonplanar rings of carbons to relieve hydrogen-hydrogen eclipsing interactions. Cyclobutane, cyclopropane, and "cycloethane" (ethene) have substantial deviations of $C-C-C$ angles from the normal values.

The Baeyer strain theory suggested that deviations of bond angles in cycloalkanes could lead to unusual properties. However, all cycloalkanes were assumed to have planar carbon rings, an assumption now known to be wholly incorrect. This part of the Baeyer theory has been replaced by the Sachse-Mohr theory of nonplanar, essentially strain-free, large carbon rings. Strain in cycloalkanes is manifested by higher than normal heats of combustion and greater reactivity in C—C bond-breaking reactions than normally observed for alkanes. Cyclopropane has higher strain than propene and undergoes many C—C cleavage reactions, as with Br_2, H_2, and H_2SO_4, which also are observed for alkenes.

Cycloheptane has a very flexible, slightly strained favorable conformation with a two-fold axis of symmetry, called the twist chair. Cyclooctane exists most favorably in the boat-chair conformation, with a minor contribution from the crown form. Ring inversion in cyclooctane has an activation energy of about 5 kcal/mole^{-1}. Cyclononane is most stable as a twist-boat-chair conformation, which has a three-fold symmetry axis and a barrier to inversion of about 6 kcal/mole^{-1}.

Cyclodecane, like the higher cycloalkanes (Dale theory), has a conformation based on trans-butane segments, with considerable hydrogen-hydrogen interactions and a barrier to inversion of about 6 kcal/mole^{-1}. Cyclotetradecane is essentially strain-free and is made up of four trans-butane segments.

The small-ring cycloalkenes are expected to be much more highly strained than the corresponding cycloalkanes. This expectation is realized with cyclopropene but not cyclobutene. trans-Cyclooctene exists as two chiral forms because the double bond cannot turn over inside the chain of carbons.

Cyclooctyne is the smallest reasonably stable cycloalkyne, although evidence has been adduced for cyclopentyne as a reaction intermediate.

A special nomenclature system is required for polycycloalkanes that takes cognizance of the sizes of bridging rings. Examples are

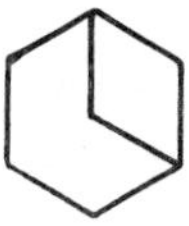

bicyclo [3.1.1] heptane

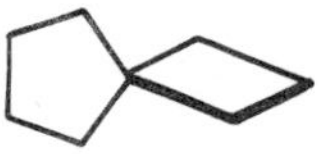

spiro [4.3] octane

Cis and trans forms of bicyclo [4.4.0] decane (cis- and trans-decalin) are known, with the trans being 2 kcal/mole^{-1} more stable.

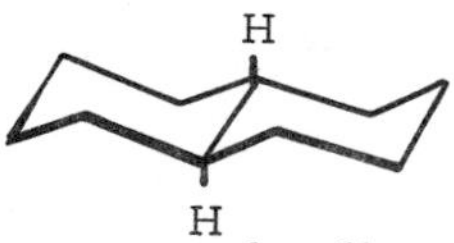

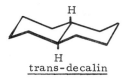

trans-decalin   cis-decalin

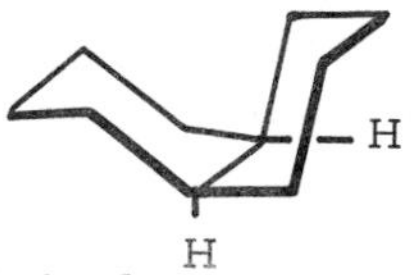

The stability of these isomers is contrary to the Baeyer theory of planar cycloalkanes. Cis- but not trans-decalin undergoes ring inversion with a barrier of about 14 kcal mole^{-1}. The decalin ring system is very important to the chemistry of many naturally occurring substances of medical importance.

Many small-ring polycyclic hydrocarbons synthesized in recent years have extraordinary strain energies but, nonetheless, they do not decompose instantly at room temperature. Among these (with their strain energies) are

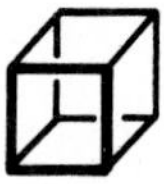

cubane

(142 kcal mole^{-1})

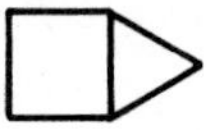

bicyclo [2.1.0] pentane

(50 kcal mole^{-1})

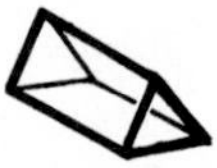

prismane

(strain energy not known)

The "propellanes" are very reactive polycyclic hydrocarbons that have two nontetrahedral carbons with all of the bonds to these carbons extending to the same side of a plane. An example is

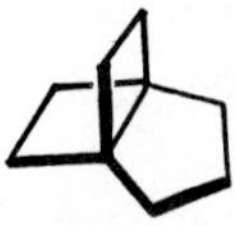

tricyclo [3.2.2.0^{1,5}] nonane

Compounds that violate Bredt's rule - that is, compounds that have a double bond at the bridgehead of a bridged polycyclic hydrocarbon - are interesting in that an angular twist of the double bond keeps the four atoms attached to the double bond from lying in a single plane. A member of this class, bicyclo [3.3.1]-1-nonene, has a strain energy of about 12 kcal mole^{-1}.

bicyclo [3.3.1]-1-nonene

ANSWERS TO EXERCISES

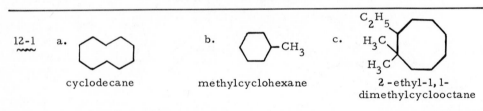

12-1   a.

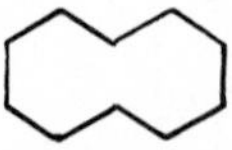

cyclodecane

b.

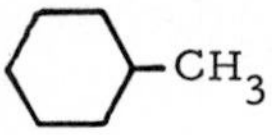

methylcyclohexane

c.

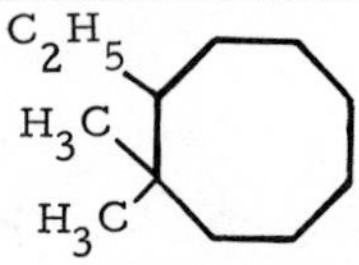

2-ethyl-1,1-dimethylcyclooctane

d. Note: In naming polysubstituted cycloalkanes, the configuration is best related to one substituent chosen as a reference (r) and designated as either cis (c) or trans (t) to the reference.

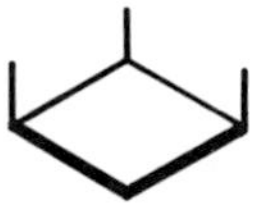

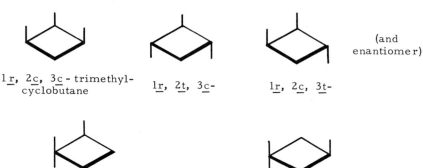

1r, 2c, 3c - trimethyl-
    cyclobutane

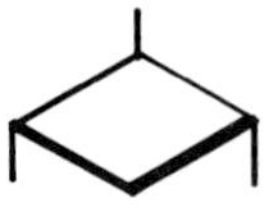

1r, 2t, 3c-

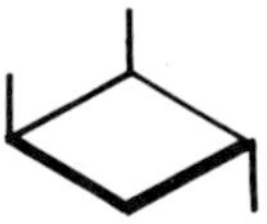

1r, 2c, 3t-

(and enantiomer)

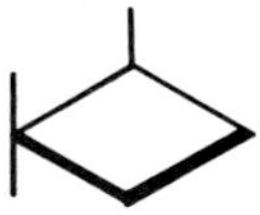

1, 1, 2-trimethylcyclobutane
(and enantiomer)

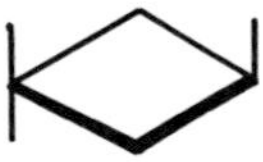

1, 1, 3-trimethylcyclobutane

e.

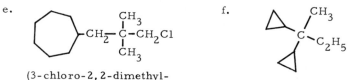

$-CH_2-\overset{\displaystyle CH_3}{\underset{\displaystyle CH_3}{C}}-CH_2Cl$

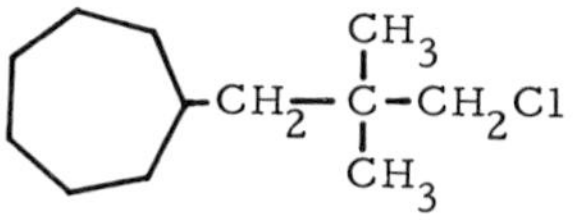

(3-chloro-2, 2-dimethyl-
propyl)cycloheptane

f.

$\overset{\displaystyle CH_3}{\underset{\displaystyle}{C}}-C_2H_5$

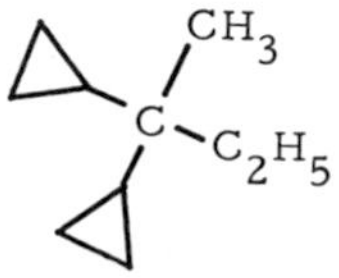

2, 2-dicyclopropylbutane

12-2   Angle strain at a carbon of planar cyclohexane = $(7.5°)^2$ x 17.5 x 6 = 5900 cal mole^{-1}. Angle strain of actual cyclohexane = $(1°)^2$ x 17.5 x 6 = 105 cal mole^{-1}.

12-3   There are 6 gauche butane segments in chair cyclohexane and 6 eclipsed C—C bonds in planar cyclohexane. These are (1, 3), (2, 4), (3, 5), (4, 6), (5, 1), and (6, 2).

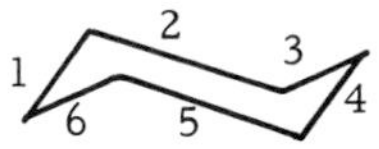

The strain energy of planar cyclohexane relative to chair cyclohexane is calculated as 6 x 5 = 30 kcal mole^{-1}. Including an angle strain of 5.9 kcal mole^{-1}, the total estimated strain energy is 36 kcal mole^{-1}.

12-4

A   B   C

D'   D   E   F (planar)

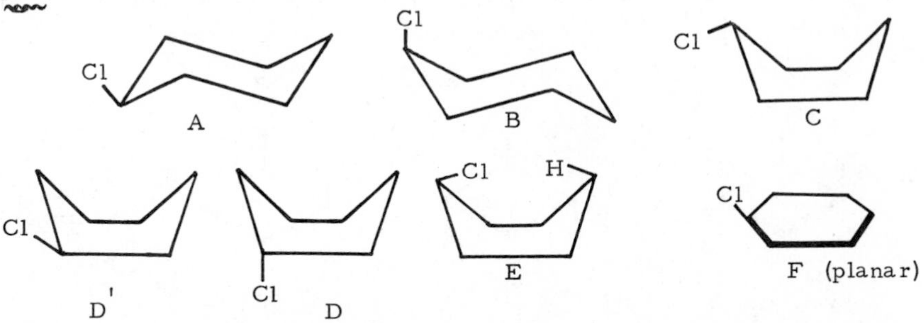

Order of decreasing stability A > B > C > D, D' >E > F. In the extreme boat forms, the chlorine is staggered with respect to the adjacent hydrogens in C, eclipsed in D, and buttressed against the 4-hydrogen in E.

12-5

a.

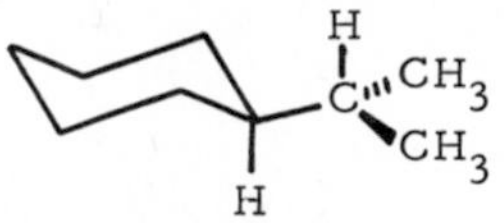

b.

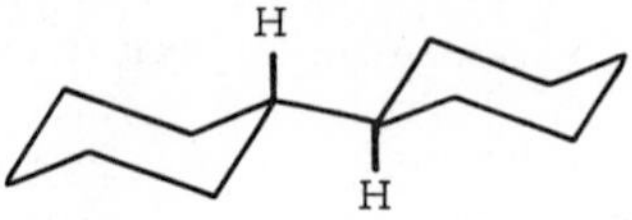

c.

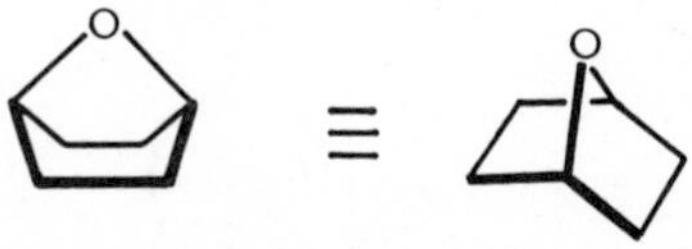

12-6 a. From the table it is clear that the distance r_{C-X} increases substantially from fluorine to iodine. The r_e and r_a also increase, but the r_e value remains close to the minimum value r_0, and the difference ($r_e - r_a$) remains almost constant at 0.4 A with halogen. Therefore, the difference in stability between axial and equatorial forms, $-\Delta G°$, is likewise expected to remain essentially constant with halogen.

b. The equilibrium should lie far to the diequatorial side because of a large unfavorable 1,3-diaxial interaction between the two axial iodine substituents.

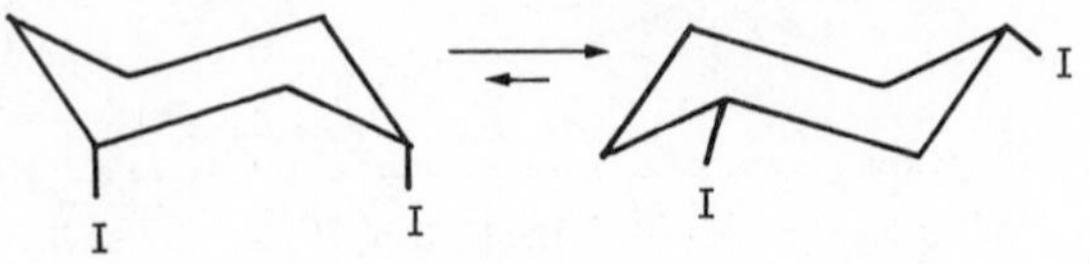

I   I   I

12-7   1 ⇌ 2, $\Delta G° = +5.0 + 0.5 = +5.5$ kcal

3 ⇌ 4, $\Delta G° = +5.0 - 0.5 = +4.5$ kcal

1 ⇌ 3, $\Delta G° = +0.5$ kcal

12-8   The additivity of ΔG° values assumed in working Exercise 12-7 does not hold when the substituents interact strongly with each other, as is the case for most 1,3-diaxial substituents and 1,2-diequatorial substituents.

12-9

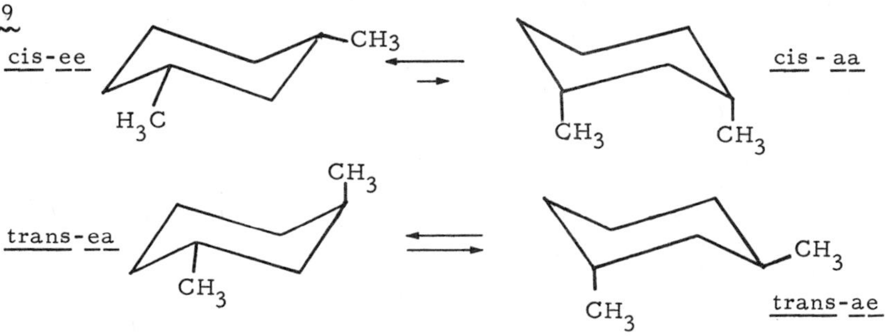

The cis isomer is more stable because it can exist in a conformation in which both methyl groups are equatorial.

12-10

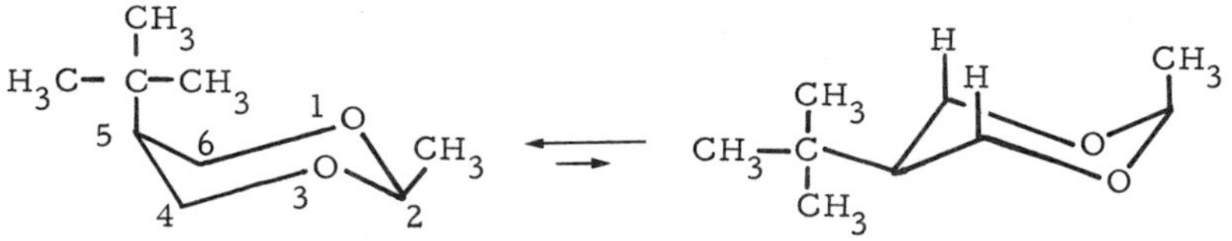

If the tert-butyl group is equatorial, then the methyl is axial and interferes with the two 4,6-diaxial hydrogens. If the tert-butyl group is axial, it interacts only with the unshared electron pairs on the 1- and 3-ring oxygen atoms. Evidently, interference by bonded atoms or groups is more severe than by nonbonding electrons. The equilibrium between the conformation of the 2,4- isomer shown is expected to lie most favorably toward the left because now these are strong axial tert-butyl-hydrogen interactions.

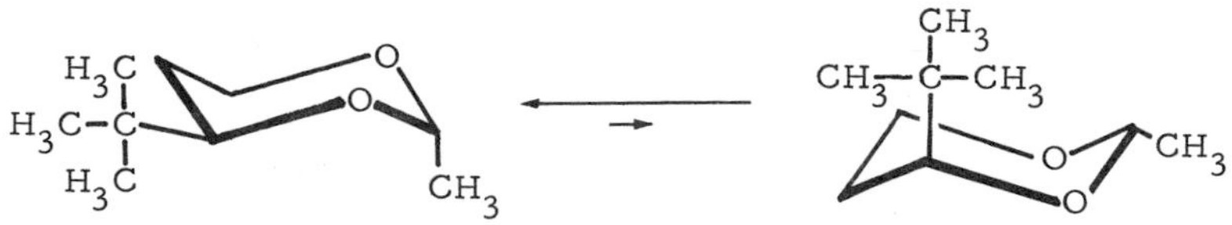

12-11   The ^{13}C spectrum of methylcyclohexane at 25° is like the spectrum at -110° (Figure 12-12) except that the ^{13}C signals for the CH_3 and ring carbons are average signals for axial and equatorial forms. The chemical shifts of the 25° spectrum are almost coincident with the major signals shown in Figure 12-12. This reflects that the major conformation at 25°, like that at -110° is the equatorial form, although the fraction of the conformation with axial methyl increases from less than 1% to about 6% on changing the temperature from -110 ° to 25 °.

12-12   a. At 30 °, the ^{19}F spectrum is virtually a single resonance corresponding to rapidly inverting chair-chair conformations. The F_e—C—F_a
coupling averages to zero, and the H—C—C—F coupling appears as an average
value of ee, ea and aa couplings.

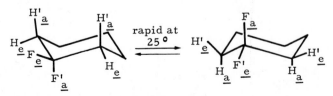

        As the temperature is lowered, ring inversion becomes slower until
at -100° it is slow enough for the equatorial and axial fluorines to appear as
separate resonances. Each is split by the other into a doublet, J_{FF} ~250 Hz.
The high-field doublet is broadened by H—C—C—F couplings of larger magnitude than the lower doublet. This is consistent with the high-field doublet
arising from axial fluorine because H_a—C—C—F_a ~15 Hz, whereas
H—C—C—F_e ~ 5.7 Hz.

        b. The ^{19}F spectrum of 1,1-difluoro-4-tert-butylcyclohexane at 25 °
resembles that of 1,1-difluorocyclohexane at -100° (Figure 12-13) because the
tert-butyl group holds the ring in a single non-inverting chair conformation.

12-13   When the \CHCl proton is axial, its nmr resonance will show a
splitting by two adjacent axial protons of J_{aa} ~ 14 Hz and a splitting by two
adjacent equatorial protons of J_{ae} ~ 3 Hz. When the   CHCl proton is
equatorial, its nmr resonance will show two sets of splittings to adjacent pairs
of protons such that J_{ae} ~ J_{ee} ~ 3 Hz.

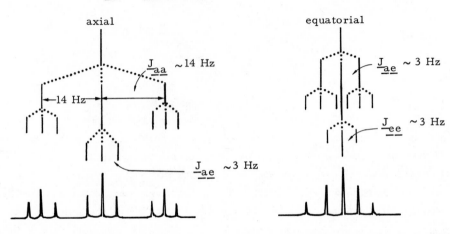

12-14

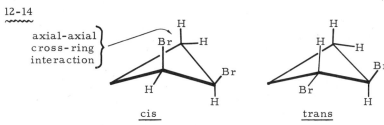

axial-axial
cross-ring
interaction

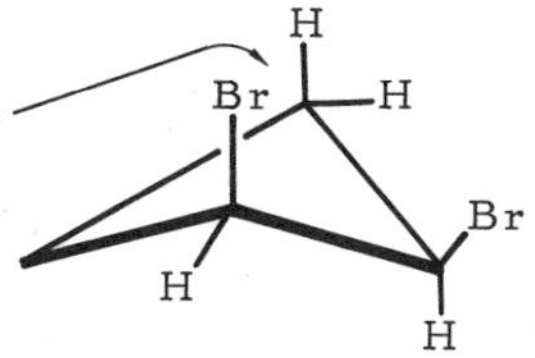

Br—H

Br

cis

H
H—H
Br
H

Br

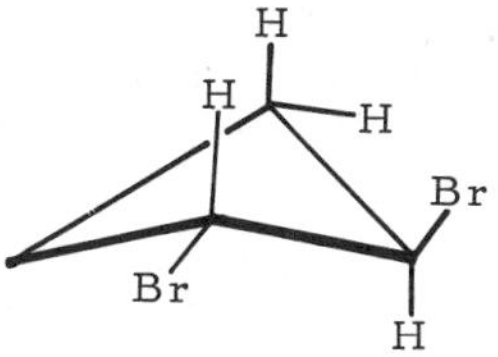

trans

The trans isomer is the more stable
isomer because there are no axial-
axial interactions.

For a similar reason, <u>cis</u>-1, 3-dibromocyclobutane is predicted to be
<u>more</u> stable than the <u>trans</u>-1, 3 isomer.

12-15   a.

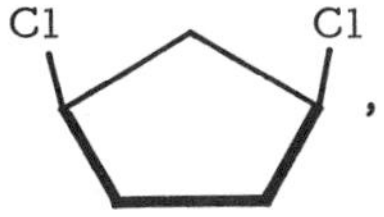

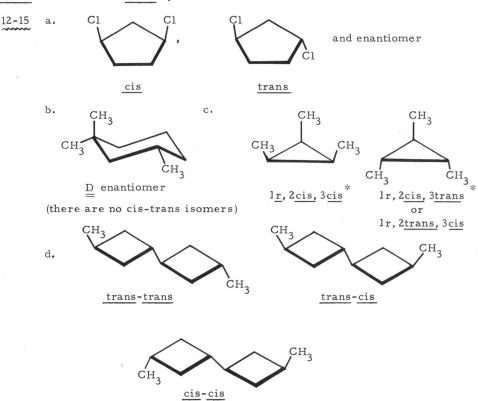

Cl   Cl   Cl

,   and enantiomer

Cl

cis   trans

b.

CH₃

CH₃

CH₃

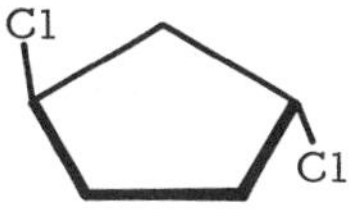

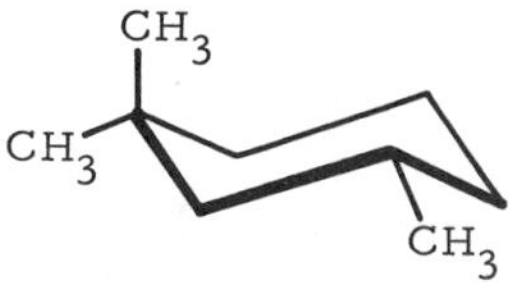

<u>D</u> enantiomer

(there are no cis-trans isomers)

c.

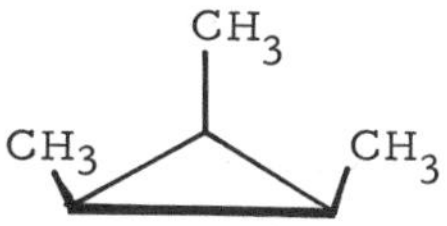

CH₃

CH₃   CH₃

1<u>r</u>, 2<u>cis</u>, 3<u>cis</u>*

CH₃

CH₃   CH₃

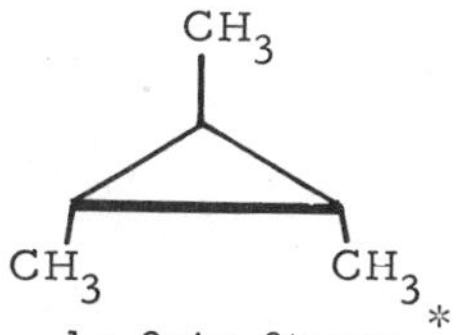

1r, 2<u>cis</u>, 3<u>trans</u>*

or

1r, 2<u>trans</u>, 3<u>cis</u>

d.

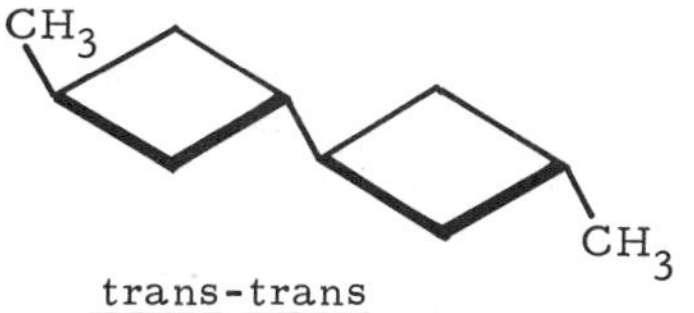

CH₃

CH₃

trans-trans

CH₃

CH₃

trans-cis

CH₃

CH₃

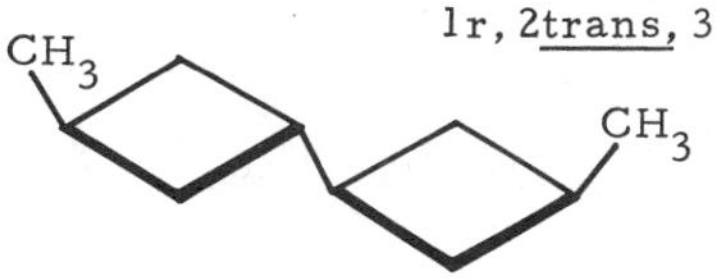

cis-cis

12-16   If the total strain in a 1-alkene is taken to be the same as in ethene, then
we can calculate ΔH° of isomerization from the data of Table 12-3 by
subtracting the total strain in the cycloalkane from that of ethene.   Thus:

* See Exercise 12-1d for an explanation of the notation <u>r</u>

$$\underline{n} \qquad \Delta \underline{H}^{\circ} \text{ for } (CH_2)_{\underline{n}} \longrightarrow CH_3(CH_2)_{\underline{n}-3}CH{=}CH_2$$

$\underline{n}$	
3	$-5.3 \text{ kcal mole}^{-1}$
4	-3.9
5	$+16$

There are other ways of obtaining the answer. For example, one can calculate ΔH for combustion of alkenes with different $\underline{n}$ values and take the difference between these and the ΔH values of combustion for the cycloalkanes as given in Table 12-3 (note that these are for formation of liquid water).

$\underline{n}$	Calcd. $\Delta \underline{H}$ of combustion of alkene (liquid water)	$\Delta \underline{H}^{\circ}$ of isomerization
2	333.3	-
3	490.2	-9.6
4	647.2	-8.7
5	804.1	+10.9

That these figures for $\Delta \underline{H}^{\circ}$ of isomerization are not the same as those given above is because the average bond energies do not give a very good calculated heat of combustion for ethene (333.3 kcal mole compared to 337.3 kcal mole by experiment).

In any case, the $\Delta \underline{H}^{\circ}$ values are such as to lead to the expectation that with $\underline{n} = 3$ or 4 the cycloalkanes should be less stable than the alkenes. The entropy effect is expected to be in the same direction, the atoms in the open-chain alkenes being generally less constrained than in the cycloalkanes. The difference is likely to be small with propene ($\underline{n} = 3$) but would increase in importance going to $\underline{n} = 4$ and 5.

12-17   The reaction $(CH_2)_{\underline{n}} + Br_2 \longrightarrow (CH_2)_{\underline{n}-2}(CH_2Br)_2$ is estimated to be exothermic by 7 kcal mole^{-1} using bond energies. If we add to this $\Delta \underline{H}^{\circ}$ value the total relief of strain on going from cycloalkane to open-chain dibromide, then: $\underline{n} = 2$, -29.4 kcal; $\underline{n} = 3$, -34.7 kcal; $\underline{n} = 4$, -33.3 kcal; $\underline{n} = 5$, -13.4 kcal; $\underline{n} = 6$, -7.4 kcal.

12-18 $\Delta \underline{H}^{\circ}$ First step: $\underline{n} = 6$, +15 kcal; $\underline{n} = 5$, +8.6 kcal; $\underline{n} = 4$, -11.3 kcal; $\underline{n} = 3$, -12.7 kcal; $\underline{n} = 2$, -7.4 kcal

$\Delta \underline{H}^{\circ}$ Second step: -22 kcal.

Reaction is feasible only for $\underline{n} = 2$, 3 and 4. It actually only occurs at a reasonable rate with $\underline{n} = 2$ and 3.

12-19 Br_2 could distinguish methylcyclopropane and 1-butene from cyclobutane; $KMnO_4$ could distinguish between methylcyclopropane and 1-butene.

12-20   1a.

The compound with axial bromine would react more rapidly because the nucleophile $I^{\ominus}$ can approach by the least-hindered "equatorial" direction.

2 a.

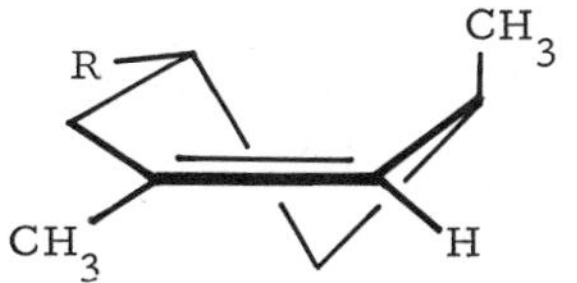

The compound with axial $\overset{\oplus}{N}(CH_3)_3$ would react more rapidly because it can achieve a favorable coplanar transition state for antarafacial E 2 elimination by having the leaving groups both axial.   The other isomer has no axial

hydrogen adjacent to $\overset{\oplus}{N}(CH_3)_3$.

b.

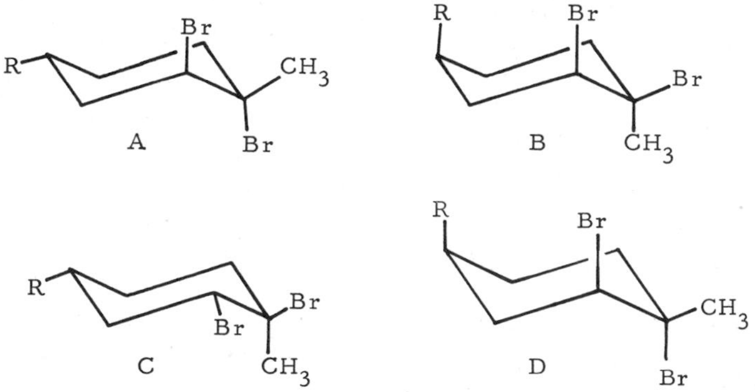

A is the product most rapidly formed because the transition state can achieve a coplanar diaxial arrangement of entering groups necessary for antarafacial addition.   Also, the bulky tert-butyl broup (R) is equatorial; B and D are excluded because R is axial; C is excluded because antarafacial addition cannot achieve a coplanar transition state when the entering groups are ee or ea.

12-21   a.   Probability of cyclization increases in dilute solution since the number of bimolecular collisions leading to addition decreases.

b.

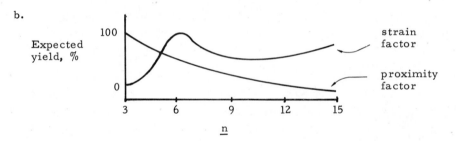

c. High yield of $(CH_2)_3$ means that favorable orientation overcomes high ring strain; with $(CH_2)_4$, opposing orientation and strain effects are almost balanced - hence low yield; for cyclohexane, yield is determined almost entirely by orientation effects; for larger rings, orientation becomes very unfavorable while strain is a relatively unimportant factor.

12-22   Notice that the axis of symmetry makes the axial-like or equatorial-like substitutions of C1 and C7 identical.

a. Position of substitution	Isomers	Total
1 (or 7)	equatorial-like, axial-like	2
2 (or 6)	equatorial-like, axial-like	2
3 (or 5)	equatorial-like, axial-like	2
4	hydrogens are equivalent	1
	7

b.   Considering chiral isomers - there are two possible chiral isomers of the twist-chair form.

mirror plane

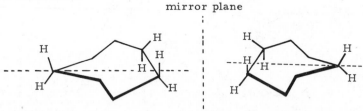

Therefore, there will be a corresponding set of monochloro-substituted isomers for each and therefore 14 isomers in all.

If you run through the same reasoning for the chair form 7 which with a plane of symmetry is meso, you will find monochlorine substituted isomers as follows: C1 a, e (achiral), C2 (C7) a, e (each chiral), C3 (C6) a, e (each chiral), C4 (C5) a, e (each chiral) ⟶ 14 total isomers.

12-23   The boat-boat conformation of cyclooctane in Figure 12-19 is unfavorable because it places two pairs of "inside" hydrogens in a 1, 5 relationship in close proximity. A diagram of a model shows this more clearly.

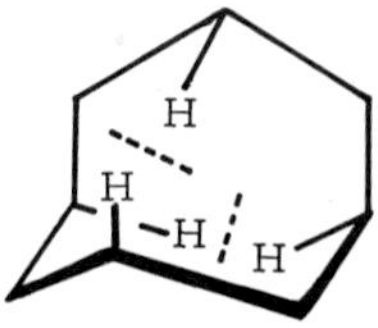

12-24 If the hydrogens on the double bond are replaced by large alkyl groups such as <u>tert</u>-butyl, the double bond will not be able to turn over and the molecule will have stable chiral forms.

12-25 a. bicyclo [3.1.0] hexane
      b. 8-chlorobicyclo [4.2.0] -2-octene
      c. 2, 3-dibromobicyclo [2.2.2] octane
      d. 4-bromo-9-methyltricyclo [5.3.0.0^{2,6}] decane
      e. tetracyclo [7.3.1.0^{3,13}.0^{7,13}] tridecane
      f. spiro [4.3] -5-octanol

12-26 If you try to construct a model of decalin with an axial-axial ring fusion, you will have discovered that it is not possible to do so without excessive strain. From this we conclude that such a molecule would be highly strained and difficult to form.

12-27

         cis   trans

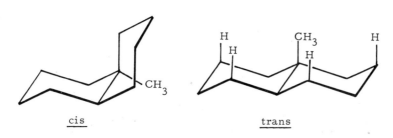

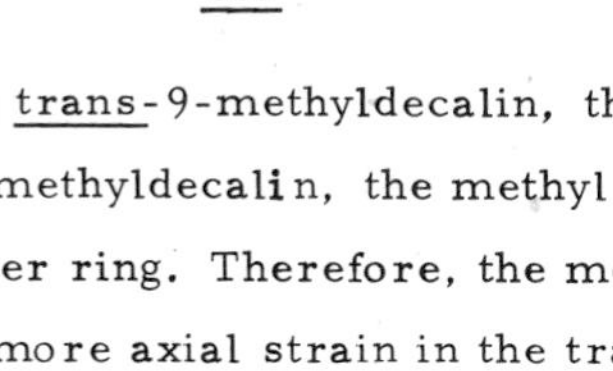

In <u>trans</u>-9-methyldecalin, the methyl group is axial to <u>both</u> rings. In cis-9-methyldecalin, the methyl group is axial to one ring and equatorial to the other ring. Therefore, the methyl group is anticipated to introduce 1. 5 kcal more axial strain in the trans isomer than in the cis isomer. Given that <u>trans</u>-decalin is 2 kcal more stable than <u>cis</u>-decalin, the net difference in energy of the 9-methyl isomers is 2 - 1. 5 = 0. 5 kcal in favor of the trans isomer. The experimental value is 0. 5 - 1 kcal.

12-28 tetracyclo [2.2.0.0^{2,6}.0^{3,5}] hexane

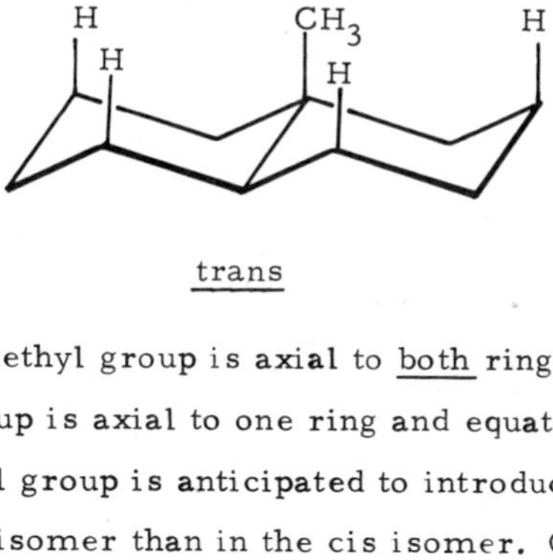

12-29

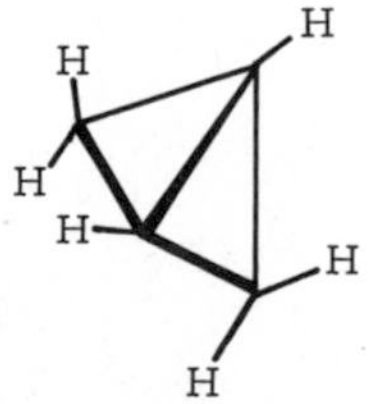

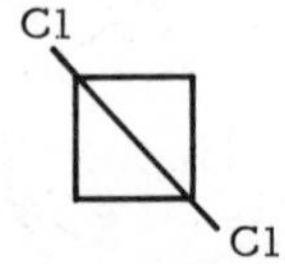

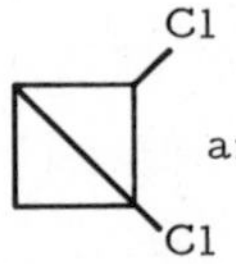

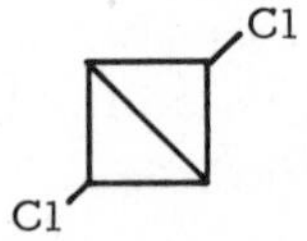

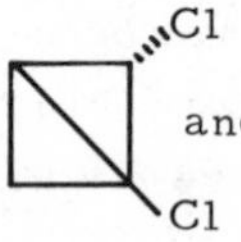

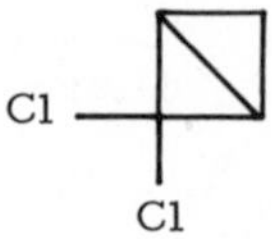

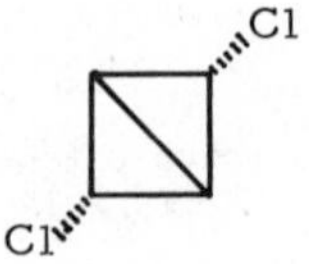

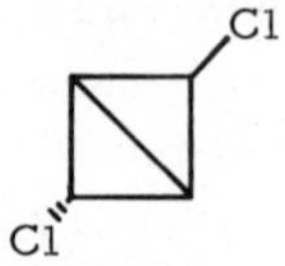

12-30   When two rings are fused together it is not possible to have a double-bonded carbon at a position common to both rings (bridgehead) unless the rings are large enough to span the planar configuration of the double bond without undue strain.   The strain is too great in bicyclo [2.2.1] heptene but is not severe in bicyclo [5.5.0] decene.

12-31   tricyclo [4.2.2^{1,4}.2^{5,8}] -1,8-dodecane

Models show that the compound is quite strained.   As stated in Exercise 12-30, Bredt's rule is not violated.   The compound is only marginally stable.

Radical addition of Br_2 is to be expected such that addition is suprafacial.   A polar (electrophilic) addition with antarafacial stereochemistry is not possible.

12-32   The answers given here are illustrative.   In many cases more than one answer is possible.

a.   methylcyclopropane

b.   cis-1,2-dimethylcyclopropane

c. $CH_3CHBrCH_2CHBrC_2H_5$,
$BrCH_2CH(CH_3)CH(C_2H_5)Br$
$CH_3CHBrCH(C_2H_5)CH_2Br$

d.

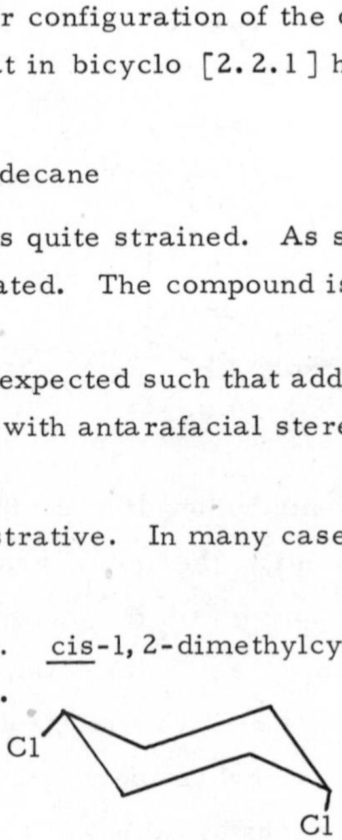

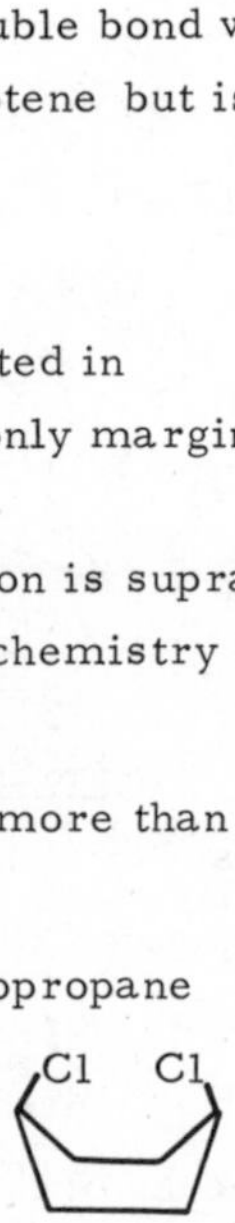

e.

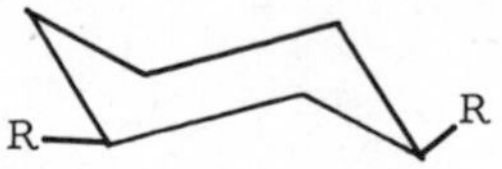

R = tert-butyl

f.

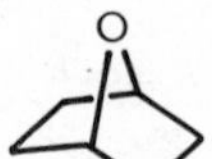

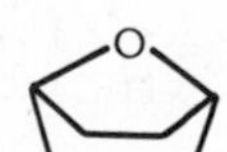

g.

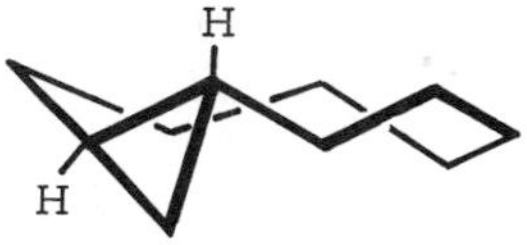

h.

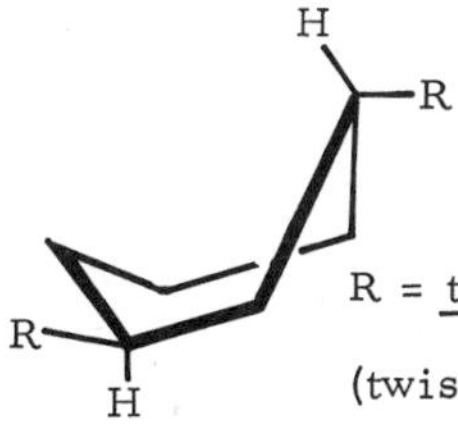

R = tert-butyl

(twist-boat form)

i.

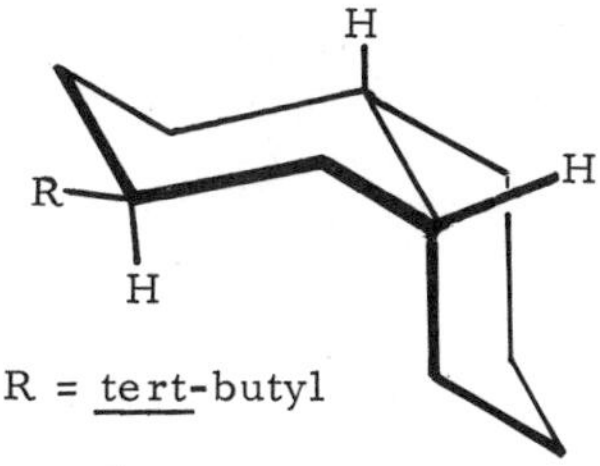

R = tert-butyl

j.

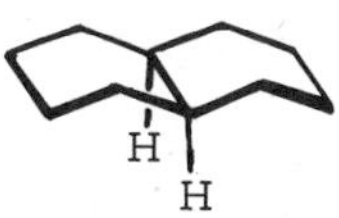

k.

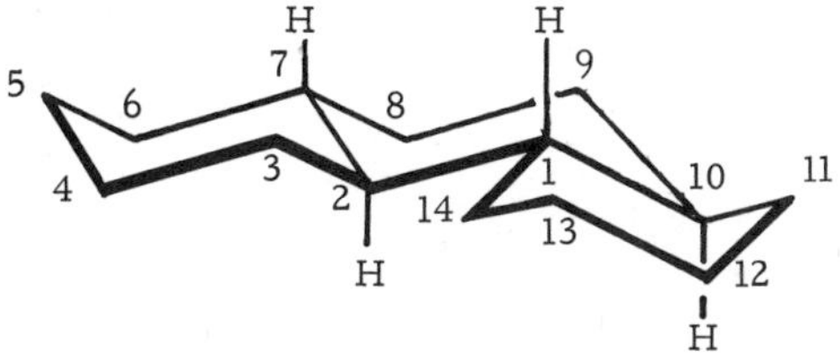

This is one of the basic steroid ring configurations (see Chapter 30).

12-33 C—C bond energy in cyclopropane may be calculated as follows, using a C—H cyclopropane bond energy of 104.7 kcal.

$(CH_2)_3 + H_2 \longrightarrow 3C + 8H$, ΔH_1^o

$CH_3CH_2CH_3 \longrightarrow 3C + 8H$, ΔH_2^o

$(CH_2)_3 + H_2 \longrightarrow CH_3CH_2CH_3$ $\Delta H_1^o - \Delta H_2^o = -37.5$ kcal

$(3x + 6 \times 104.7 + 104.2) - (2 \times 82.6 + 8 \times 98.7) = -37.5$

$x = -61.6$ kcal

a. $(CH_2)_3 \longrightarrow \cdot CH_2-CH_2-CH_2 \cdot$ $\Delta H^o = +55.6$ kcal

b. $2(CH_2)_3 \longrightarrow (CH_2)_6$ $\Delta H^o = -54$ kcal

12-34   a.   b.

H

H

cis

H

trans

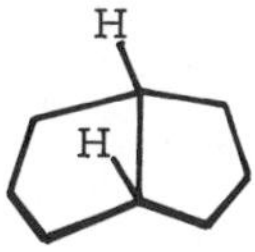

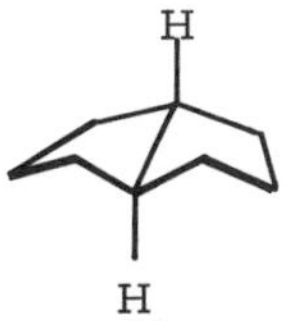

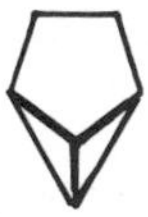

c.

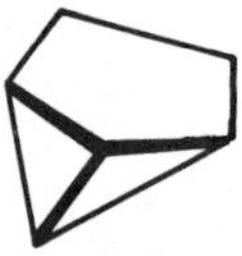

d.

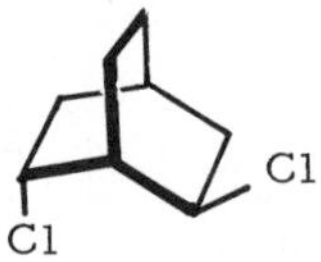

Cl
Cl

e.

<u>12-35</u>

a.

H
C(CH₃)₃

b.

c.

d.

H—
CH₃
C₆H₅
H

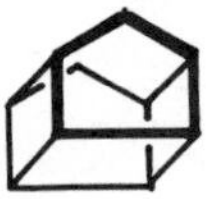

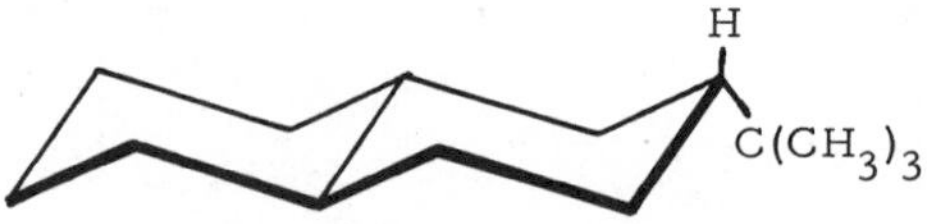

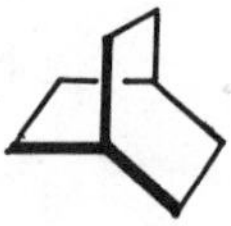

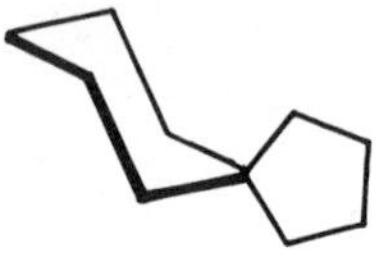

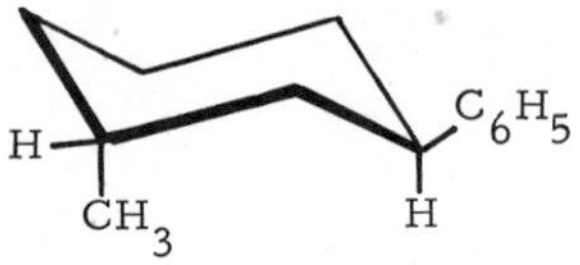

(Table 12-2 indicates phenyl is
larger than methyl; hence phenyl
is equatorial, methyl axial.)

12-36   Product is B; electrophile adds to <u>least</u> substituted carbon to give
most stable carbocation intermediate; addition is antarafacial, and the
preferred transition state places entering groups trans and diaxial; the <u>tert</u>-
butyl is equatorial.

12-37   a. <u>trans-ee</u>   b. <u>trans-ee</u>

c. 2<u>e</u>-methyl-<u>trans</u>-decalin

d.
CH₃
CH
CH₃
Cl
CH₃
≡
CH₃
''''Cl

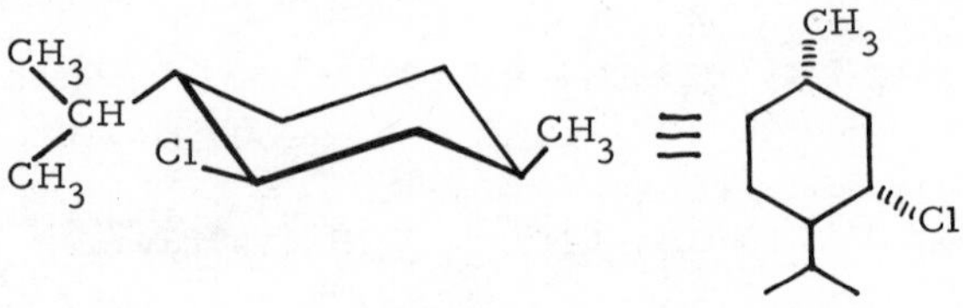

ADDITIONAL EXERCISE

12A-1 Draw the preferred conformation expected for each of the compounds:

a. <u>N</u>- methylazacyclohexane
(<u>N</u>-methylpiperidine)

b. <u>cis</u>-1-isopropyl-4-methylcyclohexane

c. <u>trans</u>-1-methyl-3-phenylcyclohexane

d. <u>cis</u>-, and <u>trans</u>-2-methyl-
cyclohexanecarbonitrile

e. <u>cis</u>-9-methyldecalin

f. CH₃''''
H
CH₃
H

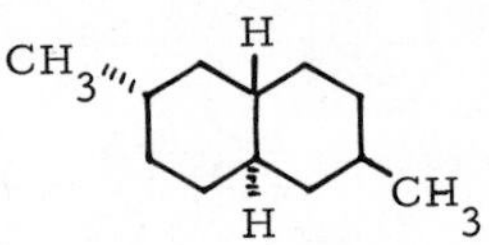

CHAPTER 13

POLYFUNCTIONAL COMPOUNDS
ALKADIENES
APPROACHES TO ORGANIC SYNTHESIS

The double bonds of conjugated polyenes normally do not act independently of one another. Electrophilic and radical additions frequently give mixtures of products by concurrent 1, 2 and 1, 4 addition:

$$CH_2{=}CH{-}CH{=}CH_2 \xrightarrow{Br_2} \begin{array}{l} \xrightarrow{\text{1, 2 addition}} BrCH_2{-}CHBr{-}CH{=}CH_2 \\[2em] \xrightarrow{\text{1, 4 addition}} BrCH_2{-}CH{=}CH{-}CH_2Br \end{array}$$

The amount of 1, 4 versus 1, 2 addition depends markedly on the reaction conditions, and expecially on whether the conditions favor kinetic or equilibrium control. Both the 1, 2 and 1, 4 products are derived from a common intermediate that is a resonance-stabilized cation for addition of electrophilic reagents, and a resonance-stabilized radical for addition of radical reagents:

$$X{-}CH_2{-}CH{=}CH{-}\overset{\oplus}{C}H_2 \longleftrightarrow X{-}CH_2{-}\overset{\oplus}{C}H{-}CH{=}CH_2$$

$$X{-}CH_2{-}CH{=}CH{-}\overset{\cdot}{C}H_2 \longleftrightarrow X{-}CH_2{-}\overset{\cdot}{C}H{-}CH{=}CH_2$$

A very important reaction of conjugated dienes is the Diels-Alder reaction, in which an alkene (dienophile) adds across the 1, 4 positions of a diene by what is known as a [4 + 2] cycloaddition. The product is a cyclohexene ring system:

Reaction occurs simply by mixing the reagents, sometimes with a solvent, usually with external heating. Catalysts seldom are required. Diels-

Alder reactions are reversible, although high temperatures usually are required for the reverse reaction. They also are highly stereospecific and give <u>reten-tion</u> of the stereochemical configurations in the diene and the dienophile. Thus if the substituents are trans in the dienophile, they will be trans in the adduct. The diene has to react in a <u>cisoid</u> conformation to give a stable adduct; there-fore, cyclic dienes usually are more reactive because they have the required <u>cisoid</u> configuration. Diels-Alder reactions are slowed by bulky substituents in the diene, and normally are facilitated by electronegative groups in the dienophile. (Table 13-1 lists some reactive dienophiles.) When the diene itself is heavily substituted with electronegative groups, such as chlorine or bromine, electron-donating groups on the dienophile facilitate the reaction.

There are several ways that the dienophile can approach the diene. The preferred approach minimizes steric interactions, but maximizes the proximity of the π centers of the diene with any π substituents in the dienophile:

(maximum π interaction between diene and dienophile)

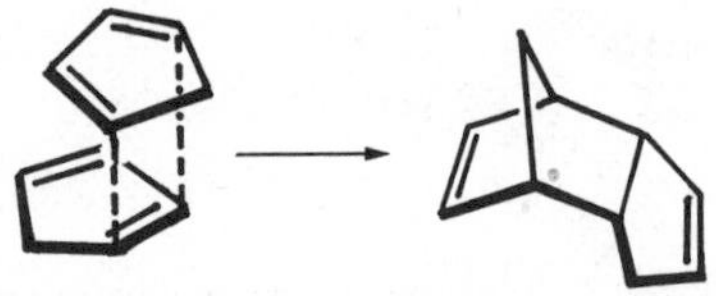

<u>endo</u> adduct (major)

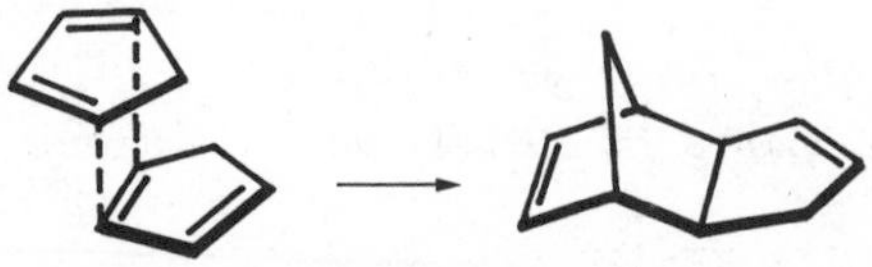

<u>exo</u> adduct (minor)

All evidence appears to indicate that the Diels-Alder and related thermal 1,4-cycloadditions to dienes are concerted reactions.

[2 + 2] Cycloaddition to conjugated dienes or even alkenes seldom occurs thermally except with perhaloalkenes or substances with cumulated double bonds, which add by stepwise mechanisms that are nonstereospecific:

Irradiation with ultraviolet light often will induce [2 + 2] cycloadditions that do not occur thermally.

Conjugated dienes are a valuable source of elastomeric polymers (rubbers). The principal monomers are 1,3-butadiene, 2-methyl-1,3-butadiene (isoprene), and 2-chloro-1,3-butadiene (chloroprene). In principle, they can give addition polymers by 1,2, cis-1,4, or trans-1,4 addition:

n x $CH_2\!=\!CH\!-\!CH\!=\!CH_2$

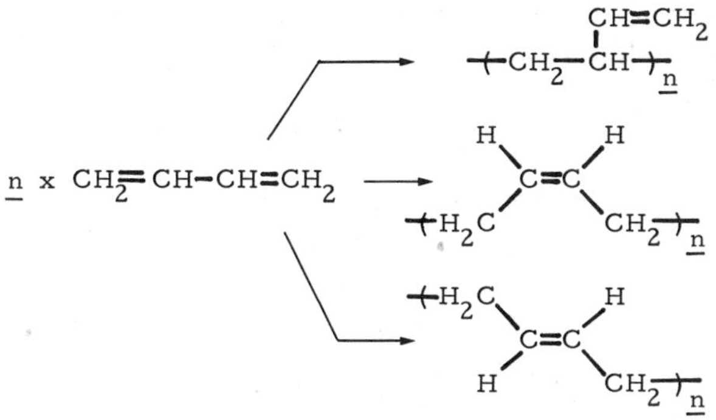

1,2 addition

cis-1,4 addition

trans-1,4 addition

Radical polymerization of chloroprene gives a mixture of atactic 1,2- and cis- and trans-1,4-polymer known as neoprene, which is valuable for its resistance to oxidants and "swelling" by chemicals. Isoprene with appropriate catalysts produces an all cis-1,4-polymer that is nearly identical to natural rubber. Butadiene copolymerizes with ethenylbenzene (styrene) to give different rubbers under different reaction conditions. The largest use for these polymers is for tire-carcass stock. (See Table 13-2 for a listing of diene polymerizations.)

Cumulated dienes, such as 1,2-propadiene, are known as allenes. They are highly reactive compounds that readily undergo electrophilic addition reactions, hydrogenation, hydroboration, cycloaddition, and base-catalyzed rearrangements. With an even number of cumulated double bonds, they can be chiral if the end carbons carry different substituents, even when the two ends are alike (a b C=C=C a b). This chirality results from restricted rotation about the cumulene bonds. Chirality also is possible for appropriately substituted diphenyls, spiranes, and related compounds in which rotation is restricted (see Table 13-3). When appropriately substituted, an odd number of cumulated double bonds can give cis-trans isomers.

The first stage in making a plan to synthesize a complex organic molecule is to conceive of suitable reactions by which the carbon skeleton of the target compound can be constructed. The second stage is to introduce the functional groups at the desired locations. The carbon framework has to be constructed to allow for functionality to be introduced at the right positions.

The preferred route will depend on many circumstances: cost and availability of starting materials, yield, ease of separation of products from unwanted by-products, number of steps, and possible need for protecting the functional groups. Lists of useful chemical transformations for planning organic syntheses are given in Tables 13-4 and 13-5.

ANSWERS TO EXERCISES

13-1   The C—H bond energy at C 3 of 1, 4-pentadiene is weaker than the methyl C—H bond energy of propene because the radical resulting from abstraction of a hydrogen atom from C 3 is stabilized by electron delocalization through <u>two</u> adjacent π bonds. There are <u>three</u> low-energy valence bond structures compared to <u>two</u> for the 2-propenyl radical.

$$CH_2{=}CH{-}CH_2{-}CH{=}CH_2 \xrightarrow[-RH]{R\cdot} CH_2{=}CH{-}\overset{\cdot}{C}H{-}CH{=}CH_2$$

$$CH_2{=}CH{-}CH{=}CH{-}\overset{\cdot}{C}H \longleftrightarrow \overset{\cdot}{C}H_2{-}CH{=}CH{-}CH{=}CH_2$$

$$CH_2{=}CH{-}CH_3 \xrightarrow[-RH]{R\cdot} CH_2{=}CH{-}\overset{\cdot}{C}H_2 \longleftrightarrow \overset{\cdot}{C}H_2{-}CH{=}CH_2$$

13-2

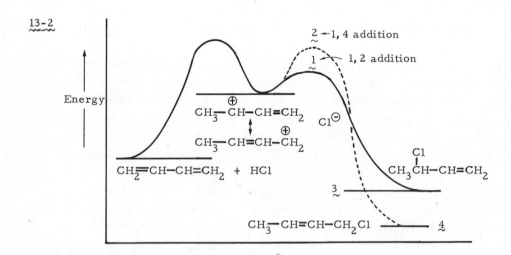

Although both products are formed by way of a common intermediate, the 1, 2-adduct is formed faster because the transition state for $Cl^{\ominus}$ attack at C 2 is lower than for attack at C 4 ($\underset{\sim}{1} < \underset{\sim}{2}$). At equilibrium, the major product is the more stable 1, 4-adduct ($\underset{\sim}{4} < \underset{\sim}{3}$).

13-3   a. 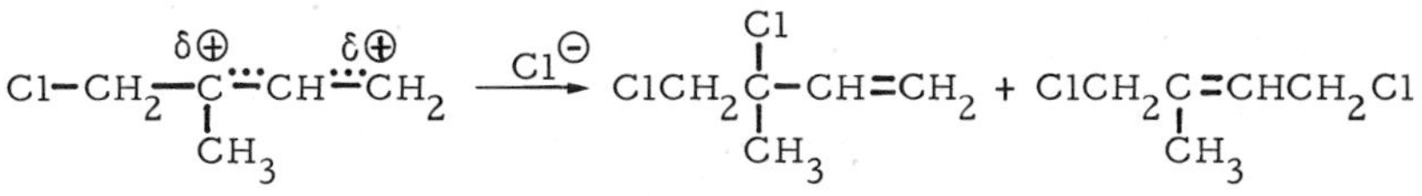 $Cl-CH_2-\overset{\delta\oplus}{C}\overset{\cdots}{\cdots}CH\overset{\delta\oplus}{\cdots}CH_2 \xrightarrow{Cl^\ominus} ClCH_2\overset{Cl}{\underset{CH_3}{\overset{|}{C}}}-CH=CH_2 + ClCH_2\underset{CH_3}{\overset{|}{C}}=CHCH_2Cl$

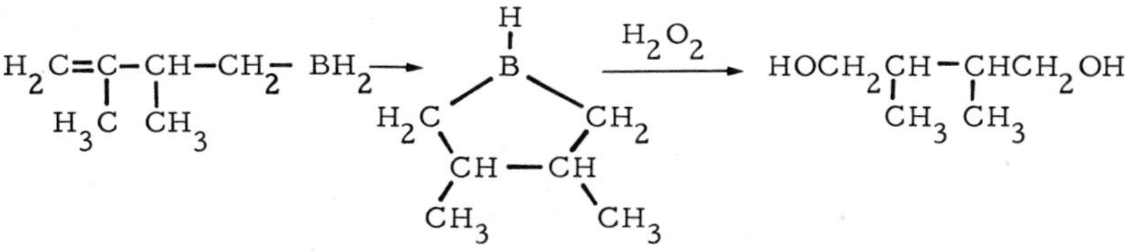

b. $H_2C=C-CH-CH_2- BH_2\longrightarrow$ $\xrightarrow{H_2O_2} HOCH_2CH-CHCH_2OH$

c. $C_6H_5S-CH_2-\overset{\frac{1}{2}\cdot}{CH}\overset{\cdots}{\cdots}CH\overset{\frac{1}{2}\cdot}{\cdots}CH_2 \xrightarrow[-C_6H_5S\cdot]{C_6H_5SH} C_6H_5S-CH_2-CH=CH-CH_3$

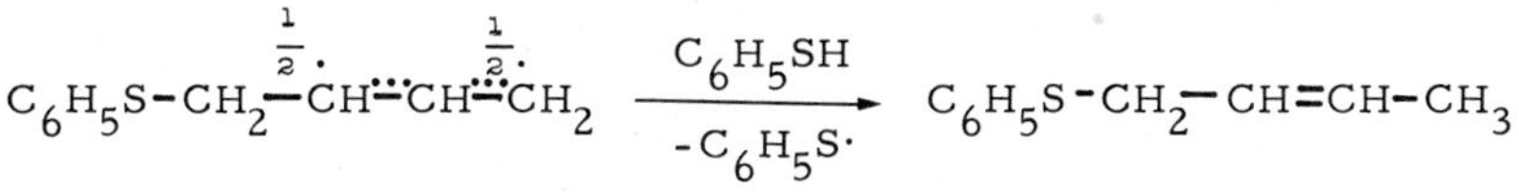

(radical additions give
high proportion of
1, 4-adduct)

d. 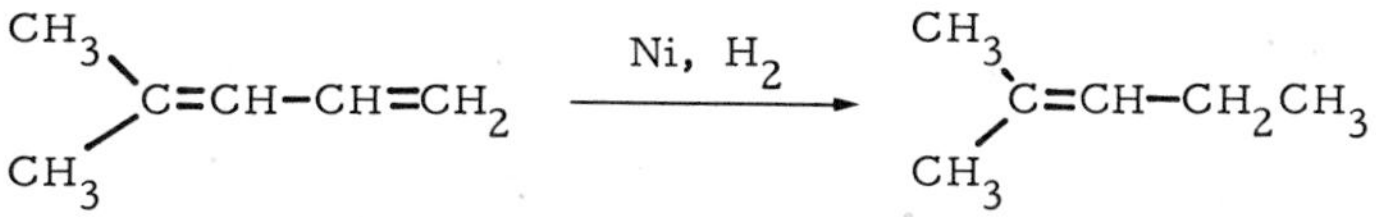 $\xrightarrow{Ni, H_2}$

(least-hindered double bond
is reduced first)

13-4

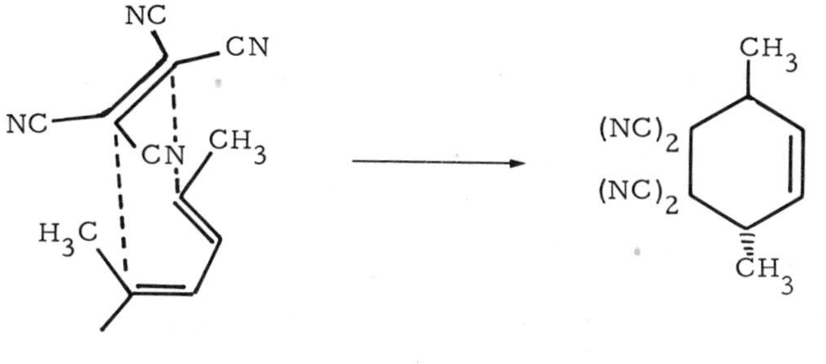

cis, trans

cis, cis-2, 4-hexadiene $\xrightarrow{(NC)_2C\equiv C(CN)_2}$

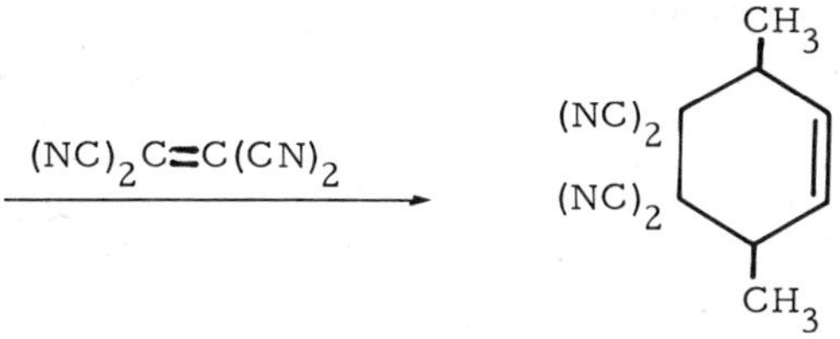

13-5

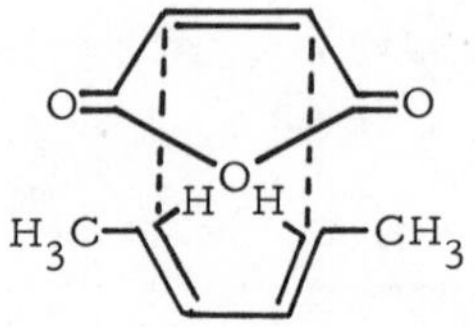

preferred orientation has
maximum interaction between
π bonds of diene and dienophile;
adduct formed is

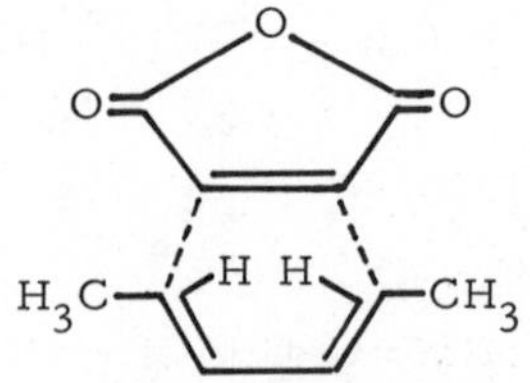

unfavorable orientation;
carbonyl interaction with
diene is negligible.

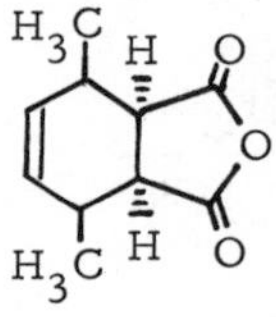

13-6

a.

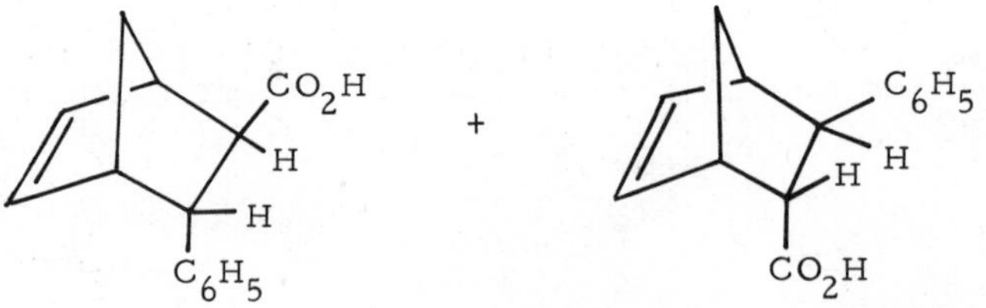

+

(a mixture of products is likely)

b.

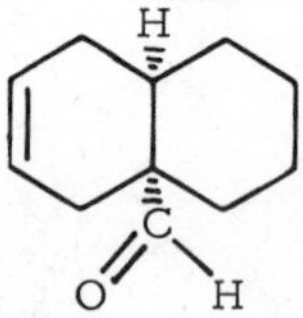

c.

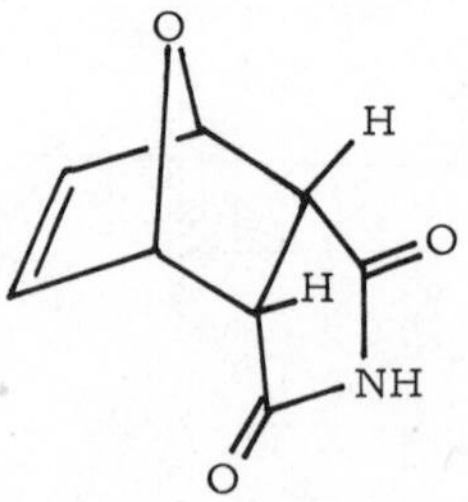

d.

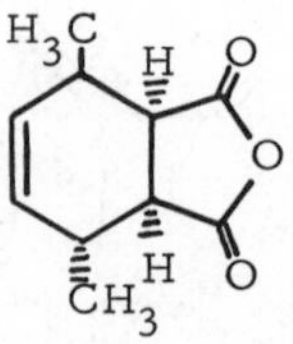

e.

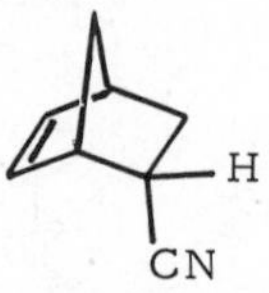

f.

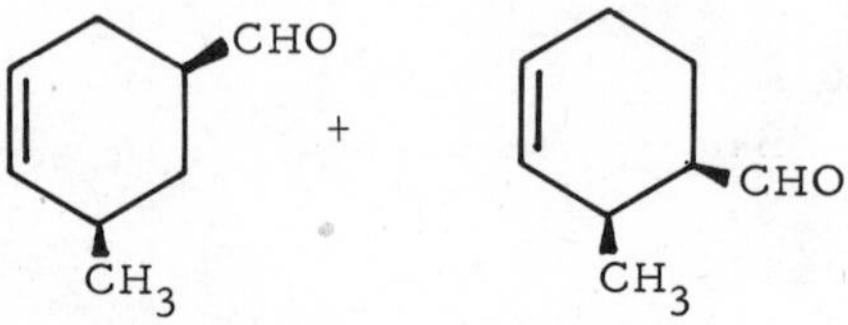

+

cis-1, 3

cis-1, 2

13-7   Cyclopentadiene (2 moles)

13-8 ΔS = -44.2 eu   for the addition of SO_2 to 1,3-butadiene at 100°

ΔS° = -67   eu   for the addition of ethene to 1,3-butadiene at 25°

The figures show that the entropy change in both reactions is large and negative; the entropy change is less favorable for the ethene-butadiene reaction by
about 23 eu. Assuming that ΔH and ΔS do not change with temperature, for
K = 1, then ΔH = $T\Delta S$, or T =~ 700° K (427° C).

13-9

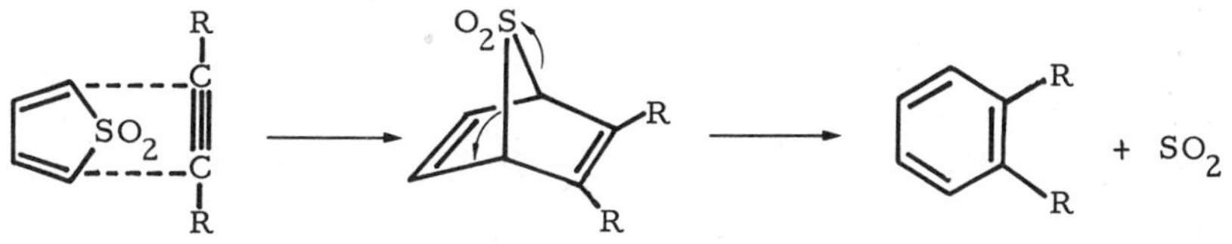

13-10

a.

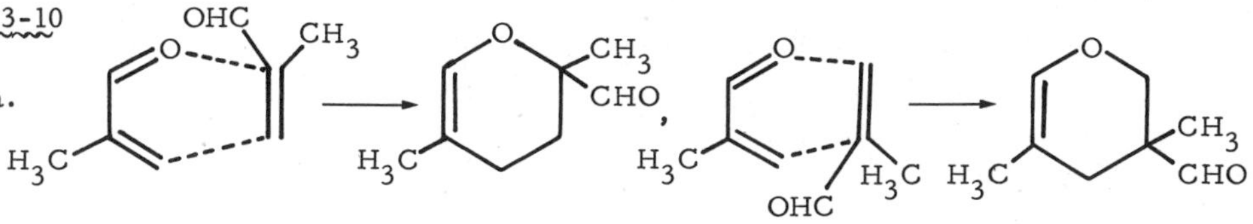

b.

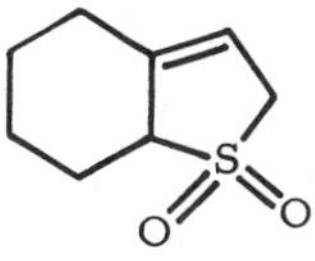

c.

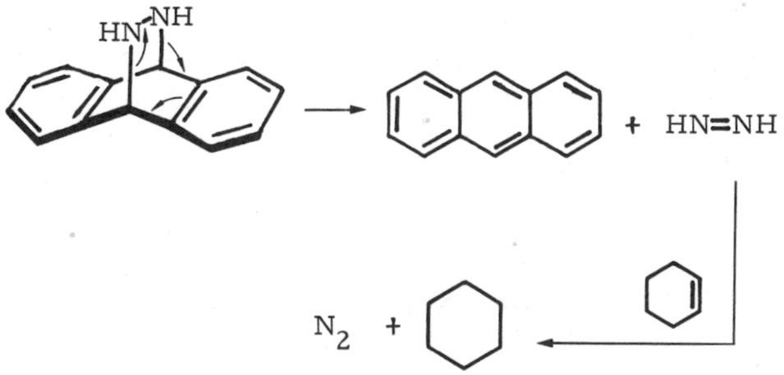

13-11

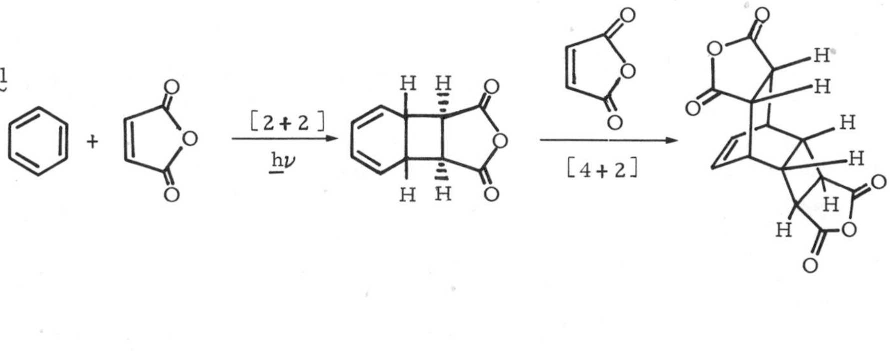

13-12

Initiation: ROOR $\longrightarrow$ 2RO·

$$RO\cdot + CH_2=CH-CH=CH_2 \longrightarrow ROCH_2\overset{\cdot}{C}H-CH=CH_2$$

$$ROCH_2CH=CH-\overset{\cdot}{C}H_2$$

Propagation by 1,2-addition:

$$ROCH_2\overset{\cdot}{C}H-CH=CH_2 + CH_2=CH-CH=CH_2 \longrightarrow$$

$$ROCH_2\underset{\underset{CH=CH_2}{|}}{C}HCH_2\overset{\delta\cdot}{CH}\cdots CH\cdots\overset{\delta\cdot}{CH_2}$$

Propagation by 1,4-addition:

$$ROCH_2CH=CH-\overset{\cdot}{C}H_2 + CH_2=CH-CH=CH_2 \longrightarrow$$

$$ROCH_2CH=CHCH_2CH_2\overset{\delta\cdot}{CH}\cdots CH\cdots\overset{\delta\cdot}{CH_2}$$

Termination by radical combination or disproportionation:

$$2R\cdot \longrightarrow R-R, \quad 2\overset{\cdot}{\underset{|}{C}}H_2-\overset{}{\underset{|}{C}}H \longrightarrow \overset{}{\underset{|}{C}}H_2\overset{}{\underset{|}{C}}H_2 + -CH=CH-$$

Gross polymer structure:

$$\underset{\underset{\underset{n}{\big\downarrow}}{}}{}\begin{matrix}CH=CH_2\\|\\-(CHCH_2-)_n\end{matrix}(CH_2-CH=CH-CH_2-)_m$$

$\Delta \underline{H}$ of polymerization = $\underline{n} \times 19.4 + \underline{m} \times 19.4$ kcal, where 19.4 kcal is $\Delta \underline{H}$ of addition $RH + \overset{}{C}=\overset{}{C} \longrightarrow R-\overset{|}{\underset{|}{C}}-\overset{|}{\underset{|}{C}}-H$ calculated from average bond energies.

13-13   The equilibrium between 1,2-butadiene, 2-butyne and 1-butyne lies in favor of 2-butyne, in the presence of a basic <u>catalyst</u> such as ethoxide ion.

$$CH_3CH_2C\equiv CH \rightleftharpoons CH_3CH=C=CH_2 \rightleftharpoons CH_3C\equiv CCH_3$$

However, in the presence of a <u>strong base</u>, such as amide ion, the equilibrium is shifted towards 1-butyne because 1-butyne is converted to the 1-butynide ion.

$$CH_3C\equiv CCH_3 \rightleftharpoons CH_3CH=C=CH_2 \rightleftharpoons CH_3CH_2C\equiv CH \xrightarrow[-NH_3]{\overset{\ominus}{N}H_2} CH_3CH_2C\equiv \overset{\ominus}{C}:$$

Ethoxide is too weak a base to convert a 1-alkyne to its conjugate base.

<u>13-14</u> $CH_3CH=CH_2 + Cl_2 + H_2O \longrightarrow CH_3\underset{\underset{Cl}{|}}{C}HCH_2Cl + CH_3\underset{\underset{OH}{|}}{C}HCH_2Cl + HCl$

100 g	125 g	130 g	40 g
2.38 moles	1.79 moles	1.16 moles	0.42 moles

Percent conversion (propene) $= \dfrac{\text{amount reacted} \cdot 100}{\text{amount present}} = \dfrac{(100-25)100}{100} = 75\%$

Percent yield (propene) $= \dfrac{\text{moles of product} \cdot 100}{\text{moles of propene reacted}} = \dfrac{(1.16+0.42)100}{1.78} = 88\%$

Percent conversion $(Cl_2) = 100\%$

Percent yield $(Cl_2) = \dfrac{\text{moles of product} \cdot 100}{\text{moles of } Cl_2 \text{ reacted}} = \dfrac{(1.16+0.42)100}{1.79} = 88\%$

<u>13-15</u>   Overall yield in 100 steps, each 99% yield $= (0.99)^{100}$ x 100 $= 37\%$

Overall yield in 100 steps, each 99.9% yield $= (0.999)^{100}$ x 100 $= 91\%$

A small difference in yield in each step translates into a very large difference in overall yield in a multistep synthesis. This illustrates the critical importance of achieving the highest possible yields in multistep reaction sequences.

<u>13-16</u>   The desired carbon skeleton can be obtained by acid-catalyzed dimerization of 2-methylpropene (Section 10-9).

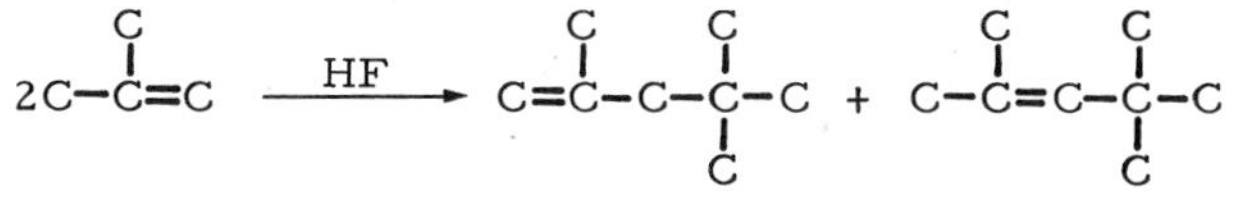

Functionality (Cl at C 2) can be achieved by addition of HCl to the alkene mixture (Section 10-3).

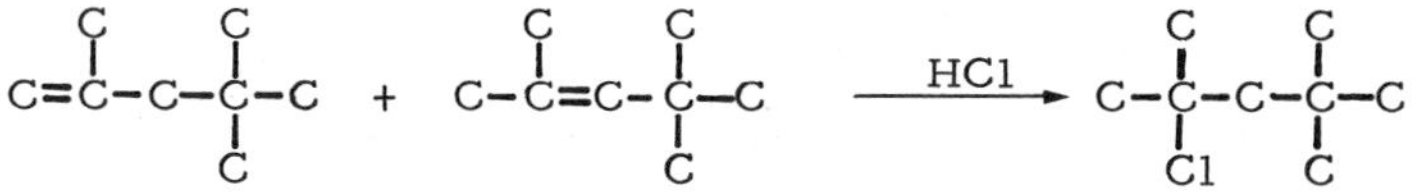

<u>13-17</u>
a.

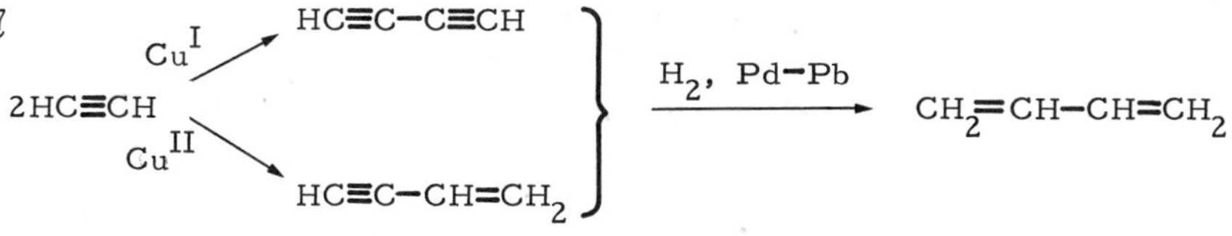

Side reactions: further Cu-catalyzed coupling of C 4 1-alkynes.

b. $H_3C-C\equiv CH$ $\xrightarrow[-NH_3]{NH_2^{\ominus}}$ $H_3C-C\equiv C:^{\ominus}$ $\xrightarrow[-Br^{\ominus}]{CH_3CH_2CH_2Br}$ $H_3CC\equiv CCH_2CH_2CH_3$

Conversion of propyne to bromopropane could be achieved by the sequence

$H_3CC\equiv CH$ $\xrightarrow{H_2,\ Pd-Pb}$ $H_3CCH=CH_2$ $\xrightarrow[ROOR]{HBr}$ $CH_3CH_2CH_2Br$;

side reactions include possible E2 reaction,

$CH_3C\equiv C:^{\ominus}$ + $CH_3CH_2CH_2Br$ $\longrightarrow$ $CH_3C\equiv CH$ + $CH_3CH=CH_2$ + $Br^{\ominus}$

c. $CH_3C\equiv CCH_3$ $\xrightarrow{H_2,\ Pd-Pb}$ $CH_3CH=CHCH_3$ $\dfrac{1.\ R_2BH}{2.\ 160^{\circ},\ 1\text{-decene}}$

$CH_3CH_2\underset{\underset{O}{\diagdown\diagup}}{CH-CH_2}$ $\xleftarrow{RCO_3H}$ $CH_3CH_2CH=CH_2$ $\longleftarrow$

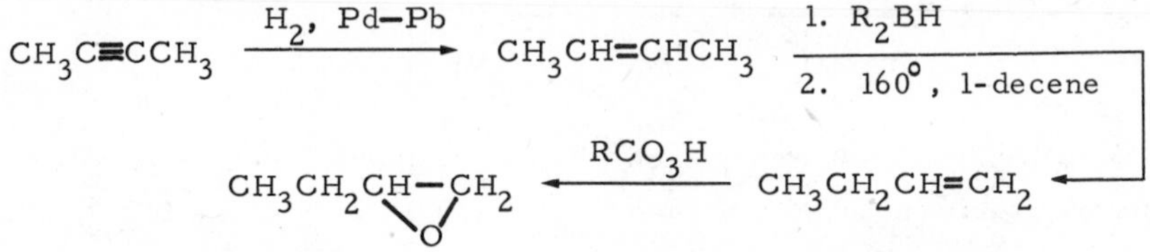

d.

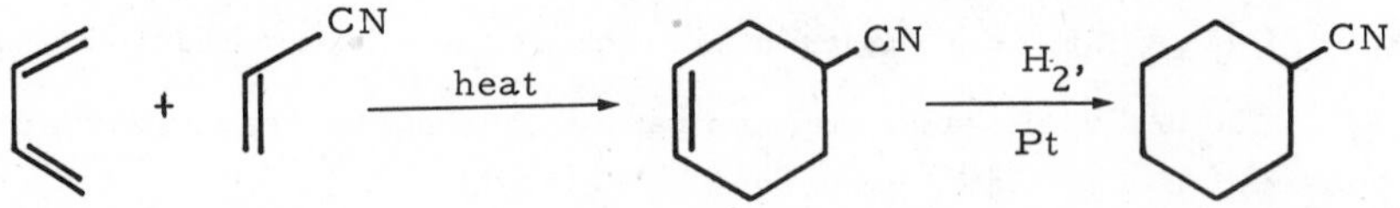

13-18

a. $(CH_3)_2C=CH_2$ $\xrightarrow[25^{\circ}]{H_2O,\ H_2SO_4\ (10\%)}$ $(CH_3)_3COH$

b. $(CH_3)_2C=CH_2$ $\xrightarrow{70\%\ HF\ in\ pyridine}$ $(CH_3)_3CF$

c. $CH_3CH=CH_2$ $\xrightarrow{Br_2,\ dilute\ NaOH,\ 25^{\circ}}$ $CH_3\underset{\underset{OH}{|}}{CH}CH_2Br$

or N-Br , H_2O

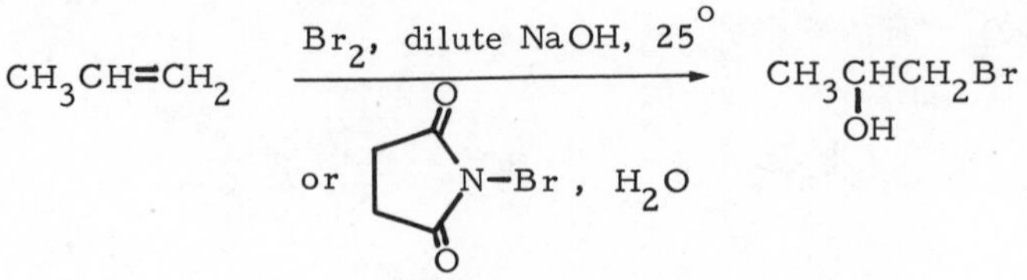

d. $(CH_3)_2C=CH_2$ $\dfrac{1.\ 60\%\ H_2SO_4,\ 70^{\circ}}{2.\ HI}$ $(CH_3)_2\underset{\underset{I}{|}}{C}CH_2C(CH_3)_3$

e. $(CH_3)_2C=CH_2$ $\xrightarrow{\text{60\% } H_2SO_4, \ 70^\circ}$

$\longrightarrow$ $CH_2=C(CH_3)CH_2C(CH_3)_3$ + $(CH_3)_2C=CHC(CH_3)_3$ $\longleftarrow$

$\xrightarrow[\text{2. Zn}]{\text{1. } O_3}$ $CH_3COCH_2C(CH_3)_3$ + lower boiling C1 - C5 carbonyl compounds

f. $(CH_3)_2C=CH_2$ + HBr $\xrightarrow[25^\circ]{\text{peroxides}}$ $(CH_3)_2CHCH_2Br$

g. $(CH_3)_2C=CH_2$ + $(CH_3)_3CH$ $\xrightarrow{\text{HF, } -25^\circ}$ $(CH_3)_2CHCH_2C(CH_3)_3$

$\xrightarrow[\text{or } SO_2Cl_2]{Cl_2, \ h\nu}$ $(CH_3)_2CHCH_2\overset{\displaystyle CH_3}{\underset{\displaystyle CH_3}{\overset{|}{\underset{|}{C}}}}-CH_2Cl$ + other monochlorination products

(about 35%)

h. $CH_3CH=CH_2$ $\xrightarrow{B_2H_6}$ $(CH_3CH_2CH_2)_3B$ $\xrightarrow[HO^\ominus]{H_2O_2}$ $CH_3CH_2CH_2OH$

i. $CH_2=C(CH_3)CH_2C(CH_3)_3$ $\xrightarrow[\text{2. } H_2O_2, \ HO^\ominus]{\text{1. } B_2H_6}$ $HOCH_2CH(CH_3)CH_2C(CH_3)_3$
(from part e)

j. $CH_2=C(CH_3)CH_2C(CH_3)_3$ $\xrightarrow[H_2O]{KMnO_4,}$ $HOCH_2C(OH)(CH_3)CH_2C(CH_3)_3$

13-19 $\langle\!\!\langle\ \rangle\!\!\rangle$—$CH_3$ $\xrightarrow[h\nu]{Br_2}$ $\langle\!\!\langle\ \rangle\!\!\rangle$—$CH_2Br$ + HBr

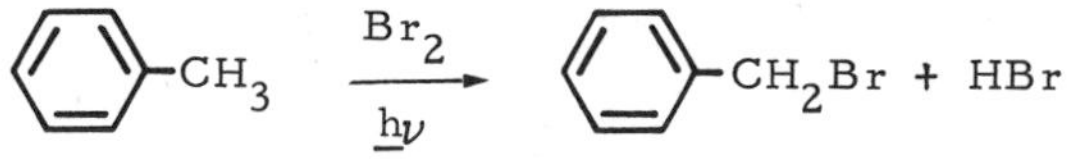

$HC\equiv CH$ $\xrightarrow[NH_3 (1), \ -33^\circ]{NaNH_2}$ $HC\equiv C:^\ominus$ $\xrightarrow[-Br^\ominus]{C_6H_5CH_2Br}$ $HC\equiv C-CH_2C_6H_5$ $\xrightarrow{NaNH_2}$

H_2, Pd—Pb $\longrightarrow$ $C_6H_5CH_2C\equiv CCH_2C_6H_5$ $\xleftarrow{C_6H_5CH_2Br}$ $^\ominus:C\equiv CCH_2C_6H_5$ $\longleftarrow$

$\longrightarrow$ cis-$C_6H_5CH_2CH=CHCH_2C_6H_5$ $\xrightarrow[\text{or } OsO_4, \ Na_2SO_3]{KMnO_4-H_2O, \ pH > 7}$

$\underset{H}{\overset{OH}{\underset{\displaystyle C_6H_5}{\overset{\displaystyle C_6H_5}{}}}}$ meso

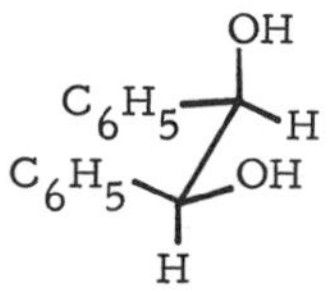

13-20

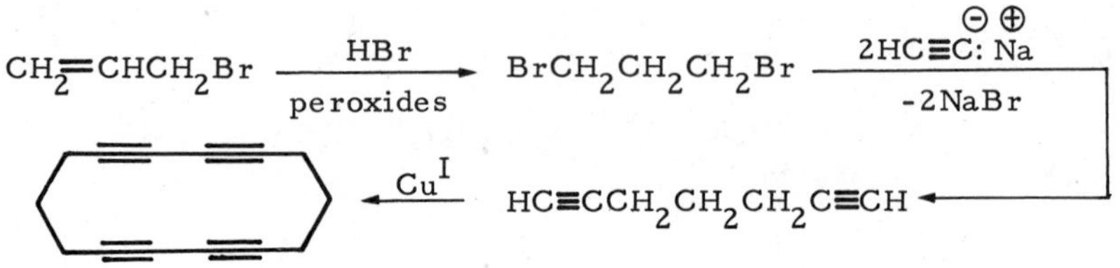

$$CH_2\!\!=\!\!CHCH_2Br \xrightarrow[\text{peroxides}]{HBr} BrCH_2CH_2CH_2Br \xrightarrow[-2NaBr]{2HC\equiv C:^{\ominus} Na^{\oplus}}$$

$$\xleftarrow{Cu^{I}} HC\equiv CCH_2CH_2CH_2C\equiv CH \longleftarrow$$

13-21 $$H_2C\!\!=\!\!CH_2 \xrightarrow{HOCl} HOCH_2CH_2Cl \xrightarrow{\overset{O}{\diagup}\diagdown, \ H^{\oplus}}$$

OCH_2CH_2Cl

1. $\overset{\oplus \ \ominus}{Na\ NH_2}$

2. OCH_2CH_2Cl

$OCH_2CH_2C\equiv CH$ $\xleftarrow[-NaCl]{HC\equiv C:^{\ominus}Na^{\oplus}}$

$OCH_2CH_2C\equiv CCH_2CH_2O$ $\xrightarrow[H_2O]{H^{\oplus}}$ $HOCH_2CH_2C\equiv CCH_2CH_2OH$

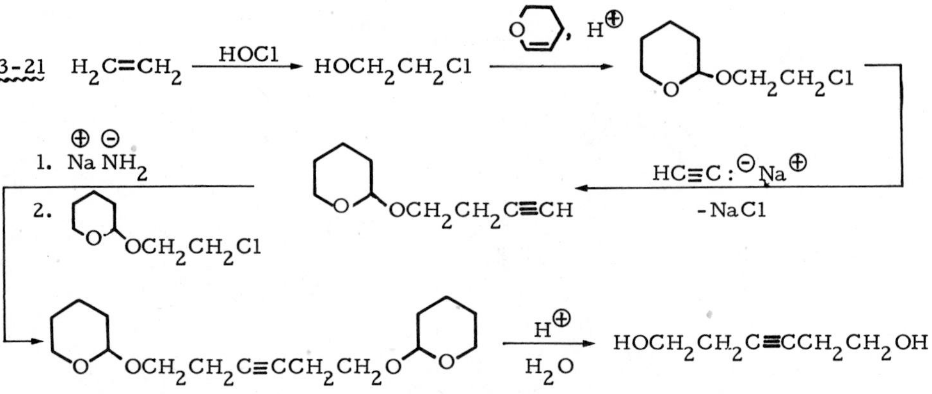

13-22   a.   The product would be very unfavorable to form because of strain inherent in the cumulated double bonds in a six-membered ring.

$$HC\equiv C\!-\!C\equiv CH \quad \xslashed{\longrightarrow} \quad HC\!\!=\!\!C\!\!=\!\!C\!\!=\!\!CH$$

b.   The product would be a cyclobutadiene derivative, and these are extremely unstable.

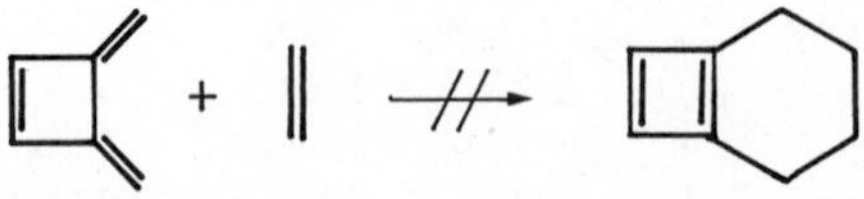

c.   The diene cannot achieve the cisoid conformation required for 1,4-cycloaddition.   Addition in the transoid manner would produce a highly strained product.

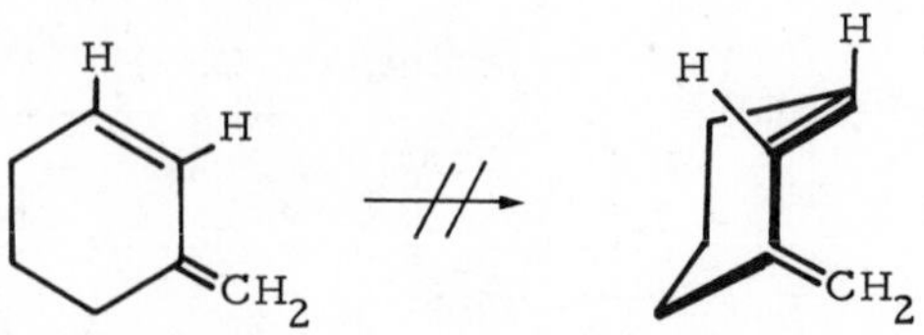

13-23 a. $CH_2=CH-CH=CH_2$ $\xrightarrow{\text{HBr, no peroxides}}$ separate from 1,4-isomer

$CH_2=CH-CH(OH)CH_3$ $\xrightarrow{\text{HBr (48\%)}}$ $CH_2=CHCHCH_3$

$CH_2=CHCH_2CH_3$ $\xrightarrow{Br_2,\ h\nu}$

(with $\underset{Br}{|}$ on product)

b.

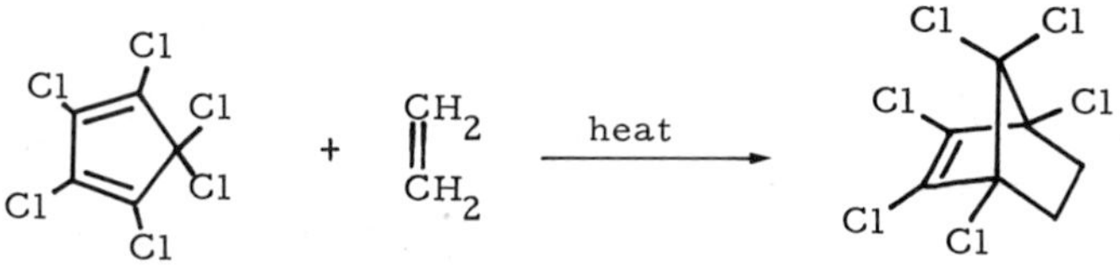

c. 1,3-cyclohexadiene + ethene $\xrightarrow{\text{heat}}$ bicyclo[2.2.2]-2-octene

d.

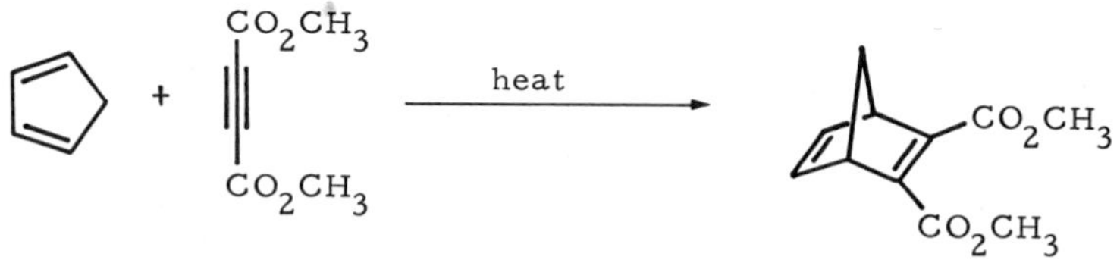

e. $CH_2=CH-CH=CH_2 + Cl_2C=CF_2$ $\xrightarrow{\text{heat}}$ $CH_2=CH-$ (cyclobutane with F, F, Cl, Cl)

f.

$\xrightarrow[\text{heat}]{\text{ROOR}}$

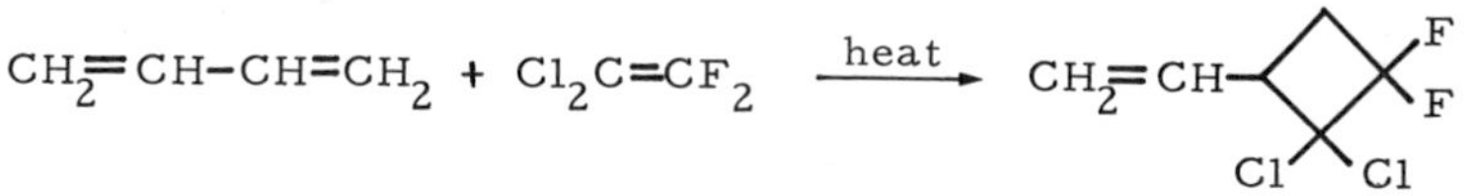

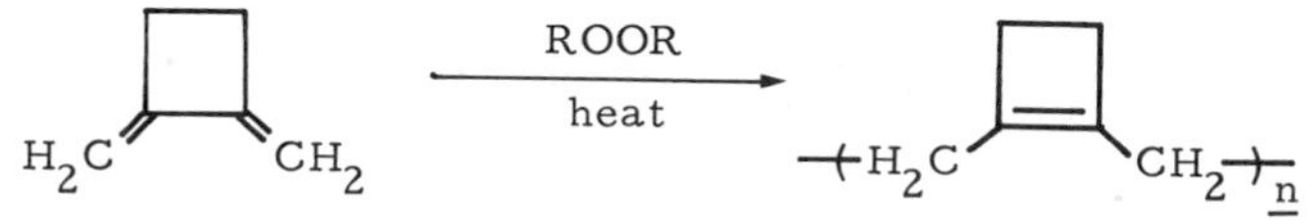

g. polymerization of 2-fluoro-1,3-butadiene

13-24 The 2-propenyl cation, $CH_2=CH-CH_2^{\oplus}$, is not formed from propadiene as easily as the isomer $CH_3-\overset{\oplus}{C}=CH_2$ even though the 2-propenyl cation is the more stable ion. The reason is that propadiene cannot form a planar 2-propenyl cation directly. Proton-transfer to the middle carbon initially gives an ion in which the two methylene groups at each end are in perpendicular planes. The full "allylic" resonance cannot be achieved until one group rotates into coplanarity with the other.

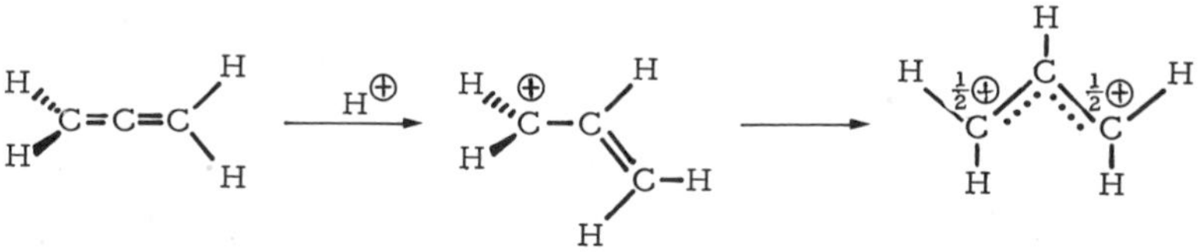

13-25

a.

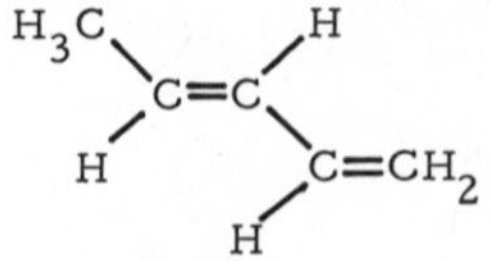

$$H_3C \quad H$$
$$C=C$$
$$H \quad C=CH_2$$
$$H$$

$$H_3C \quad CH=CH_2$$
$$C=C$$
$$H \quad H$$

trans 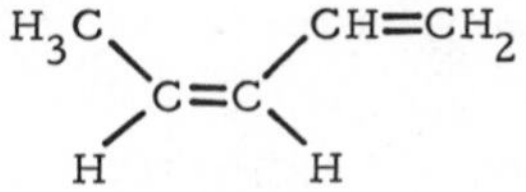   cis

b.

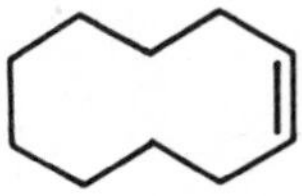

cis 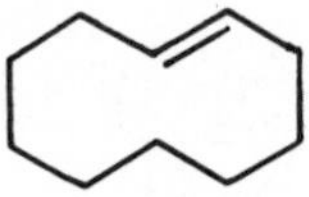   trans

c.

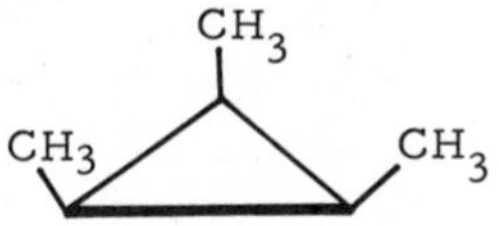

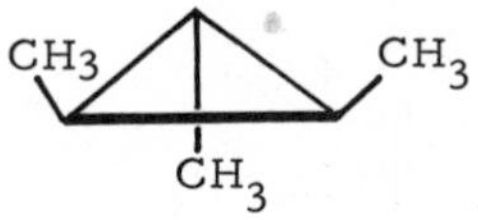

d.

trans, trans, trans 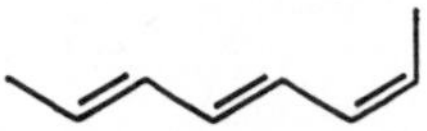   trans, trans, cis 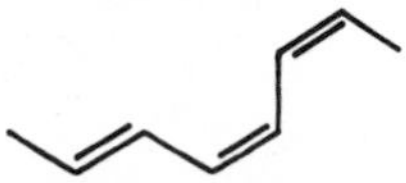   trans, cis, cis

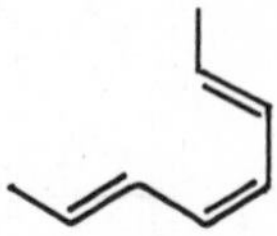

trans, cis, trans 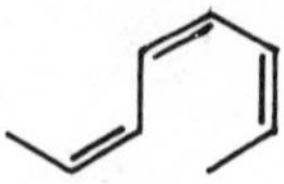   cis, cis, cis 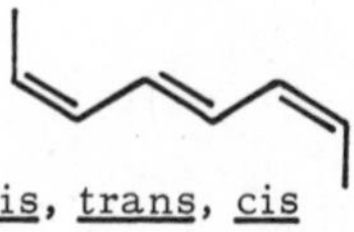   cis, trans, cis

e.

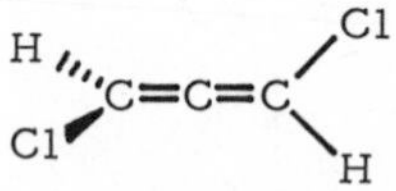

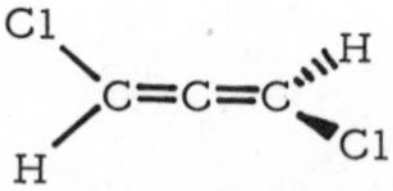

$$H \quad Cl$$
$$C=C=C$$
$$Cl \quad H$$

$$Cl \quad H$$
$$C=C=C$$
$$H \quad Cl$$

enantiomers

f.

$$H \quad H$$
$$C=C=C=C$$
$$Cl \quad Cl$$

$$Cl \quad H$$
$$C=C=C=C$$
$$H \quad Cl$$

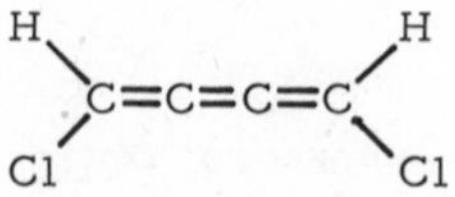

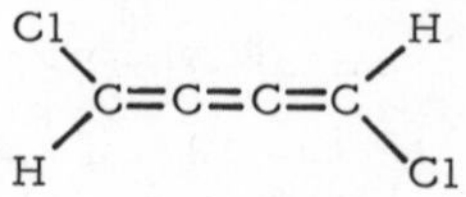

cis-trans
isomers

g.

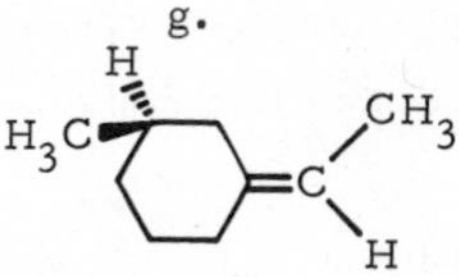

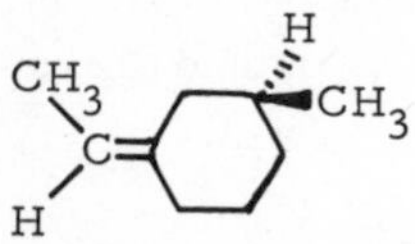

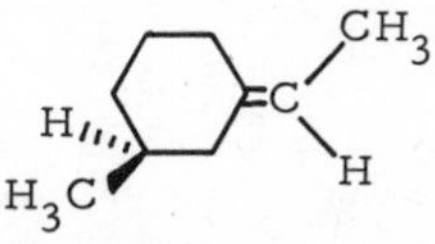

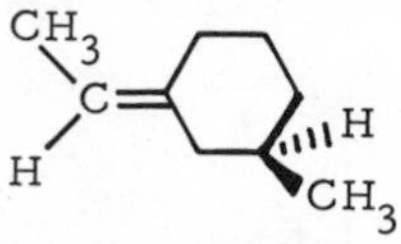

enantiomers   enantiomers

diastereomers

13-26

a. $CH_2=C=CH_2 + Cl-OH \longrightarrow \left[CH_2=\overset{\overset{\displaystyle OH}{|}}{C}-CH_2Cl \right] \longrightarrow CH_3-\overset{\overset{\displaystyle O}{||}}{C}-CH_2Cl$

(See Exercise 13-24 for an explanation of the orientation in the addition step.)

b. $CH_3CH=CH-CH=CH_2 + HCl \longrightarrow CH_3CH=CHCH(Cl)CH_3 +$

$CH_3CH_2CH(Cl)CH=CH_2 + CH_3CH(Cl)CH=CHCH_3 + CH_3CH_2CH=CHCH_2Cl$

c. $CH_2=CH-CH=CH_2 \xrightarrow{O_3}$ $\xrightarrow{Zn}$

$2CH_2O + H\overset{}{C}-\overset{}{C}H$ (with O O double-bonded)

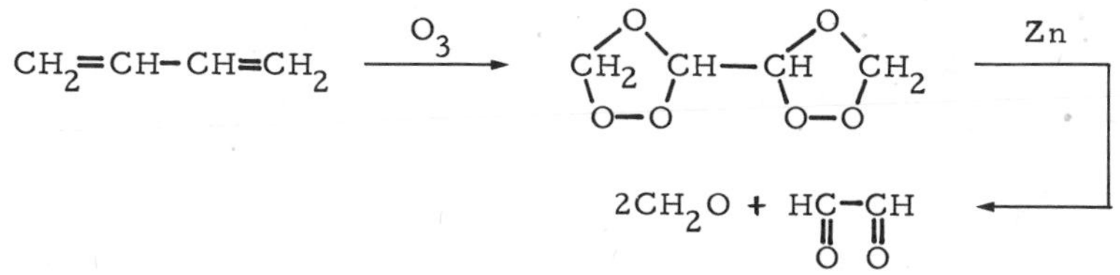

d.

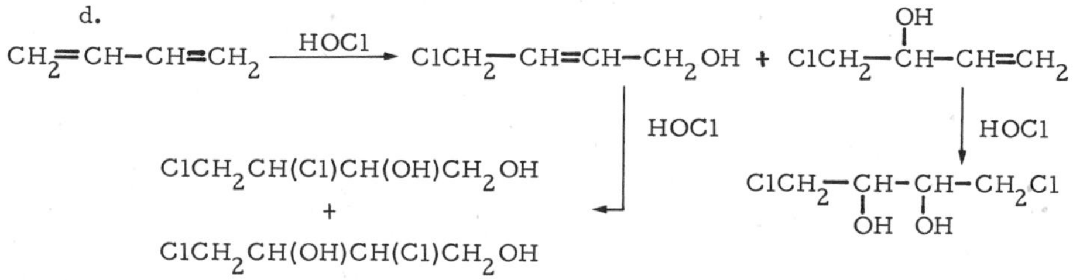

$CH_2=CH-CH=CH_2 \xrightarrow{HOCl} ClCH_2-CH=CH-CH_2OH + ClCH_2-\overset{\overset{\displaystyle OH}{|}}{CH}-CH=CH_2$

$\downarrow$ HOCl $\downarrow$ HOCl

$ClCH_2CH(Cl)CH(OH)CH_2OH$ $ClCH_2-\overset{}{CH}-\overset{}{CH}-CH_2Cl$ (OH OH)

$+$

$ClCH_2CH(OH)CH(Cl)CH_2OH$

e.

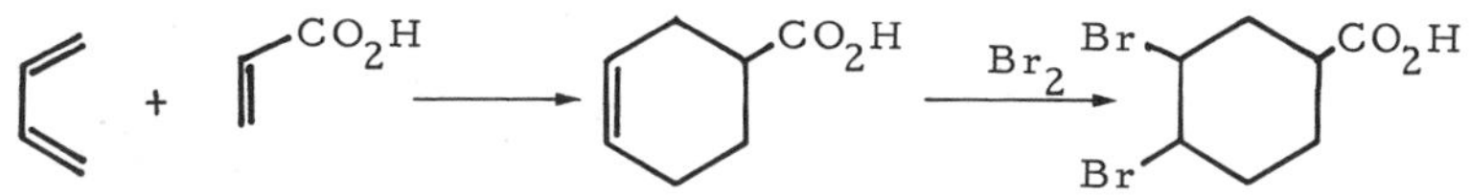

f. $CH_3CH=C=CHCH_3 \xrightarrow{ICl} CH_3CH=C\overset{\diagup I}{\diagdown CHCH_3}$ (Cl)   cis and trans

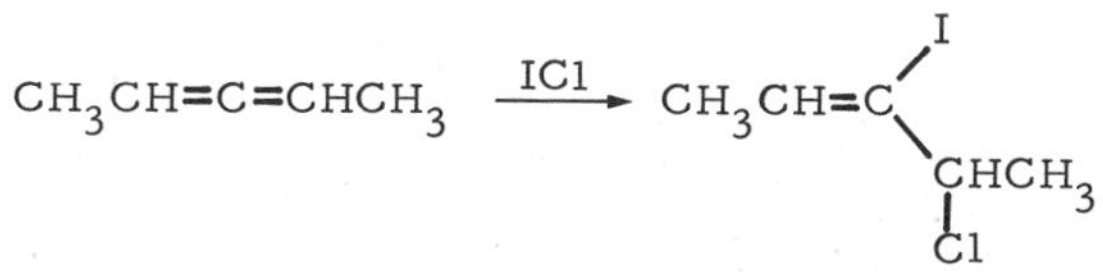

(In the case of substituted cumulenes, electrophiles add to the central carbon.)

13-27 a. chiral   b. achiral   c. achiral
       d. chiral   e. achiral   f. chiral
       g. chiral   h. chiral

13-28

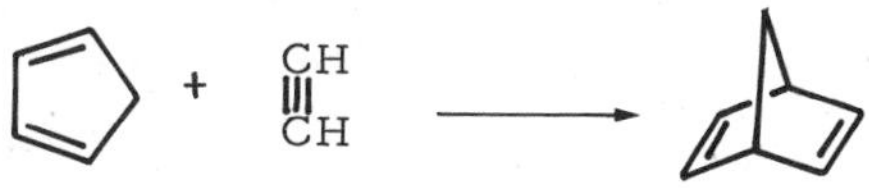

13-28 (cont.)

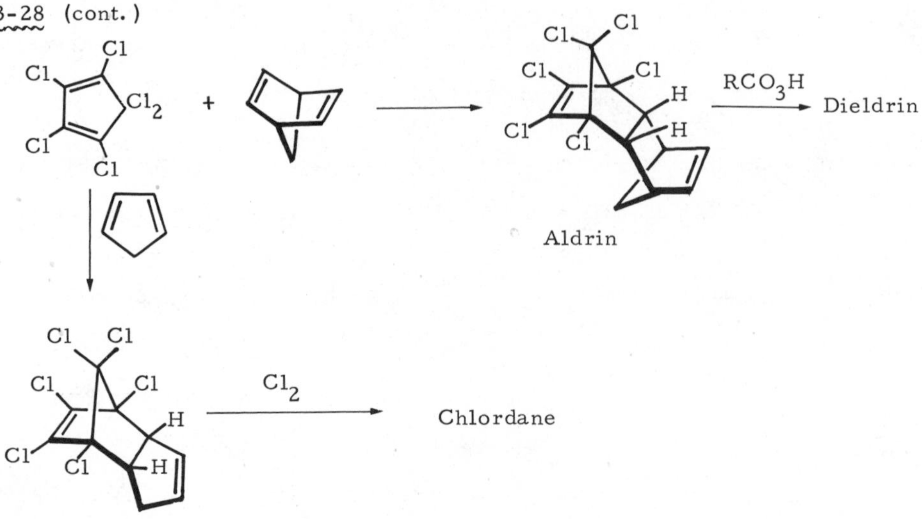

Aldrin

Chlordane

13-29

a.

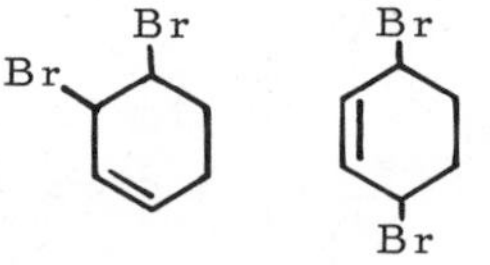

b.

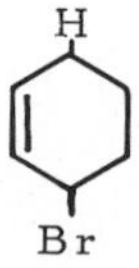

c. 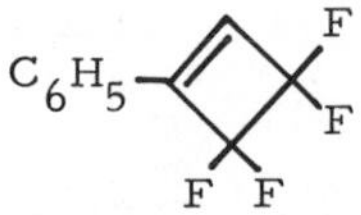 C_6H_5

d.

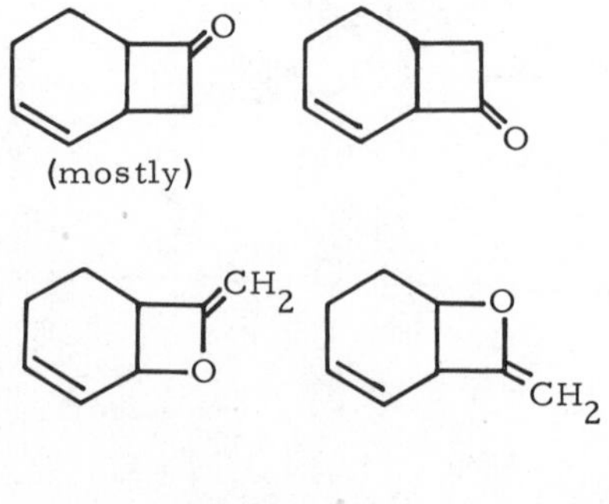

(mostly)

13-30   a.

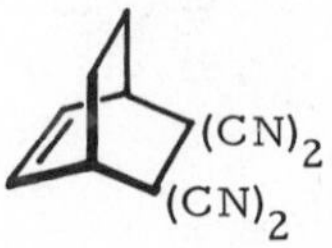

$(CN)_2$
$(CN)_2$

b.

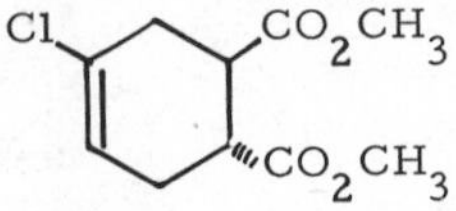

Cl CO_2CH_3

CO_2CH_3

c.

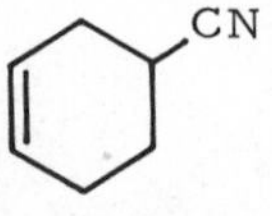

CN

d.

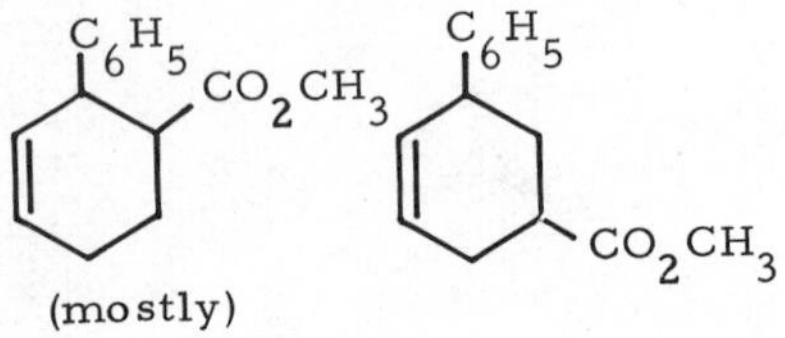

C_6H_5 CO_2CH_3

(mostly)

C_6H_5

CO_2CH_3

13-31   The orientation of addition in reactions under kinetic control always favors formation of the most stable intermediate.   The two possible intermediate cations in the addition of HCl to 3-butenyne are A and B.

$$HC{\equiv}C-CH{=}CH_2$$

$$H^{\oplus}$$

A

$$\overset{\oplus}{H_2C{=}C}-CH{=}CH_2$$

$$H_2C{=}C{=}CH-\overset{\oplus}{CH_2}$$

$$Cl^{\ominus}$$

$$CH_2{=}C(Cl)-CH{=}CH_2$$

B

$$HC{\equiv}C-\overset{\oplus}{CH}-CH_3$$

$$HC{=}\overset{\oplus}{C}{=}CH-CH_3$$

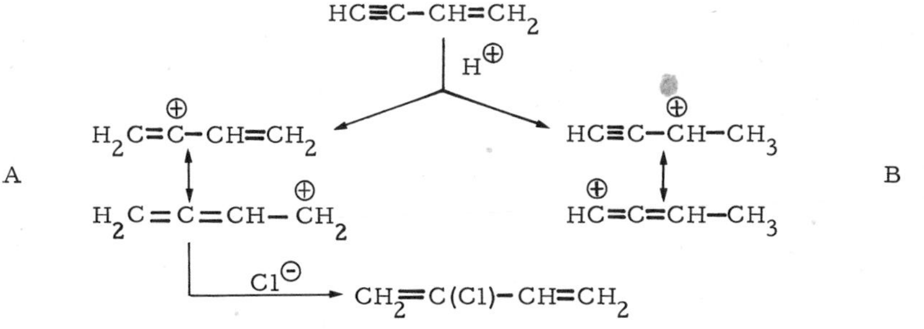

Ion A (with only one sp-hybrid carbon) is more stable than ion B (with two sp-hybrid carbons) and, accordingly, A leads to the observed product. You may recall (Section 10-5) that a likely reason why $-\overset{\oplus}{\underset{|}{C}}$ ions are more stable than $={\overset{\oplus}{C}}$ ions is because the optimum geometry for a carbocation is planar and trigonal, which can be achieved only with sp^2 hybridization at carbon.

13-32

$$R-CH_2-\overset{\overset{\textstyle Cl}{|}}{\underset{\textstyle \cdot}{C}}-CH{=}CH_2 \quad\longleftrightarrow\quad R-CH_2-\overset{\overset{\textstyle Cl}{|}}{C}{=}CH-\overset{\textstyle \cdot}{CH}_2$$

$$CH_2{=}C(Cl)CH{=}CH_2$$

$$RCH_2-\overset{\overset{\textstyle Cl}{|}}{\underset{\overset{\textstyle CH}{\underset{\textstyle \|}{CH_2}}}{C}}-CH_2-\overset{\overset{\textstyle Cl}{|}}{\underset{\overset{\textstyle CH}{\underset{\textstyle \|}{CH_2}}}{C}}\cdot$$

(1, 2)

or

$$R-CH_2\,\overset{\overset{\textstyle Cl}{|}}{\underset{\textstyle C}{\diagup}}{}\diagdown C-CH_2-CH_2-\overset{\cdot}{\underset{\overset{\textstyle |}{\textstyle Cl}}{C}}-CH{=}CH_2$$

$$\underset{\textstyle H}{}$$

(trans-1, 4)

$$R-CH_2\,\diagup\overset{\overset{\textstyle Cl}{|}}{C}\diagdown\overset{\textstyle H}{} \\ CH_2-CH_2-\overset{\cdot}{\underset{\overset{\textstyle |}{\textstyle Cl}}{C}}-CH{=}CH_2$$

(cis-1, 4)

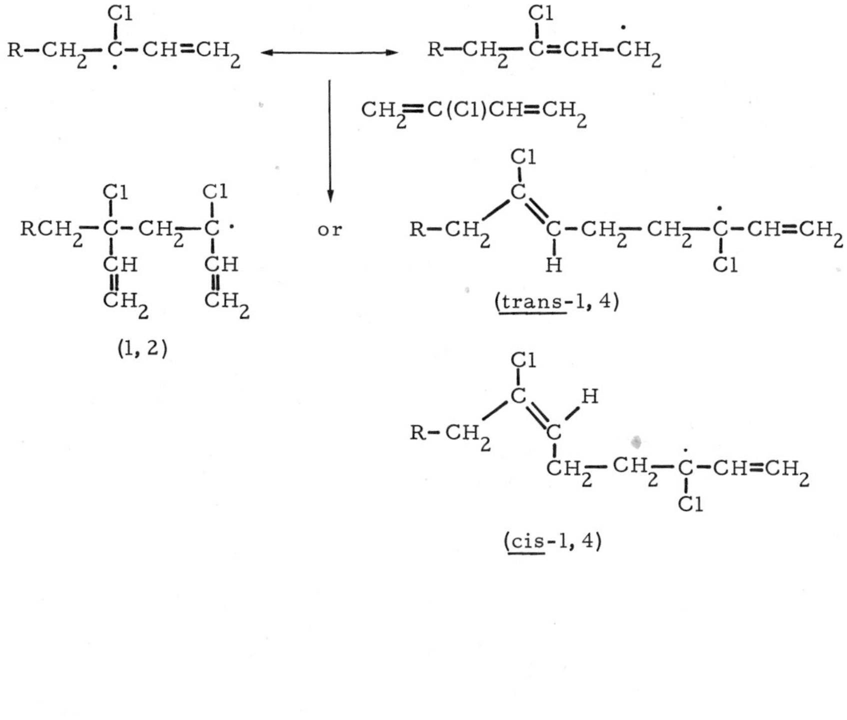

Also, similar 1, 2 and 1, 4 addition but by initial addition at the C 4 end instead of C 1:

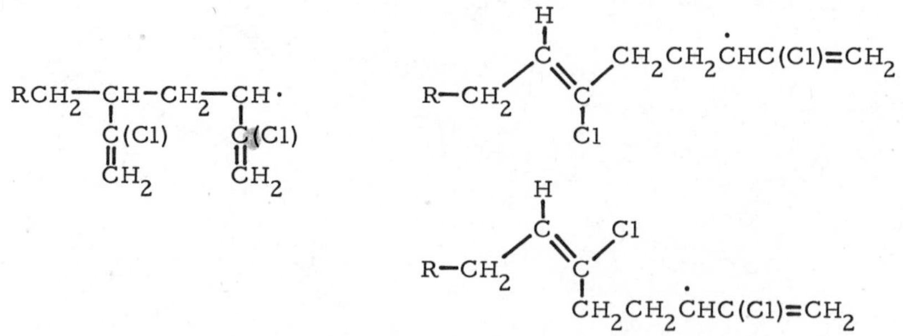

13-33

a. $CH_3CH_2C\equiv CCH_2CH_3$ $\xrightarrow{Na^{\oplus} \; {}^{\ominus}NH_2}$ $CH_3CH_2CH_2CH_2C\equiv C: {}^{\ominus} \; Na^{\oplus}$

(See Exercise 13-13)

b.

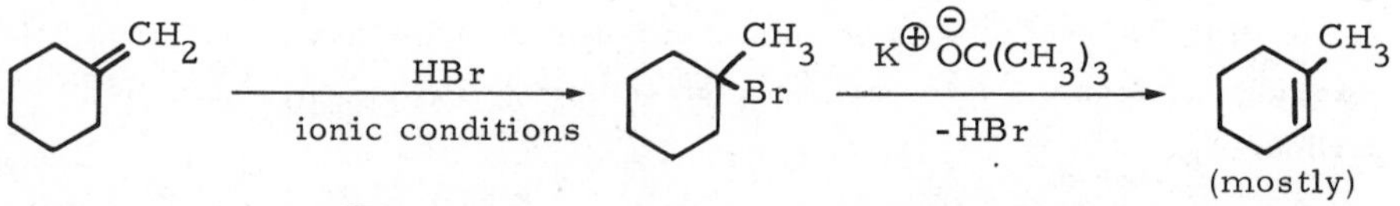

c.

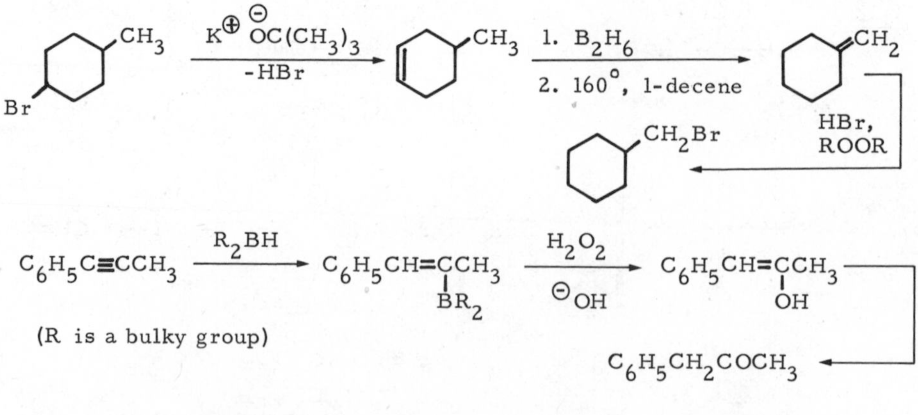

d. $C_6H_5C\equiv CCH_3$ $\xrightarrow{R_2BH}$ $C_6H_5CH=\underset{\underset{BR_2}{|}}{C}CH_3$ $\xrightarrow[{}^{\ominus}OH]{H_2O_2}$ $C_6H_5CH=\underset{\underset{OH}{|}}{C}CH_3$

$C_6H_5CH_2COCH_3$

(R is a bulky group)

e. $CH_3C\equiv CCH_3$ $\xrightarrow{R_2BH}$ $CH_3CH=\underset{\underset{BR_2}{|}}{C}CH_3$ $\xrightarrow{CH_3CH_2CO_2D}$ $CH_3CH=CDCH_3$

(R is a bulky group)   (cis)   (cis)

13-34 a.

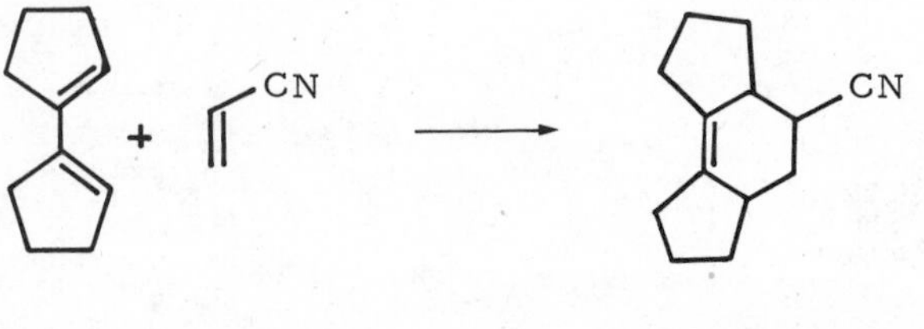

b.

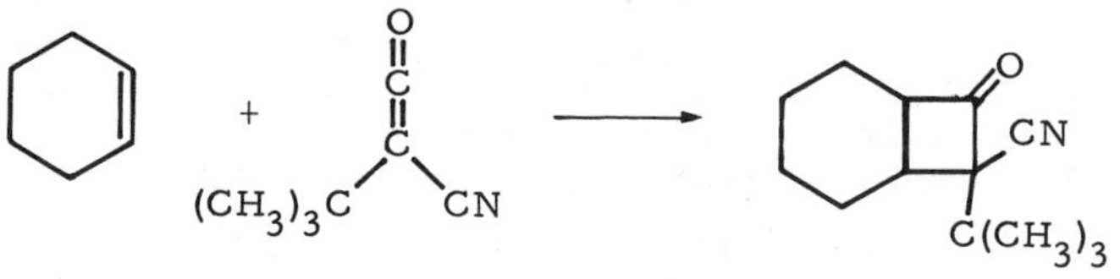

c.

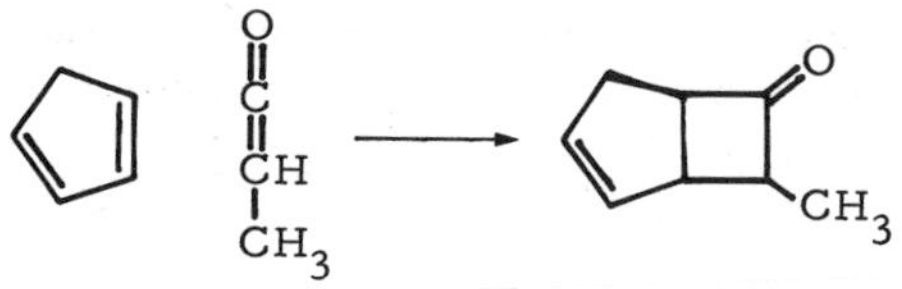

d.

$(R = CO_2H)$

13-35

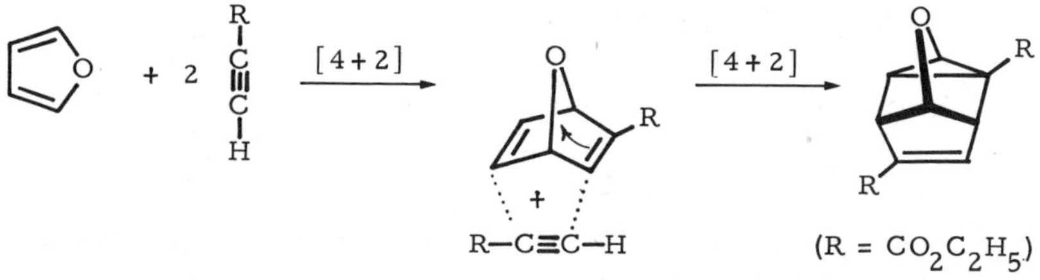

$(R = CO_2C_2H_5)$

13-36

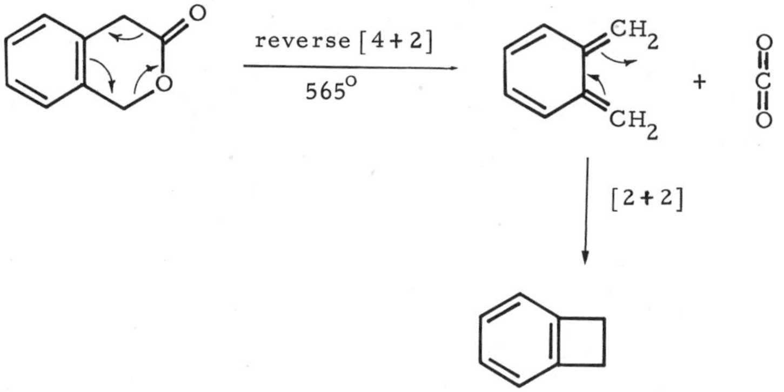

ADDITIONAL EXERCISES

(The following Exercises are designed to offer practice in devising synthetic routes using reactions discussed in Chapters 4, 8, 10, 11 and 13).

13A-1   Specify the reagents, catalysts, and reaction conditions that would be required to effect the transformation of benzene to cyclohexane, cyclohexene, and 1,3-cyclohexadiene. Note that more than one step may be required. (See Exercise 11A-1 for further practice in this type of problem).

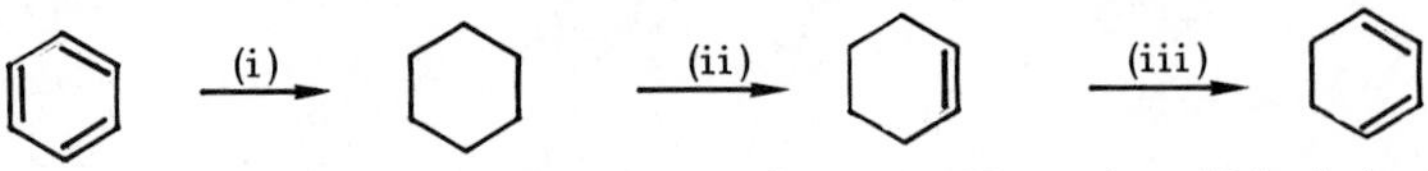

13A-2   Using any of the hydrocarbons shown in Exercise 13A-1 devise a synthesis of each compound shown below. Assume that any needed inorganic reagents are available, and that organic compounds of three carbons or less are also available.

a. $O=CH(CH_2)_4CH=O$

b. ![cyclohexane]-CN

c. ![cyclohexane]-OCH$_3$

d. ![cyclohexane]-SCH$_3$

e. ![bicyclic structure] with H and CN

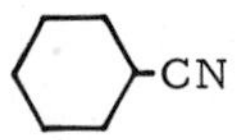

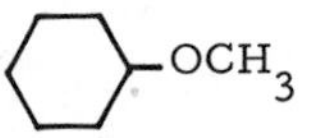

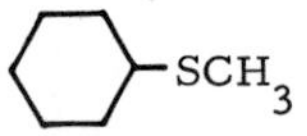

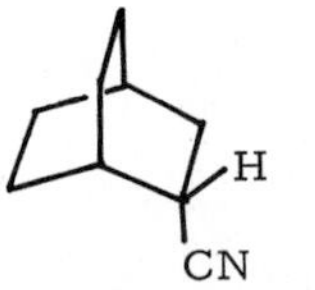

13A-3   Devise a synthesis of each compound shown starting with methylbenzene (toluene) and any other needed reagents. Specify reaction conditions and catalysts as closely as possible.

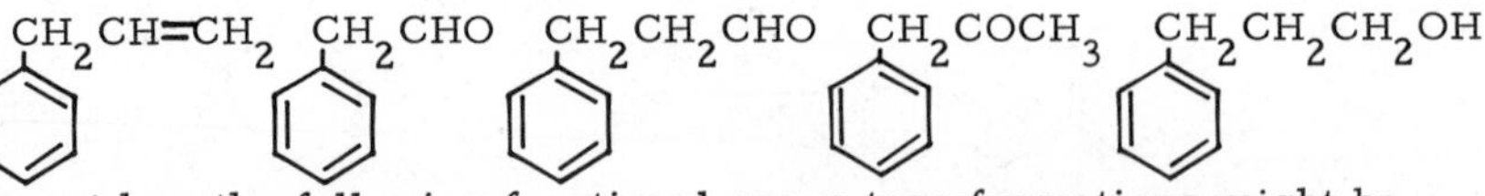

13A-4   Suggest how the following functional group transformations might be accomplished.

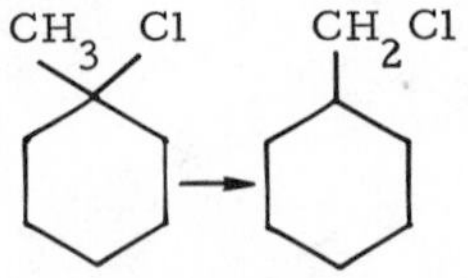

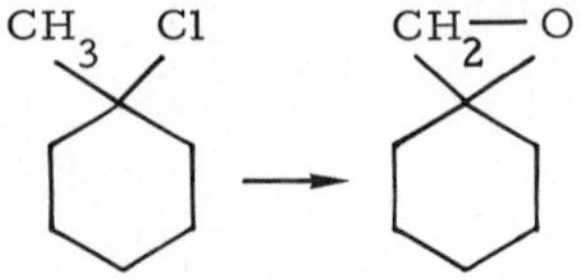

c.   2-butyne to meso-2, 3-butanediol

d.   2-butyne to 2L, 3D, -3-chloro-2-butanol and enantiomer

ORGANOHALOGEN AND ORGANOMETALLIC COMPOUNDS

The types of halogen compounds described in Chapter 14 include the following:

$$-\overset{|}{\underset{|}{C}}-X \left\{ \begin{array}{lll} RCH_2X & \underline{primary} & \text{alkyl halides} \\ R_2CHX & \underline{secondary} & \text{alkyl halides} \\ R_3CX & \underline{tertiary} & \text{alkyl halides} \\ RCH{=}CHCH_2X & & \text{allylic halides} \\ \text{⟨⟩-CH}_2X & & \text{benzylic halides} \end{array} \right.$$

$$\overset{\backslash\backslash}{\underset{/}{C}}-X \quad {\equiv}C{-}X \left\{ \begin{array}{ll} RCH{=}CHX & \text{alkenyl halides} \\ RC{\equiv}CX & \text{alkynyl halides} \\ \text{⟨⟩-X} & \text{aryl halides} \end{array} \right.$$

$$X_2C{=}CX_2, \ CX_4 \qquad \text{polyhalocarbons}$$

The behavior and methods of synthesis of these substances depend on the halogen (F, Cl, Br, or I) and on the structure of the organic group. Alkyl halides are prepared by Reactions 1-9 listed in Table 14-5 in the text. Of these, radical halogenation of alkenes is well suited for the preparation of allylic halides (Section 14-3A):

$$CH_2{=}CHCH_3 + Cl_2 \xrightarrow{400^0} CH_2{=}CHCH_2Cl + HCl$$

Benzylic halides are similarly prepared from alkylbenzenes:

$$\text{⟨⟩-CH}_3 + Br_2 \xrightarrow{h\nu} \text{⟨⟩-CH}_2Br + HBr$$

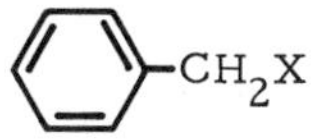

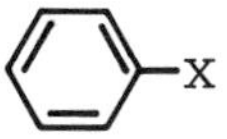

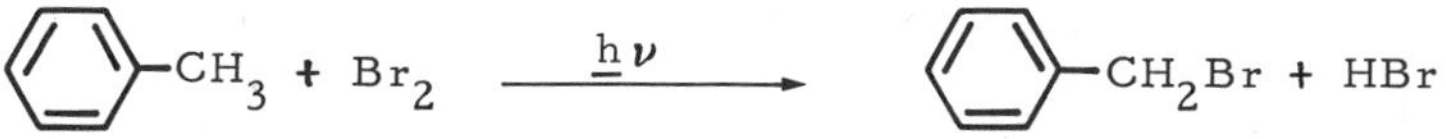

The allylic (or benzylic) hydrogen is removed selectively in radical halogenation of alkenes because less energy is required to break this bond than any of the other C—H bonds. The intermediate is a delocalized radical that reacts to form the allylic halide by abstracting a halogen atom from the halogenating reagent. The propagation steps in the chlorination of propene are

$$CH_2{=}CH{-}CH_3 + {\cdot}Cl \longrightarrow \overset{\frac{1}{2}\cdot}{CH_2}{\cdots}CH{\cdots}\overset{\frac{1}{2}\cdot}{CH_2} + HCl$$

$$\overset{\frac{1}{2}\cdot}{CH_2}{\cdots}CH{\cdots}\overset{\frac{1}{2}\cdot}{CH_2} + Cl_2 \longrightarrow CH_2{=}CH{-}CH_2{-}Cl + Cl{\cdot}$$

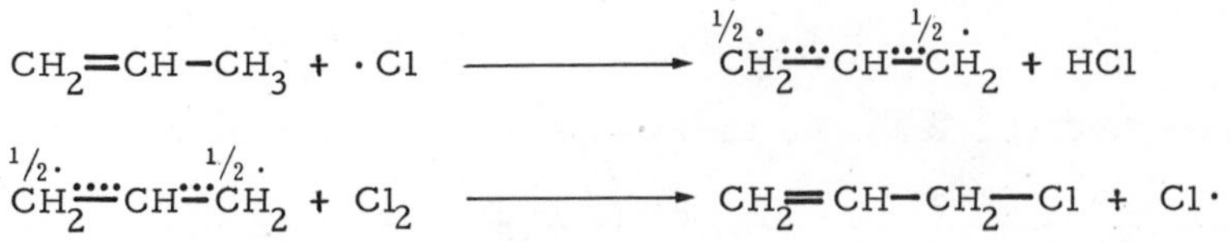

Most of the methods by which alkyl carbon-halogen bonds are made are unsatisfactory for making carbon-halogen bonds of the type $={C}{-}X$, $\equiv{C}{-}X$, or Ar—X (see Reactions 10-17 in Table 14-5).

The most important reactions of organic halides that directly involve the C—X bonds are nucleophilic displacement and elimination reactions and formation of organometallic compounds.

The behavior of various kinds of organohalogen compounds toward nucleophilic and basic reagents varies markedly with structure as summarized in Table 14-1S.

Table 14-1S

Reactivities of Organohalogen Compounds
in Displacement and Elimination Reactions

Organic halide	Reactivity		
	S_N2	E2	S_N1, E1
prim-alkyl, RCH_2X	good	fair	very poor
sec-alkyl, R_2CHX	fair	fair	fair
tert-alkyl, R_3CX	poor	good	good
allylic, $RCH{=}CHCH_2X$	good	----	good
benzylic, ⟨⟩—CH_2X	good	----	good
alkenyl, $RCH{=}CHX$	poor	fair	poor
alkynyl, $RC{\equiv}CX$	poor	----	poor
aryl, ⟨⟩—X	very poor	----	poor

Allylic and benzylic halides have high S_N reactivity, whereas alkenyl, alkynyl, and aryl halides have low S_N reactivity. For the usual S_N mechanisms, iodides are most reactive and fluorides are least reactive.

The S_N1 reactivity of allylic and benzylic halides is the result of

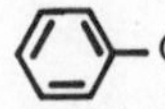

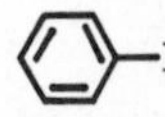

ionization to carbocations that are stabilized by electron delocalization. With
allylic halides these cations can lead to mixtures of isomeric products in
solvolysis reactions:

$$(CH_3)_2C{=}CHCH_2Cl \longrightarrow \left[(CH_3)_2C{=}CH{-}\overset{\oplus}{CH_2} \longleftrightarrow (CH_3)_2\overset{\oplus}{C}{-}CH{=}CH_2\right]$$

$$(CH_3)_2C{=}CH{-}CH_2OH + (CH_3)_2\underset{OH}{C}{-}CH{=}CH_2$$

(via H_2O, $-H^{\oplus}$)

   Aryl halides react with nucleophiles under rather specific conditions.
The halogen is displaced readily provided the ring carries groups such as
NO_2, CN, NO, $N_2^{\oplus}$, and other strongly electron-attracting groups. These
groups function to facilitate halogen displacement when in the ortho (2-) or
para (4-) or both positions relative to the halogen. The mechanism of displacement is stepwise and involves addition of the nucleophile to the ring carbon
carrying the halogen. Subsequent elimination of halide ion gives the substitution product (or its conjugate acid). This is an example of an addition-
elimination mechanism.

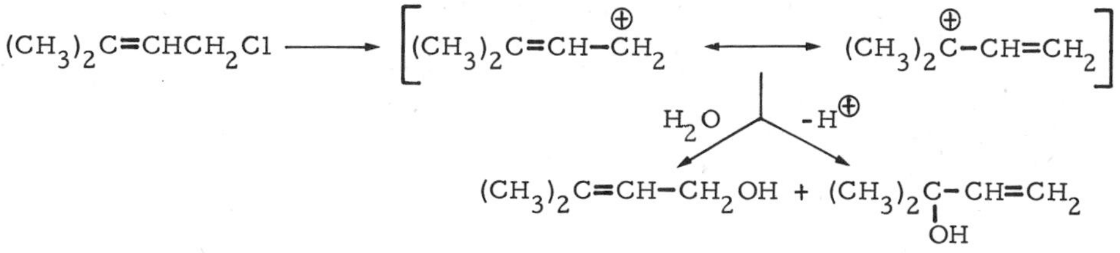

Addition-elimination mechanism of substitution

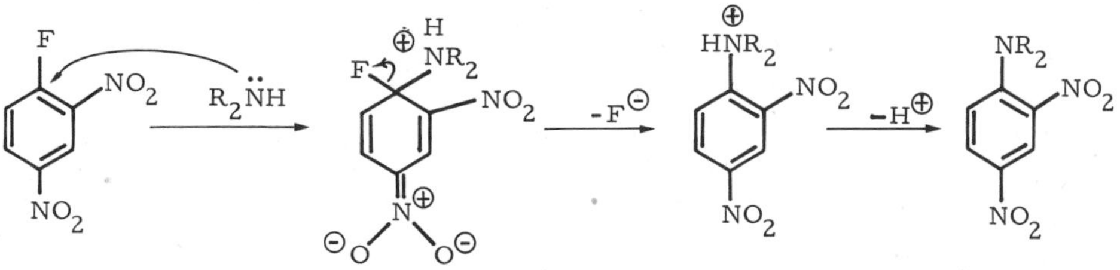

The electron-pair-accepting groups stabilize the intermediate through
substantial contributions of valence-bond structures such as 15, which of course
is only one of several contributing structures.
   Substitution reactions of aryl halides occur when there are no activating
substituents present, provided that the nucleophile also is a very strong base.
The weaker the base the higher the temperature required to effect reaction.
The base causes an E2 reaction by way of an intermediate carbanion, which
eliminates halide ion. However, the product of elimination is a highly reactive
intermediate (aryne), which rapidly adds available nucleophiles. This is an

example of an _elimination-addition_ mechanism.

Elimination-addition mechanism of substitution

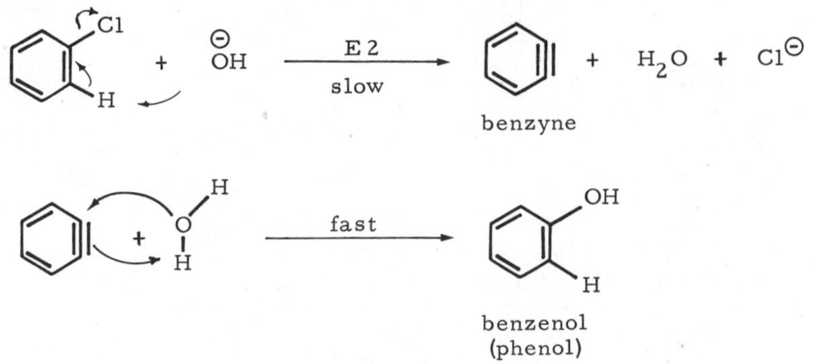

benzyne

benzenol
(phenol)

The most commonly used methods of forming aryne intermediates include the following:

1. _aryl halide with metal amides_

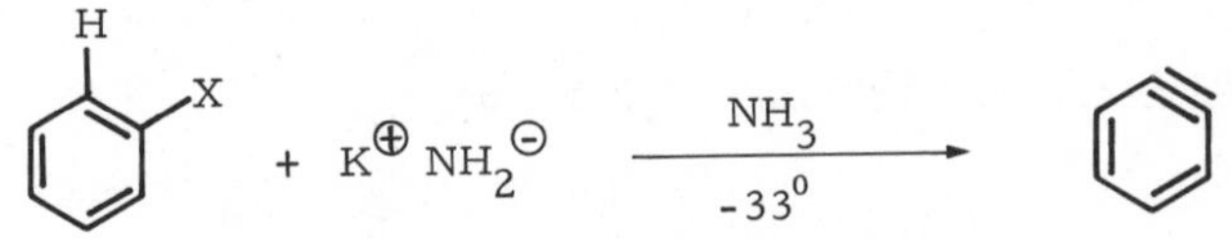

X = Cl, Br, or I

2. _from 2-haloarylmagnesium or aryllithium derivatives_

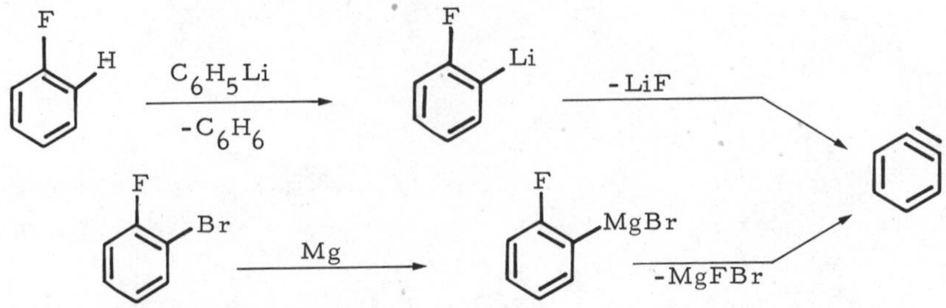

3. _thermal decomposition reactions_

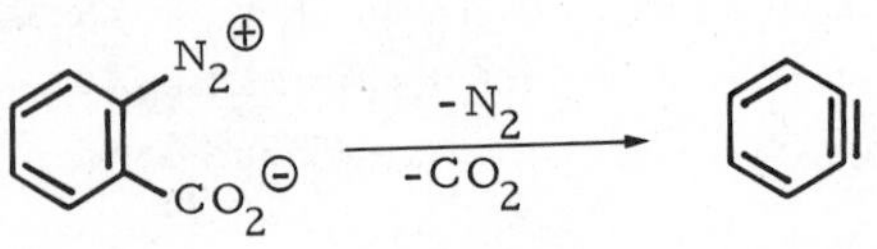

Aryne intermediates are particularly useful for the synthesis of polycyclic compounds by cycloaddition reactions. When arynes are intermediates in substitution reactions, the possibility of formation of rearrangement products always must be considered:

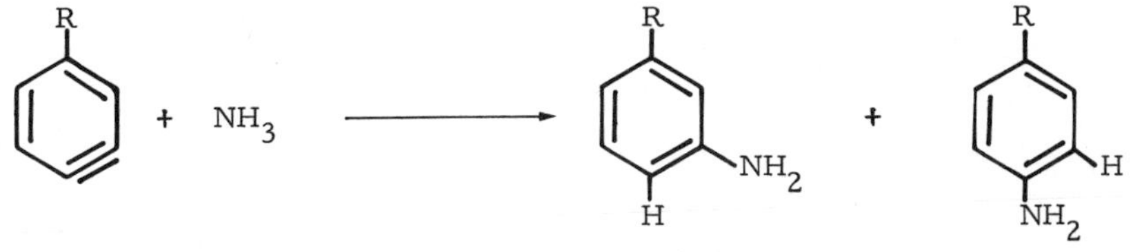

Polyhalogen compounds of alkanes and alkenes have many practical uses as solvents, polymer intermediates, refrigerants, fire retardants, propellants, and so on. The fluorocarbons require special preparative methods because the methods of Table 14-5 usually are not satisfactory. Chlorofluorocarbons are made most often from chloroalkanes using antimony halides:

$CHCl_3 + SbF_3 \longrightarrow CHF_2Cl \xrightarrow{700-900^0} \underline{n}(CF_2\!\!=\!\!CF_2) \longrightarrow \underline{(CF_2\!\!-\!\!CF_2)}_{\underline{n}}$

Teflon

Polyhalomethanes are less reactive in S_N displacement reactions than alkyl halides, but they react by 1,1- or α-elimination in the presence of strong bases, organolithium reagents, and with some active metals. The products are unstable carbene intermediates (Table 14-2):

1. $Br_3CH + {}^{\ominus}OC(CH_3)_3 \longrightarrow Br_3C{:}{}^{\ominus} + HOC(CH_3)_3$

   $Br_3C{:}{}^{\ominus} \longrightarrow Br_2C{:} + Br^{\ominus}$ (dibromocarbene)

2. $CH_2Cl_2 + CH_3Li \xrightarrow{-CH_4} \underset{\displaystyle Cl}{\overset{\displaystyle Li}{CHCl}} \xrightarrow{-LiCl} {:}CHCl$

   (chlorocarbene)

3. $CH_2I_2 + Zn \xrightarrow{Cu} ZnI_2 + {:}CH_2$ (methylene)

Carbenes are like carbocations in being powerful electrophilic species.   They
react rapidly with even very weak nucleophiles, and they also can rearrange
to more stable substances.   Their value in synthesis is for the preparation of
cyclopropanes and of products derived therefrom:

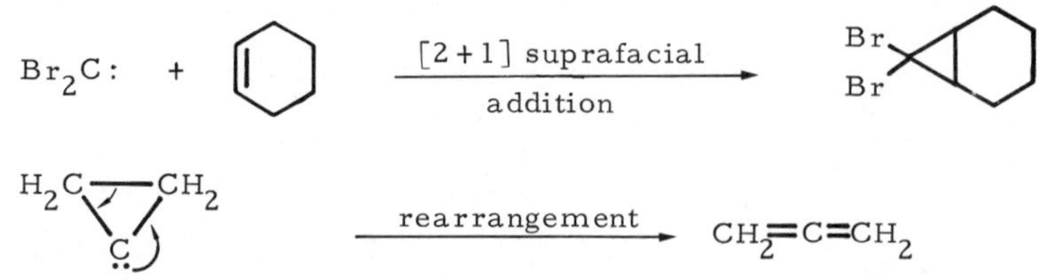

$$Br_2C: \; + \quad \bigcirc \quad \xrightarrow[\text{addition}]{[2+1] \text{ suprafacial}} \quad \overset{Br}{\underset{Br}{\rlap{\diagdown}}} \triangleleft\!\!\bigcirc$$

cyclopropylidene $\qquad \xrightarrow{\text{rearrangement}} \quad CH_2{=}C{=}CH_2$

cyclopropylidene $\qquad\qquad\qquad\qquad\qquad$ 1, 2-propadiene

Organometallic compounds are conveniently prepared directly or
indirectly from organohalogen compounds.   A summary of the most useful
methods is given in Table 14-7 in the text.

Organometallic reagents of electropositive metals (Li, Na, K, Mg) are
extremely reactive.   They seldom are isolated or stored for long periods, but
are prepared as they are needed.   The carbon metal bonds are partially ionic
and react (often violently) to transfer carbon as a nucleophile to electrophiles
such as acids, halogens, and carbonyl compounds:

$$\overset{\delta\ominus}{R}{-}\overset{\delta\oplus}{MgX} + \overset{\delta\oplus}{H}{-}\overset{\delta\ominus}{OH} \longrightarrow RH + MgXOH$$

$$\overset{\delta\ominus}{R}{-}\overset{\delta\oplus}{Li} + Br{-}Br \longrightarrow RBr + LiBr$$

$$\overset{\delta\ominus}{R}{-}\overset{\delta\oplus}{Li} + \overset{\delta\oplus}{CH_2}{=}\overset{\delta\ominus}{O} \longrightarrow RCH_2{-}OLi$$

Organomagnesium compounds (Grignard reagents) and organolithium
compounds are used extensively in organic synthesis.   Their most important
reactions are additions to the polar double bonds of carbonyl groups:

$$\overset{\delta\ominus}{R}{-}\overset{\delta\oplus}{MgX} + \overset{\delta\oplus}{\underset{\diagup}{\diagdown}C}{=}\overset{\delta\ominus}{O} \longrightarrow R{-}\overset{|}{\underset{|}{C}}{-}OMgX \xrightarrow{H^{\oplus}} R{-}\overset{|}{\underset{|}{C}}{-}OH$$

The C—O—metal compounds so formed must be hydrolyzed, preferably with a
weak acid such as $NH_4^{\oplus}$, to release the organic product.   Depending on the
nature of the carbonyl compound, these reactions provide synthetic routes to

alcohols, ketones, and carboxylic acids:

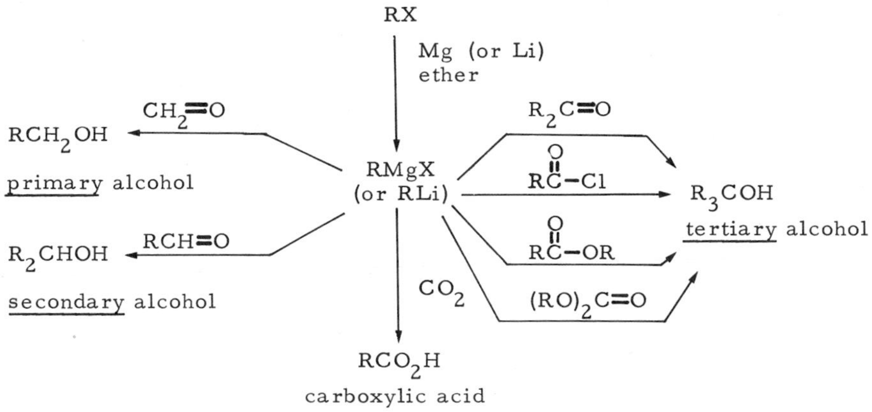

Ketones are prepared best from acid chlorides and <u>selective</u> reagents such as RCdCl and RCu. Addition of alkyllithiums to carboxylic acids also gives ketones:

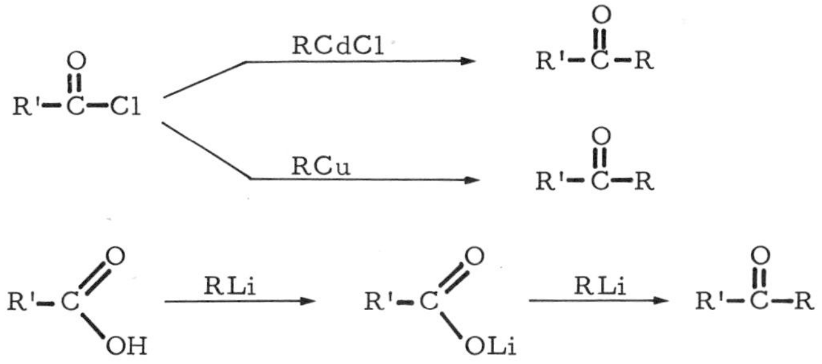

Highly branched alkyl groups on the Grignard reagent or the carbonyl component often lead to poor yields of adducts because of competing reactions (enolization and reduction, see Section 14-12A). This complication can be largely avoided by using the corresponding lithium reagents, which are more reactive and usually give good yields of adducts.

Another synthetic use of organometallic compounds is in 1, 4 addition to conjugated unsaturated carbonyl compounds. Grignard reagents may give mixtures of products; organocopper compounds tend to give exclusively 1, 4 addition.

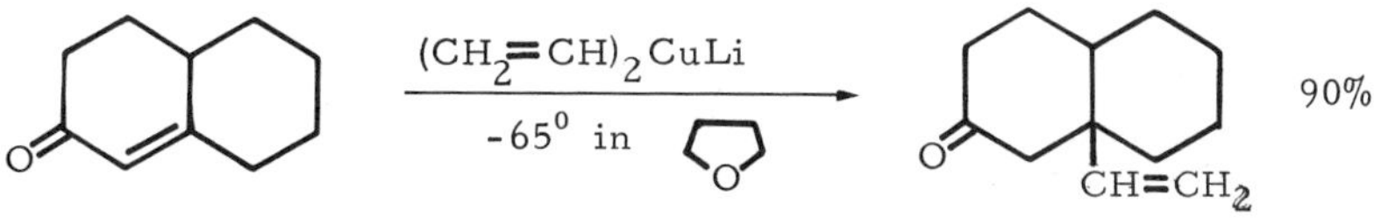

Grignard reagents (RMgX) react with oxygen, sulfur, and halogens to give ROH, RSH, and RX, respectively.

ANSWERS TO EXERCISES

14-1 C_4H_7Br

Infrared	band position cm^{-1}	structural inference
	3000 (w)	$\equiv C-H$
	1630 (m)	$C=C$
	940 (s)	$=CH_2$

Nmr	δ ppm	area	$\underline{J}$ Hz	mult.	structural inference
	1.8	3	small	complex	$CH_3C=$
	3.9	2	-	singlet	$BrCH_2-$
	4.8	1	small	complex	$-CH=$
	5.0	1	small	complex	$-CH=$

Structure:

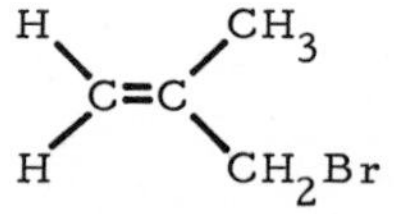

2-methyl-3-bromopropene

$C_5H_8Br_2$ Infrared indicates $-\overset{\mid}{\underset{\mid}{C}}-H$ from band at 2900 cm^{-1}. No other functional groups are evident from spectrum.

Nmr	δ ppm	area	mult.	structure
	0.8	4	singlet	$(CH_2)_2$
	3.5	4	singlet	$(CH_2Br)_2$

Structure:

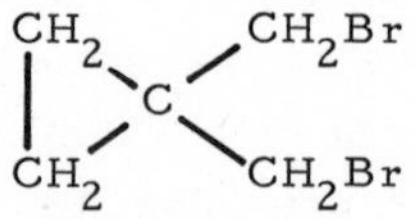

1,1-dibromomethyl-cyclopropane

14-2 a. The reaction is:

$$2K^{\oplus}I^{\ominus} + (CH_3O)_2SO_2 \longrightarrow 2CH_3I + 2K^{\oplus} + SO_4^{2\ominus}$$

and occurs readily because dimethyl sulfate has a good leaving group (Section 8-7C):

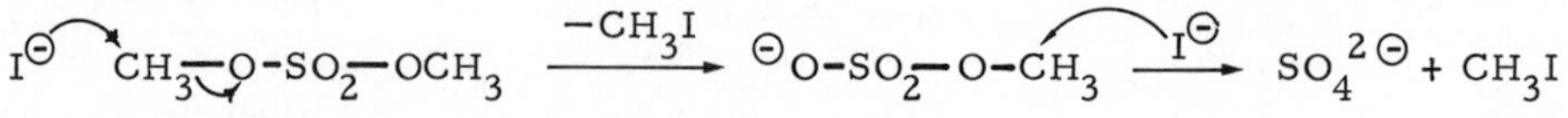

$$I^{\ominus} \quad CH_3-O-SO_2-OCH_3 \xrightarrow{-CH_3I} {}^{\ominus}O-SO_2-O-CH_3 \xrightarrow{I^{\ominus}} SO_4^{2\ominus} + CH_3I$$

Methanol does not react with potassium iodide because $^{\ominus}$OH is a poor leaving group.

b. The sulfuric acid protonates the alcohol and converts a poor leaving group ($\overset{\ominus}{O}$H) into a good leaving group (OH_2).

$$ROH \quad \xrightarrow[\substack{-HSO_4^{\ominus}}]{H_2SO_4} \quad ROH_2^{\oplus} \quad \xrightarrow[\substack{-H_2O}]{Br^{\ominus}} \quad RBr$$

c. The reaction would be

$$CH_3CH_2OH + NaCl \longrightarrow CH_3CH_2Cl + NaOH$$

and would not occur to any significant extent because chloride ion cannot displace OH from an alcohol unless a strong acid is present to convert —OH to —$\overset{\oplus}{O}H_2$, which is a better leaving group. Under those conditions, the acid would be the poison rather than ethyl chloride.

d. tert-Butyl bromide gives the most alkene with Na$\overset{\oplus}{}$ $\overset{\ominus}{O}C_2H_5$ in ethanol (E2), and on solvolysis in 60% aqueous ethanol (S$_N$1). Of the two reagents, Na$\overset{\oplus}{}$ $\overset{\ominus}{O}C_2H_5$ in ethanol would give the most alkene from either bromide by an E2 reaction.

14-3   Initiation: $>$N—Br $\longrightarrow$ Br·   (details unspecified)

Propagation:

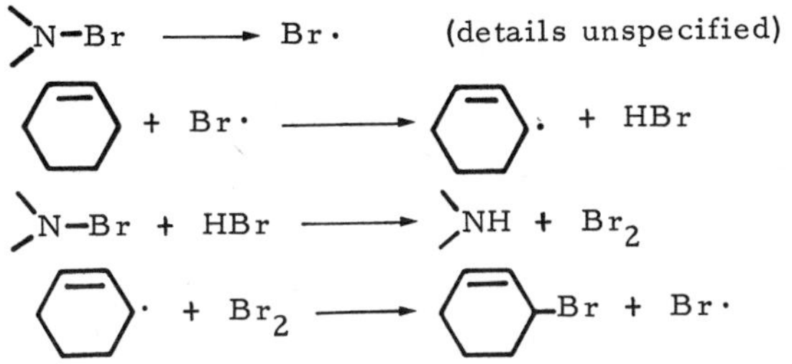

Byproducts expected are 1,2-dibromocyclohexane and 3,6-dibromocyclohexene. The 1,2-dibromide is the product of addition; the 3,6-dibromide.. disubstitution.

14-4

a. $C_6H_5CH=CHCH_2OH \quad \xrightarrow[\text{or } SOCl_2]{HCl \text{ conc.}} \quad C_6H_5CH=CHCH_2Cl$

(rearrangement is possible to $C_6H_5\underset{\underset{Cl}{|}}{C}HCH=CH_2$)

b. $HC\equiv CCH_3 \quad \xrightarrow[h\nu]{NBS} \quad HC\equiv CCH_2Br$

c. $\xrightarrow[\text{or } Cl_2, \underline{h}\nu]{SO_2Cl_2, \underline{h}\nu}$ (Cl$_2$ addition to the double
                                                                                        bond is also possible)

or $\diagdown$N—Cl compounds

d. $CH_3CH=CHCH=CHCH_3$ $\xrightarrow{HCl}$ $CH_3CH(Cl)CH=CHCH_2CH_3$

(byproduct is $CH_3CH=CH-CH(Cl)CH_2CH_3$)

e. $(CH_2=CH)_2CH_2$ $\xrightarrow[\text{or } Cl_2, \underline{h}\nu]{SO_2Cl_2, \underline{h}\nu}$ $(CH_2=CH)_2CHCl$

or $\diagdown$N—Cl compounds

14-5   Both chlorides ionize readily and reversibly through a common intermediate carbocation:

At equilibrium, the same chloride mixture will be obtained from either chloride by way of the common intermediate.

14-6   a.   Initiation: $Br_2 \xrightarrow{\underline{h}\nu} 2Br\cdot$

Propagation: $C_6H_5CH_3 + Br\cdot \longrightarrow C_6H_5CH_2\cdot + HBr$

$C_6H_5CH_2\cdot + Br_2 \longrightarrow C_6H_5CH_2Br + Br\cdot$

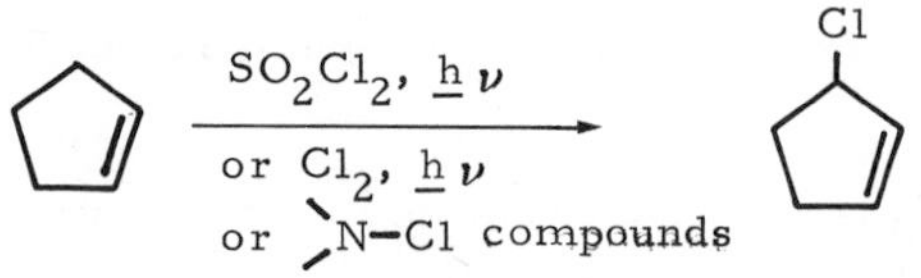

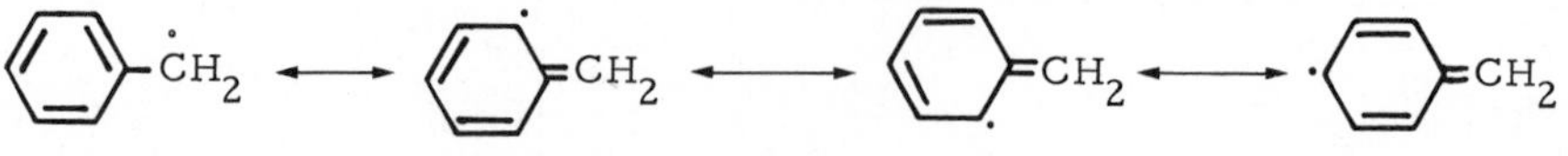

b. $C_6H_5CH_2\cdot + Br_2 \longrightarrow C_6H_5CH_2Br + Br\cdot$ $\Delta\underline{H}^o = -8.6$ kcal
        (+46.4)   (-55)

$C_6H_5CH_2\cdot + Br_2 \longrightarrow$ $+ Br\cdot$ $\Delta\underline{H}^o = +24.4$ kcal

(+38)   (+46.4)   (-5)   (-55)

stabilization   stabilization

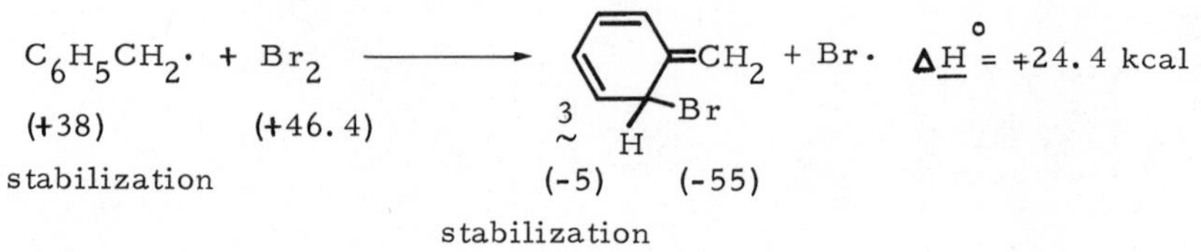

Reaction at a ring carbon to give a triene $\underset{\sim}{3}$ is energetically unfavorable and is therefore unlikely to occur. The reason is that $\underset{\sim}{3}$ is relatively unstable - having lost all but 5 kcal of stabilization associated with a phenyl group.

14-7   a. Abstraction of a hydrogen atom (by Cl·) from propadiene gives a radical in which the odd electron is delocalized. Subsequent reaction of this delocalized radical with Cl_2 takes place to give 3-chloropropyne.

$$CH_2\!\!=\!\!C\!\!=\!\!CH_2 \xrightarrow[-HCl]{Cl\cdot} CH_2\!\!=\!\!C\!\!=\!\!\overset{\cdot}{C}H \longleftrightarrow \overset{\cdot}{C}H_2\!\!-\!\!C\!\!\equiv\!\!CH \xrightarrow[-Cl\cdot]{Cl_2} ClCH_2C\!\!\equiv\!\!CH$$

b. The lack of S_N1 reactivity of $ClCH_2COCH_3$ is due to the instability of a carbocation center adjacent to an electronegative carbonyl group.

$$CH_3COCH_2Cl \xrightarrow{-Cl^{\ominus}} CH_3\!\!-\!\!\overset{\overset{\ddot{O}}{\|}}{C}\!\!-\!\!\overset{\oplus}{C}H_2 \longleftrightarrow CH_3\!\!-\!\!\overset{\overset{\ddot{O}^{\oplus}}{|}}{C}\!\!=\!\!CH_2$$

unfavorable   structure with positive
                              oxygen

c. Racemization accompanies ionization of 3-chloro-1-butene to give an achiral butenyl cation. In a weakly nucleophilic solvent, the cation reacts more rapidly with $Cl^{\ominus}$ to reform 3-chloro-1-butene than it reacts with solvent to give products.

$$CH_3CH(Cl)CH\!\!=\!\!CH_2 \xrightarrow{-Cl^{\ominus}} CH_3\overset{\frac{1}{2}\oplus}{CH}\cdots CH\cdots\overset{\frac{1}{2}\oplus}{CH_2} \xrightarrow{Cl^{\ominus}} \begin{array}{l}\text{racemic}\\\text{chlorides}\end{array}$$

chiral   achiral

$$\downarrow \text{solvent}$$

racemic products

14-8   3-Chloropropyne resembles 3-chloropropene in having high S_N1 and S_N2 reactivity. The activation is provided by the triple bond in the same way that 3-chloropropene is activated by the double bond. The S_N1 reaction of 3-chloropropyne occurs by way of a delocalized carbocation.

$$HC\!\!\equiv\!\!C\!\!-\!\!CH_2\!\!-\!\!Cl \xrightarrow{-Cl^{\ominus}} HC\!\!\equiv\!\!C\!\!-\!\!\overset{\oplus}{C}H_2 \longleftrightarrow HC\!\!=\!\!C\!\!=\!\!\overset{\oplus}{C}H_2$$

14-9   a. NaI in CH_3COCH_3 involves an S_N2 displacement and

$$C_6H_5CH\!\!=\!\!CHCl \sim C_6H_5C\!\!\equiv\!\!CCl < C_6H_5CH_2CH_2Cl < C_6H_5C\!\!\equiv\!\!CCH_2Cl$$

b. $AgNO_3$ in C_2H_5OH promotes S_N1 reaction, and the order of reactivity is the same as in part a.

14-10

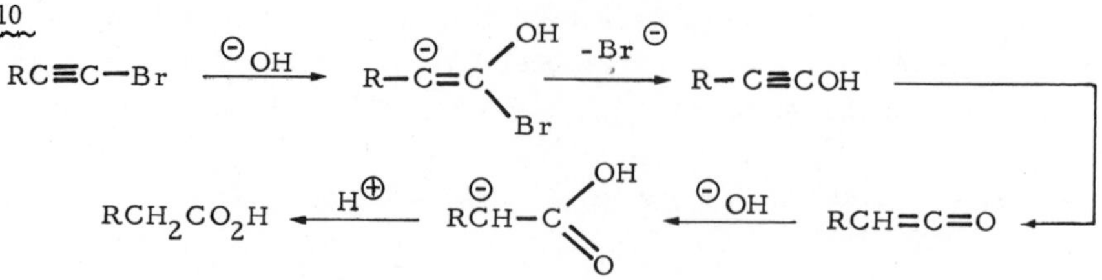

$$RC\equiv C-Br \xrightarrow{\ominus OH} R-\underset{\underset{Br}{|}}{C}=C\overset{OH}{\diagup} \xrightarrow{-Br^{\ominus}} R-C\equiv COH \longrightarrow$$

$$RCH_2CO_2H \xleftarrow{H^{\oplus}} RCH-\overset{OH}{\underset{O}{C}} \xleftarrow{\ominus OH} RCH=C=O \longleftarrow$$

14-11

a.

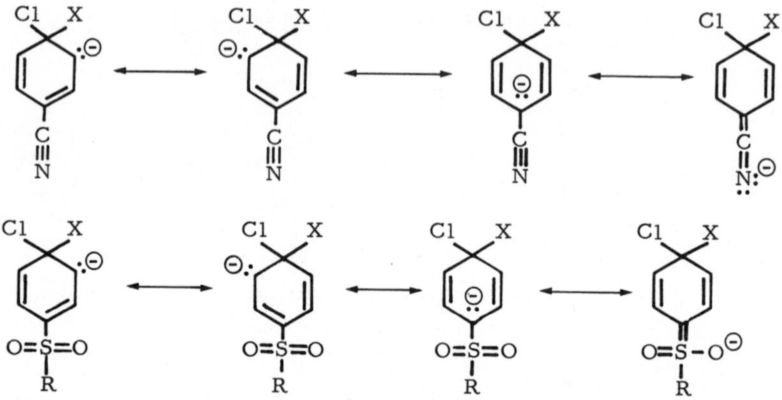

b.　Methyl groups ortho to the activating group decrease reactivity.　The reason is a steric one - the methyl groups force the activating group to turn out of coplanarity with the ring thereby reducing electron-delocalization in the anion intermediate.

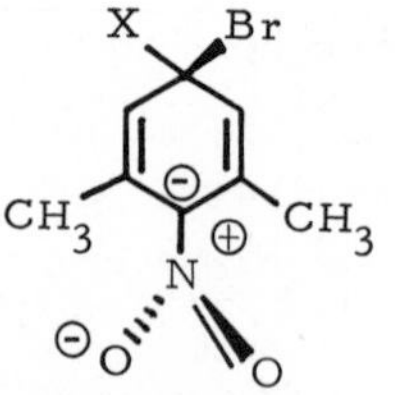

Steric hindrance to resonance of NO_2 group with ring. Hindrance (and hence decrease in reactivity) is less for the $C\equiv N$ group because of its linear structure.

14-12　　The nitro substituent is a better activating group than $\overset{\oplus}{N}(CH_3)_3$ because the anion intermediate is delocalized through the substituent (some charge is assumed by the oxygens).　The $\overset{\oplus}{N}(CH_3)_3$ group stabilizes the anionic charge mainly by electrostatic attraction - not by delocalization of charge.

14-13　　The order of activation is CH_3O < H < CF_3.　The methoxy group is electron-donating by a resonance mechanism which is deactivating relative to hydrogen.　The CF_3 group is strongly electron-withdrawing and stabilizes adjacent anionic charge - hence the CF_3 group is activating relative to hydrogen.

14-14   If halide elimination (Equation 14-4) were slow, then a reactivity
sequence similar to that observed for alkyl halides would be expected.   The
opposite sequence,   F >> Cl ~ Br ~ I,   suggests that attack of the nucleophile
is the slow step (Equation 14-3).   The high electron-attracting properties of
fluorine promote the attack of a nucleophile at a faster rate than with Cl, Br
or I.   See also Exercise 14-15.

14-15

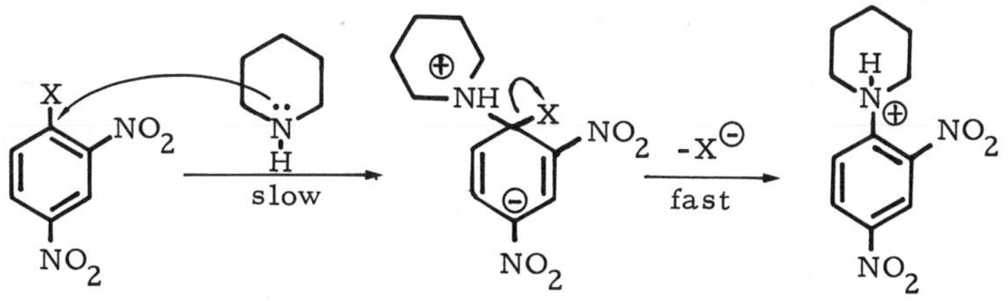

14-16

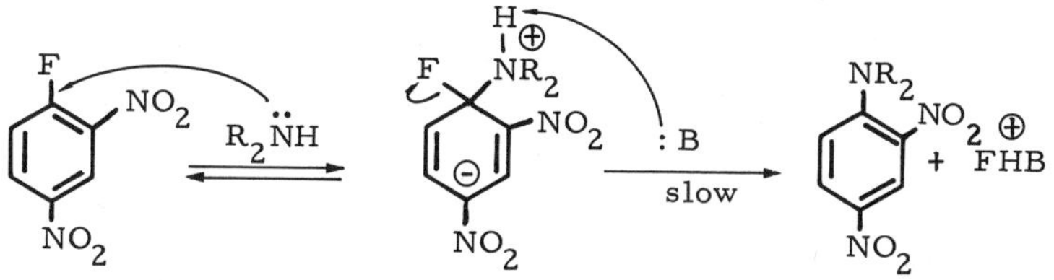

If the second step is slow, then the base probably assists in the elimination
of HF, as shown above.

14-17
a.

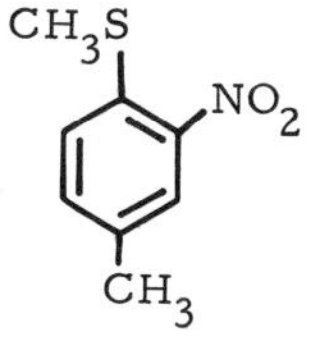

b.

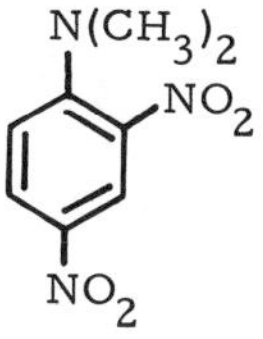

c.

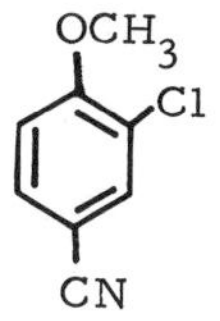

14-18   The additional π bond in benzyne is localized and does not interact
(overlap) with the kekule-type π bonds.   This is shown more clearly by an
atomic-orbital model.

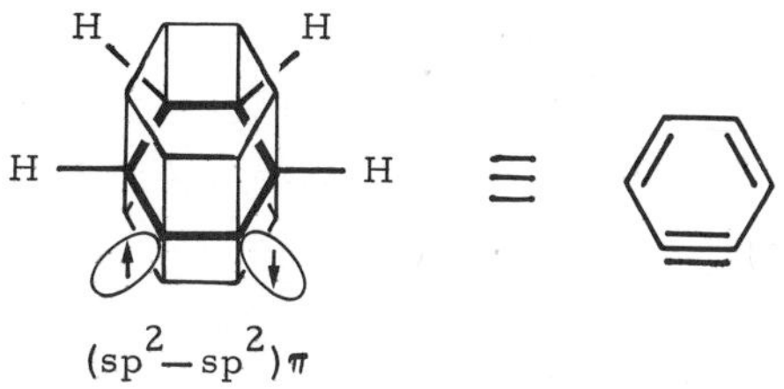

$(sp^2 - sp^2)\pi$

14-19

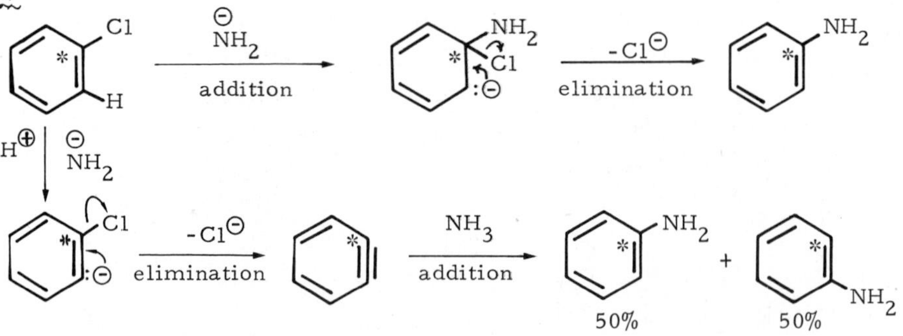

14-20

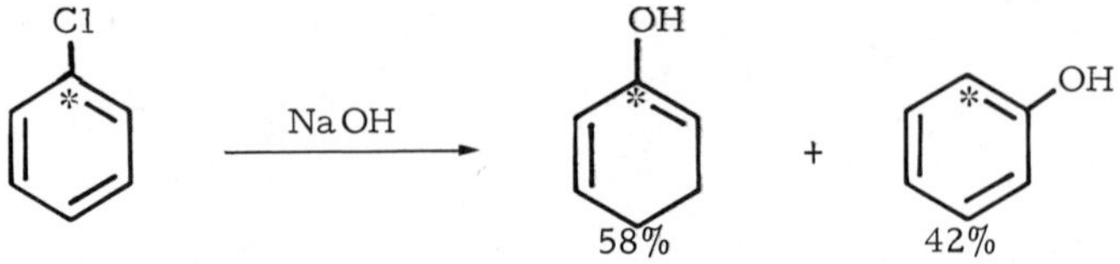

Assuming all the benzenol-2-^{14}C (phenol-2-^{14}C) arises by an elimination-addition sequence, the amount of phenol formed by this mechanism is 84%; by direct substitution 16%.

At lower temperatures ($240°$), the amount of direct substitution is expected to **increase**. Decreasing concentration of NaOH would not change the ratio of direct substitution to aryne formation because both reactions are second order in HO$^{\ominus}$.

14-21 a. 2,6-Dimethylchlorobenzene is unreactive with KNH$_2$ in liquid NH$_3$ because there are no **ortho** hydrogens and therefore no possibility of forming a benzyne intermediate.

b. The **ortho** hydrogens in fluorobenzene are acidic and readily exchange in deuterated solvents and strong base. Aniline is not formed because F$^{\ominus}$ is a poor leaving group and is not eliminated from the intermediate aryl anion.

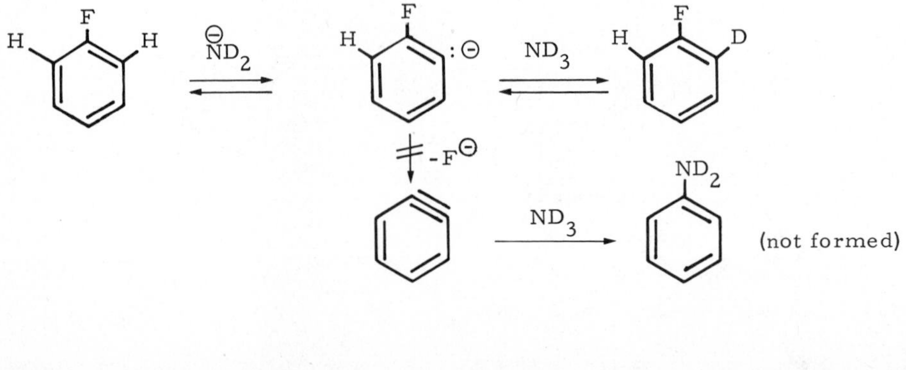

14-22

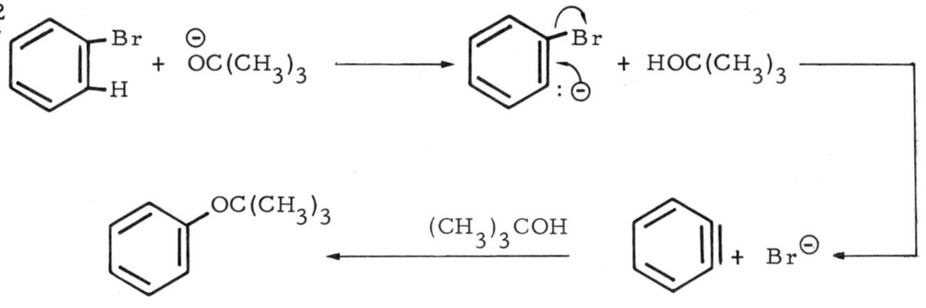

Dimethyl sulfoxide (DMSO) is a better solvent than _tert_-butyl alcohol for the above reaction because the alkoxide anion is not solvated in DMSO and is therefore more reactive. 4-Bromo-1-methylbenzene gives a mixture of _tert_-butyl 4-methylphenyl ether and _tert_- butyl 3-methylphenyl ether.

14-23 a.

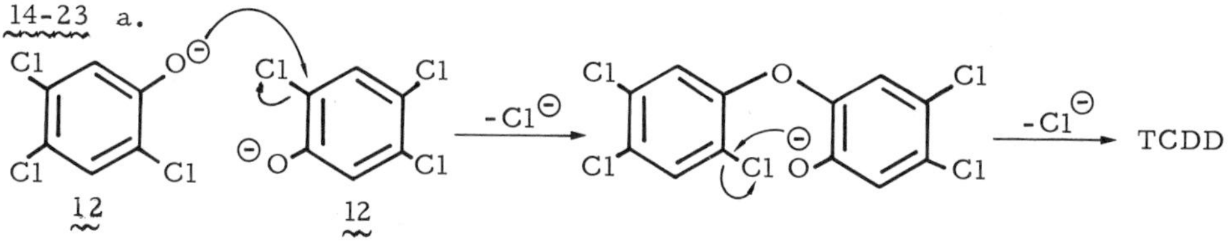

Alternatively, an aryne mechanism (elimination-addition) might operate:

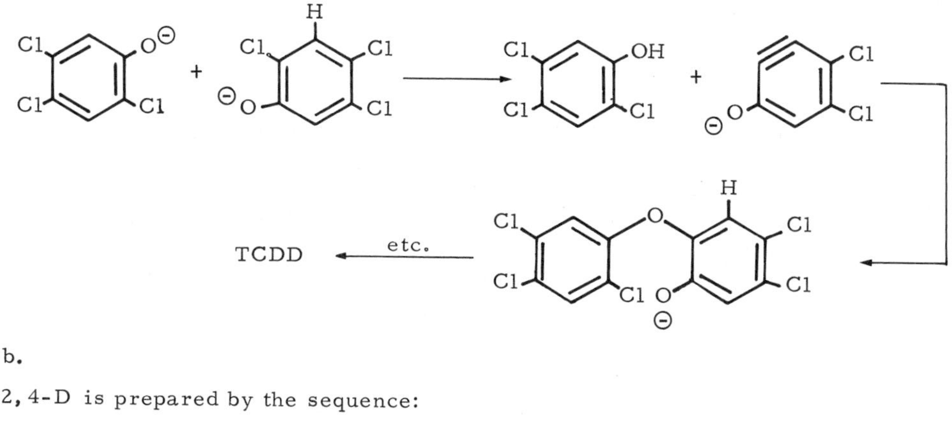

b.

2, 4-D is prepared by the sequence:

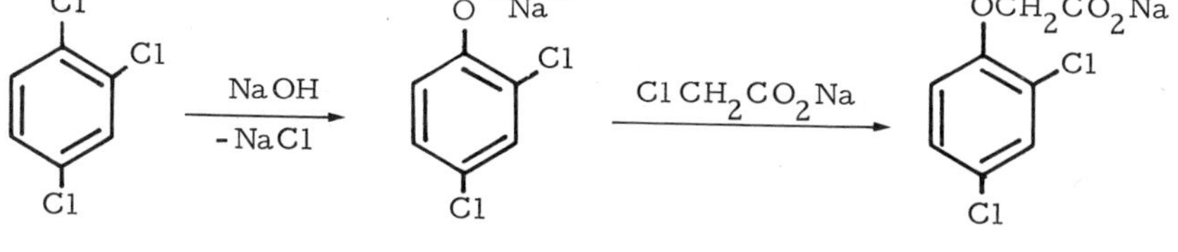

It is unlikely that the preparation of 2, 4-D would lead to TCDD or any analogs by an addition-elimination sequence because the chlorines in the intermediate benzenolate anion are not activated (they are _meta_ to each other). An aryne mechanism at high temperatures could possibly give a dichlorodibenzodioxin analog,

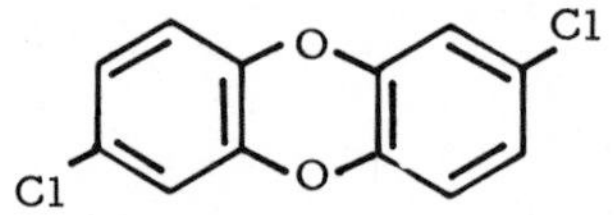

14-24 $CHBr_3$ + $K^{\oplus} \; {}^{\ominus}OC(CH_3)_3 \longrightarrow$ KBr + $HOC(CH_3)_3$ + $:CBr_2$

a.

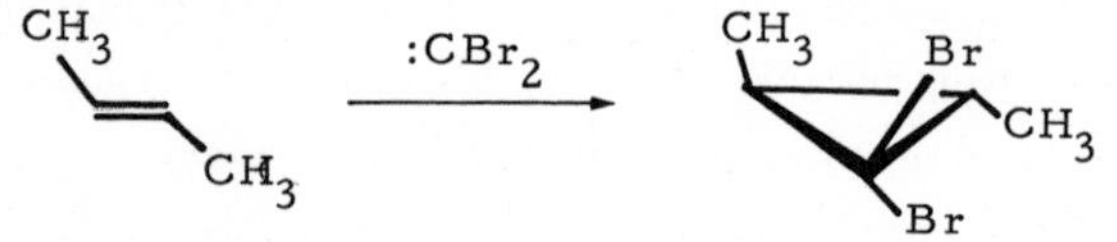

b.

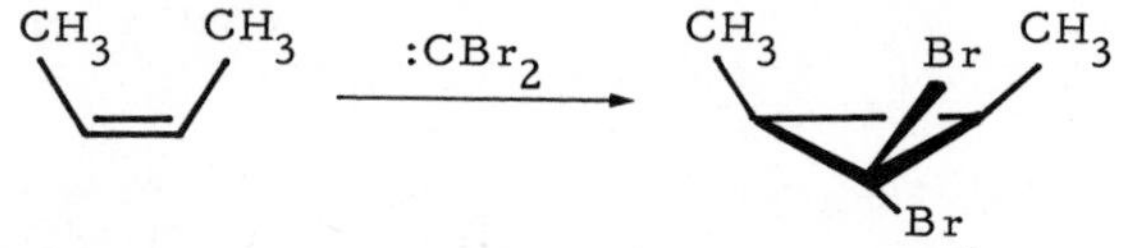

14-25

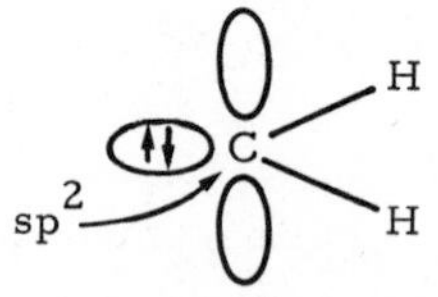

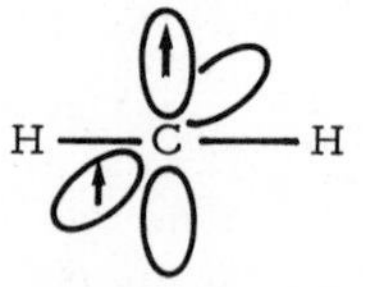

singlet: $(1s - 2sp^2)\,\sigma$ C—H
$\underline{n}^2$ in $2sp^2$ orbital
vacant 2p orbital
HCH = 120^0

triplet: $(1s - 2sp)\,\sigma$ C—H
$\underline{n}^1$ in $2p_y$ orbital
$\underline{n}^1$ in $2p_z$ orbital
HCH = 180^0

14-26 a. $2C_6H_5CH_2CH_2CH_3$ + $MgSO_4$ + $MgBr_2$

b. $CH_3C{\equiv}CCH_2CH_2CH_3$ + $MgBr_2$

c. $CH_2{=}CHCH_2CH_2CH{=}CH_2$ + LiCl

d. $CH_3CH_2CH_2CH_2OCH_3$ + Mg(Cl)Br

e. $CH_3CH_2CH_2CH_2CH_2CH_2CH_2CH_3$ + NaBr + $CH_3CH_2CH_2CH_3$
    + $CH_3CH_2CH{=}CH_2$

14-27 CBr_4 + $CH_3Li \longrightarrow CH_3Br$ + Br_3CLi

$Br_3CLi \longrightarrow Br_2C:$ + LiBr

14-27 (cont.)

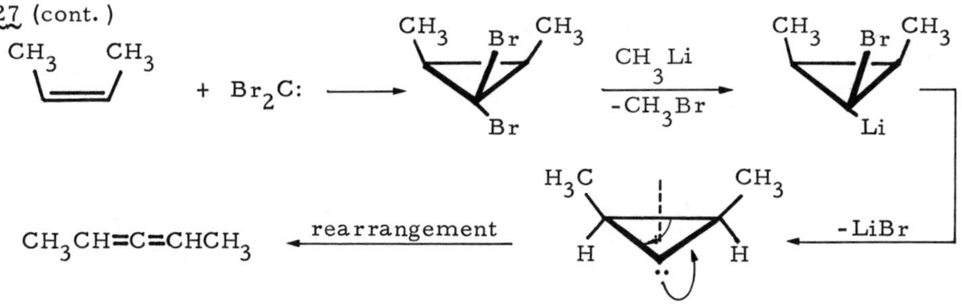

CH$_3$ CH$_3$ + Br$_2$C: $\longrightarrow$

$\xrightarrow[-CH_3Br]{CH_3Li}$

$\xrightarrow{-LiBr}$

CH$_3$CH=C=CHCH$_3$ $\xleftarrow{\text{rearrangement}}$

14-28   a.   C$_6$H$_5$CH=CH$_2$ $\xrightarrow{\text{:CCl}_2}$ $\underset{CH_2}{\overset{C_6H_5CH}{\diagdown}}CCl_2$

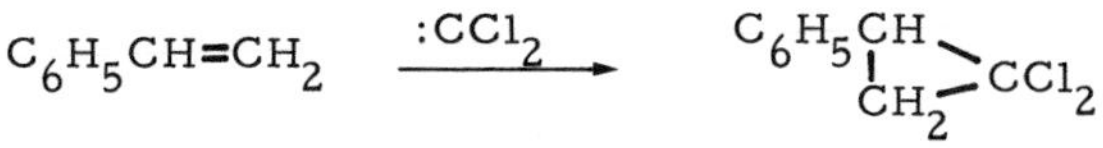

Dichlorocarbene can be prepared from one of the following reactions.   (See also Table 14-2.)

C$_6$H$_5$HgCCl$_2$Br $\xrightarrow{\text{heat}}$ Cl$_2$C: + C$_6$H$_5$HgBr

Cl$_3$CH + $^\ominus$OC(CH$_3$)$_3$ $\longrightarrow$ Cl$_2$C: + Cl$^\ominus$ + HOC(CH$_3$)$_3$

Cl$_3$CCO$_2$$^\ominus$Na$^\oplus$ $\xrightarrow{\text{heat}}$ Cl$_2$C: + CO$_2$ + NaCl

b. $\xrightarrow[\text{tetrahydrofuran}]{\text{Mg}}$ $\xrightarrow{-\text{Mg(F)Br}}$

[4 + 2]

c.   C$_6$H$_5$CHBr-$\overset{\overset{O}{\|}}{C}$-CHBrC$_6H_5$ $\xrightarrow[-\text{HOC(CH}_3)_3]{^\ominus\text{OC(CH}_3)_3}$ C$_6$H$_5$$\underset{Br}{\overset{\ominus}{C}}$:$\,$$\overset{O}{\overset{\|}{C}}$$\underset{Br}{CH}$-C$_6H_5$

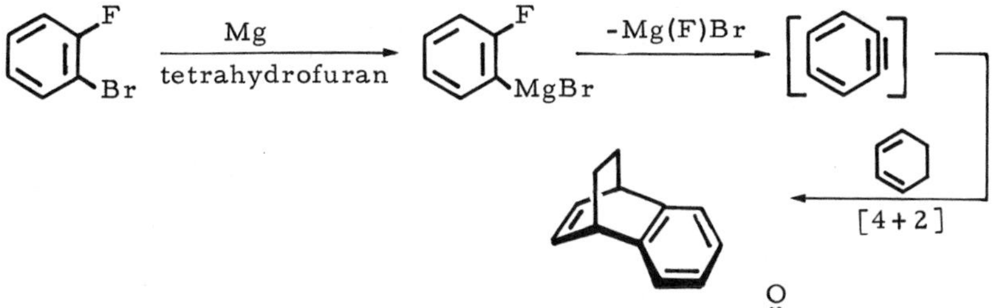

$\xleftarrow[\text{by E 2}]{-\text{HBr}}$ C$_6$H$_5$$\overset{O}{\diagup}$H $\xleftarrow{-\text{Br}^\ominus}$

H$_5$C$_6$ C$_6$H$_5$ Br   C$_6$H$_5$

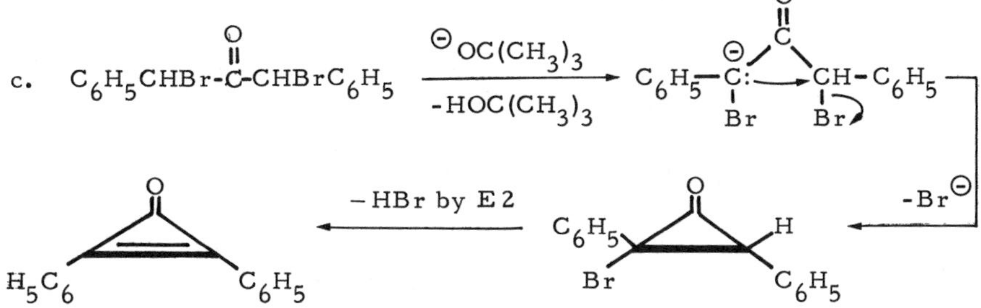

14-29

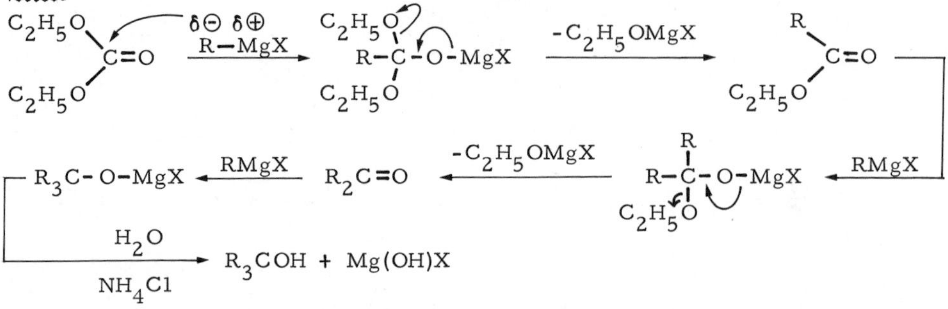

14-30

a. $C_6H_5MgBr + C_6H_5CHO \longrightarrow (C_6H_5)_2CHOMgBr \xrightarrow[NH_4Cl]{H_2O} (C_6H_5)_2CHOH$

b. 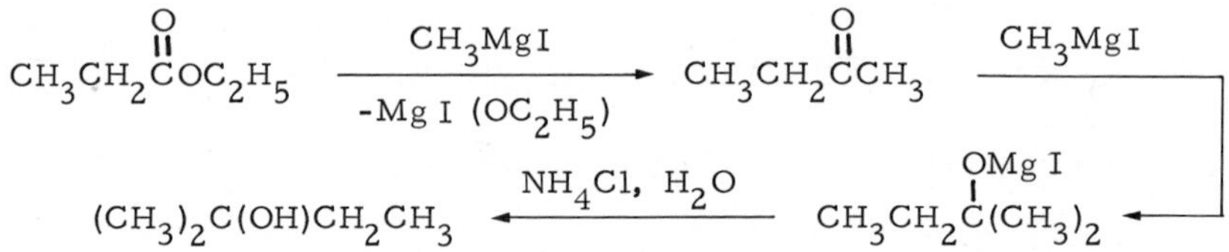 $CH_3CH_2\overset{O}{\overset{\|}{C}}OC_2H_5 \xrightarrow[-MgI(OC_2H_5)]{CH_3MgI} CH_3CH_2\overset{O}{\overset{\|}{C}}CH_3 \xrightarrow{CH_3MgI}$

$(CH_3)_2C(OH)CH_2CH_3 \xleftarrow{NH_4Cl,\ H_2O} CH_3CH_2\overset{OMgI}{\overset{|}{C}}(CH_3)_2$

c. $Cl-\overset{O}{\overset{\|}{C}}-OC_2H_5 \xrightarrow[-Mg(Cl)Br]{CH_3CH_2MgBr} CH_3CH_2\overset{O}{\overset{\|}{C}}OC_2H_5 \xrightarrow[-MgBr(OC_2H_5)]{CH_3CH_2MgBr}$

$(CH_3CH_2)_3COMgBr \xleftarrow{CH_3CH_2MgBr} (CH_3CH_2)_2C=O$

$\xrightarrow{NH_4Cl,\ H_2O} (CH_3CH_2)_3COH$

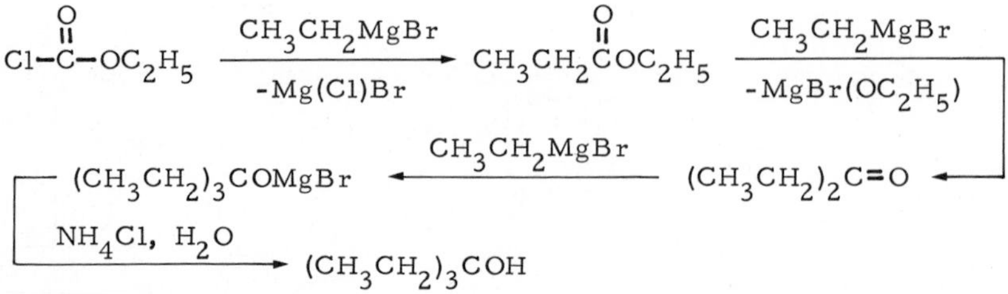

d. 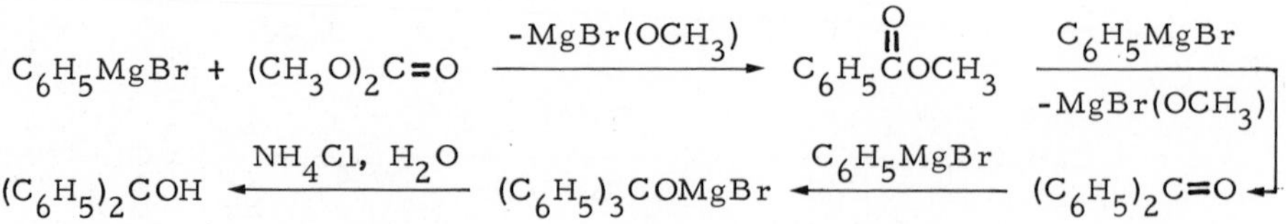 $C_6H_5MgBr + (CH_3O)_2C=O \xrightarrow{-MgBr(OCH_3)} C_6H_5\overset{O}{\overset{\|}{C}}OCH_3 \xrightarrow[-MgBr(OCH_3)]{C_6H_5MgBr}$

$(C_6H_5)_2COH \xleftarrow{NH_4Cl,\ H_2O} (C_6H_5)_3COMgBr \xleftarrow{C_6H_5MgBr} (C_6H_5)_2C=O$

e.

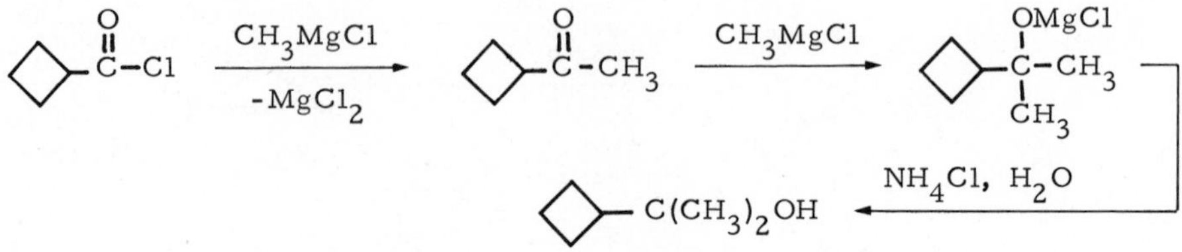

14-31

a.

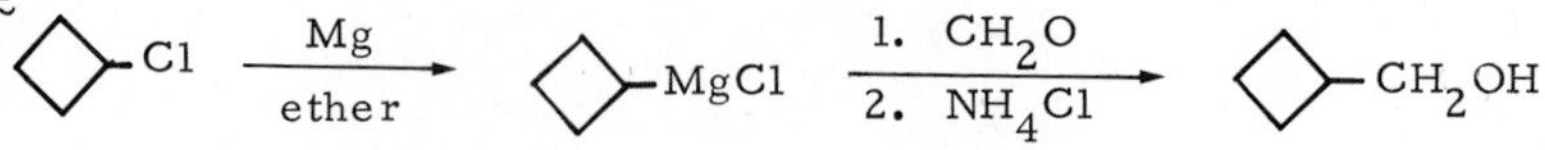

b.

$$2CH_3I \xrightarrow[\text{ether}]{Mg} 2CH_3MgI \xrightarrow[\text{2. } NH_4Cl]{\text{1. } CH_2=CHCOCH_3} CH_2=CHC(CH_3)_2OH$$

(A better route to the product would be from $CH_2=CHCH_2Cl$ and CH_3COCH_3.)

c. $C_2H_5Cl \xrightarrow[\text{ether}]{Mg} C_2H_5MgCl \xrightarrow[-C_2H_6]{C_6H_5C\equiv CH} C_6H_5C\equiv CMgCl \xrightarrow[\text{2. } NH_4Cl]{\text{1. } CH_2O}$

$$C_6H_5C\equiv CCH_2OH \longleftarrow$$

d. $C_2H_5Cl \xrightarrow[\text{ether}]{Mg} C_2H_5MgCl \xrightarrow[\text{2. } NH_4Cl]{\text{1. } CH_3CHO} C_2H_5CH(OH)CH_3$

14-32

addition product	enolization products	reduction products
a. $(CH_3)_3C\underset{OH}{C}(CH_3)_2$	$CH_3COCH_3 +$ $(CH_3)_3CH$	$(CH_3)_2CHOH +$ $(CH_3)_2C=CH_2$
b. $(C_6H_5)_2\underset{OH}{C}CH_2CH_3$	enolization is not possible	$(C_6H_5)_2CHOH +$ $CH_2=CH_2$
c. $C_6H_5CH=CHC(OH)(C_6H_5)_2$ (1,2 addition is expected)	enolization is not favorable for the $=CH-\overset{O}{\overset{\|}{C}}-$ grouping	reduction is only possible by forming benzyne, a very unfavorable product.

14-33 $\overset{\delta\oplus \; \delta\ominus}{RC\equiv N} \xrightarrow{CH_3MgI} \left[RC\overset{CH_3}{\underset{\|}{=}}N-MgI \right] \xrightarrow[-MgI(OH)]{H_2O} R-\overset{CH_3}{\underset{HO \;\; H}{\overset{\|}{C}}}-N-H$

$$R-CO-CH_3 \xleftarrow{-NH_3}$$

14-34

a. $(CH_3)_3CMgCl + CO_2 \longrightarrow (CH_3)_3C-C\overset{OMgCl}{\underset{O}{\diagdown}} \xrightarrow{H_2SO_4} (CH_3)_3CC\overset{OH}{\underset{O}{\diagdown}}$

b. $-CO_2H \xrightarrow[-CH_4]{CH_3Li}$ cyclohexyl$-\overset{O}{\overset{\|}{C}}-OLi \xrightarrow{CH_3Li}$ cyclohexyl$-\overset{OLi}{\underset{CH_3}{\overset{\|}{C}}}-OLi$

cyclohexyl$-\overset{O}{\overset{\|}{C}}-CH_3 \xleftarrow{H_2SO_4}$

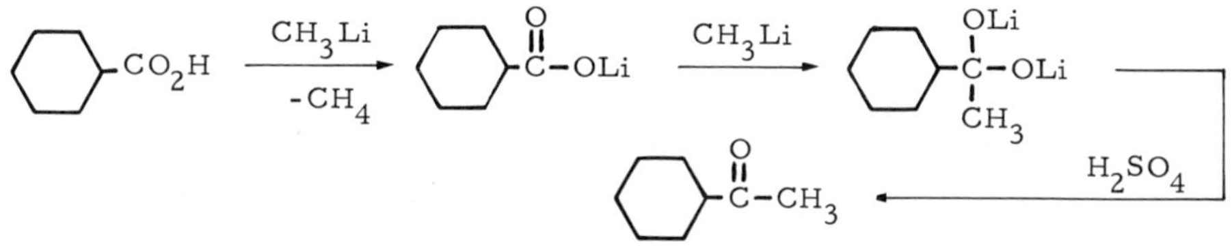

c. $C_6H_5\overset{\overset{O}{\|}}{C}Cl$ $\xrightarrow[-MgCl_2]{CD_3MgCl + CdCl_2}$ $C_6H_5\overset{\overset{OCdCl}{|}}{\underset{Cl}{C}}-CD_3$ $\xrightarrow{-CdCl_2}$ $C_6H_5\overset{\overset{O}{\|}}{C}CD_3$

d. $CH_3\overset{\overset{O}{\|}}{C}Cl$ $\xrightarrow[-LiI, \ -CuCl]{CD_3Li + CuI}$ $CH_3\overset{\overset{O}{\|}}{C}CD_3$

14-35

a. ⬡–Cl $\xrightarrow[\text{ether}]{Mg}$ ⬡–MgCl $\xrightarrow[\text{2. } NH_4Cl]{1.\{ \begin{matrix} CdCl_2 + \\ CH_3COCl \end{matrix}}$ ⬡–$\overset{\overset{O}{\|}}{C}$–CH_3

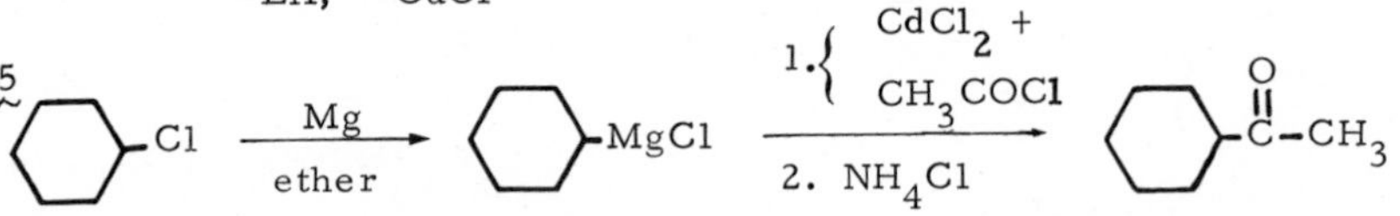

b. $C_6H_5C\equiv CH$ $\xrightarrow[-C_2H_6]{C_2H_5MgCl}$ $C_6H_5C\equiv CMgCl$ $\xrightarrow[\text{2. } H_2SO_4]{\text{1. } CO_2}$ $C_6H_5C\equiv CCO_2H$

c. $HC\equiv CH$ $\xrightarrow{NaOCl}$ $ClC\equiv CH$ $\xrightarrow[-C_2H_6]{C_2H_5MgCl}$ $ClC\equiv CMgCl$ ⎤

or

$HC\equiv CNa$ $\xrightarrow[\text{2. } H^{\oplus}]{\text{1. } CO_2}$ $HC\equiv CCO_2H$ $\xrightarrow[\text{2. } Cl_2]{\text{1. } NaNH_2}$ $ClC\equiv CCO_2H$ $\xleftarrow[\text{2. } H_2SO_4]{\text{1. } CO_2}$

14-36

a. $C_2H_5CH_2CH_2COCH_3$ (by 1,4-addition) and $CH_2=CHC(OH)(CH_3)CH_2CH_3$ (by 1,2-addition)

b.   +

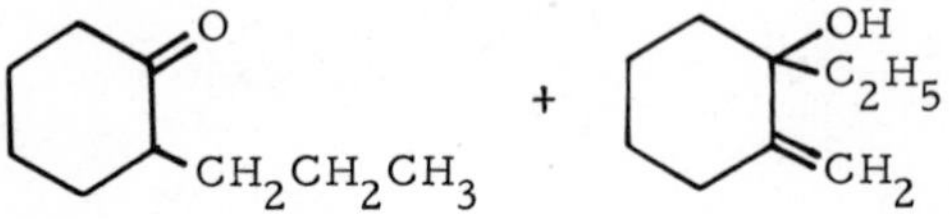

c.   d.

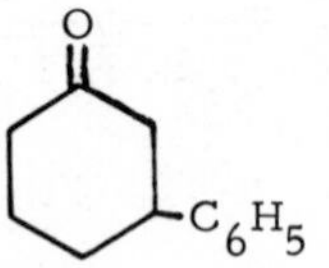

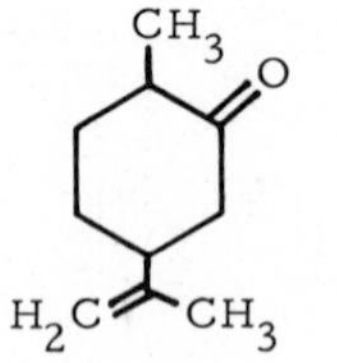

14-37 $CH_2=CHCOCH_3 + RH \longrightarrow CH_2=CHCR(OH)CH_3$ $\Delta\underline{H}^{\circ}_1$

$CH_2=CHCOCH_3 + RH \longrightarrow RCH_2CH_2COCH_3$ $\Delta\underline{H}^{\circ}_2$

Of the two addition reactions, the one corresponding to $\Delta\underline{H}^{\circ}_2$ is more exothermic than the one corresponding to $\Delta\underline{H}^{\circ}_1$ because a carbon-carbon double bond is weaker than a carbon-oxygen double bond. From Table 4-3, we can estimate:

$$\Delta\underline{H}^{\circ}_1 - \Delta\underline{H}^{\circ}_2 = [+93.5 \ (C=O \rightarrow C-O) - 110.6 \ (OH)] - [+63.2 \ (C=C \rightarrow C-C) - 98.7]$$
$$= [-17.1] - [-35.5] \ = 18.4 \text{ kcal}$$

14-38

a. $C_6H_5C\equiv CH \xrightarrow{NaNH_2} C_6H_5C\equiv CNa \xrightarrow{D-2-bromobutane} C_6H_5C\equiv C-\overset{\overset{\displaystyle CH_3}{|}}{\underset{\underset{\displaystyle C_2H_5}{|}}{C}}-H$

b. $CH_2{=}CH_2 \xrightarrow{ICl} Cl-CH_2-CH_2-I \xrightarrow[-HI]{KOH} ClCH{=}CH_2$

c. $(C_6H_5)_2CH_2 \xrightarrow{Br_2,\ h\nu} (C_6H_5)_2CHBr \xrightarrow{Br_2,\ h\nu} (C_6H_5)_2CBr_2$

d. $3KI + H_3PO_4 \xrightarrow{-K_3PO_4} 3HI \xrightarrow{C_2H_5OH} C_2H_5I + H_2O$

e. $(CH_3)_2C(OH)C_2H_5 \xrightarrow{48\%\ HBr} (CH_3)_2CBrC_2H_5$

14-39 2-Methyl-2-butanol first dissolves in the aqueous acid by the reaction:

$$ROH + H_3O^\oplus \rightleftharpoons RO\overset{\oplus}{H}_2 + H_2O$$

The second phase appears when RBr is formed by the reaction:

$$RO\overset{\oplus}{H}_2 + Br^\ominus \longrightarrow RBr + H_2O$$

The alkyl bromide, unlike the alcohol, is insoluble in aqueous acid. The simplest way to tell which layer is the alkyl bromide is to add a drop from one layer to a beaker of water. If the drop dissolves, it is the aqueous layer. If the drop is insoluble in water, it is the product (RBr) layer.

14-40

a.

cis-2-butenol   trans-2-butenol

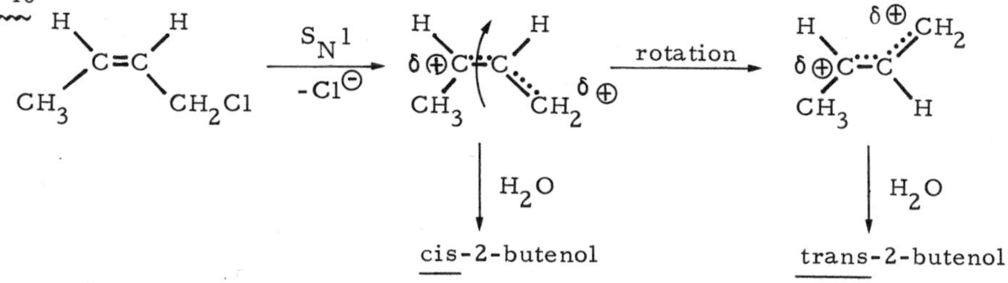

If the intermediate carbocation is sufficiently long-lived for rotation to occur prior to nucleophilic attack, then a mixture of cis and trans-2-butene will result.

b. An S_N2 mechanism will give only cis-2-butenol.

14-41 $\left.\begin{array}{l} (CH_3)_2C(Cl)CH{=}CH_2 \\ \text{or} \\ (CH_3)_2C{=}CHCH_2Cl \end{array}\right\} \xrightarrow{H_2O} \begin{array}{l} (CH_3)_2C(OH)CH{=}CH_2 \qquad A \\ + \\ (CH_3)_2C{=}CHCH_2OH \qquad B \end{array}$

Product B is expected to be more stable than A because it is primary (stronger C—O bond) and the double bond is more highly substituted. Under equilibrium control, B will then predominate. The products of kinetic control are determined by relative rates of attack of H_2O at the two positive carbons of the intermediate cation,

$$(CH_3)_2 \overset{\delta\oplus}{C} \cdots CH \cdots \overset{\delta\oplus}{CH_2} .$$

For steric reasons, the primary carbon should be preferred. However, the tertiary carbon is likely to carry more of the charge, and should then be attacked preferentially. No clear-cut prediction is possible as to the major product under kinetic control, but, in fact, a substantial amount of A is formed.

14-42   The intermediate anion formed in the reaction of a nucleophile $X:^{\ominus}$ with 2-chloropyridine is better stabilized by the ring nitrogen than in 3-chloropyridine.

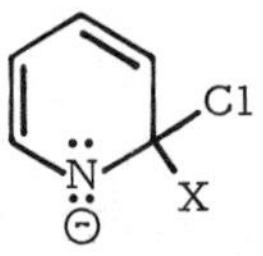

important valence-bond form with negative charge on electronegative nitrogen

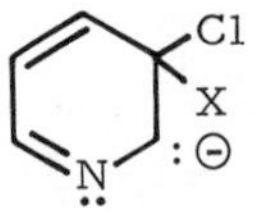

no valence-bond structures can be drawn in which the charge resides on nitrogen

14-43   Two different mechanisms are involved:
2-chloropyridine reacts by addition-elimination,

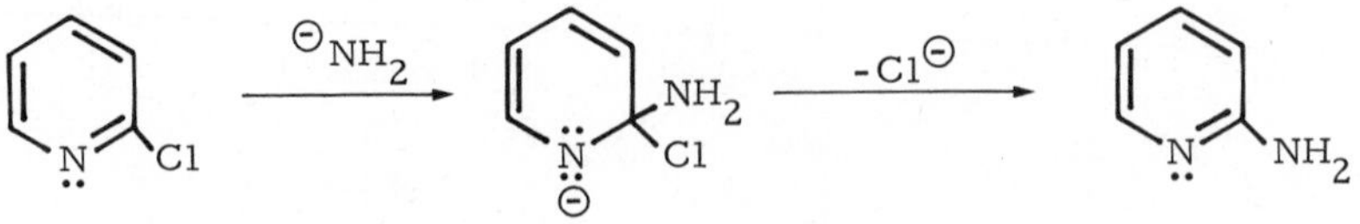

3-chloropyridine reacts by elimination-addition,

65%   35%

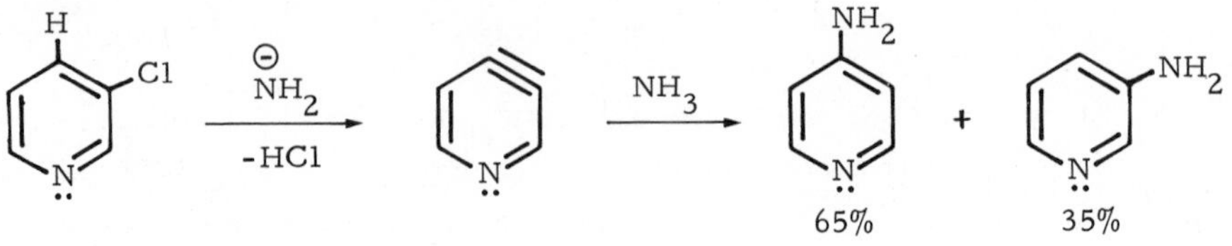

14-44
a.
b.
c.

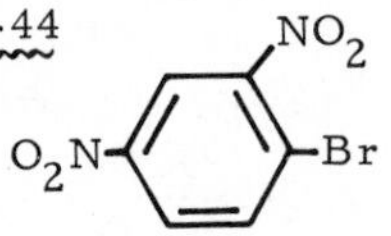

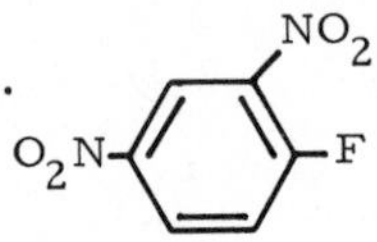

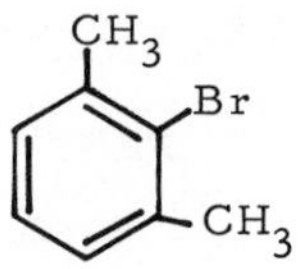

d.

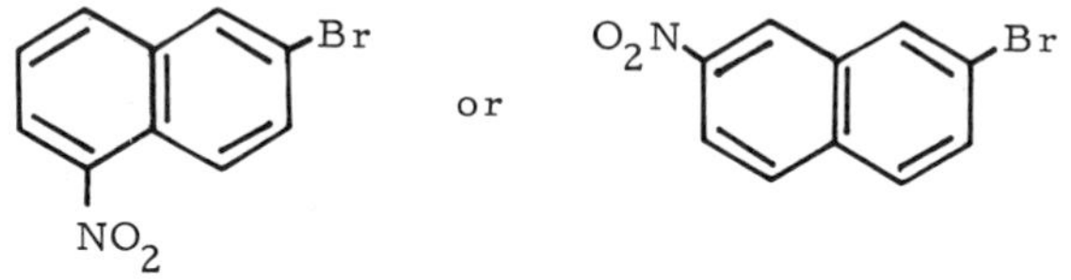

or

(The charge in the anionic intermediate formed on reaction with $^\ominus OC_2H_5$ cannot be accommodated by the NO_2 group.)

14-45   The same 1-naphthalyne intermediate is formed from each of the halonaphthalenes.   The nucleophilic reagent adds to the naphthalyne to give a mixture of 1- and 2-substituted naphthalenes.   The product mixture is therefore independent of the nature of the starting halogen,

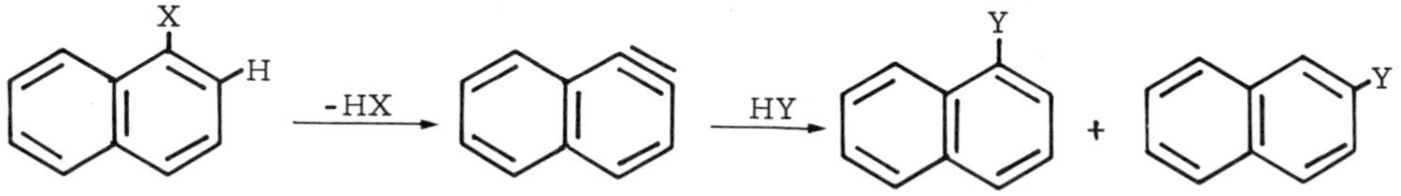

14-46   a.

$C_6H_5CH_2Cl + AgNO_3 + C_2H_5OH \longrightarrow C_6H_5CH_2OC_2H_5 + \underline{AgCl} + HNO_3$

b.

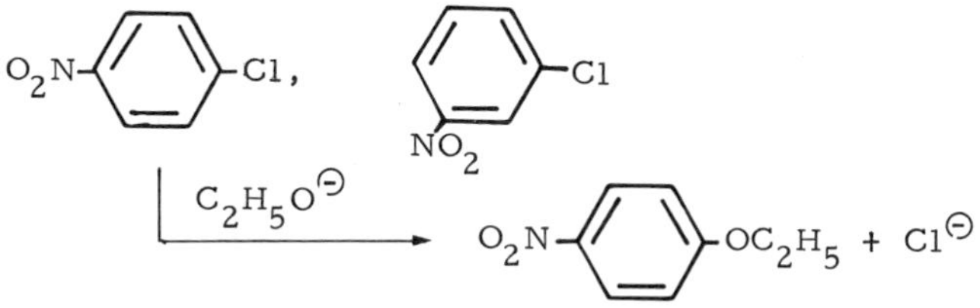

c.

d. $HC{\equiv}CCH_2Br + NaI \longrightarrow HC{\equiv}CCH_2I + NaBr$

14-47

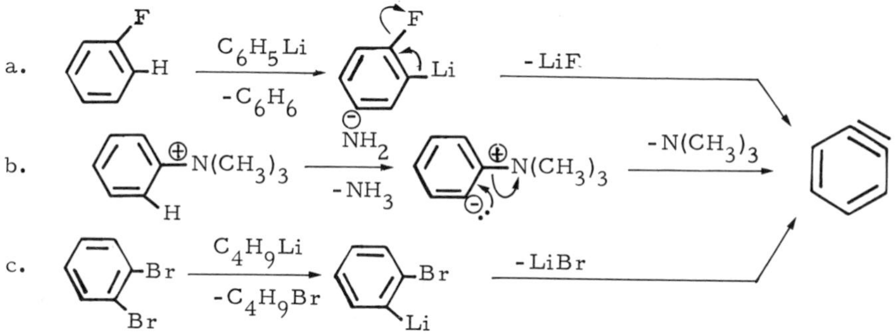

a.

b.

c.

14-48

a. C_6H_5Br $\xrightarrow[-KBr, \ -NH_3]{KNH_2}$

$C_6H_5C\equiv CH$ $\xrightarrow[-NH_3]{KNH_2}$ $C_6H_5C\equiv C^{\ominus}K^{\oplus}$

$\xrightarrow{\quad}$ (structure with $C\equiv CC_6H_5$ and $:^{\ominus}$)

$\downarrow$ NH_3, $-NH_2^{\ominus}$

$C_6H_5C\equiv CC_6H_5$

b.   (structure: ring with $(CH_2)_2NHCH_3$ and Cl) $\xrightarrow[-C_6H_6]{C_6H_5Li}$ (structure: ring with $(CH_2)_2\overset{Li}{N}CH_3$ and Cl)

$\xrightarrow[-C_6H_6, \ -LiCl]{C_6H_5Li}$

(structure with $(CH_2)_2\overset{Li}{\underset{\cdot\cdot}{N}}CH_3$)

$\longrightarrow$ (indoline structure with N–Li, CH_3)

$\xrightarrow[-Li^{\oplus}]{H^{\oplus}}$ (indoline structure with N, CH_3)

c.   (structure: ring with $N_2^{\oplus}$ and $CO_2^{\ominus}$) $\xrightarrow[-CO_2]{-N_2}$ (benzyne structure)

$\xrightarrow{[4+2]}$ (bicyclic structure with φ, O)

$\xleftarrow[\text{reverse} \ [4+1]]{-CO}$ (naphthalene structure with φ)

(φ = phenyl)

14-49

a.   (tricyclic structure with O)

b. $(C_6H_5)_4C$

c.   (biphenylene structure) $+ N_2 + SO_2$

14-50   See Exercise 14-11b.

14-51 $C_4H_{10}O_3 + CH_3MgI \longrightarrow CH_4$ (84.1 ml, 740 mm, 25°)

(0.1776 g or 0.00168 moles)

Volume of methane at standard conditions $= \dfrac{740 \times 84.1 \times 273}{760 \times 298}$

Moles of methane $= \dfrac{740 \times 273 \times 84.1}{760 \times 298 \times 22.4 \times 10^3} = 0.00335$

Hence, one mole of A gives two moles of methane. Compound A must then

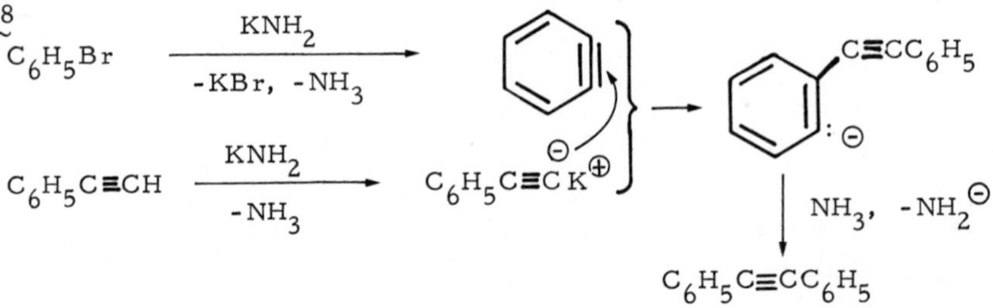

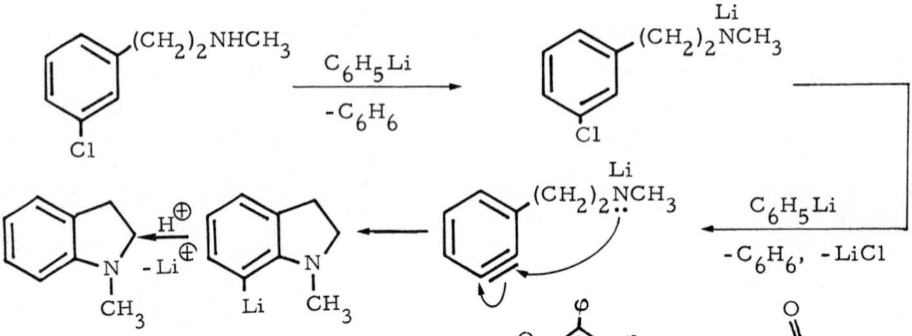

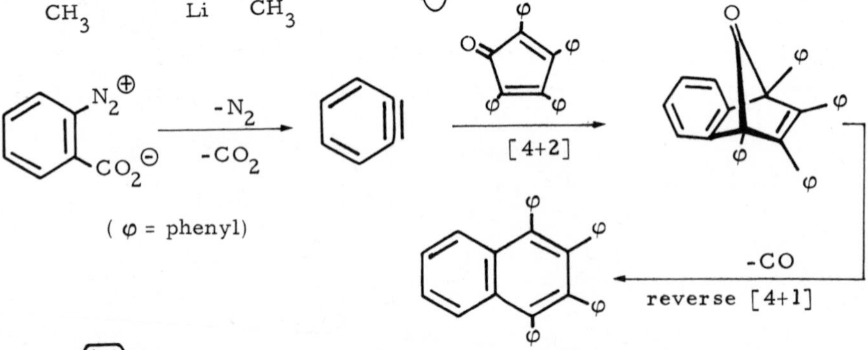

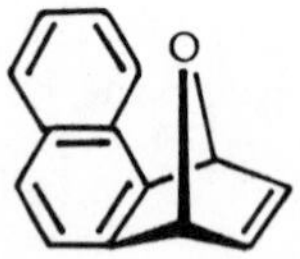

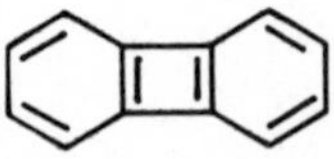

have <u>two</u> active hydrogens. A possible structure is:

$$HOCH_2CH_2OCH_2CH_2OH$$

14-52 a. $CH_3MgCl + ICl \longrightarrow CH_3I + MgCl_2$

b. $C_6H_5Li + CH_3OH \longrightarrow C_6H_6 + LiOCH_3$

c. $CH_3Li + HC\equiv CH \longrightarrow CH_4 + LiC\equiv CH$

d. $C_6H_5Li + CuI \longrightarrow C_6H_5Cu + LiI$

14-53

a. $(CH_3)_2C=CHCH_2CH_2Br \xrightarrow{\text{Mg, ether}} (CH_3)_2C=CHCH_2CH_2MgBr \longrightarrow$

$\underset{\underset{CH_2CH_2CH=C(CH_3)_2}{|}}{CH_3C=CHCO_2C_2H_5} \xleftarrow[\text{2. } H_2O]{\text{1. } CH_3C\equiv CCO_2C_2H_5} (CH_3)_2C=CHCH_2CH_2Cu \xleftarrow[-80°]{CuI}$

b. $C_2H_5Cl \xrightarrow{Mg} C_2H_5MgCl \xrightarrow[-C_2H_6]{C_6H_5C\equiv CH} C_6H_5C\equiv CMgCl \xrightarrow{CO_2}$

$\xleftarrow{H_2O} \underset{\underset{CH_3}{|}}{C_6H_5C\equiv C\overset{OLi}{\underset{|}{C}}-OMgCl} \xleftarrow{CH_3Li} C_6H_5C\equiv CCO_2MgCl \longleftarrow$

$\longrightarrow C_6H_5C\equiv CCOCH_3$

14-54 a. The reaction in (2) is a straightforward metalation followed by carbonation:

$(CH_3)_3C-Cl + 2Li \xrightarrow[\substack{\text{ether} \\ -40°'}]{-LiCl} (CH_3)_3C-Li \xrightarrow[\text{2. } H^{\oplus}]{\text{1. } CO_2} (CH_3)_3CCO_2H$

At higher temperatures ($35°$) the <u>tert</u>-butyllithium is evidently reactive enough to attack both the solvent and any unreacted chloride in E2 elimination.

$(CH_3)_3C-Li + CH_3CH_2OCH_2CH_3 \longrightarrow (CH_3)_3CH + CH_2=CH_2 + LiOC_2H_5$

$(CH_3)_3C-Li + (CH_3)_3C-Cl \longrightarrow (CH_3)_3CH + (CH_3)_2C=CH_2 + LiCl$

The ethene produced from the solvent reacts further with the alkyllithium by nucleophilic addition:

$(CH_3)_3CLi + CH_2=CH_2 \longrightarrow (CH_3)_3CCH_2CH_2Li$

and carbonation gives 4,4-dimethylpentanoic acid (Reactions 1 and 3).

$$(CH_3)_3CCH_2CH_2Li \xrightarrow[\text{2. } H^{\oplus}]{\text{1. } CO_2} (CH_3)_3CCH_2CH_2CO_2H$$

b. Methyllithium does not react similarly. It is less reactive than tert-butyllithium and does not attack the ether solvent at room temperature. Carbonation would give only ethanoic acid.

c. Since methyllithium is reasonably stable in diethyl ether, then 3,3-dimethylbutyllithium, $(CH_3)_3CCH_2CH_2Li$, is expected to be stable (both are primary carbanionic reagents). Further addition to ethene is not then expected and 6,6-dimethylheptanoic acid is not likely to be a significant product on carbonation.

$$(CH_3)_3CCH_2CH_2Li \xrightarrow{CH_2=CH_2} (CH_3)_3C(CH_2)_4Li \xrightarrow{CO_2}$$

$$(CH_3)_3C(CH_2)_4CO_2H$$
$$\text{(not observed)}$$

14-55 a. $H-C(CH_3)_2OMgI \longrightarrow (CH_3)_2CHOH$

b.   No adduct is expected (Section 12A). Products of reduction and enolization will be formed.

$$(CH_3)_2CHCH(OH)CH(CH_3)_2 + \text{butenes} + \text{butane}$$

c. $C_2H_5 \underset{S}{\overset{\parallel}{C}}-SH$

d. $CH_3CH_3 + H_2N-MgBr$

14-56

a. $(CH_3)_3CCl \xrightarrow[\text{ether}]{Mg} (CH_3)_3CMgCl \xrightarrow{D_2O} (CH_3)_3CD$

b. $CH\equiv CH + NaNH_2 \longrightarrow CH\equiv CNa + NH_3$

$CH\equiv CNa + CH_3I \longrightarrow CH\equiv C-CH_3 + NaI$

$CH_3C\equiv CH + C_2H_5MgBr \longrightarrow CH_3C\equiv CMgBr + C_2H_6$

$CH_3C\equiv CMgBr \xrightarrow[\text{2. } H_2O]{\text{1. } CO_2 \text{ (slow)}} CH_3C\equiv C-CO_2H$

c. $(CH_3)_4C + Cl_2 \xrightarrow{h\nu} (CH_3)_3CCH_2Cl$

14-56 c. (cont.)

$(CH_3)_3CCH_2Cl \xrightarrow[\text{ether}]{Mg} (CH_3)_3CCH_2MgCl \xrightarrow{I_2} (CH_3)_3CCH_2I$

d. 1. $(CH_3)_2CHMgCl + CH_3COCH_3 \longrightarrow (CH_3)_2CH-\underset{\underset{OH}{|}}{\overset{\overset{CH_3}{|}}{C}}-CH_3$

2. $(CH_3)_2CHCO_2C_2H_5 + 2\ CH_3MgI \longrightarrow (CH_3)_2CH\underset{\underset{OH}{|}}{\overset{\overset{CH_3}{|}}{C}}-CH_3$

3. $(CH_3)_2C=C(CH_3)_2 \xrightarrow{HCl} (CH_3)_2\underset{\underset{Cl}{|}}{C}-CH(CH_3)_2$

$(CH_3)_2\underset{\underset{Cl}{|}}{C}-CH(CH_3)_2 \xrightarrow[\text{ether}]{Mg} (CH_3)_2\underset{\underset{MgCl}{|}}{C}-CH(CH_3)_2$

$(CH_3)_2\underset{\underset{MgCl}{|}}{C}-CH(CH_3)_2 \xrightarrow[\text{2. } H_2O]{\text{1. } O_2} (CH_3)_2\underset{\underset{OH}{|}}{C}-CH(CH_3)_2$

e. —Br $\xrightarrow[\text{ether}]{Mg}$ —MgBr

3 —MgBr $+ CH_3O-\overset{\overset{O}{\|}}{C}-OCH_3 \longrightarrow$ $\left(\text{}\right)_3$—C-OH

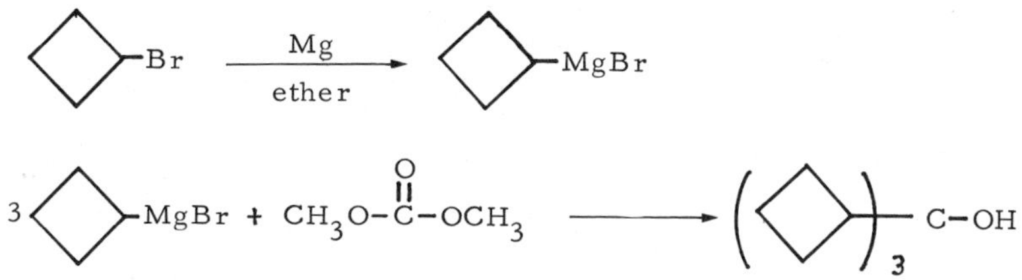

f. $(CH_3)_3CCH_2Cl \xrightarrow[\text{ether}]{Mg} (CH_3)_3CCH_2MgCl$

$(CH_3)_3CCH_2MgCl + \underset{O}{CH_2-CH_2} \longrightarrow (CH_3)_3CCH_2CH_2CH_2OH$

g. $CH_3CH=CH_2 \xrightarrow[400°]{Cl_2} CH_2=CHCH_2Cl$

$CH_2=CHCH_2Cl \xrightarrow[\text{ether}]{Mg} CH_2=CHCH_2MgCl$

$CH_2=CHCH_2MgCl + CH_2=CHCH_2Cl \longrightarrow CH_2=CHCH_2CH_2CH=CH_2$

14-57

a.

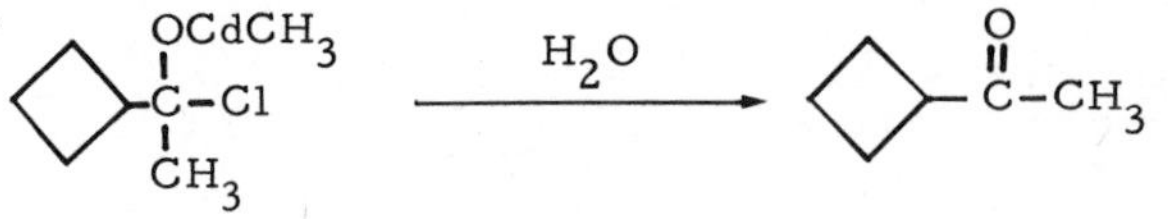

$\xrightarrow{\text{H}_2\text{O}}$

b. $C_6H_5-\underset{\underset{Li}{|}}{\overset{\overset{CH_3}{|}}{C}}-C\equiv N \xrightarrow{CO_2} C_6H_5-\underset{\underset{CO_2Li}{|}}{\overset{\overset{CH_3}{|}}{C}}-C\equiv N \xrightarrow{\text{H}_2\text{O}} C_6H_5-\underset{\underset{CO_2H}{|}}{\overset{\overset{CH_3}{|}}{C}}-C\equiv N$

c. $HC\equiv CCH_2MgBr \xrightarrow{HC\equiv CCH_2Br} HC\equiv CCH_2CH_2C\equiv CH \quad + $
$\qquad\qquad\qquad\qquad\qquad\qquad\qquad\qquad HC\equiv CCH_2CH=C=CH_2$

d. $(CH_3)_3CMgBr \xrightarrow[-70^0]{O_2} (CH_3)_3C-O-O-H \quad$ <u>tert</u>-butyl hydroperoxide

e.

f. $CH_2Cl_2 \xrightarrow[-C_4H_{10}]{C_4H_9Li} LiCHCl_2 \xrightarrow{-LiCl} :CHCl \longrightarrow$

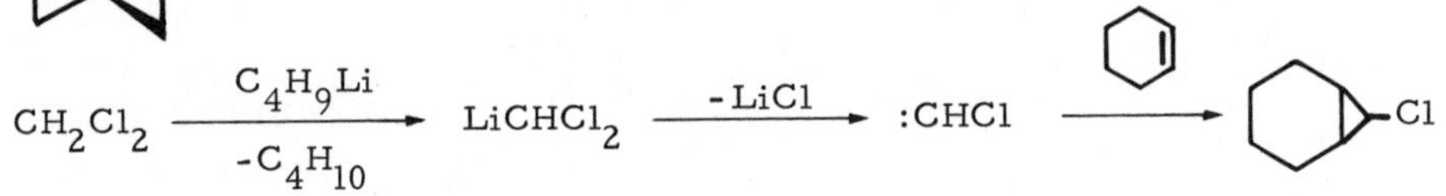

14-58   a. Addition of Grignard reagent to the carbonyl bond is unlikely because the bulky <u>tert</u>-butyl group will render reaction very slow. Most likely products are those of enolization and reduction.

$$[\,(CH_3)_2CH]_2C=O \;+\; (CH_3)_3CMgBr$$

reduction $\diagup$ $\diagdown$ enolization

$[(CH_3)_2CH]_2CHOH \;+\; (CH_3)_2C=CH_2 \qquad (CH_3)_3CH$

b. Addition will not occur for steric reasons. Reduction and enolization products are not possible. Therefore, there will be no reaction.

c. Reaction will not stop at the ketone stage. Therefore, there will be some tertiary alcohol formation, $CH_3(CH_2)_2C(OH)(CH_3)_2$.

d. A Grignard reagent is unlikely to add to a C=N bond in the direction indicated by the proposed reaction product. The expected course is:

$$\overset{\delta\oplus}{CH_3}\overset{\delta\ominus}{CH=N}CH_3 + \overset{\delta\ominus}{CH_3}-\overset{\delta\oplus}{MgI} \longrightarrow (CH_3)_2CH-N(CH_3)MgI \xrightarrow{\text{H}_2\text{O}}$$

$$(CH_3)_2CHNHCH_3 + Mg(OH)I \longleftarrow$$

e. The Grignard reagent from 2-bromoethyl ethanoate is not expected to be stable. It will eliminate rapidly to form ethene as follows:

$$2 \; CH_3\overset{\overset{\displaystyle O}{\|}}{C}-O-CH_2-CH_2-MgBr \longrightarrow 2 \; CH_2{=}CH_2 + (CH_3CO_2)_2Mg + MgBr_2$$

f. 1-Pentene is only one of several products to be expected in the reaction of 2-propenylmagnesium bromide with bromoethane. Other products are:

$$CH_2{=}CHCH_2MgBr + H{-}CH_2{-}CH_2{-}Br \xrightarrow{-MgBr_2} CH_2{=}CHCH_3 + CH_2{=}CH_2$$

$$CH_2{=}CHCH_2MgBr + CH_3CH_2Br \longrightarrow CH_2{=}CHCH_2Br + CH_3CH_2MgBr$$

$$CH_2{=}CHCH_2MgBr + CH_2{=}CHCH_2Br \xrightarrow{-MgBr_2} (CH_2{=}CHCH_2)_2$$

14-59

a. $(CH_3)_2CH-\overset{\overset{\displaystyle OH}{|}}{\underset{\underset{\displaystyle C_6H_5}{|}}{C}}-CH(CH_3)_2 + C_6H_6$ and starting ketone (from enolization)

b. $(CH_3)_3C-\overset{\overset{\displaystyle OH}{|}}{\underset{\underset{\displaystyle CH_3}{|}}{C}}-C(CH_3)_3$ (no reduction or enolization products are possible)

c.   no

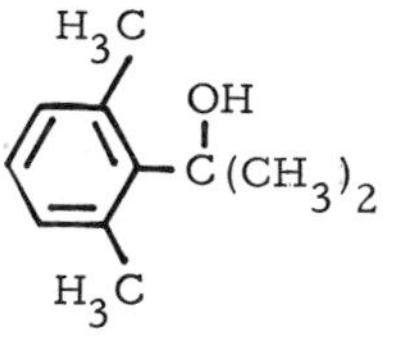

only CH_4 from enolization

14-60

a. $HC{\equiv}CH \xrightarrow{NaOCl} ClC{\equiv}CH \xrightarrow[NH_3, \, -33^0]{NaNH_2} ClC{\equiv}CNa \xrightarrow[-NaBr]{Br(CH_2)_4Br}$

$$ClC{\equiv}C(CH_2)_4C{\equiv}CCl \xleftarrow[-NaBr]{ClC{\equiv}CNa} ClC{\equiv}C(CH_2)_4Br \longleftarrow$$

b. $ClC{\equiv}CH \xrightarrow{C_2H_5MgBr} ClC{\equiv}CMgBr \xrightarrow[2. \; NH_4Cl, \; H_2O]{1. \; cyclohexanone}$

(see part a)

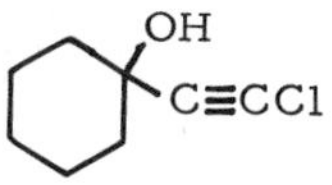

c. $HC\equiv CH$ $\xrightarrow[{-C_2H_6}]{C_2H_5MgCl}$ $HC\equiv CMgCl$ $\xrightarrow[{2.\ NH_4Cl,\ H_2O}]{1.\ 2\text{-butanone}}$ $\underset{\overset{|}{C\equiv CH}}{CH_3\overset{|}{C}(OH)C_2H_5}$

d. $HC\equiv CH$ $\xrightarrow[{NH_3,\ -33^0}]{NaNH_2}$ $HC\equiv CNa$ $\xrightarrow{CH_3I}$ $HC\equiv CCH_3$ $\xrightarrow[{NH_3,\ -33^0}]{NaNH_2}$

$CH_2\!\!=\!\!CHCH_2C\equiv CCH_3$ $\xleftarrow[{-NaBr}]{CH_2\!\!=\!\!CHCH_2Br}$ $NaC\equiv CCH_3$ $\longleftarrow$

14-61   A second-order rate (first order in each reagent) is consistent with a
reaction in which the slow step is the attack of $(CH_3)_2Mg$ on the ketone.

$$CH_3\!\!-\!\!Mg\!\!-\!\!CH_3 + O\!\!=\!\!CR_2 \xrightarrow{\text{slow}} CH_3\!\!-\!\!Mg\!\!-\!\!O\!\!-\!\!\overset{\overset{\displaystyle CH_3}{|}}{C}R_2$$

As the reaction proceeds, deviation from second-order kinetics occurs
because the products complex with unreacted Grignard reagent, and both the
initial product, $CH_3MgOCR_2CH_3$, and complexed product can react with the
ketone in a similar way to the reaction of $(CH_3)_2Mg$ with ketone. There are,
therefore, several concurrent reactions involving different CH_3—Mg reagents
in the later stages. A fuller account of this is given by E. C. Ashby,
J. Laemmle, and H. M. Neumann, Accounts of Chemical Research, 7, 272
(1974).

14-62 Methylene dichloride and methyllithium is a source of chlorocarbene,
which then cycloadds to the cyclopentadienide anion.

CH_2Cl_2 $\xrightarrow[{-CH_4}]{CH_3Li}$ $Li\!-\!CHCl_2$ $\xrightarrow{-LiCl}$ $:CHCl$

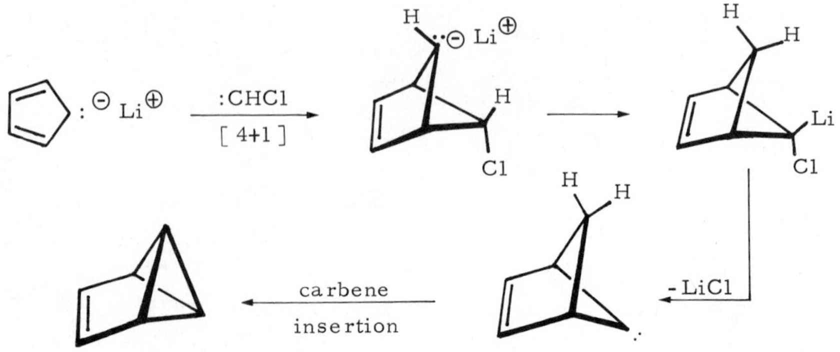

14-63   The initial adduct of a Grignard reagent and an amide is stable because spontaneous elimination of $XMgN(CH_3)_2$ does not occur.

$$\underset{\substack{\| \\ O}}{R'C}-N(CH_3)_2 + RMgX \longrightarrow R'-\underset{\substack{| \\ R}}{\overset{\substack{OMgX}}{C}}-N(CH_3)_2$$

$$R'-\underset{\substack{| \\ R}}{\overset{\substack{OH}}{C}}-N(CH_3)_2 \quad \xleftarrow{H_2O} \quad \cancel{\rightarrow} -XMgN(CH_3)_2$$

$$-HN(CH_3)_2 \searrow$$

$$R'-CO-R$$

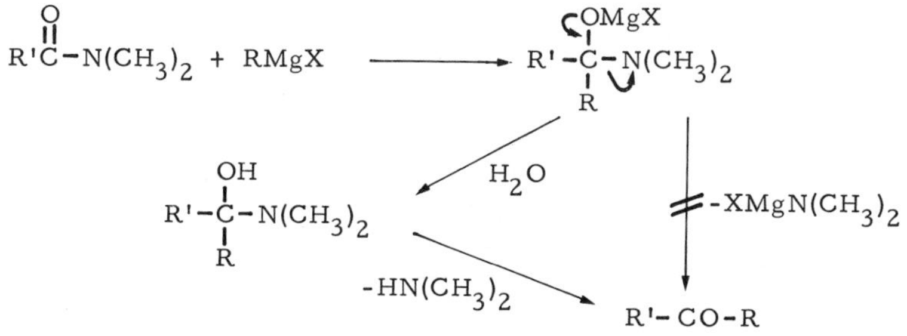

This is consistent with the $^{\ominus}N(CH_3)_2$ group being a strongly basic and very poor leaving group. Less basic groups, $^{\ominus}OCH_3$, $Cl^{\ominus}$, are better leaving groups, and eliminate readily as $XMg(OCH_3)$ and $XMgCl$.

14-64   The ΔH° values for the reaction

$$CH_3CH_3 + CH_3CH{=}O \longrightarrow CH_3CH_2CH(CH_3)OH$$

and reactions a - e are, respectively: -4.0 kcal, -19.4 kcal, +10.6 kcal, -38 kcal, -46 kcal (includes release of -27 kcal of strain energy), -10.7 kcal. Clearly, the most favorable reactions energetically are c and d. The equilibrium constant for both reactions is likely to be large.

14-65 $C_3H_5Br_3 \quad \xrightarrow{CH_3Li} \quad \triangleright{-}Br + CH_2{=}CHCH_2Br$

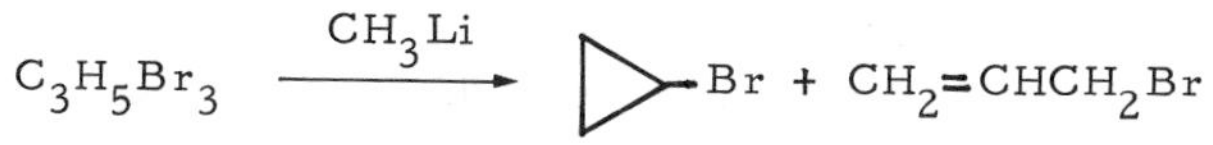

The nmr spectrum of $C_3H_5Br_3$ suggests the groupings $CH_2{-}CH$ and $CH_2{-}CH_2$ in order to give a one-proton triplet and a two-proton triplet resonance. The structure of $C_3H_5Br_3$ is therefore $BrCH_2CH_2CHBr_2$, and is consistent with chemical-shift data and the splitting patterns. The reaction with methyllithium follows.

$$BrCH_2CH_2CHBr_2 \quad \underset{-CH_3Br}{\overset{CH_3Li}{\longrightarrow}} \quad BrCH_2CH_2CH(Li)Br \quad \xrightarrow{-LiBr} \quad BrCH_2CH_2\overset{..}{C}H$$

(halogen-metal exchange)

$$BrCH_2CH{=}CH_2 \quad + \quad Br{-}\underset{\substack{\diagdown \\ CH_2}}{\overset{\substack{CH_2 \diagup \\ }}{CH}} \Big| \quad \xleftarrow{\text{rearrangement}}$$
by C-H insertion

(α C-H insertion)   (β C-H insertion)

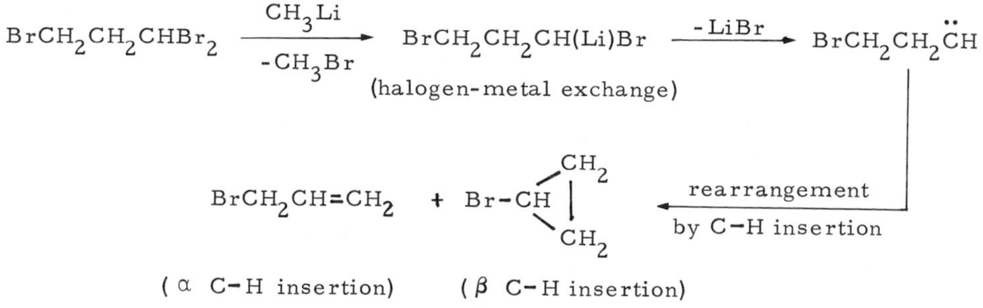

14-66   1. The conditions $(Ag^{\oplus})$ would promote ionization of 1-chloro-2-butene rather than either S_N2 or S_N2' reaction. The product mixture is obtained from the intermediate butenyl cation.

$$CH_3CH=CHCH_2Cl \xrightarrow[-AgCl]{Ag^{\oplus}} CH_3\overset{\tfrac{1}{2}\,\oplus}{CH}\cdots\overset{\tfrac{1}{2}\,\oplus}{CH}\cdots CH_2 \xrightarrow{CH_3CO_2H}$$

$$CH_3CH=CHCH_2O_2CCH_3 \;+\; CH_3\underset{O_2CCH_3}{CH}CH=CH_2 \longleftarrow$$

$$\text{A}\quad(65\%)\qquad\qquad \text{B}\quad(35\%)$$

2. The products are the result of mixed S_N1, S_N2 or S_N2' mechanism. That 36% of reaction product results from a process zeroth order in nucleophile means that 36% of reaction is S_N1. The remaining 64% must be S_N2 or S_N2' processes, both of which would be first order in nucleophile. The proportions of products from each mechanism would be

$$S_N1 \qquad\qquad A:B = 65:35 \qquad \text{where } A + B = 36$$

$$S_N2 \qquad\qquad A \;\;\text{(only)}$$

$$S_N2' \qquad\qquad B \;\;\text{(only)}$$

Solving for A and B derived by S_N1 reaction,

$$A = 23.4\% \qquad \text{and} \qquad B = 12.6\%$$

Thus, the amount of A derived by S_N2 reaction $= 85 - 23.4 = 61.6\%$
The amount of B derived by S_N2' reaction $\quad= 15 - 12.6. = 2.4\%$

From which we conclude that the S_N2' is not a major pathway under these reaction conditions.

3. The conditions (good nucleophile in poor ionizing solvent) favor S_N2 rather than ionization. Since no rearrangement is observed, the S_N2' mechanism is not operating here.

$$CH_3CH=CHCH_2Cl + \overset{\ominus}{O_2}CCH_3 \xrightarrow{S_N2} CH_3CH=CHCH_2O_2CCH_3 + \overset{\ominus}{Cl}$$

$$CH_3CH(Cl)CH=CH_2 + \overset{\ominus}{O_2}CCH_3 \xrightarrow{S_N2} CH_3CH(O_2CCH_3)CH=CH_2 + \overset{\ominus}{Cl}$$

CHAPTER 15

ALCOHOLS AND ETHERS

The physical properties of alcohols are influenced by hydrogen-bonding, which tends to associate molecules. The boiling points, therefore, are higher than those of hydrocarbons with similar molecular weight, and the solubilities in nonpolar solvents are lower.

Some important synthetic procedures for alcohols are summarized in Table 15-2. The reactions of alcohols can be classified into three main types, depending on which bonds are involved:

the $O \mathbin{\vert} H$, the $C \mathbin{\vert} O$, or the $R \mathbin{\vert} C(OH)$ bonds.

Reactions that involve only the O—H bond include those in which the alcohol behaves as an acid, a base, and a nucleophile. The important examples of each type follow:

1. Acidic properties - alkoxide salts

$$C_2H_5OH + \overset{\oplus\ominus}{KOH} \rightleftharpoons C_2H_5O^{\ominus}K^{\oplus} + H_2O \qquad \text{(Section 15-4A)}$$

potassium
ethoxide

2. Basic properties - oxonium salts

$$C_2H_5OH + H^{\oplus}BF_4^{\ominus} \rightleftharpoons C_2H_5-\overset{\overset{\textstyle H}{\vert}}{O}H^{\oplus} + BF_4^{\ominus} \qquad \text{(Section 15-4B)}$$

ethyloxonium
fluoroborate

3. Nucleophilic properties

a. <u>ether formation</u> by way of alkoxide ions and alkyl halides or sulfates

$$RO^{\ominus} + R'X \longrightarrow ROR' + X^{\ominus} \qquad \text{(Section 15-4C)}$$

The strongly basic reaction conditions of this S_N2 displacement also can promote E2 elimination of RX to an alkene.

Ether formation through S_N1 reactions when the alcohol is the solvent also can be included here:

$$ROH + R'X \xrightarrow{S_N 1} ROR' + HX$$

b. <u>ester formation</u> - Alcohols behave as nucleophiles toward carboxylic acids and related compounds to give carboxylic esters.

$$ROH + CH_3-\overset{\overset{\displaystyle O}{\|}}{C}-X \rightleftharpoons CH_3-\overset{\overset{\displaystyle O}{\|}}{C}-OR + HX \qquad \text{(Section 15-4D)}$$

$$X = \text{halogen, OH, or acyl } (-\overset{\overset{}{\underset{\underset{\displaystyle O}{\|}}{C}}}{}-R')$$

The displacement of X takes place by an addition-elimination sequence:

$$ROH + CH_3-\overset{\overset{\displaystyle O}{\|}}{C}-X \rightleftharpoons CH_3-\overset{\overset{\displaystyle O^{\ominus}}{|}}{\underset{X}{C}}-\overset{\displaystyle \oplus}{O}\overset{H}{\underset{R}{<}} \rightleftharpoons CH_3-\overset{\overset{\displaystyle O-H}{|}}{\underset{X}{C}}-OR \rightleftharpoons$$

$$CH_3-\overset{\overset{\displaystyle O}{\|}}{C}-OR + HX \rightleftharpoons$$

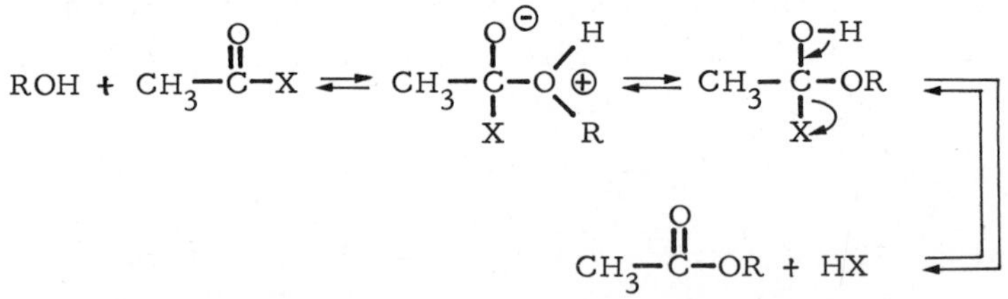

Compounds that are less reactive than acyl halides require an acidic catalyst to make the carbonyl carbon more receptive to addition by nucleophilic alcohol, and also to make X a better leaving group (H_2O).

Esterification of alcohols with acids does not have a very favorable equilibrium constant. The extent of conversion to ester is poor if the alcohol or ester is hindered. Tertiary alcohols do not form esters by this <u>addition-elimination</u> process.

c. <u>acetal and ketal formation</u> - Alcohols behave as nucleophiles toward aldehydes and ketones to give, first, hemiacetals (or hemiketals), which seldom are isolated, and then acetals or ketals. <u>These reactions, like esterification, are acid-catalyzed and reversible, and involve addition-elimination steps.</u> They work best with unhindered primary alcohols and unhindered carbonyl compounds:

$$R'_2C{=}O + ROH \rightleftharpoons R'_2C\overset{\displaystyle OH}{\underset{\displaystyle OR}{<}} \xrightarrow{\underset{\displaystyle ROH}{\longleftarrow}} R'_2C\overset{\displaystyle OR}{\underset{\displaystyle OR}{<}} + H_2O$$

(Section 15-4E)

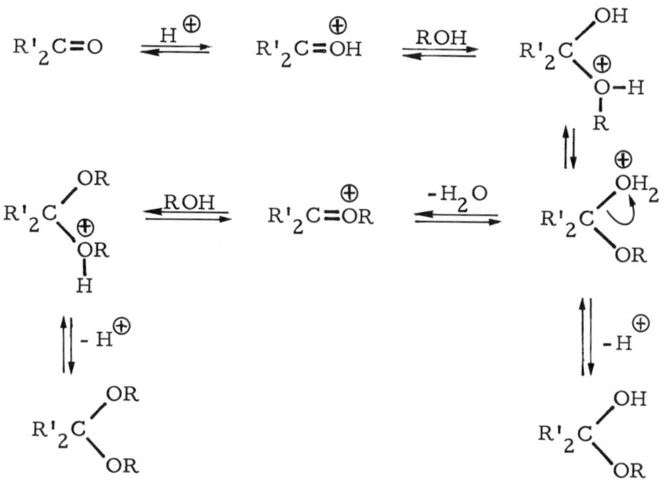

acetal or ketal   hemiacetal or hemiketal

Reactions involving the C—O bond usually occur only when the —OH function is converted into a better leaving group, such as $-\overset{\oplus}{O}H_2$, sulfate ester, $-OSO_3H$, sulfonate ester, $-O-SO_2R$, or a phosphate ester, $-OPO_3H$. Departure of the leaving group can occur as the result of S_N1, S_N2, E1, or E2 processes, depending on the reaction conditions of the alkyl groups present. The overall result of esterification and either the E1 and E2 reaction is <u>dehydration</u> of the alcohol:

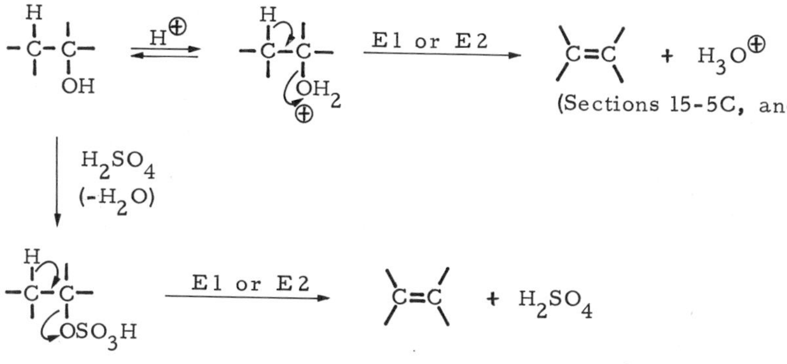

(Sections 15-5C, and 15-5E)

(Section 15-5B)

Tertiary alcohols, or other alcohols that form relatively stable carbocations, can be expected to react readily to give substitution and/or elimination products under acidic conditions:

$$\text{tert-ROH} + \text{HBr} \rightleftharpoons \text{tert-R}\overset{\oplus}{\text{O}}\text{H}_2 \rightleftharpoons \text{tert-R}^{\oplus}$$

$$\text{Br}^{\ominus} \swarrow \qquad \searrow -\text{H}^{\oplus}$$

$$\text{tert-RBr} \qquad \text{alkene}$$

(Section 15-5D)

Alkyl halides are formed by either S_N1 or S_N2 processes using hydrogen halides (HCl, HBr, HI) or acid halides ($SOCl_2$, PCl_3, PBr_3, PI_3):

$$\text{ROH} + \text{HCl} \rightleftharpoons \overset{\oplus}{\text{ROH}}_2 + \text{Cl}^{\ominus} \longrightarrow \text{RCl} + \text{H}_2\text{O}$$

(Section 15-5A)

Tertiary alcohols readily react with HCl and HBr. Primary alcohols require an electrophilic catalyst (ZnX_2) to give halides from HCl and HBr. Acid halides of inorganic acids, $SOCl_2$, PCl_3, and so on, work best with primary alcohols to give primary halides.

Reactions involving the R—C(OH) or H—C(OH) bonds are mostly oxidation reactions:

primary $RCH_2OH \xrightarrow{[O]} RCHO \xrightarrow{[O]} RCO_2H$ (Section 15-6)

secondary $R_2CHOH \xrightarrow{[O]} R_2C=O$

tertiary $R_3COH \xrightarrow{[O]}$ fragmentation

Common laboratory oxidizing agents are potassium permanganate (in acidic or basic solution) and potassium or sodium dichromate (in acidic solution).

Under acidic conditions, oxidation gives first an unstable ester of the inorganic oxyacid which, in a subsequent slow step, undergoes an elimination-type reaction:

$$\text{H–C–O–H} + \text{H–O–Metal} \rightleftharpoons \text{H–C–O–Metal} + \text{H}_2\text{O}$$

$$\downarrow$$

$$\text{C=O} + \text{H}_3\text{O}^{\oplus} + \text{Metal (reduced)}$$

Other oxidations involve hydride transfer (or the equivalent) from the alpha carbon to the oxidant. Permanganate usually functions in this way, as does the pyridine nucleotide, $NAD^{\oplus}$:

$$MnO_4^{\ominus} + H-\overset{|}{\underset{|}{C}}-O^{\ominus} \longrightarrow HMnO_4^{2\ominus} + \overset{\diagdown}{\underset{\diagup}{C}}=O$$

$$NAD^{\oplus} + H-\overset{|}{\underset{|}{C}}-O-H \xrightarrow{\text{enzyme}} NADH + \overset{\diagdown}{\underset{\diagup}{C}}=O + H^{\oplus}$$

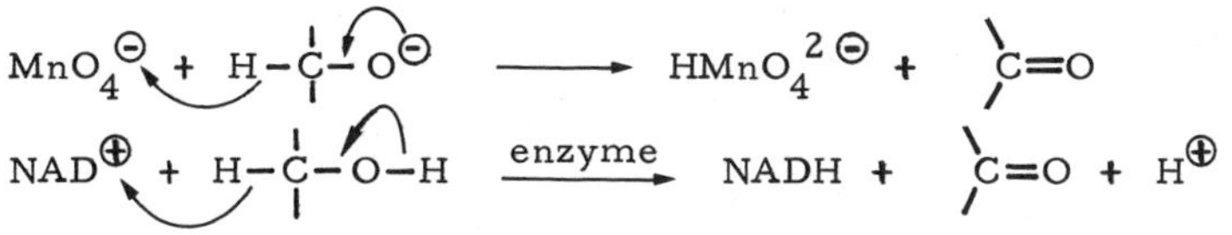

Protection of the OH function of alcohols is sometimes necessary in synthetic work and can be achieved by converting the −OH group to an ether, ester, or acetal function, which is constructed so it can be removed easily when desired:

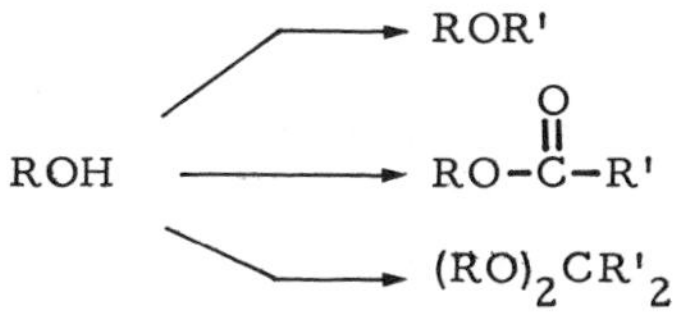

$$ROH \begin{array}{l} \longrightarrow ROR' \\ \\ \longrightarrow RO-\overset{\overset{\displaystyle O}{\|}}{C}-R' \\ \\ \longrightarrow (RO)_2CR'_2 \end{array} \qquad \text{(Section 15-9)}$$

The simple alkyl ethers R−O−R may be prepared by the reactions summarized in Table 15-4. Ethers are unreactive except in radical reactions or when converted to oxonium salts by strong acids. Cleavage of simple alkyl ethers is effected by hydrogen bromide or hydrogen iodide:

$$R-O-R + HBr \longrightarrow R-\overset{\overset{\displaystyle H}{|}}{\underset{\oplus}{O}}-R + Br^{\ominus} \begin{array}{l} \overset{S_N2}{\longrightarrow} ROH + RBr \\ \\ \underset{S_N1}{\longrightarrow} ROH + R^{\oplus} + Br^{\ominus} \end{array}$$

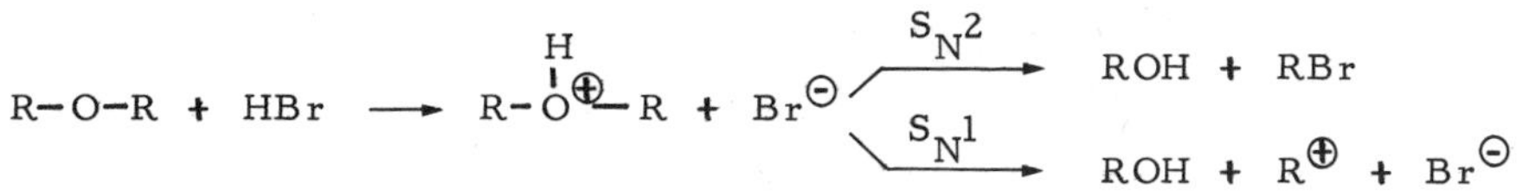

Whether the R−O−R' linkages will be cleaved by HX to give ROH + R'X or RX + R'OH depends on the structures of the attached alkyl groups. We can expect the C−O bonds to those groups to cleave most readily that are the more easily displaced in S_N2 reactions, or that readily form carbocations in S_N1 reactions:

$$\text{(benzene ring)}-\overset{\overset{\displaystyle H}{|}}{\underset{\oplus}{O}}-CH_3 \xrightarrow[S_N2]{Br^{\ominus}} \text{(benzene ring)}-OH + CH_3Br$$

$$CH_2{=}CH-CH_2\overset{\overset{\displaystyle H}{|}}{\underset{\oplus}{O}}CH_3 \xrightarrow{S_N1} CH_2{=}CH-\overset{\oplus}{C}H_2 + CH_3OH$$

$$\downarrow Br^{\ominus}$$

$$CH_2{=}CH-CH_2Br$$

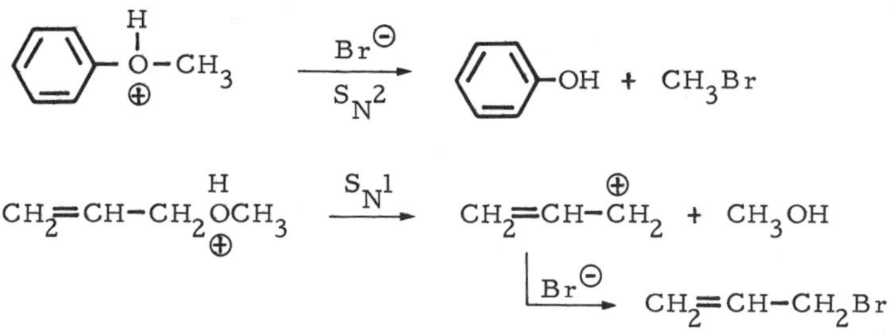

Oxacyclopropanes (oxiranes) are strained cyclic ethers with three-membered rings. They readily are cleaved by strong nucleophiles, particularly in the presence of acids:

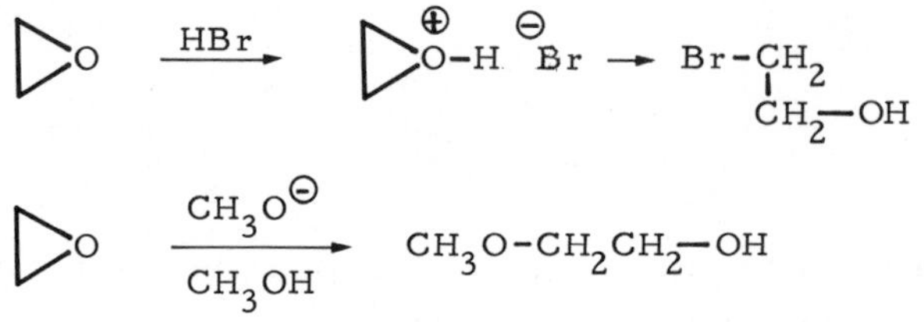

(Section 15-11D)

Oxacyclopropanes are prepared by oxidative additions to alkenes:

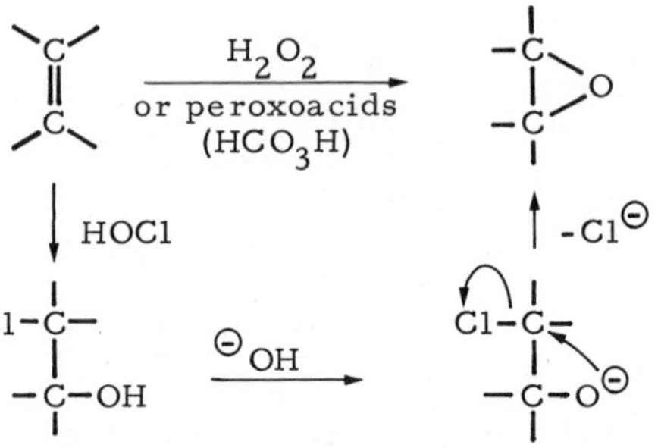

(Section 15-11C)

Many cyclic polyethers form stable complexes with metal cations and this property often permits making solutions of ionic compounds in nonpolar solvents. Because the anion of the salt usually is not strongly solvated in such solutions, it may show greatly enhanced S_N2 reactivity.

ANSWERS TO EXERCISES

15-1   a. 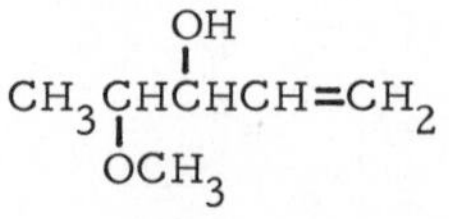
$$CH_3\underset{\underset{OCH_3}{|}}{\overset{\overset{OH}{|}}{C}H}CHCH=CH_2$$
b. 4-ethoxy-2-cyclopentenol

15-2   The cis diol is more volatile than the trans diol because it forms reasonably strong intramolecular hydrogen bonds and is not therefore strongly associated with neighboring molecules. The trans diol is sterically unable to form stable intramolecular hydrogen bonds but it is appreciably associated through intermolecular hydrogen bonds.

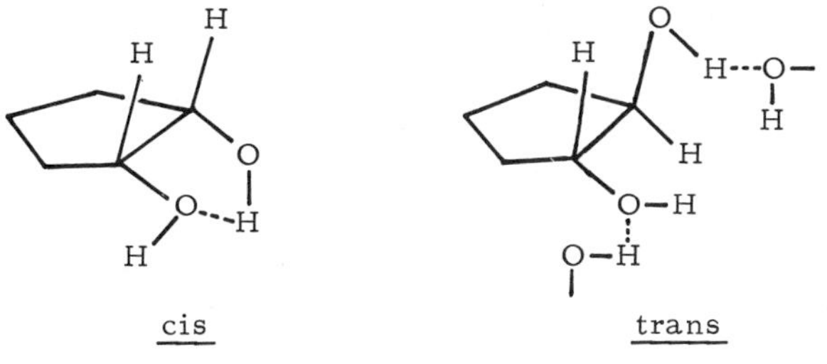

cis   trans

$\underset{\sim\sim\sim}{15\text{-}3}$ a. In very dilute solution, both diols would show sharp infrared bands due to unassociated hydroxyl groups (3600 cm^{-1}); 1,2-butanediol would probably also show a band due to intramolecularly associated hydroxyl groups (3500 cm^{-1}). This type of intramolecular H-bonding would be unfavorable for the trans-1,2-cyclobutanediol.

b. In moderately concentrated solution, the 3600 cm^{-1} bands present in very dilute solution of both compounds would diminish in intensity and new, intense broad bands would appear (3300 cm^{-1}) indicative of intermolecularly associated hydroxyl groups. The 3500 cm^{-1} band in the spectrum of 1,2-butanediol would not disappear.

c. As pure liquids, the infrared spectra of both compounds would show intense broad bands around 3300 cm^{-1} arising from intermolecularly associated hydroxyl groups. In addition, the spectrum of 1,2-butanediol would probably show an intense sharp band at 3500 cm^{-1} due to intramolecular hydrogen bonding.

$\underset{\sim\sim\sim}{15\text{-}4}$ a. The nmr spectrum of compound C_4H_6O shows four types of protons in the ratio 1:2:2:1. The infrared spectrum shows that one proton is that of an OH group (broad band centered at 3350 cm^{-1}) and another is an alkynic hydrogen (strong sharp band at 3300 cm^{-1} and weak band at 2200 cm^{-1}). The alkynic hydrogen shows in the nmr spectrum as a triplet at 2.1 ppm and the OH proton as a broad signal at 4.3 ppm. The triplet at 3.8 ppm is clearly that of the $\underline{CH_2}$—OH methylene protons. The correct structure for compound A is, therefore,

$$HC{\equiv}CCH_2CH_2OH$$

The signal at 2.4 ppm is that of the β-CH$_2$ group which appears as a sextet of lines due to spin coupling with the α-methylene protons and with the alkynic proton. The resonance due to the latter is accordingly split into a triplet.

b. The nmr spectrum of compound $C_3H_8O_2$ shows three different types of protons in the ratio of $1:4:3$. One of these (corresponding to the signal at 4.3 ppm) is a hydroxyl proton since the infrared spectrum clearly shows the presence of a hydroxyl group (sharp band at 3600 cm^{-1}, broad band near 3400 cm^{-1}). There are no other features in the infrared spectrum to indicate the presence of other functional groups. However, the sharp singlet at 3.3 ppm in the nmr spectrum is indicative of an $-OCH_3$ group. A likely structure for compound B is, therefore,

$$CH_3O-CH_2-CH_2-OH \qquad \text{2-methoxyethanol (methyl cellosolve)}$$

The complex of lines at 3.5 ppm corresponds to the four protons of the methylene groups.

15-5   Because the addition of 5% water produces no significant change in the ethanol nmr spectrum but contributes an additional resonance due to the water protons, the OH protons of water and ethanol cannot be exchanging rapidly.

Addition of 30% causes rapid exchange of the OH protons - as evidenced by the appearance of a <u>single</u> resonance for both types of hydroxyl groups, and by the disappearance of spin-coupling between the methylene and hydroxyl protons of ethanol (see Section 9-10E).

$$CH_3CH_2OH' + HOH \rightleftharpoons CH_3CH_2OH + H'OH$$

15-6   Several routes are possible. Only one representative scheme is given here.

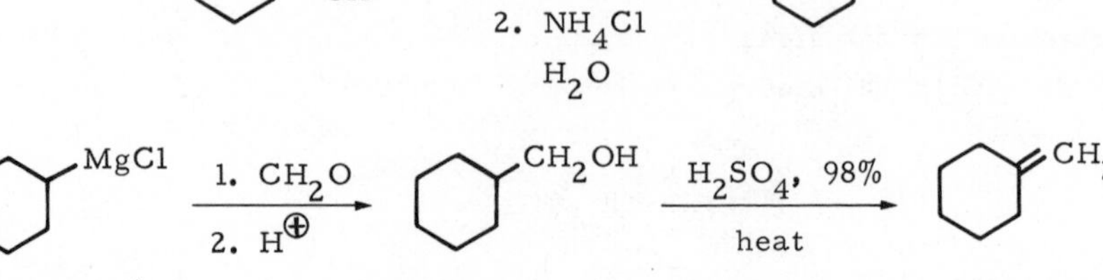

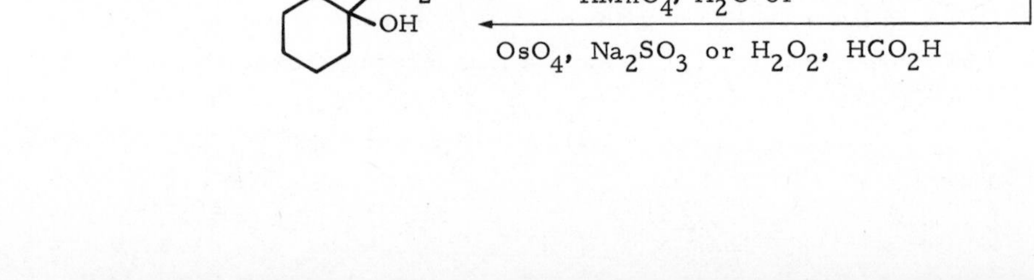

c.

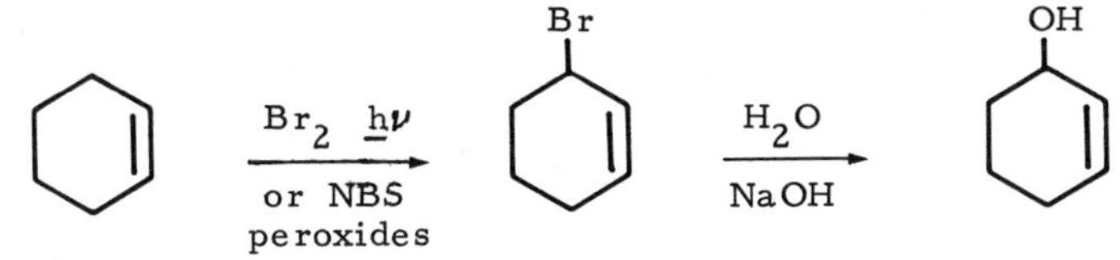

$$\text{Br}_2 \xrightarrow{h\nu} \text{ or NBS} \text{ peroxides}$$

$$\xrightarrow[\text{NaOH}]{\text{H}_2\text{O}}$$

d.

(from part a)

1.

2. $\text{H}^{\oplus}$

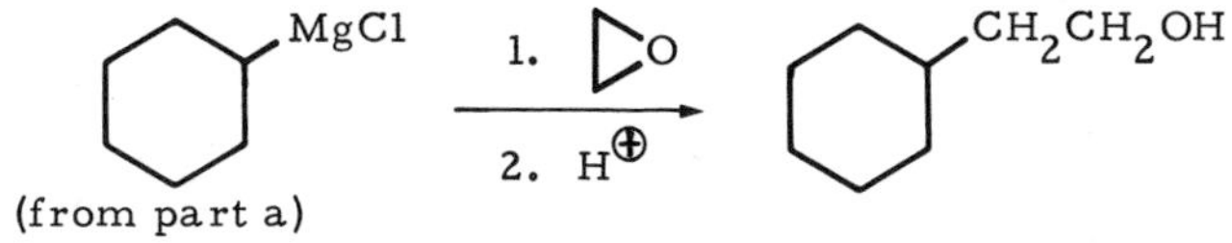

15-7   The order of basicity of alcohols should parallel their order of acidity in the gas phase. That is to say, the larger the alcohol, the more basic it will be because equilibria of the type $\text{ROH} + \text{R'OH}_2^{\oplus} \rightleftharpoons \text{ROH}_2^{\oplus} + \text{R'OH}$ will favor the ion with the larger (more polarizable) R group.

15-8

a. $\text{CH}_3\text{CH}_2\text{OH} \xrightarrow[-\text{H}_2]{\text{NaH}} \text{CH}_3\text{CH}_2\text{ONa} \xrightarrow[-\text{NaI}]{\text{CH}_3\text{I}} \text{CH}_3\text{CH}_2\text{OCH}_3$

b. $\text{CH}_2{=}\text{CHCHO} \xrightarrow[\text{2. NH}_4\text{Cl}]{\text{1. CH}_3\text{MgI}} \text{CH}_2{=}\text{CHCH(CH}_3)\text{OH} \xrightarrow[-\text{H}_2]{\text{NaH}} \xrightarrow[-\text{NaI}]{\text{C}_2\text{H}_5\text{I}}$

$$\text{CH}_2{=}\text{CHCH(CH}_3)\text{OC}_2\text{H}_5 \longleftarrow$$

c.

$$\xrightarrow[\substack{\text{or 1. B}_2\text{H}_6 \\ \text{2. H}_2\text{O}_2}]{\text{H}_2\text{SO}_4 - \text{H}_2\text{O}}$$

$$\xrightarrow[-\text{H}_2]{\text{NaH}} \xrightarrow[-\text{NaI}]{\text{CH}_3\text{I}}$$

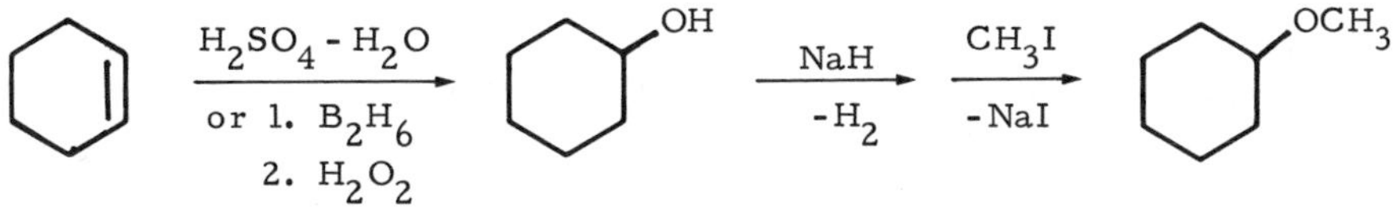

15-9   If the methanol were labelled as $\text{CH}_3{}^{18}\text{OH}$ the $\text{S}_\text{N}2$ mechanism of esterification would lead to <u>unlabelled</u> ester. The ^{18}O would be lost as $\text{H}_2{}^{18}\text{O}$. The addition-elimination mechanism would lead to labelled ester and unlabelled water.

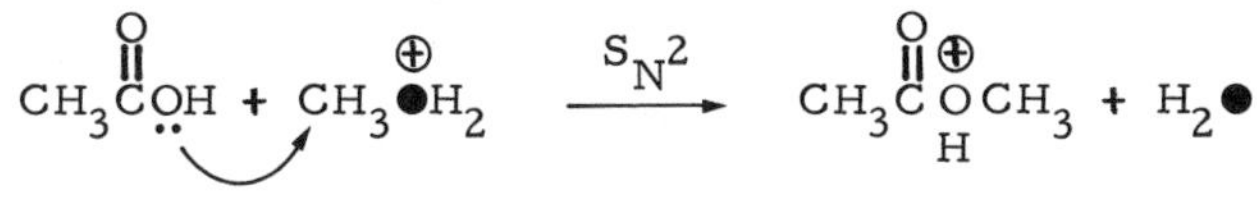

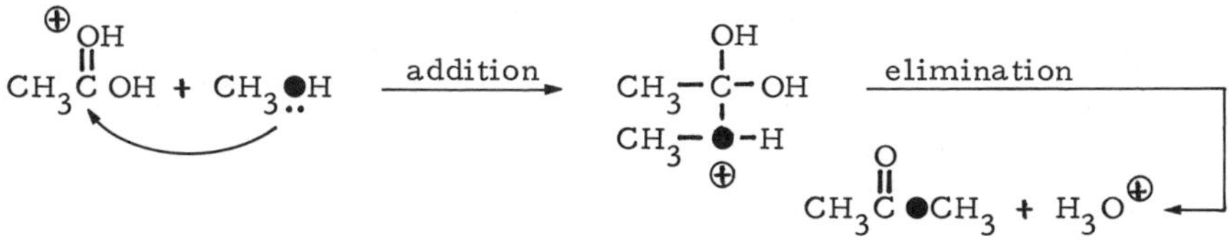

15-10   The tertiary alcohol readily dehydrates to 2-methylpropene on heating in a strong acid. Unless special precautions are taken the 2-methylpropene escapes as a gas.

$$(CH_3)_3COH \; \overset{H^{\oplus}}{\rightleftharpoons} \; (CH_3)_3\overset{\oplus}{C}OH_2 \; \overset{-H_2O}{\rightleftharpoons} \; (CH_3)_3C^{\oplus} + H_2O \quad \overset{-H_3O^{\oplus}}{\longrightarrow}$$

$$(CH_3)_2C=CH_2$$

The tert-butyl ethanoate will likewise eliminate by an El mechanism on heating with strong acid.

$$CH_3CO_2C(CH_3)_3 \; \overset{H^{\oplus}}{\longrightarrow} \; CH_3CO_2H + (CH_3)_2C=CH_2 + H^{\oplus}$$

15-11   From the comments in Exercise 15-10, the temperature of esterification of ethanoic acid by 2-methylpropene must be kept low to prevent the product ester from eliminating by the reverse reaction. Water must be excluded in order to keep the ester from hydrolyzing by S_N1-El processes. The steps in ester formation follow:

$$(CH_3)_2C=CH_2 \; \overset{H^{\oplus}}{\rightleftharpoons} \; (CH_3)_3C^{\oplus} \; \overset{H\ddot{O}_2CCH_3}{\rightleftharpoons} \; CH_3\overset{O}{\overset{\|}{C}}\overset{\oplus}{O}C(CH_3)_3 \; \overset{-H^{\oplus}}{\rightleftharpoons}$$

$$\overset{}{\underset{H}{}}$$

$$CH_3CO_2C(CH_3)_3$$

15-12

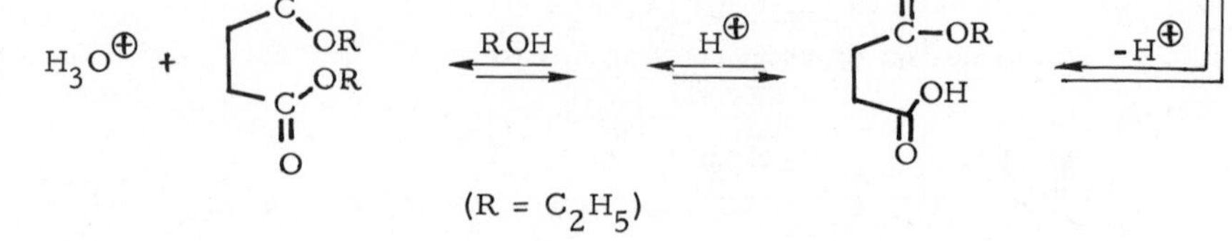

$$(R = C_2H_5)$$

The equilibrium is favorable to diester formation in the presence of excess acid because the acid converts the water formed to $H_3O^{\oplus}$. The ester will be a weaker base than H_2O.

15-13
a.

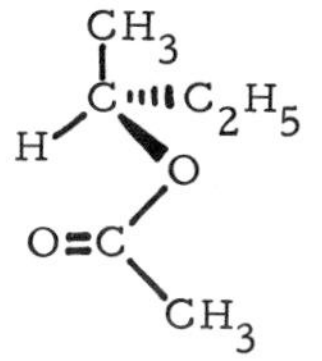

b. $CH_3\overset{O}{\overset{\|}{C}}-O-CH_2CH=CH_2$

c. $CH_2CO_2CH_3$
   $|$
   $CH_2CO_2CH_3$

15-14   The net reaction may be written as

$$CH_3CH_2OH + CH_3CO_2^{\ominus} \rightleftharpoons CH_3CO_2CH_2CH_3 + HO^{\ominus}$$

and is not likely to be energetically favorable or thermoneutral ($\underline{K} \leq 1$) because of the resonance stabilization of $CH_3CO_2^{\ominus}$. Furthermore, the reaction is unlikely to occur at a reasonable rate because the first step involves loss of the resonance stabilization of $CH_3CO_2^{\ominus}$.

15-15   One of the C—O bonds in a hemiacetal or hemiketal is a potential carbonyl group:

$$-O-\overset{|}{\underset{|}{C}}-OH \longrightarrow -OH + \overset{\diagdown}{\underset{\diagup}{C}}=O$$

If we write a similar conversion for the simple sugars, we see that it involves a rearrangement to an open-chain structure; in reverse, the cyclic structure arises by the addition of one of the hydroxyl groups on the carbon chain to the carbonyl carbon.

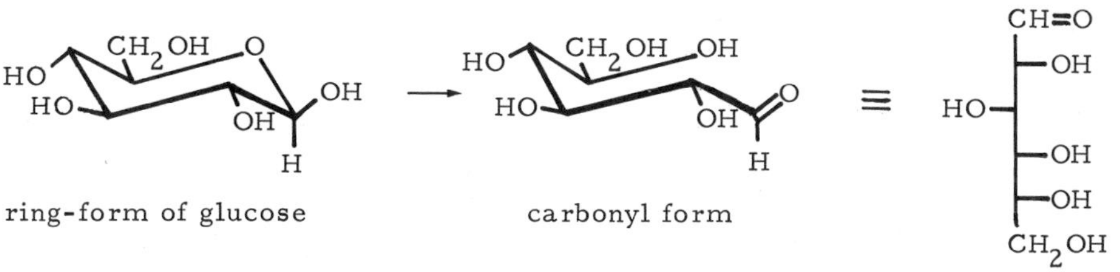

ring-form of glucose   carbonyl form

projection formula of
$2\underline{D}$, $3\underline{L}$, $4\underline{D}$, $5\underline{D}$, 6-pentahydroxyhexanal

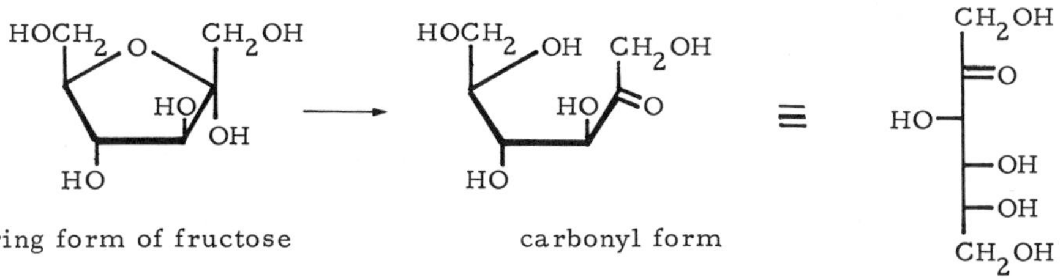

ring form of fructose   carbonyl form

projection formula of
1, $3\underline{L}$, $4\underline{D}$, $5\underline{D}$, 6-pentahydroxy-2-hexanone

D-ribose is $2\underline{D}$, $3\underline{D}$, $4\underline{D}$, 5-tetrahydroxypentanal

15-16

$$CH_3-\overset{O}{\underset{}{C}}-H \; + \; \overset{\ominus}{OCH_3} \; \rightleftharpoons \; CH_3-\overset{O^\ominus}{\underset{H}{C}}-OCH_3 \; \xrightarrow{CH_3OH} \; CH_3-\overset{OH}{\underset{H}{C}}-OCH_3 \; + \; \overset{\ominus}{OCH_3}$$

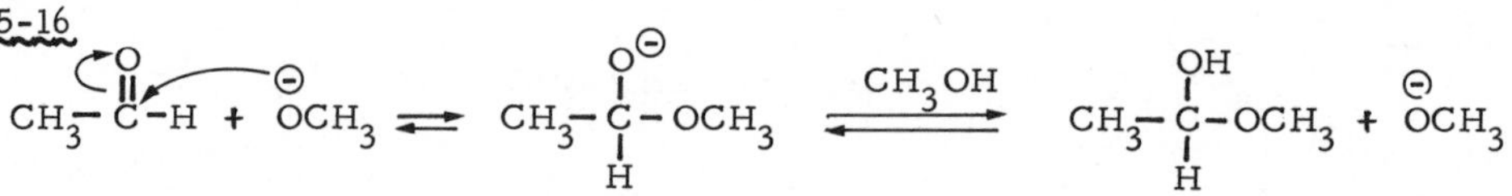

It is not possible to convert 1-methoxyethanol to 1,1-dimethoxyethane under basic conditions because $^\ominus OH$ is a poor leaving group.

$$CH_3-\overset{OH}{\underset{H}{C}}-OCH_3 \; + \; \overset{\ominus}{OCH_3} \; \not\rightarrow \; CH_3-\overset{OCH_3}{\underset{H}{C}}-OCH_3 \; + \; \overset{\ominus}{OH}$$

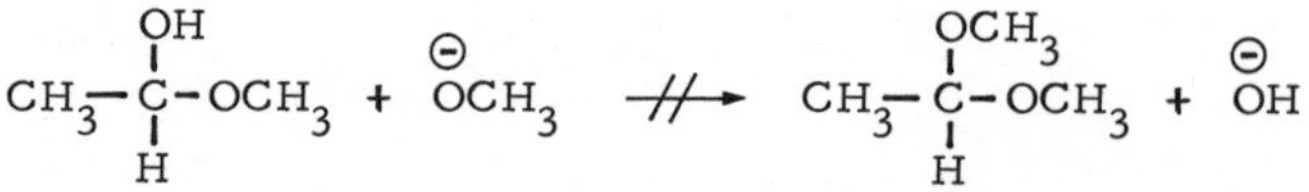

15-17   If ^{18}O-labelled methanol were used to convert a hemiacetal to an acetal, the label will be lost as $H_2^{18}O$ if a displacement mechanism operates, but will be incorporated into the acetal by an elimination-addition mechanism.

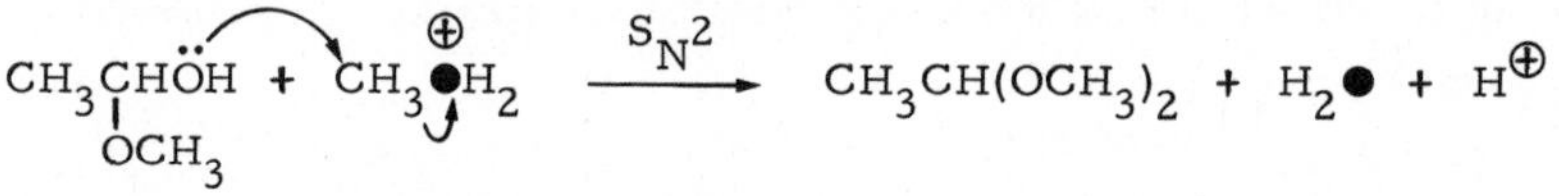

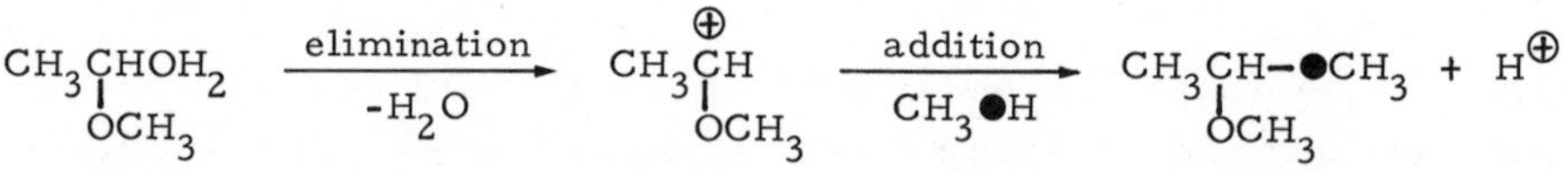

$$CH_3\underset{OCH_3}{\overset{}{CH\ddot{O}H}} \; + \; CH_3\overset{\oplus}{OH_2} \; \xrightarrow{S_N2} \; CH_3CH(OCH_3)_2 \; + \; H_2\bullet \; + \; H^\oplus$$

$$CH_3\underset{OCH_3}{\overset{}{CHOH_2}} \; \xrightarrow[-H_2O]{elimination} \; CH_3\underset{OCH_3}{\overset{\oplus}{CH}} \; \xrightarrow[CH_3\bullet H]{addition} \; CH_3\underset{OCH_3}{\overset{}{CH-\bullet CH_3}} \; + \; H^\oplus$$

15-18
1. $H-\overset{OCH_3}{\underset{OCH_3}{C}}-OCH_3 \; \underset{}{\overset{H^\oplus}{\rightleftharpoons}} \; H-\overset{CH_3O}{\underset{CH_3O}{C}}-\overset{H}{\underset{CH_3}{O^\oplus}} \; \xrightarrow{-CH_3OH} \; H-\overset{OCH_3}{\underset{OCH_3}{C^\oplus}}$

$$H-\overset{O}{\underset{}{C}}-OCH_3 \; \underset{-H^\oplus}{\overset{-CH_3OH}{\longleftarrow}} \; H-\overset{OCH_3}{\underset{OCH_3}{C}}-\overset{\oplus}{OH_2} \; \overset{H_2O}{\longleftarrow}$$

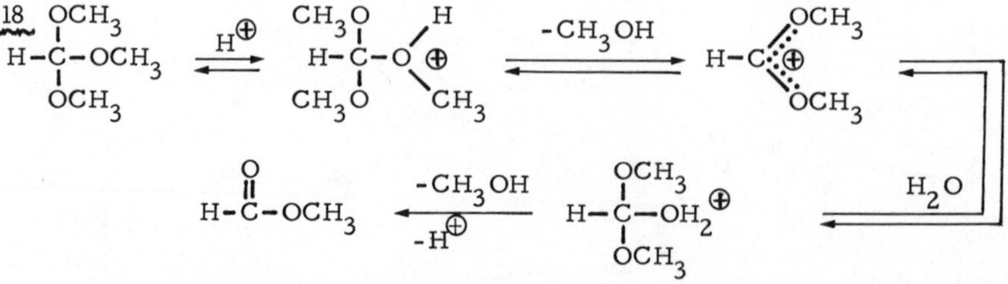

Notice that the hydrolysis in reaction 1 produces methanol which is utilized in reaction 2.

2. $C_6H_5\overset{O}{\underset{}{C}}CH_3 \; \overset{H^\oplus}{\rightleftharpoons} \; C_6H_5\overset{\overset{\oplus}{OH}}{\underset{}{C}}CH_3 \; \xrightarrow{CH_3OH} \; C_6H_5-\overset{\overset{\oplus}{OH_2}}{\underset{CH_3}{C}}-OCH_3 \; \xrightarrow{-H_2O}$

$$C_6H_5C(OCH_3)_2CH_3 \quad \underset{-H^\oplus}{\overset{CH_3OH}{\longleftarrow}}$$

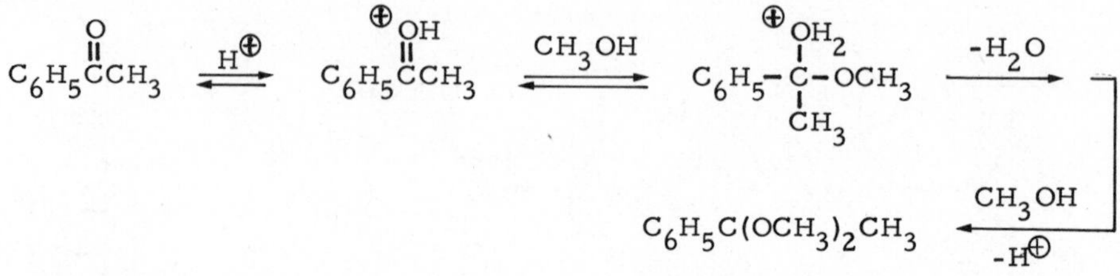

The water produced in reaction 2 is utilized in the hydrolysis of reaction 1.

15-19

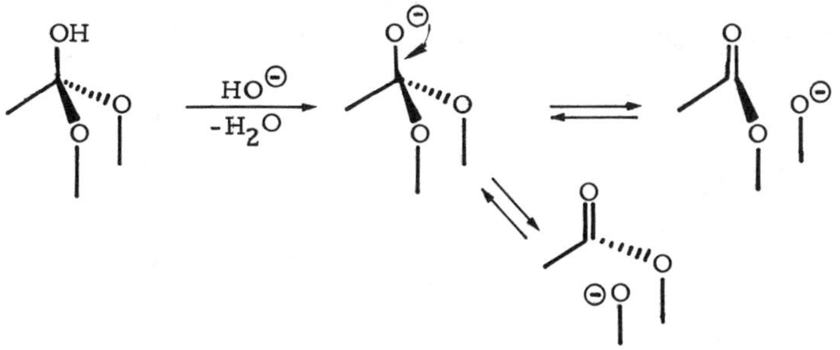

The C—O bonds to the bridgehead carbon are likely to break under basic conditions because the product has a stable ester linkage. This cleavage will lead to rearrangement of the caged structure.

15-20 S_N2 reaction with inversion is the desired process and will be achieved with $SOCl_2$ in pyridine. The chlorosulfate ester is formed first; then the displacement step is attack of $Cl^{\ominus}$ on the chlorosulfite ester. If base (pyridine) is not present, the product of the first step is HCl, which is a poor nucleophile, and which promotes S_N1 reaction with racemization. (See also Exercise 15-57 for a fuller explanation.) The combination HCl and $ZnCl_2$ also promotes racemization by S_N1 reaction.

15-21   We can write a sequence of steps that initially involve formation of a phosphite ester, analogous to the formation of chlorosulfite ester in the $SOCl_2$ + ROH reaction. The displacement step to form RBr probably occurs on the ester intermediate.

$$ROH + PBr_3 \longrightarrow RO—PBr_2 + HBr$$
$$\text{phosphite ester}$$

$$HBr + \overset{}{\underset{}{\text{(pyridine)}}} N \longrightarrow \overset{}{\underset{}{\text{(pyridine)}}} \overset{\oplus}{N}-H \ Br^{\ominus}$$

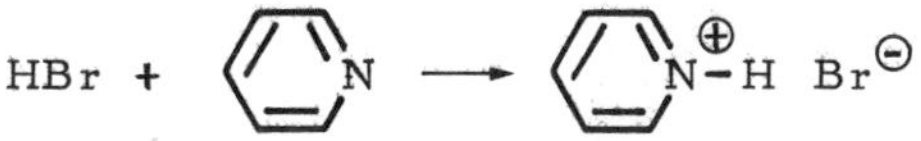

$$Br^{\ominus} + R-O—PBr_2 \xrightarrow{S_N2} RBr + O{=}PBr_2$$

The remaining P-Br bonds also are reactive and are replaced to give RBr and phosphite salts.

15-22   The Grignard reaction proceeds normally:

$$CH_3CH{=}CHCHO + CH_3MgI \rightarrow CH_3CH{=}CHCH(CH_3)OMgI \xrightarrow{10\% \ H_2SO_4}$$

$$CH_3CH{=}CHCH(CH_3)OH \longleftarrow$$

Traces of residual acid in the crude ether extract catalyzes the dehydration of the product alcohol and ether formation:

$$CH_3CH=CHCH(CH_3)OH \xrightarrow{H^\oplus} CH_3CH=CHCH(CH_3)\overset{\oplus}{O}H_2$$

$$\downarrow -H_2O$$

$$CH_3\overset{\frac{1}{2}\oplus}{C}H\cdots CH\cdots\overset{\frac{1}{2}\oplus}{C}HCH_3$$

$$CH_3CH=CH\overset{\overset{\displaystyle CH_3}{|}}{C}HOH \xleftarrow{-H^\oplus} \quad \xrightarrow{-H^\oplus}$$

$$\overset{-H^\oplus}{\swarrow} \qquad \searrow$$

$$(CH_3CH=CH\overset{\overset{\displaystyle CH_3}{|}}{CH}\!\!\rightarrow_2 O \qquad\qquad CH_3CH=CH-CH=CH_2$$

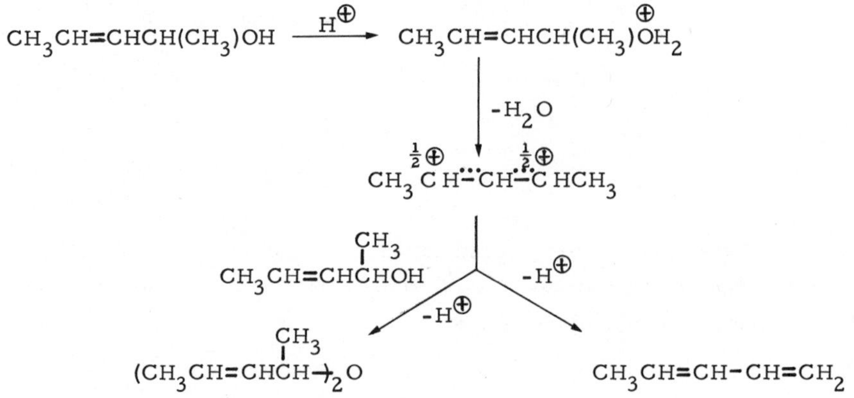

To avoid ether and diene formation the organic extract from acid hydrolysis must be neutralized by washing with dilute base (5% Na_2CO_3).

<u>15-23</u>

a. cyclohexene

b.

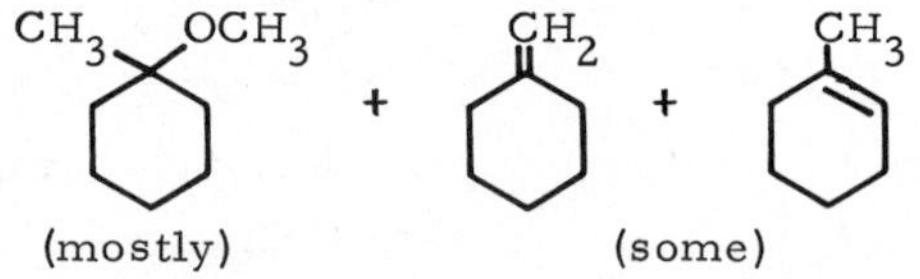

$$\underset{\text{(mostly)}}{\overset{CH_3\quad OCH_3}{\bigcirc}} \quad + \quad \overset{CH_2}{\bigcirc} \quad + \quad \underset{\text{(some)}}{\overset{CH_3}{\bigcirc}}$$

c. $CH_2=\overset{\overset{\displaystyle}{|}}{\underset{\overset{\displaystyle}{CH_3}}{C}}-CH=CH_2$

<u>15-24</u> $(CH_3)_3COH \xrightarrow[-HSO_4^\ominus]{H_2SO_4} (CH_3)_3C\overset{\oplus}{O}H_2 \xrightarrow{-H_2O} (CH_3)_3\overset{\oplus}{C}$

$$\downarrow :N\equiv C-CH_3$$

$(CH_3)_3CNH\overset{\overset{\displaystyle O}{||}}{C}CH_3 \xleftarrow{-H^\oplus} (CH_3)_3C-\underset{\oplus}{N}H=\overset{\overset{\displaystyle OH}{|}}{C}-CH_3 \xleftarrow{H_2O} (CH_3)_3C-\overset{\oplus}{N}\equiv C-CH_3$

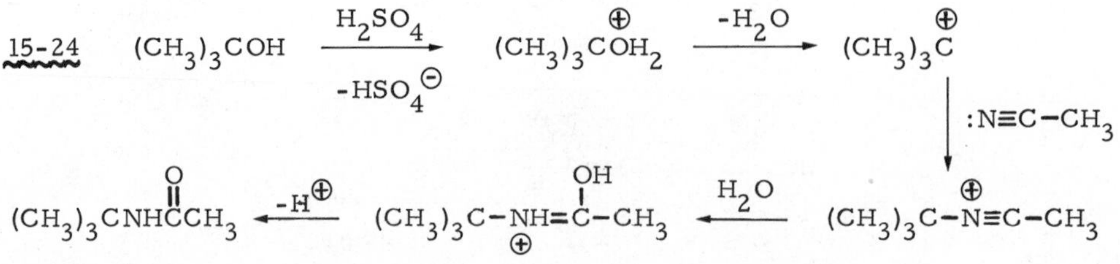

The Ritter reaction is more fully described in Section 24-3B.

<u>15-25</u>   By analogy with the discussion of ester formation coupled to ATP hydrolysis (Equations 15-5 and 15-6) we will assume that the carbodiimide, $RN=C=NR$, is converted by the carboxylic acid to an acyl intermediate that is

more reactive towards the alcohol, CH_3OH, than is the acid itself.

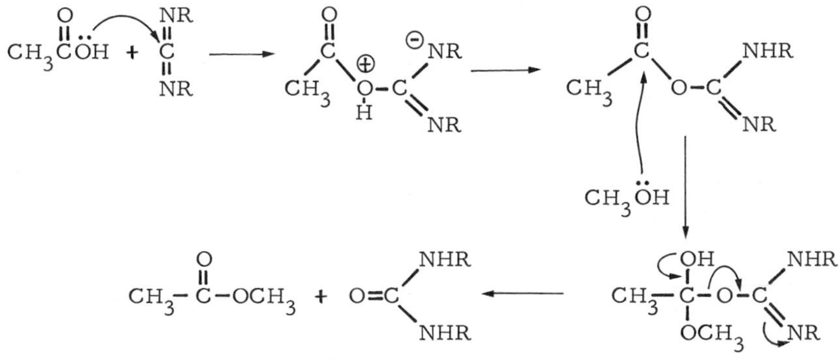

(See also Section 25-7C.)

15-26   a.   Esters are stabilized by electron delocalization of the type

$$-\overset{|}{\underset{\cdot\cdot}{O}}-C\!=\!\overset{\cdot\cdot}{O} \longleftrightarrow -\overset{\cdot\cdot}{O}\!=\!\overset{|}{C}-\overset{\cdot\cdot}{\underset{\cdot\cdot}{O}}:$$

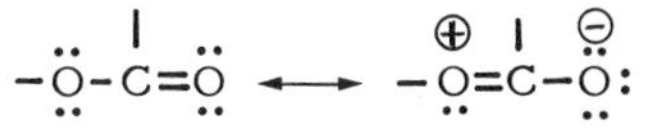

This stabilization is lost on formation of a trialkoxyalkane. Energetically, therefore, addition to an ester carbonyl has $\underline{K} < 1$.

b.   The position of equilibrium in ketal formation is unfavorable according to the reaction:

$$R_2C\!=\!O + 2CH_3OH \rightleftharpoons R_2C(OCH_3)_2 + H_2O \qquad \underline{K} < 1$$

but when coupled to the favorable process of trimethoxymethane hydrolysis:

$$HC(OCH_3)_3 + H_2O \rightleftharpoons HCO_2CH_3 + 2CH_3OH \qquad \underline{K} > 1$$

then conversion of a ketone to a ketal becomes favorable.

$$HC(OCH_3)_3 + R_2C\!=\!O \rightleftharpoons R_2C(OCH_3)_2 + HCO_2CH_3 \qquad \underline{K} > 1$$

The steps involved in this reaction are given in Exercise 15-18.

c.   Treatment of chloroform with sodium ethoxide in anhydrous ethanol gives triethoxyethane by way of dichlorocarbene.

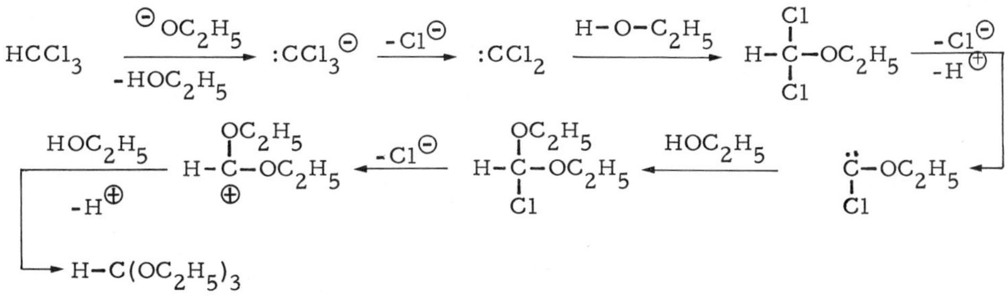

15-27   The first reaction can be described as a nucleophilic attack of the alcohol at phosphorus.

$$CH_3CH_2\overset{..}{\underset{..}{O}}H + {}^{\ominus}O-\overset{O}{\underset{\underset{O}{\ominus}}{\overset{\|}{P}}}-O-\overset{O}{\underset{\underset{O}{\ominus}}{\overset{\|}{P}}}-O-\overset{O}{\underset{\underset{O}{\ominus}}{\overset{\|}{P}}}-OR \longrightarrow CH_3CH_2-\bullet-\overset{O}{\underset{\underset{O}{\ominus}}{\overset{\|}{P}}}-OR + HP_2O_7^{3\ominus}$$

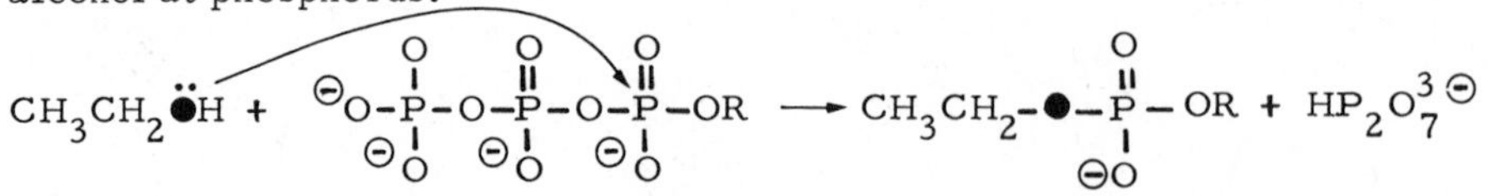

The second reaction may be described as a nucleophilic attack of ethanoic acid at carbon and would best proceed by attack of $CH_3CO_2^{\ominus}$ on the ethyl phosphate derivative.

$$CH_3\overset{O}{\overset{\|}{C}}-O^{\ominus} + CH_3CH_2-\bullet-\overset{O}{\underset{\underset{O}{\ominus}}{\overset{\|}{P}}}-OR \longrightarrow CH_3\overset{O}{\overset{\|}{C}}-OCH_2CH_3 + \bullet-\overset{O}{\underset{\underset{O}{\ominus}}{\overset{\|}{P}}}-OR$$

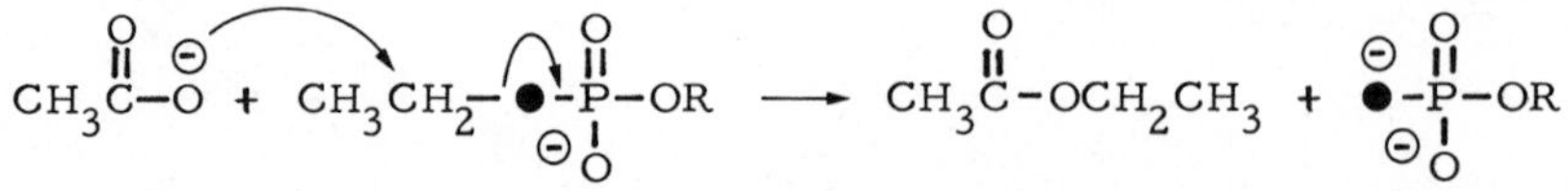

The above sequence appears less likely than the sequence of Equations 15-5 and 15-6 because S_N2 attack of $CH_3CO_2^{\ominus}$ is likely to be a slower reaction than attack of ethanol on the acyl phosphate. The roles assigned for the OH groups of the alcohol and acid are reasonable, but not as likely to give a favorable reaction rate. Labelling with ${}^{18}O$ would distinguish between the two mechanisms. By Equations 15-5 and 15-6, the ester from $R^{18}OH$ would incorporate the label. By the alternate mechanism, the ester would be unlabelled. Alternatively, one could use a stereochemical test to see if inversion of configuration in the ethyl derivative occurs (see for example Exercise 8-10).

15-28   Carbon changes its oxidation state from 0 to +2 in the elimination step.

15-29   In esterification, the slow step normally is the addition of the alcohol to the carbonyl carbon of the acid. The more hindered the alcohol or the acid the slower will be reaction. Thus, the relative rate is C3 > C11 because the OH at C3 is equatorial and less hindered than the OH at C11, which is axial and hindered by the methyls at C9 and C13.

   In chromic acid oxidation, the slow step is cleavage of the carbon-hydrogen of $H-\overset{|}{\underset{|}{C}}-OH$. The less hindered the C-H bond, the faster will be reaction. Thus, the relative rate is C11 > C3 because the C-H at C11 is equatorial and less hindered than the C-H at C3, which is axial.

15-30
a.

$$\langle 0 \rangle-OH \longrightarrow \langle +2 \rangle{=}O + 2e^{\ominus} + 2H^{\oplus}$$

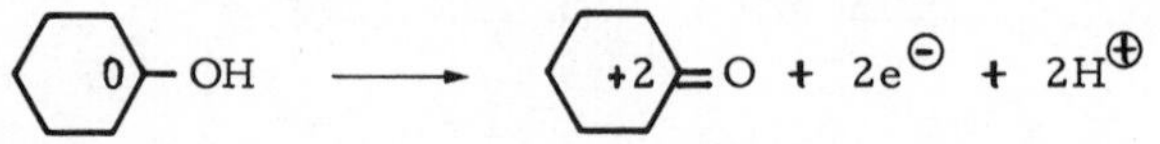

$$\overset{+7}{Mn}O_4^{\ominus} + 4H^{\oplus} + 3e^{\ominus} \longrightarrow \overset{+4}{Mn}O_2 + 2H_2O$$

One mole of cyclohexanol requires 2/3 mole of $MnO_4^{\ominus}$.

b. $\overset{-1}{C_6H_5CH_2OH}$ + H_2O $\longrightarrow$ $\overset{+3}{C_6H_5CO_2H}$ + $4H^{\oplus}$ + $4e^{\ominus}$

One mole of phenylmethanol requires 4/3 mole of $MnO_4^{\ominus}$.

15-31 $C_6H_5{-}\overset{O}{\overset{\|}{C}}{-}H$ + $\overset{\ominus}{OH}$ $\rightleftharpoons$ $C_6H_5{-}\overset{O^{\ominus}}{\overset{|}{\underset{H}{C}}}{-}OH$

$C_6H_5{-}\overset{O^{\ominus}}{\overset{|}{\underset{H}{C}}}{-}OH$ + $\overset{\ominus}{O}{-}Mn{\equiv}O$ $\longrightarrow$ $C_6H_5\overset{O}{\overset{\|}{C}}{-}OH$ + $\overset{\ominus}{O}{-}\overset{O}{\underset{OH}{Mn}}{-}O^{\ominus}$

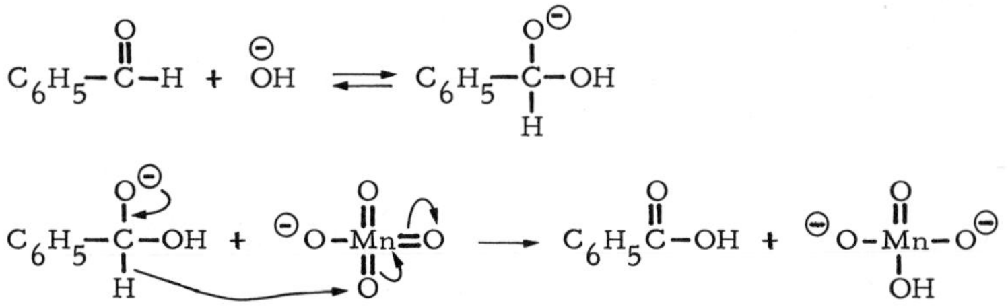

15-32 The mechanism of oxidation of alcohols in <u>acid</u> solution probably involves intermediate permanganic ester formation which leads to ^{18}O-labeled ketone.

R_2CHOH $\xrightarrow{H^{\oplus}}$ $R_2\overset{\oplus}{C}HOH_2$ + $H\overset{\cdot\cdot}{\bullet}{-}MnO_3$ $\xrightarrow{-H_3\overset{\oplus}{O}}$ $R_2\overset{}{\underset{H}{C}}{-}\bullet{-}MnO_3$ $\longrightarrow$

$R_2C{=}\bullet$ + $MnO_3^{\ominus}$ + $H_3O^{\oplus}$ $\xleftarrow{H_2O}$

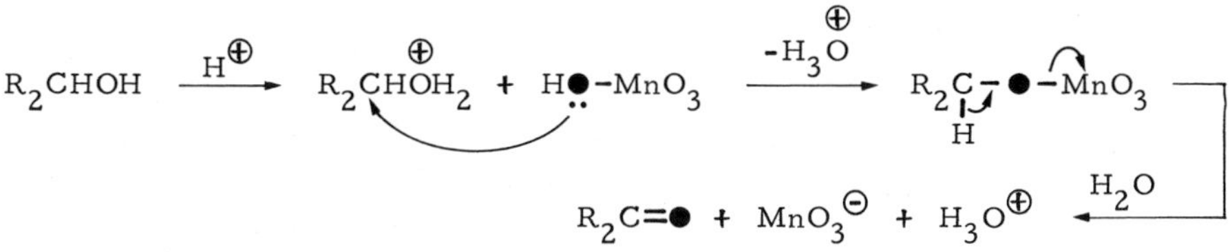

The reaction in <u>basic</u> solution occurs on the alkoxide anion rather than the neutral alcohol. Attack of $MnO_4^{\ominus}$ on the α-hydrogen of the anion thereby leads to <u>unlabelled</u> ketone.

R_2CHOH + $\overset{\ominus}{OH}$ $\rightleftharpoons$ $R_2CHO^{\ominus}$ + H_2O

$R_2C\overset{O^{\ominus}}{\underset{H}{\diagup}}$ + $Mn\bullet_4^{\ominus}$ $\longrightarrow$ $R_2C{=}O$ + $HMn\bullet_4^{2\ominus}$

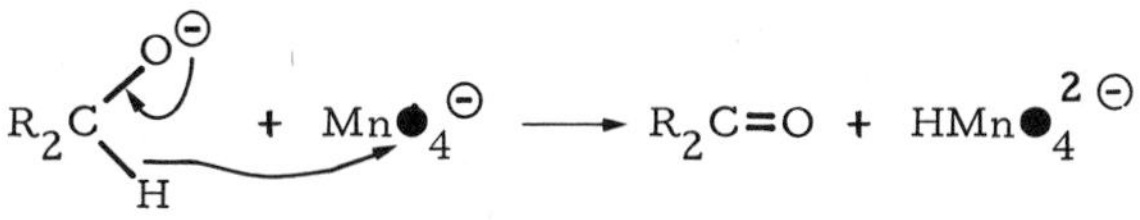

15-33 Because the reaction is significantly slower for the deuterium-labelled alcohol in acid solution, the slow step in the reaction must be the cleavage of the C$-$H (or C$-$D) bond (i.e., <u>second equation</u> in Exercise 15-31).

15-34

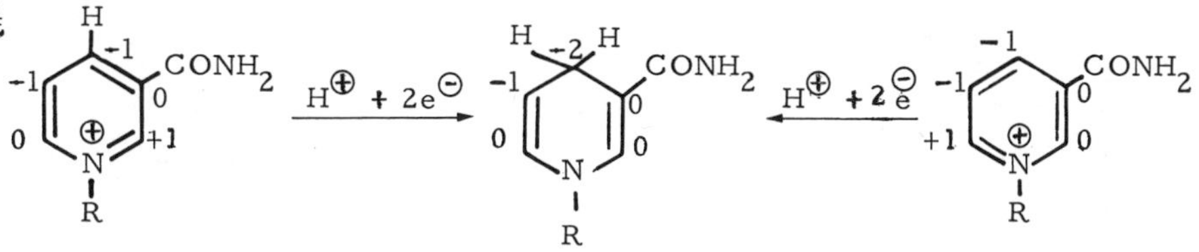

Thus, reduction is associated with the 4, 2 (and 6) positions of the pyridinium ring.

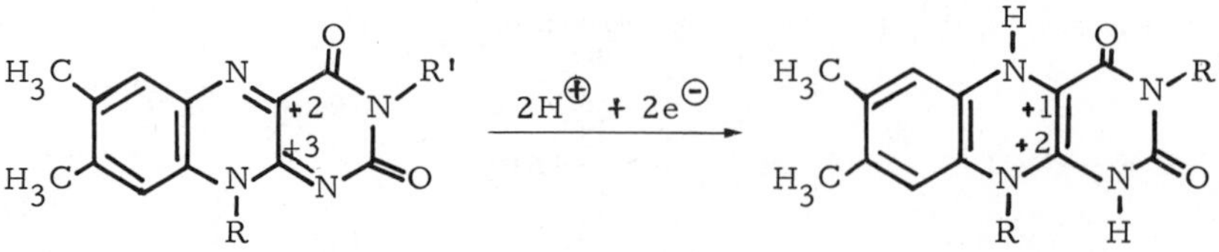

In FAD reduction, each of the two ring carbons indicated above undergoes a net change of +1 in oxidation state.

15-35

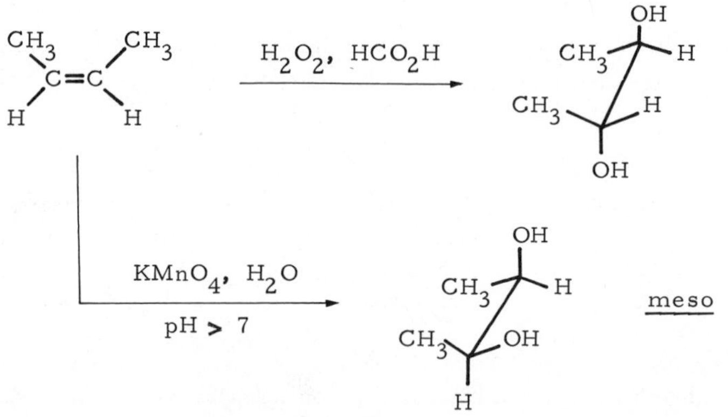

15-36

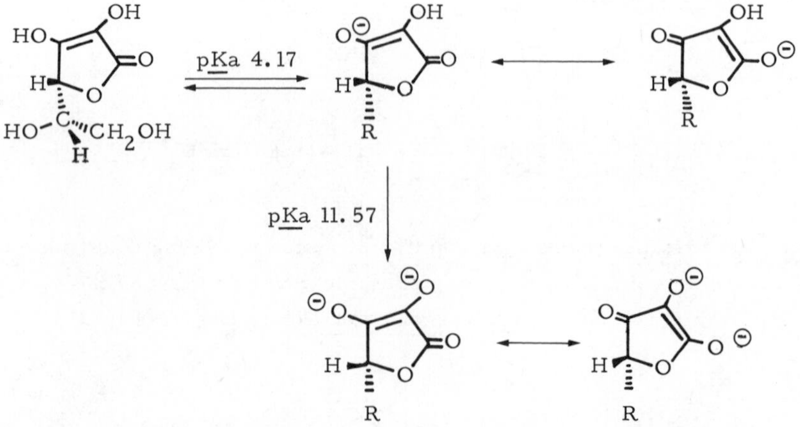

Ascorbic acid is an alkene-diol whereas 1,2-cyclopentanediol is an alkane-diol. The alkenediol is the stronger acid because its conjugate base is a delocalized (resonance-stabilized) anion.

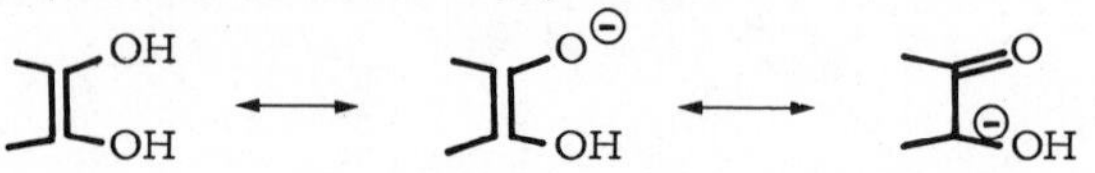

15-37 C_6H_5—CH_2—OCH_3 + H_2 $\xrightarrow{Pt}$ C_6H_5—CH_3 + $HOCH_3$

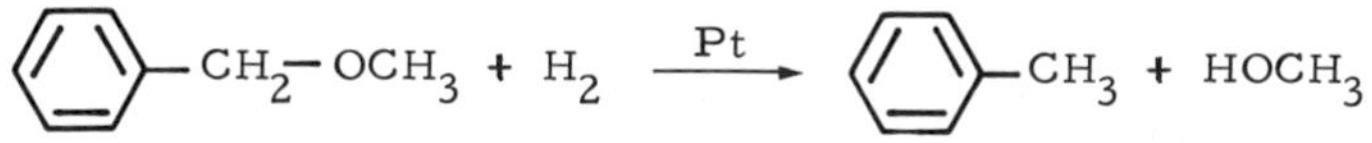

R—OCH_3 + H_2 $\longrightarrow$ R—H + $HOCH_3$

Hydrogenation of a phenylmethyl ether is more facile than hydrogenation of an alkyl ether because the C—O bond is significantly weaker in $C_6H_5CH_2$—OCH_3 than in R—OCH_3. The low C—O bond strength is associated with the stability of the benzyl radical $C_6H_5\dot{C}H_2$. (See Section 26-4D.)

15-38

a. $HOCH_2CH_2CHO$ $\xrightarrow{HOCH_2CH_2OH,\ H^{\oplus}}$ $HOCH_2CH_2CH\begin{smallmatrix}O\\ \\O\end{smallmatrix}$ $\xrightarrow{KMnO_4}$

$\xrightarrow{2CH_3MgCl}$ HO_2CCH_2CHO $\xleftarrow{H^{\oplus}}$ $HO_2CCH_2CH\begin{smallmatrix}O\\ \\O\end{smallmatrix}$

$\longrightarrow$ $ClMgO$—$\overset{O}{\overset{\|}{C}}CH_2\underset{CH_3}{CHOMgCl}$ $\xrightarrow{H^{\oplus}}$ $HO_2CCH_2\underset{CH_3}{CHOH}$

Alternatively: [pyran] + $HOCH_2CH_2CHO$ $\xrightarrow{H^{\oplus}}$ [tetrahydropyran]OCH_2CH_2CHO

$HO_2C\overset{O}{\overset{\|}{C}}CH_2\overset{O}{\overset{\|}{C}}CH_3$ $\xleftarrow{H_2CrO_4}$ $HOCH_2CH_2CH(CH_3)OH$ $\xleftarrow[2.\ H^{\oplus}]{1.\ CH_3MgCl}$

$\xrightarrow{NaBH_4}$ $HO_2CCH_2CH(OH)CH_3$ (See Section 16-4E)

b. CH_2=$CHCHO$ $\xrightarrow[NH_4^{\oplus}(catalyst)]{HC(OCH_3)_3}$ CH_2=$CHCH(OCH_3)_2$ $\xrightarrow{KMnO_4,\ H_2O}$

$HOCH_2CH(OH)CHO$ $\xleftarrow{H^{\oplus}}$ $\underset{HO\ \ \ OH}{CH_2CHCH(OCH_3)_2}$

c. HO—[cyclobutane]—CH_2Br $\xrightarrow{[pyran]}$ [tetrahydropyran]—O—[cyclobutane]—CH_2Br $\xrightarrow[\substack{2.\ (CH_3)_2C=O \\ 3.\ H^{\oplus}}]{1.\ Mg}$

HO—[cyclobutane]—$CH_2\underset{CH_3}{\overset{OH}{\underset{|}{\overset{|}{C}}}}CH_3$

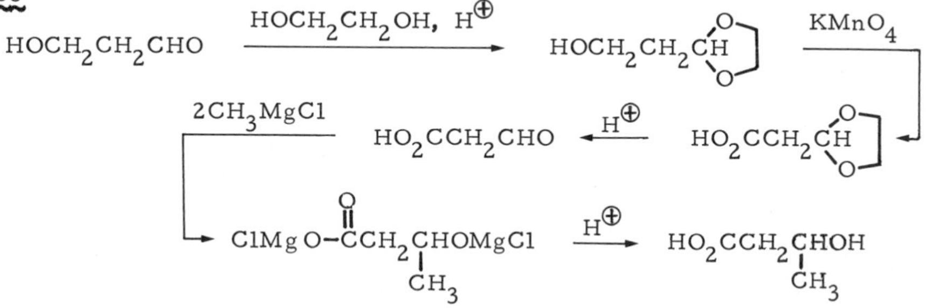

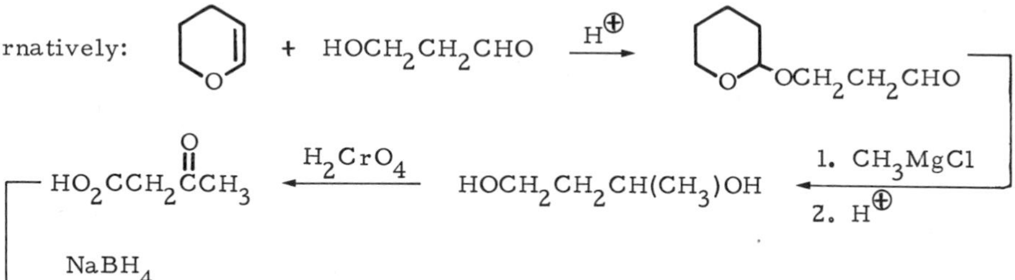

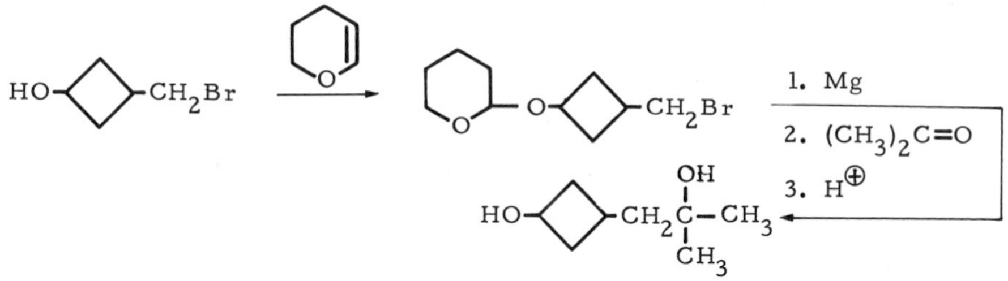

15-39

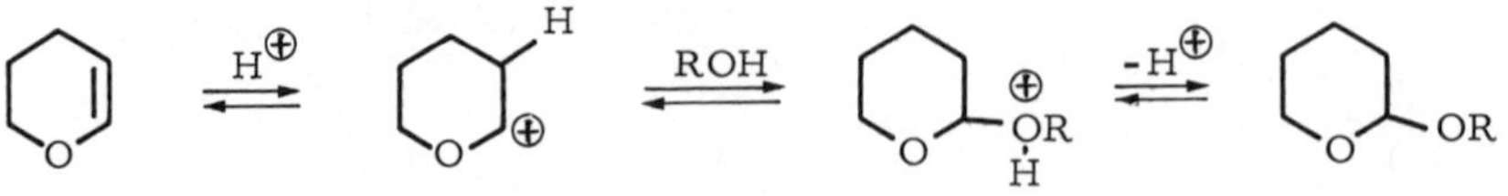

16

The higher reactivity of 16 relative to cyclohexene is because of the stabilization of the carbocation intermediate by the adjacent ether oxygen,

$$R-\overset{..}{\underset{..}{O}}-\overset{\oplus}{C}H- \longleftrightarrow R-\overset{\oplus}{\underset{..}{O}}=CH-$$

Ethoxyethene would be a comparably useful reagent for the protection of OH groups because, like 16, it possesses the structural feature $O-\underset{|}{C}=C$.

15-40 a. $CH_2{=}CHCH_2I + CH_3OH$ b. $CH_3CH_2I + O{=}CHCH_3$

c. $(CH_3)_3CCH_2OH + CH_3I$ d. $(CH_3)_3CI + (CH_2)_2C{=}CH_2 + HOCH_3$

e. $I(CH_2)_4OH + I(CH_2)_4I$ f. 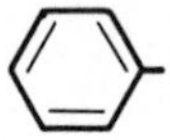 —$OH + CH_3I$

15-41 a. Methyl iodide and dimethyl sulfate are less reactive than trimethyl-oxonium salts towards nucleophiles because $I^{\ominus}$ and $^{\ominus}OSO_2OCH_3$ are poorer leaving groups than CH_3OCH_3. Therefore, whereas $(CH_3)_3O^{\oplus}$ will react with weak nucleophiles such as neutral alcohols, the less reactive CH_3I and $CH_3OSO_2OCH_3$ require stronger nucleophiles such as $RO^{\ominus}$ formed under basic conditions.

b.

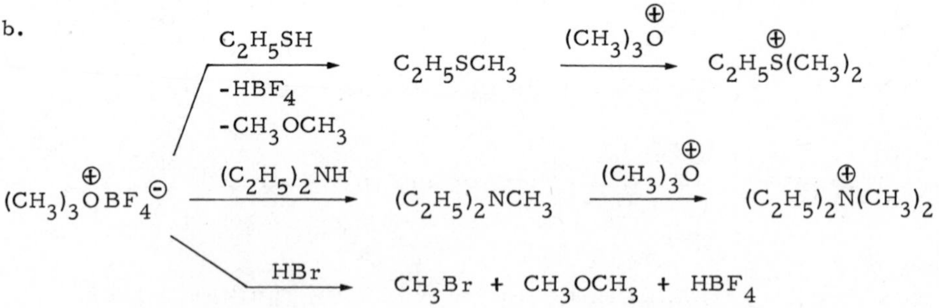

15-42 Methionine, abbreviated here as $R'SCH_3$, is a good sulfur nucleophile and can react with ATP in an S_N2 process. The initial product is S-adenosyl-methionine. (See also Exercise 8-35.)

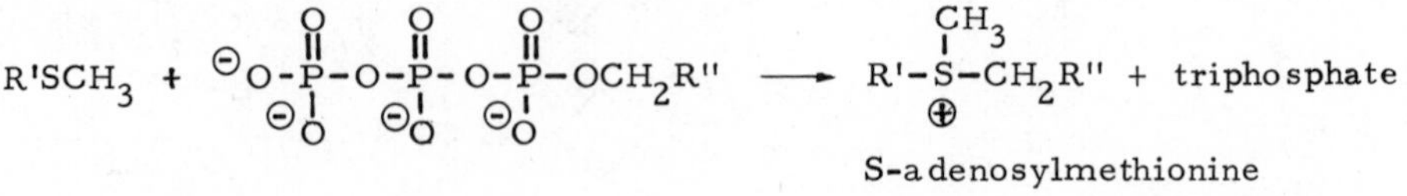

S-adenosylmethionine

The intermediate S-adenosylmethionine is a sulfonium salt. It is reactive towards nucleophiles in S_N2 processes because the methyl carbon is unhindered and there is a good leaving group (neutral sulfide, as homocysteine).

$$\text{RÖH} + \text{R'-S-CH}_2\text{R''} \longrightarrow \text{ROCH}_3 + \text{R'-S-CH}_2\text{R''} + \text{H}^{\oplus}$$

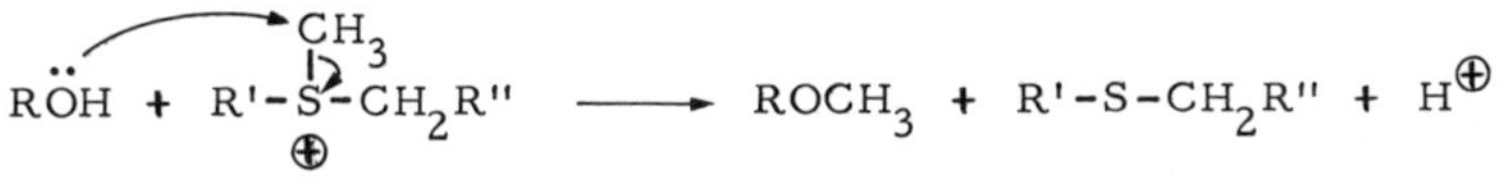

The net reaction corresponds to the methylation of ROH coupled to ATP hydrolysis.

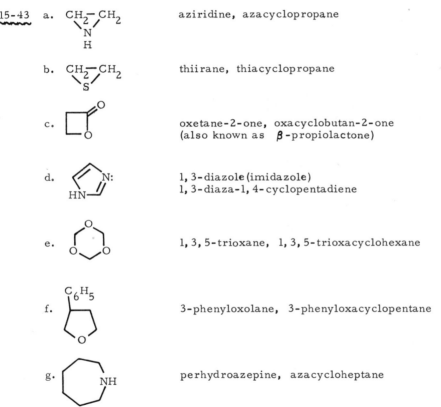

15-43   a. $\text{CH}_2\text{-CH}_2$
              N
              H

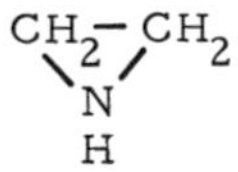

aziridine, azacyclopropane

b. $\text{CH}_2\text{-CH}_2$
      S

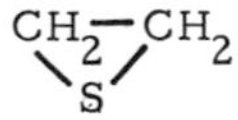

thiirane, thiacyclopropane

c.

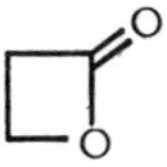

oxetane-2-one, oxacyclobutan-2-one
(also known as β-propiolactone)

d.

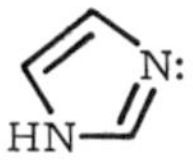

1,3-diazole (imidazole)
1,3-diaza-1,4-cyclopentadiene

e.

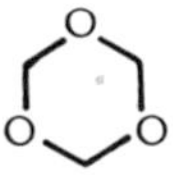

1,3,5-trioxane, 1,3,5-trioxacyclohexane

f. C_6H_5

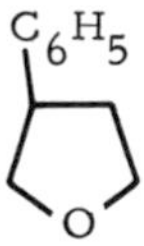

3-phenyloxolane, 3-phenyloxacyclopentane

g.   NH

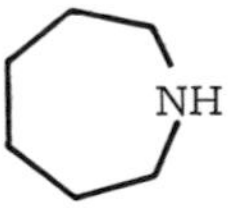

perhydroazepine, azacycloheptane

15-44   a.   trans-2-Chlorocyclohexanol; displacement occurs internally only by attack from the rear of the leaving group.

b.   The cyclopentane derivative reacts faster because the product is less strained than that from the cyclobutane analog.

c.   2-Chloro-3-buten-1-ol reacts faster than chloroethanol because the chlorine is allylic and therefore more labile. (See Section 14-3A.)

15-45

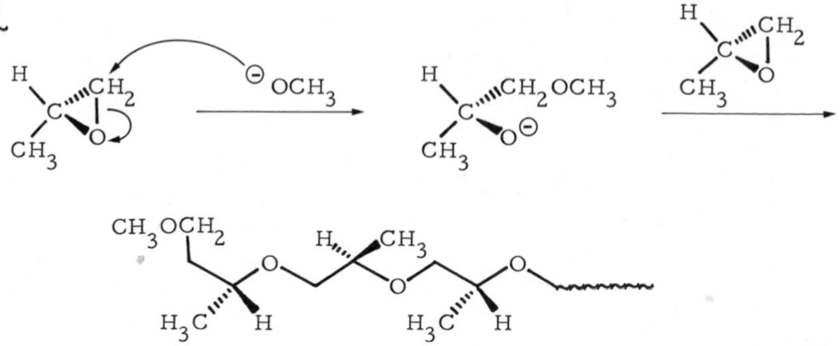

If the starting configuration is D, the polymer configuration will be D. Polymerization does not involve reactions which change the chiral carbon.

15-46   a. $HOCH_2CH_2NHCH_3$

b. CH_3CHCH_2Cl + CH_3CHCH_2OH
       | |
       OH   Cl

c.

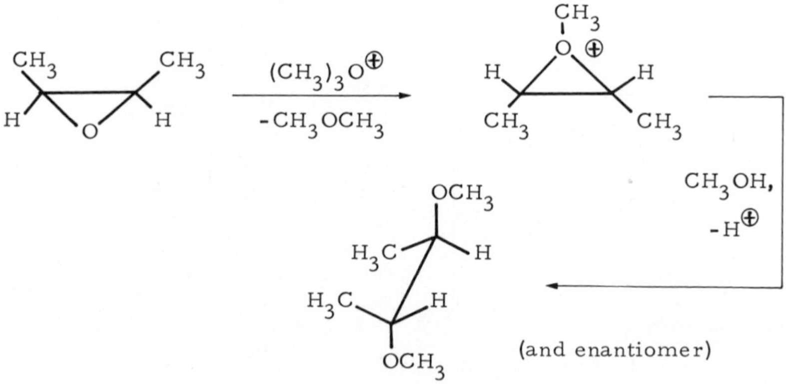

(and enantiomer)

d.

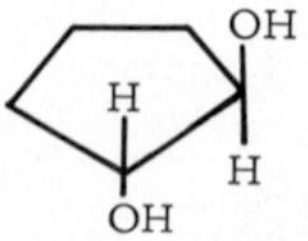

15-47   One of several possible routes is suggested below for each example:

a. $CH_2{=}CH_2$ $\xrightarrow[\text{or } O_2,\ Ag_2O]{CF_3CO_3H}$ (epoxide) $\xrightarrow[H^{\oplus}]{CH_3OH}$ $\begin{array}{cc}CH_2{-}CH_2\\ |\qquad\ |\\ CH_3O\quad OCH_3\end{array}$

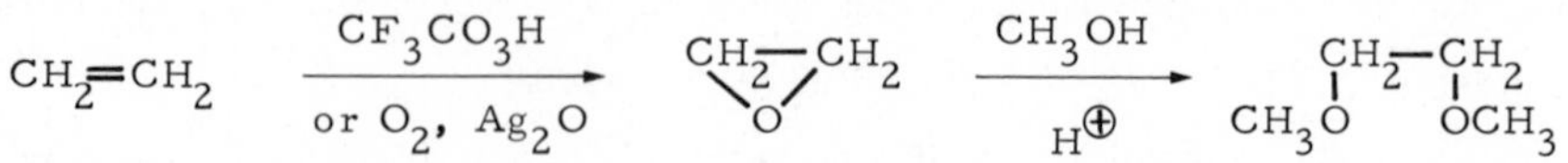

b. $CH_3CH_2CH_2OH$ $\xrightarrow[150°]{96\%\ H_2SO_4}$ $CH_3CH{=}CH_2$ $\xrightarrow[\text{benzene}]{CF_3CO_3H}$ $CH_3CH{-}CH_2$ (epoxide)

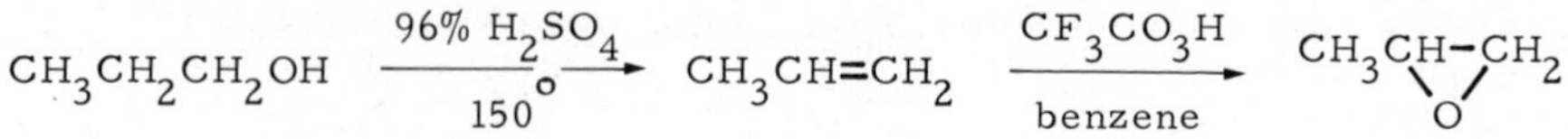

c. $CH_2\!\!=\!\!CHCH_2Cl \xrightarrow{\ NaOH\ } CH_2\!\!=\!\!CHCH_2ONa$

$CH_2\!\!=\!\!CHCH_2ONa + CH_2\!\!=\!\!CHCH_2Cl \xrightarrow{\ -NaCl\ } (CH_2\!\!=\!\!CHCH_2)_2O$

d.

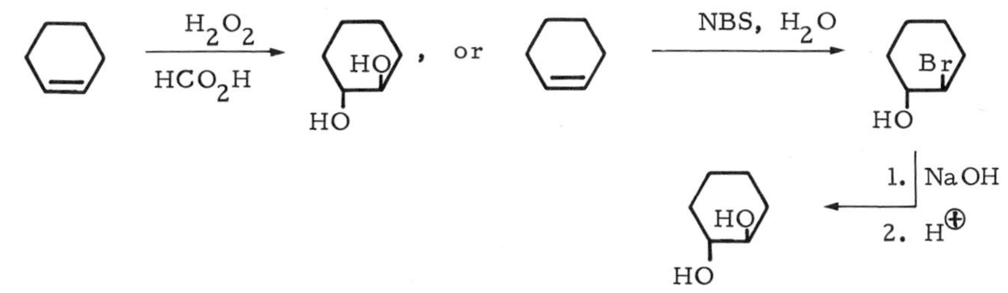

e.

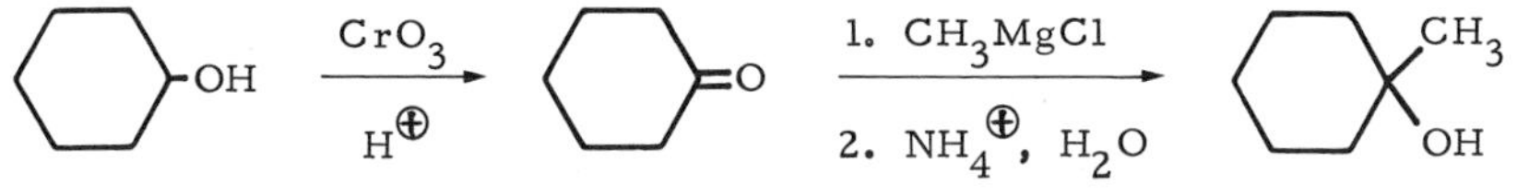

f.

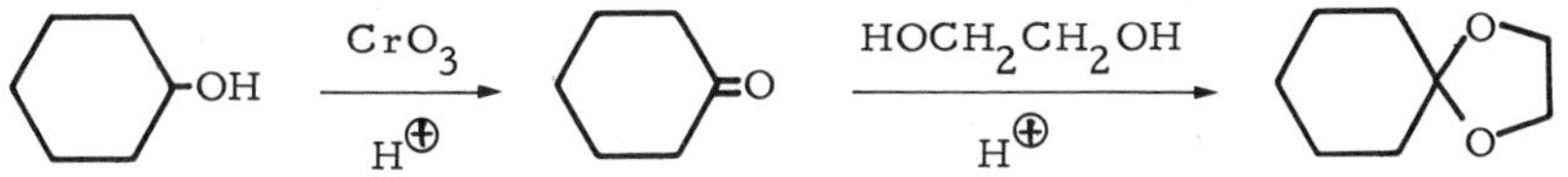

15-48. One of several possible routes is suggested:

a. $CH_3CH_2CH_2OH \xrightarrow[\text{pyridine}]{SOCl_2} CH_3CH_2CH_2Cl \xrightarrow[\text{ether}]{Mg} CH_3CH_2CH_2MgCl$

1. CH_3CHO
2. $NH_4Cl,\ H_2O$

$CH_3CH_2CH_2CHClCH_3 \xleftarrow[ZnCl_2]{HCl} CH_3CH_2CH_2CH(OH)CH_3$

b. $CH_3CH_2OH \xrightarrow[\text{fuming}]{HBr} CH_3CH_2Br \xrightarrow[\text{ether}]{Mg} CH_3CH_2MgBr$

$+ \ 2CH_3CH_2MgBr \xrightarrow{\ H_2O\ } CH_3CH_2CHCH_2CH_3$
                                                       $\quad\quad\quad\quad\quad\quad\quad\quad |$
                                                       $\quad\quad\quad\quad\quad\quad\quad\quad OH$

$CH_3CH_2CHCH_2CH_3 \xleftarrow[-HCl]{CH_3COCl}$
$\quad\quad\quad |$
$\quad\quad\quad OCOCH_3$

c. $(CH_3)_3COH$ $\xrightarrow[-H_2O]{K^{\oplus}\ ^{\ominus}OC(CH_3)_3}$ $(CH_3)_2C=CH_2$ $\xrightarrow[\text{peroxides}]{HBr}$ $(CH_3)_2CHCH_2Br$

d. $CH_3CH_2CH_2CH_2OH$ $\xrightarrow[-H_2O]{\text{hot, conc. } H_2SO_4}$ $CH_3CH_2CH=CH_2$

$CH_3CH_2\underset{\underset{OSO_3H}{|}}{C}HCH_3$ $\xleftarrow{100\% \ H_2SO_4}$ $CH_3CH_2\underset{\underset{OH}{|}}{C}HCH_3$ $\xleftarrow[70^{\circ}]{60\% \ H_2SO_4}$

e. $CH_3\underset{\underset{CH_3CH_2}{|}}{\overset{\overset{CH_3}{|}}{C}}-OH$ $\xrightarrow{48\% \ HBr}$ $CH_3\underset{\underset{CH_3CH_2}{|}}{\overset{\overset{CH_3}{|}}{C}}-Br$ $\xrightarrow[\text{ether}]{Mg}$ $CH_3\underset{\underset{CH_3-CH_2}{|}}{\overset{\overset{CH_3}{|}}{C}}-MgBr$ $\xrightarrow{CH_2O}$

$CH_3\underset{\underset{CH_3CH_2}{|}}{\overset{\overset{CH_3}{|}}{C}}-CHO$ $\xleftarrow{H_2CrO_4}$ $CH_3\underset{\underset{CH_3CH_2}{|}}{\overset{\overset{CH_3}{|}}{C}}-CH_2OH$ $\xleftarrow{}$

f. $\xrightarrow[-HCl]{CH_3-\langle\rangle-SO_2Cl}$

$\xleftarrow[C_2H_5OH]{NaCl,}$

g. $(CH_3)_3C-OH$ $\xrightarrow{HBr}$ $(CH_3)_3C-Br$ $\xrightarrow[\text{ether}]{Mg}$ $(CH_3)_3CMgBr$

$(CH_3)_3CCH_2CH_2Br$ $\xleftarrow{HBr}$ $(CH_3)_3CCH_2CH_2OH$ $\xleftarrow{CH_2-CH_2 \text{ (epoxide)}}$

$\xrightarrow[E2]{NaOC_2H_5}$ $(CH_3)_3CCH=CH_2$ $\xrightarrow[Pt]{H_2}$ $(CH_3)_3CCH_2CH_3$.

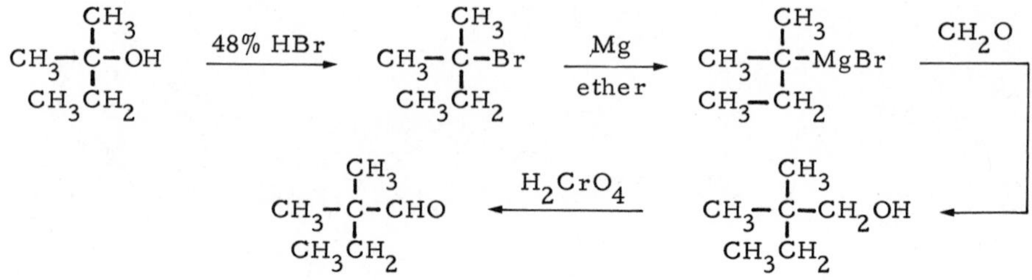

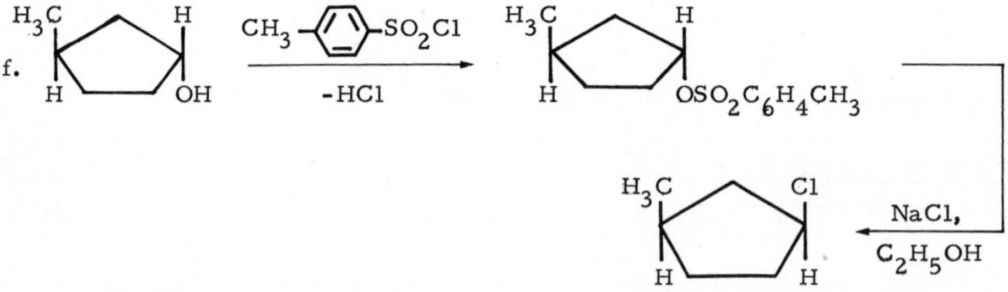

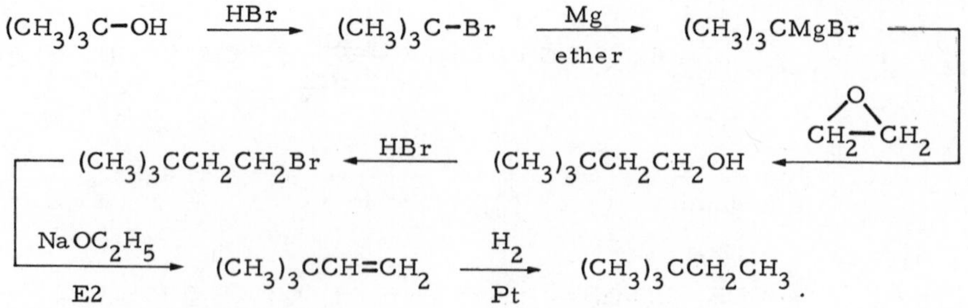

15-49

a. $(CH_3)_3COH + HCl$ $\xrightarrow[25^{\circ}]{ZnCl_2}$ $(CH_3)_3CCl + H_2O$ (rapid)

$(CH_3)_2CHCH_2OH + HCl$ $\xrightarrow[25^{\circ}]{ZnCl_2}$ no reaction unless heated

b. $CH_2{=}CH{-}CH_2CH_2OH + HCl \xrightarrow[25^\circ]{ZnCl_2} CH_2{=}CH{-}CH_2{-}CH_2Cl$ (slow)

$CH_3CH{=}CH{-}CH_2OH + HCl \xrightarrow[25^\circ]{ZnCl_2} CH_3CH{=}CH{-}CH_2Cl$ (fast)

c. $(CH_3)_2CHCH_2CH_2OH \xrightarrow[\text{heat}]{H_2SO_4} (CH_3)_2CHCH{=}CH_2$

no reaction with 2, 2-dimethyl-1-propanol (neopentyl alcohol)

d. $CH_3CH_2CH_2CH_2{-}O{-}SO_3H$ is a very strong acid while

$CH_3CH_2{-}O{-}SO_2{-}O{-}CH_2CH_3$ is essentially neutral.

e. $CH_3C\overset{O}{\underset{Cl}{\diagdown}} + C_2H_5OH \longrightarrow CH_3{-}C\overset{O}{\underset{OC_2H_5}{\diagdown}} + HCl$

no reaction under comparable conditions with $ClCH_2\overset{O}{\overset{\|}{C}}OH$

f. $CH_3{-}\overset{^{18}O}{\overset{\|}{C}}{-}OCH_3 \xrightarrow{H^\oplus, H_2O} CH_3{-}\overset{^{18}O}{\overset{\|}{C}}{-}OH + CH_3OH$

$CH_3{-}\overset{O}{\overset{\|}{C}}{-}^{18}OCH_3 \xrightarrow{H^\oplus, H_2O} CH_3{-}\overset{O}{\overset{\|}{C}}{-}OH + CH_3{}^{18}OH$

The presence or absence of ^{18}O in the methanol would determine the position of the ^{18}O-label in the methyl ester.

g. $CH_3{-}\overset{CH_3}{\underset{H}{\overset{|}{\underset{|}{C}}}}{-}\overset{H}{\underset{H}{\overset{|}{\underset{|}{C}}}}{-}O{-}CrO_3H \xrightarrow{\text{fast}} (CH_3)_2CHCHO + H_3\overset{\oplus}{O} + H\overset{\ominus}{C}rO_3$

No corresponding reaction occurs for tert-butyl chromate ester because there is no α-H.

h. $\underset{OH}{\overset{|}{C}H_2}{-}\underset{OH}{\overset{|}{C}H}{-}CH_3 \xrightarrow[HCl]{ZnCl_2} \underset{OH}{\overset{|}{C}H_2}{-}\underset{Cl}{\overset{|}{C}H}{-}CH_3$

much slower reaction with 1, 3-propanediol

i. CH_3CH-CH_2 $\xrightarrow{HCl}$ CH_3CH-CH_2OH
         $\diagdown O \diagup$ $\underset{Cl}{|}$

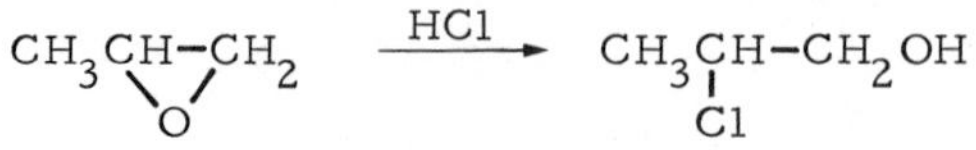

Ring opening of oxacyclobutane is less facile.

j. $CH_3-\underset{\underset{CH_3}{|}}{\overset{\overset{CH_3}{|}}{C}}-OCH_2CH_2\underset{\underset{CH_3}{|}}{\overset{\overset{CH_3}{|}}{C}}-CH_3$ $\xrightarrow[\text{fast}]{HBr}$ $(CH_3)_3CBr + (CH_3)_3CCH_2CH_2OH$

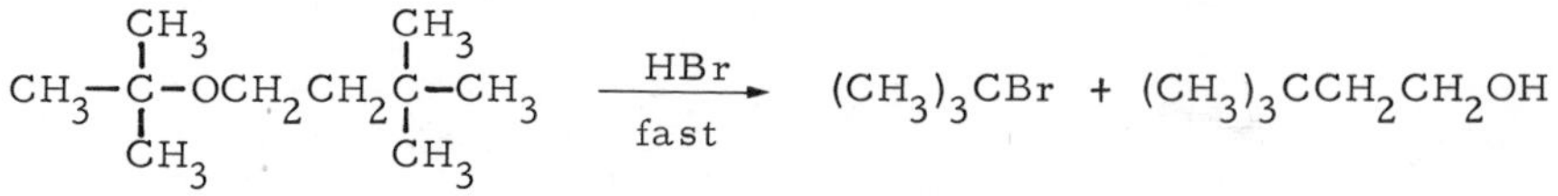

Cleavage of the ether linkage in dineopentyl ether would not occur under comparable conditions owing to steric hindrance of neopentyl groups to S_N2 attack.

15-50

$CH_3CH_2CH_2CH_2CH_2OH$ — reacts with Lucas reagent ($ZnCl_2$ + HCl) only on heating

$CH_3CH_2CH_2\underset{\underset{OH}{|}}{C}HCH_3$ — reacts with Lucas reagent slowly at room temperature

$(CH_3)_2\underset{\underset{OH}{|}}{C}CH_2CH_3$ — reacts immediately with Lucas reagent at room temperature

$CH_3CH=CHCH_2CH_2OH$ — decolorizes $KMnO_4$; and reacts with Lucas reagent only on heating

$HC\equiv C-CH_2CH_2CH_2OH$ — gives precipitate with $AgNO_3$; and reacts with Lucas reagent only on heating

$(CH_3CH_2CH_2CH_2)_2O$ — no reaction with Lucas reagent

$CH_3\overset{\overset{O}{||}}{C}OCH_2CH_2CH_2CH_2CH_3$ — reacts with sodium hydroxide solution to give sodium acetate and 1-pentanol, identified as above

15-51   Structure of A is $(CH_3)_3COOH$
            Structure of B is $(CH_3)_3COOC(CH_3)_3$

Reaction with hydrogen

A $(CH_3)_3COOH$ $\xrightarrow{H_2, Ni}$ $(CH_3)_3COH + H_2O$

B $(CH_3)_3COOC(CH_3)_3$ $\xrightarrow[Ni]{H_2}$ $2(CH_3)_3COH$

Reaction with acylating agents

A $(CH_3)_3COOH$ $\xrightarrow[\text{-HCl}]{CH_3COCl}$ $(CH_3)_3COO-\overset{\overset{\displaystyle O}{\|}}{C}-CH_3$

There is no reaction with B because there is no hydroxyl group.

Reaction with methylmagnesium iodide

A $(CH_3)_3COOH$ $\xrightarrow{CH_3MgI}$ $(CH_3)_3COOMgI + CH_4$

$(CH_3)_3COOMgI$ $\xrightarrow{CH_3MgI}$ $(CH_3)_3COMgI + CH_3OMgI$ $\xrightarrow{H_2O}$

$(CH_3)_3COH + CH_3OH$ ◄——┘

B $(CH_3)_3COOC(CH_3)_3$ $\xrightarrow{CH_3MgI}$ $(CH_3)_3COMgI + CH_3OC(CH_3)_3$

$\downarrow H_2O$

$(CH_3)_3COH$

Polymerization of chloroethene (vinyl chloride)

$(CH_3)_3COOC(CH_3)_3$ $\xrightarrow{\text{heat}}$ $2(CH_3)_3CO\cdot$

$(CH_3)_3CO\cdot + CH_2{=}CHCl$ $\longrightarrow$ $(CH_3)_3COCH_2\overset{\cdot}{C}HCl$

$(CH_3)_3COCH_2\overset{\cdot}{C}HCl + CH_2{=}CHCl \rightarrow (CH_3)_3COCH_2CH(Cl)CH_2\overset{\cdot}{C}HCl$ etc.

Thermal decomposition of B

$(CH_3)_3COOC(CH_3)_3$ $\xrightarrow{\text{heat}}$ $2(CH_3)_3CO\cdot$

$CH_3-\overset{\overset{\displaystyle CH_3}{|}}{\underset{\underset{\displaystyle CH_3}{|}}{C}}-O\cdot$ $\longrightarrow$ $CH_3-\overset{\overset{\displaystyle O}{\|}}{C}-CH_3 + CH_3\cdot$

$CH_3\cdot + CH_3\cdot$ $\longrightarrow$ CH_3CH_3

Thermal decomposition of B in air

$CH_3\cdot + O_2$ $\longrightarrow$ $CH_3OO\cdot$

$CH_3OO\cdot + H{-}R$ $\longrightarrow$ $CH_3O_2H + R\cdot$ (HR is any hydrogen donor)

CH_3O_2H $\longrightarrow$ $CH_2O + H_2O$

Two possible alternative structures

$$
\text{A} \quad CH_3O-\underset{\underset{CH_3}{|}}{\overset{\overset{CH_3}{|}}{C}}-OH
\qquad\qquad
\text{B} \quad (CH_3)_2CHOCH_2CH_2OCH(CH_3)_2
$$

(You should try to work out how these alternative structures for A and B would behave in the given set of reactions.)

15-52

a.

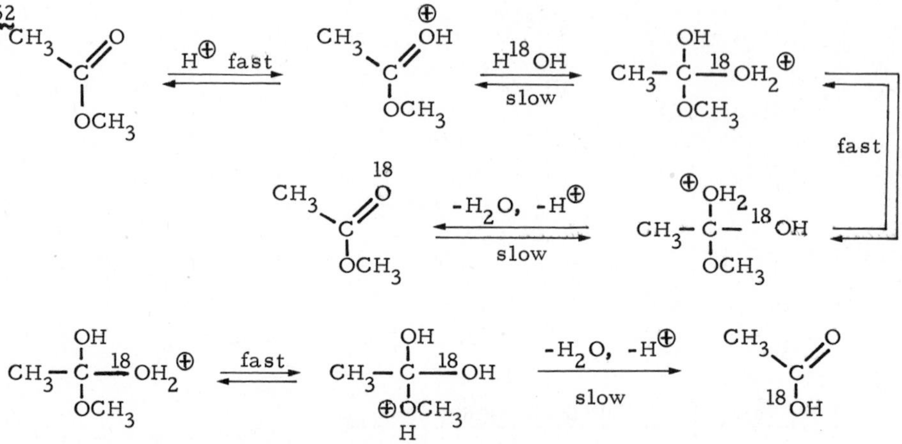

b. If methyl ethanoate were to behave as a very weak base (much weaker than water), then the sulfuric acid catalyst would protonate water exclusively and the methyl ester not at all. Acid catalysis would not then be observed.

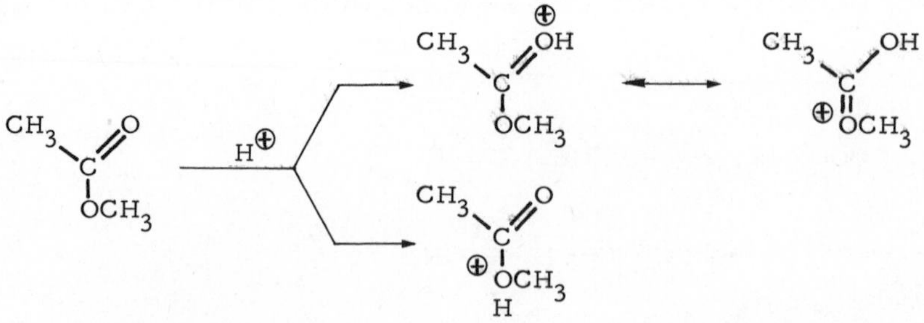

A proton is more likely to add to the carbonyl oxygen than to the methoxyl oxygen because of the possibility of stabilization of the protonated form by electron delocalization involving two nearly equivalent electron-pairing schemes.

c. At high acid concentrations, both the methyl ester and the water present would be extensively protonated. While protonation of the ester has a catalytic effect, protonation of water has a retarding effect because the effective water concentration is lowered by conversion to $H_3O^{\oplus}$ which has no nucleophilic properties.

15-53 See mechanism given in Section 11-7D.

15-54

$$CHBr_3 \xrightarrow[-(CH_3)_3COH]{K^\oplus \ ^\ominus OCH(CH_3)_3} \ :\!CBr_3^\ominus \longrightarrow :\!CBr_2 + Br^\ominus$$

$$(CH_3)_3CO^\ominus + :\!CBr_2 \longrightarrow (CH_3)_3C-O-\ddot{C}-Br + Br^\ominus$$

$$(CH_3)_3C-O-\ddot{C}-Br \longrightarrow (CH_3)_3C^\oplus + CO + Br^\ominus$$

$$(CH_3)_3C^\oplus + (CH_3)_3CO^\ominus \longrightarrow (CH_3)_3COC(CH_3)_3$$

or

$$(CH_3)_3CMgCl + R\overset{O}{\underset{\|}{C}}-O-OC(CH_3)_3 \longrightarrow R\overset{O}{\underset{\|}{C}}-OMgCl + (CH_3)_3COC(CH_3)_3$$

or

$$2(CH_3)_3CCl \xrightarrow[\text{ether}]{Ag_2CO_3} (CH_3)_3COC(CH_3)_3 + CO_2 + 2\,AgCl$$

15-55 In E2 elimination from $CH_3CH_2\underset{Cl}{C}(CH_3)_2$ and from $CH_3CD_2\underset{Cl}{C}(CH_3)_2$, proportionately more elimination toward the methyl group than towards the methylene is to be expected for the deuterium-labelled compound relative to the unlabelled compound. This is because the rate at which the products are formed is determined by the rate at which the C—H (or C—D) bond is broken, and since C—D bonds are broken more slowly than C—H bonds, elimination in the direction of the CD_2 group will be slower.

In E1 elimination, the same argument holds even though the slowest step in the reaction does not involve the C—H or C—D bonds directly. However, the product-forming steps do involve these bonds and the deuterium isotope effect will influence the product distribution as described for E2 elimination.

15-56 The BF_3 serves as an electrophilic catalyst that complexes with the oxygen of the oxacyclopropane ring and activates it towards nucleophiles. The ether, R_2O, serves as the nucleophile and attacks a ring carbon to give an oxonium salt. Alkylation of R_2O then gives the symmetrical oxonium salt, $R_3O^\oplus$.

$$\text{ClCH}_2\overset{\triangle}{}\text{O} + R_2\overset{\oplus}{\text{O}}:\overset{\ominus}{\text{BF}}_3 \;\rightleftharpoons\; \text{ClCH}_2\overset{\triangle}{}\overset{\oplus}{\text{O}}-\overset{\ominus}{\text{BF}}_3 + R_2\overset{..}{\overset{..}{\text{O}}}:$$

$$\text{ClCH}_2\underset{\underset{\ominus \text{BF}_3}{\overset{|}{\text{O}}}}{\overset{|}{\text{CHCH}}}_2\text{OR} + R_3\overset{\oplus}{\text{O}} \;\longleftarrow\; \overset{R_2\overset{..}{\text{O}}:}{}\; \text{ClCH}_2\underset{\underset{\ominus \text{BF}_3}{\overset{|}{\text{O}}}}{\overset{|}{\text{CH}}}-\text{CH}_2-\overset{\oplus}{\text{OR}}_2$$

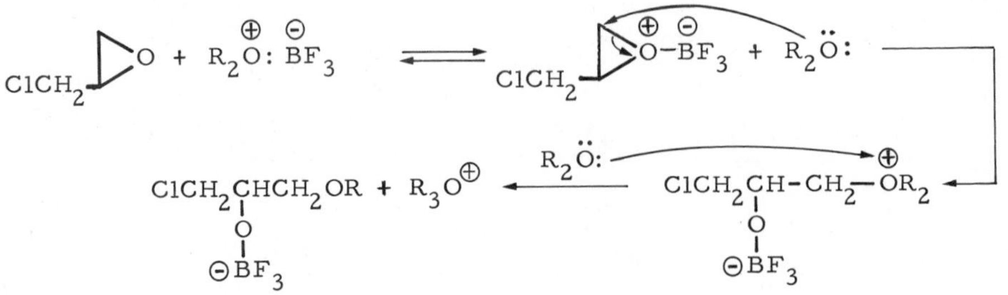

15-57 a. The first step is formation of the chlorosulfite ester of unchanged configuration. The displacement step follows, and proceeds with inversion to give the L-configuration.

$$\text{Cl}^{\ominus} + \underset{H}{\overset{C_6H_5}{\underset{}{\overset{}{\text{C}}}}}\overset{\text{}}{\underset{\text{O}-\text{S}-\text{Cl}}{\text{}}}\cdots\text{CH}_3 \;\longrightarrow\; \text{Cl}-\underset{H}{\overset{C_6H_5}{\overset{|}{\text{C}}}}\cdots\text{CH}_3 + \text{SO}_2 + \text{Cl}^{\ominus}$$

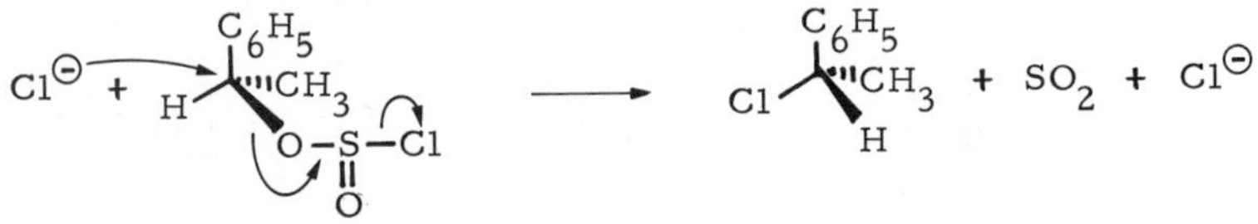

b. In the presence of base, the reaction is S_N2 because the pH is high and chloride ion concentration is high.

$$\text{CH}_3\text{CH=CHCH}_2\text{OH} + \text{SOCl}_2 + R_3\text{N} \;\longrightarrow\; \text{CH}_3\text{CH=CHCH}_2\text{OS(O)Cl} + R_3\overset{\oplus}{\text{N}}\text{H}\overset{\ominus}{\text{Cl}}$$

$$\text{Cl}^{\ominus} + \text{CH}_3\text{CH=CHCH}_2\text{OS(O)Cl} \;\xrightarrow{S_N2}\; \text{CH}_3\text{CH=CHCH}_2\text{Cl} + \text{SO}_2 + \overset{\ominus}{\text{Cl}}$$

Without an equivalent of base, the reaction mixture is strongly acidic and promotes S_N1 reaction, with rearrangement.

$$\text{CH}_3\text{CH=CHCH}_2\text{OH} + \text{SOCl}_2 \;\longrightarrow\; \text{CH}_3\text{CH=CHCH}_2\text{OS(O)Cl} + \text{HCl} \;\xrightarrow{S_N1}$$

$$\text{SO}_2 + \text{HCl} + \text{Cl}^{\ominus} + \left[\text{CH}_3\overset{\oplus}{\text{CH}}-\text{CH=CH}_2 \;\longleftrightarrow\; \text{CH}_3\text{CH=CH}\overset{\oplus}{\text{CH}}_2\right]$$

$$\longrightarrow\; \text{CH}_3\text{CH(Cl)CH=CH}_2 + \text{SO}_2 + \text{HCl}$$

15-58 Besides lowering the freezing point, ethylene glycol reduces the volatility of water through association by hydrogen bonding. Evaporation of the water from radiators filled with water-ethylene glycol mixtures is therefore reduced.

CHAPTER 16

CARBONYL COMPOUNDS I.   ALDEHYDES AND KETONES.
ADDITION REACTIONS OF THE CARBONYL GROUP

The chemical, physical, and spectral properties of carbonyl compounds indicate that the bond is polarized in the sense $\overset{\delta\oplus}{C}=\overset{\delta\ominus}{O}$. The dipole moment of carbonyl bonds is about 2.7 debye. Carbonyl compounds associate with hydroxylic solvents through hydrogen bonding:

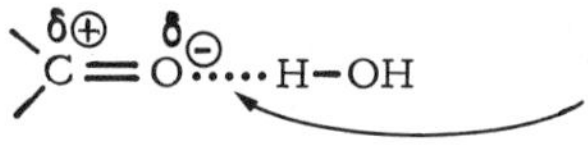

$$\overset{\delta\oplus}{C}=\overset{\delta\ominus}{O}\cdots H-OH \qquad \text{hydrogen bond}$$

The principal spectral characteristics of aldehydes and ketones follow:

infrared $C=O$ stretch $\sim$ 1720 cm^{-1} (strong) (Table 16-3)

$C-H$ stretch in $-CHO \sim$ 2720 cm^{-1} (medium)

ultraviolet $C=O$ $n \to \pi *$ $\sim$ 275 - 295 nm (weak)

$\pi \to \pi *$ $\sim$ 180 - 190 nm (strong)

nmr   aldehyde proton chemical shift (- CHO) δ = 9 - 10 ppm

α - hydrogens $CH-C=O$ $\delta \sim$ 2.1 ppm

mass spectra   Molecular ion observable

Major fragments are α -cleavage ions:

$$R-CO{+}R \longrightarrow R\overset{\oplus}{C}O + \underset{\ominus}{e} + R^{\cdot}$$

Rearrangements may occur if there is a γ - hydrogen

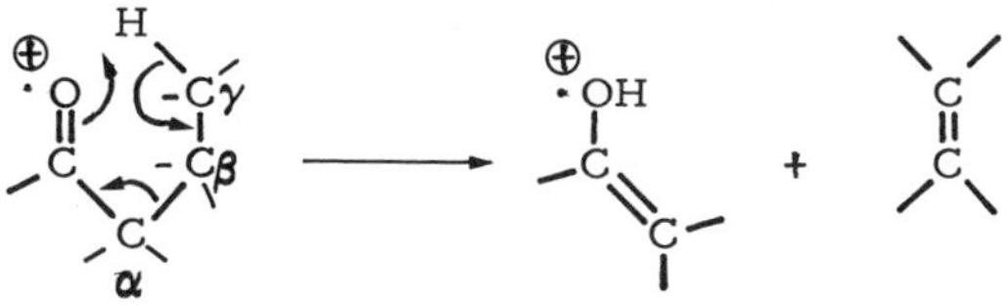

The carbonyl bond is both stronger and more reactive in ionic reactions than a carbon-carbon double bond. Addition reactions are common and the direction of addition is such that the carbon becomes bonded to a nucleophile

(Nu) and the oxygen becomes bonded to an electrophile (E):

$$\diagdown C=O \ + \ Nu-E \longrightarrow Nu-\overset{|}{\underset{|}{C}}-O-E$$

Various reagents that add to the carbonyl bond of aldehydes and ketones are listed in Tables 16-4 and 16-5.

Addition may require either an acidic or a basic catalyst. Acidic catalysts generally function to activate the carbonyl group towards nucleophilic attack:

$$\diagdown C=O \ \underset{}{\overset{H^{\oplus}}{\rightleftharpoons}} \ \diagdown C=\overset{\oplus}{O}H \ \underset{}{\overset{Nu^{\ominus}}{\rightleftharpoons}} \ Nu-\overset{|}{\underset{|}{C}}-OH$$

Basic catalysts normally function by generating a more reactive attacking nucleophile. Thus, for HCN,

$$HCN \ \xrightarrow{\ ^{\ominus}OH\ } \ H_2O \ + \ ^{\ominus}CN$$

$$\diagdown C=O \ + \ ^{\ominus}CN \longrightarrow \ ^{\ominus}O-\overset{|}{\underset{|}{C}}-CN \ \xrightarrow{\ H_2O\ } \ HO-\overset{|}{\underset{|}{C}}-CN \ + \ ^{\ominus}OH$$

Aldehydes are more reactive than ketones and generally have more favorable equilibrium constants for addition. Unfavorable equilibria with ketones may be the result of steric hindrance, but may also be because the ketone $C=O$ bond is 13 kcal mole^{-1} stronger than the $C=O$ bond of methanal and 3 kcal mole^{-1} stronger than other aldehyde $C=O$ bonds (see Table 16-1).

Aldehydes undergo a number of reactions that usually are not observed with ketones. These include polymerization, whereby chains or rings of alternating carbon and oxygen atoms are formed (Section 16-4B):

$$\underline{n}\,CH_2=O \ \underset{}{\overset{H^{\oplus}}{\rightleftharpoons}} \ HO-CH_2-O-CH_2-O-CH_2-O(CH_2-O)_{\underline{n}-3}H$$

$$\text{polyoxymethylene}$$

Aldehydes behave as reducing agents because they are oxidized readily to carboxylic acids.

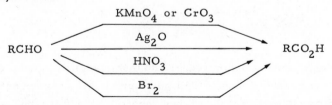

$$RCHO \xrightarrow{\substack{KMnO_4 \ \text{or} \ CrO_3 \\ Ag_2O \\ HNO_3 \\ Br_2}} RCO_2H$$

Another important reaction of aldehydes is self oxidation and reduction under the influence of base, which is known as the Cannizzaro reaction (Section 16-4E). This reaction is restricted to aldehydes with no α hydrogens:

$$2\,C_6H_5CHO \xrightarrow{\text{NaOH}} C_6H_5CO_2H + C_6H_5CH_2OH$$

Ketones are not oxidized as readily as aldehydes. However, ketones can be oxidized to esters with peroxyacids (Section 16-7).

$$R_2C=O \xrightarrow{R'CO_3H} RCO_2R$$

Both aldehydes and ketones can be reduced to alcohols or hydrocarbons by a variety of reagents (Section 16-4E, 16-5, 16-6, and Table 16-6):

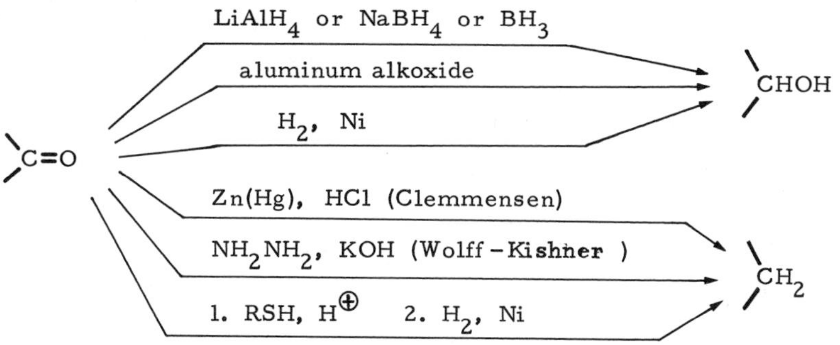

Carbonyl groups can be protected in organic reactions by conversion to a ketal or acetal function (Section 16-8). A carbonyl group masked as a ketal is stable to base but not to acid.

Commonly used methods of preparation of aldehydes and ketones are listed in Tables 16-7 and 16-8.

ANSWERS TO EXERCISES

16-1

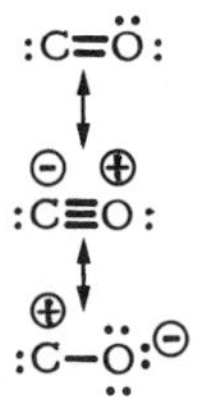

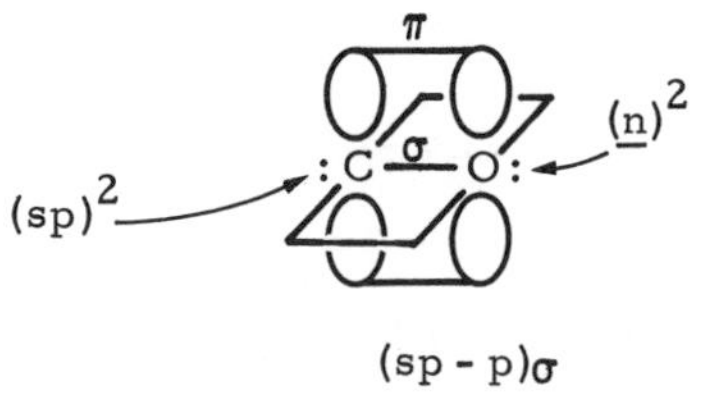

The bond energy for carbon monoxide is higher than for other carbonyl compounds because the carbon-oxygen multiple bond in carbon monoxide has

more than four bonding electrons. It has some triple-bond character. How-
ever, the contribution of the dipolar valence bond structure $:C\overset{\ominus}{\equiv}\overset{\oplus}{O}:$ must nearly
cancel the contribution of $:\overset{\oplus}{C}-\overset{\ominus}{\underset{..}{O}}:$ and the normal polarity of C-O σ bonds as
$\overset{\delta\oplus}{C}-\overset{\delta\ominus}{O}$ because the dipole moment is very small (0.13 Debye).

16-2   <u>Zero dipole moment</u>   <u>Nonzero dipole moment</u>

        (opposed dipoles)

    if ring carbons   if ring carbons
are planar   are nonplanar

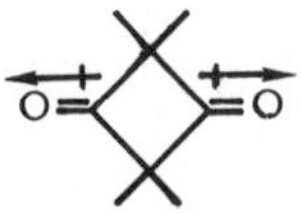

(See Section 12-3F)

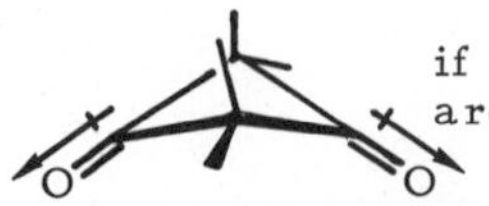

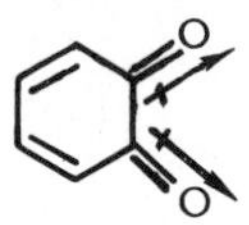

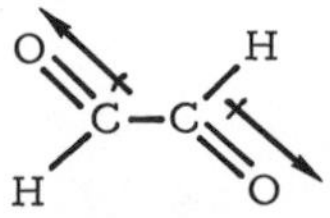

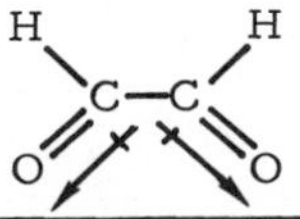

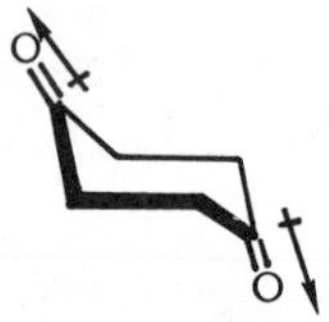

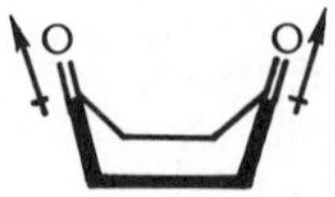

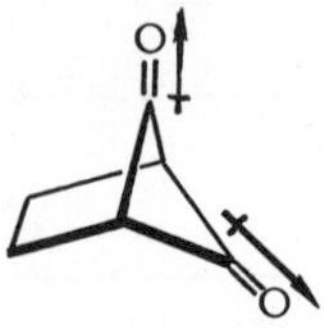

━━━ + represents a polarized bond ⊖·····⊕

    if in s-trans   if in s-cis
conformation   conformation

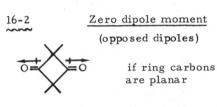

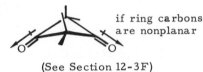

O=C=C=C=O

    if in chair   if in boat
conformation   conformation

16-3   The propagation steps would be:

$$CH_2{=}O + Br\cdot \longrightarrow Br-CH_2-\overset{.}{O} \qquad \Delta\underline{H}_1 = +12.5 \text{ kcal}$$

$$BrCH_2-\overset{.}{O} + HBr \longrightarrow Br-CH_2-O-H + Br\cdot \qquad \Delta\underline{H}_2 = -23.1 \text{ kcal}$$

and

$$CH_2{=}CH_2 + Br\cdot \longrightarrow Br-CH_2-\overset{.}{C}H_2 \qquad \Delta\underline{H}_1' = -5.8 \text{ kcal}$$

$$BrCH_2-\overset{.}{C}H_2 + HBr \longrightarrow Br-CH_2-CH_3 + Br\cdot \qquad \Delta\underline{H}_2' = -11.2 \text{ kcal}$$

Even though both reactions are exothermic overall, the first step in radical addition to methanal is quite endothermic and will therefore be very slow. The reactivity of methanal will be <u>low</u> towards radical addition compared to ethene. The <u>position</u> of equilibrium in radical addition will be the same as in polar addition because equilibrium is not dependent on reaction mechanism but only on the relative energies of the reactants and products.

16-4   a. $CH_3COCCl_3 > CH_3COCH_3$;   electron-withdrawing effect of the CCl_3 group increases the electropositive nature of the carbonyl carbon making it more susceptible to nucleophilic attack.

b. $CH_3COCH_3 > (CH_3)_3CCHO$;   the bulk of the <u>tert</u>-butyl group hinders addition of a nucleophile to the carbonyl carbon.

c. $CH_3COCO_2CH_3 > CH_3COCH_2CO_2CH_3$;   electron-withdrawing effect of the ester group $-CO_2CH_3$ is larger than the group $-CH_2CO_2CH_3$.

d. $CH_3COCOCH_3 > CH_3COCH_3$;   electron-withdrawing effect of $-COCH_3$ is larger than $-CH_3$.

e. $CH_3COCN > CH_3COCH_3$;   electron-withdrawing effect of $-CN$ is larger than $-CH_3$.

f. $CH_2{=}C{=}O$ > cyclobutanone;   according to the arguments presented in Section 16-1C, the reactivity of cyclopropanone and cyclobutanone is associated with the relief of angle strain in the ring bonds on carbonyl addition. By this reasoning, if a double bond is viewed as a two-membered ring, the angle strain in ketene ring bonds is $(120° - 0°) = 120°$, which is more than the angle strain in cyclobutanone ring bonds $(120° - 90°) = 30°$. Accordingly, ketene carbonyl is more reactive.

g.   ;   bicyclo [2.1.1]-5-hexanone is both a 5-ring and a 4-ring ketone; addition to a 5-ring ketone contributes nothing to the relief of angle strain but introduces unfavorable eclipsing strain between non-bonded atoms. The CH_2 bridge also interferes with exo substituents. For these reasons the bicyclic ketone is less reactive.

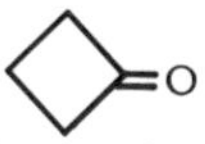

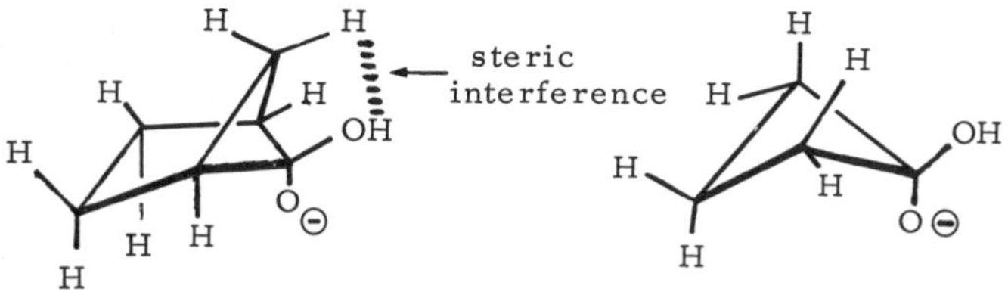

more highly strained adduct

16-5   A,   1710 cm^{-1},   ketone   RCOR
       B,   1720 cm^{-1},   ~ 2700 cm^{-1} (weak)   aldehyde   RCHO
       C,   1680 cm^{-1},   ~ 3400 cm^{-1} (broad, N—H)   amide   RCONH$_2$
       D,   1730 cm^{-1},   ~ 1220 cm^{-1} (C—O)   ester   RCO$_2$R
       E,   1800 cm^{-1},   acyl halide   RCOX
       F,   1720 cm^{-1},   3100 cm^{-1} (broad)   acid   RCO$_2$H

16-6   Ketone A is CH$_3$COCH$_3$
       Ketone B is CH$_3$COCH(CH$_3$)CH$_2$CH$_2$CH$_3$
           or   CH$_3$COCH(CH$_3$)CH(CH$_3$)$_2$
       Fragmentation patterns for B are:

               MW 100   m/e 58   MW 42

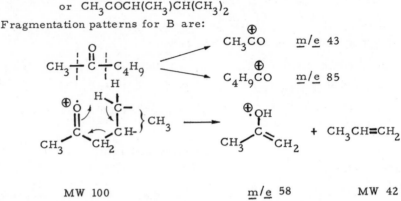

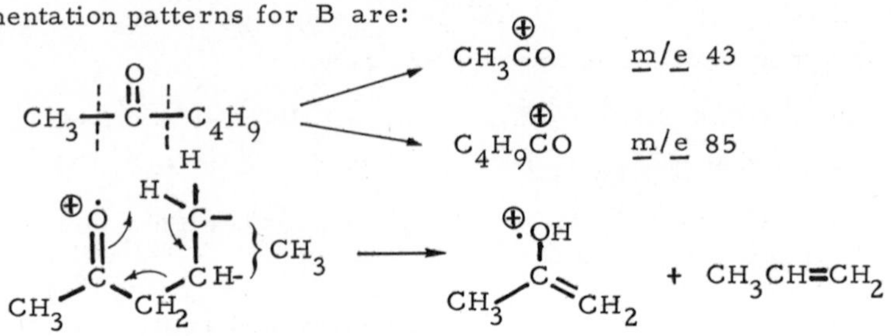

The structure of the parent hydrocarbon is:

       CH$_3$CCH$_2$CH$_2$CH$_2$CH$_3$ CH$_3$CCH$_2$CH(CH$_3$)$_2$
          ‖   or   ‖
       CH$_3$CCH$_3$ CH$_3$CCH$_3$

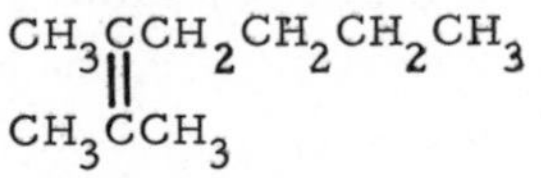

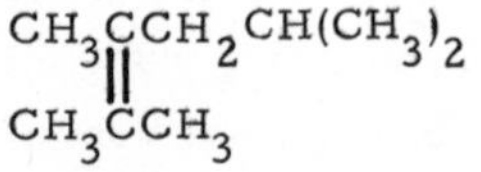

16-7   C$_5$H$_{10}$O: Infrared indicates carbonyl group (1720 cm^{-1}). Nmr spectrum
shows the presence of two kinds of hydrogen in the ratio 2:3; the triplet-quartet
pattern is characteristic of ethyl (CH$_3$CH$_2$—). Structure consistent with spectra
is:

                   CH$_3$CH$_2$COCH$_2$CH$_3$

       C$_4$H$_8$O: Infrared indicates carbonyl group (1720 cm^{-1}). Nmr spec-
trum shows the presence of three kinds of hydrogen in ratio of 1:1:6 with
chemical shifts characteristic of aldehyde, —CH and CH$_3$ hydrogens. The
doublet aldehyde signal and doublet methyl is consistent with the structure:

                   CH$_3$
                       ⟍
                        CHCHO
                       ⁄
                   CH$_3$

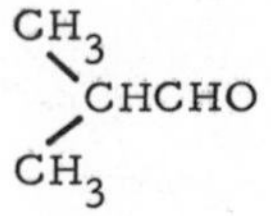

The $-\overset{|}{\underset{|}{C}}H$ proton is split into seven lines by the 6 methyl protons, and each line is further split by the aldehyde proton.

C_3H_6O: Nmr spectrum shows two types of hydrogen in the ratio of 1:2; the triplet and quintet could arise by spin-spin splitting of two equivalent protons by four equivalent protons. The infrared spectrum indicates that the oxygen function is <u>not</u> present as a carbonyl group nor as a hydroxyl group. These facts are consistent with the structure:

$$
\begin{array}{ccc}
CH_2 & \!\!-\!\! & CH_2 \\
| & & | \\
CH_2 & \!\!-\!\! & O
\end{array}
$$

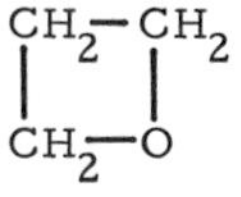

C_4H_6O: Nmr spectrum shows four distinct kinds of hydrogen in the ratio of 1:1:1:3 with chemical shifts characteristic of aldehyde (9.5 ppm), $=CH$ ($\sim$ 6.5 ppm) and $=C-CH-$ (1.9 ppm) protons. The infrared spectrum shows the presence of a carbonyl function (1690 cm^{-1}) and a carbon-carbon double bond (1650 cm^{-1}). A structure which accounts for the spectra is:

$$CH_3CH=CHCHO \quad (\underline{cis} \text{ or } \underline{trans} \text{ or both})$$

The complexity of the alkenyl (vinyl) resonance arises from spin-spin coupling of the vinyl protons with each other, with the aldehyde proton and with the methyl protons.

<u>16-8</u> a. Nmr; the spectrum of $CH_3COCH_2CH_3$ would show a methyl singlet near 2.1 ppm for CH_3CO and a triplet-quartet pattern for the ethyl group; the spectrum of $CD_3COCH_2CH_3$ would be the same except for the absence of the singlet resonance near 2.1 ppm.

b. Uv; the conjugated ketone, 2-cyclohexenone, would show strong absorption in the ultraviolet around 217 nm due to $\pi-\pi^*$ transition; the nonconjugated ketone would show a comparable absorption at shorter wavelengths, below 200 nm.

c. Ir; the ketone $CH_2=CHCH_2COCH_3$ would show strong carbonyl absorption in the infrared around 1720 cm^{-1}; this band would not appear in the spectrum of $CH_2=CHCH_2CCl_2CH_3$.

d. Nmr; propanal, CH_3CH_2CHO, would be distinguished from $CH_3CH_2COCH_3$ by the resonance of the aldehyde proton, which would appear as a triplet around 9.5 ppm; infrared would also distinguish propanal from the aldehyde C—H stretching band near 2710 cm^{-1}.

e. Nmr; the triplet-quartet splitting pattern for the ethyl group in $CH_3CH_2COCH_2CH_2CD_3$ would appear as a singlet (slightly broad) **centered** around 2.4 ppm for $CD_3CH_2COCH_2CH_2CH_3$; there would also be no methyl resonance around 1.0 ppm in the spectrum of $CH_3CH_2COCH_2CH_2CD_3$.

16-9 a. See Exercise 15-16; b. and c. See pp 622-623; d. See pp 616-617 (text)

16-10 The two equilibria are:

$$R_2C=O + HCN \rightleftharpoons HOCR_2CN \qquad \underline{K}_1$$

$$R_2C=O + CN^\ominus + H_2O \rightleftharpoons HOCR_2CN + HO^\ominus \qquad \underline{K}_2$$

and the question is whether $\underline{K}_2 > \underline{K}_1$ or $\underline{K}_1 > \underline{K}_2$. Notice that the equilibrium expressed by $\underline{K}_2$ can be written as the sum of the two equilibria:

$$R_2C=O + HCN \rightleftharpoons HOCR_2CN \qquad \underline{K}_1$$

$$CN^\ominus + H_2O \rightleftharpoons HCN + HO^\ominus \qquad \underline{K}_3$$

Thus,
$$\underline{K}_2 = \underline{K}_1 \cdot \underline{K}_3$$

Now $\underline{K}_3$ is an unfavorable equilibrium because water is a weaker acid (p$\underline{K}$ 14) than hydrogen cyanide (p$\underline{K}_a$ 9.31), or $^\ominus$OH is a stronger base than $CN^\ominus$, and the position of equilibrium lies in favor of H_2O and $CN^\ominus$. That is, $\underline{K}_3 << 1$.

Since $\underline{K}_2 = \underline{K}_1 \cdot \underline{K}_3$

then $\underline{K}_3 = \underline{K}_2 / \underline{K}_1 << 1$

or $\underline{K}_2 << \underline{K}_1$

In words, the equilibrium for cyanohydrin formation is less favorable with aqueous sodium cyanide than with HCN. The function of the $C\equiv N^\ominus$ in the HCN addition is to act as catalyst. It is consumed in the addition step but is then regenerated by the reaction

$$HCN + HO^\ominus \longrightarrow H_2O + CN^\ominus$$

16-11
a. $$R_2C=O + CN^\ominus \xrightarrow[\underline{k}_1]{slow} R_2\overset{O^\ominus}{\underset{|}{C}}-CN \xrightarrow[\underline{k}_2]{HCN,\ fast} R_2\overset{OH}{\underset{|}{C}}-CN + CN^\ominus$$

$(R = CH_3)$

If the slow step is attack of $CN^\ominus$ at carbonyl group, the rate law becomes:

$$\text{rate} = \underline{k}_1 [CH_3COCH_3][CN^\ominus]$$

or $\qquad\text{rate} = \underline{k}_1 \underline{K}[HCN][OH^\ominus][CH_3COCH_3]$

where $\underline{K}$ is equilibrium constant for the formation of $CN^\ominus$

$$\underline{K} = \frac{[C\equiv N^\ominus]}{[HCN][OH]}$$

b. If $\underline{k}_2$ is the slow step:

$$\text{rate} = \underline{k}_2 [R_2\overset{\overset{O^\ominus}{|}}{C}CN][HCN]$$

Also, if $R_2\overset{\overset{O^\ominus}{|}}{C}\text{-}CN$ is formed in a rapidly established equilibrium, then:

$$\underline{k}_1 [R_2C=O][CN^\ominus] = \underline{k}_{-1}[R_2\overset{\overset{O^\ominus}{|}}{C}\text{-}CN]$$

or $\qquad\text{rate} = \dfrac{\underline{k}_2\underline{k}_1}{\underline{k}_{-1}} [R_2C=O][CN^\ominus][HCN]$

16-12

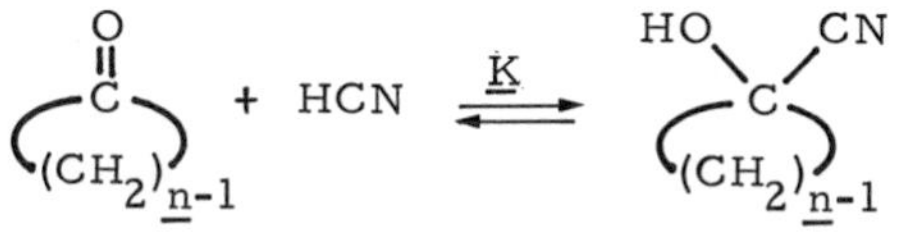

$$\underline{K}_{n=6} = 1000 \ \underline{K}_{n=5}$$

Addition to the carbonyl group of the five-ring ketone is less favorable than to the six-ring ketone because the substituents are eclipsed with neighboring hydrogens in the five-ring adduct but are staggered in the six-ring adduct. Equilibrium should favor $\underline{n} = 4$ over $\underline{n} = 5$ because there is some relief in the angle strain of the ring bonds on changing hybridization from sp^2 to sp^3 for $\underline{n} = 4$, but not for $\underline{n} = 5$.

16-13

a. $(C_6H_5)_3P: + CH_3CH_2\text{—}Br \xrightarrow[S_N2]{C_6H_6 \text{ (solvent)}} (C_6H_5)_3\overset{\oplus}{P}CH_2CH_3 \ \underset{Br^\ominus}{}$

$(C_6H_5)_3\overset{\oplus}{P}\text{-}CH_2CH_3 + C_6H_5:Li \longrightarrow (C_6H_5)_3\overset{\oplus}{P}\text{-}\overset{\ominus}{C}HCH_3 + C_6H_6 + LiBr$
$\underset{Br^\ominus}{}$

b. $(C_6H_5)_3\overset{\oplus}{P}\text{-}\overset{\ominus}{C}HCH_3 \longrightarrow \longrightarrow (C_6H_5)_3P=O + CH_3CH=C\diagup^{CH_2CH_2CH_3}_{\diagdown CH_3}$
$\begin{array}{c} | \ | \\ O=CCH_2CH_2CH_3 \\ | \\ CH_3 \end{array}$
$\qquad\qquad\qquad\qquad\qquad\qquad\qquad\qquad \underline{\text{cis}}$ and $\underline{\text{trans}}$

16-14

a. $ClCH_2OCH_3 \xrightarrow{Ar_3P:} Ar_3\overset{\oplus}{P}CH_2OCH_3 \;\overset{\ominus}{Cl} \xrightarrow[-ArH,\; -LiCl]{ArLi} Ar_3\overset{\oplus}{P}-\overset{\ominus}{C}HOCH_3$

$$\bigcirc\!\!=\!\!CHOCH_3 + Ar_3PO \longleftarrow \bigcirc\!\!=\!\!O$$

b. $Br(CH_2)_4Br \xrightarrow{Ar_3P:} Ar_3\overset{\oplus}{P}(CH_2)_4\overset{\oplus}{P}Ar_3$

$$\xrightarrow{2\,ArLi} Ar_3\overset{\oplus}{P}\overset{\ominus}{C}H(CH_2)_2\overset{\ominus}{C}\overset{\oplus}{H}PAr_3$$

$$\downarrow 2\,C_6H_5CHO$$

$$C_6H_5CH=CHCH_2CH_2CH=CHC_6H_5 + 2\,Ar_3PO$$

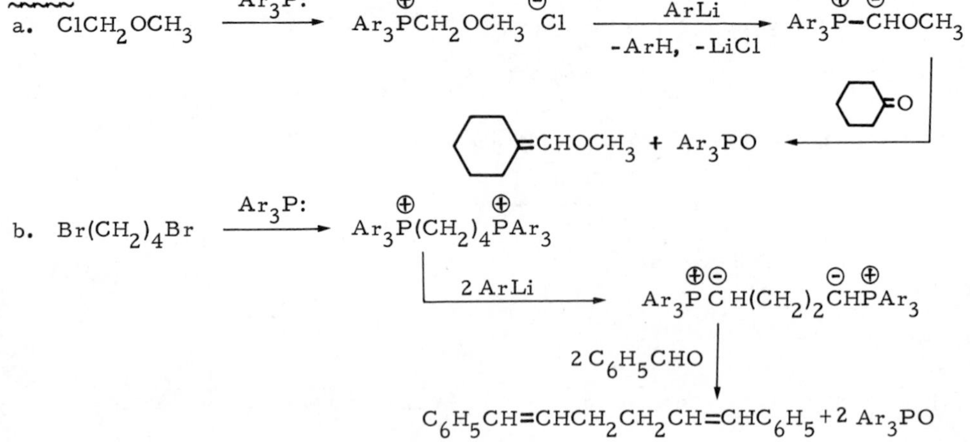

16-15 a. $:\overset{\ominus}{CH_2}-\overset{\oplus}{N}\equiv N: \longleftrightarrow CH_2=\overset{\oplus}{N}=\overset{\cdot\cdot}{N}:^{\ominus}$

Both structures must be important based on the dipole moment of diazomethane. The opposing moments of the two valence-bond forms lead to a small resultant moment. Low polarity is also manifest in the high volatility of diazomethane (b.p. $-23°$).

b.

$$\underset{:N=N:}{\overset{CH_2}{\diagup\diagdown}} \longleftrightarrow \underset{:N\equiv N^{\oplus}}{\overset{\ominus:CH_2}{\diagup}} \longleftrightarrow \underset{^{\oplus}N\equiv N:}{\overset{CH_2:^{\ominus}}{\diagdown}}$$

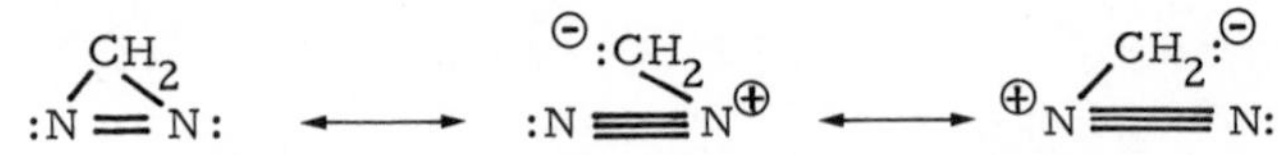

Although ylid valence-bond forms can be written for 1,2-diazacyclopropene, they are grossly distorted forms of diazomethane. Accordingly, they would be of high energy and would not contribute to the structure of 1,2-diazacyclopropene which, although thermodynamically unstable, is actually quite unreactive toward polar reagents (see Section 24-7C).

c. For valence-bond structures such as $\overset{}{\underset{}{>}}C=\overset{|}{P}-$ and $\overset{}{\underset{}{>}}C=\overset{\cdot\cdot}{\underset{|}{S}}-$ to contribute significantly to ylid structure, the sulfur and the phosphorus atoms each must have at least five low-lying atomic orbitals available for bonding with carbon. Five orbitals are required to accommodate a share of ten electrons around sulfur or phosphorus. Figure 6-4 shows that the valence orbitals of sulfur and phosphorus are 3s, 3p (three), and 3d (five). In contrast, nitrogen has only four low-lying valence orbitals, 2s, 2p (three), and cannot then form more than four bonds to carbon. Therefore, ylid structures such as $>C=N(CH_3)_3$ are not feasible.

16-16

a. 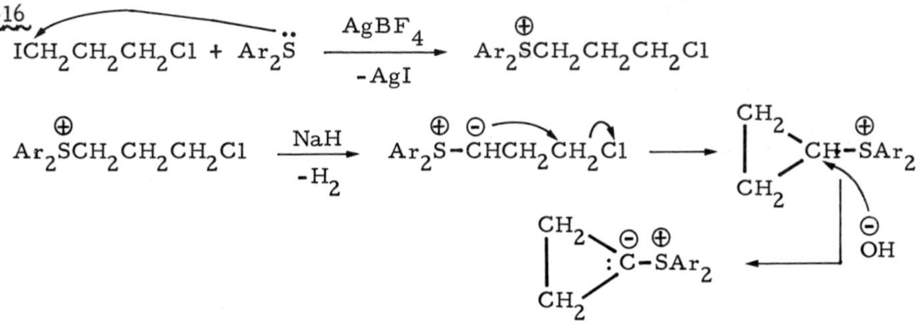 $ICH_2CH_2CH_2Cl + Ar_2\overset{\cdot\cdot}{S} \xrightarrow[-AgI]{AgBF_4} Ar_2\overset{\oplus}{S}CH_2CH_2CH_2Cl$

$Ar_2\overset{\oplus}{S}CH_2CH_2CH_2Cl \xrightarrow[-H_2]{NaH} Ar_2\overset{\oplus}{S}-\overset{\ominus}{C}HCH_2CH_2Cl \longrightarrow$

16-17   Equilibrium for hydration (and for carbonyl addition in general) is influenced unfavorably by large substituent groups on the carbonyl carbon.   The smallest, least-hindered carbonyl compound is methanal $CH_2{=}O$, and equilibrium of hydration is accordingly favorable.   Cyclopropanone hydration is also favored because of steric effects; in this case the argument is relief of angle strain on hydration (see Section 16-1C).   Electron-withdrawing groups, CCl_3 and $RC{=}O$, also favor hydration; evidently an electron-withdrawing group attached to the relatively positive carbon of a carbonyl bond, as in $Cl_3C-\overset{\delta\oplus}{\underset{|}{C}}\overset{\delta\ominus}{=}O$ is destabilizing relative to the adduct $Cl_3C-\underset{|}{C}(OH)_2$.

16-18

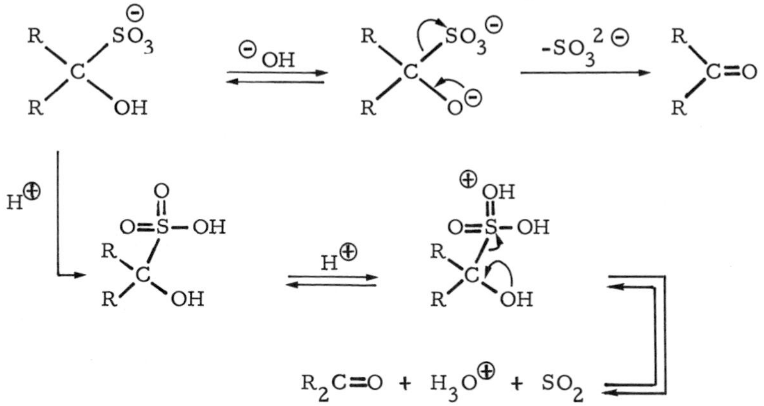

16-19   Addition of $NaHSO_3$ to a mixture of cyclohexanone and cyclohexanol precipitates the ketone as a crystalline adduct,

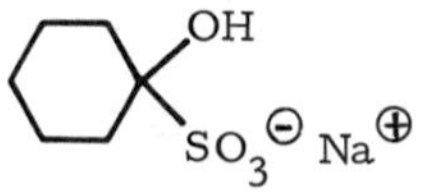

The cyclohexanol can be extracted from the mixture with ether and subsequently recovered by distillation. The hydrogen sulfite addition compound can be decomposed with acid to give cyclohexanone.

16-20

$$CH_2{=}O \xrightarrow{\ominus OH} HOCH_2{-}O^\ominus \xrightarrow{CH_2{=}O} HOCH_2OCH_2{-}O^\ominus \xrightarrow{\underline{n}CH_2{=}O}$$

$$HOCH_2O(CH_2O)_{\underline{n}}CH_2O^\ominus$$

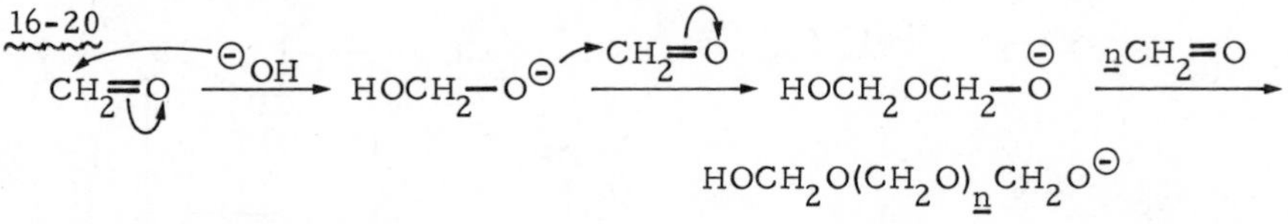

Base-catalyzed polymerization would not produce any 1,3,5-trioxacyclohexane because the ring-closure step is an unfavorable displacement reaction in which $^\ominus OH$ would be the leaving group.

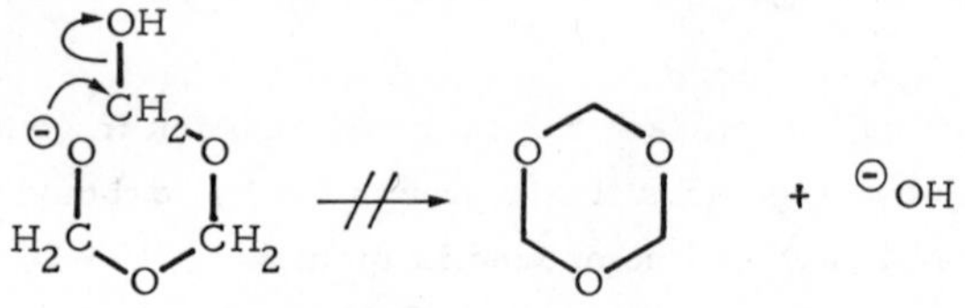

16-21

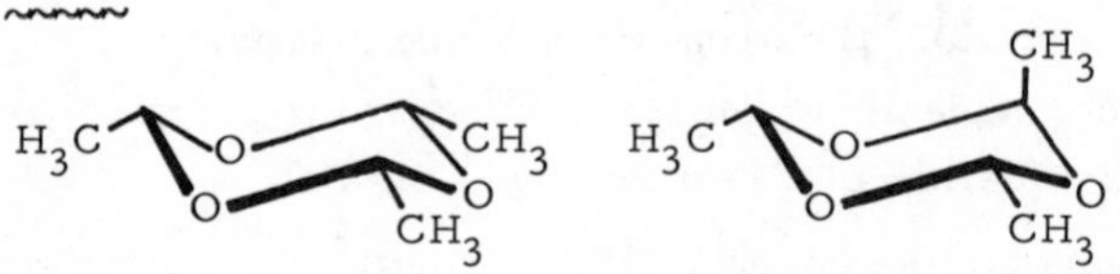

16-22 A reagent that would convert the terminal hydroxyl groups into a less reactive, less acid-base sensitive function would stabilize the polymer. Suitable reagents would be acylating or methylating reagents, such as ethanoic anhydride or dimethyl sulfate, which would convert $-OH$ to $-O-\overset{O}{\overset{\|}{C}}-CH_3$ or $-OCH_3$. For methylation:

$$HOCH_2O(CH_2O)_{\underline{n}}CH_2OH \xrightarrow[NaOH]{(CH_3O)_2SO_2} CH_3OCH_2O(CH_2O)_{\underline{n}}CH_2OCH_3$$

16-23 The two equilibria that involve RNH_2 may be written as follows:

$$R\overset{..}{N}H_2 + \overset{\oplus}{H_3O} \rightleftharpoons \overset{\oplus}{RNH_3} + H_2O$$

$$R\overset{..}{N}H_2 + (CH_3)_2C{=}O \overset{K}{\rightleftharpoons} (CH_3)_2C{=}NR + H_2O$$

In strong acid, most of the RNH_2 will be converted to $\overset{\oplus}{RNH_3}$. Therefore, the conversion of RNH_2 to $(CH_3)_2C=NR$ will be small. The equilibrium <u>constant</u> K will remain unchanged. Only the <u>position</u> of equilibrium (percent conversion) will change with pH.

The equilibrium, $(CH_3)_2C=NR + H_3O^\oplus \underset{}{\overset{K'}{\rightleftharpoons}} (CH_3)_2C=\overset{\oplus}{N}HR + H_2O$,

will be far less favorable than addition of a proton to RNH_2. As a result, very little $(CH_3)_2C=\overset{\oplus}{N}HR$ will be formed in strong acid.

16-24   a.   Only two position isomers are possible, one from N-substitution and one from C-substitution. You might suppose that there could be configurational isomers of C-substitution, one with the substituent "axial", and the other with the substituent "equatorial". However, these structures are equivalent in all respects, although you may have to make models to convince yourself of this.

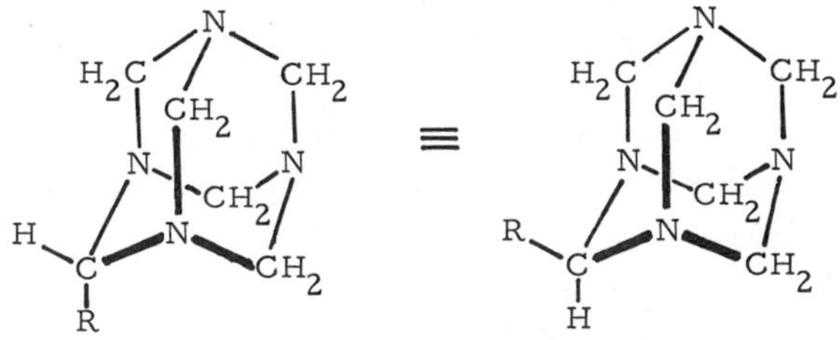

b.   Two (see Exercise 16-21 for analogous example).

c.   2, 4, 6-trimethyl-1, 3, 5-triazacyclohexane.

16-25

	Ketone	RNH_2	Product	Isomers
a.		H_2NOH		-
b.	$C_6H_5COCH_3$	$H_2NNHCONH_2$	}	cis-trans
c.		H_2NNH—NO$_2$, NO$_2$		-

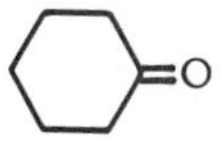

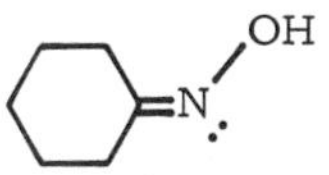

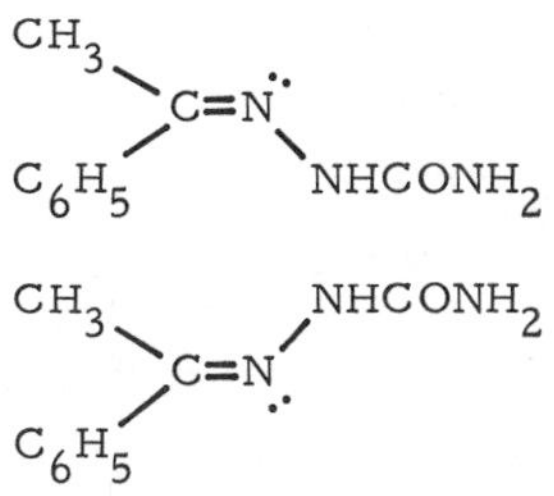

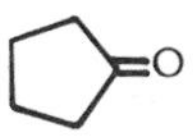

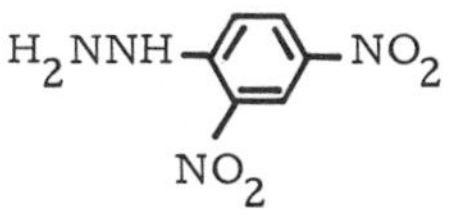

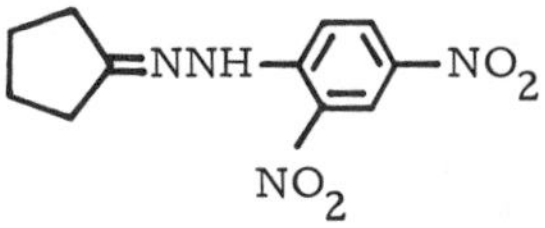

16-25 (cont.)

Ketone	RNH_2	Product	Isomers

d. $CH_3COCH_2CH_3$ $H_2NC_2H_5$

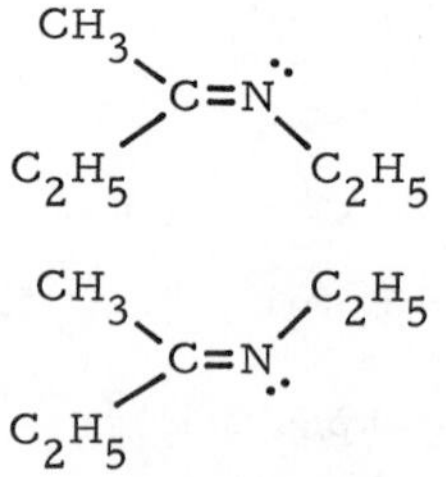

$$CH_3\diagdown \atop C_2H_5 \diagup C=N \diagdown \atop \diagup C_2H_5\ddot{}$$

$$CH_3\diagdown \atop C_2H_5 \diagup C=N\ddot{} \diagdown C_2H_5$$

$\Bigg\}$ cis-trans

Configuration at double-bonded nitrogen makes cis-trans isomers possible in Parts b and d. However, the cis, trans isomers of type d are rapidly interconverted at ordinary temperatures.

16-26 $CH_3CH=NCH_3 \rightleftharpoons CH_2=CH-NHCH_3$

$\Delta H = +16.3$ kcal; assuming $\Delta S = 0$ and neglecting possible resonance effects. The more stable isomer is therefore ethylidenemethanamine, $CH_3CH=NCH_3$. Some correction for electron-delocalization may be necessary because N-methylethenamine is expected to be stabilized to some extent by contributions from a dipolar valence-bond structure:

$$CH_2=CH-\ddot{N}HCH_3 \longleftrightarrow \overset{\ominus}{\ddot{C}}H_2-CH=\overset{\oplus}{N}HCH_3$$

16-27

+ H_2O

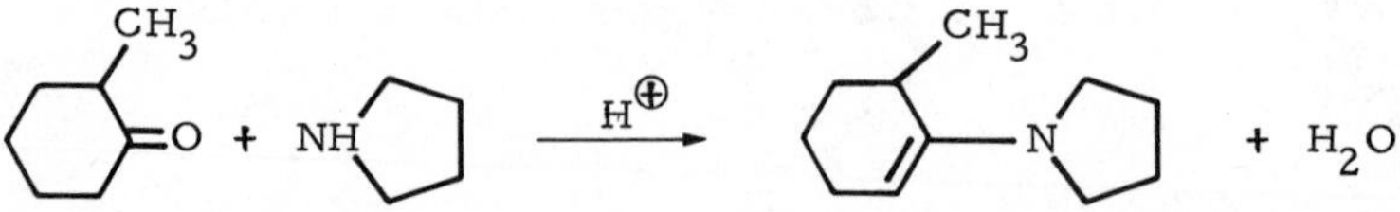

(acid catalyst required)

A possible byproduct is:

16-28

$$\underset{\underset{\displaystyle O}{\|}}{CH_3CH_2\overset{\displaystyle O}{C}CH_3} + HCl + CH_3OH \longrightarrow CH_3CH_2-\overset{\displaystyle OCH_3}{\underset{\displaystyle Cl}{C}}-CH_3 + H_2O$$

$$CH_3CH_2-\overset{\displaystyle OCH_3}{\underset{\displaystyle SCH_3}{C}}-CH_3 \xleftarrow[-HCl]{CH_3SH}$$

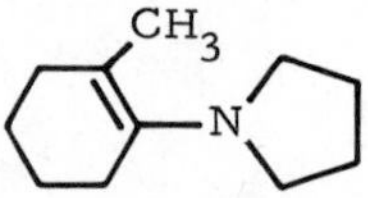

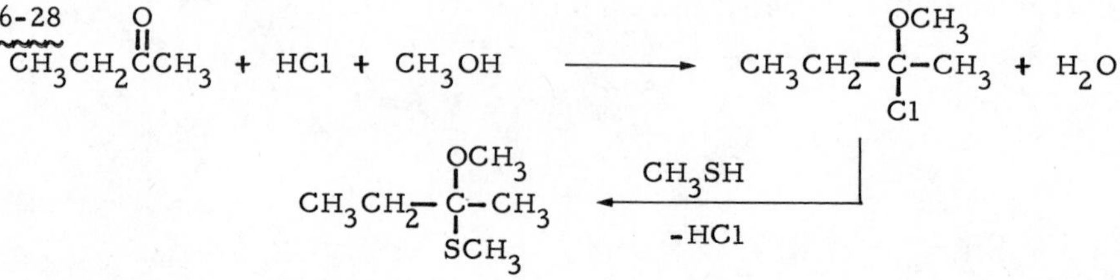

16-29   The following equilibria are probably involved:

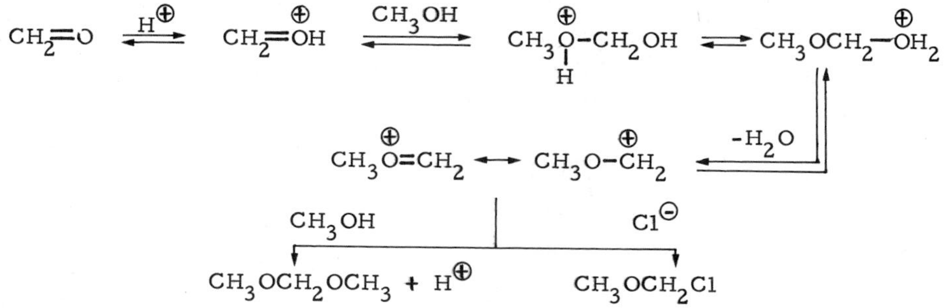

$$CH_2{=}O \; \underset{}{\overset{H^{\oplus}}{\rightleftarrows}} \; CH_2{=}\overset{\oplus}{OH} \; \underset{}{\overset{CH_3OH}{\rightleftarrows}} \; CH_3\underset{H}{\overset{\oplus}{O}}{-}CH_2OH \rightleftarrows CH_3OCH_2{-}\overset{\oplus}{OH}_2$$

$$CH_3\overset{\oplus}{O}{=}CH_2 \longleftrightarrow CH_3\overset{\oplus}{O}{-}CH_2 \;\overset{-H_2O}{\longleftarrow}$$

$$\underset{CH_3OCH_2OCH_3 + H^{\oplus}}{\overset{CH_3OH}{\downarrow}} \qquad \underset{CH_3OCH_2Cl}{\overset{Cl^{\ominus}}{\downarrow}}$$

Good conversions to the chloroether are achieved with methanal in methanol saturated with dry HCl gas. The ion $^{\oplus}CH_2Cl$ is not likely to be formed in preference to $^{\oplus}CH_2OCH_3$ and consequently CH_2Cl_2 is not obtained as a product.

16-30

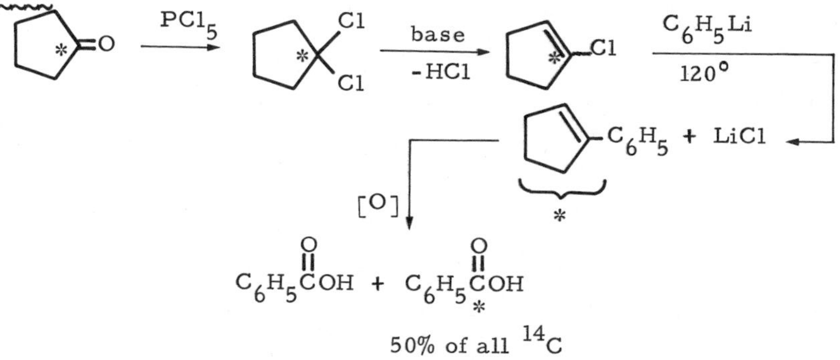

50% of all ^{14}C

Mechanism of reaction with phenyllithium:

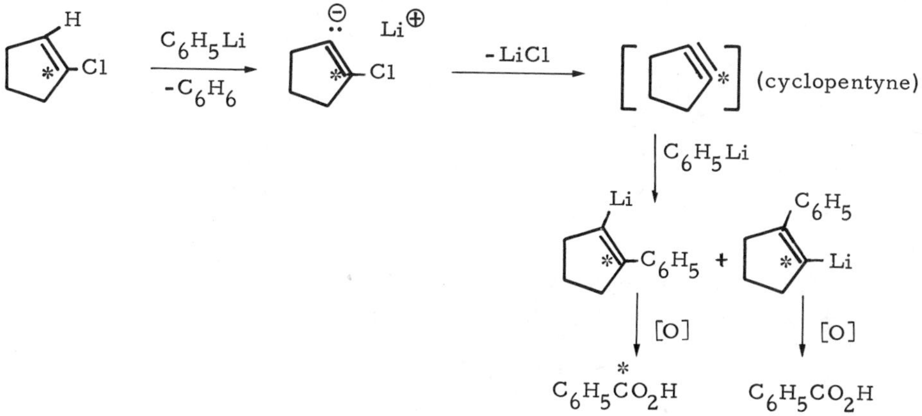

<u>16-31</u>   The following mechanisms are undoubtedly oversimplified but they illustrate reasonable pathways.

a. $\overset{\delta\oplus}{C}=\overset{\delta\ominus}{O}$ + $\overset{\oplus}{PCl_4}$ ⟶ $-\overset{|}{\underset{|}{C}}-O-\overset{\oplus}{PCl_4}$ $\xrightarrow[-PCl_5]{\overset{\ominus}{PCl_6}}$ $-\overset{Cl}{\underset{|}{C}}-O-PCl_4$

$-\overset{Cl}{\underset{|}{\underset{}{C}}}\overset{Cl}{-O-PCl_3}$ ⟶ $-\overset{Cl}{\underset{|}{C}}-Cl$ + $OPCl_3$

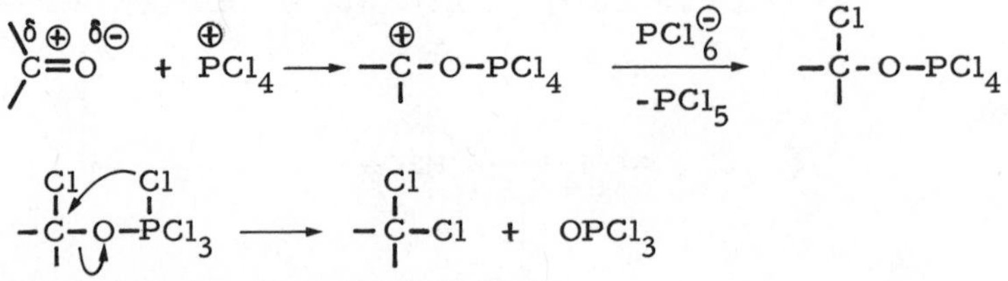

probably closely analogous
to S_N reactions of thionyl

chloride and an alcohol
(Section 15-5A)

b. $\overset{\delta\oplus}{C}=\overset{\delta\ominus}{O}$ + SF_4 ⟶ $-\overset{\oplus}{\underset{|}{C}}-O-\overset{F}{\underset{\ominus}{SF_3}}$ ⟶ $-\overset{F}{\underset{|}{C}}-O-\overset{F}{\underset{\underset{F}{|}}{S}}F$ ⟶ $-\overset{F}{\underset{|}{C}}-F$ + OSF_2

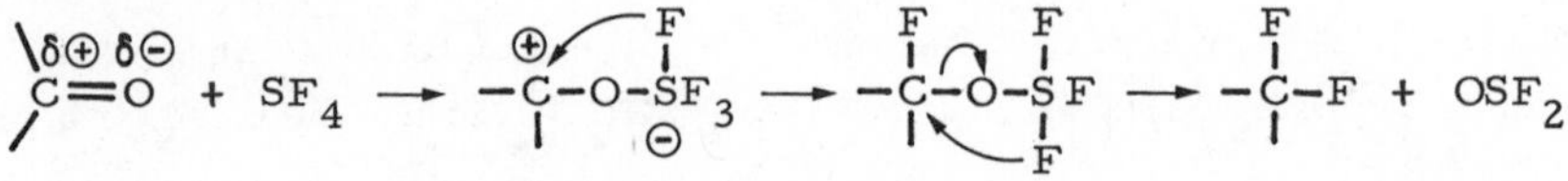

Also likely to be analogous to S_N reactions of thionyl chloride and alcohols (Section 15-5A)

<u>16-32</u>   a.   Reduction of C=C can be achieved selectively over C=O using borane (Table 16-6).

$CH_2{=}$⟨cyclohexanone⟩${=}O$ $\xrightarrow[\text{2. } H^\oplus]{\text{1. } BH_3}$ $CH_3{-}$⟨cyclohexanone⟩${=}O$

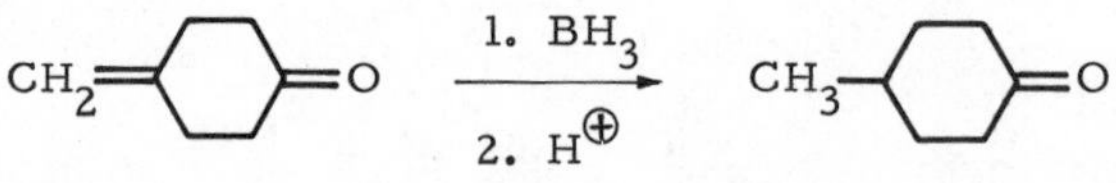

b.   Reduction of a carboxylic acid carbonyl in preference to a ketone carbonyl can be achieved with borane (Table 16-6).

$O{=}$⟨cyclohexane⟩${-}CO_2H$ $\xrightarrow{BH_3}$ $O{=}$⟨cyclohexane⟩${-}CH_2OH$

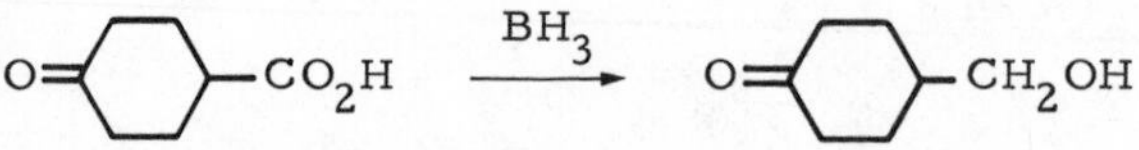

c.   Reduction of an aldehyde carbonyl in preference to a carboxylic acid carbonyl can be achieved with $NaBH_4$ (Table 16-6).

$HO_2CCH_2CH_2CH{=}O$ $\xrightarrow[\text{2. } H^\oplus]{\text{1. } NaBH_4}$ $HO_2CCH_2CH_2CH_2OH$

d.   Reduction of an aldehyde carbonyl in the presence of carbon-halogen bonds can be achieved with $NaBH_4$ (Table 16-6).

$Cl_3CCH{=}O$ $\xrightarrow[\text{2. } H^\oplus]{\text{1. } NaBH_4}$ Cl_3CCH_2OH

16-33

$$2CH_2{=}O \xrightarrow{\text{NaOH}} HCO_2H + CH_3OH$$

$$(CH_3)_3C{-}CH{=}O \xrightarrow{\text{NaOH}} (CH_3)_3C{-}CO_2H + (CH_3)_3C{-}CH_2OH$$

$$(CH_3)_3C{-}CH{=}O + CH_2{=}O \xrightarrow{\text{NaOH}} (CH_3)_3C{-}CO_2H + CH_3OH$$

or

$$(CH_3)_3C{-}CH_2OH + HCO_2H$$

Methanal, used in excess, is expected to <u>reduce</u> 2, 2-dimethylpropanal. Addition to the methanal carbonyl is much more favorable (for steric reasons) than to the 2, 2-dimethylpropanal carbonyl. Therefore, the hydroxide base is converted to $HOCH_2{-}O^{\ominus}$ which is the reducing agent; it reduces free methanal readily and 2, 2-dimethylpropanal more slowly.

$$CH_3{-}\underset{\underset{CH_3}{|}}{\overset{\overset{CH_3}{|}}{C}}{-}C\overset{H}{\underset{O}{\diagup}} + H{-}\underset{|}{\overset{OH}{C}}H{-}O^{\ominus} \longrightarrow CH_3{-}\underset{\underset{H_3C}{|}}{\overset{\overset{H_3C}{|}}{C}}{-}\underset{\underset{H}{|}}{\overset{\overset{H}{|}}{C}}{-}OH + HCO_2^{\ominus}$$

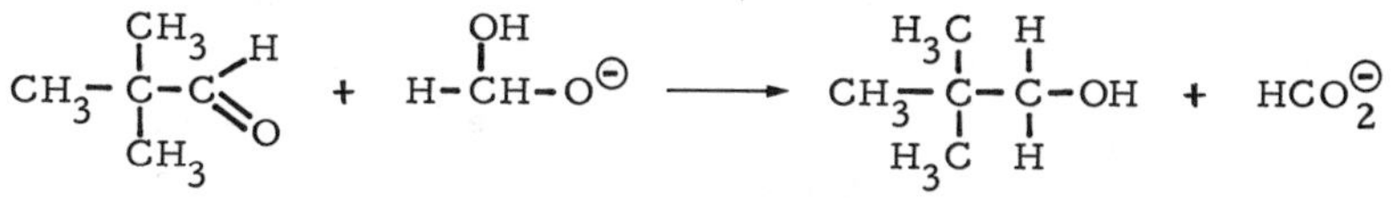

<u>16-34</u>   The slow step in the reaction must be the breaking of the aldehyde C–H bond because $\underline{k}_D < \underline{k}_H$. The slow step must also involve the equivalent of $[H^{\oplus}]$, $[MnO_4^{\ominus}]$ and $[ArCHO]$. The mechanism must also show the addition of $MnO_4^{\ominus}$ to the carbonyl group to account for the ^{18}O-labelling data. Likewise, addition of H_2O to the carbonyl must not be significant. A mechanism which accommodates these facts follows.

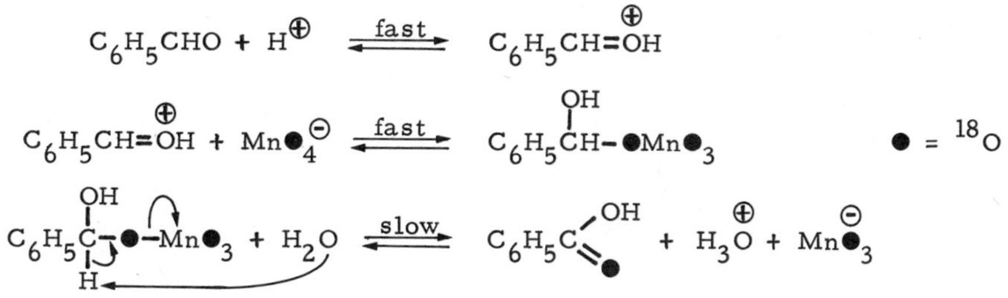

$$C_6H_5CHO + H^{\oplus} \underset{\longleftarrow}{\overset{\text{fast}}{\longrightarrow}} C_6H_5CH{=}\overset{\oplus}{O}H$$

$$C_6H_5CH{=}\overset{\oplus}{O}H + Mn{\bullet}_4^{\ominus} \underset{\longleftarrow}{\overset{\text{fast}}{\longrightarrow}} C_6H_5\underset{|}{\overset{OH}{C}}H{-}{\bullet}Mn{\bullet}_3 \qquad {\bullet} = {}^{18}O$$

$$C_6H_5\underset{\underset{H}{|}}{\overset{\overset{OH}{|}}{C}}{\bullet}{-}Mn{\bullet}_3 + H_2O \underset{\longleftarrow}{\overset{\text{slow}}{\longrightarrow}} C_6H_5C\overset{OH}{\underset{\bullet}{\diagup}} + H_3\overset{\oplus}{O} + Mn{\bullet}_3^{\ominus}$$

16-35

Initiation $\begin{cases} RO{-}OR \longrightarrow 2RO{\cdot} \\[4pt] CH_3CHO + RO{\cdot} \longrightarrow CH_3\overset{\bullet}{C}O + ROH \end{cases}$

Propagation $\begin{cases} CH_3CH{=}CH_2 + CH_3\overset{\bullet}{C}O \longrightarrow CH_3\overset{\bullet}{C}H{-}CH_2{:}\overset{\overset{O}{\|}}{C}CH_3 \\[6pt] CH_3\overset{\bullet}{C}HCH_2COCH_3 + CH_3CHO \to CH_3CH_2CH_2COCH_3 + CH_3\overset{\bullet}{C}O \end{cases}$

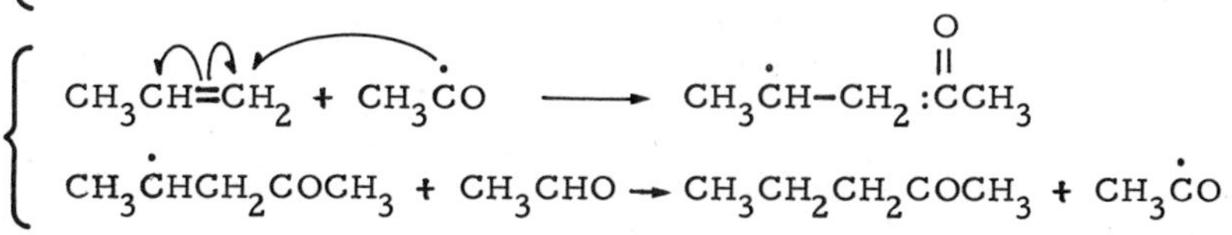

16-36

Initiation
$$
\begin{cases}
RO-OR \longrightarrow 2RO\cdot \\
(CH_3)_2CHCH_2CHO + RO\cdot \longrightarrow (CH_3)_2CHCH_2\overset{\cdot}{C}O + ROH
\end{cases}
$$

Propagation
$$
\begin{cases}
(CH_3)_2CHCH_2\overset{\cdot}{C}O \longrightarrow (CH_3)_2CH\overset{\cdot}{C}H_2 + CO \qquad \Delta \underline{H}_1 = +4.3 \text{ kcal} \\
(CH_3)_2CH\overset{\cdot}{C}H_2 + (CH_3)_2CHCH_2CHO
\end{cases}
$$

$$\Delta \underline{H}_2 = -12 \text{ kcal}$$

$$(CH_3)_2CHCH_3 + (CH_3)_2CHCH_2\overset{\cdot}{C}O$$

The bond energies used to calculate $\Delta \underline{H}$ of propagation were C–C, 82.6 kcal; CO, 257.3 kcal; C=O, 179 kcal; RCH_2–H, 98 kcal; $R\overset{O}{\overset{\|}{C}}$–H, 86 kcal. The questionable step is the decarbonylation step, $\Delta \underline{H}_1 = +4.3$ kcal. There is a significant activation energy for this dissociation which means that the rate of dissociation and hence the efficiency of the chain reaction increases with increasing temperature. In fact, as noted, most aldehydes lose carbon monoxide only at temperatures in excess of $120°$.

16-37   The first stage in the reaction is the hydroxylation of the double bond by either OsO_4 or $KMnO_4$ whereby the oxidant is reduced to $Os(VI)$ or $Mn(V)$

$$
\underset{/}{\overset{\backslash}{}}C=C\underset{\backslash}{\overset{/}{}} + OsO_4 \text{ (or } KMnO_4) \xrightarrow{H_2O} \overset{\mid}{\underset{HO}{-C-}}\overset{\mid}{\underset{OH}{C-}}
$$

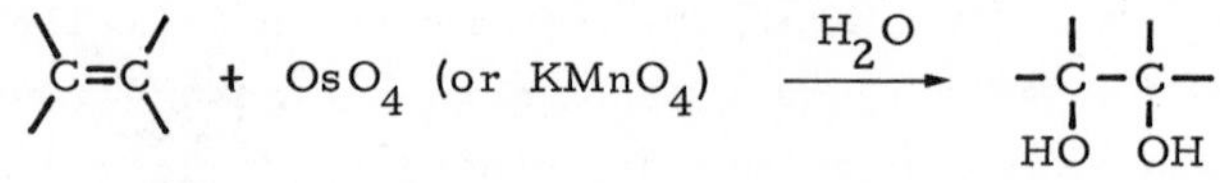

Neither oxidant reacts further with the diol, which is subsequently converted to carbonyl by periodate. However, the periodate regenerates the OsO_4 or the $KMnO_4$ so that it is available to hydroxylate more alkene.

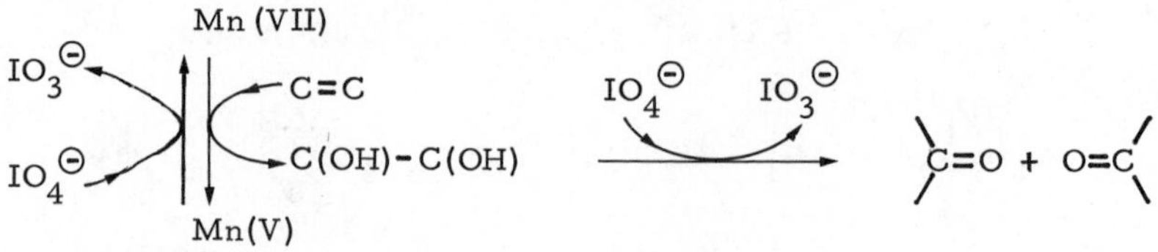

16-38

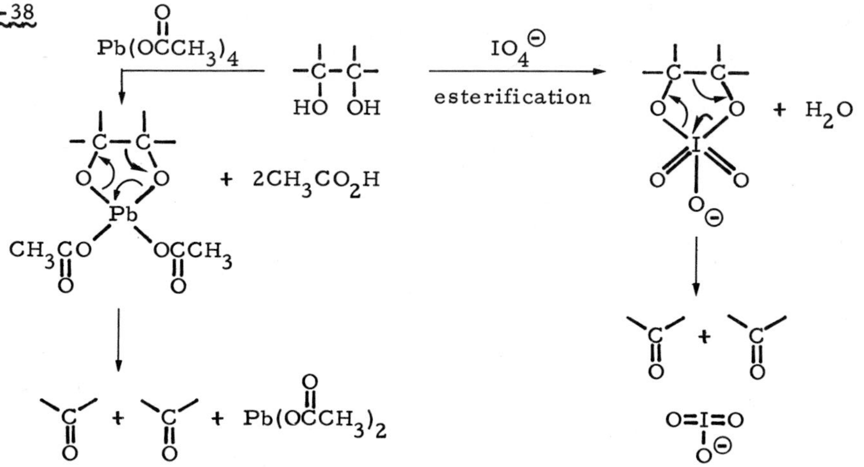

16-39 Resonance stabilization of the carboxyl function is lost on reduction to
R CHO, as the following structures indicate:

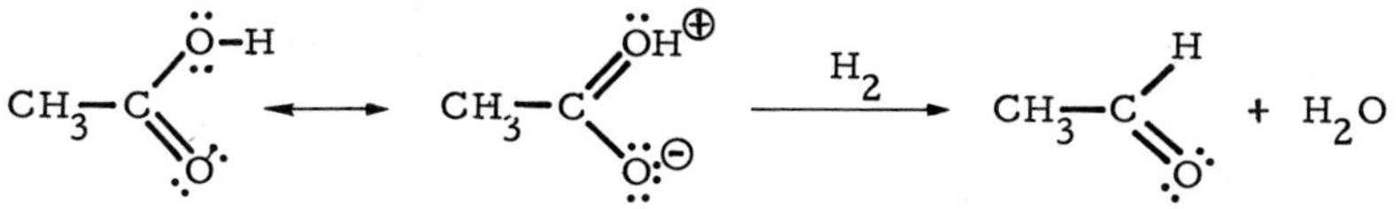

Related stabilization of CH_3COCl is apparently unimportant because of the low
tendency for Cl to participate in double bond formation. Resonance stabilization of CH_3CONH_2 is expected to be comparable to CH_3CO_2H, and $\Delta \underline{H}$ for
reduction should then be more positive than calculated from bond energies.

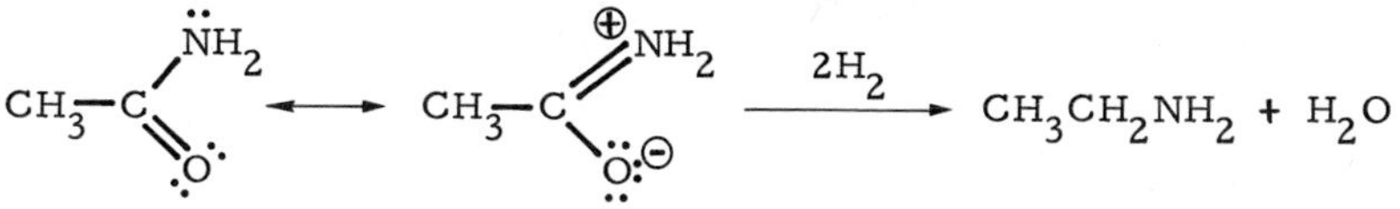

16-40

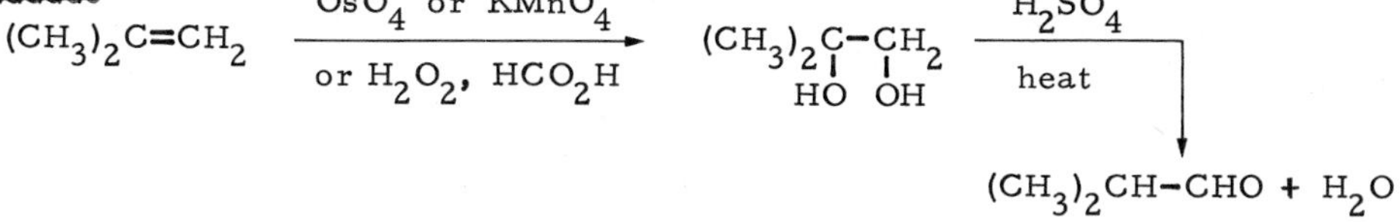

Some butanone may be formed by the process:

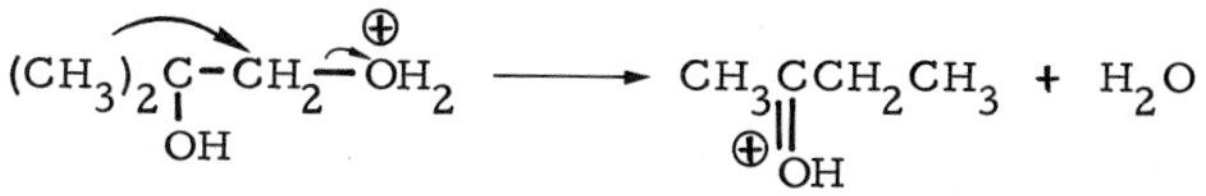

but methyl migration is less favored than hydride migration, because H_2O is
much more difficult to have leave from a primary carbon than from a tertiary

carbon.

16-41 $CH_3CH(OH)CH_2OH$ $\xrightarrow{H_2SO_4}$ CH_3CH_2CHO + CH_3COCH_3
                                                    (mostly)   (some)

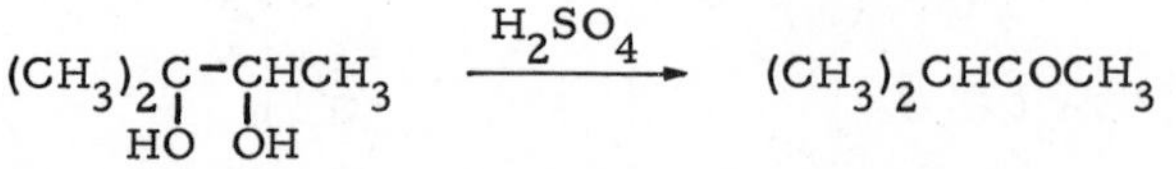

$(CH_3)_2\underset{\underset{HO}{|}}{C}-\underset{\underset{OH}{|}}{C}HCH_3$ $\xrightarrow{H_2SO_4}$ $(CH_3)_2CHCOCH_3$

16-42 $(CH_3)_2\underset{O}{C-C}(CH_3)_2$ $\underset{}{\overset{H^{\oplus}}{\rightleftharpoons}}$ $(CH_3)_2\underset{\underset{H^{\oplus}}{\overset{|}{O}}}{C-C}(CH_3)_2$ $\rightleftharpoons$ $CH_3-\underset{\underset{OH}{|}}{\overset{\overset{CH_3}{|}}{C}}-\underset{\underset{CH_3}{|}}{\overset{+}{C}}-CH_3$

$\quad\quad\quad\quad\quad\quad\quad\quad\quad\quad\quad\quad\quad\quad\quad\quad\quad\quad\quad\quad\quad\quad\quad\quad$ CH$_3$:
$\quad\quad\quad\quad\quad\quad\quad\quad\quad\quad\quad$ $CH_3-\underset{O}{\overset{||}{C}}-C(CH_3)_3$ $\xleftarrow{-H^{\oplus}}$ rearr.

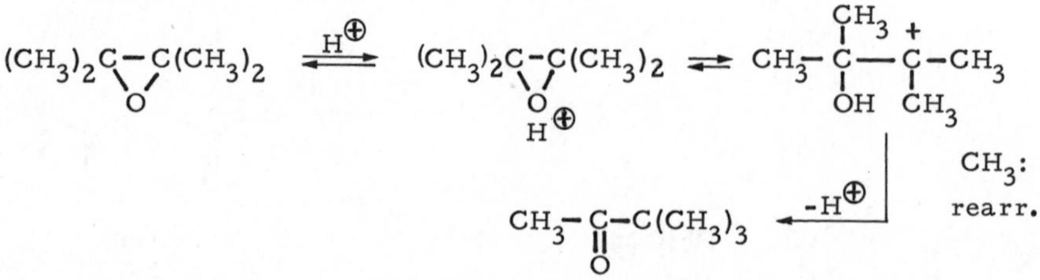

16-43   It is necessary to avoid carbocation formation. One possibility is to convert the diol to the corresponding dichloride with $SOCl_2$ or cold conc. HCl and follow this by a double E2 reaction to form 2,3-dimethyl-1,3-butadiene. Actually, the easiest route is by dehydration over aluminum oxide at high temperatures (Section 15-5C).

16-44   Of the two products, $(C_6H_5)_2CHCHO$ and $C_6H_5CH_2COC_6H_5$, the ketone is the more stable because it has a stronger C≡O bond and less steric hindrance. The product of equilibrium control is therefore $C_6H_5CH_2COC_6H_5$. However, the equilibrium product is not formed as rapidly as 1,1-diphenylethanal because its formation requires phenyl migration in a primary carbocation (which is much less easily formed than the tertiary carbocation).

$$C_6H_5-\underset{\underset{HO}{|}}{\overset{\overset{C_6H_5}{|}}{C}}-\underset{\underset{OH}{|}}{C}H_2$$

$H^{\oplus}$ ⇌   ⇌ $H^{\oplus}$

$C_6H_5-\underset{\underset{OH}{|}}{\overset{\overset{H_5C_6}{|}\oplus}{C}}-CH_2$ + H_2O $C_6H_5-\underset{\underset{OH}{|}}{\overset{\overset{C_6H_5}{|}}{C}}-\overset{\oplus}{C}H_2$ + H_2O

$-H^{\oplus}$ | C_6H_5:
rearr.   H: | $-H^{\oplus}$
                                                rearr:

$C_6H_5-\underset{O}{\overset{||}{C}}-CH_2C_6H_5$ $C_6H_5-\underset{\underset{H}{|}}{\overset{\overset{C_6H_5}{|}}{C}}-CH=O$

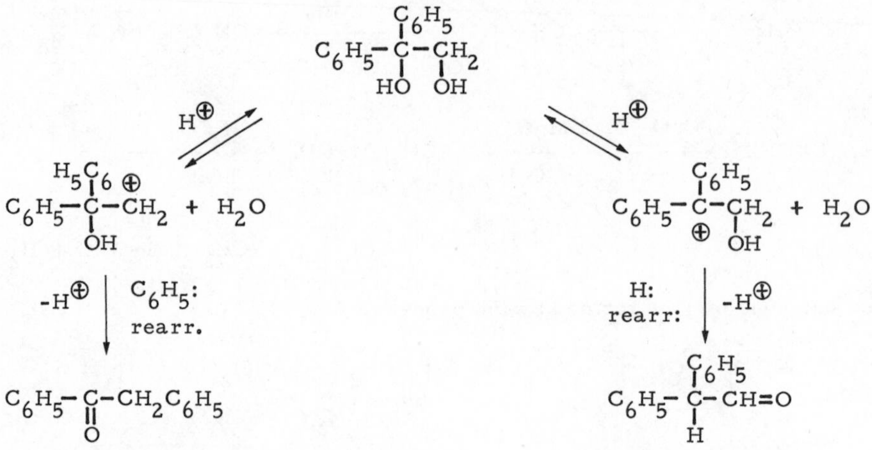

(equilibrium product)   (kinetic product)

16-45   The possible reactions are:

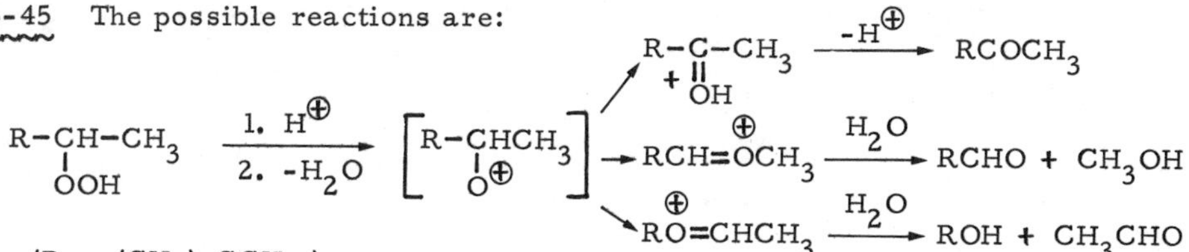

$$R-CH-CH_3 \xrightarrow[\text{2. } -H_2O]{\text{1. } H^{\oplus}} \left[R-CHCH_3 \right]$$

(R = $(CH_3)_3CCH_2^-$)

Prediction of the product ratio is difficult, but an important consideration may be the favored conformation. Thus:

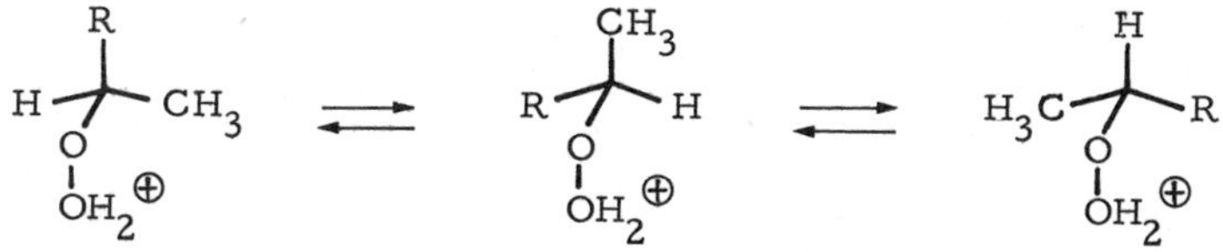

Of these, the first should be the most favorable, and if R moves over as OH_2 leaves then the R migration product should be favored. As, in fact, it is.

16-46

a.

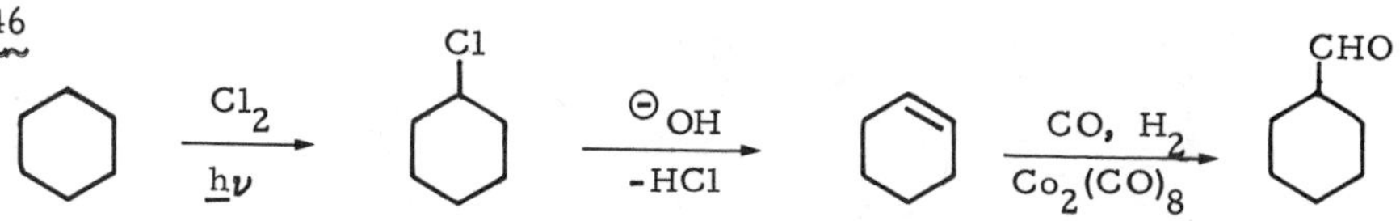

b.

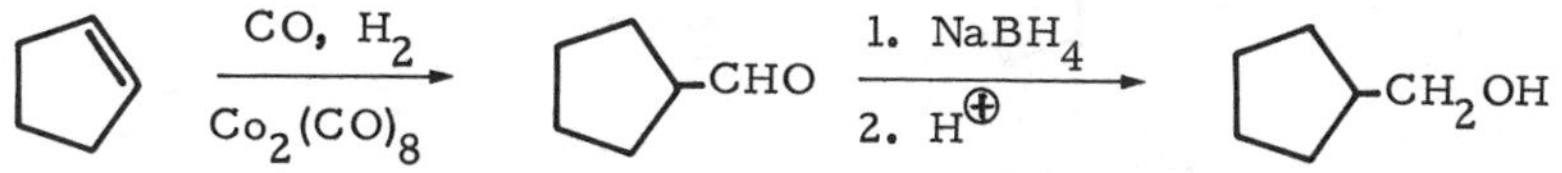

c. $C_6H_5COCH_2Br \xrightarrow[\text{base}]{(CH_3)_2S=O} C_6H_5COCHO + (CH_3)_2S + (HBr)$

(Section 16-9B)

d. $RCH=CH_2 + HBr \xrightarrow[\text{peroxides}]{h\nu \text{ or}} RCH_2CH_2Br \xrightarrow[\text{base, } -HBr]{(CH_3)_2S=O} RCH_2CHO$

(R = C_5H_{11})

e.

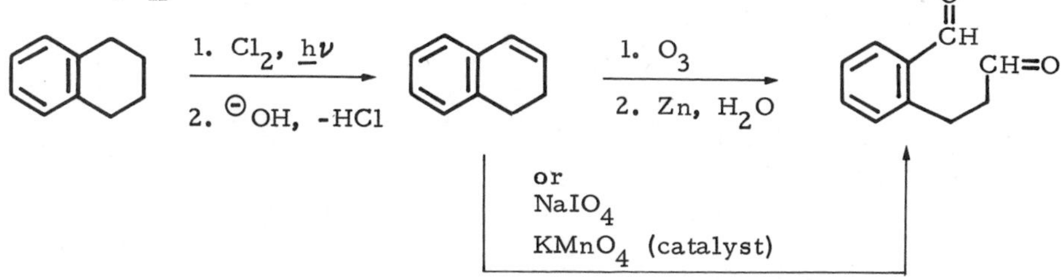

f.

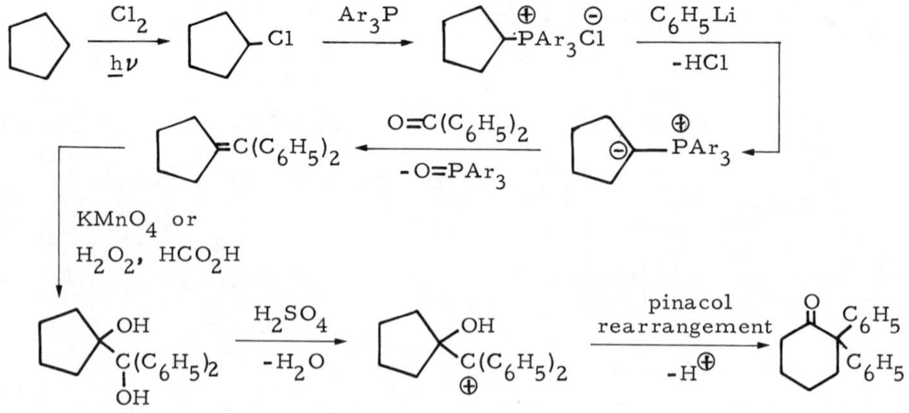

$$\text{(cyclopentane)} \xrightarrow[h\nu]{Cl_2} \text{(chlorocyclopentane)} \xrightarrow{Ar_3P} \overset{\oplus}{PAr_3}Cl^{\ominus} \xrightarrow[-HCl]{C_6H_5Li}$$

$$\xrightarrow[-O=PAr_3]{O=C(C_6H_5)_2} =C(C_6H_5)_2 \qquad \overset{\ominus}{}-\overset{\oplus}{PAr_3}$$

$$\xrightarrow[H_2O_2, HCO_2H]{KMnO_4 \text{ or}} \overset{OH}{\underset{OH}{C(C_6H_5)_2}} \xrightarrow[-H_2O]{H_2SO_4} \overset{OH}{\underset{\oplus}{C(C_6H_5)_2}} \xrightarrow[-H^{\oplus}]{\text{pinacol rearrangement}} \overset{O}{\underset{C_6H_5}{C_6H_5}}$$

16-47

a.

$$CH_3 \xrightarrow{RBH_2} CH_3 \overset{H}{\underset{B-R}{}} \longrightarrow CH_3 \overset{}{B-R} \xrightarrow{CO}$$

$$CH_3 \overset{O}{\underset{R}{B}} \longleftarrow CH_3 \overset{\ominus}{\underset{R}{B}} \overset{\oplus}{C=O} \quad \text{...}$$

$$CH_3 \overset{O}{\underset{R}{B}} \longleftarrow CH_3 \overset{\oplus}{\underset{\ominus}{B}} \overset{O}{\underset{R}{}}$$

$$\xrightarrow{H_2O_2} CH_3 \overset{}{=O}$$

(1)b.

$$\overset{C}{C-}\overset{}{C}\equiv C \xrightarrow[\text{(R is bulky)}]{RBH_2} \overset{C}{C-}\overset{R}{C-C-BH} \xrightarrow{C-C\equiv C} \overset{C}{C-}\overset{R}{C-C-B-C-C-C}$$

$$\overset{C}{C-}\overset{R}{C-}\overset{O}{C-B-C-C-C-C} \xleftarrow{\text{rearr.}} \overset{C}{C-}\overset{R}{C-}\overset{\ominus}{C-}\overset{}{B}\overset{}{C-C-C} \xleftarrow{CO}$$

$$\xrightarrow[\text{rearr.}]{} \overset{C}{C-}\overset{}{C-}\overset{}{\underset{O\text{---}BR}{C-C}-C-C-C} \xrightarrow{H_2O_2} \overset{C}{C-}\overset{}{C-C-}\overset{O}{C}-C-C-C$$

(2)b.

$$C=C-C-C-C=C \xrightarrow{2R_2BH} R_2B-C-C-C-C-C-C-BR_2$$

$$OHC-C-C-C-C-C-C-CHO \xleftarrow[2. H_2O]{1. 2CO}$$

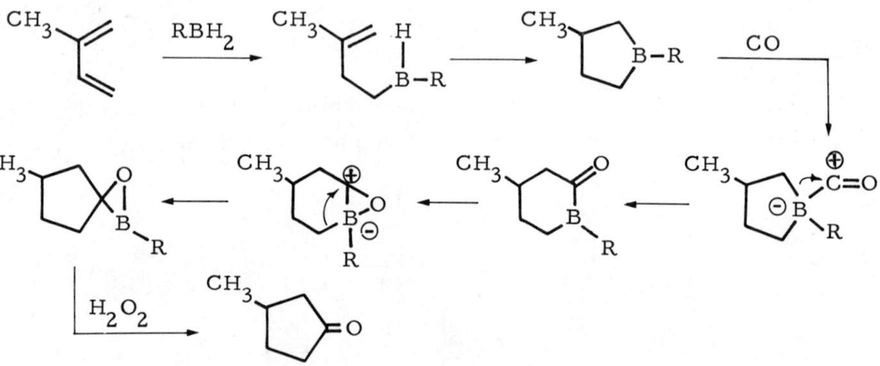

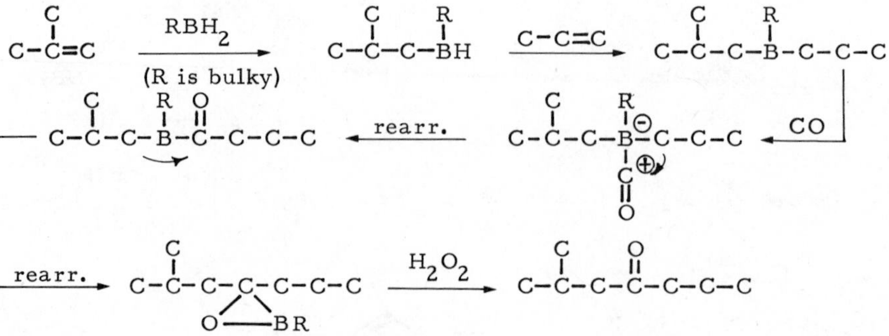

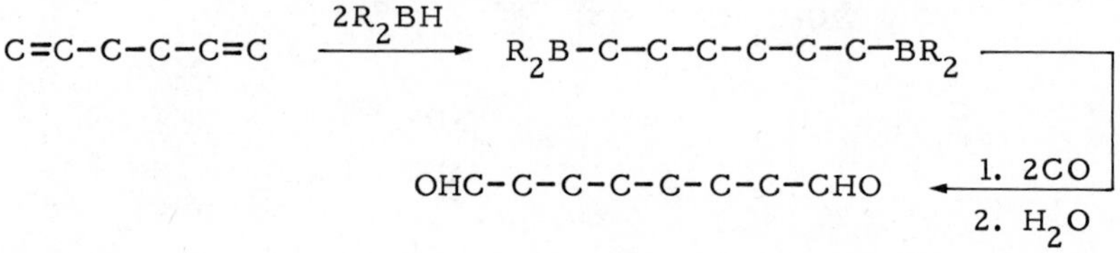

16-48

a. $(CH_3)_2CHCH_2CH_2OH$ $\xrightarrow{\text{HBr}}$ $(CH_3)_2CHCH_2CH_2Br$ $\xrightarrow[\text{E2}]{\text{base}}$

$(CH_3)_2CHCHO$ $\xleftarrow[\text{KMnO}_4]{\text{NaIO}_4}$ $(CH_3)_2CHCH=CH_2$ $\longleftarrow$

(H_2SO_4 dehydration of $(CH_3)_2CHCH_2CH_2OH$ causes rearrangement
to $(CH_3)_2C=CHCH_3$; for this reason, the E2 reaction even with an
additional step is preferred.)

b. [cyclobutane]$-CO_2H$ $\xrightarrow[\text{2. H}^{\oplus}]{\text{1. 2CH}_3\text{Li}}$ [cyclobutane]$-\overset{\overset{\text{O}}{\|}}{C}-CH_3$

c. [cyclopentanone ring]$=O$ $\xrightarrow[\text{2. H}^{\oplus}]{\text{1. LiAlH}_4}$ [cyclopentane ring with H and OH] $\xrightarrow[-\text{H}_2\text{O}]{\text{H}_2\text{SO}_4}$ [cyclopentene] $\xrightarrow[\substack{\text{or NaIO}_4, \\ \text{KMnO}_4}]{\substack{\text{1. O}_3 \\ \text{2. Zn}}}$ [ring]$\overset{\text{CHO}}{\underset{\text{CHO}}{}}$

d. [cyclobutane]$=CH_2$ $\xrightarrow[\text{KMnO}_4]{\text{NaIO}_4}$ [cyclobutanone]$=O$ $\xrightarrow[\text{KOH}]{\text{H}_2\text{NNH}_2}$ [cyclobutane]

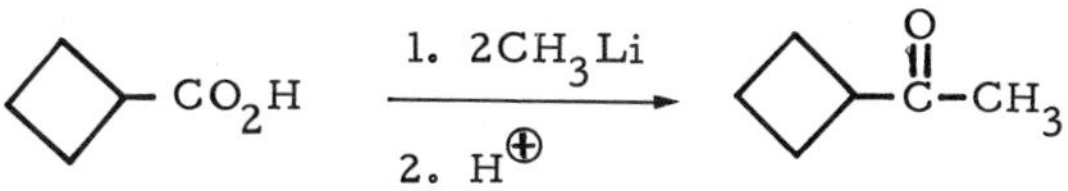

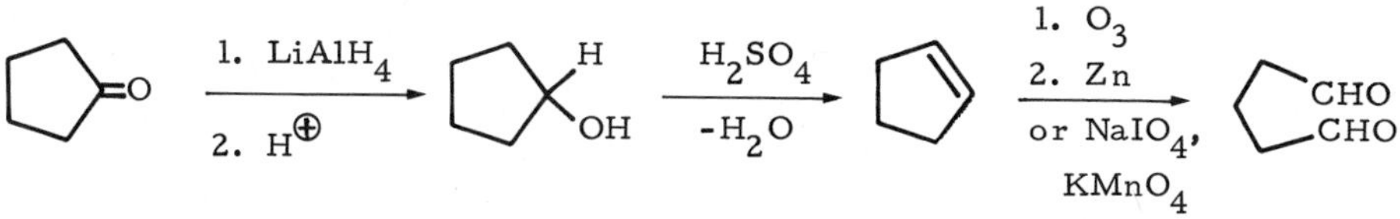

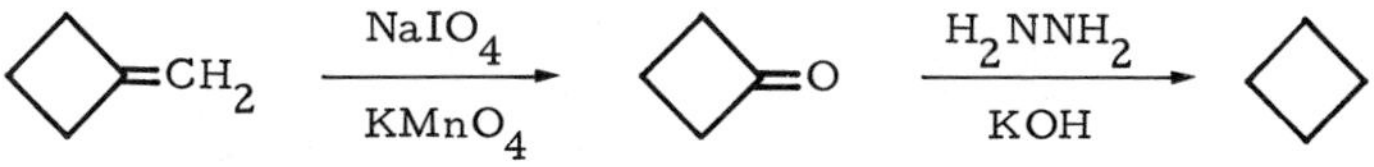

e. Cl_3CCHO $\xrightarrow{[O]}$ Cl_3CCO_2H
    $\downarrow$
    $\xrightarrow{\text{NaBH}_4}$ Cl_3CCH_2OH
    $\left.\right\}$ $\xrightarrow[\text{H}^{\oplus}]{\text{esterification}}$ $Cl_3C\overset{\overset{\text{O}}{\|}}{C}OCH_2CCl_3$

f. [cyclopentanone]$=O$ $\xrightarrow{\text{HCN}}$ [cyclopentane]$\overset{\text{OH}}{\underset{\text{CN}}{}}$ $\xrightarrow[-\text{H}_2\text{O}]{\text{H}_2\text{SO}_4}$ [cyclopentene]$-CN$ $\xrightarrow{\text{H}_2\text{SO}_4}$ [cyclopentene]$-CO_2H$

16-49

a. $H-\overset{\overset{\text{O}}{\|}}{C}-\overset{\overset{\text{O}}{\|}}{C}-H$ $\xrightarrow{\ominus\text{OH}}$ $H-\overset{\overset{\curvearrowleft\text{O}}{\|}}{C}-\overset{\overset{\text{O}^{\ominus}}{|}}{\underset{H}{C}}-OH$ $\longrightarrow$ $H-\overset{\overset{\text{HO}}{|}}{\underset{H}{C}}-\overset{\overset{\text{O}}{\|}}{C}-O^{\ominus}$

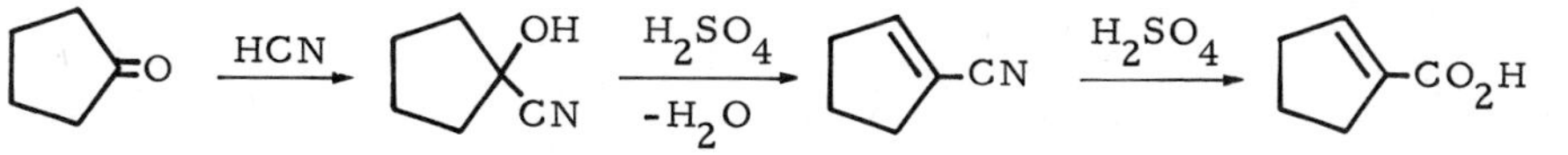

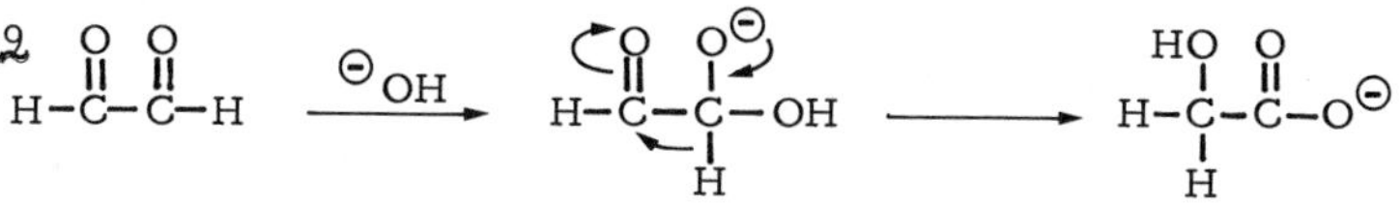

(internal Cannizarro reaction)

b. $\overset{..}{N}H_2OH + \overset{\ominus}{O}H$ $\rightleftharpoons$ $:\underset{\ominus}{\overset{H}{N}}-\overset{..}{O}H$

16-49 b. (cont.)

$$CH_3 \atop CH_3 \!\!\! \diagdown \!\! C{=}O \; + \; {:}NHOH \xrightarrow{\text{slow}} CH_3{-}\underset{\underset{CH_3}{|}}{C}{-}O^{\ominus} \xrightarrow{H^{\oplus}} CH_3{-}\underset{\underset{CH_3}{|}}{\overset{\overset{:NH-OH}{|}}{C}}{-}OH \longrightarrow$$

$$(CH_3)_2C{=}\ddot{N}OH \xleftarrow{-H_2O} CH_3{-}\underset{\underset{CH_3}{|}}{\overset{\overset{\ominus:\ddot{N}-OH}{|}}{C}}{-}OH_2^{\oplus} \longleftarrow$$

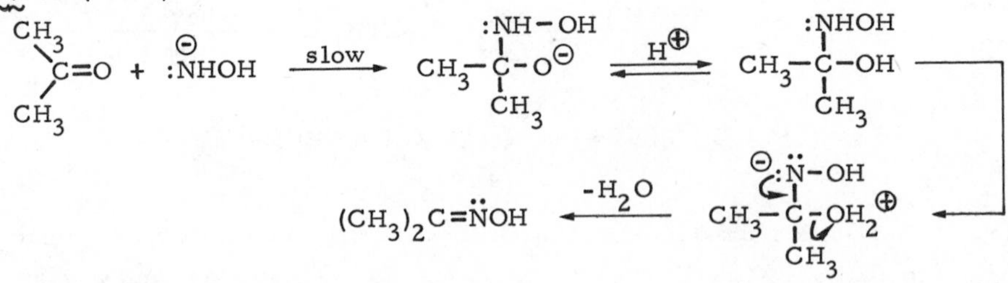

(a condensation of carbonyl compounds with RNH_2 derivatives which in this case is <u>base</u> catalyzed)

c. $CH_3{-}\underset{\underset{Br}{|}}{\overset{\overset{H}{|}}{C}}{-}\underset{\underset{OH}{|}}{\overset{\overset{H}{|}}{C}}{-}CH_3 \xrightarrow[-AgBr]{Ag^{\oplus}} CH_3{-}\underset{\oplus}{\overset{\overset{H}{|}}{C}}{-}\underset{\underset{OH}{|}}{\overset{\overset{H}{|}}{C}}{-}CH_3 \longrightarrow CH_3{-}CH_2{-}\underset{\underset{+\,OH}{\|}}{\overset{}{C}}CH_3$

$$CH_3\underset{\underset{O}{\|}}{C}CH_2CH_3 \xleftarrow{-H^{\oplus}}$$

d. $CH_2{=}O \xrightarrow[-H_2O]{NH_3} CH_2{=}NH \xrightarrow{NH_3} H_2N{-}CH_2{-}NH_2 \xrightarrow[-H_2O]{CH_2{=}O}$

$$\underset{HOCH_2}{}\overset{CH_2OH}{\underset{\begin{smallmatrix}\\ \end{smallmatrix}}{N}} \quad \xleftarrow{3\,CH_2{=}O} \quad \overset{NH}{N} \quad \xleftarrow[CH_2O]{NH_3} \quad \overset{CH_2}{\underset{CH_2}{\|}}N{-}\underset{CH_2}{NH_2}$$

$$\xrightarrow[-H_2O]{H^{\oplus}} \quad \xrightarrow[-H^{\oplus}]{NH_3} \quad \text{etc.}$$

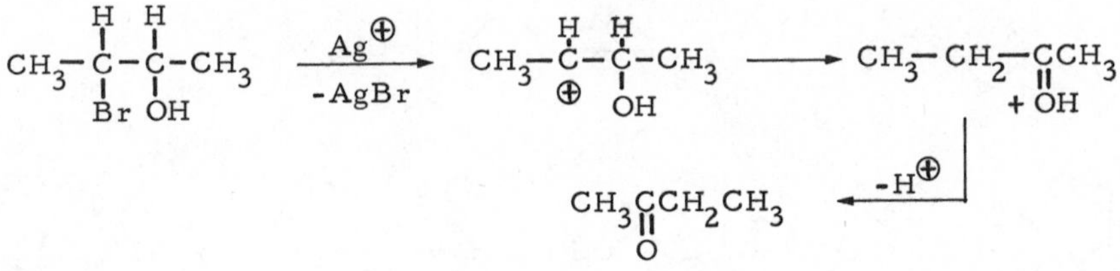

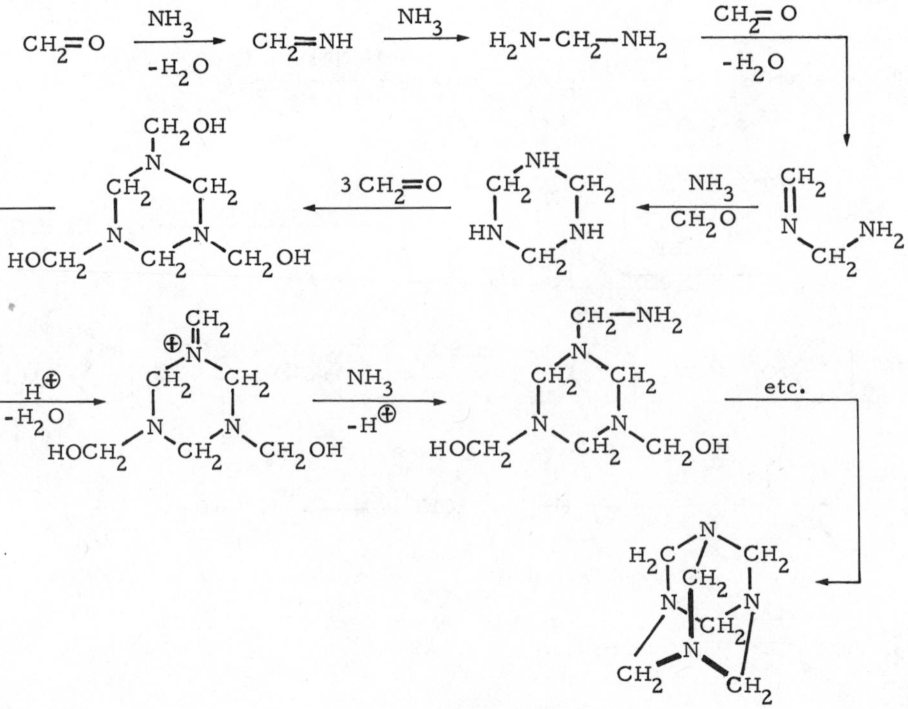

b. $(CH_3)_3CCCH_2CH_3$ $\underset{}{\overset{\ominus OH}{\rightleftharpoons}}$ $(CH_3)_3CC=CHCH_3$ $\xrightarrow[H_2O]{KMnO_4}$

$(CH_3)_3CCO_2H + CH_3CHO$

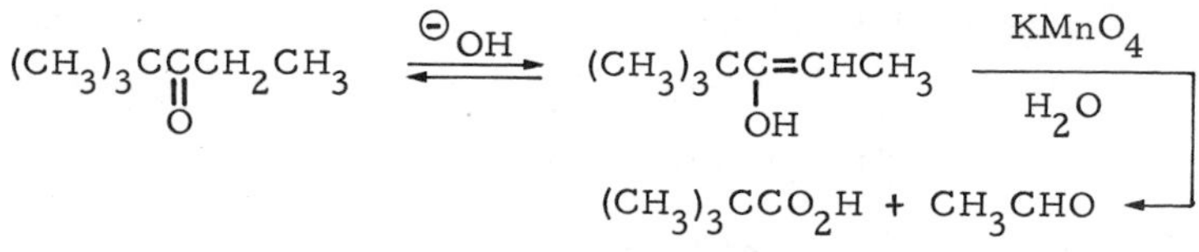

Oxidation will proceed in the direction indicated above. Oxidation under basic conditions will not lead to fission of the $(CH_3)_3C-C$ bond since this α-carbon has no enolizable hydrogens.

c. $CH_3-\underset{\underset{Cl}{|}}{\overset{\overset{CH_3}{|}}{C}}-\underset{\underset{Cl}{|}}{CH_2}$ $\xrightarrow[\substack{-CH_3OH \\ -Cl^{\ominus}}]{\ominus OCH_3}$ $(CH_3)_2C=CH_2Cl + (CH_3)_2C=CHOCH_3$

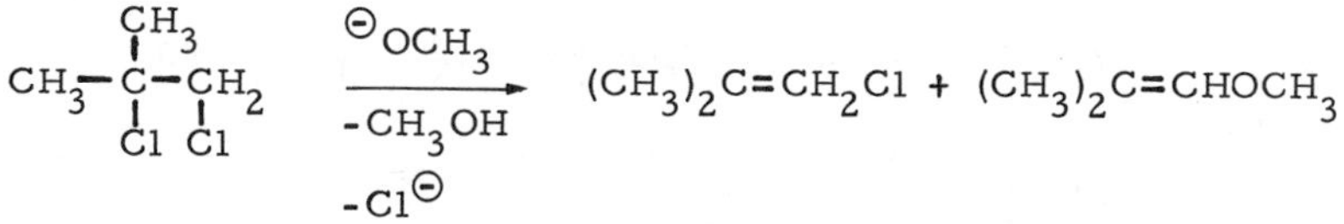

Elimination is expected to be the faster reaction under the reaction conditions.

d. The major product would be propanoic acid formed by migration of hydride to electron-deficient oxygen (see Section 16-7).

$$\underset{}{\overset{O}{\underset{}{Ar\overset{||}{C}OOH}}} + RCH=O \longrightarrow \underset{}{\overset{OH}{\underset{}{R\overset{|}{C}H}}}-O-O-\overset{O}{\overset{||}{C}}-Ar \xrightarrow{H^{\oplus}} \underset{}{\overset{OH}{\underset{}{R\overset{|}{C}H}}}-O\overset{\oplus OH}{\overset{\frown}{O}}-\overset{||}{C}-Ar \longrightarrow$$

$$\underset{\underset{H}{|}}{\overset{\overset{OH}{|}}{R-\overset{\oplus}{C}-O}} + ArCO_2H$$

$HO-CH=O-R$ $\xrightarrow{-H^{\oplus}}$ $O=CH-OR$ (minor)

$R-\overset{OH}{\overset{\oplus}{C}}=OH$ $\xrightarrow{-H^{\oplus}}$ $RC\overset{O}{\underset{OH}{\diagup}}$ (major)

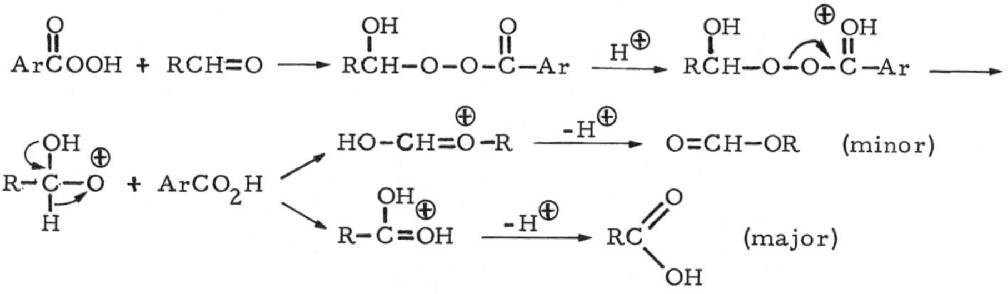

That hydride migrates preferentially in the Baeyer-Villiger reaction, but not so preferentially with alkyl hydroperoxides (see discussion of Exercise 16-45), may indicate that the free $RCH(OH)-O^{\oplus}$ ion is not formed as indicated in Section 16-7. Instead, a cyclic transition state may be involved which has a _different_ configuration of R and H _relative to the developing positive oxygen_ than is the case for the acid-catalyzed decomposition of alkyl hydroperoxides.

Baeyer-Villiger

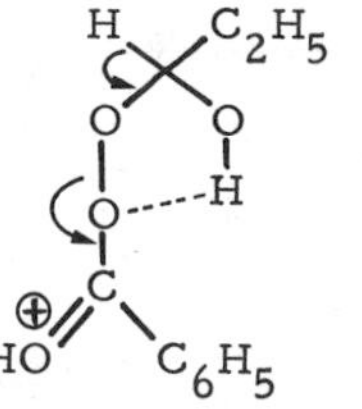

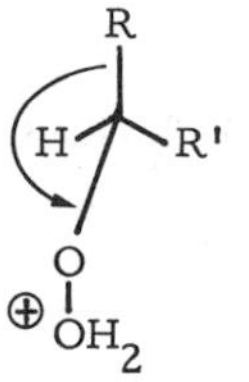

alkyl hydroperoxide

e. Sodium borohydride is expected to reduce the aldehyde carbonyl but not the carboxyl carbonyl.

$\underset{\underset{CO_2H}{|}}{CHO}$ $\xrightarrow{NaBH_4}$ $\underset{\underset{CO_2H}{|}}{CH_2OH}$

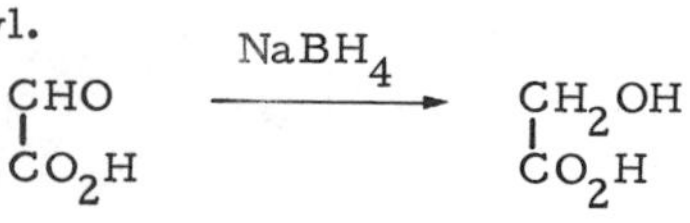

f. The reaction is possible, but the more rapid process would be the self-oxidation-reduction of methanal (see Section 16-4E, Cannizarro reaction).

$$2CH_2{=}O + NaOH \longrightarrow HCO_2Na + CH_3OH$$

Another possible reaction of $(CH_3)_2C{=}O + CH_2{=}O + NaOH$ is discussed in Section 17-3.

16-51 $MnO_4^{\ominus} + H{-}\overset{\overset{\displaystyle O}{\|}}{C}{-}O^{\ominus} \xrightarrow{\text{slow}} HMnO_4^{2\ominus} + CO_2$

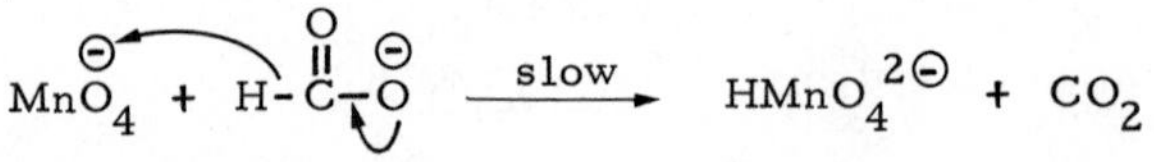

Here hydride is transferred in something like the same way as in the Cannizzaro reaction:

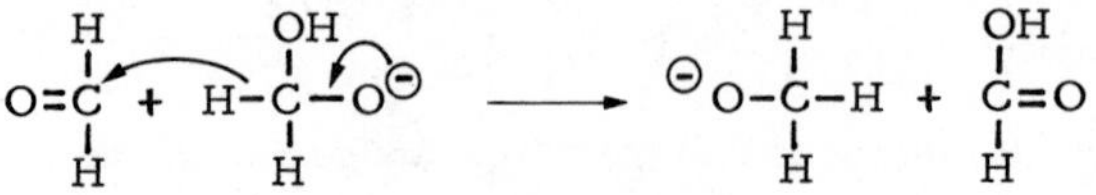

$$O{=}\overset{\overset{\displaystyle H}{|}}{\underset{\underset{\displaystyle H}{|}}{C}} + H{-}\overset{\overset{\displaystyle OH}{|}}{\underset{\underset{\displaystyle H}{|}}{C}}{-}O^{\ominus} \longrightarrow {}^{\ominus}O{-}\overset{\overset{\displaystyle H}{|}}{\underset{\underset{\displaystyle H}{|}}{C}}{-}H + \overset{\overset{\displaystyle OH}{|}}{\underset{\underset{\displaystyle H}{|}}{C}}{=}O$$

16-52 $CHCl_3 + OH^{\ominus} \rightleftharpoons :CCl_3^{\ominus} + H_2O$

$$(CH_3)_2C{=}O + :CCl_3^{\ominus} \rightleftharpoons (CH_3)_2\overset{\overset{\displaystyle \ominus}{}}{\underset{\underset{\displaystyle CCl_3}{|}}{C}}{-}O \xrightarrow[-OH^{\ominus}]{H_2O} (CH_3)_2\underset{\underset{\displaystyle CCl_3}{|}}{C}{-}OH$$

16-53

$CH_3CH_2CH_2Br \xrightarrow[-NaBr]{NaC{\equiv}CH} CH_3CH_2CH_2C{\equiv}CH \xrightarrow[\text{2. HCHO}]{\text{1. } CH_3MgCl}$
                                                                              A

$\xrightarrow{Ar_3P}$ $CH_3CH_2CH_2C{\equiv}CCH_2Br \xleftarrow{PBr_3} CH_3CH_2CH_2C{\equiv}CCH_2OH$
                          C   B

$CH_3CH_2CH_2C{\equiv}CCH_2\overset{\oplus}{P}Ar_3 \overset{\ominus}{Br} \xrightarrow[\text{2. } O{=}CH(CH_2)_8CO_2C_2H_5]{\text{1. } C_2H_5ONa}$
              D

$\xrightarrow[Pd-Pb]{H_2}$ $CH_3CH_2CH_2C{\equiv}CCH{=}CH(CH_2)_8CO_2C_2H_5$
                          E (cis and trans)

$\underset{\underset{\displaystyle H}{}}{\overset{\overset{\displaystyle C_3H_7}{}}{}}C{=}C\underset{\underset{\displaystyle H}{}}{\overset{\overset{\displaystyle CH{=}CH(CH_2)_8CO_2C_2H_5}{}}{}} \xrightarrow{LiAlH_4}$

$\underset{\underset{\displaystyle H}{}}{\overset{\overset{\displaystyle C_3H_7}{}}{}}C{=}C\underset{\underset{\displaystyle H}{}}{\overset{\overset{\displaystyle CH{=}CH (CH_2)_8CH_2OH}{}}{}}$

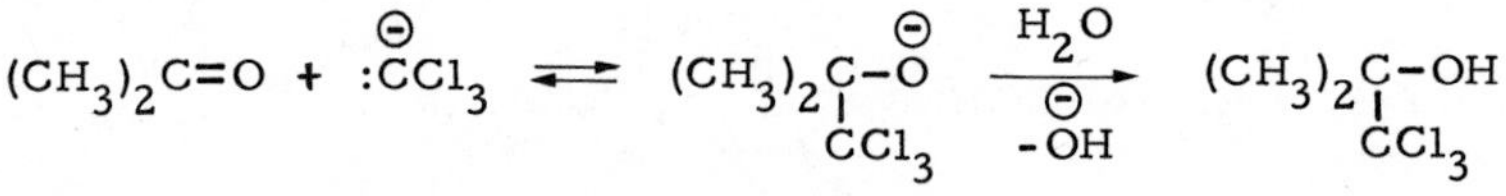

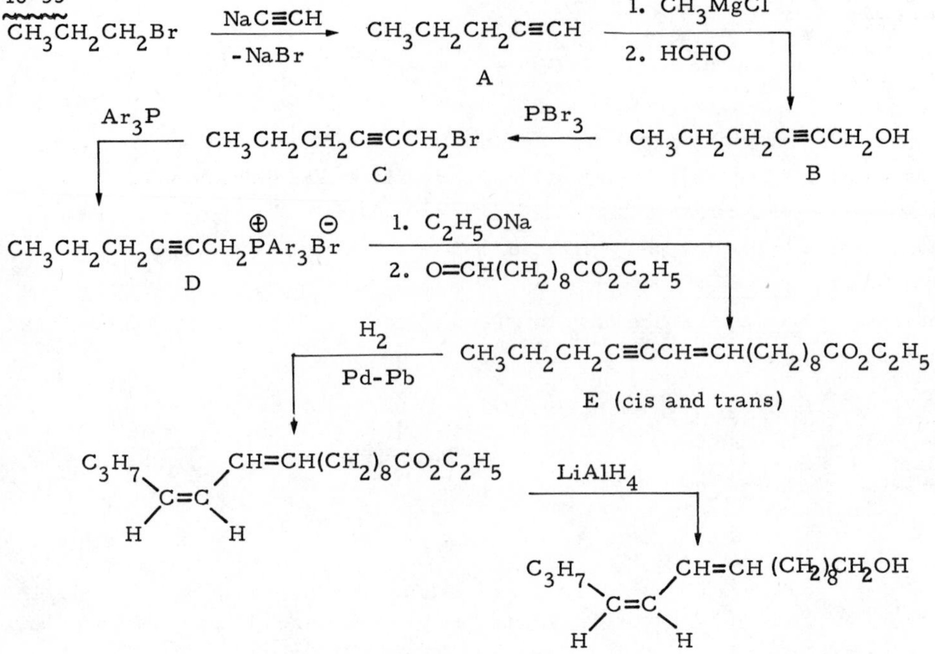

CARBONYL COMPOUNDS II.
ENOLS AND ENOLATE ANIONS.   UNSATURATED
AND POLYCARBONYL COMPOUNDS

Much of Chapter 17 is concerned with the <u>acidic properties</u> of aldehydes
and ketones.   With bases, an <u>alpha</u> hydrogen may be removed as a proton to
form an enolate anion:

$$R-\overset{\overset{O}{\parallel}}{C}-CH_3 \quad \xrightarrow[(-H_2O)]{^{\ominus}OH} \quad \left[R-C\overset{O}{\underset{CH_2:^{\ominus}}{\diagdown}} \longleftrightarrow R-C\overset{O^{\ominus}}{\underset{CH_2}{\diagdown}} \right] \quad \sim \quad \left[R-C\overset{O}{\underset{CH_2}{\diagdown}} \right]^{\ominus}$$

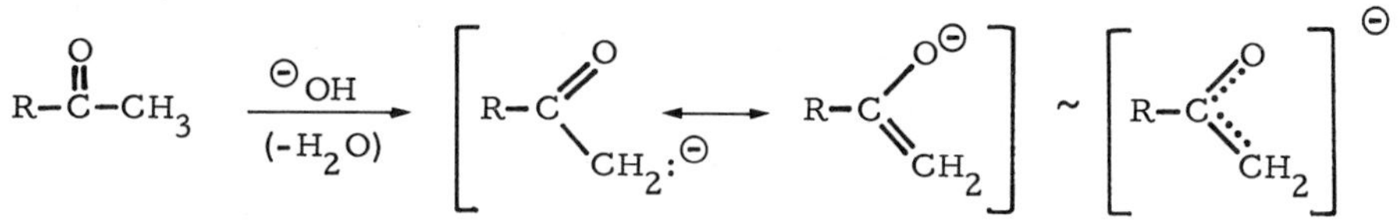

The ion formed is <u>ambident,</u> meaning it has two reactive sites because the
negative charge is shared by oxygen and carbon.   Addition of a proton to the
oxygen gives the <u>enol</u>; addition to carbon gives the <u>ketone</u>:

$$R-C\overset{O}{\underset{CH_3}{\diagdown}} \quad \xleftarrow[-OH^{\ominus}]{H_2O} \quad \left[R-C\overset{O}{\underset{CH_2}{\diagdown}} \right]^{\ominus} \quad \xrightarrow[-OH^{\ominus}]{H_2O} \quad R-C\overset{OH}{\underset{CH_2}{\diagdown}}$$

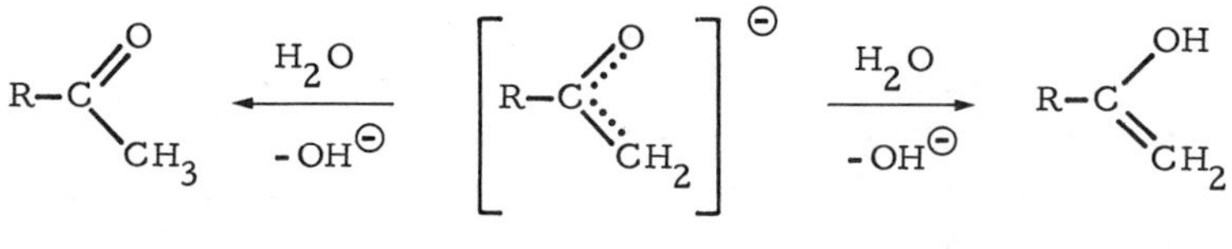

The equilibrium between a simple ketone and its enol is not established
rapidly because <u>proton transfers to and from carbon are slow.</u>

The acidity of aldehydes and ketones varies widely with structure.
Electronegative atoms or groups on the α carbon increase the acidity of the
α hydrogens (Table 17-1).   The more the enolate ion is stabilized by
electron delocalization, the more acidic is the parent ketone, and the more the
ketone-enol equilibrium favors the enol form (Table 17-2).

Acid-catalyzed formation of enols results from the sequence:

$$R-C\overset{O}{\underset{CH_3}{\diagdown}} \quad \underset{\longleftarrow}{\overset{H^{\oplus}}{\longrightarrow}} \quad R-C\overset{\overset{\oplus}{OH}}{\underset{CH_3}{\diagdown}} \quad \underset{H^{\oplus}}{\overset{H_2O}{\rightleftarrows}} \quad R-C\overset{OH}{\underset{CH_2}{\diagdown}}$$

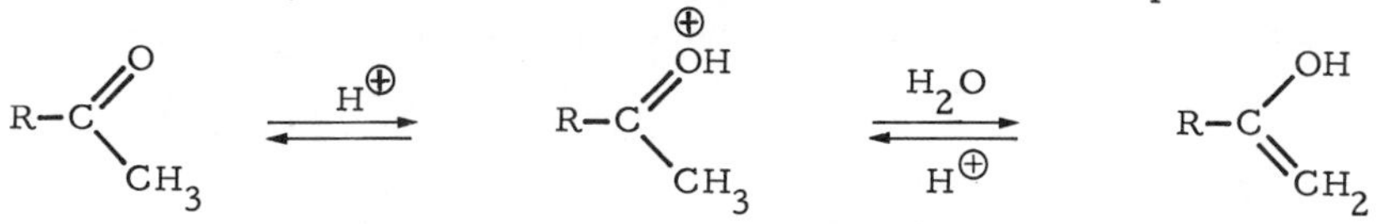

Some reactions that proceed through intermediate formation of enols or their anions are summarized as follows.

A. <u>Halogenation of ketones</u> (Section 17-2A)

$$R-\overset{O}{\overset{\|}{C}}-CH_3 \xrightarrow{Cl_2} R-\overset{O}{\overset{\|}{C}}-CH_2Cl$$

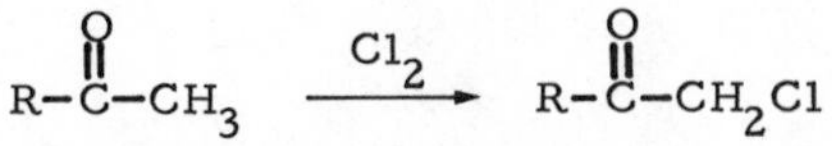

The slow step is formation of the enol or enolate. Hence the rate at which
product is formed usually is <u>independent</u> of the concentration and nature of the
halogen:

$$R-\overset{O}{\overset{\|}{C}}-CH_3 \xrightarrow[H^{\oplus} \text{ or } {}^{\ominus}OH]{\text{slow}} R-C\overset{OH}{\underset{CH_2}{\diagdown}}$$

or

$$\xrightarrow[\text{fast}]{Cl_2} R-\overset{O}{\overset{\|}{C}}-CH_2Cl$$

$$\left[R-C\overset{O}{\underset{CH_2}{\diagdown}} \right]^{\ominus}$$

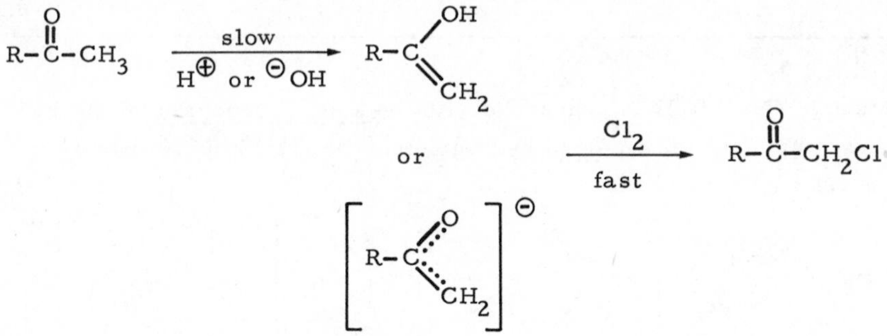

B. <u>Haloform reaction</u> (Section 17-2B)

$$R-\overset{O}{\overset{\|}{C}}-CH_3 \xrightarrow[{}^{\ominus}OH]{Cl_2} R-\overset{O}{\overset{\|}{C}}-CCl_3 \xrightarrow{{}^{\ominus}OH} R-\overset{O}{\overset{\|}{C}}-O^{\ominus} + HCCl_3$$

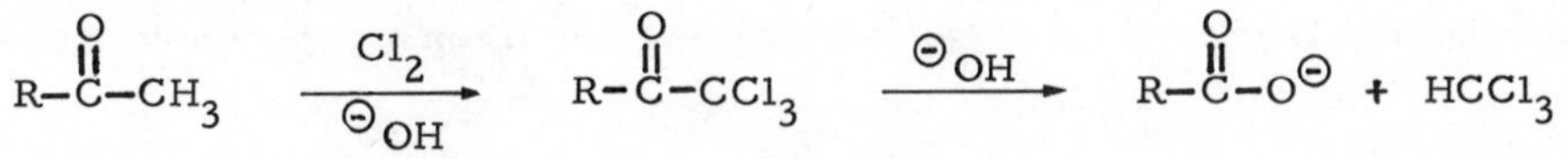

Halogenation of ketones in <u>basic</u> solution generally leads to polyhalogenation
because, once the first halogen is introduced, the second is introduced much
more rapidly. With methyl ketones, this leads ultimately to cleavage of a
carbon-carbon bond and formation of a carboxylic acid and haloform.

C. <u>Aldol addition</u> (Section 17-3)

$$2\,R-\overset{O}{\overset{\|}{C}}-CH_3 \xrightarrow{\text{base}} R-\overset{OH}{\underset{CH_3}{\overset{|}{\underset{|}{C}}}}-CH_2-\overset{O}{\overset{\|}{C}}-R$$

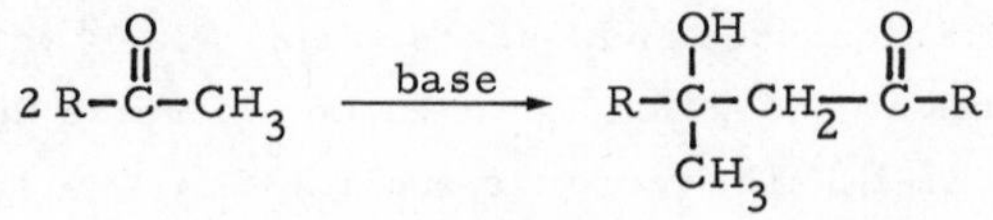

Enolate anions can behave as either carbon or oxygen nucleophiles. In the
presence of a reactive carbonyl compound, such as $CH_2{=}O$, nucleophilic
attack by the carbon of the enolate anion occurs on the carbonyl carbon. A new

C-C bond thus is formed and a β-hydroxy aldehyde or ketone results:

aldol-addition product

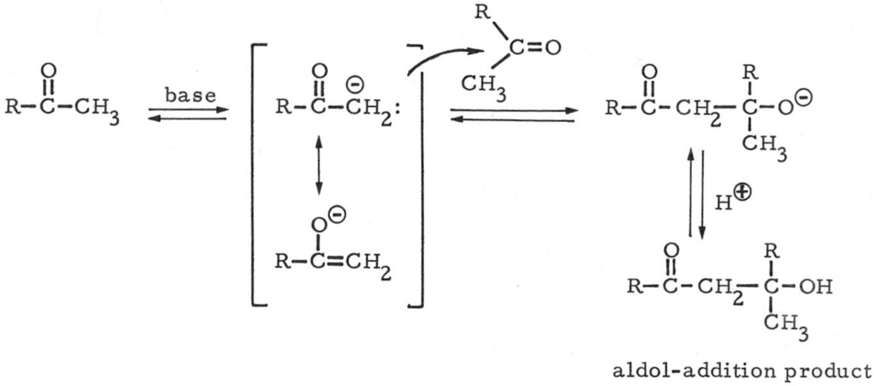

The reaction is easily reversible. The position of equilibrium depends on the nature of the carbonyl compound. Aldehydes are better acceptors of nucleophiles than are ketones because they have less steric hindrance and weaker C=O bonds. Crossed aldol reactions occur when the acceptor has no α hydrogen:

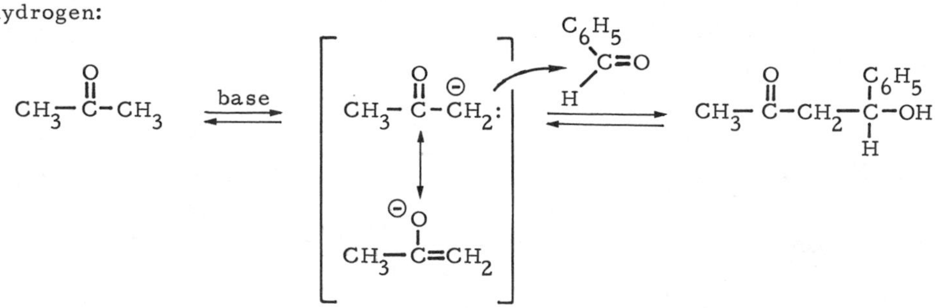

Dehydration of aldol-addition products occurs when the aldol is heated in acidic or basic solution:

$$\underset{\text{OH}}{\text{R}\overset{|}{\text{C}}\text{HCH}_2\text{CHO}} \xrightarrow{\text{H}^{\oplus} \text{ or } {}^{\ominus}\text{OH}} \text{RCH=CHCHO} + \text{H}_2\text{O}$$

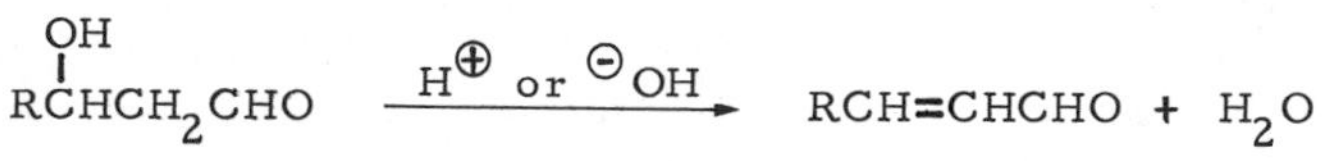

When a conjugated aldehyde or ketone serves as the acceptor in an aldol addition, the intermediate aldol normally cannot be isolated because it readily dehydrates under the reaction conditions:

$$\text{C}_6\text{H}_5\text{CHO} + \text{CH}_3\text{COCH}_3 \xrightarrow{{}^{\ominus}\text{OH}} \text{C}_6\text{H}_5\text{CH=CHCOCH}_3 + \text{H}_2\text{O}$$

D. Alkylation of enols (Section 17-4A)

$$\left[\begin{array}{c} R-\overset{\overset{\displaystyle O}{\|}}{C}-CH_2\!:^{\ominus} \\ \updownarrow \\ R-\overset{\overset{\displaystyle O^{\ominus}}{|}}{C}-CH_2 \end{array}\right] + R'X \longrightarrow R-\overset{\overset{\displaystyle O}{\|}}{C}-CH_2R' + X^{\ominus}$$

C-alkylation

OR

$$R-\overset{\overset{\displaystyle O^{\ominus}}{|}}{C}=CH_2 + X^{\ominus}$$

O-alkylation

Proportions of each process depend on the reaction conditions and the reactivity of R'X.

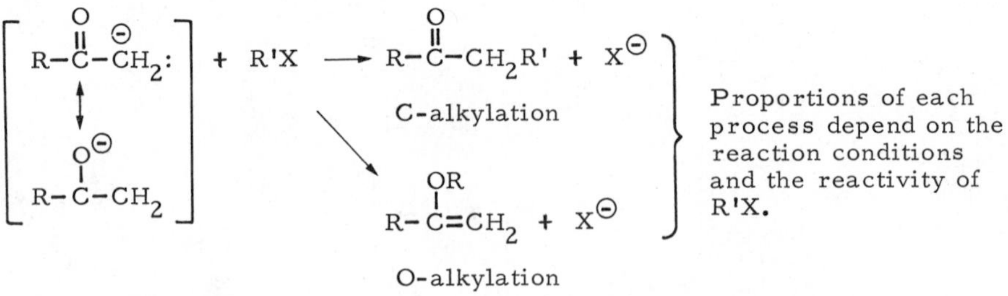

E. Alkylation of enamines (Section 17-4B)

$$\left[\begin{array}{c} R-C\!\!\begin{array}{c} \nearrow NR_2'' \\ \searrow CH_2 \end{array} \longleftrightarrow R-C\!\!\begin{array}{c} \nearrow \overset{\oplus}{N}R_2 \\ \searrow CH_2\!:^{\ominus} \end{array} \end{array}\right] \xrightarrow{\ R'X\ } R-C\!\!\begin{array}{c} \nearrow \overset{\oplus}{N}R_2'' \\ \searrow CH_2R' \end{array} + X^{\ominus}$$

$$\Big\downarrow H_2O$$

$$R-C\!\!\begin{array}{c} \nearrow\!\!\overset{\displaystyle O}{} \\ \searrow CH_2R' \end{array} + HNR_2''$$

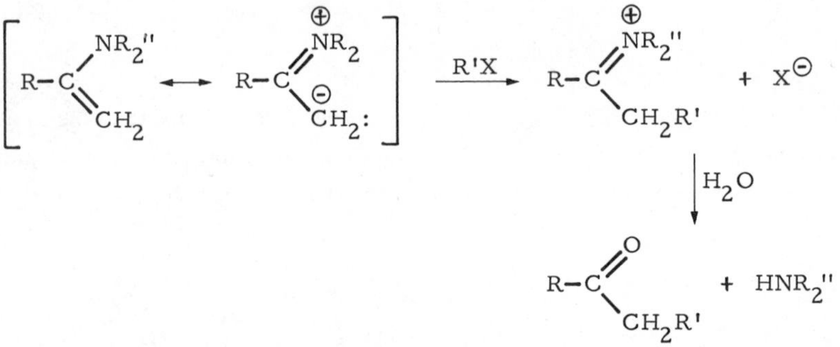

The reaction of enamines with alkyl compounds followed by hydrolysis achieves the same result as alkylating the corresponding ketone directly, but under milder conditions. For this reason alkylation of enamines may be the preferred route.

F. Conjugate and Michael-type additions (Section 17-5B)

Addition to the C≡C bond of α, β-unsaturated ketones may occur under either acidic or basic conditions. The orientation of addition of electrophiles such as HCl, and nucleophiles such as $Na^{\oplus\ominus}OC_2H_5$, suggests that enols or enolate anions are intermediates:

$$R-CH=CH-\overset{\overset{\displaystyle O}{\|}}{C}-CH_3 \xrightleftharpoons{H^{\oplus}} R-CH=CH-\overset{\overset{\displaystyle \overset{\oplus}{O}H}{\|}}{C}-CH_3 \longleftrightarrow R-\overset{\oplus}{C}H-CH=\overset{\overset{\displaystyle OH}{|}}{C}-CH_3$$

$$R-\underset{\underset{\displaystyle Cl}{|}}{C}H-CH_2\!-\overset{\overset{\displaystyle O}{\|}}{C}CH_3 \longleftarrow R-\underset{\underset{\displaystyle Cl}{|}}{C}H-CH=\overset{\overset{\displaystyle OH}{|}}{C}-CH_3 \xleftarrow{Cl^{\ominus}}$$

enol

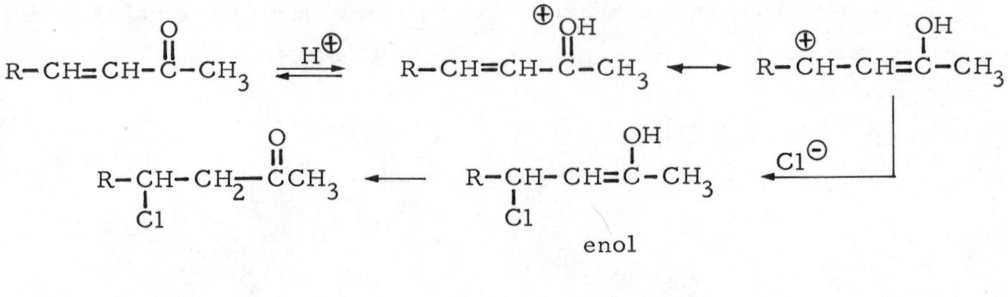

The Michael addition involves carbon-carbon bond formation by addition of an enolate anion to an aldehyde or a ketone with a conjugated double bond:

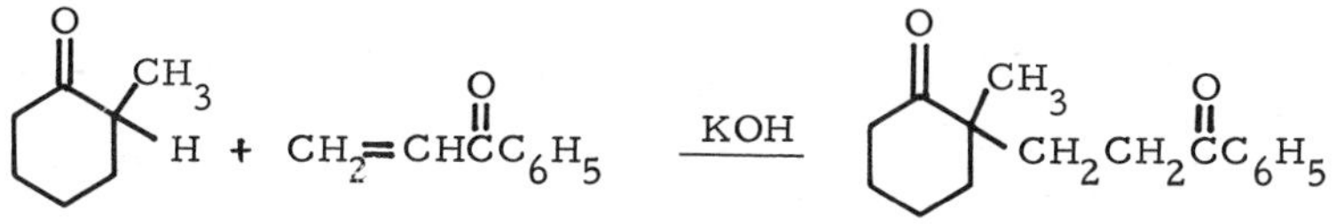

Ketenes have the structural unit >C=C=O and may be regarded as anhydrides of carboxylic acids having an α hydrogen, $H-\overset{|}{\underset{|}{C}}-CO_2H$. Ketenes are highly reactive substances, which with nucleophilic reagents - alcohols, amines - are converted to derivatives of carboxylic acids. In the absence of polar reagents, they undergo [2 + 2] cycloaddition reactions with themselves, or with suitably active carbon-carbon double bonds, to give ketene dimers or cyclobutanones.

1, 2-Dicarbonyl compounds usually are rather reactive to nucleophilic reagents. With strong base, ethanedial and diphenylethanedione undergo rearrangement to hydroxyethanoic and hydroxydiphenylethanoic acid, respectively. The latter reaction is unusual in that it is a carbon-skeleton rearrangement that occurs under basic conditions, in striking contrast to the usual acid-induced carbocation rearrangements.

1, 3-Dicarbonyl compounds, which have the structural unit $-CO-\overset{|}{\underset{|}{C}H}-\overset{|}{C}O$, usually exist largely as the enol, $-CO-\overset{|}{C}=\overset{|}{C}(OH)$. These enols are relatively strong acids - stronger than alcohols but weaker than carboxylic acids. The salts of these enols may undergo either C- or O-alkylation. Many 1, 3-dicarbonyl compounds form very stable, relatively nonpolar complexes with dipositive or tripositive metal cations such as $Cu^{(II)}$, $Be^{(II)}$, $Ni^{(II)}$, $Cr^{(III)}$, and so on.

1, 4-Dicarbonyl compounds are especially useful for preparation of unsaturated heterocyclic cyclopentane rings with one oxygen, sulfur, or nitrogen in the ring.

1, 2, 3-Tricarbonyl compounds readily undergo addition of water to the central carbonyl group, and some lose carbon monoxide with Lewis-acid catalysts.

ANSWERS TO EXERCISES

17-1   Nitromethane and ethanenitrile are acids, albeit weak acids, because their conjugate bases are relatively stable anions. Stabilization is largely due to electron delocalization whereby the negative charge is transferred to a more electronegative atom than carbon.

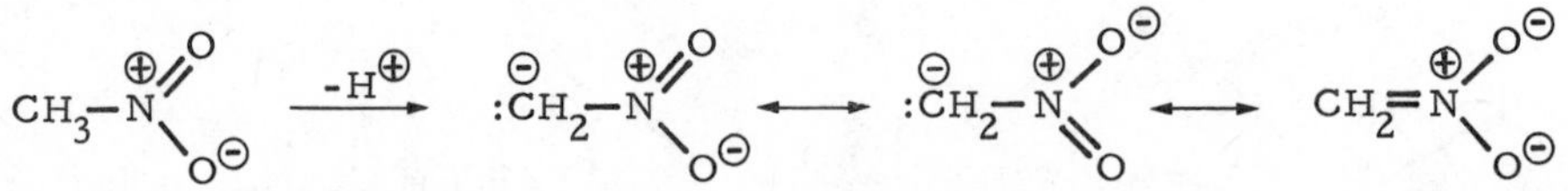

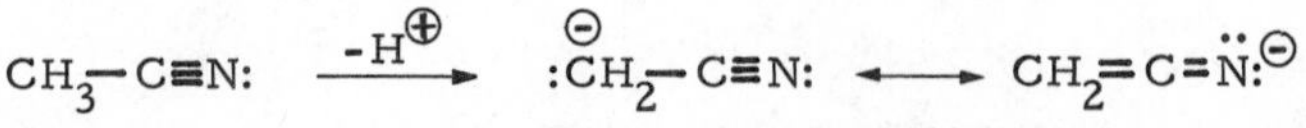

Methyl ethanoate is a weaker acid than 2-propanone because the neutral ester is significantly stabilized by electron-delocalization of the type:

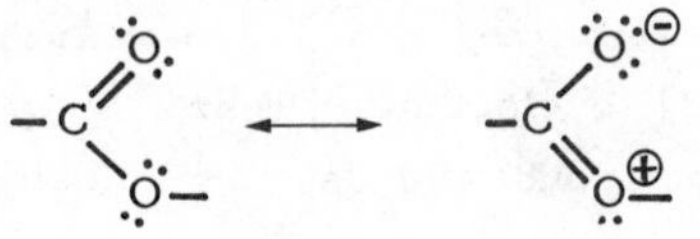

and this stabilization is lost in the conjugate base.

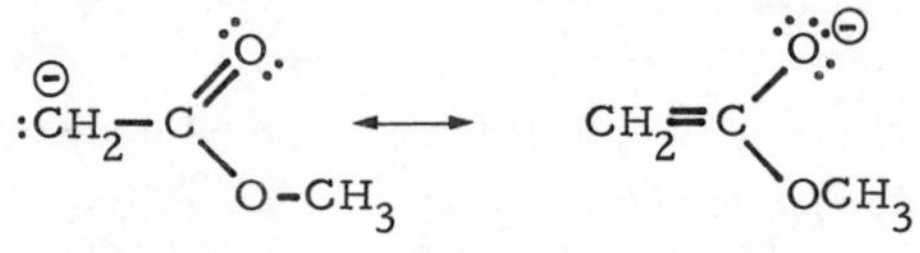

2-Propanone is not similarly stabilized as the neutral acid.

17-2
a.

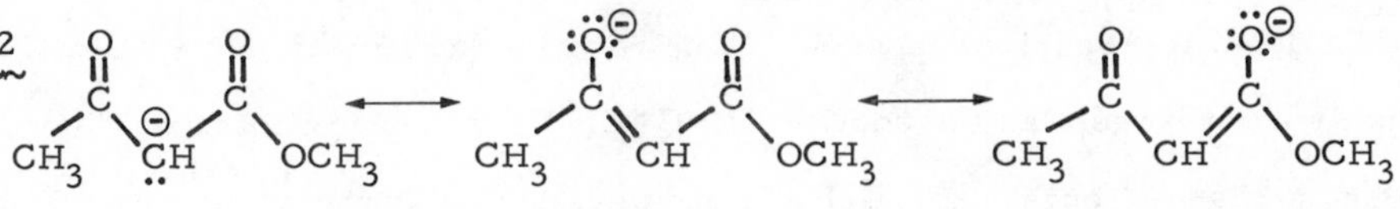

b.

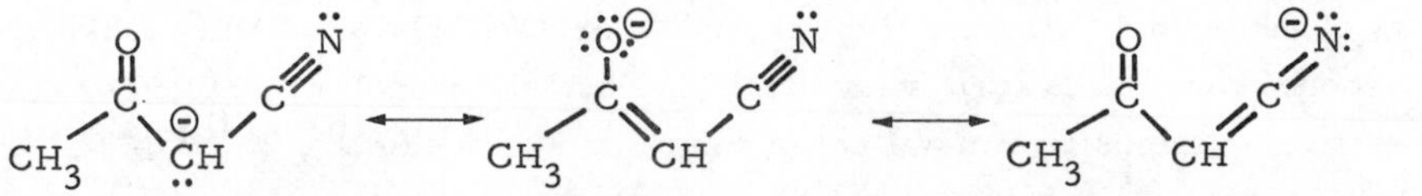

c.

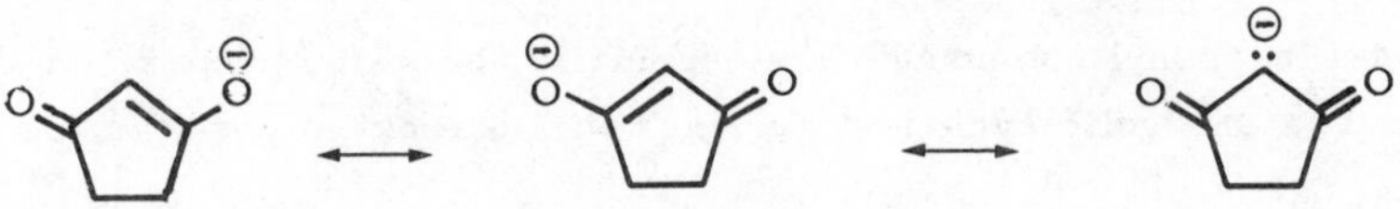

d.

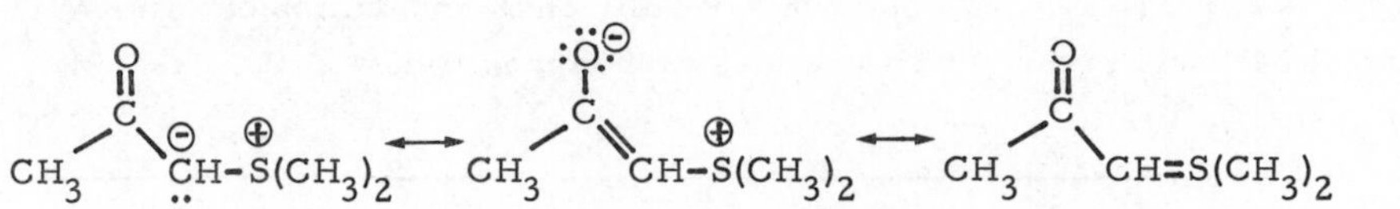

17-3   Racemization of a ketone in which the chiral center is α to the carbonyl group occurs by way of the enol or enolate anion. Therefore, acidic or basic reagents which catalyze enolization of ketones will also catalyze racemization.

| chiral ketone | achiral enol |

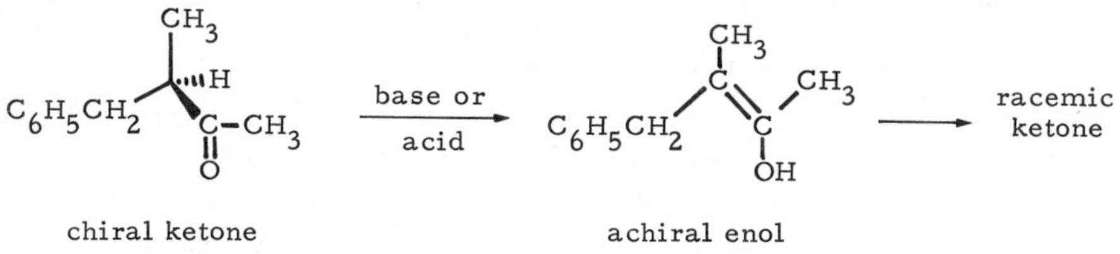

17-4   a.   The spectrum shows the presence of both keto and enol forms, $CH_3COCH_2COCH_3$ and $CH_3C(OH)=CHCOCH_3$.

δ ppm	assignment	δ ppm	assignment
2	CH_3 of enol	5.5	=CH of enol
2.2	CH_3 of ketone	15	=C—OH of enol
3.6	CH_2 of ketone		

The broad OH resonance of the enol form is at low field because the OH proton is strongly hydrogen-bonded to the neighboring carbonyl,

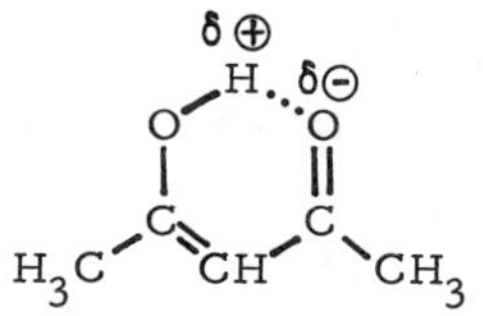

which reduces the electron density around (i.e., deshields) the proton.

b.   The rate of _intramolecular_ proton transfer between the two oxygen atoms must be rapid because the two-methyl groups in the enol form appear as a _single_ resonance at 2 ppm.   Interconversion of enol to ketone must be slow because separate sharp resonances are observable for both forms.

17-5.   a.   The enolate anion is more reactive than the enol because it is the stronger nucleophile.

b.   The difference in enthalpy of reaction at oxygen relative to carbon is the difference in enthalpy of the two products,

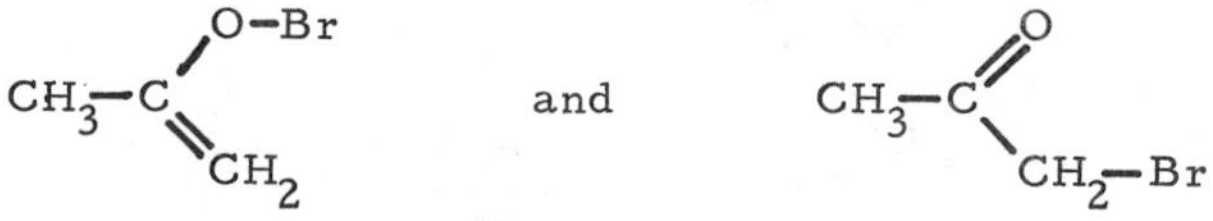

This difference, which is the heat of isomerization, is calculated from bond energies to be 50.3 kcal in favor of CH_3COCH_2Br.   The reaction is therefore 50.3 kcal more exothermic at carbon than at oxygen.   The unfavorable nature

of the reaction at oxygen is because the $O-Br$ and $C=C$ bonds are weaker than $C-Br$ and $C=O$ bonds, respectively.

17-6   The slow step in halogenation of ketones is the enolization step in which a proton is removed from carbon of the neutral ketone (base catalysis) or from carbon of the conjugate acid of the neutral ketone (acid catalysis). Since $C-D$ bonds are less easily broken than $C-H$ bonds, then halogenation of CD_3COCD_3 is anticipated to be (and indeed is) slower than halogenation of CH_3COCH_3.

17-7   a. $\nu = (6 \times 10^{-9} + 5.6 \times 10^{-4} \times 10^{-7} + 7 \times 10^{-7}) \times 1$

$= 7.06 \times 10^{-7}$ moles/l. sec.

b.   At pH 5, $[H^{\oplus}] = 10^{-5}$, $\dfrac{[H^{\oplus}][CH_3CO_2^{\ominus}]}{[CH_3CO_2H]} = 1.75 \times 10^{-5}$

Let x be $[CH_3CO_2^{\ominus}]$.

$\dfrac{10^{-5}x}{1-x} = 1.75 \times 10^{-5}$; $x = 0.636$; $1-x = 0.364$

$\nu = (6 \times 10^{-9} + 5.6 \times 10^{-4} \times 10^{-5} + 1.3 \times 10^{-6} \times 0.364 + 7 \times 10^{-9}$

$+ 3.3 \times 10^{-6} \times 0.636 + 3.5 \times 10^{-6} \times 0.364 \times 0.636)$

$= 3.4 \times 10^{-6}$ moles/l. sec.

c.   In the absence of CH_3CO_2H and $CH_3CO_2^{\ominus}$ the rate becomes:

$$\nu = \underline{k} + \underline{k}_{H_3O^{\oplus}} [\overset{\oplus}{H_3O}] + \underline{k}_{OH^{\ominus}} [\overset{\ominus}{OH}]$$

In acid solution, the rate becomes $\nu = \underline{k} + \underline{k}_{H_3O^{\oplus}} [\overset{\oplus}{H_3O}]$; while in basic solution the rate is $\nu = \underline{k} + \underline{k}_{OH}^{\ominus} [\overset{\oplus}{OH}]$. Thus, a plot of rate versus $[\overset{\oplus}{H_3O}]$ should be a straight line of slope $\underline{k}_{H_3O^{\oplus}}$ and intercept $\underline{k}$ at low pH. Similarly, a plot of rate versus $[\overset{\ominus}{OH}]$ should be a straight line of slope $\underline{k}_{OH}^{\ominus}$ and intercept $\underline{k}$ at high pH.

With CH_3CO_3H and $CH_3CO_2^{\ominus}$ present but at constant pH, the rate becomes:

$$\nu = \underline{k}' + \underline{k}_{CH_3CO_2H} [CH_3CO_2H] + \underline{k}_{CH_3CO_2^{\ominus}} [CH_3CO_2^{\ominus}]$$
$$+ \underline{k}'' [CH_3CO_2H][CH_3CO_2^{\ominus}]$$

By measuring the rate at constant pH (i. e., at a constant buffer ratio $[CH_3CO_2^{\ominus}]/[CH_3CO_2H]$) but varying concentrations of CH_3CO_2H and $CH_3CO_2^{\ominus}$, contributions of the rate terms in CH_3CO_2H and $CH_3CO_2^{\ominus}$ may be evaluated.

d. At equilibrium, the rate of the forward reaction (ketone $\rightarrow$ enol) is equal to the rate of back reaction (enol $\rightarrow$ ketone).

At equilibrium, if the concentration of enol is $1.5 \times 10^{-4}\%$, or $1\underline{M}$, then the concentration of acetone is $\dfrac{100 \times 1}{1.5 \times 10^{-4}} = 6.7 \times 10^5$. Hence, rate of forward or back reaction in neutral water (pH 7) is:

$$7.06 \times 10^{-7} \times 6.7 \times 10^5 = 0.473 \text{ moles/l. sec.}$$

e. One explanation has it that a stepwise process takes place in which CH_3CO_2H is complexed with the $C=O$ group and the $C-H$ bond is broken by attack of $CH_3CO_2^{\ominus}$.

$$CH_3COCH_3 + CH_3CO_2H \underset{}{\overset{fast}{\rightleftharpoons}} \underset{CH_3\overset{O\cdots HO_2CCH_3}{\overset{\|}{C}}CH_3}{} \xrightarrow[slow]{CH_3CO_2^{\ominus}} CH_3-\underset{\overset{OH}{\,}}{C}=CH_2$$

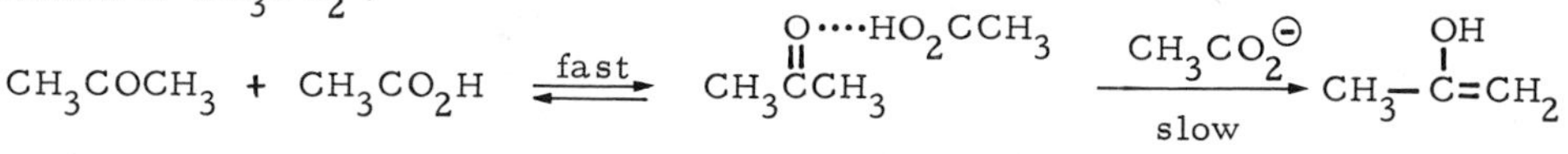

<u>17-8</u>   When the equilibrium between a ketone and its enol favors the enol form, then the rate of halogenation is determined by the reaction of the halogen with the enol rather than the rate of enol formation. From Table 17-2 we see that diketones of the type $RCOCH_2COR$ exist extensively in the enol form; another extreme example is 2,4-cyclohexadienone which, at equilibrium, exists entirely as phenol. Such examples generally react with halogens at a rate determined by the expression $\underline{k}$ [enol] [halogen] .

<u>17-9</u>   a. Under conditions of kinetic control, the major product of halogenation of an unsymmetrical ketone is derived from the enol or enolate that is formed most rapidly. Of the two enols formed from 2-butanone in acidic solution, the kinetically preferred enol is $CH_3CH=C(OH)CH_3$, which gives $CH_3CH_2BrCOCH_3$ with bromine. This enol is also the more stable enol because it is more highly substituted than the isomeric enol $CH_3CH_2C(OH)=CH_2$ (see Section 11-3).

b. $C_6H_5CH_2COCH_3 + Br_2 \xrightarrow{H^{\oplus}} C_6H_5CHBrCOCH_3$

The most rapidly formed enol is expected to be the more stable enol, which is the conjugated enol, $C_6H_5CH=C(OH)CH_3$.

<u>17-10</u>   Reaction is initially slow because little or no catalyst is present. The reaction becomes progressively more rapid because the acidic product, HBr, catalyzes the enolization step which controls the rate.

<u>17-11</u>   B $CH_3CHBrCHBrO_2CCH_3$

$\underset{\sim}{\text{C}}$ $CH_3CBr=CHO_2CCH_3$ and $CH_3CBrCH\overset{\overset{H}{|}}{\underset{O_2CCH_3}{\diagdown}}OCH_3$

Potassium ethanoate is required as the basic catalyst for the formation of the enol ester $\underset{\sim}{\text{A}}$. The steps involve enolization, as shown in Section 17-1A, and esterification of the enol by ethanoic anhydride, as in Section 15-4D. (See also Exercise 15-12.) Formation of $\underset{\sim}{\text{B}}$ is a straightforward electrophilic addition of bromine to the enol ester double bond (Section 10-3A). The adduct, $\underset{\sim}{\text{B}}$, is very S_N1 reactive because of the stabilizing influence of the enol oxygen on the adjacent carbocation (Section 8-7B).

$$CH_3CHBr-CHBr-O-CO-CH_3 \xrightarrow[S_N1]{CH_3OH} CH_3CHBr-\overset{\oplus}{CH}-O-CO-CH_3 + \overset{\ominus}{Br}$$

$$CH_3CHBr-CH\overset{\overset{\diagup OCH_3}{}}{\underset{\diagdown O-CO-CH_3}{}} \quad \xleftarrow[-H^{\oplus}]{} \quad \xrightarrow[CH_3OH]{} \quad CH_3CBr=CHO_2CCH_3 \quad \xleftarrow[-H^{\oplus}]{}$$

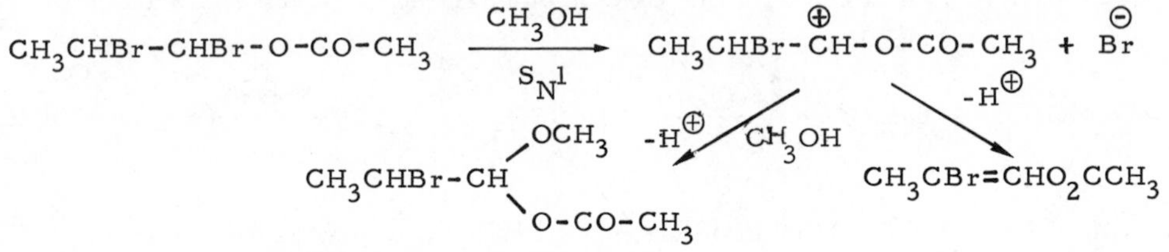

Subsequent acid hydrolysis to produce $CH_3CHBrCHO$ is the reverse of acetal formation (Section 15-4E) and the reverse of ester formation (Section 15-4D).

17-12   The hypochlorite first oxidizes ethanol to ethanal,

$$CH_3CH_2OH + NaOCl \xrightarrow{-NaOH} CH_3CH_2OCl \xrightarrow[-HCl]{NaOH}$$
$$CH_3CHO + NaCl + H_2O$$

The next step is chlorination of ethanal (hypochlorite is the source of chlorine by the reaction $H-O-Cl + Cl^{\ominus} \rightleftharpoons HO^{\ominus} + Cl_2$).

$$CH_3CHO + 3Cl_2 \longrightarrow Cl_3CCHO + 3HCl$$

Cleavage by the haloform reaction then gives chloroform

$$Cl_3CCHO + H_2O \longrightarrow Cl_3CCH(OH)_2 \xrightarrow{H_2O} Cl_3\overset{\ominus}{C} + HCO_2H + H_3O^{\oplus}$$
$$\downarrow$$
$$Cl_3CH + HCO_2H + H_2O$$

17-13   The reaction proceeds by addition of hydroxide to the carbonyl carbon. Subsequent cleavage of a C—C bond occurs only if a reasonably stable carbon anion ($:CCl_3^{\ominus}$) is formed. The alternate $:CH_3^{\ominus}$ is an exceedingly poor leaving group as the anion is not stabilized by electronegative atoms or groups.

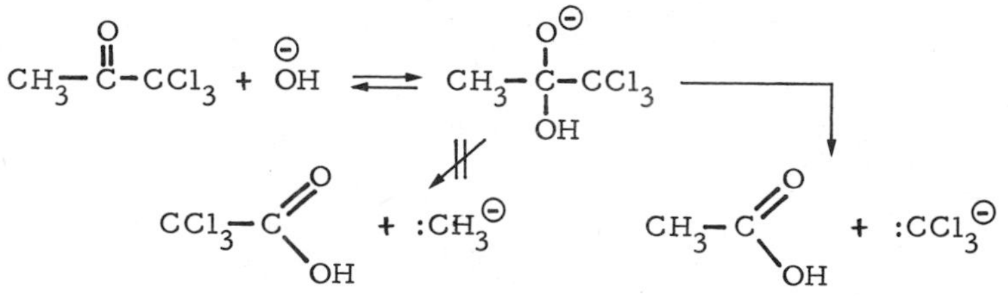

17-14   The steps in the Haller-Bauer reaction are:

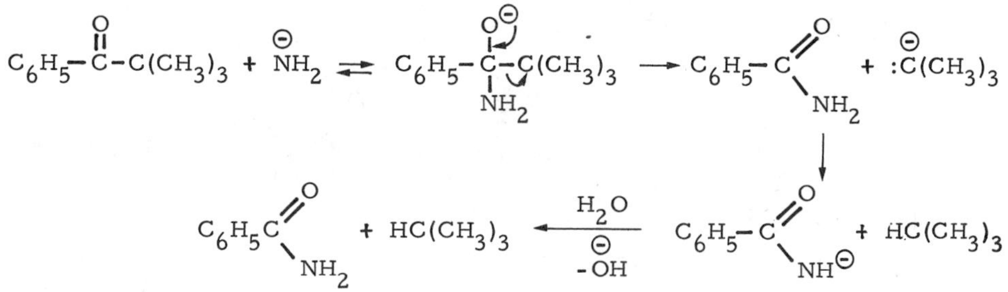

Although one might expect a phenyl anion, $C_6H_5:^{\ominus}$, to be more stable than an alkyl anion, $:CR_3^{\ominus}$, because sp^2 carbon orbitals support an electron pair better than sp^3 carbon orbitals (see Section 11-8B), cleavage occurs to form the conjugated (more stable) amide.

17-15

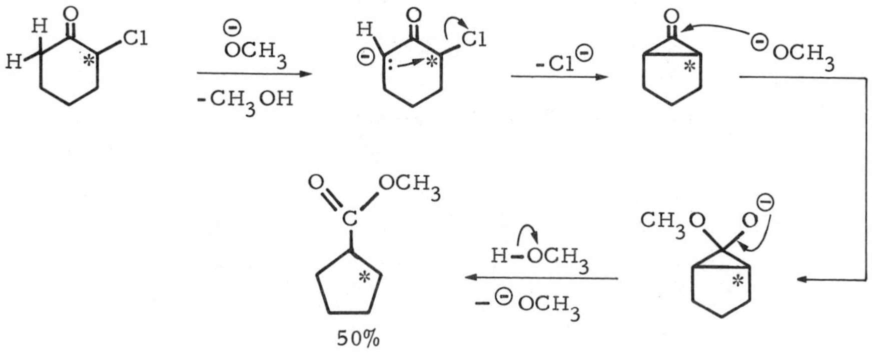

50%

and

50%

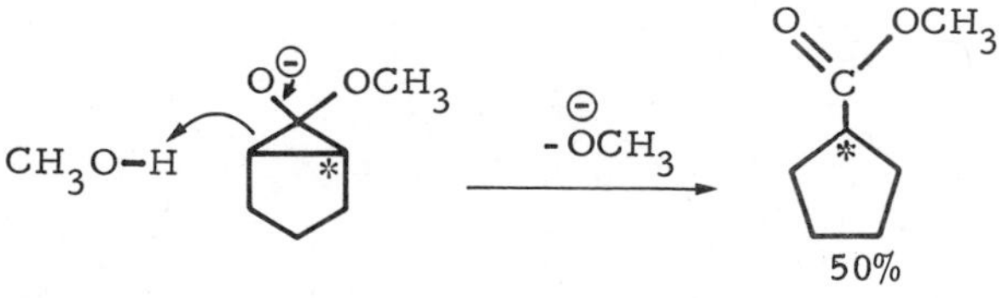

The alternate scheme would produce only

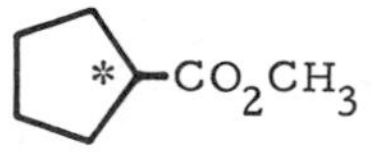

Since the two ring carbons at C2 and C5 are equivalent, they will contain 50% ^{14}C total, or 25% ^{14}C per carbon.

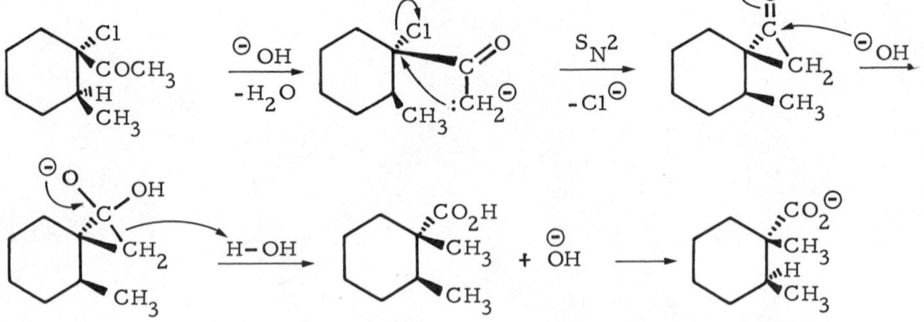

17-16 $CH_3CHO + \overset{\ominus}{OD} \underset{k_{-1}}{\overset{k_1}{\rightleftharpoons}} \overset{\ominus}{CH_2}CHO + HDO \underset{D_2O}{\overset{k_{-1}}{\rightleftharpoons}} D-CH_2CHO + \overset{\ominus}{OD}$

$CH_3CHO + \overset{\ominus}{CH_2}CHO \underset{k_{-2}}{\overset{k_2}{\rightleftharpoons}} CH_3\underset{\underset{O^{\ominus}}{|}}{CH}CH_2CHO$

The first step is the slow step because, if the reverse were the case, the product would possess C—D bonds. That is to say, deuterium exchange by the equilibrium $\underset{k_{-1}}{\overset{k_1}{\rightleftharpoons}}$ is slow compared to addition (k_2).

The rate expression for reaction in which the first step is slow is:

$$\text{Rate} = k_1 [CH_3CHO][\overset{\ominus}{OD}]$$

At very low concentrations of ethanal, the second step becomes very slow because its rate is second order in ethanal. The kinetics of reaction will change from first order to second order in ethanal as the concentration of ethanal decreases, and the first step becomes rapid compared to the second step. The product will also possess more C—D bonds as the rate of the second step decreases relative to the first. The rate expression can be written as:

$$\text{Rate} = \frac{k_1 k_2}{k_{-1}} [CH_3CHO]^2 [OD^{\ominus}] / \left(1 + \frac{k_2}{k_{-1}} [CH_3CHO]\right)$$

At moderate concentrations of CH_3CHO, the term $(1 + \frac{k_2}{k_{-1}} [CH_3CHO])$ approaches $\frac{k_2}{k_{-1}} [CH_3CHO]$ and the rate $= k_1 [CH_3CHO][OD^{\ominus}]$.

At low concentrations of CH_3CHO, the term approaches unity and the

rate $= \dfrac{k_1 k_2}{k_{-1}} [CH_3CHO]^2 [OD^{\ominus}]$.

17-17

$2CH_3CH_2CHO \xrightarrow{HO^{\ominus}} CH_3CH_2\overset{\overset{\displaystyle OH}{|}}{C}HCHCHO$
                                                        $\underset{\underset{\displaystyle CH_3}{|}}{}$

$(CH_3)_3CCHO \xrightarrow{HO^{\ominus}}$ no reaction

$(CH_3)_3CCHO + CH_3CH_2CHO \xrightarrow{HO^{\ominus}} (CH_3)_3C\overset{\overset{\displaystyle OH}{|}}{C}HCHCHO$
                                                                       $\underset{\underset{\displaystyle CH_3}{|}}{}$

17-18   Because the apparatus of Figure 17-2 does not permit equilibrium to be established between 2-propanone and the aldol adduct 11 the system will in principal continue to produce 11 until all the 2-propanone is used up. However, should some of the basic catalyst escape from the thimble into the boiler, equilibrium will be rapidly established, and no more 11 will be produced than the equilibrium concentration. Barium hydroxide is preferred over sodium hydroxide because it is less soluble and therefore less likely to "leak" into the boiler.

17-19   The aldol condensation occurs in the column packed with glass beads, and the conversion of methanal is good because a large excess of 2-propanone is present. When the mixture drips back into the flask, the base catalyst is neutralized by the carboxylic acid present. Neutralization removes the catalyst and prevents reversal of the reaction. The adduct $(CH_3)_2C(OH)CH_2OCH_3$ is not formed in significant amounts because the enolate anion, $CH_3COCH_2^{\ominus}$ adds to the carbonyl group of methanal much faster than to the carbonyl group of 2-propanone.

17-20
a. $CH_3CH(OH)CH_2CHO$ (major), and possibly $CH_3CH(OH)CH_2COCH_3$
     Addition occurs to the less-hindered carbonyl of ethanal.

b. CH_3COCH_3 (reversal of aldol addition)

c. HCO_2H, $(CH_3)_3CCH_2OH$ (Cannizzaro reaction; see Exercise 16-33)

d. $HOCH_2-C(CH_3)_2CHO$

17-21

a. $CH_2=O + HN(CH_3)_2 \longrightarrow \overset{\ominus}{O}CH_2\overset{\oplus}{N}H(CH_3)_2 \longrightarrow HOCH_2\overset{\oplus}{N}(CH_3)_2 \longrightarrow$

$$CH_2\overset{\oplus}{=}N(CH_3)_2 \xleftarrow[-H_2O]{H^{\oplus}}$$

$$C_6H_5COCH_3 \underset{}{\overset{H^{\oplus}}{\rightleftharpoons}} C_6H_5C(OH)=CH_2 \xrightarrow{\overset{\oplus}{C}H_2=N(CH_3)_2}$$

$$C_6H_5COCH_2CH_2\overset{\oplus}{N}H(CH_3)_2 \longleftarrow C_6H_5\overset{\overset{\oplus}{O}H}{\overset{\|}{C}}-CH_2CH_2N(CH_3)_2$$

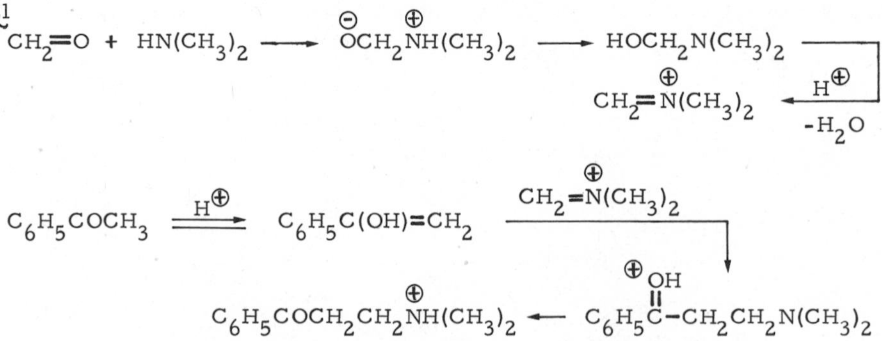

b.   Product from Part a $\xrightarrow{CH_3I} \xrightarrow{Ag_2O} \xrightarrow[-N(CH_3)_3]{heat} C_6H_5COCH=CH_2$

(See Sections 8-8 and 23-9D)

17-22   The acidity of the α-hydrogen of β-halo ketones is pronounced because of stabilization of the enolate anion by the carbonyl and bromine substituents. Thus E2 elimination under basic conditions is facile:

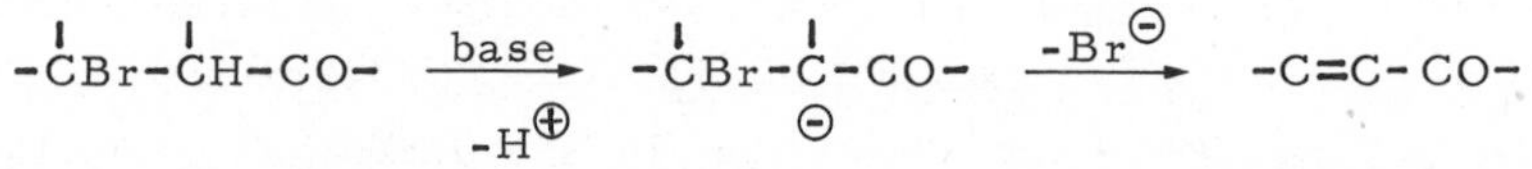

$$-\overset{|}{\underset{|}{C}}Br-\overset{|}{\underset{|}{C}}H-CO- \xrightarrow[-H^{\oplus}]{base} -\overset{|}{\underset{|}{C}}Br-\overset{|}{\underset{\ominus}{C}}-CO- \xrightarrow{-Br^{\ominus}} -C=C-CO-$$

If there is no α-hydrogen, β-halo ketones do not eliminate readily.

17-23

$$CH_3COCH_3 \xrightarrow{H^{\oplus}} CH_3\overset{\overset{\oplus}{O}H}{\overset{\|}{C}}CH_3 \xrightarrow{-H^{\oplus}} CH_3\overset{\overset{OH}{|}}{C}=CH_2 \xrightarrow{O=C(CH_3)_2}$$

$$\xrightarrow[]{-H_2O, -H^{\oplus}} \; CH_3-\overset{O}{\overset{\|}{C}}-CH_2-\overset{\overset{\oplus}{O}H_2}{\overset{|}{C}}(CH_3)_2 \xleftarrow{H^{\oplus}} CH_3-\overset{\overset{\oplus}{O}H}{\overset{\|}{C}}-CH_2-\overset{\overset{O^{\ominus}}{|}}{C}(CH_3)_2$$

$$CH_3COCH=C(CH_3)_2$$

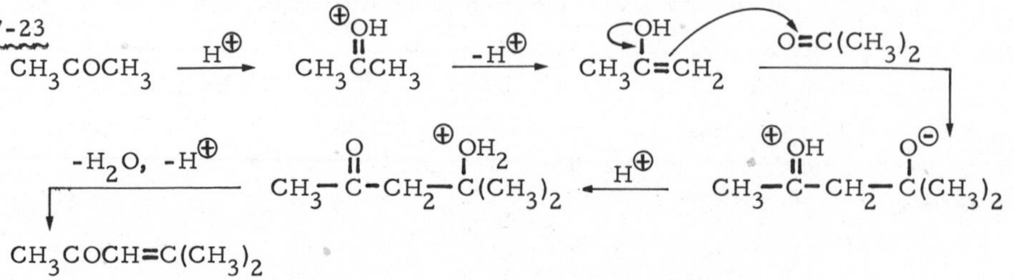

A similar sequence of enolization of mesityl oxide and addition of the enol to $(CH_3)_2C=O$ gives phorone.

17-24   Base-catalyzed dehydration of 2-butanol is not observed because the C—H acidity is too low and elimination of $^{\ominus}OH$ is not favorable. In contrast the base-catalyzed dehydration of 3-hydroxybutanal is relatively fast because the C—H acidity is enhanced by the carbonyl group and elimination of $^{\ominus}OH$ is made favorable by formation of a __conjugated__ system of double bonds.

$$HO-\overset{|}{\underset{|}{C}}-\overset{|}{\underset{|}{C}}H-C=O \xrightarrow{-H^{\oplus}} HO-\overset{|}{\underset{|}{C}}-\overset{\ominus}{\underset{|}{C}}-C=O \xrightarrow{-OH^{\ominus}} -\overset{|}{C}=\overset{|}{C}-C=O$$

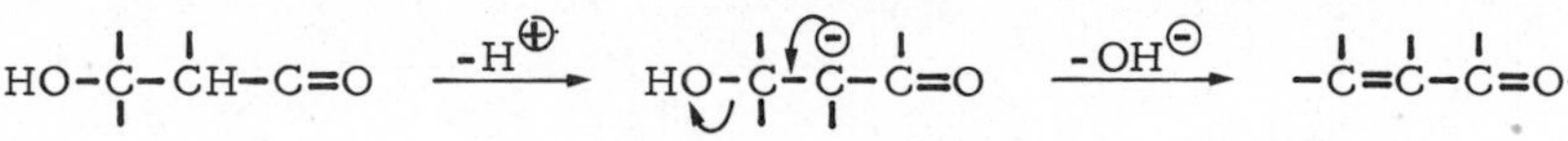

17-25

a. $2CH_3CHO$ $\xrightarrow{\text{base}}$ $CH_3CH(OH)CH_2CHO$ $\xrightarrow[\text{or LiAlH}_4]{\text{NaBH}_4}$ $H^{\oplus}$

$CH_3CH(OH)CH_2CH_2OH$ ⟵

b. $CH_3CH(OH)CH_2CHO$ $\xrightarrow[-H_2O]{H^{\oplus}}$ $CH_3CH=CHCHO$ $\xrightarrow{\text{LiAlH}_4}$ $H^{\oplus}$

(from Part a)

$CH_3CH=CHCH_2OH$ ⟵

c. $2(CH_3)_2C=O$ $\xrightarrow{\text{base}}$ $(CH_3)_2C(OH)CH_2COCH_3$ $\xrightarrow[-H_2O]{H^{\oplus}}$

$\xrightarrow[\text{HCl}]{\text{Zn-Hg}}$ $(CH_3)_2CHCH_2COCH_3$ $\xleftarrow[\text{Pt}]{H_2}$ $(CH_3)_2C=CHCOCH_3$

$(CH_3)_2CHCH_2CH_2CH_3$

d. $2CH_3CH_2CHO$ $\xrightarrow{\ominus OH}$ $CH_3CH_2CH(OH)CH(CH_3)CHO$ $\xrightarrow[\text{2. H}_2\text{, Ni}]{\text{1. CH}_3\text{SH}}$

$CH_3CH_2COCH(CH_3)_2$ $\xleftarrow[H^{\oplus}]{\text{K}_2\text{Cr}_2\text{O}_7}$ $CH_3CH_2CH(OH)CH(CH_3)_2$

e. $C_6H_5CHO + CH_3CHO$ $\xrightarrow[-H_2O]{\text{base}}$ $C_6H_5CH=CHCHO$ $\xrightarrow[-H_2O]{CH_3CHO}$

$C_6H_5CH_2CH_2CH_2CH_2CHO$ $\xleftarrow[\text{H}_2]{\text{Pt}}$ $C_6H_5CH=CHCH=CHCHO$

17-26

a. CH_3CHO $\xrightarrow[\text{H}_2\text{O}]{\text{NaCN}}$ $CH_3CH(OH)CN$ $\xrightarrow[-H_2O]{\text{H}_2\text{SO}_4}$ $CH_2=CHCN$

b. 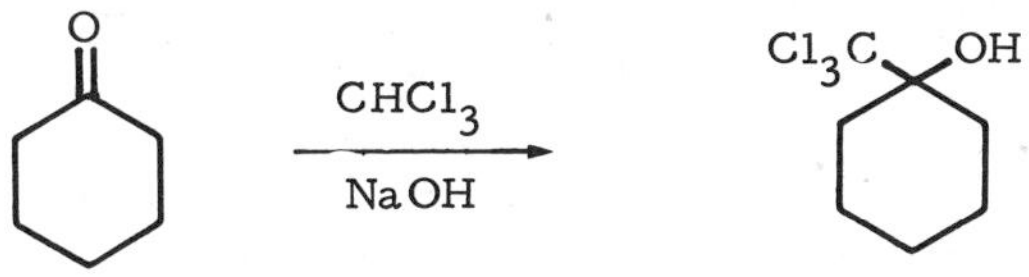 $\xrightarrow[\text{NaOH}]{\text{CHCl}_3}$ (nucleophile is $\overset{\ominus}{C}Cl_3$)

c. $(CH_3)_2CHCHO + CH_2{=}O \xrightarrow{\ominus OH} (CH_3)_2\overset{\overset{\displaystyle CH_2OH}{|}}{C}CHO \xrightarrow[\text{2. } H^{\oplus}]{\text{1. } LiAlH_4}$

$$(CH_3)_2C(CH_2OH)_2$$

d.

$+ C_6H_5CHO \xrightarrow[-H_2O]{NaOH}$

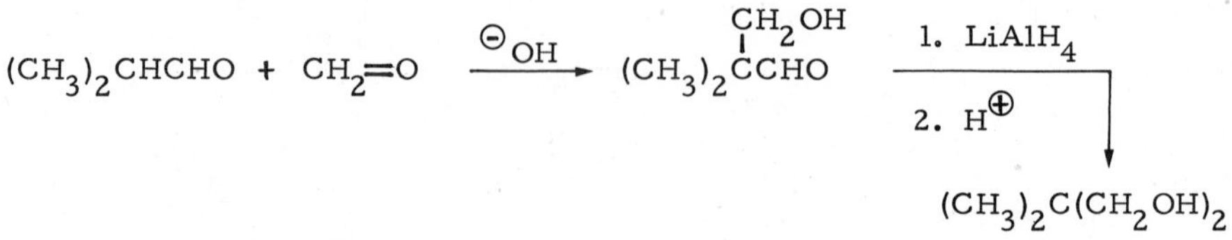

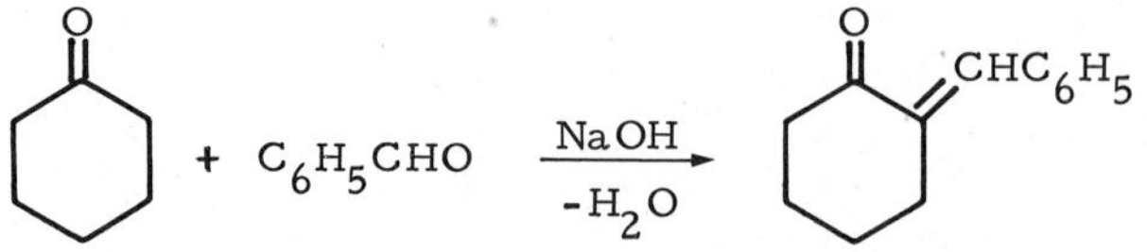

e. $C_6H_5CH_2CN + C_6H_5CHO \xrightarrow[-H_2O]{NaOH} C_6H_5C(CN){=}CHC_6H_5$

f.

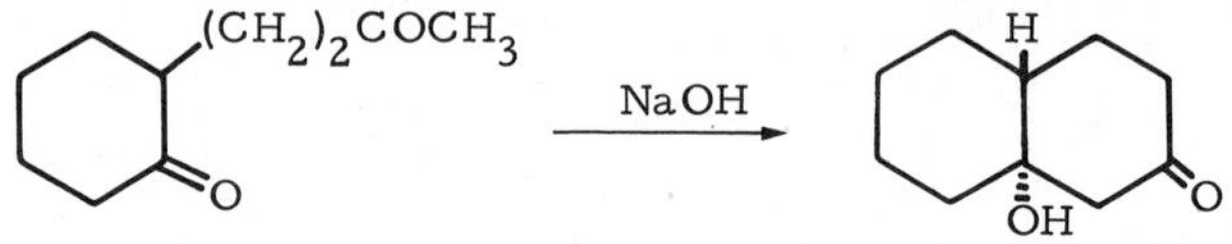

$\xrightarrow{NaOH}$

g.

$\xrightarrow{NaOH}$

$\xrightarrow[-H_2O]{H_2SO_4}$

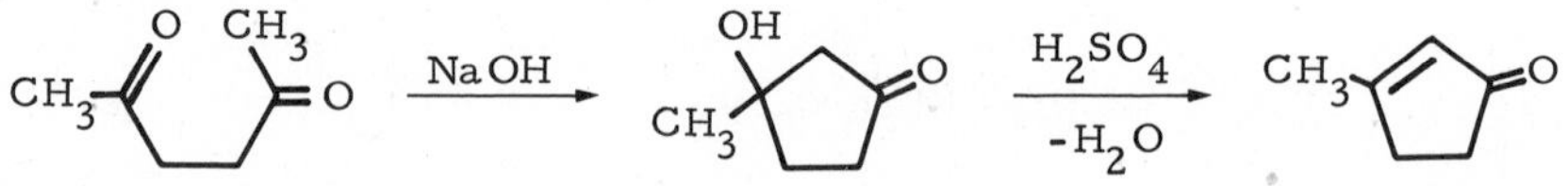

17-27 The most reactive chlorides towards nucleophiles, and hence the most likely to give O-alkylation are $C_6H_5CH_2Cl$ and $CH_2{=}CHCH_2Cl$.

17-28 a. Higher primary halides are subject to E2 elimination under the basic conditions used in the alkylation of ketones.

b. Ethanal cannot be alkylated with KNH_2 and CH_3I because aldolization occurs more rapidly than alkylation.

17-29 The most S_N2 reactive methylating agent is $(CH_3)_3\overset{\oplus}{O}\overset{\ominus}{B}F_4$. The base catalyst that might favor O-alkylation would be $NaNH_2$ in $(CH_3)_2S{=}O$, or $NaOC_2H_5$ in $(CH_3)_2S{=}O$. This ensures poor solvation of the enolate anion and high reactivity at oxygen.

17-30

a. $CH_3COCH_2CH_2CH_2Cl \xrightarrow{NaOH} CH_3CO\overset{\ominus}{C}HCH_2CH_2Cl \longrightarrow CH_3CO$

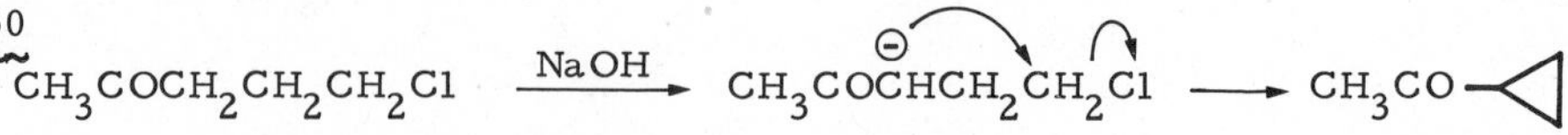

b.

$$CH_2CH_2COCH_2COCH_3$$

$\xrightarrow[\text{-HCl}]{KNH_2}$

1. $H^{\oplus}$
2. enolization

17-31

a.

$\xrightarrow[K_2CO_3]{CH_3I}$

$\xrightarrow[NaNH_2 \text{ in } (CH_3)_2S=O]{ClCH_2COC_6H_5}$

$OCH_2COC_6H_5$

b.

$$\begin{array}{c} Cl \\ Cl \end{array} + \begin{array}{c} COCH_3 \\ CH_2 \\ CO_2C_2H_5 \end{array} \xrightarrow[\substack{C_2H_5OH \\ -HCl}]{NaOC_2H_5}$$

$$\begin{array}{c} COCH_3 \\ CHCO_2C_2H_5 \\ Cl \end{array} \xrightarrow[\substack{C_2H_5OH \\ -HCl}]{NaOC_2H_5}$$

$$\begin{array}{c} COCH_3 \\ CO_2C_2H_5 \end{array}$$

c. $(C_6H_5)_3P: + BrCH_2CO_2C_2H_5 \longrightarrow (C_6H_5)_3\overset{\oplus}{P}CH_2CO_2C_2H_5 \ \overset{\ominus}{Br}$

$(C_6H_5)_3\overset{\oplus}{P}-\overset{\ominus}{C}(CH_3)CO_2C_2H_5 \xleftarrow[\text{2. NaH}]{\text{1. } CH_3I} (C_6H_5)_3\overset{\oplus}{P}-\overset{\ominus}{C}HCO_2C_2H_5 \xleftarrow{\text{NaH}}$

17-32 $R'-\overset{|}{\underset{|}{C}}-\overset{|}{C}=\overset{\oplus}{N}R_2 \xrightarrow{H_2O} R'-\overset{|}{\underset{|}{C}}-\overset{\oplus}{\underset{|}{C}}-\overset{\cdot\cdot}{N}R_2 \longrightarrow R'-\overset{|}{\underset{|}{C}}-\overset{OH}{\underset{|}{C}}-\overset{\oplus}{N}HR_2 \longrightarrow$

$R'-\overset{|}{\underset{|}{C}}-\overset{|}{C}=O + \overset{\oplus}{N}H_2R_2 \longleftarrow R'-\overset{|}{\underset{|}{C}}-\overset{\overset{\oplus}{O}H}{\underset{\|}{C}}- + NHR_2 \longleftarrow$

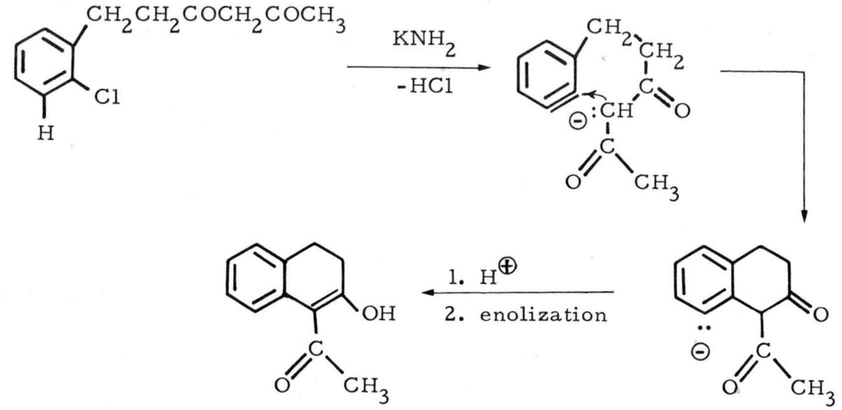

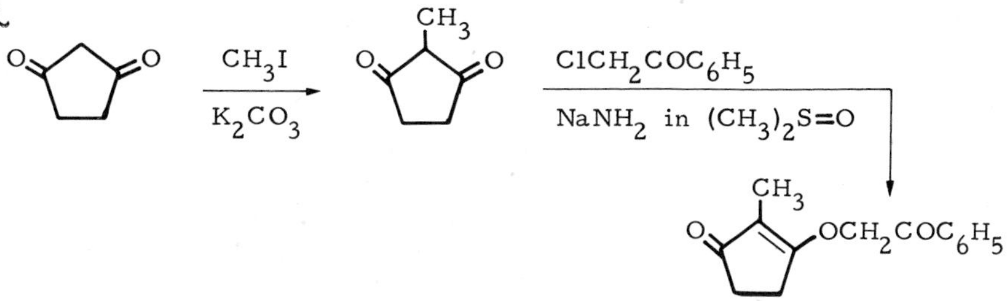

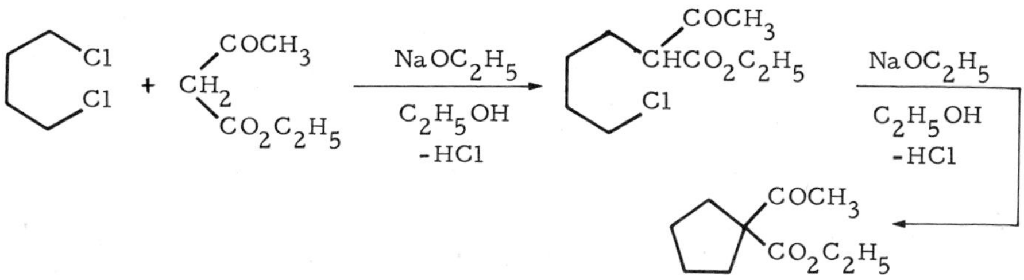

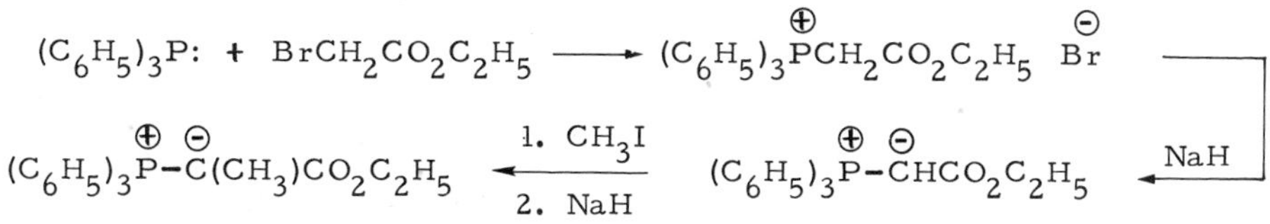

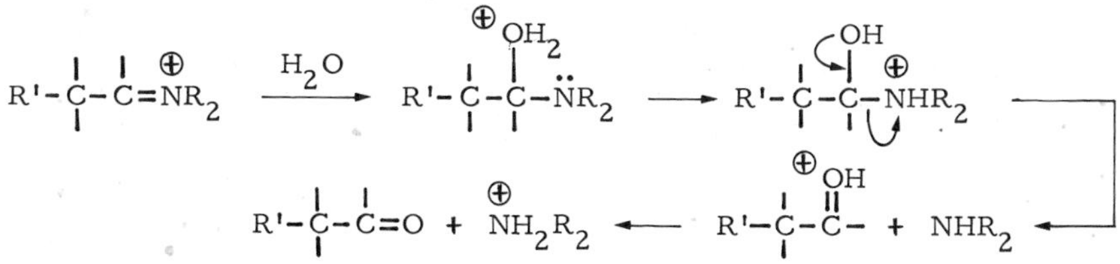

The only difference between the above scheme and enamine formation is the absence of proton elimination from the β-carbon of the imminium ion.

$$H-\overset{|}{\underset{|}{C}}-\overset{|}{C}=\overset{\oplus}{NR_2} \xrightarrow{-H^{\oplus}} -\overset{|}{C}=\overset{|}{C}-NR_2$$

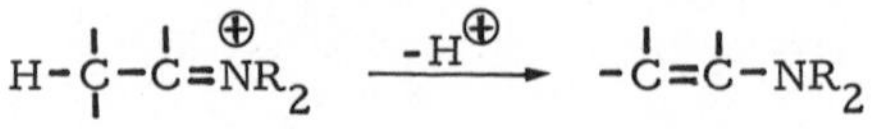

<u>17-33</u>

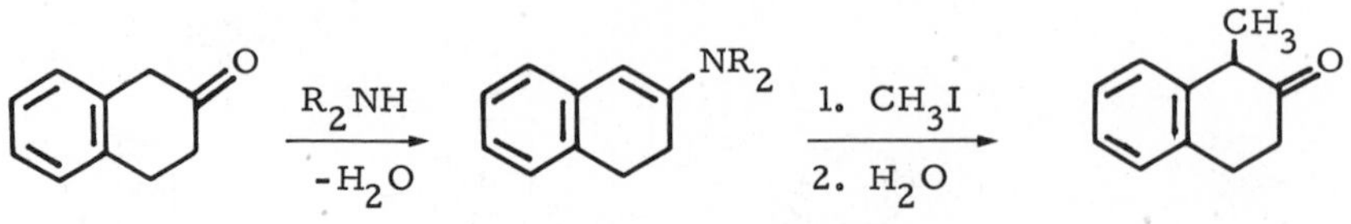

The other enamine that could be formed is

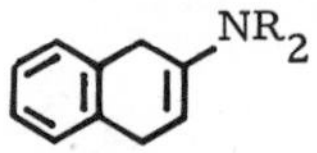

which would lead to methylation at C3. However, this product is not formed because the reaction proceeds to give the conjugated enamine in preference to the non-conjugated enamine.

<u>17-34</u>

a.

b. $(CH_3)_2CHCHO + HS(CH_2)_3SH$

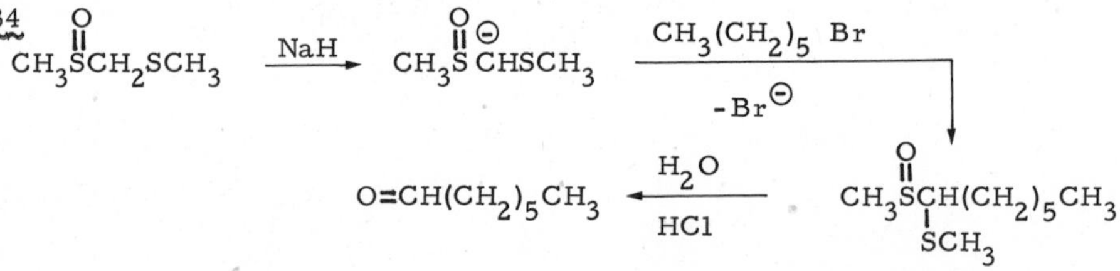

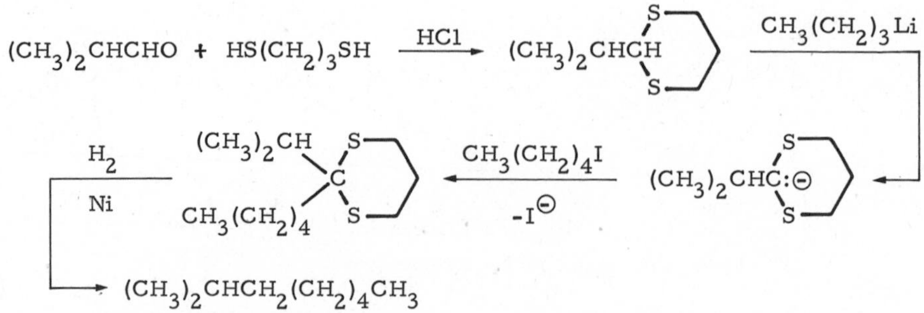

c.

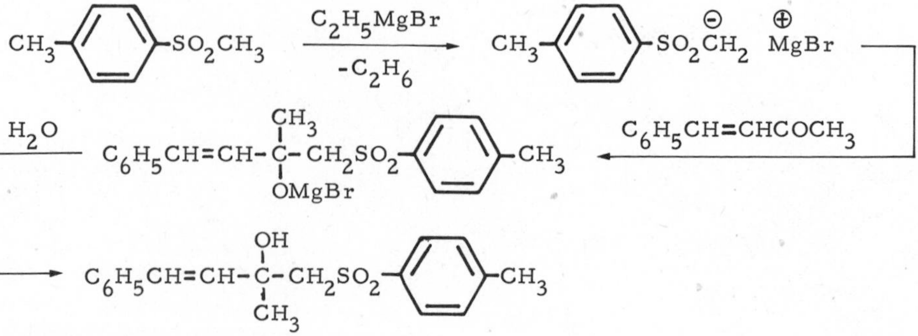

17-35 a. $C_6H_{10}O$: the spectrum indicates three types of hydrogen in the ratio 1 : 3 : 6 with chemical shifts characteristic of =CH-, CH_3CO, and CH_3-C=C protons. A suitable structure follows:

$$(CH_3)_2C=CHCOCH_3$$

The single peak at 2.1 ppm is assigned to the CH_3CO grouping while the pair of doublets near 1.8 and 2.1 ppm are assigned to the nonequivalent allylic methyl groups, each of which is split into a doublet by the ethenyl proton, $\underline{J} \sim 1$ Hz.

b. C_9H_8O: the principal groups of signals are in the ratio of 1 : 6 : 1 with chemical shifts characteristic of aldehyde, phenyl and ethenyl protons. The spectrum is that of $C_6H_5CH=CHCHO$ and is complicated by the fact that:

1. The chemical shift of the β-ethenyl proton is very close to the chemical shifts of the phenyl protons and all appear as a complex group of lines centered at 7.3 ppm.

2. The spin coupling between the α- and β- ethenyl protons is close in magnitude to the difference in their chemical shifts and gives rise to a pair of lines of unequal intensity for the α-proton (see Section 9-10K), each of which is further split by spin coupling with the aldehyde proton.

Synthesis: $2CH_3COCH_3 \xrightarrow{H^{\oplus}} (CH_3)_2C=CHCOCH_3$ (Exercise 17-23)

$C_6H_5CHO + CH_3CHO \xrightarrow{HO^{\ominus}} C_6H_5CH=CHCHO + H_2O$

17-36 a. Even though the orientation of HCl addition to propenal is opposite to that to 1,3-butadiene, the orientation in both reactions conforms to the general observation that <u>the most stable reaction intermediate is formed most rapidly.</u> For propenal, the intermediate is:

$$CH_2=CH-CH=\overset{\oplus}{O}H \longleftrightarrow \overset{\oplus}{CH_2}-CH=CH-OH,$$

which leads to $BrCH_2$-CH=CH-OH and thence to $BrCH_2CH_2CHO$. For 1,3-butadiene, the intermediate is

$$CH_2=CH-\overset{\oplus}{CH}-CH_3 \longleftrightarrow \overset{\oplus}{CH_2}-CH=CHCH_3,$$

which leads to $BrCH_2CH=CHCH_3$ and $CH_2=CHCHBrCH_3$.

b. $H_2C=CHCOCH_3 + Br_2 \longrightarrow BrCH_2CHBrCOCH_3$

1,4-Addition would not be expected because of the low O-Br bond strength.

Addition would be slower than to 1-butene because the intermediate is a less favored carbocation,

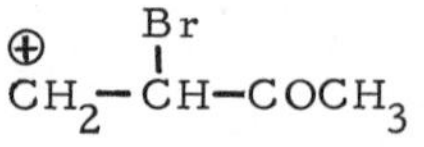

$$\overset{\oplus}{CH_2}-\overset{\overset{\displaystyle Br}{|}}{CH}-COCH_3$$

Formation of this cation rather than $BrCH_2-\overset{\oplus}{CH}-COCH_3$ would be attributed to the electron-withdrawing nature of the carbonyl group.

<u>17-37</u> Addition of $CN^{\ominus}$ to the carbonyl carbon of $CH_2{=}CHCH_2CHO$ leads to an intermediate with charge localized on <u>oxygen</u>. If addition were to occur at the carbon-carbon double bond, the charge would be localized on <u>carbon</u> - a less stable circumstance. Addition of $CN^{\ominus}$ to $CH_2{=}CHCOCH_3$ occurs at C4 because the intermediate has charge delocalized over <u>carbon</u> and <u>oxygen</u>,

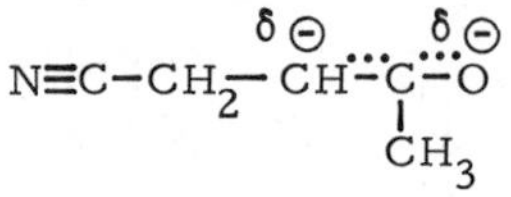

$$N{\equiv}C-CH_2-\overset{\delta\ominus}{CH}\cdots\overset{\delta\ominus}{\underset{\underset{\displaystyle CH_3}{|}}{C}}{=}O$$

<u>17-38</u> Ketene with alcohols:

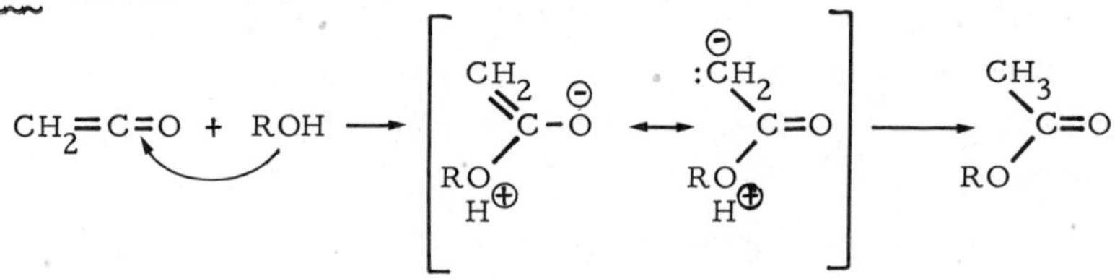

Ketene with amines:

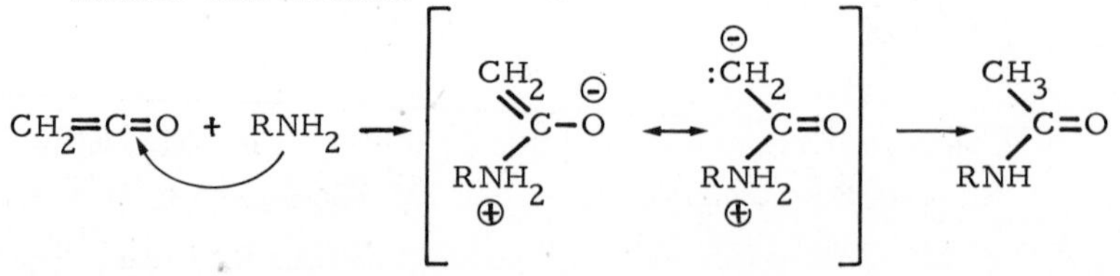

Reaction with alcohols would be accelerated by acids and bases because an equilibrium favoring formation of $RO^{\ominus}$ would be set up in the presence of a base ($ROH + \overset{\ominus}{OH} \longrightarrow RO^{\ominus} + H_2O$) while protonation of the carbonyl oxygen is expected in the presence of acids.

Reaction with amines is expected to be pH dependent. At low pH, reaction would be slow since the amine would be tied up as an amine salt. At some intermediate pH, the rate would be optimized by protonation of the carbonyl oxygen.

Actually, most acylation reactions of ketenes are quite fast even in the absence of catalysts.

17-39

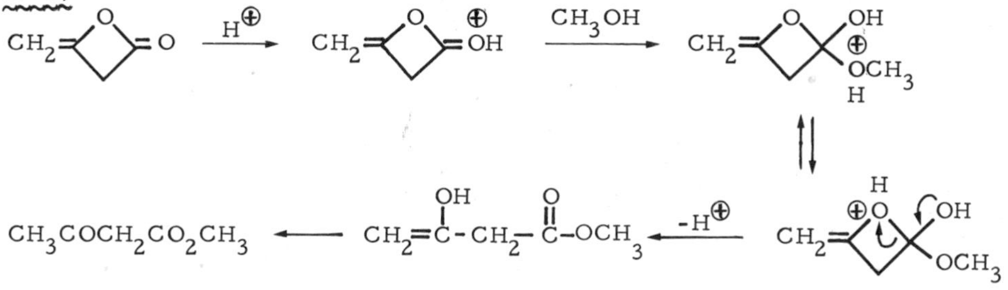

CH$_2$=C(O)(C=O) $\xrightarrow{\ H^\oplus\ }$ CH$_2$=C(O)(C=OH)$^\oplus$ $\xrightarrow{\ CH_3OH\ }$ CH$_2$=C(O)(C)(OH)$^\oplus$OCH$_3$ H

$\updownarrow$

CH$_3$COCH$_2$CO$_2$CH$_3$ $\longleftarrow$ CH$_2$=C(OH)-CH$_2$-C(=O)-OCH$_3$ $\xleftarrow{\ -H^\oplus\ }$ CH$_2$=C(O)(H)$^\oplus$(OH)(OCH$_3$)

17-40 If diketene possessed an enolic OH group, as in structures (iii), (iv), (vii) and (viii), this function could readily be detected by means of infrared and nmr.

If diketene were an α, β-unsaturated carbonyl compound, as in (iii), (vii) and (ix), this would be apparent from its ultraviolet spectrum.

The cumulated double bonds of structure (ix) could also be identified by infrared spectroscopy.

The presence of a carbonyl function could be ascertained by infrared thus eliminating structures (v), (iv) and (viii) .

The presence of a carbon-carbon double bond could be determined by infrared, and the number of ethenyl hydrogens determined by nmr. In this manner, structure (i) would remain the only structure consistent with the spectral data.

17-41 C$_6$H$_8$O$_2$: The principle groups of lines are in the ratio of $1:1:3:3$ with chemical shifts characteristic of enolic, allylic CH, allylic CH$_3$ and $-\overset{|}{\underset{|}{C}}-CH_3$ protons. The doublet near 1 ppm indicates the grouping CH$_3$-CH. A suitable structure follows:

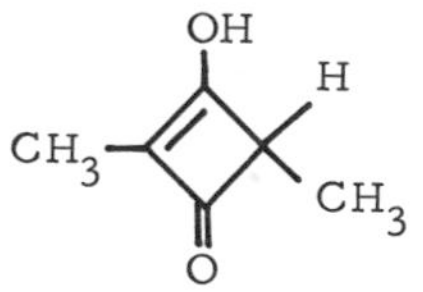

It would appear from the signals due to the CH and CH$_3$-C= protons that there is a small long-range coupling between them.

17-42 (i) Using deuterium labeled ethanedial, the mechanism predicts deuterium exchange to take place.

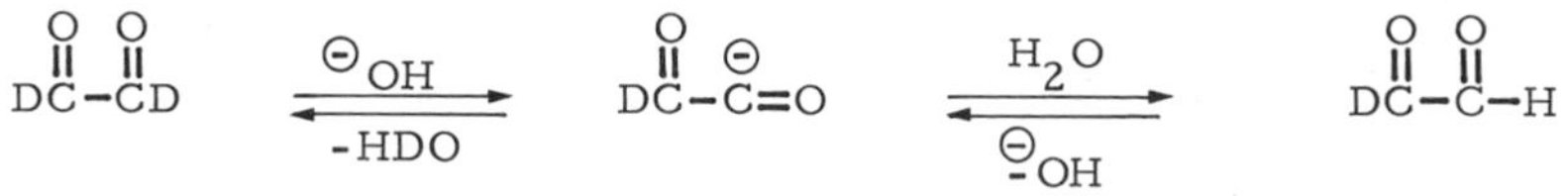

D$\overset{O}{\overset{||}{C}}$-$\overset{O}{\overset{||}{C}}$D $\underset{-HDO}{\overset{\ominus OH}{\rightleftharpoons}}$ D$\overset{O}{\overset{||}{C}}$-$\overset{\ominus}{C}$=O $\underset{\ominus OH}{\overset{H_2O}{\rightleftharpoons}}$ D$\overset{O}{\overset{||}{C}}$-$\overset{O}{\overset{||}{C}}$-H

(ii) Rearrangement in the presence of D_2O should lead to deuterium-labeled hydroxyethanoic acid, $HDC(OH)CO_2H$.

(iii) Rearrangement in the presence of other reagents which ordinarily add to ketenes should lead to derivatives of hydroxyethanoic acid.

$$H-\overset{\overset{\displaystyle OH}{|}}{C}=C=O \xrightarrow{CH_3OH} H_2\overset{\overset{\displaystyle OH}{|}}{C}CO_2CH_3$$

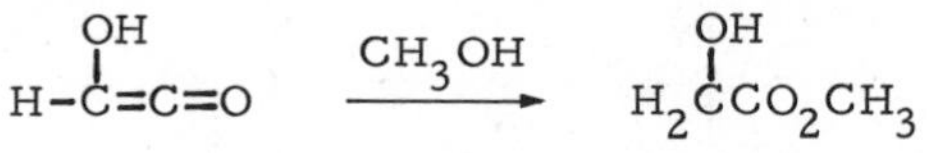

17-43

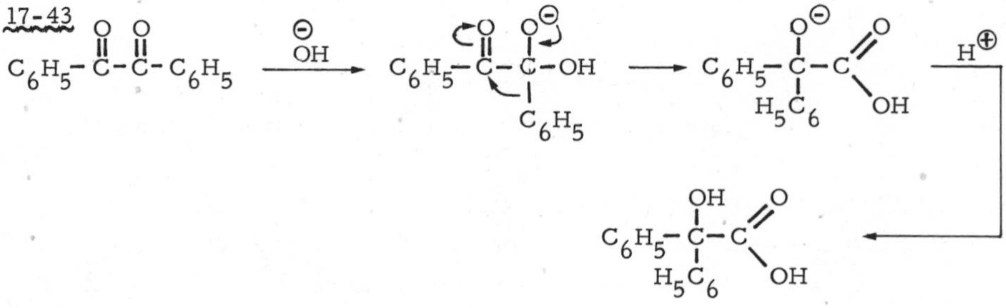

With $(CH_3)_3CO^{\ominus}K^{\oplus}$ in $(CH_3)_3COH$ the corresponding benzilic acid ester is formed in 93% yield.

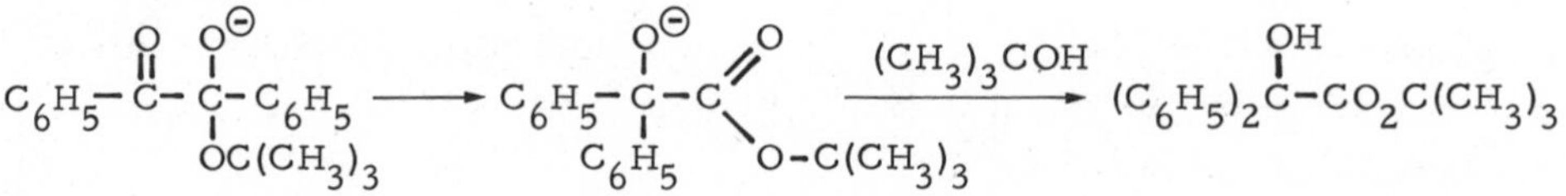

In principle, a benzilic acid-type rearrangement could occur with $CH_3COCOCH_3$. That it does not may be due in part to the fact that a methyl group does not migrate (as $CH_3:$) as readily as a phenyl group ($C_6H_5:$). Possibly more important is the fact that the Cannizzaro reaction is insignificant for carbonyl compounds having α-hydrogens because other reactions supervene.

17-44 Equilibrium favors formation of the monoenol of 1,2-cyclopentanedione because:

a. There are fewer eclipsing H⋯H interactions in the enol than there are in the diketone.

b. Electrostatic repulsions between the carbonyl groups are relieved by enol formation. In the case of 2,3-butanedione, free rotation enables the carbonyl groups to assume a more stable s-trans conformation in which interaction between the carbonyl groups is minimized.

$$\underset{CH_3}{\overset{O}{\diagdown}} C - C \underset{O}{\overset{CH_3}{\diagup}}$$

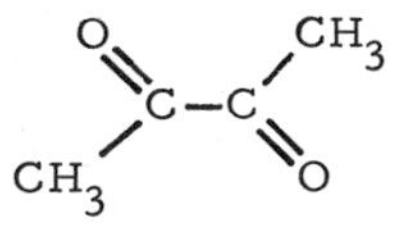

Bicyclo[2.2.1]heptan-2,3-dione, 15, is not expected to enolize owing to the prohibitive strain that would exist in the enol which would place a double bond at a bridge-head carbon.

17-45 See previous Exercise, regarding 15.

17-46 In water, the equilibrium keto $\rightleftarrows$ enol has $\underline{K} < 1$. The acidity of the keto form is expressed by the equation

$$keto \quad \overset{\underline{K}_{keto}}{\rightleftarrows} \quad enolate^{\ominus} + H^{\oplus}$$

and the acidity of the enol by

$$enol \quad \overset{\underline{K}_{enol}}{\rightleftarrows} \quad enolate^{\ominus} + H^{\oplus}$$

Hence:

$$\frac{\underline{K}_{keto}}{\underline{K}_{enol}} = \frac{enol}{keto} = \underline{K} < 1$$

or

$$\underline{K}_{keto} < \underline{K}_{enol}$$

17-47 $C_{10}H_{10}O_2$: There are four main types of hydrogen roughly in the ratio of $1:5:1:3$ with chemical shifts characteristic of enol, phenyl, $=CH-$ and CH_3CO hydrogens. The obvious structure is that of the enol form of 1-phenyl-1,3-butanedione.

$$C_6H_5-\overset{O}{\overset{\|}{C}}\diagdown_{CH_2}\overset{O}{\overset{\|}{C}}-CH_3 \quad \rightleftarrows \quad C_6H_5-\overset{O-H}{\overset{|}{C}}\diagdown_{CH}\overset{O}{\overset{\|}{C}}-CH_3$$

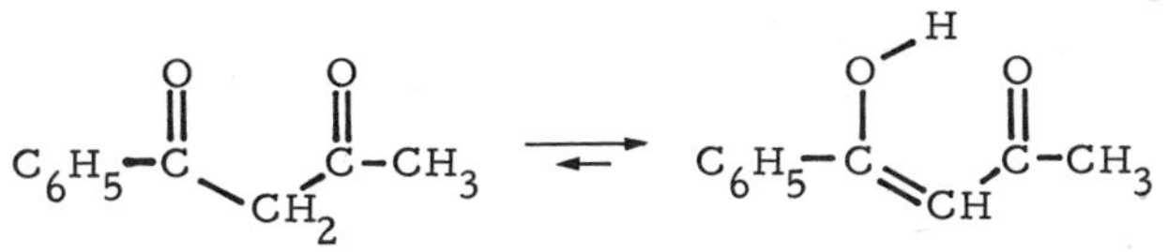

The methylene and methyl groups of the keto form can be seen as small peaks at 3.9 and 2.1 ppm respectively.

17-48

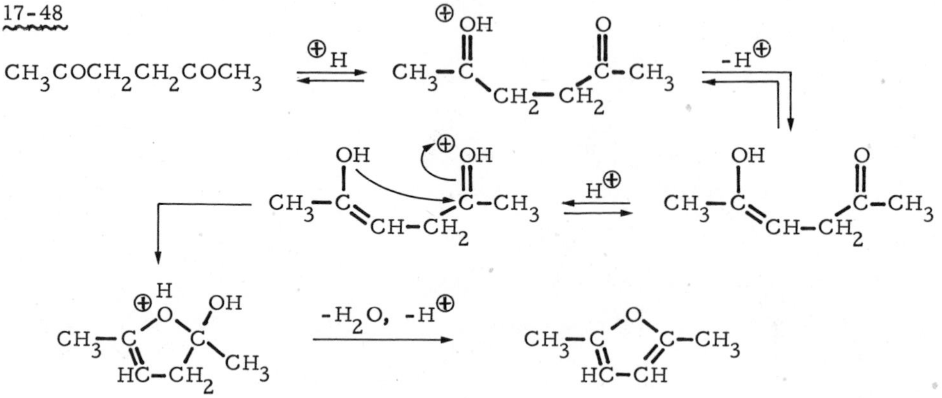

The mechanism is analogous to acid-catalyzed hemiketal formation (see Section 15-4E) and acid-catalyzed dehydration of alcohols (Section 15-5C).

17-49

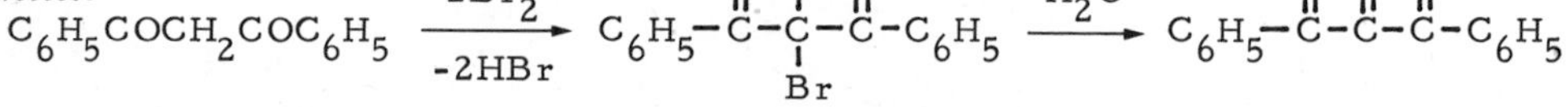

$$C_6H_5COCH_2COC_6H_5 \xrightarrow[-2HBr]{2Br_2} C_6H_5-\overset{O}{\underset{}{C}}-\overset{Br}{\underset{Br}{C}}-\overset{O}{\underset{}{C}}-C_6H_5 \xrightarrow{H_2O} C_6H_5-\overset{O}{\underset{}{C}}-\overset{O}{\underset{}{C}}-\overset{O}{\underset{}{C}}-C_6H_5$$

It would be necessary to label one of the carbonyl groups as ^{14}C or ^{13}C in order to tell which one is lost as carbon monoxide on treatment with aluminum chloride.

17-50

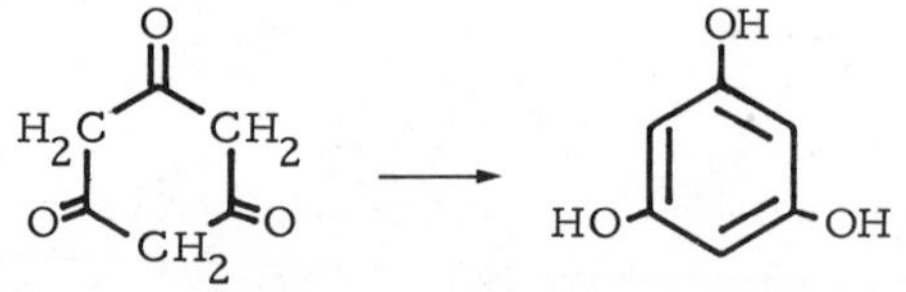

1,3,5-Cyclohexanetrione should exist exclusively in the enol form due to the large gain in resonance stabilization on forming the benzenoid ring.

17-51   a.   Only gives CH_3CO_2Na which does not react further.

   b. $(CH_3)_3CCl \xrightarrow[-HCl]{NaNH_2} (CH_3)_2C=CH_2$

   c.   In the presence of base, enolization of 3-methyl-2-butanone is more likely to occur on the methyl side.

$$(CH_3)_2CHCOCH_3 \xrightarrow[-H_2O]{\ominus OH} (CH_3)_2CHCOCH_2^{\ominus} \xrightarrow{CH_2=O} (CH_3)_2CHCOCH_2CH_2OH$$

[with excess $H_2C=O$ and base $\longrightarrow (CH_3)_2C(CH_2OH)COC(CH_2OH)_3$]

d. Aldolization and saponification (Chapter 18) would take precedence:

$$2CH_3CHO \xrightarrow{\overset{\ominus}{OH}} CH_3CH(OH)CH_2CHO$$

$$CH_3CO_2C_2H_5 + \overset{\ominus}{OH} \longrightarrow CH_3CO_2^{\ominus} + C_2H_5OH$$

e. Okay.

17-52

a. $\triangleright\!\!-COCH_3 \xrightarrow[NaOH]{Br_2} \xrightarrow{H^{\oplus}} \triangleright\!\!-CO_2H \xrightarrow[-SO_2, -HCl]{SOCl_2} \triangleright\!\!-COCl$

$\triangleright\!\!-CO_2H \xrightarrow{BH_3} \xrightarrow{H^{\oplus}} \triangleright\!\!-CH_2OH$

$\triangleright\!\!-COCl + \triangleright\!\!-CH_2OH \xrightarrow{-HCl} \triangleright\!\!-\overset{\overset{O}{\parallel}}{C}-O-CH_2\!\!-\!\triangleleft$

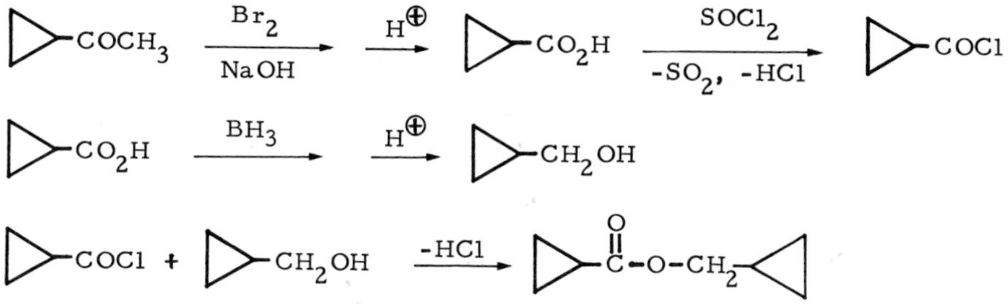

b. $CH_3COCH_3 + CH_2O \xrightarrow{\overset{\ominus}{OH}} CH_2(OH)CH_2COCH_3 \xrightarrow{H^{\oplus}} CH_2=CHCOCH_3$

c. $CH_3-\overset{\overset{H_3C}{|}}{\underset{\underset{HO}{|}}{C}}-\overset{\overset{CH_3}{|}}{\underset{\underset{OH}{|}}{C}}-CH_3 \xrightarrow{H^{\oplus}} CH_3-\overset{\overset{CH_3}{|}}{\underset{\underset{CH_3}{|}}{C}}-COCH_3 \xrightarrow[NaOH]{Br_2} \xrightarrow{H^{\oplus}} CH_3-\overset{\overset{CH_3}{|}}{\underset{\underset{CH_3}{|}}{C}}-CO_2H$

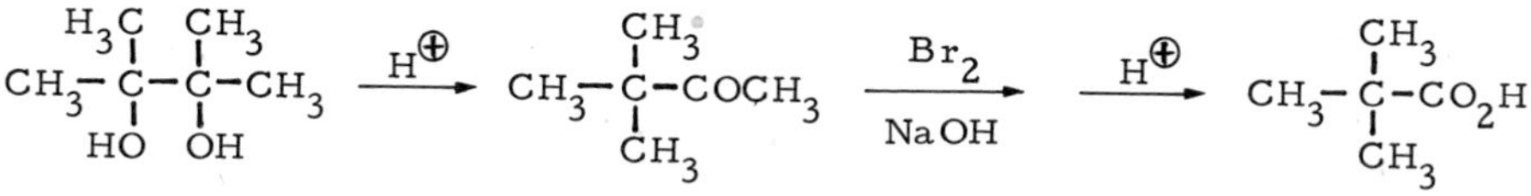

d. $CH_3CH_2COCH_2CH_3 + 4NaNH_2 + 4CH_3I \longrightarrow$

$$(CH_3)_3CCOC(CH_3)_3 + 4NaI + 4NH_3$$

e. $(CH_3CH_2)_2CHCOBr \xrightarrow{Br_2} (CH_3CH_2)_2\!-\!\overset{\overset{}{}}{\underset{\underset{Br}{|}}{C}}COBr \xrightarrow{Zn}$

$(CH_3CH_2)_2C\overset{\overset{CO}{\diagup}}{\underset{\underset{CO}{\diagdown}}{}}C(CH_2CH_3)_2 \longleftarrow (CH_3CH_2)_2C=C=O$

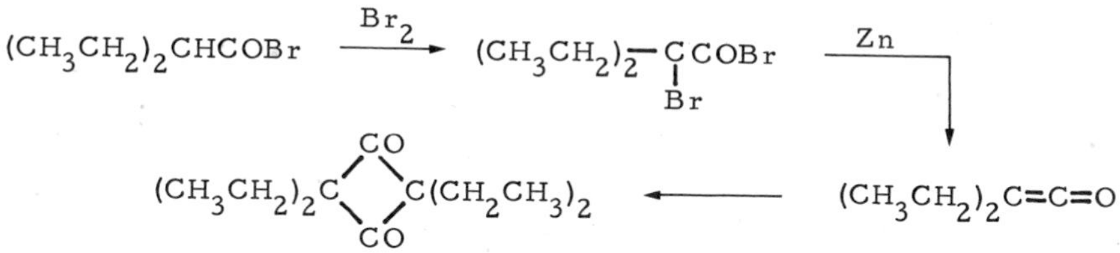

f. $2CH_3COCH_3 \xrightarrow{Ba(OH)_2} (CH_3)_2\overset{}{\underset{\underset{OH}{|}}{C}}CH_2COCH_3 \xrightarrow[I_2]{-H_2O} (CH_3)_2C=CHCOCH_3$

$\downarrow H_2, Pt$

$(CH_3)_2CHCH_2COCH_2Br \xleftarrow[OAc^{\ominus}]{Br_2} (CH_3)_2CHCH_2COCH_3$

(must run to low conversion)

g. $CH_3CHO + HN\langle$ ⟩ $\xrightarrow{H^{\oplus}}$ $CH_2=CH-N\langle$ ⟩ $\xrightarrow[\text{2. } H_2O, H^{\oplus}]{\text{1. } CH_3I}$

$CH_3CH_2CH(OH)CN \xleftarrow{HCN} CH_3CH_2CHO$

h. $CH_3COCH_3 \xrightarrow{Ba(OH)_2} (CH_3)_2\underset{\underset{OH}{|}}{C}CH_2COCH_3 \xrightarrow[-H_2O]{H^{\oplus}} (CH_3)_2C=CHCOCH_3$

$\downarrow H_2, Pt$

$\xleftarrow[-H_2O]{H^{\oplus}} (CH_3)_2CHCH_2\underset{\underset{OH}{|}}{\overset{\overset{CH_3}{|}}{C}}-CH_3 \xleftarrow{CH_3MgI} (CH_3)_2CHCH_2COCH_3$

$(CH_3)_2CHCH=C(CH_3)_2 \xrightarrow[Pt]{H_2} (CH_3)_2CHCH_2CH(CH_3)_2$

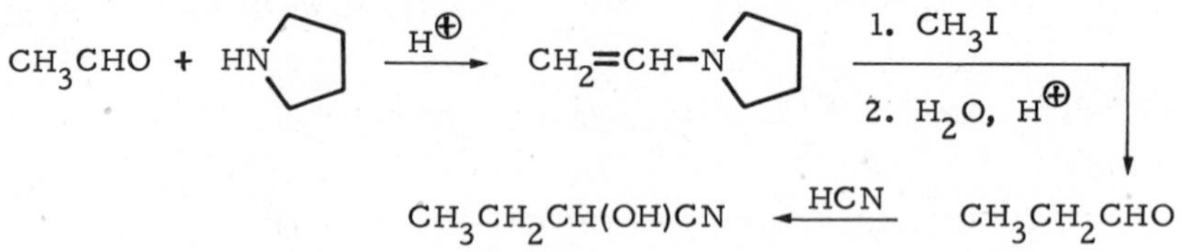

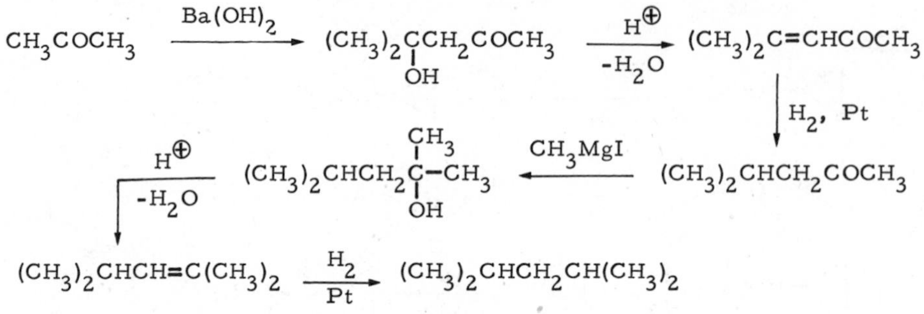

i. By way of enamine formation followed by alkylation with CH_3I, as in (g).

j. $CH_3CH_2CHO \xrightarrow[Ca(OH)_2]{\overset{\ominus}{O}H, CH_2O} CH_3\underset{\underset{CH_2OH}{|}}{\overset{\overset{CH_2OH}{|}}{C}}-CH_2OH \xrightarrow{CH_3COCl} CH_3C(CH_2O\overset{\overset{O}{||}}{C}CH_3)_3$

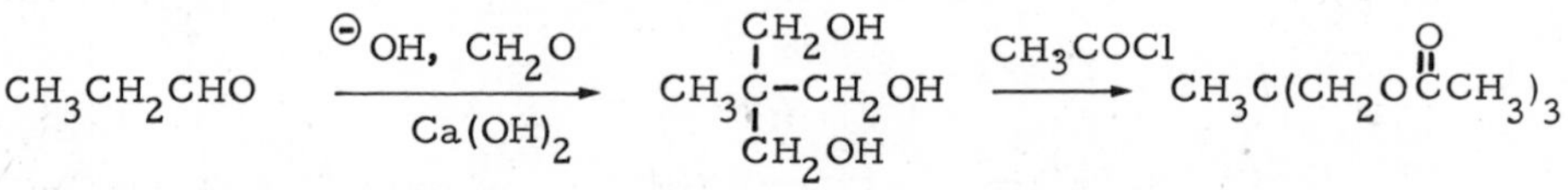

k. $(CH_3)_3CCOCH_3 + NaNH_2 + CH_3I \xrightarrow[-NaI]{-NH_3} (CH_3)_3CCOCH_2CH_3$

$\downarrow Zn, HCl$

$(CH_3)_3CCH_2CH_2CH_3$

17-53   a. Bromination: 2,4-pentanedione will react instantaneously due to its high enolic composition, formation of $Cu^{2\oplus}$ salts, etc.

b. Di-n-butyl ketone is expected to form a 2,4-dinitrophenylhydrazone (Table 16-5), but di-tert-butyl ketone is too sterically conjested to react readily.

c. Treatment of $(CH_3)_2C(OH)CH_2COCH_3$ with dilute base will cause it to cleave to 2-propanone.

d. Test for enol formation will be positive in the case of $C_6H_5COCH_2COC_6H_5$ and negative for $C_6H_5COCOC_6H_5$.

e. Ozonization of diketene will give $CH_2=O$.

f.   Treat with water then with $KMnO_4$:

$$CH_3CH=C=O \xrightarrow{H_2O} CH_3CH_2CO_2H \qquad \text{no reaction with } KMnO_4;$$

$$CH_2=CH-CH=O \qquad \text{reacts with } KMnO_4$$

g.   The enamine will react with methyl iodide to give after hydrolysis the same methylcyclopentanone as obtained by hydrolysis of the imine.

h.   Treatment with dilute base will cause elimination of HCl in the case of the β-chloro ketone.

$$ClCH_2CH_2COCH_3 \xrightarrow[-HCl]{\ominus OH} CH_2=CHCOCH_3$$

17-54   a.   Nmr and infrared spectra will show the presence of $-CO-CH_2-CO-$ and the enol $-C(OH)=CH-CO-$.

b.   Nmr will show only a singlet resonance near 1 ppm for $[(CH_3)_3C]_2CO$.

c.   Splitting patterns in nmr spectra will be different. In $(CH_3)_2C(OH)CH_2COCH_3$ only singlet resonances will be observed.

d.   As in Part a.

e.   Infrared absorption near 1950 cm^{-1} is characteristic of the cumulated system $C=C=O$.

f.   As in Part e; also, nmr resonance at 9.5 ppm identifies aldehyde.

g.   Infrared spectrum will differentiate between $C=N$ and $C=C$.

h.   Splitting patterns in nmr spectra will be different. In $CH_3CH(Cl)COCH_3$ the C4 methyl gives a doublet resonance.

17-55   a.   The ultraviolet spectrum will show the presence of the **grouping** $C=C-C=O$. Infrared and nmr spectra will further determine whether or not the grouping $CH_2=CH$ is present.

b.   A carboxyl group can be identified by the presence in the infrared of an intense hydrogen-bonded OH band (2500-3000 cm^{-1}), and by nmr (chemical shift of OH proton ~9.2 ppm).

c.   A methyl resonance in the nmr would identify the cyclobutanone. Also, the position of carbonyl absorption in the infrared spectrum of cyclic ketones varies with ring size. Cyclopentanones absorb near 1745 cm^{-1} and cyclobutanones absorb near 1780 cm^{-1}.

d. An intense OH absorption is expected in the infrared spectrum of the 1,3-diketone, but not in the spectrum of the isomeric 1,2-dicarbonyl compound.   The nmr spectrum will also clearly differentiate between the isomers by the presence or absence of enolic OH absorption.

e. Infrared will show absorption near 1950 cm^{-1} if the cumulated grouping, C=C=O, is present.

f. The presence or absence of carbonyl absorption in the infrared spectrum will differentiate between the two compounds.

g. Double-bond absorption in the infrared and resonance due to =CH— protons is expected for the infrared and nmr spectra of $CH_3COCH=CCH_3$.
                                                                       |
                                                                      OCH_3

The alternative structure, $CH_3COCHCOCH_3$, is expected to be largely enolic
                                           |
                                          CH_3
by analogy with 2,4-pentanedione and should show enolic OH absorption in its infrared and nmr spectra.

17-56   a. The keto form is present to the extent of 6 - 7%, as determined from the integrals of the methyl protons of the enol and keto forms.

b. 1. ~ 1% enol because of conjugation with C_6H_5 groups.

2. 0% enol

3. 0% enol

4. 100% enol   =

17-57

a. $(C_6H_5)_2CHCHO$ $\xrightarrow[-H^\oplus]{K_2CO_3}$ $(C_6H_5)_2\overset{\ominus}{\underset{\cdot\cdot}{C}}CHO$ $\xrightarrow[2.\ H^\oplus]{1.\ CH_2O}$ $(C_6H_5)_2\underset{CH_2OH}{\overset{|}{C}}CHO$

(Cannizarro) $(C_6H_5)_2C(CH_2OH)_2$ $\xleftarrow[CH_2O]{\ominus OH}$

b. $(CH_3)_2C=CH-N$⟨⟩ + $(CH_3)_2C=CHCH_2Br$ ⟶ $(CH_3)_2\underset{CH=N⟨⟩}{\overset{|}{C}}-CH_2CH=C(CH_3)_2$ $Br^\ominus$

$(CH_3)_2\underset{CHO}{\overset{|}{C}}CH_2CH=C(CH_3)_2$ + $H_2\overset{\oplus}{N}⟨⟩$ $Br^\ominus$ $\xleftarrow{H_2O}$

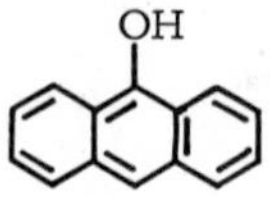

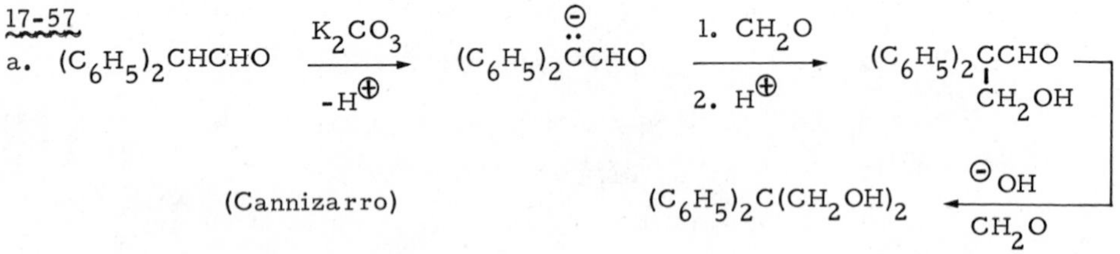

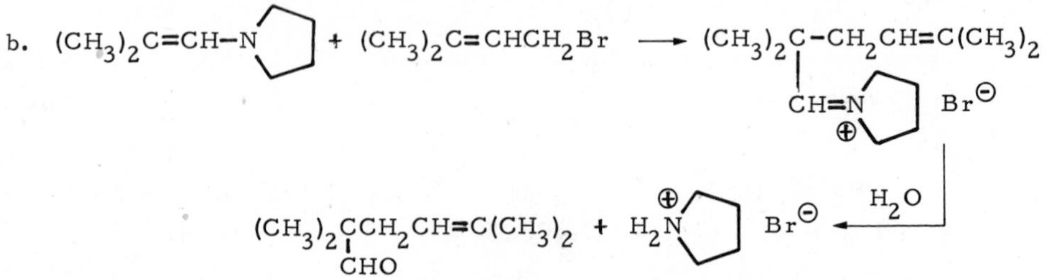

c. $C_6H_5\overset{\displaystyle\parallel}{\underset{\displaystyle O}{C}}-CH(CN)_2 \rightleftharpoons C_6H_5\underset{\displaystyle OH}{C}=C(CN)_2 \xrightarrow[\substack{or \\ (CH_3)_3O^{\oplus}}]{CH_2N_2} C_6H_5\underset{\displaystyle OCH_3}{C}=C(CN)_2$

d. $(C_6H_5)_2C=C=O \xrightarrow{LiAlH_4} (C_6H_5)_2C=CH-O-Al\diagdown \xrightarrow{H_2O}$

$(C_6H_5)_2CH-CHO \longleftarrow (C_6H_5)_2C=CHOH$

e.

f.

g. $CH_2=C=O + :\overset{\ominus}{C}H_2-\overset{\oplus}{N}\equiv N: \longrightarrow$ $\longrightarrow$ $=O + N_2$

h. $CH_2=CHOR \xrightarrow{H^{\oplus}} CH_3\overset{\oplus}{C}HOR \xrightarrow[-H^{\oplus}]{\underset{\displaystyle ClCH_2CH_2\overset{OH}{C}HCN}{}}$

$ClCH_2CH_2\underset{\displaystyle \ominus}{\overset{\displaystyle CN}{C}}-OCH(CH_3)OR \xleftarrow{\text{strong base}} ClCH_2CH_2\overset{\displaystyle CN}{C}HOCH(CH_3)OR$

$\xrightarrow{H^{\oplus}, H_2O}$ (acetal hydrolysis)

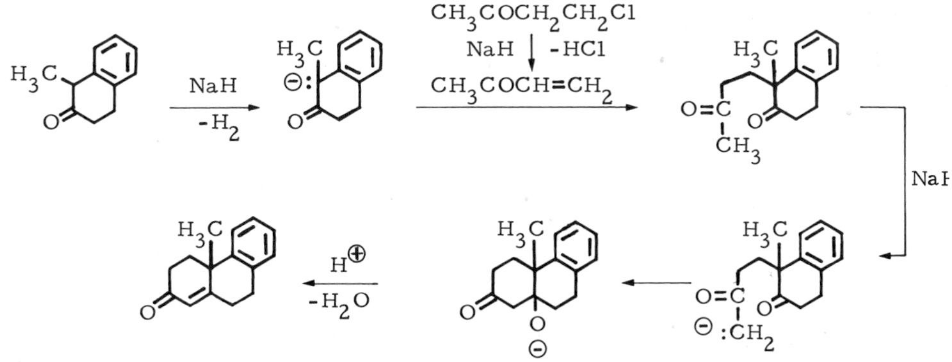

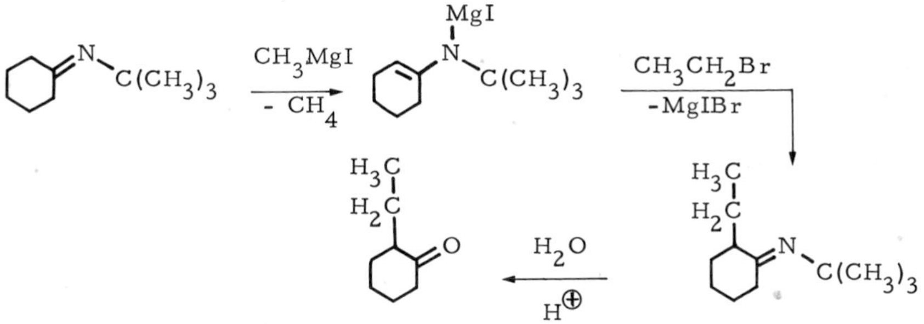

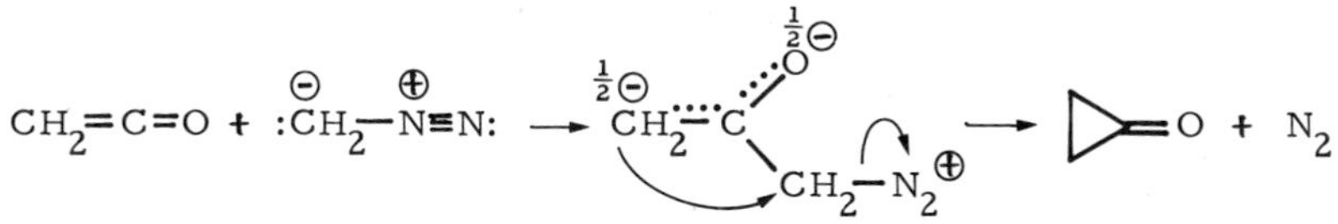

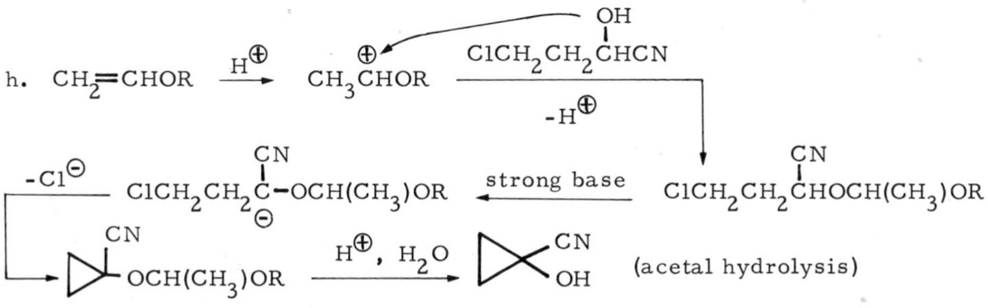

17-58   The Oppenauer oxidation employs a basic medium (aluminum alkoxides), and under these conditions, a β,γ-unsaturated ketone will rearrange to the more stable α,β-unsaturated ketone by way of enolate anion formation.

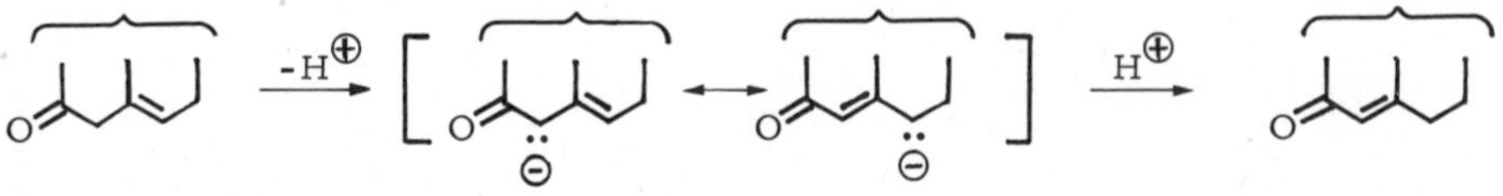

17-59 C_4H_7OCl: The spectrum shows three types of hydrogen in ratio of 2:2:3.   The single peak at 2.2 ppm is that of an unsplit methyl group, and the pair of triplets at lower fields indicates the grouping $-CH_2-CH_2-$.   The triplet near 3.7 ppm may be assigned to the grouping $Cl-CH_2-$; the singlet methyl at 2.2 ppm to CH_3CO; and the triplet near 2.9 ppm to $-CH_2CO-$.   The structure is $CH_3COCH_2CH_2Cl$.   As a β-chloro ketone, this compound may be expected to undergo rapid elimination of HCl in basic solution.

$$CH_3COCH-CH_2 \underset{-HCl}{\overset{\ominus OH}{\longrightarrow}} CH_3COCH=CH_2$$
$$\quad\quad\quad | \quad\;\; |$$
$$\quad\quad\quad H \quad Cl$$

C_4H_7OBr: There are three types of hydrogen in the ratio of 1:3:3. The single peak at 2.3 ppm may be assigned to the grouping CH_3CO ; the doublet at 1.7 ppm and quartet at 4.5 ppm to the grouping CH_3CH.   A structure consistent with the spectrum is $CH_3COCH(Br)CH_3$.   This compound is an α-bromo ketone, and is expected to be highly reactive towards nucleophilic reagents in S_N2 reactions and to rapidly enolize on the side of the α-bromoethyl group in basic solution.

17-60   The β,γ isomer has relatively acidic hydrogens at the α-carbon because the enolate anion formed by proton removal is also conjugated with the C=C.   Equilibrium between the β,γ and α,β isomers is established by way of the enolate anion, and equilibrium favors the more stable conjugated α,β isomer.

17-61

a. $C_6H_5CHO + CH_3COC_6H_5 \xrightarrow{\ominus OH} C_6H_5CH=CHCOC_6H_5$

b. $C_6H_5CHO + CH_3COCH_3 \xrightarrow{\ominus OH} C_6H_5CH=CHCOCH_3 \xrightarrow{C_6H_5CHO}$

$C_6H_5CH=CHCOCH=CHC_6H_5$

c. $C_6H_5CHO + CH_3COCH_2COCH_3 \xrightarrow{\ominus OH} C_6H_5CH=C(COCH_3)_2$

d.

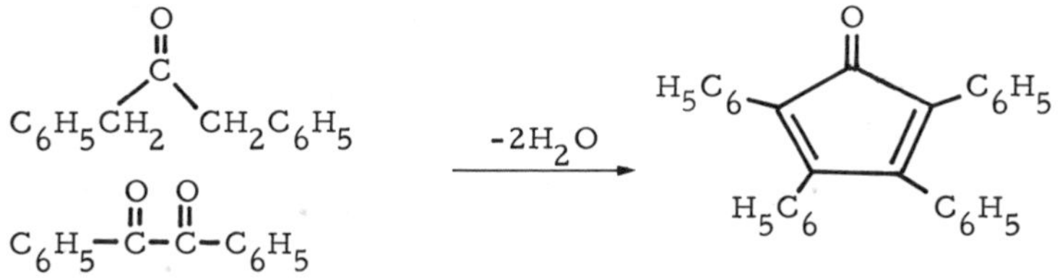

$$\text{C}_6\text{H}_5\text{CH}_2\overset{\overset{\displaystyle O}{\|}}{\text{C}}\text{CH}_2\text{C}_6\text{H}_5$$

$$\xrightarrow{\;-2\text{H}_2\text{O}\;}$$

$$\text{C}_6\text{H}_5-\overset{\overset{\displaystyle O}{\|}}{\text{C}}-\overset{\overset{\displaystyle O}{\|}}{\text{C}}-\text{C}_6\text{H}_5$$

e.

$$\xrightarrow[\text{2. Zn, H}_2\text{O}]{\text{1. O}_3}$$

$$\xrightarrow[-\text{H}_2\text{O}]{\ominus\text{OH}}$$

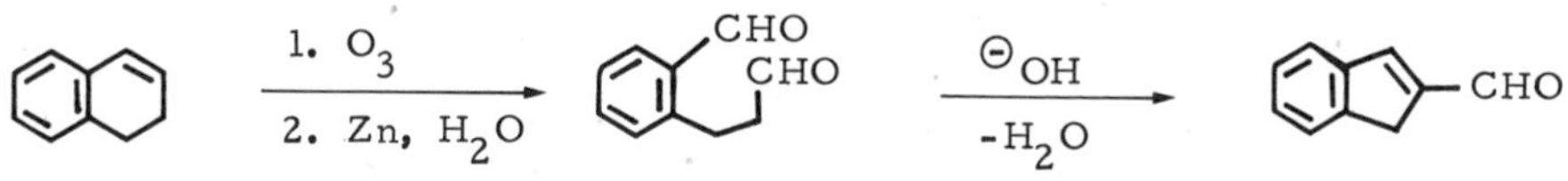

17-62 The equilibrium $\text{O=CHCH}_2\text{R} \rightleftharpoons \text{ROCH=CH}_2$ is very difficult to establish compared to $\text{O=CHCH}_3 \rightleftharpoons \text{HO-CH=CH}_2$ because C–C and C–O bonds are much less readily broken than C–H and O–H bonds. Rearrangement of an alkenyl ether might occur in the presence of an electrophilic reagent capable of donating $R^{\oplus}$ to the terminal carbon.

$$\text{ROCH=CH}_2 + \text{RX} \longrightarrow \text{R}\overset{\oplus}{\text{O}}\text{CH-CH}_2\text{R} + \text{X}^{\ominus} \longrightarrow \text{RX} + \text{O=CHCH}_2\text{R}$$

17-63 If cyclopropanone existed in the dipolar form, its nmr spectrum would be solvent dependent and exhibit low field (deshielded) resonance for the methylene protons compared to that expected of the cyclic structure.

17-64 C-alkylation, $\Delta \underline{H}^{\circ} = -4.3$ kcal;   O-alkylation, $\Delta \underline{H}^{\circ} = +23$ kcal
These are essentially the same values as in the aldol reaction.

If we wish to make the corresponding calculation for the enolate anion we need to have more information which we can get from the relative $p\underline{K}a$ values. Thus,

$$\text{CH}_3\text{COCH}_3 + \text{CH}_3\text{I} \longrightarrow \text{CH}_3\text{COCH}_2\text{CH}_3 + \text{HI} \qquad \Delta \underline{H}^{\circ} = -4.3 \text{ kcal}$$

$\Big\downarrow \Delta \underline{H}^{\circ}_1$ (ionization) $\qquad\qquad\qquad\qquad\qquad\qquad \Big\downarrow \Delta \underline{H}^{\circ}_2$ (ionization)

$$\text{H}^{\oplus} + \text{CH}_3\text{COCH}_2^{\ominus} + \text{CH}_3\text{I} \longrightarrow \text{CH}_3\text{COCH}_2\text{CH}_3 + \text{I}^{\ominus} + \text{H}^{\oplus} \Delta \underline{H}^{\circ} = ?$$

Because HI is a 29 powers of 10 stronger acid than CH_3COCH_3, its $\Delta \underline{H}^{\circ}$ of ionization (neglecting $\Delta \underline{S}$!) is perhaps -40 kcal $(-RT \ln (\underline{K}_1/\underline{K}_2) = -1.99 \times 300 \times \ln(10^{29})$ less than the $\Delta \underline{H}^{\circ}$ of ionization of CH_3COCH_3 and $\Delta \underline{H}^{\circ}_2 - \Delta \underline{H}^{\circ}_1 = 40$ kcal. On this basis, $\Delta \underline{H}^{\circ}$ for C-alkylation of the enolate anion is about -44 kcal and $\Delta \underline{H}^{\circ}$ for O-alkylation would be +23 - 40 = -17 kcal - much more favorable than +23 kcal.

CHAPTER 18

CARBOXYLIC ACIDS AND THEIR DERIVATIVES

To a very substantial degree, the chemistry of carboxylic acids involves principles that we have discussed previously in detail in connection with other compounds.

The physical properties of carboxylic acids, like those of alcohols, are influenced greatly by hydrogen bonding. Acids tend to form rather stable cyclic dimers by hydrogen bonding in nonpolar solvents. Hydrogen-bond formation strongly influences both the infrared and proton nmr spectra. Methods of synthesis of carboxylic acids, RCO_2H, are summarized in a general way in Table 18-5.

The simple alkanoic acids are not very strong acids in water ($\underline{K}_a \sim 10^{-5}$) compared to inorganic acids such as H_2SO_4, HCl, and $HClO_4$, but are about 10^{11} times stronger than alcohols. The acidity arises from greater electron delocalization in the anion relative to the acid itself. Compared to the difference in energy between an alkoxide ion and an alcohol, a carboxylate ion has about a 15 kcal mole^{-1} greater stabilization energy than the acid.

Substitution of electron-attracting substituents on the hydrocarbon chain of a carboxylic acid can greatly enhance the $\underline{K}_a$. Indeed, trifluoroethanoic acid is about as strong as HCl. This kind of enhancement of acid strengths can be understood qualitatively in terms of simple electrostatic theory.

Carboxylic acids can act as bases and accept protons from strong acids. This protonation of acids is important in accelerating esterification reactions through the reaction of acids with alcohols. Primary alcohols and unhindered alcohols give the best results in esterification reactions.

The cleansing action of long-chain carboxylate salts (as soaps) arises from their ability both to reduce the surface tension of water and to dissolve oils and grease through micelle formation. Chemical reactions often proceed at greatly different rates in soap micelles. Lipids, which are like soap molecules in having polar and nonpolar parts, tend to aggregate in bilayers. Such structures are important in cell membranes.

Acyl chlorides are formed from carboxylic acids and PCl_3, PCl_5, or $SOCl_2$.

Carboxylic acids are reduced by $LiAlH_4$ to alcohols. Special reactions are required for reduction of RCO_2H to $RCHO$ or RCH_3, and usually these reductions are better achieved indirectly.

Carboxylic acids undergo thermal decarboxylation rather easily whenever there are strong electron-attracting groups attached to the α carbon or the structural unit $Z = Y - \overset{|}{\underset{|}{C}} - CO_2H$ is present, in which Z is oxygen, nitrogen, or carbon, and Y is nitrogen or carbon. All carboxylic acids lose CO_2, when by some oxidative process they are converted to $RCO_2 \cdot$. The final products of decarboxylation, $RCO_2 \cdot \longrightarrow R \cdot + CO_2$ depend on what is available for $R \cdot$ to react with.

The α carbon of carboxylic acids can be substituted with bromine (or, less well, with chlorine) to give α-halocarboxylic acids. The halogens in these substances are very reactive in S_N2 reactions, but not in S_N1 reactions. A summary of some important reactions of carboxylic acids follows.

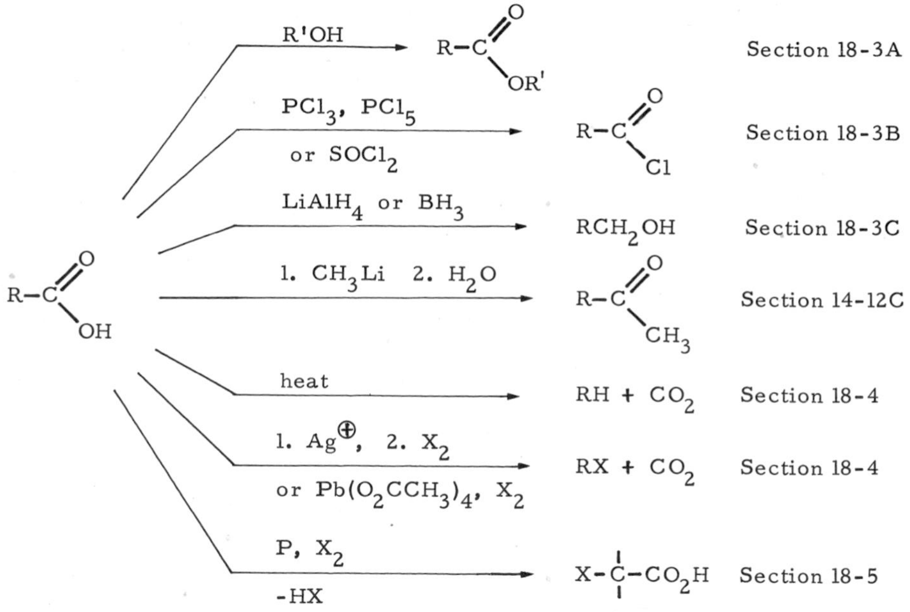

Functional derivatives of carboxylic acids, R—C (=O)—Z or RCN, can be regarded as substances that regenerate the acid on hydrolysis. Examples of such derivatives are esters, anhydrides, acyl halides, amides, nitriles, imides,

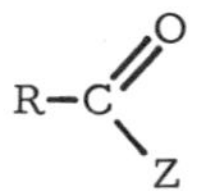

hydrazides, and hydroxamic acids:

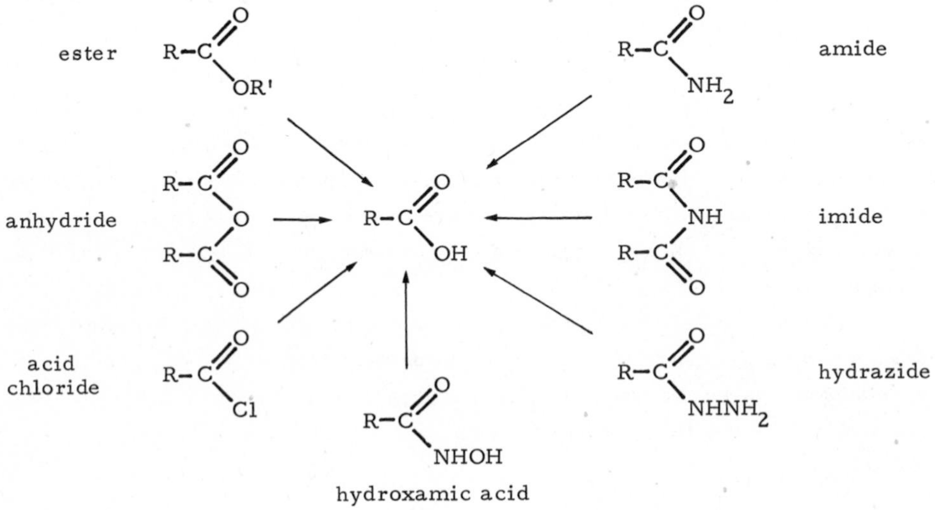

These compounds often are important as acylating agents, whereby

$RC\overset{O}{\underset{|}{\parallel}}$ is transferred from Z to some other atom. Such reactions include ester formation (and hydrolysis), ester interchange, and amide formation. Addition of R'MgX and other organometallic compounds to RCOZ gives ketones or tertiary alcohols, depending on the nature of Z. Reduction of RCOZ with $LiAlH_4$ usually gives the corresponding alcohol, RCH_2OH. If Z is NH_2, the amine, RCH_2NH_2, is the expected product; $-NHR'$ gives a secondary amine, and $-NR_2'$ gives a tertiary amine.

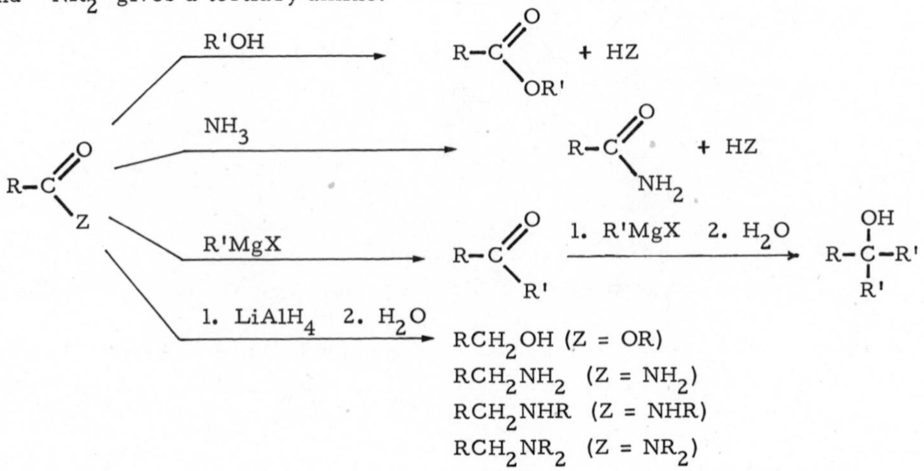

RCH_2OH (Z = OR)
RCH_2NH_2 (Z = NH_2)
RCH_2NHR (Z = NHR)
RCH_2NR_2 (Z = NR_2)

A summary of the methods ot synthesis of carboxylic acid derivatives is in Tables 18-6 and 18-7.

Carboxylic acid derivatives, especially esters, are particularly useful for forming enolate anions, $-\overset{..}{\underset{|}{C}}\overset{\ominus}{}-CO_2R'$, which are not easily formed from the carboxylic acids themselves. Such ester anions and, in particular, those derived from diethyl propanedioate (malonate) and ethyl 3-oxobutanoate (acetoacetate) can be <u>acylated, alkylated,</u> or used in <u>Michael additions.</u>

<u>Acylation of ester anions (Claisen condensation) gives 1,3-dicarbonyl compounds.</u> The equilibrium for β-keto ester formation by acylation of an anion with an ester normally is not favorable, but usually can be made so by proper selection of reagents so that either the β-keto ester or the alcohol formed in the reaction is removed by being converted to the corresponding salt.

Acylation

$$R-\overset{O}{C}\underset{Z}{\diagdown} + \overset{\ominus}{-}\underset{H}{\overset{}{C}}CO_2R' \rightleftharpoons \xrightarrow{-Z^{\ominus}} R-\overset{O}{\overset{||}{C}}-\underset{H}{\overset{|}{C}}-CO_2R' \xrightarrow[\;]{-H^{\oplus}} R-\overset{O}{\overset{||}{C}}-\overset{\ominus}{\underset{|}{C}}-CO_2R'$$

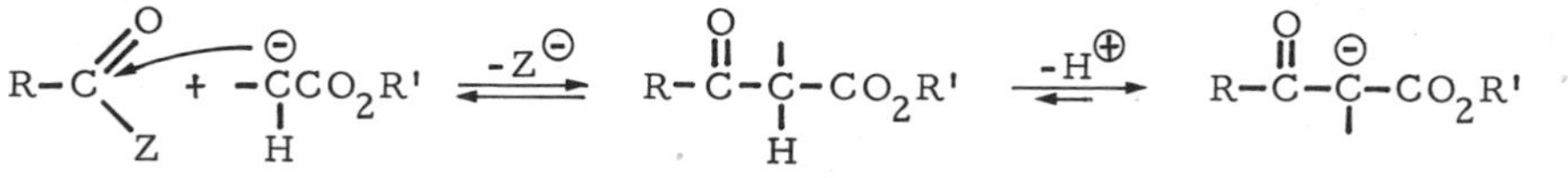

Z = -OR

Z = -Cl   (Claisen condensation)

<u>Alkylation of ester anions</u> can give either C- or O-products. <u>Hydrolysis and decarboxylation</u> of the C-alkylated esters of β-keto esters provides ketones or carboxylic acids.

Alkylation

$$RX + \overset{O}{\overset{||}{-C}}-\overset{\ominus}{\underset{|}{C}}-CO_2R' \xrightarrow{-X^{\ominus}} \overset{O}{\overset{||}{-C}}-\underset{R}{\overset{|}{C}}-CO_2R' \xrightarrow{H_2O,\ H^{\oplus}} \overset{O}{\overset{||}{-C}}-\underset{R}{\overset{|}{C}}-CO_2H$$

$$\downarrow H_2O,\ OH^{\ominus} \qquad\qquad \downarrow -CO_2$$

$$\overset{O}{\overset{||}{-C}}-OH + R-\underset{H}{\overset{|}{C}}-CO_2H \qquad\qquad \overset{O}{\overset{||}{-C}}-CHR$$

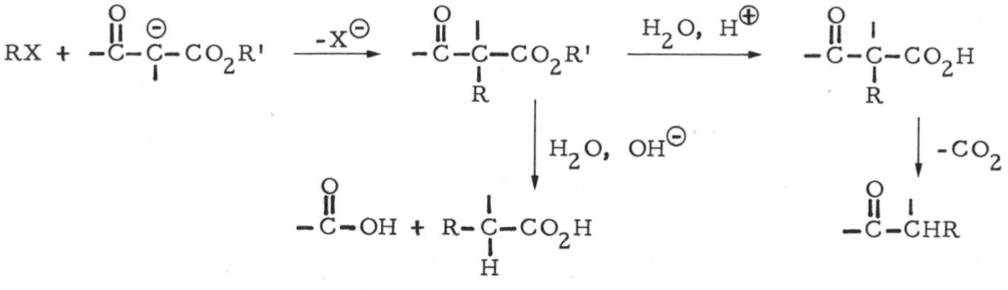

(acid synthesis)   (ketone synthesis)

The lithium salt of ethyl ethanoate, $LiCH_2CO_2C_2H_5$, undergoes <u>aldol-type addition</u> to aldehydes and ketones at -80°:

$$\diagdown C=O + \overset{\ominus}{-}\overset{|}{\underset{|}{C}}-CO_2R \longrightarrow \underset{\diagup}{\overset{O^{\ominus}}{\overset{|}{C}}}-\overset{|}{\underset{|}{C}}-CO_2R \xrightarrow{H^{\oplus}} \overset{HO}{\underset{\leftarrow}{\overset{|}{C}}}-\overset{|}{\underset{|}{C}}-CO_2R$$

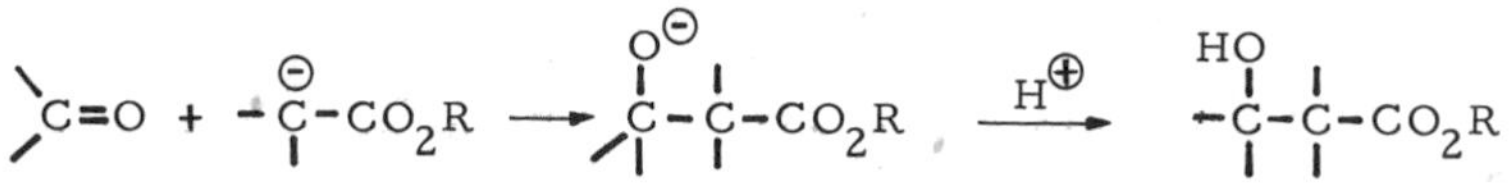

The <u>Michael</u> addition of carbanions or enamines to α,β-unsaturated carbonyl compounds is an important synthetic method for preparing 1, 5-dicarbonyl compounds:

$$-\overset{\overset{O}{\|}}{\underset{|}{C}}-\underset{|}{C}=C\diagdown + \overset{\ominus}{-\underset{|}{C}}-CO_2R \longrightarrow -\overset{\overset{O^{\ominus}}{:}}{\underset{|}{C}}\overset{}{=}\underset{|}{C}-\underset{|}{C}-\underset{|}{C}-CO_2R$$

$$\Big\downarrow H^{\oplus}$$

$$-\overset{\overset{O}{\|}}{C}-CH-\underset{|}{C}-\underset{|}{C}-CO_2R$$

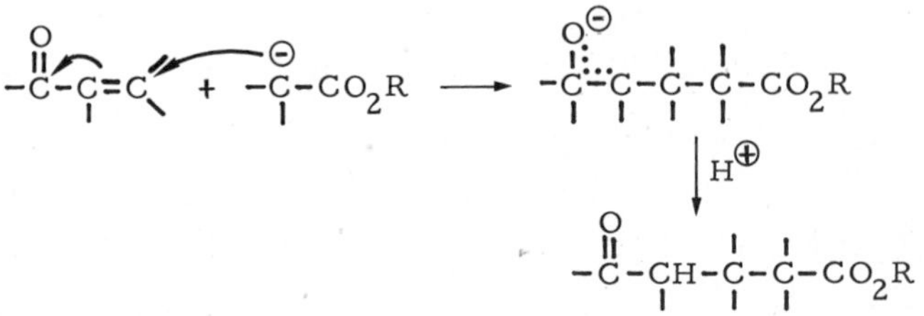

Unsaturated carboxylic acids have relatively unexceptional chemical properties when the double bond is far down the chain from the carboxyl group. The α, β and β, γ isomers of an unsaturated acid or its functional derivatives usually are interconverted easily by treatment with bases.

Additions of electrophilic reagents, such as HBr, $H_2O(H^{\oplus})$, and so on, to α, β-unsaturated acids usually goes opposite to Markownikoff's rule, as the result of "1, 4-" or "conjugate" addition. The carboxylic acid group adds easily to a double bond in the same chain under the influence of strong acid, if a five- or six-membered ring can be formed. The products, which are cyclic esters, are called lactones. Cyclizations of 4- and 5-hydroxyalkanoic acids by internal esterification also form lactones:

$$\underset{/}{\overset{\diagdown}{C}}=\underset{|}{C}-CO_2R \xrightarrow{\text{HX}} X-\underset{|}{C}-\underset{|}{CH}-CO_2R$$

$$\underset{/}{\overset{\diagdown}{C}}=C\!\!-\!\!\Big(\underset{|}{C}\Big)_{\!\underline{n}}\!\!-CO_2H \xrightarrow[-H_2O]{\text{HX}} \quad \text{or}$$

$$\underline{n} = 1 \qquad\qquad \underline{n} = 2$$

$$\overset{H^{\oplus}}{\underset{-H_2O}{\diagup}}$$

$$HO-\underset{|}{C}-\underset{|}{C}\!\!-\!\!\Big(\underset{|}{C}\Big)_{\!\underline{n}}\!\!-CO_2H$$

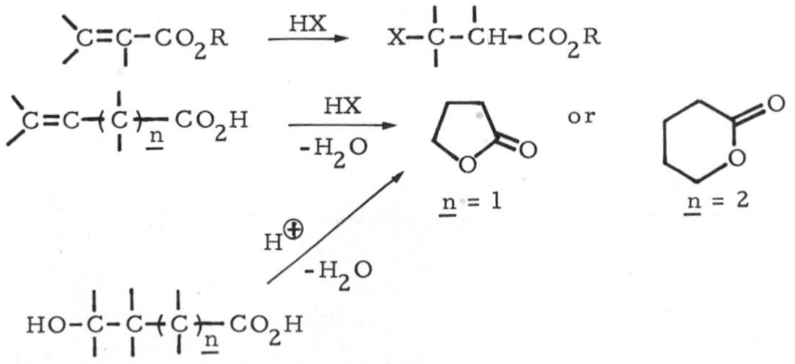

Dicarboxylic acids are of considerable industrial importance for making polymers used as plastics, fibers, surface coatings, and the like. The first ionization constant, $\underline{K}_1$, of a dicarboxylic acid is greater than that of a simple alkanoic acid, whereas the second ionization constant, $\underline{K}_2$, is smaller because of the electrostatic effect of the $-CO_2^{\ominus}$ group at the other end of the molecule.

When dicarboxylic acids are heated strongly, they tend to form cyclic anhydrides or cycloalkanones. Which product results depends on which has a five- or six-membered ring. Thus butanedioic and pentanedioic acids form cyclic anhydrides, whereas hexanedioic and heptanedioic acids form cyclopentanone and cyclohexanone, respectively.

The Dieckmann and acyloin reactions of dicarboxylic esters are important for the synthesis of medium- and large-ring compounds.

ANSWERS TO EXERCISES

<u>18-1</u>   As explained in Section 9-10E, the chemical shift of the OH proton of an alcohol varies with the degree of hydrogen bonding which in turn varies with the concentration of alcohol, particularly when dissolved in a nonpolar solvent such as carbon tetrachloride. In contrast, the OH proton of carboxylic acids does not show the same variation of chemical shift with concentration because carboxylic acids are more strongly associated by hydrogen bonding and generally associate in solution as dimers. These dimers are not easily dissociated on dilution; hence the chemical shift of the OH proton does not change appreciably on dilution.

<u>18-2</u>

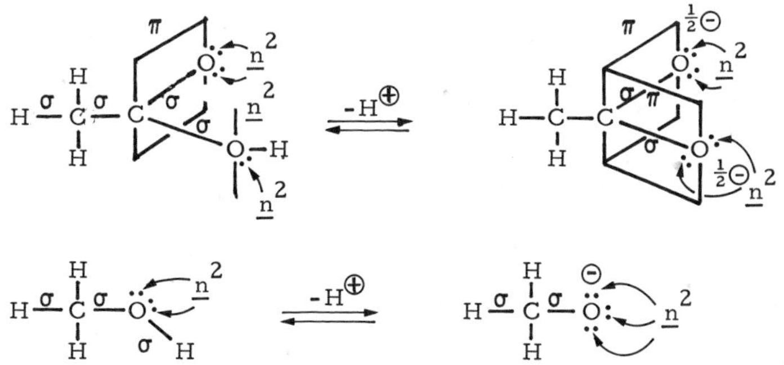

Electron delocalization is more effective in stabilizing ethanoate anion than it is for ethanoic acid because the oxygens of the anion are made equivalent by delocalization over three nuclei. The greater stabilization of the ethanoate anion than of the ethanoic acid accounts for the acidity of CH_3CO_2H. Neither ethoxide anion nor ethanol can be similarly stabilized; accordingly, ethanol is a weaker acid than ethanoic acid.

<u>18-3</u>   To predict or rationalize the absolute or relative strength of an acid, you must consider, the absolute or relative difference in energy between the acid and its anion and the effect of structural changes on that energy difference.

Three valence-bond structures for carbonic acid can be written, two of these are <u>equivalent</u> dipolar forms. Ethanoic (acetic) acid has two <u>nonequivalent</u> valence-bond structures. Accordingly, carbonic acid is expected to have a greater degree of resonance stabilization than ethanoic acid.

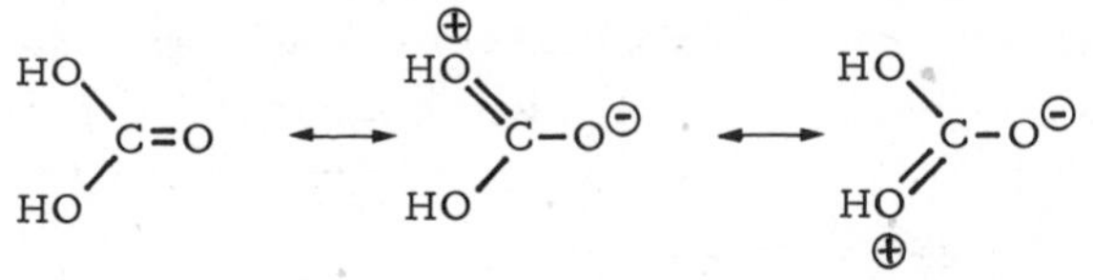

carbonic acid

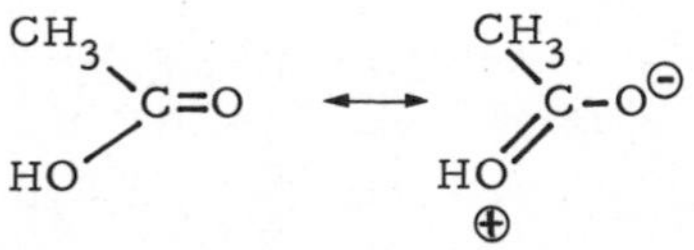

ethanoic acid

What about the corresponding anions? For hydrogen carbonate (bicarbonate) ion, we can write again three structures - two of which are equivalent, with no charge separation. The third is dipolar and relatively less favorable.

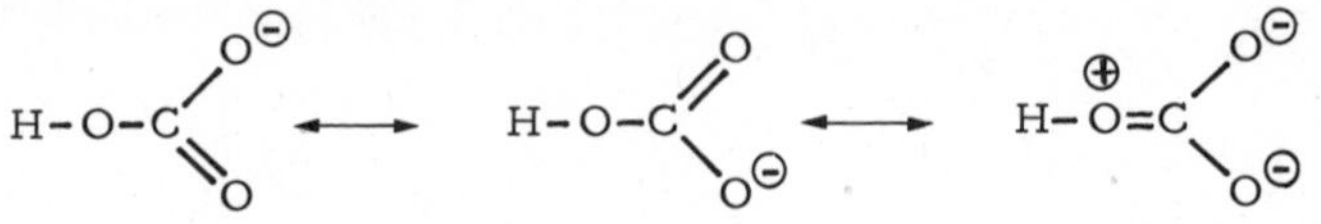

important valence-bond structures   unimportant structure

For ethanoate, there are two equivalent valence-bond structures - both important:

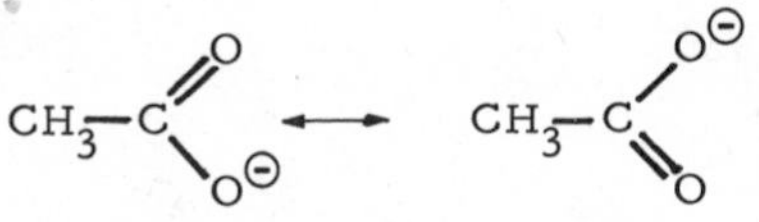

Thus, we conclude that stabilization by electron delocalization will be comparable for hydrogen carbonate and ethanoate ions. But because carbonic acid is expected to be more stabilized than ethanoic acid relative to its anion, it is a weaker acid than ethanoic acid.

18-4

a. $(CH_3)_3\overset{\oplus}{N}CH_2CO_2H$;   electron-withdrawing inductive effect.

b. $(CH_3)_3\overset{\oplus}{N}CH_2CO_2H$;   electrostatic field effect of $R_3\overset{\oplus}{N}$ is greater than that of $R_3\overset{\oplus}{N}-\overset{\ominus}{O}$.

c. CH_3CO_2H;   has fewer electron-donating alkyl groups.

d. $CH_3OCH_2CO_2H$;   as in Part a.

e. $HC\equiv CCH_2CO_2H$;   the sp-carbons of the triple bond are more electron-attracting than the sp^2 carbons of 3-butenoic acid (Section 11-8B)

18-5   If you have studied electrostatics in your courses in physics, you will
remember that the electrostatic force between a dipole such as $O^{\ominus}\!\!-\!N^{\oplus}$ and a
charged particle such as a proton falls off very much <u>more rapidly with distance</u>
than does the electrostatic force between two charges such as $N^{\oplus}$ and a proton.
Consequently, the effect of a dipole, as in acids of the type
$O^{\ominus}\!\!-\!N^{\oplus}(CH_3)_2(CH_2)_n CO_2H$ on the acid strength will decrease faster than the
effect of a charged group on the strengths of acids such as
$(CH_3)_3 N^{\oplus}(CH_2)_n CO_2H$.

If you are unfamiliar with the physics of electrostatics, the effect can be
calculated using the equations for the electrostatic energy (Section 18-2C) required
to bring charges together from a large distance (r_1) to the distances they will
have in the unionized acid. This is easiest to do by simply substituting in illustrative numbers. Let us assume the following reasonable distances:

$$\overset{\oplus}{(CH_3)_3 N}\!-\!CH_2\!-\!CO_2H \qquad\qquad \overset{\oplus}{(CH_3)_3 N}\!-\!CH_2\!-\!CH_2\!-\!CH_2\!-\!CO_2H$$

$$\vert\!\!-\!\!\text{— 3.6A —}\!\!\rightarrow\!\vert \qquad\qquad \vert\text{— 7.2A —}\vert$$

$$\text{— 8.7A —}$$

$$\overset{\overset{CH_3}{\vert}}{O^{\ominus}\!\overset{\oplus}{\!-\!N}\!-\!CH_2\!-\!CO_2H} \qquad\qquad \overset{\overset{CH_3}{\vert}}{O^{\ominus}\!\overset{\oplus}{\!-\!N}\!-\!CH_2\!-\!CH_2\!-\!CH_2\!-\!CO_2H}$$

$$\underset{\vert}{CH_3} \qquad\qquad\qquad \underset{\vert}{CH_3}$$

$$\vert\text{— 5.1A —}\vert \qquad\qquad \vert\text{— 7.2A —}\vert$$

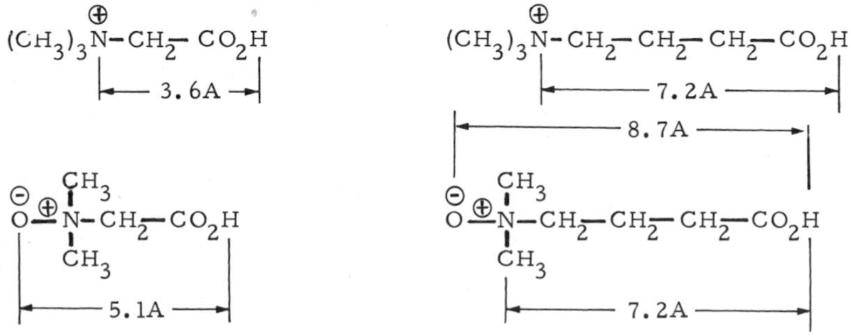

For both of the dipolar acids the $O^{\ominus}$ is assumed to be 1.5A further
away from the ionizable proton and the $N^{\oplus}$ at the same distance as for the
$(CH_3)_3 N^{\oplus}$ acids. The equation for the energy is $(e_1 e_2 / \underline{D})(1/\underline{r}_2)$, if $\underline{r}_1$ is very
large. For $(CH_3)_3 N^{\oplus}$ acids, the <u>ratios</u> of the electrostatic energies for the
short- and long-chain acids is simply $(1/3.6)/(1/7.2) = 2$. The equation is
more complicated for the dipolar acids because there is attraction between the
proton and $O^{\ominus}$ along with the repulsion between the proton and $N^{\oplus}$. The
ratios of the electrostatic energies is $(1/3.6 - 1/5.1)/(1/7.2 - 1/8.7) = 3.41$,
where the minus signs are for the attractive forces. This larger ratio shows
the simple. electrostatic theory predicts that the acid strength falls off faster
with chain length for the dipolar acids, $O^{\ominus}\!\!-\!\underset{\oplus}{N}(CH_3)_2(CH_2)_n CO_2H$, than for acids
such as $(CH_3)_3 N^{\oplus}\!\!-\!(CH_2)_n CO_2H$.

18-6   The chloro acid <u>3</u> is stronger than the acid without chlorine because of
the inductive effect of chlorine - the positive end of the C—Cl dipole being
closer to the carboxyl group than the negative end. Because of the special
structure of the chloro acid <u>4,</u> the positive end of the C—Cl dipole is <u>farther</u>
away from the carboxyl group than the negative end and <u>4</u> is thus a <u>weaker</u> acid

than 3 or the unsubstituted acid.

18-7   The effect of halogen on the acid strength of XCH_2CO_2H depends on the magnitude of the X–C dipole, which in turn is measured by the product of the electrostatic charge (q) and the distance (r) between charges ($q \cdot r$). The electrical effects of C–F bonds are less than might be expected relative to C–Cl bonds because the bond distance (r) is <u>shorter</u> for C–F than for C–Cl, even though q is larger for C–F than for C–Cl.

<u>18-8</u>   Proton transfer to the carbonyl oxygen of a carboxylic acid gives an ion in which the charge is delocalized in two equivalent valence-bond structures (Equation 18-4). If a proton is transferred to the hydroxyl oxygen no charge delocalization is possible (Equation 18-5). Thus, $RC^{\oplus}(OH)_2$ is more stable than $RC(O)O^{\oplus}H_2$.

<u>18-9</u> $CH_3CO_2H + H_2O \rightleftharpoons CH_3C(OH)_3$ $\Delta \underline{H}^{\circ} = +26$ kcal

(assuming bond energy C=O is 179 kcal and S. E. of CH_3CO_2H is 18 kcal)

$\qquad CH_3CHO + H_2O \rightleftharpoons CH_3CH(OH)_2$ $\Delta \underline{H}^{\circ} = +5$ kcal

Clearly, hydration of the acid is far less favorable than hydration of the aldehyde. However, the rate of hydration of the acid is expected to increase in the presence of sulfuric acid since protonation of the carbonyl oxygen (a rapid process) facilitates attack of water on carbon.

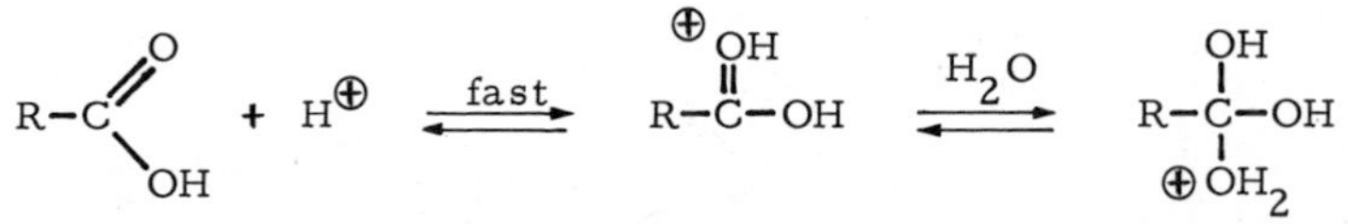

The <u>position</u> of equilibrium will be unaffected by the acid catalyst since the equilibrium constant is determined by the thermodynamic stabilities of the reactants and products and not on the reaction path by which the products are formed.

18-10   No <u>tert</u>-butyl ethanoate would be formed; the reaction product would be <u>tert</u>-butyl chloride.

$$(CH_3)_3COH + HCl \longrightarrow (CH_3)_3CCl + H_2O$$

18-11

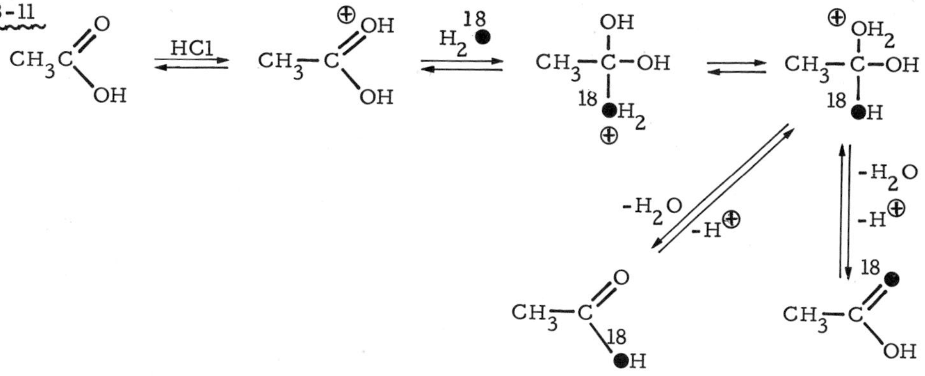

The ^{18}O label would become scrambled between water and the carbonyl and hydroxyl oxygens of the acid. Here water is acting in the same way as an alcohol molecule in esterification.

18-12 Steric congestion in the conjugate acid of 2, 4, 6-trimethylbenzoic acid is relieved in the acyl cation. The electron-donating effect of the methyl substituents stabilizes the acyl cation from 2, 4, 6-trimethylbenzoic acid. Neither steric nor electronic factors operate to favor the formation of $C_6H_5C\overset{\oplus}{=}O$ from $C_6H_5CO_2H$, and protonation of the carbonyl oxygen of benzoic acid is therefore the favored process.

18-13 If decarboxylation is a concerted, cyclic process, then 2-methyl-3-butenoic acid would give 2-butene.

18-14 Decarboxylation by a concerted, cyclic process would lead to an enol form of 3-methyl-2-butanone which would react rapidly with bromine to give 3-methyl-3-bromo-2-butanone.

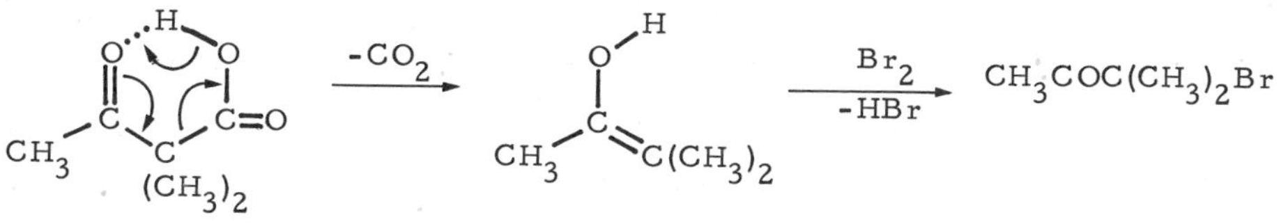

18-15

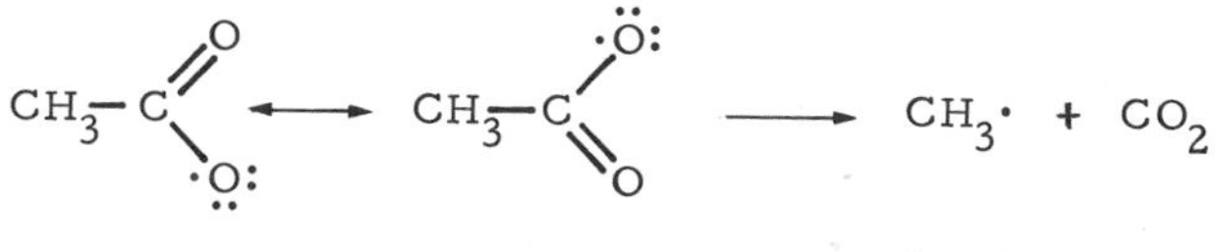

To calculate $\Delta \underline{H}$ for the above reaction, one would need to know the stabilization energy of the carboxylate radical in addition to the data given in Table 4-3.

18-16   The $\Delta \underline{H}^{\circ}$ value for the process $R \cdot + H_2O \longrightarrow RH + \cdot OH$ in the vapor phase is calculated to be +12 kcal.   Hydrocarbon formation by this path is not then energetically favorable.

18-17   Radicals formed in the Kolbe electrolysis can be oxidized to carbocations by electron transfer to the electrode.

$$RCO_2 \cdot \longrightarrow R \cdot + CO_2$$

$$R \cdot \longrightarrow R^{\oplus} + e$$

Oxidation is a competing reaction pathway to dimer formation and is (not surprisingly) significant when the carbocation $R^{\oplus}$ is relatively stable (eg. tert-butyl).   The reaction products are those expected from the interaction of the carbocations with nucleophiles.

$$R^{\oplus} + H_2O \xrightarrow{\;-H^{\oplus}\;} ROH$$

$$R^{\oplus} + RCO_2^{\ominus} \longrightarrow RCO_2R$$

18-18 $-\overset{|}{\underset{|}{Pb}}-O-\overset{O}{\overset{\|}{C}}-CH_3 + RCO_2H \longrightarrow -\overset{|}{\underset{|}{Pb}}-O-\overset{O}{\overset{\|}{C}}-R + CH_3CO_2H$

$-\overset{|}{\underset{|}{Pb}}-\overset{O}{\overset{\|}{C}}-OR \longrightarrow -\overset{|}{\underset{|}{Pb}} \cdot + \cdot \overset{O}{\overset{\|}{C}}-OR \longrightarrow CO_2 + R \cdot$

$$R \cdot + Cu(II) \longrightarrow R^{\oplus} + Cu(I)$$

$$Cu(I) + -\overset{|}{\underset{|}{Pb}} \cdot \longrightarrow Cu(II) + Pb(II)$$

If $R = HO_2C(CH_2)_6$, then $HO_2C(CH_2)_6^{\oplus} \xrightarrow{\;-H^{\oplus}\;} HO_2C(CH_2)_4CH=CH_2$.

18-19 $CH_3CH_2CO_2H + PBr_3 \longrightarrow CH_3CH_2COBr + POBr + HBr$

$\qquad CH_3CH_2COBr \longrightarrow CH_3CH=C(Br)OH$

$CH_3CH=C(Br)OH + Br_2 \longrightarrow CH_3CHBrCOBr + HBr$

$CH_3CHBrCOBr + CH_3CH_2CO_2H \longrightarrow CH_3CHBrCO_2H + CH_3CH_2COBr$

The electron-withdrawing properties of bromine has the effect of enhancing the acidity of the α-hydrogens of propanoyl bromide relative to the parent acid. The increased acidity leads to greater rate of enolization - and hence a greater rate of bromination.

18-20   Reaction at low concentrations of $\overset{\ominus}{O}H$ is a slow intramolecular S_N2 displacement of bromine by the carboxylate anion followed by rapid hydrolysis of the cyclic ester (α-lactone) intermediate.

$$CH_3-\underset{\underset{Br}{|}}{\overset{\overset{H}{|}}{C}}-\overset{O^{\ominus}}{\overset{|}{C}}=O \xrightarrow[-Br^{\ominus}]{\text{slow}} CH_3-C\overset{O}{\underset{\underset{H}{|}}{\triangleleft}}C=O \xrightarrow[H_2O]{\text{fast}} CH_3\underset{\underset{OH}{|}}{\overset{\overset{H}{|}}{C}}-CO_2^{\ominus}$$

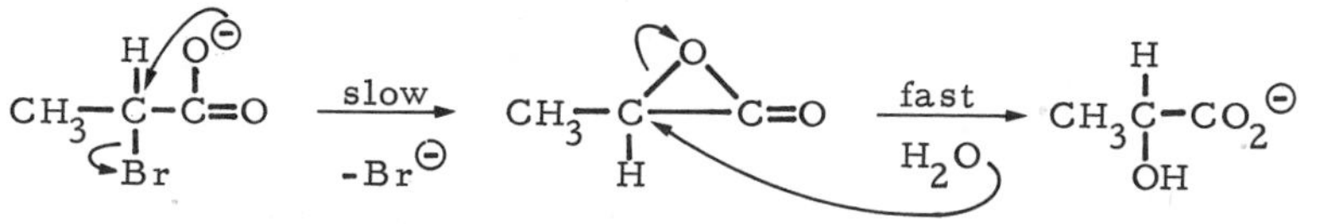

The configuration is unchanged at C2 because both reactions lead to inversion
at C2, which amounts to net retention. At high-concentrations of $^{\ominus}$OH the
rate becomes dependent on $[OH^{\ominus}]$ and the configuration at C2 is inverted
because the reaction becomes one of direct S_N2 displacement of bromine by
$^{\ominus}$OH.

$$CH_3-\underset{\underset{Br}{|}}{\overset{\overset{H}{|}}{C}}-CO_2^{\ominus} + \overset{\ominus}{OH} \longrightarrow CH_3-\underset{\underset{H}{|}}{\overset{\overset{OH}{|}}{C}}-CO_2^{\ominus} + Br^{\ominus}$$

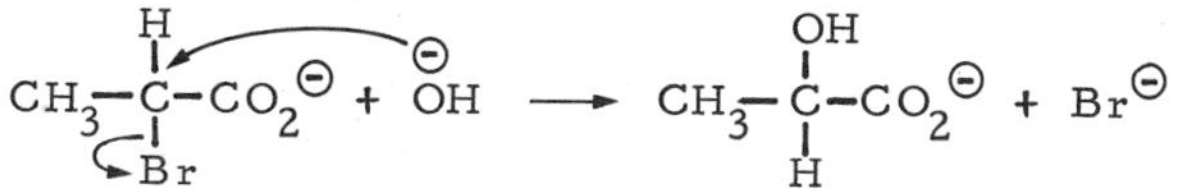

18-21 Volatility decreases in the order $CH_3CO_2CH_2CH_3 > CH_3CO_2H >$
$(CH_3CO)_2O > CH_3CONH_2$. In part, the order correlates with increasing
molecular weight. Thus ethanoic anhydride (MW 102) is less volatile than
ethanoic acid (MW 60). But the ester (MW 88) is heavier yet more volatile
than either the acid or the amide (MW 76), which means that another factor
must influence volatility besides weight. Association of molecules through
hydrogen bonding accounts for the differences, and hydrogen bonding arises when
there are polar bonds (C=O) and hydrogen bonds to electronegative elements
(N—H, O—H).

$$CH_3-C\overset{O\cdots H-O}{\underset{O-H\cdots O}{\diagup \diagdown}}C-CH_3 \qquad\qquad CH_3-C\overset{O}{\underset{\underset{H}{\overset{|}{N-H}\cdots O=C}}{\diagup}}\overset{CH_3}{\underset{\underset{H}{\overset{|}{N-H}\cdots}}{\diagdown}}$$

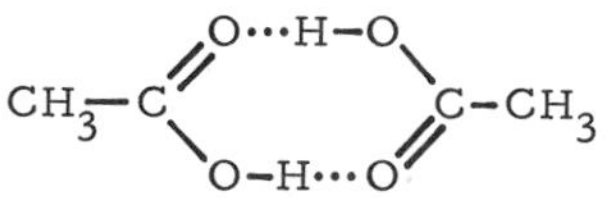

(dimer)   (polymer)

18-22 Attack of a nucleophile at carbon of a carboxylate anion is less favorable
than attack at the carbonyl carbon of a carboxylate ester because the product, in
the case of the carboxylate anion, would be a dinegative ion with loss of all of
the stabilization energy of the carboxylate anion.

$$R-CO_2^{\ominus} + CH_3O^{\ominus} \rightleftharpoons R-\underset{\underset{OCH_3}{|}}{\overset{\overset{\overset{\ominus}{O}}{||}}{C}}-O^{\ominus}$$

large SE   SE = 0

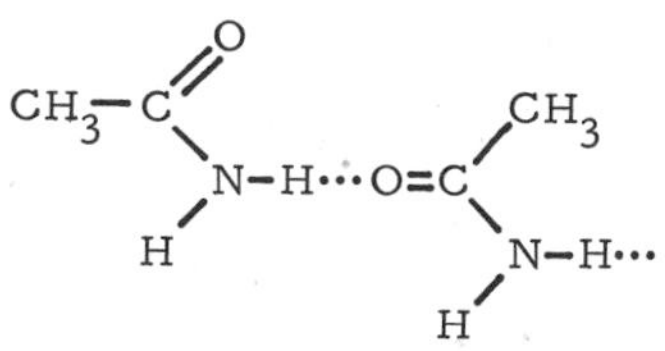

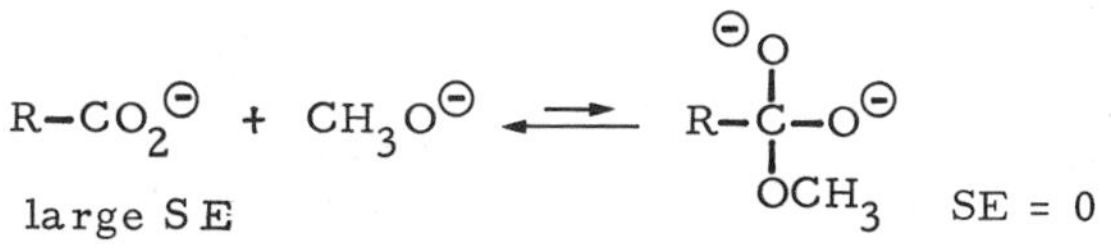

$$RCO_2CH_3 + CH_3O^{\ominus} \rightleftharpoons R-\overset{\overset{\displaystyle O^{\ominus}}{|}}{\underset{\underset{\displaystyle OCH_3}{|}}{C}}-OCH_3 \quad SE = 0$$

smaller SE

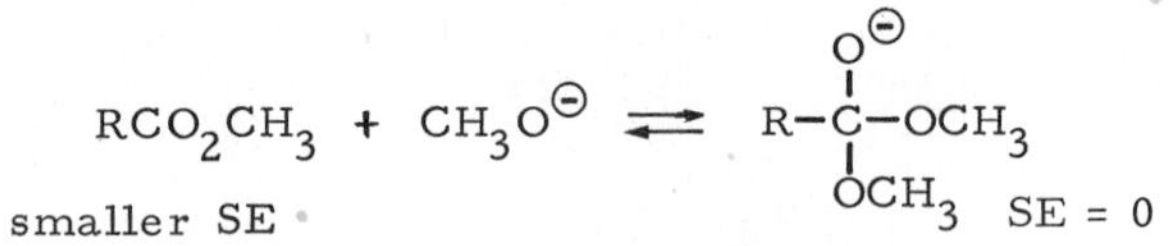

18-23   a. $C_2H_5OH + \overset{\ominus}{OR} \rightleftharpoons C_2H_5O^{\ominus} + HOR$

$$C_2H_5-\overset{\displaystyle O}{\underset{\displaystyle OCH_3}{\overset{\diagup\!\diagup}{C}}} + C_2H_5O^{\ominus} \rightleftharpoons C_2H_5-\overset{\overset{\displaystyle O^{\ominus}}{|}}{\underset{\underset{\displaystyle OCH_3}{|}}{C}}-OC_2H_5 \rightleftharpoons C_2H_5-\overset{\overset{\displaystyle O}{||}}{C}-OC_2H_5 + CH_3O^{\ominus}$$

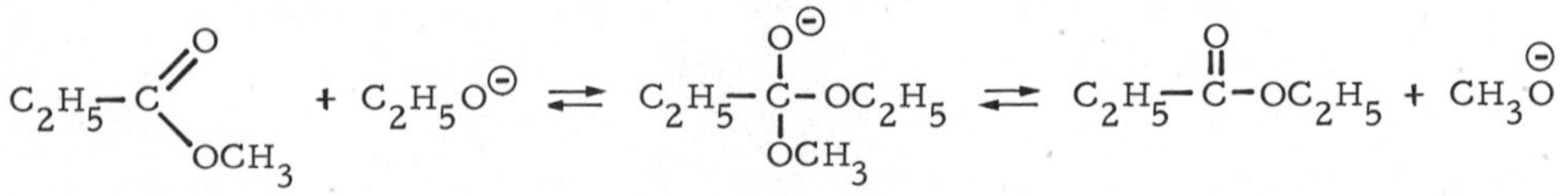

b. If, in the steps of Part a, $\overset{\ominus}{OR}$ is $\overset{\ominus}{OCH_3}$, then methoxide functions as a true catalyst in that it is consumed in one step but regenerated in another. If it is present in catalytic (i.e., small) amounts, its function is merely to convert ethanol to ethoxide ion. In excess of ethanol, it doesn't matter what base is used as a catalyst provided it converts ethanol measureably to ethoxide, and is regenerated in a subsequent step.

c. Configuration of the D-2-butyl group would be unaffected because the reaction does not involve the chiral carbon of the alcohol.

18-24   See Exercise 18-25a.

18-25
a. $C_6H_5CO_2CH_3 + H^{\oplus} \rightleftharpoons C_6H_5\overset{\overset{\displaystyle \oplus OH}{||}}{C}-OCH_3 \xrightarrow{C_2H_5OH} C_6H_5-\overset{\overset{\displaystyle OH}{|}}{\underset{\underset{\displaystyle \underset{\displaystyle H}{\oplus OC_2H_5}}{|}}{C}}-OCH_3$

$C_6H_5CO_2C_2H_5 \underset{-CH_3OH}{\overset{-H^{\oplus}}{\rightleftharpoons}} C_6H_5-\overset{\overset{\displaystyle OH}{|}}{\underset{\underset{\displaystyle OC_2H_5}{|}}{C}}\!\!\overset{\displaystyle H}{\underset{\oplus}{-OCH_3}}$

b. $CH_3COCl + C_2H_5OH \rightleftharpoons CH_3\overset{\overset{\displaystyle O^{\ominus}}{|}}{\underset{\underset{\displaystyle \underset{\displaystyle H}{\oplus OC_2H_5}}{|}}{C}}-Cl \xrightarrow{-HCl} CH_3CO_2C_2H_5$

c. $(CH_3CO)_2O + H^{\oplus} \rightleftharpoons CH_3-\overset{\overset{\displaystyle \oplus OH}{||}}{C}-OCOCH_3 \xrightarrow{CH_3OH} CH_3-\overset{\overset{\displaystyle OH}{|}}{\underset{\underset{\displaystyle \underset{\displaystyle H}{\oplus OCH_3}}{|}}{C}}-OCOCH_3$

$$CH_3CO_2CH_3 + CH_3CO_2H \xleftarrow{-H^{\oplus}}$$

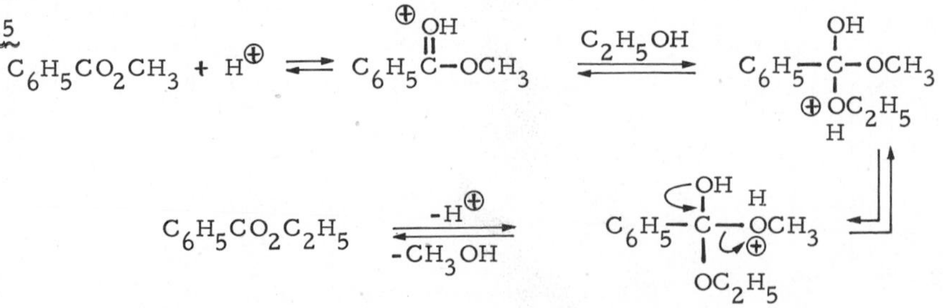

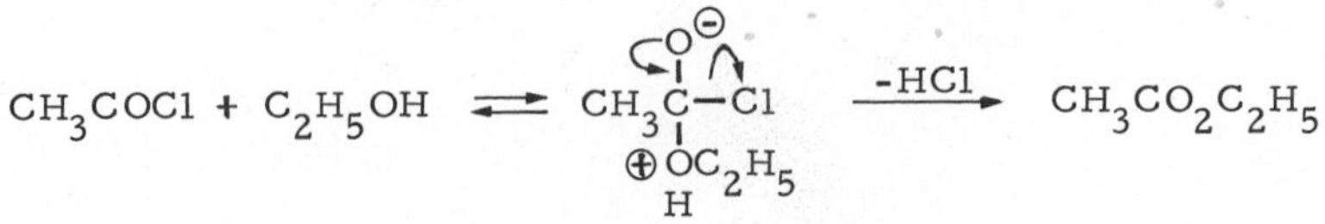

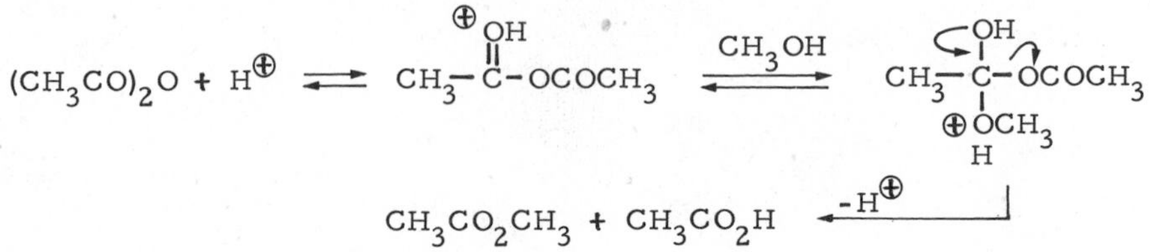

d. $CH_3CONH_2 + H^{\oplus} \rightleftharpoons CH_3\overset{\overset{\oplus}{\underset{\|}{OH}}}{C}-NH_2 \left(or\ CH_3\overset{O}{\underset{\|}{C}}-\overset{\oplus}{N}H_3 \right) \xrightarrow{H_2O}$

$CH_3CO_2H + NH_4^{\oplus} \rightleftharpoons CH_3\overset{\overset{OH}{\underset{|}{}}}{\underset{\underset{OH}{|}}{C}}-\overset{\oplus}{N}H_3 \rightleftharpoons CH_3\overset{\overset{OH}{\underset{|}{}}}{\underset{\underset{\oplus}{\overset{\oplus}{O}H_2}}{C}}-NH_2$

e. $CH_3CONH_2 + \overset{\ominus}{O}H \rightleftharpoons CH_3\overset{\overset{O}{\underset{\|}{}}}{\underset{\underset{O-H}{|}}{C}}-NH_2 \rightleftharpoons CH_3\overset{\overset{O}{\underset{\|}{}}}{\underset{\underset{O\ominus}{|}}{C}}-\overset{\oplus}{N}H_3 \longrightarrow$

$CH_3\overset{O}{\underset{\|}{C}}-\overset{\ominus}{O} + NH_3$

f. $CH_3COCl + NH_3 \rightleftharpoons CH_3\overset{\overset{\ominus}{\underset{\|}{O}}}{\underset{\underset{\oplus}{\overset{}{N}H_3}}{C}}-Cl \rightleftharpoons CH_3CONH_2 + HCl \xrightarrow{NH_3}$

$CH_3CONH_2 + NH_4Cl$

g. $CH_3CO_2CH_3 + CH_3NH_2 \rightleftharpoons CH_3\overset{\overset{O}{\underset{\|}{}}}{\underset{\underset{\oplus}{\overset{}{H_2}NCH_3}}{C}}-OCH_3 \rightleftharpoons CH_3\overset{\overset{OH}{\underset{|}{}}}{\underset{\underset{HNCH_3}{|}}{C}}-OCH_3$

$CH_3-\overset{O}{\underset{\|}{C}}-NHCH_3 + CH_3OH$

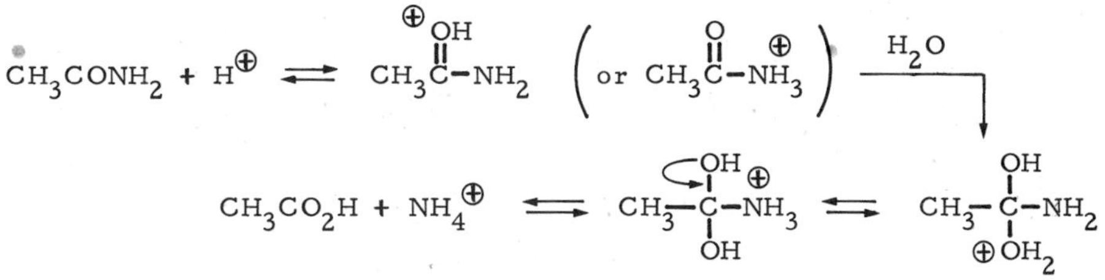

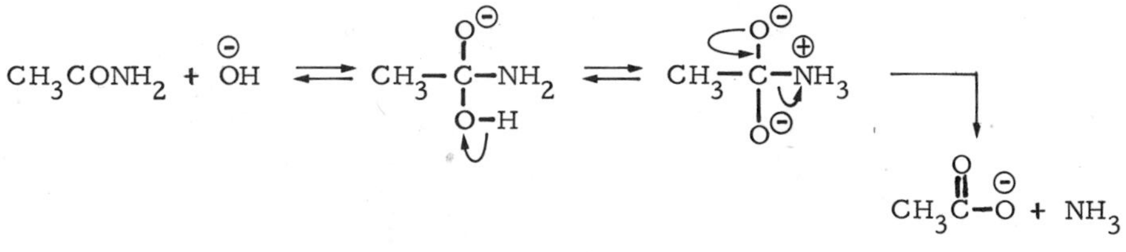

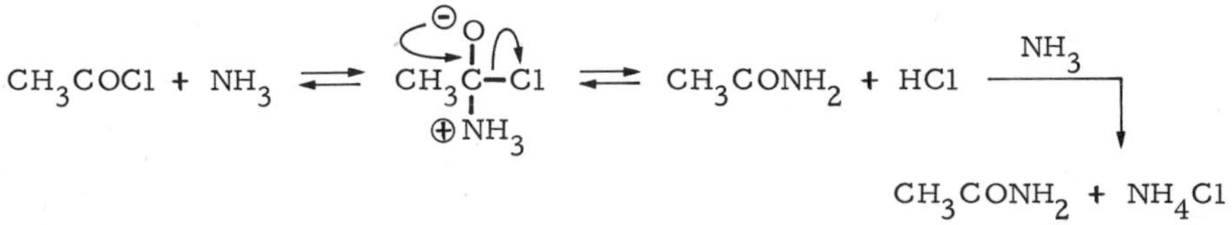

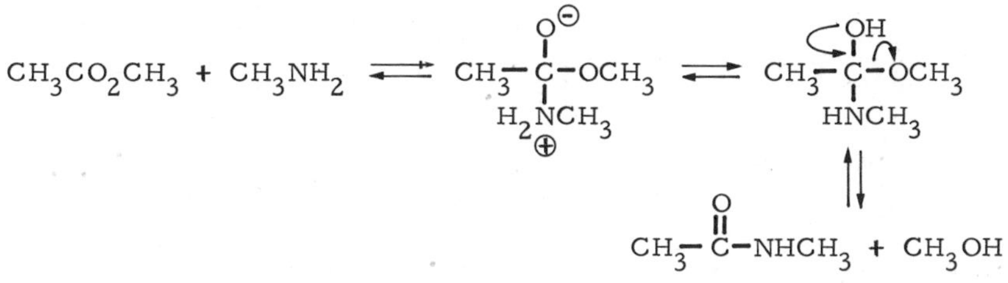

18-26   For the ^{18}O label to appear exclusively as the hydroxyl oxygen of the carboxyl group means that cleavage of the acyl-oxygen bond must have occurred and not S_N2-attack at the $-CH_2-O-$ with alkyl-oxygen bond cleavage.

$\begin{matrix} CH_2-C=O \\ | \quad\quad | \\ CH_2-O \end{matrix} \xrightarrow{H^{\oplus}} \begin{matrix} CH_2-C=O \\ | \quad\quad | \\ CH_2-\overset{\oplus}{O}H \end{matrix} \xrightarrow{H_2{}^{18}O} \begin{matrix} {}^{18}\overset{\oplus}{O}H_2 \\ | \\ CH_2-C-\overset{\ominus}{O} \\ | \quad\quad | \\ CH_2-OH \\ \oplus \end{matrix} \underset{H^{\oplus}}{\overset{-H^{\oplus}}{\rightleftharpoons}} \begin{matrix} {}^{18}OH \\ | \\ CH_2-C \\ | \quad\quad \diagdown O \\ CH_2OH \end{matrix}$

and not

$\begin{matrix} H \\ \diagup {}^{18} \\ \quad O: \\ \diagup \\ H \end{matrix} \quad \begin{matrix} CH_2-C=O \\ | \quad\quad | \\ CH_2-OH \\ \oplus \end{matrix} \xrightarrow{-H^{\oplus}} \begin{matrix} CH_2-CO_2H \\ | \\ H^{18}O-CH_2 \end{matrix}$

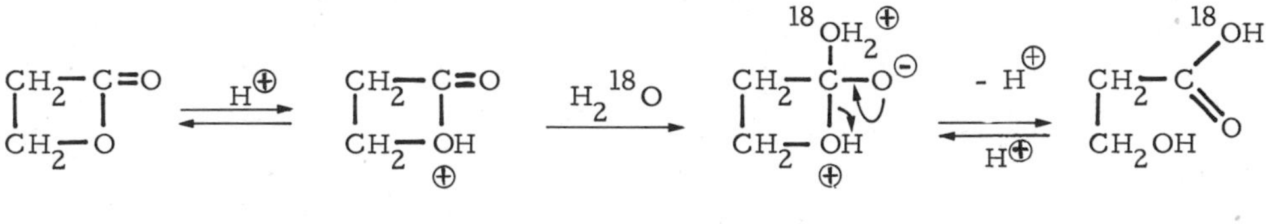

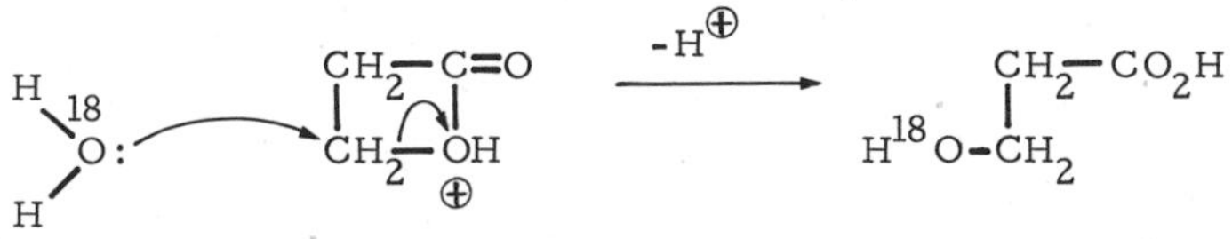

18-27

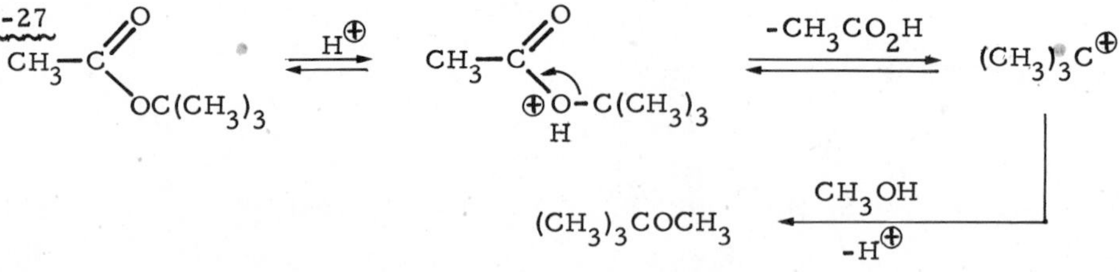

18-28   Steric hindrance to formation of the hydrolysis intermediate by attack of $^{\ominus}$OH at the carbonyl carbon is the reason for slow hydrolysis under basic conditions.   However acid-catalyzed hydrolysis is rapid by the following pathway:

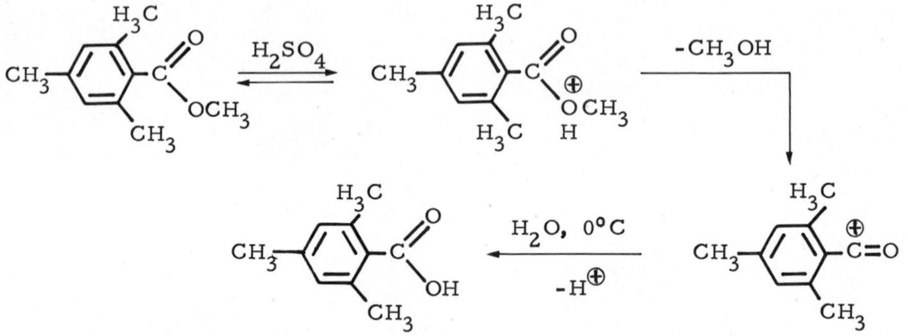

18-29   Acid derivatives react with R'MgX by the following route:

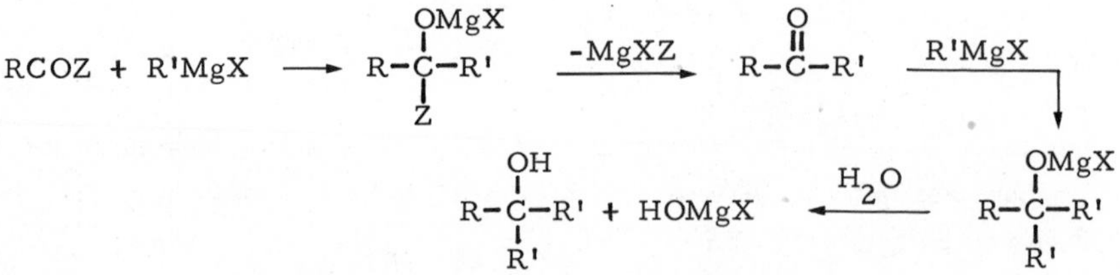

However, if the elimination step forming ketone does not proceed (as when the intermediate adduct is stable and Z is a poor leaving group) the product isolated after hydrolysis is a ketone rather than a tertiary alcohol.   In the case of amides Z is NR_2, which is a poor leaving group under almost any circumstances.

Aldehydes may be prepared from N, N-dialkylmethanamides.

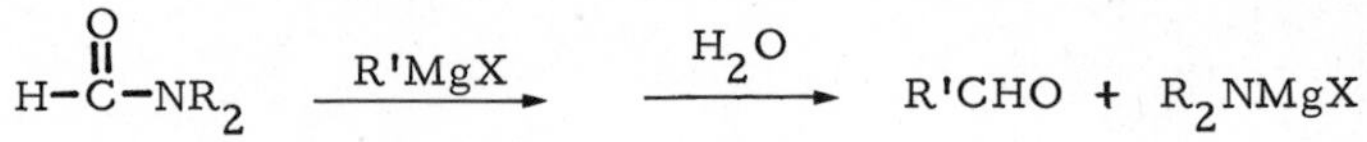

18-30   The position of a keto-enol equilibrium is determined by the relative
stability of the keto and enol forms:

$$CH_3COCH_2COX \rightleftharpoons CH_3C(OH)=CHCOX$$

Factors that increase the stability of the keto form decrease enolization. In the
case of $X=OC_2H_5$, the keto form is stabilized by the electron-delocalization of
the ester function,

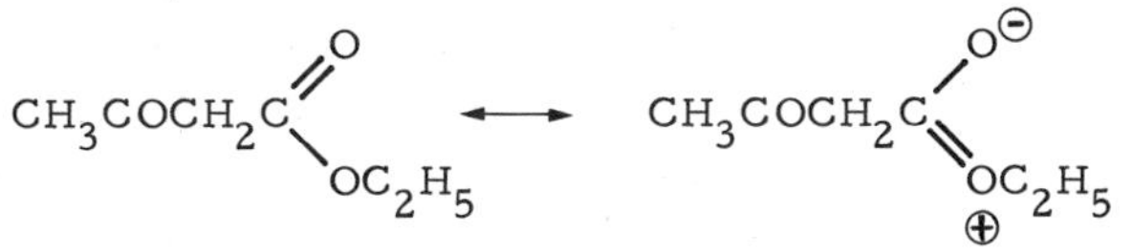

Stabilization of the enol by conjugation is lessened by the ester function because
there is "competition" for the electrons of the ester carbonyl group by the
enol $-OH$ and the ester $-OC_2H_5$ group.

        By this reasoning, one would predict very little enol in diethyl propane-
dioate, for the stabilization energy of both ester groups would be diminished on
enolization.

        On the other hand, 3-oxobutanal has nothing to lose in the form of
stabilization energy and everything to gain by enolization.

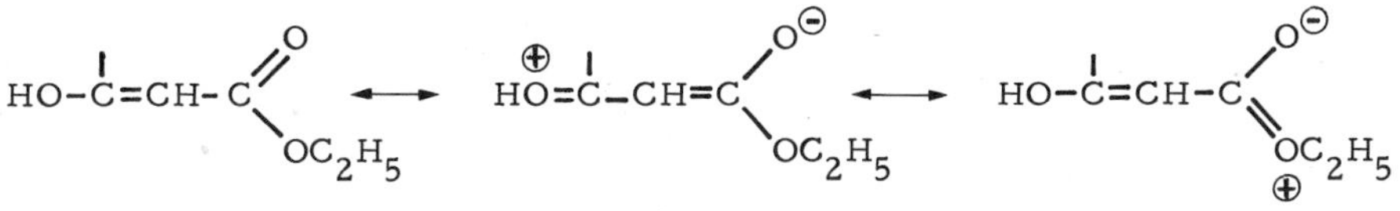

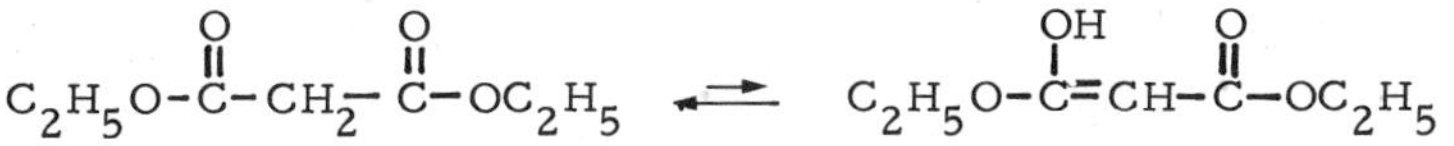

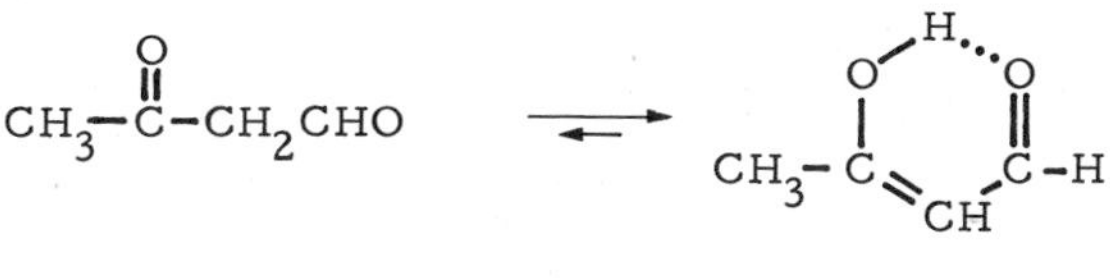

18-31

        In C- versus O-alkylation of enolate anions, it appears that O-alkyla-
tion becomes more important as the alkylating agent RX becomes more reac-

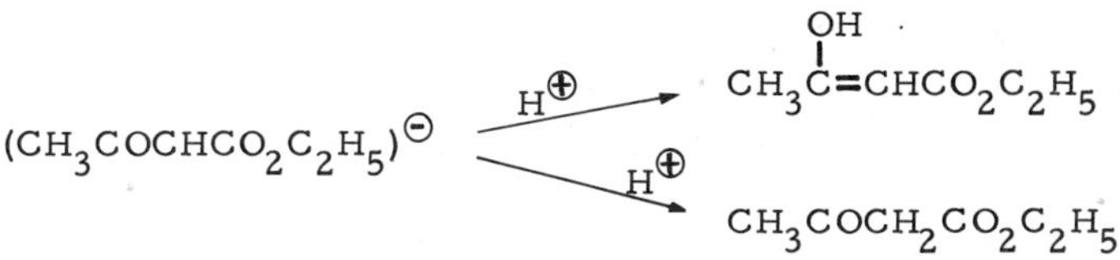

tive - reactivity depending on the ease with which the R—X bond is broken. Extrapolating this reasoning to protonation of enolate anions by <u>strong</u> acids, one would expect proton transfer from a <u>strong</u> acid would occur predominantly on oxygen to give the enol. Proton transfer to oxygen generally is much faster than proton transfer to carbon.

18-32 Peaks a, b and c in Figure 18-6 are assigned respectively to the OH, alkenyl and methyl protons of the enol form of ethyl 3-oxobutanoate. Addition of a small amount of $NaOC_2H_5$ catalyzes the interconversion of the keto and enol forms, $CH_3COCH_2CO_2C_2H_5 \rightleftharpoons CH_3C(OH)CHCO_2C_2H_5$. The interconversion is so rapid that the nmr resonances for the keto and enol forms coalesce. Thus, the a, b, c resonances disappear.

18-33

a. $2CH_3CO_2C_2H_5 \xrightarrow{NaOC_2H_5} (CH_3COCHCO_2C_2H_5)^{\ominus} Na^{\oplus} \xrightarrow{CH_3CO_2C_2H_5}$

$$CH_3COCHCO_2C_2H_5 \xleftarrow{-NaOC_2H_5} CH_3COCHCO_2C_2H_5$$
$$\underset{CH_3-\overset{|}{C}=O}{} \qquad \underset{CH_3-\overset{|}{\underset{O_{\ominus}Na^{\oplus}}{C}}-OC_2H_5}{}$$

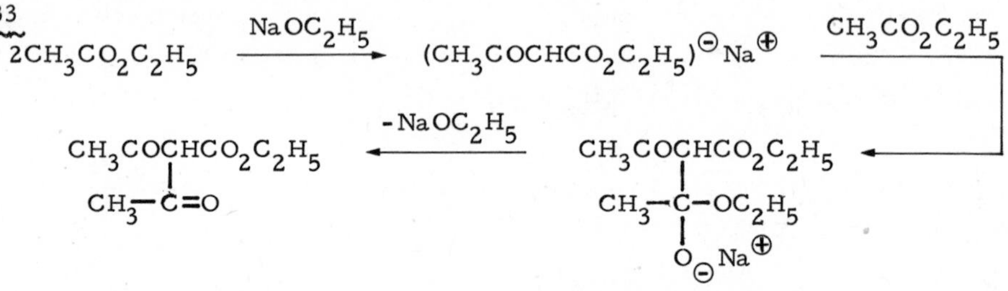

The above side reaction, which could take place in the Claisen condensation, is not likely to be important because attack of the sodium salt of $CH_3COCH_2CO_2C_2H_5$ on $CH_3CO_2C_2H_5$ is expected to be very slow owing to steric hindrance and the fact that the enolate anion $CH_3CO^{\ominus}CHCO_2C_2H_5$ is a less powerful nucleophile than the anion $^{\ominus}CH_2CO_2C_2H_5$.

b. $CH_3CO_2C_2H_5 \xrightarrow{NaOC_2H_5} \left[{}^{\ominus}CH_2C\overset{O}{\underset{OC_2H_5}{\diagup}} \longleftrightarrow CH_2{=}C\overset{O^{\ominus}}{\underset{OC_2H_5}{\diagup}} \right] \longrightarrow$

$$CH_2{=}C\overset{OCOCH_3}{\underset{OC_2H_5}{\diagup}} \xleftarrow{-OC_2H_5{}^{\ominus}} CH_2{=}C\overset{O-\overset{\ominus}{\underset{O}{\parallel}}-CH_3}{\underset{OC_2H_5}{\diagdown OC_2H_5}} \xleftarrow{CH_3CO_2C_2H_5}$$

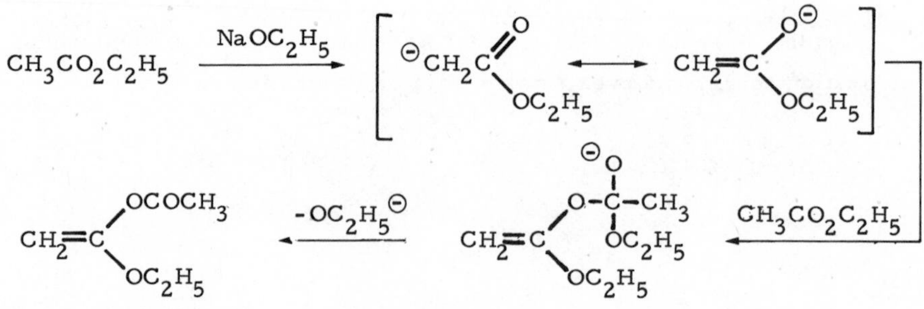

O-acylation in the Claisen condensation is not expected to be significant for the same reasons that O-alkylation of enolate anions does not occur in the aldol addition (see Section 17-3B).

18-34

1. $C_2H_5OH \rightleftharpoons C_2H_5O^\ominus + H^\oplus$ $\qquad$ $\underline{K}_1 = 10^{-18}$

2. $CH_3COCH_2CO_2C_2H_5 \rightleftharpoons CH_3CO\overset{\ominus}{C}HCO_2C_2H_5 + H^\oplus$ $\qquad$ $\underline{K}_2 = 10^{-11}$

Subtracting 1 from 2 gives:

3. $CH_3COCH_2CO_2C_2H_5 + C_2H_5O^\ominus \overset{K_3}{\rightleftharpoons} CH_3CO\overset{\ominus}{C}HCO_2C_2H_5 + C_2H_5OH$

$$\underline{K}_3 = \underline{K}_2 / \underline{K}_1 = 10^{-11}/10^{-18} = 10^7$$

$$\Delta\underline{G}^\circ = -2.303\,RT\log 10^7$$

$$= -2.303 \times 1.987 \times 298 \log 10^7 \text{ cal/mole}$$

$$= -9.5 \text{ kcal/mole}$$

18-35

a. $CH_3COCH_2CO_2C_2H_5$; $\qquad$ $CH_3CH_2CO\overset{\overset{\displaystyle CH_3}{|}}{C}HCO_2C_2H_5$

$CH_3CO\overset{\overset{\displaystyle |}{C}H}{\underset{\displaystyle CH_3}{}}CO_2C_2H_5$; $\qquad$ $CH_3CH_2COCH_2CO_2C_2H_5$

b. $CH_3COCH_2CO_2C_2H_5$

c. no Claisen reaction possible

18-36

a. $2CH_3CH_2CO_2C_2H_5 \xrightarrow{NaOC_2H_5} \left[CH_3CH_2CO\overset{}{C}(CH_3)CO_2C_2H_5\right]^\ominus \longrightarrow$

$CH_3CH_2COCH(CH_3)CO_2C_2H_5 \xleftarrow{H^\oplus}$

b. $CH_3COCH_3 \xrightarrow{NaOC_2H_5} CH_3CO\overset{\ominus}{C}H_2 \xrightarrow[-OC_2H_5^\ominus]{(CO_2C_2H_5)_2} CH_3COCH_2COCO_2C_2H_5$

c. $C_6H_5CH_2CO_2C_2H_5 \xrightarrow{NaOC_2H_5} C_6H_5\overset{\ominus}{C}HCO_2C_2H_5 \xrightarrow[-OC_2H_5^\ominus]{(C_2H_5O)_2C{=}O}$

$C_6H_5CH(CO_2C_2H_5)_2$

d. $CH_3COCH_3 \xrightarrow{NaOC_2H_5} CH_3CO\overset{\ominus}{C}H_2 \xrightarrow[-OC_2H_5^\ominus]{CH_3CO_2C_2H_5} CH_3COCH_2COCH_3$

e. $(CH_3)_3CCOCH_3$ $\xrightarrow[\text{2. } H^{\oplus}, \ C_2H_5OH]{\text{1. } Br_2, \ NaOH}$ $(CH_3)_3CCO_2C_2H_5$

$(CH_3)_3CCOCH_3$ $\xrightarrow{NaOC_2H_5}$ $(CH_3)_3CCOCH_2^{\ominus}$

$(CH_3)_3CCOCH_2^{\ominus}$ + $(CH_3)_3CCO_2C_2H_5$ $\xrightarrow{-OC_2H_5^{\ominus}}$ $(CH_3)_3CCOCH_2COC(CH_3)_3$

f. $(CH_3)_2CHCO_2C_2H_5$ $\xrightarrow{(C_6H_5)_3\overset{\oplus}{C} \ Na}$ $(CH_3)_2\overset{\ominus}{C}CO_2C_2H_5$ $\xrightarrow[\text{$-OC_2H_5^{\ominus}$}]{C_6H_5CO_2C_2H_5}$

$\begin{array}{c} (CH_3)_2C-CO_2C_2H_5 \\ | \\ COC_6H_5 \end{array}$ $\longleftarrow$

18-37 Disadvantages to using sodium hydride are:

  a. hydrogen is evolved

  b. insolubility of NaH in organic solvents

  c. unsuitable for use in hydroxylic (or protic) solvents.

Advantages are:

  a. readily available

  b. stronger base than $NaOC_2H_5$

  c. no unwanted products (like C_2H_5OH from $NaOC_2H_5$).

18-38 In principle, the reaction could go as written. However, under the particular conditions, the fastest reaction will be of $NaOC_2H_5$ with CH_3I to give $CH_3OCH_2CH_3$.

18-39
$CH_3COCH_2CO_2C_2H_5$ $\xrightarrow[\text{2. } CH_3CH_2I]{\text{1. } NaOC_2H_5}$ $\begin{array}{c} CH_3COCHCO_2C_2H_5 \\ | \\ CH_2CH_3 \end{array}$ $\xrightarrow[\text{2. } CH_3CH_2I]{\text{1. } NaOC_2H_5}$

$CH_3COCH(CH_2CH_3)_2$ $\xleftarrow[\text{2. heat, } -CO_2]{\text{1. } H^{\oplus}, \ H_2O}$ $\begin{array}{c} CH_2CH_3 \\ | \\ CH_3COCCO_2C_2H_5 \\ | \\ CH_2CH_3 \end{array}$ $\longleftarrow$

An alternative synthesis of 3-ethyl-2-pentanone by ethylation of 2-pentanone in the presence of sodium amide would be unsatisfactory because attack of $NaNH_2$ would occur mainly at the CH_3 group and thus give 3-heptanone. Also E2 is more likely with $NaNH_2$ than with the enolate anion.

18-40

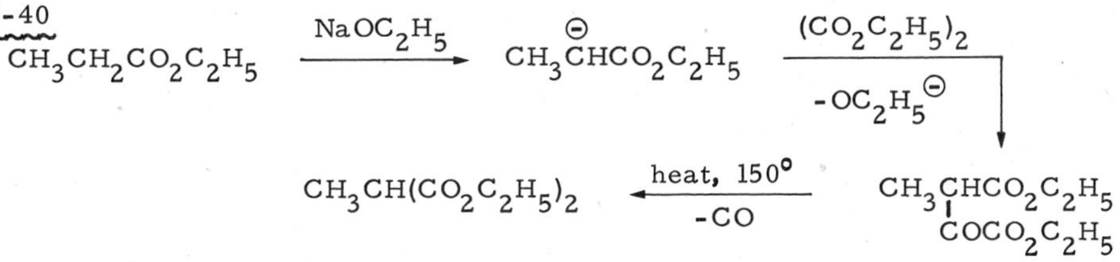

$$CH_3CH_2CO_2C_2H_5 \xrightarrow{NaOC_2H_5} CH_3\overset{\ominus}{C}HCO_2C_2H_5 \xrightarrow[-OC_2H_5^{\ominus}]{(CO_2C_2H_5)_2}$$

$$CH_3CH(CO_2C_2H_5)_2 \xleftarrow[-CO]{heat,\ 150^\circ} \underset{\underset{COCO_2C_2H_5}{|}}{CH_3CHCO_2C_2H_5}$$

18-41

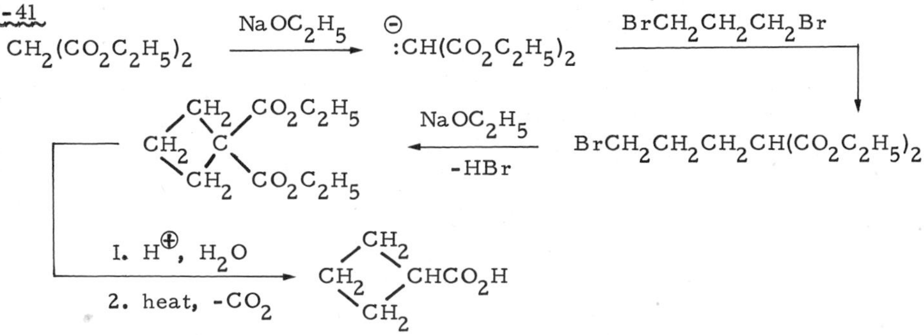

$$CH_2(CO_2C_2H_5)_2 \xrightarrow{NaOC_2H_5} {:}\overset{\ominus}{C}H(CO_2C_2H_5)_2 \xrightarrow{BrCH_2CH_2CH_2Br}$$

$$\xleftarrow[-HBr]{NaOC_2H_5} BrCH_2CH_2CH_2CH(CO_2C_2H_5)_2$$

$$\begin{array}{l} 1.\ H^{\oplus},\ H_2O \\ 2.\ heat,\ -CO_2 \end{array}$$

18-42

reverse Claisen {

hydrolysis {

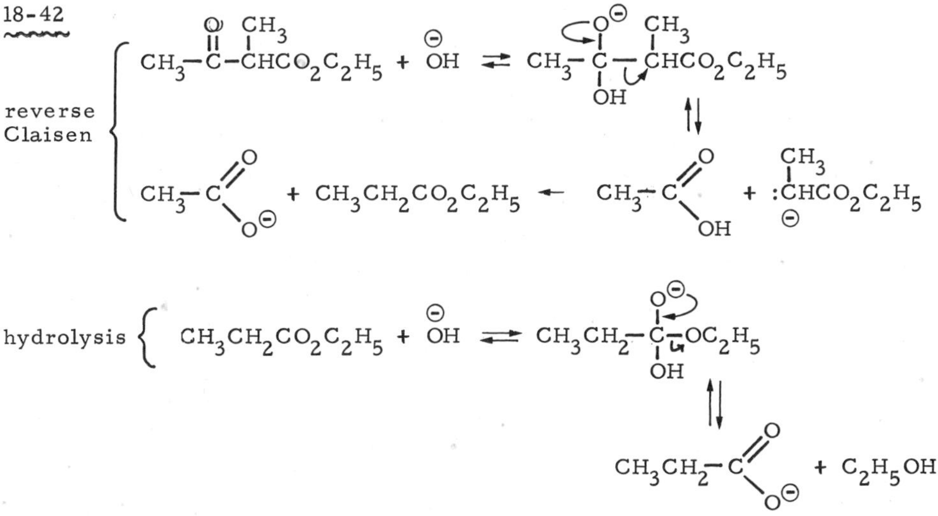

18-43 a. The ester must be added to the base if a Claisen condensation of the ester anion with neutral ester is to be avoided. By adding the ester to the base, the ester is converted almost completely to the anion, and the amount of neutral ester is kept to a minimum.

b. At equilibrium the possible Claisen and **aldol** products (after hydrolysis) are $HOC(CH_3)_2CH_2CO_2CH_3$, $CH_3COCH_2CO_2C_2H_5$, $CH_3COCH_2COCH_3$, and $(CH_3)_2C(OH)CH_2COCH_3$. The major product at equilibrium in excess base

would correspond to the strongest acid (weaker conjugate base) which is
$CH_3COCH_2COCH_3$.

c.

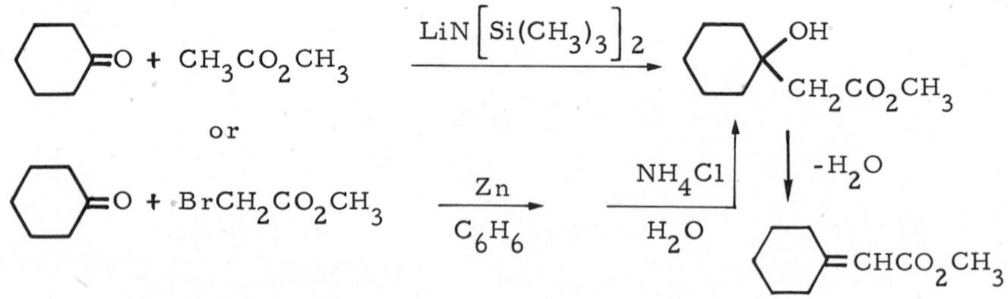

18-44

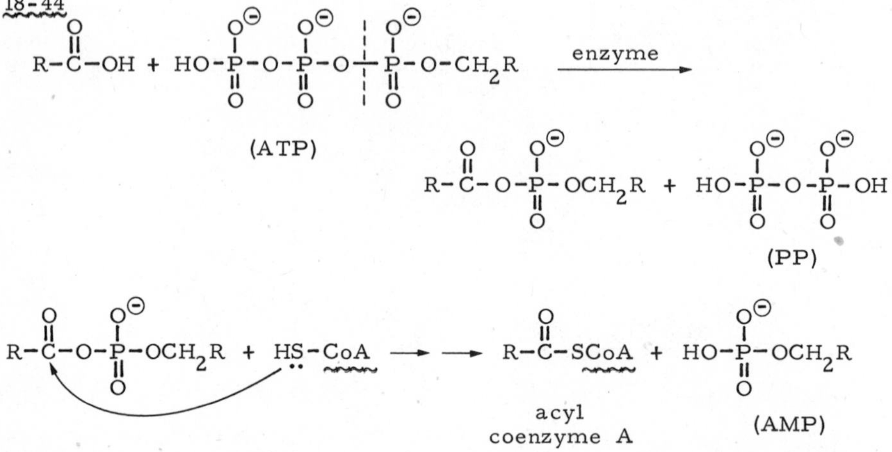

(ATP)

(PP)

acyl
coenzyme A   (AMP)

18-45

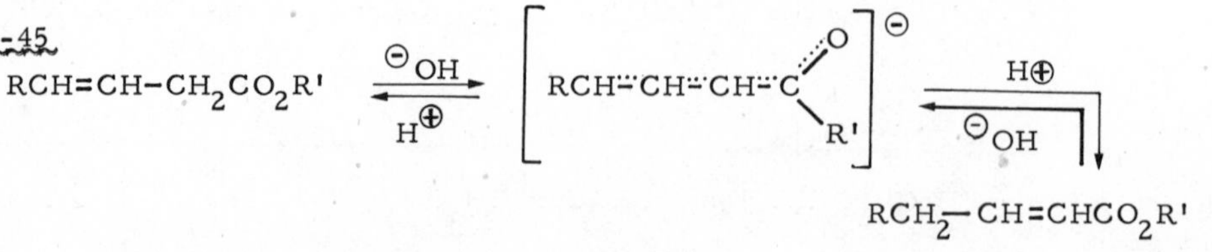

The α,β-unsaturated ester, in which the carbonyl group and double bond
are conjugated, is expected to predominate.

With the corresponding acids, carboxylate anions are formed under basic
conditions. Equilibration between the α,β- and β,γ-unsaturated anions is much
less facile than with the esters because of the difficulty in removing a proton
from a negatively charged ion. Rearrangement of a γ,δ- to an α,β-unsatura-
ted ester does not form a stabilized anion of the type $\left[-\overset{|}{C}\overset{..}{=}\overset{|}{C}\overset{..}{=}\overset{|}{C}\overset{..}{=}\overset{|}{C}\overset{..}{=}O\right]^{\ominus}$, which
is necessary if rearrangement is to take place.

18-46 Lactone formation by 3-butenoic acid in the presence of sulfuric acid by

Markownikoff addition would lead to a strained four-membered ring.

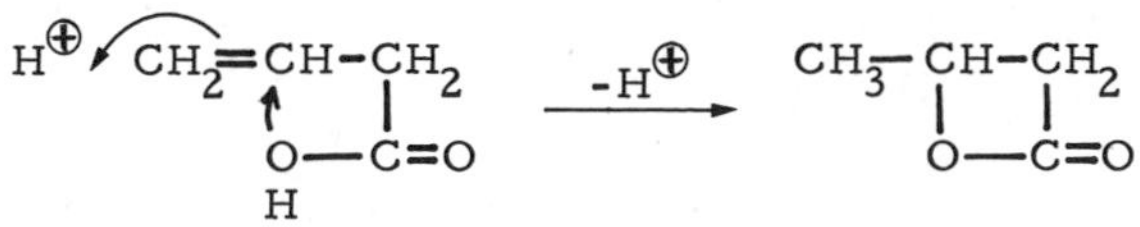

Actually, the anti-Markownikoff-product oxacyclopentan-2-one (γ-butyrolactone) is the product.

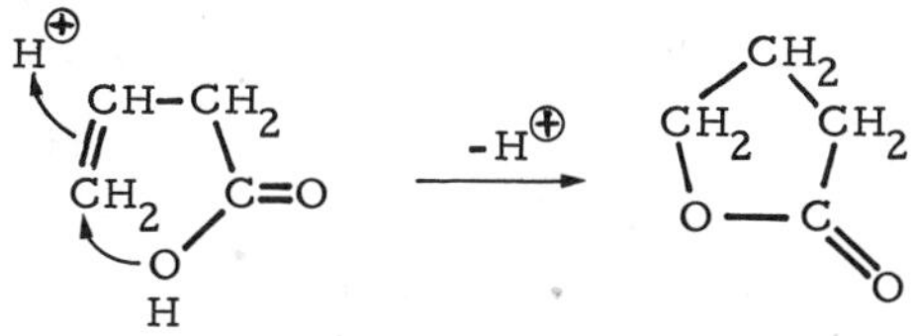

<u>18-47</u> Michael addition is unlikely because, under the basic reaction conditions, the 3-phenylpropenoate (cinnamate) anion would be formed. Attack of a carbanion at the β carbon would give a relatively poorly stabilized dianion.

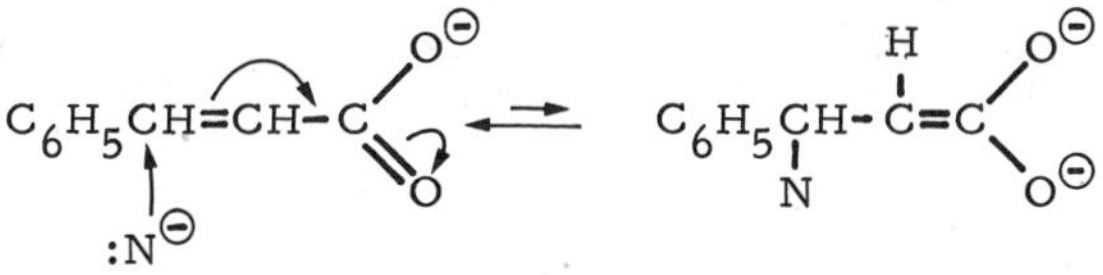

<u>18-48</u>

$$CH_3CO_2C_2H_5 \xrightarrow{NaOC_2H_5} {}^{\ominus}:CH_2CO_2C_2H_5$$

$$C_6H_5CH=C(CO_2C_2H_5)_2 + {}^{\ominus}:CH_2CO_2C_2H_5 \longrightarrow$$

$$\underset{\underset{CH_2CO_2C_2H_5}{|}}{C_6H_5CHCH(CO_2C_2H_5)_2} \xleftarrow{H^{\oplus}} \underset{\underset{CH_2CO_2C_2H_5}{|}}{C_6H_5CH-\overset{..}{\overset{\ominus}{C}}(CO_2C_2H_5)_2}$$

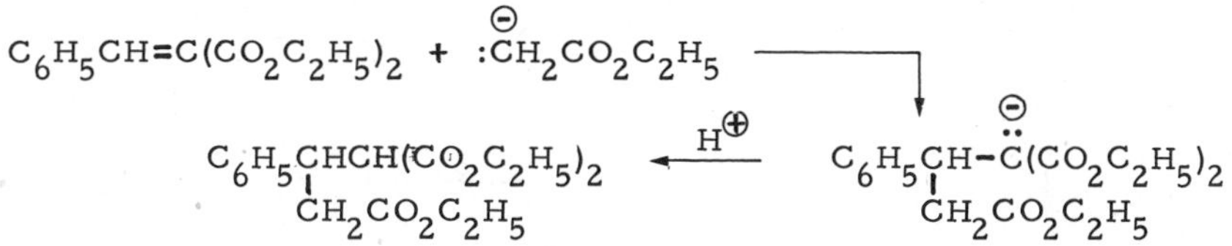

Claisen condensation of the ethyl ethanoate is likely to be a side reaction but more important will be addition of $NaOC_2H_5$ to the unsaturated ester to give the considerably stabilized propanedioate (malonate) anion.

$$C_6H_5-CH=C(CO_2C_2H_5)_2 \underset{\longleftarrow}{\overset{NaOC_2H_5}{\longrightarrow}} \underset{\underset{OC_2H_5}{|}}{C_6H_5-CH-\overset{\ominus}{\overset{..}{C}}(CO_2C_2H_5)_2}$$

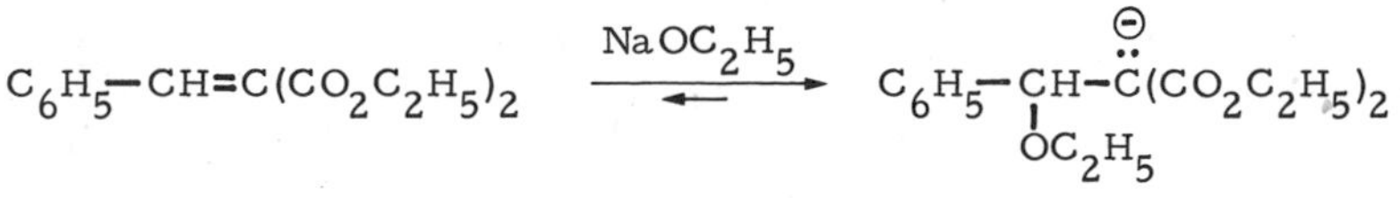

This reaction is much less likely with ethyl 3-phenylpropenoate in competition with formation of diethyl propanedioate (malonate) anion.

$$C_6H_5CH=CHCO_2C_2H_5 \xrightleftharpoons{NaOC_2H_5} C_6H_5\overset{..}{\underset{OC_2H_5}{C}H-\overset{\ominus}{C}H-CO_2C_2H_5}$$

18-49

a. $C_6H_5CH=CHCO_2C_2H_5$ + $CH_2(CO_2C_2H_5)_2$ $\xrightarrow{NaOH}$

$$C_6H_5\underset{CH_2CO_2H}{CHCH_2CO_2H} \underset{\text{2. heat, } -CO_2}{\overset{\text{1. } H^{\oplus}, H_2O}{\longleftarrow}} C_6H_5\underset{CH_2(CO_2C_2H_5)_2}{CHCH_2CO_2C_2H_5}$$

b. $C_6H_5COCH=CHC_6H_5$ + $CH_2(CO_2C_2H_5)CN$ $\xrightarrow{NaOH}$

$$C_6H_5COCH_2\underset{}{\overset{C_6H_5}{C}HCH_2CN} \underset{\text{2. heat}}{\overset{\text{1. } H^{\oplus}, H_2O}{\longleftarrow}} C_6H_5COCH_2\overset{C_6H_5}{CHCH(CO_2C_2H_5)CN}$$

c. $CH_2=CHCN$ + $CH_2(CO_2C_2H_5)_2$ $\xrightarrow{NaOC_2H_5}$ $\underset{CH(CO_2C_2H_5)_2}{CH_2CH_2CN}$

$$NCCH_2CH_2\underset{CO_2C_2H_5}{\overset{CO_2C_2H_5}{C}CH_2CH_2CN} \xleftarrow{NaOC_2H_5, \ CH_2=CHCN}$$

d.

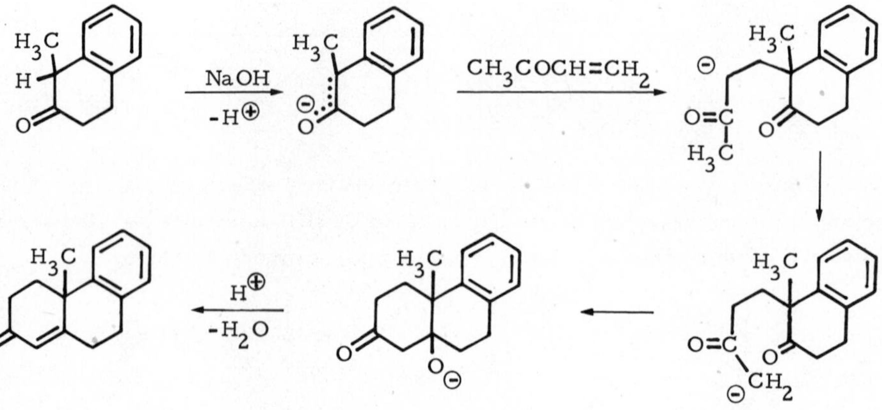

This sequence is a combination of a Michael and an aldol addition.   (See Section 30-4C.)

18-50   The proton-transfer product from the addition of N-(1-cyclohexenyl) azacyclopentane to methyl 2-methylpropenoate is <u>not</u> that shown below because steric hindrance between the substituents at the two ends of the double bond prevent its formation.

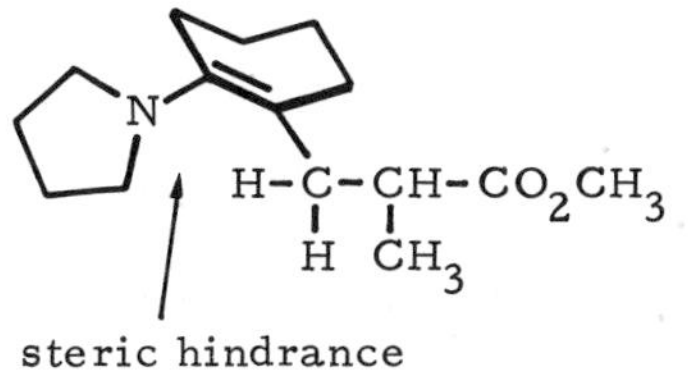

steric hindrance

In the actual product, the two ring substituents do not interfere sterically.

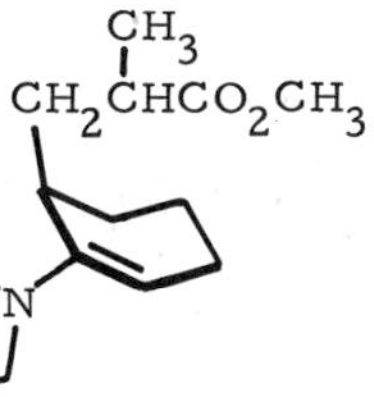

18-51

cis

trans

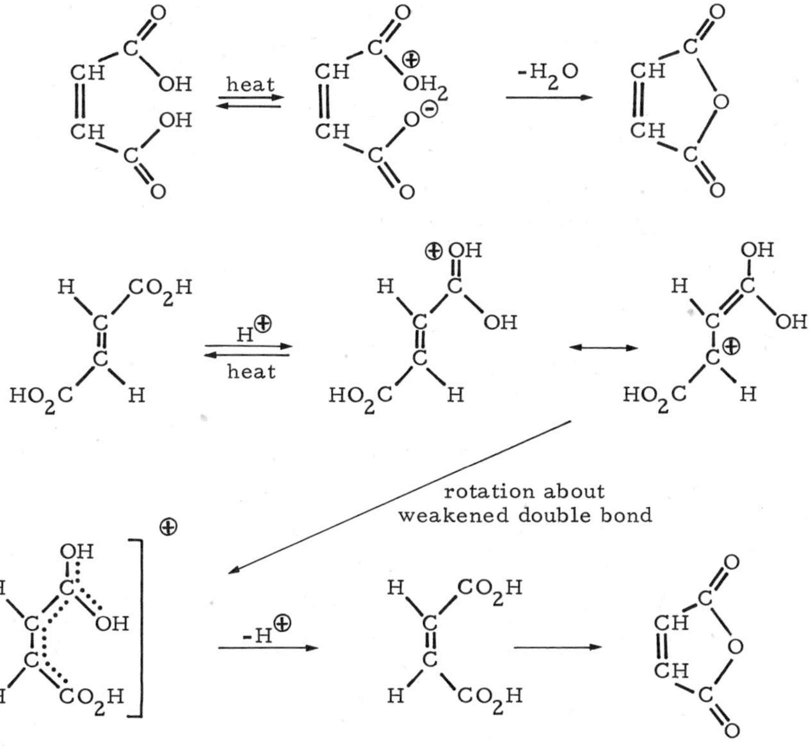

rotation about
weakened double bond

For the <u>trans</u> acid to form the anhydride it must first be isomerized to the <u>cis</u> acid, and this isomerization requires strong heating and is self-catalyzed by <u>trans</u>-butenedioic acid acting as an acid.

18-52   The following are illustrative solutions to these syntheses problems.

a. $CH_3CH_2CH_2OH$ $\xrightarrow{SOCl_2}$ $CH_3CH_2CH_2Cl$ $\xrightarrow[\text{ether}]{Mg}$ $CH_3CH_2CH_2MgCl$

$CH_3CH_2CH_2CO_2H$ $\xleftarrow[\text{2. } H^{\oplus}]{\text{1. } CO_2}$

b. $(CH_3)_3CCl$ $\xrightarrow[\text{ether}]{Mg}$ $(CH_3)_3CMgCl$ $\xrightarrow[\text{2. } H^{\oplus}]{\text{1. } CO_2}$ $(CH_3)_3CCO_2H$

c. $(CH_3)_2C=CH_2$ $\xrightarrow[\text{2. } H_2O_2]{\text{1. } B_2H_6}$ $(CH_3)_2CHCH_2OH$ $\xrightarrow{KMnO_4}$ $(CH_3)_2CHCO_2H$

d. $(CH_3)_3CCl$ $\xrightarrow[\text{ether}]{Mg}$ $(CH_3)_3CMgCl$ $\xrightarrow[\text{2. } H^{\oplus}]{\text{1. } \overset{O}{\overset{\diagup\diagdown}{CH_2-CH_2}}}$ $(CH_3)_3CCH_2CH_2OH$ —

$(CH_3)_3CCH(Br)CO_2H$ $\xleftarrow{P,\ Br_2}$ $(CH_3)_3CCH_2CO_2H$ $\xleftarrow{KMnO_4}$

e.   ⟨⟩$-CO_2H$ $\xrightarrow{Ag^{\oplus}}$ ⟨⟩$-CO_2Ag$ $\xrightarrow[-CO_2]{Br_2}$ ⟨⟩$-Br$ $\xrightarrow[\text{ether}]{Mg}$

⟨⟩$\overset{14}{}CH_2OH$ $\xleftarrow{LiAlH_4}$ ⟨⟩$\overset{14}{}CO_2H$ $\xleftarrow{^{14}CO_2}$ ⟨⟩$-MgBr$

(from $Ba^{14}CO_3$ and $H^{\oplus}$)

f. $CH_2{=}CHCH_2Cl$ + $Na\overset{\oplus\ \ominus}{C}H(CO_2C_2H_5)_2$ $\longrightarrow$ $CH_2{=}CHCH_2CH(CO_2C_2H_5)_2$ —

$CH_2{=}CHCH_2CH_2CONH_2$ $\xleftarrow[\text{2. } NH_3]{\text{1. } SOCl_2}$ $CH_2{=}CHCH_2CH_2CO_2H$ $\xleftarrow[\text{2. heat}]{\text{1. } H^{\oplus},\ H_2O}$

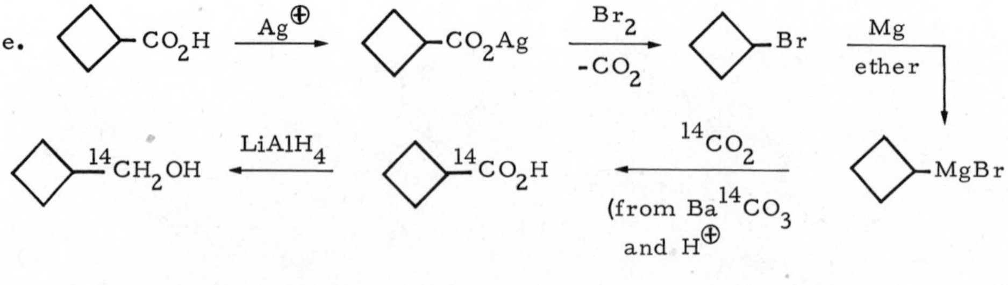

g. $(CH_3)_3CCl$ $\xrightarrow[\text{2. } CO_2]{\text{1. Mg}}$ $(CH_3)_3CCO_2H$ $\xrightarrow{LiAlH_4}$ $(CH_3)_3CCH_2OH$

$(CH_3)_3CCO_2H$ $\xrightarrow{SOCl_2}$ $(CH_3)_3CCOCl$ $\xrightarrow{(CH_3)_3CCH_2OH}$

$(CH_3)_3CCO_2CH_2C(CH_3)_3$

<u>18-53</u>

a.

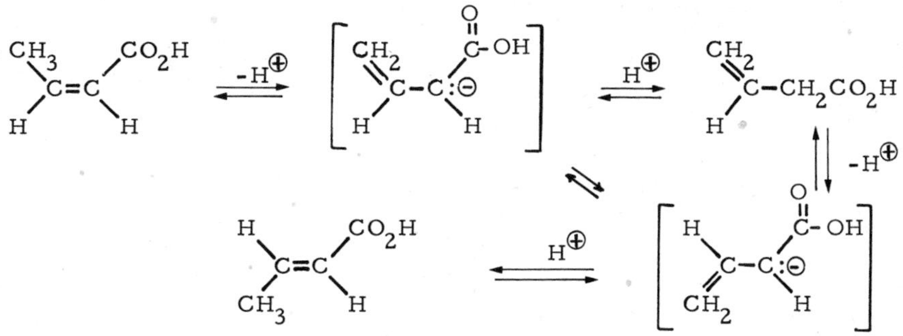

b. $CH_2{=}CHCH_2CH_2CO_2H$ $\xrightarrow{HOBr}$

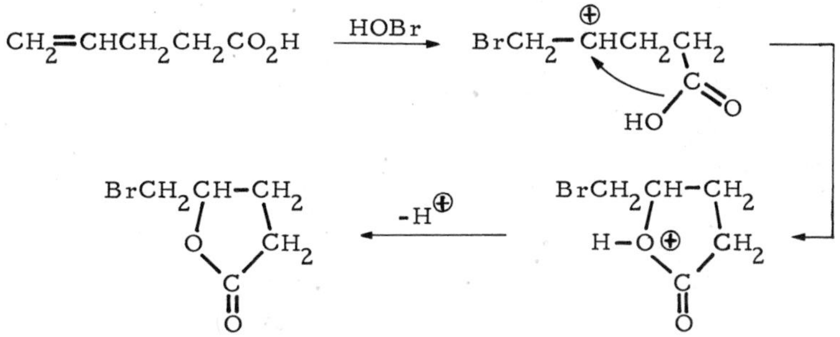

c. The order of reactivity with different R groups is consistent with acyl-oxygen not alkyl-oxygen fission.

Thus,

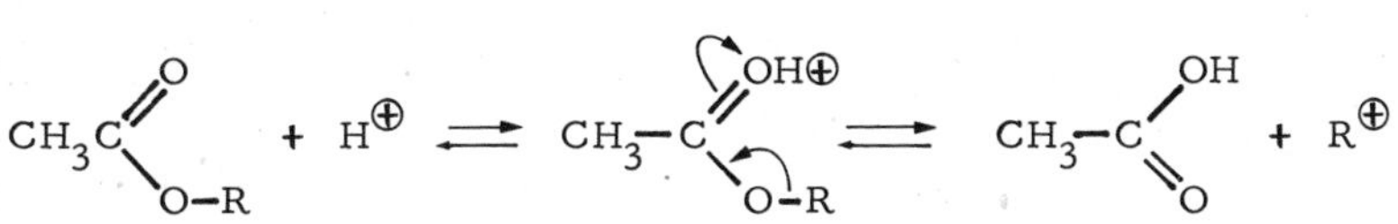

is <u>not</u> the correct mechanism.

More likely is

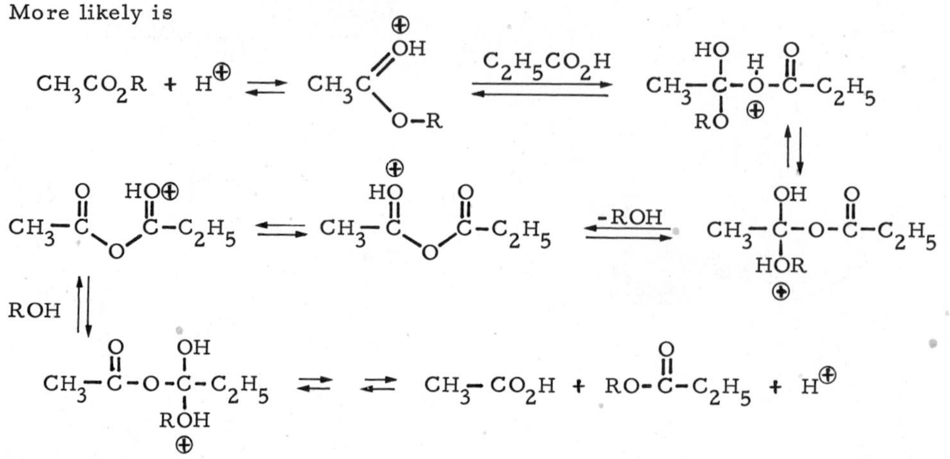

$$CH_3CO_2R + H^\oplus \rightleftharpoons CH_3C\overset{\overset{\oplus}{OH}}{\underset{O-R}{\diagup}} \xrightarrow{C_2H_5CO_2H} CH_3-\overset{HO}{\underset{RO}{C}}-\overset{H}{\underset{\oplus}{O}}-\overset{O}{C}-C_2H_5$$

$$CH_3-\overset{O}{\overset{\|}{C}}\diagdown_{O}\diagup\overset{HO^\oplus}{\overset{\|}{C}}-C_2H_5 \rightleftharpoons CH_3-\overset{\overset{\oplus}{HO}}{\overset{}{C}}\diagdown_{O}\diagup\overset{O}{\overset{\|}{C}}-C_2H_5 \xleftarrow{-ROH} CH_3-\overset{OH}{\underset{HOR}{\underset{\oplus}{C}}}-O-\overset{O}{\overset{\|}{C}}-C_2H_5$$

$$ROH \updownarrow$$

$$CH_3-\overset{O}{\overset{\|}{C}}-O-\overset{OH}{\underset{\underset{\oplus}{ROH}}{\underset{|}{C}}}-C_2H_5 \rightleftharpoons \rightleftharpoons CH_3-CO_2H + RO-\overset{O}{\overset{\|}{C}}-C_2H_5 + H^\oplus$$

The slowness of the reaction with bulky R groups is probably due to steric hindrance.

18-54   4-Bromobicyclo[2.2.2]octane-1-carboxylic acid (A) is a stronger acid than 5-bromopentanoic acid (B) possibly for two reasons. First, A has three chains over which the inductive electron-withdrawing influence of the bromine can be transmitted compared to one for B.

Br —◇— CO$_2$H   Br —⋀— CO$_2$H

(A) three paths   (B) one path

Second, rotation about the C-C bonds of B may permit assumption of conformers whereby the field effect of the bromine has an acid-weakening effect. This, of course, is not possible for A which has a rigid arrangement of carbons.

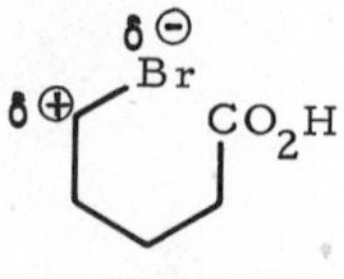

$$\overset{\delta\ominus}{\underset{\delta\oplus}{Br}} \diagup CO_2H$$

18-55

a. 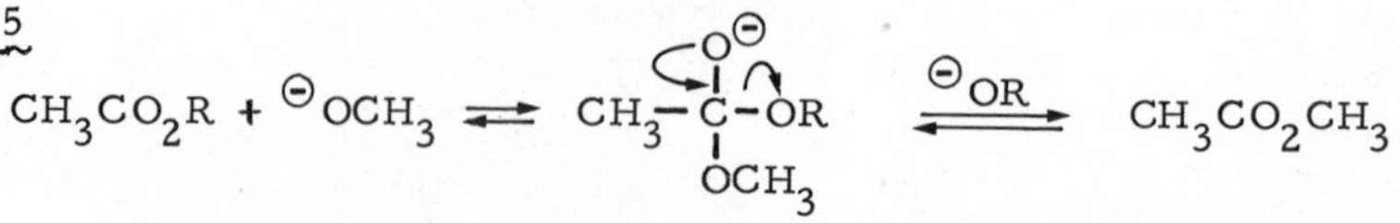 $CH_3CO_2R + {}^\ominus OCH_3 \rightleftharpoons CH_3-\overset{O^\ominus}{\underset{OCH_3}{\overset{|}{C}}}-OR \xrightarrow{{}^\ominus OR} CH_3CO_2CH_3$ (1)

When R is tert-butyl, the first step in the above reaction is extremely slow owing to steric hindrance offered by the <u>tert</u>-butyl group. When R is ethyl, hindrance is less severe and the reaction rate is faster.

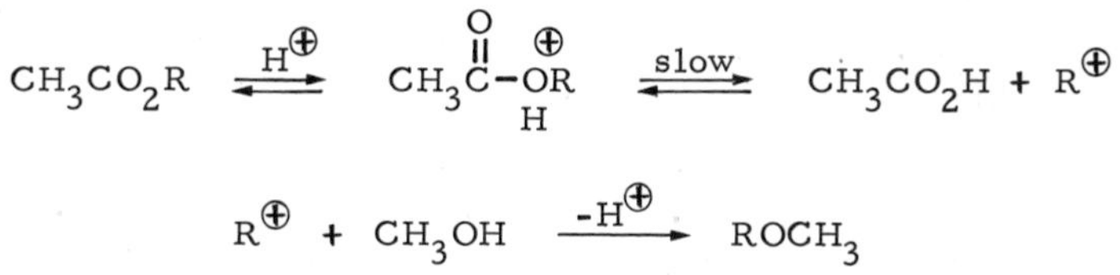

$$CH_3CO_2R \xrightarrow{H^{\oplus}} CH_3\overset{O}{\overset{\|}{C}}-\overset{\oplus}{O}R \xrightarrow{slow} CH_3CO_2H + R^{\oplus}$$

$$R^{\oplus} + CH_3OH \xrightarrow{-H^{\oplus}} ROCH_3 \qquad\qquad (2)$$

The above reaction proceeds at a rapid rate when R is <u>tert</u>-butyl since this group forms a relatively stable carbocation, $R^{\oplus}$. When R is ethyl, formation of an ethyl carbocation is very unfavorable and reaction takes a different course.

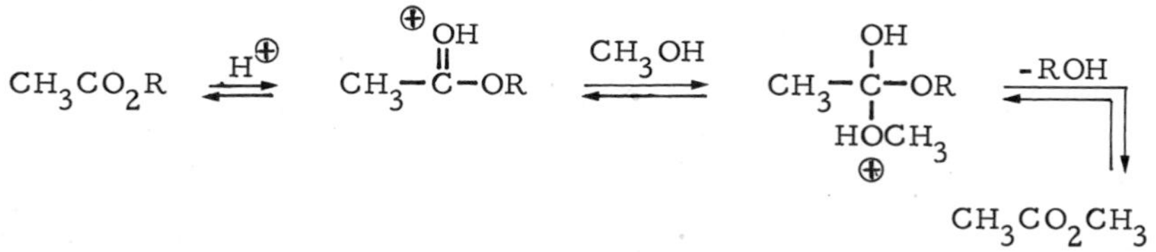

$$CH_3CO_2R \xrightarrow{H^{\oplus}} CH_3-\overset{\overset{\oplus}{O}H}{\underset{}{C}}-OR \xrightarrow{CH_3OH} CH_3-\overset{OH}{\underset{\overset{\oplus}{HOCH_3}}{C}}-OR \xrightarrow{-ROH} $$

$$CH_3CO_2CH_3 \qquad (3)$$

b.   Using ^{18}O-labeled methanol, the label would be found exclusively at the methoxyl oxygen of methyl ethanoate by the reactions of Equations (1) and (3), and entirely at the ether oxygen by Equation (2).

<u>18-56</u>

$$R-C\overset{O}{\underset{OR'}{}} + {}^{18}OH^{\ominus} \rightleftharpoons R-\overset{O^{\ominus}}{\underset{{}^{18}OH}{C}}-OR' \xrightarrow{H^{\oplus}} R-\overset{OH}{\underset{{}^{18}OH}{C}}-OR' \quad \underset{\sim}{1}$$

$$\xrightarrow{-R'OH} \quad\quad \xrightarrow{-H_2O}$$

$$R-C\overset{{}^{18}O}{\underset{OH}{}} + R-C\overset{O}{\underset{{}^{18}OH}{}} \qquad R-C\overset{{}^{18}O}{\underset{OR'}{}}$$

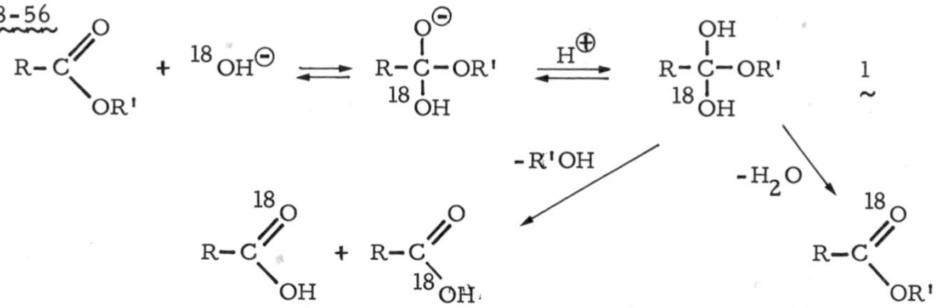

The fact that an ester, RCO_2R', is converted to $R-C\overset{{}^{18}O}{\underset{OR'}{}}$ competitively with hydrolysis in $H_2^{18}O + {}^{18}OH^{\ominus}$, means that the neutral ester intermediate $\underset{\sim}{1}$ loses $^{\ominus}OH$ in competition with loss of $^{\ominus}OR$.

<u>18-57</u>

a. $CH_3CO_2H \xrightarrow{Br_2, \ P} BrCH_2COBr$

$$CH_3CH_2OH \xrightarrow[\text{heat}]{\text{conc. } H_2SO_4} CH_2{=}CH_2 \xrightarrow{HOCl} HOCH_2CH_2Cl$$

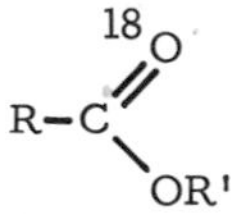

18-57 a. (cont.)

$$BrCH_2COBr + HOCH_2CH_2Cl \longrightarrow BrCH_2CO_2CH_2CH_2Cl$$

b. $(CH_3)_2CHCO_2H \xrightarrow{Br_2, P} (CH_3)_2C(Br)COBr \xrightarrow{{}^{\ominus}OCH_3}$

$$(CH_3)_2C(OCH_3)CONH_2 \xleftarrow{NH_3} (CH_3)_2C(OCH_3)CO_2CH_3$$

c. $(CH_3)_3CCH_2C(CH_3)=CH_2 \xrightarrow[\text{2. Zn, }H_2O]{\text{1. }O_3} (CH_3)_3CCH_2COCH_3$

$$(CH_3)_3CCH_2\underset{OH}{C}(CH_3)CH_2CH_3 \xleftarrow[\text{2. }H^{\oplus}]{\text{1. }CH_3CH_2MgBr}$$

d. $(CH_3)_3CCO_2H \xrightarrow{LiAlH_4} (CH_3)_3CCH_2OH \xrightarrow{HBr} (CH_3)_3CCH_2Br$

$$(CH_3)_3CCH_2CHO \xleftarrow[\text{2. }H^{\oplus}, H_2O]{\text{1. LiAlH}_4 \text{ (inverse addition)}} (CH_3)_3CCH_2CN \xleftarrow{NaCN}$$

e. $(CH_3)_2C=C(CH_3)_2 \xrightarrow[\text{pH 7}]{KMnO_4} \underset{HO\ \ OH}{CH_3-\overset{H_3C}{\underset{|}{C}}-\overset{CH_3}{\underset{|}{C}}-CH_3} \xrightarrow{H^{\oplus}} \underset{CH_3}{CH_3-\overset{H_3C}{\underset{|}{C}}-\overset{O}{\overset{||}{C}}-CH_3}$

$$\underset{H_3C\ \ OH}{CH_3-\overset{H_3C}{\underset{|}{C}}-\overset{CH_3}{\underset{|}{C}}-CH_3} \xleftarrow[\text{2. }H^{\oplus}, H_2O]{\text{1. }CH_3MgBr}$$

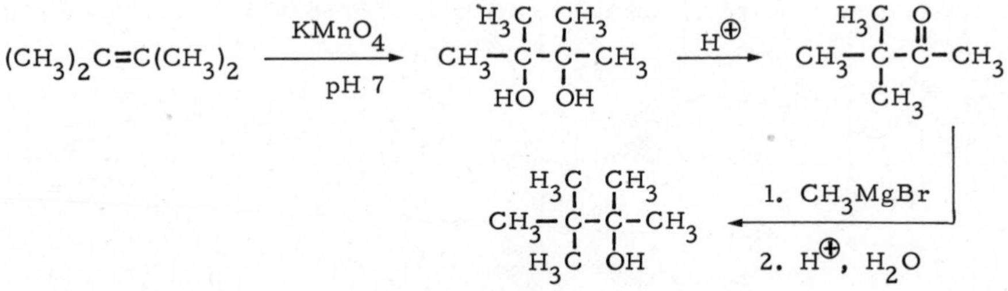

f.

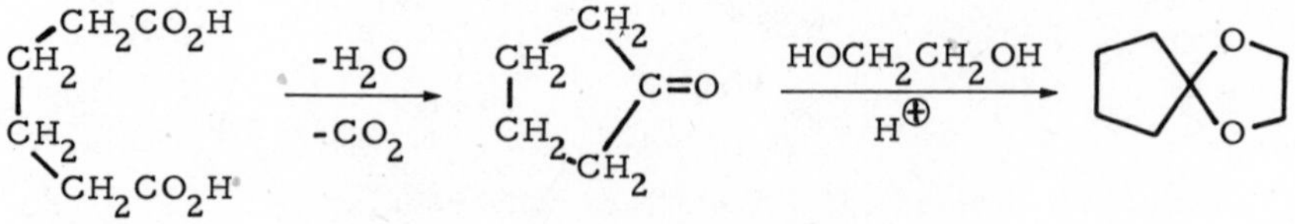

18-58 a. Heating methanoic acid with thionyl chloride will give carbon monoxide. Ethanoic acid will be converted to ethanoyl chloride.

$$HCO_2H + SOCl_2 \longrightarrow [HCOCl] + HCl + SO_2 \longrightarrow CO + 2HCl + SO_2$$

$$CH_3CO_2H + SOCl_2 \longrightarrow CH_3COCl + SO_2 + HCl$$

b. Treatment with cold, dilute base will result in solution of the acidic compound.

$$CH_3OCH_2CO_2H + NaOH \longrightarrow CH_3OCH_2CO_2Na + H_2O$$

c. Test for unsaturation with $KMnO_4$ or Br_2 in water.

$$CH_2=CHCO_2H \xrightarrow{KMnO_4} \underset{\underset{OH\ \ \ OH}{|\ \ \ \ |}}{CH_2-CH-CO_2H}$$

d. Cold ethanol will give ethyl ethanoate with ethanoyl bromide but will not react with α-bromoethanoic acid.

$$CH_3COBr + C_2H_5OH \longrightarrow CH_3CO_2C_2H_5 + HBr$$

e. Nucleophilic reagents (e.g. $^{\ominus}OCH_3$) react more rapidly with α-halo esters than with primary alkyl halides.

$$CH_3CH_2CHBrCO_2CH_3 \xrightarrow[^{\ominus}OCH_3]{fast} CH_3CH_2CH(OCH_3)CO_2CH_3$$

f. Hydrolysis in water solution will form white crystals of butanedioic acid whereas ethanoyl anhydride will give a solution of ethanoic acid.

g. The cis acid will readily form an anhydride on heating, whereas the trans acid will not do so under the same conditions.

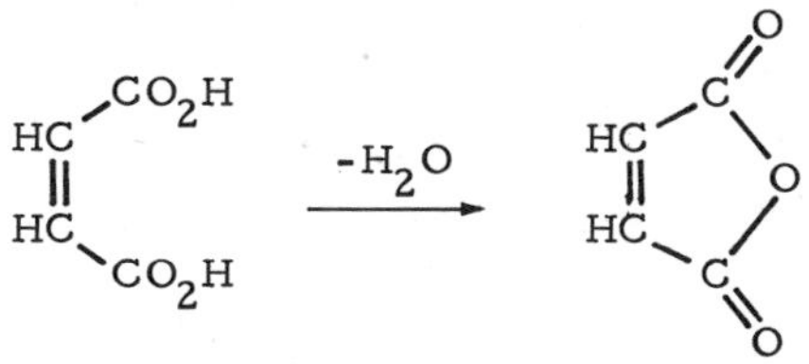

h. Precipitation of the ethyne derivative as a silver alkynide:

$$CH\equiv CCO_2CH_3 \xrightarrow{Ag^{\oplus}(NH_3)_2} AgC\equiv CCO_2CH_3$$

i. Alkali generates ammonia in the cold from the ammonium salt, but not from the amide.

$$CH_3CO_2NH_4 + NaOH \longrightarrow CH_3CO_2Na + H_2O + NH_3$$

j.   Lactone formation will occur readily with the γ,δ-unsaturated acid in the presence of strong acid:

$$CH_2{=}CHCH_2CH_2CO_2H \xrightarrow{H_2SO_4}$$

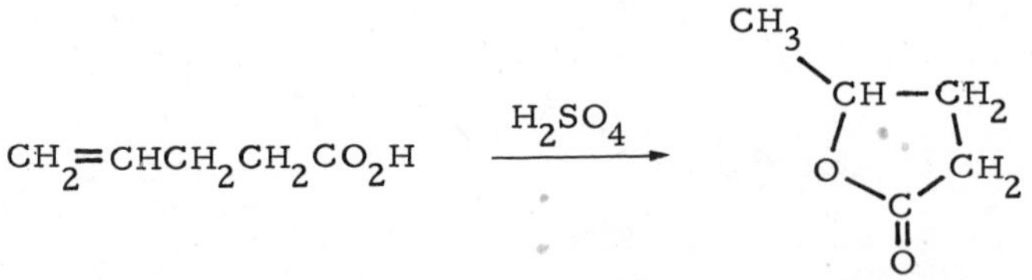

k.   The anhydride will react on heating with ethanol to give ethanoic acid and ethyl ethanoate.

$$(CH_3CO)_2O + CH_3CH_2OH \longrightarrow CH_3CO_2CH_2CH_3 + CH_3CO_2H$$

18-59   a.   nmr (CH_3- group)

b.   nmr (C_2H_5- group) and ir   (OH absorption of carboxyl group)

c.   nmr (vinyl protons) and ir   (C=C stretch)

d.   nmr (CH_3- and $-CO_2H$ protons) and ir   (OH absorption of carboxyl group)

e.   nmr ($BrCH_2-$ and $-CHBr-$ groups)

f.   nmr (CH_3CH_2- and $-CH_2CH_2-$ groups) and ir   (C=O of cyclic anhydride)

g.   ir (and ^{13}C nmr, HC = CH carbons 4 ppm <u>upfield</u> in cis acid compared to trans acid because of a steric effect. p. 460; also, H—C—C—H coupling in trans acid is <u>larger</u>, but must be measured in an unusual way, <u>cf</u> p. 320 and Exercises 9-32 and 9-33)

h.   nmr (HC≡ and $H_2C{=}CH-$) and ir   (C≡C and C=C stretch)

i.   nmr ($\overset{\oplus}{NH_4}$ vs. $-NH_2$) and ir   (H-bonded NH stretch)

j.   nmr ($CH_2{=}CH$ <u>vs</u>. $-CH{=}CH-$) and ir   ($CH_2{=}C$ stretch and stretching and bending vibration frequencies of $H_2C{=}$)

k.   nmr (C_2H_5 group), ir   (anhydride carbonyl)

18-60   Effect of <u>heat</u> on isomeric hydroxypentanoic acids:

$$2CH_3CH_2CH_2CH(OH)CO_2H \xrightarrow{-2H_2O}$$

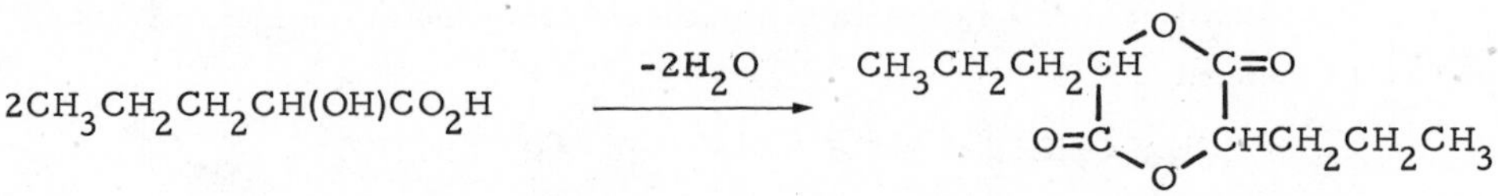

$$CH_3CH_2CH(OH)CH_2CO_2H \xrightarrow{-H_2O} CH_3CH_2CH{=}CHCO_2H$$

18-60 (cont.)

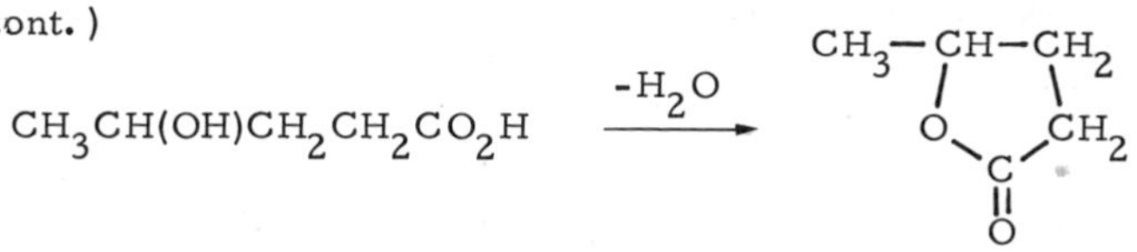

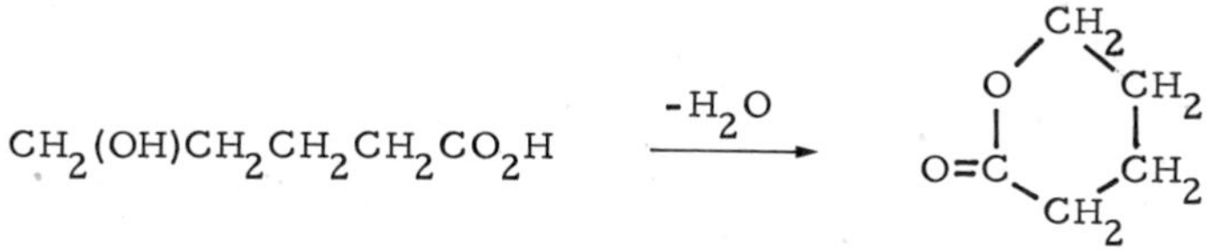

$CH_3CH(OH)CH_2CH_2CO_2H$ $\xrightarrow{-H_2O}$

$CH_2(OH)CH_2CH_2CH_2CO_2H$ $\xrightarrow{-H_2O}$

18-61 Compound A $CH_3\overset{*}{C}H(OH)CH_2CO_2H$ (C is chiral carbon)

Compound B $CH_3CH=CHCO_2H$

Compound C CH_3COCH_3

$CH_3CH(OH)CH_2CO_2H$ $\xrightarrow[-H_2O]{heat}$ $CH_3CH=CHCO_2H$

$CH_3CH(OH)CH_2CO_2H$ $\xrightarrow{CrO_3,\ H^{\oplus}}$ $CH_3COCH_2CO_2H$ $\xrightarrow{-CO_2}$

$CH_3CO_2H + CHI_3$ $\xleftarrow{NaOH,\ I_2}$ CH_3COCH_3

iodoform
(yellow)

Compound A is uniquely defined by the above description. No other hydroxybutanoic acid could be optically active and dehydrate on heating to an α,β-unsaturated acid (compound B).

18-62   a.   1,1-dicarbomethoxycyclohexane

b.   ethyl 2-isopropyl-3-oxo-butanoate

c.   N,N-dimethyl-2,2-dimethyl-3-oxopentanamide

d.   ethyl 3-hydroxypropenoate

e.   diethyl 2-oxobutanedioate

f.   diethyl 3-carbethoxy-2-oxopentanedioate

g.   methyl 1-methylcyclopentanecarboxylate

h.   2(2-propenyl)-4-pentenoic acid

i.   4,7-dimethyl-3-octanone

j.   2-propyl-4-pentenoic acid

   k.   diethyl 2-ethanoylpropanedioate

   l.   ethyl 3-hydroxyl-5-phenyl-4-pentenoate

18-63

a. $ClCH_2(CH_2)_3CH_2Cl + CH_2(CO_2CH_3)_2 \xrightarrow[-2HCl]{NaOCH_3}$

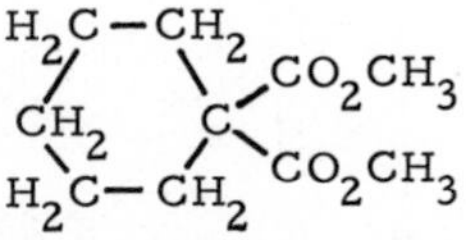

b. $CH_3COCH_2CO_2C_2H_5 + (CH_3)_2CHBr \xrightarrow[-HBr]{NaOC_2H_5} CH_3COCHCO_2C_2H_5$
$\underset{\quad\quad\quad\quad\quad\quad CH(CH_3)_2}{}$

c. $2CH_3CH_2CO_2C_2H_5 \xrightarrow{NaOC_2H_5} CH_3CH_2CO-CH-CO_2C_2H_5 \xrightarrow[2.\ CH_3I]{1.\ NaOC_2H_5}$
$\underset{\quad\quad\quad\quad\quad\quad\quad\quad\quad CH_3}{}$

$CH_3CH_2COC(CH_3)_2CON(CH_3)_2 \xleftarrow{HN(CH_3)_2} CH_3CH_2CO-\overset{\overset{CH_3}{|}}{\underset{\underset{CH_3}{|}}{C}}-CO_2C_2H_5 \longleftarrow$

d. $CH_3CO_2C_2H_5 \xrightarrow{NaOC_2H_5} {}^{\ominus}{:}CH_2CO_2C_2H_5 \xrightarrow[-OC_2H_5^{\ominus}]{HCO_2C_2H_5}$

$HOCH{=}CHCO_2C_2H_5 \rightleftharpoons H\overset{O}{\overset{||}{C}}CH_2CO_2C_2H_5$

e. $CH_3CO_2C_2H_5 \xrightarrow{NaOC_2H_5} {}^{\ominus}{:}CH_2CO_2C_2H_5 \xrightarrow[-OC_2H_5^{\ominus}]{(CO_2C_2H_5)_2}$

$C_2H_5OCOCOCH_2CO_2C_2H_5$

f. $(CH_2)_2(CO_2C_2H_5)_2 + C_2H_5O_2CCO_2C_2H_5 \xrightarrow{NaOC_2H_5}$

$C_2H_5OCOCOCHCO_2C_2H_5$
$\underset{\quad\quad\quad CH_2CO_2C_2H_5}{}$

g.

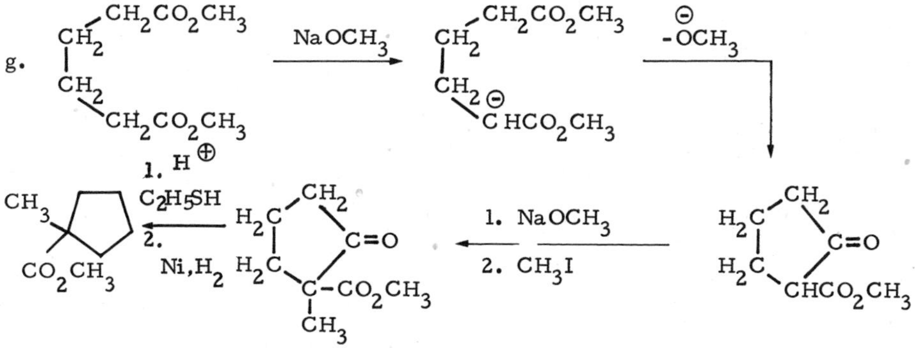

$$\text{(structure) } \begin{array}{c} CH_2CO_2CH_3 \\ CH_2 \\ | \\ CH_2 \\ CH_2CO_2CH_3 \end{array} \xrightarrow{\text{NaOCH}_3} \begin{array}{c} CH_2CO_2CH_3 \\ CH_2 \\ | \\ CH_2 \ominus \\ \overset{|}{C}HCO_2CH_3 \end{array} \xrightarrow{-OCH_3}$$

$$\begin{array}{c} CH_3 \\ \diagdown \\ CO_2CH_3 \end{array} \xleftarrow[\substack{2.\ Ni,H_2}]{\substack{1.\ H^{\oplus} \\ C_2H_5SH}} \begin{array}{c} CH_2 \\ H_2C \diagup \diagdown \\ H_2C \diagdown \diagup C=O \\ C-CO_2CH_3 \\ | \\ CH_3 \end{array} \xleftarrow[\substack{2.\ CH_3I}]{\substack{1.\ NaOCH_3}} \begin{array}{c} CH_2 \\ H_2C \diagup \diagdown \\ H_2C \diagdown \diagup C=O \\ CHCO_2CH_3 \end{array}$$

h. $CH_2(CO_2C_2H_5)_2 \xrightarrow[\substack{2.\ CH_2=CH-CH_2Cl}]{\substack{1.\ NaOC_2H_5}} CH_2=CHCH_2-CH(CO_2C_2H_5)_2 \longrightarrow$

$$\xleftarrow[\substack{1.\ H_2O,\ H^{\oplus} \\ 2.\ heat,\ -CO_2}]{} (CH_2=CH-CH_2)_2C(CO_2C_2H_5)_2 \xleftarrow[\substack{2.\ CH_2=CHCH_2Cl}]{\substack{1.\ NaOC_2H_5}}$$

$$\longrightarrow (CH_2=CH-CH_2)_2CHCO_2H$$

i. $CH_3CH_2COCH_2CO_2C_2H_5 \xrightarrow[\substack{2.\ (CH_3)_2CHCH_2CH_2Br}]{\substack{1.\ NaOH}}$

$$\begin{array}{c} CH_3 \\ | \\ CH_3CH_2CO\overset{|}{C}-CH_2CH_2CH(CH_3)_2 \\ | \\ CO_2C_2H_5 \end{array} \xleftarrow[\substack{2.\ CH_3I}]{\substack{1.\ NaOH}} \begin{array}{c} CH_3CH_2COCHCH_2CH_2CH(CH_3)_2 \\ | \\ CO_2C_2H_5 \end{array}$$

$$\xrightarrow[\substack{2.\ heat,\ -CO_2}]{\substack{1.\ H_2O,\ H^{\oplus}}} CH_3CH_2COCH(CH_3)CH_2CH_2CH(CH_3)_2$$

j. $CH_3CH_2CH_2Br + \overset{\oplus \ominus}{NaCH}(CO_2C_2H_5)_2 \xrightarrow{-NaBr} CH_3CH_2CH_2CH(CO_2C_2H_5)_2$

$$\xleftarrow[\substack{2.\ heat}]{\substack{1.\ H_2O,\ H^{\oplus}}} \begin{array}{c} CH_3CH_2CH_2\overset{|}{C}(CO_2C_2H_5)_2 \\ | \\ CH_2CH=CH_2 \end{array} \xleftarrow[\substack{CH_2=CH-CH_2Br}]{\substack{NaOH}}$$

$$\longrightarrow \begin{array}{c} CH_3CH_2CH_2\overset{|}{C}HCO_2H \\ | \\ CH_2CH=CH_2 \end{array}$$

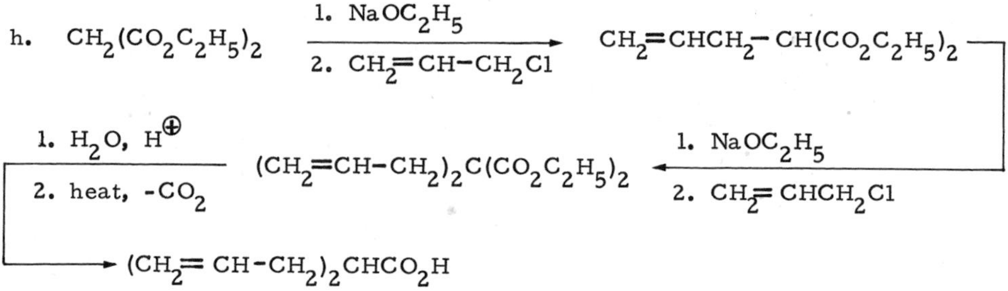

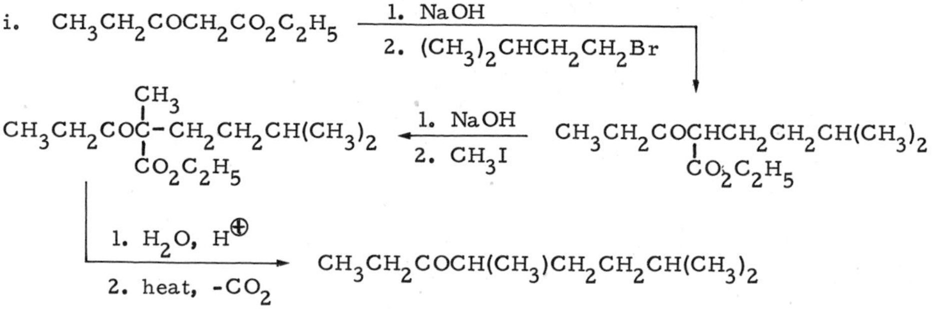

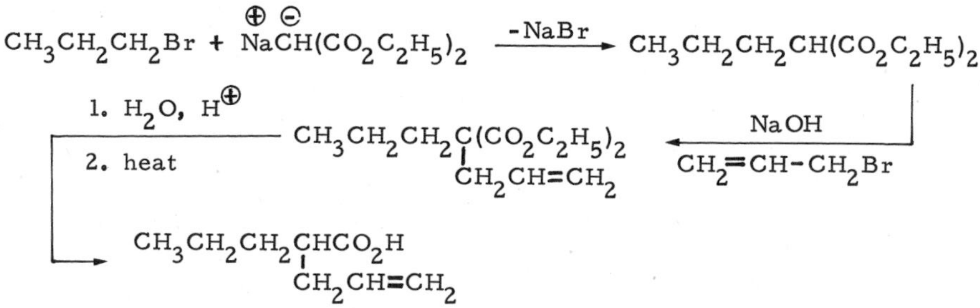

k. $CH_3COCl + \overset{\oplus}{Na}\ \overset{\ominus}{CH}(CO_2C_2H_5)_2 \xrightarrow{-NaCl} CH_3COCH(CO_2C_2H_5)_2$

l. $C_6H_5CHO + CH_3CHO \xrightarrow{\overset{\ominus}{OH}} C_6H_5CH=CHCHO$

$CH_3CO_2C_2H_5 \xrightarrow[-80°]{1.\ LiN\left[Si(CH_3)_3\right]_2} \xrightarrow{2.\ C_6H_5CH=CHCHO}$

$C_6H_5CH=CHCH(OH)CH_2CO_2C_2H_5$

<u>18-64</u>   With reference to Section 18-3C, the borane has only <u>two</u> available hydrogens - one of which is wasted to form H_2 on reaction of the borane with RCO_2H, and the other is used to reduce the acid to the aldehyde. A third B—H bond is required if reduction is to proceed to the alcohol stage.

CHAPTER 19

MORE ON STEREOCHEMISTRY

Optical activity is associated with chirality and is useful for monitor-
ing the behavior of chiral compounds. A beam of plane-polarized light propagates an electric field that oscillates in one plane and a magnetic field that
oscillates in another plane. Optical activity means that a substance causes the
plane of polarization of polarized light passing through it to be rotated to the
right (dextrorotatory) or to the left (levorotatory). The measured rotation, α ,
depends on the concentration $\underline{c}$ of the chiral substance, the path length $\underline{1}$ through
which the light passes, the nature of the solvent (if any), the temperature $\underline{t}$ and
the wavelength of the polarized light λ . The specific rotation $[\alpha]_{\lambda}^{t}$ of a chiral
substance in solution at wavelength λ is expressed by the equation $[\alpha]_{\lambda}^{t}$ =
$100\,\alpha/\underline{1}\,\underline{c}$, in which the concentration $\underline{c}$ is in grams of sample per 100 ml of
solution. The molecular rotation $[\underline{M}]_{\lambda}^{t}$ is related to the specific rotation by
the relationship $[\underline{M}]_{\lambda}^{t} = [\alpha]_{\lambda}^{t}\,\underline{M}/100$, in which $\underline{M}$ is the molecular weight.

Enantiomers, in the absence of a chiral environment, have the same
physical properties except for the sign of their optical rotations and can seldom
be separated by simple crystallization or distillation. Reliable separation
methods involve conversion of the enantiomers by a reaction with a chiral
substance to give two diastereomers. Thus an enantiomer mixture $\underset{\sim}{A}_+\ \underset{\sim}{A}_-$
can be converted by a chiral substance $\underset{\sim}{B}_+$ to the diastereomers $\underset{\sim}{A}_+\,\underset{\sim}{B}_+$ and
$\underset{\sim}{A}_-\,\underset{\sim}{B}_+$. Diastereomers have different physical properties and, in principle,
can be separated by crystallization, distillation, and chromatography. Mixtures
of enantiomeric acids can be converted to diastereomeric salts with a chiral
base, enantiomeric alcohols can be converted to esters with a chiral acid, and
so on. Another method of resolution involves selective destruction of one
enantiomer by a chiral substance. This works especially well for substances
that have one enantiomer that can be metabolized by a living organism while the
other cannot.

Enantiomeric purity is the fractional excess of one enantiomer over the
other. A racemic mixture has an enantiomeric purity of zero and a single
enantiomer a purity of one. For $\underline{n}_1$ moles of one enantiomer and $\underline{n}_2$ moles of
the other in a mixture, the purity is $(\underline{n}_1 - \underline{n}_2)/(\underline{n}_1 + \underline{n}_2)$. The enantiomeric purity

is related to the observed rotation α_{obs} by α_{obs}/α_o where α_o is the rotation of the pure enantiomer.

Many procedures have been developed to determine the enantiomeric purity when α_o is not known or the optical rotation is so small as not to be helpful. Gas-liquid chromatography and nmr spectroscopy are particularly useful. Each depends on differences in behavior of enantiomers in a chiral environment or on combination with a chiral substance. Chemical-shift differences in the nmr spectra of enantiomers often can be induced by the use of chiral-shift reagents, which usually are chelates of a rare-earth metal, such as europium, made with chiral ligands attached to the metal.

The sign of rotation of a particular enantiomer is difficult to relate to the absolute configuration(s) of the chiral center(s). Before any absolute configurations were known, the configurations of many compounds with chiral centers were assigned relative to the enantiomers of glyceraldehyde. Now that the absolute configuration of the glyceraldehyde enantiomers have been established, the absolute configurations of all compounds that have been related to these enantiomers also are known. A very simple example of relating configurations is provided by the oxidation of D-(+)-glyceraldehyde to (-)-glyceric acid:

$$
\begin{array}{ccc}
\text{CHO} & & \text{CO}_2\text{H} \\
| & & | \\
\text{H-C-OH} & \xrightarrow{\text{[o]}} & \text{H-C-OH} \\
| & & | \\
\text{CH}_2\text{OH} & & \text{CH}_2\text{OH}
\end{array}
$$

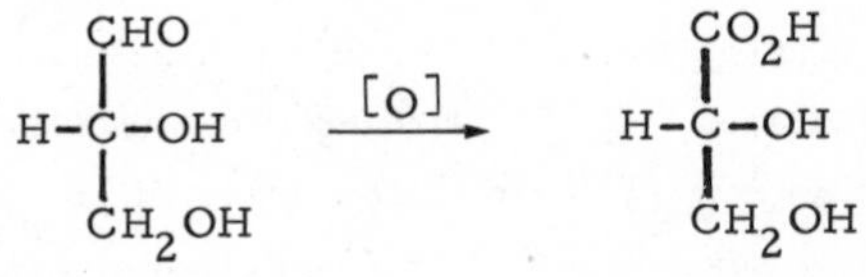

D-(+)-glyceraldehyde   D-(-)-glyceric acid

Because there is no expectation that this oxidation will change the configuration at the chiral carbon, (-)-glyceric acid can be safely assigned the D configuration. Reactions that are used for the determination of relative configurations must have known stereochemical consequences with respect to the chiral centers in the conversion of reactants to products. An example is afforded by the relationship of the configuration of (+)-lactic acid to (+)-alanine in Figure 19-5.

When more than one chiral carbon is present in a molecule, the one with the highest number is correlated with the configuration of D-(+)-glyceraldehyde to establish the absolute configuration by the carbohydrate convention. Examples are:

$$
\text{D}\quad
\begin{array}{c}
\text{CHO} \\
| \\
\text{H-C-OH} \\
| \\
\boxed{\text{H-C-OH}} \\
| \\
\text{CH}_3
\end{array}
\qquad
\text{L}\quad
\begin{array}{c}
\text{CO}_2\text{H} \\
| \\
\text{H-C-OH} \\
| \\
\boxed{\text{HO-C-H}} \\
| \\
\text{CO}_2\text{H}
\end{array}
\qquad \text{tartaric acid}
$$

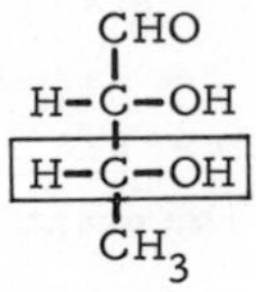

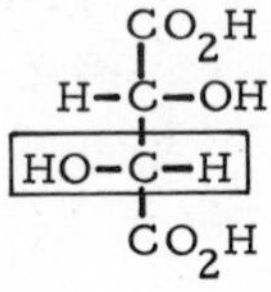

In contrast, by the <u>amino-acid conversion,</u> the <u>lowest-numbered</u> chiral carbon determines the absolute configuration:

$\underline{\underline{\text{L}}}$ serine

$$\begin{array}{c} CO_2H \\ | \\ \boxed{H_2N-\overset{|}{C}-H} \\ | \\ H-\overset{|}{C}-OH \\ | \\ CH_3 \end{array}$$

$\underline{\underline{\text{D}}}$

$$\begin{array}{c} CO_2H \\ | \\ \boxed{H-\overset{|}{C}-OH} \\ | \\ HO-\overset{|}{C}-H \\ | \\ CO_2H \end{array}$$

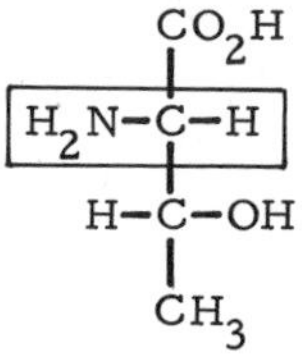

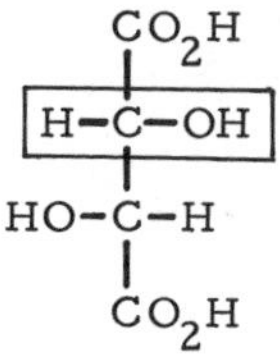

tartaric acid

This leads to a contradiction with (+)-tartaric acid, which is $\underline{\underline{\text{L}}}$ by the carbohydrate convention and $\underline{\underline{\text{D}}}$ by the amino-acid convention. One way that has been proposed to remove the ambiguity is to use subscripts $\underline{\text{g}}$ (glucose) and $\underline{\text{s}}$ (serine) so that $\underline{\underline{\text{D}}}_{\underline{\text{s}}}$ is the $\underline{\underline{\text{D}}}$ configuration by the amino-acid convention and $\underline{\underline{\text{D}}}_{\underline{\text{g}}}$ is the $\underline{\underline{\text{D}}}$ configuration by the carbohydrate convention.

Configurations can be established with much less ambiguity by the <u>R, S</u> or <u>Cahn-Ingold-Prelog convention.</u> The $\underline{\text{R}}$, $\underline{\text{S}}$ configurations are determined on the basis of rules which assign an order of preference to the substituents at a chiral center. The rules may be summarized as:

1. Atoms at the chiral center highest in atomic number have preference.

2. If the first atoms are the same, the priority goes to the substituent with a <u>second</u> atom highest in atomic number.

3. Multiple bonds are treated as if they were multiple single bonds. For example, $-HC \hspace{-3pt}=\hspace{-3pt} CH_2$ becomes
$$-CH{\overset{\textstyle CH_2}{\underset{\textstyle (C)}{<}}} \; .$$

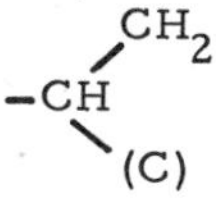

4. If the chiral center is viewed with the group of <u>lowest</u> priority on the opposite side from you, and if the other groups are arranged in a clockwise sequence with decreasing priority, configuration is $\underline{\text{R}}$; if the sequence is counter-clockwise, configuration is $\underline{\text{S}}$.

Thus $\underline{\underline{\text{L}}}$-serine is also $\underline{\text{S}}$-serine, and $\underline{\underline{\text{D}}}$-lactic acid is also $\underline{\text{R}}$-lactic acid. There is, however, no relationship between the $\underline{\underline{\text{D}}}$, $\underline{\underline{\text{L}}}$ and $\underline{\text{R}}$, $\underline{\text{S}}$ systems.

Application of the sequence rules to configuration about double bonds is useful in cases where cis-trans notation is ambiguous. First, at each end of the double bond, decide which group has the higher priority with the aid of the sequence rules for the $\underline{\text{R}}$, $\underline{\text{S}}$ system. If the highest priority groups are on the same side of the double bond, the configuration is $\underline{\text{Z}}$; if on opposite sides, it is **E**.

higher priority
group on this
end ——————————→

H   CH$_2$OH
  \\C=C/
CH$_3$ CH$_3$

higher priority
group on this
end

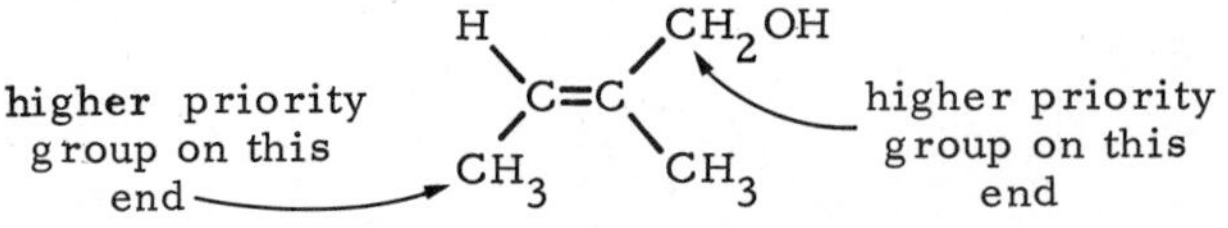

2-methyl-2(**E**)-buten-1-ol

H   CH$_3$
  \\C=C/
CH$_3$ CH$_2$OH

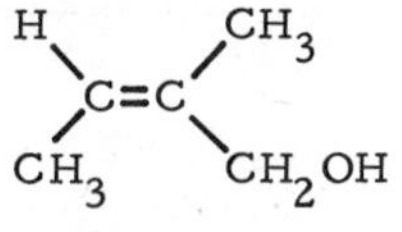

2-methyl-2(**Z**)-buten-1-ol

When there is more than one chiral center in the molecule the configuration of each can be specified by the **R**, **S** system. Examples are

CHO
|
H–C–OH
|
H–C–OH
|
CH$_3$

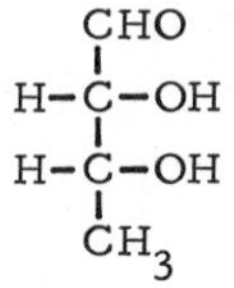

(2**R**, 3**R**)-2, 3-dihydroxybutanal

HO   Br

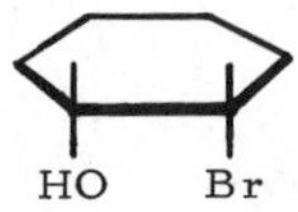

(1**R**, 2**S**)-2-bromocyclohexanol

Some molecules are identical with their mirror images (achiral) but
may behave as if they were enantiomers when placed in a chiral environment.
They are described as being <u>prochiral</u> and the prerequisite is the absence of an
<u>axis</u> of symmetry. We also can use the term prochirality with reference to
atoms within molecules, irrespective of whether the molecule is chiral or not.
Any carbon in a molecule having the substitution pattern C(AAXY), where A. X,
and Y are different,is a prochiral atom. Replacement of one of the like groups
(A) by a different group (Z) leads to a chiral center of one configuration,
whereas replacement of the other A group gives the opposite configuration:

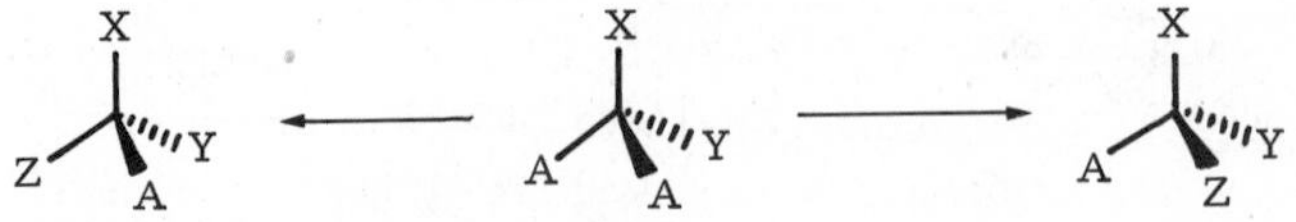

X
|
Z–C⟋⟍Y
|
A

chiral

X
|
A–C⟋⟍Y
|
A

prochiral

X
|
A–C⟋⟍Y
|
Z

chiral

An asymmetric synthesis is a synthesis in which a chiral center is
created with a preference for a particular configuration. Such a synthesis can
be the result of reaction of a nonchiral substance with chiral substances or of
reactions in a chiral environment. Asymmetric syntheses are the rule in the
reactions carried out by living systems.

Racemization can occur by many processes. Chiral biphenyl derivatives usually racemize on heating. Ketones with a chiral α carbon of the type

O=C
    |
  –C–H
    |

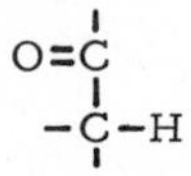

racemize by acid- or base-induced enolization. Chiral alkyl halides or alcohols where the halogen or hydroxyl is attached to the chiral carbon can racemize
by either S_N1 or S_N2 reactions. Chiral hydrocarbons are racemized by
reagents that convert the chiral carbons to carbocations or carbanions.

ANSWERS TO EXERCISES

19-1

a.

$$C_6H_5CH_2\underset{\underset{(D, L)}{NH_2}}{CHCH_3} + R*CO_2H \xrightarrow{-H_2O} C_6H_5CH_2\underset{\underset{(D, D + L, D)}{\overset{\oplus}{NH_3} \ominus O_2CR*}}{CHCH_3}$$

1. separation by recrystallization

$$\underset{*}{D}-RCO_2H + D-C_6H_5CH_2\underset{NH_2}{CHCH_3} + L-C_6H_5CH_2\underset{NH_2}{CHCH_3} \xleftarrow{\substack{\text{1. separation by} \\ \text{recrystallization} \\ \text{2. } H_3O^{\oplus}}}$$

(R*CO$_2$H is a chiral acid, such as listed in Section 19-3B)

b.

$$HO_2CCH=C=CHCO_2H + R*NH_2 \xrightarrow{-H_2O} HO_2CCH=C=CHCO_2^{\ominus} \ H_3\overset{\oplus}{N}R*$$

$$(D, L) \qquad\qquad (L) \qquad\qquad\qquad (D, L + L, L)$$

1. separation by recrystallization

$$D-HO_2CCH=C=CHCO_2H + L-HO_2CCH=C=CHCO_2H \xleftarrow{\substack{\text{1. separation by} \\ \text{recrystallization} \\ \text{2. } H_3O^{\oplus}}}$$

(R*NH$_2$ is a chiral base, such as listed in Section 19-3A.)

c.

$$L-C_6H_5\underset{CH_3}{CHOH} + D-C_6H_5\underset{CH_3}{CHOH} \xleftarrow{\substack{\text{1. separation by} \\ \text{recrystallization} \\ \text{2. } H_3O^{\oplus} \\ \text{3. ester hydrolysis}}}$$

(L, L + D, L)

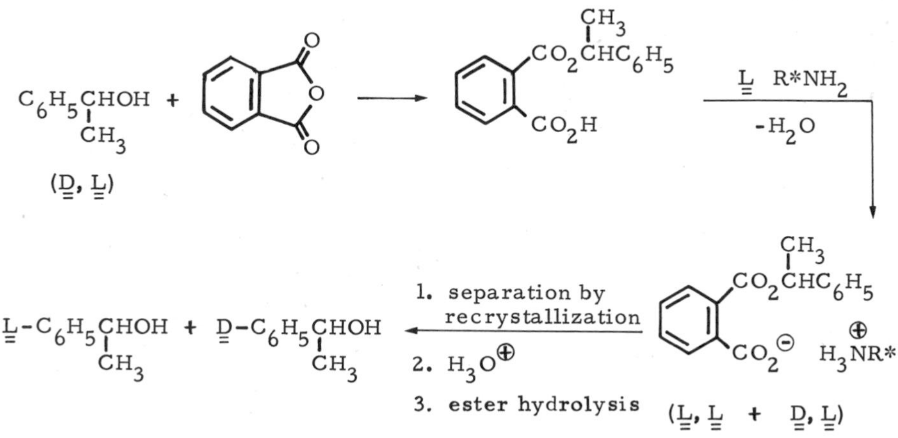

19-2
a.

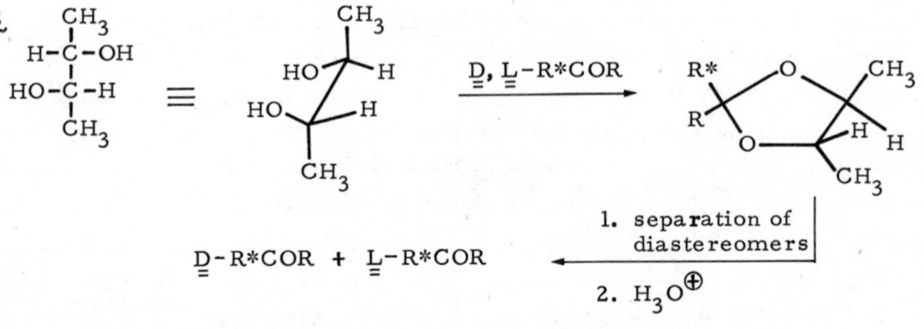

$$\underline{D}\text{-R*COR} + \underline{L}\text{-R*COR} \xleftarrow{\begin{array}{c}\text{1. separation of}\\ \text{diastereomers}\\ \hline\\ \text{2. } H_3O^{\oplus}\end{array}}$$

b.

$$\underline{D}\text{-R*COR} + \underline{L}\text{-R*COR} \xleftarrow{\begin{array}{c}\text{1. separation of diastereomers}\\ \hline\\ \text{2. } H_2O\end{array}}$$

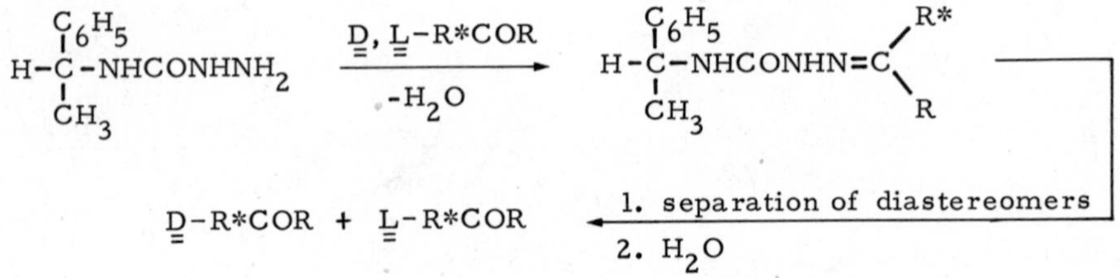

Hydrolysis conditions must be chosen to avoid possible racemization by enolization when the chiral carbon is adjacent (alpha) to the carbonyl group (see
Exercise 17-3).

19-3   Enantiomeric purity $= \dfrac{10.1}{19.34} \times 100 = 52\%$

Resolution of the (±)-acid with (-)-amine would result in the separation of the
(-)-acid. If the resolving amine is only 80% enantiomerically pure, then the
partially resolved acid will be no more than 80% as enantiomerically pure as
when resolved with 100% of one amine enantiomer. That is $80 \times 10.1 / 19.34 =$
41.6%

19-4

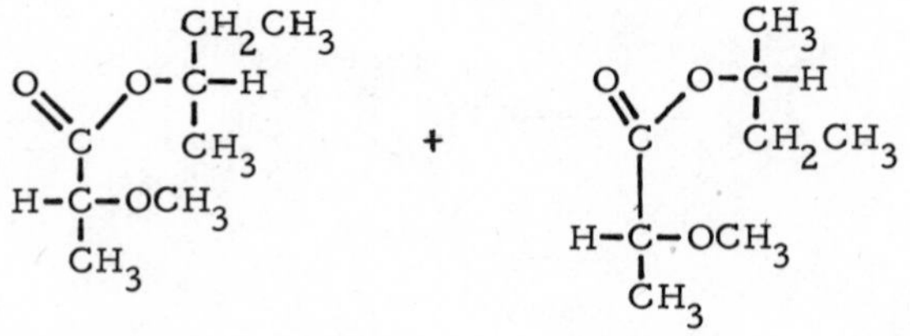

The resolution can be followed conveniently from the relative areas of
the singlet resonances of the OCH_3 protons. The chemical shifts of the
methoxyl protons are unlikely to be identical in the two diastereomers. Also,
the signals are uncomplicated by splitting and are well separated from other
resonances. Their relative intensities can therefore be measured with reasonable accuracy.

19-5   a. In the chiral amine solvent, $(+)-C_6H_5CH(NH_2)CH_3$, the racemic
alcohol, $C_6H_5CH(OH)CF_3$, is in an asymmetric environment and is chirally
solvated. The chemical shift of the ^{19}F resonance of the CF_3 group of the
(+) alcohol solvated by the (+) amine is not the same as that of the (-) alcohol
solvated by the (+) amine. Thus, two resonances are observed, each of which
is split into a doublet by spin-coupling of ^{19}F with the proton at C1. When the
solvent is racemic, only one doublet resonance is observed because there is no
preference of one type of chiral solvation over another, and all the possible
combinations of chiral solvent with chiral alcohol are in rapid equilibrium and
give rise to only one average signal.

$$(-)(-) \ + \ (+)(+) \ \rightleftharpoons \ (+)(-) \ + \ (-)(+)$$

Similar equilibration in the chiral solvent does not produce an average signal.

b. The resolution of $CF_3CH(OH)C_6H_5$ could be followed by ^{19}F nmr
using one of the chiral forms of 1-phenylethanamine as solvent. The relative
intensities of the doublet pairs for the CF_3 resonance would be a measure of
optical purity.

19-6

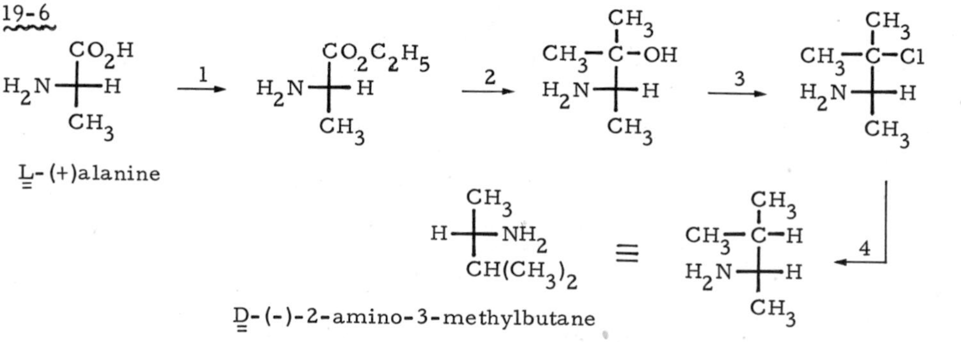

$\underline{L}$-(+)alanine

$\underline{D}$-(-)-2-amino-3-methylbutane

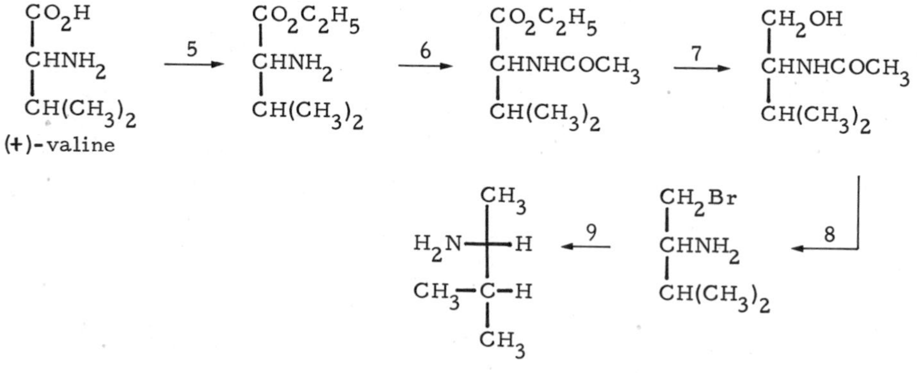

(+)-valine

$\underline{L}$-(+)-2-amino-3-methylbutane

If alanine is $\underline{\underline{L}}$ then (-)-2-amino-3-methylbutane is the $\underline{\underline{D}}$ isomer. With this information, the scheme establishes that (+)-valine must have the $\underline{\underline{L}}$ configuration in order to give $\underline{\underline{L}}$-(+)-2-amino-3-methylbutane.

19-7

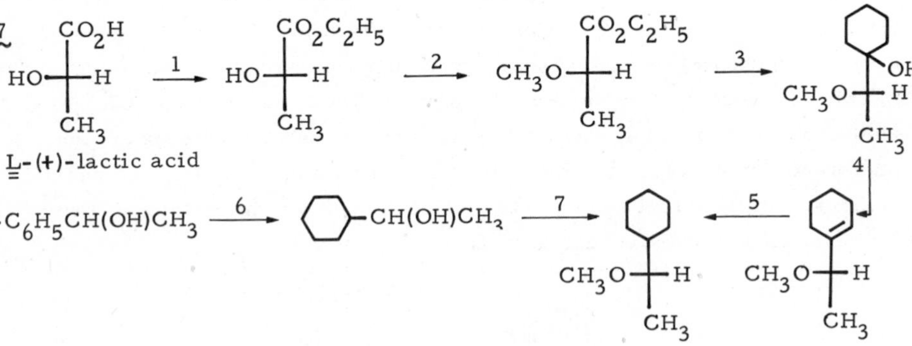

$\underline{\underline{L}}$-(+)-lactic acid

(-)-$C_6H_5CH(OH)CH_3$ $\xrightarrow{6}$ ◯-CH(OH)CH$_2$ $\xrightarrow{7}$ ...

The scheme established that (-)-phenylethanol has the $\underline{\underline{L}}$ configuration. The (+) sulfonium salt derived from the $\underline{\underline{L}}$ alcohol has the $\underline{\underline{L}}$ configuration.

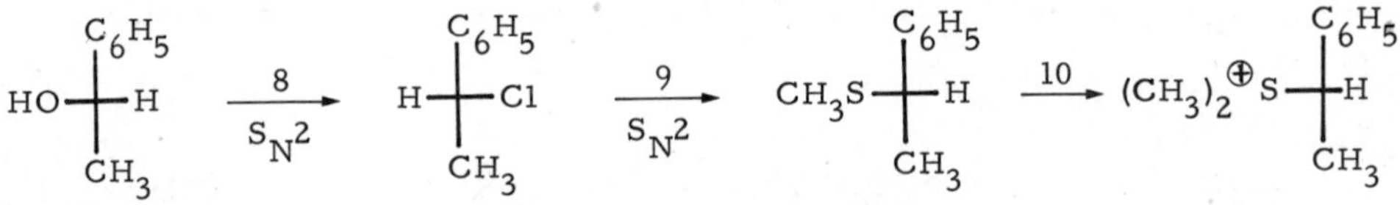

$\underline{\underline{L}}$-(-)-phenylethanol   (+)

19-8 a. $-CONH_2$ b. 2-methylcyclohexyl   c. 1-propenyl

      d. phenyl   e. chloromethyl

19-9 a. $\underline{\underline{L}}, \underline{S}$ b. $\underline{\underline{L}}, \underline{S}$ c. $2\underline{\underline{D}}, 3\underline{\underline{D}}$; $2\underline{S}, 3\underline{R}$

      d. $2\underline{\underline{L}}, 3\underline{\underline{D}}, 4\underline{\underline{D}}$; $2\underline{S}, 3\underline{R}, 4\underline{R}$ e. $\underline{S}$ f. $\underline{S}$

19-10

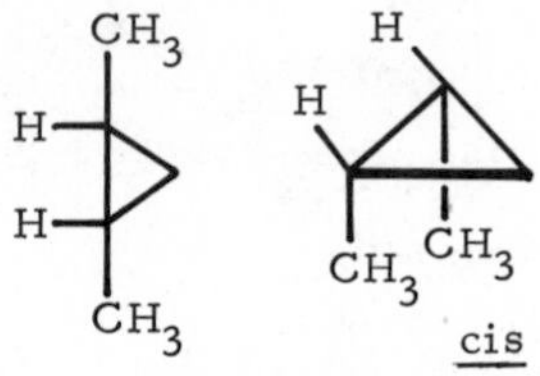

a.

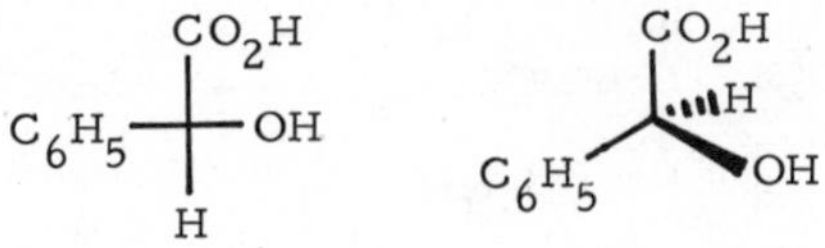

c.

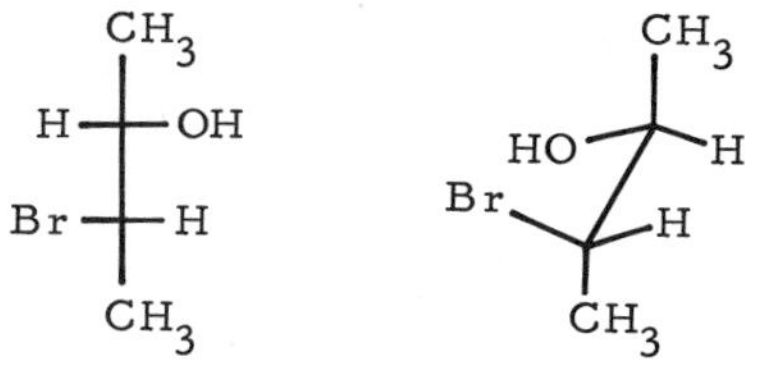

CH_3
H——OH
Br——H
CH_3

CH_3
HO——H
Br——H
CH_3

threo

d.

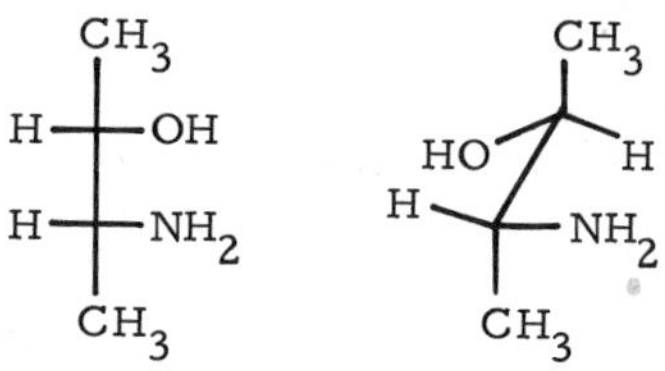

CH_3
H——OH
H——NH_2
CH_3

CH_3
HO——H
H——NH_2
CH_3

erythro

e.

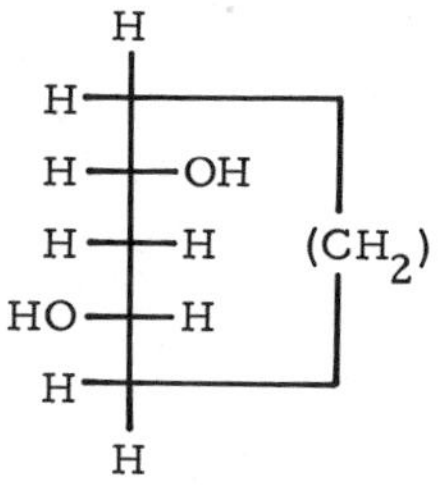

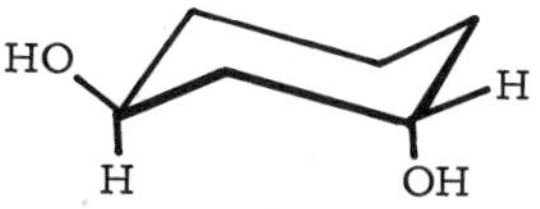

H
H——
H——OH
H——H (CH_2)
HO——H
H——
H

HO——[ring]——H
H   OH

trans

f.

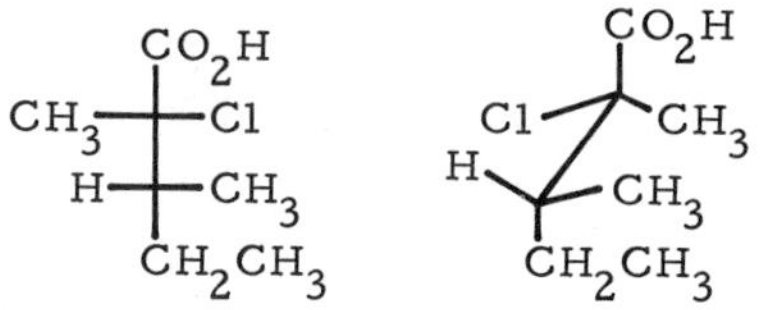

CO_2H
CH_3——Cl
H——CH_3
CH_2CH_3

CO_2H
Cl——CH_3
H——CH_3
CH_2CH_3

g.

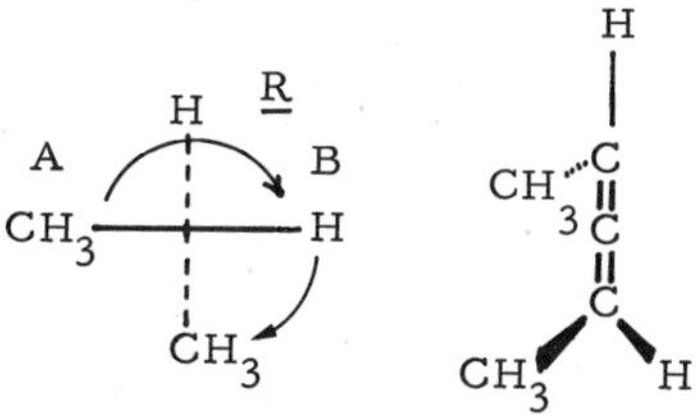

$\underline{R}$
A   H   B
CH_3————H
CH_3

H
CH_3——C
C
CH_3 H

19-11
a. 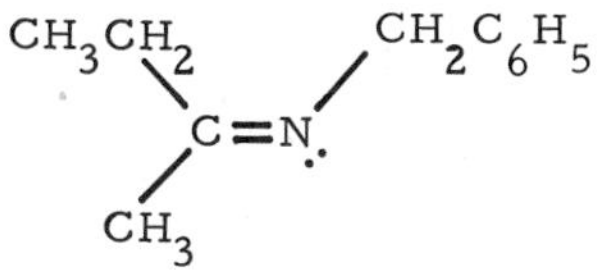 CH_3CH_2 $CH_2C_6H_5$
C=N
CH_3

b. 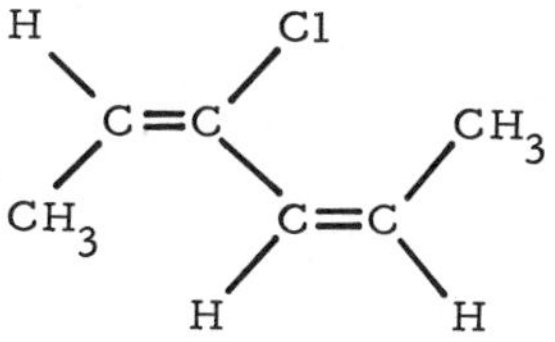   H   Cl
C=C
CH_3 CH_3
C=C
H   H

19-12   a.   (1$\underline{E}$, 3$\underline{E}$)-1, 2, 4-trichloro-1, 3-butadiene

b.   ($\underline{E}$)-2-chloro-3-cyanobutenoic acid

c.   ($\underline{Z}$)-ethylidene-2-methylcyclohexane

d.   ($\underline{Z}$)-2-methoxypentene

19-13   <u>chiral</u>   2-methylbutanedioic acid

<u>prochiral</u> ethenylbenzene, 2-butanone, glycine, 1-chloro-2-phenylethane, cis-2-butene, butanedioic acid

<u>achiral</u>   2-propanone

prochiral carbon (*): $C_6H_5\overset{*}{C}H=CH_2$, $CH_3\overset{*}{C}H=\overset{*}{C}HCH_3$, $CH_3\overset{*}{C}O\overset{*}{C}H_2CH_3$

19-13 (cont.)

prochiral carbon (*): $ClCH_2CH_2C_6H_5$, $H_2NCH_2CO_2H$, $CH_3\overset{*}{C}HCO_2H$,
$(\overset{*}{C}H_2CO_2H)_2$ $*CH_2CO_2H$

19-14. Both 2-oxobutanedioic (oxaloacetic) acid and citric acid are prochiral molecules. For the C4 carbon of 2-oxobutanedioic to appear only at C1 of 2-oxopentanedioic (2-ketoglutaric) acid, both $HO_2CCH_2COCO_2H$ and $HO_2CCH_2\underset{CO_2H}{C}(OH)CH_2CO_2H$ must function as if they were chiral in an asymmetric (enzyme) environment.

$$
\begin{array}{ccccc}
\begin{array}{c} *CO_2H \\ | \\ CH_2 \\ | \\ CO \\ | \\ CO_2H \end{array}
& \xrightarrow[\text{enzyme}]{\text{acetyl CoA}} &
\begin{array}{c} *CO_2H \\ | \\ CH_2 \\ | \\ HO-C-CO_2H \\ | \\ CH_2 \\ | \\ CO_2H \end{array}
& \xrightarrow{\text{enzyme}} &
\begin{array}{c} *CO_2H \\ | \\ C=O \\ | \\ CH_2 \\ | \\ CH_2 \\ | \\ CO_2H \end{array}
\end{array}
$$

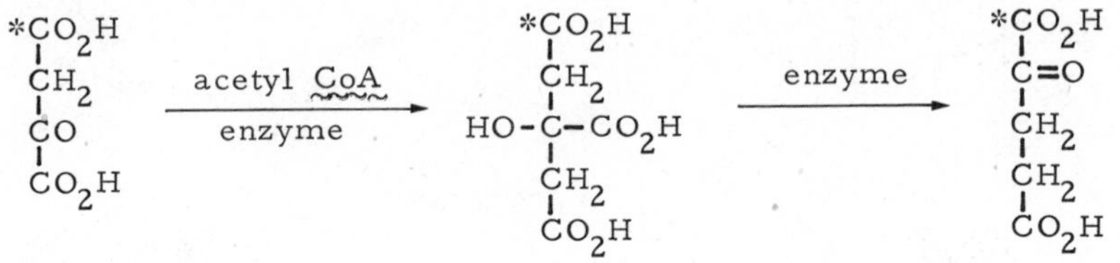

(The $-CH_2CO_2H$ groups are enantiotopic and distinguishable)

19-15

a. $\begin{array}{c} CH_3 \\ C_6H_5 - \!\!\!\! \boxed{} \!\!\!\! - OH \\ C_2H_5 - \!\!\!\! \boxed{} \!\!\!\! - H \\ C_6H_5 \end{array}$ + $\begin{array}{c} CH_3 \\ HO - \!\!\!\! \boxed{} \!\!\!\! - C_6H_5 \\ C_2H_5 - \!\!\!\! \boxed{} \!\!\!\! - H \\ C_6H_5 \end{array}$ b. $\begin{array}{c} CH_3 \\ H - \!\!\!\! \boxed{} \!\!\!\! - OH \\ H - \!\!\!\! \boxed{} \!\!\!\! - CH_3 \\ C_6H_5 \end{array}$ + $\begin{array}{c} CH_3 \\ HO - \!\!\!\! \boxed{} \!\!\!\! - H \\ H - \!\!\!\! \boxed{} \!\!\!\! - CH_3 \\ C_6H_5 \end{array}$

c. $C_6H_5 - \!\!\!\! \underset{H}{\overset{OH}{\boxed{}}} \!\!\!\! - CO - O - \!\!\!\! \underset{CH_2CH_3}{\overset{CH_3}{\boxed{}}} \!\!\!\! - H$ + $C_6H_5 - \!\!\!\! \underset{OH}{\overset{H}{\boxed{}}} \!\!\!\! - CO - O - \!\!\!\! \underset{CH_2CH_3}{\overset{CH_3}{\boxed{}}} \!\!\!\! - H$

d. $\begin{array}{c} CN \\ H - \!\!\!\! \boxed{} \!\!\!\! - OH \\ H - \!\!\!\! \boxed{} \!\!\!\! - OH \\ H - \!\!\!\! \boxed{} \!\!\!\! - OH \\ CH_2OH \end{array}$ + $\begin{array}{c} CN \\ HO - \!\!\!\! \boxed{} \!\!\!\! - H \\ H - \!\!\!\! \boxed{} \!\!\!\! - OH \\ H - \!\!\!\! \boxed{} \!\!\!\! - OH \\ CH_2OH \end{array}$ e. $\begin{array}{c} CO_2H \\ H - \!\!\!\! \boxed{} \!\!\!\! - CH_3 \\ C_6H_5 \end{array}$ + $\begin{array}{c} CO_2H \\ CH_3 - \!\!\!\! \boxed{} \!\!\!\! - H \\ C_6H_5 \end{array}$

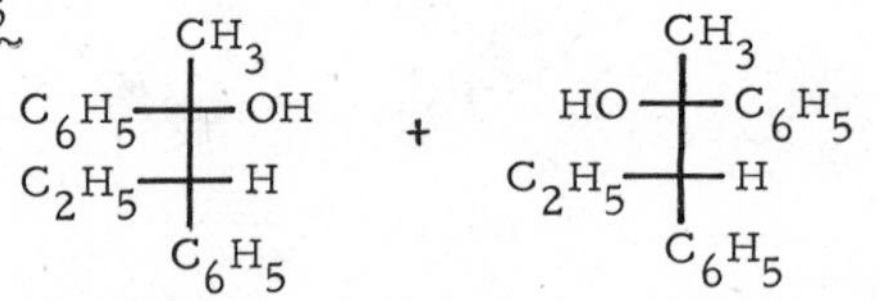

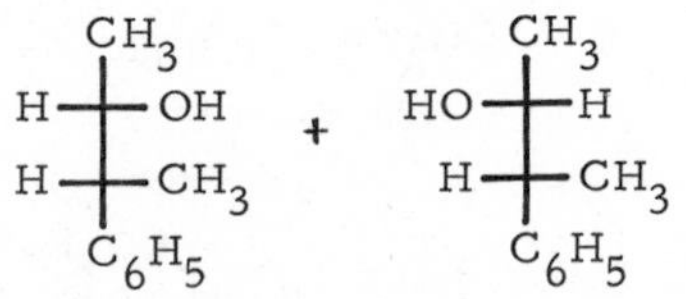

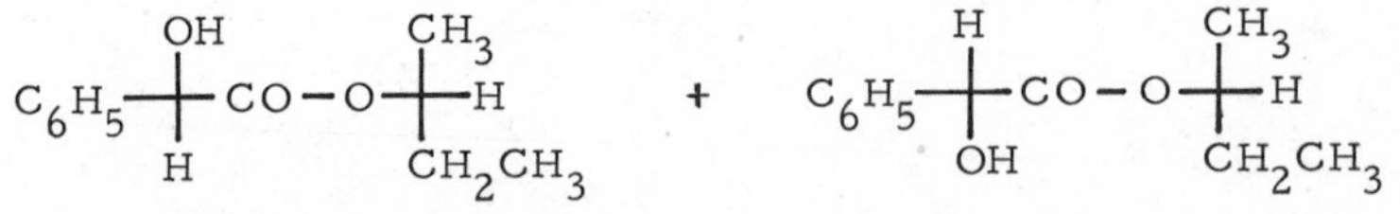

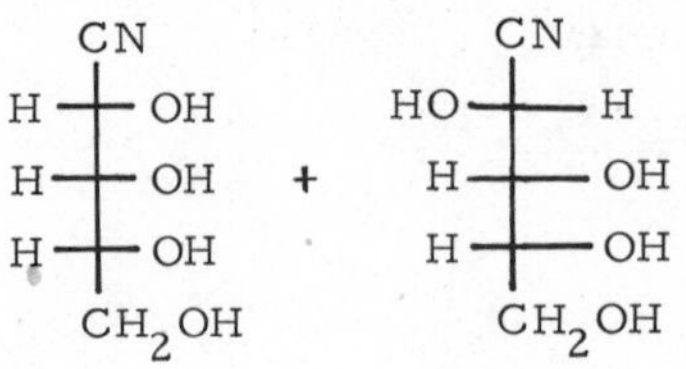

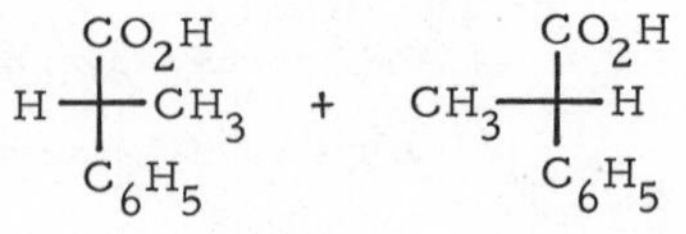

19-16

a. R_2BH $\xrightarrow{CH_3CH=CHCH_3}$

$\left(R = \vcenter{\hbox{\includegraphics{pinane}}} \right)$

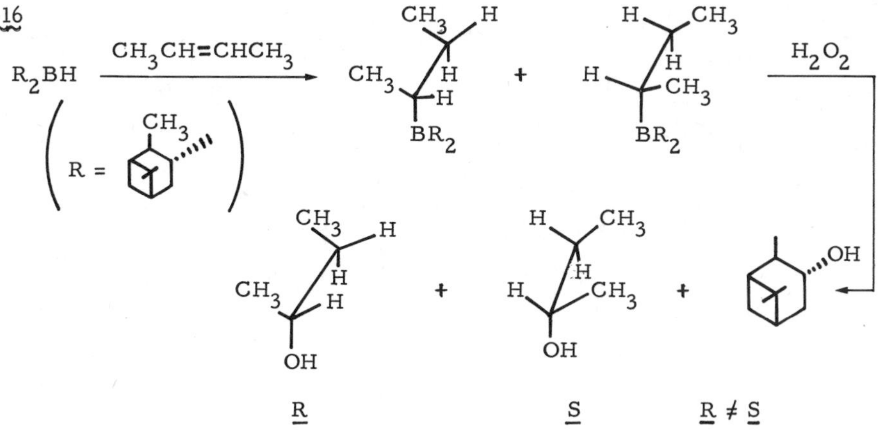

The two diastereomeric boranes are not formed in equal amounts because of the influence of the chiral pinane group on the rate of formation of diastereomers. This nonequivalence shows up as a predominance of (-)-2-butanol in the oxidation products.

b. The chiral borane 26 reacts more rapidly with one enantiomer of 3-methylcyclopentene than with the other. When less than an equimolar amount of 26 relative to alkene is used, the residual alkene is enriched (partially resolved) in the less reactive enantiomer. If equimolar amounts of reagents are used, there will be no residual alkene and therefore no resolution at this stage.

19-17   The three projections on the next page represent the same configuration.

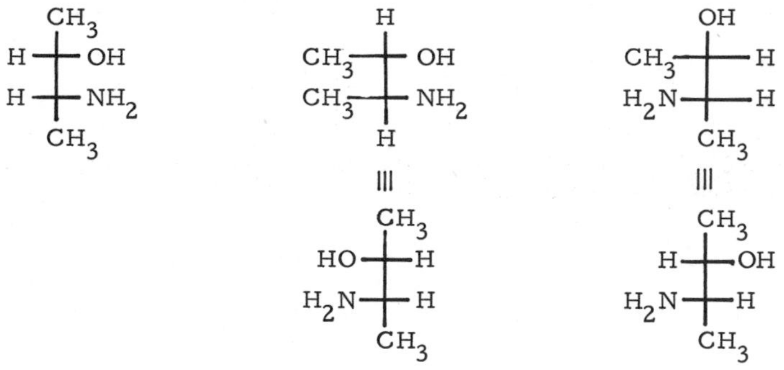

19-17 (cont.)

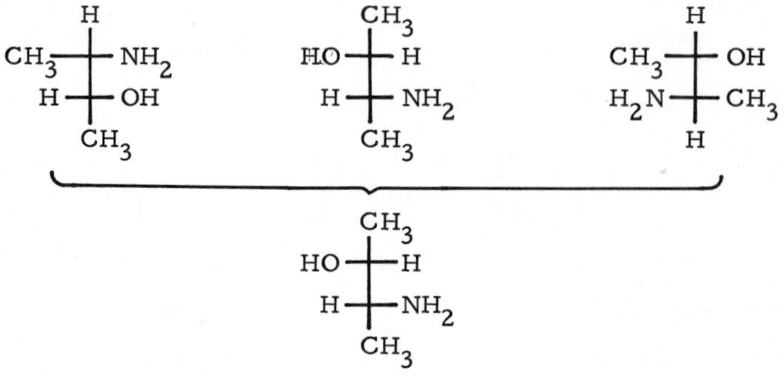

19-18
a.

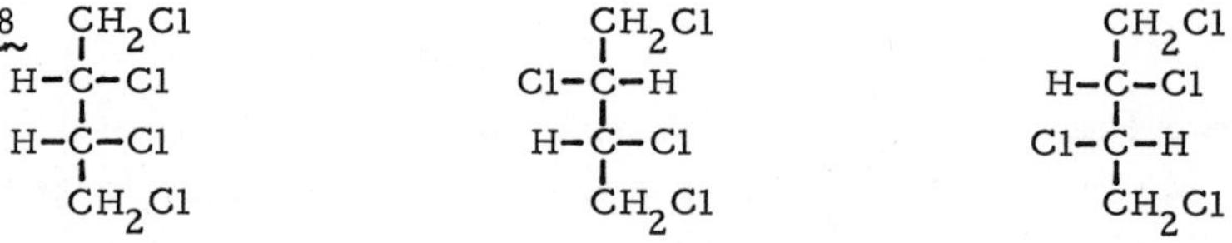

b.

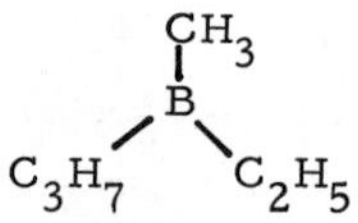

one isomer only
(planar molecule)

c.

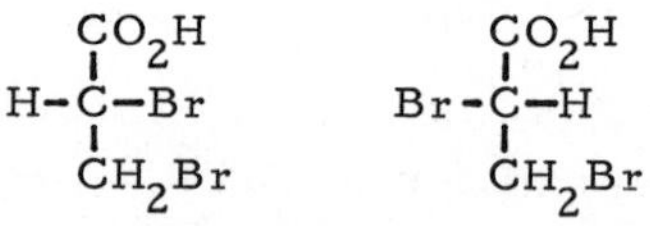

d.

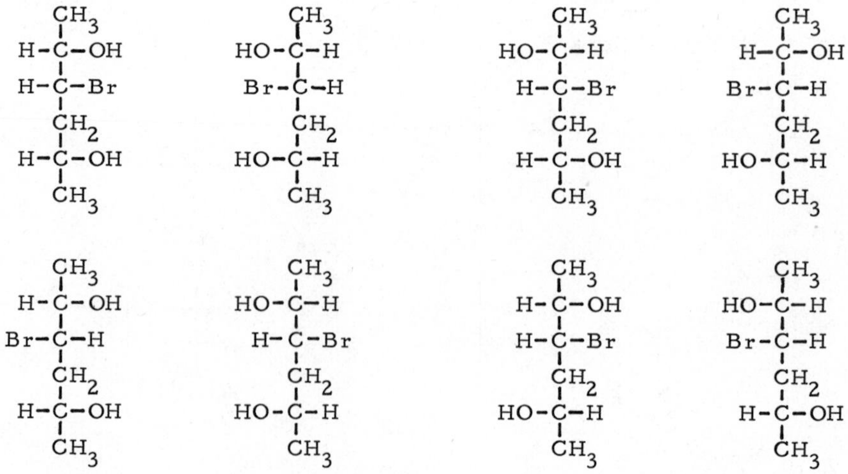

e.

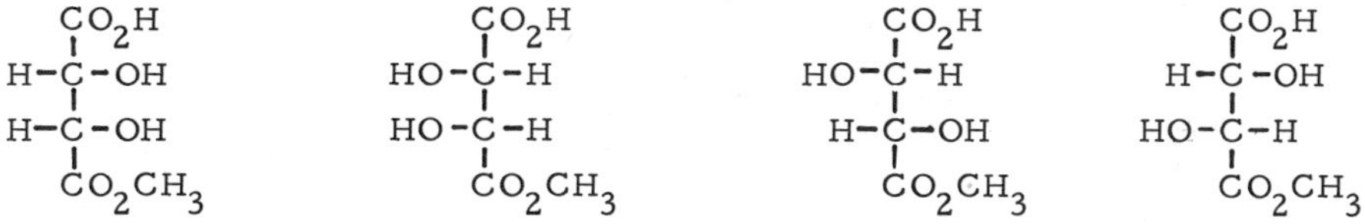

$$\begin{array}{cccc}
\text{CO}_2\text{H} & \text{CO}_2\text{H} & \text{CO}_2\text{H} & \text{CO}_2\text{H} \\
\text{H}-\text{C}-\text{OH} & \text{HO}-\text{C}-\text{H} & \text{HO}-\text{C}-\text{H} & \text{H}-\text{C}-\text{OH} \\
\text{H}-\text{C}-\text{OH} & \text{HO}-\text{C}-\text{H} & \text{H}-\text{C}-\text{OH} & \text{HO}-\text{C}-\text{H} \\
\text{CO}_2\text{CH}_3 & \text{CO}_2\text{CH}_3 & \text{CO}_2\text{CH}_3 & \text{CO}_2\text{CH}_3
\end{array}$$

f.

$$\begin{array}{cccc}
\text{CH}_2\text{CH}_3 & \text{CH}_2\text{CH}_3 & \text{CH}_2\text{CH}_3 & \text{CH}_2\text{CH}_3 \\
\text{CH}_3-\text{C}-\text{H} & \text{H}-\text{C}-\text{CH}_3 & \text{H}-\text{C}-\text{CH}_3 & \text{CH}_3-\text{C}-\text{H} \\
\text{O} & \text{O} & \text{O} & \text{O} \\
\text{C}=\text{O} & \text{C}=\text{O} & \text{C}=\text{O} & \text{C}=\text{O} \\
\text{H}-\text{C}-\text{OH} & \text{HO}-\text{C}-\text{H} & \text{H}-\text{C}-\text{OH} & \text{HO}-\text{C}-\text{H} \\
\text{CH}_3 & \text{CH}_3 & \text{CH}_3 & \text{CH}_3
\end{array}$$

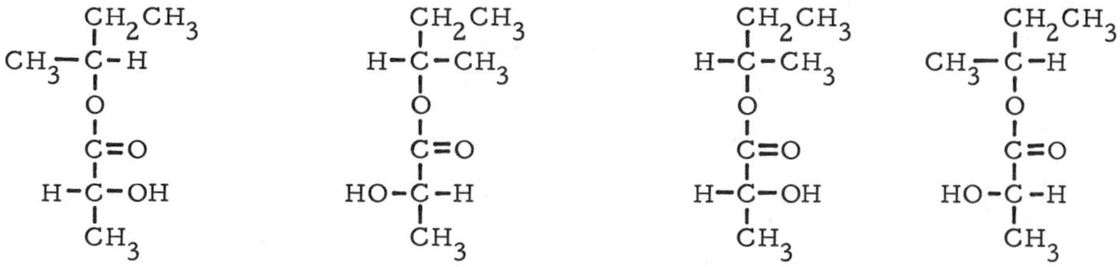

g.

, which in Fischer projection is

, which in Fischer projection is

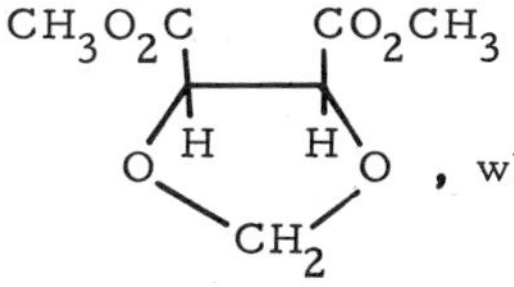

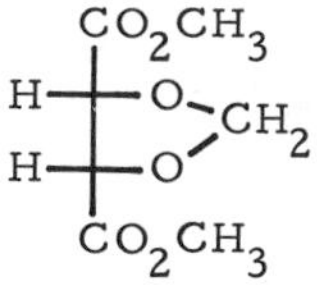

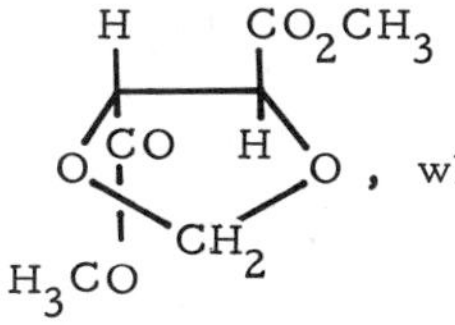

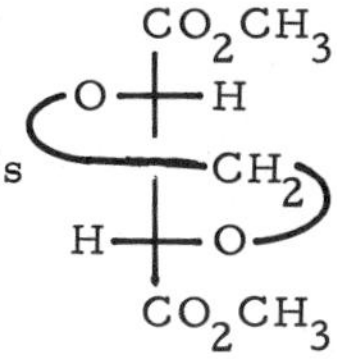

(and enantiomer)

19-19

a.

$$\begin{array}{c}
\text{CH}_3 \\
\text{H}-\text{C}-\text{OH} \\
\text{C}_2\text{H}_5 \\
\underline{\underline{\text{D}}}
\end{array}
\xrightarrow{(\text{CH}_3\text{CO})_2\text{O}}
\begin{array}{c}
\text{CH}_3 \\
\text{H}-\text{C}-\text{OCOCH}_3 \\
\text{C}_2\text{H}_5 \\
\underline{\underline{\text{D}}}
\end{array}$$

b.

$$\begin{array}{c}
(\text{CH}_3)_2\text{CH} \\
\text{CH}_3-\text{C}-\text{OH} \\
\text{CH}_2\text{CH}_2\text{CH}_3 \\
\underline{\underline{\text{D}}}
\end{array}
\xrightarrow[\text{S}_\text{N}^1]{\text{HCl}}
\begin{array}{c}
(\text{CH}_3)_2\text{CH} \\
\text{CH}_3-\text{C}-\text{Cl} \\
\text{CH}_2\text{CH}_2\text{CH}_3
\end{array}
+
\begin{array}{c}
(\text{CH}_3)_2\text{CH} \\
\text{Cl}-\text{C}-\text{CH}_3 \\
\text{CH}_2\text{CH}_2\text{CH}_3 \\
\underline{\underline{\text{D}}},\ \underline{\underline{\text{L}}}
\end{array}$$

c.

$$\begin{array}{c}
\text{CH}_2\text{OCOCH}_3 \\
\text{H}-\text{C}-\text{OH} \\
\text{CH}_2\text{OH} \\
\underline{\underline{\text{D}}}
\end{array}
\xrightarrow{\text{H}_2\text{O},\ \overset{\ominus}{\text{OH}}}
\begin{array}{c}
\text{CH}_2\text{OH} \\
\text{H}-\text{C}-\text{OH} \\
\text{CH}_2\text{OH}
\end{array}
+
\text{CH}_3\text{CO}_2^{\ominus}$$

achiral (prochiral)

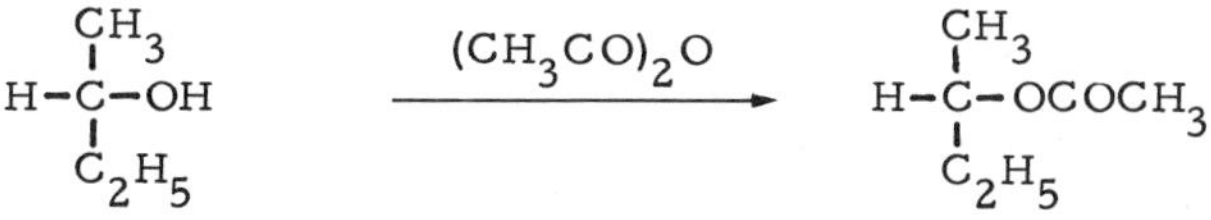

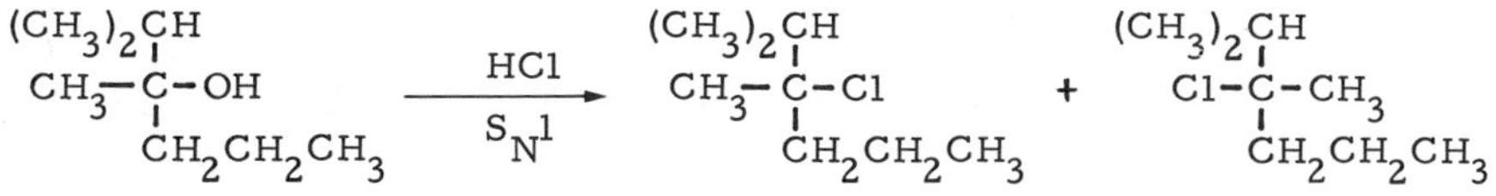

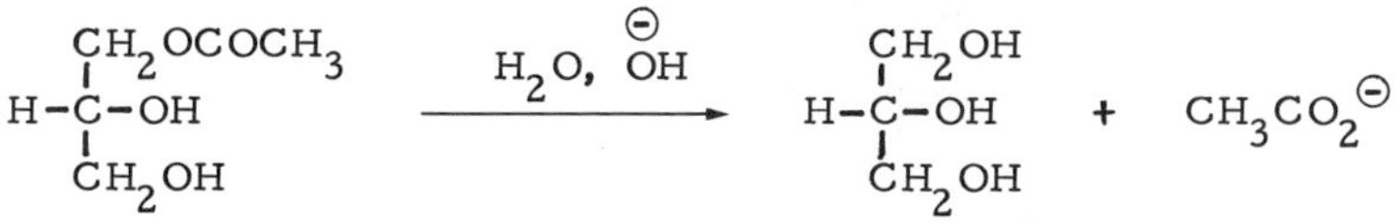

d.

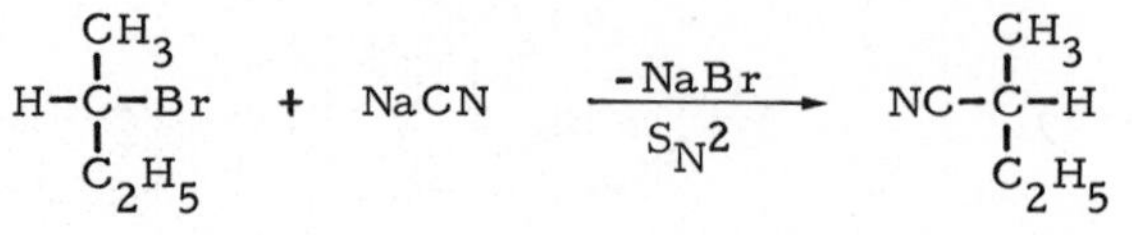

$$\begin{array}{c} CH_3 \\ | \\ H-C-Br \\ | \\ C_2H_5 \end{array} + NaCN \xrightarrow[S_N2]{-NaBr} \begin{array}{c} CH_3 \\ | \\ NC-C-H \\ | \\ C_2H_5 \end{array}$$

D   L

e.

$$\begin{array}{c} CH_3 \\ | \\ H-C-COC(CH_3)_3 \\ | \\ C_2H_5 \end{array} \xrightarrow[HO^\ominus]{Br_2} \begin{array}{c} CH_3 \\ | \\ Br-C-COC(CH_3)_3 \\ | \\ C_2H_5 \end{array} + \begin{array}{c} CH_3 \\ | \\ (CH_3)_3CCO-C-Br \\ | \\ C_2H_5 \end{array}$$

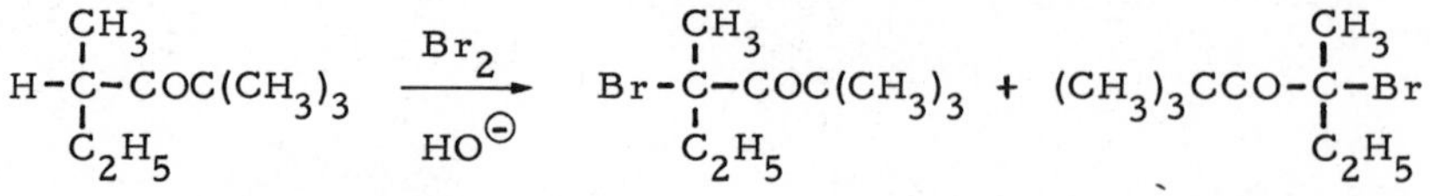

D   (formed through enol)   D, L

f.

$$\begin{array}{c} CH_3 \\ | \\ H-C-COC_2H_5 \\ | \\ C_2H_5 \end{array} \xrightarrow{CH_3MgBr} \begin{array}{c} H_3C \quad C_2H_5 \\ | \quad\;\; | \\ H-C-C-OH \\ | \quad\;\; | \\ H_5C_2 \quad CH_3 \end{array} + \begin{array}{c} H_3C \quad CH_3 \\ | \quad\;\; | \\ H-C-C-OH \\ | \quad\;\; | \\ H_5C_2 \quad C_2H_5 \end{array}$$

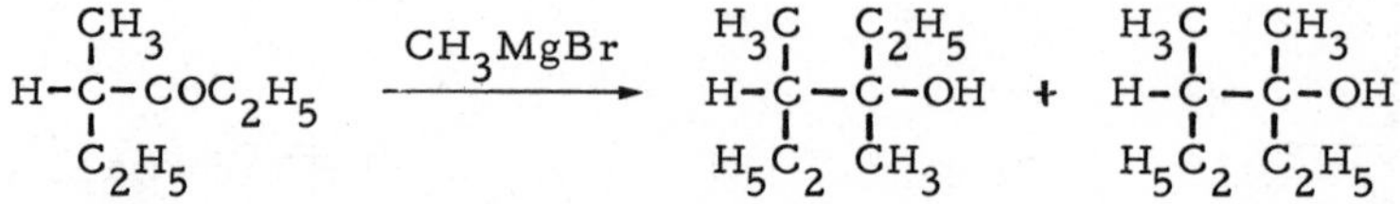

D   (not expected to be formed in equal amounts)

19-20

a.

$$\begin{array}{c} C_6H_5 \\ | \\ Br-C-H \quad R \\ | \\ H-C-Cl \quad R \\ | \\ C_6H_5 \end{array}$$

b.

$$\begin{array}{c} CH_3 \\ | \\ H-C-OH \quad S \\ | \\ H-C-D \quad R \\ | \\ CH_3 \end{array}$$

c.

$$\begin{array}{c} CO_2H \\ | \\ H-C-CH_3 \quad R \\ | \\ H-C-CH_3 \quad S \\ | \\ CO_2H \end{array}$$

d.

$$\begin{array}{c} CO_2^\ominus \\ | \\ H-C-OH \\ | \\ CH_2 \\ | \\ CO_2^\ominus \end{array}$$
R

$$\begin{array}{c} \overset{\oplus}{H_3N}-\overset{CH_3}{\underset{C_6H_5}{C}}-H \quad R \\ \\ \overset{\oplus}{H_3N}-\overset{CH_3}{\underset{C_6H_5}{C}}-H \quad R \end{array}$$

$$\begin{array}{c} CO_2^\ominus \\ | \\ H-C-OH \\ | \\ CH_2 \\ | \\ CO_2^\ominus \end{array}$$
R

$$\begin{array}{c} \overset{\oplus}{H_3N}-\overset{C_6H_5}{\underset{CH_3}{C}}-H \quad S \\ \\ \overset{\oplus}{H_3N}-\overset{C_6H_5}{\underset{CH_3}{C}}-H \quad S \end{array}$$

$$\begin{array}{c} CO_2^\ominus \\ | \\ H-C-OH \\ | \\ CH_2 \\ | \\ CO_2^\ominus \end{array}$$
R

$$\begin{array}{c} \overset{\oplus}{H_3N}-\overset{CH_3}{\underset{C_6H_5}{C}}-H \quad R \\ \\ \overset{\oplus}{H_3N}-\overset{C_6H_5}{\underset{CH_3}{C}}-H \quad S \end{array}$$

$$\begin{array}{c} CO_2^\ominus \\ | \\ H-C-OH \\ | \\ CH_2 \\ | \\ CO_2^\ominus \end{array}$$
R

$$\begin{array}{c} \overset{\oplus}{H_3N}-\overset{C_6H_5}{\underset{CH_3}{C}}-H \quad S \\ \\ \overset{\oplus}{H_3N}-\overset{CH_3}{\underset{C_6H_5}{C}}-H \quad R \end{array}$$

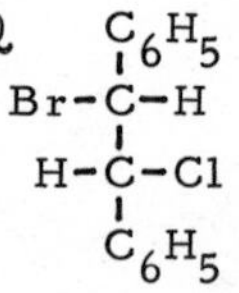

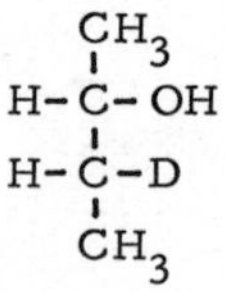

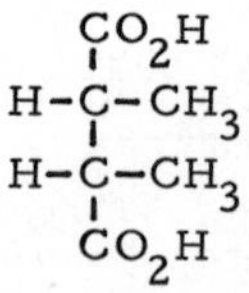

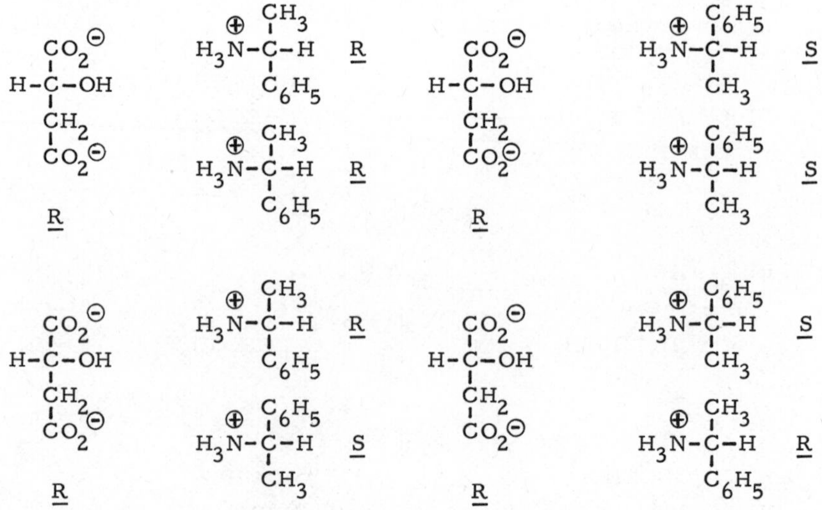

19-21

$$H_2O_2, \quad HCO_2H$$

cis   trans

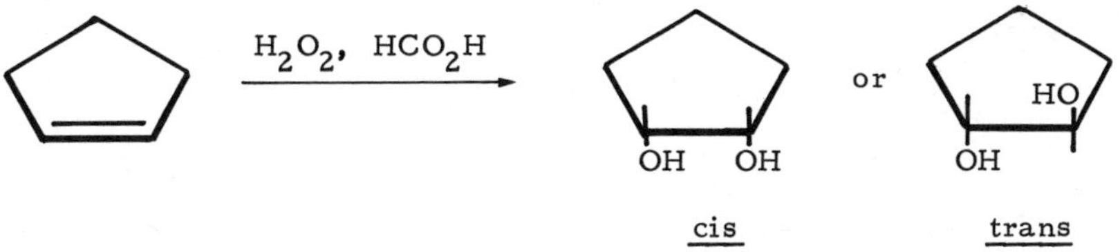

If _trans_-1, 2-cyclopentanediol is formed, it should be resolvable into optically active forms. In contrast, _cis_-1, 2-cyclopentanediol has a plane of symmetry and will not be resólvable.

19-22

$$\begin{array}{c} CH_3 \\ | \\ H-C-OH \\ | \\ CH_2 \\ | \\ CH_3 \end{array}$$

$$\begin{array}{c} CHO \\ | \\ H-C-OH \\ | \\ CH_2OH \end{array}$$

H_2 | Pt

RSH; Ni, H_2

Start here, resolve and take one isomer

$$\begin{array}{c} CH_3 \\ | \\ H-C-OH \\ | \\ CH \\ \| \\ CH_2 \end{array}$$

O_3

$$\begin{array}{c} CH_3 \\ | \\ H-C-OH \\ | \\ CHO \end{array}$$

NaBH$_4$

$$\begin{array}{c} CH_3 \\ | \\ H-C-OH \\ | \\ CH_2OH \end{array}$$

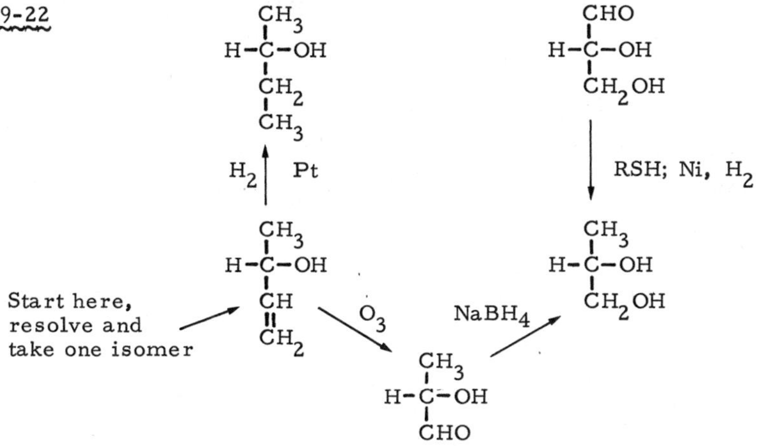

The above is an illustrative procedure which was selected to avoid breaking the H—C or —OH bonds at the chiral center. The key part of the scheme is to relate one isomer of 3-buten-2-ol to an isomer of 2-butanol and also to D-glyceraldehyde. If reduction of D-glyceraldehyde gives a 1, 2-propanediol of the same rotation as that obtained from a particular isomer of 3-buten-2-ol then that isomer is D and the hydrogenated material from it is D-2-butanol.

19-23
a.

$$\begin{array}{c} CO_2C_2H_5 \\ | \\ CHOH \\ | \\ CH_3 \end{array}$$

1. NaOH
2. H$^\oplus$

$$\begin{array}{c} CO_2H \\ | \\ CHOH \\ | \\ CH_3 \end{array}$$

resolve brucine

D-lactic acid

L-lactic acid

$C_2H_5OH, \ H^\oplus$

ethyl D-lactate + ethyl L-lactate

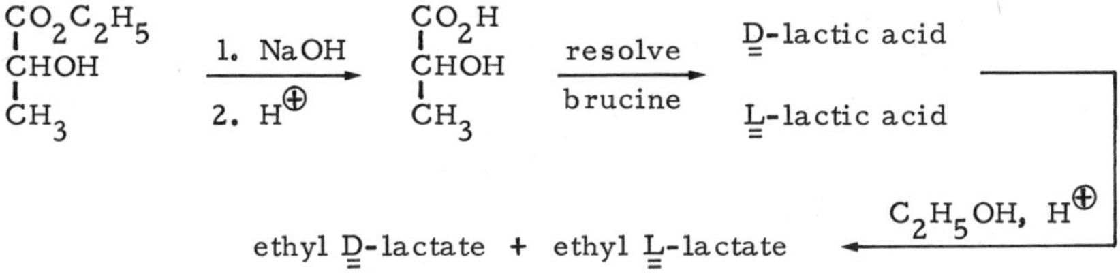

19-23 (cont.)

b.

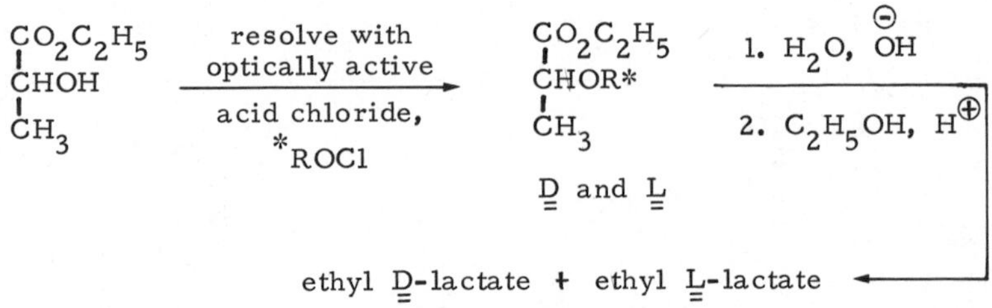

$$\begin{array}{c} CO_2C_2H_5 \\ | \\ CHOH \\ | \\ CH_3 \end{array} \xrightarrow[\substack{\text{acid chloride,} \\ *ROCl}]{\text{resolve with} \\ \text{optically active}} \begin{array}{c} CO_2C_2H_5 \\ | \\ CHOR* \\ | \\ CH_3 \\ \\ \underline{\underline{D}} \text{ and } \underline{\underline{L}} \end{array}$$

1. H_2O, $\overset{\ominus}{OH}$

2. C_2H_5OH, $H^{\oplus}$

ethyl $\underline{\underline{D}}$-lactate + ethyl $\underline{\underline{L}}$-lactate

19-24

a.

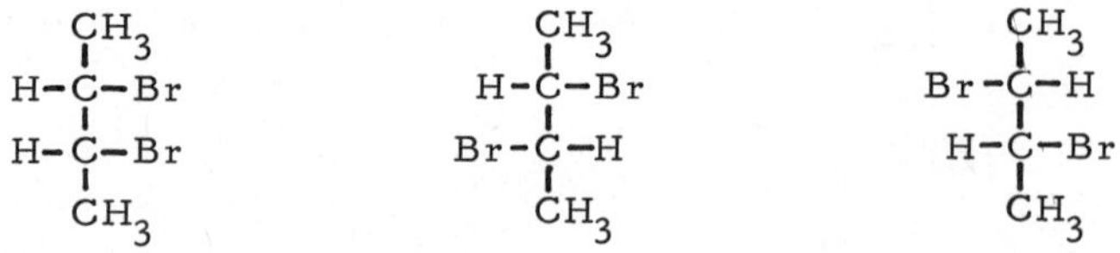

$$\begin{array}{c} CH_3 \\ | \\ H-C-Br \\ | \\ H-C-Br \\ | \\ CH_3 \end{array} \qquad \begin{array}{c} CH_3 \\ | \\ H-C-Br \\ | \\ Br-C-H \\ | \\ CH_3 \end{array} \qquad \begin{array}{c} CH_3 \\ | \\ Br-C-H \\ | \\ H-C-Br \\ | \\ CH_3 \end{array}$$

b.   Bromine addition to alkenes proceeds in a <u>trans</u> manner. The reverse reaction, debromination of a 1,2-dibromide, must also proceed in a <u>trans</u> manner. Thus, optically active 2,3-dibromobutane will give cis-2-butene, as can be seen from the mechanism given in part (c).

c.

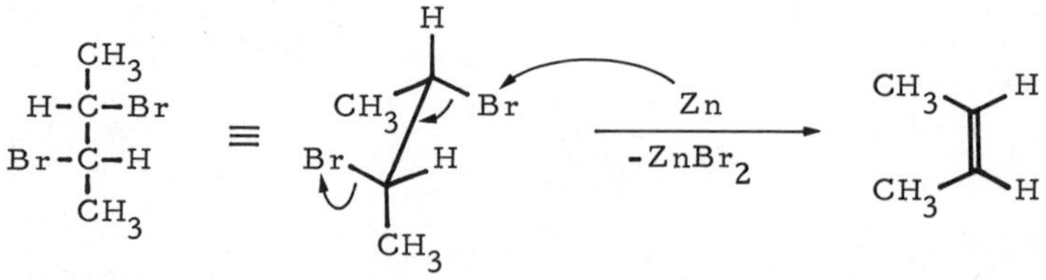

19-25   Each elimination reaction with the butene dibromides can be formulated as a <u>trans</u> 1,2-elimination, as shown below:

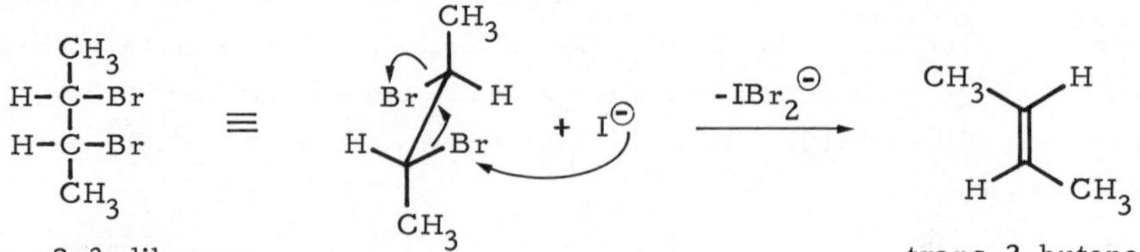

meso-2,3-dibromo-butane

trans-2-butene

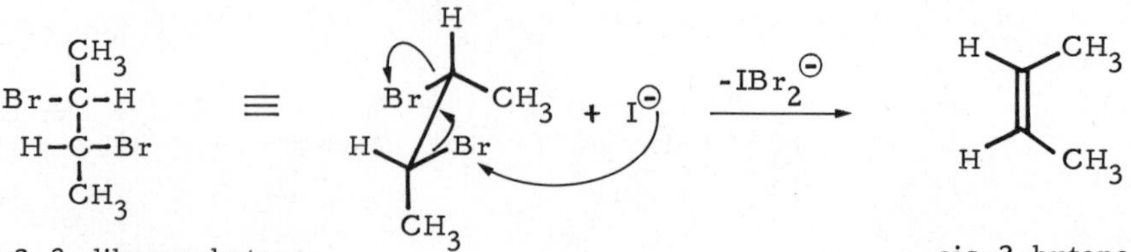

$\underline{\underline{D}}$-2,3-dibromobutane

cis-2-butene

(Similarly, the $\underline{\underline{L}}$-isomer gives <u>cis</u>-2-butene.)

With the 1, 2-dibromo-1, 2-dideuterioethane the same mechanism gives
the wrong isomer.

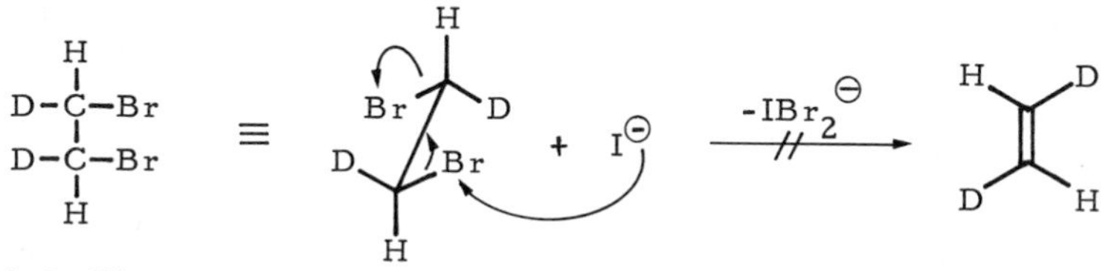

meso-1, 2-dibromo-
1, 2-dideuterioethane

In order to have something like the same elimination mechanism and obtain the
observed product we must have an inversion step somewhere. This could be the
result of an S_N2-displacement by $I^{\ominus}$ on the original dibromide followed by
elimination of IBr in place of Br_2.

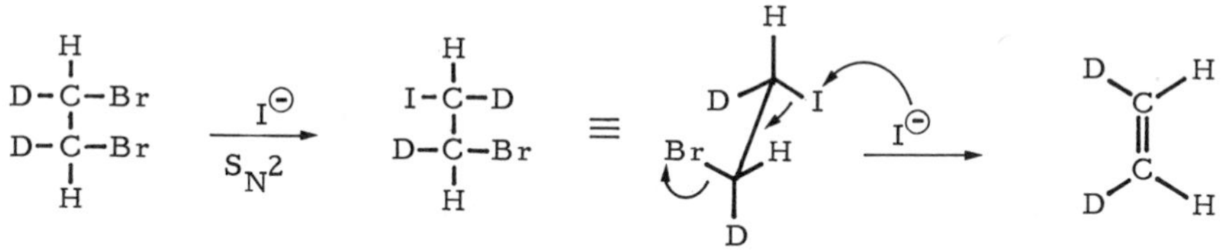

19-26   From the following detailed mechanism of the acid-catalyzed dehydration
of diastereomeric 2, 3-butanediols one may predict that the D or L forms will
give mostly 2-butanone because they lead to a transition state in which the
methyl groups offer the least hindrance to $:H^{\ominus}$ transfer - being in a trans
conformation.

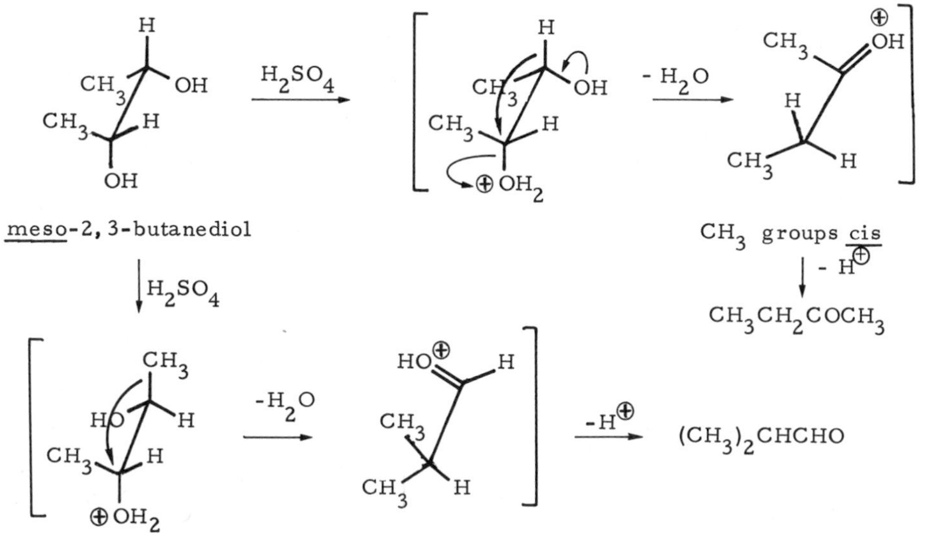

meso-2, 3-butanediol

CH_3 groups cis

$CH_3CH_2COCH_3$

$(CH_3)_2CHCHO$

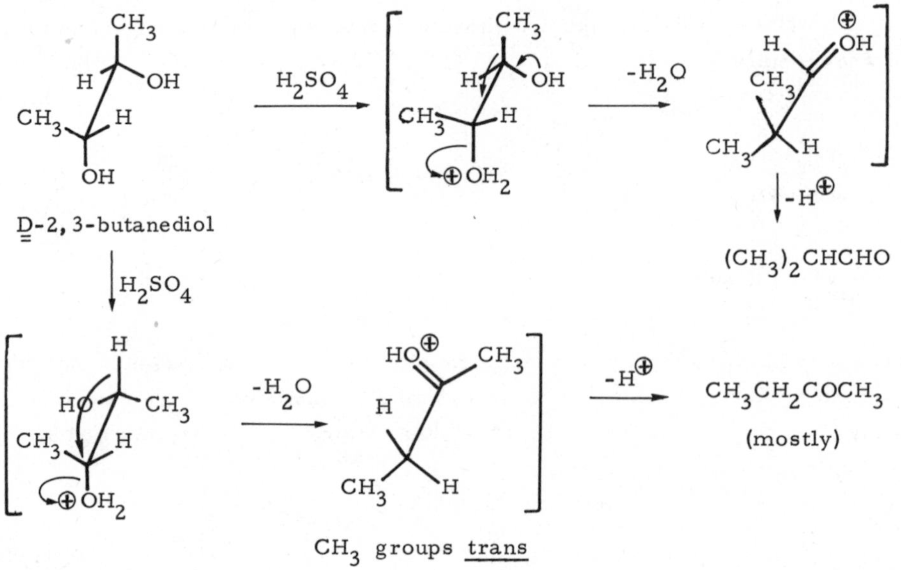

D-2,3-butanediol

(CH₃)₂CHCHO → $(CH_3)_2CHCHO$

CH₃ groups trans

19-27   If (+)-tartaric acid were transformed by reduction of either carboxyl to methyl followed by reduction of the CHOH group next to the remaining carboxyl to a methylene group, L-3-hydroxybutanoic acid would result. One would then say that (+)-tartaric acid belongs to the L series, which is contradictory to the conclusion based on reduction to D-malic acid that it belongs to the D series.

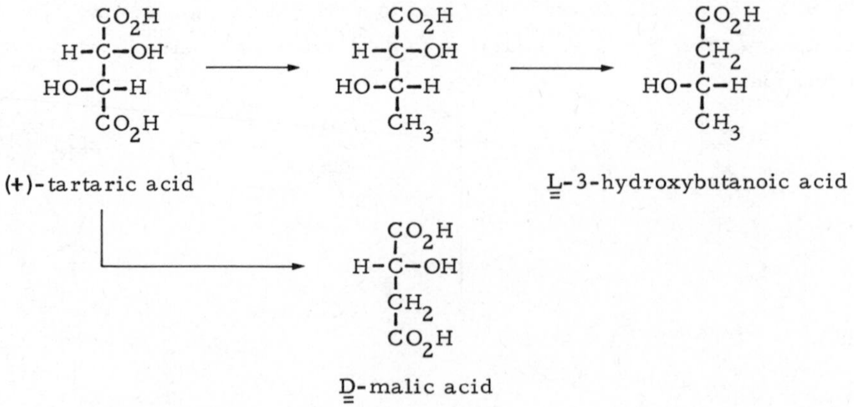

(+)-tartaric acid

L-3-hydroxybutanoic acid

D-malic acid

19-28   Racemization of A by way of enolization is precluded since the rate of racemization is not enhanced in an acidic solvent and no deuterium is observed in A when racemization is carried out in CH_3CO_2D.

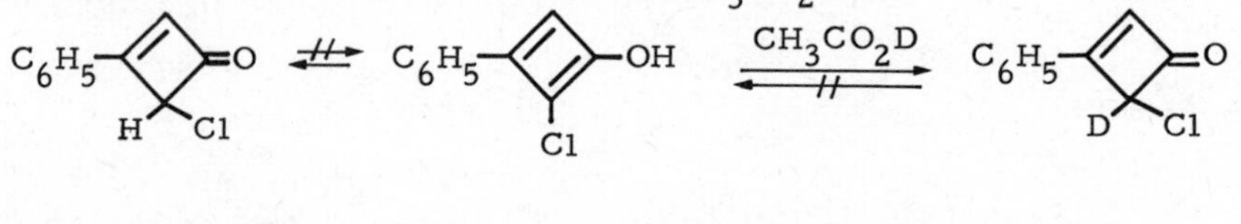

Racemization by S_N2-type chloride exchange is also not important. Racemization is most reasonably explained by a reversible ring opening reaction brought about thermally.

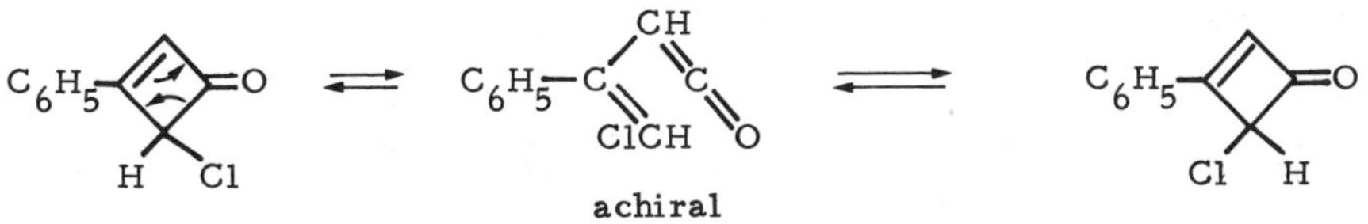

achiral

19-29   The specific rotation of a compound is given by the equation

$$[\alpha]^{t}_{\lambda} = \frac{\alpha}{1 \cdot c}$$

To verify that a measured rotation of -100° is or is not actually +260°, measurements at two or more concentrations (c) must be carried out. The correct data will mutually satisfy the above equation relating $[\alpha]$ to α, 1 and c.

19-30   The scale drawings reveal a strong "buttressing" effect of the 3- and 3'-iodines.

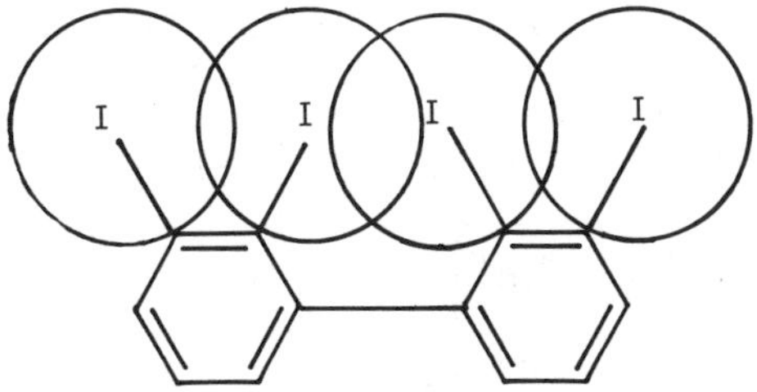

The rotation would **never pass** I against I but **rather** I against H.

This prevents the 2,2'-iodines from being bent backward in the transition state for racemization and thus slows the rate of racemization.

19-31   a.   The <u>ortho</u>-methyl substituents must interfere sufficiently to force them out of the plane of the phenanthrene ring. With one CH_3 above and one CH_3 below the ring plane, the compound has no elements of symmetry and is therefore resolvable.

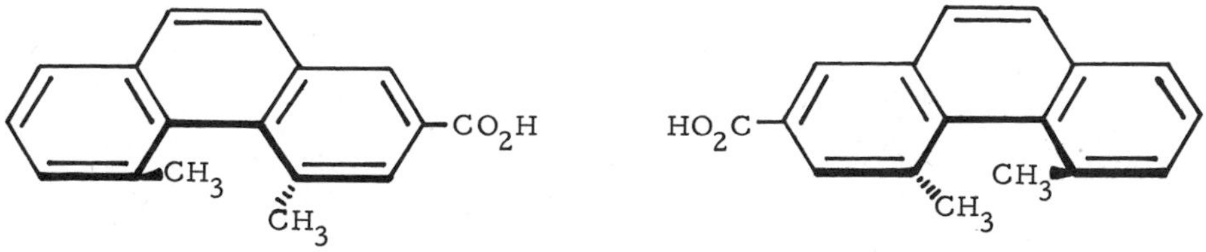

b.   Three stereoisomers are expected for <u>30</u> consisting of one pair of enantiomers and one <u>meso</u> isomer.

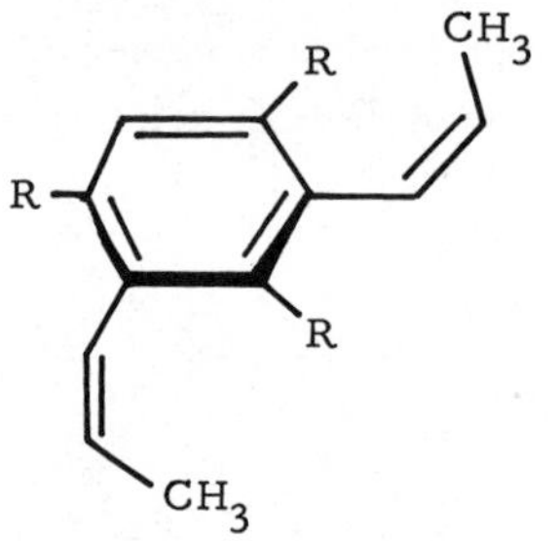

(and enantiomer)   (mes6)

19-32   (Asterisk denotes ^{14}C.)

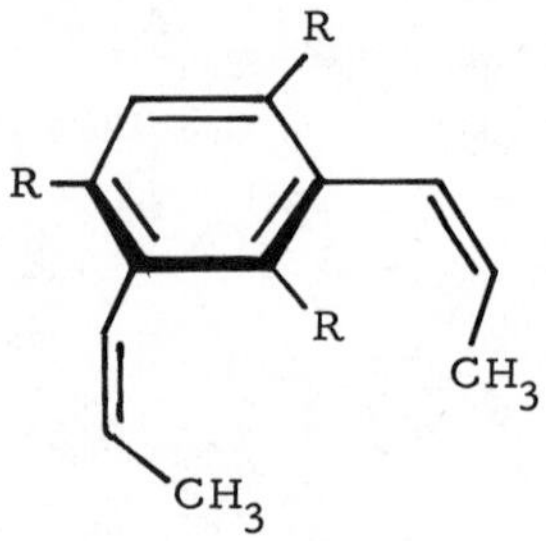

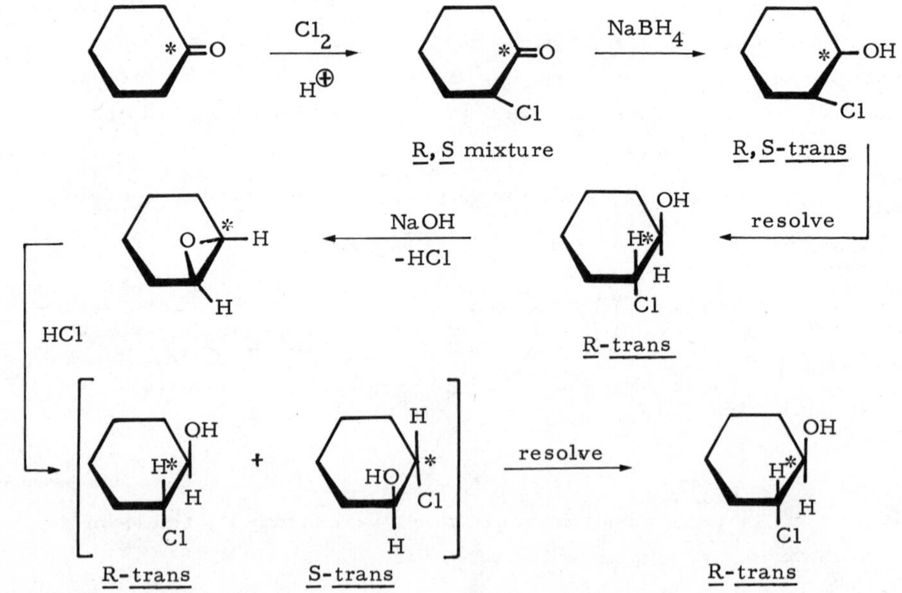

Note that the label $^{14}C = *$ is still in the same place in the R-trans final product as before conversion to the oxide, despite going through what might be regarded as a symmetrical intermediate, the oxide.

CHAPTER 20

CARBOHYDRATES

Carbohydrates are polyhydroxy aldehydes or polyhydroxy ketones of the general formula $(C \cdot H_2O)_x$. They include substances that can be hydrolyzed to polyhydroxy aldehydes or ketones, or, in the very broadest sense, derivatives of polyhydroxy aldehydes or ketones. The monosaccharides or simple sugars are called aldotetroses, ketopentoses, and so on, as befits the number of carbons in the chain, and whether there is an aldehyde or ketone carbonyl. If the carbon chain is long enough and the carbonyl group suitably located, the monosaccharides exist principally as cyclic hemiacetals or hemiketals, with five-membered oxacyclopentane rings (furanoses) or six-membered oxacyclohexane rings (pyranoses).

D-(+)-Glucose $(C_6H_{12}O_6)$ is an especially important aldohexose, which, when written in the open-chain aldehyde form, is predicted to have $2^4 = 16$ possible stereoisomers. Only one corresponds to D-(+)-glucose. However, formation of the cyclic hemiacetal with a six-membered ring results in the creation of two additional isomers, called anomers or anomeric forms and designated as α - and β -D-glucose.

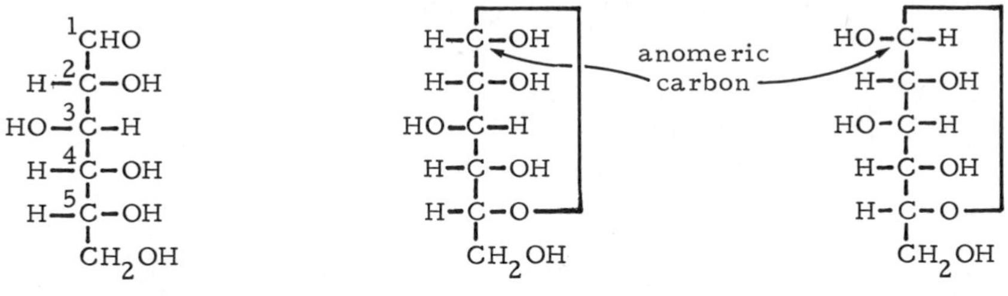

D-glucose	α -D-glucose	β -D-glucose
(aldehyde form)		

The configurations of these isomers are shown in Fischer projections. The α anomer is the isomer having the same configuration at the anomeric carbon as at the carbon that determines whether the monosaccharide is D or L (C5 for glucose; the anomeric carbon is always C1 in aldoses.)

The configurations of the four chiral carbons of the aldehyde form of glucose were related to the configuration of D-(+)-glyceraldehyde by Emil Fischer. All of the 16 possible stereoisomers of glucose occur naturally or have been synthesized. A key reaction in determining aldose configurations is controlled oxidation of the end groups on the chain to give dicarboxylic acids, which may be meso and hence achiral. For example, oxidation of the first of the two following chiral aldotetroses gives the achiral meso-tartaric acid, whereas oxidation of the second gives the chiral D-tartaric acid:

```
    CHO                    CO2H                    CHO                    CO2H
     |                      |                       |                      |
  H-C-OH      [O]        H-C-OH                 HO-C-H       [O]        HO-C-H
     |         --->         |                       |          --->        |
  H-C-OH                 H-C-OH                  H-C-OH                  H-C-OH
     |                      |                       |                      |
   CH2OH                  CO2H                    CH2OH                   CO2H

 D-erythrose              meso                  D-threose                  D
```

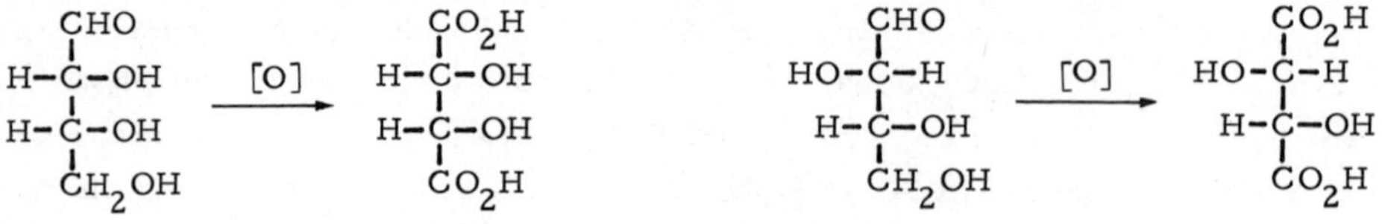

The Wohl degradation achieves the conversion of an aldose to the next lower aldose with loss of C1. For an aldotetrose, a Wohl degradation allows the configuration of the highest-numbered chiral carbon to be compared with that of glyceraldehyde:

```
    CHO
     |
  H-C-OH              Wohl                CHO
     |            ------------->           |
  H-C-OH          degradation           H-C-OH
     |                                     |
   CH2OH                                 CH2OH
```

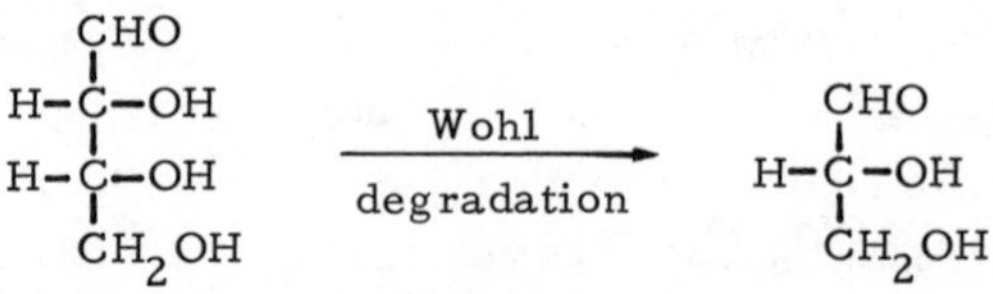

The Kiliani-Fischer cyanohydrin synthesis converts the - CHO group to - CHOH-CHO, thereby transforming a tetrose to a pentose, pentose to hexose, and so on. Because a new chiral center is created, two stereoisomers are expected to be formed. For example,

```
    CHO                                    CHO                    CHO
     |                                      |                      |
  H-C-OH       Kiliani-Fischer          H-C-OH                 HO-C-H
     |        cyanohydrin synthesis        |                      |
  H-C-OH      ------------------->      H-C-OH        +        H-C-OH
     |                                      |                      |
   CH2OH                                 H-C-OH                  H-C-OH
                                            |                      |
                                          CH2OH                  CH2OH
```

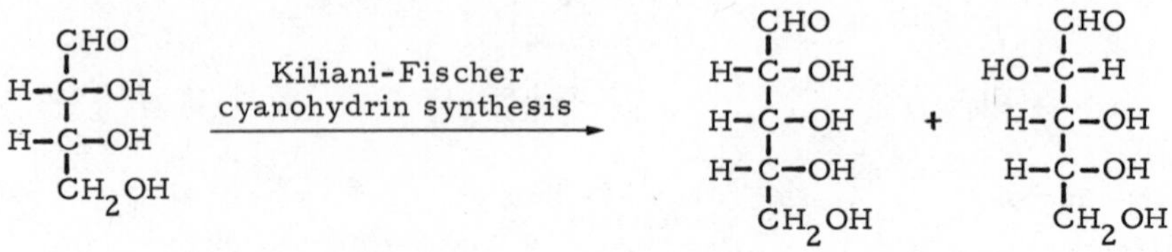

Glucose is most stable in the cyclized form of the α and β anomers, but because of their hemiacetal structures, they are converted readily to the aldehyde form (and to each other) by acidic or basic catalysts. As a result, glucose shows many aldehyde reactions - it forms an oxime, reduces Fehling's and Tollen's reagents, and so on. With many reagents, such as hydroxy compounds and amines, glucose gives products known as O- or N-glucosides, which have cyclic structures and can exist in two anomeric configurations. With methanol and acid, glucose forms two O-glucosides; both are pyranosides and differ only in the configuration at C1, the anomeric carbon. In the chair form of the oxacyclohexane ring of glucose, the substituents at C2, C3, C4, and C5 occupy equatorial positions. The β forms of the O- and N-glucosides of D-glucose normally are the most stable because the O or N substituent is equatorial:

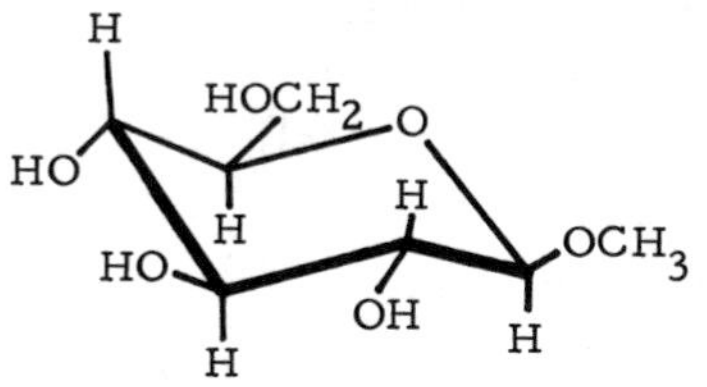

methyl β -D-glucopyranoside   methyl α -D-glucopyranoside
(a β -O-glycoside)   (an α -O-glycoside)

These derivatives of glucose are acetals and under the mildly alkaline conditions used with Fehling's and Tollen's reagents do not revert to the aldehyde
form.   Therefore such derivatives do not act as reducing agents and are called
nonreducing sugar derivatives.

Interconversion of sugar anomers, such as α- and β-D-glucose under
the influence of acid or base catalysts, is called mutarotation. Carbon-13 nmr
spectroscopy is a powerful tool for studying anomeric equilibria.

The size of the oxide ring of a monosaccharide usually can be determined by examination of the periodate oxidation products of its O-methyl
glycoside.

D-Glucose combines with two molecules of phenylhydrazine (phenyl-
diazane) to give D-glucose phenylosazone, in which C2 has been oxidized to
the carbonyl level. The aldohexose, D-mannose, and the ketohexose, D-fructose, give the same phenylosazone as D-glucose, which shows that the
configurations of C3, C4, and C5 of D-glucose, D-mannose, and D-fructose
are the same.

Many important sugar derivatives are O-glycosides. An example is
glucovanillin, which is a β -O-glycoside of D-glucose and vanillin, 3-methoxy-
4-hydroxybenzenecarbaldehyde. The vanillin is connected to C1 of the glucose
at the 4—OH position. The nonsugar component of such glycosides are called
aglycones.

The ribonucleosides and deoxyribonucleosides are N-glycosides of

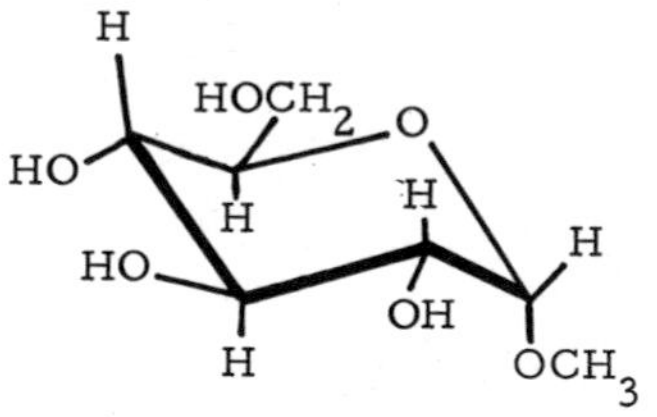

ribose and 2-deoxyribose and nitrogen bases such as adenine.   Nucleotides are various phosphate esters of the nucleosides.

Combination of monosaccharides with one another by glycoside linkages gives disaccharides, trisaccharides, and so on.   Oligosaccharides are such substances with less than ten monosaccharide units.   Polysaccharides have more than ten monosaccharide units.   The combination of two monosaccharides can occur so that one acts as an aglycone for the other by having the oxygen of a hydroxyl group at a carbon other than C1 (usually C4) attached to C1 of the other sugar.   This combination normally gives a reducing sugar, because the saccharide acting as the aglycone has a hemiacetal or hemiketal group.   Lactose, cellobiose, and maltose are such disaccharides.   Cellobiose is 4-O-β-D-glucopyranosyl-β-D-glucopyranose.

Sucrose (common table sugar) has a different kind of glycoside connection in that the two monosaccharides, D-glucose and D-fructose, are linked together in a 1,1'-glycosidic linkage, with the anomeric hydroxyl of each sugar effectively acting as the aglycone for the other.   The result is acetal formation for the glucose and ketal formation for the fructose.   Disaccharides of this type are nonreducing sugars because they have no hemiacetal or hemiketal groups.   Sucrose also is interesting in that the fructose is bound in the furanose form.   Therefore sucrose is β-D-fructofuranosyl-α-D-glucopyranoside.

It often is possible to distinguish between α and β glycosidic linkages by testing the ease of hydrolysis with enzymes that show a preference for one configuration over the other.   Thus the enzyme emulsin works better with β linkages, whereas maltase is more effective with α linkages.

Cellulose is a polysaccharide made up of long chains of β-D-glucose units linked between C1 of one glucose and C4 of the next.   Cotton, jute, flax, and hemp are natural cellulose fibers.   Rayon, cellulose acetate, cellulose acetate butyrate, and cellulose nitrate are derivatives of cellulose that have important commercial uses.

Starch is found in two varieties.   Amylose is a polysaccharide with the same structure as cellulose except that the anomeric carbon has the α configuration.   Amylopectin has a similar structure except that some branching is present involving the C6 hydroxyls (that is, 1,6-glycosidic linkages).   Dextrin is a complex mixture of partially hydrolyzed starch molecules.   Cyclohexaamylose is a cyclic, doughnut-shaped, partially hydrolyzed starch.   Its cavity can act as a host for proper-sized nonpolar molecules, and the molecules held in the cavity often show different reactivities from those outside.

Chitin, which constitutes the shells of insects and crustaceans, is like cellulose except that the 2-hydroxyl is replaced by a 2-N-ethanoylamide group.   The important blood anticoagulant, heparin, is an O-sulfate ester and N-sulfamido derivative of a complex polysaccharide comprised of $\underline{D}$-glucuronic acid, $\underline{L}$-iduronic acid, and α-2-deoxy-2-amino-$\underline{D}$-glucose. The pectins and plant gums also are largely polyuronic acids. The hemicelluloses are constituents of plant cell walls, which are largely made up of pentose units.

Vitamin C is a lactone, which, because of its considerable acidity arising from the presence of an enediol group, is called ascorbic acid. It has the $\underline{L}$ configuration of its single chiral carbon.

Carbohydrates are formed in green plants by photosynthesis by the overall reaction $x\,CO_2 + \underline{x}\,H_2O \longrightarrow (C \cdot H_2 O)_x + x\,O_2$. The light energy is used to oxidize water to oxygen, reduce $\overline{NADP}^{\oplus}$ to NADPH, and convert ADP to ATP. Carbon dioxide is actually "fixed" in a dark reaction with an enzyme-mediated carboxylation of a ketopentose diphosphate ($\underline{D}$-ribulose 1,5-diphosphate). Cleavage of the six-carbon acid so formed gives two molecules of 3-phospho-$\underline{D}$-glyceric acid, which subsequently are converted to $\underline{D}$-fructose. The $\underline{D}$-ribulose 1,5-diphosphate is regenerated for combination with another molecule of carbon dioxide by an amazing series of reactions in which sugars of different sizes combine and fragment until two pentoses emerge (Calvin cycle).

Metabolic energy is gained by converting carbohydrates to carbon dioxide and water - the reverse of photosynthesis. Some of the energy is converted to heat, but fully 40% is stored for other purposes by producing ATP from ADP.

The first part of the metabolism, glycolysis, breaks glucose down in several steps to two molecules of 3-phospho-$\underline{D}$-glyceric acid. This sequence consumes ATP but generates NADH, which provides energy for ATP synthesis when oxidized by O_2. The 3-phospho-$\underline{D}$-glyceric acid then is converted to 2-oxopropanoic (pyruvic) acid in a nonoxidative process that produces ATP from ADP.

The 2-oxopropanoic acid is oxidatively cleaved to carbon dioxide and ethanoyl (acetyl) CoA with the production of NADH from $NAD^{\oplus}$. The ethanoyl CoA then enters the "citric acid cycle" and is oxidized to CO_2 through a succession of intermediates, the production of which at almost every stage results in energy storage through reduction of $NAD^{\oplus}$ to NADH or FAD to $FADH_2$, or through a phosphorylation reaction. Additional conversions of ADP to ATP result from coupling oxygen into the overall process by indirect oxidation of NADH to $NAD^{\oplus}$ and $FADH_2$ to FAD. The overall result is the production of 36 molecules of ATP from ADP and inorganic phosphate per molecule of glucose oxidized to CO_2 and water.

The metabolic oxidation of long-chain carboxylic acids (fatty acids) involves removing two carbons at a time from the carboxylate end of the molecule in the form of ethanoyl CoA. The ethanoyl CoA so formed enters the citric cycle and is converted to CO_2 and water.

ANSWERS TO EXERCISES

20-1   a.   The four possible D-aldopentoses are:

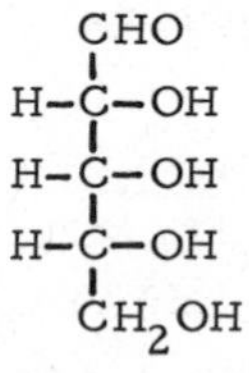

$$
\begin{array}{cccc}
\text{CHO} & \text{CHO} & \text{CHO} & \text{CHO} \\
\text{H-C-OH} & \text{HO-C-H} & \text{H-C-OH} & \text{HO-C-H} \\
\text{H-C-OH} & \text{H-C-OH} & \text{HO-C-H} & \text{HO-C-H} \\
\text{H-C-OH} & \text{H-C-OH} & \text{H-C-OH} & \text{H-C-OH} \\
\text{CH}_2\text{OH} & \text{CH}_2\text{OH} & \text{CH}_2\text{OH} & \text{CH}_2\text{OH} \\
1 & 2 & 3 & 4
\end{array}
$$

Oxidation of 2 and 4 would give optically active 2, 3, 4-trihydroxypentane-dioic acids.

$$
2 \xrightarrow{\text{HNO}_3}
\begin{array}{c}
\text{CO}_2\text{H} \\
\text{HO-C-H} \\
\text{H-C-OH} \\
\text{H-C-OH} \\
\text{CO}_2\text{H}
\end{array}
\qquad
4 \xrightarrow{\text{HNO}_3}
\begin{array}{c}
\text{CO}_2\text{H} \\
\text{HO-C-H} \\
\text{HO-C-H} \\
\text{H-C-OH} \\
\text{CO}_2\text{H}
\end{array}
$$

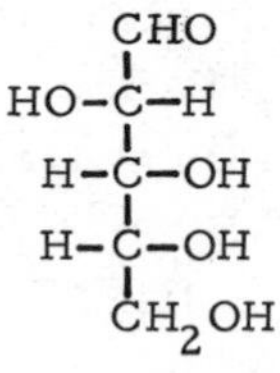

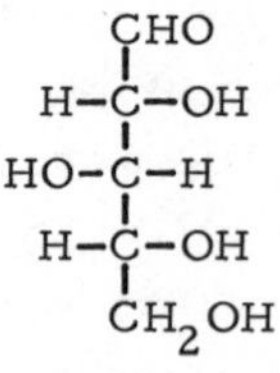

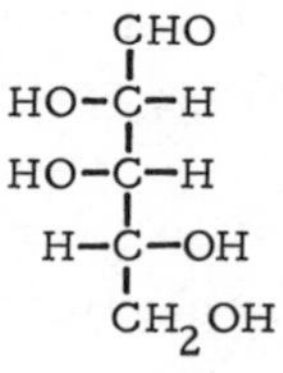

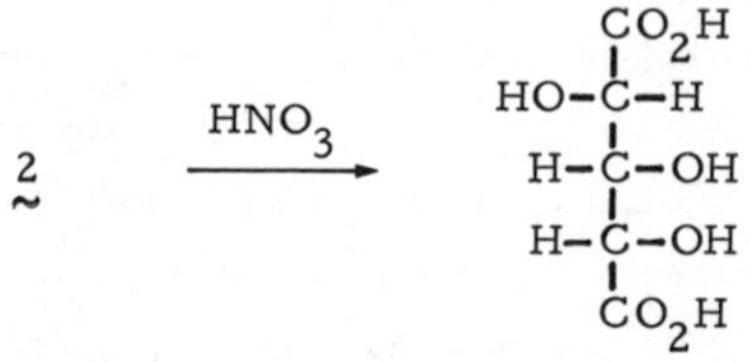

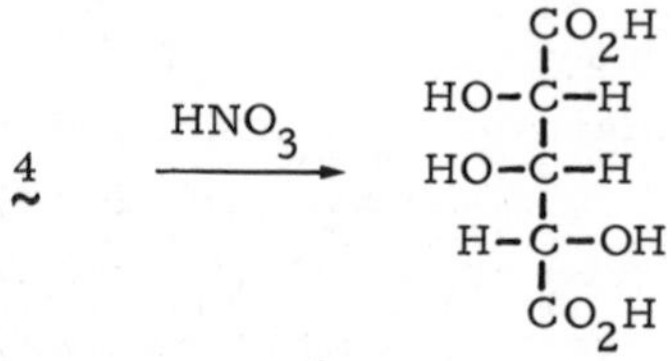

Therefore, either 2 or 4 could be D-arabinose.

b.   Because D-glucose and D-mannose can be prepared from D-arabinose, they must differ only in the configuration of C2.

c.   The possible configurations of glucose and mannose are:

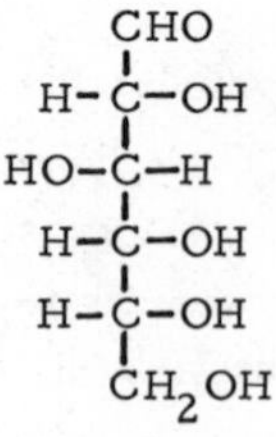

$$
\begin{array}{cccc}
\text{CHO} & \text{CHO} & \text{CHO} & \text{CHO} \\
\text{H-C-OH} & \text{HO-C-H} & \text{H-C-OH} & \text{HO-C-H} \\
\text{HO-C-H} & \text{HO-C-H} & \text{HO-C-H} & \text{HO-C-H} \\
\text{H-C-OH} & \text{H-C-OH} & \text{HO-C-H} & \text{HO-C-H} \\
\text{H-C-OH} & \text{H-C-OH} & \text{H-C-OH} & \text{H-C-OH} \\
\text{CH}_2\text{OH} & \text{CH}_2\text{OH} & \text{CH}_2\text{OH} & \text{CH}_2\text{OH} \\
5 & 6 & 7 & 8
\end{array}
$$

Configuration 7 may be rejected since oxidation would lead to an optically inactive dicarboxylic acid. Therefore, 5 or 6 represents glucose or mannose. Oxidation of either would give an optically active dicarboxylic acid. The configuration of D-arabinose is therefore represented by 2 . L-Arabinose is the enantiomer of 2 .

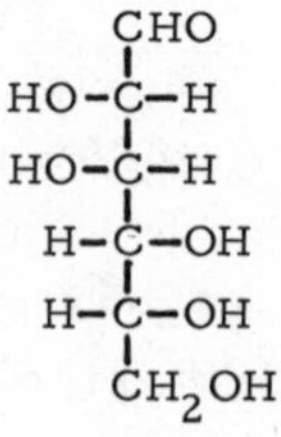

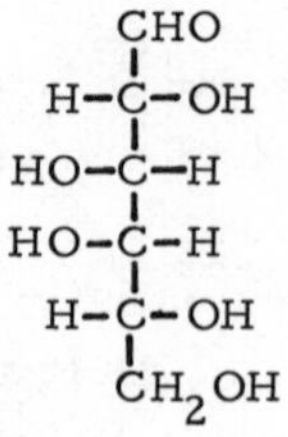

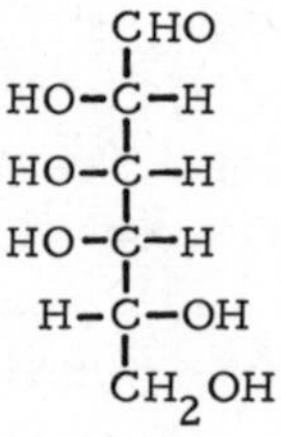

d.   The γ-lactones from glucaric and manaric acids may have the following configurations:

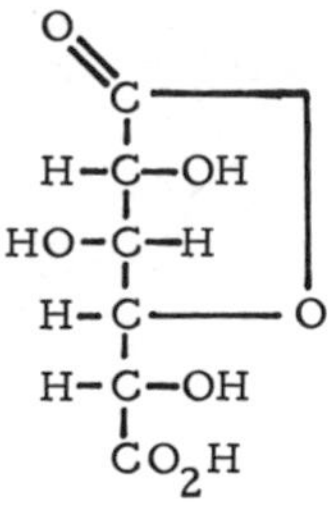

$$
\begin{array}{cccc}
\text{9} & \text{10} & \text{11} & \text{12}
\end{array}
$$

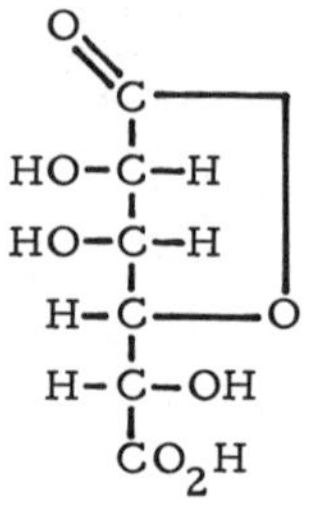

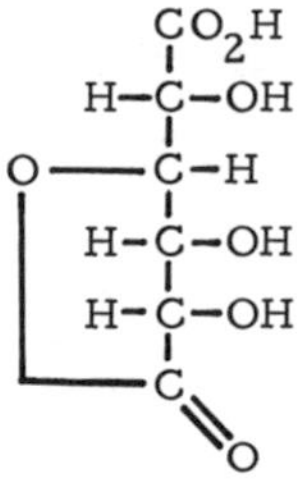

Note that   10   and   12   are identical and must therefore represent the lactone from D-mannaric acid.   D-Mannose must then have the configuration 6 , and by elimination, D-glucose must have configuration 5.

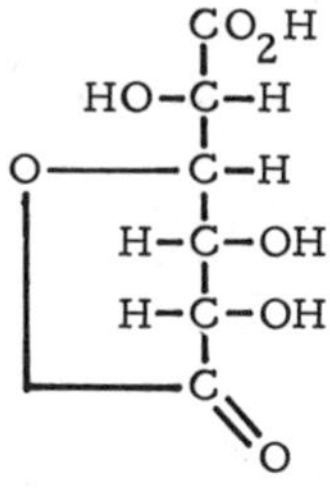

20-2   Starting with the information in Part (2), we may conclude that the configurations at C4 and C5 of D-galactose are L and D, respectively.   The optically active pentose of Part (1) may have one of the following configurations, 1   or   2 .   However,   1   is excluded because oxidation would lead to an optically inactive dicarboxylic acid.   Configuration at C3 of D-galactose is therefore L.

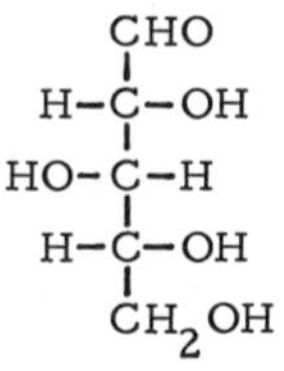

$$
\begin{array}{cc}
\text{CHO} & \text{CHO} \\
\text{H-C-OH} & \text{HO-C-H} \\
\text{HO-C-H} & \text{HO-C-H} \\
\text{H-C-OH} & \text{H-C-OH} \\
\text{CH}_2\text{OH} & \text{CH}_2\text{OH} \\
\text{1} & \text{2}
\end{array}
$$

D-Galactose must have one of the following configurations,   3   or   4 .

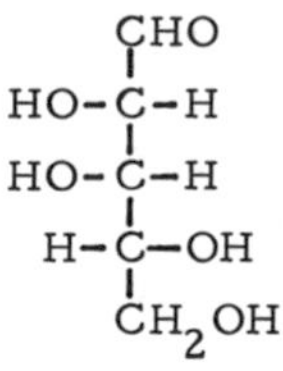

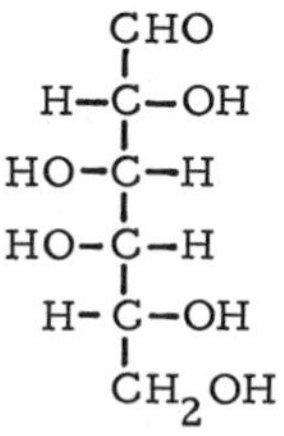

$$
\begin{array}{cc}
\text{CHO} & \text{CHO} \\
\text{H-C-OH} & \text{HO-C-H} \\
\text{HO-C-H} & \text{HO-C-H} \\
\text{HO-C-H} & \text{HO-C-H} \\
\text{H-C-OH} & \text{H-C-OH} \\
\text{CH}_2\text{OH} & \text{CH}_2\text{OH} \\
\text{3} & \text{4}
\end{array}
$$

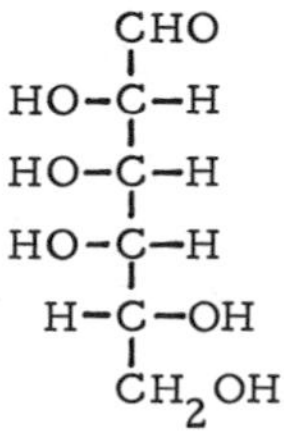

Mechanism of the Wohl degradation:

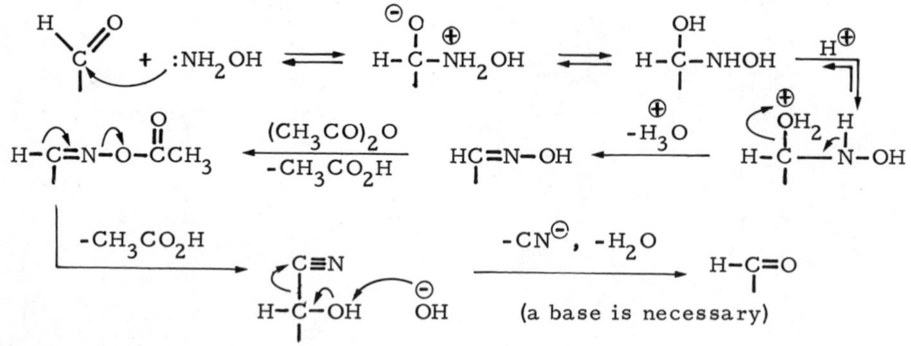

<u>20-3</u>   a. different (enantiomers)   b. same   c. all three are different

<u>20-4</u>

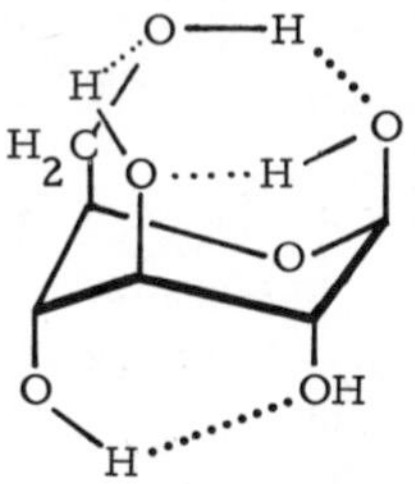

        Hydrogen-bonding between the <u>cis</u>-1, 3-diaxial hydroxyl groups and the
CH$_2$OH group is possible and would stabilize the all-axial conformation. Hydrogen bonding between <u>trans</u>-1, 2- and <u>cis</u>-1, 3-diol groups is not a factor that
contributes much to the stabilization of the all-equatorial conformer.

<u>20-5</u>   The chair conformations of α- and β-$\underline{\underline{D}}$-ribopyranose are:

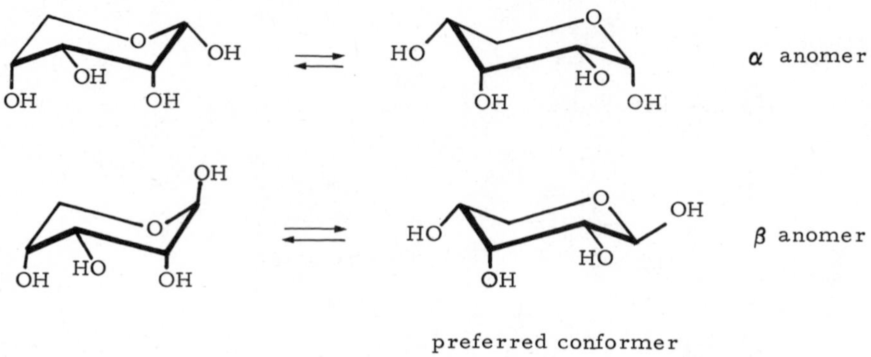

preferred conformer

The chair conformations of α-, and β-D-idopyranose are:

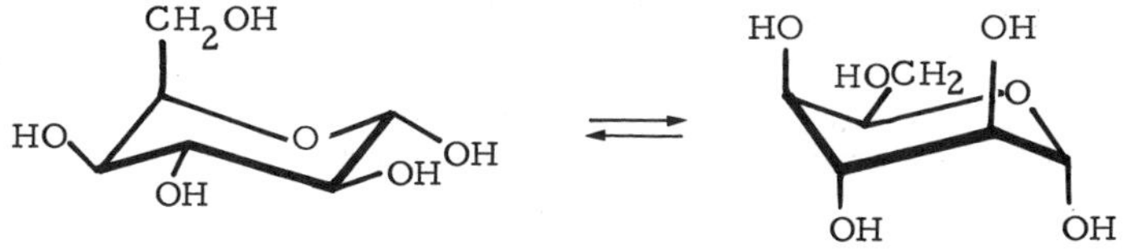

preferred conformer

α anomer

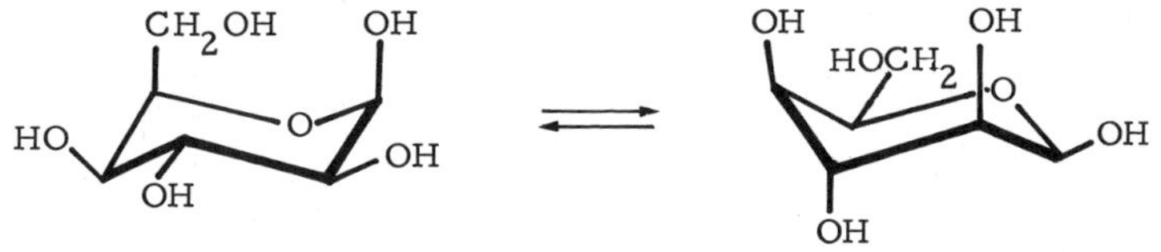

β anomer

20-6

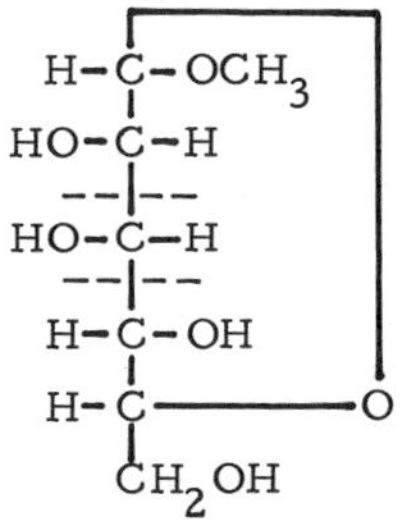

methyl α-D-mannopyranose

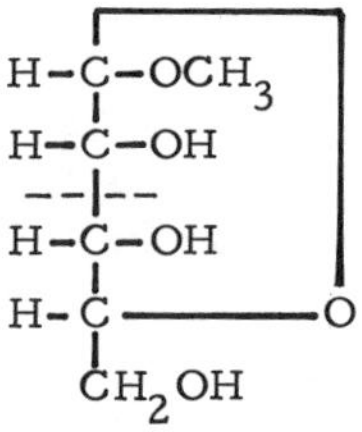

methyl α-D-ribofuranose

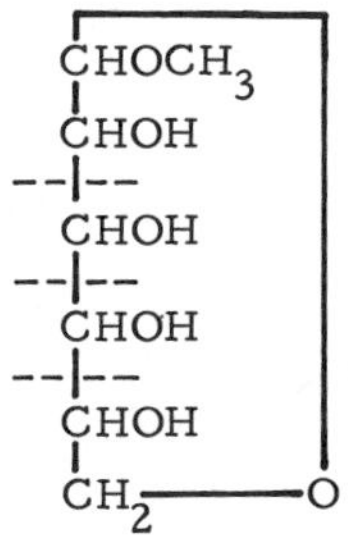

methyl glycoside of an aldohexose

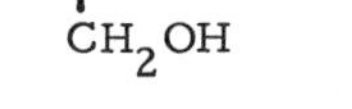

methyl glycoside of an aldohexose

20-7   D-Arabinose must have the same configuration at C3 and C4 as
D-ribose since both sugars give the same phenylosazone. The configuration at
both C3 and C4 must be D since the following transformations give rise to
meso-tartaric acid.

$$
\begin{array}{c}
\text{CHO} \\
\text{CHOH} \\
\text{H-C-OH} \\
\text{H-C-OH} \\
\text{CH}_2\text{OH}
\end{array}
\quad \xrightarrow{\text{Ruff degradation}} \quad
\begin{array}{c}
\text{CHO} \\
\text{H-C-OH} \\
\text{H-C-OH} \\
\text{CH}_2\text{OH}
\end{array}
\quad \xrightarrow{\text{HNO}_3} \quad
\begin{array}{c}
\text{CO}_2\text{H} \\
\text{H-C-OH} \\
\text{H-C-OH} \\
\text{CO}_2\text{H}
\end{array}
$$

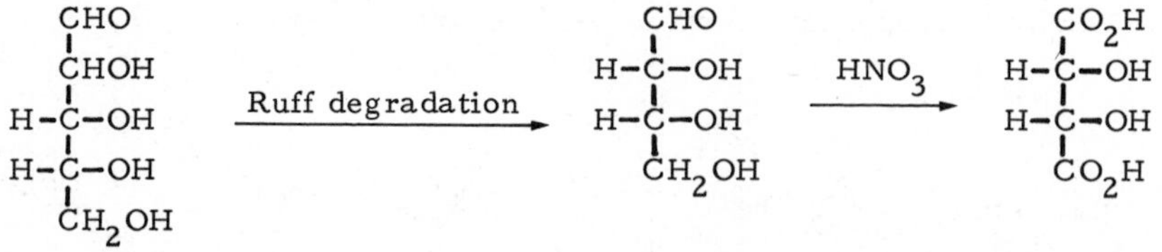

D-arabinose   D-erythrose   meso-tartaric acid

The configuration at C2 of D-ribose must also be D since reduction leads to optically _inactive_ ribitol. D-Arabinose therefore has the L configuration at C2.

$$
\begin{array}{c}
\text{CHO} \\
\text{H-C-OH} \\
\text{H-C-OH} \\
\text{H-C-OH} \\
\text{CH}_2\text{OH}
\end{array}
\quad \xrightarrow{\text{H}} \quad
\begin{array}{c}
\text{CH}_2\text{OH} \\
\text{H-C-OH} \\
\text{H-C-OH} \\
\text{H-C-OH} \\
\text{CH}_2\text{OH}
\end{array}
\qquad
\begin{array}{c}
\text{CHO} \\
\text{HO-C-H} \\
\text{H-C-OH} \\
\text{H-C-OH} \\
\text{CH}_2\text{OH}
\end{array}
$$

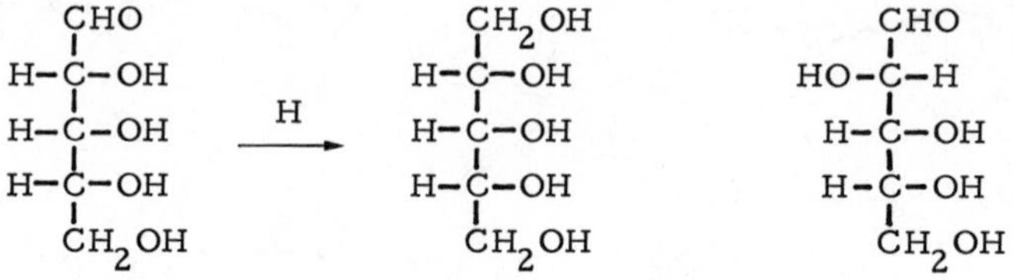

D-ribose   ribitol   D-arabinose

20-8

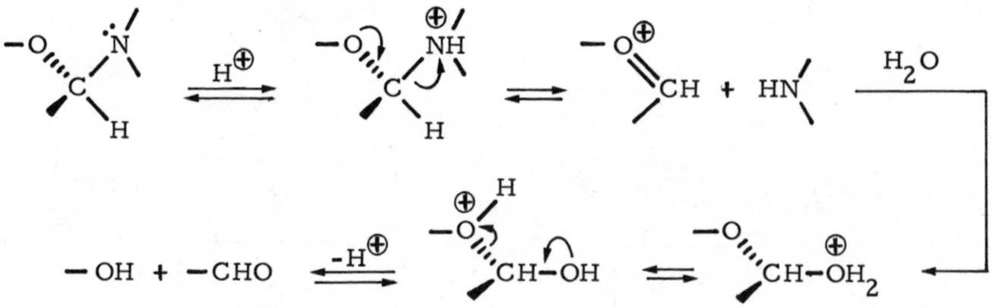

Adenosine would be expected to hydrolyze _less_ readily than N-methyl-α-ribosylamine because the adenine nitrogen is much less basic than the methanamine nitrogen (Section 23-7D).

20-9   a. 25, 26, and 27 are reducing sugars; each has a free anomeric carbon of an aldose.

        b. 24 α-glucosyl connected to β-fructosyl; 25 both α; 26 both β; 27 β-galactosyl to C4 of β-glucose.

        c. 24 $\longrightarrow$ D-glucose + D-fructose

            25 $\longrightarrow$ D-glucose

            26 $\longrightarrow$ D-glucose

            27 $\longrightarrow$ D-galactose and D-glucose

20-10

a.

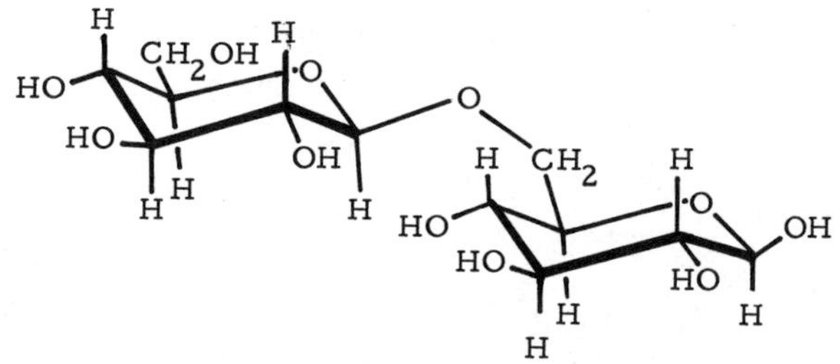

b.

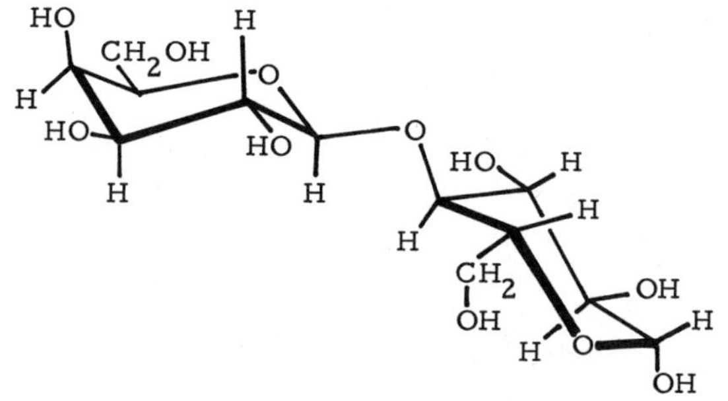

c.

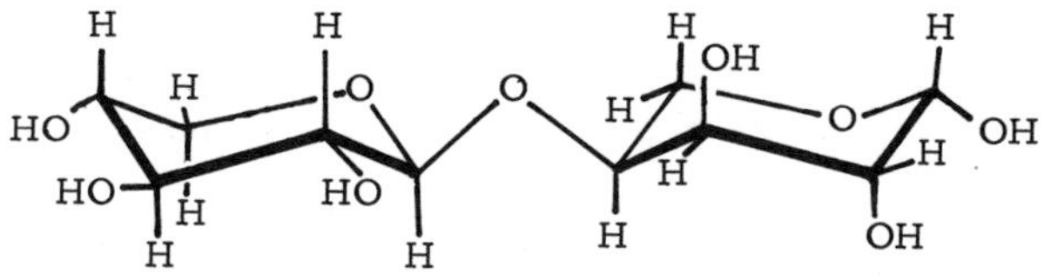

d.

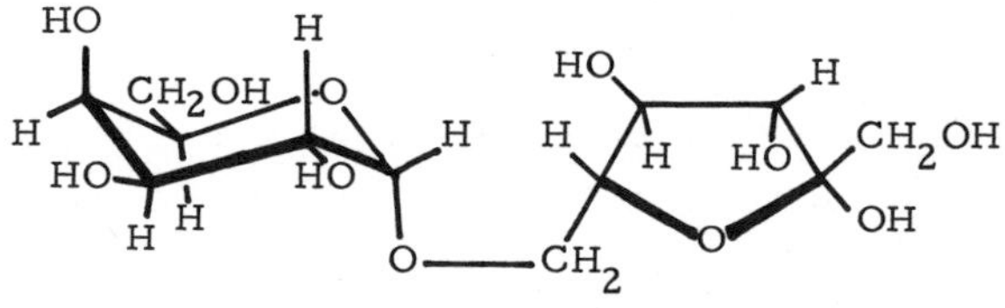

20-11   a.   Anomeric configuration is α , and the component sugar is D-glucose.

        b.   The oxide bridge between the glucose units does not involve both anomeric carbons.

        c.   At least the glucose units must be present in the pyranose form; the evidence for this comes from the formation of 2, 3, 4, 6-tetra-O-methyl-D-glucopyranose on methylation of maltose followed by hydrolysis. Glucoside formation may involve the C4 or C5 hydroxyl of the other glucose unit since either would give rise to 2, 3, 6-tri-O-methyl-D-glucose.

d.   Bromine oxidation, methylation and hydrolysis give a tetramethyl-
D-gluconic acid which forms a γ-lactone; this means that C4 of the second
glucose unit of maltose is involved in glucoside formation.

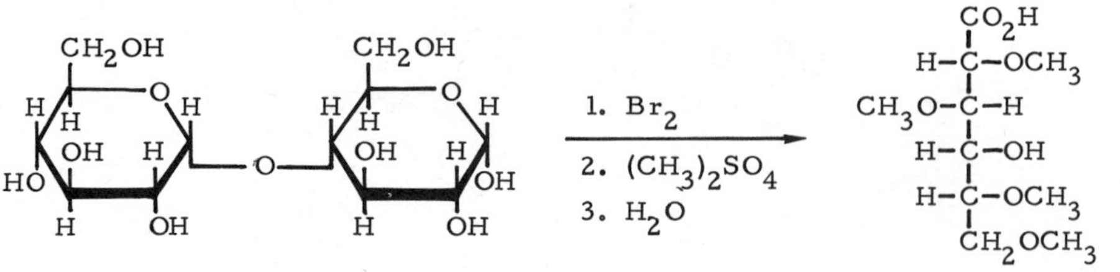

maltose   tetramethyl-D-gluconic acid

20-12   The configuration of the anomeric carbon of glucose is β in cellobiose.

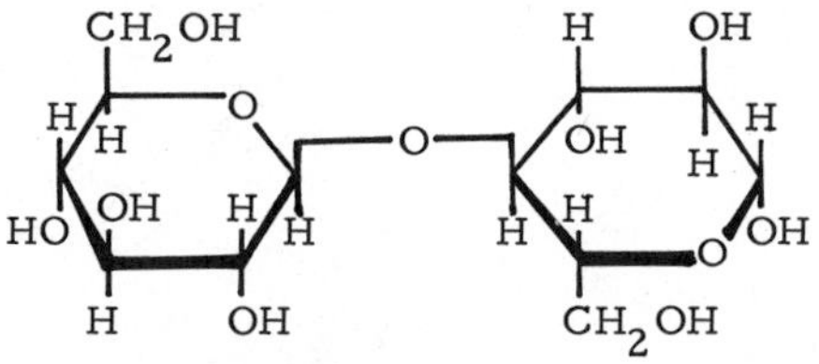

cellobiose

20-13   a.   Lactose must be composed of D-glucose and D-galactose - the
configuration of the galactosidic link being β.

b.   Galactoside formation involves only one of the two anomeric carbons.

c.   The oxide bridge linking the two sugars is between C1 of galactose
and one of the glucose carbons at other than C1.

d.   The galactose ring of lactose must be a pyranose ring. The oxide
bridge to glucose may be linked to either C4 or C5 of glucose.

e.   Since bromination of lactose followed by methylation and hydrolysis
leads to tetra-O-methyl-1, 4-gluconolactone, the oxide bridge in lactose must
be between C1 of galactose and C4 of glucose.   The structure of lactose
follows.

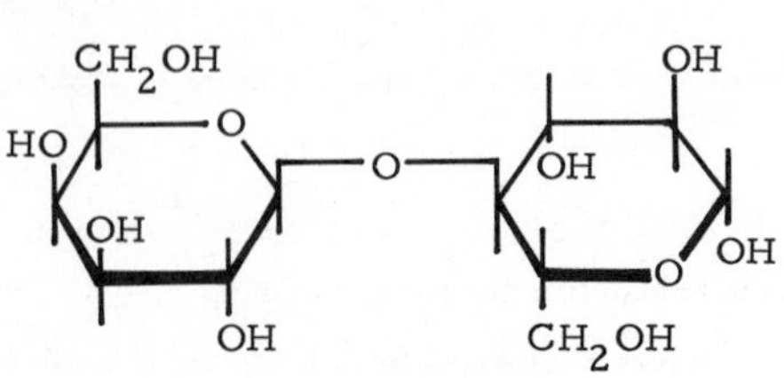

20-14   The C—O bonds in β-D-glucoside rings (as in cellulose) can be all equatorial. This makes for a more stable structure (less steric interaction) than one with α-D̲-glucoside rings (as in starch) in which the α-glucoside link has to be axial if the other C—O bonds are equatorial.

20-15

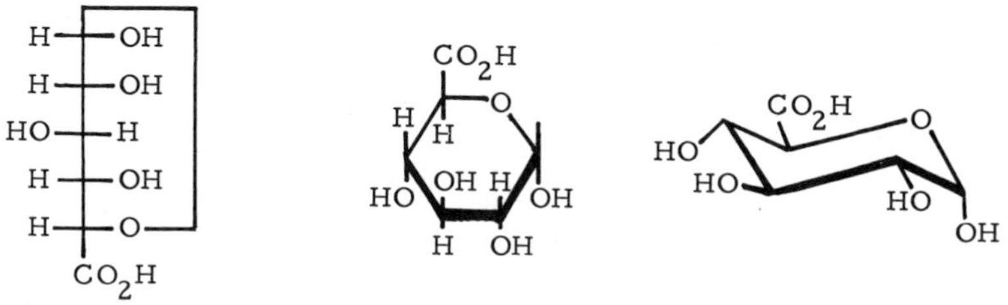

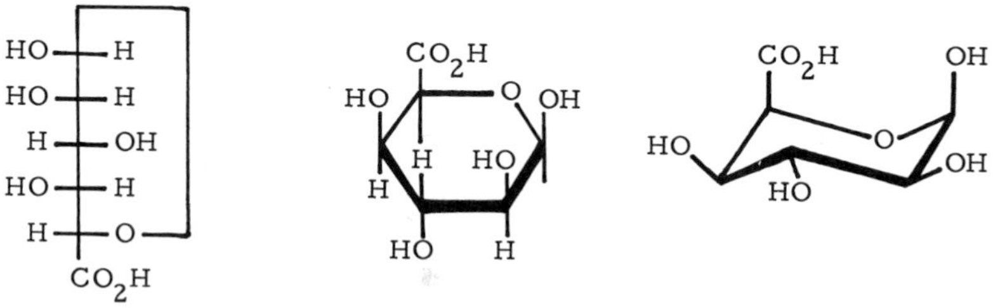

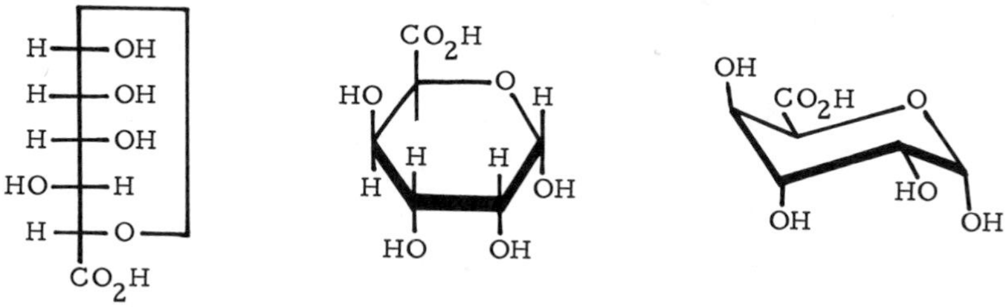

20-16   The most important factor is that the enediol structure provides the opportunity for electron-delocalization involving the 3 -OH electron-donating oxygen and the electron-attracting lactone carbonyl. In the alternative structures, no similar conjugation is possible.

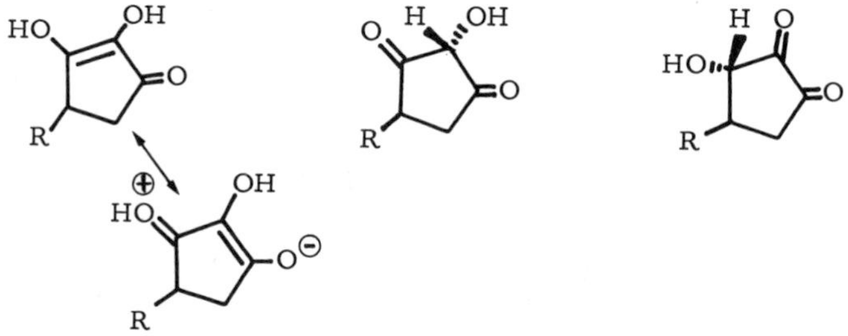

20-17

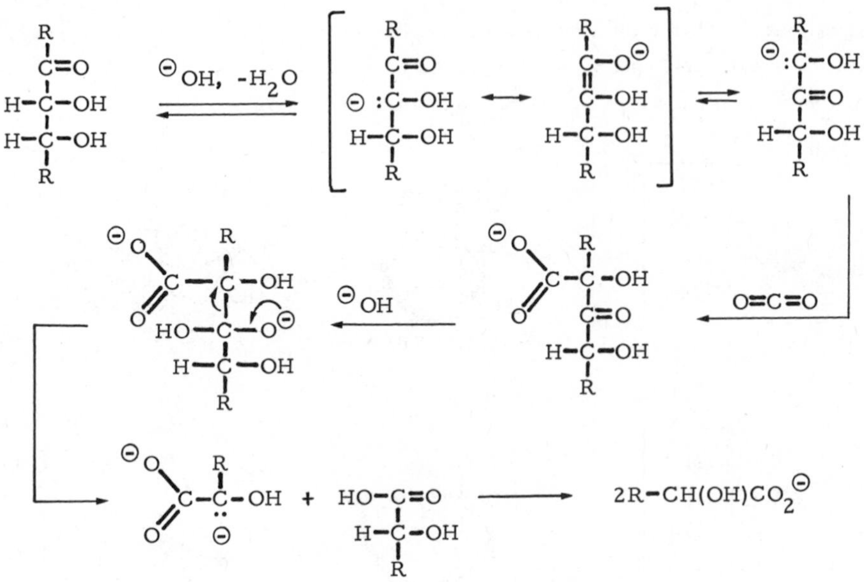

R
|
C=O
|
H-C-OH $\xrightarrow{\ominus OH, -H_2O}$
|
H-C-OH
|
R

$\left[\begin{array}{ccc} \begin{array}{c} R \\ | \\ C=O \\ | \\ \ominus :C-OH \\ | \\ H-C-OH \\ | \\ R \end{array} & \longleftrightarrow & \begin{array}{c} R \\ | \\ C-O^{\ominus} \\ \| \\ C-OH \\ | \\ H-C-OH \\ | \\ R \end{array} \end{array}\right]$
$\rightleftharpoons$
$\begin{array}{c} R \\ | \\ \ominus :C-OH \\ | \\ C=O \\ | \\ H-C-OH \\ | \\ R \end{array}$

$\begin{array}{c} \ominus O \\ \diagdown \\ C- \\ \diagup \\ O \end{array} \begin{array}{c} R \\ | \\ C-OH \\ | \\ HO-C-O^{\ominus} \\ | \\ H-C-OH \\ | \\ R \end{array}$
$\xleftarrow{\ominus OH}$
$\begin{array}{c} \ominus O \\ \diagdown \\ C- \\ \diagup \\ O \end{array} \begin{array}{c} R \\ | \\ C-OH \\ | \\ C=O \\ | \\ H-C-OH \\ | \\ R \end{array}$
$\xleftarrow{O=C=O}$

$\longrightarrow$
$\begin{array}{c} \ominus O \\ \diagdown \\ C- \\ \diagup \\ O \end{array} \begin{array}{c} R \\ | \\ C-OH \\ | \\ \ominus \end{array}$
$+$
$\begin{array}{c} HO-C=O \\ | \\ H-C-OH \\ | \\ R \end{array}$
$\longrightarrow$
$2R-CH(OH)CO_2^{\ominus}$

The reactions can be classified as base-catalyzed hydroxy-ketone rearrangement, $-CHOH-CO- \rightarrow -CO-CHOH$, followed by an aldol-type addition with CO_2 as the acceptor molecule. This is followed by a reverse Claisen condensation.

20-18   The equilibrium $CH_2=C(OH)CO_2H \rightleftharpoons CH_3COCO_2H$ strongly favors the keto form, and the energy gained by forming the ketone outweighs the energy lost in the conversion $ADP \rightarrow ATP$.

20-19   glucose $+ 9O_2 \rightarrow 6CO_2 + 6H_2O\,(\underline{l})$ $\Delta \underline{H}^{\circ} = -670$ kcal

$4\,H_2O\,(\underline{l}) + 6\,CO_2 \rightarrow 2CH_3COCO_2H + 8O_2$ $\Delta \underline{H}^{\circ} = +2 \times 280$ kcal

neglecting heats of solution:

glucose $(\underline{aq}) + O_2 \rightarrow 2CH_3COCO_2H\,(\underline{aq}) + 2H_2O\,(\underline{l})$ $\Delta H^{\circ} = -110$ kcal

20-20   The energy difference between these two substances is principally in the difference between a ketone $C=O$ and an aldehyde $C=O$. Table 4-3 gives ketone 179 kcal and aldehyde as 176 kcal or 3 kcal in favor of the ketonic isomer.

20-21   Two moles of ATP are formed in the net conversion

glucose $\longrightarrow 2CH_3COCO_2H$

20-22   The difference is in the resonance energy of $R-\overset{\overset{O}{\|}}{C}-OR$ and $R-\overset{\overset{O}{\|}}{C}-SR$. Esters are stabilized by resonance in accord with the valence-bond structures

$$\overset{\displaystyle\overset{O}{\|}}{R-C-OR} \longleftrightarrow \overset{\displaystyle\overset{O^{\ominus}}{|}\,\oplus}{R-C=OR}.$$ Corresponding structures for sulfur are less impor-

tant because sulfur is more reluctant to form double bonds. SE for ethanoic

acid is about 18 kcal (Section 18-2A). The sulfur compound is unlikely to be

more than half of that figure.

20-23   The answer here is the same as for Exercise 19-14.

20-24   This is a reverse Claisen-type condensation as observed for

$CH_3COCH_2CO_2C_2H_5 + HOC_2H_5 \longrightarrow 2CH_3CO_2C_2H_5$ (Section 18-3B). It

could be achieved in water solution with a mildly basic catalyst - not so basic as

to convert all of the $RCOCH_2COSCoA$ to $RCOCH^{\ominus}COSCoA$, but basic enough

to convert HSCoA to $^{\ominus}S\,CoA$ (thiols are stronger acids but also more nucleo-

philic than alcohols). The equation is

$$RCOCH_2COSCoA + \overset{\ominus}{SCoA} \rightleftharpoons \underset{\underset{SCoA}{|}}{R-\overset{\overset{O^{\ominus}}{|}}{C}-CH_2COSCoA} \longrightarrow R-\overset{\overset{O}{\|}}{C}SCoA + \overset{\ominus}{CH_2COSCoA}$$

$$CH_3COSCoA \xleftarrow{\;H^{\oplus}\;}$$

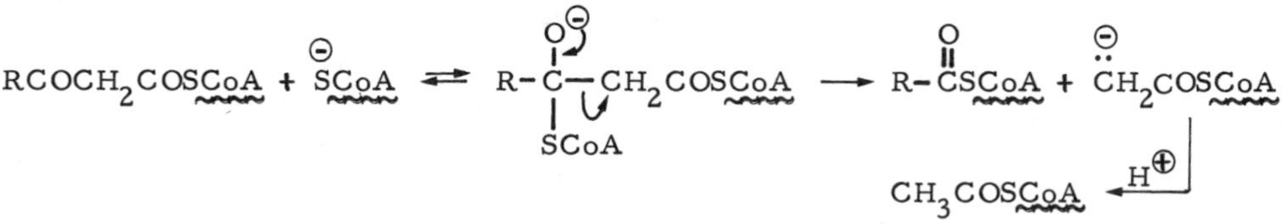

20-25   Knoop's discovery showed that fatty acids are degraded biochemically
two carbons at a time by " β oxidation".

no β -carbon to oxidize to a ketone

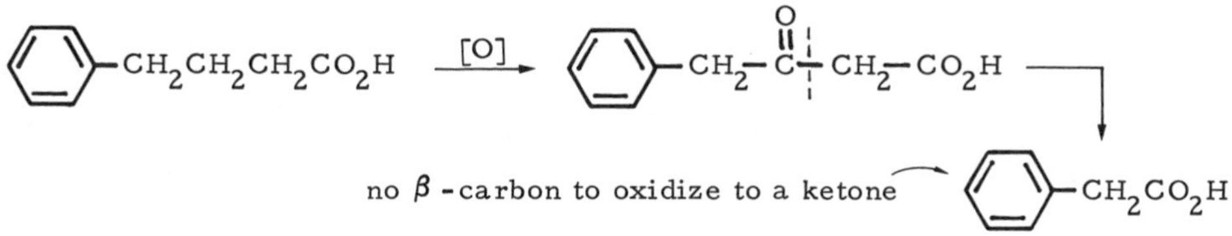

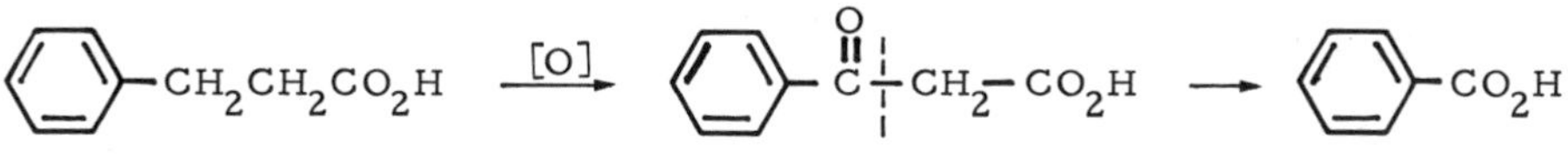

two carbons lost

20-26   This exercise was evolved while watching the champion Russian weight
lifter in the 1976 Olympics.

a.   225 kg raised 2 meters = 225 x 2 x 2.3 cal

                               = 1.035 kcal

   with 50% efficiency   = 2.070 kcal

$$ATP + H_2O \longrightarrow ADP + H_2PO_4^{\ominus} \qquad \Delta G^{\circ} = -7 \text{ kcal}$$

   hence,   2.07/7 = 0.30 mole ATP required = 150 g.

b.   Glucose oxidized at 40% efficiency to replenish ATP $= \dfrac{2.070}{0.40 \times 686}$

$$= 0.0075 \text{ mole}$$

$$= 1.12 \text{ g}$$

20-27   The pentose in question is an aldopentose (reduces Tollen's reagent) and has four hydroxyl groups available for ester formation. Because the phenylosazone is optically inactive it must have no asymmetric carbon atom. This necessitates rejection of all straight-chain structures for this sugar. The only structure which fits all of the data is

$$
\begin{array}{c}
\text{CHO} \\
| \\
\text{H}-\text{C}-\text{OH} \\
| \\
\text{HO}-\text{C}-\text{CH}_2\text{OH} \\
| \\
\text{CH}_2\text{OH}
\end{array}
$$

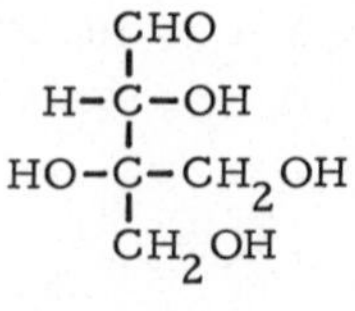

This sugar is known as apiose and occurs in parsley.

20-28

$$
\begin{array}{c}
\text{CH}_2\text{OH} \\
| \\
\text{H}-\text{C}-\text{OH} \\
| \\
\text{HO}-\text{C}-\text{H} \\
| \\
\text{H}-\text{C}-\text{OH} \\
| \\
\text{H}-\text{C}-\text{OH} \\
| \\
\text{CHO}
\end{array}
\quad = \quad
\begin{array}{c}
\text{CHO} \\
| \\
\text{HO}-\text{C}-\text{H} \\
| \\
\text{HO}-\text{C}-\text{H} \\
| \\
\text{H}-\text{C}-\text{OH} \\
| \\
\text{HO}-\text{C}-\text{H} \\
| \\
\text{CH}_2\text{OH}
\end{array}
\quad \xrightarrow{\text{C}_6\text{H}_5\text{NHNH}_2} \quad
\begin{array}{c}
\text{HC}=\text{NNHC}_6\text{H}_5 \\
| \\
\text{C}=\text{NNHC}_6\text{H}_5 \\
| \\
\text{HO}-\text{C}-\text{H} \\
| \\
\text{H}-\text{C}-\text{OH} \\
| \\
\text{HO}-\text{C}-\text{H} \\
| \\
\text{CH}_2\text{OH}
\end{array}
$$

$\Big\downarrow$ Na, Hg

$$
\begin{array}{c}
\text{CH}_2\text{OH} \\
| \\
\text{H}-\text{C}-\text{OH} \\
| \\
\text{HO}-\text{C}-\text{H} \\
| \\
\text{H}-\text{C}-\text{OH} \\
| \\
\text{H}-\text{C}-\text{OH} \\
| \\
\text{CH}_2\text{OH}
\end{array}
$$

D-sorbitol

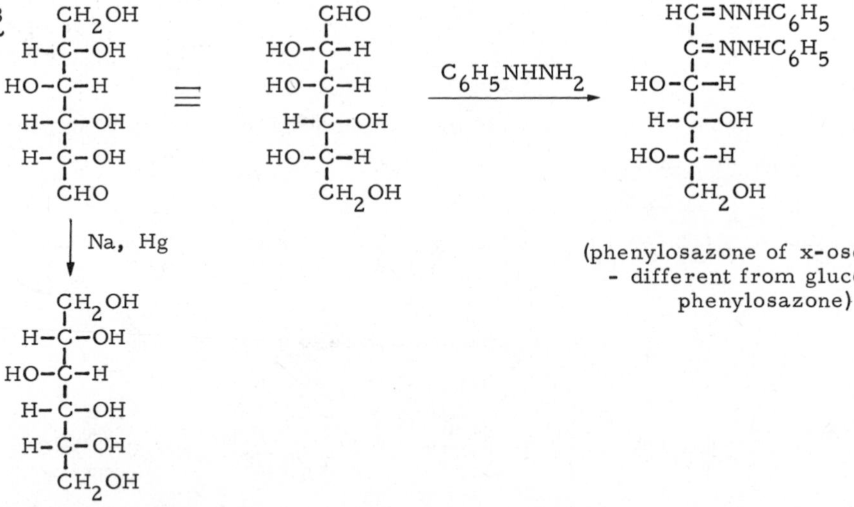

(phenylosazone of x-ose
 - different from glucose
   phenylosazone)

20-29   Treatment of an amide with NaOCl solution generally leads to loss of the amide carbon and formation of an amine (Hofmann reaction, Sec. 23-12E). In the case of the amide of D, $\text{C}_3\text{H}_7\text{O}_3 \cdot \text{CONH}_2$, this reaction leads to glyceraldehyde; the initially formed NH_2 group must have been hydrolyzed. Working

backwards we have,

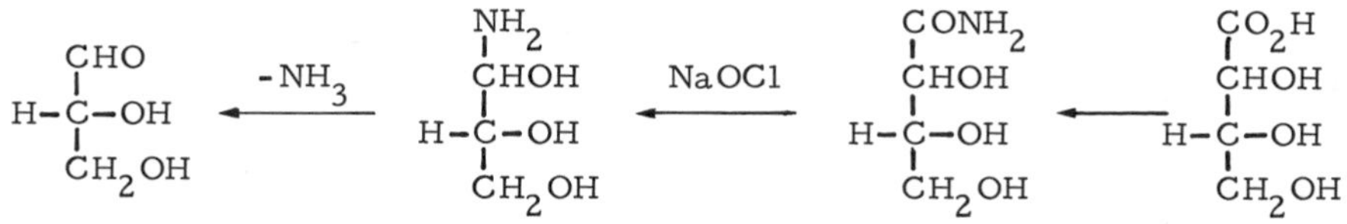

$$
\underset{\text{D-glyceraldehyde}}{\begin{array}{c}\text{CHO}\\ |\\ \text{H-C-OH}\\ |\\ \text{CH}_2\text{OH}\end{array}}
\xleftarrow{\text{-NH}_3}
\begin{array}{c}\text{NH}_2\\ |\\ \text{CHOH}\\ |\\ \text{H-C-OH}\\ |\\ \text{CH}_2\text{OH}\end{array}
\xleftarrow{\text{NaOCl}}
\begin{array}{c}\text{CONH}_2\\ |\\ \text{CHOH}\\ |\\ \text{H-C-OH}\\ |\\ \text{CH}_2\text{OH}\end{array}
\xleftarrow{}
\underset{\text{D}}{\begin{array}{c}\text{CO}_2\text{H}\\ |\\ \text{CHOH}\\ |\\ \text{H-C-OH}\\ |\\ \text{CH}_2\text{OH}\end{array}}
$$

The configuration at C2 of compound D is determined by the transformations D $\xrightarrow{[O]}$ B $\xrightarrow{[H]}$ C. Compound B is a reducing sugar with three free hydroxyl groups (forms triacetate) and its potential aldehyde group may be reduced to inactive C.

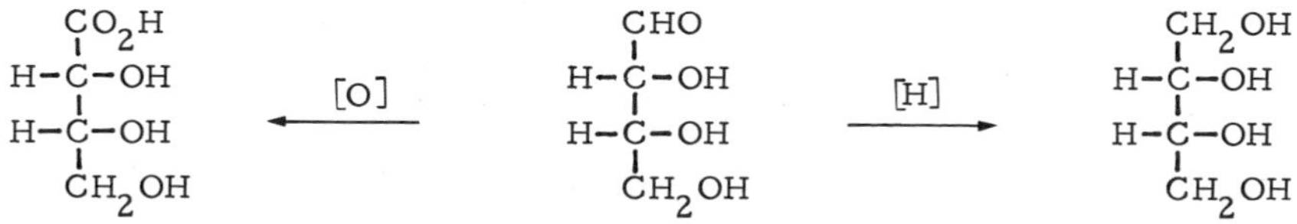

$$
\underset{\text{D}}{\begin{array}{c}\text{CO}_2\text{H}\\ |\\ \text{H-C-OH}\\ |\\ \text{H-C-OH}\\ |\\ \text{CH}_2\text{OH}\end{array}}
\xleftarrow{[O]}
\underset{\text{B}}{\begin{array}{c}\text{CHO}\\ |\\ \text{H-C-OH}\\ |\\ \text{H-C-OH}\\ |\\ \text{CH}_2\text{OH}\end{array}}
\xrightarrow{[H]}
\underset{\substack{\text{C}\\ \text{(inactive)}}}{\begin{array}{c}\text{CH}_2\text{OH}\\ |\\ \text{H-C-OH}\\ |\\ \text{H-C-OH}\\ |\\ \text{CH}_2\text{OH}\end{array}}
$$

Compound B is formed from A by acid hydrolysis of an 'acetal' link. Compound A is a methyl glycoside. The most likely configuration for A and B follow:

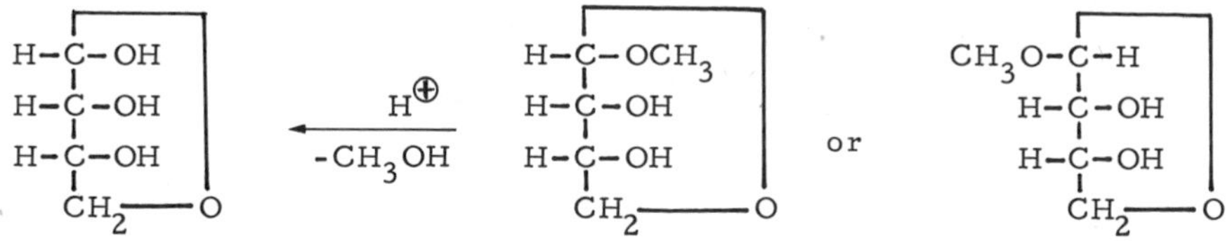

$$
\underset{\substack{\text{B}\\ \text{(or }\beta\text{-anomer)}}}{\begin{array}{c}\text{H-C-OH}\\ |\\ \text{H-C-OH}\\ |\\ \text{H-C-OH}\\ |\\ \text{CH}_2\text{—O}\end{array}}
\xleftarrow[\text{-CH}_3\text{OH}]{\text{H}^{\oplus}}
\underset{\substack{\text{A}\\ (\alpha\text{-methyl glycoside})}}{\begin{array}{c}\text{H-C-OCH}_3\\ |\\ \text{H-C-OH}\\ |\\ \text{H-C-OH}\\ |\\ \text{CH}_2\text{—O}\end{array}}
\quad\text{or}\quad
\underset{\substack{\text{A}\\ (\beta\text{-methyl glycoside})}}{\begin{array}{c}\text{CH}_3\text{O-C-H}\\ |\\ \text{H-C-OH}\\ |\\ \text{H-C-OH}\\ |\\ \text{CH}_2\text{—O}\end{array}}
$$

20-30

a.

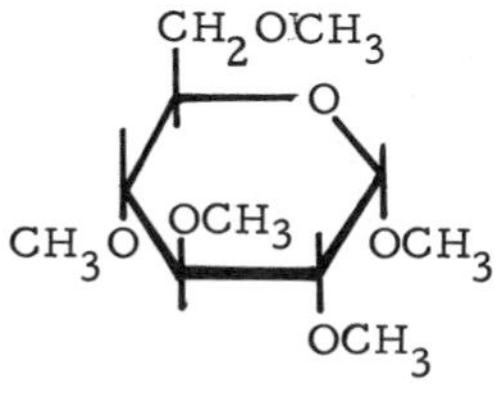

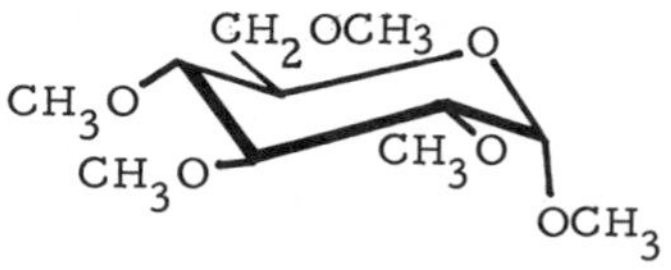

20-30 (cont.)

b.

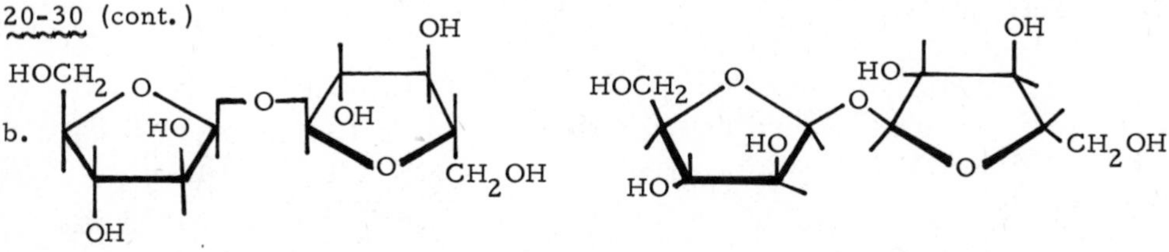

D   L

c.

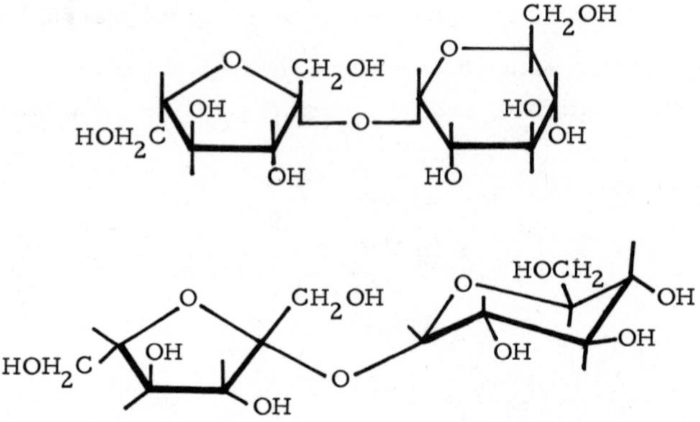

20-31   The reaction of D-glucose with 2-propanone is expected to involve only cis pairs of hydroxyl groups.

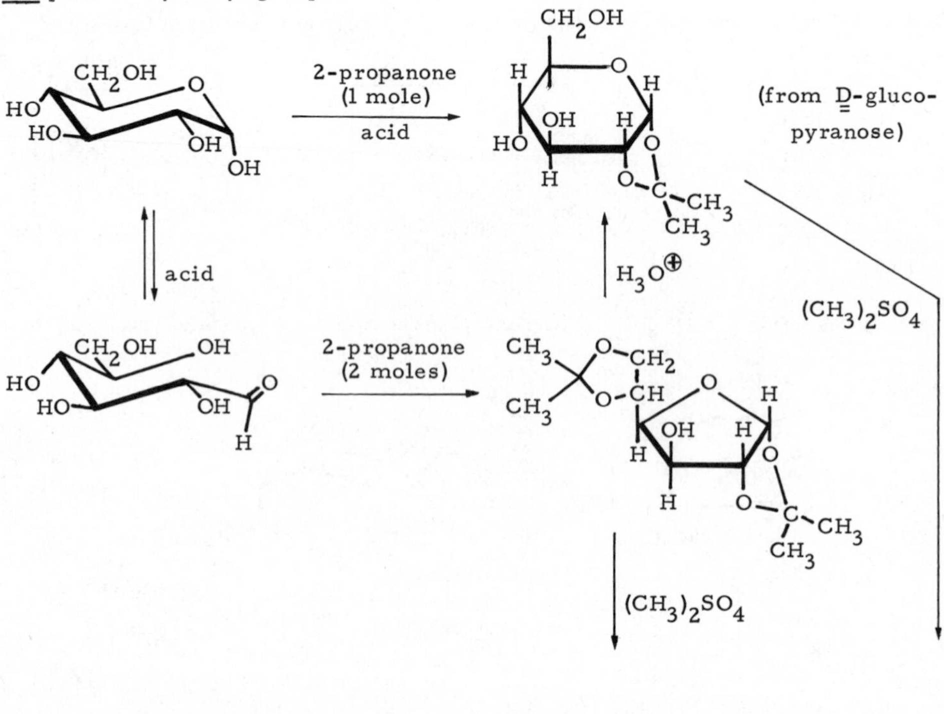

20-31 (cont.)   (cont.)   (cont.)

$$H_3O^{\oplus}$$

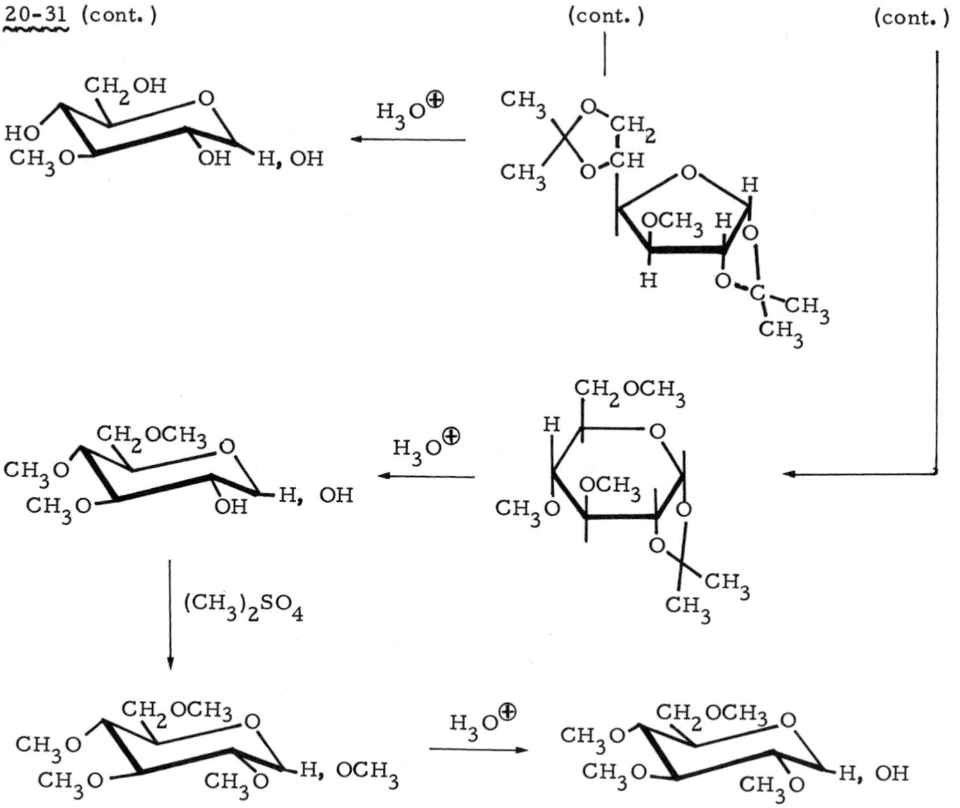

20-32

a. α-D-Glucofuranose $\xrightarrow[\text{HCl}]{\text{acetone (one mole)}}$

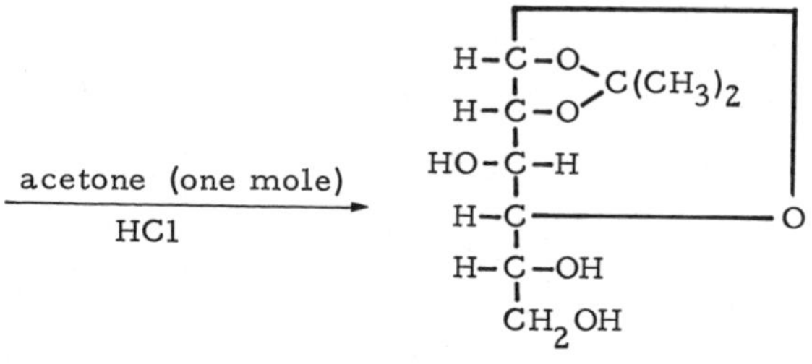

(A)

b.   (A) $\xrightarrow{\text{NaIO}_4}$ c.   (B) $\xrightarrow{\text{Na}^{14}\text{CN}}$

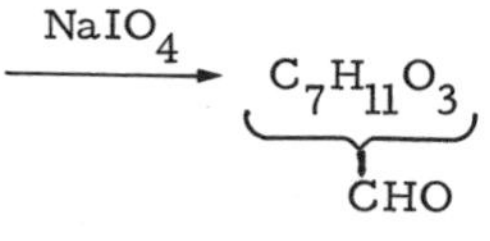

(B)   (C)   (D)

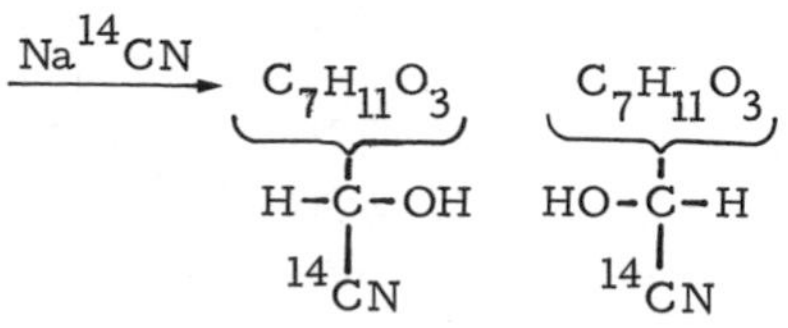

20-32 (cont.)

d. 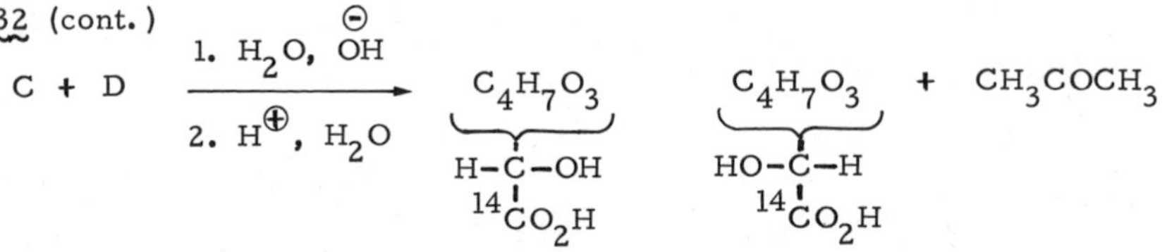 C + D $\xrightarrow[\text{2. } H^{\oplus}, H_2O]{\text{1. } H_2O, \ \overset{\ominus}{OH}}$

$C_4H_7O_3$
$$\begin{array}{c} H-\overset{|}{\underset{|}{C}}-OH \\ {}^{14}CO_2H \end{array}$$
(E)

$C_4H_7O_3$
$$\begin{array}{c} HO-\overset{|}{\underset{|}{C}}-H \\ {}^{14}CO_2H \end{array}$$
(F)

$+ \ CH_3COCH_3$

e. E + F $\xrightarrow[-H_2O]{\text{heat}}$

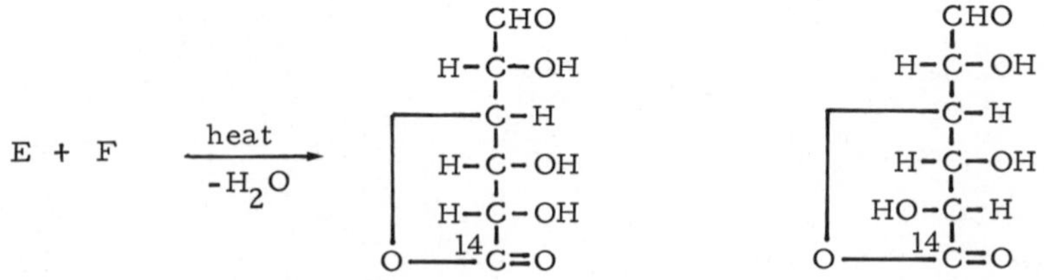

(G)

(H)

f. 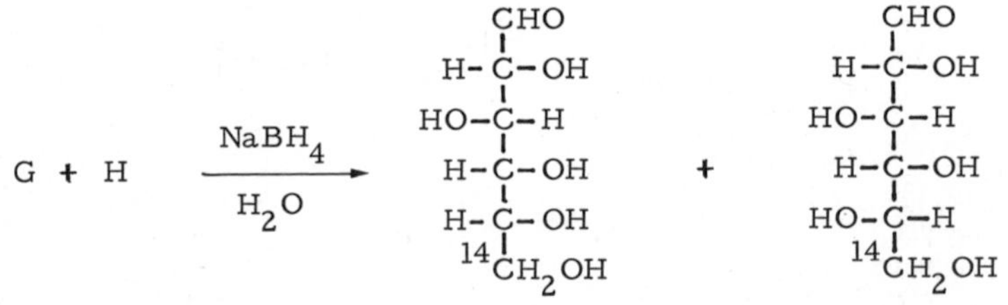 G + H $\xrightarrow[H_2O]{NaBH_4}$

$$\begin{array}{c} CHO \\ H-\overset{|}{\underset{|}{C}}-OH \\ HO-\overset{|}{\underset{|}{C}}-H \\ H-\overset{|}{\underset{|}{C}}-OH \\ H-\overset{|}{\underset{|}{C}}-OH \\ {}^{14}CH_2OH \end{array}$$
D-glucose-6-^{14}C

$+$

$$\begin{array}{c} CHO \\ H-\overset{|}{\underset{|}{C}}-OH \\ HO-\overset{|}{\underset{|}{C}}-H \\ H-\overset{|}{\underset{|}{C}}-OH \\ HO-\overset{|}{\underset{|}{C}}-H \\ {}^{14}CH_2OH \end{array}$$
(I)

20-33

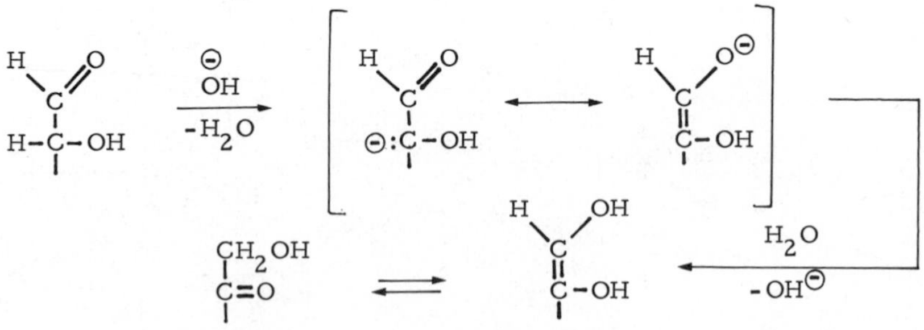

D-Glucose in the presence of hydroxide ion is expected to equilibrate to a mixture of D-glucose, D-mannose and D-fructose provided that carbonyl formation proceeds no further than C2 (Section 20-2D).

20-34   The glycosidic linkage cannot be α-1,4 as in maltose. The sugar part is a disaccharide of D-glucose. The linkage is deduced as 1,6 from the exhaustive methylation studies.   The structure follows.

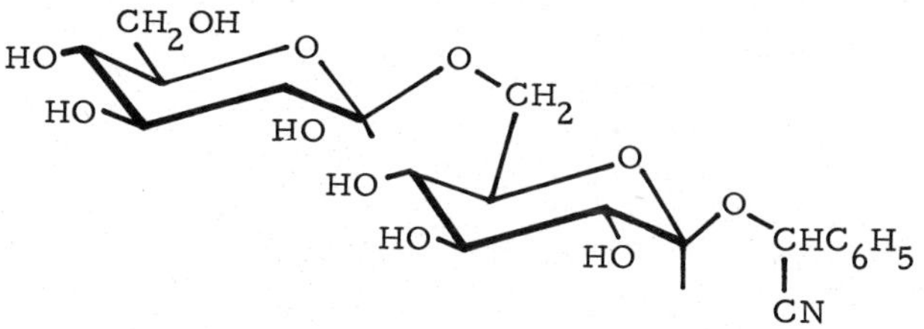

amygdalin

This compound is a controversial anticancer drug known as "Laetrile." It has not been approved for clinical use by the United States Food and Drug Administration but has been approved by a few States.

CHAPTER 21

THE RESONANCE AND MOLECULAR-ORBITAL METHODS.
PERICYCLIC REACTIONS

The most useful qualitative procedures for application of quantum-mechanical principles to chemical bonding are the molecular-orbital (MO) and resonance (or valence-bond, VB) methods. Each of these has its roots in a particular approximate way of calculating bonding energies and, in the simplest form, each starts from the same atomic-orbital model.

The MO procedure mixes atomic orbitals to obtain an optimum set of molecular orbitals - optimum meaning that the AOs provide a set of molecular orbitals of minimum energy from the particular set of atomic orbitals. The appropriate number of electrons then is included in these orbitals to give the electronic configuration.

The VB procedure starts with different electronic configurations (VB structures) corresponding to different ways of pairing the electrons in the atomic orbitals. Mixing of these electronic configurations produces different possible energy states, the lowest of which corresponds to the normal state of the molecule and is called the resonance hybrid. The energy of the resonance hybrid usually is substantially lower than that of a single VB structure, especially when there are two or more reasonably equivalent VB structures. A set of simple rules permits assessment of the importance of different VB structures to a given hybrid.

Neither the MO nor the VB method should be regarded as "correct". Neither can be used to calculate bonding energies accurately and neither, in its simplest qualitative form, accounts for important properties of some kinds of systems. However, when both predict the same result, that result is usually correct. When the predictions differ, a higher level of theory may be required.

Both the VB and MO methods account for the unusual stability and low reactivity of benzene. Both methods also give a good account of the structures and properties of 1,3-butadiene, propenal, and the 2-propenyl (allyl) carbocation, radical, and carbanion.

Stabilization energies (SE) of conjugated polyenes are the differences between experimental heats of reactions and those calculated on the basis of bond energies. An SE value is a net quantity, which is a composite of electron delocalization, strain, and steric effects. The delocalization energy (DE) is the difference in energy calculated for a localized structure and a delocalized structure with the same geometry. If steric and strain effects are small, as for benzene, then DE $\cong$ SE.

Carbon-carbon bond lengths can be used as a test for important electron delocalization provided proper account is taken of how the states of hybridization of the carbons involved influence bond lengths. Thus the C—C bond distance between two sp^3 carbons normally is about 1.54 A and between two sp^2 carbons about 1.48 A. A carbon-carbon bond distance of 1.47 A represents a very significant shortening from 1.54 A, but not from 1.48 A, so that use of the proper reference distance is very important. The carbon-carbon bond distance in benzene is about 1.40 A, which is close to the average between the 1.34 A distance of a normal C—C and the 1.48 A C—C bond distance between two sp^2 carbons, and can be taken to reflect either the contribution of two equivalent Kekule structures in the VB method or the strong π bonding suggested by the MO method.

The MO method predicts that monocyclic conjugated systems will be stable only when they possess 2, 6, 10, 16, or 4n + 2 electrons, because this number just suffices to fill the bonding molecular orbitals. This is the basis of the 4n + 2 rule deduced by Hückel. Conjugated monocyclic systems with 4n electrons are much less stable because degenerate pairs of bonding molecular orbitals are incompletely filled with one unpaired electron in each, in accord with Hund's rule. The resulting electronic configuration is a triplet, and is expected to lead to reactivity similar to that of a diradical.

The simple VB method does not account for the 4n + 2 rule. The rule does not apply to acyclic polyenes and should be applied with caution to polycyclic conjugated systems.

Molecular orbital theory provides a means of rationalizing reactions occurring by cyclic transition states (pericyclic reactions). To evaluate whether a given thermal pericyclic reaction meets the conditions imposed by the molecular orbital theory, it is necessary to determine first if the orbital arrangement is a Hückel system (no nodes, or an even number of nodes) and second if there are 4n + 2 or 4n participating electrons. Hückel transition states are favored with 4n + 2 electrons, and Mobius transition states are favored with 4n electrons.

Pericyclic reactions that are unfavorable thermally often can be caused to occur photochemically. Concerted reactions that proceed thermally and photochemically may have opposite stereochemical consequences.

The [2 + 2] cycloadditions to conjugated dienes or alkenes, which involve perhaloalkenes or substances with cumulated double bonds, often occur by radical mechanisms:

$$
\begin{array}{c}
CF_2 \\
\| \ 2 \\
CF_2
\end{array}
\ + \
\begin{array}{c}
CH_2 \\
\| \ 2 \\
CH_2
\end{array}
\ \xrightarrow[\text{nonconcerted}]{\text{heat}} \
\begin{array}{c}
F_2C\!-\!CH_2 \\
| \quad\quad | \\
F_2C\!-\!CH_2
\end{array}
$$

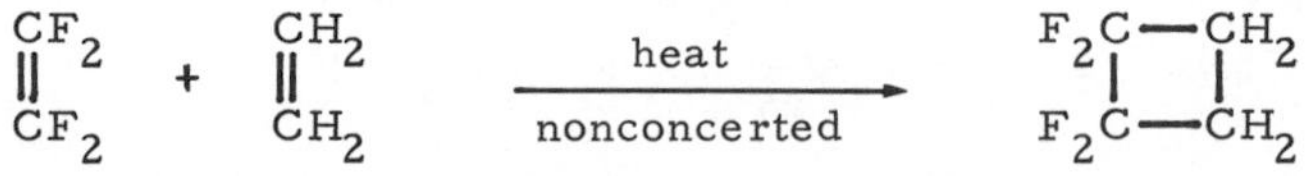

These stepwise reactions normally are not stereospecific.

ANSWERS TO EXERCISES

21-1   Naphthalene can be represented by three different Kekulé-type structures. Neither of the other two structures can be drawn even as one Kekulé structure. (Try various combinations of bonds.)

21-2
$$4C\ (\underline{s}) + 3H_2\ (\underline{g}) \longrightarrow C_4H_6\ (\underline{g}) \qquad \Delta\underline{H}_f^\circ\ (exp) = 26.3\ kcal$$
for 1, 3-butadiene

$$4C\ (\underline{g}) + 3H_2\ (\underline{s}) \longrightarrow C_4H_6\ (\underline{s}) \qquad \Delta\underline{H}^\circ = -653.8\ kcal$$
from bond energies

$$4C\ (\underline{s}) \longrightarrow 4C\ (\underline{g}) \qquad \Delta\underline{H}^\circ = 4 \times 171.3\ kcal$$

Net $$4C\ (\underline{s}) + 3H_2\ (\underline{g}) \longrightarrow C_4H_6\ (\underline{g}) \qquad \Delta\underline{H}_f^\circ\ (calc) = -653.8 + 4 \times 171.3$$
$$= 31.4\ kcal$$

Hence the stabilization energy is

$$\Delta\underline{H}_f^\circ\ (calc) - \Delta\underline{H}_f^\circ\ (exp) = 31.4 - 26.3 = 5.1\ kcal$$

21-3   Using the same approach as in Exercise 21-2, the stabilization energy of propenal is calculated to be 7.7 kcal.

21-4

π -electron energy = $2(\alpha + 1.41\beta)$
$= 2\alpha + 2.82\beta$

π -electron energy = $2(\alpha + \beta)$
$= 2\alpha + 2\beta$

3-chloro-cyclohexene

4-chloro-cyclohexene

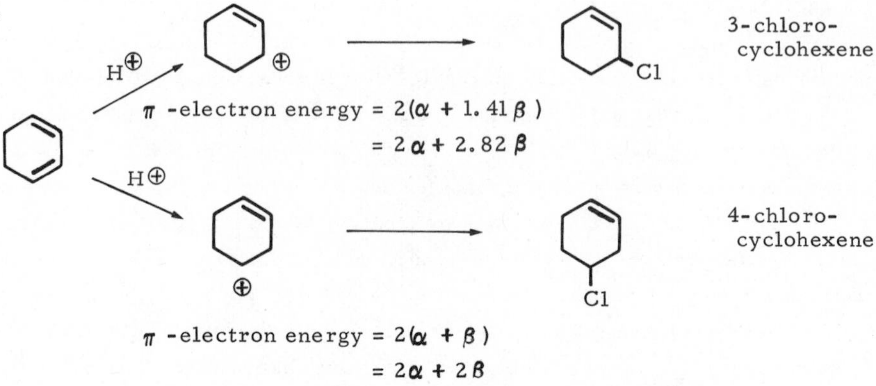

By the MO method the 2-cyclohexenyl ion is more stable than the 3-cyclohexenyl ion by $2(\alpha + 1.41\beta) - 2(\alpha + \beta) = 0.82\beta$. The dominant product expected under conditions of kinetic control is therefore 3-chlorocyclohexene.

By the VB method the 2-cyclohexenyl ion is a hybrid of two equivalent structures,

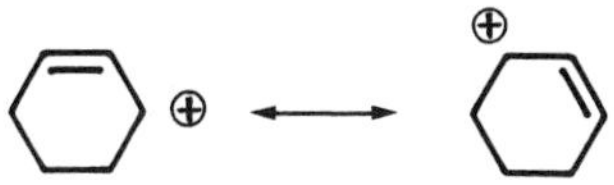

Similar resonance stabilization is not possible for the 3-cyclohexenyl ion.

21-5

Atomic orbital model:
(planar - all angles $120°$)

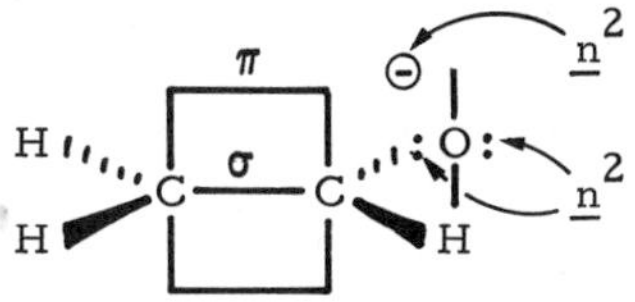

By the VB method the ion is a hybrid of the following structures:

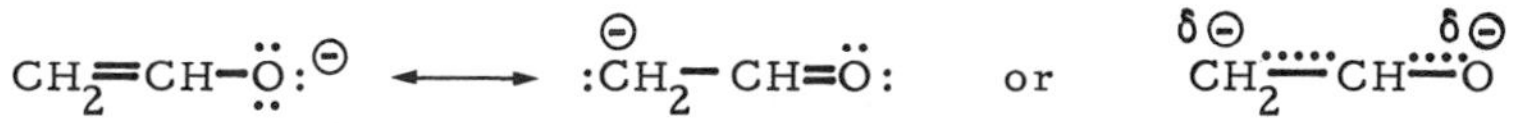

By the MO method the ion is a delocalized structure in which the π -molecular orbitals are a combination of three parallel 2p atomic orbitals containing a total of four electrons, two each in the two lowest energy MOs.

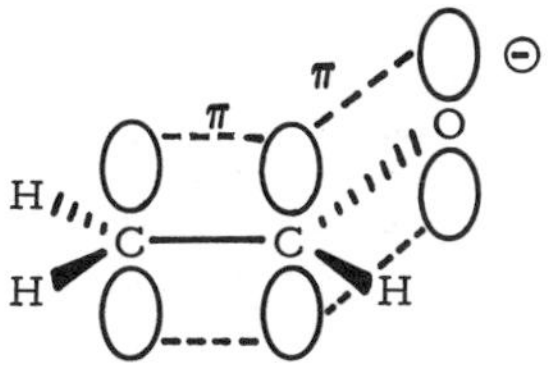

21-6   From Figure 21-9 and the discussion in Section 21-5B, the lowest energy MO of the 2-propenyl system, when filled with two electrons, gives an electron distribution in which the center carbon is neutral (i.e., $\frac{1}{2}\oplus CH_2^{\cdots}CH^{\cdots}CH_2^{\frac{1}{2}\oplus}$). In the 2-propenyl radical and anion, one and two additional electrons, respectively, must occupy the NBMO. This orbital contributes zero amplitude (zero electron density) at C2. Therefore, electron density is distributed symmetrically between C1 and C3. That is,

$\frac{1}{2}\cdot CH_2^{\cdots}CH^{\cdots}CH_2^{\frac{1}{2}\cdot}$ and $\frac{1}{2}\ominus CH_2^{\cdots}CH^{\cdots}CH_2 \frac{1}{2}\ominus$.

21-7   Reactivity depends on the energy of the intermediate cation relative to the neutral alkene. The smaller the energy gap between them the more reactive is the alkene. Thus, despite the stabilization energy of 1,3-butadiene, the intermediate $CH_2\!\!=\!\!CHCHCH_2\overset{\oplus}{B}r$ lies closer to 1,3-butadiene than $\overset{\oplus}{C}H_2CH_2Br$ does to ethene. This can be seen from the following crude MO calculation derived from Figures 21-10 and 21-9.

$$CH_2\!\!=\!\!CH-CH\!\!=\!\!CH_2 \xrightarrow{Br\overset{\oplus}{}} \overset{Br}{\underset{\ddot{}}{CH_2}}\!\!-\!\!\overset{\delta\oplus}{CH}\!\!\cdots\!\!\overset{\delta\oplus}{CH}\!\!\cdots\!\!CH_2 \quad \text{(localization of 2e in C—Br bond)}$$

π energy $2(\alpha+1.62\,\beta) + 2(\alpha+0.62\,\beta)$ $\qquad 2\alpha + 2(\alpha+1.41\,\beta)$

net change in π energy $= -1.66\,\beta$

$$CH_2\!\!=\!\!CH_2 \xrightarrow{Br\overset{\oplus}{}} Br\!:\!CH_2\!\!-\!\!\overset{\oplus}{C}H_2 \quad \text{(localization of 2e in C—Br bond)}$$

energy $2(\alpha+\beta)$ $\qquad\qquad 2\alpha$

net change in π energy $= -2\,\beta$

Thus, the energy gap is larger for ethene $(2\,\beta)$ than it is for butadiene $(1.66\,\beta$).

21-8

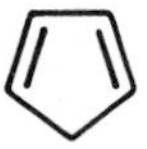

$$+ 13/2\,O_2 \longrightarrow 5CO_2 + 3H_2O\ (\underline{l}) \qquad \Delta\underline{H}^{\circ}_{-exp} = -707.7\ kcal$$

$$3H_2O\ (\underline{l}) \longrightarrow 3H_2O\ (\underline{g}) \qquad\qquad \Delta\underline{H}^{\circ} = 3 \times 10\ kcal$$

Hence, the heat of combustion of 1,3-pentadiene to CO_2 and H_2O, all in the vapor state is $-707.7 + 30 = -677.7$ kcal. From bond energies of Table 4-3, the heat of combustion is calculated to be -679.1 kcal. Therefore, the stabilization is $-677.7 + 679.1 = 1.4$ kcal. This value is subject to more uncertainty than a similar calculation for 1,3-butadiene because of the uncertain strain effects of introducing two double bonds in a five-membered ring. Any strain effects will reduce the stabilization energy calculated from the heat of combustion.

21-9   a. The calculated heat of combustion of benzene to liquid water is -828 kcal. The stabilization energy of benzene therefore is $828 - 789 = 39$ kcal.

b. If we assume stronger C—H and C—C bonds in benzene than normal (112 kcal for $\equiv$C—H and 90.6 kcal for $\equiv$C—C$\equiv$), the calculated heat of combustion is -726 kcal, and the stabilization energy becomes negative by 63 kcal!

21-10   a. The strain energy of the four-membered ring in biphenylene decreases the stabilization energy to less than twice that of benzene.

b. Azulene is less stable than naphthalene and therefore has a higher heat of combustion. Only two Kekulé-type VB structures can be written for azulene, three can be written for naphthalene. That is to say, resonance stabilization is more important for naphthalene than for azulene.

21-11

a.   3.35 A   1.421 A   (overlapping layers of
                                  chicken-wire)

b. sp^2

c. 33.3%

d.

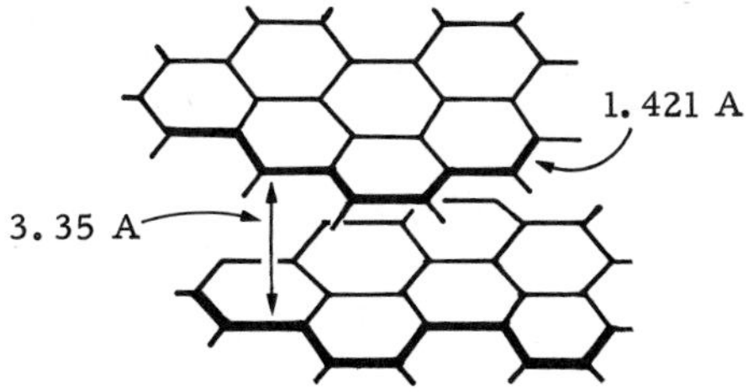

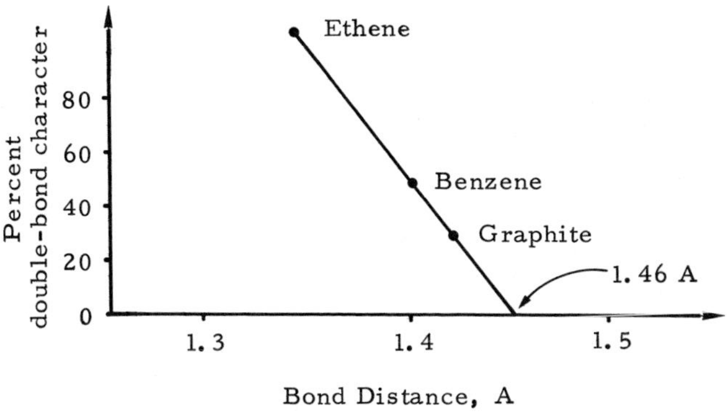

e. 1.46 A; agrees reasonably well with the single bond distance between
    sp^2—sp^2 carbons (Table 21-3), which suggests that to whatever degree reso-
    nance is important in 1,3-butadiene it does not influence the bond length
    significantly.

21-12

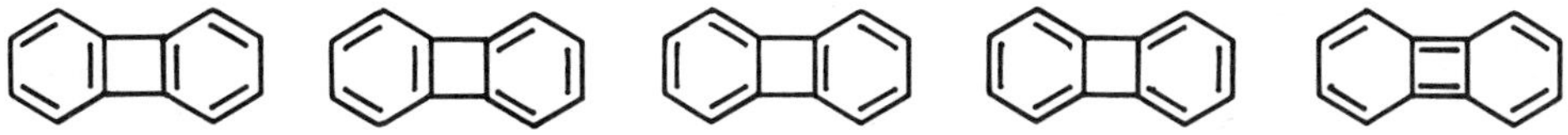

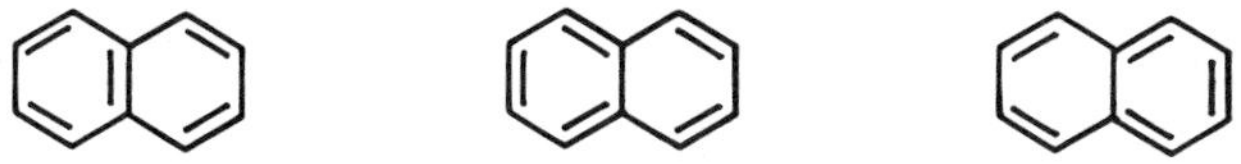

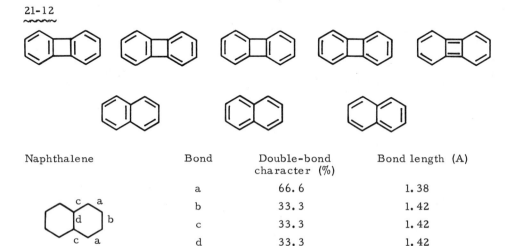

Naphthalene	Bond	Double-bond character (%)	Bond length (A)
	a	66.6	1.38
	b	33.3	1.42
	c	33.3	1.42
	d	33.3	1.42

Bond (a) should be attacked preferentially by ozone since it has the

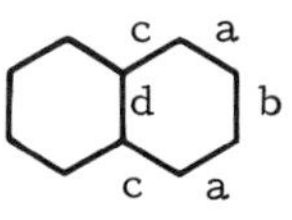

most double-bond character.

Biphenylene:

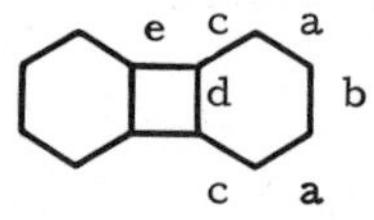

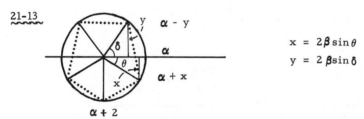

Bond	Double-bond character (%)	Bond length (A)
a	60	1.39
b	40	1.41
c	40	1.41
d	40	1.41
e	20	1.43

Bond (a) should be the most reactive towards ozone.

21-13

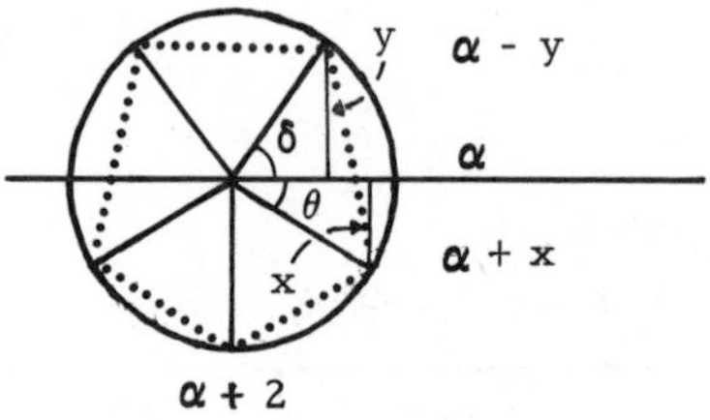

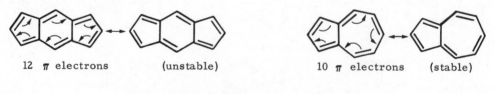

$$x = 2\beta \sin \theta$$
$$y = 2\beta \sin \delta$$

The internal angle of a regular pentagon is $(\delta + \theta) = 360/5 = 72°$. Also, $2\theta + 72 = 180°$, or $\theta = 18°$ and $\delta = 72 - 18 = 54°$. Hence:

$$x = 2\beta \sin 18 \qquad \text{and} \qquad y = 2\beta \sin 54$$
$$= 0.62\beta \qquad\qquad\qquad = 1.62\beta$$

and $\qquad \alpha + x = \alpha + 0.62\beta \qquad \alpha - y = \alpha - 1.62\beta$

21-14   The $4\underline{n} + 2$ rule applies to the following:

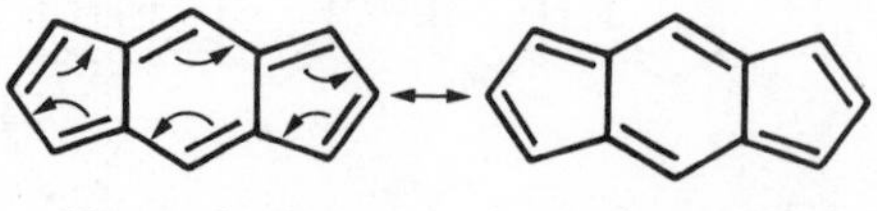

12 π electrons   (unstable)   10 π electrons   (stable)

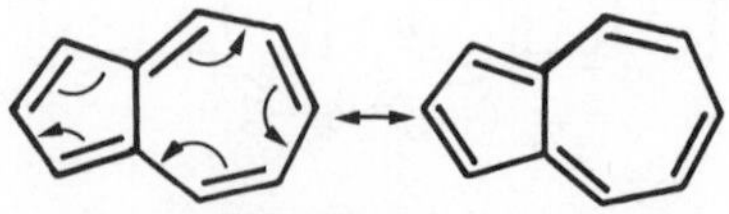

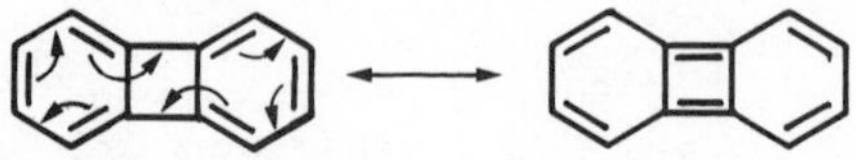

12 π electrons (unstable if there is extensive conjugation between the two benzenoid rings)

The $4\underline{n} + 2$ rule does not apply to:

formal bond

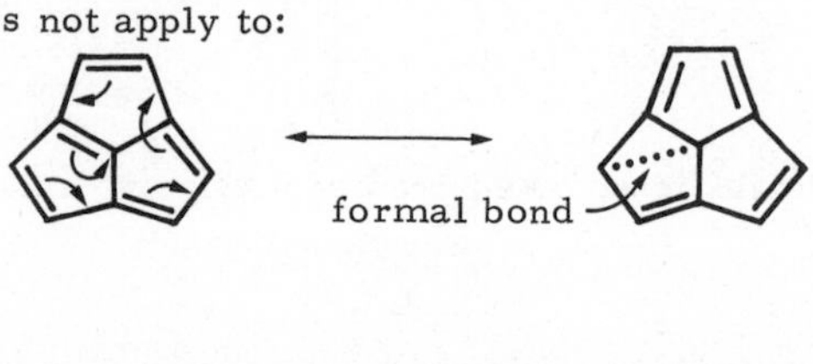

21-15

Energy levels

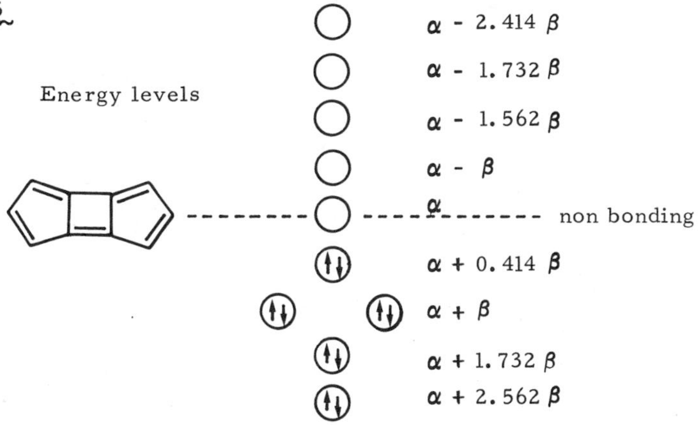

$\alpha - 2.414\ \beta$

$\alpha - 1.732\ \beta$

$\alpha - 1.562\ \beta$

$\alpha - \beta$

α non bonding

$\alpha + 0.414\ \beta$

$\alpha + \beta$

$\alpha + 1.732\ \beta$

$\alpha + 2.562\ \beta$

Total π -electron energy $= 10\ \alpha + 13.4\ \beta$

Delocalization energy $= (10\alpha + 13.4\beta) - (10\ \alpha + 10\beta)$

$= 3.4\ \beta$

21-16. $C_8H_8{}^{2\ominus}$ energy levels:

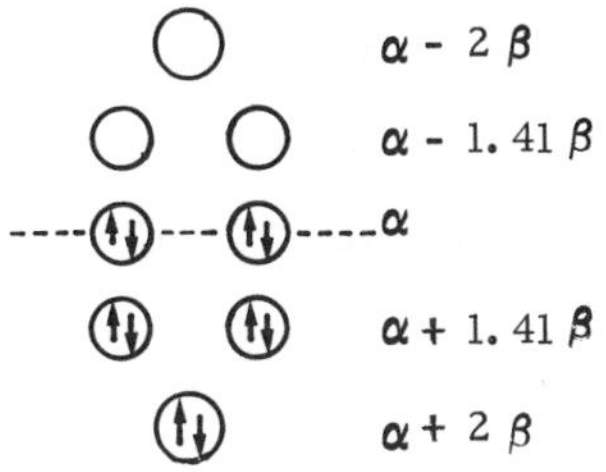

$\alpha - 2\ \beta$

$\alpha - 1.41\ \beta$

α

$\alpha + 1.41\ \beta$

$\alpha + 2\ \beta$

Total π -electron energy

of $C_8H_8{}^{2\ominus}$ is

$10\ \alpha + 9.64\ \beta$

Total π -electron energy of planar localized C_8H_8 $= 8\ \alpha + 8\beta$

Total π -electron energy of localized $C_8H_8 + 2e$ $= 10\ \alpha + 6\beta$

Hence, the delocalization energy of cyclooctatetraene dianion is

$(10\ \alpha + 9.64\ \beta) - (10\ \alpha + 6\beta) = 3.64\ \beta$

21-17   Valence-bond structures are as follows where the electrons can be paired or unpaired:

$$CH_2{=}C\begin{smallmatrix}CH_2\\\\CH_2\end{smallmatrix} \longleftrightarrow CH_2{-}C\begin{smallmatrix}CH_2\\\\CH_2\end{smallmatrix} \longleftrightarrow CH_2{-}C\begin{smallmatrix}CH_2\\\\CH_2\end{smallmatrix}$$

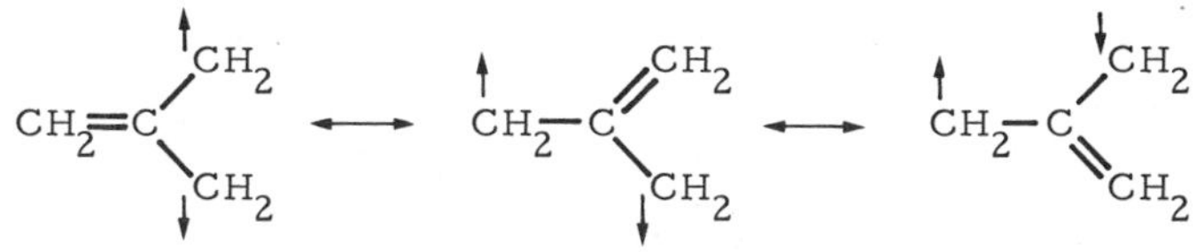

However, the VB theory makes clear that each form only has one π -bond.

The molecular orbital energies and configuration are:

$\alpha - 1.73\,\beta$

α

$\alpha + 1.73\,\beta$

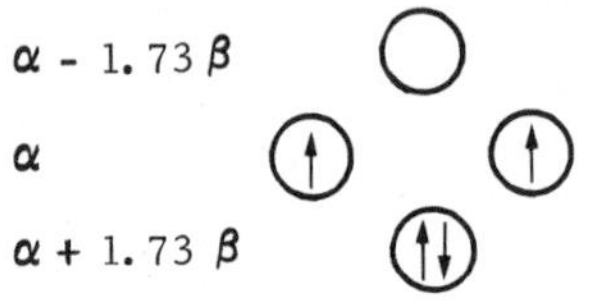

A triplet ground state is predicted by MO method. The VB method does not predict the triplet ground state.

21-18	a	b	c	d	e
DE cation	2β	$1.24\,\beta$	3β	$0.82\,\beta$	$1.46\,\beta$
DE anion	0	$2.48\,\beta$	$2.1\,\beta$	$0.82\,\beta$	$1.46\,\beta$

21-19   The two possible adducts are:

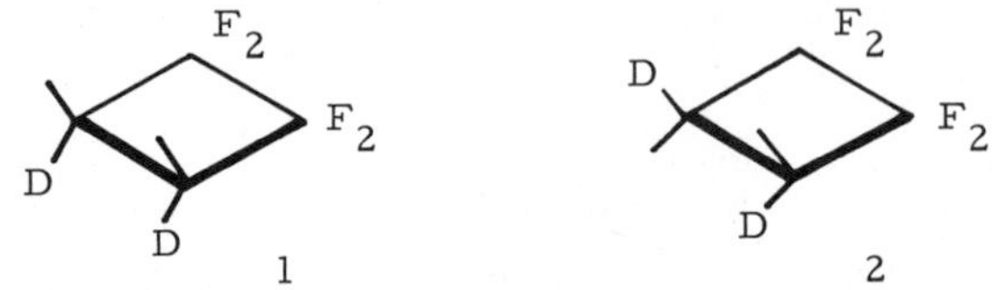

a.   Both 1 and 2 would be anticipated as products because the diradical could undergo rotation about the CHD–CHD bond and thus lose memory of the cis orientation of the D's in the starting ethene.

b.   1 only

c.   There are two possible transition states depending on which double bond undergoes antarafacial addition.   One leads to 1 and the other to 2.

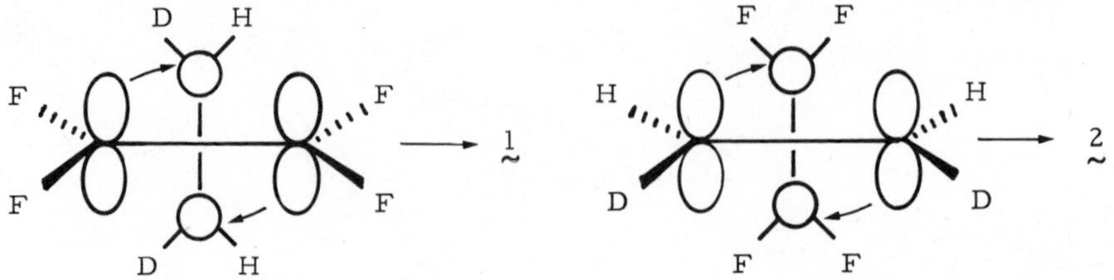

21-20   The Hückel π-electron energy for the transition state corresponding to 39 is approximately $2(\alpha + 2\beta) + 4(\alpha + 1.41\,\beta) + 2\alpha = 8\alpha + 9.64\beta$.   The Möbius π-electron energy for 40 is approximately $4(\alpha + 1.84\beta) + 4(\alpha + 0.77\,\beta) = 8\alpha + 10.44\beta$.   The difference in energy between the two transition states is therefore $(8\alpha + 10.44\beta) - (8\alpha + 9.64\beta) = 0.8\beta$.   The Möbius transition state is favored.

<u>21-21</u>

Hückel cyclic transition state
with 6 electrons

or

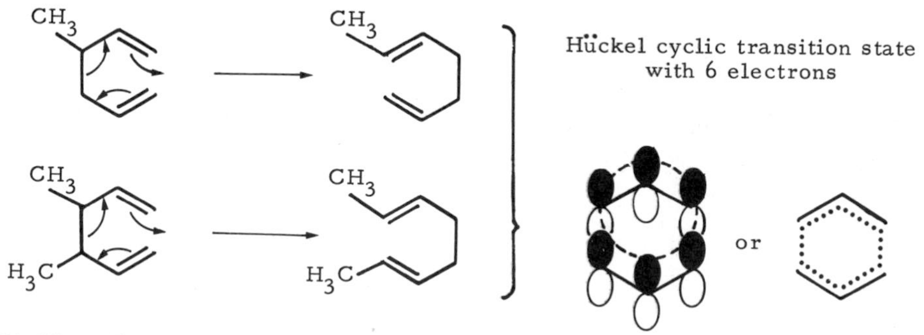

<u>21-22</u>   For steric reasons, disrotatory ring-opening of the bicyclic rings is
preferred. Electronically, only the 6-electron system can undergo disrotatory
cleavage easily.

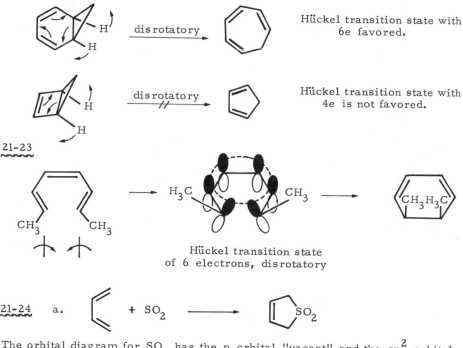

Hückel transition state with
6e favored.

Hückel transition state with
4e is not favored.

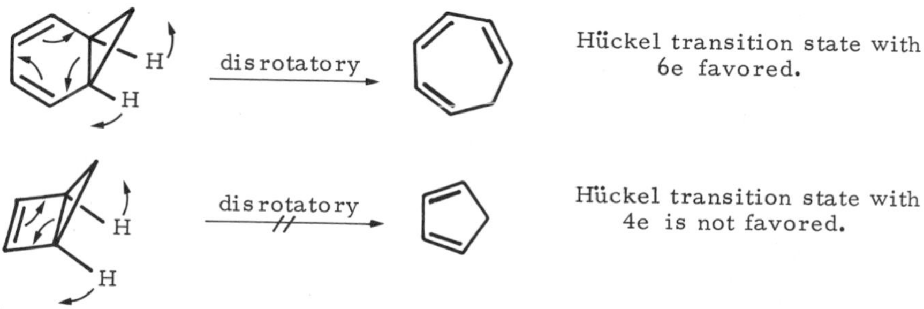

21-23

Hückel transition state
of 6 electrons, disrotatory

<u>21-24</u>   a. $\begin{array}{c}\end{array}$ + SO_2 $\longrightarrow$ SO_2

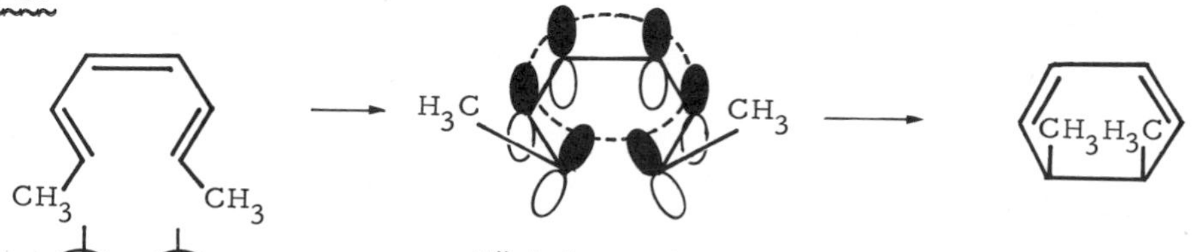

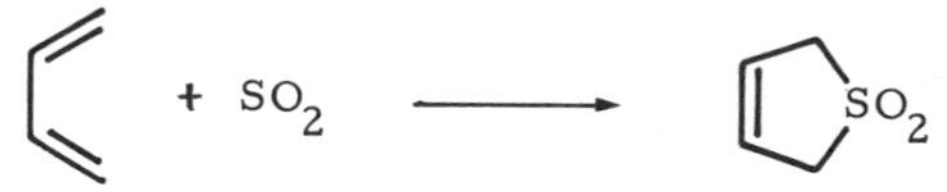

The orbital diagram for SO_2 has the p orbital "vacant" and the $\underline{sp}^2$ orbital
doubly filled:

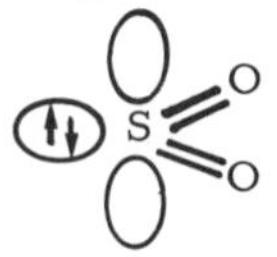

The transition state for cycloaddition to butadiene can be formulated using both the $\underline{p}$ and $\underline{sp}^2$ orbitals of SO_2 as follows.

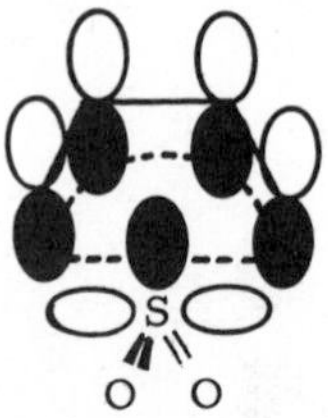

Overlap of the diene orbitals with the $\underline{filled}$ $\underline{sp}^2$ gives a Hückel array and 6 electrons (allowed).

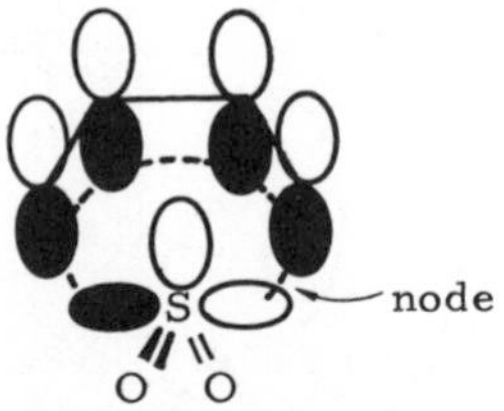

Overlap of the vacant $\underline{p}$ orbital with the diene orbitals gives a Möbius array and 4 electrons (allowed).

b. $\begin{array}{c} CH_2 \\ \| \\ CH_2 \end{array}$ + SO_2 $\longrightarrow$ $\triangleright SO_2$

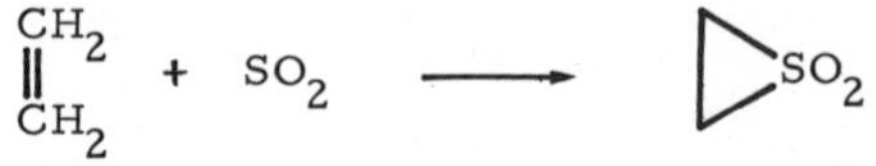

This reaction is interesting because the only stable orbital arrays one can write involve the following:

and

node

vacant p
sp^2

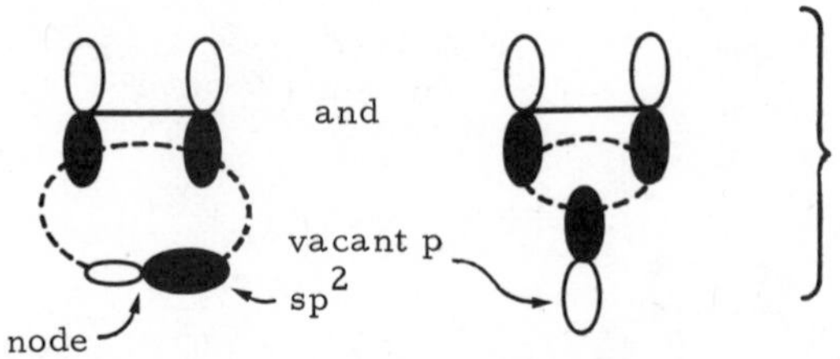

Möbius, 4e, stable   Hückel, 2e, stable

which means a transition state with the geometry shown:

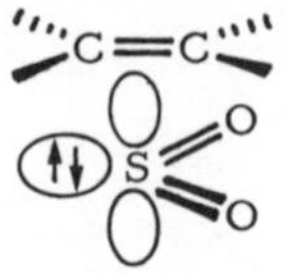

a.

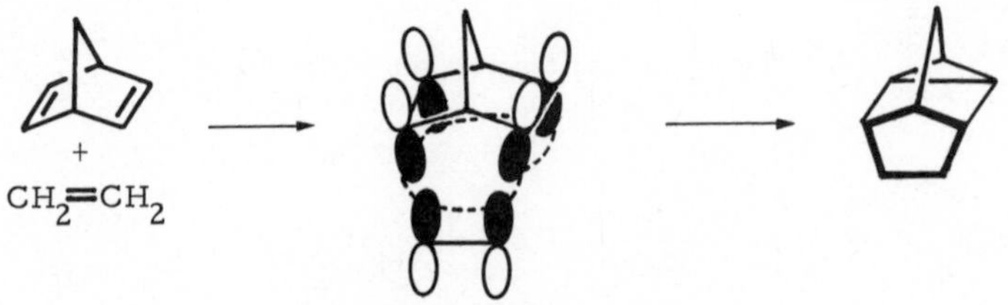

Hückel transition state
of 6 electrons (favorable)

b.

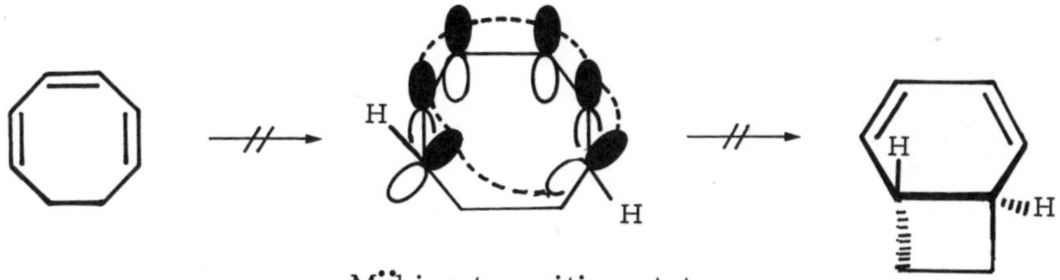

Möbius transition state
of 6 electrons (unfavorable)

trans

c.

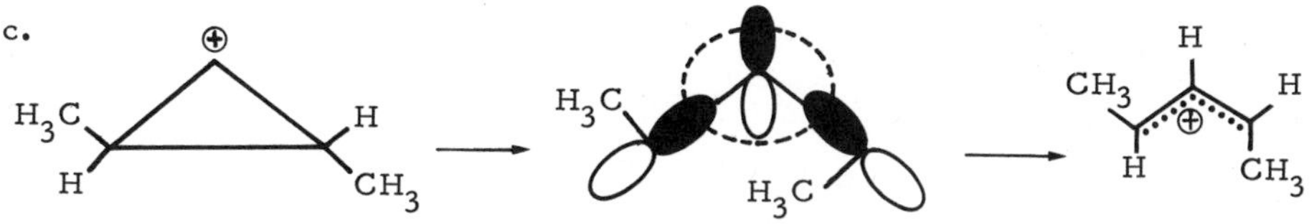

Hückel transition state of
2 electrons, favored

21-26

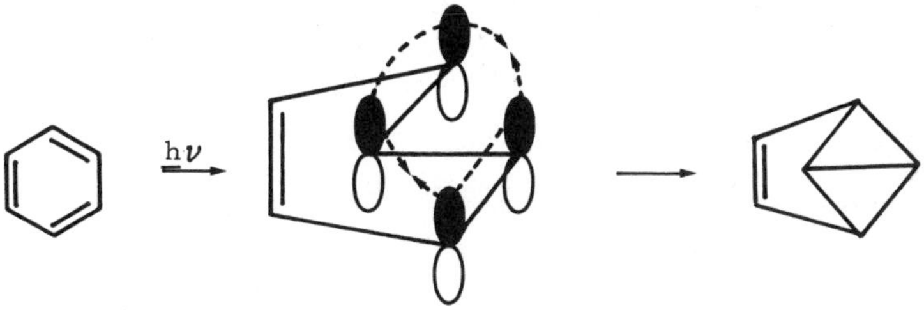

Hückel transition state of 4 electrons.
Unfavorable thermally, but photochemically allowed.
Reverse reaction would not occur thermally by a concerted mechanism.

21-27

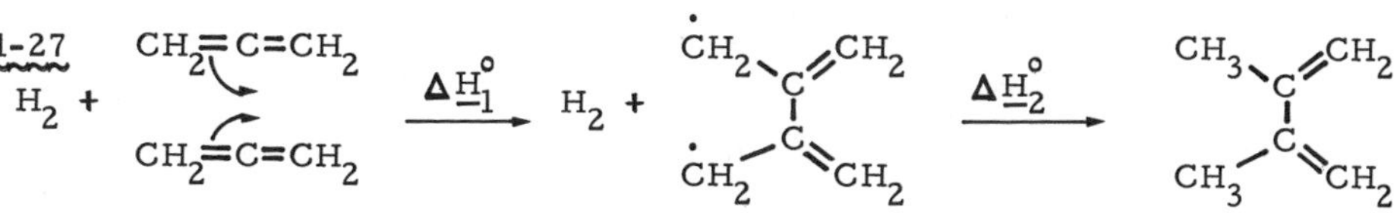

$\underline{50}$, DE = 1.64 β

$$\Delta \underline{H}^{o}_{net} = \Delta \underline{H}^{o}_{\underline{1}} + \Delta \underline{H}^{o}_{\underline{2}} = \Delta \underline{H}^{o}_{\underline{1}} - 104 \text{ kcal (H–H)} + 2 \times 89 \text{ kcal (CH)}$$

$$-80 \text{ kcal} = \Delta \underline{H}^{o}_{\underline{1}} - 74 \text{ kcal}, \qquad \Delta \underline{H}^{o}_{\underline{1}} = -6 \text{ kcal (for } \underline{50})$$

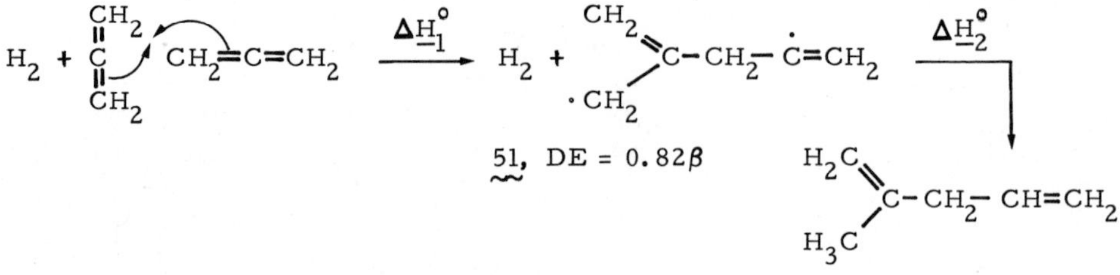

$$H_2 + \underset{CH_2}{\overset{CH_2}{C}} \underset{CH_2=C=CH_2}{\overset{\Delta H_1^o}{\longrightarrow}} H_2 + \underset{\cdot CH_2}{\overset{CH_2}{C}} C-CH_2-\dot{C}=CH_2 \xrightarrow{\Delta H_2^o}$$

$$\underset{51, \ DE = 0.82\beta}{} \qquad \underset{H_3C}{\overset{H_2C}{C}} C-CH_2-CH=CH_2$$

$$\Delta H^o_{net} = -76 = \Delta H_1^o - 104 + 89 + 105 = \Delta H_1^o - 90 \qquad \Delta H_1^o = 14 \text{ kcal}$$
$$\text{(for } 51\text{)}$$

$$\underset{CH_2=C=CH_2}{\overset{CH_2=C=CH_2}{}} \longrightarrow \underset{CH_2-C}{\overset{CH_2-C}{\overset{CH_2}{\diagdown}}} \underset{CH_2}{\overset{\diagup CH_2}{}} \xrightarrow{\Delta H_1^o, \ H_2} \underset{CH_2-CH=CH_2}{\overset{CH_2-CH=CH_2}{}}$$

$$52, \ DE = 0$$

$$\Delta H^o_{net} = -72 = \Delta H_1^o - 104 + 210 = \Delta H_1^o - 106 \qquad \Delta H_1^o = 34 \text{ kcal}$$
$$\text{(for } 52\text{)}$$

The predicted and observed order of stability is $50 > 51 > 52$.

21-28   a.   Unimportant since it violates rule that positions of all nuclei must be the same in the contributing structures.

b.   As in Part a.

c.   Unimportant.

d.   Neither form is a very favorable structure - the structure with positive oxygen is relatively unfavorable.

e.   Unimportant. Nitrogen would have to use 3d orbitals in forming $\overset{\oplus}{CH_2}-CH=NH_3$. Hence the right-hand structure would have a very high energy.

f.   Unimportant for steric reasons. A double bond at a bridgehead carbon of a bicyclic structure would be very highly strained.

g.   Unimportant since one form represents a very distorted 2-propenyl cation and would not significantly contribute to structure of cyclopropyl cation.

h.   Likely to be of considerable importance since it involves resonance stabilization of a primary cation. The resonance postulated here is very like that of the 2-propenyl cation.

i.   Reasonable on grounds of electronegativity. Less reasonable on steric grounds because the favored geometry for right-hand form would have

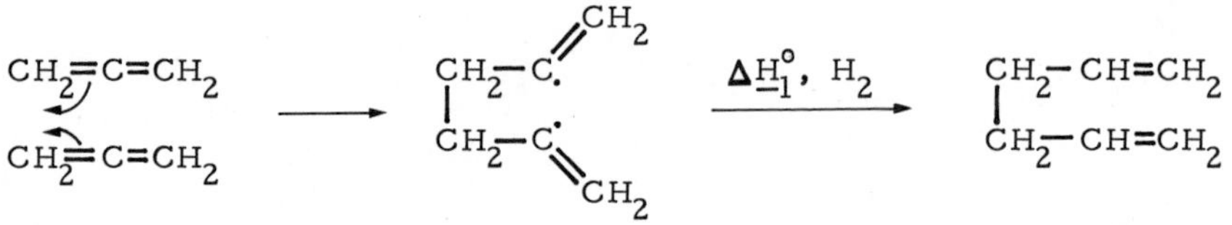

two fluorines, the carbon and the oxygens lying in one plane.

21-29

$$H-C\equiv C-C\equiv C-H; \qquad \begin{matrix} C-CH \\ \| \quad \| \\ C-CH \end{matrix} \; ; \quad \begin{matrix} C=CH \\ \| \quad | \\ C=CH \end{matrix}$$

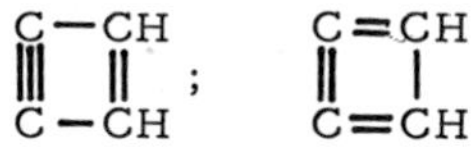

The first written structure is the most favorable geometric configuration since it is a linear structure in which the bond angles have their normal values. Both of the cyclic structures would be highly strained since formation of them would require severe distortion from preferred bond angles. The resonance energy of the alkadiyne structure would be greater than that of 1,3-butadiene since rotation about the central bond does not effect the possibility of electron pairing across this bond.

$$HC\equiv C-C\equiv CH \quad \longleftarrow \quad H\overset{.}{C}=C=C=\overset{.}{C}H$$

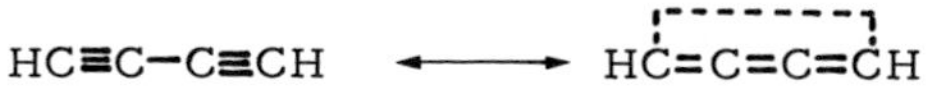

21-30

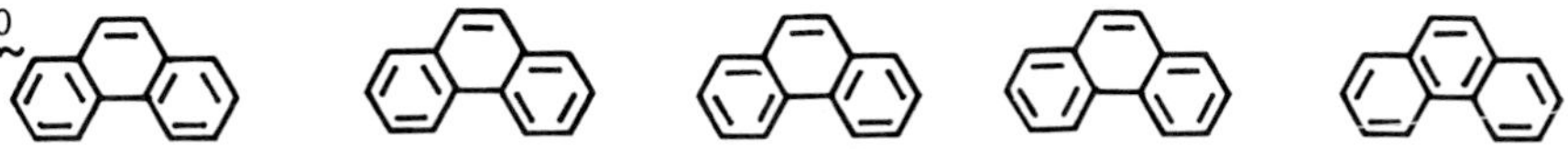

(1)   (2)   (3)   (4)   (5)

The five resonance structures of phenanthrene suggest that the 9,10 bond has 80% double-bond character whereas all other bonds have 60% (or less) double-bond character. The 9,10 bond is therefore more like an alkene double bond than a benzene ring bond, and it is not surprising that bromine adds to phenanthrene largely across the 9,10 positions.

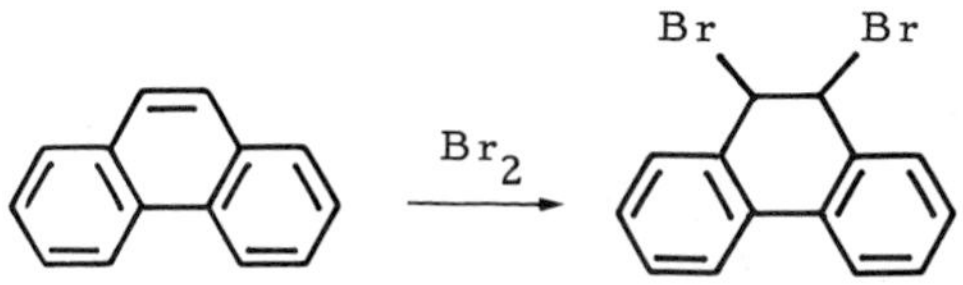

SE = 118.7   SE = 2 x 43.2

$\Delta \underline{H}° =$ 63.2 + 46.1 - 2 x 68

+ 118.7 - 2 x 43.2

= +5.6 kcal

21-31   a. $CH_2=CH-CH_2Cl$;   ionization is favored by a gain in resonance energy of 2-propenyl cation.

b. $CH_2=CH-\underset{\underset{Cl}{|}}{CH}-CH=CH_2$;   the same carbonium ion is formed from both halides. Therefore, the difference in ease of ionization will depend mainly on the difference in the stabilities of the halides. The secondary halide in which the two double bonds are not conjugated is less stable than the primary halide and therefore ionizes more readily.

21-32

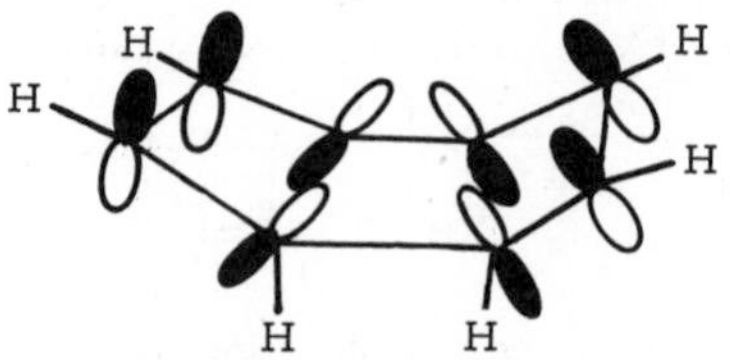

This atomic orbital model has a pair of parallel p orbitals on each double bond but these pairs of orbitals on adjacent double bonds are not oriented well for effective electron delocalization.   A view down one of the σ single bonds shows

parallel   not parallel

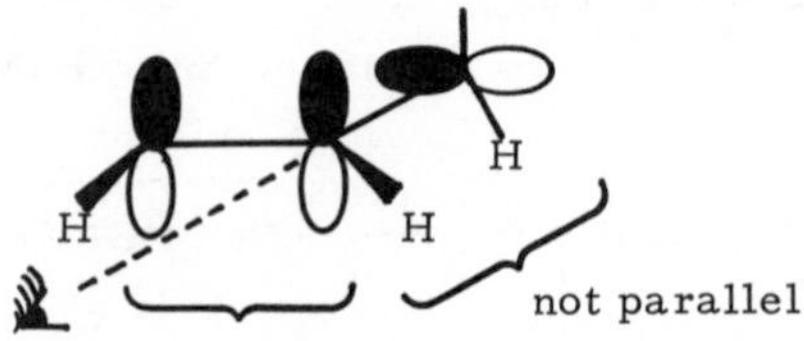

21-33   The problem with 53 is that there is a very substantial repulsion between the hydrogens in the middle of the ring which forces the double bonds to twist out of the planar configuration which would allow two __equivalent__ valence-bond structures even with the favorable 10π electrons.   With azulene 54, the extra C—C bond and different configurations of the double bonds allow for a stable planar structure with two equivalent valence-bond structures.

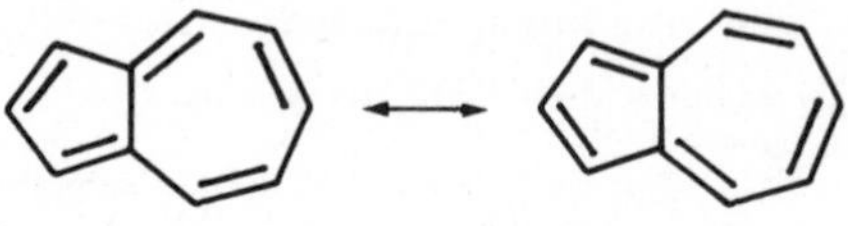

The CH$_2$ bridged structure 55 will be seen from models to be not quite planar, but despite one's first impression it is also seen to be surprisingly strain free and quite capable of having two equivalent resonance structures.

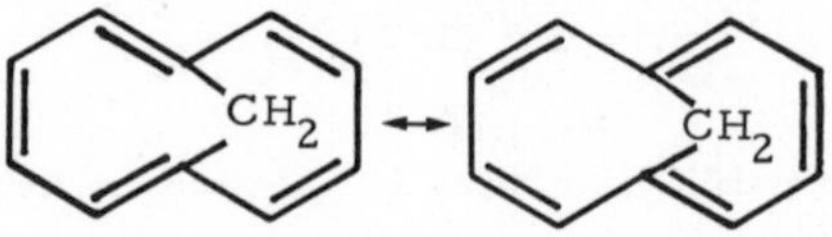

When syntheses expected to give 53 are performed the product is __cis__-bicyclo [4. 4. 0] -2, 4, 7, 9-decatetraene apparently formed by disrotatory Hückel electrocyclic ring closure (Section 21-10E).

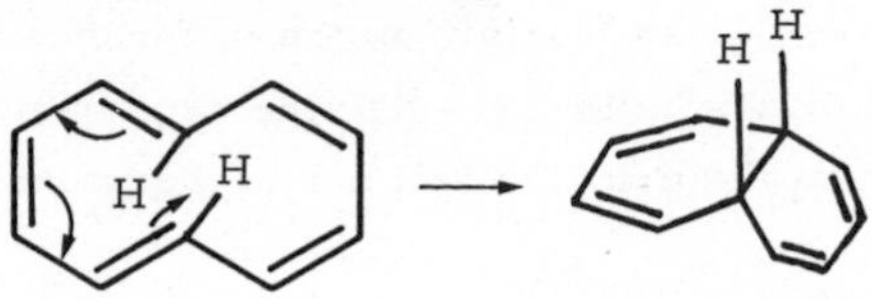

A corresponding ring closure of $\underset{\sim}{54}$ would give a strained ring.

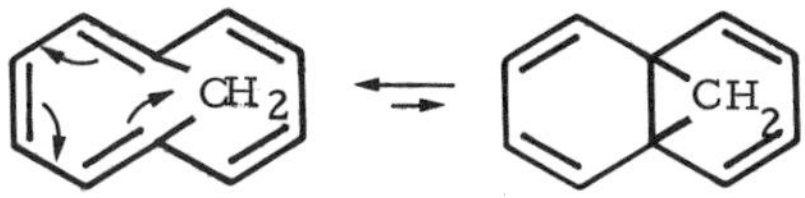

$\underset{\sim\sim\sim\sim}{21\text{-}34}$

a.

most important

$\underset{\sim\sim\sim\sim}{21\text{-}35}$ Molecules with some resonance stability according to Hückel's rule include $\underline{a}$ (2π electrons), $\underline{b}$ (6π electrons), $\underline{c}$ (6π electrons), and $\underline{d}$ (2π electrons corresponding to OH). The rule predicts $\underline{f}$ would have no resonance stabilization; the rule does not apply to $\underline{e}$ as written, but there are two valence-bond structures of $\underline{e}$ which by the $4n + 2$ rule should have a degree of stabilization.   That $\underline{e}$ is not a known compound may be a consequence of the considerable angle strain that it would be expected to have.

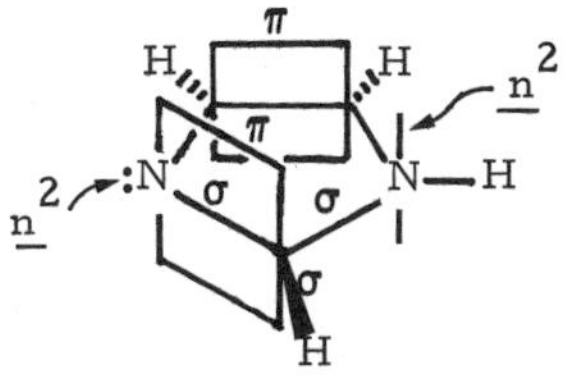

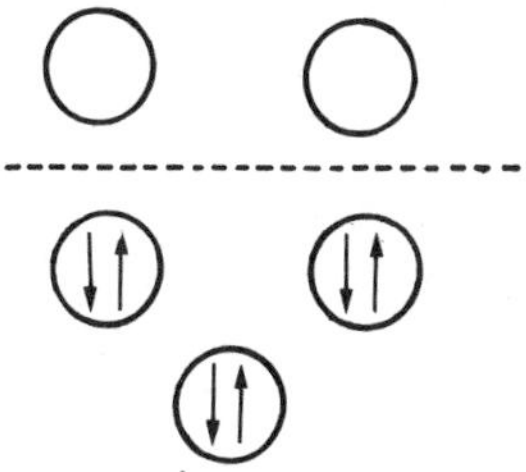

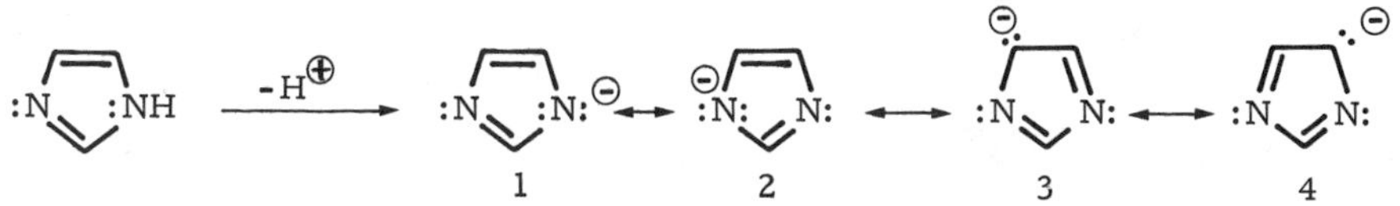

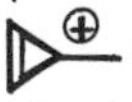

By the rule as formulated at the end of Section 21-9A these should correspond to substantial stabilization.

$\underset{\sim\sim\sim\sim}{21\text{-}36}$ a.   The adduct would be a cyclobutadiene derivative which, by Hückel's rule, is an unstable $4n$ electron system.

b.   The dianion is a delocalized structure in which all four oxygens are equivalent and the ring is a $4n + 2$ electron system.

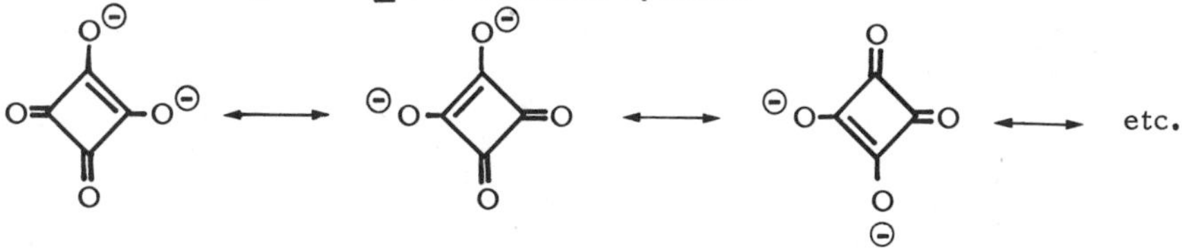

21-36   b.   (cont.)

or

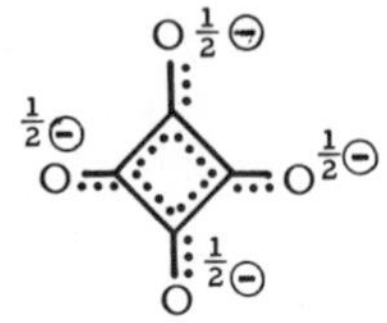

c.   2,4-Cyclopentadienone approaches a 4n electron system within the five-membered ring by virtue of the polarity of the carbonyl.   This is seen best in the VB structure:
As such it would be unstable.

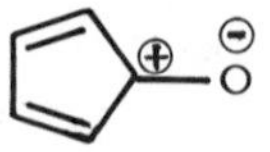

21-37

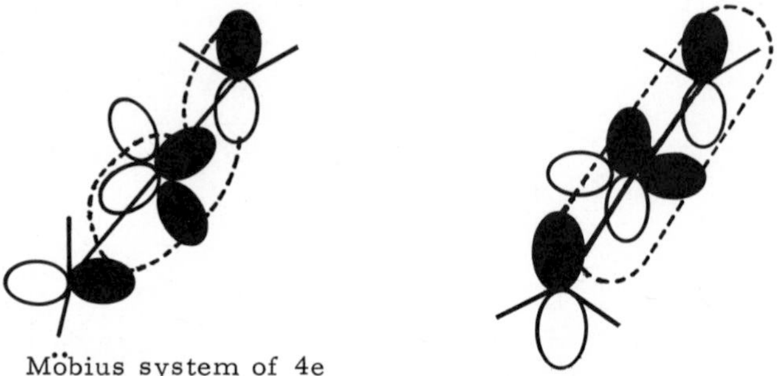

Möbius system of 4e
(stable)

Hückel system of 4e (unstable)

21-38
a.

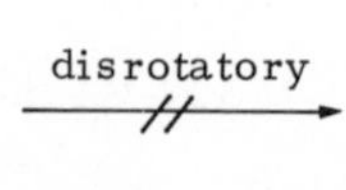

b.

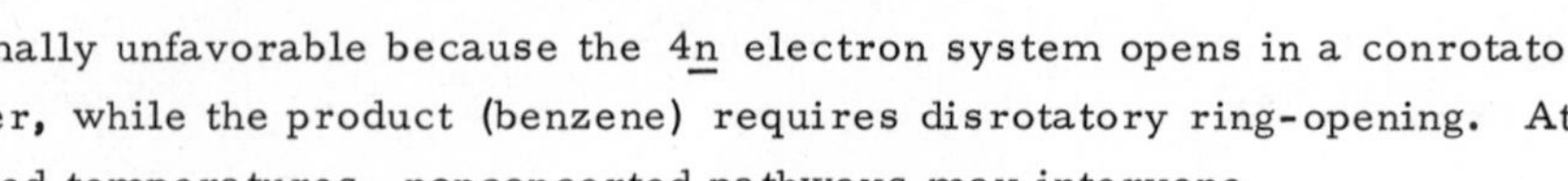

$\xrightarrow{\text{electrocyclic} \atop \text{ring opening}}$ [4 + 2] cycloaddition

21-39

$\xrightarrow{\text{disrotatory}} ✗$

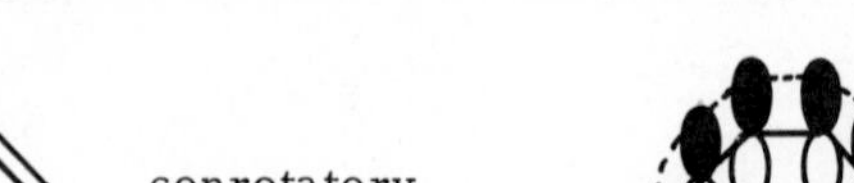

Thermally unfavorable because the 4n electron system opens in a conrotatory manner, while the product (benzene) requires disrotatory ring-opening.   At elevated temperatures, nonconcerted pathways may intervene.

21-40

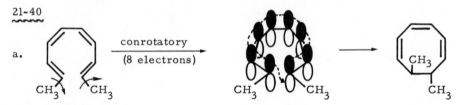

a. $\xrightarrow[\text{(8 electrons)}]{\text{conrotatory}}$

b.    $\xrightarrow[\text{(6 electrons)}]{\text{disrotatory}}$ $\longrightarrow$

cis

c. $C_6H_5 \diagdown \diagup CD_3$ $\xrightarrow[\text{(4 electrons)}]{\text{conrotatory}}$

H_3C C_6H_5

or

C_6H_5 C_6H_5

CH_3 CD_3

d.

C_6H_5

IIII H

IIII H

C_6H_5

$\xrightarrow[\text{rotatory}]{\text{con-}}$

cis

C_6H_5

C_6H_5

cis, trans

H C_6H_5

H

H_5C_6 H

21-41

a.

Hückel transition state
2 electrons - favorable

b.   Möbius transition state
4 electrons - favorable

c.   Hückel transition state
4 electrons - unfavorable

Möbius transition state is
not physically possible.

21-42

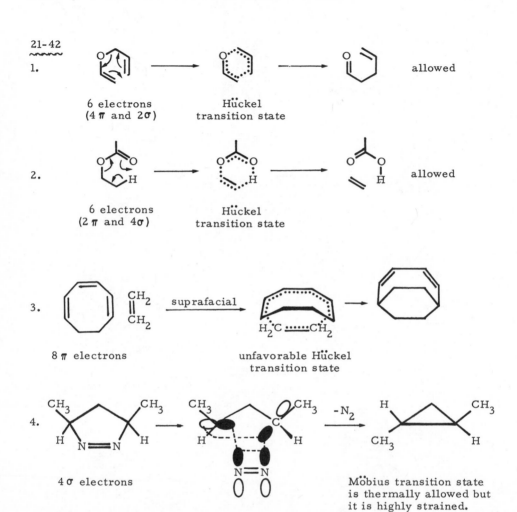

1.

6 electrons
(4 π and 2σ)

Hückel
transition state

allowed

2.

6 electrons
(2 π and 4σ)

Hückel
transition state

allowed

3.   suprafacial

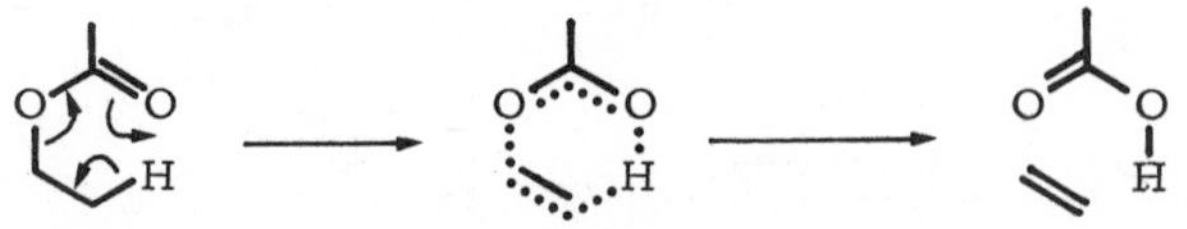

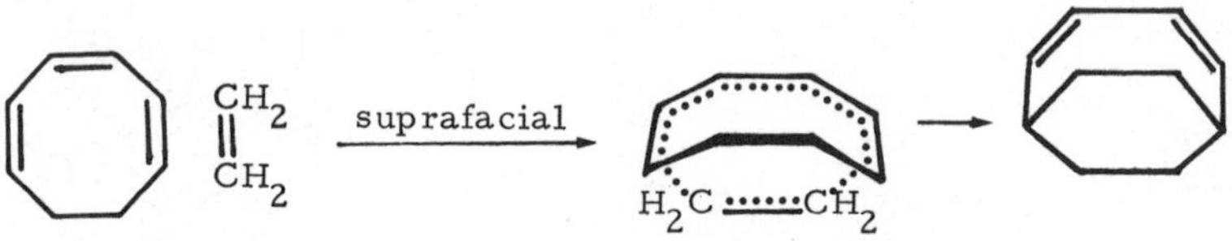

8 π electrons

unfavorable Hückel
transition state

4.

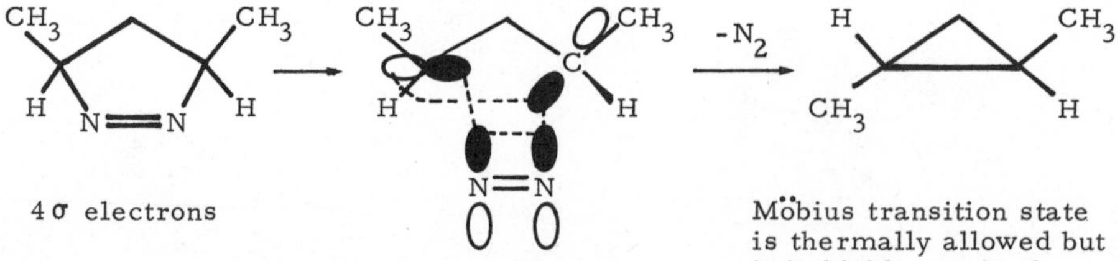

4 σ electrons

$-N_2$

Möbius transition state
is thermally allowed but
it is highly strained.

5.

$HO_2C-C\equiv C-CO_2H$

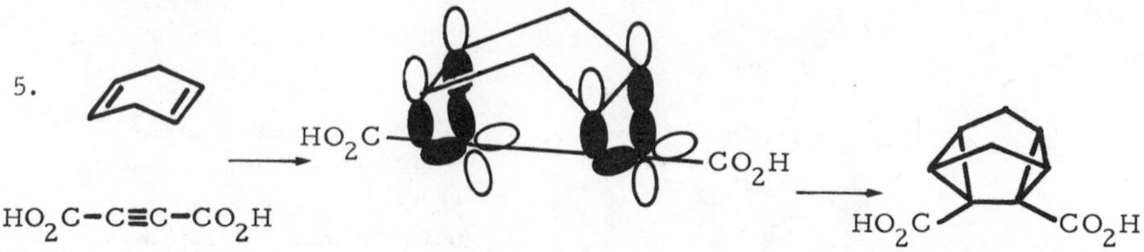

8 π electrons

Hückel transition state

thermally disallowed
photochemically allowed

21-43   a.

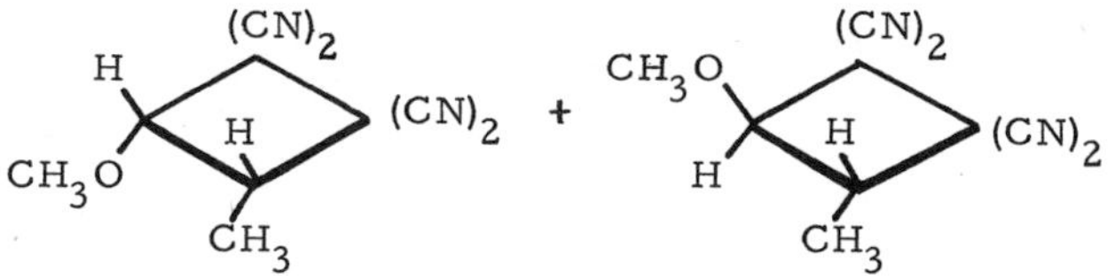

b.   Facts 1 and 2 indicate that the reaction is stepwise and involves ionic intermediates. Fact 3 shows that reaction is reversible from the intermediate (see Section 21-11).

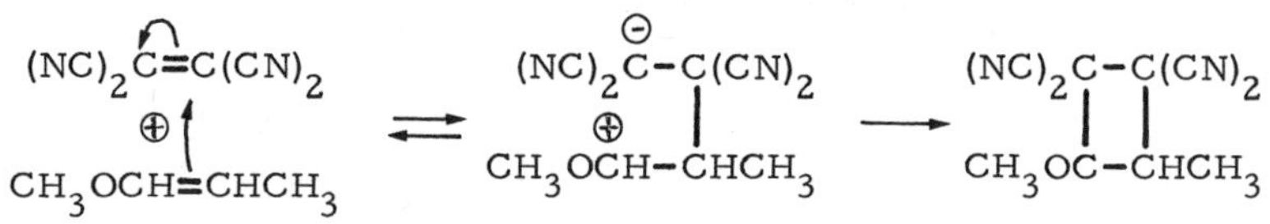

d.   The dipolar intermediate can be trapped by methanol.

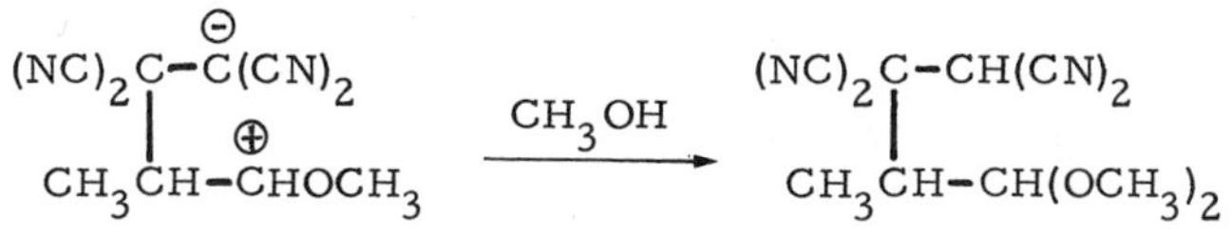

The suggested possible alternative product would form only if the cycloaddition went through the dipolar intermediate,

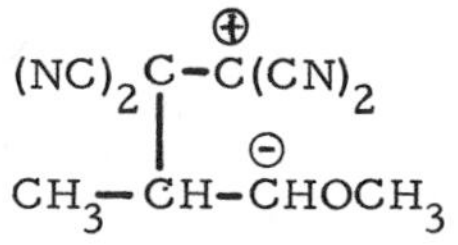

which is very unlikely because $-CN$ is not expected to stabilize a positive carbon and $-OCH_3$ is not likely to be very effective at stabilizing a negative carbon.

CHAPTER 22

ARENES. ELECTROPHILIC AROMATIC SUBSTITUTION

Arenes are cyclic or polycyclic conjugated unsaturated hydrocarbons related to benzene. They have many special chemical and spectroscopic properties. Their remarkable stability is associated with π-electron delocalization. This stability is manifest in the way that these compounds react with electrophiles by substitution rather than addition:

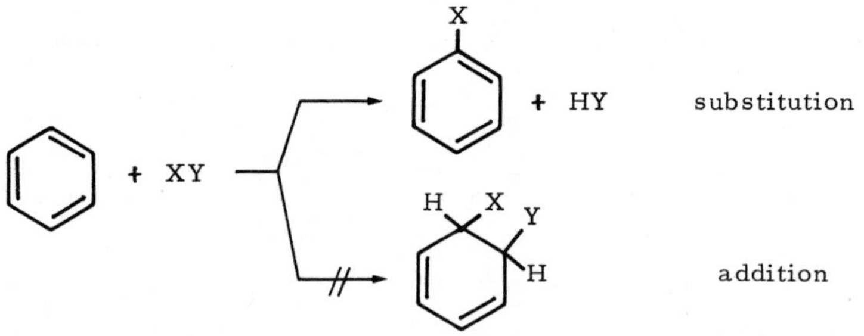

The mechanism of electrophilic substitution involves first addition of the electrophile $X^{\oplus}$, then elimination of a proton $H^{\oplus}$ from the same ring position:

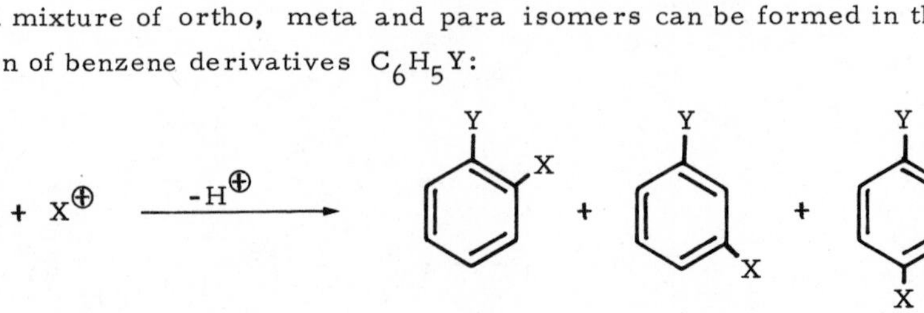

A mixture of ortho, meta and para isomers can be formed in the substitution of benzene derivatives C_6H_5Y:

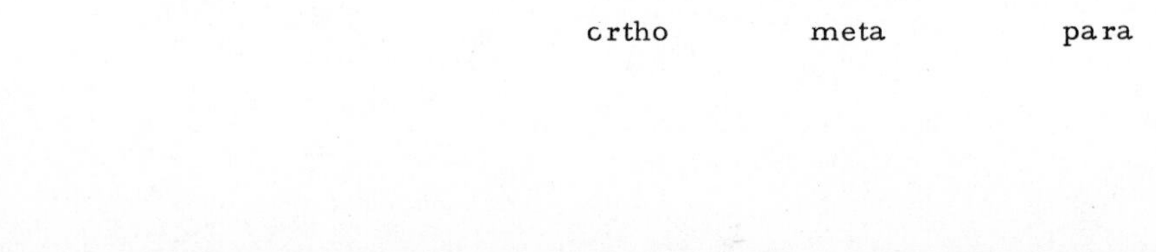

The actual distribution of isomers formed depends on the nature of the Y group, and to some extent on $X^{\oplus}$. Groups that can stabilize the transition state (or intermediate cation) by electron donation (e.g., alkyl and $\ddot{Y}$-type groups) invariably lead to ortho-para substitution. Moreover, substituents of this type generally make the ring more reactive than that of benzene, provided they are not strongly electronegative (e.g., halogens), in which case they deactivate the ring and retard substitution. If the substituents are strongly electronegative and incapable of electron donation by resonance, they both deactivate the ring and direct the entering group to the meta position.

ortho-para directing

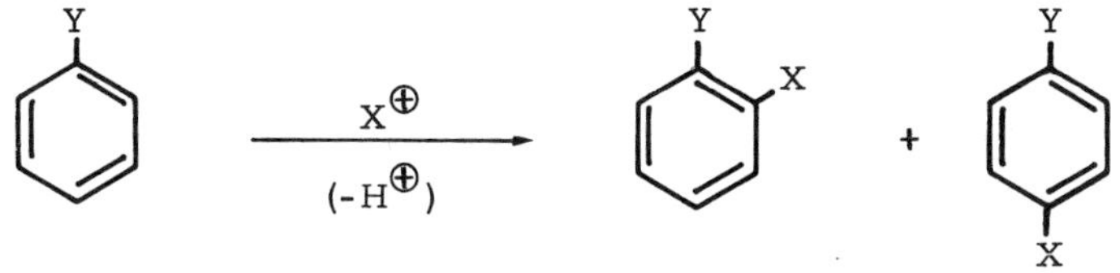

Y = alkyl, $-OH$, $-O^{\ominus}$, $-SCH_3$, $-NH_2$, $-Cl$, $-Br$, $-I$, $-F$, $-NO$, etc.

meta-directing

$Z = -NH_3^{\oplus}$, $-N_{\oplus}\diagup^{O}_{\diagdown O^{\ominus}}$, $-S(CH_3)_2^{\oplus}$, $-C\diagup^{O}_{\diagdown CH_3}$, $-C\equiv N$

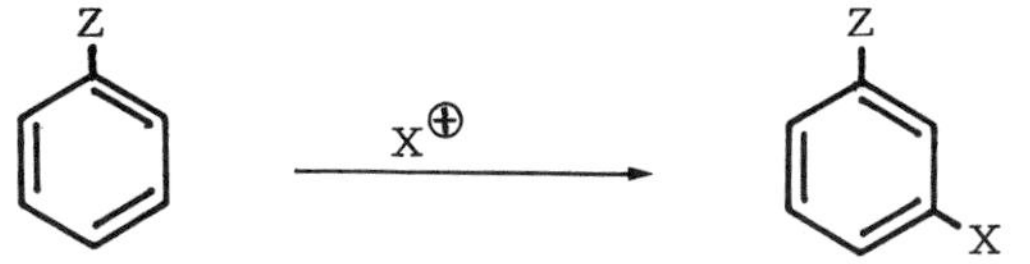

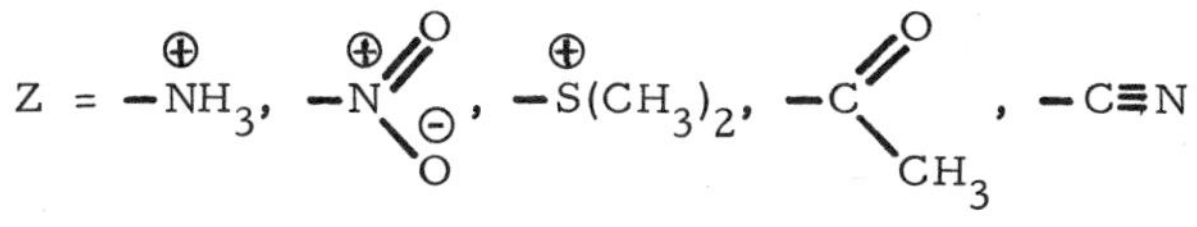

Steric effects also are important. The amount of ortho substitution decreases if the group already present is large, or if the electrophilic substituting agent is large. In contrast, the more reactive the substituting agent, the less selective it is between the ortho, meta, and para positions.

The most common types of electrophilic substitution reactions of benzene are summarized in Figure 22-7. From the standpoint of organic synthesis, the most important substitution reactions are nitration, halogenation, alkylation, acylation, and sulfonation. Catalysts may be required for speedy reaction. Their function may be to create a more reactive electrophile from the given reagents. Thus, in nitration in mixed acids (HNO_3 and H_2SO_4), the sulfuric acid generates the substituting agent, $NO_2^{\oplus}$, from nitric acid according to the equation:

$$HNO_3 + H_2SO_4 \longrightarrow NO_2^{\oplus} + H_2O + HSO_4^{\ominus}$$

Metal halide catalysts function similarly in halogenation, alkylation and acylation.

$$Br_2 + FeBr_3 \longrightarrow Br^{\oplus}\!\cdots\!FeBr_4^{\ominus}$$

$$RCl + AlCl_3 \longrightarrow R^{\oplus}\!\cdots\!AlCl_4^{\ominus}$$

$$RCOCl + AlCl_3 \longrightarrow RCO^{\oplus}\!\cdots\!AlCl_4^{\ominus}$$

Alkylation reactions are complicated by several features:

a. polysubstitution; introduction of one alkyl substituent activates the ring towards substitution of another.

b. rearrangement; the alkylating species, $R^{\oplus}$, may rearrange to a more stable carbocation, $R'^{\,\oplus}$, and will then lead to mixtures of isomeric products.

c. rearrangement; the equilibrium product of polyalkylation may be isolated rather than the kinetic product - depending on the amount of catalyst used and the reaction time.

Acylation reactions do not suffer from the same limitations as alkylation, and the two step acylation-reduction sequence,

$$ArH \xrightarrow[AlCl_3]{RCOCl} ArCOR \xrightarrow[Pt]{H_2} ArCH_2R$$

may be preferable to the one-step alkylation sequence,

$$ArH \xrightarrow[AlCl_3]{RCH_2Cl} ArCH_2R$$

for the synthesis of $ArCH_2R$ if rearrangement and polysubstitution are to be avoided.

Naphthalene behaves like benzene in showing a wide variety of electrophilic substitution reactions. Anthracene and phenanthrene, however, show increased tendency to react by addition, but only at the 9,10 positions. Addition reactions of benzene occur with difficulty, but are important in the commercial production of cyclohexane (by catalytic hydrogenation) and hexachlorocyclohexane (by photochlorination).

Reduction of benzene and its derivatives to nonconjugated dienes occurs in metal-ammonia solutions and has wide application for the reduction of aromatic rings in organic synthesis. Oxidation of arenes is not a routine laboratory reaction - rather, it is accomplished on an industrial scale for the production of cis-butenedioic (maleic) anhydride (from benzene) and phthalic anhydride (from 1, 2-dimethylbenzene and naphthalene). Some industrial uses of benzene derivatives are summarized in Figures 22-9 and 22-10.

The azulenes and annulenes are nonbenzenoid conjugated cyclic polyenes with aromatic properties. Azulene is very polar for a hydrocarbon and has the larger ring somewhat positive and the other correspondingly negative. Annulenes have unusual nmr proton shifts and often are configurationally very mobile. The large annulenes, $\geq$ [26]annulene, appear not to have equivalent valence-bond structures. Cyclooctatetraene is nonaromatic and undergoes many reactions that indicate that it is in equilibrium with a bicyclic isomer formed by an electrocyclic interconversion reaction.

ANSWERS TO EXERCISES

22-1   a. 1-Methylnaphthalene, 2-methylnaphthalene.

b. 1-Methylanthracene, 2-methylanthracene, 9-methylanthracene.

c. 1-Methylphenanthrene, 2-methylphenanthrene, 3-methylphenanthrene, 4-methylphenanthrene, 9-methylphenanthrene.

22-2   1, 2-dimethylbenzene (ortho-xylene) - 2 products;
1-chloro-2, 3-dimethylbenzene; 1-chloro-3, 4-dimethylbenzene.

1, 3-dimethylbenzene (meta-xylene) - 3 products;
1-chloro-2, 6-dimethylbenzene; 1-chloro-2, 4-dimethylbenzene; 1-chloro-3, 5-dimethylbenzene.

1, 4-dimethylbenzene (para-xylene) - 1 product; 1-chloro-2, 5-dimethylbenzene.

22-3   a. diphenylmethyl chloride or diphenylchloromethane (benzhydryl chloride); b. phenyldichloromethane (benzal chloride); c. phenyltrichloromethane (benzotrichloride); d. 4-methylbiphenyl or 4-methylphenylbenzene; e. 3-nitro-1-(2-propenyl)benzene or meta-nitro-(2-propenyl)benzene; f. 3-(2, 4-dichlorophenyl)-2-propen-1-ol.

22-4   Both compounds are derivatives of benzene since both have bands at 3000 cm^{-1} ($\equiv$C-H stretch), and 1600 cm^{-1} and 1500 cm^{-1} (C$\equiv$C stretch). Compound A (top) has a single band at 800 cm^{-1} and an absorption pattern in the region 2000-1650 cm^{-1} characteristic of a para-substituted benzene derivative. Compound A must therefore be 1-chloro-4-methylbenzene (para-chlorotoluene). Compound B (bottom) has two bands (690 and 770 cm^{-1})

typical of meta-substituted benzene derivatives; the absorption pattern at
$2000 - 1650$ cm^{-1} is also consistent with B as 1-chloro-3-methylbenzene (meta-chlorotoluene).

22-5   The u.v. absorption spectrum of benzenamine (aniline) shifts to much
shorter wavelengths in acid solution as the result of anilinium ion formation,
the anilinium ion not having an unshared electron pair on nitrogen to stabilize
the excited state.

     Resonance stabilization of the excited state of benzenolate ion is more
significant than for benzenol (phenol) because it would have no charge separation. Therefore, benzenolate ion absorbs at longer wavelengths.

22-6   a.   The alpha (and beta) methylene protons are predicted to have about
the same chemical shifts in 1,2- and 1,4-hexamethylenebenzene. But the
gamma methylene protons of 1,4-hexamethylenebenzene are expected to be
strongly shielded by the aromatic ring current and to give an nmr signal at high
fields relative to the corresponding gamma protons of 1,2-hexamethyleneben-
zene.

     b.   The chemical shift of the protons of cyclooctatetraene is typically
that of alkenyl protons, =CH-. The ring has no aromatic character, and no
aromatic ring current, and hence no paramagnetic deshielding effect on the
ring protons.

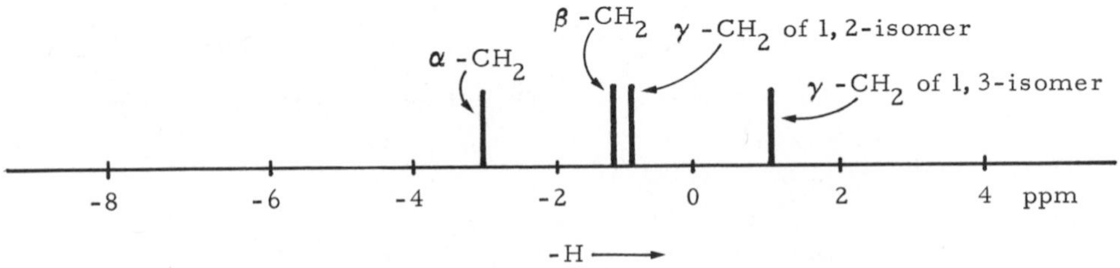

22-7   a. C_8H_{10}:   two types of hydrogen in ratio of 2:3 with chemical shifts
typical of aryl and arylalkyl protons. A structure consistent with the spectrum
is that of 1,4-dimethylbenzene (para-xylene).

     b. C_8H_7OCl: there are two principal types of hydrogen that are in the
ratio 4:3 and have chemical shifts typical of aryl and arylalkyl or acyl protons.
The spectral data, in conjunction with the molecular formula, suggests the compound be either 2-, 3- or 4-chlorophenylethanone. The symmetrical quartet
of lines centered at 7.5 ppm with $J > 7$ Hz identifies the compound as 1-
(4-chlorophenyl)ethanone (para-chloroacetophenone). CH_3-⟨⟩-COCl is also a
possible structure.

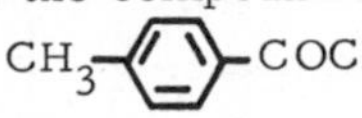

c. $C_9H_{10}O_2$: the three principal types of hydrogen are in the ratio of 5:2:3. The 3-4 line pattern of the upfield resonances is characteristic of an ethyl group; this is indicated to be an ethyl ester because of the shift to low fields (~4.4 ppm) of the CH_2 resonance. The data indicate the compound to be ethyl benzoate, $C_6H_5CO_2C_2H_5$.

Note the general resemblance of the resonances of the aromatic protons around 7.5 ppm to those of nitrobenzene in Figure 22-5.

d. C_9H_{12}: the spectrum is very similar to that of para-xylene in (a) except the hydrogens are in the ratio of 1:3. The spectrum is that of 1,3,5-trimethylbenzene (mesitylene).

22-8   a. $C_6H_6 + Cl_2 \longrightarrow C_6H_5Cl + HCl$ $\Delta\underline{H}° = -27.3$ kcal

b. $C_6H_6 + Cl_2 \longrightarrow C_6H_5Cl_2$ $\Delta\underline{H}° = -1$ kcal

(assuming stabilization energy of 1,2-dichloro-3,5-cyclohexadiene to be 3 kcal and of benzene to be 43 kcal)

22-9   We could study the reaction in the presence of a very large excess of bromine or an oxidizing agent. If the addition product has any appreciable lifetime, it should be possible to trap it by addition to its double bonds. If we were to use the following deuterated benzene:

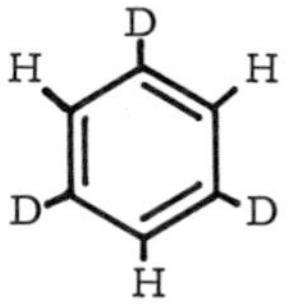

Addition-elimination should lead to quite preferential formation of 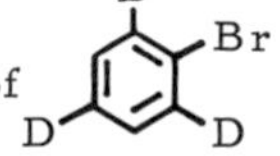 (primary isotope effect).

If addition-elimination were important we might expect in presence of CH_3OH or the like, formation of 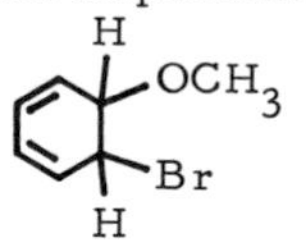 which should lead to 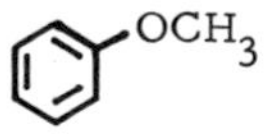 instead of bromobenzene.

22-10   A strong acid with a nonnucleophilic anion such as HBF_4 facilitates formation of $NO_2^{\oplus}$ from ethanoyl nitrate.

In the case of hydrochloric acid, the chloride ion is nucleophilic and destroys the ethanoyl nitrate by the reaction:

$$CH_3-\overset{O}{\overset{\|}{C}}-O-NO_2 + HCl \longrightarrow CH_3-\overset{O}{\overset{\|}{C}}-\overset{\oplus}{\underset{H}{O}}-NO_2 + Cl^{\ominus} \longrightarrow CH_3-\overset{O}{\overset{\|}{C}}-Cl + HNO_3$$

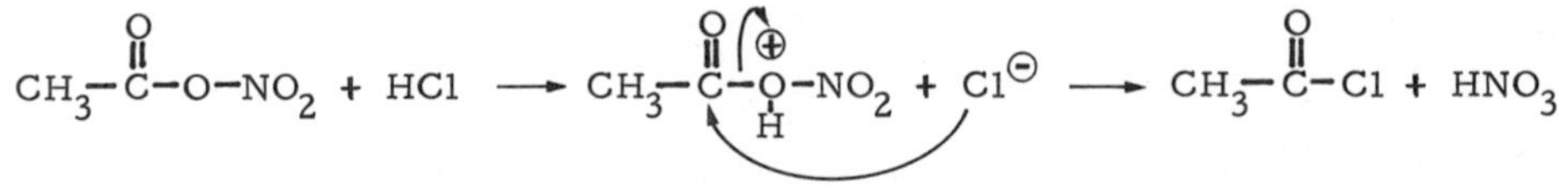

22-11   For benzene, methylbenzene and ethylbenzene to undergo nitration at a rate independent of the concentration of arene means that the <u>slow</u> step in the nitration of these compounds does not involve attack of $NO_2^{\oplus}$ on the arene.   The slow step must then be formation of nitronium ion (or some other activated species of the same type).

$$HNO_3 \xrightarrow[CH_3NO_2]{slow} NO_2^{\oplus} \xrightarrow[C_6H_6]{fast} C_6H_5NO_2$$

Equimolal mixtures of nitrobenzene and nitromethylbenzenes (nitrotoluenes) would <u>not</u> be formed on nitration of an equimolar mixture of benzene and methylbenzene.   The amount of each nitro compound formed is proportional to the rate of its formation from $NO_2^{\oplus}$ and the arene, even though each of these rates is faster than the measured overall rate of nitration.

22-12   A strong acid converts HOCl or HOBr into more effective halogenating agents, $H_2OCl^{\oplus}$ or $Cl^{\oplus}$, $H_2OBr^{\oplus}$ or $Br^{\oplus}$.

$$HOCl + HX \rightleftharpoons H_2\overset{\oplus}{O}Cl + X^{\ominus} \rightleftharpoons H_2O + Cl^{\oplus} + X^{\ominus}$$

If $X^{\ominus}$ is $Cl^{\ominus}$, the reaction tends to be reversible; nucleophilic chloride reacts with positive halogen species.

22-13   (See Figure 10-4) for electronegativity order $Br^{\oplus} > H_2OBr^{\oplus} > HOBr$ $ClBr > Br_2 >> Br_3^{\ominus} >> HBr$.

22-14   Aluminum chloride gives no chlorobenzene in the reaction of bromine with benzene.   This is the result one would anticipate if the function of $AlCl_3$ is to polarize and weaken the Br—Br bond.   Even if ionization equilibria are involved, little chlorination would be expected to occur in competition with bromination because of the greater bond strength Al—Cl relative to Al—Br.

$$Br_2 + AlCl_3 \rightleftharpoons Br^{\oplus}(AlCl_3Br)^{\ominus} \rightleftharpoons BrCl + AlCl_2Br$$

(BrCl is a brominating agent, being polarized $\overset{\delta\oplus}{Br} - \overset{\delta\ominus}{Cl}$.)

Formation of $Cl^{\oplus}(AlCl_2Br_2)^{\ominus}$ would be much less favorable.

22-15   a.   Bromine and iodine react according to the following equation:

$$Br_2 + I_2 \rightleftharpoons 2\,IBr$$

Apparently, the IBr functions as a catalyst in the bromination of benzene by assisting in the breaking of the Br—Br bond.

$$C_6H_6 + Br_2 \xrightarrow{IBr} C_6H_6 \cdot \cdot \overset{\oplus}{Br} \cdot \cdot \overset{\ominus}{Br} - IBr \xrightarrow{slow} C_6H_6\overset{\oplus}{Br} + IBr_2^{\ominus}$$

$$C_6H_5Br + HIBr_2 \xleftarrow{fast}$$

b.   The second molecule of bromine is thought to be involved in much the same way as IBr is involved in the bromination of benzene (see (a) above). It assists in the breaking of the Br—Br bond in the bromination step, thereby removing $Br^{\ominus}$ as $Br_3^{\ominus}$.

$$ArH + Br_2 \xrightarrow{Br_2} ArH \cdot \cdot \overset{\delta\oplus}{Br} \cdot \cdot \overset{\delta\ominus}{Br} \cdot \cdot Br_2 \xrightarrow{slow} Ar\overset{\oplus}{H}Br + Br_3^{\ominus}$$

$$ArBr + HBr_3 \xleftarrow{fast}$$

In 50% aqueous CH_3CO_2H, the kinetic order in bromine drops to first order probably because the solvent now replaces bromine in breaking the Br—Br bond.

$$ArH + Br_2 \xrightarrow{H_2O} ArH \cdot \cdot \cdot \overset{\delta\oplus}{Br} - \overset{\delta\ominus}{Br} \xrightarrow[slow]{CH_3CO_2H} Ar\overset{\oplus}{H}Br + \overset{\ominus}{Br} \cdot \cdot \cdot HO_2CCH_3$$

$$ArBr + HBr \xleftarrow{fast} CH_3CO_2H$$

22-16 $(CH_3)_2CHOH + BF_3 \rightleftharpoons (CH_3)_2CH-\overset{\oplus}{\underset{H}{O}}-\overset{\ominus}{BF_3}$

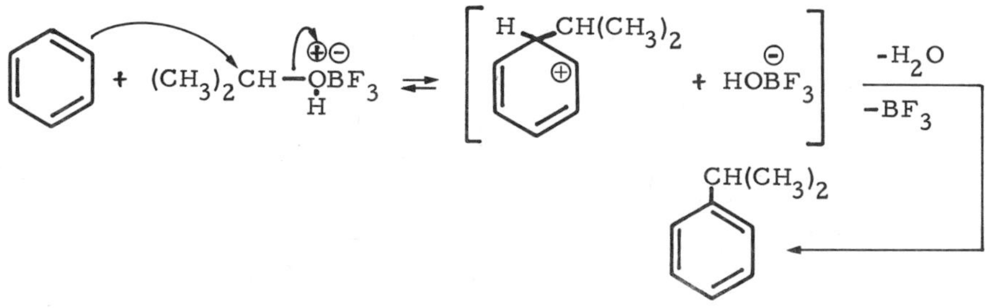

<u>22-17</u>   In the presence of excess HF , BF_3 (or $H^{\oplus} \, BF_4^{\ominus}$), <u>salt formation</u> of the dimethylbenzenes becomes significant.

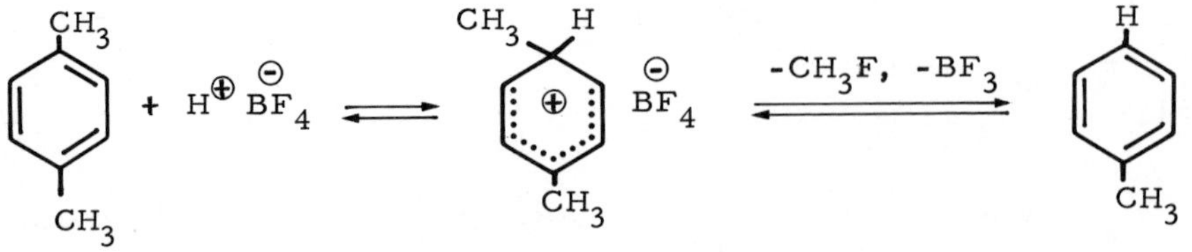

The most stable salt of the dimethylbenzenes is 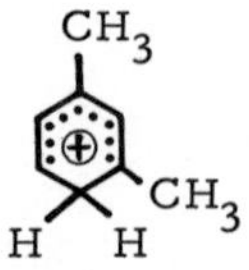   because two methyl groups are located on the positions of maximum positive charge.

<u>22-18</u>   a.   Rearrangement precedes substitution:

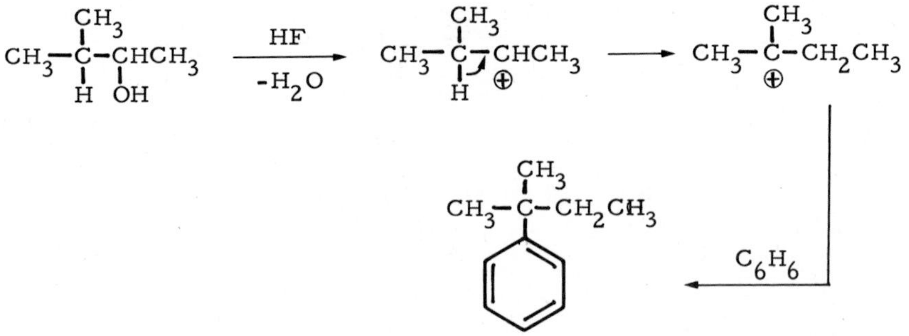

     b.   The bicyclic compound cannot ionize to form a stable tertiary carbon cation because the bonds to the positive carbon cannot become planar. Because no ionization occurs, no electrophilic substitution occurs.

     c.   Rearrangement accompanies ionization in each case.

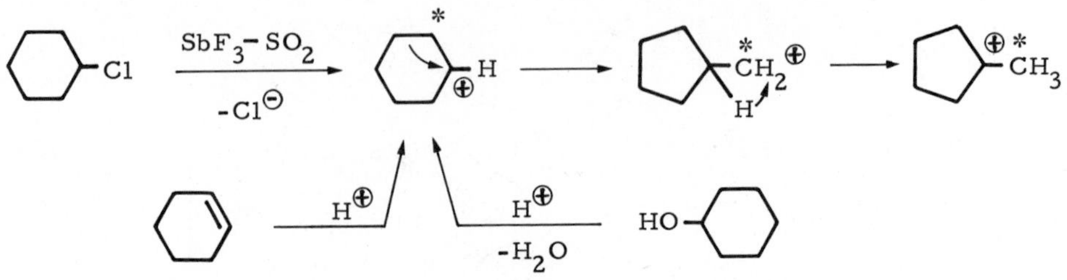

22-19   The first step is a Friedel-Crafts acylation catalyzed by $AlCl_3$.

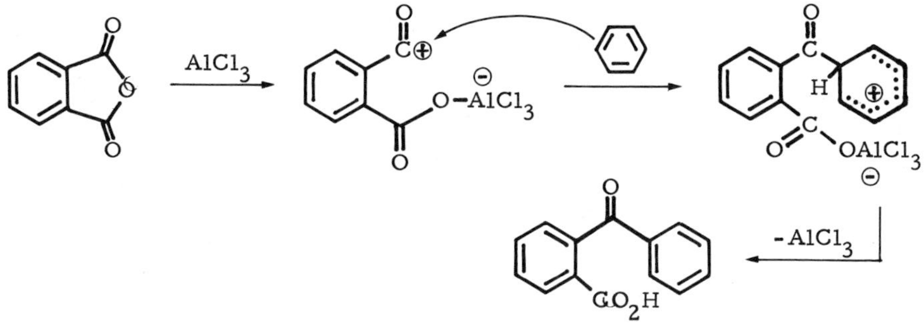

The second step is also a Friedel-Crafts acylation but requires a stronger catalyst because the carbonyl substituent deactivates the ring to further substitution, and the water formed as a product must be removed by the SO_3.

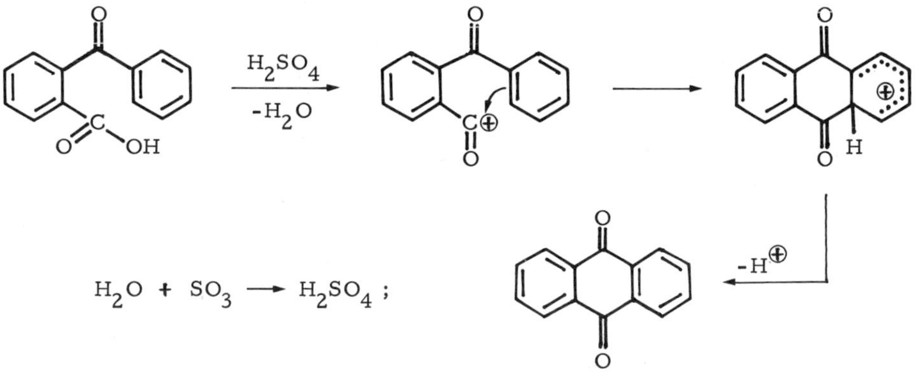

$$H_2O + SO_3 \longrightarrow H_2SO_4 \; ;$$

22-20

a. $C_6H_6 + C_6H_5COCl \xrightarrow[\text{nitrobenzene}]{AlCl_3} C_6H_5COC_6H_5 \xrightarrow{Zn, \; HCl}$

$C_6H_5CH_2C_6H_5$

$C_6H_6 + C_6H_5CH_2Cl \xrightarrow[\text{nitrobenzene}]{AlCl_3} C_6H_5CH_2C_6H_5$

(not easy to control)

b. $C_6H_5CH_3 + CH_3COCl \xrightarrow[\text{nitrobenzene}]{AlCl_3} \text{para-}CH_3CO-C_6H_4-CH_3$

$\underline{\text{para}}-CH_3CH_2-C_6H_4-CH_3 \xleftarrow{Zn, \; HCl}$

22-21

a. $CH_2{=}O + HCl \rightleftharpoons Cl-CH_2-OH \xrightarrow{ZnCl_2} \overset{\oplus}{Cl}CH_2\cdots\overset{\ominus}{O}-ZnCl_2$ over H

$\langle\!\!\!\!\!\bigcirc\!\!\!\!\!\rangle + ClCH_2-\overset{\oplus}{\underset{H}{O}}-\overset{\ominus}{Z}nCl_2 \rightleftharpoons$ $+ HO\overset{\ominus}{Z}nCl_2 \xrightarrow{-ZnCl_2}$

$\langle\!\!\!\!\!\bigcirc\!\!\!\!\!\rangle-CH_2Cl + H_2O$

b. $ClCH_2OCH_3 + SnCl_4 \rightleftharpoons ClCH_2-\overset{\oplus}{O}-CH_3 \rightleftharpoons ClCH_2\cdots\overset{\oplus}{O}CH_3$ with $\ominus SnCl_4$ and $\ominus SnCl_4$

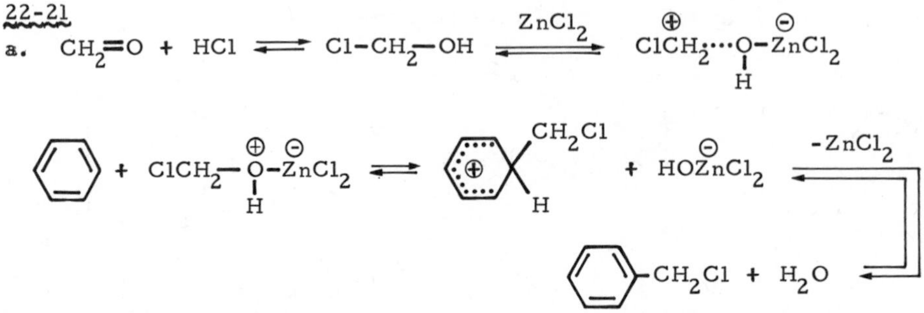

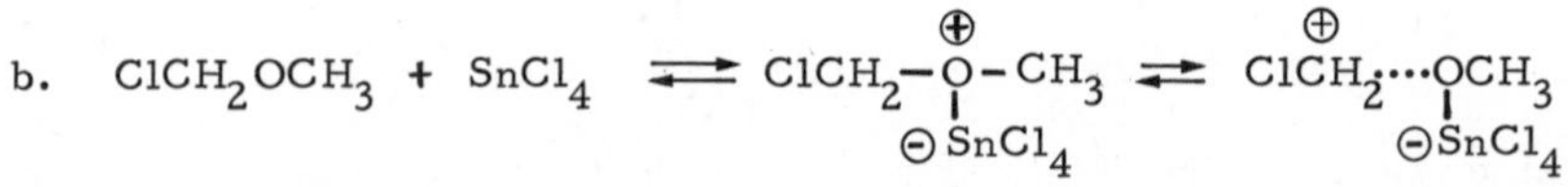

(and continues as in Part a)

22-22 $H-C{\equiv}N + ZnCl_2 + HCl \rightleftharpoons H-\overset{\oplus}{C}{=}\overset{Cl\ \overset{\ominus}{Z}nCl_2}{N}-H \longleftrightarrow H-\overset{\oplus}{\underset{\ominus}{C}}-\overset{Cl\ \overset{\ominus}{Z}nCl_2}{N}-H$

$\langle\!\!\!\!\!\bigcirc\!\!\!\!\!\rangle-CHCl(NH_2) \xrightarrow{-ZnCl_2}$ $CHCl-NH\overset{\ominus}{Z}nCl_2 \quad C_6H_6$

$\xrightarrow{-HCl} \langle\!\!\!\!\!\bigcirc\!\!\!\!\!\rangle-CH{=}NH \xrightarrow{H_2O} \langle\!\!\!\!\!\bigcirc\!\!\!\!\!\rangle-CHO$

22-23

$\underset{\substack{\text{branched-chain}\\\text{alkane}}}{\overset{\overset{\textstyle C}{|}}{C-CH-C}} \xrightarrow[-HCl]{Cl_2,\ h\nu} \overset{\overset{\textstyle C}{|}}{\underset{\underset{\textstyle Cl}{|}}{C-C-C}} \xrightarrow[AlCl_3,\ -HCl]{C_6H_6} \overset{\overset{\textstyle C}{|}}{C-C-C}$ with phenyl

22-24

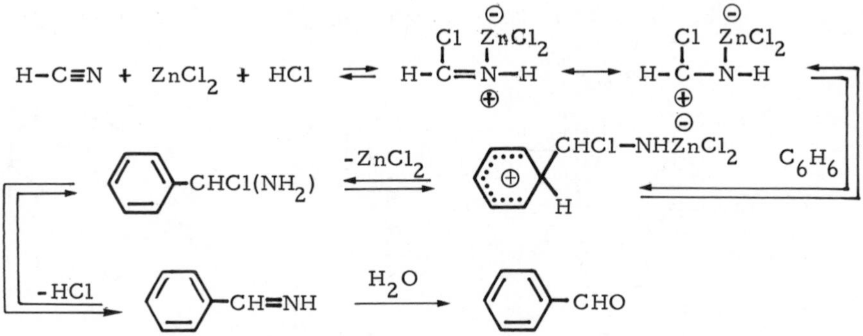

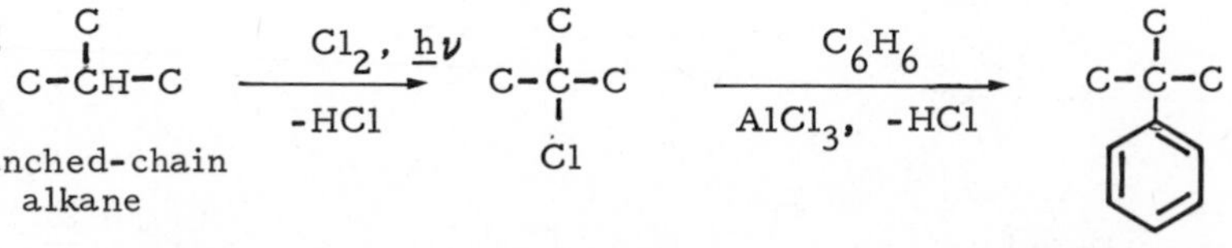

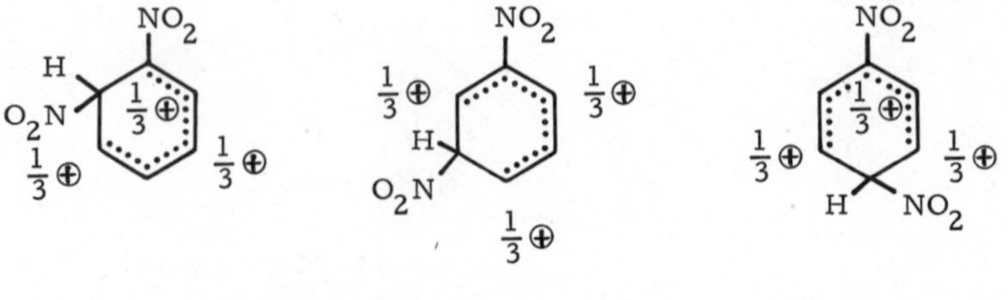

All three intermediates have a less favorable charge distribution than in the nitration of benzene itself because the substituent (NO_2) is strongly electron-attracting on an already electron-deficient ring system.   Nitrobenzene therefore reacts less readily than benzene in nitration.   Of the three intermediates, the lowest in energy is the <u>meta</u> intermediate because the positive charge in the ring is furthest from the nitro substituent. Similar reasoning applies to the CF_3 and CHO substituents which also are electron-withdrawing. Substituents that are electron-donating (e.g., $\ddot{N}H_2$) stabilize the intermediates for ortho and para substitution.

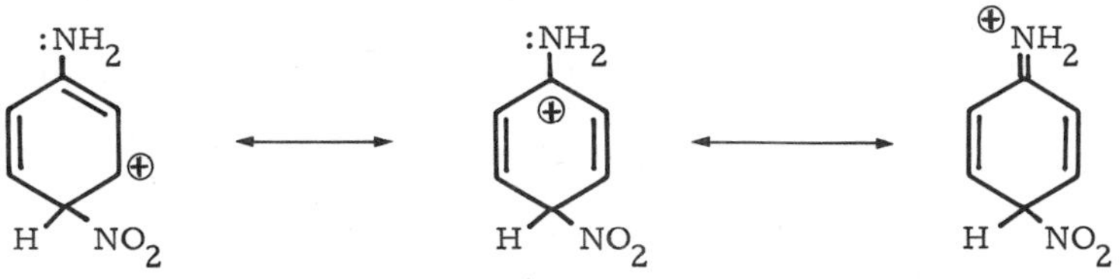

The $ClCH_2$ group is deactivating because of the electron-withdrawing halogen but the effect is not strong enough to divert orientation from <u>ortho-para</u> (as in alkyl) to <u>meta</u>.

<u>22-25</u> The least reactive brominating agent (Br_2) is the more selective (no <u>meta</u> product). It is also the largest species and gives mostly para substitution over ortho substitution. The more reactive species $(H_2OBr^{\oplus})$ is less selective (more meta product); and gives almost a 2:1 or a statistical distribution of ortho-para substitution products. In the gas phase, $Br^{\oplus}$ may be expected to be totally nonselective and add to the ring carbons at equal rates to give the statistical distribution (2:2:1) of ortho, meta and para ions.

<u>22-26</u>

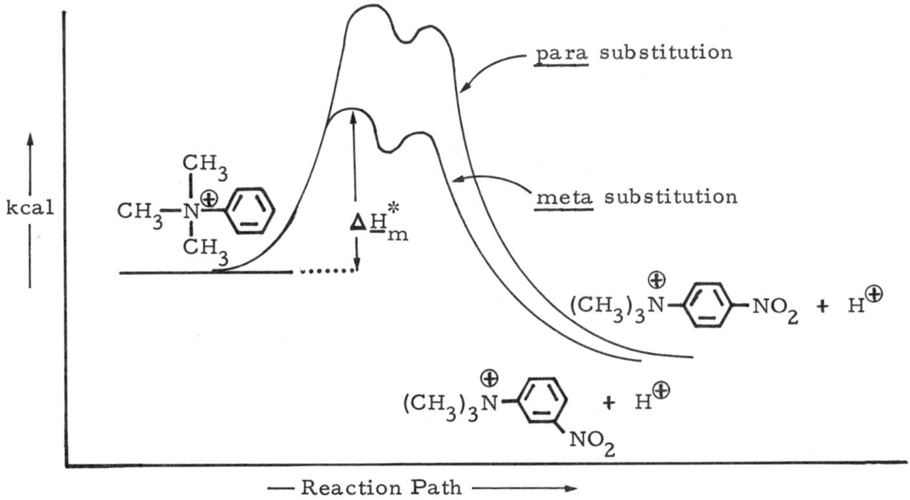

$$\Delta H^*_m < \Delta H^*_p$$

22-27

a. [benzene ring]—CH₂NO₂   ortho-para substitution with deactivation

b. [benzene ring]—S̈CH₃   ortho-para substitution with activation

c. [benzene ring]—N̈=O   ortho-para substitution with activation

d. [benzene ring]—P(CH₃)₂ with O δ⊖ and δ⊕   meta substitution with deactivation

22-28

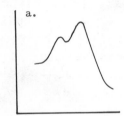 a.

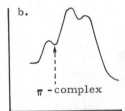 b.
π -complex

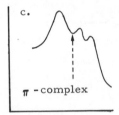 c.
π -complex

mechanism:

[reaction scheme: nitro-substituted cyclohexadiene with CH₃CO—O group and isopropyl group] —H₂SO₄→ [protonated intermediate CH₃C(OH)⊕] —-CH₃CO₂H→ [NO₂-substituted cation ⊕]

[arrow down]

[1-methyl-4-isopropyl-NO₂ product] ←-H⊕— [cyclohexadiene cation with NO₂, H and isopropyl] ←

22-29

a. O₂N—[biphenyl with arrows]

(deactivation)

b. CH₃—[benzene ring]—CO₂H with arrows

(activation from CH₃ ,
deactivation from —CO₂H)

c. —CO$_2$H

   CH$_3$ ◄···this position is hindered ·············►

   (orientation is controlled
   by the activating substituent)

d.   Br—◄—

                              Br

   (deactivation)

e.   F—◄—

   (minor)   OCH$_3$

   (orientation is controlled
   by the activating substituent)

f.   CH$_3$—◄—

                              CH$_3$

   (activation)

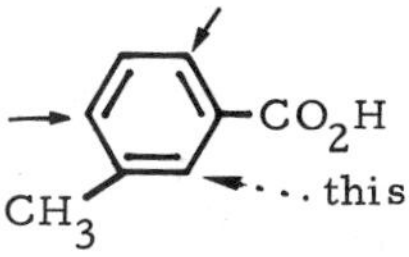

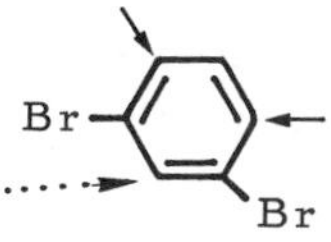

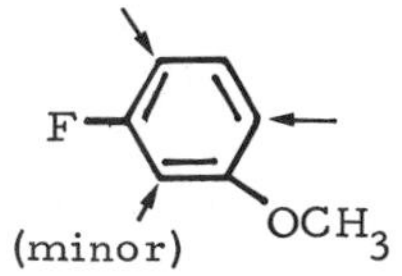

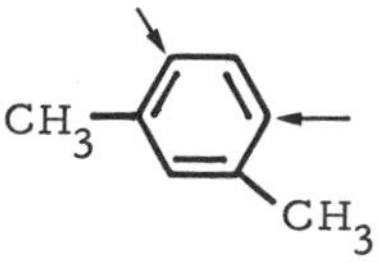

22-30

         4-Nitromethylbenzene (10%) is formed by attack at C4.
         4 (8%) is formed by attack at C3.
         3 (41%) is formed by attack at C2.
         5 and 6 (41%) are formed by attack at C1.

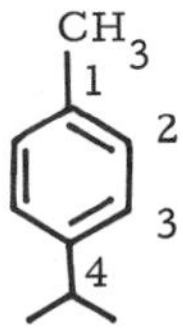

Relative rates of attack at C1 : C2 : C3 : C4 $= 41 : \dfrac{8}{2} : \dfrac{41}{2} : 10$

$$= 41 : 4 : 20.5 : 10 \qquad 10 : 1 : 5 : 2.5$$

If ipso substitution is neglected and we assume that the products 3 and 4 are derived from attack at C3 and C2 respectively, then the relative rates of attack at C2 : C3 $= \dfrac{8}{2} : \dfrac{82}{2} = 1 : 10$, which is approximately twice the rate calculated with the ipso correction.

22-31   See Exercises 21-22 and 21-30.

22-32   If acylation of naphthalene in the 2-position in nitrobenzene is the result of thermodynamic control, then 1-ethanoylnaphthalene under the same experimental conditions   would isomerize to the 2-isomer.

22-33   a. 1-methyl-4-bromonaphthalene and possibly some 1-methyl-2-bromonaphthalene

         b. 2-methyl-1-nitronaphthalene

         c. 5-nitro-2-naphthalenecarboxylic acid and 8-nitro-2-naphthalenecarboxylic acid

22-34   No Kekulé-type resonance structures can be written for acenaphthylene in which the 1, 2-bond is anything but a double bond. One would therefore predict this bond to be as reactive as an alkene double bond, as indeed it is.

22-35   The most stable intermediate anion in the reduction of methoxybenzene has negative charge further from the substituent (OCH_3).   In the reduction of sodium benzoate, the intermediate is more stable when the charge is adjacent to the substituent $(CO_2^{\ominus})$

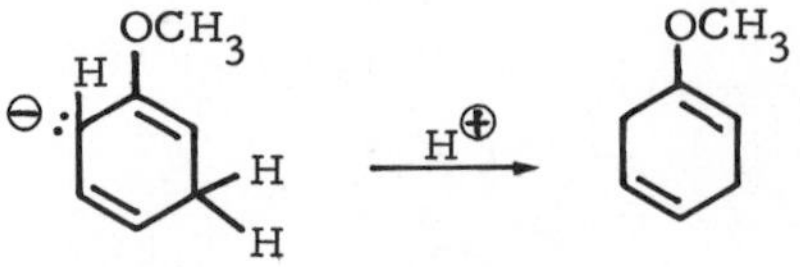

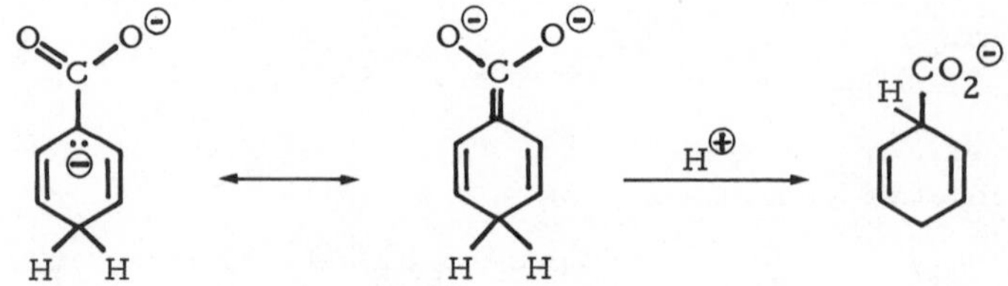

22-36

a.

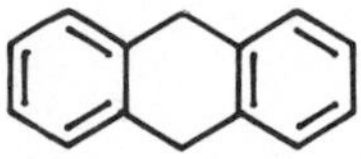

b.

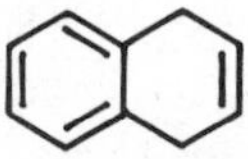

c.

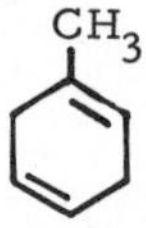

22-37   Over-reduction occurs if the nonconjugated 1,4-diene is equilibrated with the conjugated 1,3-diene, which is rapidly reduced further.   To prevent this, ethanol is added to rapidly and irreversibly protonate the anion.   Ethanol also prevents any stronger base (e.g., $NH_2^{\ominus}$) from forming which might reform the cyclohexadienyl anion.

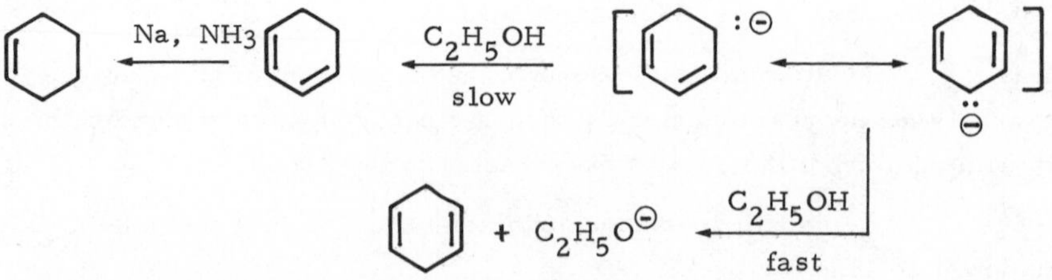

22-38   Loss in stabilization energy on cycloaddition to benzene is 43.2 kcal; loss in SE on 9,10-cycloaddition to anthracene is 26.6 kcal; on 1,4-cycloaddition, 33.5 kcal.   Clearly 9,10-cycloaddition is preferable.

22-39

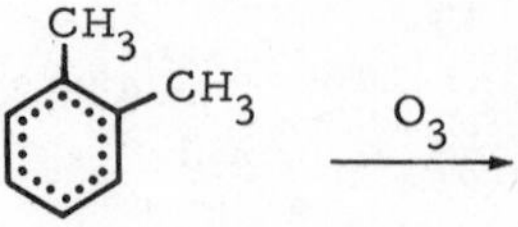

O_3 →   OHCCHO, $CH_3COCOCH_3$,   OHCCOCH$_3$

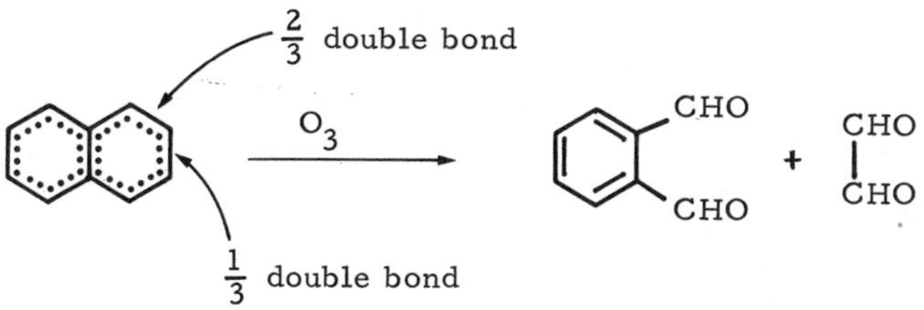

$-\frac{2}{3}$ double bond

O_3

CHO
CHO

+

CHO
CHO

$\frac{1}{3}$ double bond

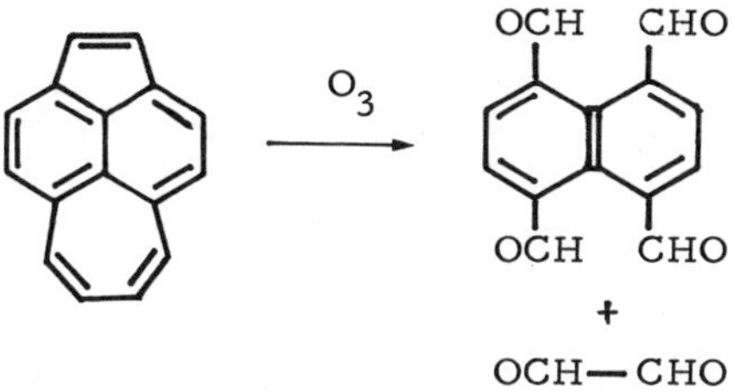

O_3

OCH   CHO

OCH   CHO

+

OCH—CHO

22-40

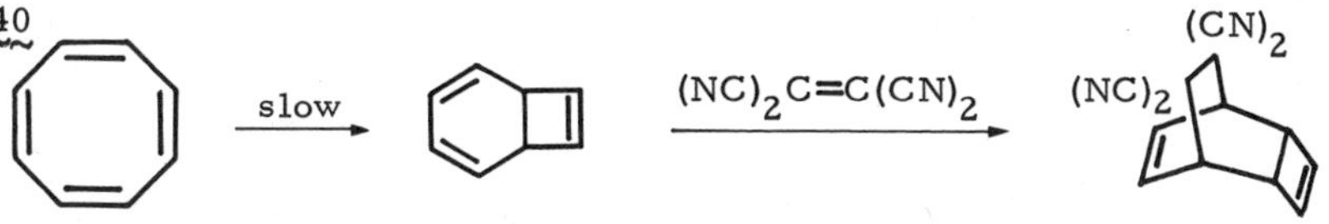

slow

$(NC)_2C=C(CN)_2$

$(CN)_2$

$(NC)_2$

The rate of addition of tetracyanoethene to the bicyclooctatriene is the rate-determining step at low concentrations of tetracyanoethene. At high concentrations, addition occurs faster than the bicyclooctatriene is formed; therefore the slow step becomes the isomerism of cyclooctatetraene and the rate is independent of the concentration of tetracyanoethene.

22-41   a.   with ethanoic acid

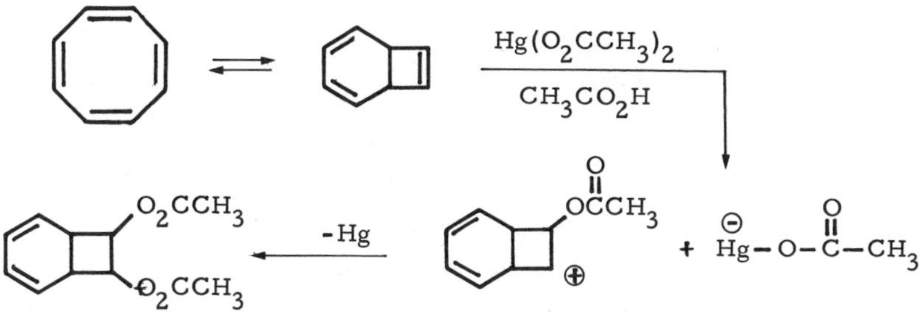

$Hg(O_2CCH_3)_2$

CH_3CO_2H

O_2CCH_3

O_2CCH_3

$-Hg$

$\overset{O}{\overset{\|}{OCCH_3}}$

$\oplus$

$+$ $\overset{\ominus}{Hg}-O-\overset{O}{\overset{\|}{C}}-CH_3$

b. <u>with water</u>

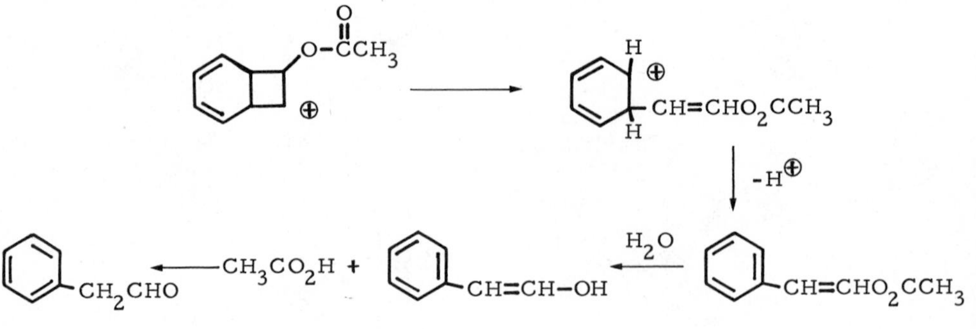

c. <u>with methanol</u>

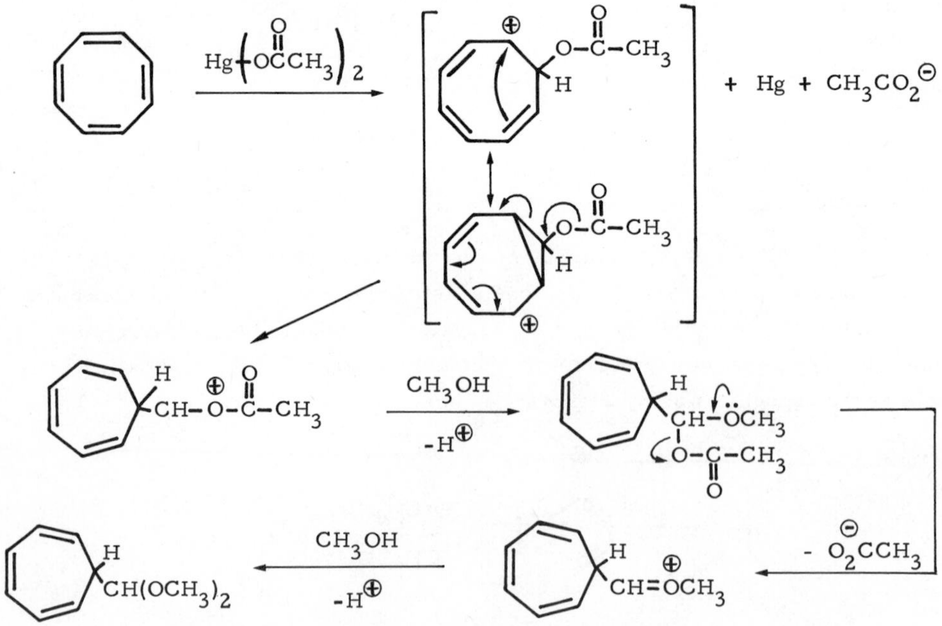

22-42 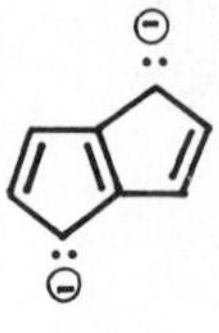   The dianion of pentalene corresponds to a pair of <u>ortho</u>-fused cyclopentadiene anions. The system has 10 π electrons and has the proper number for stability in accord with the $4\underline{n} + 2$ rule. It is therefore expected to be more stable than pentalene which has only 8 π electrons.

22-43

a.   or   aromatic $(4\underline{n} + 2)$ polyene

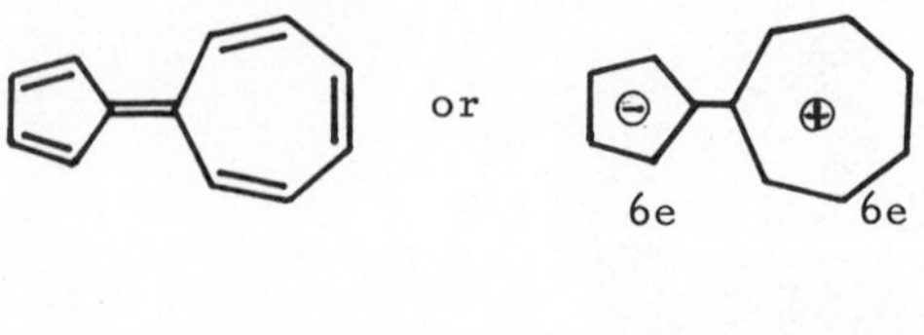

b.   [10] annulene is in principle an aromatic ($4\underline{n} + 2$) polyene, if planar. But it cannot assume a planar ring without strain because of H···H repulsion.

c.   Aceplieadylene is a stable compound.   It is not a ($4\underline{n} + 2$) polyene by the definition of Section 21-9.   The two benzenoid rings are aromatic.   The other rings (5 and 7) are alkene-like.

d. 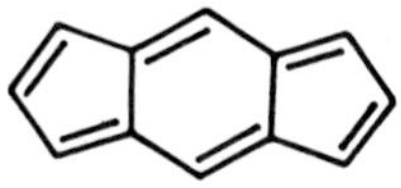   non-aromatic $4\underline{n}$ polyene

22-44 C_8H_{10}

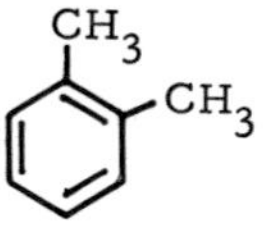

1 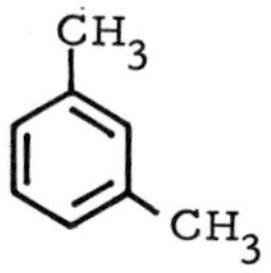   2 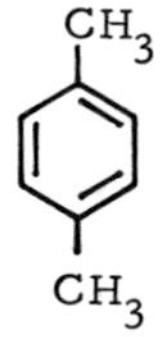   3 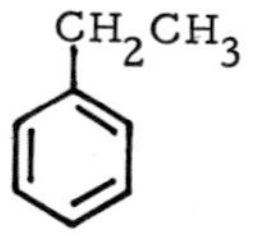   4

1    $\xrightarrow{NO_2^{\oplus}}$

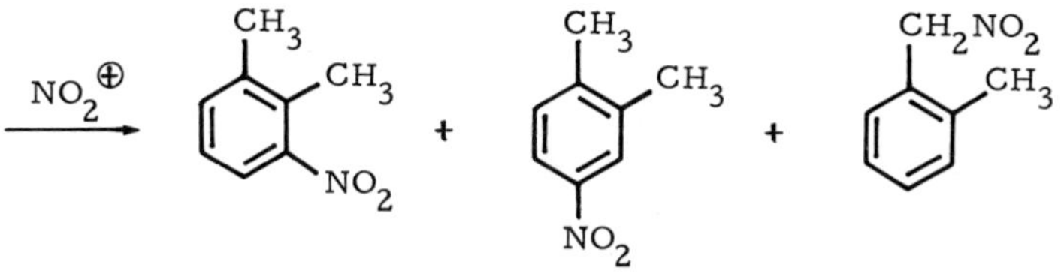

2,3-dimethyl-
nitrobenzene

3,4-dimethyl-
nitrobenzene

(2-methylphenyl)-
nitromethane

2    $\xrightarrow{NO_2^{\oplus}}$

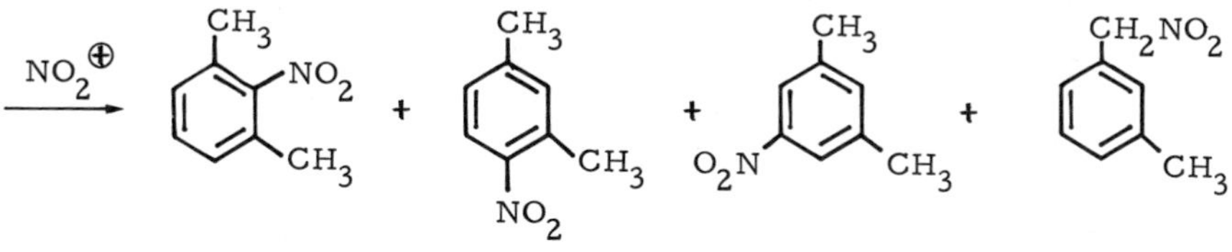

2,6-dimethyl-
nitrobenzene

2,4-dimethyl-
nitrobenzene

3,5-dimethyl-
nitrobenzene

(3-methylphenyl)-
nitromethane

3   $\xrightarrow{\text{NO}_2^{\oplus}}$

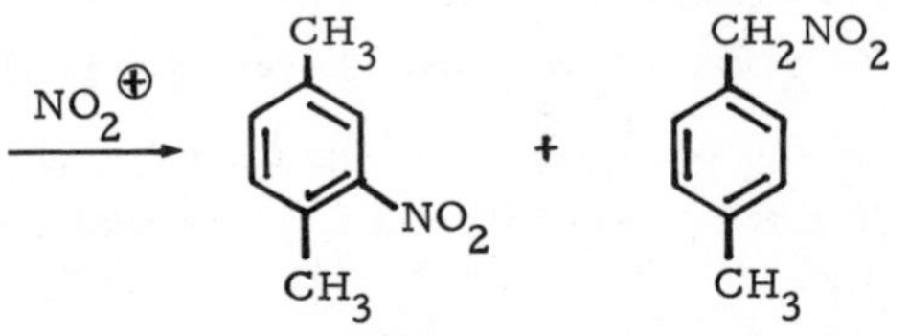

2,5-dimethyl-
nitrobenzene

(4-methylphenyl)-
nitromethane

4   $\xrightarrow{\overset{\oplus}{\text{NO}_2}}$

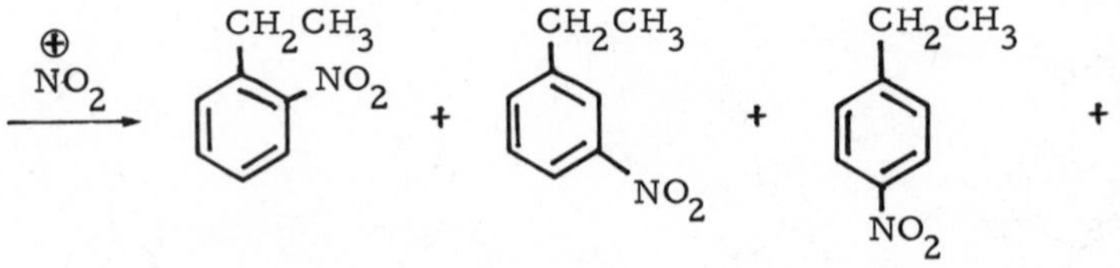

2-ethyl-
nitrobenzene

3-ethyl-
nitrobenzene

4-ethyl-
nitrobenzene

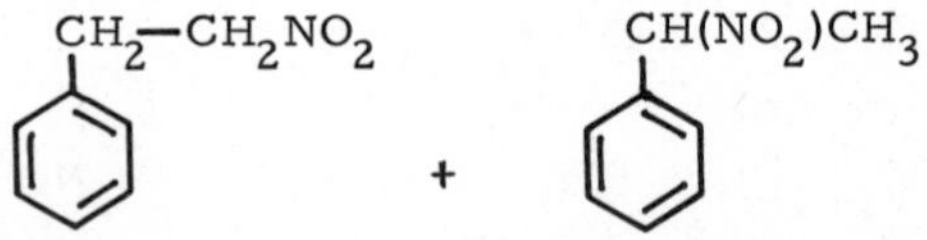

+

2-phenyl-1-nitro-
ethane

1-phenyl-1-
nitroethane

22-45

a. $C_8H_{10} \equiv CH_3\text{—}\langle\bigcirc\rangle\text{—}CH_3$

b. $C_6H_3Br_3 \equiv Br\text{—}\langle\bigcirc\rangle\text{—}Br$

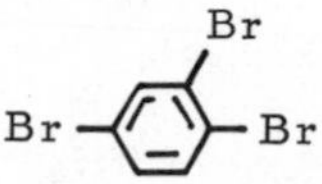

c. $C_6H_3Br_2Cl \equiv Br\text{—}\langle\bigcirc\rangle\text{—}Cl$

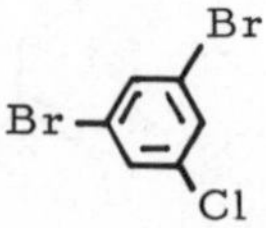

d. $C_8H_8(NO_2)_2 \equiv$ 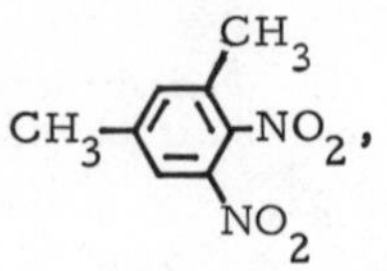 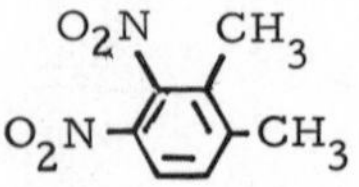   ,   or

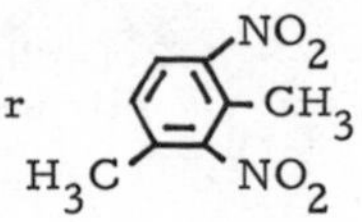

22-46

a.

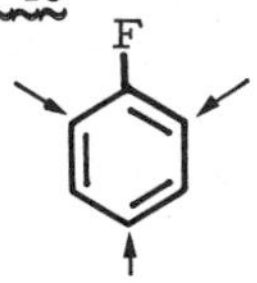

(see Table 22-6)

b.

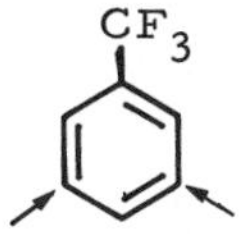

(see Table 22-6)

c.

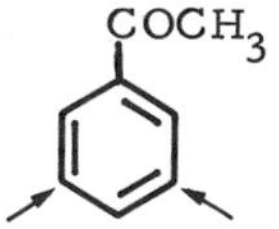

(see Table 22-6)

d.

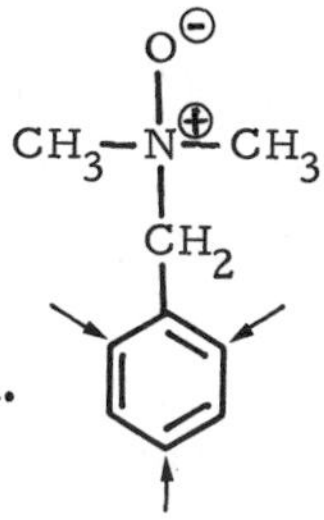

(ortho, para-orientation; hard to evaluate reactivity)

e.

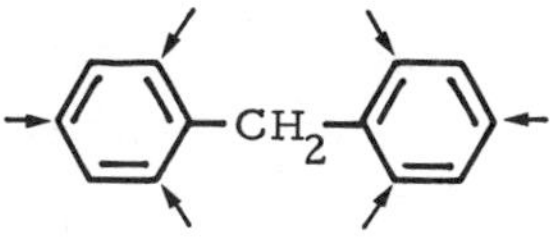

(ortho, para-orienting with activation)

f.

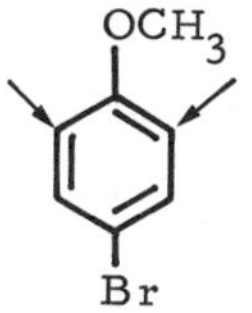

(ortho, para-orientation of methoxyl overshadows that of bromine Rate > benzene)

g.

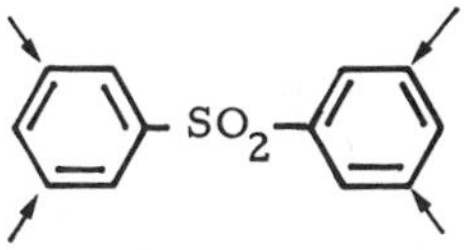

(Electron withdrawing effect of the sulfone group leads to meta-orientation. Rate < benzene)

h.

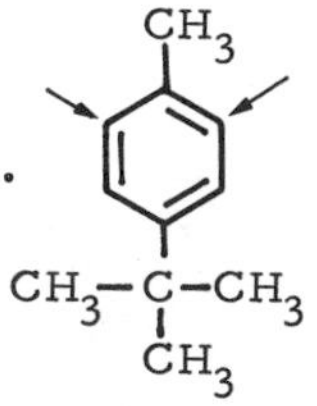

(Steric effect of tert-butyl group directs entering group ortho to methyl. Rate > benzene)

i.

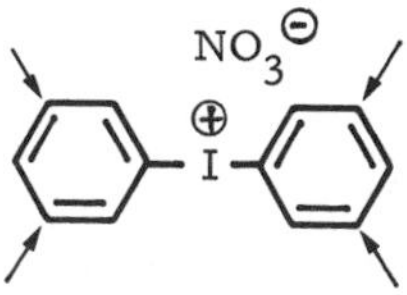

(see Table 22-6)

j.

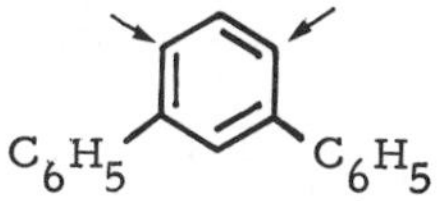

(Conjugative effect of two phenyl groups tends to lead to substitution ortho to one phenyl and para to the other. For steric reasons, substitution ortho to both groups is expected to be less significant.)

k. 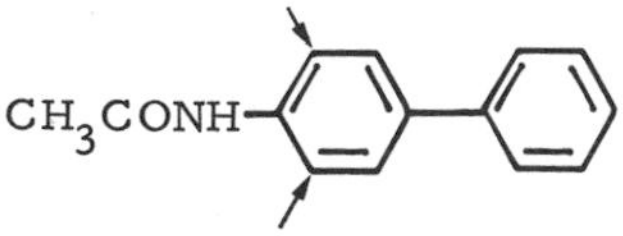 CH_3CONH-

(see Table 22-6)

The effect of the CH_3CONH- group on reactivity of the right-hand ring is expected to be smaller than in the ring to which it is attached.

22-47   Conjugation of the amino group of aniline with the π electrons of the ring activates the ortho and para positions towards electrophilic reagents. Activation is strong enough that bromination does not stop at the mono- or dibromoanilines but proceeds all the way to 2,4,6-tribromoaniline. In the case of nitration in the presence of strong acids, aniline is largely protonated, and the $\overset{\oplus}{N}H_3$ group deactivates the ring towards electrophilic attack. Nitration therefore occurs at the least deactivated (i.e., meta) position.

22-48   The orientation of NO_2, CN and $CH=CHNO_2$ groups is predicted on a simple basis to be meta. Comparison of the resonance structures for meta substitution with those of para substitution leads to this conclusion: para substitution in each case leads to an unfavorable charge distribution - adjacent positively charged atoms in the case of the NO_2 group; positively charged nitrogen in the case of CN; and positively charged carbon adjacent to electron-withdrawing NO_2 group in the case of $CH=CHNO_2$. These predictions are correct except for the $-CH=CH-NO_2$ group which is actually ortho, para-directing with deactivation. Apparently, the contribution of

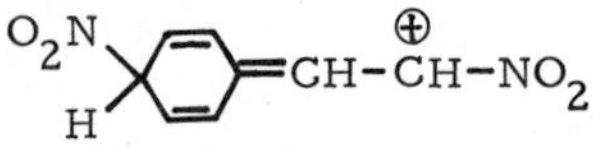

$=CH-\overset{\oplus}{C}H-NO_2$, although not large, is sufficient to lead to ortho,-

para-orientation, albeit with deactivation.

22-49

a. C_6H_6 $\xrightarrow{Br_2,\ FeBr_3}$ C_6H_5Br $\xrightarrow[H_2SO_4]{HNO_3}$ para-$NO_2-C_6H_4-Br$

(Separation from ortho isomer is assumed feasible.)

b. C_6H_6 $\xrightarrow[AlCl_3]{(CH_3)_2CHCl}$ $C_6H_5CH(CH_3)_2$ $\xrightarrow[heat]{H_2SO_4\ (conc.)}$

$-SO_3H$ $\xleftarrow[H_2SO_4]{HNO_3}$ para-$(CH_3)_2CH-C_6H_4-SO_3H$

c. C_6H_6 $\xrightarrow[AlCl_3]{(CH_3)_3CCl}$ $(CH_3)_3C-C_6H_5$ $\xrightarrow[ZnCl_2]{HCN,\ HCl}$

d. $+$ $\xrightarrow[nitrobenzene]{AlCl_3}$ $-\overset{O}{\overset{\|}{C}}-CH_2CH_2CO_2H$

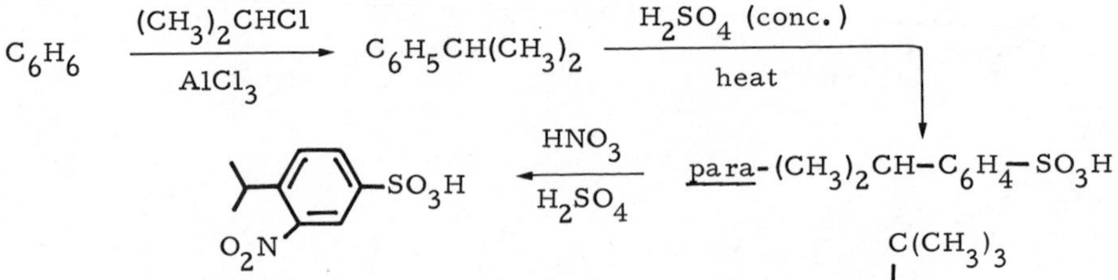

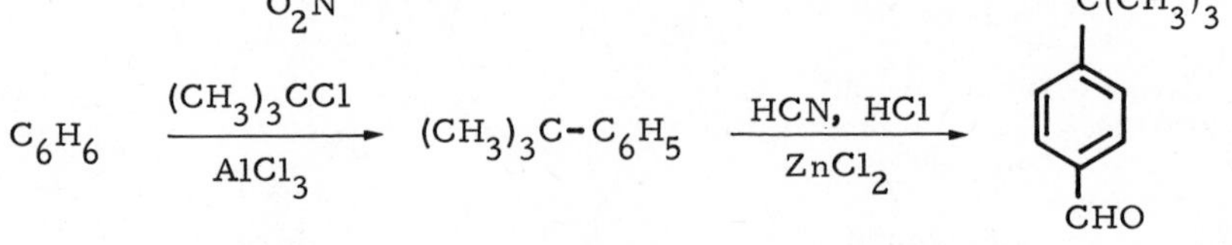

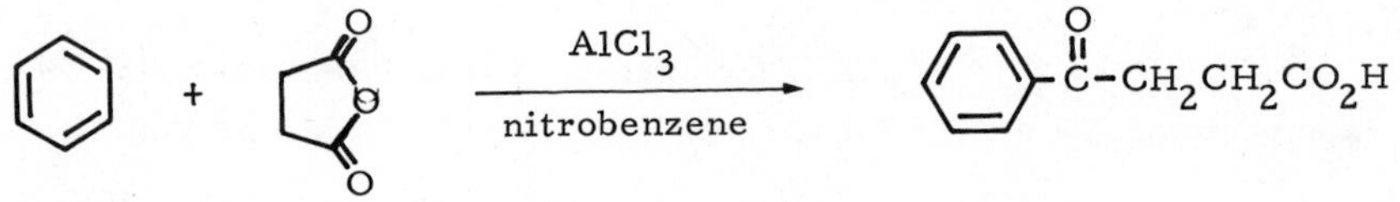

e.

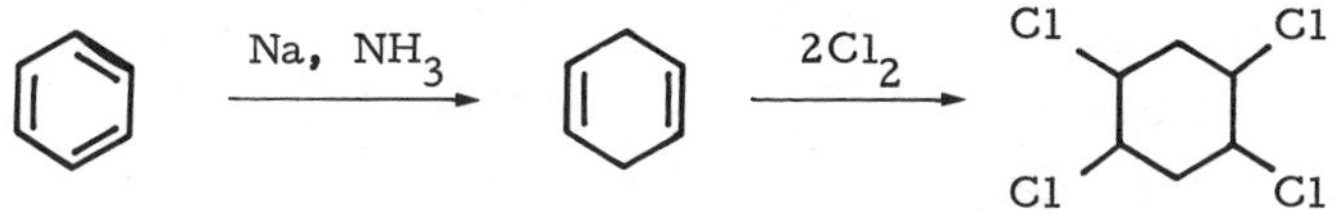

$$\text{(benzene)} \xrightarrow{\text{Na, NH}_3} \text{(1,4-cyclohexadiene)} \xrightarrow{2Cl_2} \text{(tetrachlorocyclohexane with Cl at four positions)}$$

22-50   a. Strong acid promotes the formation of $NO_2^{\oplus}$ by the reactions

$$HO-NO_2 + H_2SO_4 \longrightarrow H_2\overset{\oplus}{O}-NO_2 +. HSO_4^{\ominus}$$

$$H_2\overset{\oplus}{O}-NO_2 \longrightarrow H_2O + NO_2^{\oplus}$$

Nitrate ions reverse the formation of $NO_2^{\oplus}$ by the process

$$H_2\overset{\oplus}{O}-NO_2 + NO_3^{\ominus} \rightleftharpoons 2HONO_2$$

   b. The solvent is useful because it dissolves the reagents yet is inert. The NO_2 group deactivates the ring to substitution by the acylating agent.

   c. See Section 14-6 regarding substitution and addition reactions of aryl halides. See also Section 10-6 and Section 23-10.

22-51

a. 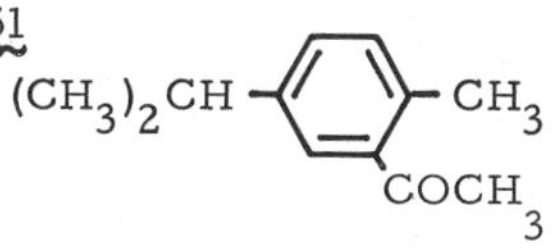 $(CH_3)_2CH$—⟨ring with CH_3 and $COCH_3$ substituents⟩

b. 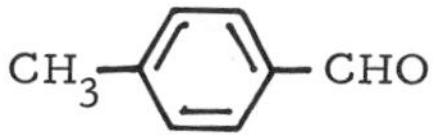 CH_3—⟨ring⟩—CHO

c. 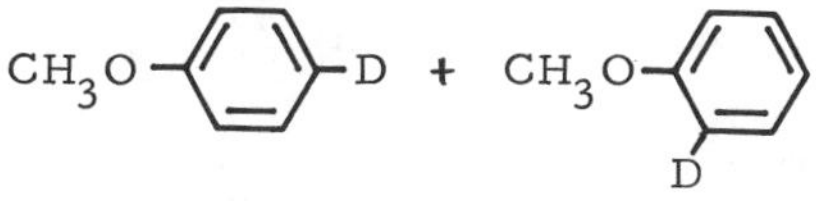 CH_3O—⟨ring⟩—D + CH_3O—⟨ring with D⟩

d. F—⟨ring⟩—H

e. 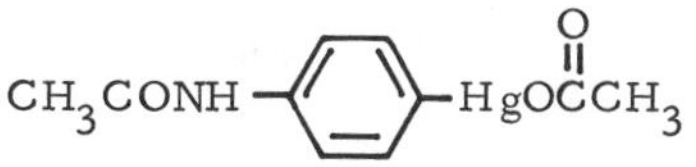 CH_3CONH—⟨ring⟩—$Hg\overset{\overset{O}{\|}}{O}CCH_3$, and some ortho isomer

f. 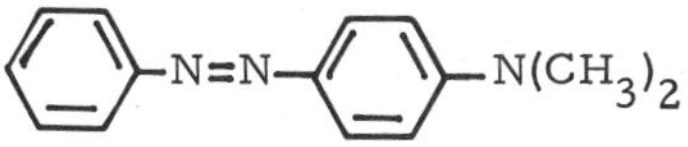   ⟨ring⟩—$N=N$—⟨ring⟩—$N(CH_3)_2$ (See Section 23-10C.)

g.

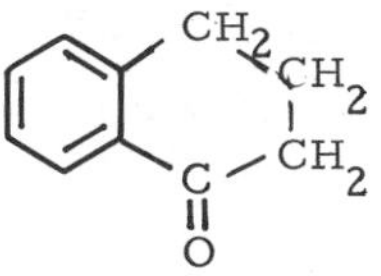

h.

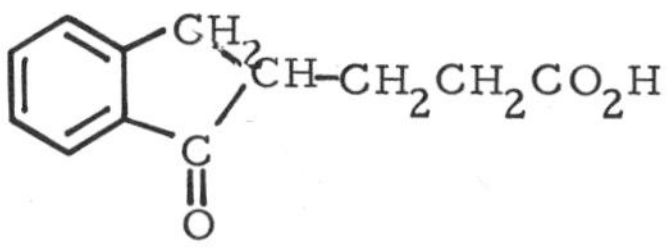

i. 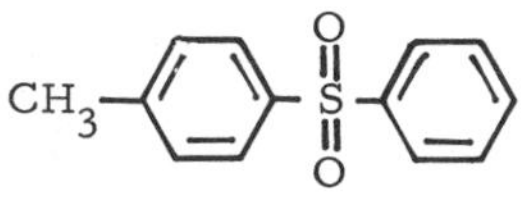 CH_3—⟨ring⟩—$\overset{\overset{O}{\|}}{\underset{\|}{S}}$—⟨ring⟩

22-52

a.

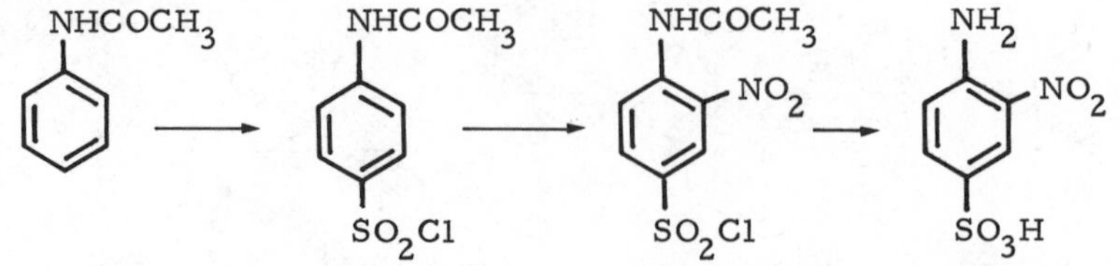
NHCOCH3
NHCOCH3
SO2Cl
A
NHCOCH3
NO2
SO2Cl
B
NH2
NO2
SO3H
C

b.

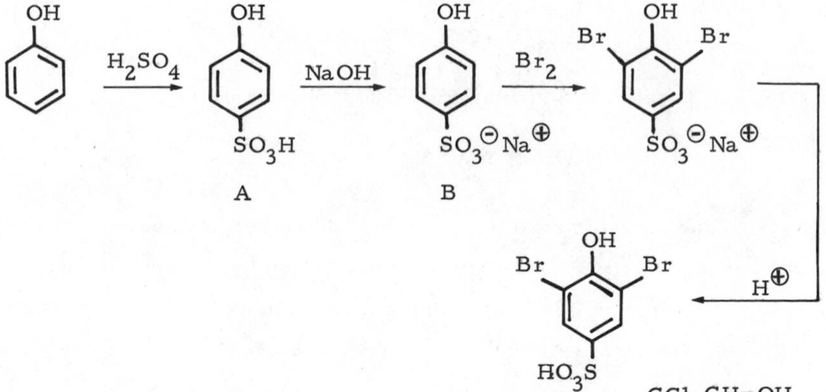
OH
H2SO4
OH
SO3H
A
NaOH
OH
SO3⊖ Na⊕
B
Br2
OH
Br
Br
SO3⊖ Na⊕
H⊕
OH
Br
Br
HO3S

22-53

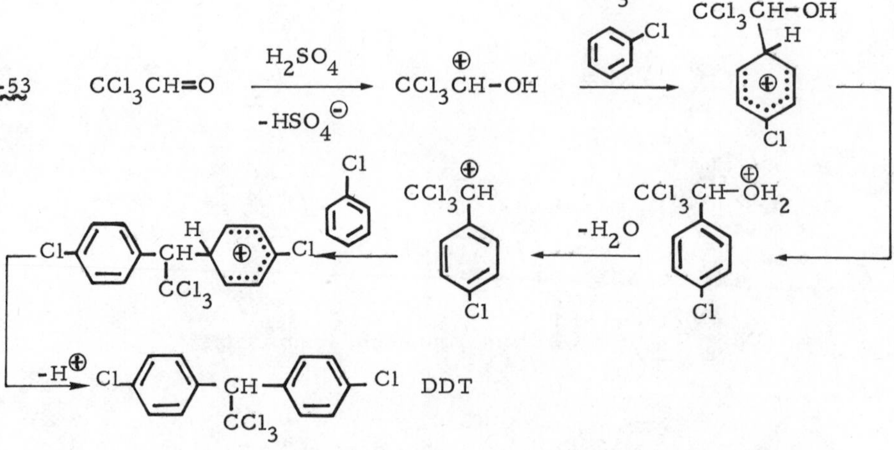
CCl3CH=O
H2SO4
-HSO4⊖
CCl3CH-OH
Cl
CCl3CH-OH
H
Cl
CCl3CH-OH2
Cl
-H2O
CCl3CH
Cl
Cl
Cl
CH
H
CCl3
Cl
-H⊕
Cl
CH
CCl3
Cl
DDT

22-54

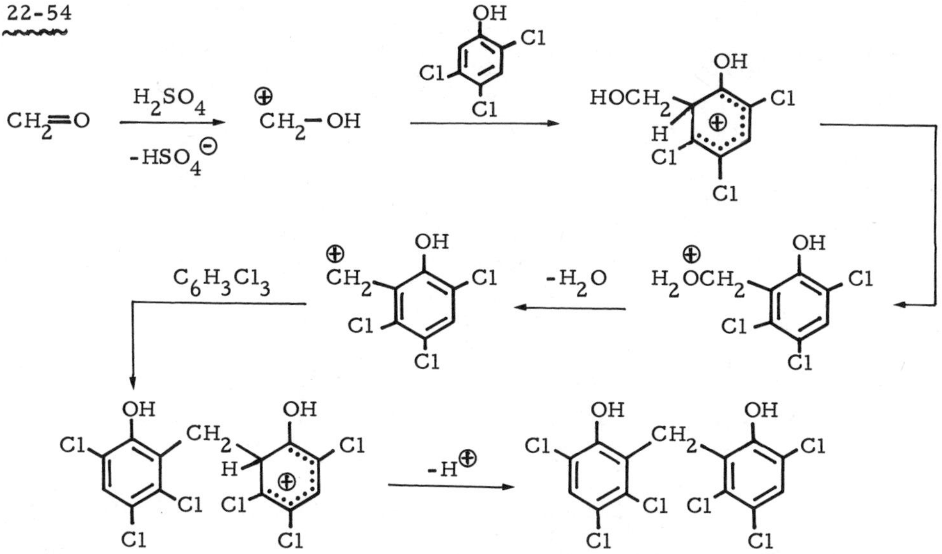

22-55   This is probably an electrophilic substitution in which the substituting agent is $HO^{\oplus}$.

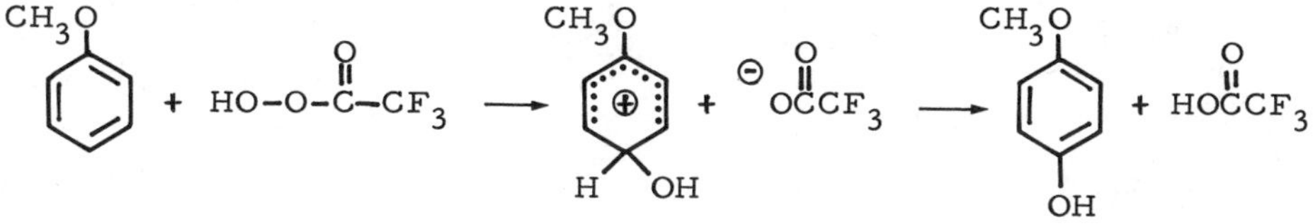

Fluorobenzene would be less reactive than methoxybenzene but would give ortho and para products.

22-56   a. $(CH_3CO)_2O + HNO_3 \longrightarrow CH_3CO_2NO_2 + CH_3CO_2H$

$CH_3CO_2NO_2 \longrightarrow CH_3CO_2^{\ominus} + NO_2^{\oplus}$

$C_6H_6 + NO_2^{\oplus} \xrightarrow{\quad -H^{\oplus} \quad} C_6H_5NO_2$

b. (See Exercise 22-11)

Slow step is the rate of formation of $NO_2^{\oplus}$. The subsequent steps are fast.

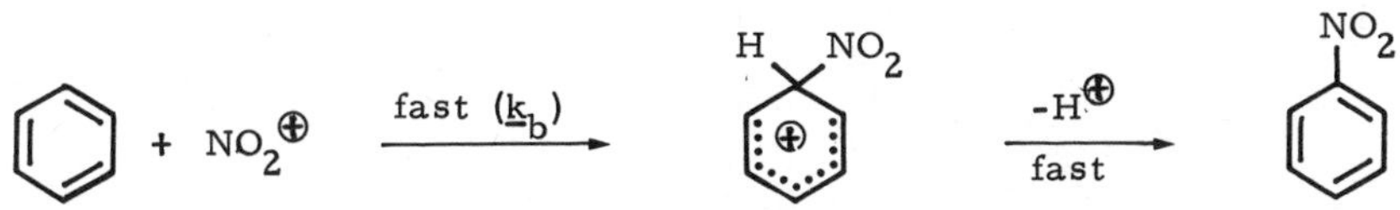

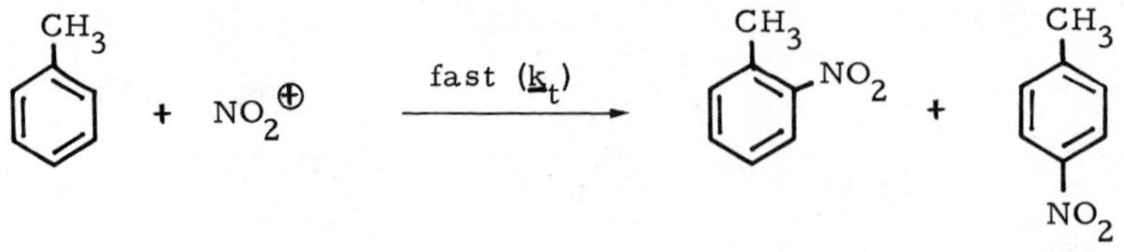

$$\frac{k_t}{k_b} = \frac{[\text{nitromethylbenzenes}]}{[\text{nitrobenzene}]} = 25$$

22-57

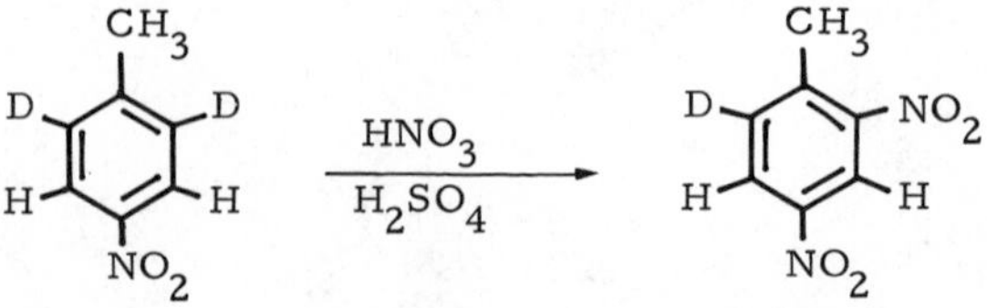

a. The fact that 2, 6-dideuterio-4-nitromethylbenzene nitrates at the same rate as benzene means that the C–H bond is not appreciably weakened in the transition state of the slow step in reaction. This step must then be attack of $NO_2^{\oplus}$ at aromatic carbon.

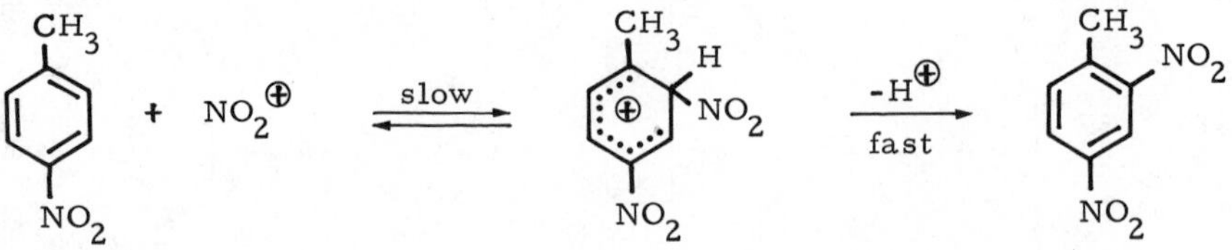

b. Rates of nitration of C_6D_6 of C_6H_6 would be the same, and independent of concentration of arene (zeroth order).

$$\nu = k[CH_3CO_2NO_2][C_6D_6]^0$$

22-58   a. Naphthalene + Cl_2 ⟶ 1, 4-dichloro-1, 4-dihydronaphthalene (Loss in stabilization energy is 33 kcal.)

Phenanthrene + Cl_2 ⟶ 9, 10-dichloro-9, 10-dihydrophenanthrene (Loss in stabilization energy is 28 kcal, assuming the SE of biphenyl remains in the product.)

Addition to phenanthrene should then be a more favorable reaction than addition to naphthalene.

b.

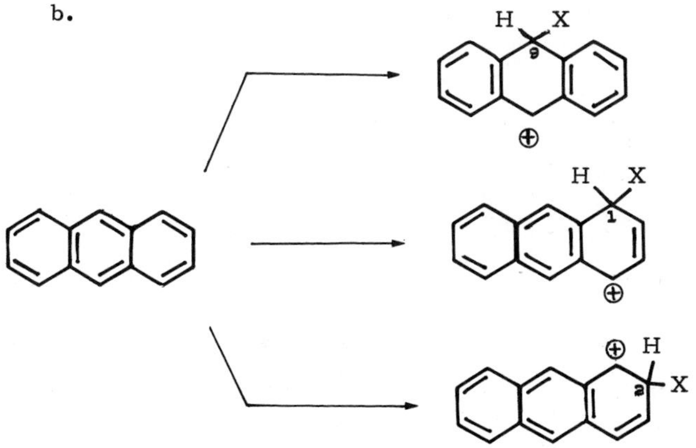

Substitution of anthracene at the 9-position is predicted to be more favorable than at the 1- or 2-positions because the intermediate retains the stabilization energy of two benzene rings (2 x 38 kcal) rather than of one naphthalene ring (71 kcal).

22-59   Structure B is the observed product. It retains the stabilization of two benzene rings, which is greater than one naphthalene ring A and the polyene system of C.

22-60   The phenyl group has an electron-withdrawing (-I) inductive effect and, at the same time, can contribute electrons by a conjugation or resonance effect. Accordingly, the ortho and para positions of biphenyl are activated towards electrophilic attack (as a result of the conjugation effect of C_6H_5) and deactivated at the meta position (as a result of the inductive effect of C_6H_5).

Biphenyl is more reactive than 2,2'-dimethylbiphenyl in nitration - possibly because the methyl groups of 2,2'-dimethylbiphenyl interfere with each other sufficiently to force the rings to lie, on the average, in different planes, thus destroying inter-ring conjugation.

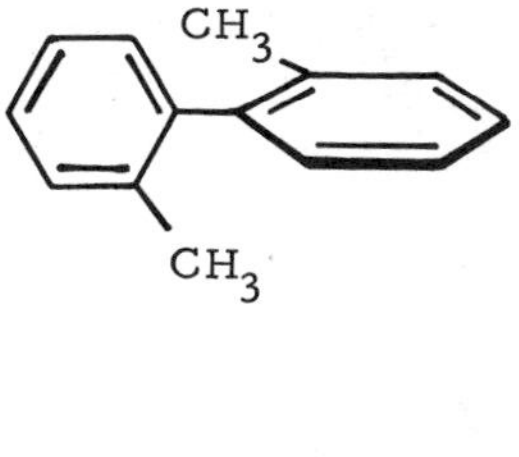

CHAPTER 23

ORGANONITROGEN COMPOUNDS. I. AMINES

When the hydrogens of ammonia (NH_3) are replaced by one or more alkyl (or aryl) groups, compounds of structure RNH_2, R_2NH, and R_3N, are formed, which are known as alkanamines or arenamines. Not surprisingly, they behave somewhat like ammonia. In fact, the structural feature $\overset{\cdot\cdot}{\underset{/}{N}}$—H, like the $-\overset{\cdot\cdot}{O}$—H group of alcohols, imparts basic and nucleophilic properties to the nitrogen and acidic properties to the N—H bond. We also see two types of hydrogen bonding, depending on whether the amine nitrogen donates or accepts the hydrogen:

$$\overset{\cdot\cdot}{\underset{/}{N}}\text{—H---:X} \qquad \text{or} \qquad \text{X—H---:}\underset{\backslash}{\overset{/}{N}}\text{—}$$

(X is F, O, N)

Self-association through hydrogen bonding reduces the volatility and increases water-solubility relative to hydrocarbons of similar weight and structure. However, $\overset{\cdot\cdot}{\underset{/}{N}}$—H----:N bonds are weaker than O—H----:O bonds.

<u>Spectroscopic Properties</u> (Section 23-5) follow:

<u>Infrared spectra</u>	N—H stretch, $3300 - 3500$ cm^{-1} (w) <u>prim.</u> 2 bands <u>sec.</u> 1 band <u>tert.</u> none
	N—H bond, 1600 cm^{-1} (m) C—N stretch, 1300 cm^{-1} (m) (arenamines only)
<u>Ultraviolet spectra</u>	C—N: $\xrightarrow{\underline{n} \to \sigma^*}$ C—$\overset{\cdot}{N}\cdot$ ~ 230 nm (not very useful for identification purposes)
<u>Nmr spectra</u>	δ_{N-H} ~ 0.8 ppm (variable due to H-bonding and chemical exchange) δ_{N-C-H} ~ 2.7 ppm
<u>Mass spectra</u>	A nitrogen-containing substance can be recognized by the odd mass number for the molecular ion M^+ and other odd-electron ions (Table 23-2).

α -Cleavage pattern for alkanamines:

$$R-\overset{|}{\underset{|}{C}}-\ddot{N}\diagup \longrightarrow R-\overset{|}{\underset{|}{C}}-\overset{\oplus}{\underset{\cdot}{N}}\diagup \longrightarrow R\cdot \ + \ \diagup C=\overset{\oplus}{N}\diagdown$$

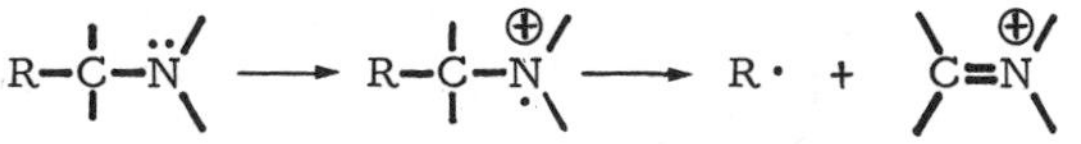

The stereochemistry of alkanamines is unusual in that the bonds to nitrogen are pyramidal yet are rapidly inverting, $^{\text{\tiny III}}$N: $\rightleftharpoons$:N$^{\text{\tiny III}}$. Consequently, configuration at the amine nitrogen usually is not stable. That is to say, configurational isomers of the RS or EZ type are not separately stable. For example, the following equilibria are established rapidly:

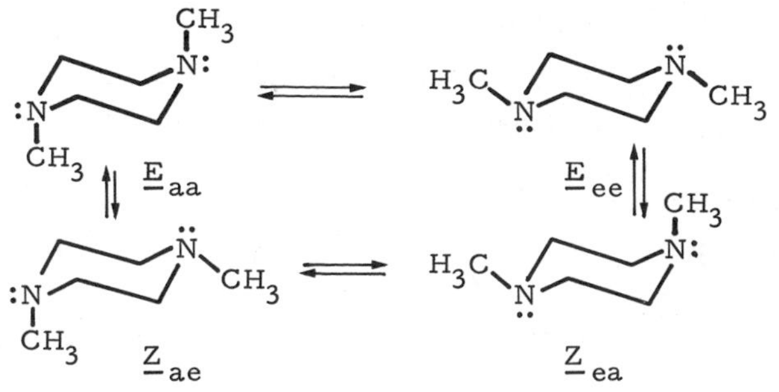

Chemical properties. Alkanamines are weak bases and therefore dissolve in acids to form water-soluble ammonium salts:

$$R\ddot{N}H_2 \ + \ HCl \longrightarrow R\overset{\oplus}{N}H_3\overset{\ominus}{Cl}$$

Arenamines are even weaker bases than alkanamines because of the loss of resonance stabilization when the amine nitrogen is protonated (Section 23-7C):

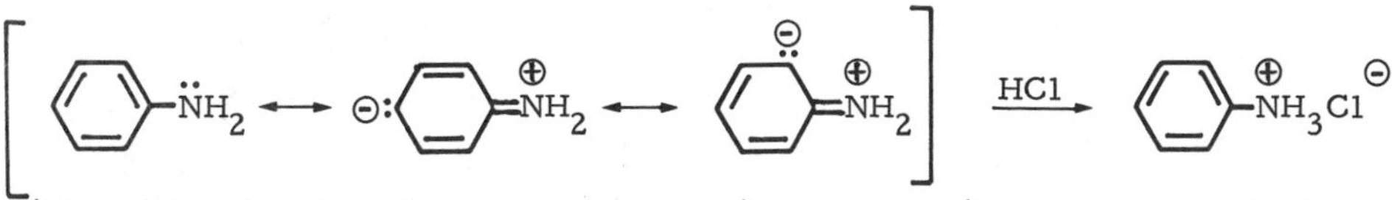

(Note delocalization of nitrogen lone pair)   (Notice that all electrons on nitrogen are localized.)

Generally, any resonance or inductive effect that operates to withdraw electrons from the amine nitrogen will stabilize the neutral amine and decrease its base strength (and its nucleophilicity). Likewise, any resonance or inductive effect that operates to stabilize the amine salt will increase base strength (and nucleophilicity) of the amine.

Primary and secondary alkanamines are very weak acids ($\underline{K}_a \sim 10^{-33}$)

and will give up their protons only to salts of _weaker_ acids (e. g. , some hydro-carbon acids):

$$R_2\overset{..}{N}H + C_6H_5\overset{\ominus \oplus}{Li} \longrightarrow R_2\overset{..}{N}:\overset{\ominus}{}\overset{\oplus}{Li} + C_6H_6$$

The amine salts formed are strongly basic and are useful reagents in synthesis to generate carbanions from compounds with C—H acidities greater than the amine N—H acidity:

$$-\overset{|}{\underset{|}{C}}-H + R_2N:^{\ominus}Li^{\oplus} \rightleftharpoons -\overset{|}{\underset{|}{C}}:^{\ominus} Li^{\oplus} + R_2\overset{..}{N}H$$

A large number of amine reactions depend on the nucleophilic proper-ties of the amine nitrogen. The most important ones include

1. Nucleophilic attack at carbonyl groups

a. carboxylic acid derivatives (acylation, amide formation)

$$R'\overset{O}{\overset{||}{C}}-X + R\overset{..}{N}H_2 \longrightarrow R'\overset{O}{\overset{||}{C}}-NHR + HX$$

(Sections 23-9A, 23-13C, 24-3)

b. aldehydes and ketones (imine and enamine formation)

$$R_2C{=}O + R\overset{..}{N}H_2 \rightleftharpoons R_2C{=}NR + H_2O \quad \text{(Sections 23-9B, 16-4C)}$$

$$\overset{\backslash}{\underset{/}{}}CH-\overset{|}{C}{=}O + R_2\overset{..}{N}H \rightleftharpoons \overset{\backslash}{\underset{/}{}}C{=}\overset{|}{C}-\overset{..}{N}R_2 \quad \text{(Sections 23-9B, 16-4C)}$$

2. Nucleophilic attack on alkyl and aryl halides

a. alkylation

$$RX + R\overset{..}{N}H_2 \longrightarrow R_2\overset{\oplus}{N}H_2 \overset{\ominus}{X} \quad \text{(Sections 23-9D, 23-13B)}$$

b. arylation

$$ArX + R\overset{..}{N}H_2 \longrightarrow Ar\overset{\oplus}{N}H_2R \overset{\ominus}{X} \quad \text{(Sections 23-9E, 14-6B, 14-6C)}$$

3. Sulfonation

$$RSO_2-Cl + R\overset{..}{N}H_2 \longrightarrow RSO_2-\overset{..}{N}HR + HCl \quad \text{(Section 23-9C)}$$

4. Amines with nitrous acid

a. Primary alkanamines react with HONO to give products derived from a carbocation intermediate according to the following scheme:

$$R\ddot{N}H_2 \xrightarrow{HONO} R-\overset{\oplus}{N}H_2-N=O \xrightarrow{-H^{\oplus}} R\ddot{N}H-\ddot{N}=O \longrightarrow R\ddot{N}=\ddot{N}-OH$$

$$R^{\oplus} \xleftarrow{-N_2} R-\overset{\oplus}{N}\equiv N \xleftarrow{-H_2O} R-\ddot{N}=\ddot{N}-\overset{\oplus}{O}H_2 \xleftarrow{H^{\oplus}}$$

rearrange / | \ H₂O

$$R^{\oplus} \qquad alkene \quad RX \qquad X^{\ominus} \qquad ROH$$

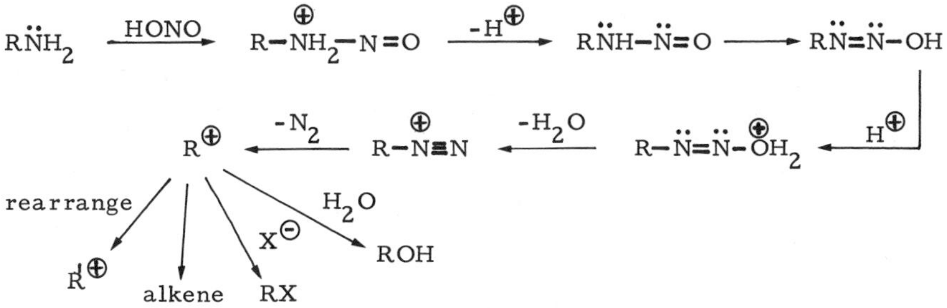

b. Primary arenamines, $Ar\ddot{N}H_2$, react similarly except that the arenediazonium salts, $Ar-\overset{\oplus}{N}\equiv N$, are more stable than alkanediazonium salts, $R-\overset{\oplus}{N}\equiv N$. The diazonium group, $-\overset{\oplus}{N}\equiv N$, easily can be replaced by other groups (OH, halogen, CN, H, NO_2) in a displacement reaction that usually, but not always, requires a cuprous catalyst:

$$Ar-\overset{\oplus}{N}\equiv N + X^{\ominus} \xrightarrow{Cu(I)} Ar-X + N_2 \qquad (Table\ 23-4)$$

$$Ar-\overset{\oplus}{N}\equiv N + H_2O \xrightarrow{heat} Ar-OH + N_2 + \overset{\oplus}{H}$$

The sequence $ArH \longrightarrow ArNO_2 \longrightarrow Ar\ddot{N}H_2 \longrightarrow Ar\overset{\oplus}{N}_2 \xrightarrow{CuX} ArX$ is an indirect method of putting a substituent, X, on an arene ring. This method is a useful alternative to direct electrophilic substitution (Sections 22-4C and 22-4D). These and other reactions of arenediazonium salts are given in Table 23-4 and Sections 23-10B and 23-10C.

c. Secondary amines with nitrous acid lead to N-nitroso compounds:

$$R_2\ddot{N}H \xrightarrow{HONO} R_2\overset{\oplus}{N}HNO \xrightarrow{-H^{\oplus}} R_2\ddot{N}-\ddot{N}=O$$

N-nitrosoarenamines rearrange to C-nitroso compounds when heated with acids. This is one example of a general rearrangement of the type $Ar-\ddot{N}\overset{R}{\underset{Y}{\diagdown}} \xrightarrow{H^{\oplus}}$

$Y-Ar-\ddot{N}HR$ (see Section 23-10D):

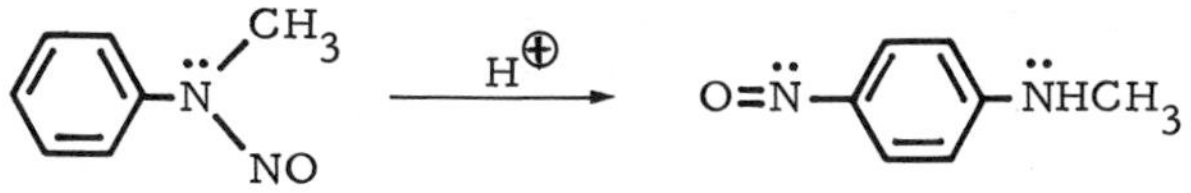

5. Oxidation of amines depends on the degree of substitution at nitrogen and the

nature of the oxidants. For instance, with peroxy compounds,

$$R_3N: \xrightarrow{H_2O_2} R_3\overset{\oplus}{N}-\overset{\ominus}{O} \quad \text{amine oxide}$$

$$R_2NH \xrightarrow{H_2O_2} R_2NOH \quad \text{azanol (hydroxylamine)}$$

$$RNH_2 \xrightarrow{H_2O_2} RNO_2 \quad \text{nitro compound}$$

Arenamines and CrO_3 or $KMnO_4$ lead to complex products (aniline black) and quinones (Section 23-11D).

6. <u>Arenamines as Carbon Nucleophiles</u>. The enamine structure $\overset{\diagdown}{C}=\overset{|}{C}-\overset{..}{N}H_2$ $\longleftrightarrow \overset{\ominus}{\underset{|}{C}}-C=\overset{\oplus}{N}H_2$ of aromatic amines makes the ring carbons relatively nucleophilic at the ortho and para carbons. Substitution of electrophiles occurs readily at these positions. Examples are halogenation (Section 23-9F), diazo coupling (Section 23-10C), and rearrangement of amines of the type $Ar-N\overset{R}{\underset{Y}{\diagup}}$ (Section 23-10D).

   <u>Methods of amine synthesis</u> are tabulated in Tables 23-6 and 23-7. The most important methods involve

   1. alkylation or arylation of a simpler amine or ammonia (Sections 23-9D, 23-9E, 23-12C)

   2. reduction methods (Sections 23-12B, 23-12C)

   3. synthesis from amides by hydrolysis, reduction, or rearrangement

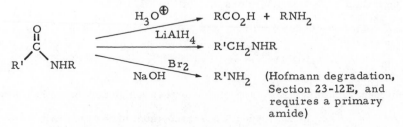

(Hofmann degradation, Section 23-12E, and requires a primary amide)

ANSWERS TO EXERCISES

23-1   a. <u>N</u>, <u>N</u>-dimethylethenamine (dimethylvinylamine)

   b. <u>N</u>-ethyl-<u>N</u>-methylbenzenamine (<u>N</u>-ethyl-<u>N</u>-methylaniline)

   c. aminoethanoic acid (glycine)

   d. 1,4-diaminobutane

e. 1,4-diazabenzene (pyrazine, 1,4-diazine)

f. cyclohexyltrimethylammonium iodide

g. 1,1,4,4-tetramethyldiazoniacyclohexane chloride

23-2   The chemical shift of the NH proton of $(C_2H_5)_2NH$ will change markedly with concentration when dissolved in a solvent capable of hydrogen bonding with the NH proton (e.g., ethers). If an alcohol is used as solvent, only an average resonance of OH and NH protons may be observed, the chemical shift varying with the concentration. The protons of the ethyl groups will not be similarly affected and will be clearly distinguishable from the NH proton.

Addition of D_2O will result in formation of $(C_2H_5)_2ND$ which will cause the N—H resonance to diminish in intensity relative to the C—H resonances.

23-3   Compound a, $C_8H_{11}N$: The infrared spectrum indicates that a primary amino group is present (two bands ~ 3350 cm^{-1} and 3250 cm^{-1}). Also, the spectrum shows that the compound is a monosubstituted benzene derivative (see Section 22-3A). The nmr spectrum shows four kinds of hydrogen in the ratio of $5:1:2:3$ with chemical shifts characteristic of the groups C_6H_5, CH, NH_2 and CH_3. The quartet at 3.9 ppm and the doublet at 1.3 ppm are consistent with the grouping $-\overset{|}{\underset{|}{C}}H-CH_3$. Compound A therefore has the following structure:

$$C_6H_5\overset{\overset{\displaystyle CH_3}{|}}{C}HNH_2$$

1-phenylethanamine (1-phenylethylamine)

Compound b, $C_8H_{11}N$, appears to be a secondary amine (one band at about 3300 cm^{-1} in its infrared spectrum) and a monosubstituted benzene derivative (see Section 22-3A). The four kinds of hydrogen in its nmr spectrum are in the ratio of $5:2:3:1$ with chemical shifts corresponding to the groups C_6H_5, CH_2, CH_3 and NH. The following structure for B is consistent with the spectral data:

$$C_6H_5CH_2NHCH_3$$

N-methylphenylmethanamine   (N-methylbenzylamine)

23-4   Odd mass for $\underline{M}^+$ suggests a compound with an odd number of nitrogens, probably one nitrogen. Even mass for fragment ion suggests that it too has one nitrogen. Because 73 - 58 = 15 (or CH_3), the fragment ion corresponds to loss of CH_3 from $\underline{M}^+$. A likely structure is:

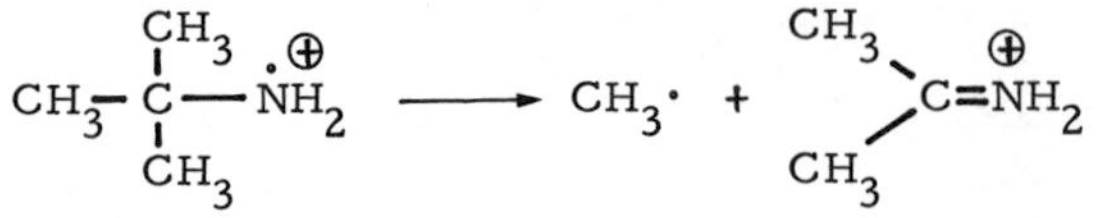

$$
\underset{\underset{CH_3}{|}}{\overset{\overset{CH_3}{|}}{CH_3-C}}\!\!-\!\!\overset{.\oplus}{NH_2} \longrightarrow CH_3\cdot \; + \; \underset{CH_3}{\overset{CH_3}{>}}C\!=\!\overset{\oplus}{NH_2}
$$

$$\underline{m/e}\ 73 \qquad\qquad\qquad\qquad \underline{m/e}\ 58$$

23-5   Molecular ion has mass 87. The ions 87, 72 and 30 must have odd nitrogen content; 57 is nitrogen free. Ion 30 is almost certainly $CH_2\overset{\oplus}{=}\overset{\oplus}{NH_2}$, while ion 72 corresponds to loss of methyl from $\underline{M}^+$. Ion 57 corresponds to loss of $\cdot CH_2NH_2$ from $\underline{M}^+$.

The nmr evidence suggests that 9-proton signal at 0.9 ppm is $(CH_3)_3C$; the 2-proton signals at 2.3 and 1.3 ppm correspond to $-CH_2-NH_2$, in that order. The most likely structure and fragmentation pattern is:

$$
(CH_3)_3CCH_2NH_2 \xrightarrow{-e} \underset{\underline{M}^+,\ 87}{(CH_3)_3CCH_2\overset{.\oplus}{NH_2}}
$$

$$
\xrightarrow{-\overset{.}{C}H_3}\ (CH_3)_2C\underset{\overset{\oplus}{NH_2}}{\overset{\overset{\cdot}{CH_2}}{\diagup|}}\ \ \underline{m/e}\ 72
$$

$$
\xrightarrow{-\overset{.}{C}(CH_3)_3}\ CH_2\overset{\oplus}{=}NH_2\ \ \underline{m/e}\ 30
$$

$$
\xrightarrow{-\overset{.}{C}H_2NH_2}\ (CH_3)_3\overset{\oplus}{C}\ \ \underline{m/e}\ 57
$$

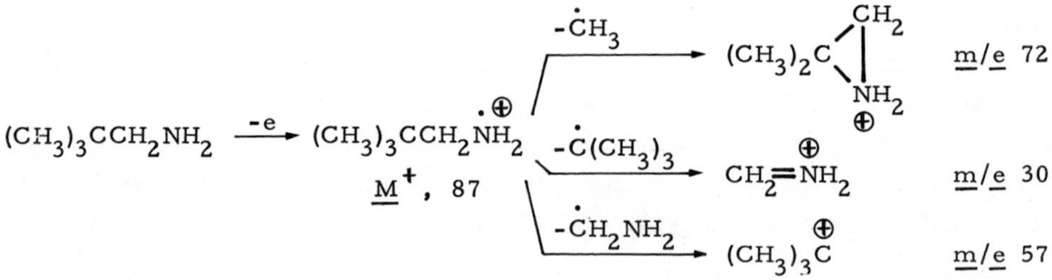

23-6   If we accept that the preferred bond angles to pyramidal nitrogen are like the HNH angle in ammonia (107°), then the change in angle strain in going to the <u>planar</u> inversion transition state of $(C_2H_5)_3N$ is proportional to $(120° - 107°)^2 = (13°)^2$ per CNC angle (See Exercise 12-2, p. 449). In contrast the corresponding change in angle strain for just the ring CNC angle of 1-ethylaza cyclopropane would be proportional to $(120°- 60°)^2 - (107°- 60°)^2 \sim (37.3)^2$. On this basis the transition state for inverting the configuration at the ring nitrogen has more angle strain than that in $(C_2H_5)_3N$.

Figure 21-3 can be adapted to represent a <u>planar</u> system $-\overset{..}{\underset{..}{O}}-\overset{..}{N}<$ where the π MO's are a combination of two parallel p orbitals, one on oxygen and one on nitrogen. Four electrons fill these two MO's, and because this requires that the antibonding orbital be filled, the net π-bonding is zero (or less). Electron repulsion will be severe, and hence the transition state for inversion at nitrogen in $\underset{\sim}{2}$ is of high energy.

$$
:\!O\!-\!N\diagdown R \;\rightleftharpoons\; :\!O\!-\!N\!-\!R \;\rightleftharpoons\; :\!O\!-\!N\diagdown R
$$

$$\underset{\sim}{2} \qquad\qquad \text{unfavorable} \qquad\qquad \underset{\sim}{2}$$

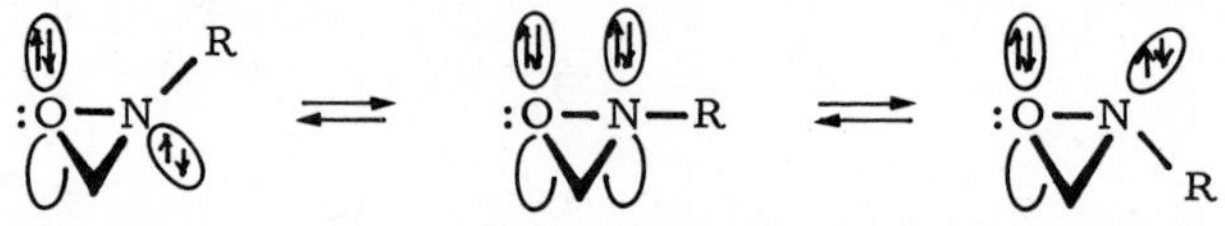

23-7   The nmr spectrum of 1, 2, 2-trimethylazacyclopropane (Fig. 23-7) may be explained as follows:

At room temperature, resonance at 133 Hz corresponds to the N—CH$_3$ protons; the two lines at 63 and 70 Hz correspond to nonequivalent methyl groups at C2, and the lines at 50 and 92 correspond to nonequivalent protons at C3.

As the temperature is raised inversion of configuration at nitrogen becomes more rapid and, at 110°, is sufficiently rapid that the methyl groups at C2 become magnetically equivalent and the protons at C3 become magnetically equivalent. Thus, the peaks at 63 and 70 Hz coalesce to a single line, and the peaks at 50 and 92 Hz similarly coalesce.

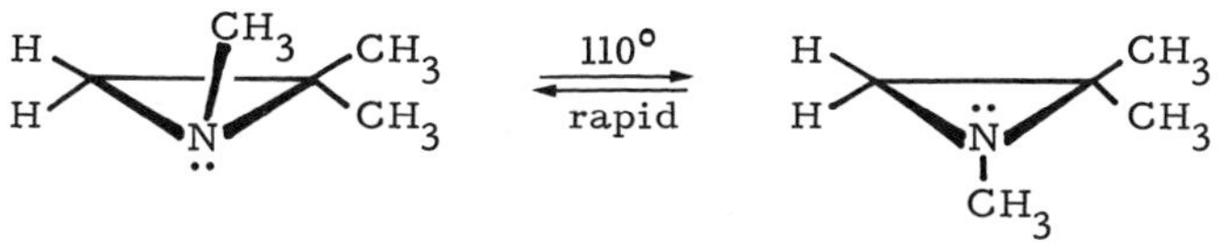

23-8   The spectral changes reflect the following conformational changes:

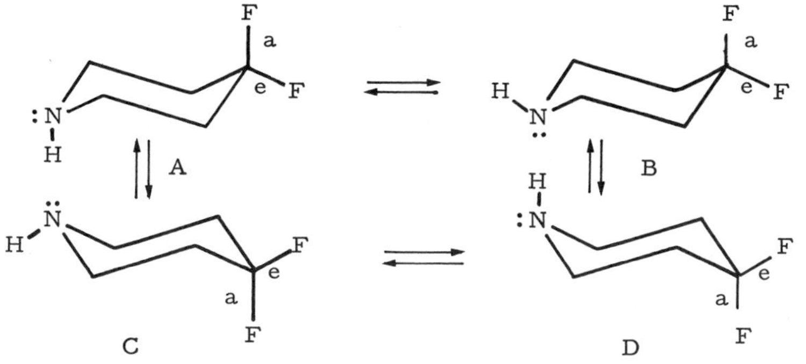

At 25°, ring inversion (and also nitrogen inversion) make the two fluorines indistinguishable as far as the nmr experiment is concerned. There is therefore only one ^{19}F chemical shift, and this signal is split into a 1:4:6:4:1 quintet by the neighboring four protons, – CH$_2$—CF$_2$—CH$_2$– ($\underline{J}_{HF}$ is small, so lines are narrowly spaced). As the temperature is lowered, the conformational changes become slower and the ^{19}F resonance broadens as distinguishable chemical shifts for nonequivalent fluorines become apparent. By -90°, the spectrum is consistent with two nonequivalent conformations A (or C) and B (or D) in the ratio of 75:25 (or 25:75). There are now two overlapping quartets in which $\underline{J}$ is large (235 Hz) because each conformation has two nonequivalent fluorines (F$_{axial}$ and F$_{equatorial}$) which couple to give a quartet in which $\underline{J}_{FF}$ = 235 Hz, and the ^{19}F shifts for conformation A (or C) are not the same as the shifts for B (or D). Schematically:

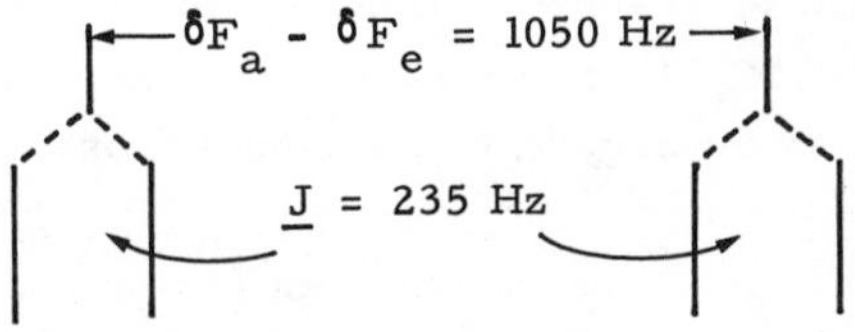

$$\delta F_a - \delta F_e = 1050 \text{ Hz}$$

$$\underline{J} = 235 \text{ Hz}$$

75%

$$\delta F_a - \delta F_e = 700 \text{ Hz}$$

$$\underline{J} = 235 \text{ Hz}$$

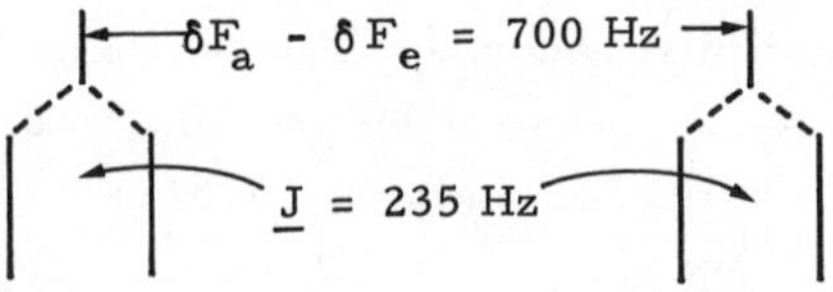

25%

23-9   a. $(CH_3)_3N: + HBF_4 \xrightarrow{CD_3NO_2} (CH_3)_3\overset{\oplus}{N}H \ \overset{\ominus}{B}F_4$

δ 2.7 ppm
(singlet)

δ 3.5 ppm for methyl
resonance

(doublet because of coupling
to NH)

b. A solution of $(CH_3)_3\overset{\oplus}{N}H \ \overset{\ominus}{B}F_4$ in the presence of a base, $(CH_3)_3N:$, undergoes rapid exchange of the NH proton.

$$(CH_3)_3\overset{\oplus}{N}H + \boxed{(CH_3)_3\overset{..}{N}} \longrightarrow (CH_3)_3\overset{..}{N} + \boxed{(CH_3)_3\overset{\oplus}{N}}H$$

Because of the rapid exchange, there is no visible H–C–N–H coupling. Hence doublet at 3.5 ppm becomes a singlet. The rapid exchange reaction also averages the methyl proton chemical shifts for the ammonium salt and the amine. The actual shift observed depends on the relative amount of salt to amine. As more amine is added the **shift** observed will be closer to that of the pure amine. This is expressed in the following equation

$$\delta_{obs} = \underline{n}_{salt}\,\delta_{salt} + \underline{n}_{amine}\,\delta_{amine}$$

where the $\underline{n}$'s are the respective mole fractions.

23-10   a. $(CH_3CH_2)_3N$ Electron-donating (alkyl) groups on nitrogen increase basicity.

b. $(CH_2)_5NH$ Additional eclipsing strain and angle strain are introduced in the conjugate acid of $(CH_2)_2NH$.

c. $CH_3CH_2CH_2NH_2$ Electron-withdrawing (CF_3) groups decrease basicity.

d. $\overset{\ominus}{O_2}CCH_2CH_2NH_2$ Field effect makes it easier to transfer a proton to a nitrogen near a negative charge than to
one near a positive charge.

23-11

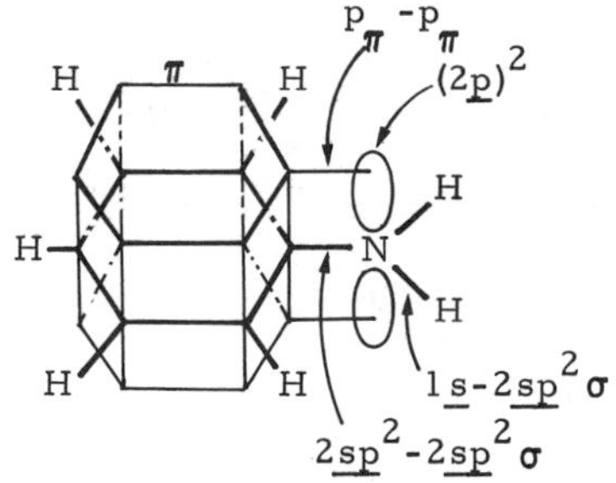

benzenamine   benzenammonium ion

Benzenamine (aniline) is a weaker base than most saturated amines
because, on forming the conjugate acid, the hybridization of nitrogen changes,
which results in a loss of stabilization arising from delocalization of the unshared
electron pair on nitrogen over the orbitals of the aromatic ring.   No similar
loss in resonance energy occurs with saturated amines.

23-12   On comparing the proton-transfer equilibria of an amidine with a
saturated amine RNH_2, it is evident that a resonance-stabilized cation can be
formed only on proton addition to the imine nitrogen.

$$R-C\overset{NH}{\underset{NH_2}{\big<}} + H^{\oplus} \rightleftharpoons R-C\overset{\overset{\oplus}{NH_2}}{\underset{NH_2}{\big<}} \longleftrightarrow R-C\overset{NH_2}{\underset{\overset{\oplus}{NH_2}}{\big<}}$$

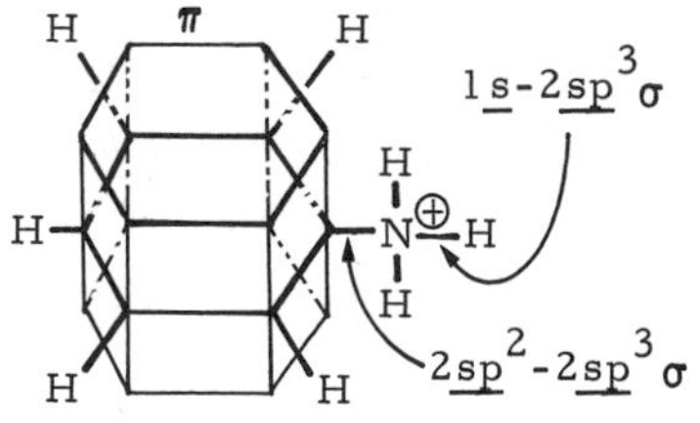

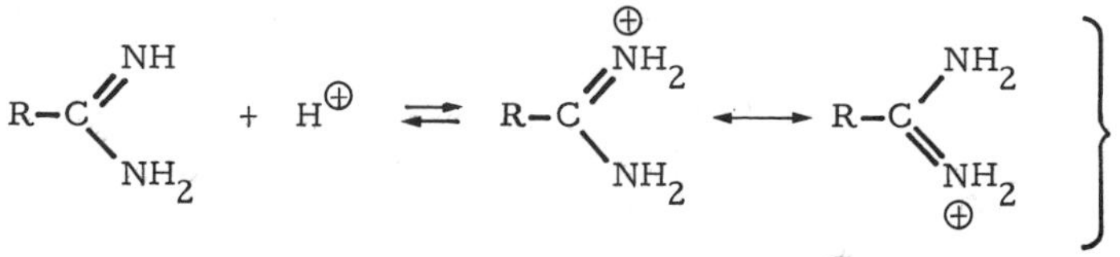

charge equally
shared by two
nitrogens

$$R-NH_2 + H^{\oplus} \rightleftharpoons R-\overset{\oplus}{NH_3}$$

A similar stabilization of the conjugate acid of RNH_2 is not possible.   Therefore, the equilibrium constant (base strength) of the amidine is greater than
that of a saturated amine.

23-13   Order of base strengths is:

$$\text{A} \quad > \quad \text{B} \quad > \quad \text{C}$$

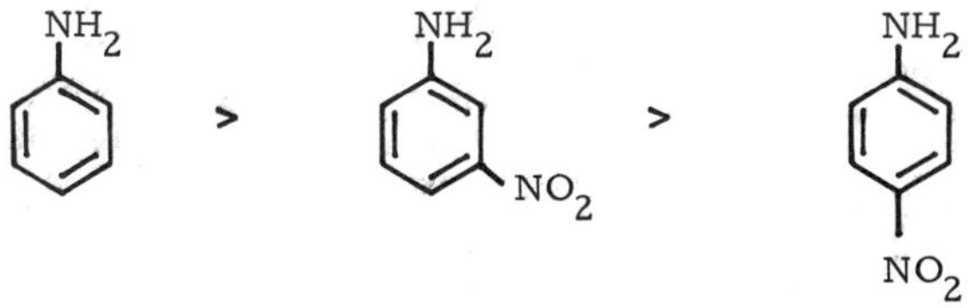

A   B   C

B is weaker than A because of the electron-withdrawing inductive effect of a nitro group. This effect also operates in C but might be thought as less impor-tant than in B because the NO_2 group is more distant from the basic amino group. However, C is actually a weaker base than B because of the stabilizing resonance effect of a nitro group in the 4-position on the neutral base. This is apparent in the following valence-bond structures for the base. This stabiliza-tion is lost on forming the conjugate acid.

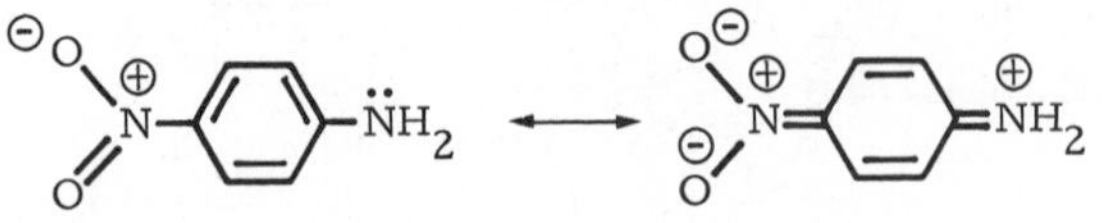

23-14   In evaluating relative basicities (or acidities), you should look for effects that would likely stabilize one conjugate acid more than the other, or one neutral base more than the other.

a. $\underline{K} < 1$; resonance stabilization is important for the para-substitu-ted amine but not for the meta-substituted amine. (See also Exercise 23-13.)

b. $\underline{K} > 1$; the monocation retains some of the stabilization associ-ated with delocalization of electron pair on nitrogen; the dication has no such stabilization.

c. $\underline{K} > 1$; methyl groups increase basicity.

d. $\underline{K} > 1$; two ortho bromines have a stronger inductive effect than an ortho-para pair; this effect is base-weakening, or acid strengthening.

e. $\underline{K} < 1$; the diazonium monocation is better resonance stabilized than is the 4-ethoxybenzenamine.

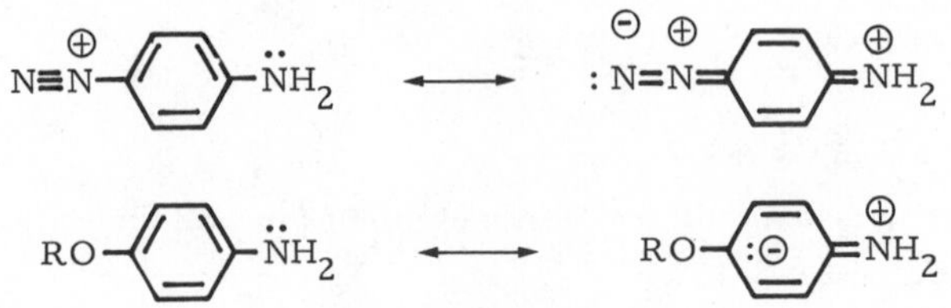

23-15   a. Proton transfer to the ring nitrogen would lead to a reactive diene in which all the aromatic stabilization of the $6\ \pi$ electron system is lost. Proton transfer to carbon would give a conjugated immonium ion with similar loss in aromatic stabilization.

b. The conjugate acid resulting from proton transfer to imidazole is

better stabilized (relative to the neutral base) than is the conjugate acid of
pyrimidine. Therefore, imidazole is the stronger base.

Notice the two equivalent VB structures.

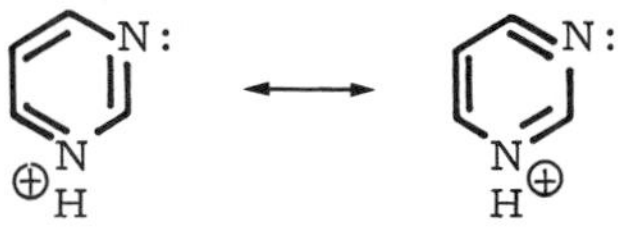

Nitrogens are not equivalent.

   c. $\overset{\oplus}{C(NH_2)_3}$ is a very weak acid because it is strongly resonance-
stabilized. Three equivalent VB structures can be written (Section 23-7D)
showing that the charge is distributed equally over three equivalent nitrogens.
Proton transfer to a base would mean loss of substantial resonance energy.

23-16   For caffeine $H^{\oplus}$, $pK_a = 10.61$; hence $pK_b = 14 - 10.61 = 3.39$, and
$pK_b = -\log K_b$, or $K_b = 4.07 \times 10^{-4}$.

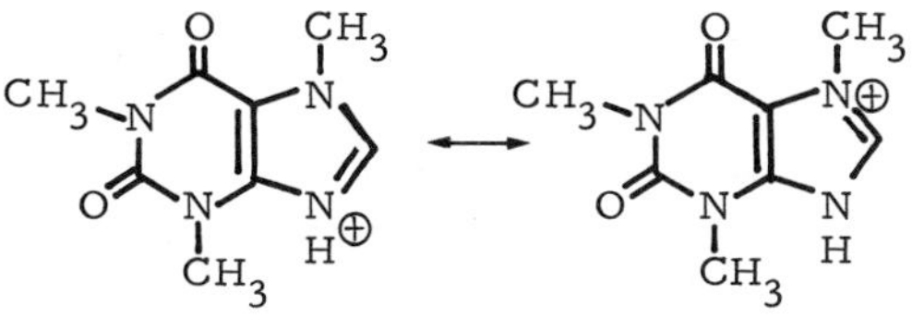

is expected
to be
preferred over

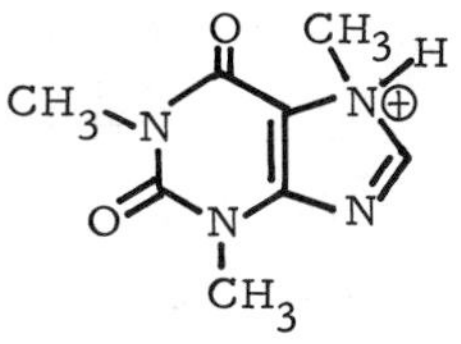

(See Exercise 23-15b.)

23-17

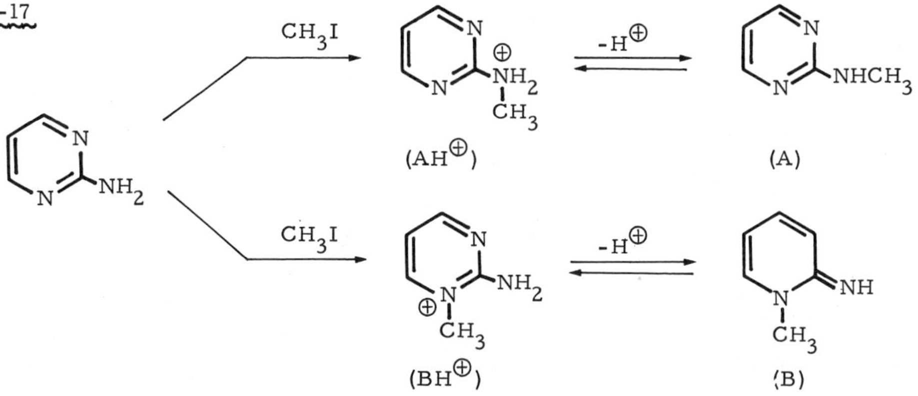

$(AH^{\oplus})$ (A)

$(BH^{\oplus})$ (B)

A is formed by methylation at the 2-amino nitrogen. B is formed by methylation at a ring nitrogen. A is a weak base because the neutral base enjoys more
resonance stabilization than the conjugate acid, $AH^{\oplus}$. B is a much stronger

base because its conjugate acid, $BH^{\oplus}$, enjoys aromatic stabilization, which is not present in B. At neutral pH, the ring nitrogens are the most nucleophilic and the major methylation product is B. At high pH, the NH_2 nitrogen is converted to $-NH^{\ominus}$ (stabilized by $-\overset{|}{N}=\overset{|}{C}-NH \longleftrightarrow \overset{|}{N}-\overset{|}{C}=NH$) making it more nucleophilic than the ring nitrogens and, under these conditions, methylation gives A.

23-18   We can write three different conjugate acids, depending on the site of protonation. The preferred site is that which leads to the most stable conjugate acid.

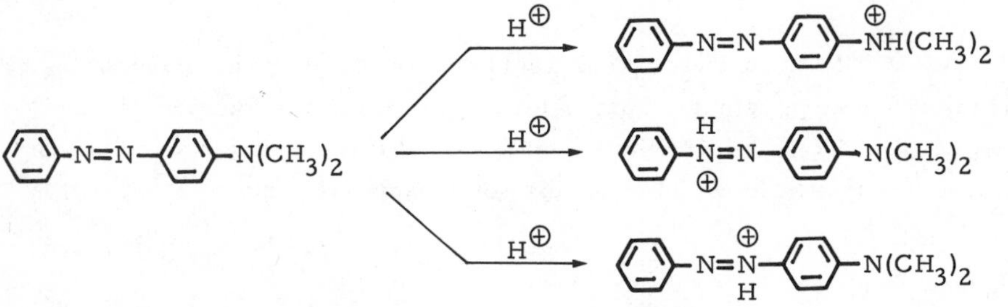

Of the three structures, A, B, and C, B is preferred because only in B can the charge be distributed profitably from the azo to the amino nitrogen.

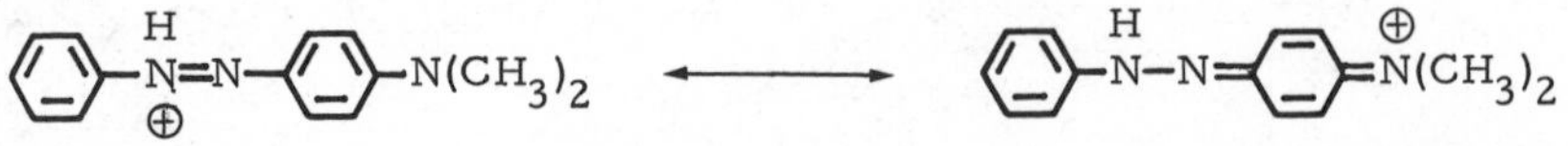

This predicts that one azo nitrogen is more basic than the amino nitrogen even though the basicity of $C_6H_5N(CH_3)_2$ is 7 - 8 powers of ten greater than that of $C_6H_5N=NC_6H_5$.

23-19 a. The conjugate base from imidazole is more strongly stabilized by resonance than is the conjugate base from pyrrole.

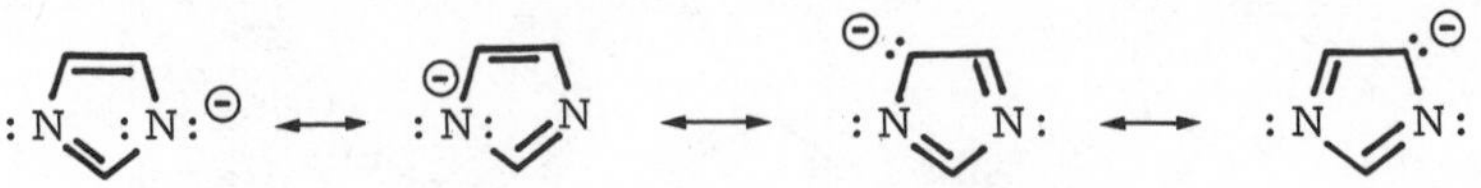

equivalent pair   equivalent pair

b. Benzenamine would be expected to be a stronger acid than cyclohexanamine because the conjugate base from benzenamine is a delocalized anion whereas that from cyclohexanamine is not.

23-20

a. $C_6H_5NHCONHC_6H_5$ b. $C_6H_5SO_2NH(CH_2)_4NHSO_2C_6H_5$

c. $CH_3CH_2CH=NCH_2CH_3$

d. 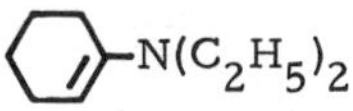 $-N(C_2H_5)_2$

e.

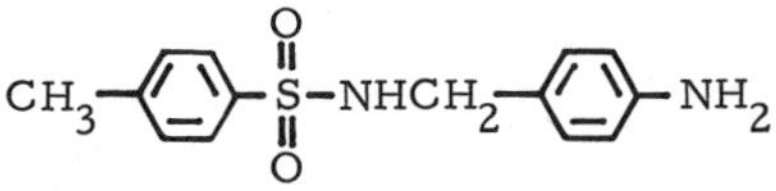

23-21

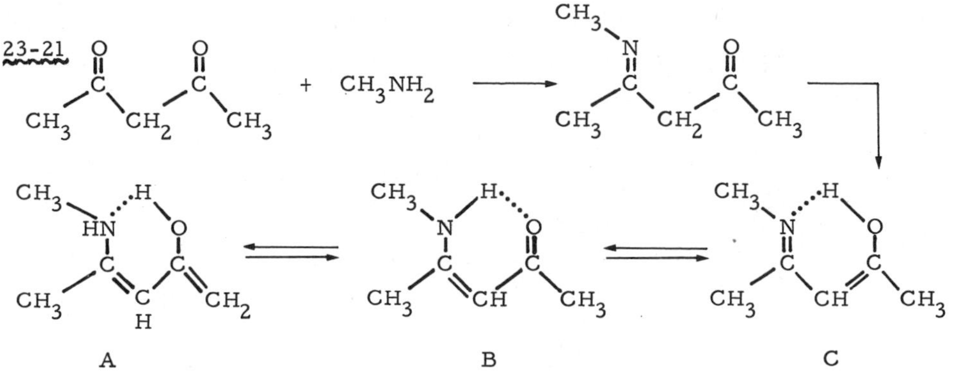

Structures A, B and C all have internal hydrogen bonding.

23-22

a. $C_3H_5-NH_2 + C_6H_5SO_2Cl + 2NaOH \longrightarrow C_3H_5\overset{\ominus}{N}-\overset{O}{\underset{O}{\overset{||}{\underset{||}{S}}}}-C_6H_5 \overset{Na^{\oplus}}{} + NaCl + 2H_2O$

$$C_3H_5NHSO_2C_6H_5 + NaCl \xleftarrow{\quad HCl \quad}$$

$C_3H_5NH_2$ has to be a primary amine and could be $CH_2=CHCH_2NH_2$ or $-NH_2$. Structures such as $CH_3CH=CH-NH_2$ are not allowed because the element $-\overset{|}{C}=\overset{|}{C}-NH_2$ is not stable.

b.   One amine function is secondary, the other is tertiary.

$$(CH_3)_2NCH_2CH_2NHCH_3 + C_6H_5SO_2Cl + NaOH \xrightarrow[-H_2O]{-NaCl}$$

$$Cl^{\ominus}(CH_3)_2\overset{\oplus}{\underset{H}{N}}CH_2CH_2\overset{CH_3}{\underset{|}{N}}SO_2C_6H_5 \xleftarrow{\quad HCl \quad} (CH_3)_2NCH_2CH_2\overset{CH_3}{\underset{|}{N}}SO_2C_6H_5$$

23-23

a. $CH_2=CH_2 \xrightarrow{Cl_2} ClCH_2CH_2Cl \xrightarrow[-2HCl]{2NH_3} H_2NCH_2CH_2NH_2$

b. $CH_2=CH_2 \xrightarrow[\text{or } O_2, Ag]{RCO_3H} \overset{O}{CH_2-CH_2} \xrightarrow{NH_3} HOCH_2CH_2NH_2$

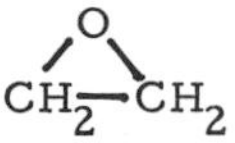

c. C_6H_5Cl $\xrightarrow[\text{NH}_3, \ -33°]{\text{NaNH}_2}$ $C_6H_5NH_2$

23-24   A mixture of primary, secondary and tertiary amine may be expected from the reaction of 1-bromobutane with excess butanamine. Some tetrabutyl-ammonium bromide may also be formed. On treatment of the product with butanedioic (succinic) anhydride, the following reactions would take place.

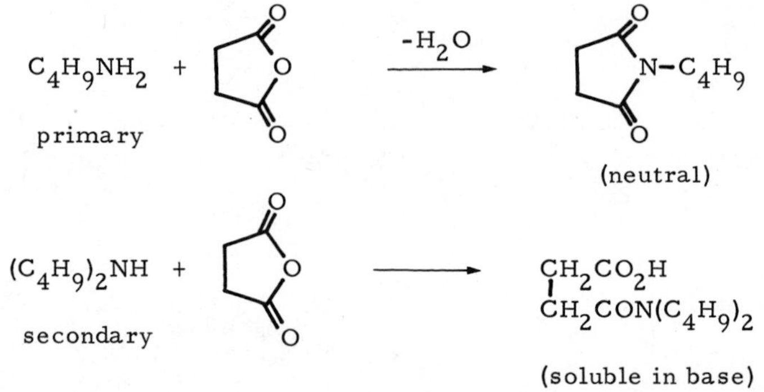

$C_4H_9NH_2$ + [structure] $\xrightarrow{-H_2O}$ [structure] $N-C_4H_9$

primary   (neutral)

$(C_4H_9)_2NH$ + [structure] $\longrightarrow$ $\underset{CH_2CON(C_4H_9)_2}{CH_2CO_2H}$

secondary   (soluble in base)

The unreacted tertiary amine may be isolated by extraction with acid; the product derived from $(C_4H_9)_2NH$ by extraction with base; and the product from $C_4H_9NH_2$ would remain as the neutral component. Each amine may be regenerated by the following reactions:

$(C_4H_9)_3\overset{\oplus}{N}H$ $\xrightarrow[-H_2O]{\overset{\ominus}{O}H}$ $(C_4H_9)_3N$

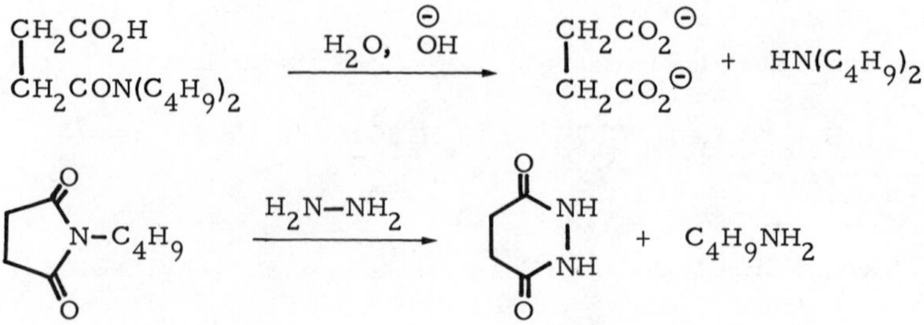

$\underset{CH_2CON(C_4H_9)_2}{CH_2CO_2H}$ $\xrightarrow{H_2O, \ \overset{\ominus}{O}H}$ $\underset{CH_2CO_2^{\ominus}}{CH_2CO_2^{\ominus}}$ + $HN(C_4H_9)_2$

[structure] $N-C_4H_9$ $\xrightarrow{H_2N-NH_2}$ [structure] $\underset{NH}{\overset{NH}{}}$ + $C_4H_9NH_2$

23-25   An ambident structure can be written for the sulfonamide anion, and on the basis of the VB structures, O- and N-alkylation would both appear possible.

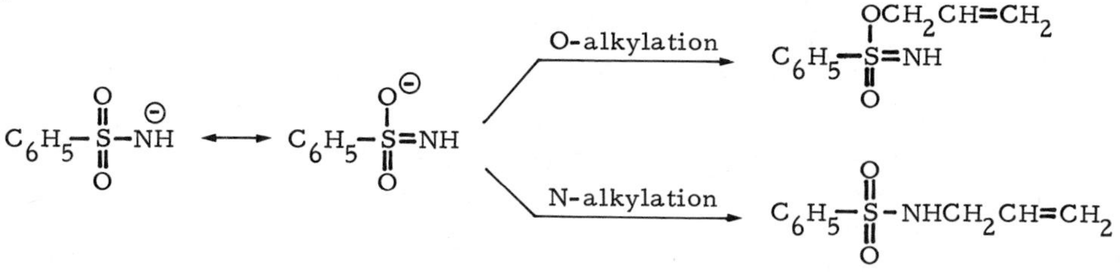

However, the resonance depicted above does not involve $3\underline{p}$ sulfur orbitals but $3\underline{d}$ sulfur orbitals, and we cannot arbitrarily assume that because π bonding or resonance involving $3\underline{d}$ orbitals is important that O-alkylation of a sulfonamide anion will occur on oxygen to give a thermodynamically favorable product. In general, we would expect that an S=O bond would be substantially more favorable than an S=NH bond.

23-26

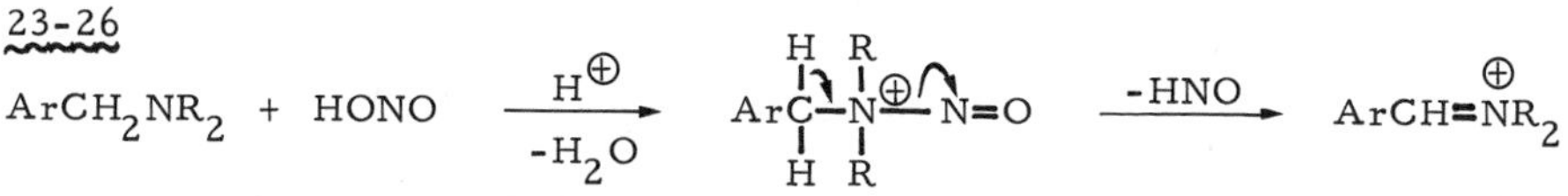

23-27

The VB structures show that the electron pair on nitrogen of the dimethylamine group is delocalized and less available for bonding to a proton than in $(CH_3)_2NH$.

b.   The VB structures show that the N—N bond has double-bond character. Rotation about the N—N bond is slow (or prevented) at ordinary temperatures, which makes the methyl groups nonequivalent (one is $\underline{E}$ to the oxygen, the other $\underline{Z}$). The fact that the stabilization energy would be lost at the middle of a rotation between the planar states ($\Delta \underline{H}^*$ for this process is 23 kcal) indicates that the SE of $\underline{N}$-nitrosoamines (in the planar state) is about 23 kcal. The separate methyl resonances in the nmr spectrum coalesce on heating as the rotation about the N—N bond becomes rapid.

23-28   a. Assuming $\overset{\oplus}{N}O$ is the nitrosating agent:

$$H_2\overset{..}{N}CH_2CO_2R + \overset{\oplus}{N}O \longrightarrow O=N-\overset{\oplus}{N}H_2CH_2CO_2R \xrightarrow{-H^{\oplus}}$$

$$\left[\!\!\!- H_2\overset{\oplus}{O}-N=NCH_2CO_2R \xleftarrow{H^{\oplus}} HO-N=NCH_2CO_2R \longleftarrow O=N-NHCH_2CO_2R \right.$$

$$\left.\xrightarrow{-H_2O} :N\equiv\overset{\oplus}{N}-CH_2CO_2R \xrightarrow{-H^{\oplus}} :N\equiv\overset{\oplus}{N}-\overset{\ominus}{C}HCO_2R \longleftarrow\!\!\!\right]$$

$$:\overset{\ominus}{\underset{..}{N}}=\overset{\oplus}{N}=CHCO_2R \longleftrightarrow etc.$$

does not lose N_2 readily to give a carbocation because of the presence of the $-CO_2R$ group (see p. 817)

ethyl diazoethanoate (ethyl diazoacetate)

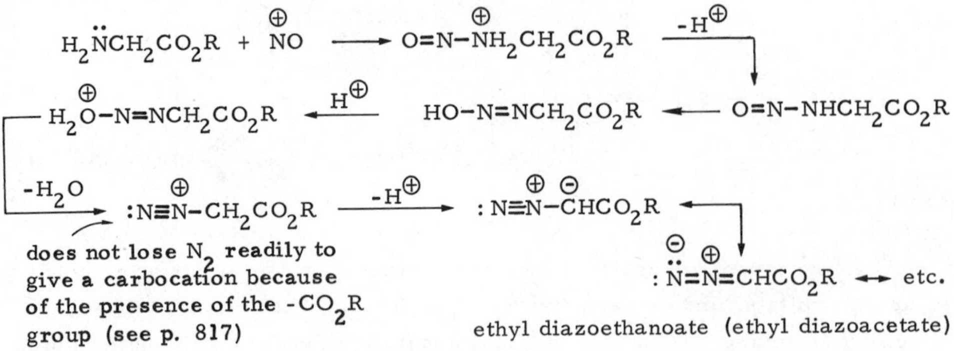

b.   The same reaction would not occur with ethyl 3-aminopropanoate because the diazonium ion formed on nitrosation, $N_2\overset{\oplus}{-}CH_2CH_2CO_2R$, cannot lose a proton to form a diazo compound that is conjugated with (stabilized by) the ester function.

23-29

a. $(CH_3)_2CHCH_2OH$; $CH_3CH(OH)CH_2CH_3$; $(CH_3)_3COH$; $(CH_3)_2C=CH_2$

$CH_3CH=CHCH_3$; $CH_2=CHCH_2CH_3$; $CH_3-\triangleleft$

chlorides corresponding to the alcohols

b.   N−NO

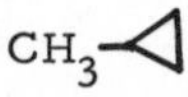

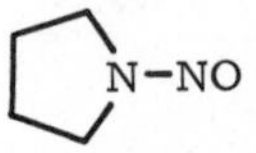

c. $CH_3CH=CHCH_2OH$; $CH_3CH(OH)CH=CH_2$; $CH_2=CH-CH=CH_2$

chlorides corresponding to the alcohols

d. $CH_3COC(CH_3)_3$ (analogous to the pinacol rearrangement)

23-30

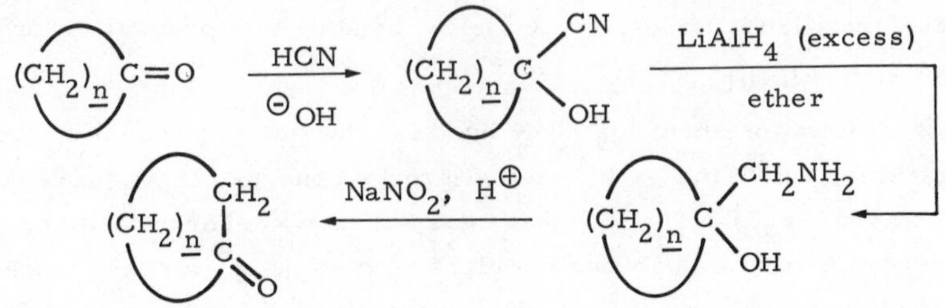

23-31

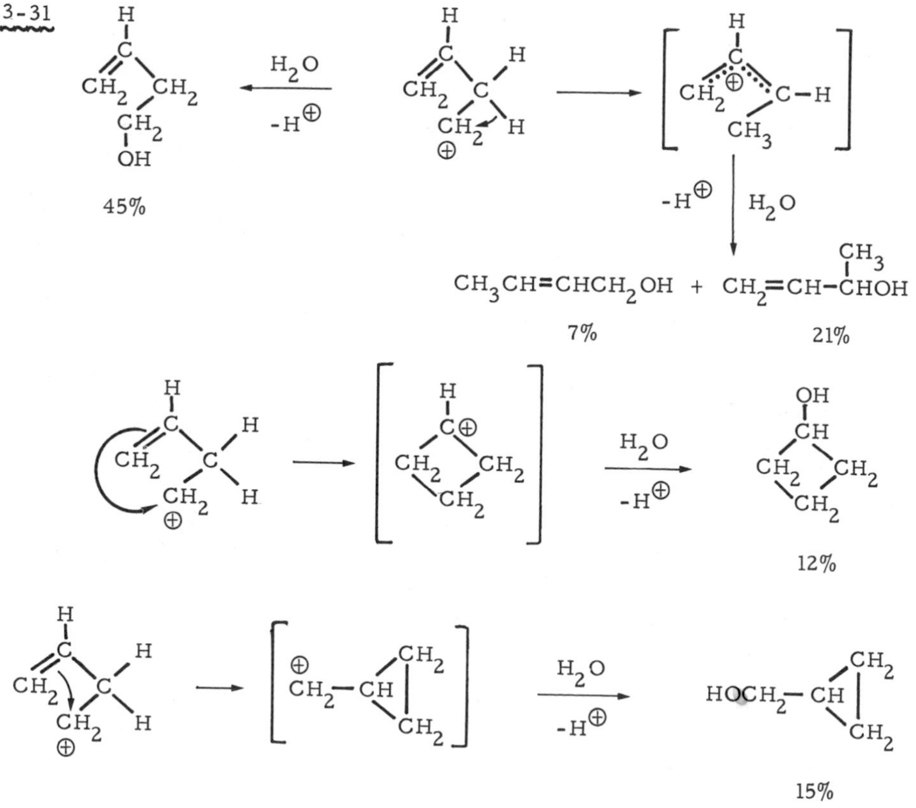

$$CH_3CH=CHCH_2OH + CH_2=CH-\overset{CH_3}{\underset{|}{C}}HOH$$

45%   7%   21%

12%

15%

23-32   Using deuterium-labeled propanamine, $CD_3CH_2CH_2NH_2$, the position and proportion of deuterium in the propene formed on nitrous-acid deamination will indicate the origin of the propene.

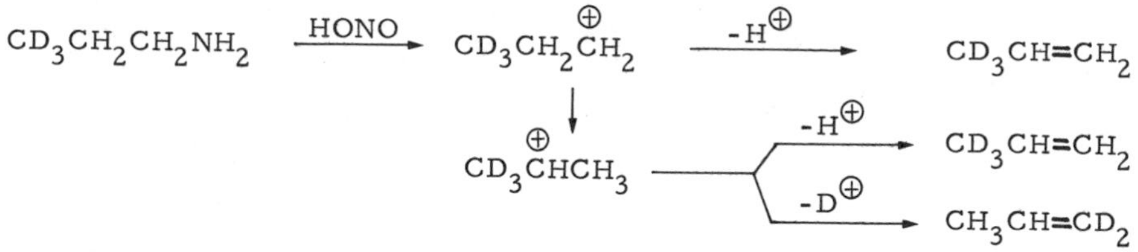

$$CD_3CH_2CH_2NH_2 \xrightarrow{HONO} CD_3CH_2\overset{\oplus}{C}H_2 \xrightarrow{-H^{\oplus}} CD_3CH=CH_2$$

$$CD_3\overset{\oplus}{C}HCH_3 \xrightarrow{-H^{\oplus}} CD_3CH=CH_2$$

$$\xrightarrow{-D^{\oplus}} CH_3CH=CD_2$$

Ignoring kinetic isotope effects, the ratio of [propyl cation]/[isopropyl cation] will be equal to $\left([CD_3CH=CH_2] - [CH_3CH=CD_2] / [CH_3CH=CD_2] \right)$.

23-33   The S_N1 mechanism (A) is:

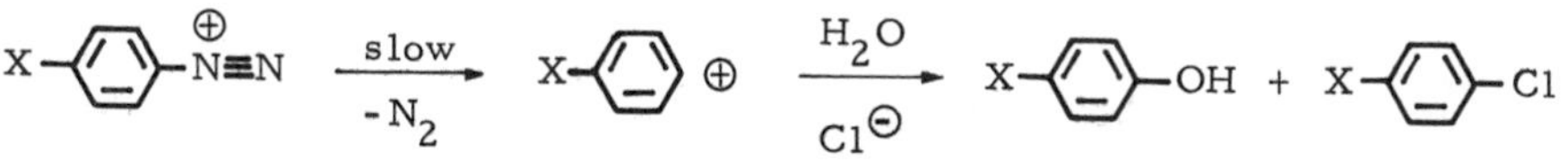

$$X-\langle\bigcirc\rangle-\overset{\oplus}{N}\equiv N \xrightarrow[-N_2]{slow} X-\langle\bigcirc\rangle^{\oplus} \xrightarrow[Cl^{\ominus}]{H_2O} X-\langle\bigcirc\rangle-OH + X-\langle\bigcirc\rangle-Cl$$

The elimination-addition (benzyne) mechanism (B) is:

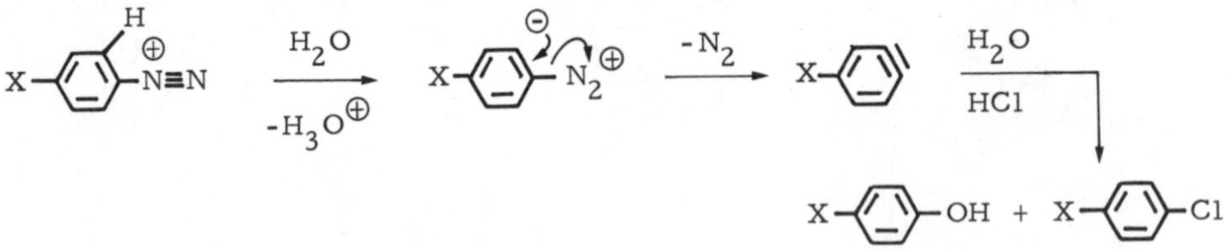

The addition-elimination mechanism (C) is:

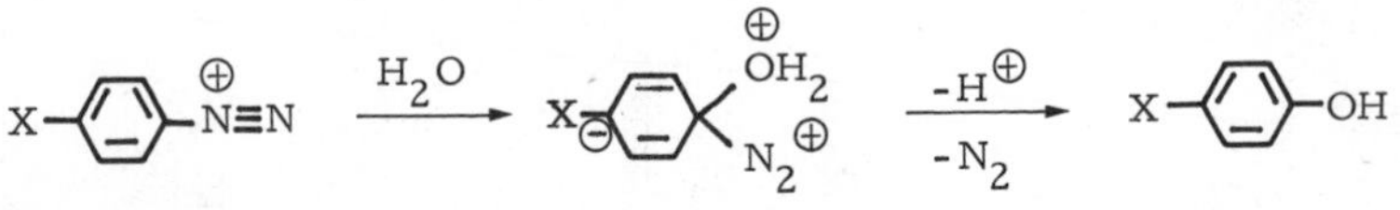

The product composition but not the rate of reaction will change with increasing $Cl^\ominus$ by mechanisms A and B. Rearrangement products and C–D bonds are expected to form by mechanism B but not A. Substituents such as CH_3O are expected to increase the rate of ionization by mechanism A, decrease rate of substitution by mechanism C. Likewise NO_2 groups should enhance rate by C but not by A. Also the rate should be independent of the nucleophile concentration (H_2O) by mechanism A but not by C. Overall, the evidence is consistent with mechanism A only.

23-34

a.

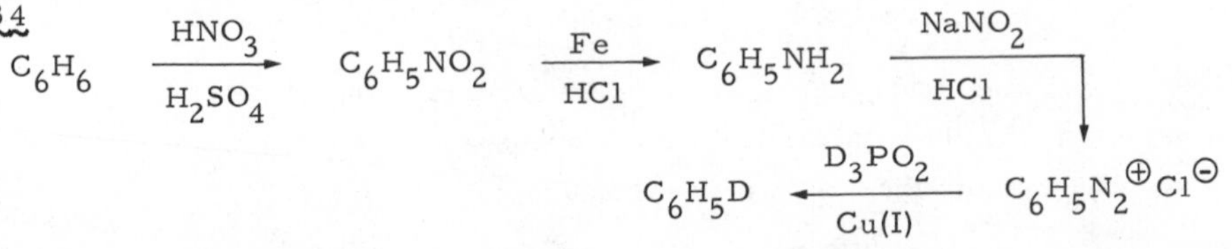

b.

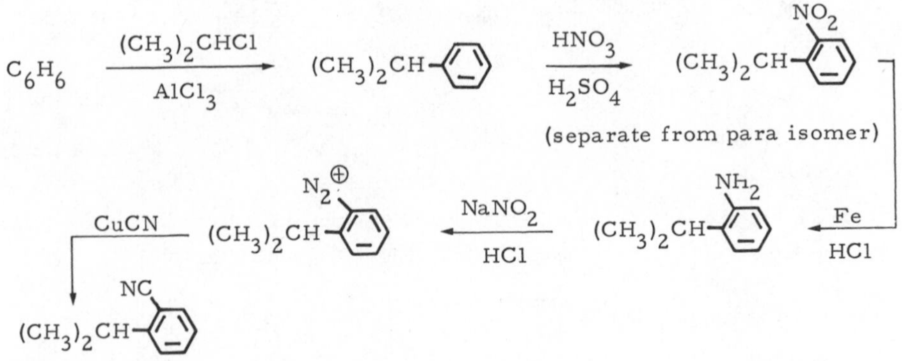

c. C_6H_6 $\xrightarrow[\text{AlCl}_3]{(CH_3)_3CCl}$ $(CH_3)_3C-C_6H_5$ $\xrightarrow[\text{H}_2SO_4]{HNO_3}$ $(CH_3)_3C-\bigcirc-NO_2$

$(CH_3)_3C-\bigcirc-OH$ $\xleftarrow[\text{3. H}_2O \quad \text{2. NaNO}_2, \text{ HCl} \quad \text{1. Fe, HCl}]{}$

d. C_6H_6 $\xrightarrow[\text{AlCl}_3]{CH_3COCl}$ $C_6H_5COCH_3$ $\xrightarrow[\text{H}_2SO_4]{HNO_3}$ $CH_3CO-\bigcirc$ NO_2

(The NO_2 group cannot be introduced first.)

$CH_3CO-\bigcirc$ $\xleftarrow[\text{3. H}_2O \quad \text{2. NaNO}_2, \text{ HCl} \quad \text{1. Fe, HCl}]{}$
OH

e. C_6H_6 $\xrightarrow[\text{H}_2SO_4]{HNO_3}$ $C_6H_5NO_2$ $\xrightarrow[\text{FeCl}_3]{Cl_2}$ $\bigcirc-NO_2$ or,
Cl

$C_6H_5NO_2$ $\xrightarrow[\text{H}_2SO_4]{HNO_3}$ $\underset{NO_2}{\overset{NO_2}{\bigcirc}}$ $\xrightarrow{(NH_4)_2S_x}$ $\underset{NO_2}{\overset{NH_2}{\bigcirc}}$ $\xrightarrow[\text{HCl}]{\text{1. NaNO}_2}$

$\underset{NO_2}{\overset{Cl}{\bigcirc}}$ $\xleftarrow{\text{2. CuCl}}$

f. C_6H_6 $\xrightarrow{Cl_2}$ C_6H_5Cl $\xrightarrow{Mg}$ C_6H_5MgCl $\xrightarrow{CO_2}$ $\xrightarrow{H^\oplus}$ $C_6H_5CO_2H$

$\underset{HO_2C}{\bigcirc-I}$ $\xleftarrow{I^\ominus}$ $\xleftarrow[\text{HCl}]{\text{NaNO}_2}$ $\xleftarrow[\text{HCl}]{Fe}$ $\underset{HO_2C}{\bigcirc-NO_2}$ $\xleftarrow[\text{H}_2SO_4]{HNO_3}$

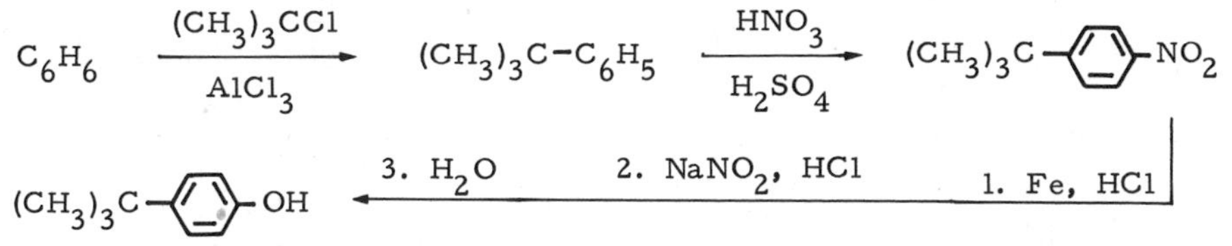

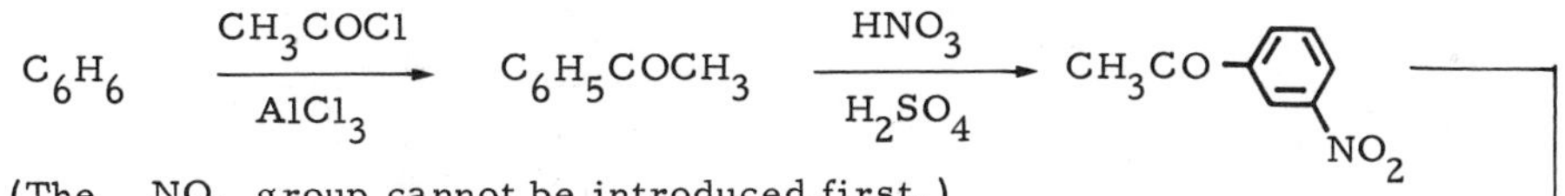

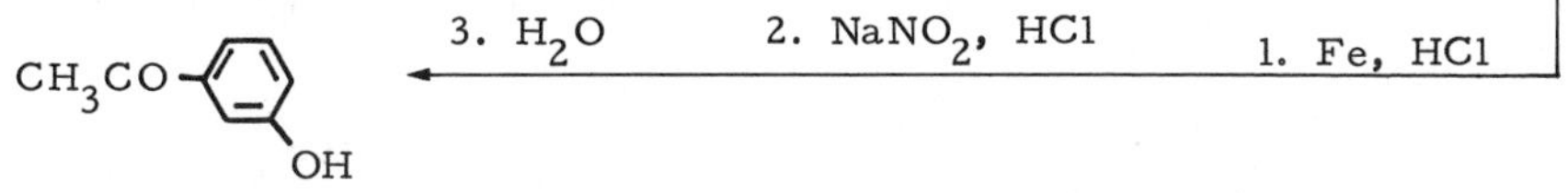

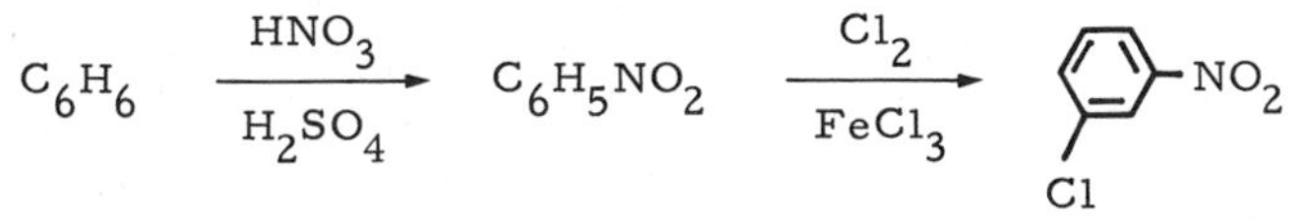

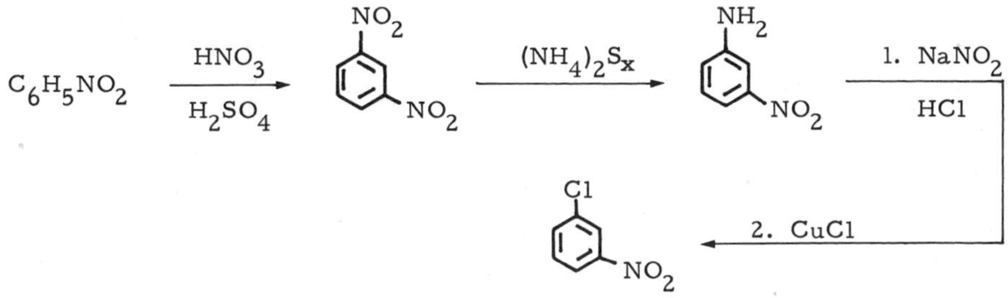

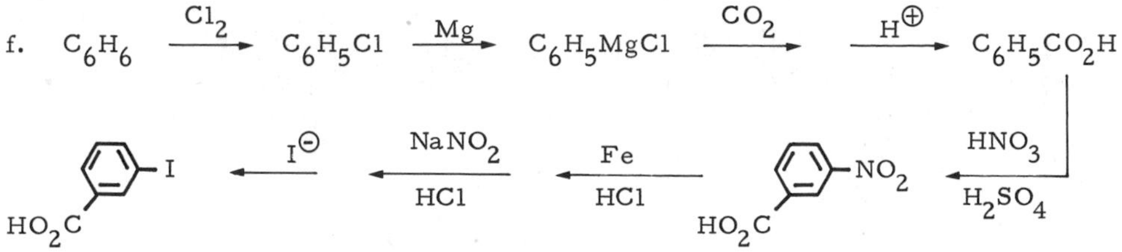

g. 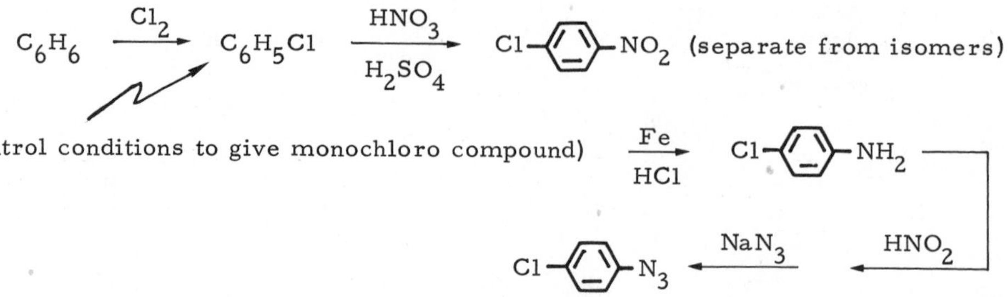 C_6H_6 $\xrightarrow{Cl_2}$ C_6H_5Cl $\xrightarrow[H_2SO_4]{HNO_3}$ Cl—⬡—NO_2 (separate from isomers)

(control conditions to give monochloro compound) $\xrightarrow[HCl]{Fe}$ Cl—⬡—NH_2

Cl—⬡—N_3 $\xleftarrow{NaN_3}$ $\xleftarrow{HNO_2}$

h. C_6H_6 $\xrightarrow[BF_3]{CH_3OH}$ CH_3—C_6H_5 $\xrightarrow[H_2SO_4]{HNO_3}$ CH_3—⬡—NO_2 (separate from

isomers) $\xrightarrow[HCl]{Fe}$ $\xrightarrow[HCl]{NaNO_2}$ CH_3—⬡—$\overset{\oplus}{N_2}$ $\xrightarrow{Na_2SO_3}$ CH_3—⬡—$NHNH_2$

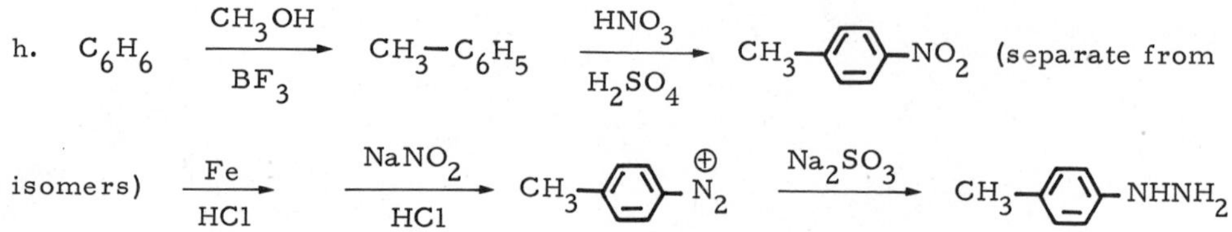

i. 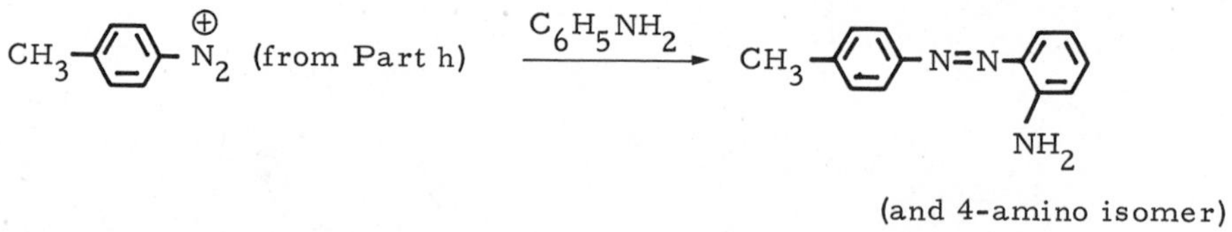 CH_3—⬡—$\overset{\oplus}{N_2}$ (from Part h) $\xrightarrow{C_6H_5NH_2}$ CH_3—⬡—N=N—⬡

NH$_2$

(and 4-amino isomer)

j. $C_6H_5NH_2$ (from Part g) $\xrightarrow[HCl]{NaNO_2}$ $C_6H_5\overset{\oplus}{N_2}\overset{\ominus}{Cl}$ $\xrightarrow[Cu(I)]{CH_2=CHCH_3}$

$C_6H_5CH_2CH(Cl)CH_3$

23-35

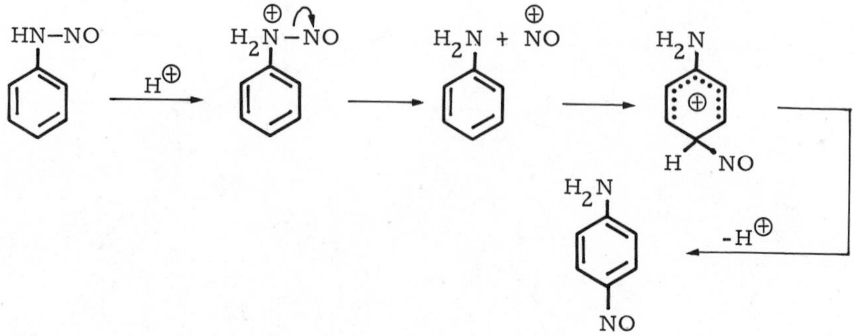

The N-chloro compound behaves similarly.

23-36   In a reaction where a mixture of AB and A'B' give rearrangement products BA and B'A' but no BA' or other crossed products, we may conclude that the rearrangement is intra rather than inter molecular.   The mechanism for the rearrangement of hydrazobenzene may be written as:

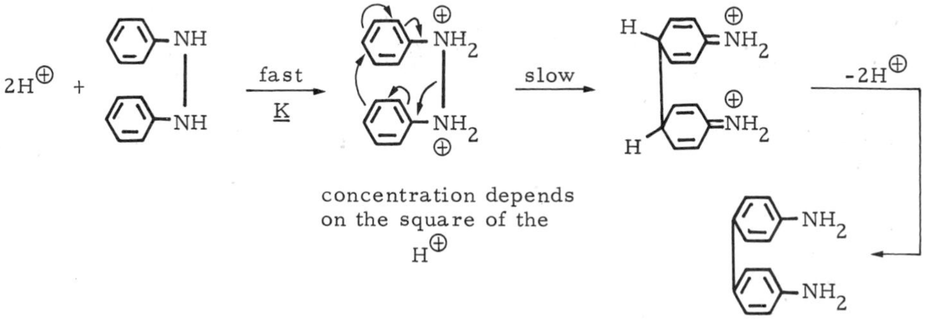
2H⊕ +
NH
NH
fast
K
NH2
NH2
slow
H
H
NH2
NH2
-2H⊕
concentration depends
on the square of the
H⊕
NH2
NH2

23-37

a. 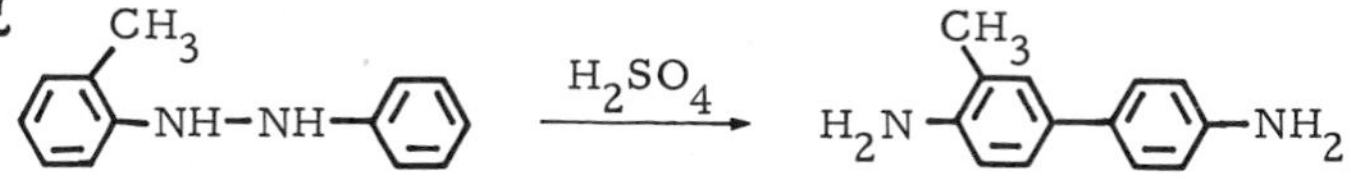
CH3
NH–NH
H2SO4
H2N
CH3
NH2

b. 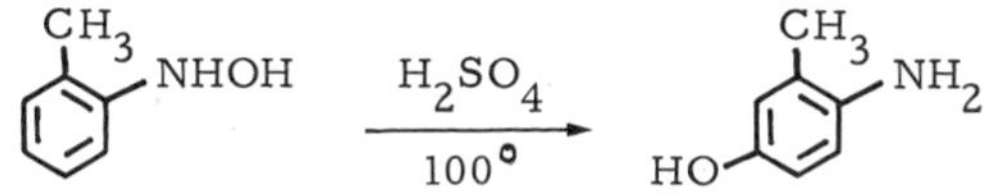
CH3
NHOH
H2SO4
100°
CH3
NH2
HO

c. 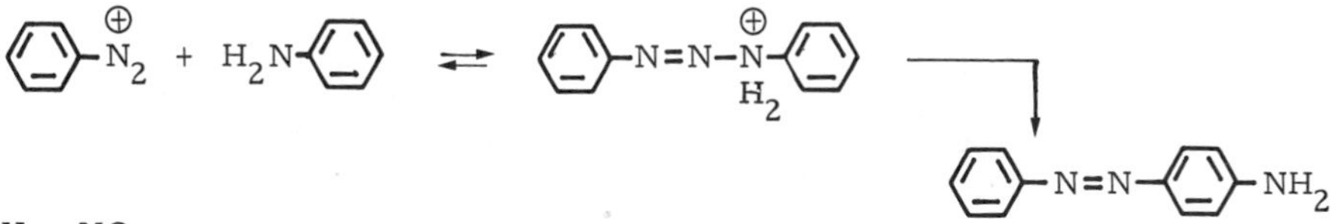
N2
+ H2N
N=N–N
H2
N=N
NH2

d. 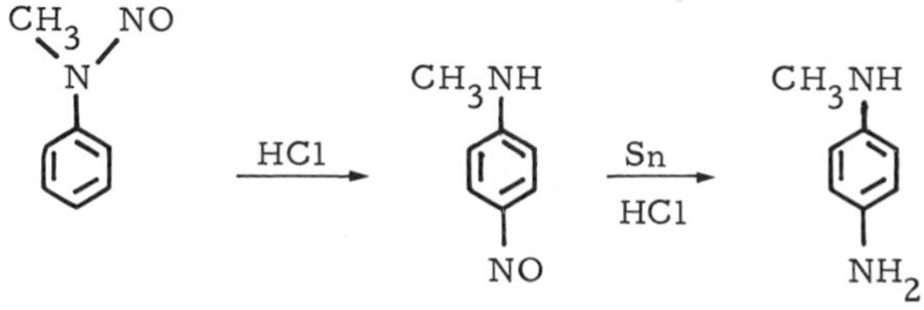
CH3
NO
N
HCl
CH3NH
NO
Sn
HCl
CH3NH
NH2

23-38 a. -3 b. +2 c. -1 d. $CH_3-\overset{0}{N}\equiv\overset{0}{N}:$ ($\oplus$ on the first N) e. -3

f. $CH_3-\overset{+1}{N}=\overset{-1}{N}-CH_3$ (with $\oplus$ on the first N bearing $O^{\ominus}$) g. $CH_3-\overset{-1}{N}=\overset{+1}{N}=\overset{-1}{N}:^{\ominus}$ ($\oplus$ on the middle N) h. +1 i. -1 j. -1

23-39 $CH_2\!\!=\!\!CHCH_2NH_2 + C_2H_5Br \xrightarrow{-HBr} CH_2\!\!=\!\!CHCH_2NHC_2H_5$ ⟶

⟶ $CH_2\!\!=\!\!CHCH_2\overset{\oplus}{\underset{\underset{O}{\ominus}}{N}}C_2H_5$ (CH_3) $\xleftarrow{H_2O_2}$ $CH_2\!\!=\!\!CHCH_2\underset{|}{\overset{CH_3}{N}}C_2H_5$ $\xleftarrow{CH_3I}{-HI}$

resolve with an optically active acid ⟶ optically active amine oxide
e.g. camphor-10-sulfonic acid
(see Sec. 19-3B)

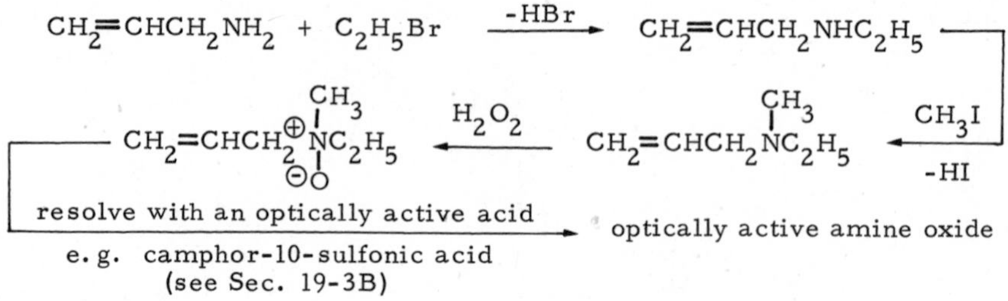

23-40

a. $CH_2\!\!=\!\!CHCH_2Br + KCN \xrightarrow[-KBr]{(CH_3)_2SO} CH_2\!\!=\!\!CHCH_2CN$ ⟶

$CH_2\!\!=\!\!CHCH_2CH_2NH_2 \xleftarrow[\text{ether}]{LiAlH_4 \text{ (excess)}}$

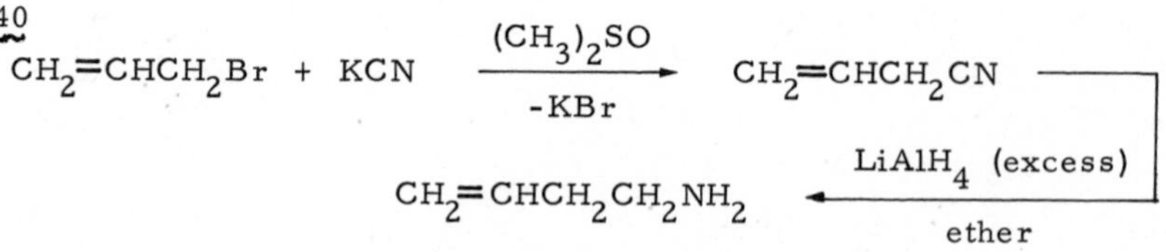

b.   cyclohexanone $\xrightarrow{H_2NOH}$ =NOH $\xrightarrow{LiAlH_4}$ —NH$_2$

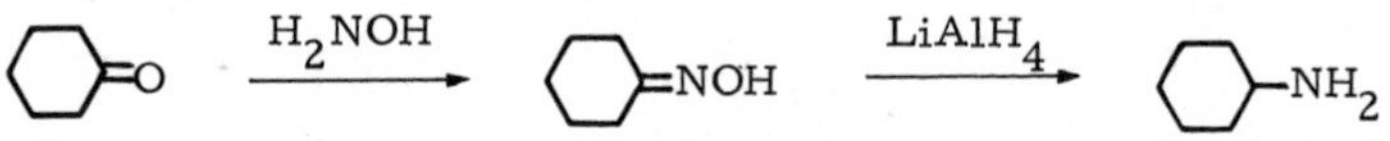

c.   Ⓐ—CO_2H $\xrightarrow{NH_3}$ Ⓐ—CO_2NH_4 $\xrightarrow[-H_2O]{heat}$ Ⓐ—$CONH_2$ $\xrightarrow{LiAlH_4}$

Ⓐ—$CO_2C_2H_5$ $\xleftarrow[H^{\oplus}]{C_2H_5OH}$ $\xrightarrow{NH_3}$ Ⓐ—CH_2NH_2

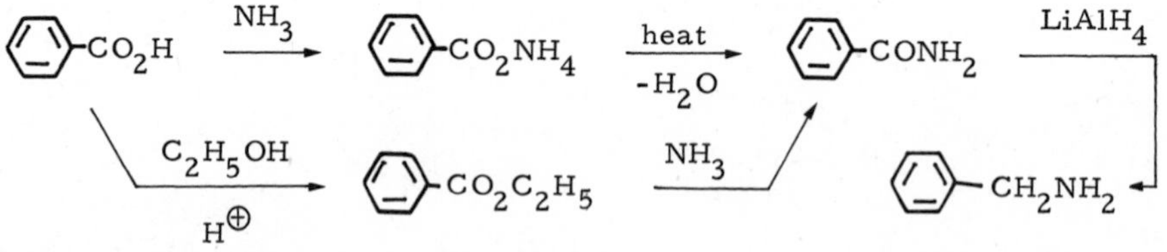

d.   Ⓐ—$CHO + CH_3NH_2 \xrightarrow{NaBH_3(CN)}$ Ⓐ—CH_2NHCH_3

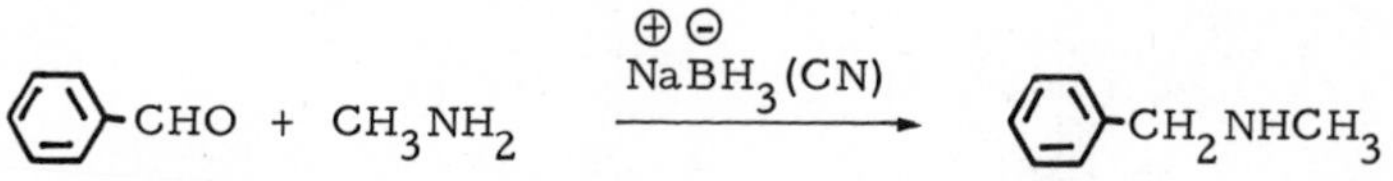

23-41   Working backwards we have:

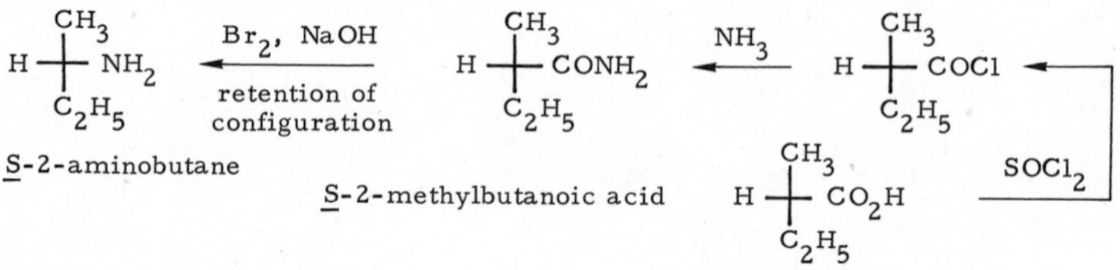

$\underset{\overset{|}{C_2H_5}}{\overset{\overset{CH_3}{|}}{H{-}\!\!-{-}NH_2}}$ $\xleftarrow[\text{retention of configuration}]{Br_2, \text{ NaOH}}$ $\underset{\overset{|}{C_2H_5}}{\overset{\overset{CH_3}{|}}{H{-}\!\!-{-}CONH_2}}$ $\xleftarrow{NH_3}$ $\underset{\overset{|}{C_2H_5}}{\overset{\overset{CH_3}{|}}{H{-}\!\!-{-}COCl}}$ ⟵

S-2-aminobutane

S-2-methylbutanoic acid $\underset{\overset{|}{C_2H_5}}{\overset{\overset{CH_3}{|}}{H{-}\!\!-{-}CO_2H}}$ $\xleftarrow{SOCl_2}$

23-42

a. $(CH_3)_2CHCH_2NH_2$ b. $C_6H_5CH_2NH_2$

c. $H_2N(CH_2)_4NH_2$ d. $C_6H_5\overset{\oplus}{N}H_3 \ \overset{\ominus}{H}SO_4 \xrightarrow{\overset{\ominus}{OH}} C_6H_5NH_2$

23-43   Cleavage of a C—N bond by a homolytic process will be easiest when the radical fragments are resonance stabilized. This circumstance arises when the C—N bond is linked to an aromatic ring as in $(C_6H_5)_3C-NR_2$. (See Section 26-4D.)

23-44

a. $(C_6H_5)_3CCl$ $\xrightleftharpoons{H_2O}$ $(C_6H_5)_3C^{\oplus} + Cl^{\ominus}$ $\xrightarrow{NH_3}$

$(C_6H_5)_3C-NH_2 + NH_4Cl$

b. $(C_6H_5)_3C-NH_2$ $\xrightarrow{CH_3CO_2H}$ $(C_6H_5)_3CNH_3^{\oplus}$ $\xrightarrow{CH_3CO_2H}$

$(C_6H_5)_3COH + H^{\oplus}$ $\xleftarrow{H_2O}$ $(C_6H_5)_3C^{\oplus} + NH_4^{\oplus} + 2CH_3CO_2^{\ominus}$

23-45   The amine function is protonated in strong acid and the ammonium group is meta-directing in electrophilic aromatic substitution. Also, the amine function is extensively oxidized with nitric acid. Protection of the amino group by acylation then permits nitration with ortho-para orientation.

23-46

a. $CH_3COCl + H_2NCH_2CH_2CH_2OH$ $\xrightarrow[-NaCl]{NaOH}$ $CH_3CONHCH_2CH_2CH_2OH$
(1 mole)

$H_2NCH_2CH_2CO_2H$ $\xleftarrow[\text{heat}]{H_3O^{\oplus}}$ $CH_3CONHCH_2CH_2CO_2H$ $\xleftarrow[H^{\oplus}]{CrO_3}$

b.

more reactive nitrogen

$\xrightarrow[\text{(1 mole)}]{CH_3COCl}$

$\xrightarrow[RCO_3H]{H_2O_2 \text{ or}}$

$\xleftarrow[\text{heat}]{H_3O^{\oplus}}$

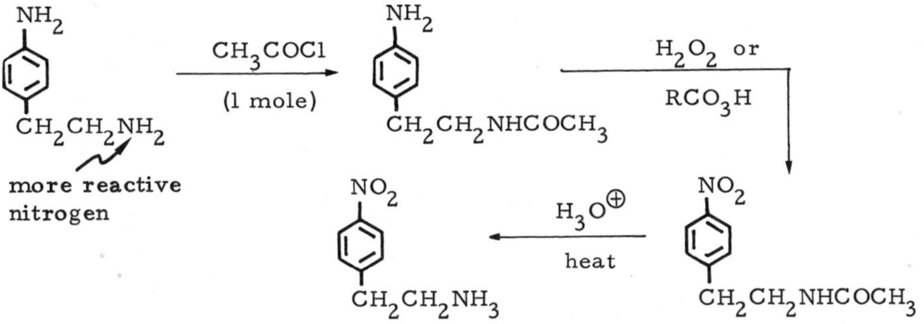

23-47

a. $(CH_3)_3CCO_2H$ $\xrightarrow[\text{2. } NH_3]{\text{1. } SOCl_2}$ $(CH_3)_3CCONH_2$ $\xrightarrow{LiAlH_4}$ $(CH_3)_3CCH_2NH_2$

b. $CH_2=CH-CH=CH_2$ $\xrightarrow{Br_2}$ $BrCH_2CH=CHCH_2Br$ (separate from 1,2-isomer)

$H_2NCH_2CH_2CH_2CH_2CH_2CH_2NH_2$ $\xleftarrow[Pt]{H_2}$ $NCCH_2CH=CHCH_2CN$ $\xleftarrow[ethanol]{KCN}$

23-48   The following answers are illustrative.

a. R_2NH or R_3N where R is a fairly large hydrocarbon group (e.g. butyl).

b. $(CH_3CH_2)4\overset{\oplus}{N}\ \overset{\ominus}{OH}$

c. $CH_3-\overset{\overset{O}{\|}}{C}-O-CH_2-\overset{\overset{NH_2}{|}}{\underset{\underset{H}{|}}{C}}-CH_2OH$

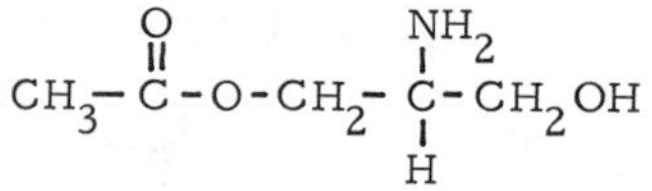

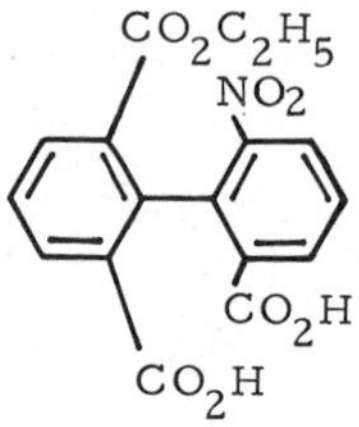

23-49   a. Compound B must be a primary amine to react with nitrous acid to give alcohols. Working backwards towards a chiral amine, the following path may be written.

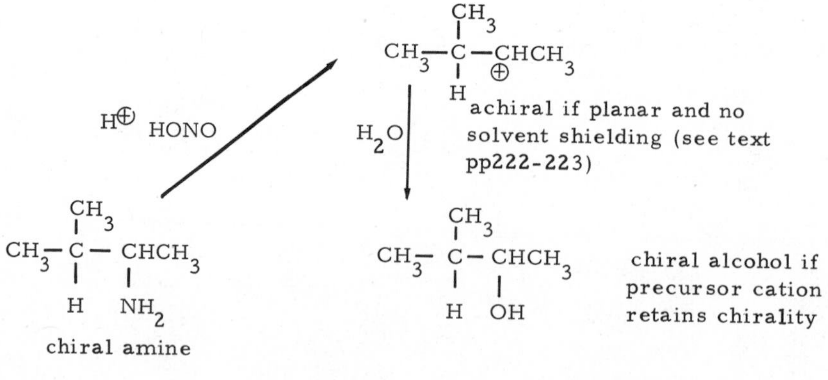

b. For compound B, $C_5H_{13}N$, to be formed from compound A, $C_5H_{11}O_2N$, with no loss in enantiomeric purity, compound A must be a nitro compound.

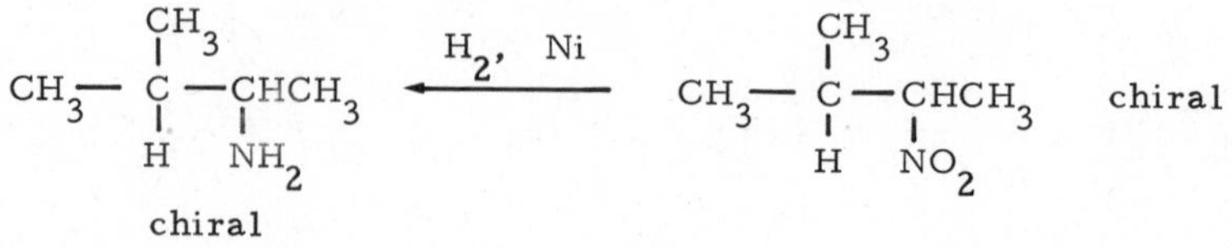

c. The fact that A is soluble in base is consistent with A being a nitro compound since such compounds are weak acids if they possess at least one α-hydrogen.

     d. Compound A racemizes in basic solution because salt formation destroys the asymmetry at the α-carbon.

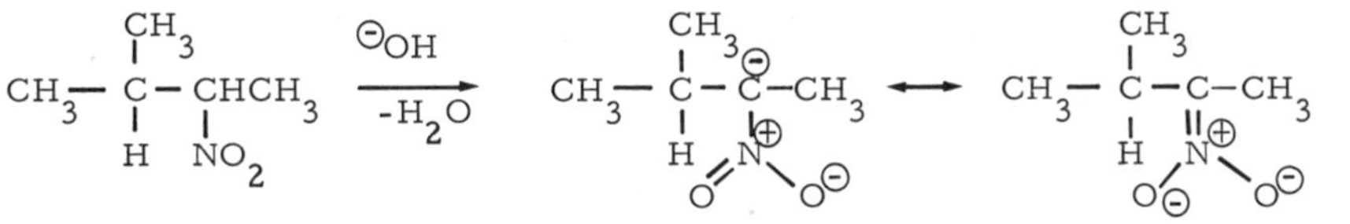

23-50   a. Inductive effect of methyl groups enhances base strength of amine; the effect of CF_3 groups decreases base strength.

     b. $C_6H_5CH_2NH_2$ is a stronger base than $p\text{-}CH_3\text{-}C_6H_4\text{-}NH_2$ because the electron pair on nitrogen is more localized.

     c. C_5H_5N is a stronger base than CH_3CN; that is, the pyridinium ion $C_5H_5NH^{\oplus}$ is a weaker acid than $CH_3C{=}NH^{\oplus}$ owing to the difference in the hybridization of nitrogen. Other things being the same for the N—H bond to be formed, the stronger the N compound as a base the smaller the s-character of the N—H σ-orbital.

     d. $H\text{-}C\overset{\displaystyle NH}{\underset{\displaystyle NH_2}{}}$ is stronger than $H\text{-}C\overset{\displaystyle O}{\underset{\displaystyle NH_2}{}}$ for the reason that protonation of the imine nitrogen leads to a resonance stabilized anion in which both nitrogens become equivalent.

     e. N-Methylazacyclopentane is a stronger base than N-methylaza-cyclopropane because, in the latter compound, adding a proton to nitrogen increases the ring strain more than for the former compound.

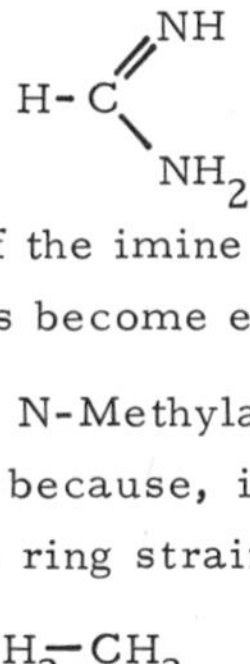

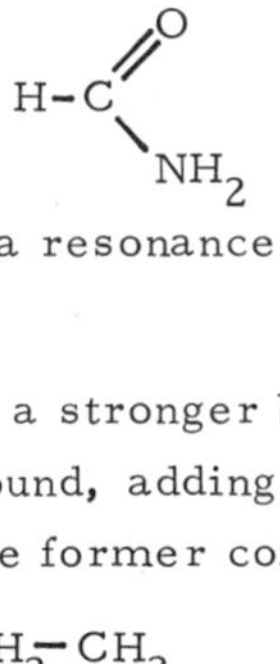

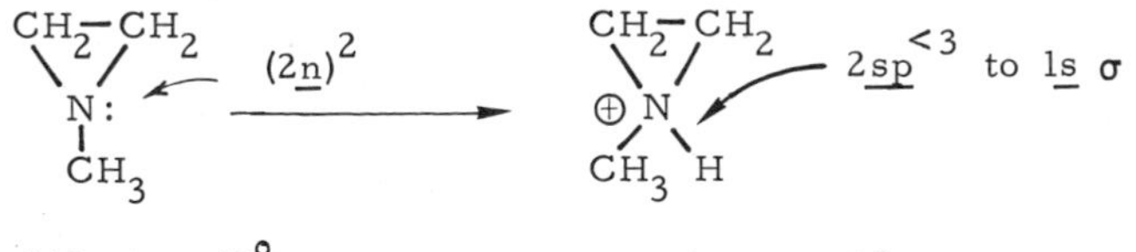

angle strain $107 - 60 = 47°$ $109.5 - 60 = 49.5°$

Remember that (to a crude approximation) angle strain depends on the square of the deviation from the normal values (see Exercise 12-2 and Exercise 23-6).

    Alternatively, because the ring C—N bonds of an azacyclopropane will have more $\underline{p}$ character than those in an azacyclopentane, we expect the N—H bond formed on protonation to have more $\underline{s}$ character, and hence the hydrogen in the conjugate acid will be easier to remove as a proton (see

Section ll-8B).

23-51

a. $(CH_3)_2CHCH_2CONH_2$ — $\xrightarrow{Br_2, \ NaOH}$

b. $(CH_3)_2CHCH_2CO_2H$ — $\xrightarrow{HN_3, \ H_2SO_4}$

c. $(CH_3)_2CHCH_2CON_3$ — $\xrightarrow{heat}$ → $(CH_3)_2CHCH_2NH_2$

d. $(CH_3)_2CHCH_2Br$ ————

e. $(CH_3)_2CH-C{\equiv}N$ $\xrightarrow{LiAlH_4}$

$\xrightarrow{H_2NNH_2}$

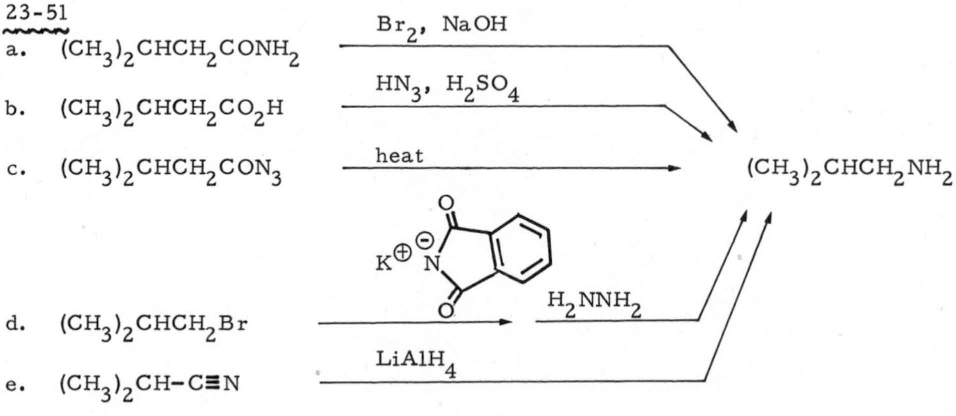

23-52   The following are representative:

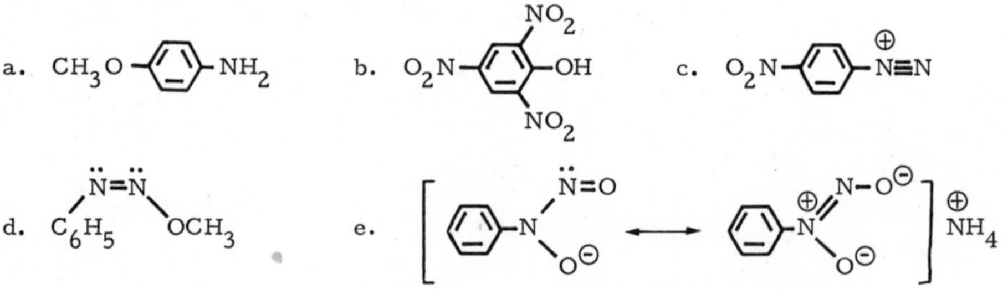

a. $CH_3O-\langle\rangle-NH_2$

b. $O_2N-\langle\rangle-OH$ with NO_2 groups

c. $O_2N-\langle\rangle-\overset{\oplus}{N}{\equiv}N$

d.

e.

23-53 $Cl-\langle\rangle-NH_2$ $\xrightarrow[HBr]{NaNO_2}$ $Cl-\langle\rangle-N_2^{\oplus} \ Br^{\ominus}$ ————

$Br-\langle\rangle-N{=}N-\langle\rangle-N(CH_3)_2$ ←———— $\langle\rangle-N(CH_3)_2$ $Br-\langle\rangle-N_2^{\oplus} \ Cl^{\ominus}$

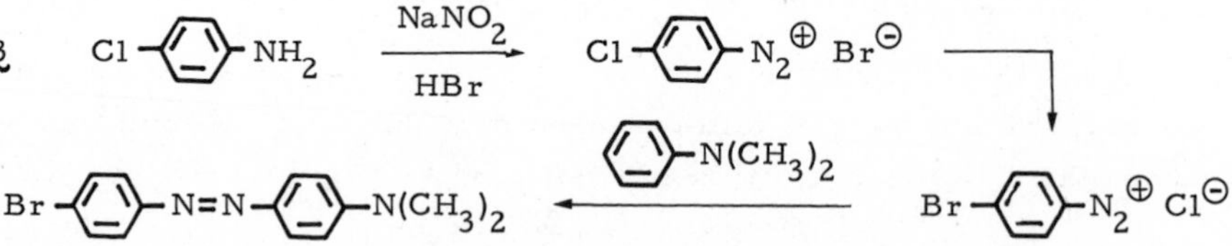

The effect of the $N_2^{\oplus}$ group activates halogen to nucleophilic substitution.

23-54 $Cu(II) + H-P(OH)_2 \longrightarrow Cu(I) + \cdot P(OH)_2 + H^{\oplus}$

Propagation $ArN_2^{\oplus} + \cdot P(OH)_2 \longrightarrow ArN_2\cdot + [\overset{\oplus}{P}(OH)_2]$

$ArN_2\cdot \longrightarrow Ar\cdot + N_2$

$Ar\cdot + H-P(OH)_2 \longrightarrow ArH + \cdot P(OH)_2$

$[\overset{\oplus}{P}(OH)_2] + H_2O \longrightarrow P(OH)_3 + H^{\oplus}$

23-55

a.

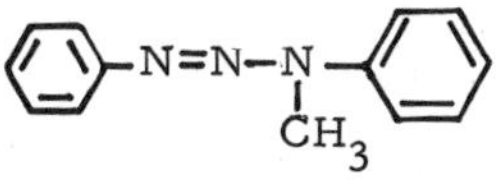

b.

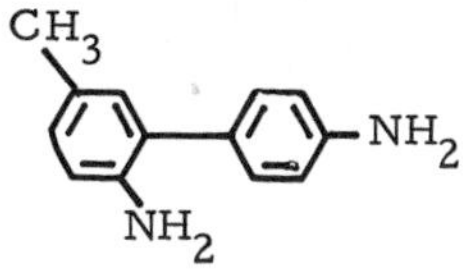

c.

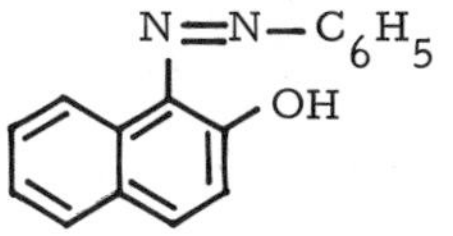

but not

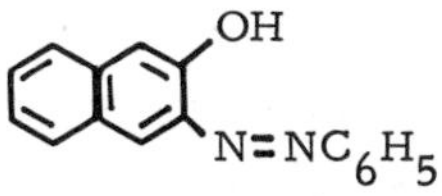

or

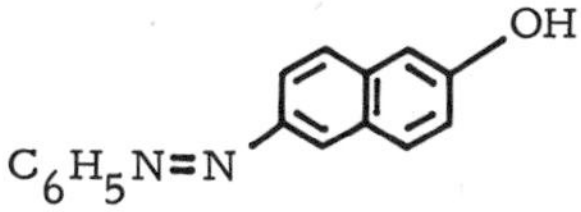

because these involve less favorable intermediates.   Thus

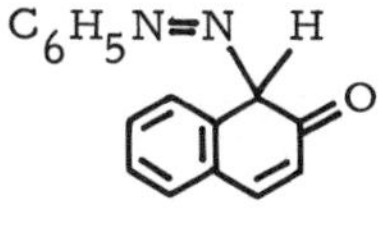

is more favorable than 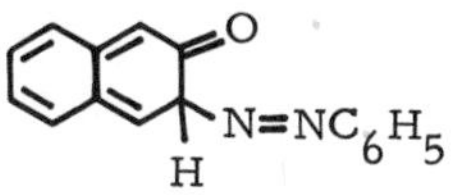   or

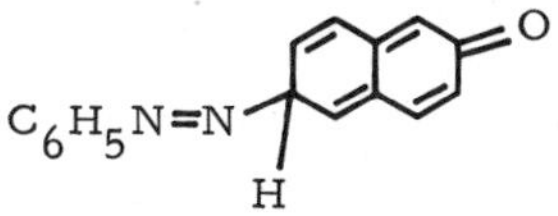

because it has an intact benzene ring with three

double bonds.

d.

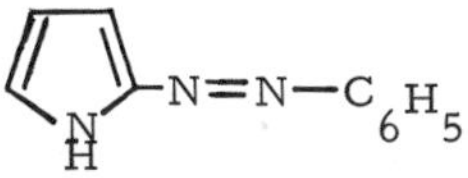

23-56   The electron pair on nitrogen is delocalized over _three_ phenyl rings in triphenylamine compared to _one_ ring in benzenamine. This extensive delocalization accounts for the low basicity and the long wavelength absorption in the electronic spectrum of triphenylamine.

N-Phenylcarbazole would be expected to be a weaker base than triphenylamine because: 1) the planarity of the carbazole ring system enhances delocalization of the nitrogen electrons even more than in triphenylamine, in which the rings are non-coplanar; and 2) because the five-membered central ring of carbazole has _six_ π electrons counting the unshared pair on nitrogen. Protonation of the unshared pair removes it from the π electron system.

CHAPTER 24

ORGANONITROGEN COMPOUNDS.

II. AMIDES, NITRILES, NITRO COMPOUNDS

Amides are "acylated" amines, either $R'\overset{O}{\overset{\|}{C}}NH_2$, $R'\overset{O}{\overset{\|}{C}}NHR$, or $R'\overset{O}{\overset{\|}{C}}NR_2$. The $-\overset{O}{\overset{\|}{C}}-$ group greatly modifies the reactivity of the nitrogen. The bonds to the amide nitrogen approach coplanarity (i.e., $\underline{sp}^2$ hybridization) to permit the maximum electron delocalization:

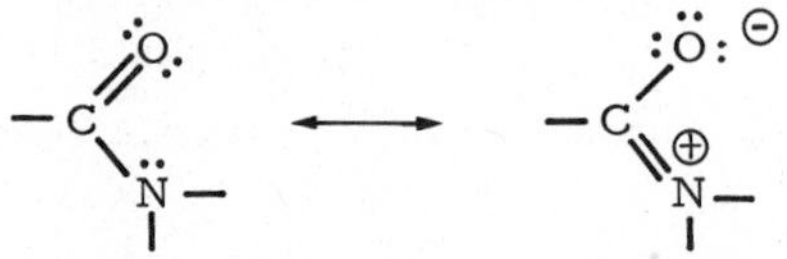

Thus amides are polar compounds with large dipole moments (3.8 D), high melting points and water solubility, and are associated extensively through hydrogen bonding:

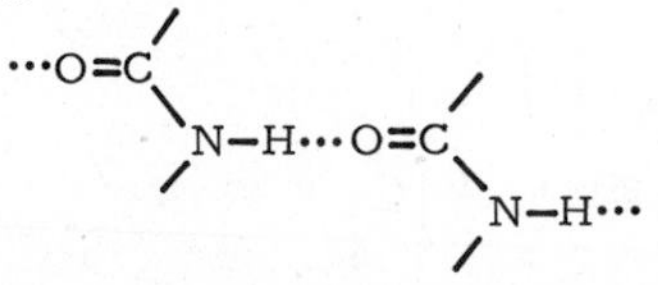

The infrared spectra of amides reflect hydrogen bonding from the position and intensity of the N–H stretch (3400 - 3500 cm^{-1}). The C=O stretching vibration absorbs at lower frequencies than most carbonyl compounds (1690 cm^{-1}). The nmr spectra of amides provide strong evidence that rotation about the C–N bond is restricted. Thus different chemical shifts are observed for similar groups (R) on nitrogen, thereby showing that they are distinguishable:

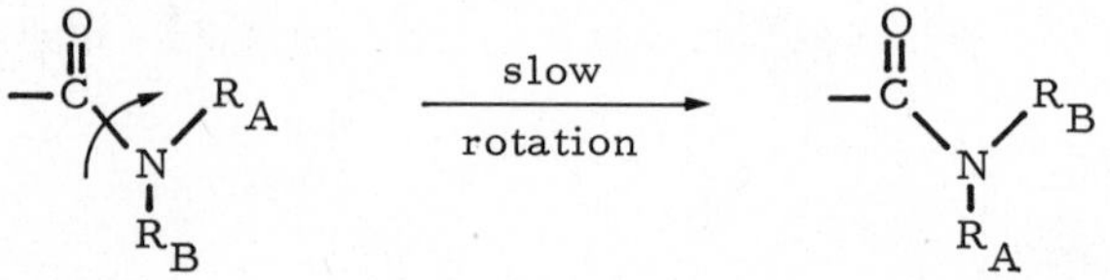

The resonance of N–H protons of primary and secondary amides are unusually broad and at much lower fields (~ 7 ppm) than amine N–H protons (~ 1 ppm). Because of resonance, the lone pair of electrons on the amide nitrogen is less available for bonding than in amines and, accordingly, amides are less basic and less nucleophilic. However, they are stronger acids than amines, again because of resonance:

$$R-\overset{\overset{\displaystyle O}{\|}}{C}-NH_2 \quad \xrightarrow{-H^{\oplus}} \quad R-\overset{\overset{\displaystyle O}{\|}}{C}-\overset{\ominus}{N}H \quad \longrightarrow \quad R-\overset{\overset{\displaystyle O^{\ominus}}{|}}{C}=NH$$

Reactions for synthesis of amides include:

1. acylation   (Sections 23-9A, 23-13C, 24-3A)

$$RCOCl + NH_3 \longrightarrow RCONH_2 + HCl$$

2. hydrolysis of nitriles   (Section 24-3B)

$$RC{\equiv}N + H_2O \text{ (or } H_2O_2) \xrightarrow{NaOH} RCONH_2$$

3. alkylation of nitriles (Ritter reaction)   (Section 24-3B)

$$ROH + R'CN \xrightarrow{H^{\oplus}} R'CONHR$$

4. Beckmann rearrangement of oximes   (Section 24-3C)

$$R_2C{=}N{-}OH \xrightarrow{H^{\oplus}} RCONHR$$

The more important reactions of amides include the following:

1. hydrolysis   (Section 24-4)

$$R'CONR_2 \xrightarrow{H^{\oplus}, \text{ or } \overset{\ominus}{O}H, \text{ or } H_2O_2} R'CO_2H + R_2NH$$

2. reduction   (Section 18-7C)

$$R'CONR_2 \xrightarrow[\text{or } H_2, \text{ Pt}]{LiAlH_4} R'CH_2NR_2$$

3. Hofmann degradation   (Section 23-12D)

$$RCONH_2 \xrightarrow{Cl_2 \text{ or } Br_2 \text{ in } NaOH} RNH_2 + CO_2$$

4. dehydration   (Section 24-5)

$$RCONH_2 \xrightarrow{P_2O_5} RCN$$

5.   nitrosation   (Section 24-4)

$$R'CONHR + HONO \longrightarrow R'CO\overset{\overset{\displaystyle NO}{|}}{N}\!-\!R + H_2O$$

General methods of <u>synthesis of nitriles</u> follow   (Section 24-5):

1.   alkylation (S_N2)

$$RX + NaCN \longrightarrow RCN + NaX$$

$$RCH_2CN \xrightarrow[\text{2. RX}]{\text{1. NaNH}_2} R_2CHCN$$

2.   arylation

$$ArNH_2 \xrightarrow{HONO} ArN_2^{\oplus} \xrightarrow{CuCN} ArCN$$

3.   carbonyl addition

$$R_2C{=}O \xrightarrow{HCN} R_2C{<}\overset{OH}{\underset{CN}{}}$$

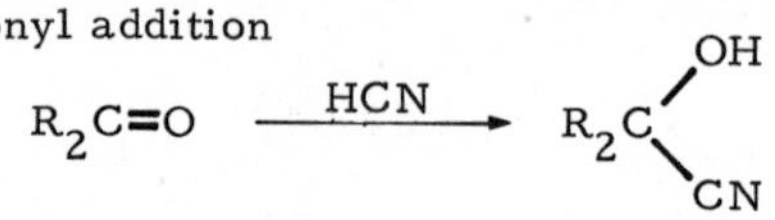

4.   Michael addition

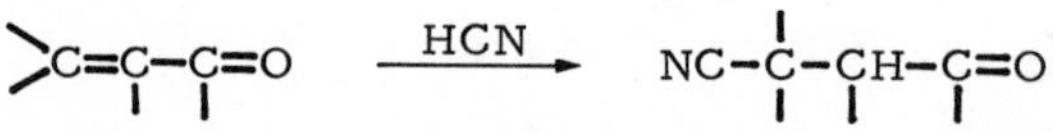

$$\overset{}{>}C{=}\underset{|}{C}{-}\underset{|}{C}{=}O \xrightarrow{HCN} NC{-}\underset{|}{\overset{|}{C}}{-}\underset{|}{CH}{-}\underset{|}{C}{=}O$$

5.   dehydration

$$\underset{H}{\overset{R}{>}}C{=}NOH \xrightarrow{-H_2O} RC{\equiv}N$$

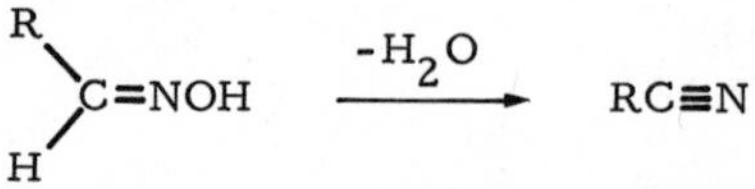

$$R{-}CONH_2 \xrightarrow{-H_2O} RC{\equiv}N$$

Nitriles are converted to amides and thence to carboxylic acids by hydrolysis.

Alkanenitriles that have <u>alpha</u> hydrogens can be alkylated under basic conditions (Section 24-5).

<u>Nitro compounds</u> are polar compounds that may be synthesized by the following reactions:

1.   from alkanes   (Section 4-6)

$$RH \xrightarrow{HNO_3} RNO_2$$

2.   from arenes   (Sections 22-4C, 24-7B)

$$ArH \xrightarrow{HNO_3} ArNO_2$$

3. from alkyl halides   (Section 23-21)

$$RX + NaNO_2 \longrightarrow RNO_2 + NaX$$

4. from amines   (Sections 23-10B and 23-11)

$$ArNH_2 \longrightarrow ArN_2^{\oplus} \longrightarrow ArNO_2$$

$$RNH_2 \xrightarrow{[O]} RNO_2$$

Polynitro compounds have wide use as high explosives. Polynitro-arenes form charge-transfer complexes with arenes that have electron-donating (alkyl) substituents.

Reduction of nitro compounds is achieved easily but the actual product obtained depends critically on the nature of the reducing agent and the pH of the medium (see Section 24-6C).

Nitroalkanes with alpha hydrogens rearrange to aci-nitro compounds in basic solution. The aci form is more acidic than the nitroalkane. The conjugate base can react by aldol and Michael addition to C=O and C=C, respectively:

aci-nitro compound

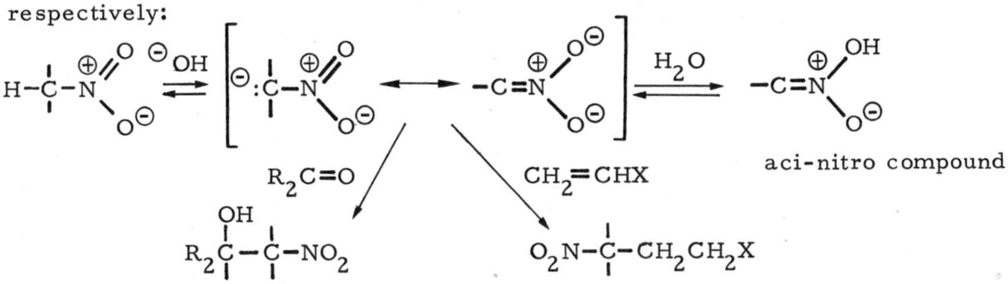

The most important organic nitrogen compounds containing N—N bonds are hydrazines, R_2N-NR_2; azo compounds, $RN=NR$; diazo compounds, $R_2C-N≡N$, and azides, $RN=N=N$. The chemistry of all compounds of this group is dominated by reactions which lead to dinitrogen, N_2. Typical examples are:

$$CH_3N=NCH_3 \xrightarrow{\text{heat}} CH_3-CH_3 + N_2$$

$$CH_2=N=N + CH_3CO_2H \longrightarrow CH_3CO_2CH_3 + N_2$$

$$CH_2=N=N + CH_3COCl \longrightarrow CH_3COCHN_2 + N_2 + CH_3Cl$$

$$C_6H_5CON_3 \longrightarrow C_6H_5NCO + N_2$$

ANSWERS TO EXERCISES

__24-1__   The stability associated with amides through electron delocalization of the type $-\overset{..}{\underset{|}{N}}-\underset{|}{C}=O \longleftrightarrow -\overset{\oplus}{\underset{|}{N}}=\underset{|}{C}-\overset{\ominus}{O}$ cannot be realized in the bicyclic amide because the bonds to the nitrogen and oxygen atoms are not coplanar.   There is essentially no overlap of the carbonyl π orbital with the lone-pair orbital on nitrogen.

__24-2__   a. (i) propenamide (acrylamide is the common name).   (ii) __N__-cyclo-hexyl-__N__-methylethanamide   (iii) __N__-phenylethanamide

b.   (i) $\text{C}_6\text{H}_5-\overset{\overset{\text{O}}{\|}}{\text{C}}-\text{NHCH}_2\text{CH}_3$ (ii) $(\text{CH}_3)_2\text{CH}\overset{\overset{\text{O}}{\|}}{\text{C}}\text{N}-\text{C}_6\text{H}_{11}$

__24-3__   Compound $\text{C}_5\text{H}_9\text{NO}_3$:   The infrared spectrum has two intense carbonyl bands around 1740 cm^{-1} and 1680 cm^{-1} indicative of carboxylate and amide carbonyl groups, respectively.   The two peaks at $3300 - 3400 \text{ cm}^{-1}$ would suggest a free and hydrogen-bonded $-\text{CO}-\underset{|}{\text{N}}-\text{H}$ grouping.   The nmr spectrum shows four types of hydrogen in the ratio $1:2:3:3$ with chemical shifts consistent with $\text{CO}\underset{|}{\text{N}}\text{H}$, $\text{CO}\underset{|}{\text{N}}-\text{CH}_2^-$, OCH_3 and CH_3CO groups.   A structure which fits with the spectral data is that of methyl ethanamidoethano-ate (methyl acetamidoacetate or methyl acetylglycine).

$$\text{CH}_3\text{CONHCH}_2\text{CO}_2\text{CH}_3$$

The CH_2 resonance is split by the NH proton.   The NH resonance is too broadened to show the triplet structure expected from splitting by the CH_2.

Compound $\text{C}_{10}\text{H}_{13}\text{NO}$:   The infrared spectrum has an amide-type carbonyl band near 1680 cm^{-1} and two bands in the region $3300 - 3400 \text{ cm}^{-1}$ due to a CONH– grouping.   The nmr spectrum has five kinds of hydrogen in the ratio $1:5:2:2:3$ with chemical shifts indicative of amide, phenyl, methylene and COCH_3 hydrogens.   A suitable structure follows:

$$\text{C}_6\text{H}_5\text{CH}_2\text{CH}_2\text{NHCOCH}_3, \quad \underline{\text{N}}\text{-(2-phenylethyl)ethanamide}$$

Resonance due to the methylene hydrogens is complicated by spin-spin splitting by protons on adjacent carbon and nitrogen atoms.

__24-4__   As an even-mass fragment ion, __m/e__ 44 retains the nitrogen from the parent amide.   A likely molecular formula is CH_2NO, which corresponds

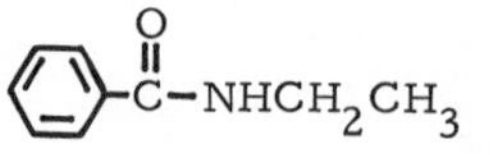

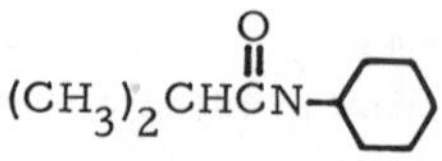

to $\overset{\oplus}{O}C-NH_2$ formed by α -cleavage of the molecular ion.

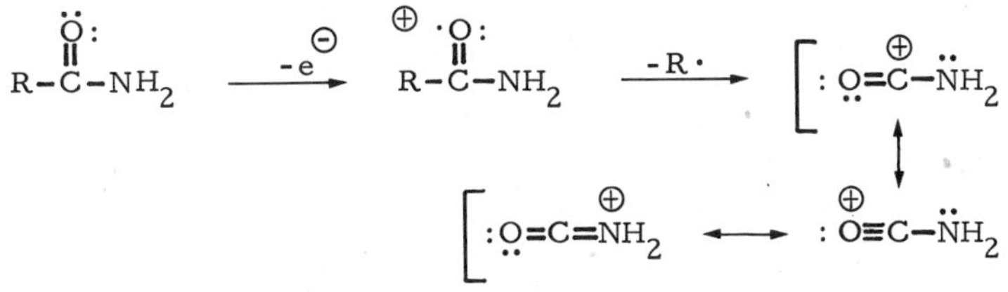

<u>24-5</u>   a.   Rotation is restricted around the C—N bond of amides because the bond has some double-bond character. Therefore, there are two isomers of <u>N</u>-methylmethanamide which have different nitrogen chemical shifts. The <u>E</u> isomer is expected to be the more stable and give the stronger nitrogen (^{15}N) signal.

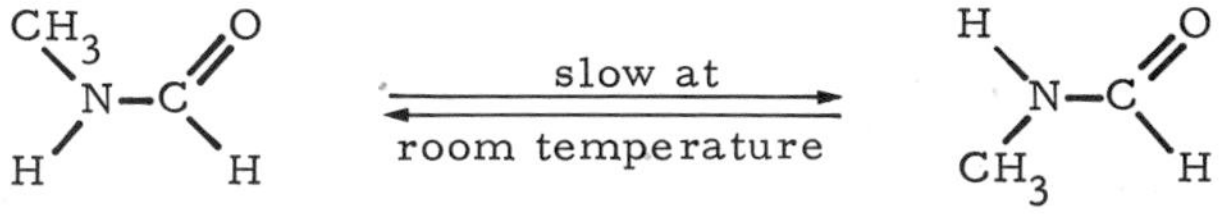

<u>Z</u> isomer	<u>E</u> isomer
	(preferred)

At $150°$, rotation around the C—N bond would be faster than the speed at which the separate nitrogen resonances can be detected - in which case only an average resonance will be observed.

b.   When the ring size is 5, 6, 7 or 8, only one stable amide con-figuration is possible (or dominant). That is the cis configuration. As the ring size increases beyond 8, the trans configuration becomes possible and dominates in rings 10 and 11. The intermediate situation is evident in the 9-membered ring in which both the <u>cis</u> and <u>trans</u> configurations are evident.

<u>cis</u>		<u>trans</u>	
<u>n</u> = 5 - 8		<u>n</u> = 10, 11, etc.	

<u>24-6</u>   If protonation of amides occurs on nitrogen, electron delocalization of the amide structure is no longer possible; free rotation about the C—N bond should then occur. In the case of <u>N</u>, <u>N</u>-dimethylmethanamide, this would make the <u>N</u>-methyl groups become magnetically equivalent and the nmr spectrum

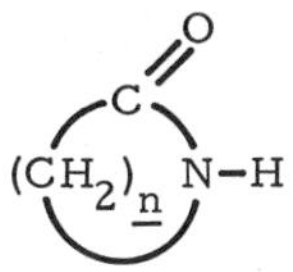

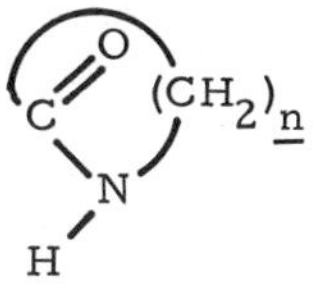

would then show a single resonance peak for the protons of both methyl groups.

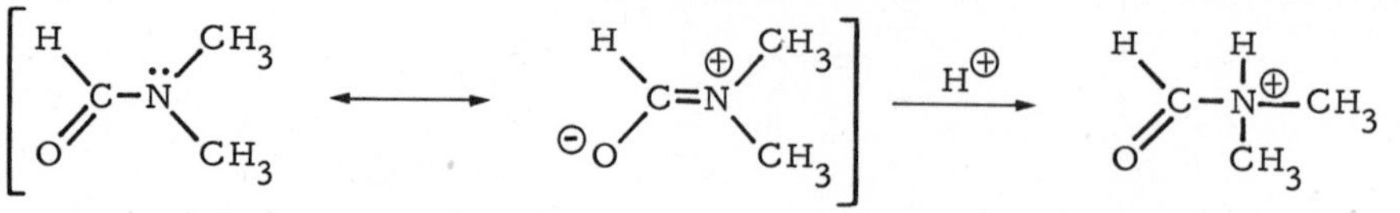

However, if protonation occurs on oxygen, electron delocalization is still possible, in fact even more favored, and the methyl groups are expected to remain nonequivalent at room temperature.

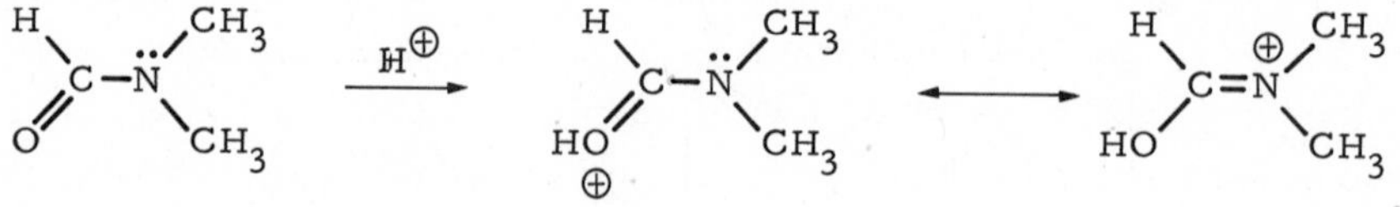

The latter circumstance is actually observed.

24-7

a.

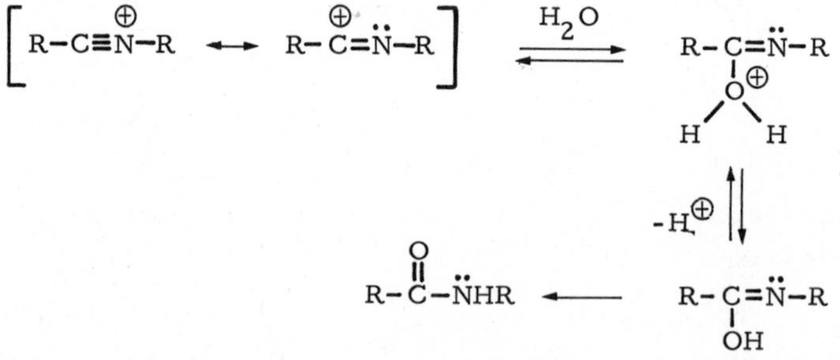

b. The following equilibria are possible:

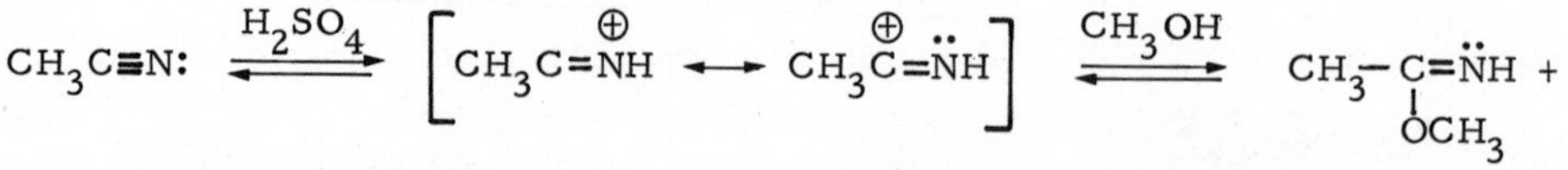

The final step to $\underline{N}$-methylethanamide, $CH_3\overset{O}{\overset{||}{C}}NHCH_3$, does not occur because enol ethers do not rearrange to carbonyl compounds as do enols

24-8

1. $HOOH + \overset{\ominus}{OH} \rightleftharpoons HOO^{\ominus} + H_2O$ $\underline{K}_A \sim 10^{-12}$

2. 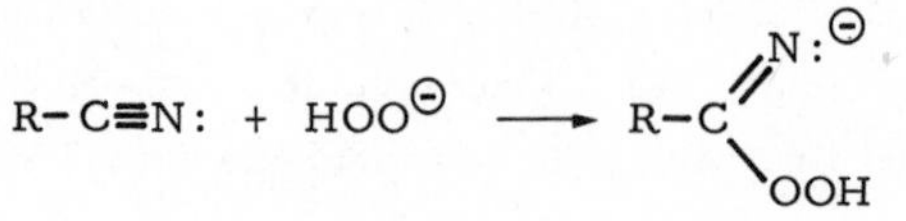 $R-C \equiv N: + HOO^{\ominus} \longrightarrow$

($HOO^{\ominus}$ is more nucleophilic than $HO^{\ominus}$.)

3. 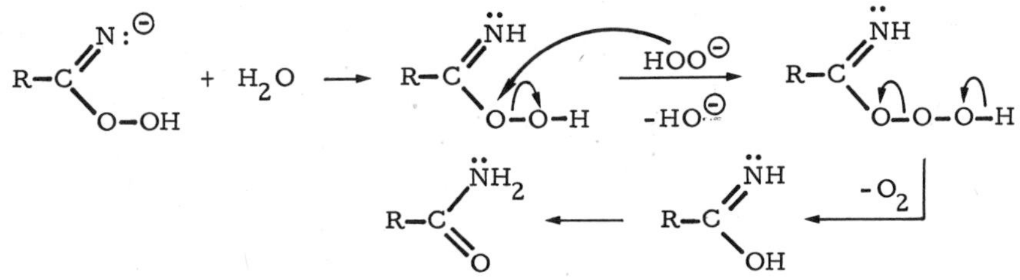 R–C(=N:⁻)(O–OH) + H₂O ⟶ R–C(=NH)(O–O–H) →[HOO⁻ / –HO⁻] R–C(=NH)(O–O–O–H)

→[–O₂] R–C(=NH)(OH) ⟶ R–C(=NH₂)(=O)

The steps by which oxygen is **evolved are** not actually known. If step 2 is slow the rate equation is:

$$\nu = \underline{k}[H_2O_2][\overset{\ominus}{O}H][RCN]$$

Using $H_2{}^{18}O_2$, the ^{18}O label is predicted by the above mechanism to turn up as the carbonyl oxygen of the amide.

24-9 $RC(=O)(OCH_3)$ →[1. 2CH₃MgCl 2. NH₄⁺] $R–C(CH_3)(CH_3)–OH$ →[H₂SO₄ / HCN] $R–C(CH_3)(CH_3)–NHC(=O)(H)$

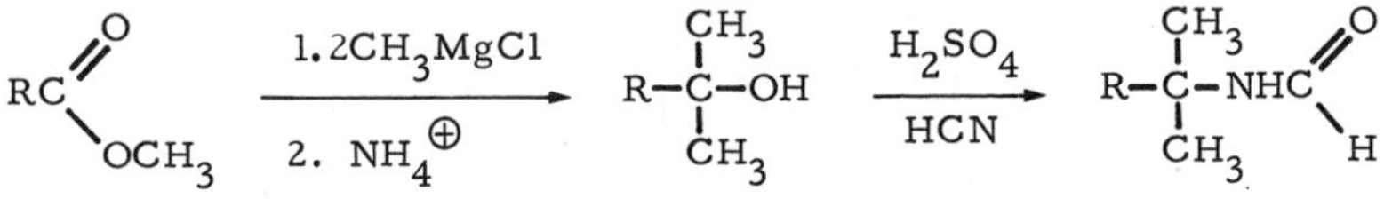

For R = $C_6H_5CH_2$, the product is N-(1,1-dimethyl-2-phenylethyl)methanamide.

24-10

a. C_6H_5NHCHO

b. 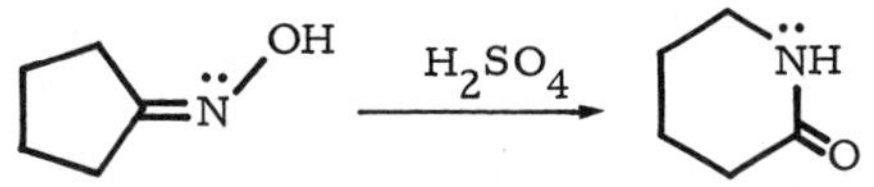   (cyclopentylidene=N–OH) →[H₂SO₄] (six-membered ring lactam, NH with C=O)

c. $CH_2{=}CH–C(=O)–NHCH_2C_6H_5$

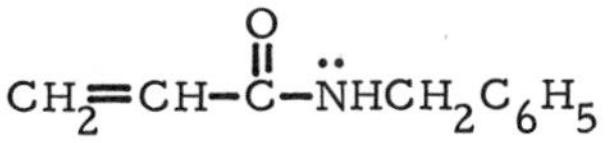

d. 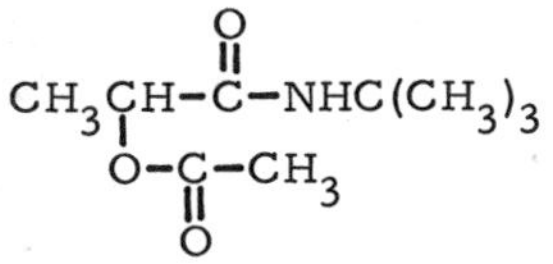 $CH_3CH(–C(=O)–NHC(CH_3)_3)(O–C(=O)–CH_3)$

24-11 RCH_2CN →[LiNH₂] $R\overset{\ominus}{C}HCN$

$RCH_2C{\equiv}N + R\overset{\ominus}{C}HCN \longrightarrow RCH_2\overset{:\overset{\ominus}{N}}{C}{-}CHR(CN) \longrightarrow RCH_2C(=\overset{\ddot N}{N}H)–\overset{\ominus}{C}R(CN) \longrightarrow$

→[H₂O, H⊕] $RCH_2C(=\overset{\ddot{H N}}{})–CHR(CN)$ →[H₂O] $RCH_2C(=O)–CHR(CN)$

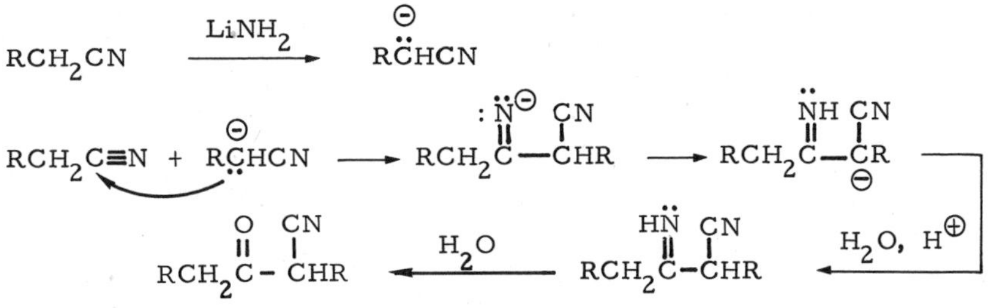

Synthesis of large-ring ketones:

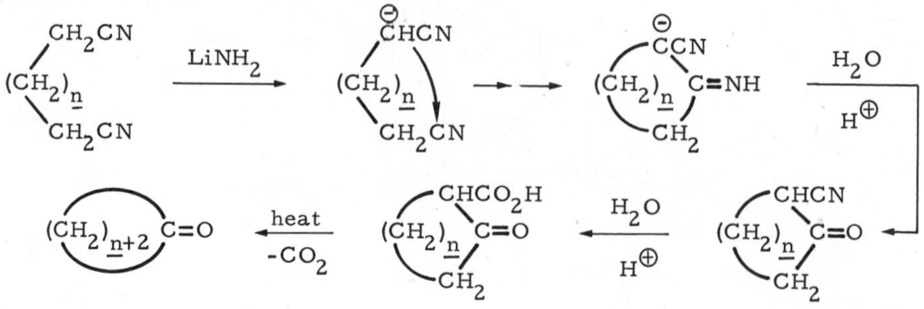

24-12

a. $(CH_3)_3CCl$ $\xrightarrow[\text{ether}]{\text{Mg}}$ $(CH_3)_3CMgCl$ $\xrightarrow[\text{2. } H^\oplus]{\text{1. } CO_2}$ $(CH_3)_3CCO_2H$ ──────

$(CH_3)_3CCN$ $\xleftarrow[-H_2O]{POCl_3}$ $(CH_3)_3CCONH_2$ $\xleftarrow[\text{2. } NH_3]{\text{1. } SOCl_2}$

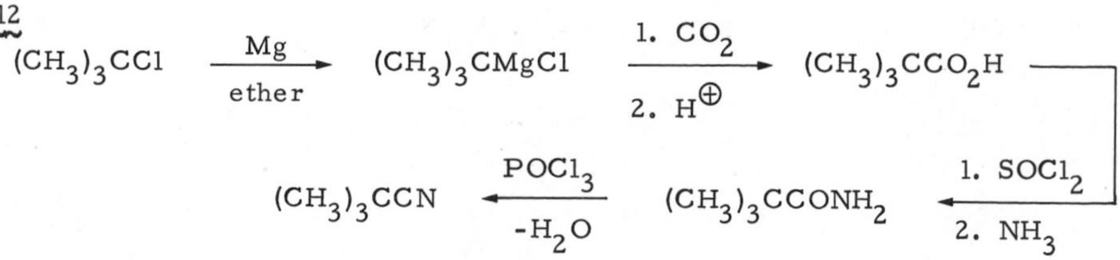

or

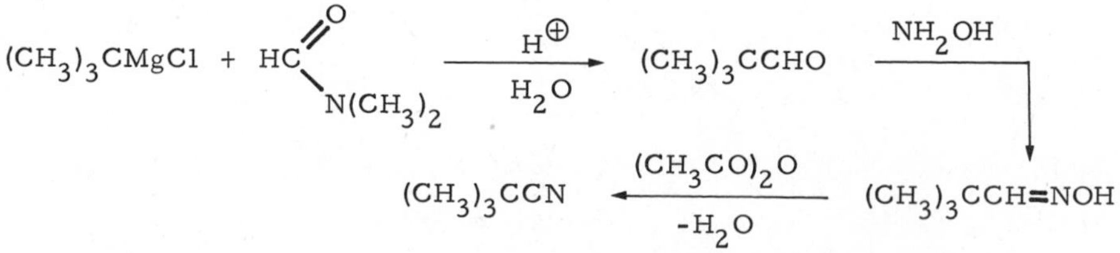

$(CH_3)_3CMgCl$ + $HC\overset{O}{\underset{N(CH_3)_2}{\diagup\!\!\!\diagdown}}$ $\xrightarrow[H_2O]{H^\oplus}$ $(CH_3)_3CCHO$ $\xrightarrow{NH_2OH}$

$(CH_3)_3CCN$ $\xleftarrow[-H_2O]{(CH_3CO)_2O}$ $(CH_3)_3CCH{=}NOH$

b. $CH_2{=}CHCH_2Br$ $\xrightarrow[(CH_3)_2SO]{KCN}$ $CH_2{=}CHCH_2CN$ $\xrightarrow{NaOH}$

$CH_3CH{=}CHCN$

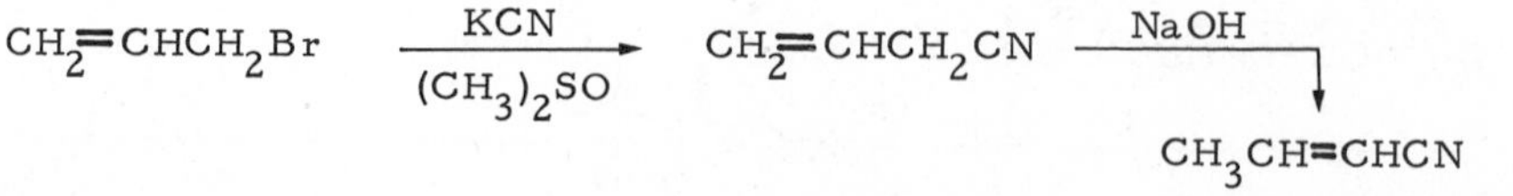

(see Sec. 18-9A)

c. CH_3CHO $\xrightarrow[]{HCN, \overset{\ominus}{OH}}$ $CH_3{-}\overset{OH}{\underset{H}{\overset{|}{C}}}{-}CN$ $\xrightarrow{H^\oplus, H_2O}$ $CH_2{=}CHCO_2H$

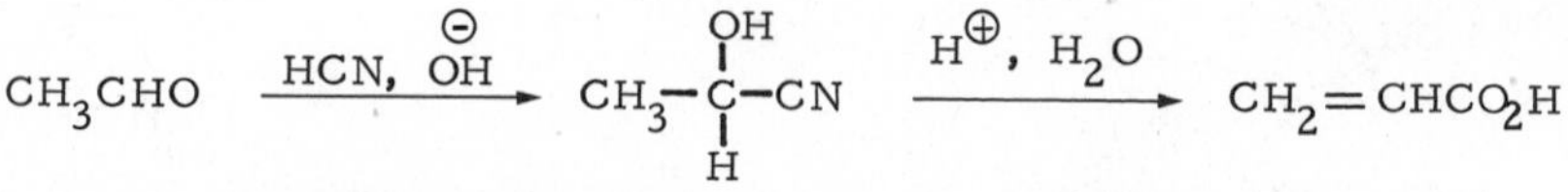

24-13   Compound $HC_3(CN)_5$ has the following structure:

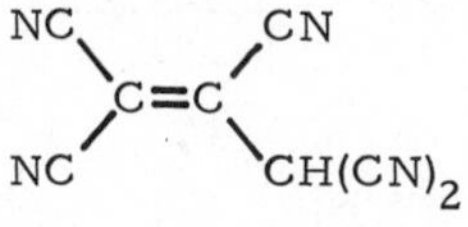

It is a strong acid because of the considerable stabilization energy of the anion
formed. The charge can be delocalized over all five cyano groups. The

compound probably is formed by the following sequence:

$$CH_2(CN)_2 \xrightarrow[-H_2O]{\ominus OH} :\overset{\ominus}{C}H(CN)_2$$

$$(NC)_2C\!\!=\!\!C(CN)_2 + :\overset{\ominus}{C}H(CN)_2 \longrightarrow (NC)_2\overset{\ominus}{C}\!-\!\underset{\underset{\displaystyle CH(CN)_2}{|}}{\overset{\overset{\displaystyle CN}{|}}{C}}\!-\!CN \xrightarrow{-CN^{\ominus}}$$

$$\text{anion} \xleftarrow{\ominus OH} (NC)_2C\!\!=\!\!C\!\!\overset{CN}{\underset{CH(CN)_2}{\big<}}$$

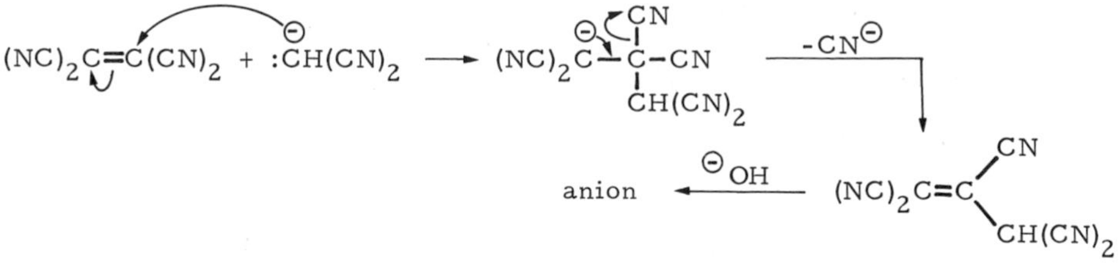

24-14   a.   Conversion of methylbenzene to 3,5-dinitromethylbenzene may be carried out by the sequence.

$$CH_3\!-\!C_6H_5 \longrightarrow 4\text{-}CH_3\!-\!C_6H_4\!-\!NO_2 \xrightarrow{[2H]} 4\text{-}CH_3\!-\!C_6H_4\!-\!NH_2$$

$$4\text{-}CH_3\!-\!C_6H_4\!-\!NHCOCH_3 \xleftarrow{CH_3COCl}$$

Dinitration of the amide followed by deamination analogous to the reactions shown in Section 24-6B for N-(4-methylphenyl)ethanamide would give the desired product.

b.

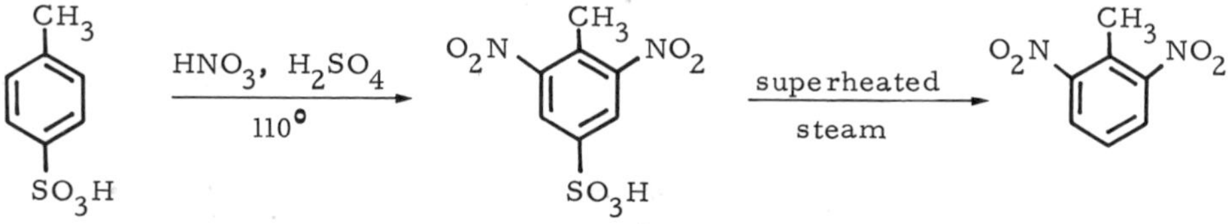

c.

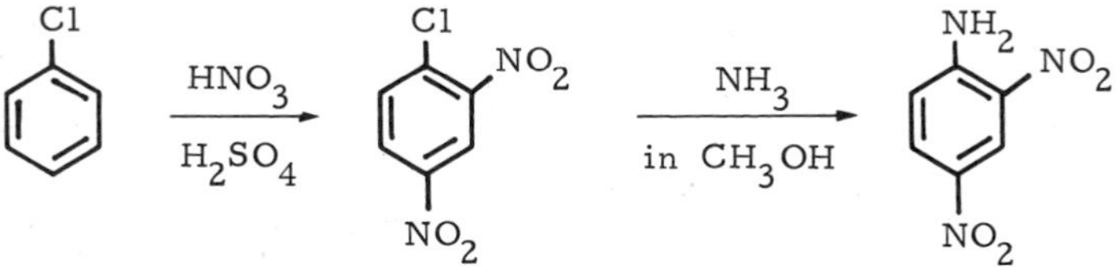

d.

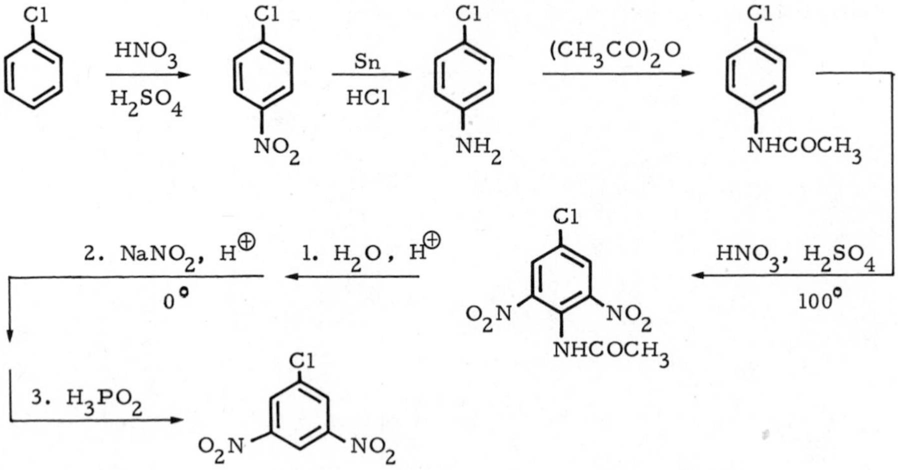

e.

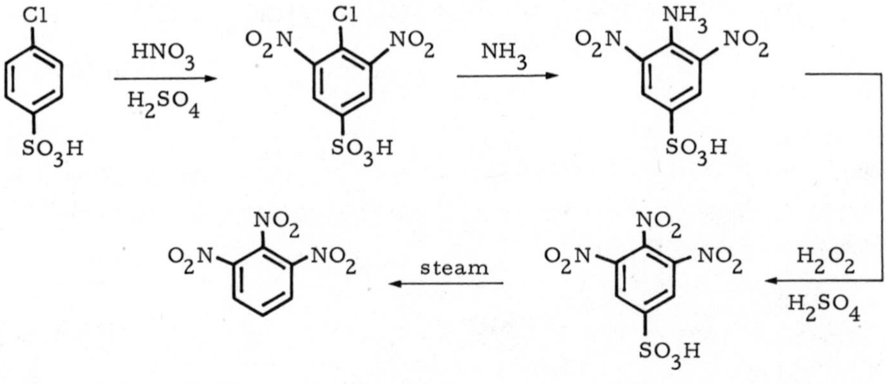

24-15   Orange ⟶ blue-green represents change from charge-transfer complex of tetracyanoethene with benzene to that of anthracene.   The color rapidly fades as the tetracyanoethene adds across the 9,10-positions of anthracene.

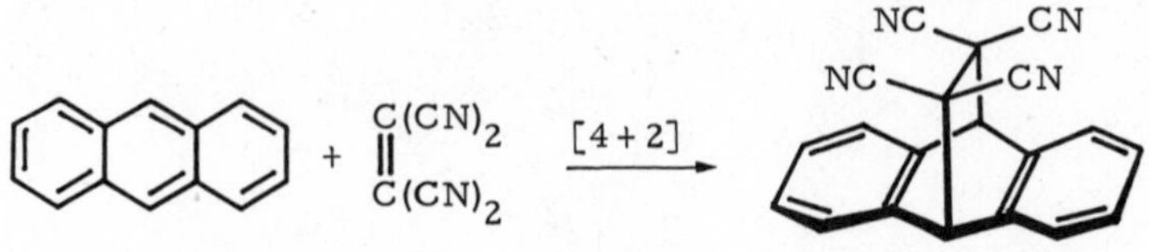

orange charge transfer
complex

Diels-Alder adduct
blue-green

24-16   Because 1,3,5-trinitrobenzene forms a charge-transfer complex with 1,3,5-trimethylbenzene, its dipole moment should not be the same in the hydrocarbon solvent as in CCl_4.

24-17   Reduce the nitro compound with a suitable mild reducing agent $(SnCl_2)$ and dissolve out the resulting amine in acid or put in any kind of a substance that forms a "tighter" complex than does anthracene. Alumina in a chromatographic column is expected to adsorb trinitrobenzene more strongly than hydrocarbons.

24-18

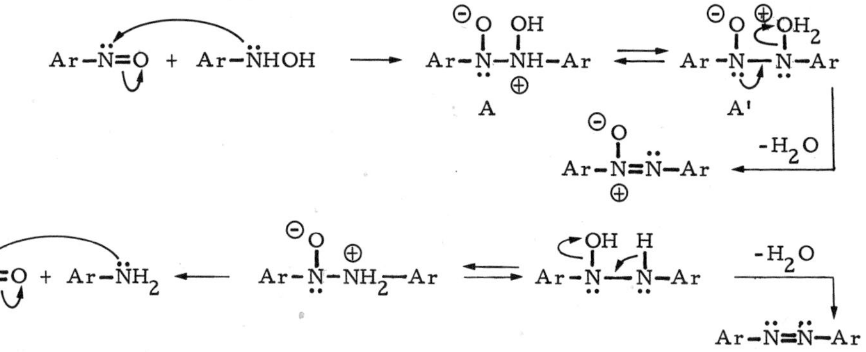

Because proton transfers between electronegative atoms (oxygen) generally are very rapid, the oxygen of the nitroso compound would become equivalent to the oxygen of the azanol in the intermediates (A and A') formed in the reaction ArNO with Ar'NHOH. This would lead to a mixture of ArN(O)=NAr' and ArN=N(O)Ar'.

24-19   The double bond of nitroethene is expected to be much less reactive towards electrophilic reagents than a normal carbon-carbon double bond. However, towards nucleophilic reagents it should be very considerably more reactive. It is an excellent dienophile in the Diels-Alder reaction.

24-20

a. $CH_3NO_2 + CH_2{=}O \xrightarrow{\ominus OH} HOCH_2CH_2NO_2$

b. $HOCH_2CH_2NO_2 \xrightarrow[-H_2O]{H^\oplus} CH_2{=}CHNO_2$

c. $CH_3NO_2 + 3CH_2{=}O \xrightarrow{\ominus OH} (HOCH_2)_3CNO_2 \xrightarrow[-H_2O]{3 HONO_2} (O_2NOCH_2)_3CNO_2$

d. $(CH_3)_2CHNO_2 + CH_2{=}O \xrightarrow{\ominus OH} (CH_3)_2\underset{CH_2OH}{C}NO_2 \xrightarrow{LiAlH_4} (CH_3)_2\underset{CH_2OH}{C}NH_2$

e. CH_3NO_2 + 3 $CH_2{=}CHCN$ $\xrightarrow{\ominus OH}$ $(NCCH_2CH_2)_3CNO_2$

f. $(CH_3)_2CHNO_2$ + $CH_2{=}CHNO_2$ $\xrightarrow{\ominus OH}$ $O_2NCH_2CH_2C(CH_3)_2NO_2$ ⌐

$H_2NCH_2CH_2C(CH_3)_2NH_2$ $\xleftarrow{LiAlH_4}$

24-21

a. $(CH_3)_3COH$ $\xrightarrow[HCN]{H_2SO_4}$ $(CH_3)_3C{-}NHCHO$ $\xrightarrow[H^{\oplus}]{H_2O}$ $(CH_3)_3C{-}NH_2$ ⌐

$(CH_3)_3C{-}NO_2$ $\xleftarrow[H_2SO_4]{H_2O_2}$

b. $(CH_3)_3CCO_2H$ $\xrightarrow[2.\ NH_3]{1.\ SOCl_2}$ $(CH_3)_3CCONH_2$ $\xrightarrow[NaOH]{Br_2}$ $(CH_3)_3C{-}NH_2$ ⌐

$(CH_3)_3C{-}NO_2$ $\xleftarrow[H_2SO_4]{H_2O_2}$

24-22

a. $(CH_3)_2NNH_2$ + HONO $\xrightarrow{-H_2O}$ $\left[(CH_3)_2\ddot{N}{-}\ddot{N}H{-}\ddot{N}{=}O\right.$

$\xleftarrow{-N_2}$ $\left[(CH_3)_2\overset{\oplus}{\ddot{N}}{-}N{\equiv}N{:} \longleftrightarrow (CH_3)_2\overset{\oplus}{N}{=}N{=}\overset{\oplus}{\overset{\ominus}{\ddot{N}}{:}}\right]$ $\xleftarrow{-OH^{\ominus}}$ $\left.(CH_3)_2\ddot{N}{-}\ddot{N}{=}\ddot{N}{-}OH\right]$

$\longrightarrow (CH_3)_2N^{\oplus}$ $\xrightarrow{-H^{\oplus}}$ $CH_2{=}N{-}CH_3$ $\xrightarrow[H^{\oplus}]{H_2O}$ $CH_2{=}O$ + $H_3\overset{\oplus}{N}{-}CH_3$

$\searrow^{H_2O}_{-H^{\oplus}}$ $(CH_3)_2N{-}OH$

b. $(CH_3)_2N{-}\overset{H}{\underset{}{N}}{-}CH_3$ + HONO $\xrightarrow{-H_2O}$ $(CH_3)_2N{-}\overset{CH_3}{\underset{}{N}}{-}N{=}O$

c. $(CH_3)_2N{-}N(CH_3)_2$ $\xrightarrow[-OH^{\ominus}]{HONO}$ $(CH_3)_2N{-}\overset{CH_3}{\underset{\underset{N=O}{|}}{\overset{\oplus}{N}}}{-}CH_2{-}H$ (see Sec. 23-10A) ⌐

$-HNO$

$(CH_3)_2N{-}\overset{CH_3}{\underset{}{N}}{-}N{=}O$ $\xleftarrow[-H_2O]{HONO}$ $(CH_3)_2N{-}\overset{CH_3}{\underset{}{N}}H$ $\xleftarrow[-CH_2{=}O]{H_2O}$ $(CH_3)_2N{-}\overset{CH_3}{\underset{}{\overset{\oplus}{N}}}{=}CH_2$

d. $CH_3-\overset{\overset{O}{\|}}{C}-NH-NH_2 \xrightarrow[\text{NaOH}]{Br_2} CH_3-\overset{\overset{O}{\|}}{C}-\underset{\underset{Br}{|}}{N}-NH_2 \xrightarrow{-HBr} CH_3\overset{\overset{O}{\|}}{C}-N=NH \longrightarrow$

$$CH_3\overset{\overset{O}{\|}}{C}-H + N_2 \longleftarrow$$

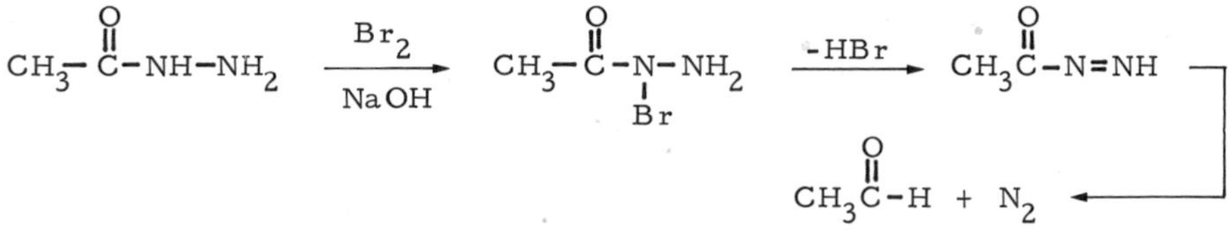

<u>24-23</u>   Order of reactivity predicted is e >>a > b ~~ c> >d. (e) is expected to be especially reactive since decomposition leads to two very stable substances, benzene and nitrogen. (a) is reactive because relatively stable benzyl radicals are formed.

(b) is less reactive than (a) but more so than (d) because a tertiary radical is produced; (c) is expected to be about like (b) because the radicals formed are tertiary and because radicals are normally non-planar there should be no strain forming the radical center at a bridgehead carbon. With (d), unstable methyl radicals would be formed.

<u>24-24</u> $2\ CH_3COCH_3 + HCN \rightleftharpoons 2\ CH_3-\underset{\underset{CH_3}{|}}{\overset{\overset{CN}{|}}{C}}-OH \xrightarrow[-2\ H_2O]{H_2NNH_2}$

$$CH_3-\underset{\underset{CH_3}{|}}{\overset{\overset{CN}{|}}{C}}-N=N-\underset{\underset{CH_3}{|}}{\overset{\overset{CN}{|}}{C}}-CH_3 \xleftarrow[OH^\ominus]{Br_2} CH_3-\underset{\underset{CH_3}{|}}{\overset{\overset{CN}{|}}{C}}-NH-NH-\underset{\underset{CH_3}{|}}{\overset{\overset{CN}{|}}{C}}-CH_3$$

a. $(CH_3)_2\underset{}{\overset{\overset{CN}{|}}{C}}-N=N-\overset{\overset{CN}{|}}{C}(CH_3)_2 \xrightarrow[\text{carbon solvent}]{\text{a perfluoro-}} (CH_3)_2\overset{\overset{CN}{|}}{C}-\overset{\overset{CN}{|}}{C}(CH_3)_2 + N_2$

b. $(CH_3)_2\overset{\overset{CN}{|}}{C}-N=N-\overset{\overset{CN}{|}}{C}(CH_3)_2 \xrightarrow{Br_2,\ CCl_4} 2(CH_3)_2\overset{\overset{CN}{|}}{C}Br + N_2$

<u>24-25</u> $:\overset{\ominus}{CH_2}-N\overset{\oplus}{\equiv}N: \longleftrightarrow CH_2\overset{\oplus}{=}N=\overset{..}{N}:{}^{\ominus}$

$$CH_3-C\overset{O}{\underset{O-H}{\diagdown}} + :\overset{\ominus}{CH_2}-N\overset{\oplus}{\equiv}N: \longrightarrow CH_3-C\overset{O}{\underset{O-CH_3}{\diagdown}} + N_2$$

<u>24-26</u>

a. $CH_3C\overset{O}{\underset{Cl}{\diagdown}} + :\overset{\ominus}{CH_2}-N\overset{\oplus}{\equiv}N: \longrightarrow CH_3-\overset{\overset{\ominus}{O}}{\underset{\underset{Cl}{|}}{C}}-\underset{\underset{H}{|}}{\overset{\overset{H}{|}}{C}}-N\overset{\oplus}{\equiv}N: \xrightarrow{-HCl}$

$$CH_3-C\overset{O}{\underset{CH=N=N:}{\diagdown}}{}^{\oplus\ \ominus}$$

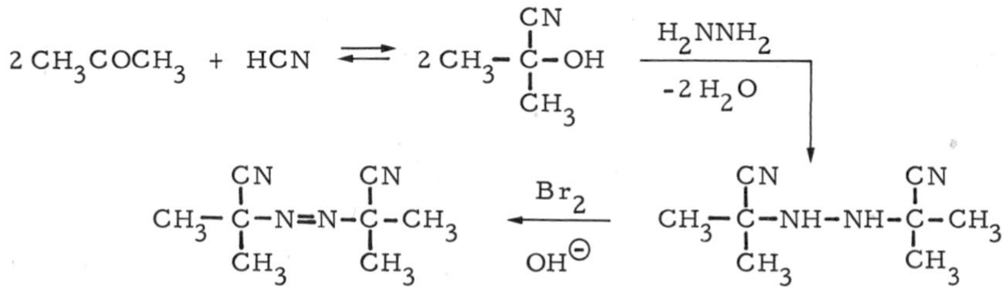

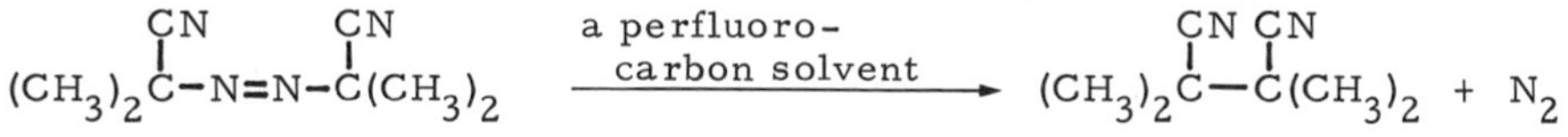

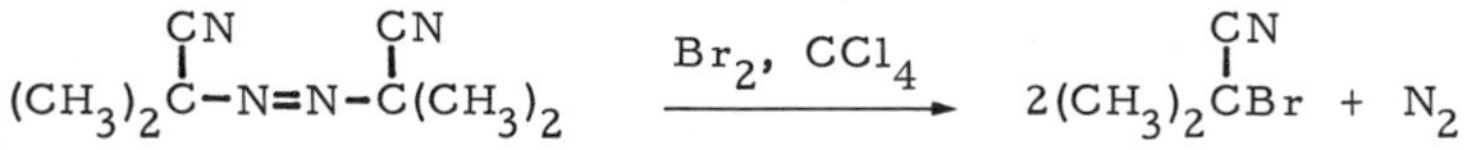

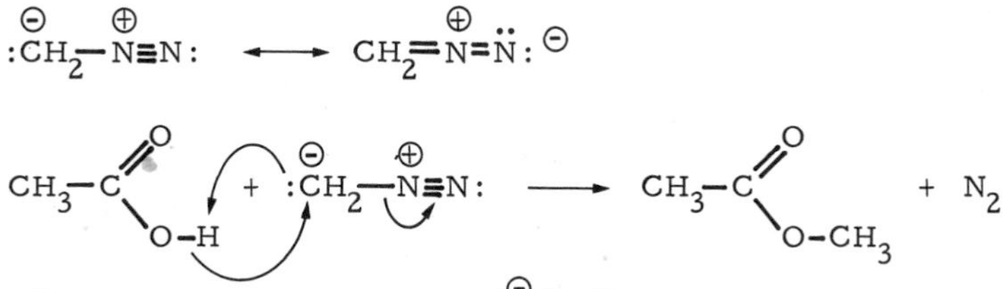

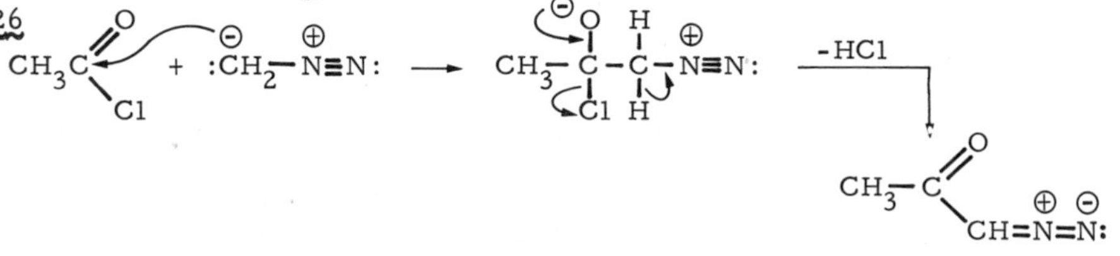

24-26   a. (cont.)

$$CH_2N_2 + HCl \longrightarrow CH_3Cl + N_2$$

Analogous to nucleophilic addition reactions of carbonyl compounds (Sec. 16-4A)

b.

-HCl

$$CH_3\overset{O}{\overset{\|}{C}}-CH_2Cl \xrightarrow[-N_2]{S_N2} CH_3\overset{O}{\overset{\|}{C}}-CH_2\overset{\oplus}{N}{\equiv}N + \overset{\ominus}{Cl}$$

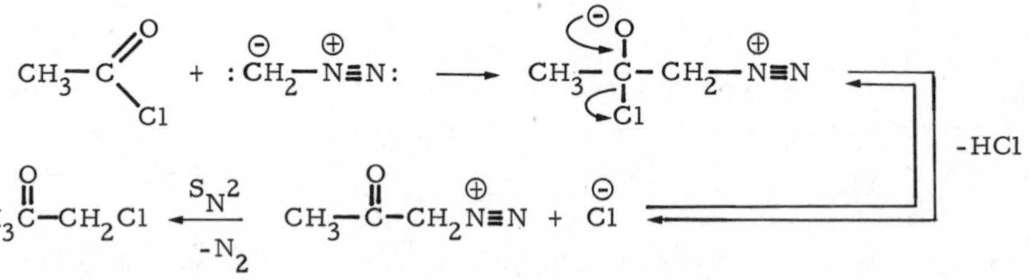

c.

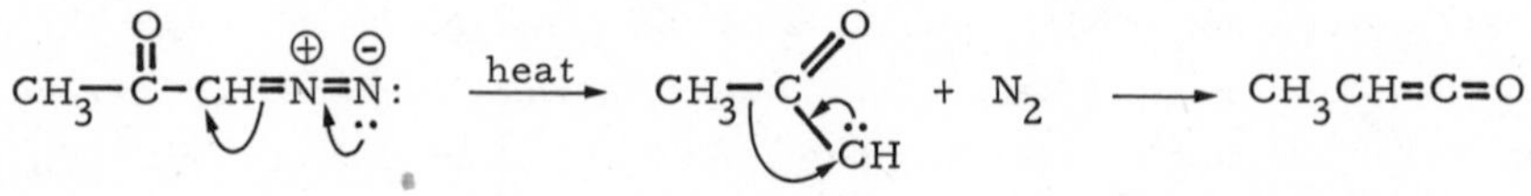

d.

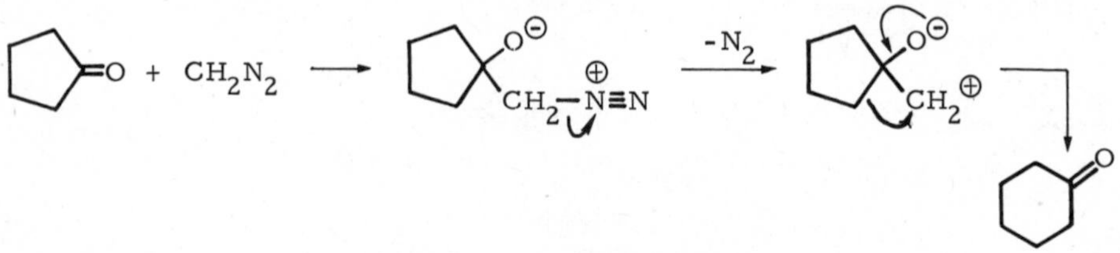

e.

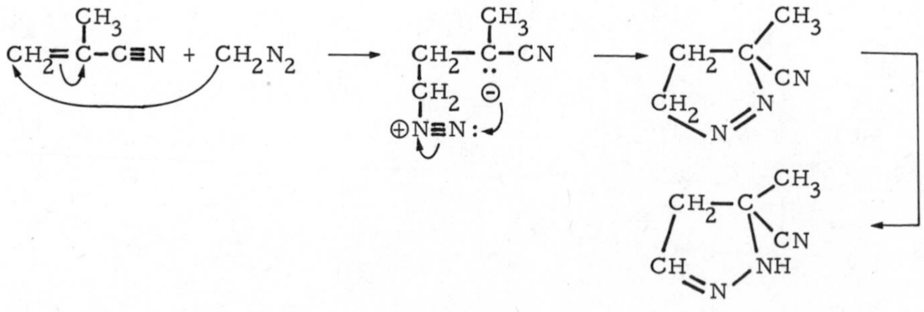

more stable product

24-27

a.

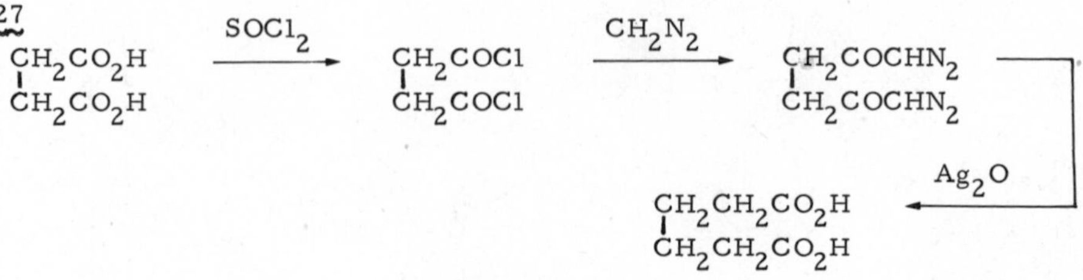

b. $(CH_3)_2C=C=O$ + CH_2N_2 $\longrightarrow$ $(CH_3)_2C=C\underset{CH_2-N\equiv N}{\overset{O^{\ominus}}{\diagdown}}$ $\xrightarrow{-N_2}$ $CH_3-\underset{CH_2}{\overset{CH_3}{\underset{|}{\overset{|}{C}}}}\diagdown C=O$

c. $CH\equiv C-CO_2CH_3$ + CH_2N_2 $\longrightarrow$ $\left[\underset{CH_2N_2}{\overset{CH=\overset{\ominus}{C}-CO_2CH_3}{\overset{\oplus}{|}}} \longrightarrow \underset{H_2C\diagdown_N\diagup NH}{\overset{CH=C-CO_2CH_3}{|}} \right]$

$\underset{CH\diagdown_N\diagup :NH}{\overset{CH=C-CO_2CH_3}{|}}$ $\Big\vert$

24-28 $R_2C=O$ + NH_3 $\longrightarrow$ $\left[R_2C\underset{NH_2}{\overset{OH}{\diagup}} \xrightarrow{-H_2O} R_2C=NH \right]$

NH_2Cl + NH_3 $\longrightarrow$ $:NH$ + NH_4Cl

$R_2C=NH$ + $:NH$ $\longrightarrow$ $R_2C\underset{NH}{\overset{NH}{\diagup}}\Big\vert$

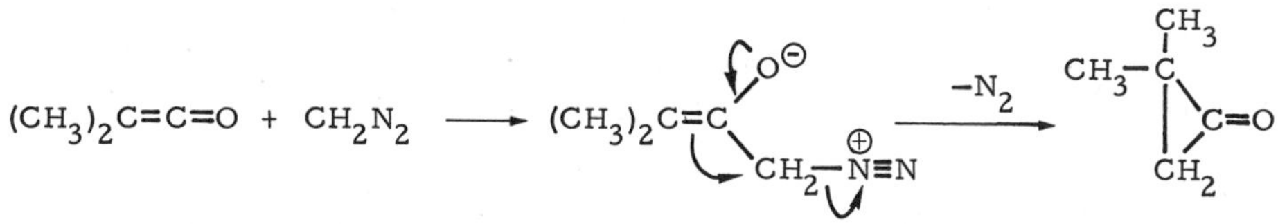

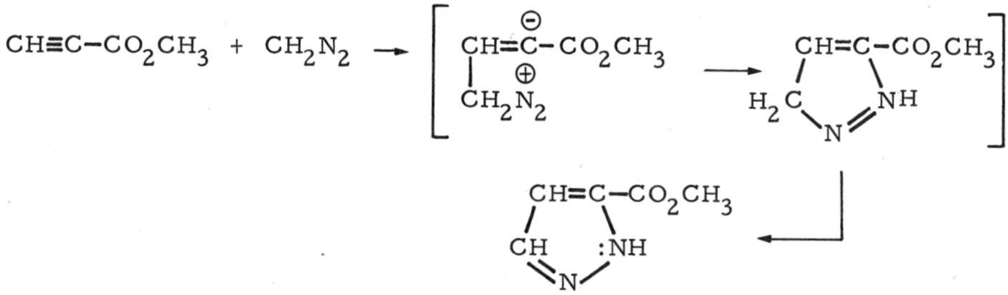

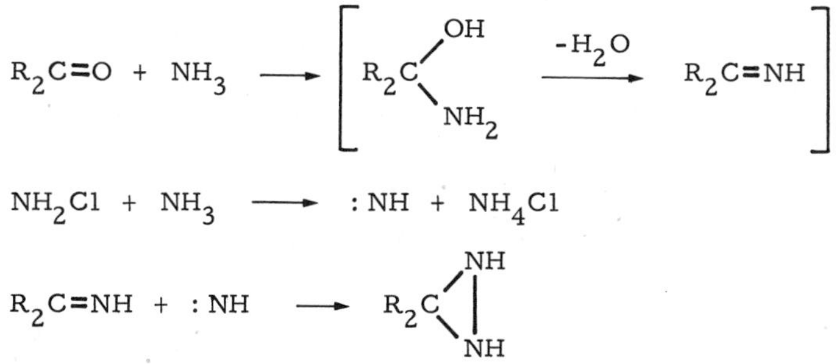

24-29   Diazacyclopropene, unlike diazomethane, cannot add a proton to carbon without first breaking the ring bonds. Protonation must necessarily occur at nitrogen, and this is unlikely to be a favorable process presumably because of the high s-character of the unshared electron pair and the ring strain involved in forming sp^3-hybridized nitrogen.

24-30

a. $CH_2=CHCH_2CH_2OH$ $\xrightarrow[\text{-HCl}]{CH_3-\langle\rangle-SO_2Cl}$ $CH_2=CHCH_2CH_2OSO_2-\langle\rangle-CH_3$

$CH_2=CHCH_2CH_2NH_2$ $\xleftarrow{LiAlH_4}$ $CH_2=CHCH_2CH_2N_3$ $\xleftarrow{NaN_3}$

b. $\underset{CO_2H}{\overset{CO_2H}{\diamondsuit}}$ $\xrightarrow[H_2SO_4]{HN_3}$ $\underset{NH_2}{\overset{NH_2}{\diamondsuit}}$ $\xrightarrow[-2\,NH_3]{H_2O}$ $\diamondsuit=O$

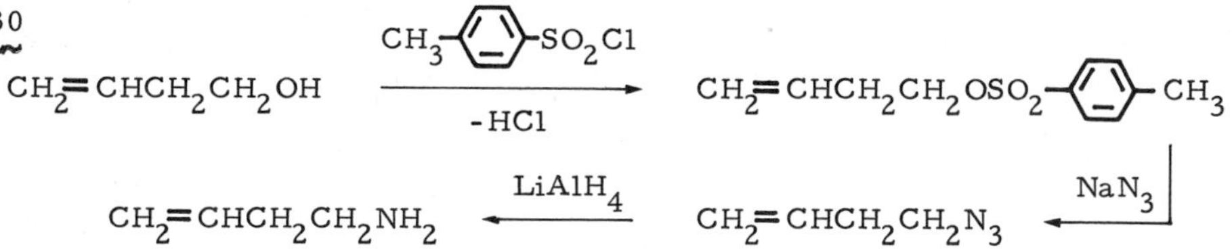

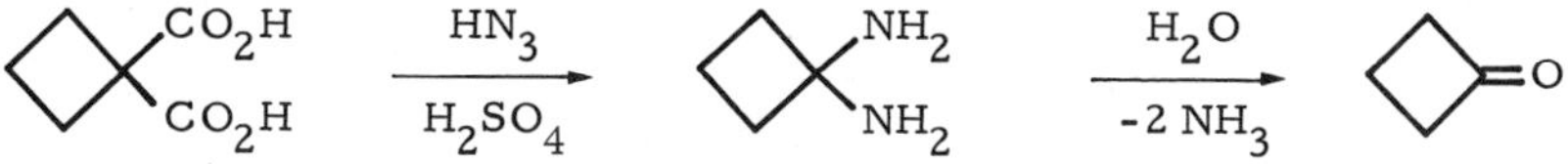

24-31

a. $(CH_3)_2CHCHO$ $\xrightarrow[-H_2O]{H_2NOH}$ $(CH_3)_2CHCH\!=\!NOH$ $\xrightarrow[-H_2O]{P_2O_5}$ $(CH_3)_2CHCN$

b. $CH_3NO_2 + 3\,CH_2O$ $\xrightarrow{NaOH}$ $(HOCH_2)_3CNO_2$ $\xrightarrow{(CH_3CO)_2O}$

$$(CH_3CO_2CH_2)_3CNO_3$$

c. $C_6H_5CN + (CH_3)_3COH$ $\xrightarrow{H_2SO_4}$ $C_6H_5\overset{\displaystyle O}{\overset{\|}{C}}\!-\!NC(CH_3)_3$

24-32

Compound	m.p. °C	b.p. °C	H_2O	$(C_2H_5)_2O$	acid	base
				Solubility		
A. octanamine	-	175	-	+	+	-
B. N-butylbutanamine	-	159	-	+	+	-
C. N, N-dipropylpropan-amine (tri-propylamine)	-	157	-	+	+	-
D. N, N-dimethylethanamide	-20	175	+	+	+	+
E. 1-nitrobutane	-	151	-	+	-	+
F. 2-nitro-2-methylbutane	-	150	-	+	-	-

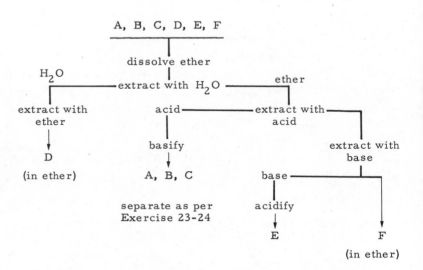

24-33   a.   Reaction with nitrous acid will give evolution of nitrogen in the case of the primary amine.

b.   Heating in basic solution will give ammonia for the primary amide.

c.   The ethynylamine will dissolve in acid and will also give a pre-cipitate as the silver salt on mixing in silver nitrate solution.

d.   Treatment with iodide ion will produce iodine from the N-chloro-amine.

e.   Acid hydrolysis and heat will produce carbon dioxide gas from the carbamate.

f.   As in Part a.

g.   Hydrolysis with dilute base will give ammonia from the primary amide.   The amine is also more strongly basic.

24-34   a.   Infrared would show N—H bands in the region $3300 - 3500 \text{ cm}^{-1}$ for the primary amine.

b.   Strong infrared absorption of NO_2 group near 1580 and 1375 cm^{-1} would distinguish the nitro compound from the amide.

c.   Strong $\equiv$C—H absorption at 3300 cm^{-1} would appear in the infra-red spectrum of the 1-alkyne.

d.   The nmr spectrum of the salt under non-exchanging conditions would show splitting of the methylene resonance by the N—H protons.

e.   The nmr spectrum of the carbamic ester would show a low field singlet resonance for the OCH_3 protons ($\delta \sim 3.5$ ppm).

f.   The nmr of 2-methoxyethanamine would show $\delta \sim 3.5$ ppm for the OCH_3 protons.

g.   As in Part f.

24-35   a.   Infrared would indicate the presence of a primary or secondary amine from the N—H absorption bands in the region $3300 - 3500 \text{ cm}^{-1}$.

b.   Whether a primary, secondary or tertiary amide could be deter-mined by infrared (N—H bands in the region $3400 - 3500 \text{ cm}^{-1}$).   The nmr would clearly differentiate between the compounds since the amide proton, —NH–C=O, would appear at characteristically low magnetic fields.

c.   Strong bands near 1580 cm^{-1} and 1375 cm^{-1} would identify the nitroalkane.

d.   A nitrile band in the region $2000 - 2300 \text{ cm}^{-1}$ and a carbonyl band near 1740 cm^{-1} in the infrared would identify the cyanoketone.   Its nmr spec-trum is likely to show some enolic OH resonance.

24-36   Compound A must be a diamide since two carbon atoms are lost on treatment of A with Br_2 in NaOH (Hofmann reaction).

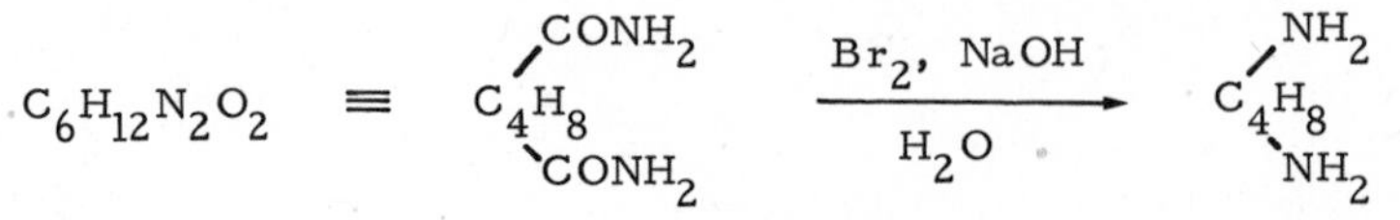

$$C_6H_{12}N_2O_2 \equiv C_4H_8 \underset{CONH_2}{\overset{CONH_2}{<}} \xrightarrow[H_2O]{Br_2, \ NaOH} C_4H_8 \underset{NH_2}{\overset{NH_2}{<}}$$

A   D

Compound D must be a diamine which, on reaction with nitrous acid, deaminates and undergoes a pinacol-type rearrangement. Working backwards, the following reactions may be written.

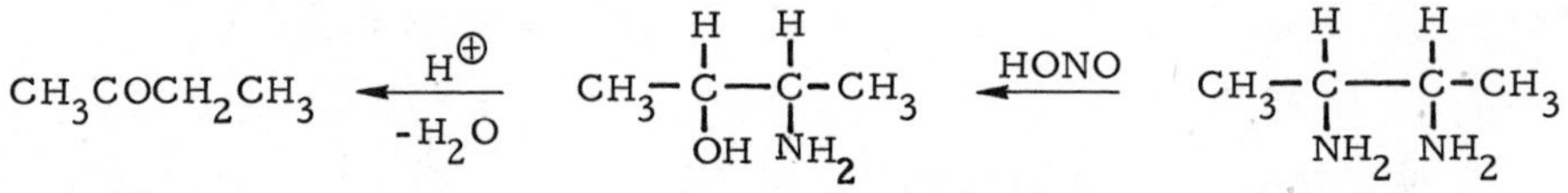

$$CH_3COCH_2CH_3 \xleftarrow[-H_2O]{H^{\oplus}} CH_3-\underset{\underset{OH}{|}}{\overset{\overset{H}{|}}{C}}-\underset{\underset{NH_2}{|}}{\overset{\overset{H}{|}}{C}}-CH_3 \xleftarrow{HONO} CH_3-\underset{\underset{NH_2}{|}}{\overset{\overset{H}{|}}{C}}-\underset{\underset{NH_2}{|}}{\overset{\overset{H}{|}}{C}}-CH_3$$

Accordingly, compound A is the diamide of 2,3-dimethylbutanedioic acid (2,3-dimethylsuccinic acid) which can exist in chiral forms and which reacts with nitrous acid to give compound B.

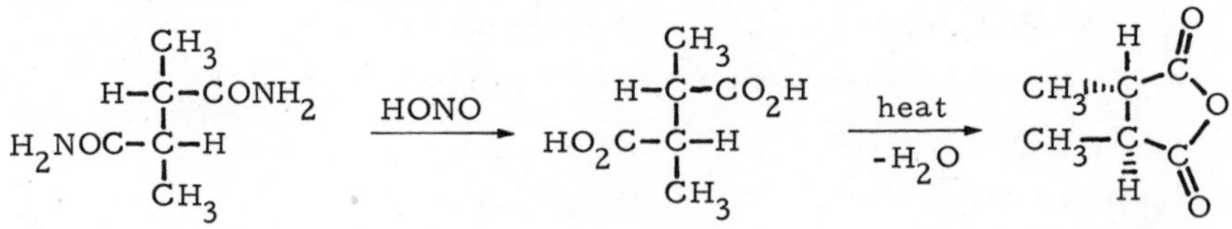

A   B   C

trans

On heating, B converts to trans-3,4-dimethyl-2,5-dioxooxacyclopentane (trans-2,3-dimethylsuccinic anhydride), C.

24-37
a. $C_6H_5CH_2CO_2C_2H_5 \xrightarrow[P]{Br_2} C_6H_5CHBrCO_2C_2H_5 \xrightarrow{AgNO_2}$

$C_6H_5CH(NO_2)CO_2H \xleftarrow{H_2O} C_6H_5CH(NO_2)CO_2C_2H_5$

b. $C_6H_5CH_2CO_2H \xrightarrow{SOCl_2} C_6H_5CH_2COCl \xrightarrow{CH_2N_2}$

$C_6H_5CH_2CH_2CO_2H \xleftarrow[-N_2]{Ag_2O \ (moist)} C_6H_5CH_2COCHN_2$

27-38

a. C_6H_6 $\xrightarrow[H_2SO_4]{HNO_3}$ $C_6H_5NO_2$ $\xrightarrow[HCl]{Fe}$ $C_6H_5NH_2$ $\xrightarrow{(CH_3CO)_2O}$

$C_6H_5NHCOCH_3$

b. 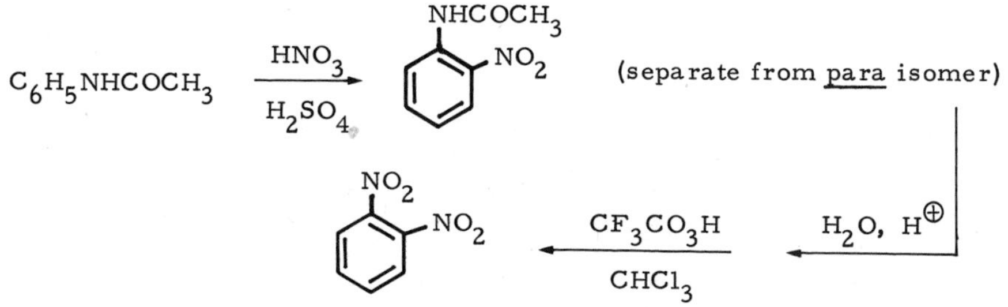 $C_6H_5NHCOCH_3$ $\xrightarrow[H_2SO_4]{HNO_3}$

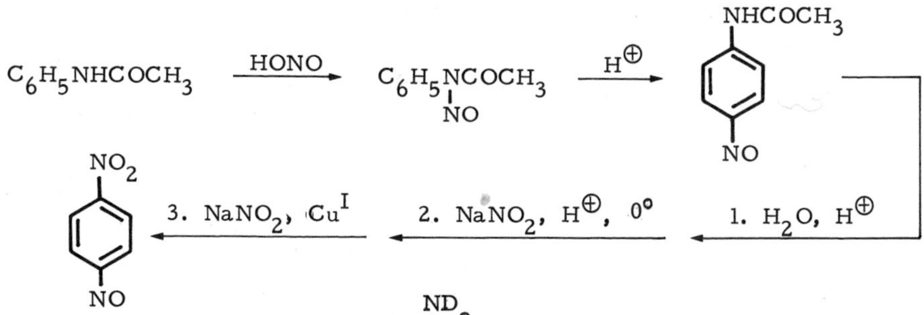

(benzene ring with NHCOCH$_3$ and NO$_2$)   (separate from para isomer)

(benzene ring with NO$_2$ and NO$_2$) $\xleftarrow[CHCl_3]{CF_3CO_3H}$ $\xleftarrow{H_2O, H^{\oplus}}$

c. $C_6H_5NHCOCH_3$ $\xrightarrow{HONO}$ $C_6H_5\underset{NO}{\overset{}{N}}COCH_3$ $\xrightarrow{H^{\oplus}}$ (benzene ring with NHCOCH$_3$ and NO)

(benzene ring with NO$_2$ and NO) $\xleftarrow{3.\ NaNO_2,\ Cu^I}$ $\xleftarrow{2.\ NaNO_2,\ H^{\oplus},\ 0^{\circ}}$ $\xleftarrow{1.\ H_2O,\ H^{\oplus}}$

d. $C_6H_5NHCOCH_3$ $\xrightarrow[D^{\oplus}]{D_2O}$ (benzene ring with ND$_2$, D, D, D) $\xrightarrow[2.\ H_3PO_2,\ Cu(I)]{1.\ NaNO_2,\ DCl,\ D_2O,\ 0^{\circ}}$

(benzene ring with D, D, D)

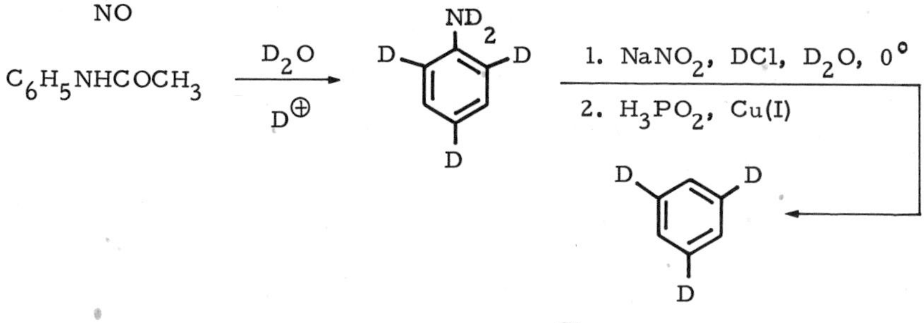

e. 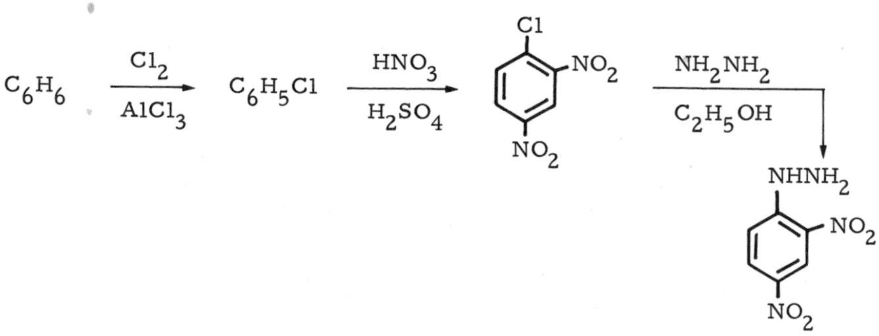 C_6H_6 $\xrightarrow[AlCl_3]{Cl_2}$ C_6H_5Cl $\xrightarrow[H_2SO_4]{HNO_3}$ (benzene ring with Cl, NO$_2$, NO$_2$) $\xrightarrow[C_2H_5OH]{NH_2NH_2}$

(benzene ring with NHNH$_2$, NO$_2$, NO$_2$)

24-39   a. Solubility in basic solutions:

$$C_6H_5CH_2NO_2 + NaOH \longrightarrow C_6H_5\overset{\ominus}{C}HNO_2 \quad Na^{\oplus} + H_2O$$

b. Hydrolysis in basic solution will liberate ammonia:

$$C_6H_5CONH_2 + NaOH \longrightarrow C_6H_5CO_2Na + NH_3$$

c. Coupling of benzenamine with a diazonium salt:

$$ArN_2^{\oplus} + C_6H_5NH_2 \xrightarrow{\text{pH 9}} ArN=N-C_6H_5NH_2$$

or treatment with bromine water to give tribromobenzenamine.

e. Diazotization of 4-methylbenzenamine will liberate N_2 on warming.

$$CH_3-C_6H_4-NH_2 \xrightarrow{\text{NaNO}_2, \text{ HCl, } H_2O} CH_3-C_6H_4-OH + N_2$$

f. Treatment of the N-nitroso compound with acid will convert it to the <u>para</u>-nitroso compound.

$$C_6H_5-N\underset{NO}{\overset{CH_3}{<}} \xrightarrow[\text{ether, } 25°]{\text{HCl}} 4- O=N-C_6H_4-NHCH_3$$

24-40

a. C_6H_6 $\xrightarrow[H_2SO_4]{HNO_3}$ $\xrightarrow{(NH_4)_2 S_x}$

b. bromobenzene $\xrightarrow[H_2SO_4]{HNO_3}$ 4-bromonitrobenzene (separate from isomers) $\xrightarrow[NH_4Cl]{Zn}$

$\xleftarrow[H_2SO_4, \ 0°]{K_2Cr_2O_7}$

c. $C_6H_5-CH_3$ $\xrightarrow[H_2SO_4]{HNO_3}$ 2,4-dinitromethylbenzene $\xrightarrow{(NH_4)_2 S_x}$

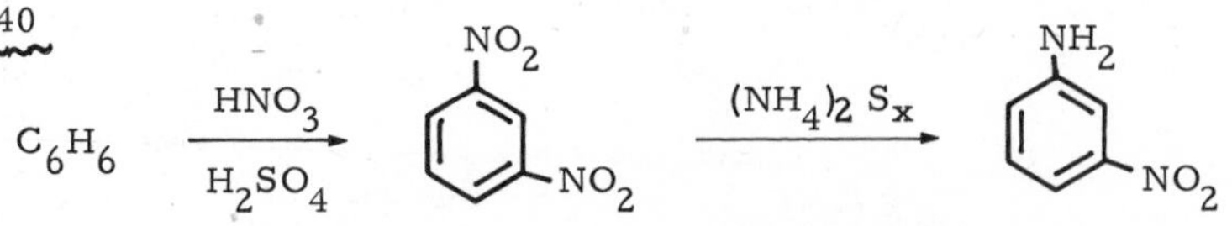

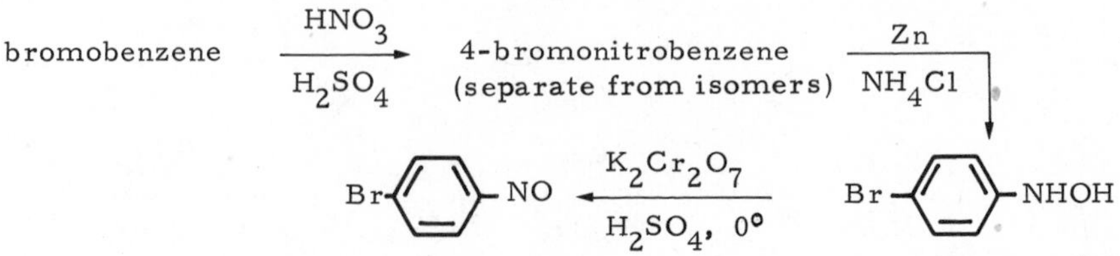

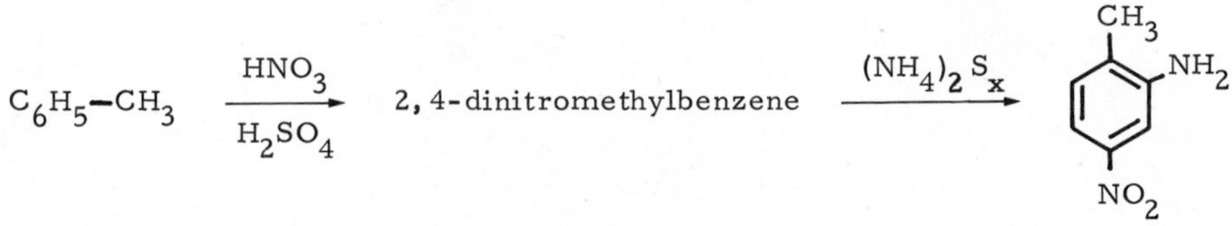

d. Chlorobenzene $\xrightarrow{\text{HNO}_3, \text{ H}_2\text{SO}_4}$ 4-chloronitrobenzene (separate from

isomers) $\xrightarrow{\text{Sn, HCl}}$ 4-chlorobenzenamine

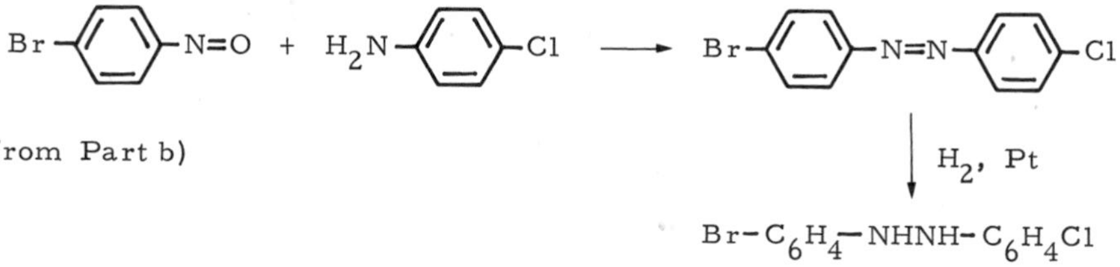

$$\text{Br}-\!\!\!\bigcirc\!\!\!-\text{N=O} + \text{H}_2\text{N}-\!\!\!\bigcirc\!\!\!-\text{Cl} \longrightarrow \text{Br}-\!\!\!\bigcirc\!\!\!-\text{N=N}-\!\!\!\bigcirc\!\!\!-\text{Cl}$$

(from Part b)

$\downarrow$ H_2, Pt

$$\text{Br}-\text{C}_6\text{H}_4-\text{NHNH}-\text{C}_6\text{H}_4\text{Cl}$$

e. nitrobenzene $\xrightarrow{\text{Sn, HCl}}$ benzenamine $\xrightarrow{(\text{CH}_3\text{CO})_2\text{O}}$ $\text{C}_6\text{H}_5\text{NHCOCH}_3 \rightharpoondown$

4-nitrobenzenamine $\xleftarrow{\text{H}_2\text{O}}$ N-(4-nitrophenyl)ethanamide $\xleftarrow{\text{HNO}_3, \text{ H}_2\text{SO}_4}$
(separate from isomers)

nitrobenzene $\xrightarrow{\text{Zn, NH}_4\text{Cl}}$ $\text{C}_6\text{H}_5\text{NHOH}$ $\xrightarrow[\text{H}_2\text{SO}_4, \ 0]{\text{K}_2\text{Cr}_2\text{O}_7}$ $\text{C}_6\text{H}_5\text{NO}$

$$\text{O}_2\text{N}-\!\!\!\bigcirc\!\!\!-\text{NH}_2 + \text{O=N}-\!\!\!\bigcirc \xrightarrow[-\text{H}_2\text{O}]{\text{azabenzene}} \text{O}_2\text{N}-\!\!\!\bigcirc\!\!\!-\text{N=N}-\!\!\!\bigcirc$$

f. methylbenzene $\xrightarrow[\text{(separate from isomers)}]{\text{1. HNO}_3, \text{ H}_2\text{SO}_4 \quad \text{2. Zn, NH}_4\text{Cl}}$ $\text{CH}_3-\!\!\!\bigcirc\!\!\!-\text{NHOH}$

$$\text{CH}_3-\!\!\!\bigcirc\!\!\!-\overset{\ominus}{\underset{\oplus}{\text{N=N}}}\!\!-\overset{O}{}\!\!-\bigcirc \xleftarrow[\text{NaOH}]{\text{C}_6\text{H}_5\text{NO}}$$

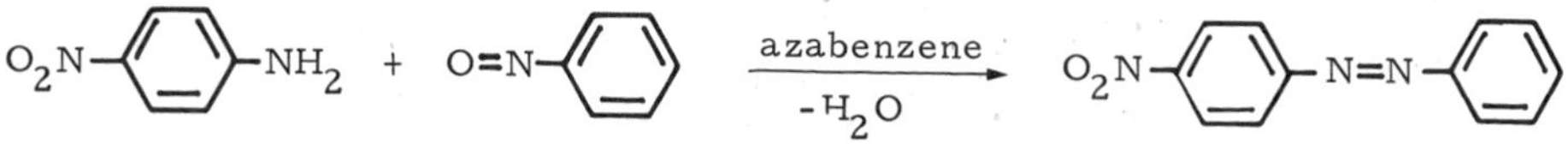

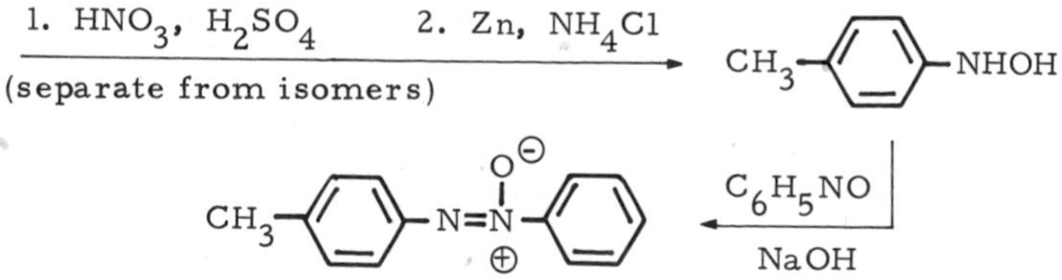

CHAPTER 25

AMINO ACIDS, PEPTIDES, PROTEINS,
ENZYMES AND NUCLEIC ACIDS

Natural α -amino acids are polyfunctional molecules of structure $H_2NCHRCO_2H$, in which the configuration at the chiral carbon is almost invariably $\underline{L}$. The R group varies from alkyl and aryl to a functionalized carbon chain (see Table 24-1).

In acid solution (pH ~ 0), amino acids are converted to the ammonium cations, $H_3\overset{\oplus}{N}CHRCO_2H$. At higher pH values, the acidic sites, $H_3\overset{\oplus}{N}$- and CO_2H, are neutralized until, at high pH, the amino acids completely convert to anions. The acid strength of a carboxyl group is higher than that of the ammonium group, that is $p\underline{K}_{CO_2H} < p\underline{K}_{NH_3^{\oplus}}$, so that neutralization occurs in the following order:

$$H_3\overset{\oplus}{N}CHRCO_2H \xrightarrow{\ -H^{\oplus}\ } H_3\overset{\oplus}{N}CHRCO_2^{\ominus} \xrightarrow{\ -H^{\oplus}\ } H_2NCHRCO_2^{\ominus}$$

Acidic or basic substituents in the R group will add further to the number of ionizable sites in the molecule.

For each amino acid there is a pH called the isoelectric point at which the number of cationic sites equals the number of anionic sites. At this pH, there is no net migration of the amino acid in an electric field, and the solubility also is at a minimum. Crystalline amino acids exist predominantly in the dipolar form, $H_3\overset{\oplus}{N}CHRO_2^{\ominus}$. As a result, the solid amino acids have salt-like properties, high melting points, and low solubilities in organic solvents.

Detection of amino acids is facilitated by the ninhydrin color test. The blue-violet product formed between the amino acid and the ninhydrin reagent is the same for all amino acids (except proline and hydroxyproline) and the color intensity can be related quantitatively to the amino acid concentration. Separation methods commonly used in amino-acid analysis include paper chromatography, thin layer chromatography, and ion-exchange chromatography.

Amino and carboxyl functions of amino acids show the expected reactions of esterification, imine formation, acylation, and diazotization, but

the products may not be stable indefinitely and often react further. Amino
esters cyclize to azlactones or diketopiperazines; imines may decarboxylate,
rearrange and hydrolyze to α-keto acids.

The synthesis of α-amino acids can be achieved by the following types
of reactions:

1. $ClCHRCO_2H \xrightarrow[\text{2. }H^{\oplus}]{\text{1. }NH_3\text{ (excess)}} H_2NCHRCO_2H$ 　　　nucleophilic
displacement

2. $RCHO + NH_3 + HCN \longrightarrow H_2NCHRCN \xrightarrow[H_2O]{H^{\oplus}} H_2NCHRCO_2H$ 　Strecker
synthesis

3.

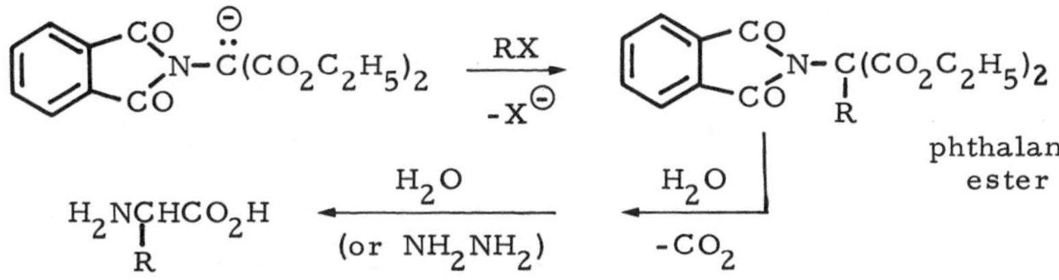

phthalamidomalonic
ester synthesis

Laboratory syntheses of this type give $\underline{D}, \underline{L}$ products which must be resolved if
the separate enantiomers are desired.

Polymerization of amino acids through amide-bond formation leads to
peptides and proteins, $\underline{n}(H_2NCRCO_2H) \longrightarrow (NHCHRCO)_{\underline{n}} + \underline{n}H_2O$. Proteins are high molecular-weight polymers while peptides are smaller molecules
of the same general type. The exact distinction between peptides and proteins is
not very sharp, but almost all proteins exist in specific hydrated conformations
that can be destroyed irreversibly on heating (denaturation), whereas peptide
conformations are formed reversibly. For these reasons, synthetic peptides
can be prepared that are identical in all respects with the natural peptide, but it
is questionable whether many proteins could be synthesized that would be identical to their natural forms.

The accepted notation for polyamide structure has the amino acid with
the free amino group written on the left (N-terminus) and the amino acid with
the free carboxyl group written on the right (C-terminus). Three-letter abbre-
viations are commonly used for each amino acid residue (see Table 25-1).

Phe-Met   is

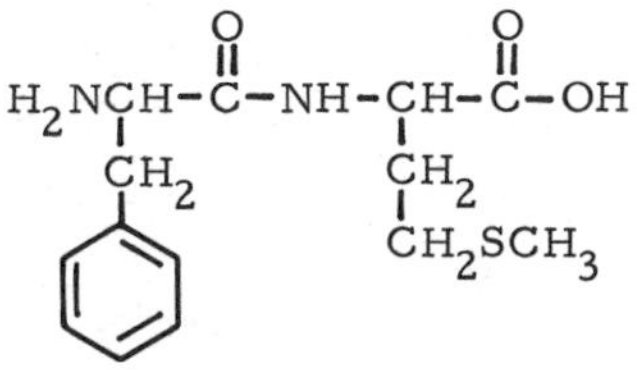

phenylalanylmethionine

The structures of polypeptides are determined by the following steps. First, the total amino-acid composition is found by complete acid hydrolysis of all the amide bonds. Second, the N-terminal and C-terminal amino acids are identified. Third, the peptide is partially cleaved in different ways by selective hydrolysis and the composition and sequence of the fragments determined. Finally, by matching the overlaps of the sequences, the composition of the original peptide is deduced from the compositions of the smaller peptides.

Reagents and methods for determining the N-terminal residue include:

1.   2,4-dinitrofluorobenzene (Sanger reagent), ArF $\simeq$ O_2N—⟨ ⟩—F, NO_2

$$Ar-F + H_2NCHRCO-\boxed{peptide} \xrightarrow{\text{(-HF)}} ArNHCHRCO-\boxed{peptide}$$

$$\downarrow H_3O^{\oplus}$$

$$ArNHCHRCO_2H + \text{amino acids}$$
(separated and identified)

2.   dansyl chloride, $ArSO_2Cl$ = ⟨⟨ ⟩⟩—SO_2Cl , $(CH_3)_2N$—⟨⟨ ⟩⟩

$$ArSO_2Cl + H_2NCHRCO-\boxed{peptide} \longrightarrow ArSO_2NHCHRCO-\boxed{peptide} + HCl$$

$$H_3O^{\oplus}\downarrow$$

$$ArSO_2NHCHRCO_2H + \text{amino acids}$$
(separated and identified)

3.   Edman degradation, $XY = C_6H_5N{=}C{=}S$,   phenylisothiocyannate

$$XY + H_2NCHRCO-\boxed{peptide} \xrightarrow{\text{addition}} HXYNHCHRCO-\boxed{peptide}$$

$$\downarrow H_2O$$

$$\xleftarrow[\text{cyclization}]{\text{(-H}_2\text{O)}} HXYCHRCO_2H + \boxed{peptide}$$

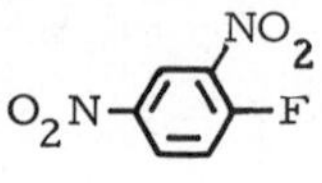

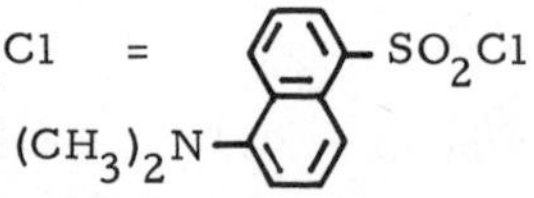

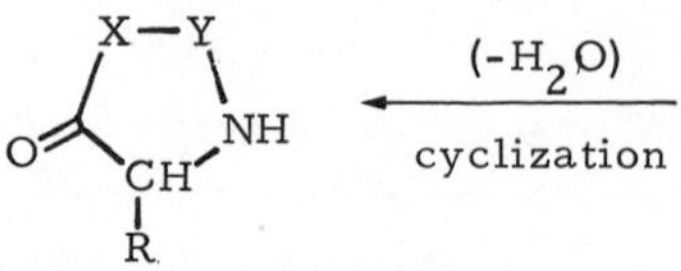

(separated and identified)

The Edman degradation is of value because it can be used progressively to sequence the chain from the N-terminal end.

Carboxypeptidase A is an enzyme that catalyzes the hydrolysis of peptide bonds (a proteinase), but specifically, at the bond linking the C-terminal acid to the chain. This enzyme is used to sequence the chain from the C-terminal end.

To find the amino-acid sequence in a large polypeptide, it is broken down selectively into smaller parts. Reagents for this purpose include several proteinases, namely,

Trypsin, which selectively cleaves peptide bonds to the basic amino acids, Lys and Arg.

$$-\text{Arg-Gly-} \xrightarrow{\text{trypsin, } H_2O} -\text{Arg} + \text{Gly-}$$

Chymotrypsin, which selectively cleaves peptide bonds to the aromatic amino acids, Phe, Tyr, and Trp.

$$-\text{Tyr-Gly-} \xrightarrow{\text{chymotrypsin, } H_2O} -\text{Tyr} + \text{Gly-}$$

In peptide synthesis, reactive functional groups not involved in peptide-bond formation must be protected in order to achieve a specific amino-acid sequence. Also, to achieve efficient coupling between the amino acids, either the carboxyl or the amino function has to be suitably activated. Most often the carboxyl group is activated, and the chain is built up from the C-terminal end. Each time a new amino acid is coupled with the growing chain, the protecting group at the N-terminal end must be removed. The steps are: (1) protection of functional groups in side chains and at nonreacting N- and C- termini; (2) activation of carboxyl of N-protected amino acid; (3) coupling of the amino acid to the free NH_2 of peptide chain; (4) removal of the protecting group on the N-terminus.

The principal reagents used to protect functional groups in peptide synthesis are listed in Table 25-2 together with brief descriptions of how they are introduced and how they are removed. In the coupling step, it is imperative that racemization be minimized and that the yield is maximized. The most useful methods to activate the carboxyl function employ azides, $RCON_3$, anhydrides, RCO-O-COR', and aryl esters, RCO-OAr. A related method utilizes N, N-dicyclohexylcarbodiimide, $C_6H_{11}N=C=NC_6H_{11}$, as a condensing agent. This type of reagent functions by reacting rapidly with the free acid to form an adduct which acylates the free amino group of the peptide chain,

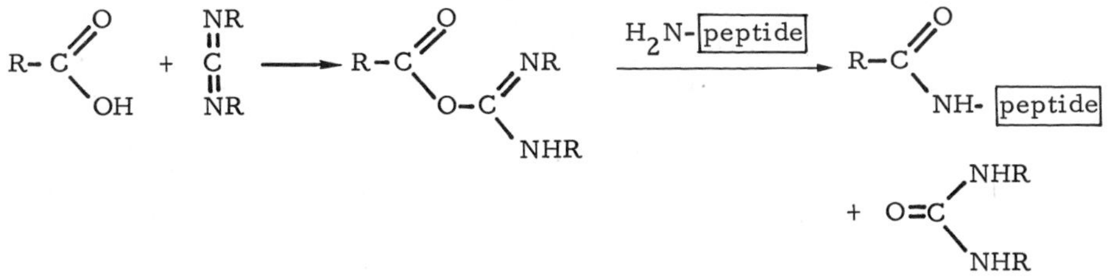

Solid-phase peptide synthesis is an elegant modification of the classical procedures for peptide syntheses that optimizes the yield by minimizing material
losses through handling. This is achieved by anchoring the peptide chain to be
lengthened to an insoluble polystyrene resin. The resin is best made porous
enough so that it can be penetrated by the reagents used for the peptide synthesis. The entire operation of multistep synthesis can be programmed to run
automatically.

Peptides and proteins can be separated and purified by physical
methods such as ultracentrifugation, gel filtration, ion-exchange chromatography,
electrophoresis and affinity chromatography.

The structure of a protein is described in terms of four levels of
complexity. The primary structure refers to the sequence of amino acids in the
polypeptide chains; the secondary structure refers to the way the chains are
coiled or pleated through local $N-H \cdots O=C$ hydrogen bonding to form the α -
helix or β -pleated sheet arrangements; the tertiary structure is the folding of
the coiled chains through long-range formation of disulfide bridges, binding of
metals or other prosthetic groups, Van der Waal's interactions, hydrogen
bonding of side chains, and attractive forces between ionic groups. The quaternary structure involves association of individual, coiled and folded protein
molecules (or subunits) into a complex "packet" of protein, as we illustrate
for hemoglobin (Section 25-8B) and keratin (Figure 25-17). Loss of the three-
dimensional structure (denaturation) generally results in the loss of biological
activity.

A number of important proteins (see Table 25-3) have metal atoms
associated, one way or the other, with the peptide chains which are especially
important to their appointed functions. Such groups are often called prosthetic
groups. Cytochrome c, hemoglobin, and myoglobin possess the same prosthetic group, called heme. The heme group is necessary for the biological
activity of these proteins and functions to transmit electrons (cytochrome c)
and to bind reversibly molecular oxygen (hemoglobin and myoglobin).

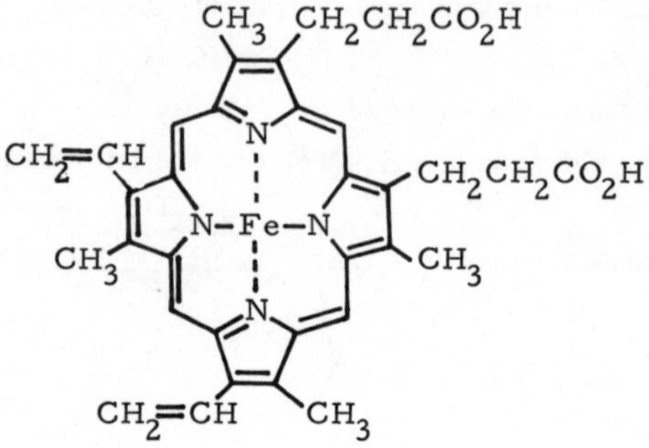

heme

In enzyme-catalyzed reactions, the enzyme first complexes with the reactant (<u>substrate</u>) at a region within the enzyme called the <u>active site.</u> It is thought that either the enzyme of the substrate may undergo conformational changes in forming the enzyme-substrate complex (induced-fit theory) and that the complex is held together by non-covalent interactions between the enzyme and the substrate, $E + S \rightleftharpoons ES$.

After complex formation, the enzyme reacts with the substrate - in some instances, at least, by binding covalently with it; $ES \longrightarrow ES'$. The reaction is completed with the release of products and regeneration of the enzyme; $ES' \longrightarrow P + E$. Some particularly well-studied enzymic reactions include peptide hydrolysis by carboxypeptidase A and chymotrypsin. It appears that enzymes can achieve what are normally base-catalyzed reactions at near neutrality through a chain of proton transfers.

Many enzymes operate in conjunction with small organic molecules called <u>coenzymes</u> or <u>cofactors</u>. Some important coenzymes are: $NAD^{\oplus}$, FAD, pyridoxal, thiamine, ATP, and coenzyme A.

The biosynthesis of proteins is dictated and regulated by the genes - the nucleoproteins containing the nucleic acids DNA and RNA.

The generation of the molecules of life before life existed on earth is believed to have taken place under the influence of solar radiation in the atmosphere of methane, ammonia, and water that is postulated to have existed in pre-biological times. The key intermediates are expected to have been methanal and hydrogen cyanide. From these substances, and ammonia, it is possible to conceive of the formation of amino acids, purine and pyrimidine bases of DNA and RNA, sugars, and, from there, to biopolymers, proteins, nucleic acids and carbohydrates.

ANSWERS TO EXERCISES

25-1   Isoleucine, threonine, cystine, hydroxylysine, and hydroxyproline each have two chiral centers. The possible stereoisomers of isoleucine with the <u>L</u> configuration at C2 are given here in projection.

$$
\begin{array}{cc}
\text{CO}_2\text{H} & \text{CO}_2\text{H} \\
\text{H}_2\text{N}\!-\!\!\!-\!\text{H} & \text{H}_2\text{N}\!-\!\!\!-\!\text{H} \\
\text{H}_3\text{C}\!-\!\!\!-\!\text{H} & \text{H}\!-\!\!\!-\!\text{CH}_3 \\
\text{CH}_2\text{CH}_3 & \text{CH}_2\text{CH}_3
\end{array}
$$

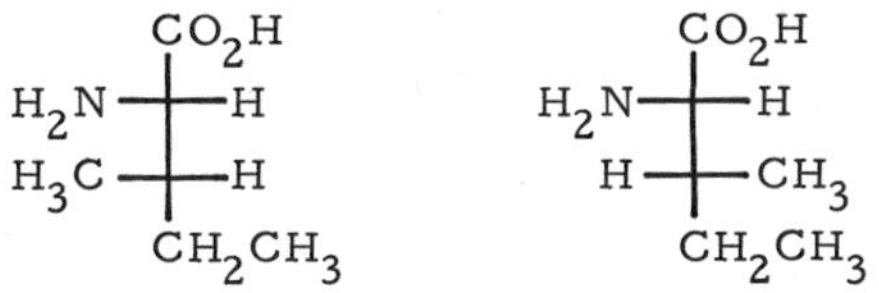

25-2   <u>Acidic amino acids:</u> aspartic, glutamic.

   <u>Basic amino acids:</u> lysine, hydroxylysine, arginine.

Notice that histidine may be classified as both an acidic and a basic amino acid because the imidazole ring is both weakly acidic and weakly basic (Section 23-7D, and Exercise 23-19). Arginine has the most strongly basic nitrogen (Section 23-7D). The amino acid with the least basic amino nitrogen function is probably tryptophan, which is unstable in acid. (Compare with pyrrole, Exercise 23-15.)

25-3 The entire plot would be displaced towards higher pH. That is to say, if we were to start at pH 0 and add base, the concentration of the monocation $\overset{\oplus}{H_3}N(CH_2)_5CO_2H$ would become significant at a <u>higher</u> pH than for glycine. This is because the CO_2H group is less acidic in the hexanoic acid ($NH_3^{\oplus}$ group is more distant). The isoelectric point would be $(4.43 + 10.75)/2 = 7.59$, and the concentration of monoanion $H_2N(CH_2)_5CO_2^{\ominus}$ would be significant at higher pH than $H_2NCH_2CO_2^{\ominus}$. Also, the <u>width</u> of the pH range over which the dipolar ion is stable is narrower for 6-aminohexanoic acid than for glycine. (Compare $10.75 - 4.43 = 6.32$ with $9.60 - 2.35 = 7.25$.)

25-4 Isoelectric point, pI, is defined as the pH at which $[H_2NCHRCO_2^{\ominus}] = [\overset{\oplus}{H_3}NCHRCO_2H]$.

On adding Equations 25-1 and 25-2 we get:

$$p\underline{K}_A + p\underline{K}'_A = 2pH + \log_{10}\frac{[\overset{\oplus}{H_3}NCH_2CO_2H][\overset{\oplus}{H_3}NCH_2CO_2^{\ominus}]}{[\overset{\oplus}{H_3}NCH_2CO_2^{\ominus}][H_2NCH_2CO_2^{\ominus}]} = 9.60 + 2.35$$

and because the log term reduces to zero at the pI,

$$p\underline{K}_A + p\underline{K}'_A = 2p\underline{I} = 9.60 + 2.35$$

or

$$p\underline{I} = (9.60 + 2.35)/2$$

25-5 a. The order of stability of the monocations is:

and the order of stability of the neutral forms is:

The rationale is that C (or F) is least stable because of the incompatibility of moderately basic functions coexisting with moderately acidic functions. Regarding D and E (or A and B), the most stable form is the one corresponding to the <u>least</u> acidic $\overset{\oplus}{N}H_3$ group (the most strongly basic NH_2 group) which is at C6 - not C2.

b.

$$HO_2CCH_2CH_2\underset{\overset{|}{\overset{\oplus}{N}H_3}}{C}HCO_2H \xrightleftharpoons{\quad p\underline{K}_a \ 2.19 \quad} HO_2CCH_2CH_2\underset{\overset{|}{\overset{\oplus}{N}H_3}}{C}HCO_2^{\ominus} \xrightleftharpoons{\quad p\underline{K}_a \ 4.25 \quad}$$

$$\overset{\ominus}{O}_2CCH_2CH_2\underset{\overset{|}{NH_2}}{C}HCO_2^{\ominus} \xrightleftharpoons{\quad p\underline{K}_a \ 9.67 \quad} \overset{\ominus}{O}_2CCH_2CH_2\underset{\overset{|}{\overset{\oplus}{N}H_3}}{C}HCO_2^{\ominus}$$

p$\underline{I}$ (glutamic acid) = (2.19 + 4.25)/2 = 3.22

$$\underset{\overset{|}{\overset{\oplus}{N}H_3}}{H_2N=}\overset{\overset{NH_2}{|}}{\underset{H}{C}}N(CH_2)_3\underset{}{C}HCO_2H \xrightleftharpoons{\quad p\underline{K}_a = 2.17 \quad} \underset{\overset{|}{\overset{\oplus}{N}H_3}}{H_2N=}\overset{\overset{NH_2}{|}}{\underset{H}{C}}N(CH_2)_3CHCO_2^{\ominus} \xrightleftharpoons{}$$

$$HN=\overset{\overset{NH_2}{|}}{\underset{H}{C}}N(CH_2)_3\underset{\overset{|}{NH_2}}{C}HCO_2^{\ominus} \xrightleftharpoons{\quad 12.48 \quad} \overset{\oplus}{H_2N}=\overset{\overset{NH_2}{|}}{\underset{H}{C}}N(CH_2)_3\underset{\overset{|}{NH_2}}{C}HCO_2^{\ominus} \xrightleftharpoons{\quad p\underline{K}_a \ 9.04 \quad}$$

p$\underline{I}$ (arginine) = (9.04 + 12.48)/2 = 10.76

<u>25-6</u>

a. C=O at 1710 cm^{-1}, $\overset{\oplus}{N}$-H at 3100 - 2600 cm^{-1}

b. $-CO_2^{\ominus}$ at 1600 cm^{-1}, N-H at 3500 - 3300 cm^{-1}

<u>25-7</u> $H_2N\underset{\overset{|}{CH_3}}{C}HCO_2H$ + 2D$_2$O (excess) $\xrightleftharpoons{}$ $D_2N\underset{\overset{|}{CH_3}}{C}HCO_2D$ + 2H$_2$O

The observed nmr spectrum would be of $D_2NCH(CH_3)CO_2D$ and would show a three-proton doublet about 1 ppm for the methyl group and a 1-proton quartet below 3.00 ppm for the methine (CH) group, and a singlet resonance near 5 ppm for H$_2$O and HDO.

25-8

(partial structure for ninhydrin)

$$\text{(partial structure)} \quad C=O + H_2\ddot{N}CH_2CO_2H \rightleftharpoons$$

$$C=NHCH_2CO_2H \xleftarrow{-H_2O} \quad \xrightarrow{H^{\oplus}}$$

$$-H^+$$

$$\xrightarrow{-CO_2} \quad CH-N=CH_2 \xrightarrow{H_2O} \quad CH-NH-CH_2-O-H$$

$$\xleftarrow{-H^{\oplus}} \quad C=N-CH \xleftarrow[\text{more ninhydrin}]{-H_2O} \quad CHNH_2 \xleftarrow[\text{reduction step}]{-CH_2O}$$

$$C=O$$

blue

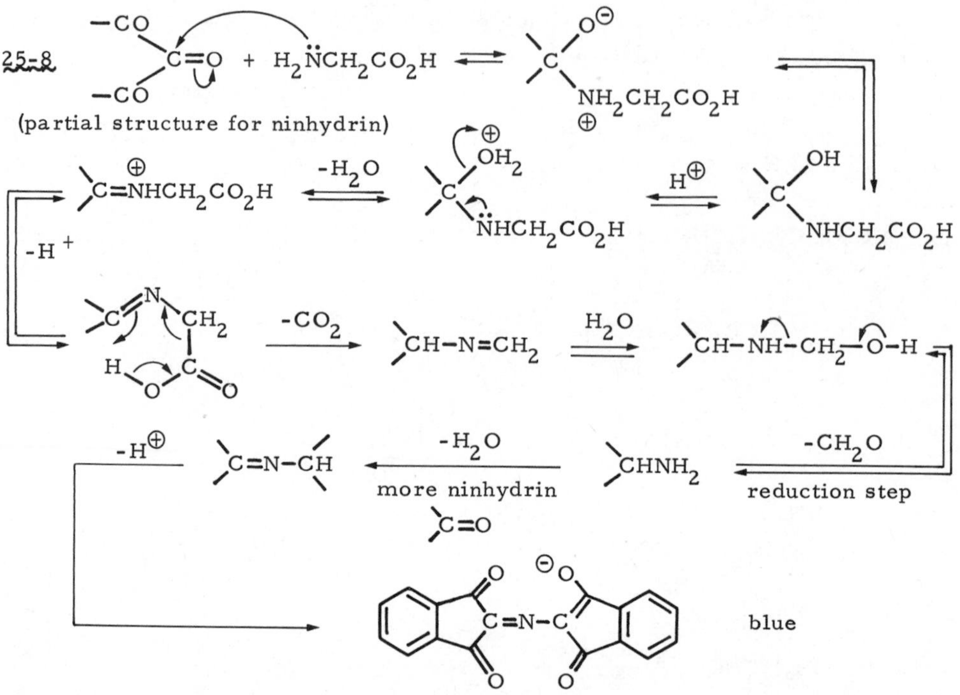

By this sequence, the color is expected to be independent of the character of the amino acid used because only the amino nitrogen of the amino acid is incorporated into the color salt. Neither ammonia nor methanamine should give the color because a reduction step is required in the sequence to give the intermediate $CHNH_2$. Neither CH_3NH_2 nor NH_3 can do more than form an imine derivative with ninhydrin.

25-9　　Arginine is a strongly basic amino acid and exists in the dipolar form

$$H_2NCHCO_2^{\ominus}$$

at much higher pH (basic conditions) than does the

$$(H_2N)_2\overset{\oplus}{C}-\underset{H}{N}(CH_2)_3$$

dipolar form of glutamic acid

$$H_3\overset{\oplus}{N}CHCO_2^{\ominus}$$
$$(CH_2)_2CO_2H$$

. That is to say, the carboxyl function of glutamic acid is more acidic than the $(H_2N)_2\overset{\oplus}{C}-\overset{H}{N}-$ function of arginine, and therefore dissociates more completely (elutes) at higher pH (less basic conditions) than does arginine.

25-10　　Radical polymerization of styrene follows:

$$CH_2=CHAr \xrightarrow{\cdot OR} RO-CH_2-\overset{\cdot}{C}HAr \xrightarrow{CH_2=CHAr} RO CH_2\overset{}{C}HCH_2\overset{\cdot}{C}HAr$$
$$\xleftarrow{n\ CH_2=CHAr}$$

$$\text{Ar Ar Ar}$$

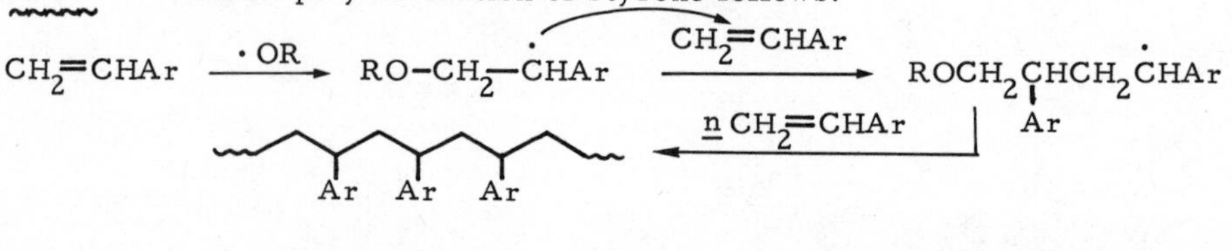

When polymerization is carried out in the presence of 2-10% of 1,4-diethenyl-
benzene, then some of the aryl groups in the growing polymer chain will possess
4-ethenylphenyl groups. Cross-linking occurs as the 4-ethenylphenyl group of
one chain adds to the growing end of another. Ultimately, this leads to a three-
dimensional network of polymer chains interconnected by $1,4-C_6H_4$ groups.

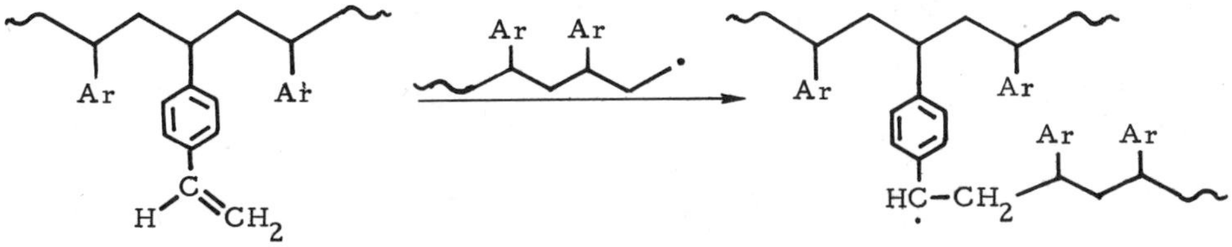

Treatment of the polymer with $H_2SO_4-SO_3$ introduces a para-HSO_3
group into some of the phenyl group

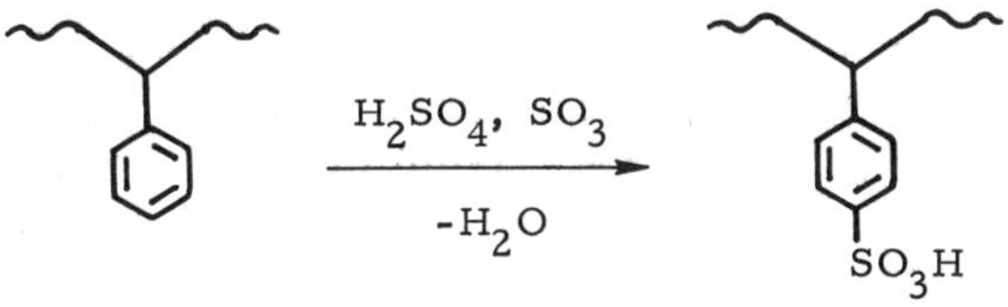

25-11   A resin that contains cationic groups, such as resin-$\overset{\oplus}{N}H_3$ $\overset{\ominus}{O}H$.

25-12

a.

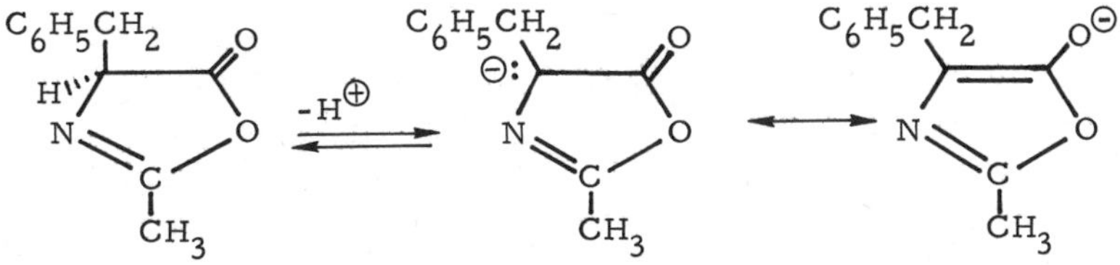

b.   The acidic hydrogen is the one that was originally the alpha hydrogen of phenylalanine. It is acidic because the conjugate base is resonance
stabilized.

c.   The chiral configuration of azlactones from α-amino acids is lost
on forming the conjugate base. Thus, the azlactone is racemized easily in the
presence of a base (e.g., ethanoate) and on hydrolysis of the azlactone, the
product amino acid will be racemic.

25-13   The nitrosation steps are the same for both compounds and would
produce the two diazonium ions A and B.

A $\overset{\oplus}{N_2}CH_2CO_2H \xrightarrow{-H^{\oplus}} \overset{\oplus}{N_2}CH_2CO_2^{\ominus}$

B $\overset{\oplus}{N_2}CH_2CO_2C_2H_5 \xrightarrow{-H^{\oplus}} N_2CHCO_2C_2H_5$

Deprotonation of B gives the diazoester $N_2CHCO_2C_2H_5$, but loss of a proton from A to give N_2CHCO_2H is not expected because the carboxyl hydrogen is more acidic than the <u>alpha</u> hydrogen. The product is not stabilized as is ethyl diazoethanoate (see Exercise 23-28) and loses nitrogen readily, possibly by formation of an unstable α lactone.

$$\overset{\oplus}{N_2}-CH_2CO_2^{\ominus} \xrightarrow{-N_2} \underset{O}{CH_2-C=O} \xrightarrow{H_2O} HOCH_2CO_2H$$

25-14

a. $HO_2C-(CH_2)_2-\overset{O}{\overset{||}{C}}-CO_2H \xrightarrow[HCN]{NH_3} HO_2C-(CH_2)_2-\overset{NH_2}{\underset{CO_2H}{\overset{|}{C}}}-CN \xrightarrow[H^{\oplus}]{H_2O}$

$HO_2C(CH_2)_2\underset{NH_2}{\overset{|}{C}}HCO_2H \xleftarrow[-CO_2]{heat} HO_2C-(CH_2)_2-\overset{NH_2}{\underset{CO_2H}{\overset{|}{C}}}-CO_2H$

b. $CH_3-\overset{CH_3}{\overset{|}{C}}H-CH_2-OH \xrightarrow[ZnCl_2]{HCl} (CH_3)_2CHCH_2Cl$

(phthalimide-$N-\overset{\ominus}{C}(CO_2R)_2$)

$-Cl^{\ominus}$

(phthalimide-$N-\overset{CH_2CH(CH_3)_2}{\underset{CO_2R}{\overset{|}{C}}}-CO_2R$) $\xleftarrow[-CO_2]{\overset{1.\ H_3O^{\oplus}}{2.\ heat}}$

(phthalimide-$N-\overset{|}{C}HCO_2H$, $CH_2CH(CH_3)_2$) $\xrightarrow{H_2NNH_2} H_2N\overset{|}{C}HCO_2H$ $CH_2CH(CH_3)_2$ $+$ (phthalhydrazide NH-NH)

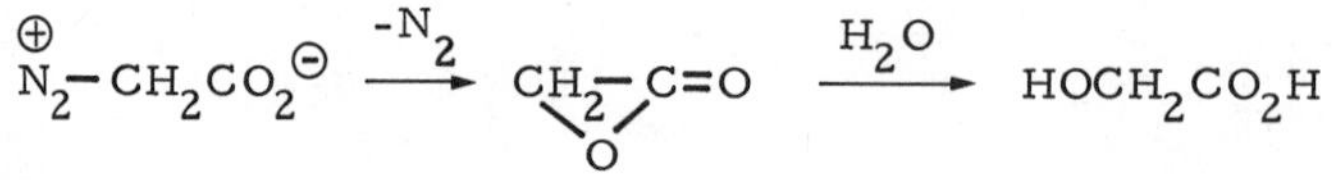

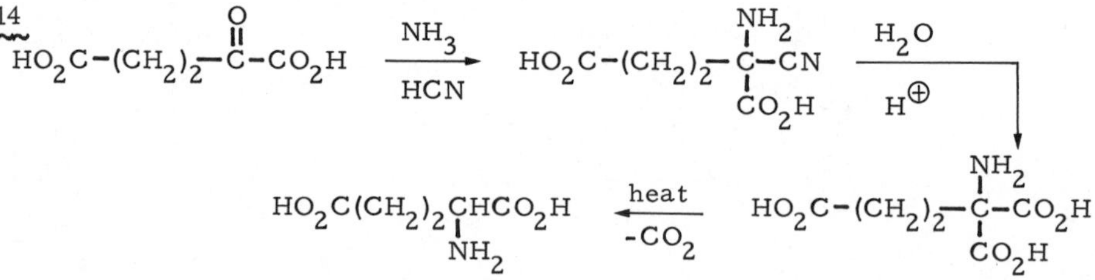

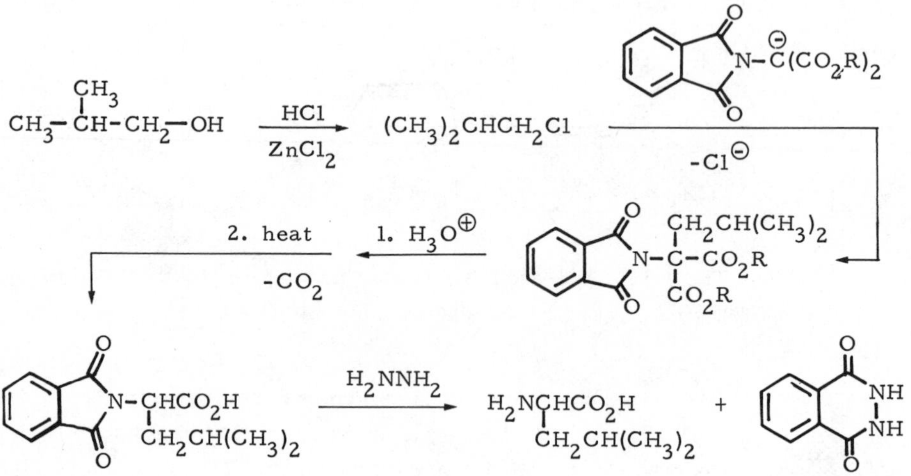

c. 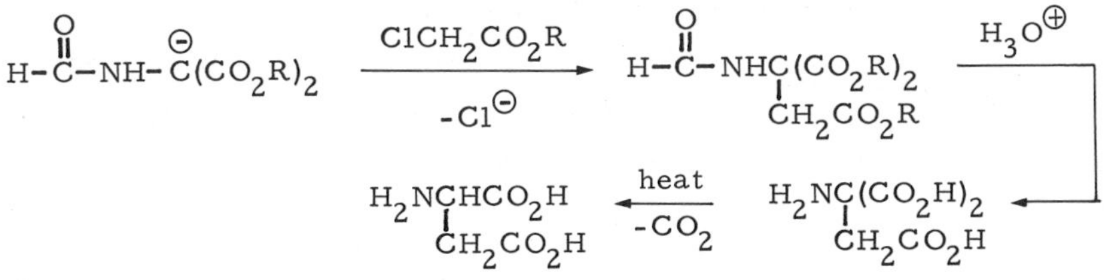
$$H-\overset{O}{\overset{||}{C}}-NH-\overset{\ominus}{\overset{|}{C}}(CO_2R)_2 \xrightarrow[-Cl^{\ominus}]{ClCH_2CO_2R} H-\overset{O}{\overset{||}{C}}-NH\overset{|}{\underset{CH_2CO_2R}{C}}(CO_2R)_2 \xrightarrow{H_3O^{\oplus}}$$

$$\underset{\overset{|}{CH_2CO_2H}}{H_2NCHCO_2H} \xleftarrow[-CO_2]{heat} \underset{\overset{|}{CH_2CO_2H}}{H_2N\overset{|}{C}(CO_2H)_2}$$

25-15

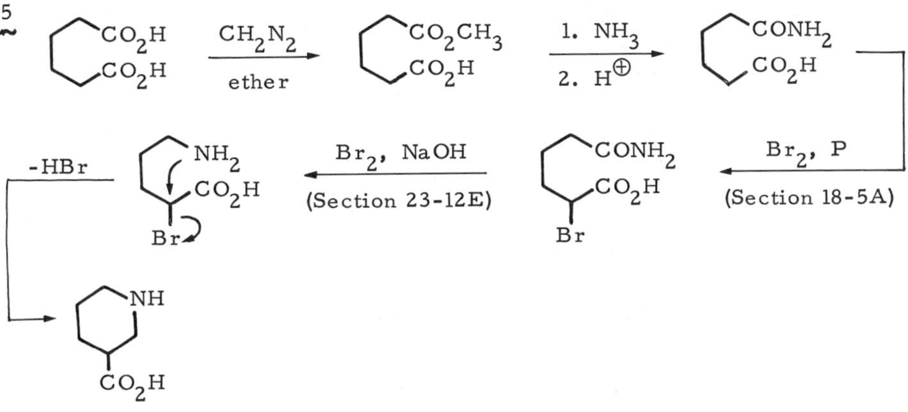

25-16   See Figure 25-9

25-17
$$\underset{\overset{|}{\underset{\overset{|}{NH_2}}{(CH_2)_4}}}{H_2NCH_2CONHCHCONH\overset{\overset{CH_3}{|}}{C}HCO_2H}$$

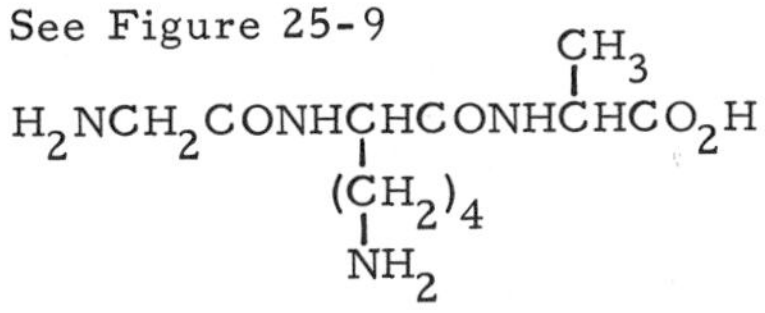

Arylation of the NH_2 group of glycine and the terminal NH_2 group of lysine would occur.

25-18   Eisenine $\xrightarrow{H_2O}$ 2 glutamic acid + alanine + ammonia

Given that **alanine** is the C-terminal acid, the only possible sequence is Glu-Glu-Ala. However, there is no free NH_2 group in the terminal amino acid because the peptide does not react with 2,4-dinitrofluorobenzene. The peptide must then be a cyclic amide. There is only one free carboxyl group in eisenine and NH_3 is given off on hydrolysis, which means that a second carboxyl is present as an amide function, $-\overset{O}{\overset{||}{C}}-NH_2$. The most likely structure is:

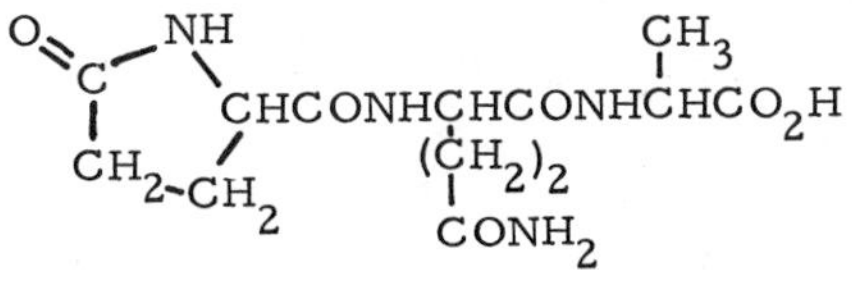

25-19   From (2) we may conclude that the peptide eledoisin is either cyclic
or that both the N-terminal and C-terminal groups are amide functions. From
(3) we know that the C-terminal residue is methionine, present as an amide,
and that the sequence from the C-terminal end is:

(Ala, Asp, Glu, Lys, Pro, Ser)– Phe – Ile – Gly– Leu– Met NH$_2$

The two possible sequences for Lys– Q are:

Lys– Ala– Asp– Phe– Ile – Gly– Leu– Met NH$_2$

Lys– Asp –Ala– Phe– Ile – Gly– Leu– Met NH$_2$

The correct sequence for Lys – Q can be deduced from the information that L
on trypsin hydrolysis gives a tripeptide of sequence Asp– Ala– Phe, with Asp
as the N-terminal group. We can now write the sequence for eledoisin as:

(Ser, Pro, Glu)– Lys– Asp– Ala– Phe– Ile– Gly– Leu– Met NH$_2$

Since partial acid hydrolysis gives dipeptides Ser– Lys and Pro – Ser, we know
the sequence must be Pro – Ser– Lys, and the total sequence is:

Glu – Pro –Ser– Lys– Asp– Ala– Phe– Ile – Gly– Leu– Met NH$_2$

Since there is no free N-terminal amino group, the N-terminal amino acid,
glutamic acid, is probably present as a cyclic amide.

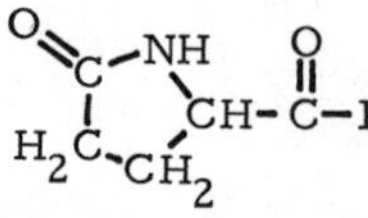

25-20   There is no free C-terminal carboxyl in the hexapeptide. It is
probably present as –CONH$_2$. The two peptides from chymotrypsin hydrolysis,
taken together with the Edman degradation and carboxypeptidase information,
establish the partial sequence:

Arg – Trp –Gly(Lys, Pro, Val)

The partial hydrolysis information gives a sequence for the larger of the two
peptides as:

Gly – Lys –Pro – Val NH$_2$

and the hexapeptide as:

Arg– Trp– Gly– Lys –Pro– Val NH$_2$

25-21   Racemization occurs by way of azlactone intermediates. See
Exercise 25-12.

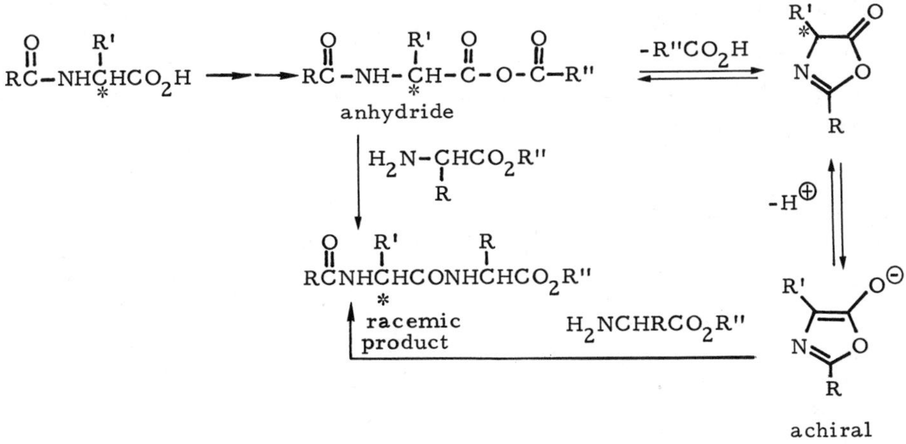

25-22   The yield of the desired polypeptide after 100 steps would be
$(0.99)^{100}$ x 100% = 36.6%.

25-23   Order of reactivity correlates with efficacy of X as a leaving group
in the reaction:

RCOX + H_2N-[peptide]-Y $\longrightarrow$ RCONH-[peptide]-Y + HX

Based on pK_a values of HX (Table 24-1) and steric factors, the order predicted
is: 3 > 6 > 2 > 4 > 5. The carbonic anhydride (1) is comparable to 3 and 6.

One of the problems with good coupling reagents 1, 3 and 6 is that
they are also good at forming azlactones and thence lead to racemization. Strong
acids, HX, are undesirable byproducts, and 4-nitrophenol is not easily
separated from a water-soluble peptide.

25-24   Route 1 gives a yield of 0.8 x 0.8 x 0.8 x 100 = 51%
Route 2 gives a yield of 0.8 x 0.8 x 100 = 64%

25-25

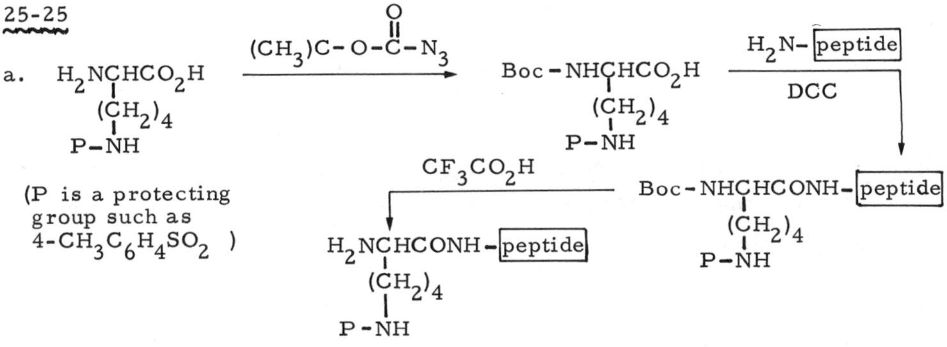

Parts b, c and d are similar to Part a except that the side-chain carboxyl of aspartic acid and the side-chain hydroxyl of serine must be suitably protected. RCO_2H can be protected as $RCO_2C_2H_5$, and ROH as $ROCH_2-C_6H_5$. See Table 25-2.

<u>25-26</u>   The abbreviations used here correspond to those of Section 23-13C and Figure 25-10.

a. Glu $\xrightarrow[H_2SO_4]{C_2H_5OH}$ $\overset{\oplus}{H_3N}-\underset{\underset{CO_2C_2H_5}{(CH_2)_2}}{CHCO_2C_2H_5}$ HO—⟨ ⟩—NO₂ ester exchange

$\overset{\oplus}{H_3N}-\underset{\underset{CO_2C_2H_5}{(CH_2)_2}}{CH-CO_2C_6H_4-NO_2}$

Boc—N₃

Boc—NH—$\underset{\underset{CO_2C_2H_5}{(CH_2)_2}}{CHCO_2-C_6H_4-NO_2}$

Notice that the ester group at C1 is the more reactive since it is activated by the $\overset{\oplus}{H_3N}$ group.

$H_2NCH_2CO_2C_2H_5$ → Boc—Glu—Gly—$\overset{O-C_2H_5}{\overset{\|}{OC_2H_5}}$ $\xrightarrow[H_2O]{\overset{\oplus}{H}}$ Glu—Gly

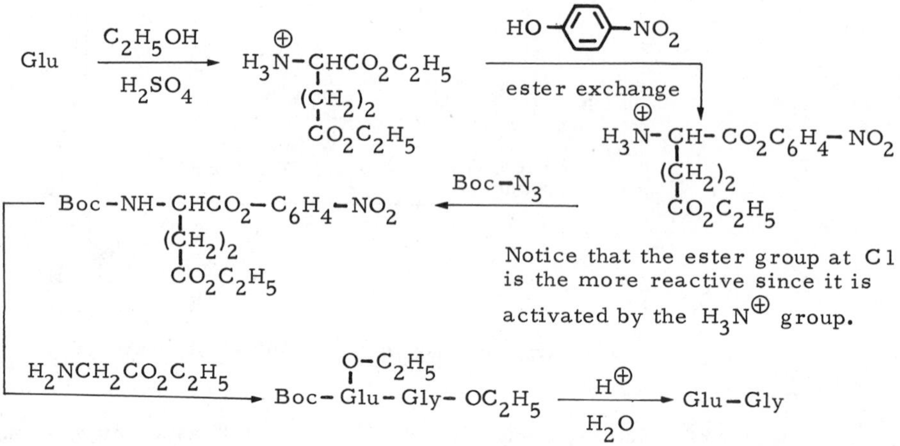

b. Ala $\xrightarrow{Z-Cl}$ Z—Ala $\xrightarrow[DCC]{Val-OC_2H_5}$ Z—Ala—Val OC₂H₅ $\xrightarrow[CH_3CO_2H]{HBr}$

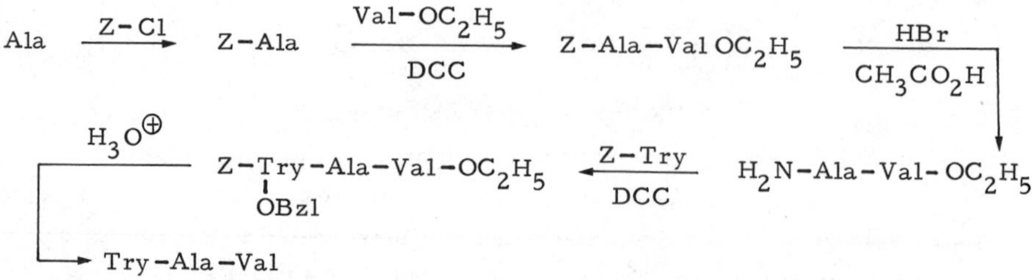

$\overset{\oplus}{H_3O}$ → Z—Try—Ala—Val—OC₂H₅ $\xleftarrow[DCC]{Z-Try}$ H₂N—Ala—Val—OC₂H₅
OBzl

→ Try—Ala—Val

25-27   polysaccharide-OH + CH₂—CH-CH₂Cl $\xrightarrow{(-HCl)}$

polysaccharide—O-CH₂-CH-CH₂

polysaccharide—OH

polysaccharide—O-CH₂CH-CH₂—O—polysaccharide
OH

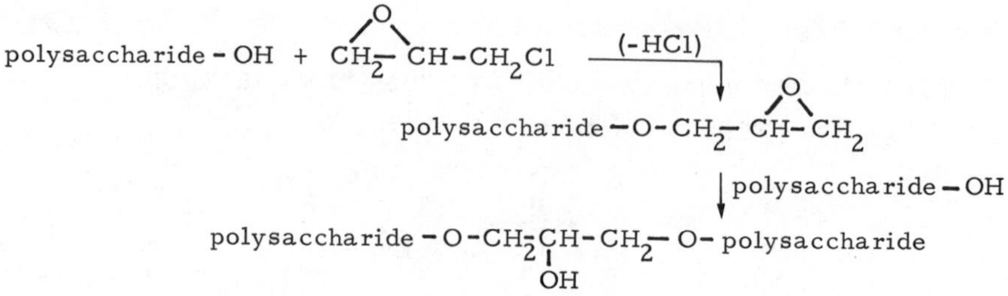

<u>25-28</u>   Let M be the molecular weight of hemoglobin. Assume that there is at least one tryptophan residue per molecule of hemoglobin. Then the number of moles of hemoglobin in 100 g is given by 1.48/204, where 204 is the molecu-

lar weight of tryptophan.   Then:

$$100/M = 1.48/204$$
$$M = 13,800$$

In 100 g of hemoglobin there are 0.355 g of iron, or 0.355/56 g atoms of iron, or 0.355/56 moles of hemoglobin, if there is at least one iron per mole of hemoglobin.   Hence:

$$M \times 0.355/55.8 = 100$$
$$M = 15,700$$

These analyses indicate that the molecular weight of hemoglobin is in the range 13,800 - 15,700, or any multiple thereof depending on whether there are 1, 2, 3 or 4 etc. g. atoms of iron or moles of tryptophan per mole of hemoglobin. Actually there are four iron atoms and the molecular weight is indicated to be in the range 55,200 to 62,400.

25-29   A hydrolytic enzyme that reacts exclusively (or more rapidly) with either $\underline{D}$- or $\underline{L}$-peptides may be used for the resolution of $\underline{D}, \underline{L}$-amino acids. For example:

$$\underline{D}, \underline{L}\text{-alanine} \xrightarrow{(CH_3CO)_2O} \underline{D}, \underline{L}\text{-}CH_3CONHCH(CH_3)CO_2H \xrightarrow[\text{enzyme}]{\text{hydrolytic}}$$

$$CH_3CO_2H + \underline{\underline{L}\text{-}H_2NCH(CH_3)CO_2H + \underline{D}\text{-}CH_3CONHCH(CH_3)CO_2H}$$

separate

25-30   Papain has a potential but not actual −SH group.   The −SH group is masked (perhaps as a hemithioketal) but is evidently easily reformed.   Acylation of the −SH group precedes the hydrolytic step in papain-catalyzed hydrolytic peptide cleavage.

$$\text{Enzyme−SH} + \text{RCONHR'} \longrightarrow \text{Enzyme−S−}\overset{\overset{O}{\|}}{C}\text{−R} + H_2NR'$$

$$\text{Enzyme−SH} + \text{HO}\overset{\overset{O}{\|}}{C}\text{R} + H_2NR' \xleftarrow{H_2O}$$

25-31   The decarboxylation step is favorable because the products are relatively stable species.   The electron delocalization that stabilizes the decarboxylated thiamine adduct is shown in the following VB structures.

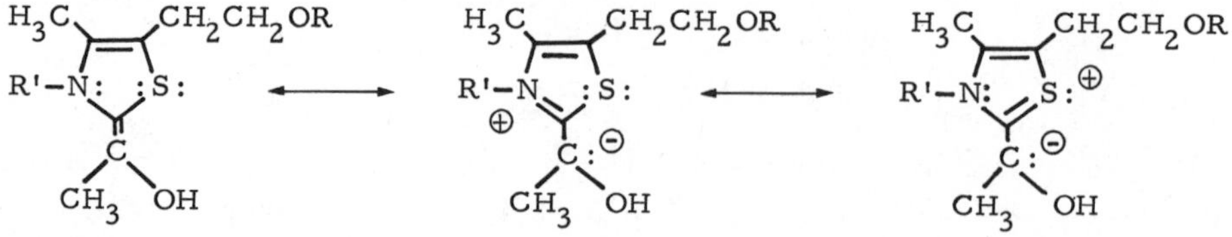

25-32

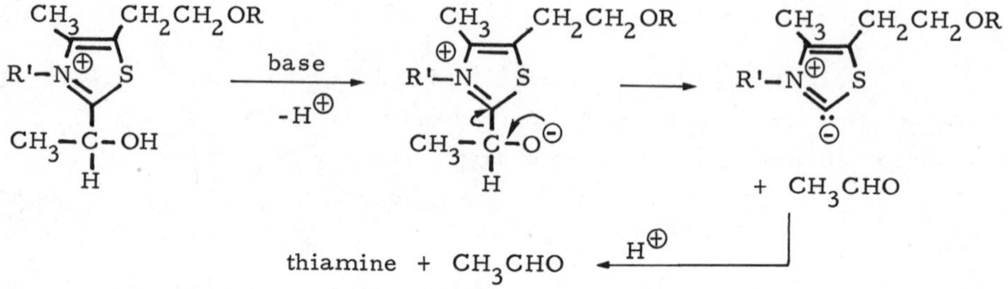

thiamine + CH_3CHO

The above is a reverse aldol-type reaction.

25-33   The equilibria in question follow:

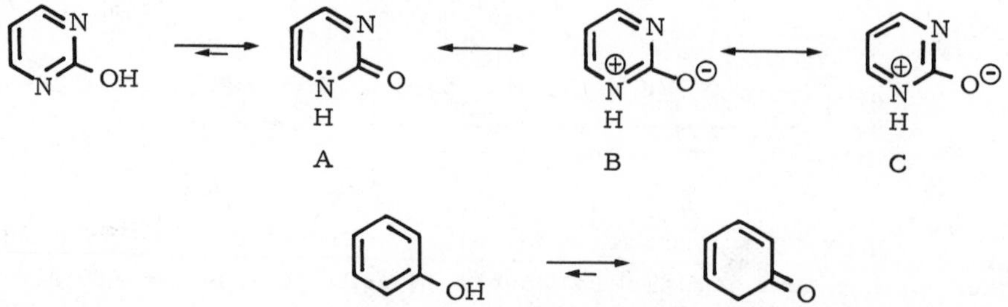

Atomic-orbital models indicate that the lactam form can enjoy aromatic-type stabilization whereas aromaticity is destroyed in the cyclohexadienone. The VB structures A ⟷ B ⟷ C best show this as an amide-style resonance overlaid on the conjugated pyrimidine ring system.

25-34

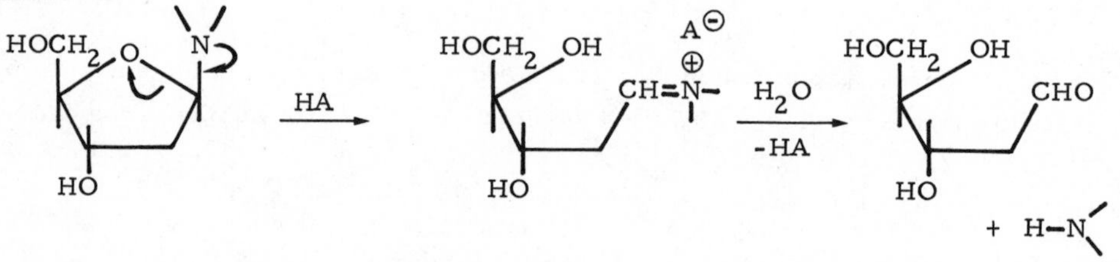

Hydrolysis is catalyzed by acids.

25-35

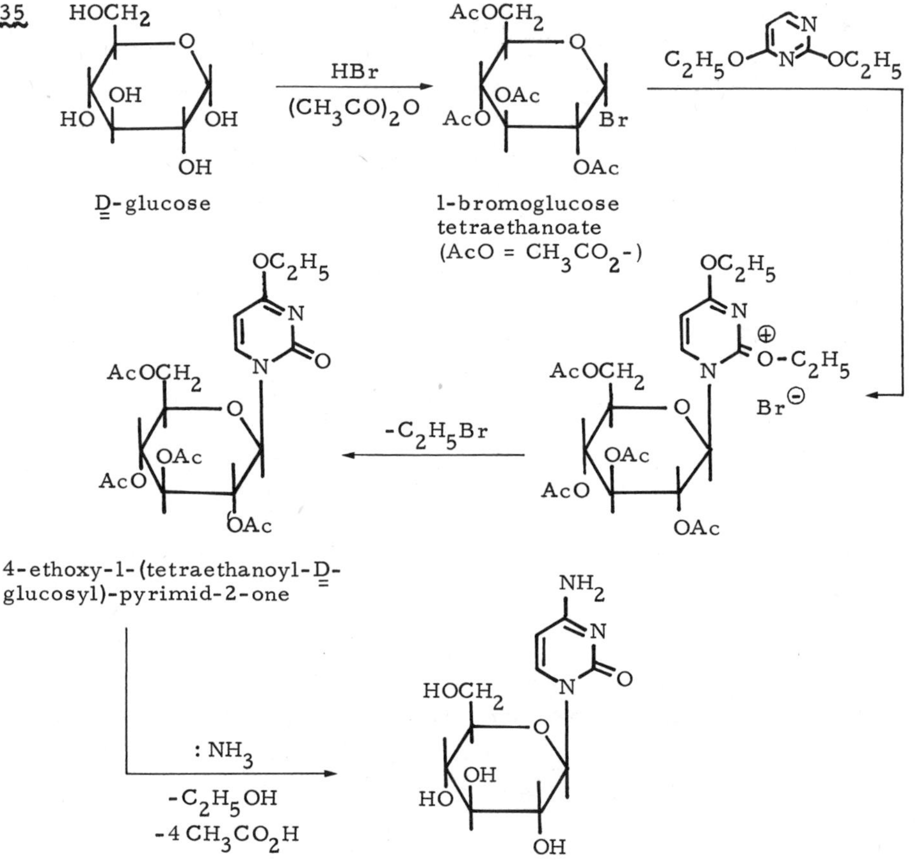

D-glucose

$\xrightarrow[\text{(CH}_3\text{CO)}_2\text{O}]{\text{HBr}}$

1-bromoglucose
tetraethanoate
$(AcO = CH_3CO_2-)$

$-C_2H_5Br$

4-ethoxy-1-(tetraethanoyl-D-
glucosyl)-pyrimid-2-one

: NH_3

$-C_2H_5OH$
$-4\,CH_3CO_2H$

The reaction does not yield 6-ethoxy-1-(tetraethanoyl-D-glucosyl)-pyrimid-2-one
because the ethoxy groups reduce the basicity (and nucleophilicity) of nitrogen
of the 2,4-diethoxypyrimidine. Steric hindrance may also be important.

25-36

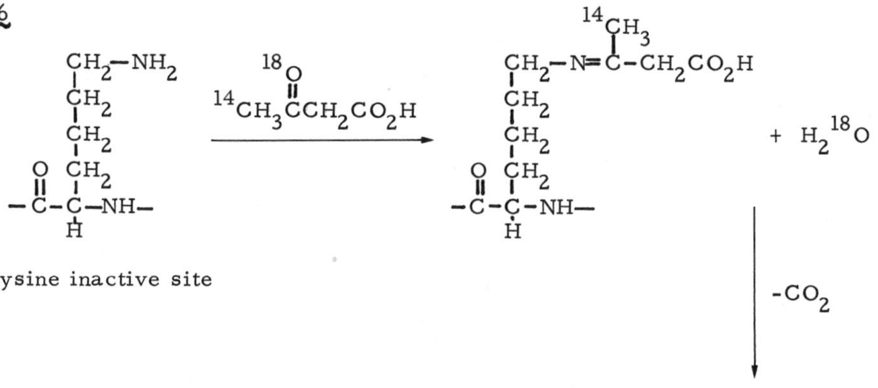

lysine inactive site

$^{14}CH_3\overset{^{18}O}{\overset{\|}{C}}CH_2CO_2H$

$+ H_2^{18}O$

$-CO_2$

(continued)

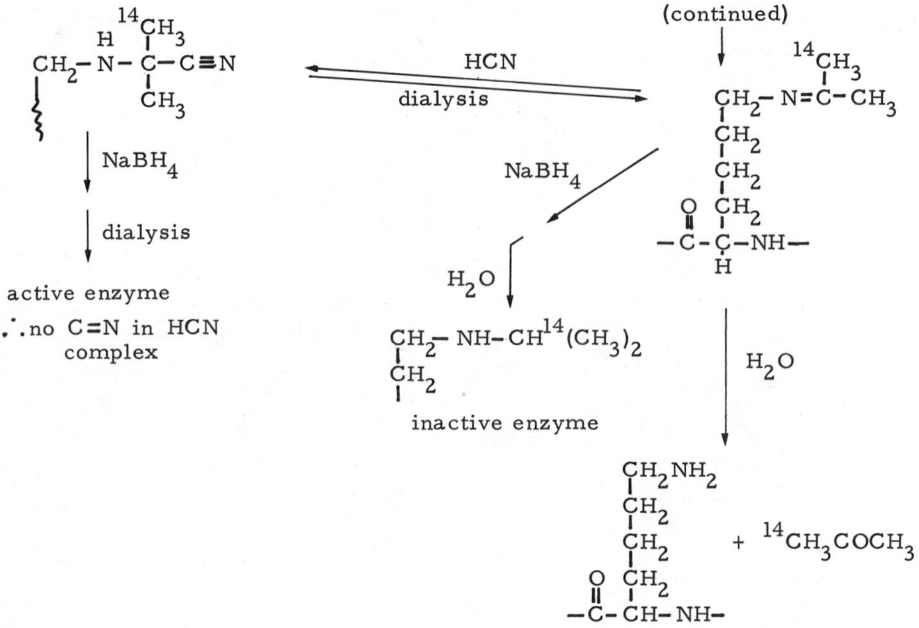

$$\underset{\overset{|}{\overset{?}{\underset{?}{}}}}{CH_2} - \overset{H}{\underset{|}{N}} - \overset{\overset{14}{C}H_3}{\underset{\underset{CH_3}{|}}{C}} - C \equiv N \quad \underset{\text{dialysis}}{\overset{HCN}{\rightleftharpoons}} \quad$$

(continued)

$$\downarrow$$

$$CH_2 - N = \overset{\overset{14}{C}H_3}{C} - CH_3$$

$$\underset{\underset{-C-C-NH-}{\overset{O}{\parallel}}}{\overset{CH_2}{\underset{\overset{|}{CH_2}}{\underset{|}{CH_2}}}}$$

$$\downarrow NaBH_4$$

$$\downarrow \text{dialysis}$$

active enzyme

∴ no C≡N in HCN
    complex

$NaBH_4$

H_2O

$$CH_2 - NH - CH^{14}(CH_3)_2$$
$$\overset{|}{CH_2}$$

inactive enzyme

H_2O

$$\underset{-C-CH-NH-}{\overset{O}{\parallel}}\overset{CH_2 NH_2}{\underset{\overset{|}{CH_2}}{\underset{\overset{|}{CH_2}}{\overset{|}{CH_2}}}} \quad + \quad {}^{14}CH_3COCH_3$$

25-37   a. Since there are five distinct resonances (one of them of double
height) for a tryptophan carbon there must be six tryptophan residues in lysozyme (in agreement with the structure shown in Figure 25-15). b. On reduction of
the S-S bonds, the lysozyme chain unfolds to some extent and allows the molecule more mobility. As a result, the six tryptophan residues, on a time-average
basis, are in much more similar environments than in the achiral enzyme.
Therefore the chemical shifts are more nearly the same.

CHAPTER 26

MORE ON AROMATIC COMPOUNDS.
ARYL OXYGEN COMPOUNDS.
SIDE-CHAIN DERIVATIVES.

There are many aspects of the chemistry of aromatic compounds that set them apart from aliphatic compounds, and it is appropriate to briefly review the main points here. Aromatic compounds owe their special stability to electron delocalization, which contributes on the order of $10 - 15$ kcal mole^{-1} of stabilization for each double bond of a Kekulé structure. This stabilization leads to modified chemical reactivity as summarized below.

1. Almost all single bonds to aromatic ring carbons, whether C—H, C-halogen, C—O, or C—N are <u>stronger</u> than corresponding single bonds to $\underline{sp}^3$ hybrid carbon. Aryl-X bonds are less easily broken in the sense $Ar \overset{\shortmid}{\cdot}{:}_{\neg}X$ or $Ar \overset{\shortmid}{|} :X$, and aryl radicals, $Ar\cdot$, and aryl cations, $Ar^{\oplus}$, are difficult to generate and are very reactive intermediates. Cleavage in the sense $Ar: \overset{\shortmid}{|} H$ is easier than in $C_{sp^3} - \overset{\shortmid}{|} H$, and Ar—H bonds are more acidic than R—H bonds.

2. Most of the useful substitution or replacement reactions on arene rings do not occur by simple radical S_N1 or S_N2 substitution mechanisms but involve <u>addition-elimination</u> or, much less commonly, <u>elimination-addition</u> (benzyne-type) reactions. Examples of addition-elimination include the very important nitration, halogenation, alkylation, and so on, electrophilic substitution reactions (Section 22-4 to 22-7 and Section 23-10C) and nucleophilic displacement reactions of activated aromatic halides (Section 14-6B). Examples are

<u>Electrophilic substitution</u>

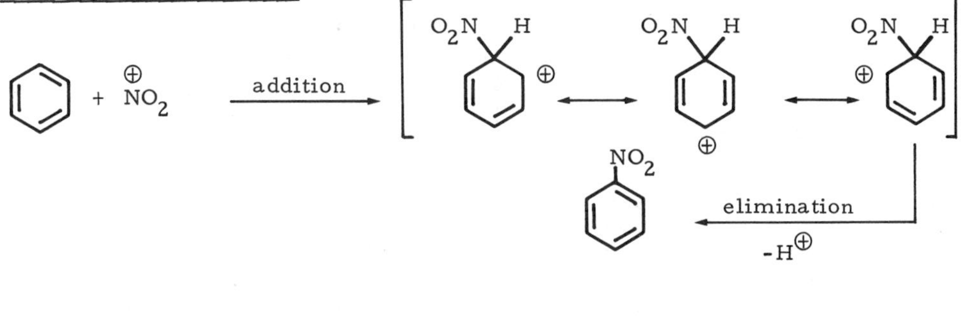

Nucleophilic substitution

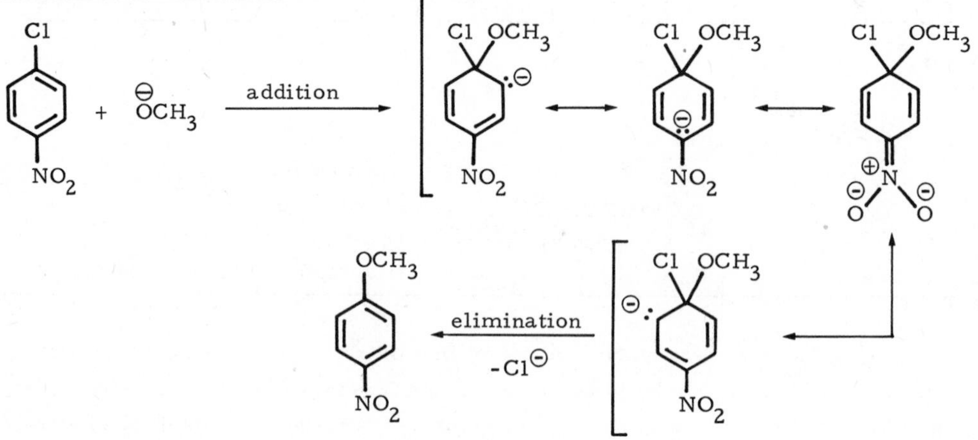

A few authenticated examples of nucleophilic substitution by way of aryl radicals and aryl cations are known, of which the most common involve replacement of the diazonium group in $Ar-N_2^{\oplus}$ in the presence of cuprous salts (Section 23-10B).

$$ArN_2^{\oplus} + Cu(I) \longrightarrow Ar\cdot + N_2 + Cu(II)$$

$$Ar\cdot + Cu(II) \longrightarrow Ar^{\oplus} + Cu(I)$$

$$Ar^{\oplus} + X^{\ominus} \longrightarrow ArX$$

Aryl-halogen bonds can also be activated by electron transfer, and conversion of aryl halides to the corresponding organometallic compounds probably proceeds through the following steps (Section 14-10A):

$$Ar-Cl \xrightarrow{e^{\ominus}} Ar-Cl^{\ominus} \longrightarrow Ar\cdot + Cl^{\ominus}$$

$$Ar\cdot \xrightarrow{e^{\ominus}} Ar:^{\ominus} \xrightarrow{Li^{\oplus}} Ar-Li$$

3. Important electron delocalization effects usually are observed with arenes carrying substituents having a p orbital on the first atom attached to the ring. Examples of substituent groups with different numbers of electrons in such a p orbital are: <u>vacant</u>, $-CH_2^{\oplus}$ and $-B(CH_3)_2$; <u>one electron</u>, $-\overset{\cdot}{C}H_2$ and $-\ddot{\overset{\cdot}{O}}\cdot$; <u>two electrons</u>, $-\ddot{C}H_2^{\ominus}$, $-\ddot{N}H_2$, $-\ddot{O}H$, and $-\ddot{\underset{\cdot\cdot}{F}}:$. The arene ring can interact with such substituents, provided that the p orbital in the substituent is essentially parallel to the p orbital of the first carbon of the ring.

Electrons are delocalized from the ring to the substituent p orbital, or from the substituent p orbital into the ring. In valence-bond notation, we can write:

$$\underset{}{\bigcirc}\text{-}\overset{\oplus}{\text{CH}}_2 \longleftrightarrow \overset{\oplus}{\bigcirc}\text{=CH}_2 \longleftrightarrow \oplus\bigcirc\text{=CH}_2 \longleftrightarrow \underset{\oplus}{\bigcirc}\text{=CH}_2$$

$$\bigcirc\text{-}\overset{..}{\text{CH}}_2 \longleftrightarrow \bigcirc\text{=CH}_2 \longleftrightarrow \cdot\bigcirc\text{=CH}_2 \longleftrightarrow \bigcirc\text{=CH}_2$$

$$\bigcirc\text{-}\overset{..}{\text{NH}}_2 \longleftrightarrow \overset{\ominus}{\underset{\oplus}{\bigcirc}}\text{=NH}_2 \longleftrightarrow \ominus\text{:}\overset{\oplus}{\bigcirc}\text{=NH}_2 \longleftrightarrow \underset{\ominus}{\overset{\oplus}{\bigcirc}}\text{=NH}_2$$

$$\bigcirc\text{-}\overset{..}{\underset{..}{\text{O}}}\text{:}^{\ominus} \longleftrightarrow \overset{\ominus}{\bigcirc}\text{=O} \longleftrightarrow \ominus\text{:}\bigcirc\text{=O} \longleftrightarrow \underset{\ominus}{\bigcirc}\text{=O}$$

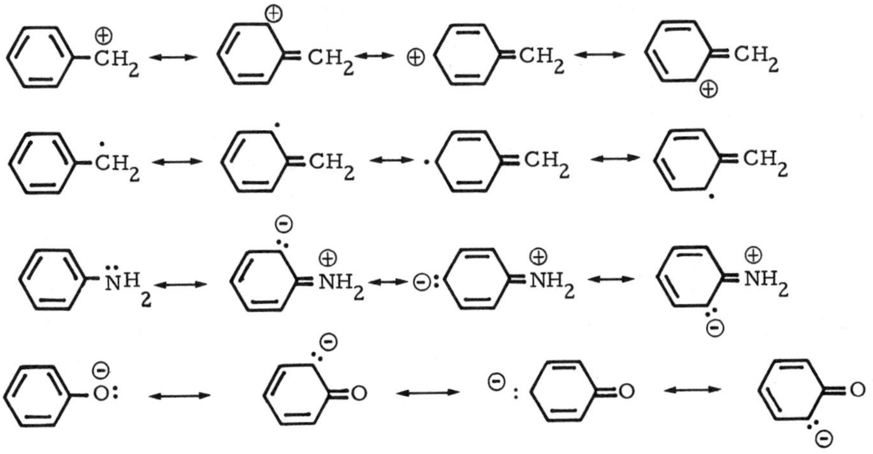

Electron delocalization of this type has been mentioned in connection with

    a.   The ease of forming the $C_6H_5CH_2^{\oplus}$ cation from phenylmethyl halides (Sections 14-3C and 26-4B).

    b.   The ease of radical-chain halogenation of alkylbenzenes,

$C_6H_5CH_3 + Cl_2 \longrightarrow C_6H_5CH_2Cl + HCl$ (Section 14-3A).

    c.   The low basicity of arenamines compared to alkylamines (Section 23-7C).

    d.   The high acidity of arenols compared to saturated alcohols (Sections 15-7A, 17-1A, 17-1D, 26-1).

    e.   The orientation of electrophiles to the ortho and para positions of benzenamine, benzenol, and similarly activated compounds (Sections 22-5, 23-9F, and 26-1E).

    f.   The activation of aryl halides toward nucleophiles by ortho and para electron-withdrawing substituents (Section 14-6B).

    4.   Replacement of a ring CH in an arene by nitrogen to give substances such as azabenzene (pyridine), 1,3-diazabenzene (pyrimidine), or 1-azanaphthalene (quinoline) does <u>not</u> cause loss of much aromatic character.

     azabenzene   1,3-diazabenzene   1-azanaphthalene

     (pyridine)   (pyrimidine)   (quinoline)

The stabilization energy is still high (Table 21-1) and the basicities of the heter-

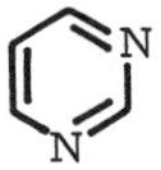

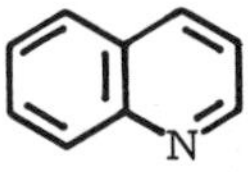

onitrogens are low (Section 23-7D), as expected for nitrogens forming $\underline{sp}^2$ bonds to protons. The differences that you will observe in behavior between arene and azarene derivatives usually can be rationalized in a straightforward way, if you remember that the nitrogen in the ring is more electron-attracting (electronegative) than carbon.

5. Addition reactions, such as catalytic hydrogenation, halogen addition, cycloaddition, are difficult and require abnormal conditions compared to the comparable addition reactions of polyenes.

6. Oxidative degradation of arenes can be accomplished, but under much more drastic conditions than for alkenes.

Substituents modify the reactivity of aromatic structures according to their nature. The effects of oxygen and carbon substitution on the ring are summarized below for arenols and side-chain compounds. Arenols are stable enols; the stability arises because the "ene" is part of an aromatic ring with $4\underline{n} + 2$ electrons in it. Arenols can be made by hydrolysis of aryl halides or arenesulfonic acids under rather drastic conditions with sodium or potassium hydroxide. Industrially, the favored reactions are by oxidation of alkylarenes and cationic rearrangement of the corresponding hydroperoxides,

$$ArCR_2O{-}OH \xrightarrow{\ -OH^{\ominus}\ } \left[ArCR_2O^{\oplus}\right] \longrightarrow R_2C{\overset{\oplus}{=}}O{-}Ar \xrightarrow{\ H_2O\ } R_2C{=}O \ + \ HOAr.$$

Arenols are about 10^8 times stronger acids than saturated alcohols, and they show considerable association by hydrogen bonding. Arenoxide ions react with alkyl halides, usually to give O-substitution products. With 2-propenyl halides, C-alkylation may result, especially in nonpolar solvents. O-2-Propenyl aryl ethers usually rearrange on strong heating to C-propenylarenols (Claisen rearrangement - Section 26-1E). A summary of reactions follows.

ArOH $\longrightarrow$ ArO$^{\ominus}$ + H$^{\oplus}$ Acidic properties (Sec. 26-1C)

$$
\text{ArO}^{\ominus}
\begin{cases}
\xrightarrow{\text{RX, }-X^{\ominus}} \text{ArOR} \\
\xrightarrow{\text{RX, }-X^{\ominus}} \text{R}{-}\text{Ar}{-}\text{OH} \\
\xrightarrow{\text{CO}_2} \text{HO}_2\text{C}{-}\text{Ar}{-}\text{OH} \\
\xrightarrow{\text{Br}_2} \text{Br}{-}\text{Ar}{-}\text{OH} \\
\xrightarrow{\text{Cl}_3\text{CH, }^{\ominus}\text{OH}} \text{ArCHO}
\end{cases}
$$

Nucleophilic properties; O-alkylation and O-acylation (Section 26-1C)

Nucleophilic properties; C-alkylation and related ring substitution with electrophilic reagents (Section 26-1E).

ArOH $\xrightarrow{\text{H}_2,\ \text{Ni}}$ cyclohexanols   (Section 26-1 F)

ArOH $\xrightarrow{[O]}$ quinones   (Sections 26-1G, 26-1H, 26-2 and 26-3)

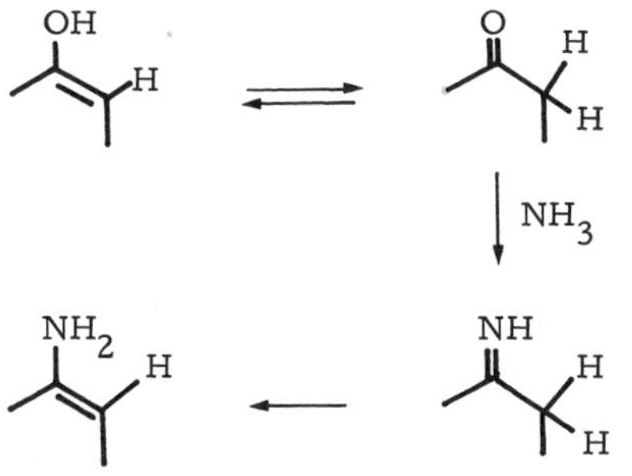

Keto-enol tautomerism
(Section 26-1 D)

Bucherer reaction
(Section 26-1 D)

Regarding the above reactions: (1) It is difficult to run substitution reactions on arenols wherein the Ar—O bond is broken. With a strong acid (HX), ethers such as Ar—O—R are cleaved on the alkyl side to give ArOH and RX. Exceptions involve addition-elimination with activating nitro groups present (Section 14-6B) or establishment of keto-enol equilibria, as in the Bucherer reaction, where arenols are converted to arenamines in a reversible reaction. (2) The ring carbons of arenols are activated toward electrophiles by the OH group and become very specially reactive if the OH is converted to $-O^{\ominus}$ by a base. Among the useful electrophilic reagents are bromine, carbon dioxide, trichloromethane and carbonyl compounds. (3) Arenols usually can be hydrogenated to alicyclic alcohols. (4) Oxidation of arenols by one-electron oxidizing agents appears to form first an arenoxy radical,

$$\text{ArOH} \xrightarrow{-\text{H}\cdot} \text{ArO}\cdot$$

Some such radicals are quite stable, but most readily dimerize, or disproportionate. Oxidation of benzenol with chromic acid forms the quinone, 1,4-benzenedione.

2-Substituted arenols often have rather strong intramolecular hydrogen bonds to the 2-substituent, which substantially alter their physical properties. Similar behavior has been encountered previously with the enol forms of 1,3-diones and has a pronounced effect on the proton nmr shift of the 2,4-pentanedione.

Many arene polyols are known. Some occur naturally and others can be synthesized by reactions similar to those used for arenols. Oxidation of 1,2- and 1,4-arenediols usually produces 1,2- or 1,4-arenediones, which commonly are known as quinones. With polycyclic arenes, the quinoid arrangement of bonds may extend across several rings, as with 3, 10-pyrenedione:

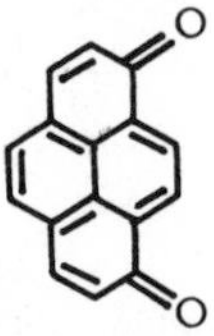

3, 10-pyrenedione

Interconversion of arenediols and arenediones by oxidation-reduction reactions occurs by way of an anion radical called a "semiquinone":

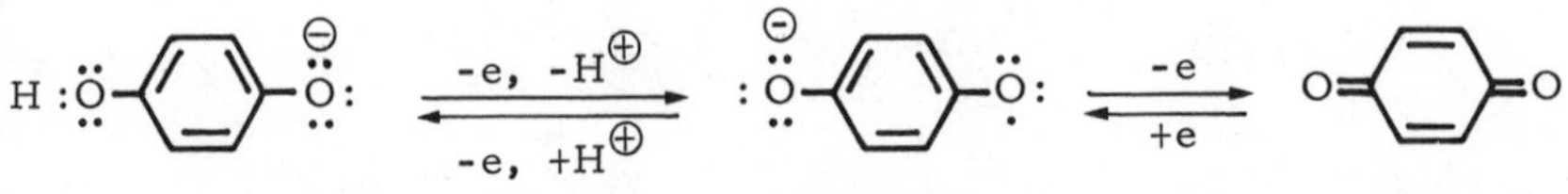

semiquinone   quinone

Similar oxidation-reduction processes are important in the chain of biochemically important reactions, which have as the overall result: NADH + $1/2 \, O_2$ + $H^{\oplus}$ —→ $NAD^{\oplus}$ + H_2O. The quinone, coenzyme Q(CoQ), is an important participant in these reactions.

Reduction of light-activated silver bromide to metallic silver in the conventional photographic process can be achieved with 1, 4-benzenediol (hydroquinone) in slightly alkaline solution.

$$HO-\!\!\!\!\bigcirc\!\!\!\!-OH \;+\; 2AgBr \xrightarrow{2OH^{\ominus}} O=\!\!\!\!\bigcirc\!\!\!\!=O \;+\; 2\,Ag^0 \;+\; 2\,Br^{\ominus} \;+\; 2\,H_2O$$

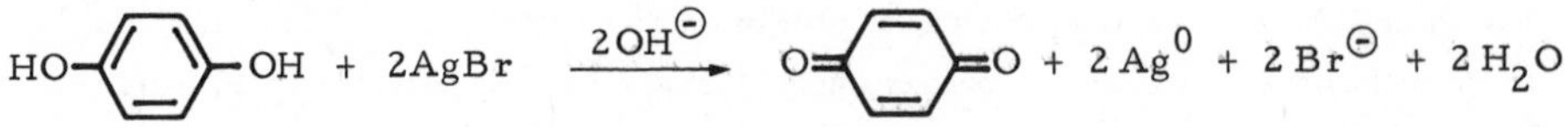

Vitamin K_1 is a 1, 4-naphthalenedione derivative that is important in blood clotting (an antihemorrhagic factor).

Quinones, like other α, β-unsaturated compounds, undergo addition reactions in the 1, 4 manner. The products can enolize and form an arenol. For example,

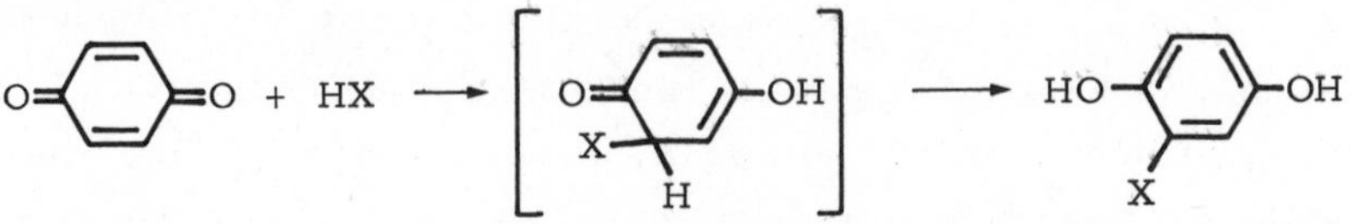

1, 4 addition product
(unstable)

Cyclobutenediones, which can be regarded as quinones of the cyclobutadiene ring system, resist reduction to cyclobutadienediols, which emphasizes the lack of aromatic stability of the cyclobutadiene and its derivatives.

Tropolones are hydroxycycloheptatrienones that exhibit many of the properties of arenols. The tropolone ring system occurs in many natural products; most notably in colchicine, a substance isolated from the autumn crocus. Colchicine is used for the treatment of acute gout.

We have defined aromatic side-chain compounds as alkylarenes, arenecarboxylic acids, arenecarbaldehydes, arylmethanols, and substances derived from them. Many such side-chain derivatives can be synthesized by taking advantage of the unusual stability of arylmethyl cations, anions, and radicals. Examples include conversion of alkylarenes to arenecarboxylic acids by vigorous oxidation, and to arylmethyl (benzyl) halides by radical chlorination.

Another useful preparation of arylmethyl chlorides involves chloromethylation, which is the introduction of a $-CH_2Cl$ group by treatment of an arene with methanal and hydrogen chloride in the presence of zinc chloride. A summary follows of reactions by which side-chain compounds may be prepared.

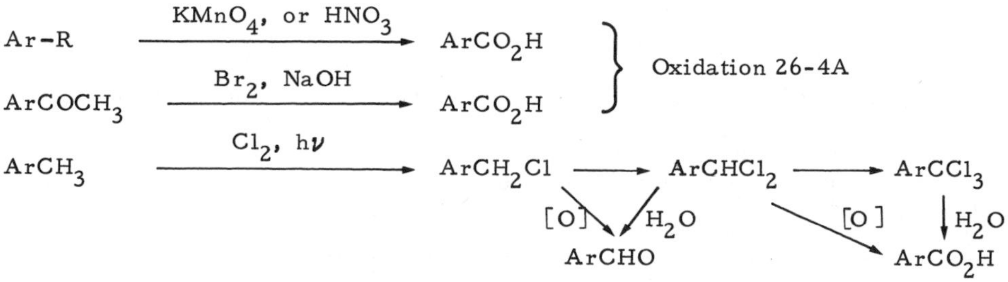

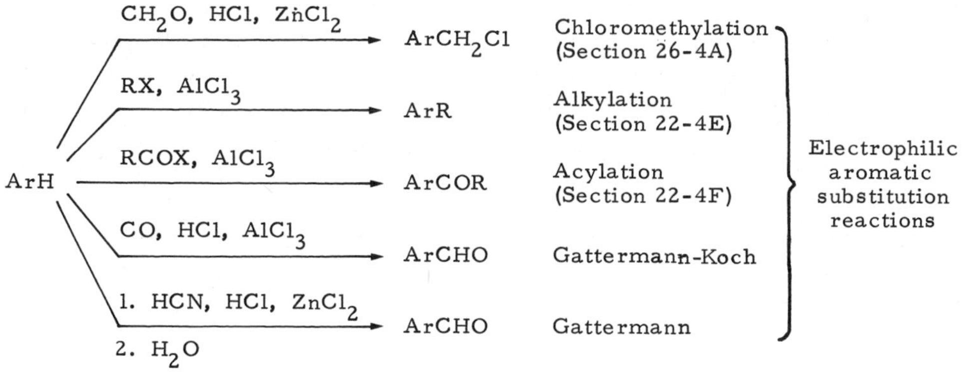

The unusual stability of arylmethyl cations, anions, and radicals resulting from electron delocalization, which we have emphasized repeatedly in this and other chapters, is greatly accentuated by having more aryl groups on the methyl carbon. Triarylmethyl cations, anions, and radicals often can be prepared and isolated at room temperature using relatively mild reagents. The triphenylmethyl radical, $(C_6H_5)_3C\cdot$, is in equilibrium with a dimer in solvents such as benzene, and this dimer is not hexaphenylethane, but is

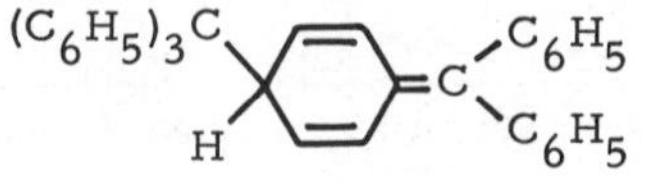

The benzoin condensation (Section 26-4E) results from heating an arenecarbaldehyde with sodium cyanide in ethanol. The products have the structure ArCHOHCOAr and are useful synthetic intermediates.

Many aromatic side-chain derivatives have important pharmaceutical, industrial, perfume, biochemical, or insecticidal utility.

The Hammett relationship (Section 26-6) correlates with considerable usefulness and precision the reactivity and equilibria of meta and para, but not ortho, substituted benzene derivatives with a simple two-parameter equation

$$\log k/k_0 \ (\text{or} \ \log K/K_0) \ = \ \sigma\rho$$

The parameter σ is the substituent constant, which is independent of the reaction involved and depends on the nature of the substituent and its position on the ring. The reaction constant, ρ , is independent of σ , but depends on the type of reaction, solvent, temperature, concentrations, and so on.

Hammett relationships are not very useful for correlating effects of substituents in other than aromatic systems, (1) unless the substituents are located on a fairly rigid part of the molecule, far enough away from the reacting center so as not to have sizeable steric effects at the reacting center, or (2) unless more parameters are used to describe the substituent properties. The Hammett equation tends to underestimate the influence of the substituent on a reaction where the substituent can have a large electron delocalization effect.

ANSWERS TO EXERCISES

26-1

a.	4-nitrobenzenol	Stronger resonance stabilization of the conjugate base from electron-accepting substituent in the 4 position.
b.	2,6-dimethyl-4-nitrobenzenol	Methyl groups in 3,5 positions distort nitro group from coplanarity with the ring and thereby reduce resonance stabilization of conjugate base.

c.   3-methoxybenzenol

CH_3O group has electron-accepting inductive effect (acid strengthening) but electron-donating resonance effect (acid weakening). From the 3 position, the acid strengthening effects are larger.

d.   4-azabenzenol

Resonance involving the electronegative ring nitrogen is important in stabilizing the conjugate base.

26-2

$\Delta H° = +20$ kcal

SE = 48   SE = $\underline{x}$

The conversion $CH_2{=}CHOH \longrightarrow CH_3CH{=}O$ has $\Delta H° = -15$ kcal. The corresponding conversion for benzenol has $\Delta H°$ equal to $-15 + 48 + \underline{x} = 20$, where $\underline{x}$ is the stabilization of the dienone ring. Hence $\underline{x} = 13$ kcal.

SE = 79.5 + 5   SE = 43.2 + 13

The corresponding conversion for 2-naphthalenol has

$$\Delta H° = 79.5 + 5 - (43.2 + 13) - 15 = +13.3 \text{ kcal}$$

Whereas the isomerization of both phenols is unfavorable, isomerization of 2-naphthalenol is more favorable than that of benzenol.

26-3

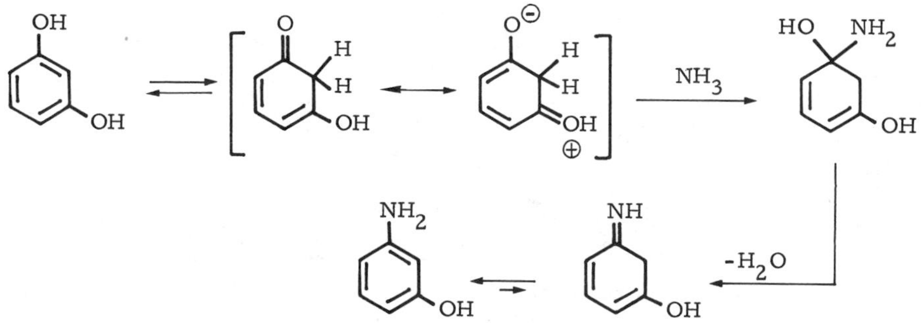

Benzenol is not expected to react similarly because of the much smaller stability expected for the keto form in equilibrium with benzenol.

26-4   The products of bromination of benzenol in various solvents indicate that reaction involves attack on the phenoxide ion rather than benzenol. This we may conclude from the following: In water solution, since dissociation of benzenol is significant, bromination proceeds at a reasonable rate. Di- and tribromination subsequently occur rapidly because of the even greater degree of dissociation of mono- and dibromobenzenols relative to benzenol and the powerful activating effect of the $O^{\ominus}$ group. However, in nonpolar solvents, the rate of monobromination of benzenol is slowed by the decrease in dissociation as is also the rate of di- and tribromination for similar reasons. The products isolated are, therefore, 2- and 4-bromobenzenols.

26-5   See Section 22-8A for reasons for preferential formation of 1-substitution in the first stage of the reaction. The substituted ring of 1-bromo-2-naphthalenol is deactivated (by the inductive effect) to further bromine substitution relative to the same ring in 2-naphthalenol. For this reason, a second substituent enters the other ring at the 6-position rather than enter the same ring at the 3-position (which is a sort of "super-para" position to the substituent).

26-6

26-7

Benzenol is not expected to undergo reaction with CO_2 with the same facility as 1,3-benzenediol for the reason that with benzenol the key reaction intermediate analogous to   1   is expected to have relatively low order of

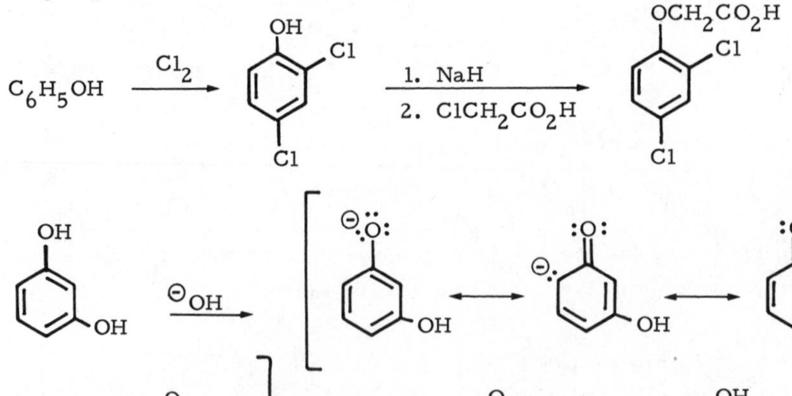

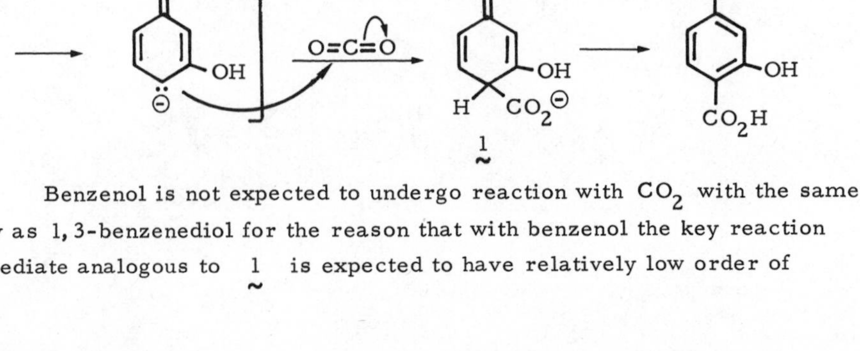

stability (see also Exercise 26-3). Formation of the 4-isomer occurs exclusively probably because of steric hindrance at the 2-position.

26-8   If the rings are linked at the 4-position, the 2,6-di-tert-butyl groups are equivalent and different from the tert-butyl group at C4. There will be only two nmr signals for the tert-butyl groups in the ratio of 2:1.

If the rings are linked at the 2-position, all three tert-butyl groups are nonequivalent and will give rise to three nmr singlets in the ratio of 1:1:1.

26-9

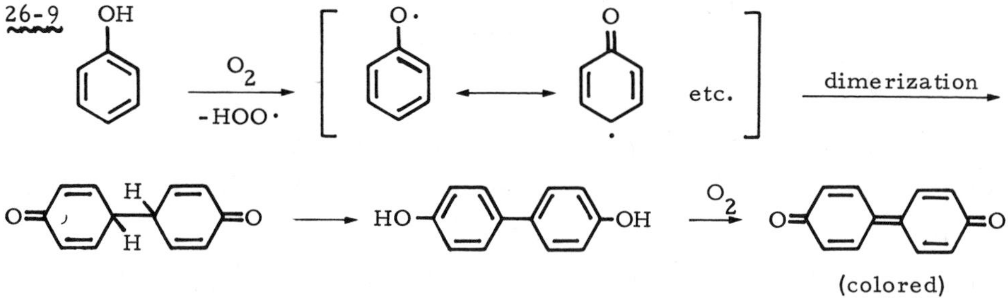

Repetition of the preceding with 4,4'-dihydroxybiphenyl and more benzenol could lead to a highly colored conjugated terphenyl quinoid structure.

26-10   The ready decarboxylation of gallic acid can be explained as arising from the stability of the keto intermediate in 2 and relative ease of formation of the pyrogallol anion therefrom:

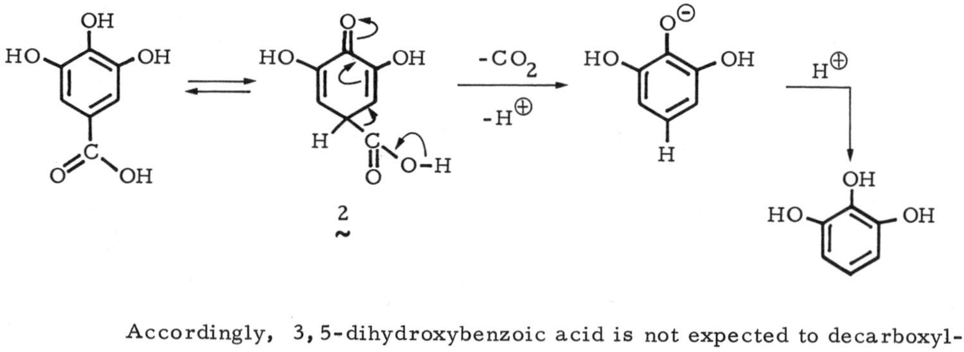

Accordingly, 3,5-dihydroxybenzoic acid is not expected to decarboxylate as readily as its 2,6-isomer since a similar reaction sequence cannot be written for this substance.

26-11

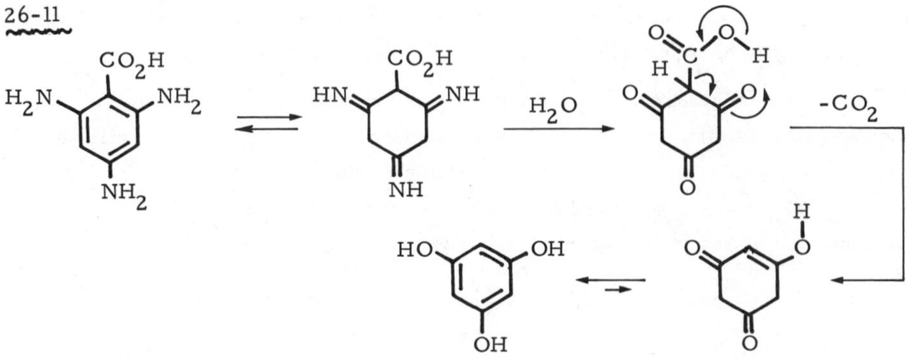

If 2,4,6-triaminobenzoic acid is in equilibrium with the tri-imino tautomeric form, hydrolysis of the imino compound to the triketo compound is to be expected.

26-12   See Section 26-5 for the synthesis of vanillin.

26-13

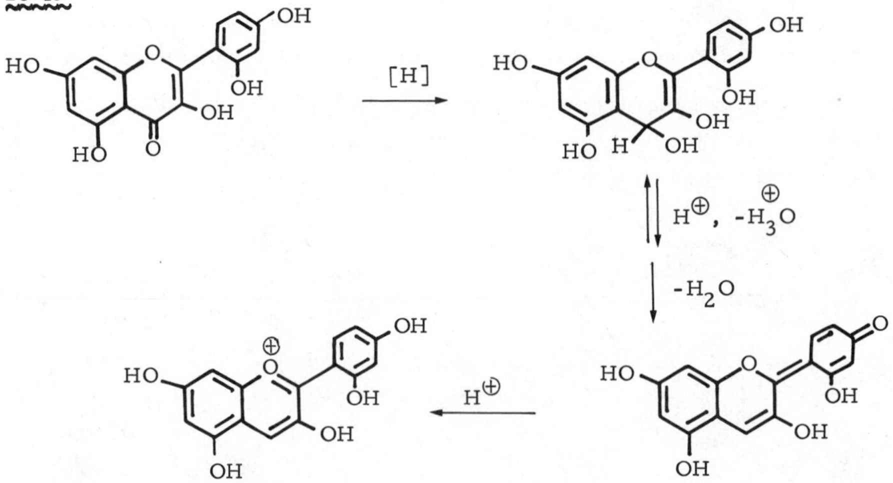

26-14

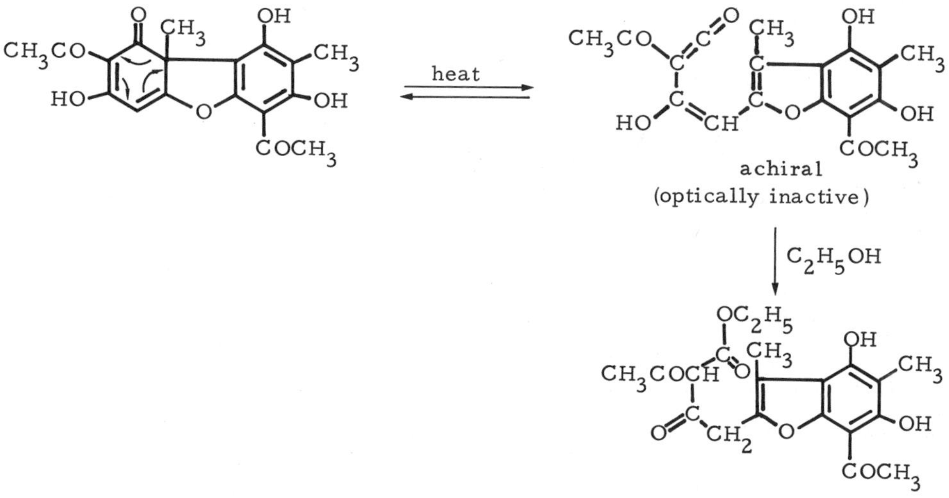

achiral
(optically inactive)

C_2H_5OH

26-15   cis-2, 2'-Dimethylbiphenyldione (least stable quinone for steric reasons, and gains stabilization energy of two phenyl rings on reduction to the
hydroquinone); 4, 4'-biphenyldione; 1, 4-benzenedione; 1, 4-naphthalenedione;
9, 10-anthracenedione (has the least to gain on reduction and is thus the most
stable quinone).

26-16

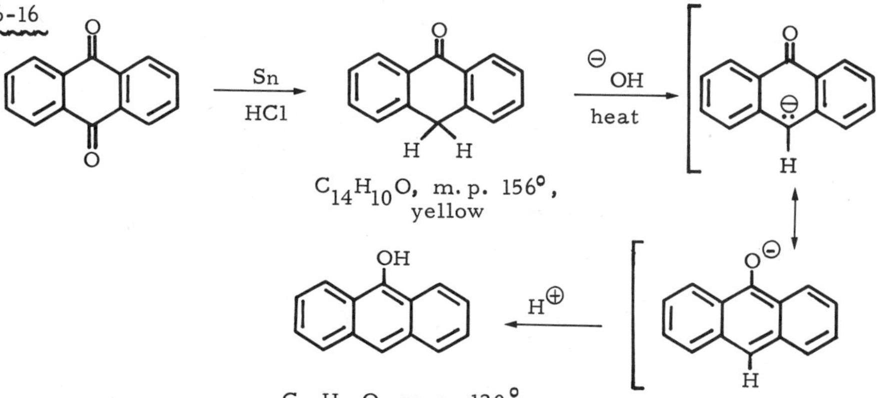

$C_{14}H_{10}O$, m.p. 156°,
yellow

$C_{14}H_{10}O$, m.p. 120°,
brown-yellow

If we assume the SE of 9-anthracenol is 113 + 5 kcal (113 for the
anthracene, 5 for the OH), then $\Delta \underline{H}°$ for rearrangement is 43.2 + 43.2 -
(113 + 5) + 15 = -17 kcal. (This ignores some SE for the carbonyl form of
$C_{14}H_{10}O$ but includes 15 kcal for the keto-enol interconversion (text p. 1288).

26-17

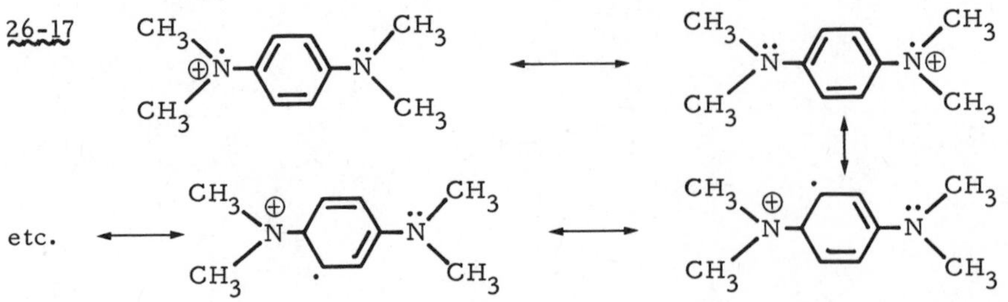

Octamethyl-1,4-benzenediamine does not form a cation radical that is stabilized to the same extent as the tetramethyl-1,4-benzenediamine cation radical because steric interference by the methyl groups inhibits resonance of the type shown in the preceding by distorting the molecular configuration such that the orbitals with unshared and odd electrons on nitrogen cannot overlap with the π electrons of the aromatic ring.

26-18

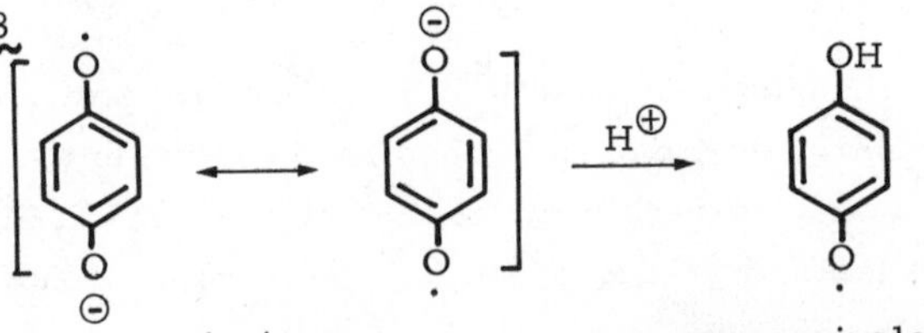

semiquinone

no equivalent
resonance structures
possible involving the OH group

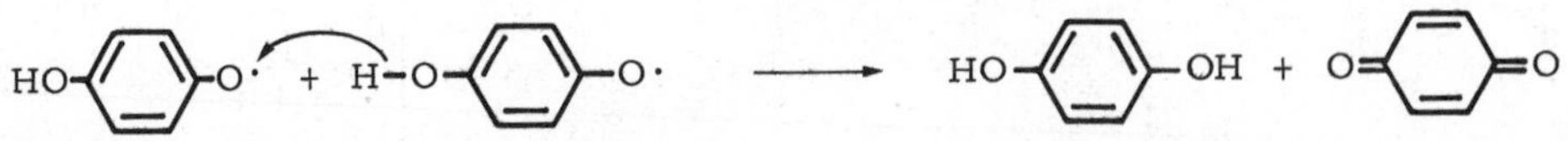

26-19   The methoxy groups of CoQ are electron donating by resonance and more polar than the methyl groups of plastoquinone. On this basis, we would expect that the reduction potential of CoQ would be lower than that of plasto-quinone because CoQ should be less ready to accept electrons in the π system. Also, CoQ should be more polar and more soluble in polar solvents than plasto-quinone. Reduction of the two quinones will give the corresponding 1,4-benzene-diols. The presence of the methoxy groups which are electron attracting by induction and electron donating by resonance makes prediction of the relative acidity of the benzenediols difficult and this is likely to be important to their biological functioning. On balance, it would seem that the $CoQH_2$ would be the weaker acid - especially if hydrogen bonding were to stabilize the phenolic hydroxyls.

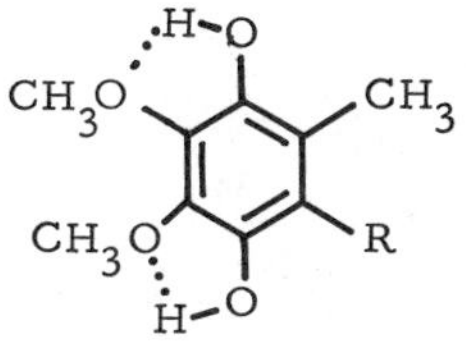

26-20   1,4-Benzenedione is a reactive dienophile in (4 + 2) cycloaddition reactions. Natural quinones are less reactive because the ring substituents obstruct the formation of cycloadducts. Ring substituents also stabilize double bonds (see Section 11-3) and addition of nucleophilic agents will be less favorable than for 1,4-benzenedione. Probably most important is the fact that with a fully substituted 1,4-benzenedione there is no possibility of additions which lead to 1,4-benzenediols, as illustrated in Section 26-2D.

26-21 $2 F_2C{=}CFCl$

This reaction is facilitated by strong electron-attracting effect of the fluorines.

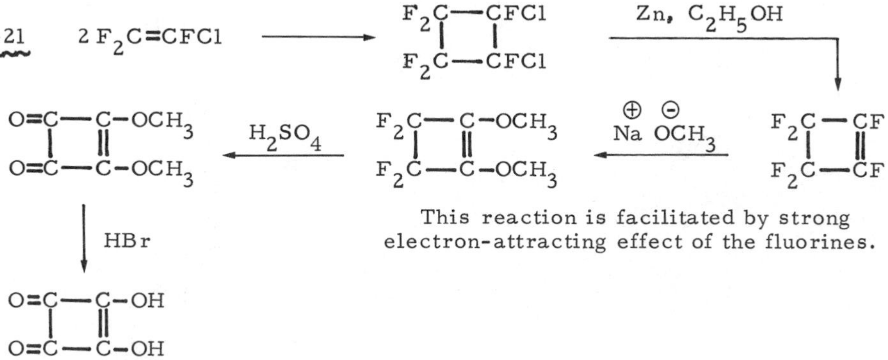

26-22   Dipolar VB structures contribute significantly to the structure of tropone with the result that the carbonyl oxygen is more strongly electronegative and therefore more strongly basic than that of an ordinary ketonic oxygen.

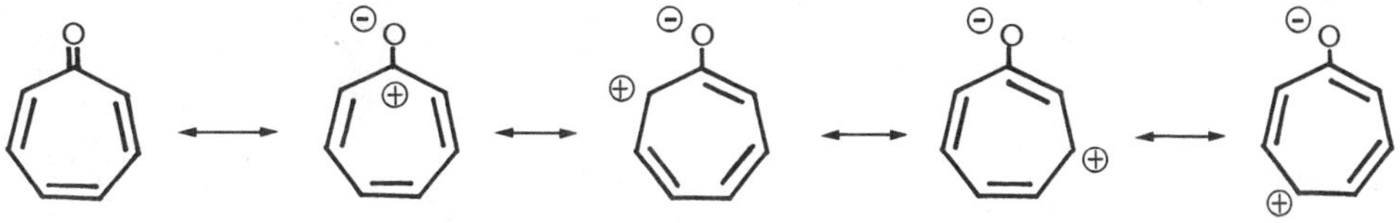

26-23   In the nitration of tropolone, the favored positions of attack by $NO_2^{\oplus}$

would be those highest in electron density, namely 3, 5 and 7.

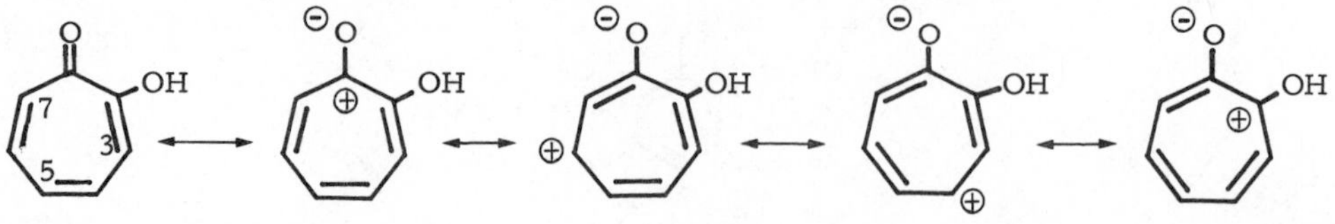

26-24   Since the benzotropylium ion is likely to be less stable than the tropy-
lium ion, it should react more readily with water to form an alcohol. It is less
stable because acquisition or loss of electrons by inter-ring delocalization will
result in a loss in stability of the cation (see Section 21-9B).

The point of attack of water would be as shown below, such that aromaticity of the phenyl ring is preserved.

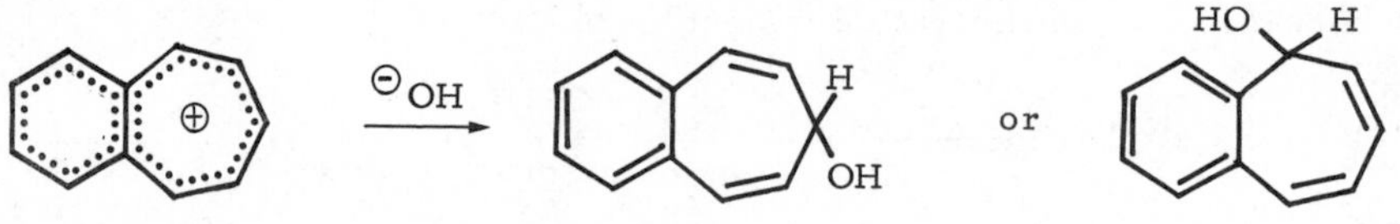

26-25

a. 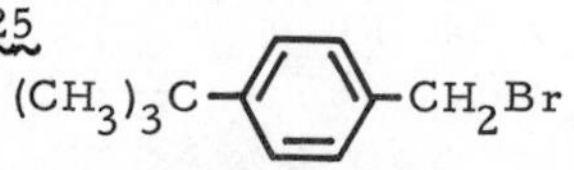 $(CH_3)_3C$—⬡—CH_2Br

b. 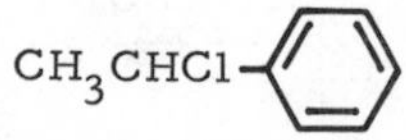 CH_3CHCl—⬡

c. 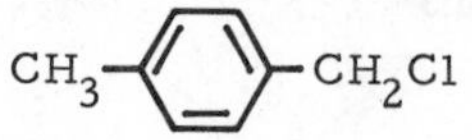 CH_3—⬡—CH_2Cl

d. 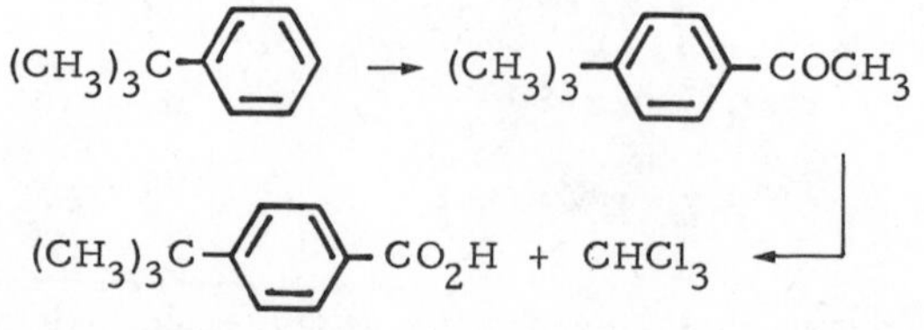 $(CH_3)_3C$—⬡ → $(CH_3)_3C$—⬡—$COCH_3$

$(CH_3)_3C$—⬡—CO_2H + $CHCl_3$ ⟵

26-26

a.   Methylbenzene (toluene) $\xrightarrow{HNO_3,\ H_2SO_4}$ 4-nitromethylbenzene   (and

some ortho isomer) $\xrightarrow{2\,Cl_2,\ h\nu}$ O_2N—⬡—$CHCl_2$ —

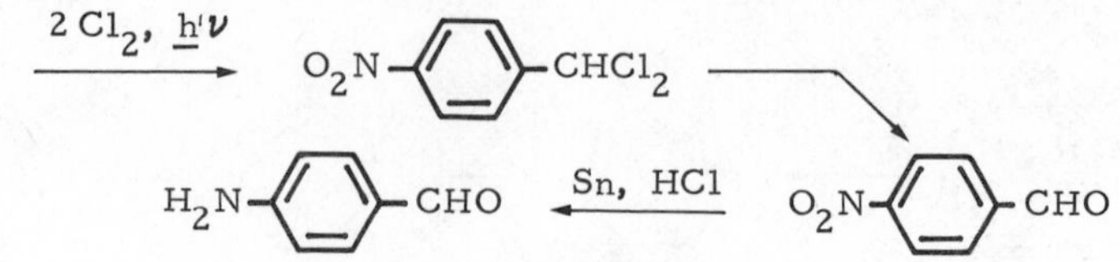

H_2N—⬡—CHO $\xleftarrow{Sn,\ HCl}$ O_2N—⬡—CHO

b.   Phenanthrene $\xrightarrow{1.\ O_3,\ 2.\ H_2O_2,\ NaOH}$

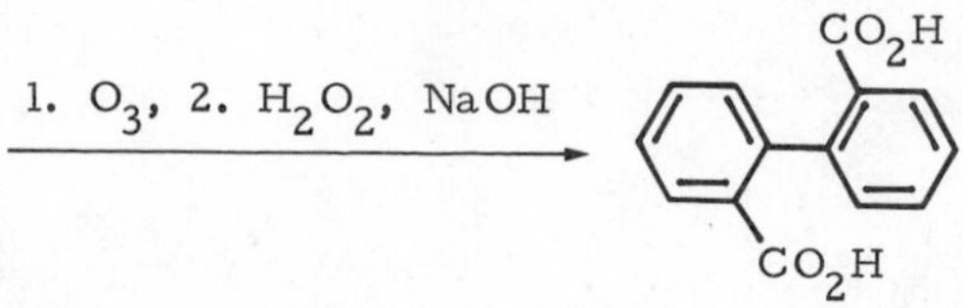

c.

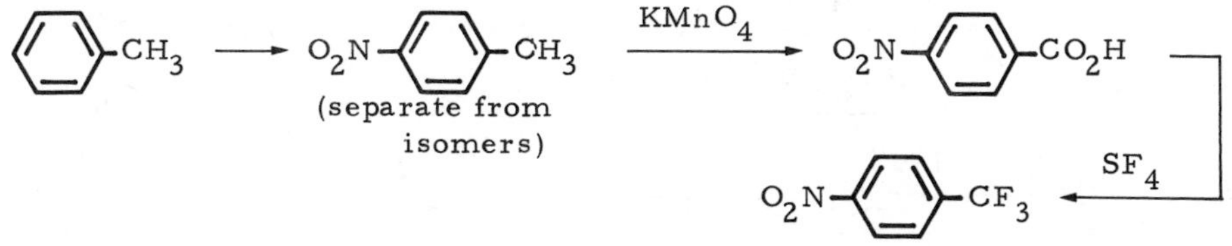

$$\text{C}_6\text{H}_5\text{-CH}_3 \longrightarrow \text{O}_2\text{N-}\underset{\substack{\text{(separate from}\\\text{isomers)}}}{\text{C}_6\text{H}_4}\text{-CH}_3 \xrightarrow{\text{KMnO}_4} \text{O}_2\text{N-C}_6\text{H}_4\text{-CO}_2\text{H}$$

$$\text{O}_2\text{N-C}_6\text{H}_4\text{-CF}_3 \xleftarrow{\text{SF}_4}$$

d. 1,2-dimethylbenzene

$$\xrightarrow{\text{HNO}_3} \quad \xrightarrow[\text{AlCl}_3]{1.\ \text{C}_6\text{H}_6} \quad \xrightarrow[\text{SO}_3]{2.\ \text{H}_2\text{SO}_4}$$

(See Exercise 22-19.)

$$\xleftarrow{3.\ \text{SF}_4}$$

e. methylbenzene

$$\xrightarrow{\text{CH}_2\text{O, HCl, ZnCl}_2} \text{CH}_3\text{-C}_6\text{H}_4\text{-CH}_2\text{Cl} \xrightarrow[(\text{CH}_3)_2\text{SO}]{\text{KCN}}$$

$$\text{CH}_3\text{-C}_6\text{H}_4\text{-CH}_2\text{CN} \longleftarrow$$

f. methylbenzene

$$\xrightarrow[\text{dark}]{\text{Cl}_2,\ \text{FeCl}_3} \text{CH}_3\text{-C}_6\text{H}_4\text{-Cl} \xrightarrow[2.\ \text{H}_2\text{O}]{1.\ \text{Cl}_2,\ h\nu} \text{O=CH-C}_6\text{H}_4\text{-Cl}$$

g. methylbenzene

$$\xrightarrow{\text{SO}_3}$$

(and some para)

$$\xrightarrow{\text{NaOH}}$$

$$\xrightarrow[\text{NaOH}]{\text{CHCl}_3}$$

h. benzene $\xrightarrow[\text{BF}_3]{\text{CH}_2\text{=CH}_2}$ ethylbenzene $\xrightarrow[\text{Cu}_2\text{Cl}_2]{\text{CO, HCl, AlCl}_3}$ 4-ethylbenzene-

carbaldehyde $\xrightarrow{\text{LiAlH}_4}$ $\text{HOCH}_2\text{-C}_6\text{H}_4\text{-CH}_2\text{CH}_3$

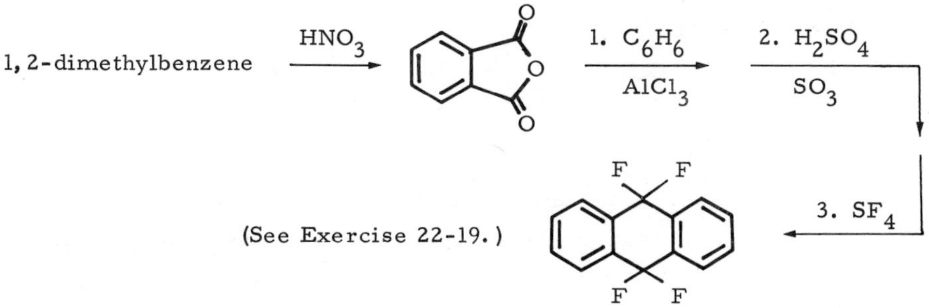

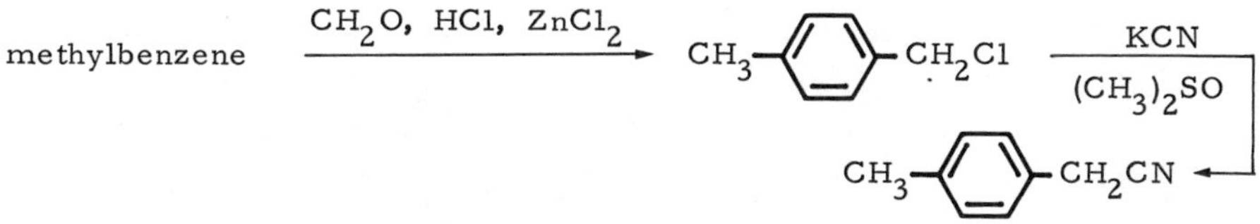

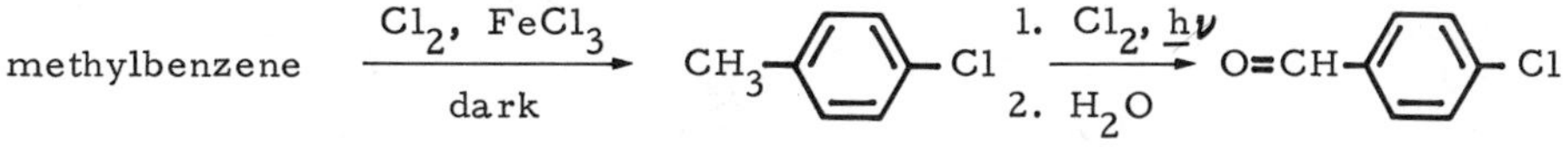

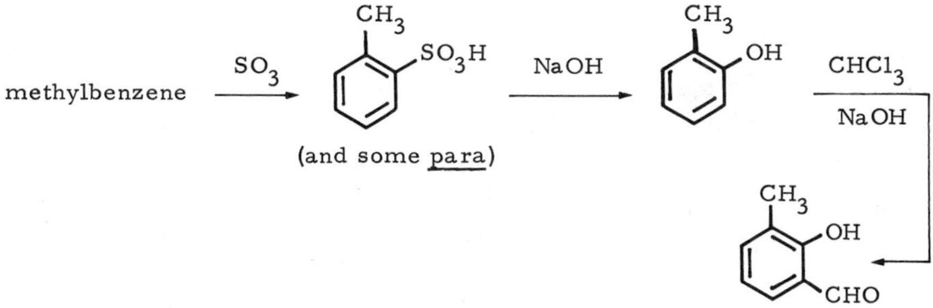

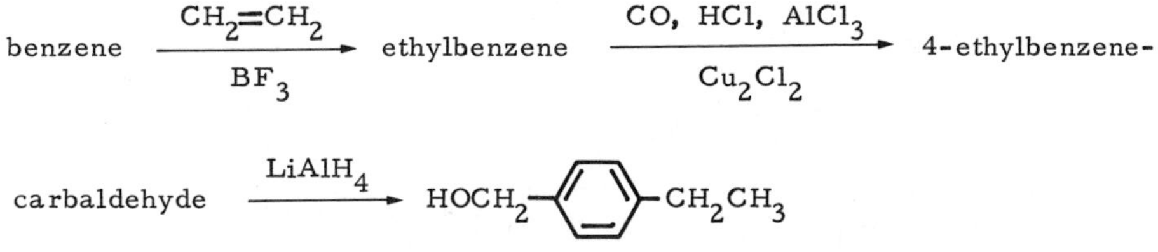

i.   benzene $\xrightarrow{\text{HNO}_3,\ \text{H}_2\text{SO}_4}$ 1,3-dinitrobenzene $\xrightarrow{(\text{NH}_4)_2\text{S}_x}$ 3-nitroben-

zenamine $\xrightarrow{\text{1. HNO}_2,\ \text{HCl, 0}^\circ\qquad \text{2. Cu}_2\text{Cl}_2}$ 3-nitrochlorobenzene

$\xrightarrow{\text{Fe, HCl}}$ 3-chlorobenzenamine $\xrightarrow{\text{1. HNO}_2,\ \text{HCl, 0}^\circ\qquad \text{2. H}_2\text{O}}$

3-chlorobenzenol $\xrightarrow{\text{1. NaH}\qquad\text{2. CH}_3\text{CH}_2\text{I}}$ 3-chloroethoxybenzene

$\xrightarrow{\text{CO, AlCl}_3}$

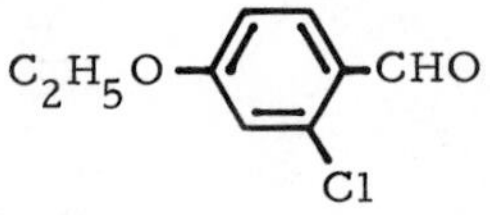

26-27   a.   A <u>decrease</u> in energy of ionization in the order $7 > 8 > 9$ means a
<u>gain</u> in stability of the cation relative to the chloride in the order $9 > 8 > 7$. We
expect that a phenyl ring will be less able to stabilize a cationic charge than a
naphthyl group because there are two more locations for the charge to be spread
over in a naphthyl group. Consequently, if the difference in steric hindrance
between RCl and $R^{\oplus}$ is not large (as expected for 7 and 8), then the ionization
energy should decrease from 7 to 8. For 8 vs. 9, the situation is more complex because there will be substantial hindrance from C8 and the hydrogen
attached to it.

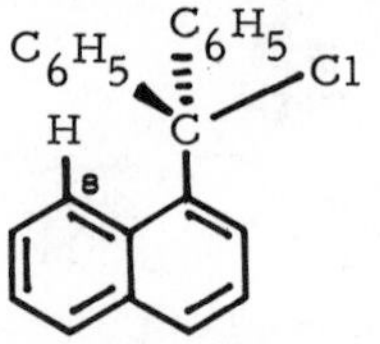

This crowding will tend to make both RCl and $R^{\oplus}$ corresponding to 9 less
stable; RCl because of simple crowding (see Section 8-7B), but $R^{\oplus}$ because it
will be more difficult to achieve delocalization of the $\oplus$ charge as the result of
interferences tending to inhibit formation of a planar cation.

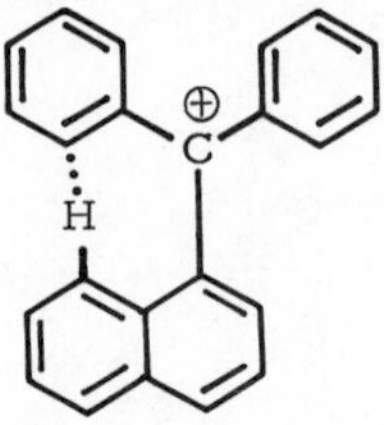

Because these steric effects on RCl and $R^{\oplus}$ of 9 will tend to
more or less balance out, probably the most important effect making the ioniza-

tion of $\underset{\sim}{9} > \underset{\sim}{8}$ is the electron delocalization effect which favors electrophilic substitution of naphthalene at the 1- relative to the 2-position as described in detail in Section 22-8.

   b.   2, 3, 6, 7-Bibenzotropyl alcohol $\underset{\sim}{11}$ will form a more stable cation than 9-fluorenol $\underset{\sim}{10}$ for the reason that cycloheptatrienyl cations are more stable than cyclopentadienyl cations (Section 21-9B). The effect is to some degree reduced by the benzo groups but by no means eradicated.

   c. $(C_6H_5)_3COH + H_2SO_4 \rightleftharpoons (C_6H_5)_3C^{\oplus} + H_2O + HSO_4^{\ominus}$

   $H_2SO_4 + H_2O \rightleftharpoons H_3O^{\oplus} + HSO_4^{\ominus}$

   $(C_6H_5)_3COH + 2H_2SO_4 \rightleftharpoons \underbrace{(C_6H_5)_3C^{\oplus} + H_3O^{\oplus} + 2HSO_4^{\ominus}}$

   4 moles of particles

<u>26-28</u>   The geometry of 9-phenylfluorene fixes two aromatic rings in a planar configuration; this has the effect of enhancing the acidity of the 9-proton since the corresponding anion is better stabilized than that from triphenylmethane, which is a more flexible, less planar molecule. Also, cyclopentadienide ions are expected to have enhanced stability (Section 21-9B).

<u>26-29</u>   a.   The radical derived from 3-phenyl-1-propene is better stabilized than that from methylbenzene because of delocalization of the odd electron, which gives the radical some of the stability of <u>both</u> phenylmethyl and 2-propenyl radicals:

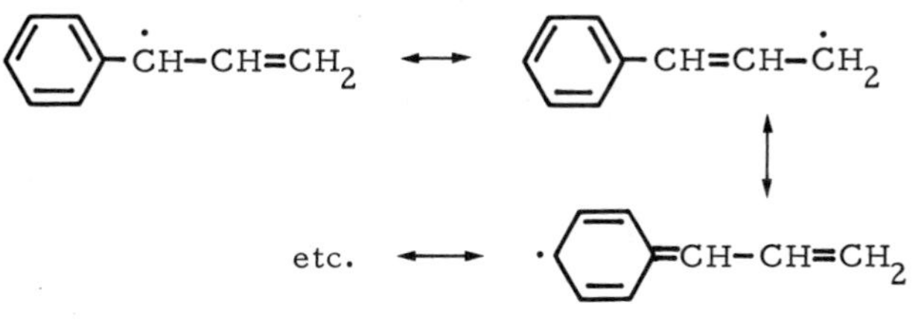

Generally, the better stabilized the radical, the easier it is formed.

   b.   The same radical is to be expected on hydrogen abstraction from 1-phenyl-1-propene and 3-phenyl-1-propene. However, the radical is not expected to be formed as easily from 1-phenyl-1-propene as from 3-phenyl-1-propene because the conjugated isomer is the more stable isomer $(\Delta H_1^{\circ} > \Delta H_2^{\circ})$.

<u>26-30</u>   a.   Since the reaction conditions approach S_N1, the compound $p\text{-}CH_3OC_6H_4CH_2Br$ would be most reactive since the $p\text{-}CH_3O$ group would better

stabilize the phenylmethyl-type carbocation intermediate.

$$\text{p-CH}_3\text{OC}_6\text{H}_4\text{CH}_2\text{Br} + \text{H}_2\text{O} \xrightarrow{\text{acetone}} \text{p-CH}_3\text{OC}_6\text{H}_4\text{CH}_2\text{OH} + \text{HBr}$$

b.   The stronger acid will be the more reactive (i.e., triphenyl-methane) since this is the compound that forms the more stable carbanion.

$$(C_6H_5)_3CH + C_6H_5Li \longrightarrow (C_6H_5)_3C:^{\ominus} + C_6H_6 + Li^{\oplus}$$

c. $(C_6H_5)_3C-N=N-C(C_6H_5)_3$ will decompose more readily than $C_6H_5)_2CH-N=N-CH(C_6H_5)_2$ because loss of nitrogen will generate the more stable $(C_6H_5)_3C\cdot$ radicals (see Section 24-7B).

d.   The steric requirements of the two compounds are about the same and cannot then be the deciding factor in reactivity differences. However, since the N—N bond is much weaker than a C—C bond, the N—N compound is expected to dissociate much more readily.

$$(C_6H_5)_2N-N(C_6H_5)_2 \longrightarrow 2(C_6H_5)_2N\cdot$$

e.   Diphenylethanoyl peroxide is more reactive because it decomposes to form the more stable **phenylmethyl free radical.**

$$(C_6H_5CH_2CO_2)_2 \xrightarrow{\;\;-2CO_2\;\;} 2\,C_6H_5CH_2\cdot$$

f.   Ketones will be stabilized best by groups attached to the carbonyl group that stabilize a positive charge. The more stable the ketone, the less reactive it should be towards addition reagents. Hence, diphenylmethanone is expected to be _less_ reactive towards $NaBH_4$ than 1,3-diphenyl-2-propanone because of resonance of the following type:

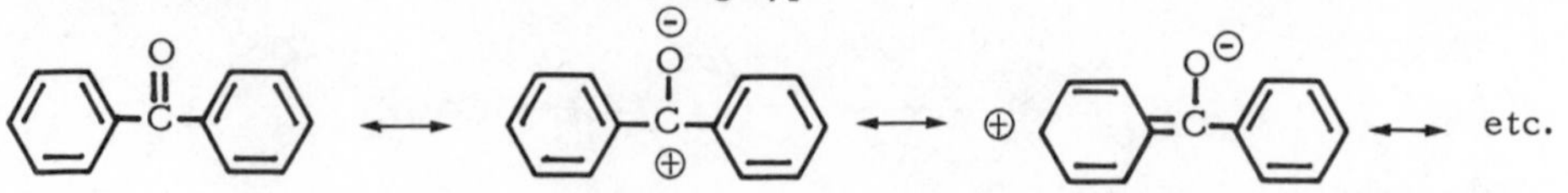

26-31   Base-induced enolization:

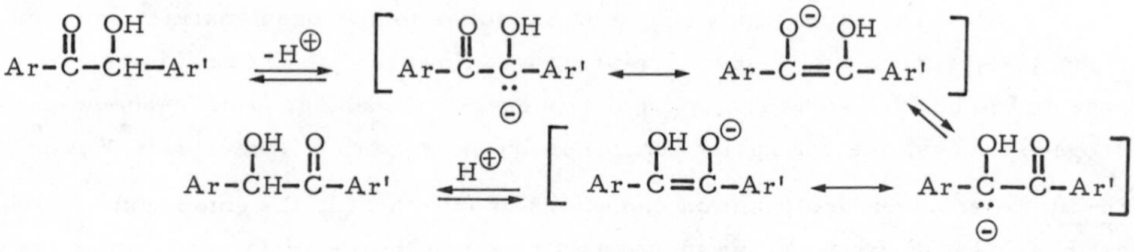

By this mechanism, rearrangement in D_2O as solvent would result in formation of C—D bonds faster than rearrangement if the second step is slow (unlikely) and at the same rate as rearrangement if the first step is the slow step. A Cannizzaro-type reaction:

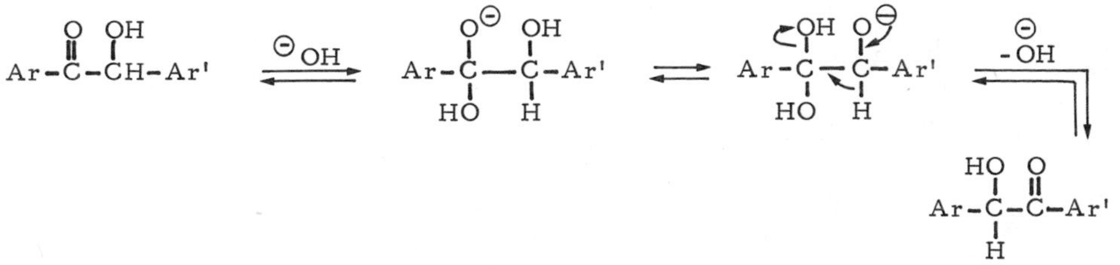

By this mechanism, rearrangement in D_2O would not result in C—H to C—D exchange.

26-32

a. 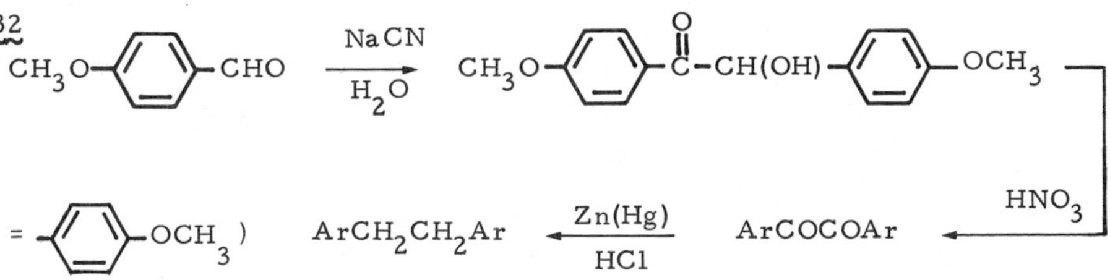 2 CH_3O—⬡—CHO $\xrightarrow[H_2O]{NaCN}$ CH_3O—⬡—$\overset{O}{\overset{||}{C}}$—CH(OH)—⬡—$OCH_3$ ⟶

$(Ar = $ —⬡—OCH_3 $)$ $ArCH_2CH_2Ar$ $\xleftarrow[HCl]{Zn(Hg)}$ $ArCOCOAr$ $\xleftarrow{HNO_3}$

b.

⬡ $\xrightarrow[ZnCl_2]{CH_2=O,\ HCl}$ ⬡—CH_2Cl $\xrightarrow[H_2SO_4]{HNO_3}$ ⬡(NO_2)—CH_2Cl

(separate isomers, considerable para formed)

$(CH_3)_2S=O$

CH_3COCH_3 / NaOH ⟶ ⬡(NO_2)—CHO $\xleftarrow{CrO_3}$ ⬡(NO_2)—CH_2OH $\xleftarrow{H_2O}$

⬡(NO_2)—CH=CH—$\overset{O}{\overset{||}{C}}$—$CH_3$

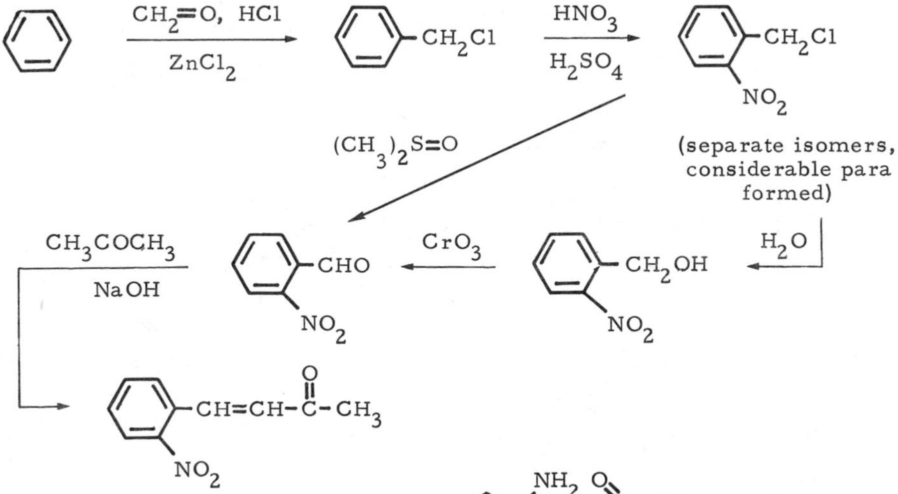

c. Product from Part b $\xrightarrow{1.\ Fe,\ HCl}$ ⬡(NH_2)(—CH=CH—$\overset{O}{\overset{||}{C}}$—$CH_3$) $\xrightarrow[heat]{-H_2O}$ ⬡⬡(N)—CH_3

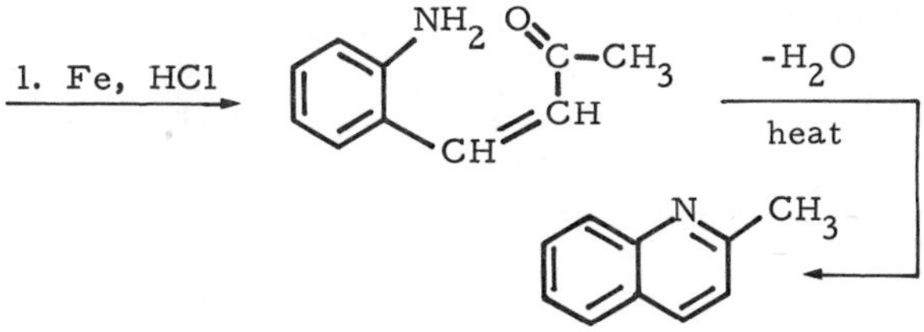

d.   Benzencarbaldehyde $\xrightarrow[\text{H}_2\text{O}]{\text{NaCN}}$ benzoin $\xrightarrow[\text{CH}_3\text{CO}_2\text{H}]{\text{HNO}_3}$ benzil $\begin{array}{l}\text{1. NaOH}\\[4pt]\text{2. H}^{\oplus}\end{array}$

(benzophenone)   diphenylmethanone $\xleftarrow[\text{KMnO}_4]{}$ benzilic acid

26-33   Mechanism is related to imine formation (Section 16-4C).

26-34 $\log (\underline{K}_{meta}/\underline{K}_o) = \sigma\ _{meta}$

$\log (2.51 \times 10^{-4}/6.76 \times 10^{-5}) = \sigma\ _{meta} = +0.570$

$\log (2.82 \times 10^{-4}/6176 \times 10^{-5}) = \sigma\ _{para} = +0.620$

26-35   a.   Substituents in the meta position influence reactivity mainly by their inductive effects. In the para position, resonance effects usually predominate. Because fluorine is electron-accepting by induction but electron-donating by resonance, these effects are opposed in the para position and almost cancel, giving σ -F$_{para}$ as nearly zero. In contrast, σ -F$_{meta}$ is large and positive, reflecting its electron-attracting inductive effect.

b.   The rationale is the same as in Part a except that the resonance effect is relatively stronger for methoxy and dominates over induction in the para position. Hence the negative para σ constant.

c.   Inductive effects fall off with distance. Therefore σ -meta > σ -para for the $-\overset{\oplus}{N}(CH_3)_3$ group, which is capable only of an inductive effect. In the case of $-N_2^{\oplus}$, both resonance and inductive effects are possible in the para position, both of which operate to withdraw electrons.

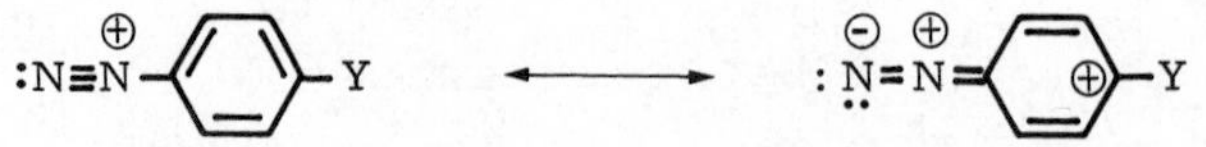

d.   The CF_3 group has an electron-withdrawing inductive effect and therefore has both σ -meta and σ -para positive. However, it is also electron-withdrawing by a hyperconjugative resonance effect, which operates through the ortho and para positions (see Section 8-7B for discussion of hyperconjugation of CH_3 group):

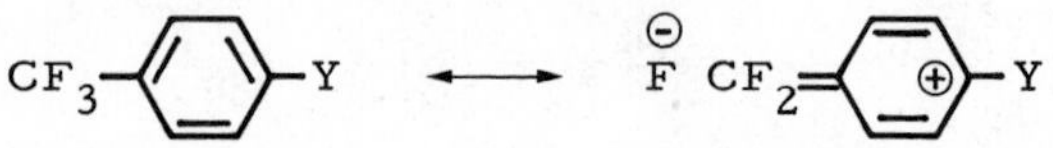

26-36   a. σ-meta, +0.5 to +0.6;   σ-para, +0.55 to +0.65.

The CN group is electron withdrawing by resonance and induction, but less so than the nitro group.

b. σ-meta, +0.2 to +0.3;   σ-para, +0.15 to 0.25.

The inductive effect (-I) of $-\overset{\oplus}{N}(CH_3)_3$ is expected to be much reduced but still dominant in $-CH_2\overset{\oplus}{N}(CH_3)_3$.

c. σ-meta, +0.3;   σ-para, +0.10.

The inductive effect (-I) would be expected to operate strongly in meta position. The fluorines would outweigh the resonance effect of the oxygen as in $-OCH_3$ to give a net electron-withdrawing effect in the para position.

d. σ-meta, -0.35;   σ-para, -0.20.

Resonance effects should be relatively unimportant relative to the electrostatic effect of the negatively charged group.

26-37   The rigidity of the structure would permit the effect of the substituent R on the acid strength and rate of ester hydrolysis to be uncomplicated by steric and entropy effects. A Hammett-type relationship is therefore expected (and found).

26-38   The large difference in ρ-values for reactions 11 and 12 of Table 26-7 reflects a change in reaction mechanism. In the presence of a strong base reaction 11 is an S_N2 reaction and is not influenced strongly by substituent effects (ρ is small and negative). However, an S_N1 reaction prevails in 48% aqueous ethanol (reaction 12), and this type of reaction is strongly accelerated by electron-supplying substituents (ρ is large and negative).

26-39   A change from a polar solvent (e.g., H_2O) to a less polar solvent (e.g., CH_3OH) will depress the degree of ionization of benzoic acid and substituted benzoic acids. At the same time, the degree of dissociation is expected to be increasingly sensitive to substituent effects as the dielectric constant and polarity of the solvent decreases (i.e., $\rho > +1$). The associated decrease in solvation energy is partially made up by increasing ion stability provided by electron-attracting substituents.

26-40   The further the reaction site is from the substituent, the more insensitive the reaction becomes to substituent effects (small ρ values). Since the carboxyl group of substituted phenylethanoic acids is insulated from the substi-

tuted phenyl ring by a methylene group, the ρ value for ionization is expected
to be smaller than for the corresponding benzoic acids. In 4-phenylbutanoic
acids $(R-C_6H_4CH_2CH_2CH_2CO_2H)$, the ρ value is expected to approach <u>zero</u>.

The slow step in saponification reactions of ethyl benzoates is attack
of $^{\ominus}OH$ at the ester carbonyl group. The more electropositive the carbonyl
carbon, the more rapid is the reaction. Thus, electron-attracting substituents
accelerate reaction (i.e., ρ is positive). The ρ -value for saponification is
larger than ρ for ionization of the benzoic acids because in saponification the
reaction site is one atom closer to the substituent than in ionization.

<u>26-41</u> $\log \underline{K}_m - \log 1.3 \times 10^{-10} = 0.71 \times 2.11$

$\log \underline{K}_p - \log 1.3 \times 10^{-10} = 0.78 \times 2.11$

$\underline{K}_m = 2.4 \times 10^{-9}$; $\underline{K}_p = 3.4 \times 10^{-9}$

The observed dissociation constant for 4-nitrobenzenol is about twenty
times larger than calculated; that for 3-nitrobenzenol is only about four times
larger than calculated. The large discrepancy in calculated and observed
values for 4-nitrobenzenol is the result of resonance interaction between the
electron-withdrawing NO_2 group and the electron-donating $O^{\ominus}$ group at the
ends of the conjugated system.

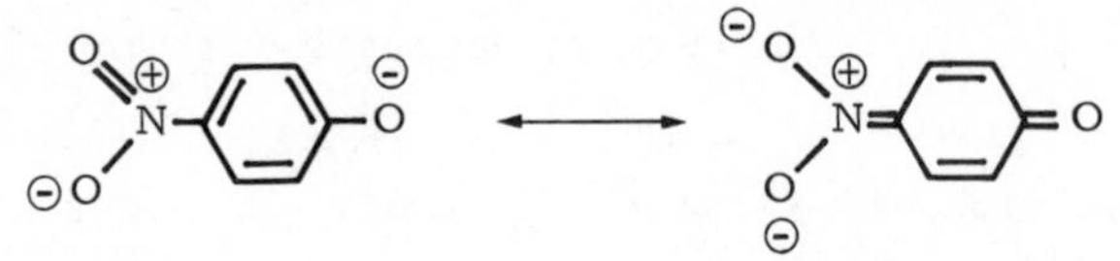

Contribution of the quinoid form to the hybrid is so significant that 4-nitroben-
zenol is a much stronger acid than may be calculated on the basis of the normal
substituent effect of a nitro group.

26-42   Relative rates $C_6H_5CH_2Cl : \underline{p}-CH_3C_6H_4CH_2Cl : \underline{p} \ CH_3O-C_6H_4CH_2Cl$
: $\underline{p}-NO_2C_6H_4CH_2Cl$ is $1 : 1.14 : 1.23 : 0.55$ in water at $30°$ in the presence of
base, and $1 : 2.3 : 3.8 : 0.020$ in 48% ethanol at $30°$.

The greater spread in rates in 48% ethanol indicates a change of
mechanism on changing conditions. In the presence of base, the reaction is
S_N2 - in 48% ethanol it becomes S_N1. Under S_N1 conditions, the substituent
effects are larger because reaction demands a supply of electrons to the reaction
site to stabilize the carbonium ion intermediate.

26-43   a. Solubility in dilute base (benzenol positive, cyclohexanol negative).

b. The acid will dissolve but not the ester in $NaHCO_3$ - both will dissolve in NaOH solution.

c.   The 1,4- but not the 1,3-benzenediol is easily oxidized to a quinone.

d.   Hot base converts tropolone to benzoic acid. 1,4-Benzendiol is converted to 1,4-benzenedione by mild oxidizing agents while tropolone is not; **tropolone forms a copper chelate, 1,4-benzenediol does not.**
e.   1,4-Anthracenedione has one reactive double bond and should therefore be a good dienophile. Diels-Alder adducts with dienes (e.g., anthracene) should form readily. 9,10-Anthracenedione will not react similarly.

26-44

a.

b.

c.   Benzenol $\xrightarrow{\text{HNO}_3}$ 4-nitrobenzenol $\xrightarrow{\text{ClCH}_2\text{CO}_2\text{H}}$

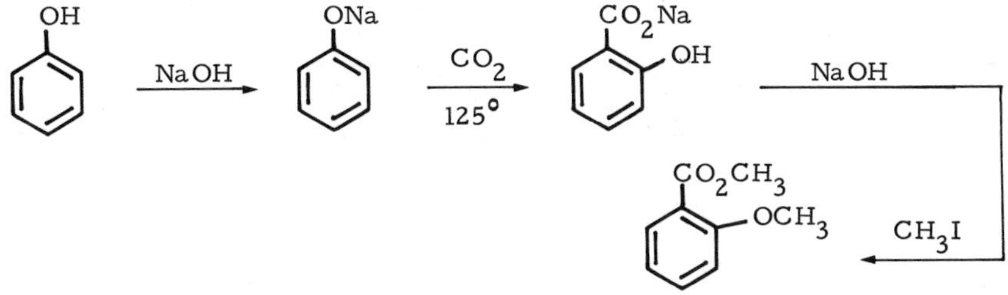

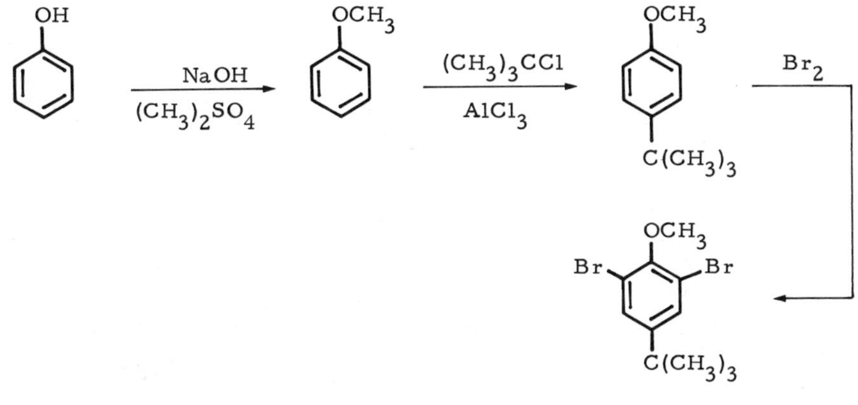

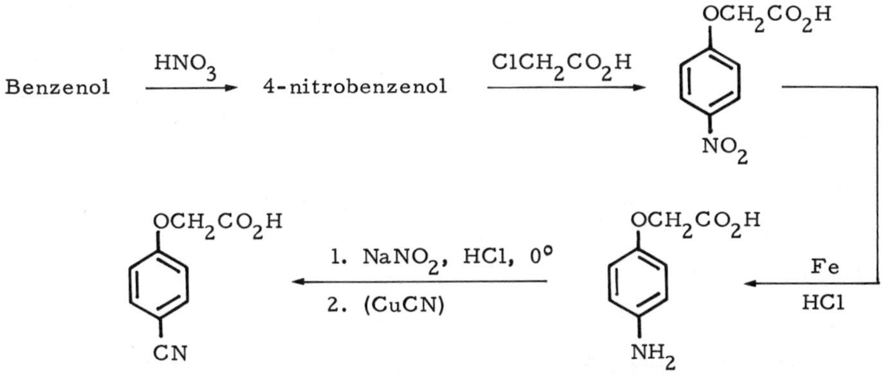

d. Benzenol ⟶

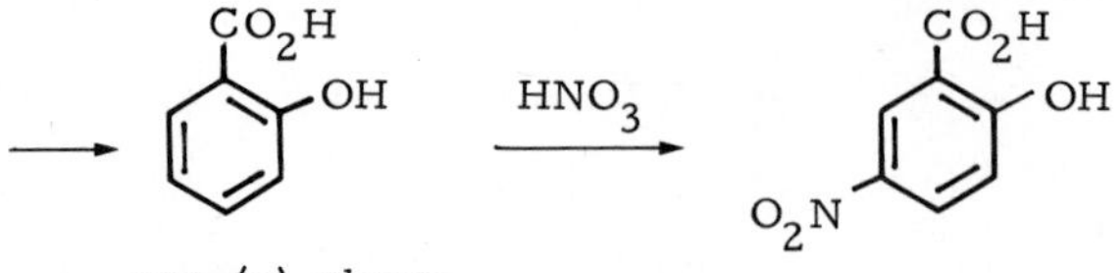

see (a) above

e. Naphthalene $\xrightarrow[\text{SO}_3]{\text{H}_2\text{SO}_4}$ 2-naphthalenesulfonic acid $\xrightarrow{\text{NaOH}}$ 2-naphtha-

lenol ⟶ 2-naphthalenamine (Bucherer reaction, Section 26-1D)

f.

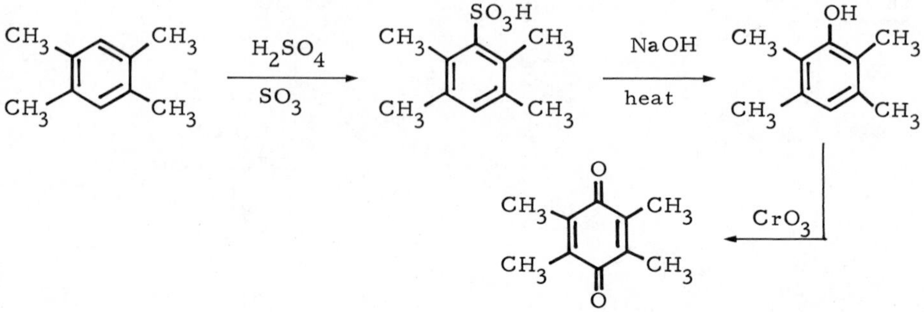

g.

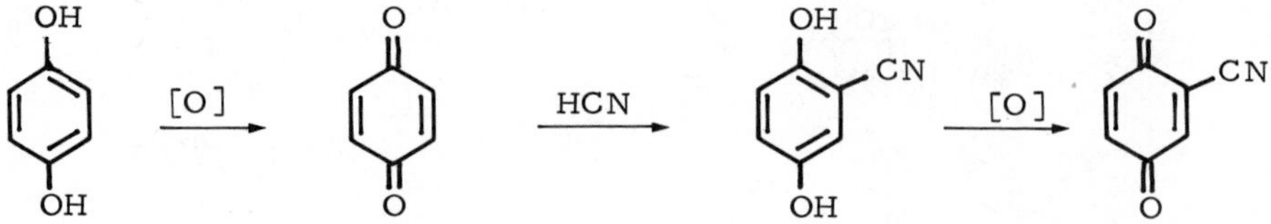

h.

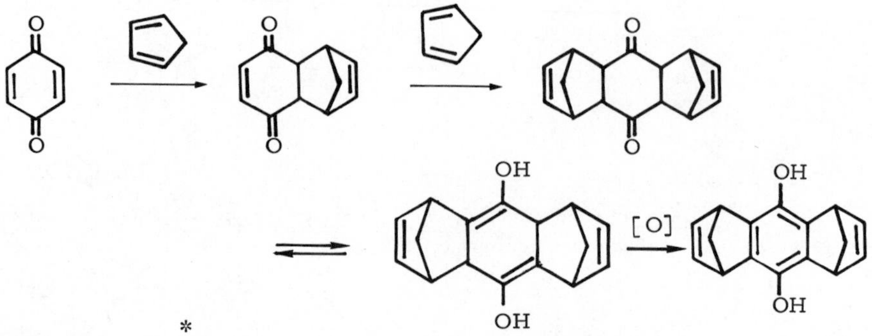

26-45

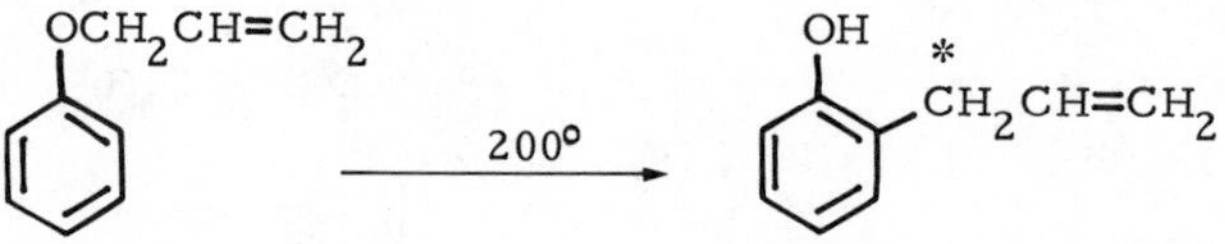

A dissociation-recombination mechanism cannot be involved since otherwise a mixture of 2-(2-propenyl-1-^{14}C)benzenol and 2-(2-propenyl-3-^{14}C)benzenol would be obtained. The actual ^{14}C-distribution in the product indicates concerted pericyclic rearrangement mechanism.

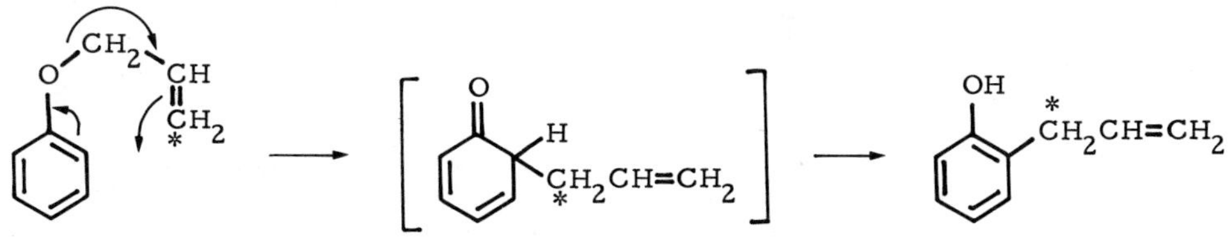

If the ortho positions are blocked, rearrangement to the para position occurs by a stepwise intramolecular process.

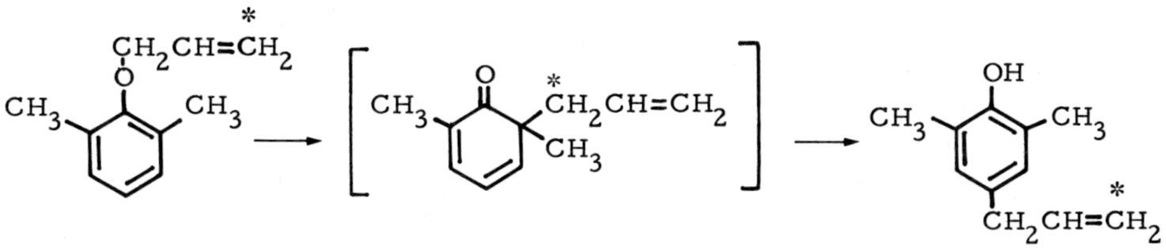

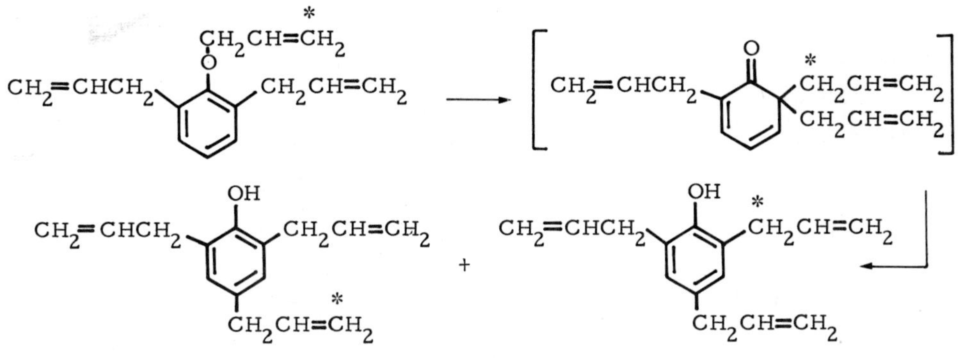

26-46   The following answers are representative:

a.   4-nitrobenzenol

b.   2,4-dichlorobenzenol

c.

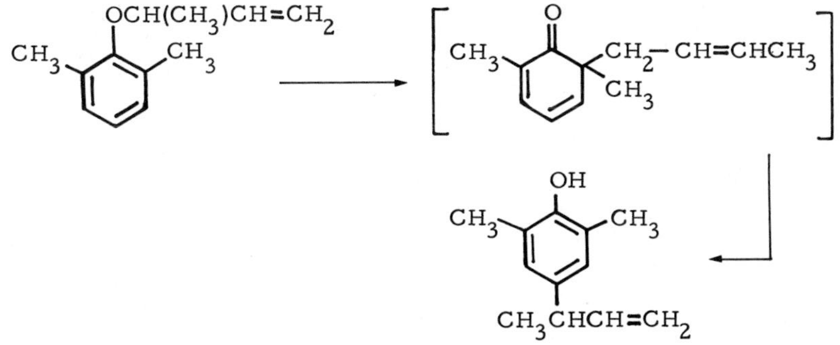

d.   9,10-anthracenedione;   e.   tetrachloro-1,4-benzenedione

f.

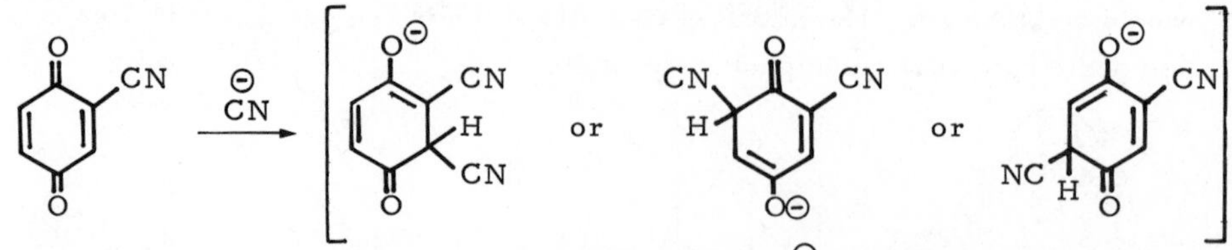

of these only the first has $-O^{\ominus}$ conjugated with $-CN$
and therefore it should be favored and the product
2,3-dicyano-1,4-benzenediol

26-47

a. 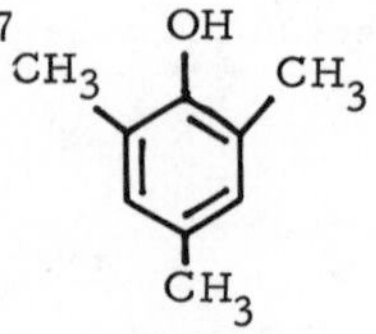   coupling w ould
not occur

b.

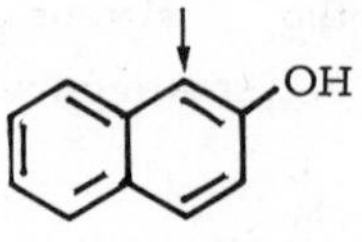

c.

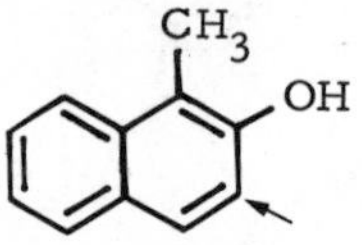

d.

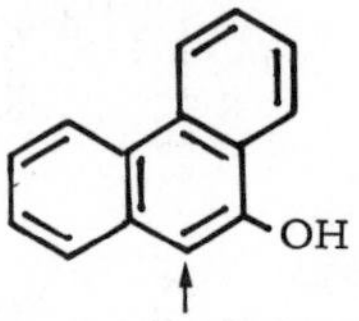

26-48

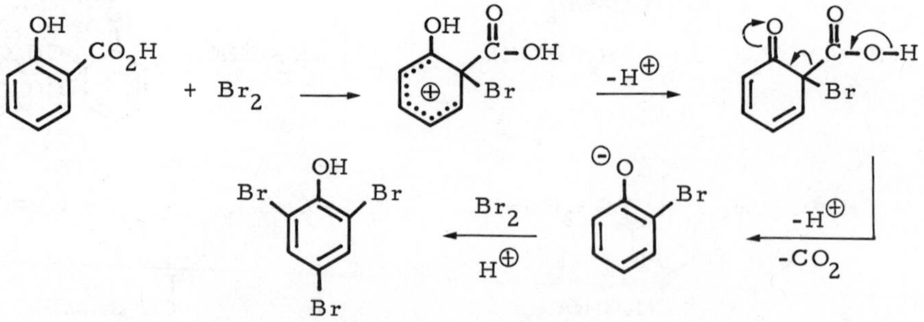

The ready bromodecarboxylation of 2-hydroxybenzoic acid may be
attributed to the intermediate formation of a β-keto acid, which is a type of
compound known to decarboxylate readily. The 3- and 4-hydroxybenzoic acids
are not therefore likely to bromodecarboxylate as readily as the ortho isomer
because no β-keto acid can be formed. However, the para-isomer could form
an α, β-unsaturated γ-keto acid which would be expected to decarboxylate fairly
readily.

26-49   The intermediate dichlorocarbene as the attacking electrophile will
give 15 by attack at the ipso or C4 position. Because loss of $CH_3^{\oplus}$ is not

possible, 15 is stable.

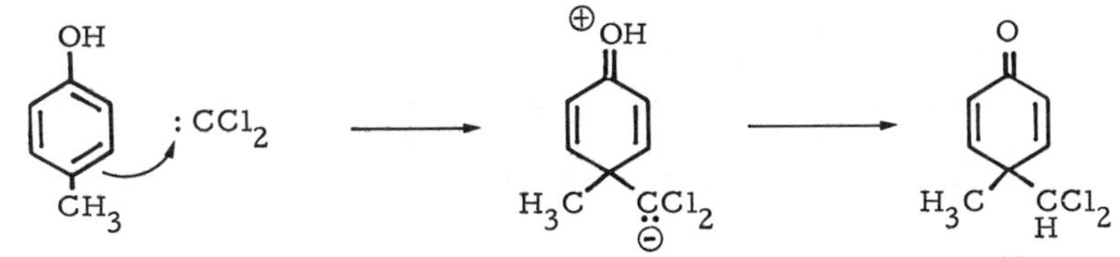

26-50 $(CH_3)_2C{=}O + H^\oplus \rightleftharpoons \left[(CH_3)_2C{\overset{\oplus}{=}}OH \longleftrightarrow (CH_3)_2\overset{\oplus}{C}{-}OH \right]$

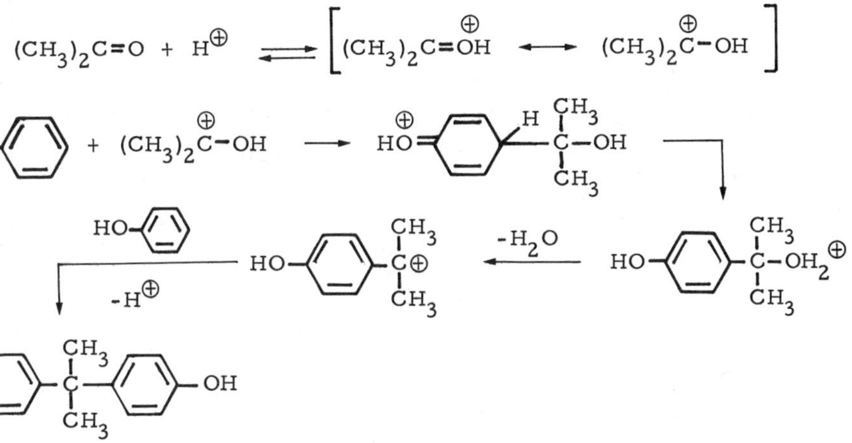

26-51

a. Benzene $\xrightarrow[\text{FeCl}_3]{\text{Cl}_2}$ chlorobenzene $\xrightarrow[250^\circ]{\text{NaOH}}$ benzenol $\xrightarrow{\text{Na}_2\text{Cr}_2\text{O}_4}$

1, 4-benzenediol $\xleftarrow{[2H]}$ 1, 4-benzenedione $\longleftarrow$
(hydroquinone)

b. Benzene $\longrightarrow$ benzenol (see Part A) $\xrightarrow{\text{HNO}_3}$ 4-nitrobenzenol (and

some 2- isomer) $\xrightarrow{\text{Fe, HCl}}$ 4-aminobenzenol (Rodinal)

(also can be obtained from C_6H_5NHOH, Section 23-10D)

c. Benzene $\xrightarrow{\text{Cl}_2,\ \text{AlCl}_3}$ chlorobenzene $\xrightarrow{\text{HNO}_3,\ \text{H}_2\text{SO}_4}$ 4-chloronitro-

benzene $\xrightarrow{\text{HN(C}_2\text{H}_5)_2}$ 4-nitro-N, N-diethylbenzenamine $\xrightarrow{\text{Fe, HCl}}$

4-amino-N, N-diethylbenzenamine

(or from reduction of the 4-nitroso derivative of N, N-diethylbenzenamine; see
Section 23-10B)

d.   Benzene ⟶ 4-aminobenzenol (see Part b) $\xrightarrow{CH_3I}$ 4-$\underline{N}$-methyl-
aminobenzenol (Metol);

or by reduction of rearrangement product of $\underline{N}$-nitroso-$\underline{N}$-methylbenzenamine
(Section 23-10D) followed by HNO_2 in H_2O.

e.   4-Aminobenzenol (see Part b) $\xrightarrow[\text{2. } H^\oplus]{\text{1. } ClCH_2CO_2H}$ $\underline{N}$-(4-hydroxyphenyl)-

aminoethanoic acid (Glycin)

f.   Benzenol (see Part a) $\xrightarrow{HNO_3, \ H_2SO_4}$ 2, 4-dinitrobenzenol $\xrightarrow{Fe, \ HCl}$

2, 4-diaminobenzenol (Amidol) ⟵┘

26-52   The oxidizing agent could be 1, 4-benzenedione (para-benzoquinone).

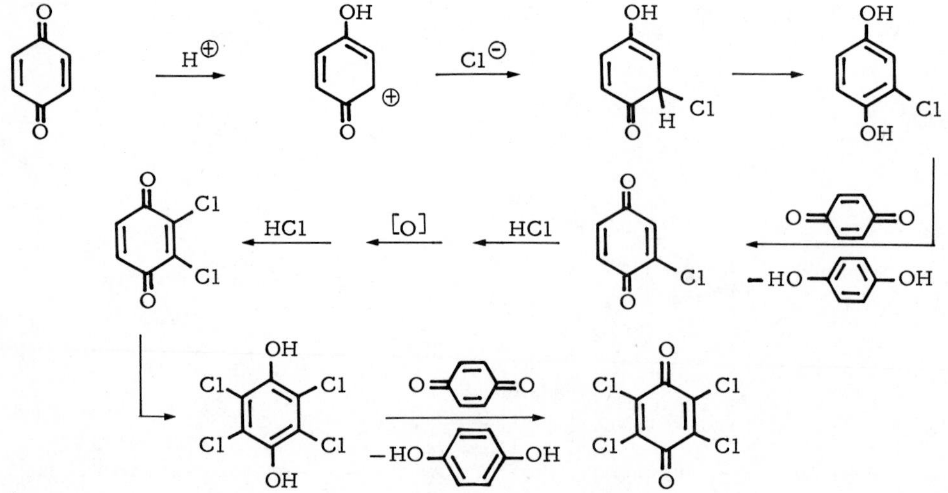

26-53   Benzenol forms 4-nitrosobenzenol ($C_6H_5O_2N$) by a straightforward
electrophilic substitution reaction.   1, 4-Benzenedione forms the same product
by a condensation-rearrangement sequence, as follows:

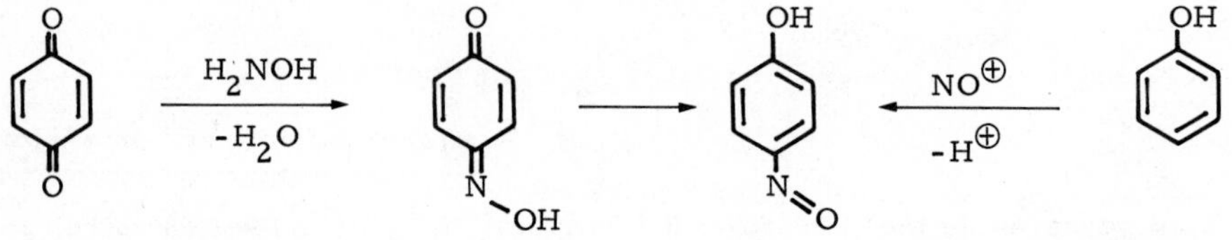

26-54

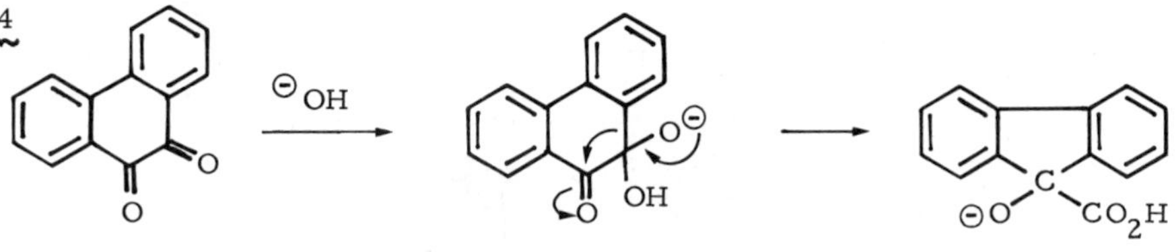

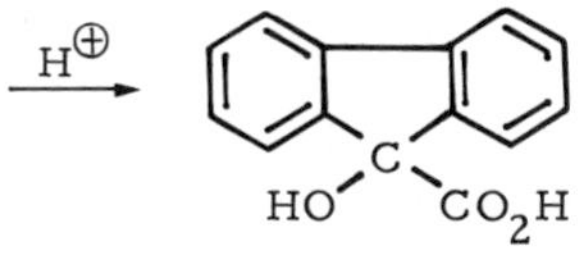

Benzilic-acid-type rearrangement of 9,10-phenanthrenedione would appear to be possible with the loss of some stabilization energy.

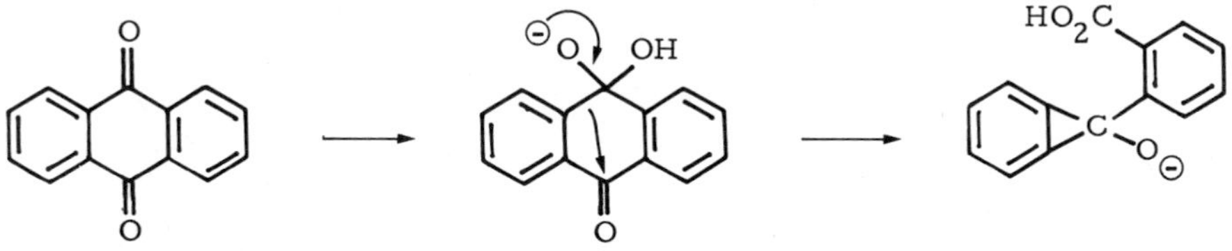

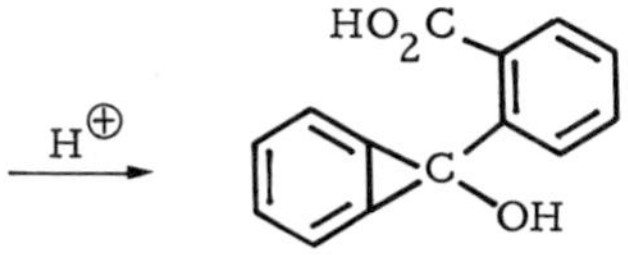

A similar rearrangement of 9,10-anthracene-dione would be highly unlikely since a highly strained compound would be formed.

26-55

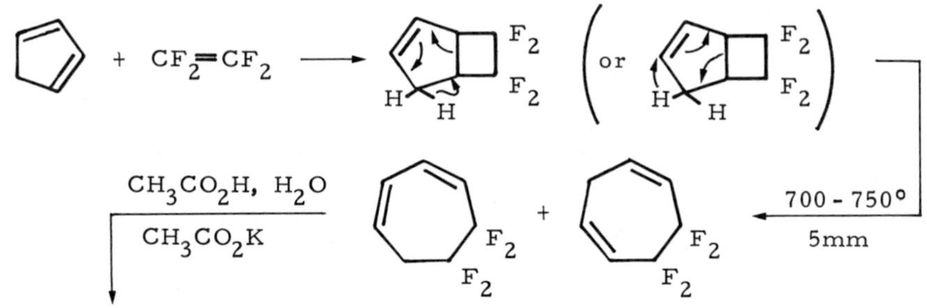

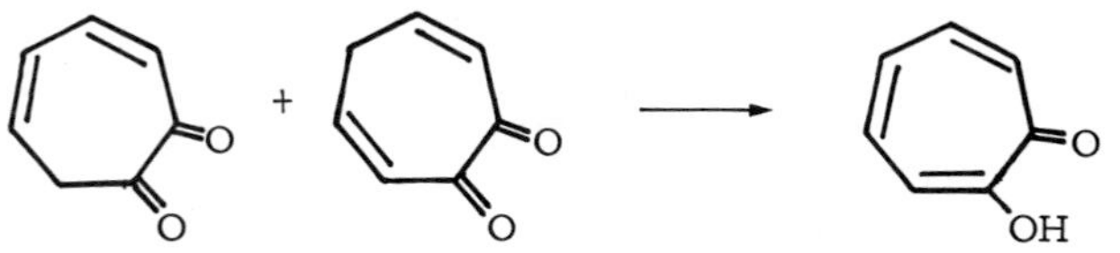

26-56   The properties of 3- and 4-hydroxy-2,4,6-cycloheptatrienone would be expected to be considerably modified compared to tropolone, mainly because of the proximity of the functional groups in tropolone which allows for strong hydrogen-bonding.

26-57

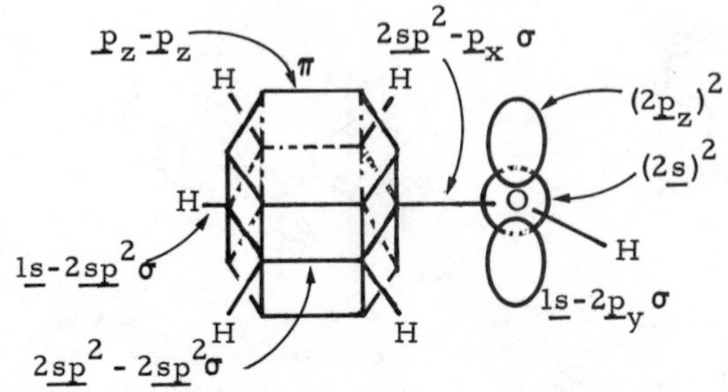

The atomic orbital model suggests that only the unshared electron pair on oxygen that occupies the p_z orbital would be delocalized over the ring. This places the direction of the OH bond coplanar with ring to achieve maximum overlap.

26-58   The reverse benzil-benzilic acid rearrangement in the case of 16 may be written as follows:

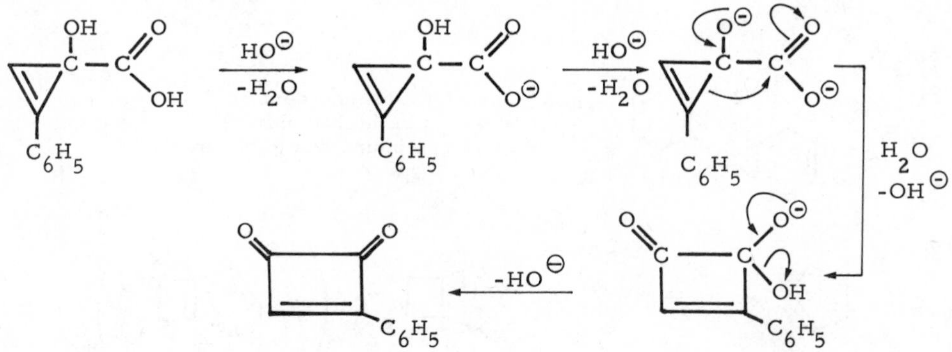

The possible reasons for why this rearrangement occurs are: (1) decrease in strain energy, and (2) gain in resonance stabilization associated with the cyclo- butenedione ring which formally can be regarded as a $4n + 2$ cyclic Hückel ring.

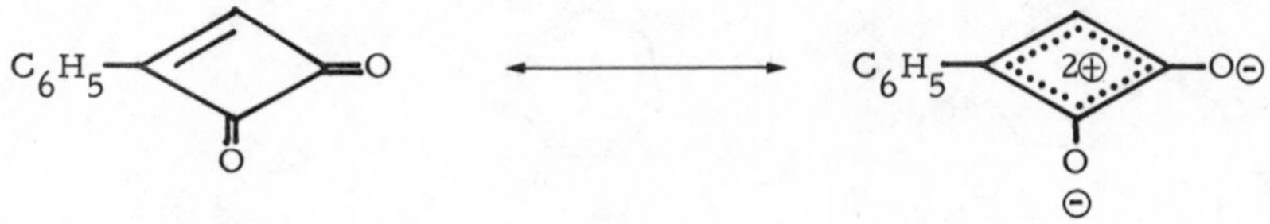

CHAPTER 27

MORE ON SPECTROSCOPY.
IMPORTANT, LESS COMMON, SPECTROSCOPIC METHODS

27-1   Given that the uncertainty in the life time of the rotational conforma-
tion $\underset{\sim}{4}$, $\underset{\sim}{5}$ or $\underset{\sim}{6}$ is 10^{-10} sec. then, from Equation 27-1, the uncertainty in the
chemical-shift difference between $\underset{\sim}{4}$ and $\underset{\sim}{5}$ (or $\underset{\sim}{6}$) is $\Delta \nu = 10^{10}/2\pi$ Hz, which is
a very large chemical shift! For separate resonances to be observed for each
rotational conformation, the chemical shift difference would have to be larger
than $10^{10}/2\pi$ Hz.

27-2   The ^{19}F nmr spectrum of 1, 2-difluorotetrachloroethane at $-120°$ is
composite of the <u>separate</u> resonances for conformations A (A') and B. At
higher temperatures rotation is faster about the C—C bond with the result that
the resonances coalesce at $-33°$.

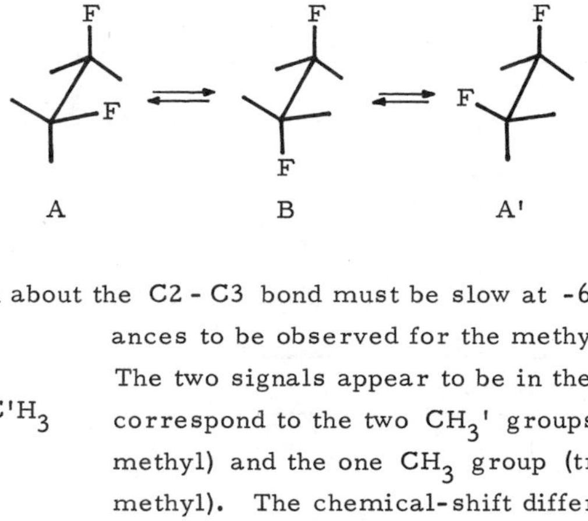

(Chlorines are
not shown.)

A   B   A'

27-3   Rotation about the C2 - C3 bond must be slow at $-64°$ for two resonances to be observed for the methyl protons at C3.
The two signals appear to be in the ratio of 2:1 and
correspond to the two CH_3' groups (gauche to the Cl
methyl) and the one CH_3 group (trans to the Cl
methyl). The chemical-shift difference at $-64°$ is
10 Hz, which gives an estimate of the rate of rotation

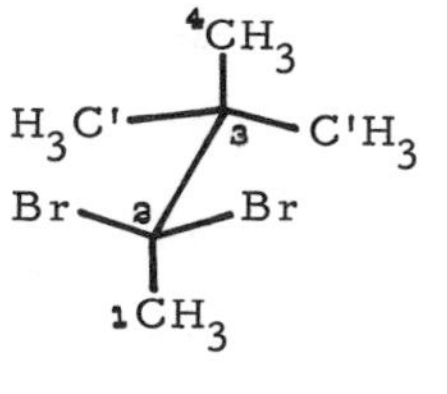

at $-33°$ as $\dfrac{1}{2\pi \cdot 10} = 0.016$ sec.

27-4   The lower limit for the lifetime of the conformational isomers of
$CH_3CH_2CH_2I$ is $1/2 \times 0.35 \times 10^9 = 4.5 \times 10^{-10}$ sec.

27-5   a. The OH resonance is split into a 1:2:1 triplet through spin-coupling with the CH_2 protons.

b.   Exchange is rapid in 100% ethanol. The uncertainty in the chemical shifts of the triplet resonance in 10% ethanol is 5 Hz + 5 Hz = 10 Hz, and the approximate lifetime for the observed spin interactions is $1/(2\pi \times 10) = 0.016$ sec.   In 100% ethanol, the lifetime of an exchanging proton would have to be <u>less</u> than 0.016 sec.

c.   Under conditions where exchange is not so rapid that all evidence of splitting between CH_2OH disappears - nor so slow that sharp splitting is evident, then both the OH and the CH_2 resonances will be broad (as in 100% ethanol).   The CH_3 resonances will not be similarly affected by exchange because the methyl protons are coupling only to the nonexchanging CH_2 protons.

27-6   The two transitions labeled 3 and 4 in Figures 27-5 and 27-6 are of equal intensities because the populations of the energy states are very nearly the same (see page 299 of text).

27-7   The energy levels are set up as follows:

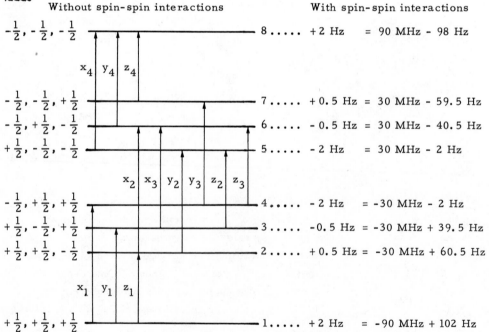

Without spin-spin interactions

With spin-spin interactions

$-\frac{1}{2}, -\frac{1}{2}, -\frac{1}{2}$ ———— 8 ..... + 2 Hz   = 90 MHz - 98 Hz

x_4 | y_4 | z_4

$-\frac{1}{2}, -\frac{1}{2}, +\frac{1}{2}$ ———— 7 ..... + 0.5 Hz = 30 MHz - 59.5 Hz

$-\frac{1}{2}, +\frac{1}{2}, -\frac{1}{2}$ ———— 6 ..... - 0.5 Hz = 30 MHz - 40.5 Hz

$+\frac{1}{2}, -\frac{1}{2}, -\frac{1}{2}$ ———— 5 ..... - 2 Hz   = 30 MHz - 2 Hz

x_2 | x_3 | y_2 | y_3 | z_2 | z_3

$-\frac{1}{2}, +\frac{1}{2}, +\frac{1}{2}$ ———— 4 ..... - 2 Hz   = -30 MHz - 2 Hz

$+\frac{1}{2}, -\frac{1}{2}, +\frac{1}{2}$ ———— 3 ..... -0.5 Hz = -30 MHz + 39.5 Hz

$+\frac{1}{2}, +\frac{1}{2}, -\frac{1}{2}$ ———— 2 ..... + 0.5 Hz = -30 MHz + 60.5 Hz

x_1 | y_1 | z_1

$+\frac{1}{2}, +\frac{1}{2}, +\frac{1}{2}$ ———— 1 ..... + 2 Hz   = -90 MHz + 102 Hz

numbers in last column are calculated as the <u>net</u> of the figures given in the Table <u>following</u>.

Energy levels are:

Correction for spin-spin interactions

(If 100 and 60 pair have the same sign, the interaction is positive; if opposite sign, it is negative.)

8    $\frac{3}{2}$ (60 MHz) $- \frac{1}{2}$(100 + 60 + 40) Hz $+\frac{1}{4}$ (5 + 3 + 0) = 2 Hz

7    $\frac{1}{2}$ (60 MHz) $- \frac{1}{2}$(100 + 60 - 40) Hz $+\frac{1}{4}$ (5 - 3 + 0) = 0.5 Hz

6    $\frac{1}{2}$ (60 MHz) $- \frac{1}{2}$(100 - 60 + 40) Hz $+\frac{1}{4}$ (-5 + 3 + 0) = -0.5 Hz

5    $\frac{1}{2}$ (60 MHz) $- \frac{1}{2}$(-100 + 60 + 40) Hz $+\frac{1}{4}$ (-5 - 3 + 0) = -2 Hz

4   $-\frac{1}{2}$ (60 MHz) $- \frac{1}{2}$(-100 + 60 + 40) Hz $+\frac{1}{4}$ (-5 - 3 + 0) = -2 Hz

3   $-\frac{1}{2}$ (60 MHz) $+ \frac{1}{2}$(100 - 60 + 40) Hz $+\frac{1}{4}$ (-5 + 3 + 0) = -0.5 Hz

2   $-\frac{1}{2}$ (60 MHz) $+ \frac{1}{2}$(100 + 60 - 40) Hz $+\frac{1}{4}$ (+5 - 3 + 0) = 0.5 Hz

1   $-\frac{3}{2}$ (60 MHz) $+ \frac{1}{2}$(100 + 60 + 40) Hz $+\frac{1}{4}$ (+5 + 3 + 0) = 2 Hz

The allowed upward transitions involve the change of any <u>one</u> given nucleus from the +1/2 to the -1/2 state. The transitions are labeled in the figure as x for the nucleus with the 100-Hz chemical shift, y for the 60-Hz shift and z for the 40-Hz shift.

The transition energies are calculated by subtracting the energy of the lower level from the energy of the upper level. Without spin-spin interaction, the x_1 transition energy will be the energy of Level 4 [-1/2(60 MHz) + 1/2(-100 + 60 + 40) Hz = -30 MHz] - the energy of Level 1 [-3/2(60 MHz) + 1/2(100 + 60 + 40) Hz = -90 MHz + 100 Hz] or 60 MHz - 100 Hz. Proceeding in this way, the transitions without spin-spin interaction will be

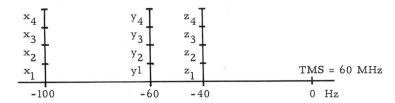

If we now take into account spin-spin interactions and, as before, subtract the energy of the lower level from that of the upper level for each transition, we have:

x_1 = -104 Hz, x_2 = -101 Hz, x_3 = -99 Hz   and x_4 = -96 Hz

$$y_1 = y_2 = -62.5 \text{ Hz} \quad \text{and} \quad y_3 = y_4 = -57.5 \text{ Hz}$$

$$z_1 = z_2 = -41.5 \text{ Hz} \quad \text{and} \quad z_3 = z_4 = -38.5 \text{ Hz}$$

Plotting we have:

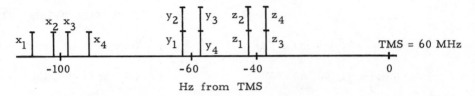

Hz from TMS

Because the chemical shifts are relatively large with respect to the magnitudes of the spin-spin interactions (see Section 9-10K), each of the transitions will be approximately equally probable. We therefore expect a spectrum of the following appearance with respect to line positions and intensities.

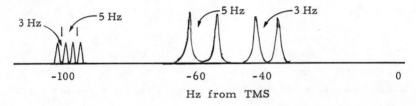

Hz from TMS

     This way of calculating spectral line positions may seem unduly complex compared to the approach used in Section 9-10J. However, it is the starting point for all analyses of spectra in which there are second-order splitting or line intensities different from simple expectations. The procedure for handling such situations is to set up the energy levels just as we have done here and then perform a quantum-mechanical "mixing" of those states which have the same net magnetic quantum numbers (such as Levels 2, 3, and 4 or 5, 6, and 7). The mixing procedure has analogy in the mixing of electronic states in the resonance and molecular-orbital methods described in Section 21-2A.

27-8   The C2 - C3 bond is the weakest bond in 3,3-dimethyl-2-butanone and is therefore cleaved preferentially on irradiation. It is the weakest bond because it forms the most stable radical fragments.

     For the photodissociation of 2-propanone (see Section 28-2A) $CH_3 \cdot$ and $CH_3CO \cdot$ radicals are formed which give CO, CH_3CH_3 and $CH_3COCOCH_3$. There will be sorting of magnetic states of $CH_3CO \cdot$ and $\cdot CH_3$, and CIDNP is expected.

27-9   The spacings in the vibrational fine structure are 0.27 ev which amounts to $23.06 \times 0.27 = 6.2$ kcal.   Ground-state vibrational energy differences for ethyne molecules are on the order of 2150 cm^{-1} for the C≡C bond (or 4651 nm) = 6.15 kcal by Equation 9-2 (text, p. 269).   The sizes of the spacings are clearly of the right magnitude to belong to the ground state.   However, at ordinary temperatures (say, 300° K) virtually all of the ethyne molecules have to be in the lowest vibrational state.   An energy difference of 6.15 kcal between the lowest and next highest vibrational state corresponds to an equilibrium constant (calculated from $\underline{\Delta G}° = -RT \ln \underline{K}$) of 3×10^{-5} or 99.996% in the lowest vibrational state.   If vibrational structure is observed in the spectrum, it must be almost all for transitions from the lowest vibrational state of the ground state to various vibrational levels of the excited state.

27-10   a.   The mean lifetime in the Mössbauer experiment is 9.9×10^{-8} sec which from Equation 27-1 corresponds to a $\Delta \nu = 1/(2\pi \underline{\Delta}\underline{t})$. $1/(2\pi \times 9.9 \times 10^{-8})$ $= 1.61 \times 10^6$ Hz and this from Sections 9-3 and 9-4 can be converted to kcal by

$$\frac{\overset{\text{(kcal/nm)}}{28,600} \times \overset{\text{(Hz)}}{1.61 \times 10^6}}{\underset{\text{(nm/Hz)}}{3 \times 10^8 \times 10^9}} = 1.53 \times 10^{-7} \text{ kcal}$$

The ratio of this to 14,400 ev (= 3.32×10^5 kcal) $= \dfrac{1.53 \times 10^{-7} \text{ kcal}}{3.32 \times 10^5 \text{ kcal}}$

$$= 4.62 \times 10^{-13}$$

(an incredible precision in energy)

b. $\Delta \nu$ as calculated by the method of Part a = $1/(2\pi \times 1.5 \times 10^{-8})$ = 1.06×10^7 Hz, while λ (589.3 nm) = $3 \times 10^{17}/589.3 = 5.09 \times 10^{14}$ Hz.   The uncertainty in λ is

$$\frac{1.06 \times 10^7 \times 589.3}{5.09 \times 10^{14}} = 1.2 \times 10^{-5} \text{ nm}$$

27-11   a.   No $\underline{M}+29$ or $\underline{M}+41$ peaks are expected because the hydrocarbon is an alkane and does not add ions $C_2H_5^{\oplus}$ or $C_3H_5^{\oplus}$.

b.   The secondary ion would be most likely.

c.   Each peak is separated from its neighbor by 14 mass units corresponding to loss of a CH_2 group.   The spectrum therefore shows the sequence of C—C bond cleavages along a straight (continuous) hydrocarbon chain.

27-12 $BrCH_2CH_2Br \xrightarrow{-e} BrCH_2CH_2Br^{\oplus}$ $(\underline{m/e} = 188)$ $\xrightarrow{-Br\cdot}$

$BrCH_2CH_2Br$ + [cyclopentyl cation with Br] $\longleftarrow$ $BrCH_2CH_2CH_2CH_2\overset{..}{Br}:$ $\longleftarrow$ $\overset{Br\oplus}{CH_2\!\!-\!\!CH_2}$

$\underline{m/e}$ 136

$\underline{m/e}$ 108

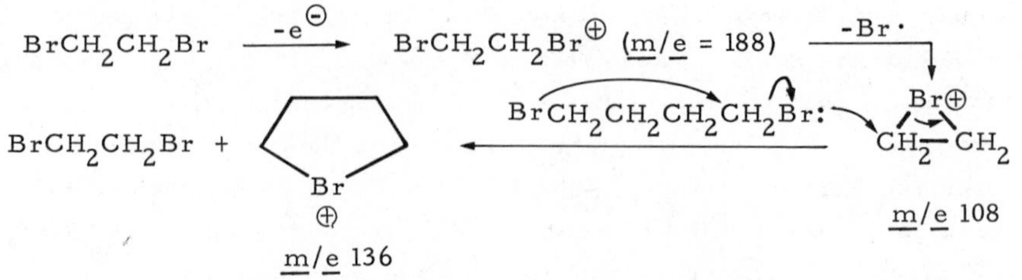

27-13   Derivative spectrum represents <u>eight</u> equally-spaced absorption peaks.

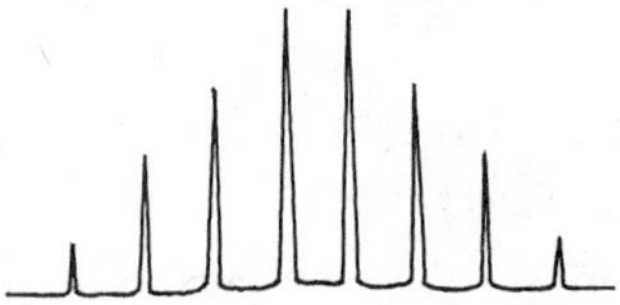

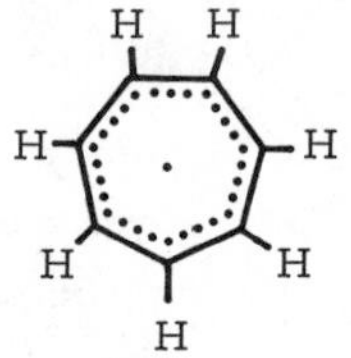

cycloheptatrienyl
radical

A possible isomeric structure follows:

     The above structure can be ruled out because the spectrum shows that the odd electron must be equally distributed over seven carbons, which would not be the case with the phenylmethyl radical.

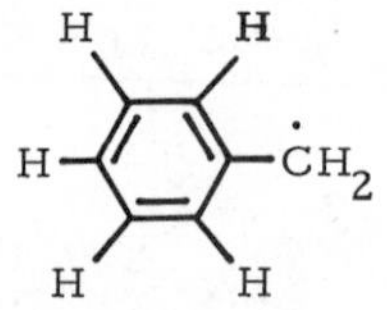

27-14   a.   The radical anion from diphenylmethanone and sodium is some-times called a ketyl and, dissolved in ethers, gives a bright blue solution sensitive to oxygen and protic substances.

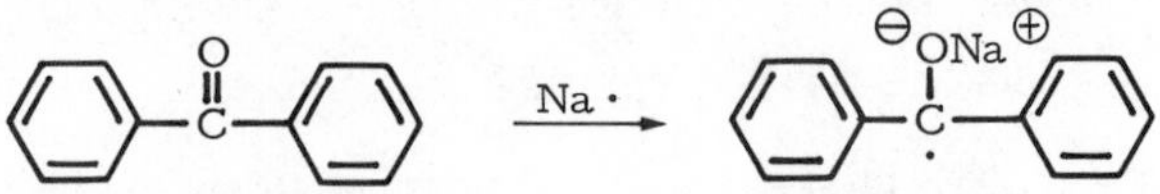

[Reaction scheme: diphenylmethanone + Na· → sodium ketyl]

The $\overset{\ominus}{-O}\overset{\oplus}{Na}$ bond must not be highly dissociated to $-\overset{\ominus}{O}$ + $\overset{\oplus}{Na}$ because if it were, fast exchange of sodiums would occur between different ketyl anions and the electron-sodium splittings would disappear.

     b.   When excess diphenylmethanone is added, exchange occurs with transfer of Na· from the ketyl to the ketone.   The electron associated with a particular sodium must be transferred <u>with that sodium</u> in order to obtain the electron-sodium splitting.   Thus,

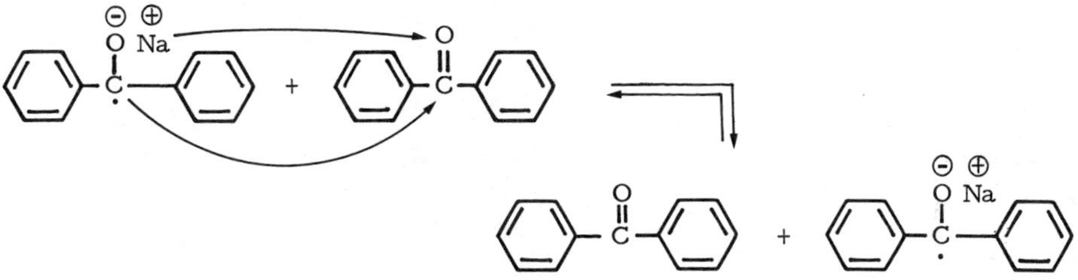

This exchange will wipe out the ring proton

1,2-Dimethoxyethane is expected to solvate $Na^{\oplus}$ much better than oxacyclopentane

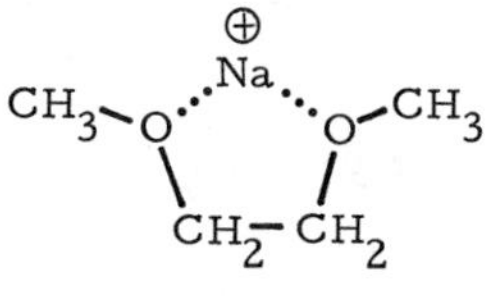

because it is bidentate and therefore increases the extent of ionic dissociation of sodium naphthalenide.

CHAPTER 28

PHOTOCHEMISTRY

28-1   No. A relatively stable upper state may well exist but it may never be reached from the ground state if it is displaced relative to the ground state such that dissociation always occurs on excitation.

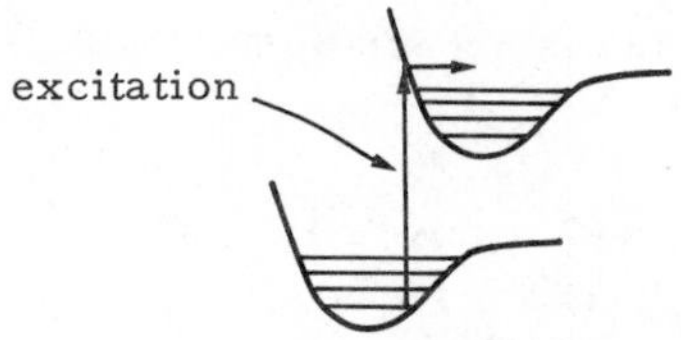

28-2

ground state   excited singlet states   triplet state

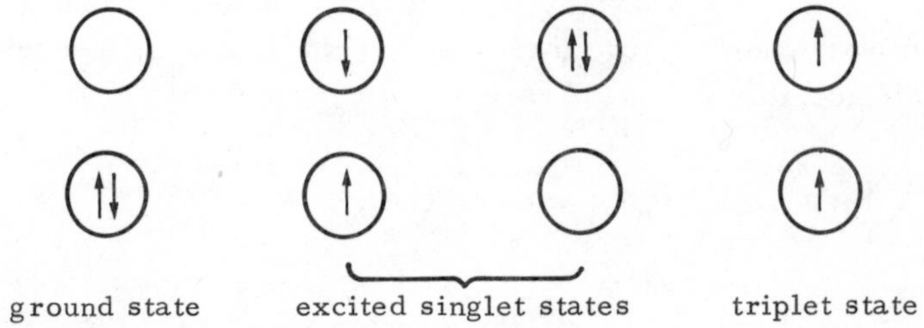

The nonplanar state should be more stable because the interelectronic repulsion should be less (the electrons would be farther apart if in perpendicular p-orbitals on each carbon).

28-3   Fluorescence is the result of a <u>radiative</u> decay of an excited <u>singlet</u> state to the ground state. The excited singlet state is a sort of <u>diradical</u> type of state in which there are electrons in unfilled orbitals. Such states are expected to be highly reactive and to be destroyed by dimerization, reaction with oxygen, radical inhibitors, etc. We should therefore expect that all those factors which would <u>increase</u> the tendency for the excited singlet state to react would tend to quench fluorescence.

Clearly phosphorescence will be subject to the same influences; the differences which would be expected being the difference in reactions and reactivity of singlet <u>vs.</u> triplet states.

28-4   If the molecules are essentially all in their lowest vibrational state
and the temperature is increased, more of them go into higher states which will
of course change (slightly) the energies of the possible transitions in Figure 28-1
thus giving a temperature effect (even if small) on the spectrum. The size and
direction of the effect will clearly depend on the relative shapes of the potential
energy curves and the spacing of the energy levels in the ground and excited
states.

28-5   The Franck-Condon principle states that the motions of the atoms do
not change on electronic excitation. This means that no great change in kinetic
energy of vibration can occur. Therefore, if the potential energy curves and
energy-level spacings of ground and excited singlet states are identical, an
$0 \longrightarrow 0^*$ or a $1 \longrightarrow 1^*$ transition will be much more probable than a $0 \longrightarrow 1^*$,
$1 \longrightarrow 2^*$, etc., because the 0 and 0^*, the 1 and 1^*, etc. states will have the
same kinetic energy over the whole range of each vibrations while this will not
be so for 0 and 1^*, the 1 and 2^*, etc. states.

28-6

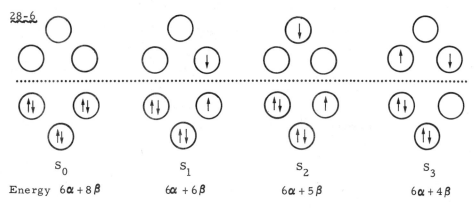

S_0	S_1	S_2	S_3
Energy $6\alpha + 8\beta$	$6\alpha + 6\beta$	$6\alpha + 5\beta$	$6\alpha + 4\beta$

The three excited singlet states differ in energy from the ground state by 2β,
3β and 4β (40, 60 and 80 kcal), and from each other by β or 20 kcal. The
λ_{max} values for the $S_0 \longrightarrow S_1$, $S_0 \longrightarrow S_2$, and $S_0 \longrightarrow S_3$ transitions are respectively:

$$\frac{28,600}{40} = 715 \text{ nm}, \quad \frac{28,600}{60} = 477 \text{ nm}, \quad \text{and} \quad \frac{28,600}{80} = 358 \text{ nm}$$

The difference in λ_{max} for the three absorptions bands is therefore 715 - 477 =
238 nm, and 477 - 358 = 119 nm. The calculated difference in λ_{max} is probably
more meaningful than the calculated λ_{max} for a single transition.

28-7   (a) At high light intensities, the radical concentrations are higher and
the quantum yields should be less. (b) The light acts to produce $Cl\cdot$ and as

long as one quantum produces only Cl· then Φ should be independent of the wavelength.   (c) Oxygen is expected to reduce Φ somewhat by diverting the R· radicals (R· + O_2 ⟶ RO_2·) and providing new chain termination reactions (2 RO_2· ⟶ ROOR + O_2, etc.).   (d) Alkenes may reduce Φ by forming relatively more stable radicals by addition or allylic substitution thereby slowing chain propagation and favoring termination.

28-8   If the reaction were $CH_3COCH_3 \xrightarrow{h\nu} 2\,CH_3\cdot + CO$, then I_2 should react with CH_3· to give CH_3I but not change the CO formation.   That I_2 reduces CO formation argues for $CH_3COCH_3 \xrightarrow{h\nu} CH_3\cdot + CH_3CO\cdot$ and $CH_3CO\cdot + I_2 \longrightarrow CH_3COI + I\cdot$ as well as $CH_3\cdot + I_2 \longrightarrow CH_3I + I\cdot\cdot$.

28-9 $(CH_3)_3C-\overset{\overset{\displaystyle O}{\|}}{C}-C(CH_3)_3 \xrightarrow{h\nu} (CH_3)_3C\cdot + \cdot\overset{\overset{\displaystyle O}{\|}}{C}-C(CH_3)_3$

$(CH_3)_3C-\overset{\overset{\displaystyle O}{\|}}{C}\cdot \longrightarrow (CH_3)_3C\cdot + CO$

$2(CH_3)_3C\cdot \longrightarrow (CH_3)_3C-C(CH_3)_3$ combination

$\longrightarrow (CH_3)_2C{=}CH_2 + (CH_3)_3CH$
    disproportionation

Norrish type II cleavage requires a hydrogen at the γ position relative to the carbonyl group.   There is no γ carbon in di-tert-butyl ketone.

28-10

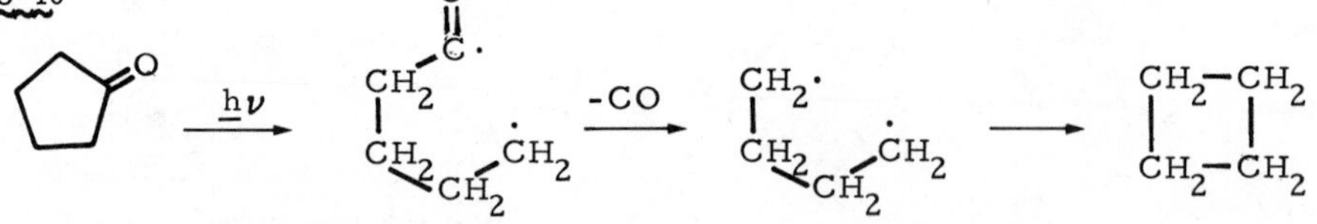

By a stepwise process:

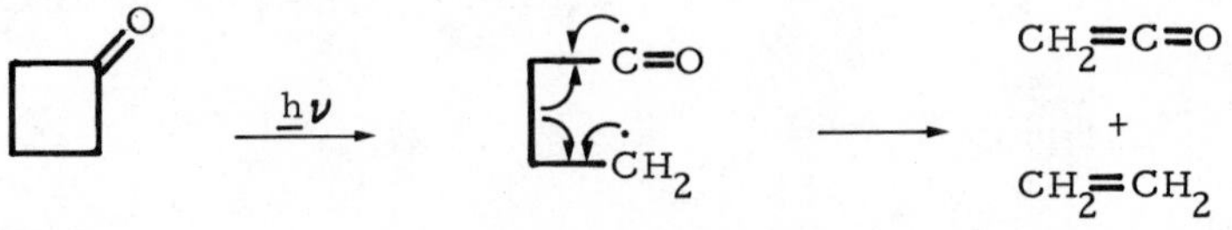

28-11   Norrish type II cleavage is expected:

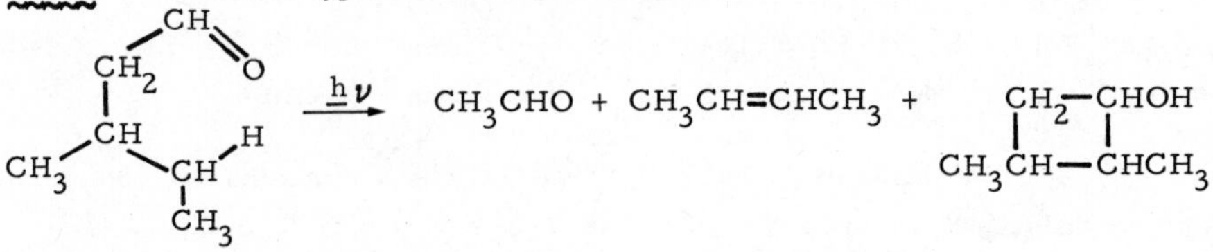

and

$$CH_3CH_2 \overset{CH_2}{\underset{CH_2}{\overset{|}{\underset{|}{CH}}}} \overset{CH}{\underset{CH_2}{\diagdown}} \overset{=O}{\underset{H}{\diagup}} \xrightarrow{h\nu} CH_3CHO + CH_3CH_2CH=CH_2 \longrightarrow \overset{CH_2}{\underset{CH_3CH_2CH}{\overset{|}{\underset{|}{}}}}\overset{CHOH}{\underset{CH_2}{}}$$

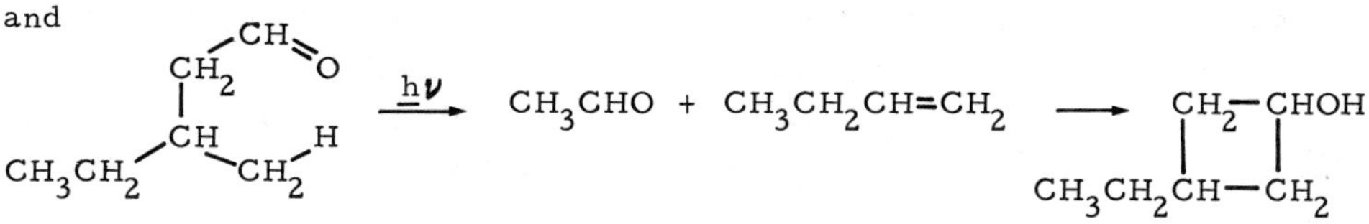

28-12

$$(C_6H_5)_2C=O \xrightarrow{h\nu} (C_6H_5)_2\overset{*}{C}=O$$

$$(C_6H_5)_2\overset{*}{C}=O + CH_3-\overset{H}{\underset{CH_3}{\overset{|}{\underset{|}{C}}}}-OH \longrightarrow (C_6H_5)_2-\overset{\cdot}{C}-OH + CH_3-\overset{\cdot}{\underset{CH_3}{\overset{|}{\underset{|}{C}}}}-OH$$

fairly stable   less stable
radical   radical

$$(CH_3)_2\overset{\cdot}{C}-OH + O_2 \longrightarrow (CH_3)_2\overset{|}{\underset{OO\cdot}{C}} - OH$$

$$(CH_3)_2\overset{|}{\underset{OO\cdot}{C}}-OH + (C_6H_5)_2\overset{\cdot}{C}-OH \longrightarrow (CH_3)_2\overset{|}{\underset{OOH}{C}}-OH + (C_6H_5)_2C=O$$

$$(CH_3)_2\overset{|}{\underset{OOH}{C}}-OH \longrightarrow (CH_3)_2C=O + H_2O_2$$

With this scheme $\Phi \approx 1$ for 2-propanone and H_2O_2 formed. The key step here is oxidation of $(C_6H_5)_2\overset{\cdot}{C}-OH$ back to benzophenone instead of benzopinacol formation. This may also occur by:

$$(C_6H_5)_2\overset{\cdot}{C}-OH \xrightarrow{O_2} (C_6H_5)_2\overset{|}{\underset{OO\cdot}{C}OH} \xrightarrow{[H]} (C_6H_5)_2\overset{|}{\underset{OOH}{C}OH} \longrightarrow$$

$$(C_6H_5)_2C=O + H_2O_2$$

28-13   The scheme outlined adds up to

$$B + \tfrac{1}{2}AH_2 + \underline{h\nu} \longrightarrow \tfrac{1}{2}(BH)_2 + \tfrac{1}{2}A$$

and therefore Φ for formation of A should be 0.5. If $\cdot$AH were formed with AH_2 being optically active 2-butanol, the alcohol should become racemic since the $CH_3-\overset{\cdot}{C}OH(CH_2CH_3)$ radical is not expected to maintain stereochemical integrity.

28-14   If $\underline{k}_t \sim \underline{k}_c$ and $\underline{k}_t^* > \underline{k}_c^*$ then trans-1,2-diphenylethene (trans-

stilbene) is destroyed faster than the cis isomer when irradiated but each is
reformed at the same rate.   Therefore, at the irradiation equilibrium the cis-
isomer is expected to be more favored than at thermal equilibrium.   Excited
(and bulky) benzophenone molecules are expected to be better able to transfer
their energy to more or less planar trans-stilbene than to sterically hindered
cis-stilbene.

28-15   a.   The reaction is an electrocyclic ring closure involving 6 π electrons.   The thermally-allowed transition state would be a Hückel orbital system
formed by disrotation, and would give 16-cis isomer.   The photochemically-
allowed transition state is the opposite of the thermal case; i.e., a Möbius
orbital system formed by conrotation to give the trans isomer.   (See Section 21-
10G.)

        b.   The trans-isomer is first photo-isomerized to the cis isomer
which then cyclizes according to Equation 28-8.

28-16

a.

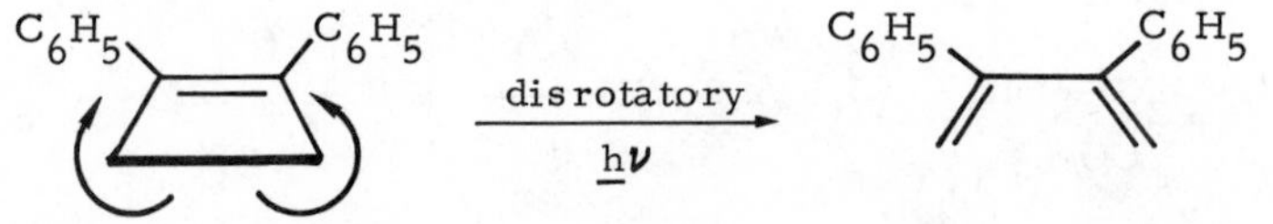

b.

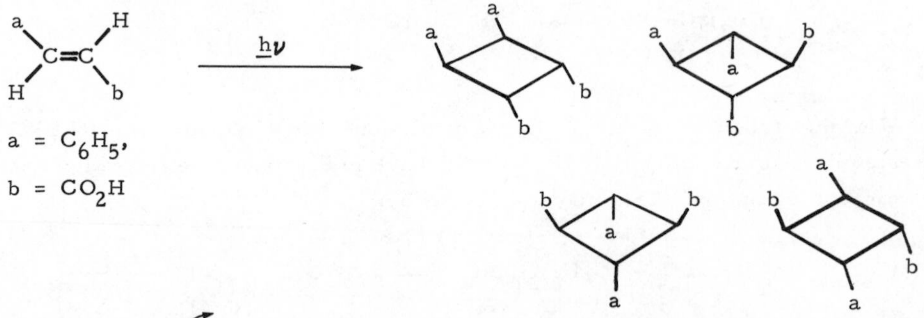

a = C_6H_5,

b = CO_2H

c.

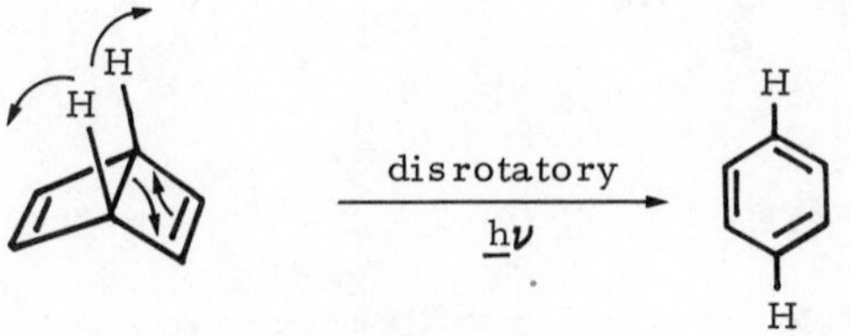

28-17   Sens $\xrightarrow{h\nu}$ Sens* (S_1) $\longrightarrow$ Sens* (T_1)

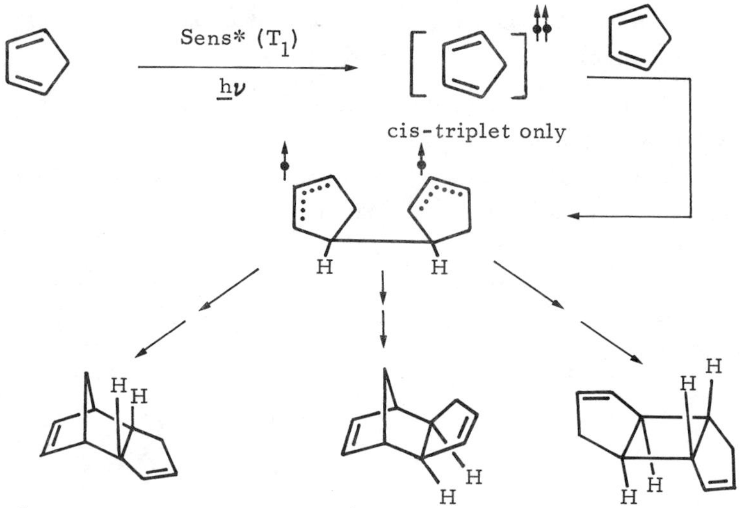

Sens* (T$_1$)

$\underline{h\nu}$

cis-triplet only

H   H

from trans   from cis   from trans

Notice that the first adduct shown is the same compound as that from a thermal [4 + 2] cycloaddition. It would be difficult to exclude the incursion of the thermal reaction during photolysis.

28-18   ArCHO $\xrightarrow{h\nu}$ ArCH=O* $\xrightarrow[\text{[2 + 2]}]{\text{R}_2\text{C=CHR}}$ cycloadducts (4 isomers possible)

28-19

a.

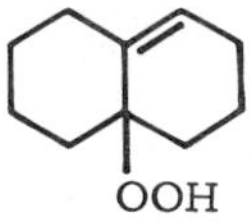

OOH

b.

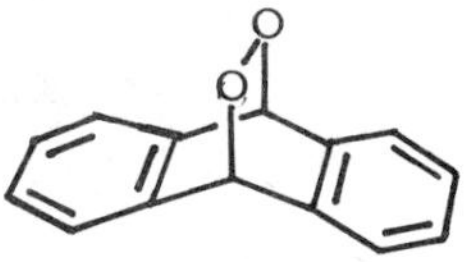

c.

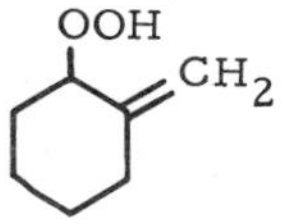

OOH

CH$_2$

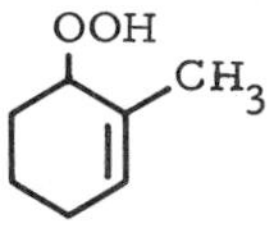

and possibly

OOH

CH$_3$

and

OOH

CH$_3$

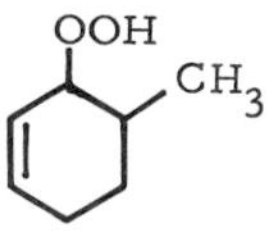

28-20   1. Photoisomerization, cis $\rightleftarrows$ trans, at one or several of the double bonds. 2. electrocyclic ring closure. 3. [2+2] cycloaddition.

28-21   Deletion of a very narrow wavelength band would have almost no effect on color perception. One would have to delete a substantial part of the wavelength band to perceive a color change, or the subtraction color.

    a. blue-green (cyan)   c. black

    b. magenta   d. very faint blue, at most

28-22   Compounds absorbing light on the blue end of the visible spectrum 400 - 480 nm, would appear to be yellow-green.

28-23   Compounds 18 and 19 are expected to have very comparable spectra since the interaction of electrons in the double bonds outside the rings and the ring electrons is not expected to lead to significant stabilization of the excited state in the manner expected for stilbene:

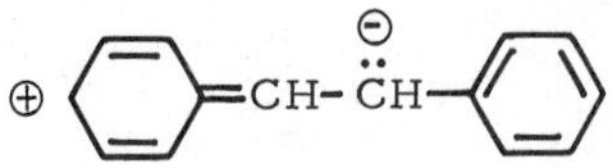

satisfactory electron-
pairing scheme for
excited state of stilbene

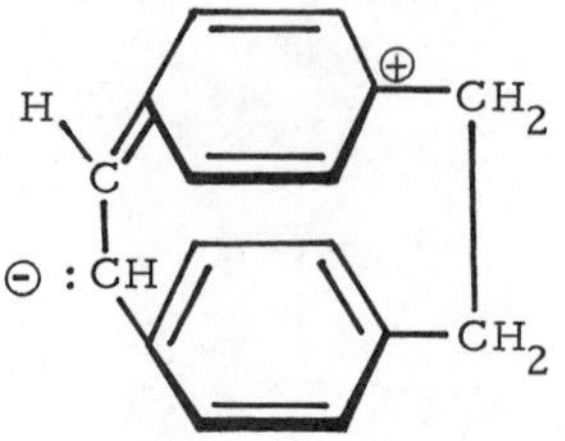

very highly strained
(unfavorable) electron-pairing
scheme for excited state of   18

28-24   The electron-pairing schemes 20a, b, c etc. must represent singlet states because they have the two non-bonding electrons on the same atoms.   This would only be possible for a singlet state or a triplet state involving a higher atomic orbital $2p, 3p,$ $2s, 3s,$ etc.

Triplet butadiene can be represented as:

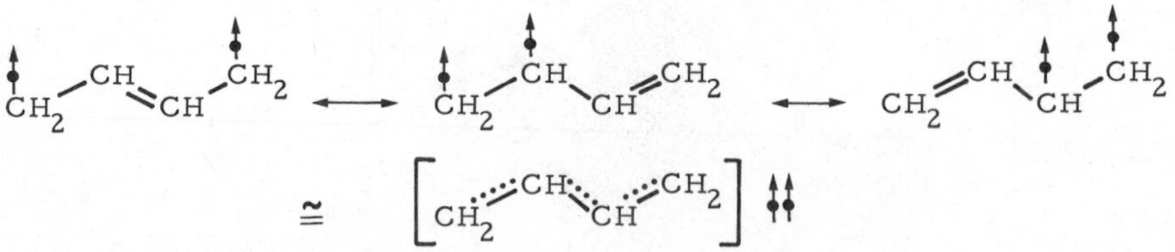

28-25 $E = \dfrac{28,600}{330} = 86.7$ kcal for trans-trans

$E = \dfrac{28,600}{310} = 92.3$ kcal for trans-cis

$E = \dfrac{28,600}{300} = 95.3$ kcal for cis-cis

Steric hindrance interferes with attainment of a planar excited state with the trans-cis and cis-cis isomers thus reducing the stabilization of the excited state which is not expected to be exactly planar for these isomers.

28-26   Successive protonations of the $-N(CH_3)_2$ groups change their effect from electron-donating to electron-attracting.

28-27

a.

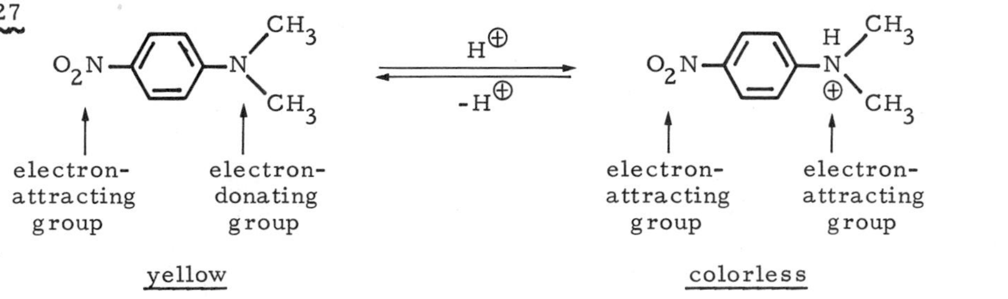

electron-
attracting
group

electron-
donating
group

yellow

electron-
attracting
group

electron-
attracting
group

colorless

b.

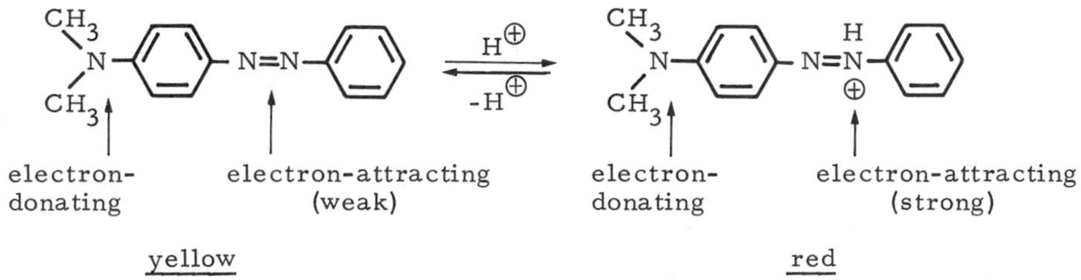

electron-
donating

electron-attracting
(weak)

yellow

electron-
donating

electron-attracting
(strong)

red

(See Exercise 23-18)

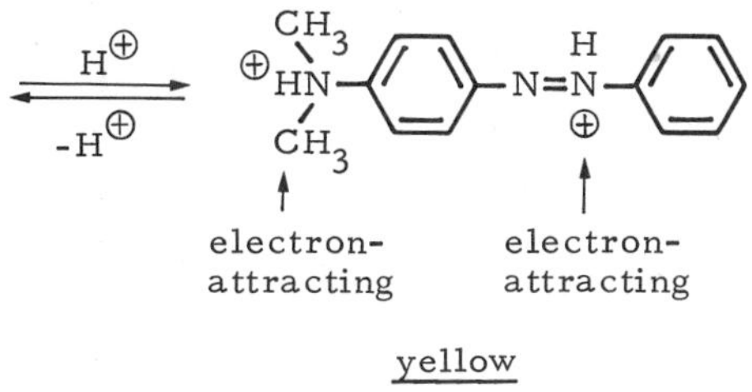

electron-
attracting

electron-
attracting

yellow

28-28

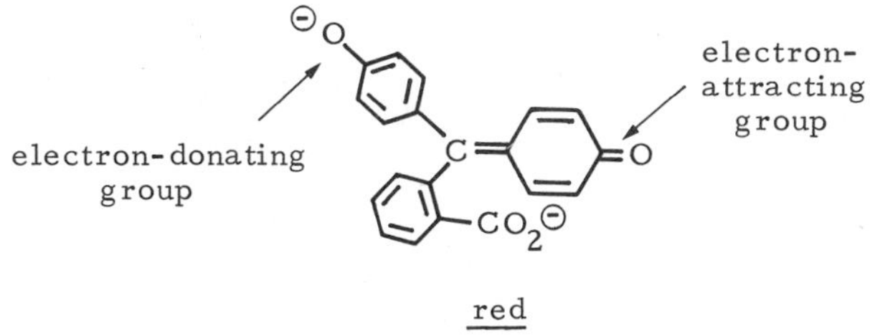

electron-donating
group

electron-
attracting
group

red

The other forms are expected to be much less highly colored.

28-29   Chromophores turn out in general to be electron-attracting groups
while auxochromes turn out to be electron-donating groups.

chromophores (strong ⟶ weak) = $-N=N-$, $-NO_2$, $-C\equiv N$,

and $-\overset{\oplus}{N}(CH_3)_3$

(The onium ions are weak chromophores.)

auxochromes (strong ⟶ weak) = $-O^{\ominus}$, $-I$ and $-CH_3$

28-30

(a) white light ⟶

clear
clear
clear

(b) green light ⟶

yellow
clear
cyan

(c) magenta ⟶

clear
magenta
clear

(d) orange ⟶

yellow (strong)
magenta (weak)
clear

28-31

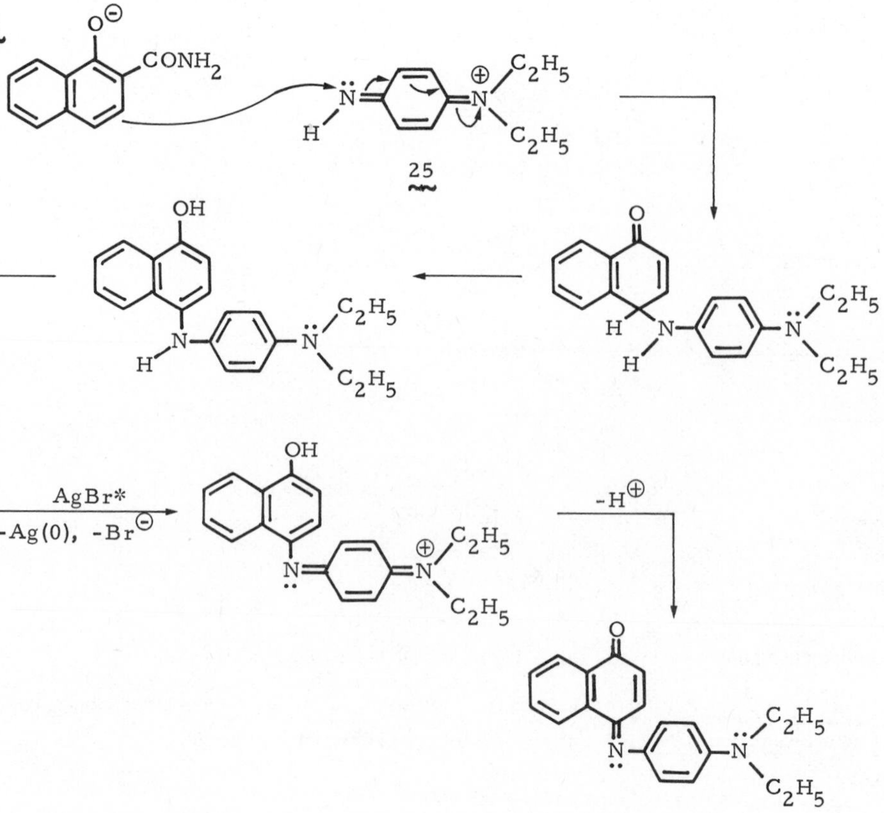

cyan dye

28-32   Addition of $: \overset{\ominus}{X}$ to 23 to give 24 rather than at a ring carbon occurs because although a C=N (relatively weak) is sacrificed, a fully aromatic ring results. With a quinone, the same type of addition would result in sacrifice of a relatively strong C=O, combined with nucleophilic attack at oxygen despite gain of a fully aromatic ring. Reaction at nitrogen is therefore more favorable. Similar attack at a =CH$_2$ position of 3,6-dimethylidene-1,4-cyclohexadiene is expected to be still more favorable.

CHAPTER 29

POLYMERS

29-1

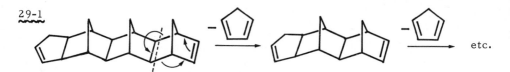

The mechanism is just the <u>reverse</u> of the Diels-Alder reaction as discussed in Sections 13-3A and 21-10A. The tetramer could <u>not</u> decompose by this route if the double bonds at the ends of the molecule were saturated as by hydrogenation, addition of water (acid catalyst), oxidation, etc.

29-2   No, because many chains will start and continue to grow. Combination of chains is not possible by the Diels-Alder mechanism. It would require 1, 2-cycloaddition (unfavorable) to link up all of the separate chains into one molecule. High molecular weight material would be favored by feeding cyclopentadiene slowly into a well-stirred polymerizing mixture - the idea being to keep the product of the concentration of polymer molecules times the concentration of cyclopentadiene as high as possible relative to the concentration of cyclopentadiene squared.

29-3

$$\overline{\underline{M}}_n = \frac{1.0 \, (C_5H_6)}{\left(\dfrac{0.3}{2} + \dfrac{0.2}{3} + \dfrac{0.15}{4} + \ldots \right)} = \frac{66.05}{0.309} = 214$$

$$\overline{\underline{M}}_w = (0.3 \times 2 + 0.2 \times 3 + 0.15 \times 4 + \ldots) C_5H_6 = 4.15 \times 66.05 = 274$$

Clearly as the molecular weight range of a polymer is narrowed $\overline{\underline{M}}_n$ approaches $\overline{\underline{M}}_w$.

Suppose one took the sample described above and determined the hydrogen absorption of 1.00 g of material - knowing there are two double bonds per polymer molecule we would calculate the molecular weight as (1.00 g x 2 moles of H_2 absorbed). Since the number of moles of hydrogen absorbed will be $\left(\dfrac{0.3}{2} + \dfrac{0.2}{3} + \dfrac{0.15}{4} + \ldots \right) \dfrac{2}{C_5H_6}$, the molecular weight calculated will be $\overline{\underline{M}}_n$.

29-4   The following syntheses are illustrative not exhaustive.

a. (i) polymerization of ethene;   (ii) hydrogenation of 1,4-polybuta-diene;   (iii) polymerization of CH_2: from diazomethane;   (iv) polymerization of cyclopropane (no practical procedures seem to be available).

b. polymerization of 1-methylazacyclopropane (N-methylaziridine)

c. polymerization of 2-butene (not easy), or 1,4-polymerization and hydrogenation of $CH_3CH=\overset{\overset{\displaystyle CH_3}{|}}{C}—\overset{\overset{\displaystyle CH_3}{|}}{C}=CHCH_3$

d. polymerization of $\overset{\overset{\displaystyle CH_2}{|}}{\underset{\underset{\displaystyle CH_2}{|}}{C}}{=}O$, or dehydration of $HOCH_2—CH_2—CO_2H$

under esterification conditions

e. addition polymerization of $CH_2{=}CH-O-\overset{\overset{\displaystyle O}{||}}{C}-CH_3$

f. hydrolysis of product of Part e

g. polymerization of $CH_2{=}$⟨benzene ring⟩$=CH_2$, or $ClCH_2$⟨benzene ring⟩$-CH_2Cl$ coupling with metals

h. 1,2,3-propanetriol (glycerol) with carbonyl dichloride (phosgene, $COCl_2$)

i. polymerization of chloral, CCl_3CHO, under basic conditions (Section 16-4B)

29-5 $-CH_2-CH_2-CH_2-CH_2-CH_2\cdot$ + $+CH_2\underset{n}{\rightarrow}$ $CH_2+CH_2\underset{m}{\rightarrow}$

growing-chain radical

$+CH_2\underset{n}{\rightarrow}$ $\overset{\displaystyle\cdot}{C}H+CH_2\underset{m}{\rightarrow}$ $+-CH_2-CH_2-CH_2-CH_2 : H$

middle-of-chain radical   polymer

$CH_2{=}CH_2$

$+CH_2\underset{n}{\rightarrow}$ $\underset{\underset{\displaystyle CH_2-CH_2\cdot}{|}}{C}H+CH_2\underset{m}{\rightarrow}$ ⟶ branched polymer

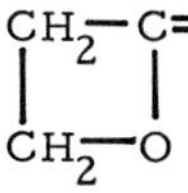

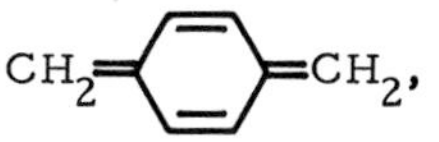

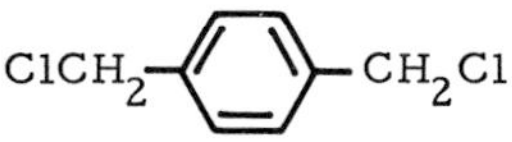

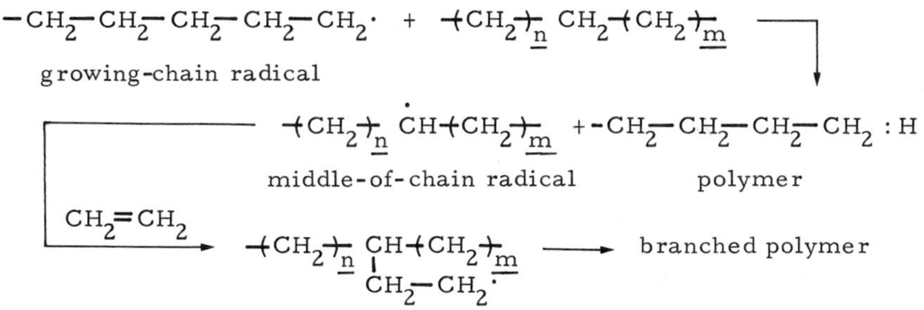

Other mechanistic routes are possible such as combination of growing chain radicals and middle-of-chain radicals.

Branches will decrease the density and lower T_m because with branches the chains will not be able to pack together well and the crystallinity will thus be reduced.

29-6   Let $+CH_2 \frac{1}{n}$ = RH

$$RH + \text{radical initiator} \longrightarrow R\cdot$$

$$R\cdot + Cl_2 \longrightarrow RCl + Cl\cdot$$
$$Cl\cdot + RH \longrightarrow R\cdot + HCl$$

} chain formation of chlorinated polymer

$$R\cdot + SO_2 \longrightarrow R-\overset{\overset{O}{\|}}{\underset{\underset{O}{\|}}{S}}\cdot$$

$$R-\overset{\overset{O}{\|}}{\underset{\underset{O}{\|}}{S}}\cdot + Cl_2 \longrightarrow R-SO_2Cl + Cl\cdot$$

$$Cl\cdot + RH \longrightarrow R\cdot + HCl$$

} chain formation of chlorosulfonated polymer

More or less random chlorination and chlorosulfonation is expected to reduce the crystallinity of polyethene and make it elastic if the motions of the chain are reasonably rapid (as they are expected to be for a polyethene chain).

Chlorosulfonated polyethene could be cross-linked with a diamine (or a diol).

$$2\,RSO_2Cl + NH_2CH_2CH_2NH_2 \xrightarrow[]{-2HCl} R-SO_2\overset{H}{N}-CH_2CH_2-\overset{H}{N}-SO_2-R$$

29-7   The density would be expected to decrease somewhat because the cross-linked chains would not be able to crystallize in quite the usual polyethene manner, T_g and T_m would be expected to increase. There should be no particular elastic properties conferred on the material; in fact, the chains would tend to be constrained not to move so easily. Plastic flow of the polymer would be expected to greatly reduce and the softening (also the molding) temperature increased.

29-8   a. The forces between the chains are too great - the material is viscous not rubbery.

b. Break up the hydrogen bonds between the amide groups by N-substitution, substitute some methyl groups possibly randomly along the chain; i.e., the following may well be rubbery.

$$-\overset{\overset{CH_3}{|}}{N}-\overset{\overset{O}{\|}}{C}-C_8H_{15}(CH_3)-\overset{\overset{O}{\|}}{C}-\overset{\overset{CH_3}{|}}{N}-$$

straight-chain, randomly substituted with
a methyl group

c. It should be a rubbery material something like poly-2-methylpropene (polyisobutylene) with a log T_g.

d. It would be expected to form the sodium salt, absorb water in the process and swell up.

e. It should be a highly crystalline polymer, with a very high T_m and T_g.

f. No, if the material were _trans_-1,4-polychloroprene it would be expected to be quite crystalline with T_m perhaps as great or greater than polychloroethene (polyvinyl chloride).

g.

m	n	Expected Properties
1	200	elastomeric but tending to undergo plastic flow easily - not enough crystallinity for good properties
30	200	crystalline areas produced by ester blocks should give thermoplastic elastomer properties
200	200	as for 30-200 but much stiffer because of higher proportion of ester blocks.
200	30	very stiff, large degree of crystallinity
200	1	essentially like Dacron

h. The blowing operation produces orientation and partial crystallinity of the propenenitrile segments as in the drawing of Orlon fibers, thus increasing the mechanical strength.

29-9   Silly Putty is expected to have a low T_g and a low T_m with rather weak forces between the chains so that plastic flow is possible. The elastic properties would fit for a polymer which does not crystallize when stretched, but undergoes slow plastic flow. Atactic polypropene, polyisobutylene or uncross-linked ethene-propene-hexadiene polymer would be expected to have properties close to those of Silly Putty. Irradiation of Silly Putty is expected to introduce at first a few cross-links which would reduce plastic flow and make it more conventionally elastic. Further cross-linking would finally lead to a resinous material.

29-10   a. No, because although the individual molecules would be chiral

$$H \overbrace{C-CH_2}^{CH_3}_n CH=CH_2 \qquad CH_2=CH \overbrace{CH_2-C}^{CH_3}_n H$$

chiral centers

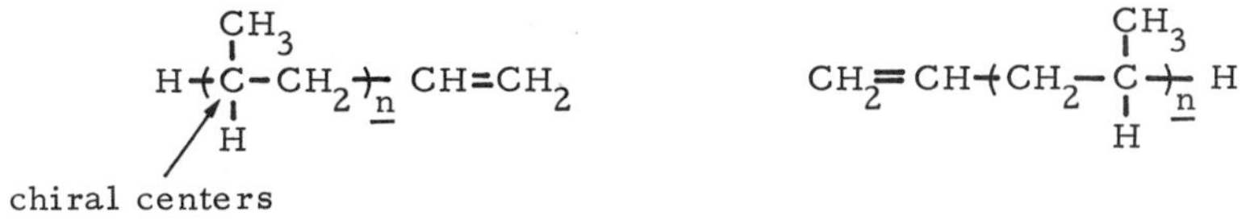

equal amounts of each chiral form would be produced from nonchiral starting materials.

b.   The chirality depends on the <u>end groups</u> being different and for a chiral center near the middle of the chain the surroundings are so nearly symmetrical that the rotation should be exceedingly small.   Theoretically yes, practically no.

$$H + \cdots\cdots -\underset{\underset{H}{|}}{\overset{\overset{CH_3}{|}}{C}}-CH_2-\underset{\underset{H}{|}}{\overset{\overset{CH_3}{|}}{C}}-CH_2-\underset{\underset{H}{|}}{\overset{\overset{CH_3}{|}}{C}}- \cdots\cdots \rightarrow CH{=}CH_2$$

For a very long chain there is "almost" a plane of symmetry here.

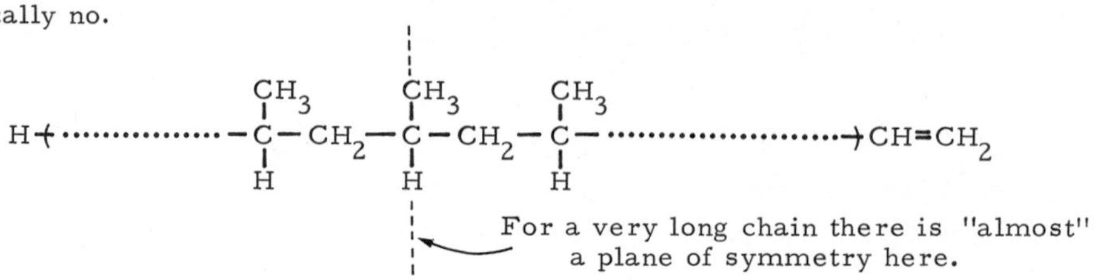

29-11   The polymer would have to be a linear polymer.

29-12 $$NH_2-\overset{\overset{O}{||}}{C}-NH_2 + \overset{\ominus}{OH} \rightleftharpoons \left[NH_2-\overset{\overset{O}{||}}{C}-\overset{\ominus}{\underset{..}{N}}H \longleftrightarrow NH_2-\overset{\overset{O^{\ominus}}{|}}{C}{=}NH \right] + H_2O$$

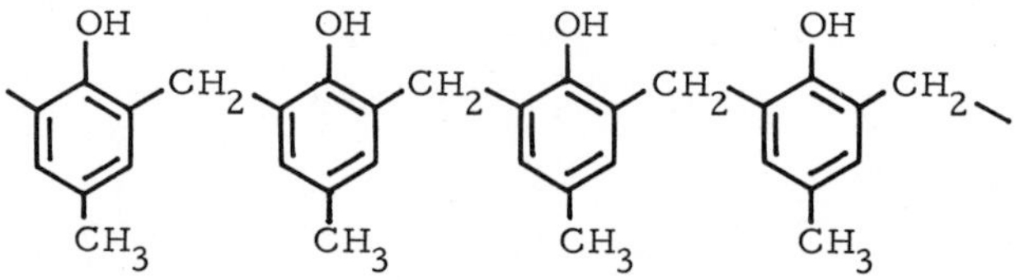

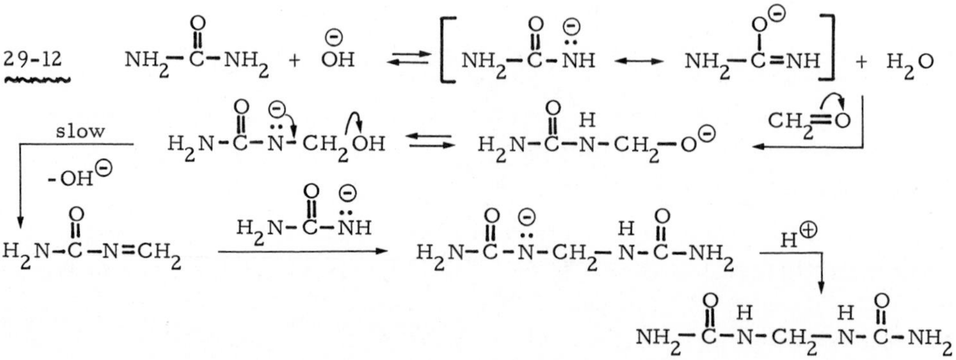

29-13

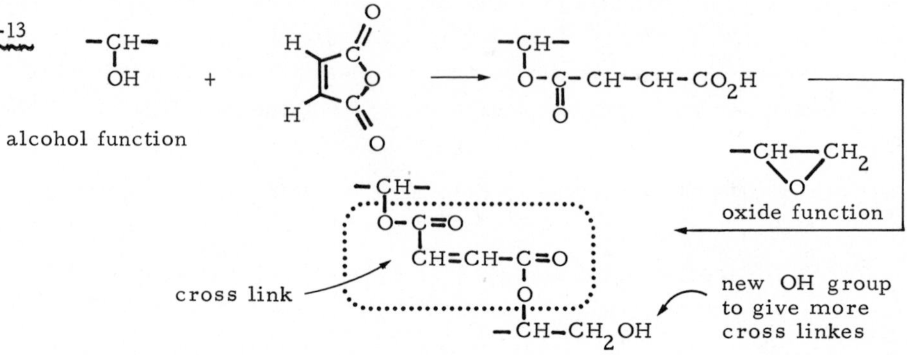

29-14

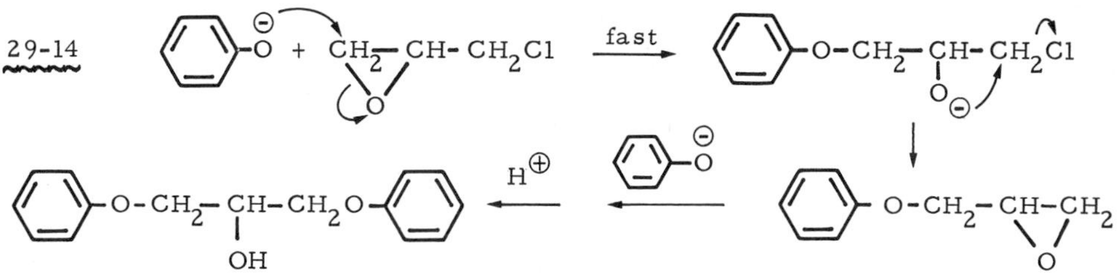

29-15   Let $CH_2=CRR'$ represent methyl 2-methylpropenoate, and $(ArCO_2)_2$* represent benzoyl peroxide labeled with ^{14}C in the ring.

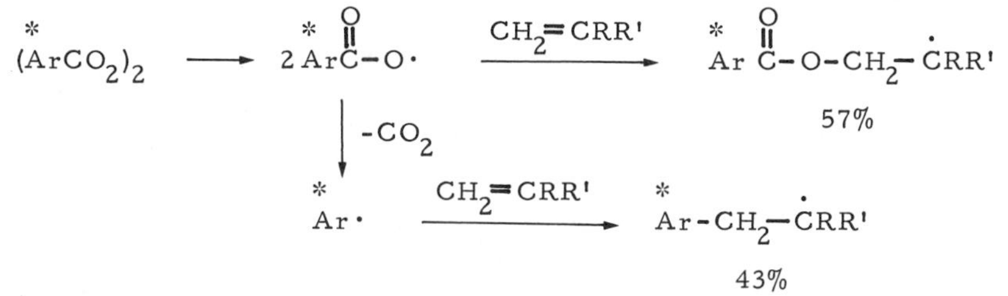

Termination by combination of radicals gives <u>two</u> initiator fragments per polymer molecule while termination by disproportionation gives only <u>one</u> initiator fragment per polymer molecule.   Therefore, 27% of the polymer molecules contain two fragments per molecule and were formed by combination.

29-16   For the growing chain end of the polymer to produce atactic material means that there is essentially no preference for attack of ethenylbenzene on either side of the α-carbon no matter what the configuration of the phenyl group once removed is.

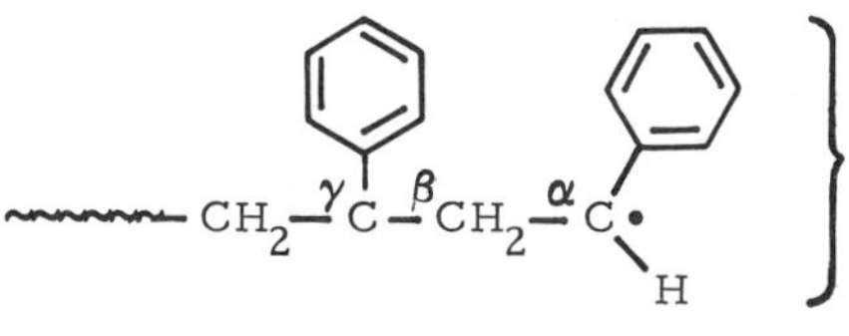

attack on ethenylbenzene can occur on either side of the α-carbon irrespective of the configuration at the γ-carbon (D or L).   There is no asymmetric induction (Section 19-10.)

29-17   The fact that no measurable amounts of periodic acid or lead tetraacetate are consumed indicates that polyvinyl alcohol very largely has the head-to-tail structure:

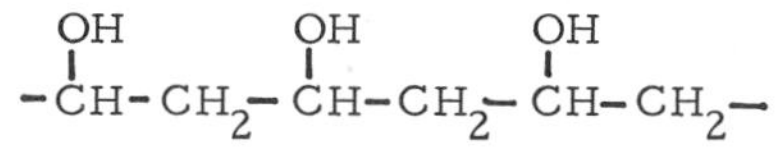

Polymer with the head-to-head structure would be rapidly cleaved.

$$-CH_2-\overset{\overset{\displaystyle OH}{|}}{CH}-\overset{\overset{\displaystyle OH}{|}}{CH}-CH_2-\ CH_2-\overset{\overset{\displaystyle OH}{|}}{CH}-\overset{\overset{\displaystyle OH}{|}}{CH}-\ \xrightarrow{[O]}\ -CH_2-\overset{\overset{\displaystyle O}{\|}}{CH}+H\overset{\overset{\displaystyle O}{\|}}{C}-CH_2-CH_2-\overset{\overset{\displaystyle O}{\|}}{CH}+H\overset{\overset{\displaystyle O}{\|}}{C}-$$

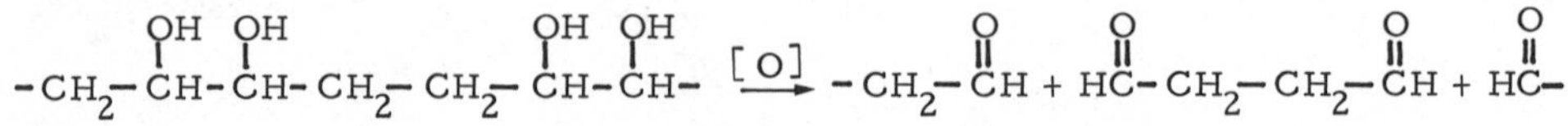

That the molecular weight of the polymer is reduced from 25,000 to 5000 by periodic acid indicates an average of four head-to-head linkages in each chain and that in the polymerization of ethenyl ethanoate (vinyl acetate) only one head-to-head structure is produced on the average for every $5000 \div (C_2H_4O) \cong 110$ molecules polymerized.

29-18   If polychloroethene contained head-to-head units on treatment with zinc it would give double bonds.

$$-\overset{\overset{\displaystyle Cl}{|}}{CH}-\overset{\overset{\displaystyle Cl}{|}}{CH}-CH_2-\ CH_2-\overset{\overset{\displaystyle Cl}{|}}{CH}-\overset{\overset{\displaystyle Cl}{|}}{CH}-CH_2-\ \xrightarrow{Zn}\ -CH=CH-CH_2-\ CH_2-CH=CH-\ CH_2-$$

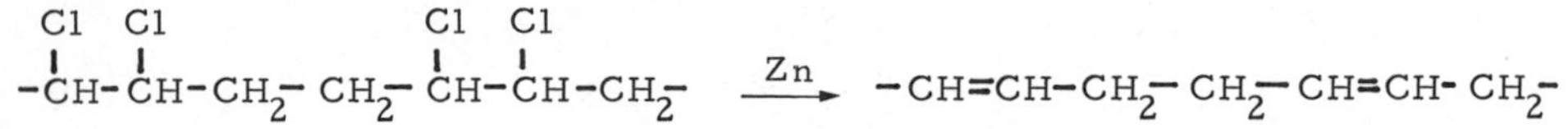

The head-to-tail structure would have the 1,3 relation between the chlorines and could lead to cyclopropane-ring formation.

$$-\overset{\overset{\displaystyle Cl}{|}}{CH}-CH_2-\overset{\overset{\displaystyle Cl}{|}}{CH}-CH_2-\overset{\overset{\displaystyle Cl}{|}}{CH}-CH_2-\overset{\overset{\displaystyle Cl}{|}}{CH}-\ CH_2-\overset{\overset{\displaystyle Cl}{|}}{CH}-CH_2-\overset{\overset{\displaystyle Cl}{|}}{CH}-CH_2-\overset{\overset{\displaystyle Cl}{|}}{CH}-CH_2-\Bigg]$$

$$-CH-CH-CH_2-\overset{\overset{\displaystyle Cl}{|}}{CH}-CH_2-CH-CH_2-CH-CH-CH_2- \xleftarrow{Zn}$$

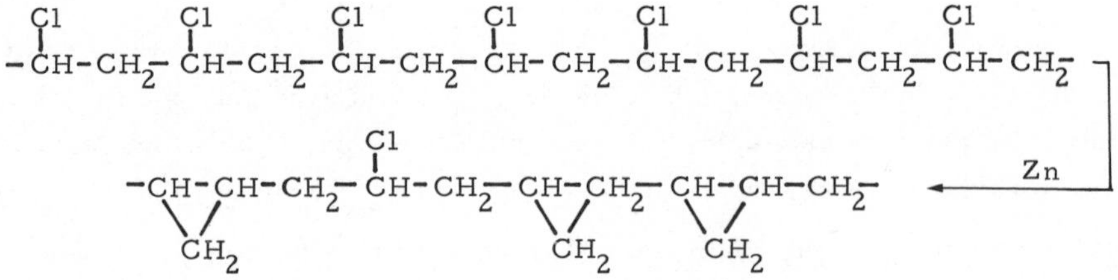

It would not be expected that all of the chlorine would be removed because, as shown in the equation, chlorine atoms could become isolated on the chain if cyclopropane rings were to form in each side of them.

29-19   Rubber and gutta percha are both head-to-tail 1,4 polymers of 2-methyl-1,3-butadiene (isoprene). Since they are different polymers, one has the cis (rubber) configuration at the double bond and the other the trans configuration.

$$\cdots CH_2-\overset{\overset{\displaystyle CH_3}{|}}{C}\diagdown_{CH}-CH_2-CH_2-\overset{\overset{\displaystyle CH_3}{|}}{C}\diagdown_{CH}\cdots \longrightarrow H\overset{\overset{\displaystyle O}{\|}}{C}CH_2CH_2\overset{\overset{\displaystyle O}{\|}}{C}CH_3$$

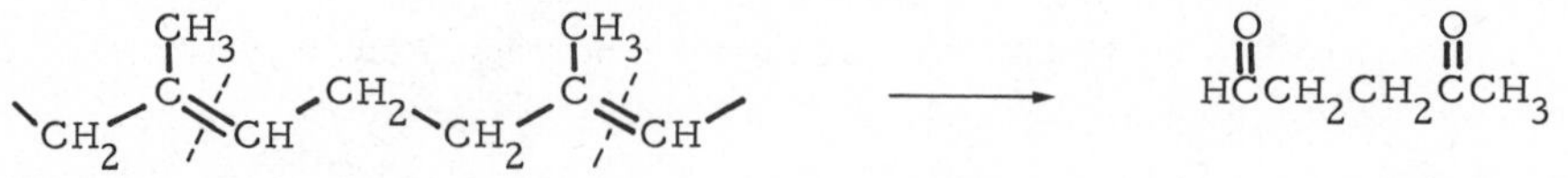

head-to-tail

Neither polymer can be a head-to-head polymer because a head-to-head structure would lead to hexane-2,5-dione on ozonolysis.

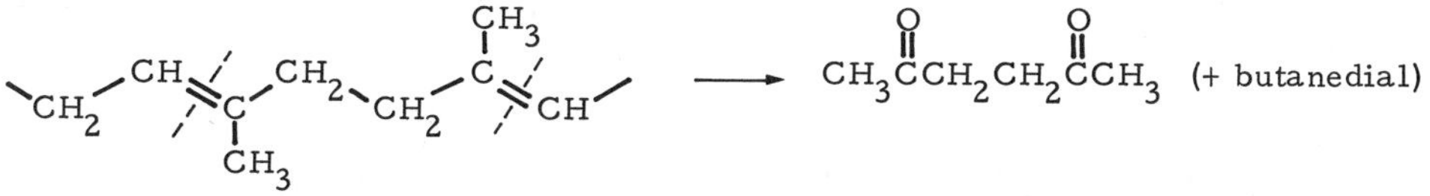

head-to-head   hexane-2,5-dione

29-20   The product is expected to be formed by the following chain mechanism in competition with formation of polyethylbenzene $(R=C_6H_5CH_2-)$.

$$C_6H_5CH=CH_2 + RS\cdot \longrightarrow C_6H_5\overset{.}{C}H-CH_2-SR$$

$$C_6H_5\overset{.}{C}H-CH_2SR + RSH \longrightarrow C_6H_5-CH_2-CH_2-SR + RS\cdot$$

$RS\cdot$ is expected to add to the double bond at the unsubstituted end which will give a stabilized benzyl-type radical.

The best conditions for the reaction will be to have absent radical traps, oxygen, etc. (to avoid interference with the preceding chain); to have a high concentration of RSH (to diminish polymerization of the $C_6H_5CH=CH_2$; and to have a relatively low initiator concentration (to diminish the importance of termination reactions).

29-21   Tetrachloromethane is involved in a chain-transfer process by the following mechanism which is in competition with polymerization and results in formation of polymer containing CCl_3 and Cl groups. Every time the growing-chain radical attacks tetrachloromethane, a polymer molecule and a $\cdot CCl_3$ radical are formed; the $\cdot CCl_3$ radical then adds to ethenylbenzene to form a new growing-chain radical.

$$\text{~~~}CH_2-\overset{.}{C}H-C_6H_5 + CCl_4 \longrightarrow \text{~~~}CH_2-\underset{Cl}{\overset{|}{C}H}-C_6H_5 + \cdot CCl_3$$

$$\cdot CCl_3 + CH_2=CHC_6H_5 \longrightarrow CCl_3-CH_2-\overset{.}{C}HC_6H_5$$
$$\overset{|\,C_6H_5CH=CH_2}{\underset{\longrightarrow}{}} \text{~~~}CH_2-\overset{.}{C}HC_6H_5$$

Although the above scheme does not affect the rate of polymerization, it leads to polymer containing chlorine, and lowers the average molecular weight at high

CCl_4 concentrations by increasing the rate of chain termination.

$$\text{\textbf{∿}CH_2-\overset{\textbf{.}}{C}H-C_6H_5} + \cdot CCl_3 \longrightarrow \begin{cases} \text{∿}CH=CH-C_6H_5 + HCCl_3 \\ \\ \text{∿}CH_2-\underset{\underset{CCl_3}{|}}{C}H-C_6H_5 \end{cases}$$

<u>29-22</u>   2-Propenyl ethanoate (allyl acetate) gives low-molecular weight polymer because of chain-transfer between the growing-chain radical and the monomer (let $CH_2=CH-CH_2OAc$ = allyl acetate).

$$\text{∿}CH_2-\overset{\textbf{.}}{C}H-CH_2OAc + CH_2=CH-CH_2OAc \longrightarrow$$

growing-chain radical

$$\text{∿} CH_2-CH_2-CH_2OAc + CH_2=CH-\overset{\textbf{.}}{C}HOAc$$

resonance-stabilized
radical

$$CH_2=CH-\overset{\textbf{.}}{C}H-OAc + CH_2=CH-CH_2OAc \longrightarrow \text{growing-chain radical}$$

The polymerization is slow because the chain-transfer radical is sufficiently stabilized so that it does not attack monomer readily and thus acts as a semi-inhibitor. With $CH_2=CH-CD_2OAc$ hydrogen removal at the α-carbon is slowed by kinetic isotope effect and more monomer polymerizes before chain transfer occurs.

29-23 $CH_2=CH-CONH_2 \longrightarrow +CH_2-\underset{\underset{CONH_2}{|}}{C}H+ \xrightarrow[Br_2]{NaOH} -CH_2-\underset{\underset{NH_2}{|}}{C}H-$

polyvinylamine

$$CH_2=CH-\overset{O}{\overset{||}{\underset{O}{\underset{||}{N}}}}\overset{CH_2}{\underset{CH_2}{\underset{C}{\big\langle}}} \longrightarrow +CH_2-\underset{\underset{OC\overset{N}{\underset{H_2C-CH_2}{}}CO}{|}}{C}H+ \xrightarrow{NH_2NH_2}$$

$$CH_2=CH-NO_2 \longrightarrow +CH_2-\underset{\underset{NO_2}{|}}{C}H+ \xrightarrow{[H]}$$

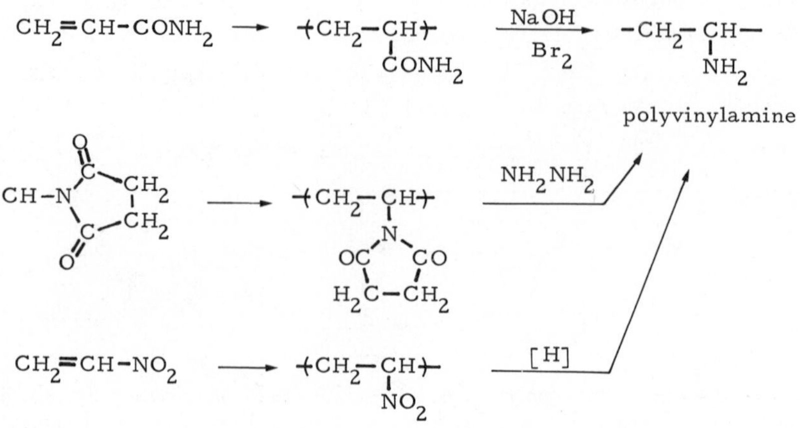

29-24

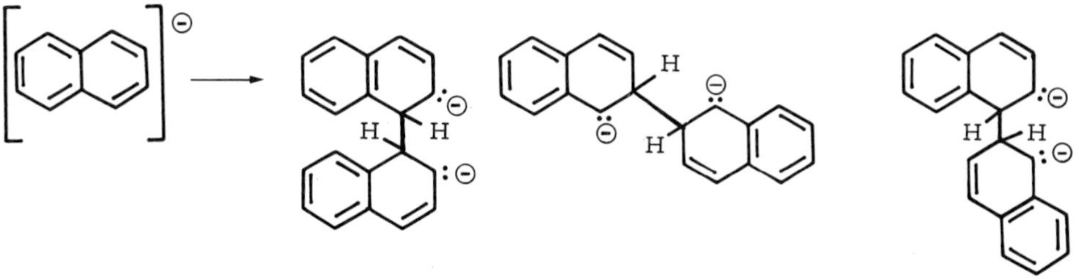

Besides the loss of resonance stabilization in one of the benzenoid rings, there is significant steric hindrance in the dimer dianion relative to the monomer radical anion. Furthermore, dimerization requires bringing together two negative ions (change repulsions).

29-25   By adding oxacyclopropane and small amounts of a protic solvent (e.g., ethanol) to a living polyethylbenzene polymer, the chains would be terminated with $-CH_2CH_2-OH$ groups.

29-26   The alternation of monomer units would lead to a high degree of regularity of this polymer particularly if it were formed stereospecifically (cis or trans) in the anhydride part.

can be cis or trans

$$- CH_2- \underset{\underset{CH_3}{|}}{\overset{\overset{CH_3}{|}}{C}} ---- \underset{\underset{CO}{|}}{\overset{\overset{H}{|}}{C}} - \underset{\underset{CO}{\diagdown}}{\overset{\overset{H}{|}}{C}} -CH_2- \underset{\underset{CH_3}{|}}{\overset{\overset{CH_3}{|}}{C}} ---$$

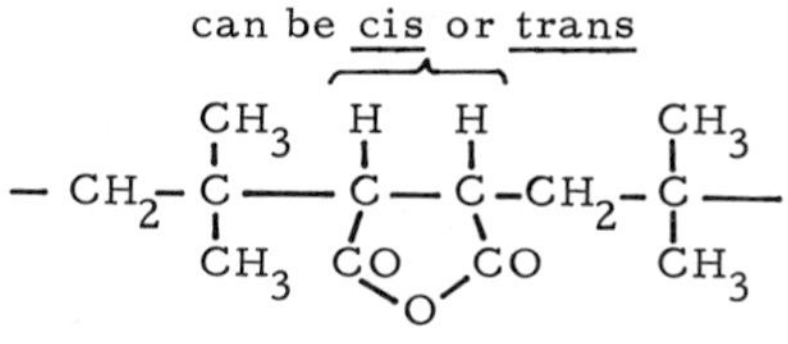

If so, with the polar groups it possesses it should be relatively highly crystalline, and reasonably high melting.

29-27   Attack on ethenylbenzene should always occur at the CH_2 group whether by a cationic, radical or anionic mechanism. As a result, the growing chain end wherever styrene is involved should be:

chain $\sim\!\!\sim\!\!\sim$ $CH_2-\underset{|}{CH}-$⟨◯⟩

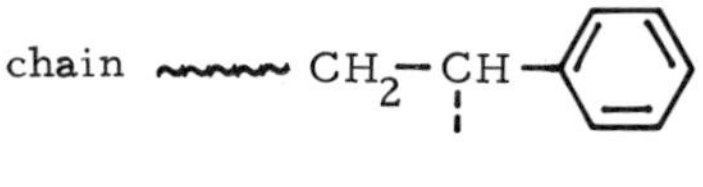

However, radical and anionic attack on propene are expected to proceed in opposite ways - a secondary radical being more favored than a primary radical while the opposite is true for a secondary anion <u>vs</u>. a primary anion.

radical attack → $\sim\sim CH_2- CH -CH_2- \overset{\cdot}{C}H-CH_3$

$\sim\sim CH_2- \underset{|}{C}H-\bigcirc$ + $CH_2\!\!=\!\!CHCH_3$

radical or anion

anion attack → $\sim\sim CH_2- \underset{\underset{CH_3}{|}}{C}H-CH-\overset{\ominus}{C}H_2$

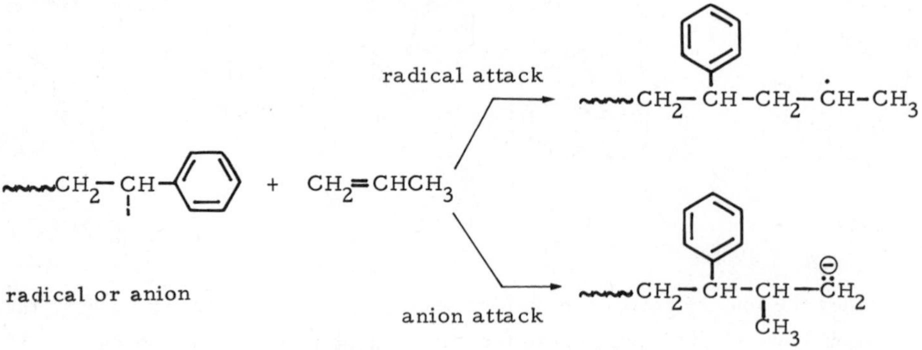

These two routes would give polymers with different structures.

<u>29-28</u>   Let $H-O-CH_2-CH_2-O\!+\!CH_2-CH_2-O\underset{\underline{n}}{\big)}CH_2-CH_2OH$ = $HO\sim\sim OH$

poly-1,2-ethanediol (polyethylene oxide)   or   A

and $N\!\equiv\!C\sim\sim C\!\equiv\!N$ represent $N\!\equiv\!CCH_2CH_2CH_2CH_2C\!\equiv\!N$

A possible route to a block polymer would be:

$N\!\equiv\!C\sim\sim C\!\equiv\!N$ $\xrightarrow{H_2O}$ $HO_2C\sim\sim CO_2H$

$\xrightarrow{[H]}$ $H_2NCH_2\sim\sim CH_2NH_2$ ⌐

$HO_2C\sim\sim CONH\sim\sim CH_2NH_2$ (nylon 66 segment)

$HO\sim\sim OH$ $\xrightarrow{\text{nylon 66 segment}}$ $HO\sim\sim O\overset{\overset{O}{\|}}{C}\sim\sim CONH\sim\sim CH_2NH_2$

$NH_2CH_2\sim\sim NHCO\sim\sim\overset{\overset{O}{\|}}{C}O\sim\sim O\overset{\overset{O}{\|}}{C}\sim\sim CONH\sim\sim CH_2NH_2$ ⌐

$\Big\downarrow HO_2C\sim\sim CONH\sim\sim$ and so on

$-A-\!\big[\text{nylon 66}\big]\underset{\underline{n}}{}-A-\!\big[\text{nylon 66}\big]\underset{\underline{m}}{}-A-\!\big[\text{nylon 66}\big]\underset{\underline{p}}{}-A-$

29-29   The first would be an elastomer with some degree of physical cross linking involving the ethenylbenzene blocks on the end of the chain.   The second where the potential elastomer part is _pendant_ to the stiff part would undergo plastic flow of the elastomer part and have little memory of the unstretched state.

29-30   Highly colored (see Section 22-3B), very low resistance to oxidative attack (see Section 22-8 and 22-10) but very highly crystalline.

29-31   1. Regularity of the units which allows for development of a high degree of crystallinity.

2. Strong hydrogen bonding between the chains.

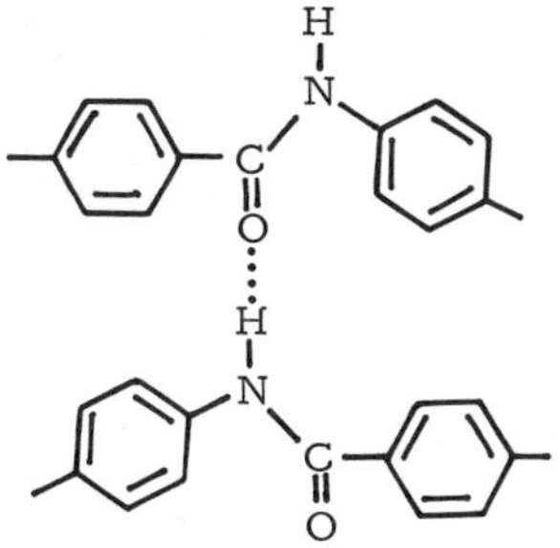

etc.

3. General stiffness of the segments, no $-CH_2-CH_2-CH_2-$ groups and restricted rotation about the $-N-C-$ bonds (Section 24-1D).

29-32   One way of altering the S-S groups would be to convert them to $-S-CH_2-CH_2-S-$ groups.

$$-S-S- \xrightarrow{[H]} 2-S-H \xrightarrow{BrCH_2CH_2Br} -S-CH_2-CH_2-S-$$

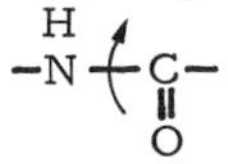

CHAPTER 30

NATURAL PRODUCTS.   BIOSYNTHESIS

30-1
a.

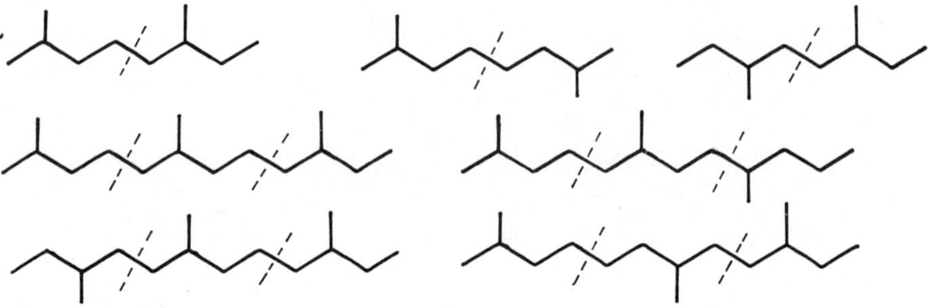

b.

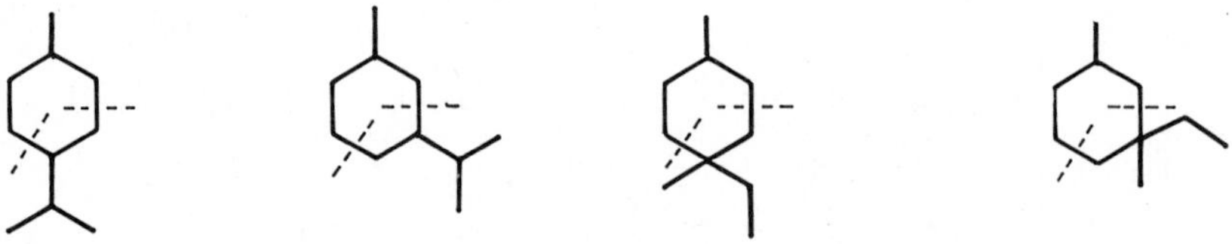

The following structures are also divisible into iso C_5 units.

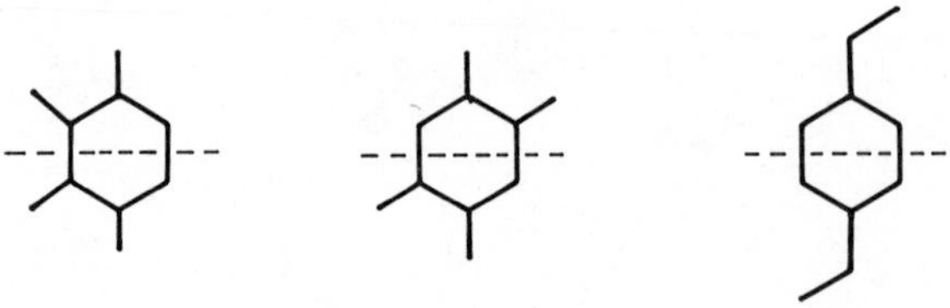

30-2   The λ_{max} of alloocimene indicates extended conjugation.   Ozonization products show presence of $(CH_3)_2C=C$ and $CH_3CH=C$.

$$\begin{matrix} CH_3 \\ \diagdown \\ C=CH-CH=CH-C=CHCH_3 \\ \diagup | \\ CH_3 CH_3 \end{matrix}$$

30-3

a.   no stereoisomers

b.   trans-trans

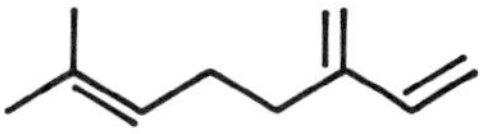

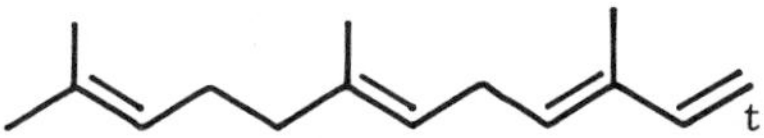

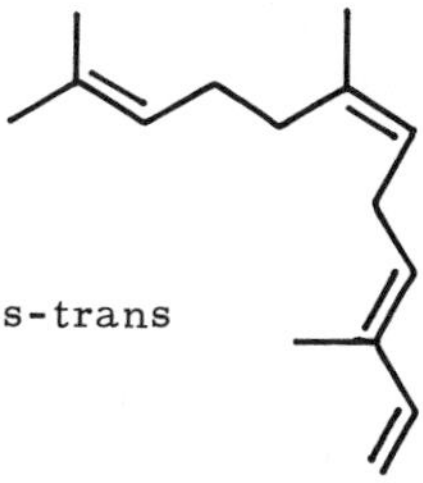

cis-trans

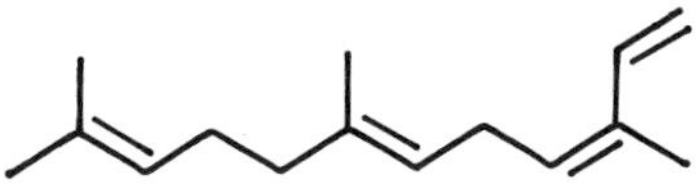

trans-cis

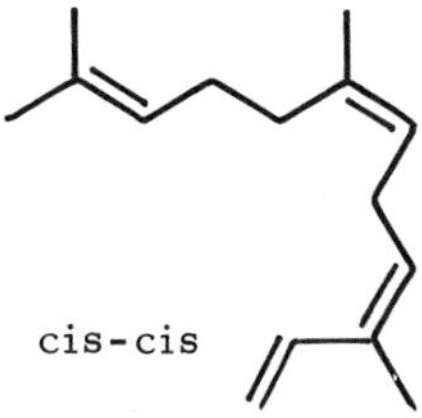

cis-cis

c.   Enantiomeric pair only.

d.

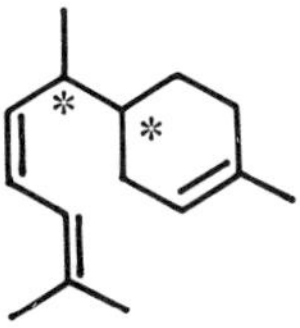

(2 enantiomeric pairs)

(2 enantiomeric pairs)

e.

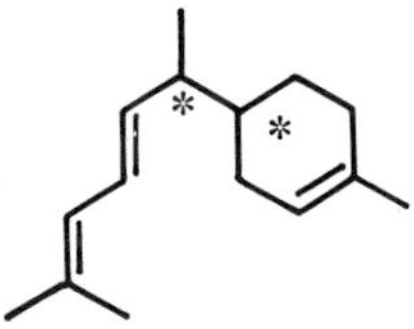

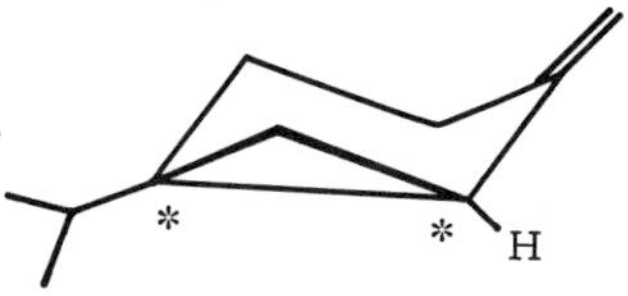

(1 enantiomeric
pair)

f.

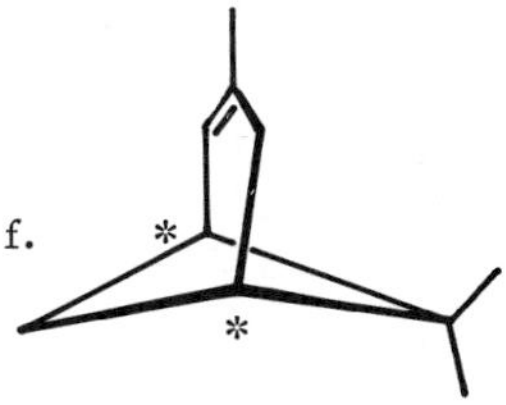

(1 enantiomeric
pair)

g.

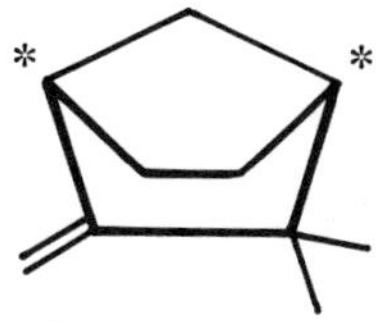

(1 enantiomeric
pair)

h.

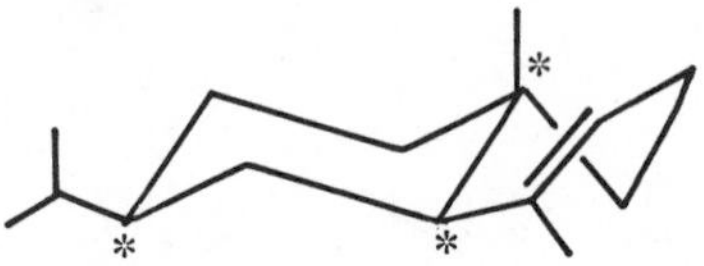

(enantiomeric pair)   (enantiomeric pair)

(enantiomeric pair)   (enantiomeric pair)

i. If the double bond in the nine-membered ring is required to be trans and
    the linkages of it to the cyclobutane ring required to be cis - then there is
    only an enantiomeric pair. If the ring double bond can be cis, there is
    another enantiomeric pair. With trans fusion to the cyclobutane (less likely)
    more enantiomeric pairs would be possible.

30-4   Racemization of camphene can be explained as the result of methyl
migration, which converts one enantiomer into the other.

        This mechanism competes with a more esoteric one where there is a
1, 3-hydrogen shift.

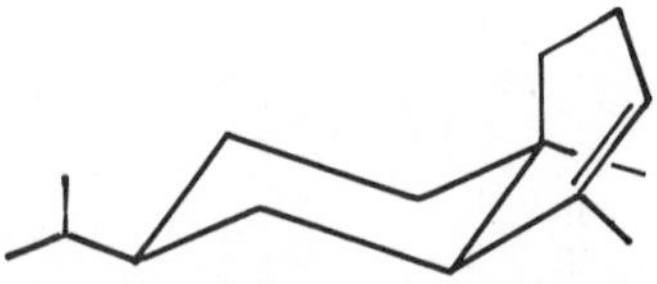

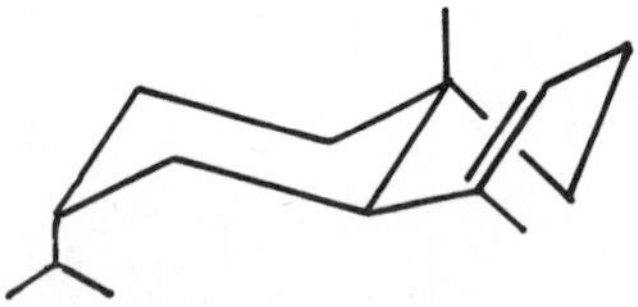

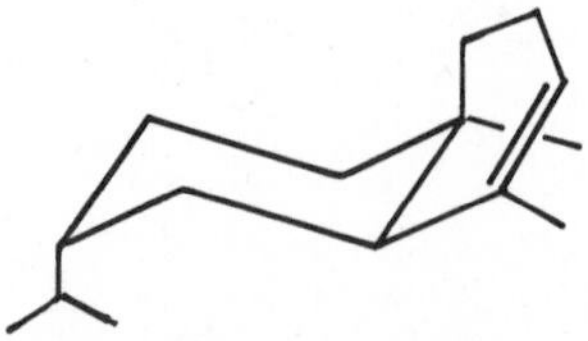

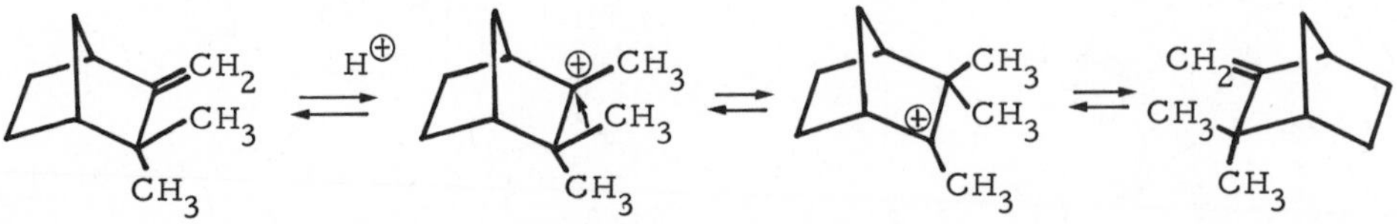

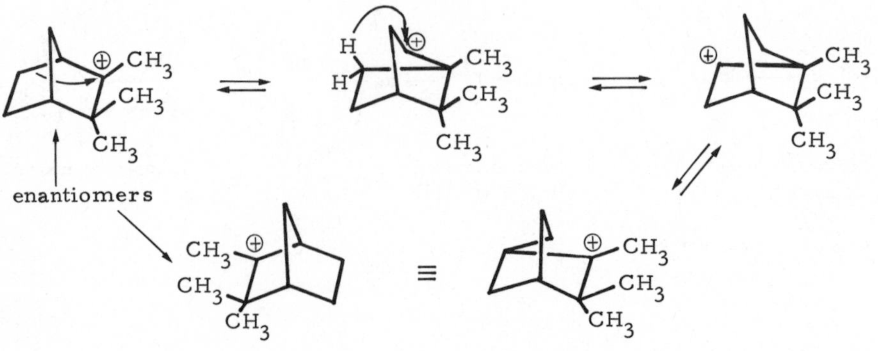

The concurrent operation of this mechanism has been demonstrated by isotopic labeling experiments [cf. J. Am. Chem. Soc., 75, 3165 (1953)].

30-5

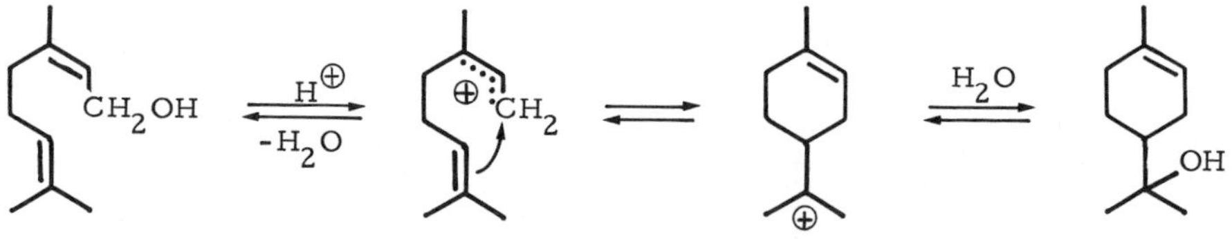

nerol α-terpineol

Geraniol is not expected to cyclize as readily as nerol because inversion of configuration at the double bond would first have to occur.

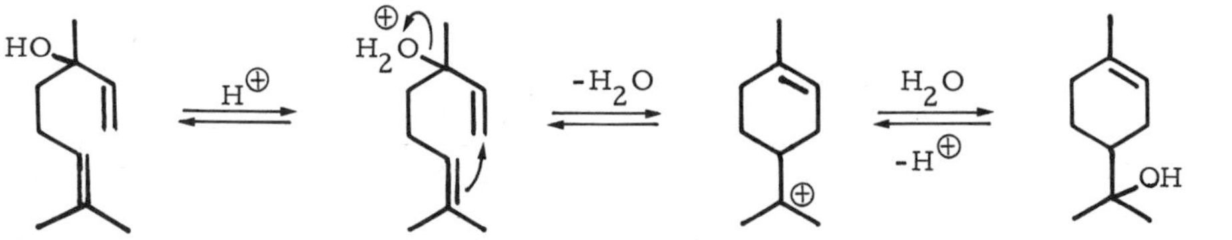

linaloöl   (chiral) α-terpineol

The above transformations are stereospecific if ring-closure is reasonably concerted and, if so, will lead to optically active α-terpineol from optically active linaloöl.

30-6

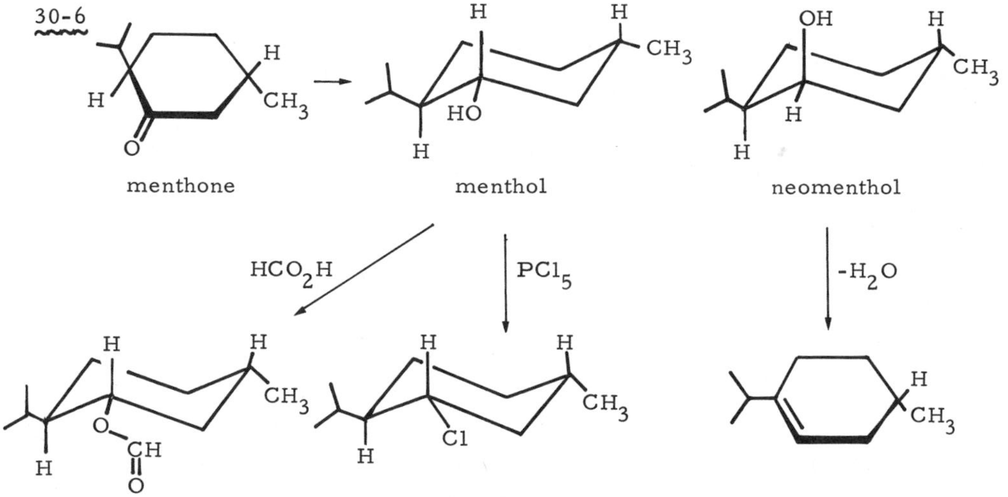

menthone   menthol   neomenthol

Trans elimination occurs very readily with neomenthol because in the favored conformation (with the isopropyl group equatorial) the H and OH are very well located for trans elimination.

30-7

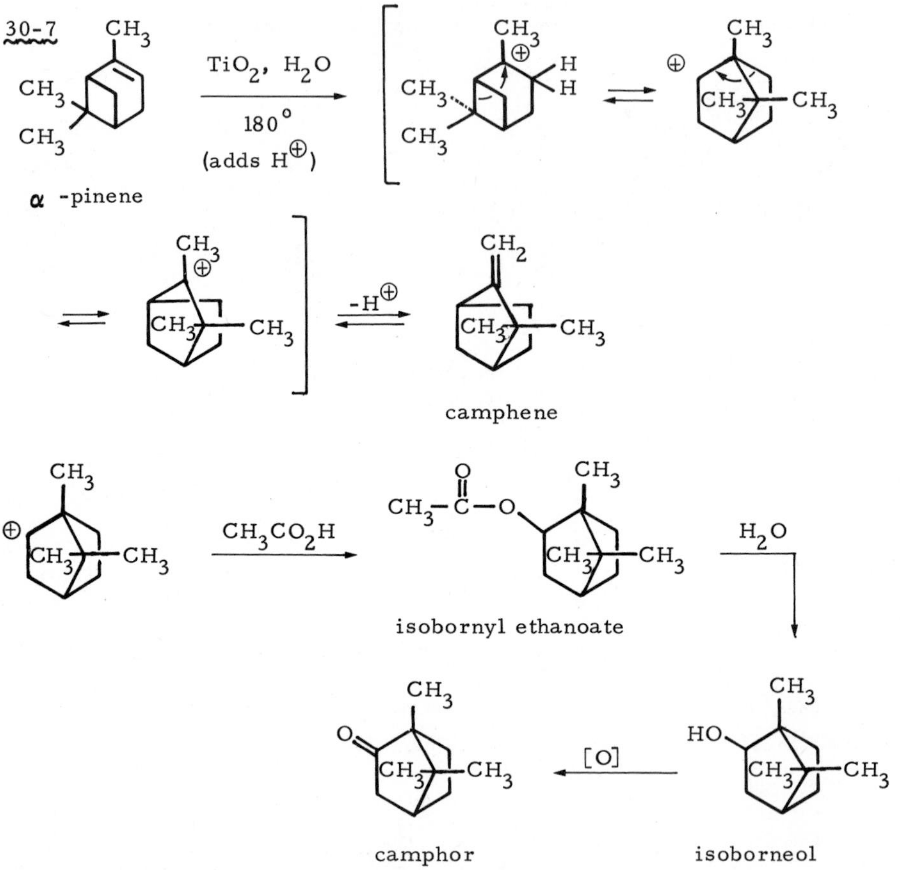

α -pinene

camphene

isobornyl ethanoate

camphor   isoborneol

30-8

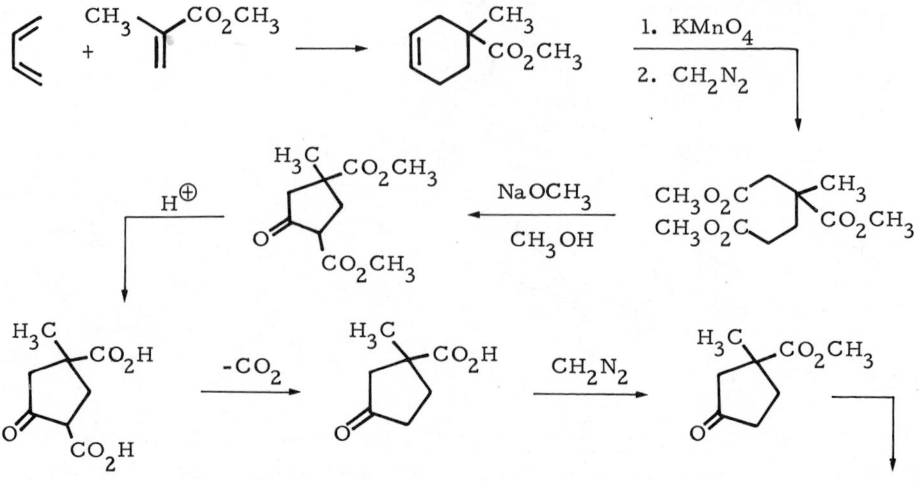

(cont.)

30-8 (cont.)   (cont.)

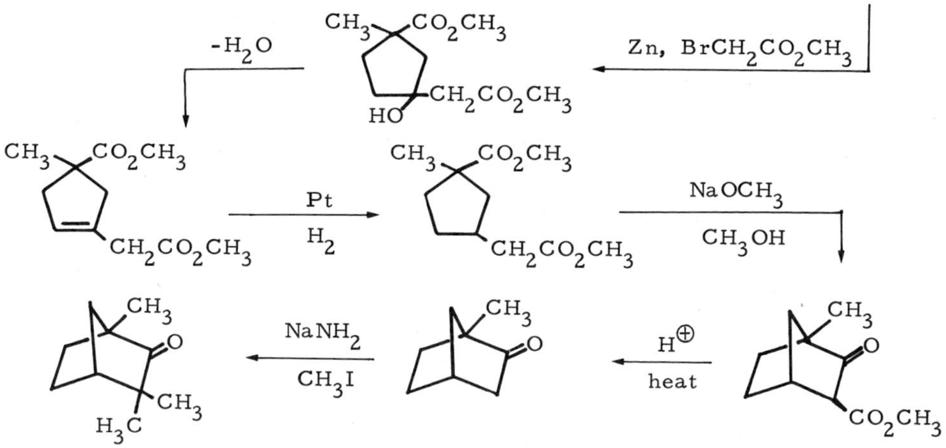

$-H_2O$ Zn, $BrCH_2CO_2CH_3$

CH$_3$ CO$_2$CH$_3$

CH$_2$CO$_2$CH$_3$

HO

CH$_3$ CO$_2$CH$_3$ CH$_3$ CO$_2$CH$_3$ NaOCH$_3$

CH$_2$CO$_2$CH$_3$ Pt / H$_2$ CH$_2$CO$_2$CH$_3$ CH$_3$OH

CH$_3$ NaNH$_2$ CH$_3$ H$^{\oplus}$ CH$_3$

O   CH$_3$I   O   heat   O

CH$_3$ CO$_2$CH$_3$

H$_3$C

fenchone
and enantiomer

30-9

(i) 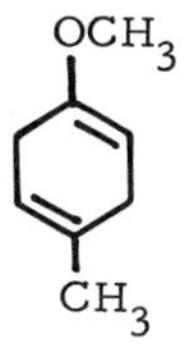 OCH$_3$ ... CH$_3$

(ii) 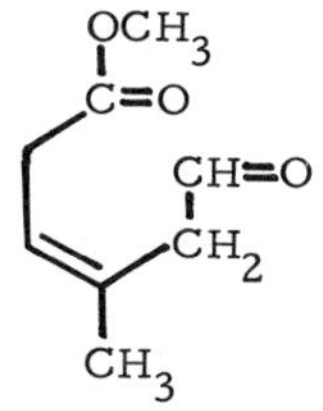 OCH$_3$ / C=O / CH=O / CH$_2$ / CH$_3$

(iii) NaBH$_4$

(iv) 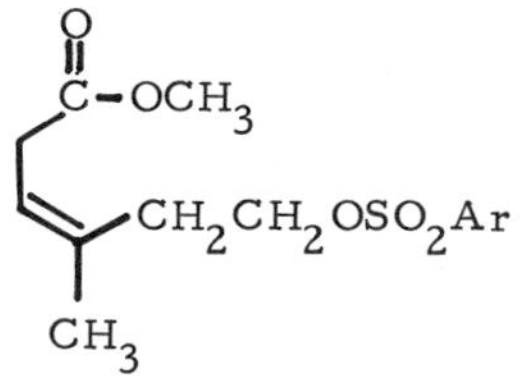 $\overset{O}{C}-OCH_3$ / CH$_2$CH$_2$OSO$_2$Ar / CH$_3$

(v) 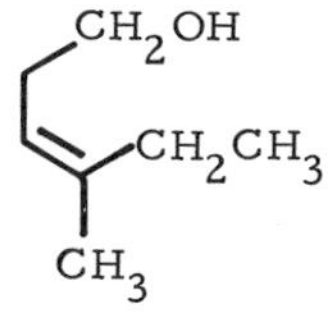 CH$_2$OH / CH$_2$CH$_3$ / CH$_3$

(vi) 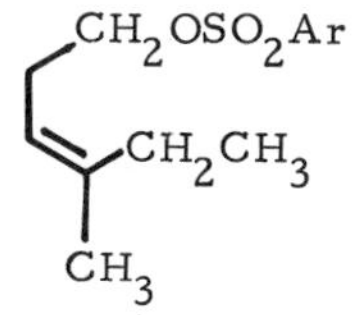 CH$_2$OSO$_2$Ar / CH$_2$CH$_3$ / CH$_3$

(vii) 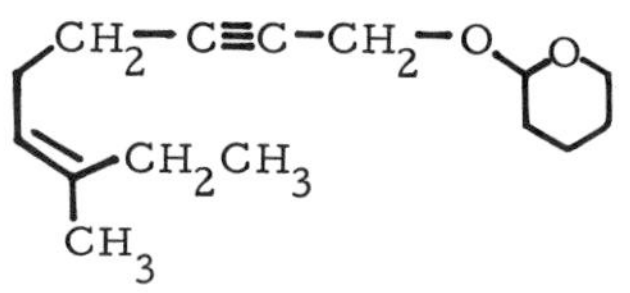 CH$_2$-C≡C-CH$_2$-O / CH$_2$CH$_3$ / CH$_3$

(viii) H$_2$O, H$^{\oplus}$

(ix) 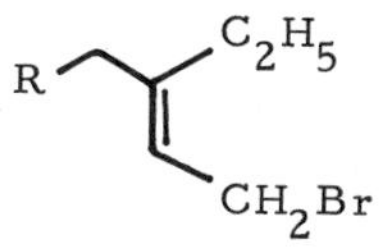 R / C$_2$H$_5$ / CH$_2$Br

(x) 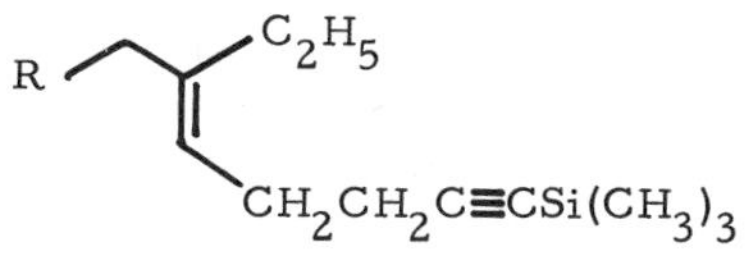 R / C$_2$H$_5$ / CH$_2$CH$_2$C≡CSi(CH$_3$)$_3$

(xi) 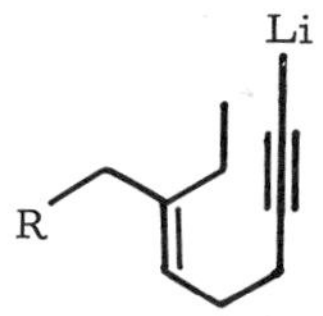 R / Li

(xii) CH$_2$=O

(xiii) LiAlH$_4$, I$_2$, NaOCH$_3$ followed by (CH$_3$)$_2$LiCu

(xiv) (CH$_3$)$_2$S=O and N, N-dicyclohexylcarbodiimide (Section 16-9B)

(xv) HCN   (xvi)

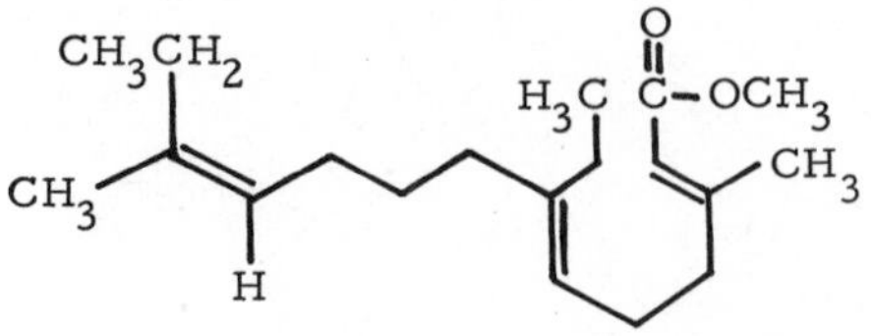

(xvii)

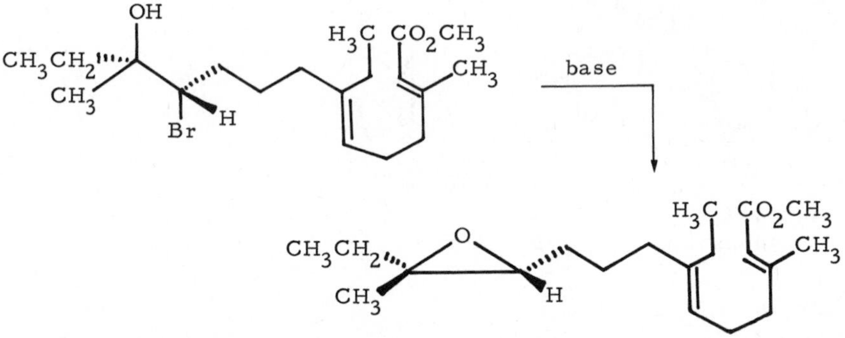

30-10   Cholic acid has only one mirror image form, however it has 11 chiral carbon atoms, 6 of which establish the stereochemistry of the ring junctions. There are, therefore, 2^5 or 32 optical isomers possible for cholic acid having the stereochemistry at the ring junctions as shown or 2^{11} (2048) conceivable optical isomers in all.

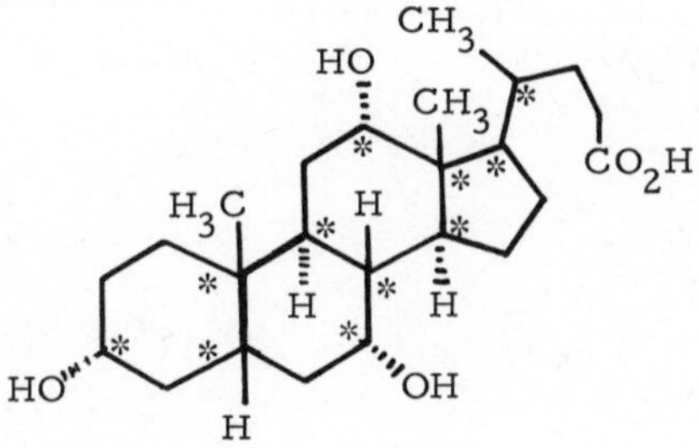

30-11

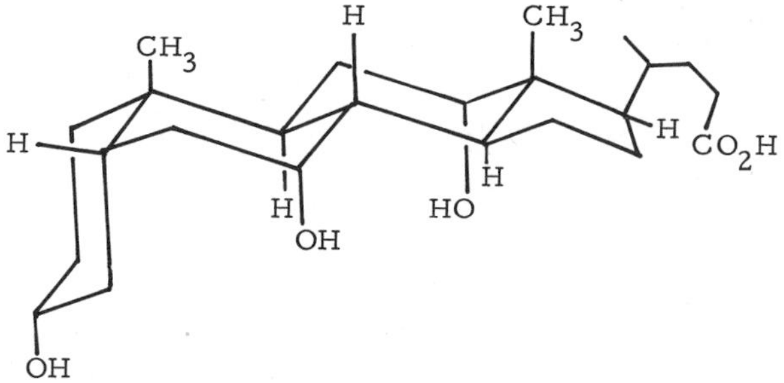

CH₃

H

CH₃

H

H CO₂H

H

HO

H

H OH

OH

OH

30-12

CH₃

HO

base

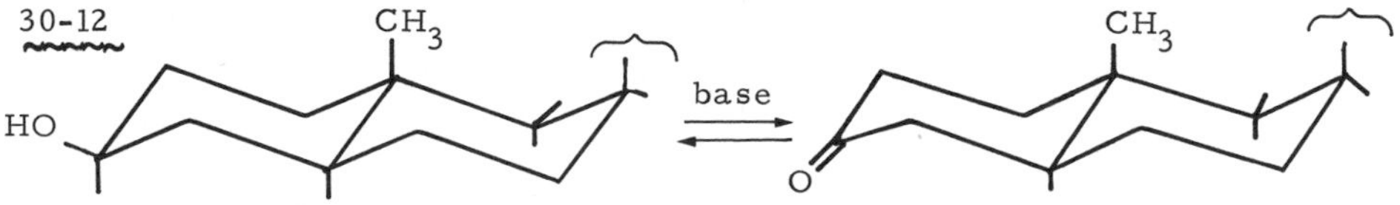

5α-cholestanol, 90%
(equatorial OH)

CH₃

O

5α-cholestanone

base

CH₃

HO

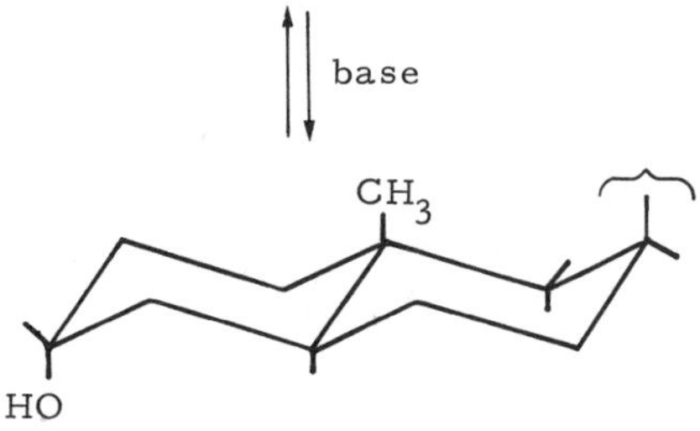

epicholestanol, 10%
(axial OH)

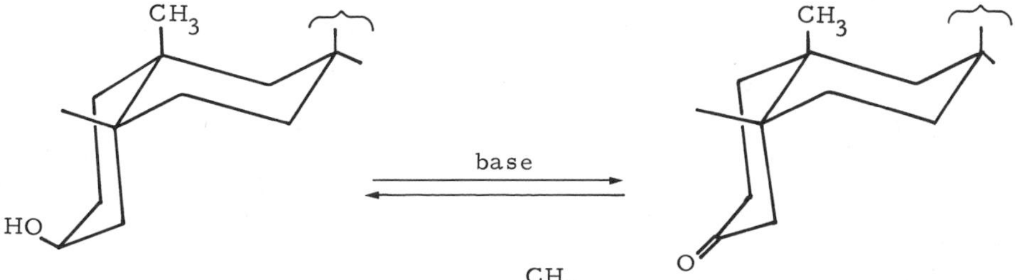

CH₃

HO

5β-cholestanol, 10%
(axial OH)

base

CH₃

O

5β-cholestanone

base

CH₃

OH

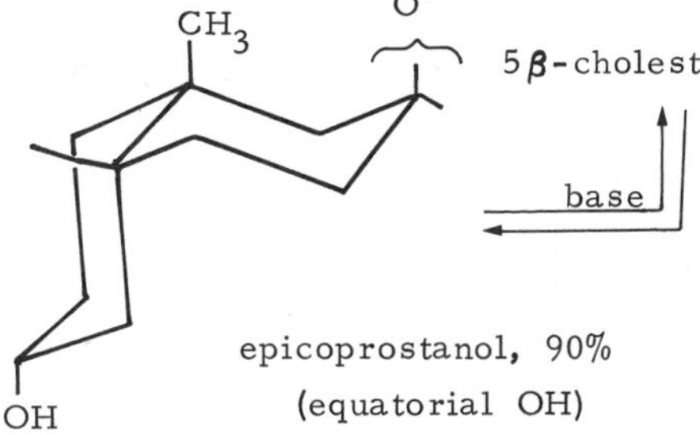

epicoprostanol, 90%
(equatorial OH)

The position of equilibrium
favors formation of the
equatorial isomer.

30-13

a.

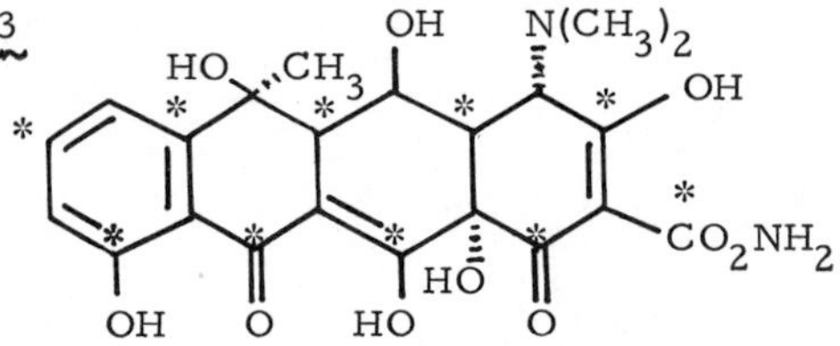

b.

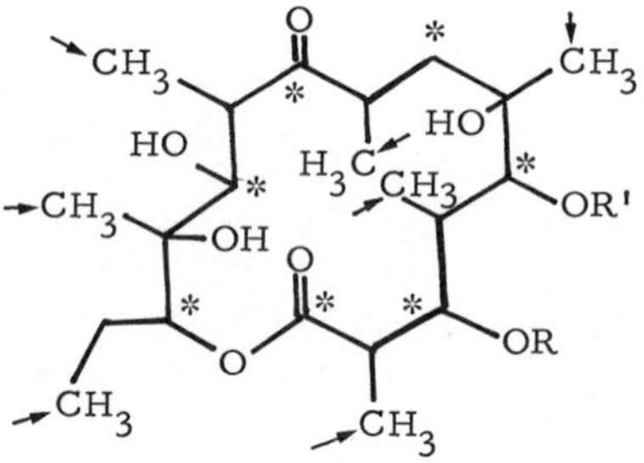

*   denotes ^{14}C

→   denotes CD_3

30-14

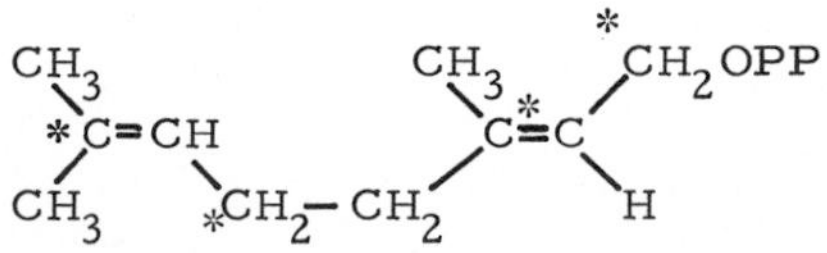

$$\underset{CH_3}{\overset{CH_3}{>}}\overset{*}{C}=CH\underset{\overset{*}{C}H_2-CH_2}{} \qquad \underset{CH_3}{\overset{CH_3}{>}}\overset{*}{C}=\overset{CH_2OPP}{\underset{H}{C}}$$

*   denotes ^{14}C

30-15   (* denotes ^{14}C)

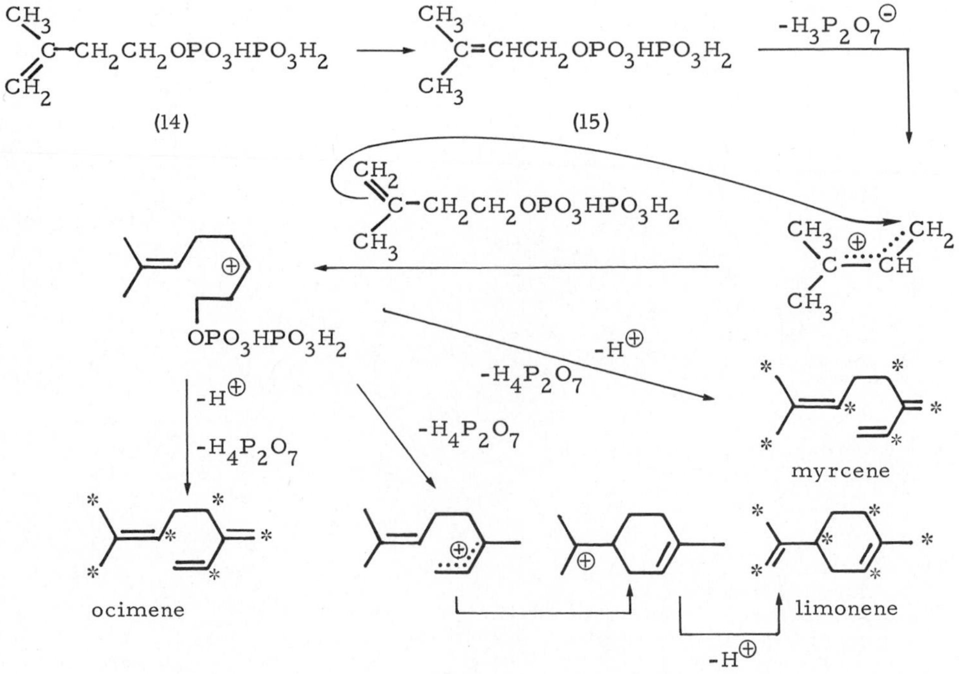

$$\underset{CH_2}{\overset{CH_3}{>}}C-CH_2CH_2OPO_3HPO_3H_2 \longrightarrow \underset{CH_3}{\overset{CH_3}{>}}C=CHCH_2OPO_3HPO_3H_2 \xrightarrow{-H_3P_2O_7^{\ominus}}$$

(14)   (15)

$$\underset{CH_3}{\overset{CH_2}{>}}C-CH_2CH_2OPO_3HPO_3H_2$$

$-H^{\oplus}$

$-H_4P_2O_7$

$-H^{\oplus}$

$-H_4P_2O_7$

$-H_4P_2O_7$

myrcene

ocimene

$-H^{\oplus}$

limonene

30-16

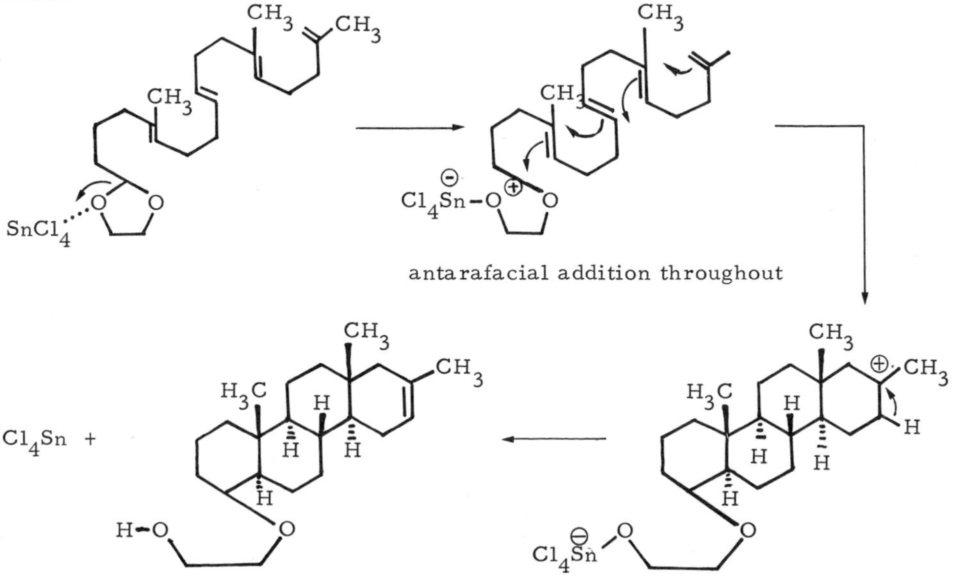

antarafacial addition throughout

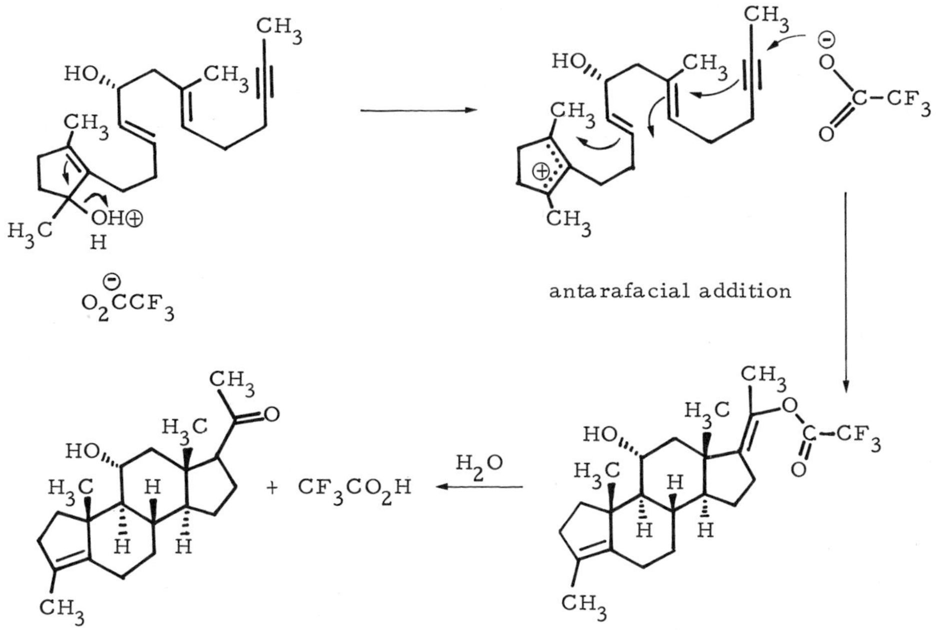

antarafacial addition

30-17   See W. E. Bachmann, W. Cole and A. L. Wilds, J. Amer. Chem. Soc., 61, 974 (1939).

30-18   See R. B. Woodward, F. Sondheimer and D. Taub, J. Amer. Chem. Soc., 73, 4057 (1951) and previous papers.

30-19   See J. E. Cole, W. S. Johnson, P. A. Robins and J. Walker, Proc. Chem. Soc., 114 (1958).

30-20   See G. Stork, E. E. van Tamelen, L. J. Friedman and A. W. Burgstahler, J. Amer. Chem. Soc., 75, 384 (1953).

30-21   See G. Stork and F. H. Clarke, Jr., J. Amer. Chem. Soc., 77, 1072 (1955).

30-22   See M. Gates and G. Tschudi, J. Amer. Chem. Soc., 78, 1380 (1956).

30-23   See R. B. Woodward, F. E. Bader, H. Bickel, A. J. Frey and R. W. Kierstead, Tetrahedron, 2, 1 (1958).

30-24   (i) $HS(CH_2)_3SH$, $H^{\oplus}$;   (ii)

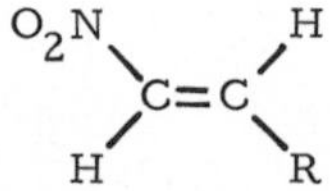

;   (iii) Fe, HCl;

(iv) HCO_2CH_3;   (v) $HOCH_2CH_2OH$, $H^{\oplus}$;   (vi) $KMnO_4$, H_2O, $NaIO_4$;

(vii) NaOH (aldol condensation);   (viii) $(CH_3CO)_2O$, $H^{\oplus}$;   (ix) $NaBH_4$;

(x) H_2O, $H^{\oplus}$;   (xi) $H^{\oplus}$, heat (dehydration of an aldol);   (xii) $NaBH_4$;

(xiii) H_2O, $H^{\oplus}$, 1 hr.   (xiv) Further acid hydrolysis for longer reaction time hydrolyzes amide group and the side-chain nitrile function.

(xv)

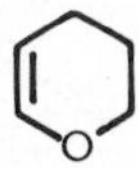

, $H^{\oplus}$;   (xvi) base (-HBr);   (xvii) H_2O, $H^{\oplus}$;

(xviii) chromatography to separate PGE_1 from the undesired stereoisomer.

CHAPTER 31

TRANSITION-METAL ORGANIC COMPOUNDS

31-1

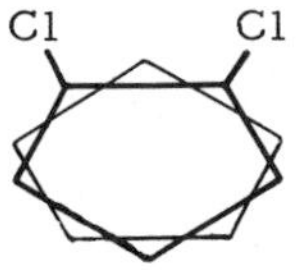

A (achiral)   B (achiral)   C (achiral)   (Fe is not shown)

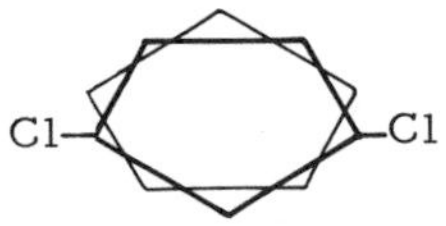

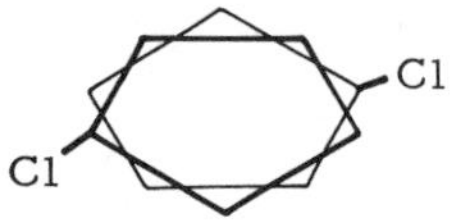

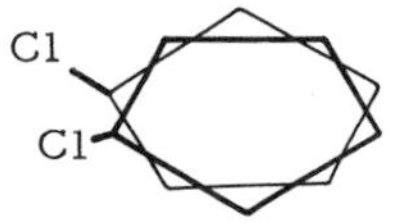

D (chiral)   E (chiral)

In brief, the achiral isomers A, B, and C would each give, respectively, five, five, and two trichloro derivatives. Chiral isomers D and E would be very difficult to distinguish by the substitution method; each would, in principle, give four trichloro derivatives.

31-2

a.

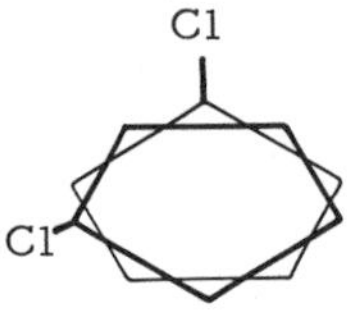

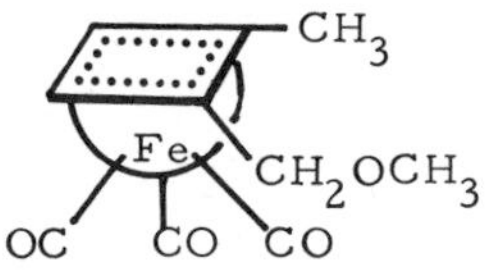

chiral -

no plane or center of symmetry

b.

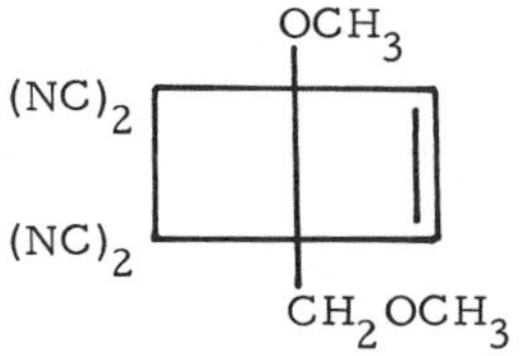

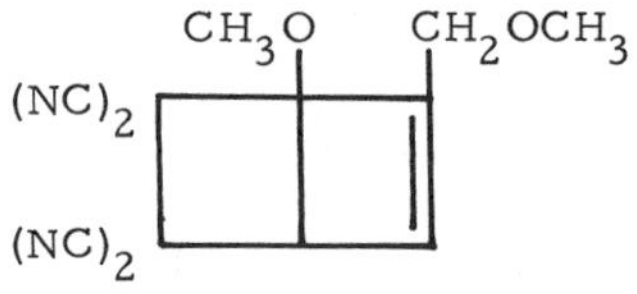

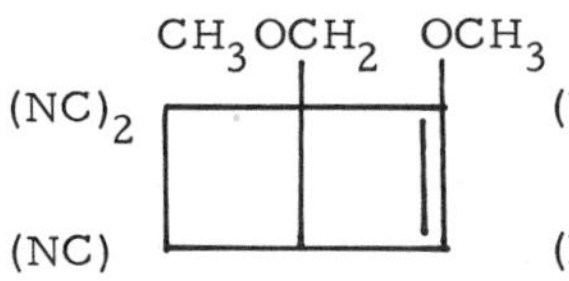

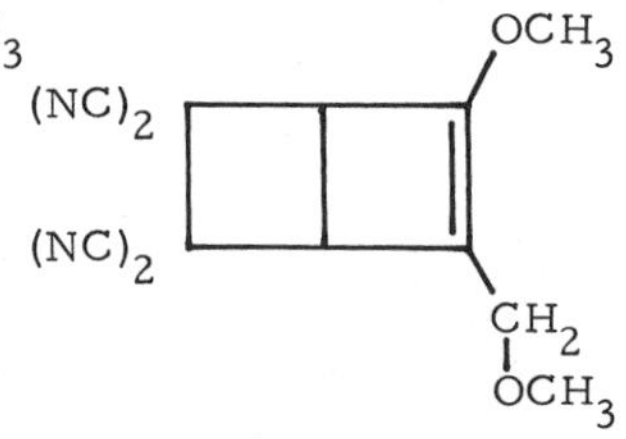

c. Because the cyclobutadiene would be achiral, cycloadducts formed from it would be optically inactive.

d.   The mixture of cycloadducts formed would suggest that the cyclo-butadiene is not a localized diene but rather a 4-electron Hückel cyclic system. The reaction therefore conforms more to a [4 + 2] cycloaddition than to a [2 + 2].

31-3   There are three nonequivalent (chemically shifted) nitrogens in 9. When formed from $^{14}N-N^{15}$, these appear as single resonances of equal intensity.

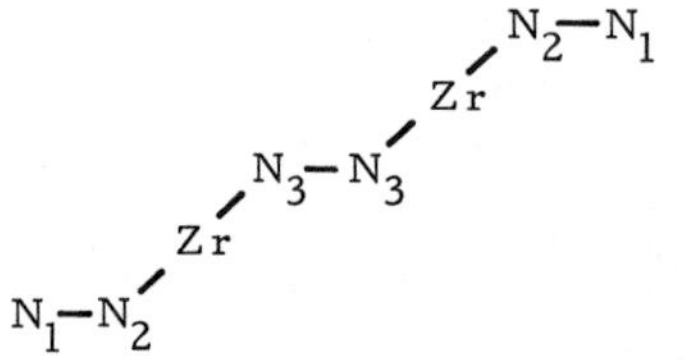

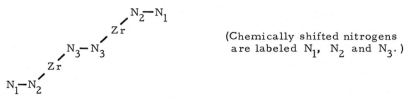

(Chemically shifted nitrogens are labeled N_1, N_2 and N_3.)

When formed with $^{15}N-N^{15}$, the N_1 and N_2 resonances become doublets because they are coupled to each other, J = 6 Hz. Coupling with the ^{14}N nucleus is not evident.

31-4

1.32 A   140°   P(C₆H₅)₃

Pt

P(C₆H₅)₃

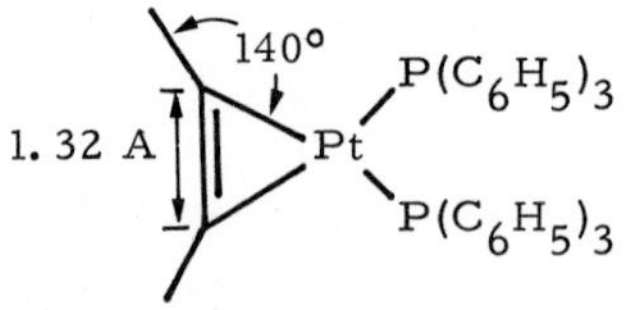

(more equivalent to llc than to lla or llb)

31-5
CH₃CH₂CH=C(CH₃)₂   ⇌ (Cp)₂ZrHCl ⇌
$$CH_3CH_2CH \overset{\overset{\displaystyle Cl}{|}}{\underset{}{-}} \overset{\overset{\displaystyle H—Zr(Cp)_2}{}}{C}(CH_3)_2$$

$$CH_3\overset{\overset{\displaystyle Zr(Cp)_2}{|}}{\underset{\overset{|}{Cl}}{C}}HCH_2CH(CH_3)_2$$ ⇌ $$H—Zr(Cp)_2 \overset{|}{\underset{|}{Cl}}$$ ⇌ $$CH_3CH_2\overset{\overset{\displaystyle Zr(Cp)_2}{|}}{\underset{\overset{|}{Cl}}{C}}HCH(CH_3)_2$$

⇅

CH₂=CHCH₂CH(CH₃)₂   CH₃CH=CHCH(CH₃)₂

H—Zr(Cp)₂
  |
  Cl
                    ⇌   (Cp)₂ZrCH₂CH₂CH₂CH(CH₃)₂
                            |
                            Cl

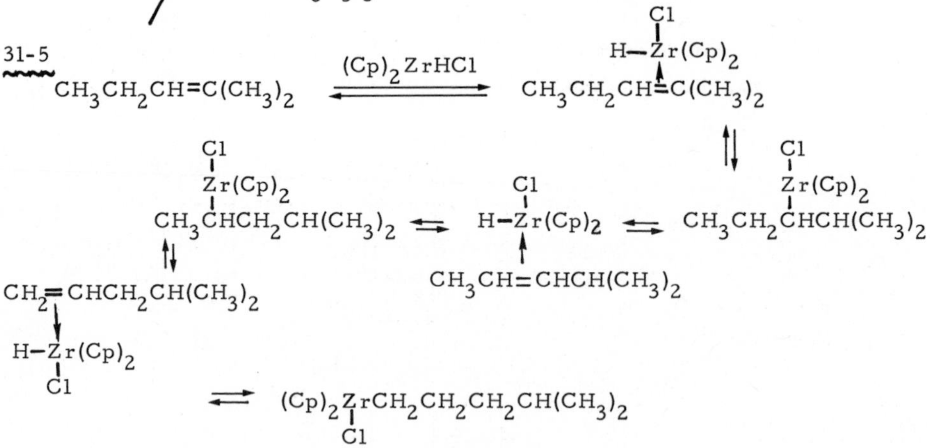

The alternate product, $$CH_3CH_2CH_2\overset{\overset{\displaystyle CH_3}{|}}{C}H\overset{}{C}H_2\overset{\overset{\displaystyle Cl}{|}}{Z}r(Cp)_2,$$ is not a significant product because the β-methyl makes it more hindered, and the orientation whereby Zr becomes σ-bonded to a tertiary carbon is kinetically unfavorable (i.e.,

$$CH_3CH_2CH_2\overset{\overset{\displaystyle C(CH_3)_2}{}}{\underset{\overset{|}{ClZr(Cp)_2}}{C}}$$ is not favored).

31-6

a.

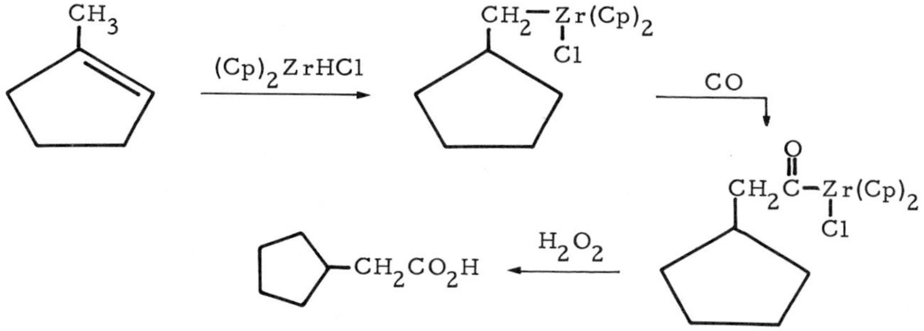

b. 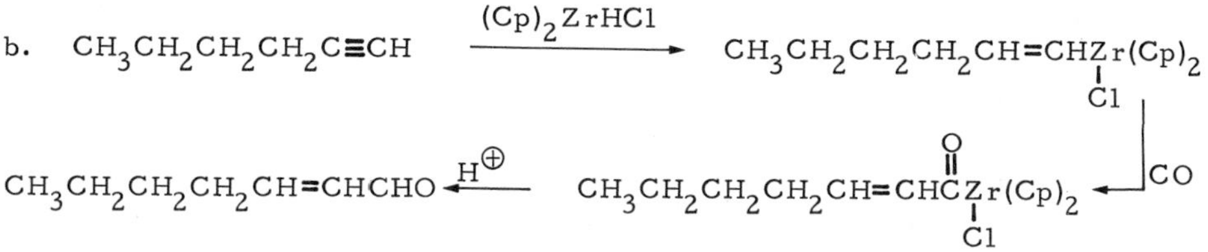 $CH_3CH_2CH_2CH_2C\equiv CH$ $\xrightarrow{(Cp)_2ZrHCl}$ $CH_3CH_2CH_2CH_2CH=CHZr(Cp)_2$ $\underset{Cl}{|}$

$CH_3CH_2CH_2CH_2CH=CHCHO \xleftarrow{H^\oplus}$ $CH_3CH_2CH_2CH_2CH=CHCZr(Cp)_2 \xleftarrow{CO}$ with C=O and $\underset{Cl}{|}$

c. $(CH_3)_3CC\equiv CH$ $\xrightarrow{(Cp)_2ZrHCl}$ $(CH_3)_3CCH=CHZr(Cp)_2$ $\underset{Cl}{|}$ $\xrightarrow{CH_3COCl}$

$(CH_3)_3CCH=CHCOCH_3$ $\underset{ClZr(Cp)_2}{|}$ $\xleftarrow{(Cp)_2ZrHCl}$ $(CH_3)_3CCH=CHCOCH_3$

$\downarrow H_3O^\oplus$

$(CH_3)_3CCH_2CH_2COCH_3$

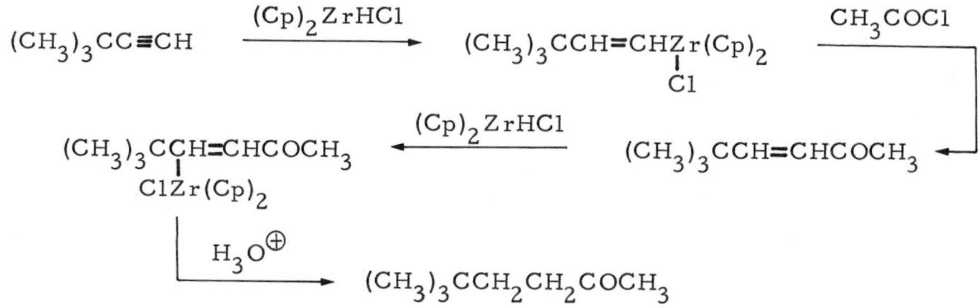

31-7   The coupling constant of 13 Hz is consistent with a preferred conformation for both 14 and 15 in which the HC—CH protons are trans. If we assume
that the two largest groups on adjacent carbons are also trans, then the most
likely configuration and conformations of 14 and 15 are:

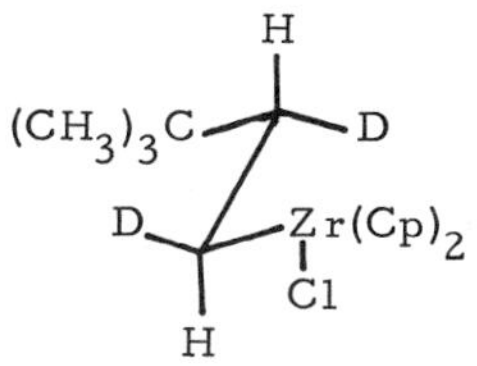

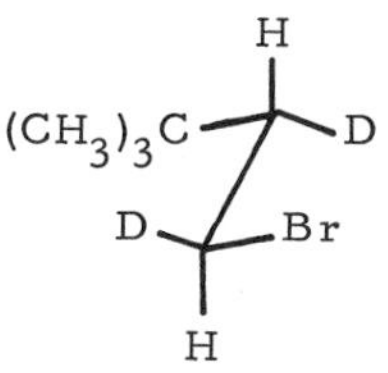

1. For 15 to be formed from 14, the metal atom must be replaced
with bromine with retention of configuration at carbon.

2. For the addition step to give 14, addition must be suprafacial.

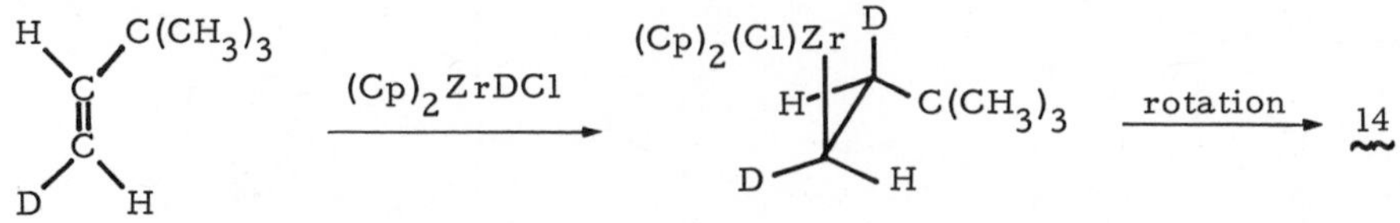

3. Replacement of the metal for deuterium in the alkenyl zirconium adduct also must have occurred with <u>no change</u> in configuration at carbon.

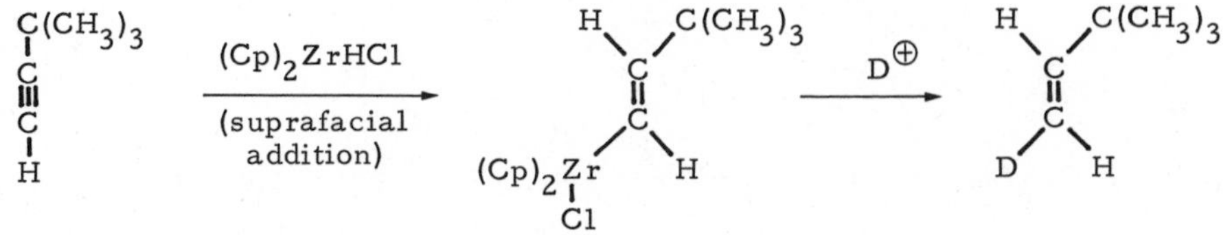

31-8

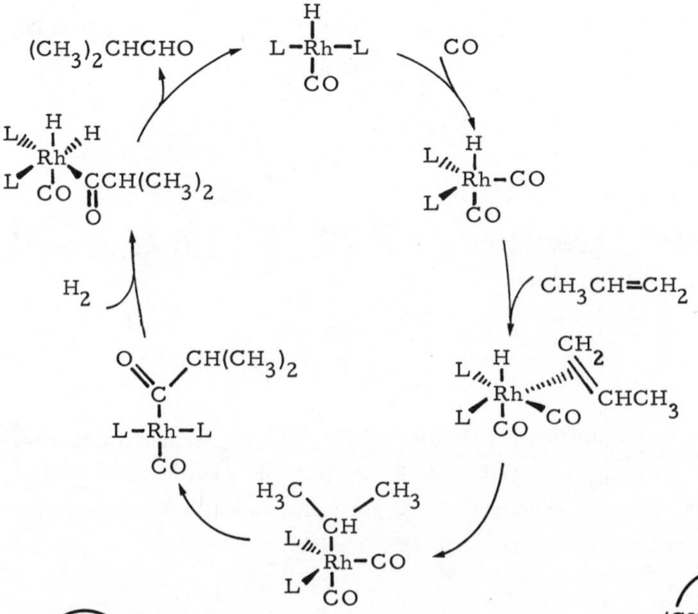

31-9   a.

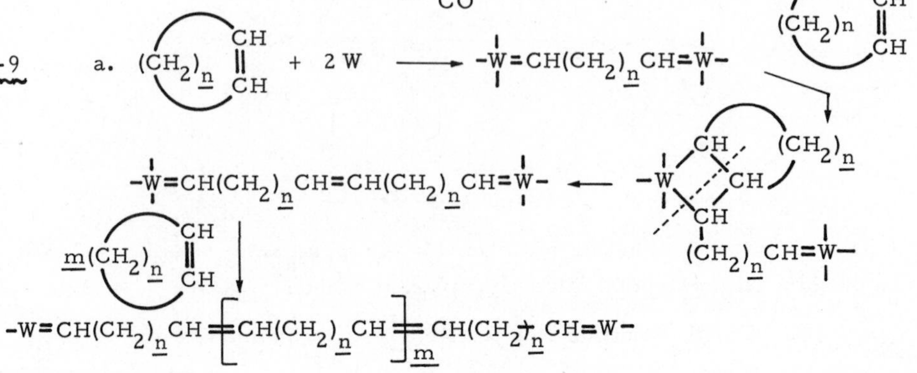

b.   If the linear polymer from Part a were to react <u>intramolecularly</u>, a large-ring polymer would be formed.

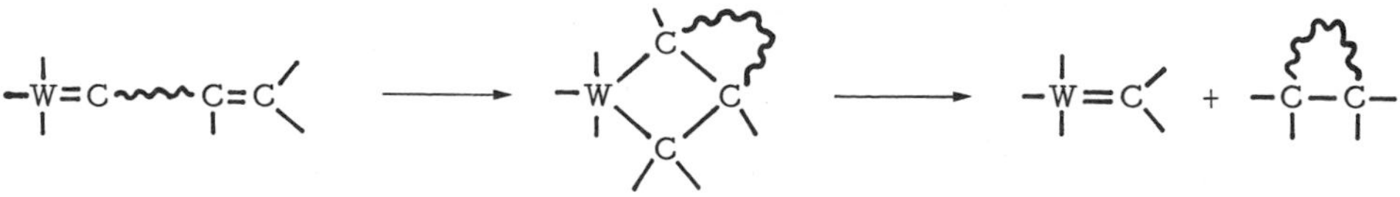

c.   If the cyclic polymers are large enough, the ring cavity may accommodate a linear chain which, if it should cyclize, gives a catenane:

Also, if the product   1   occurs then a catenane   2   can result from the twisted system.

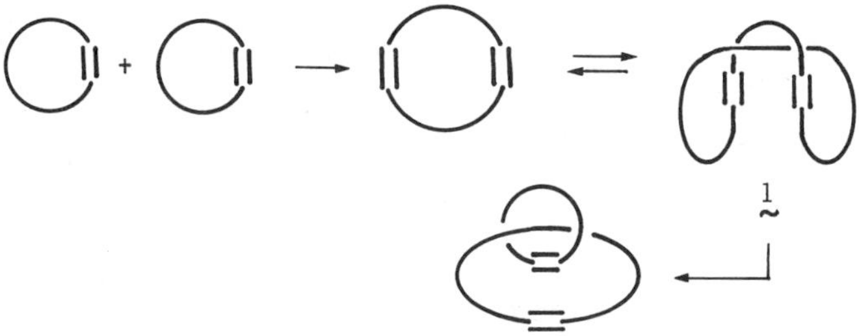

31-10

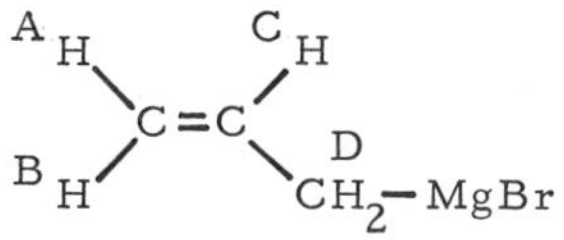

The alkenyl protons A, B and C are expected to have chemical shifts δ_A, δ_B and δ_C in the region of 350 Hz, and would show complex splitting patterns produced by couplings J_{AB}, J_{AC}, and J_{BC}. By analogy with CH_3CH_2MgBr in which δ_{CH_2} = 38 Hz, the resonance for the protons D should appear near 38 Hz as a doublet ($J_{CD} \sim$ 4-7 Hz), possibly showing smaller couplings with A and B.

The spectrum of Figure 31-4 shows a doublet at 148 Hz and a quintet at 377 Hz. This is clearly irreconcilable with the structure of 2-propenylmagnesium bromide as shown above. The achiral spectrum indicates only <u>two</u> chemically shifted sets of protons, and the splitting pattern suggests that four equivalent alkenyl protons are coupled to one other proton. This would be the case if the magnesium were in dynamic exchange between the two ends of the 2-propenyl chain.

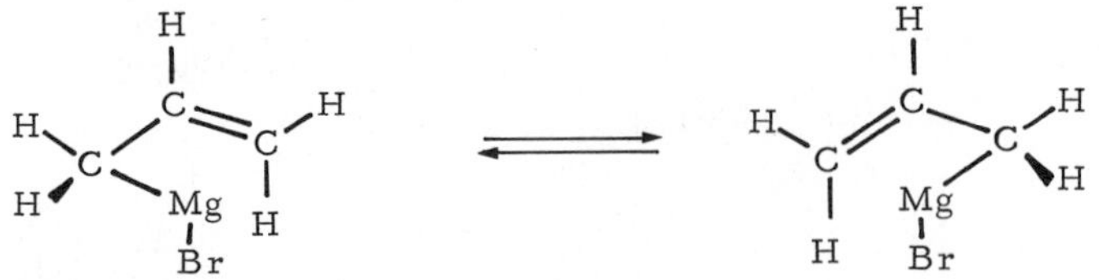

It is unlikely that the metal is a π-complex similar to the π-allyl complexes of nickel because the spectrum is too simple.

31-11   The simplification of the proton nmr spectrum of $(C_3H_5)_2Ni$, 20, on complexation with azabenzene indicates a drastic change in structure. The most reasonable and simple explanation is conversion of the π-propenyl structure of at least one of the propenyl groups to the σ-propenyl structure. A possible structure is:

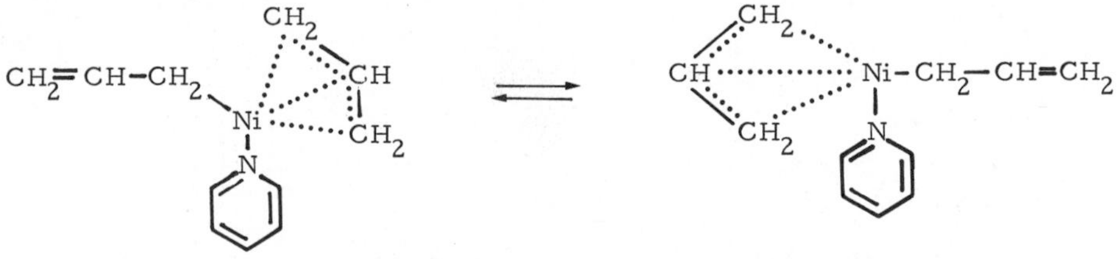

To give the simple propenyl nmr spectrum as for $CH_2{=}CH{-}CH_2MgBr$, two forms as shown would have to be in <u>rapid</u> equilibrium with each other.

31-12   If a hydride-shift mechanism were involved, the deuterium label would be scrambled by reversible transfers of the type:

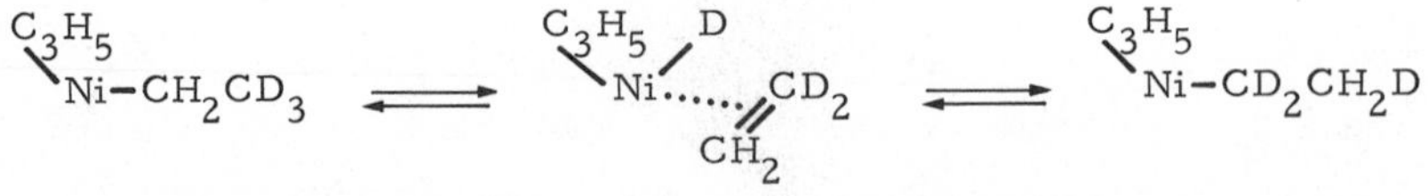

$C_3H_5 = \pi$-propenyl group

This would lead to   formation of C_3H_5D and $CH_2{=}CD_2$ but also of $C_3H_6 +$ $CD_2{=}CHD$. Because mixtures are <u>not</u> obtained, a radical decomposition is indicated.

$$C_3H_5Ni{-}CH_2CD_3 \longrightarrow Ni + C_3H_5{\cdot} + {\cdot}CH_2CD_3$$
$$C_3H_5{\cdot} + {\cdot}CH_2{-}CD_3 \longrightarrow C_3H_5D \text{ and } CH_2{=}CD_2$$

Similarly,

$$C_3H_5{-}Ni{-}CD_2CH_3 \longrightarrow Ni + C_3H_5{\cdot} + {\cdot}CD_2CH_3$$
$$C_3H_5{\cdot} + {\cdot}CD_2CH_3 \longrightarrow C_3H_6 + CD_2{=}CH_2$$

If the hydride-shift step were not reversible, then the products would be the same as by the radical mechanism. For example,

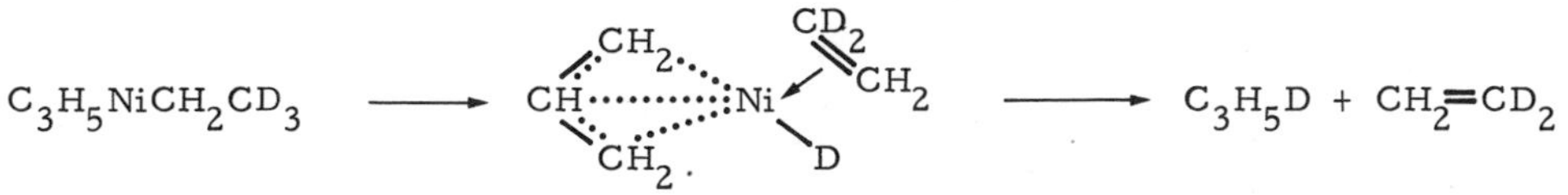

$$C_3H_5NiCH_2CD_3 \longrightarrow \text{[structure]} \longrightarrow C_3H_5D + CH_2{=}CD_2$$

31-13

$$S_N2, \text{ inversion at carbon}$$
(cf. Section 31-3B and 31-4C)

retention of
configuration
at carbon

CO
-L

$$CH_3OH$$
$$2 L$$

$$+ \ PdL_4 \ + \ HBr$$

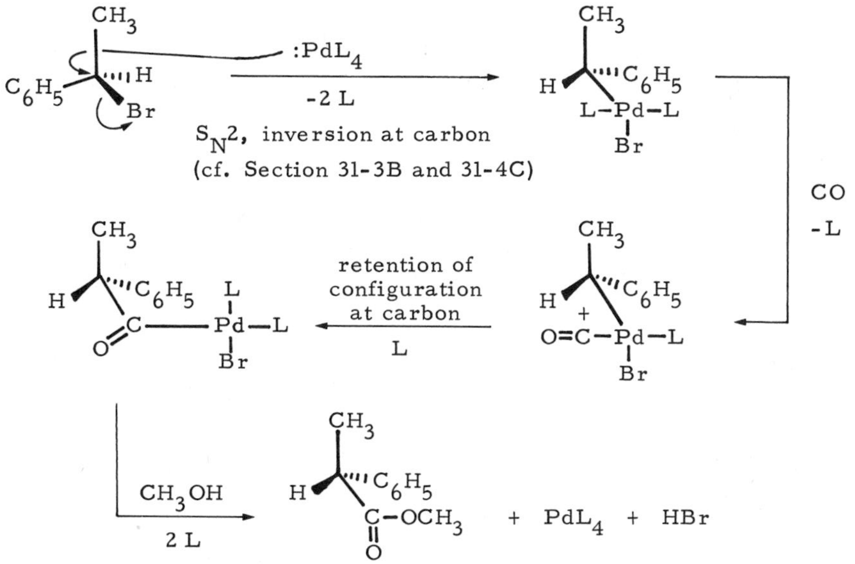

31-14   a. Your mechanism does not have to be the same as that given below, but should have the feature that the four hydrogens of the ethene are the same four hydrogens in the ethanol formed. This would not be the case if you allowed $CH_2{=}CHOH$ to be released from the Pd and rearrange in D_2O solution, because then CDH_2CHO would be formed.

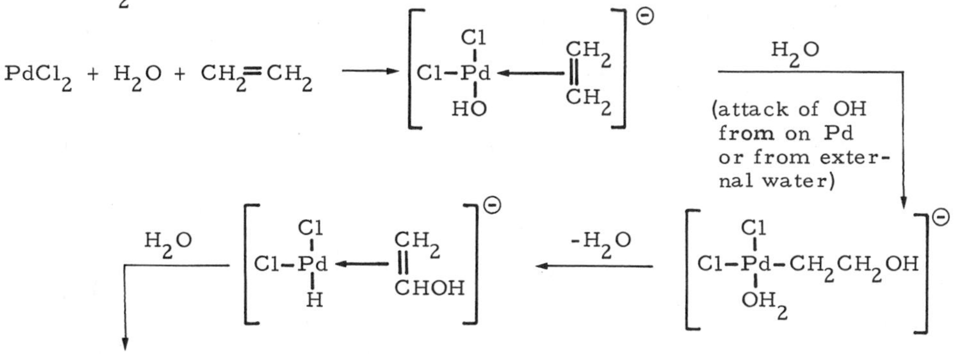

$$PdCl_2 + H_2O + CH_2{=}CH_2 \longrightarrow \left[\begin{array}{c} Cl \\ Cl-Pd \leftarrow \ \| \ ^{CH_2}_{CH_2} \\ HO \end{array}\right]^{\ominus} \xrightarrow{H_2O}$$

(attack of OH
from on Pd
or from exter-
nal water)

$$\xrightarrow{H_2O} \left[\begin{array}{c} Cl \\ Cl-Pd \leftarrow \ \| \ ^{CH_2}_{CHOH} \\ H \end{array}\right]^{\ominus} \xleftarrow{-H_2O} \left[\begin{array}{c} Cl \\ Cl-Pd-CH_2CH_2OH \\ OH_2 \end{array}\right]^{\ominus}$$

(cont.)

31-14 a. (cont.)

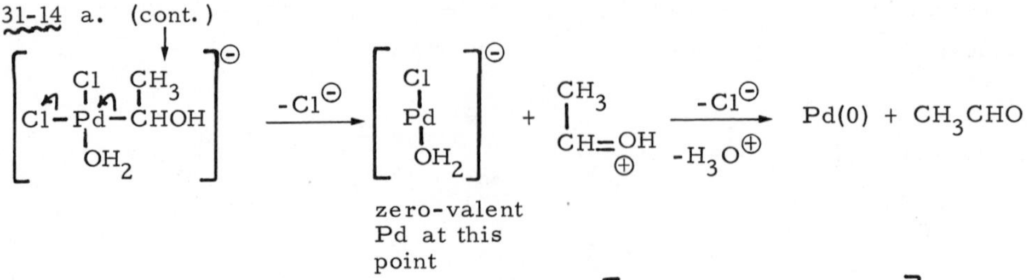

$$\left[\begin{array}{c}\text{Cl}\quad\text{CH}_3\\ \text{Cl}-\overset{|}{\underset{|}{\text{Pd}}}-\text{CHOH}\\ \text{OH}_2\end{array}\right]^{\ominus} \xrightarrow{-\text{Cl}^{\ominus}} \left[\begin{array}{c}\text{Cl}\\ \overset{|}{\underset{|}{\text{Pd}}}\\ \text{OH}_2\end{array}\right]^{\ominus} + \begin{array}{c}\text{CH}_3\\ |\\ \text{CH}=\overset{\oplus}{\text{OH}}\end{array} \xrightarrow[-\text{H}_3\text{O}^{\oplus}]{-\text{Cl}^{\ominus}} \text{Pd(0)} + \text{CH}_3\text{CHO}$$

zero-valent
Pd at this
point

b. PdCl_2 + trans-2-butene + CH_3OH $\xrightarrow{\text{CO}}$

attack by exter-
nal CH_3OH gives
inversion at carbon

$\xrightarrow[\text{at carbon}]{\text{retention}}$

$\xrightarrow{\text{CH}_3\text{OH}}$

D, L mixture

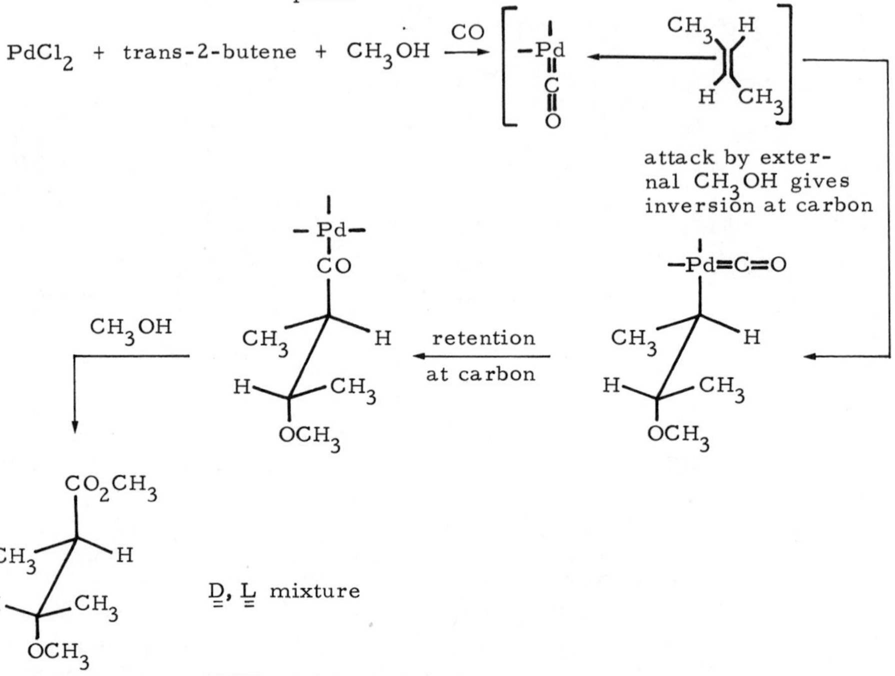

The other product could be obtained by the sequence:

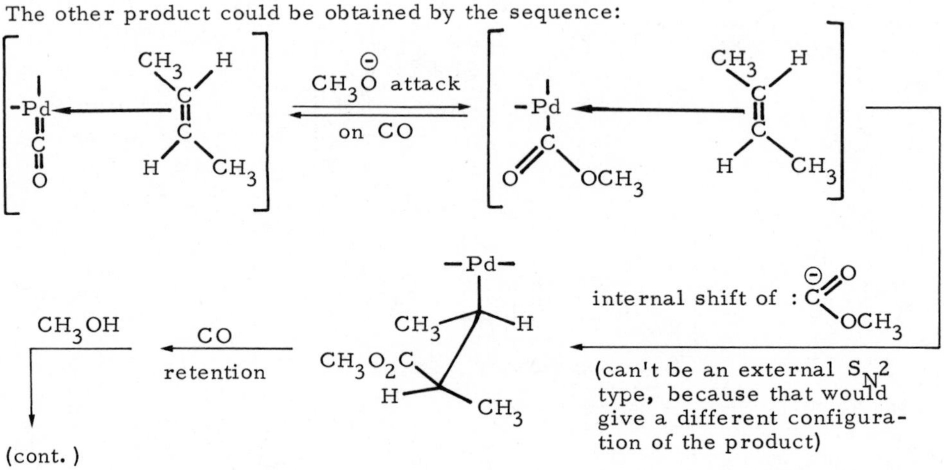

$\xrightleftharpoons[\text{on CO}]{\text{CH}_3\overset{\ominus}{\text{O}} \text{ attack}}$

internal shift of $:\overset{\ominus}{\text{C}}\overset{\diagup\text{O}}{\diagdown\text{OCH}_3}$

$\xrightarrow[\text{retention}]{\text{CO}}$ $\xleftarrow{\text{CH}_3\text{OH}}$

(can't be an external S_N2
type, because that would
give a different configuration of the product)

(cont.)

31-14 b. (cont.)

+

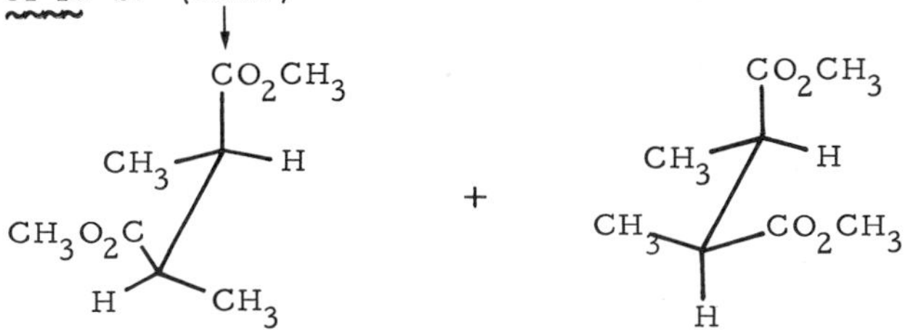

D, L mixture

ANSWERS TO ADDITIONAL EXERCISES

Chapter 1

1A-1 $C_6H_4O_2$ 1A-2   (by the principle of least structural change)

1A-3   C—K; carbon is electronegative relative to potassium.

1A-4   C—F; carbon is electropositive relative to fluorine.

1A-5   a. $CH_3Na + HCl \longrightarrow CH_4 + NaCl$ c. $CH_3Na + HOH \longrightarrow CH_4 + NaOH$
b. $CH_3Na + CH_3Cl \longrightarrow CH_3CH_3 + NaCl$ d. $CH_3Na + CH_3OH \longrightarrow CH_4 + NaOCH_3$

1A-6 CH_3CH_2OH, CH_3NH_2

1A-7   Hydrogen bonding of the type $\overset{|}{O}$—H$\cdots\overset{|}{O}$— is stronger than of the type $\overset{|}{S}$—H$\cdots\overset{|}{S}$—. Hence water is more associated and less volatile.

1A-8 $(CH_3)_3N:\cdots\cdot$H—OH hydrogen bonding accounts for water solubility.

1A-9   a, b, c, and d CH_3OH e. CH_3SNa

1A-10   a. NH_3 b. ClOH   c. Cl_3CH

1A-11 $(CH_3)_2CHCH_2CH_3$

1A-12 $(CH_3)_3CCH_2C(CH_3)_3$; 3 kinds of carbon

Chapter 2

2A-1   a. alkene, arene, ether   b. arene, alcohol, ester, amine (note that an amine function can have one, two, or three R groups on nitrogen, RNH_2, R_2NH, or R_3N)   c. arene, halide, alcohol, ketone, amine, amide   d. carboxylic acid, amine, thiol, amide.

2A-2 $(CH_3)_3N$, $(CH_3)_3NH^{\oplus}Cl^{\ominus}$

2A-3   four ways

2A-4   1. elimination (of water) 2. addition (of H_2); may also be classed as reduction 3. substitution (of Br for OH) 4. substitution (of CN for Br) 5. addition (of H_2O) 6. rearrangement

Chapter 3

3A-1   6 prim, 1 sec, 2 tert, 1 quat

3A-2 $(CH_3)_2C(Br)CH_2CH_3$, 2-bromo-2-methylbutane (<u>tert</u>-pentylbromide)

3A-3   (1-chloroethyl)cyclopropane, 1-chloro-2-ethylcyclopropane, 2-chloro-1,1-dimethylcyclopropane, 3-chloro-1,2-dimethylcyclopropane

3A-4   1,3,5-trimethylbenzene (mesitylene), 2-chloro-1,3,5-trimethylbenzene, 1-chloromethyl-3,5-dimethylbenzene

3A-5   a. $CH_3CH=C(CH_3)_2$ b. $CH_3CH=CHC(Cl)=CH_2$ c. $(CH_3)_2CH-$
d. $HC\equiv C-CH=C=CH_2$ e. $(CH_3)_2C(Cl)C\equiv CH$ f. $(CH_3)_3C-$$-CH_2CH_3$

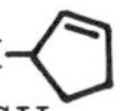

3A-6   2-methyl-1, 3-pentadiene, 4-methyl-1, 3-pentadiene, 3-methyl-1, 3-
pentadiene, 2-ethyl-1, 3-butadiene (the latter is included here although it is not
strictly a pentadiene.) Three cumulated isomers are possible.
3A-7   a. 3, 7-dimethyl-1, 3, 6-octatriene   b. 3-methylidene-7-methyl-1, 6-
octadiene   c. 3, 7, 11-trimethyl-1, 3, 6, 11-dodecatetraene   d. 1-methyl-3-(1-
methylethenyl)cyclohexene   e. 2-methyl-5-(1, 5-dimethyl-4-hexenyl)1, 3-cyclo-
hexadiene

Chapter 4

4A-1   3-bromo-3-methylpentane
4A-2   (C) $RH + \cdot CCl_3 \longrightarrow R\cdot + HCCl_3$; $R\cdot + BrCCl_3 \longrightarrow RBr + \cdot CCl_3$
4A-3   trichloromethane (E)
4A-4   17.3% 1-chloromethyl-3-methylcyclopentane; 19.1% 2-chloro-1, 3-
dimethylcyclopentane; 25.4% 1-chloro-1, 3-dimethylcyclopentane; 38.2% 2-
chloro-1, 4-dimethylcyclopentane
4A-5   (B) H_2S
4A-6   cyclohexene $\xrightarrow{Br_2}$ 1, 2-dibromocyclohexane $\xrightarrow{Br_2}$ 1, 1, 2-tribromo-
cyclohexane
4A-7   (Z) 2, 3-dimethylbutane,   (X) 2-chloro-2, 3-dimethylbutane,
(Y) 1-chloro-2, 3-dimethylbutane

4A-8 $O_3 + :\overset{\bullet}{N}: :\overset{\bullet\bullet}{\underset{\bullet\bullet}{O}}: \longrightarrow O_2 + \cdot \overset{\bullet\bullet}{\underset{\bullet\bullet}{O}} : \overset{\bullet}{N}: :\overset{\bullet\bullet}{\underset{\bullet\bullet}{O}}:$ first step
        $O + \cdot O-N=O \longrightarrow O_2 + :N=O$ second step

4A-9 $O_3 + Cl\cdot \longrightarrow O_2 + Cl-O$ first step
        $O + ClO\cdot \longrightarrow O_2 + Cl\cdot$ second step

---

Chapter 5

5A-1   There are three chiral monochloro compounds possible of formula
$C_5H_{11}Cl$. They are:

    A $CH_3CH_2CH_2CH(Cl)CH_3$ from pentane
    B $CH_3CH(Cl)CH(CH_3)_2$ from 2-methylbutane
    C $CH_3CH_2CH(CH_3)CH_2Cl$ from 2-methylbutane

        The calculated percentage composition of A formed in the monochlorin-
ation of pentane is $(4 \times 3.3 \times 100)/(6 \times 1 + 4 \times 3.3 + 2 \times 3.3) = 51\%$; A is the major
chlorination product of pentane.
        A similar calculation for B and C shows they would be formed in
comparable amounts from 2-methylbutane.
        B = $(2 \times 3.3 \times 100)/(3 \times 1 + 6 \times 1 + 2 \times 3.3 + 1 \times 4.4) = 33\%$
        C = $(6 \times 1 \times 100)/(3 \times 1 + 6 \times 1 + 2 \times 3.3 + 1 \times 4.4) = 30\%$
        Because only one major product is obtained that is also chiral, the
hydrocarbon must be pentane and A must be 2-chloropentane.
5A-2   D, L-3, 4-dimethylhexane; because addition of hydrogen can occur at
either face of the double bond with equal probability. cis-3, 4-Dimethyl-3-hexene
would give meso-3, 4-dimethylhexane.
5A-3   L-1-Amino-1-phenylethane. The reaction does not involve the chiral
center directly. Therefore, the configuration does not change.

5A-4   D, L-2,3- Dibromobutane

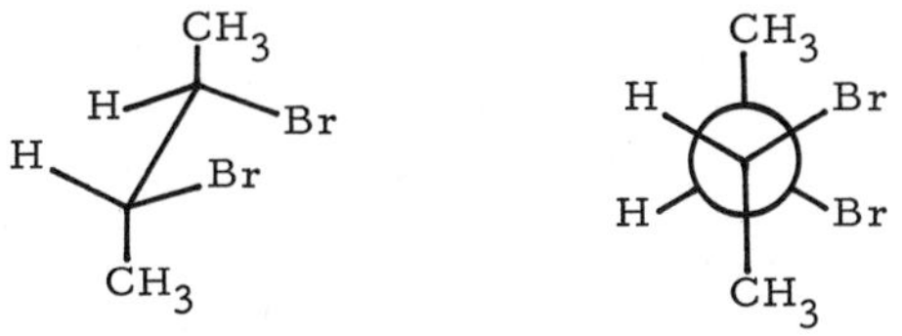

The conformations shown
assume that the methyl groups
have a greater preference for
the trans positions than the
bromines.

5A-5   Eight configurational isomers are conceivable.

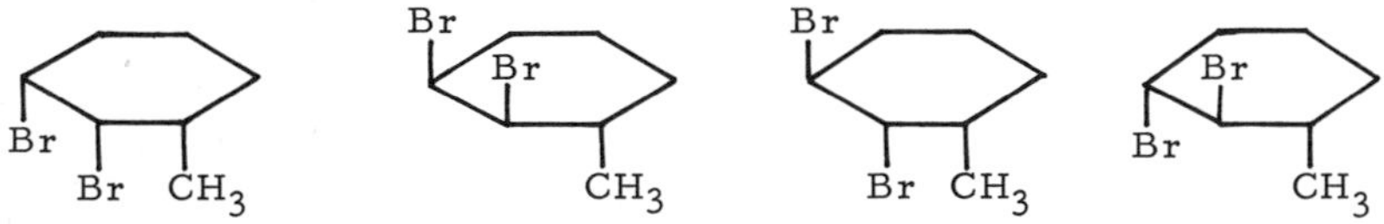

Each of the above has an enantiomeric configuration.

5A-6   D, L. The configuration shown for 2-bromobutane is D. Replacing H_a
at C3 with Br gives the L configuration at C3.

5A-7   1. Four configurational isomers are possible. That is, there would be
two pairs of enantiomers.

2 and 3. There is clearly a cis and a trans arrangement possible about
the double bond. Both 2 and 3 are achiral.

5A-8   oleic acid, two configurations, cis (c) and trans (t)

linoleic acid, four configurations, cc and tt

linolenic acid, eight configurations, ccc, cct, ttc, ttt, ctc, ctt, tcc, tct

5A-9   a. Three chiral carbons are shown, C2, C3 and C4. Therefore, eight
configurational isomers are possible. b. Inversion of configuration at C2 would
produce a diastereomer of L-streptose. The diastereomer would be chiral.

5A-10   All the double bonds in the chain have the trans configuration. There
are three chiral carbons - one in the six-membered ring and two in the five-ring.

Chapter 6

6A-1   a. $CH_3-C\overset{\oplus}{=}\overset{..}{O}: \longleftrightarrow CH_3-C\overset{\oplus}{\equiv}O:$

b.

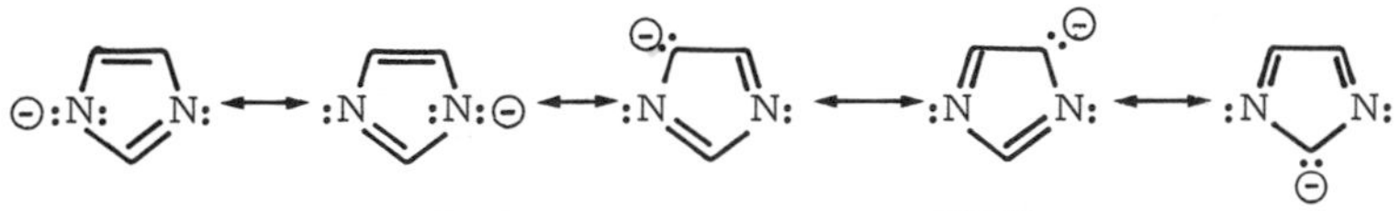

c.

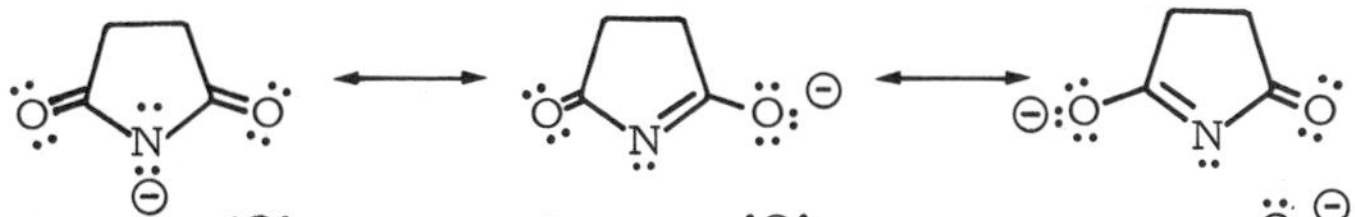

d.

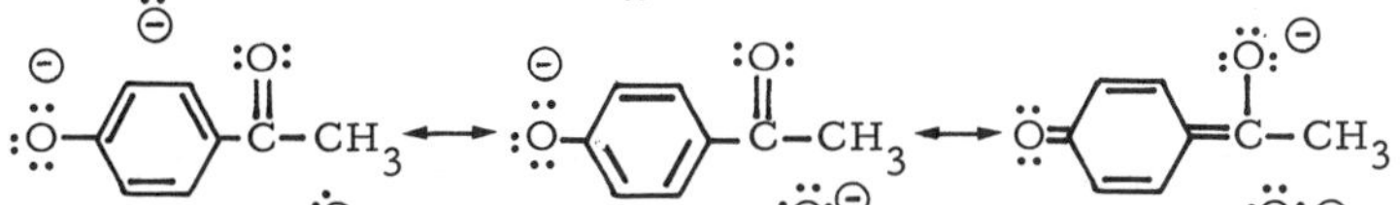

e.

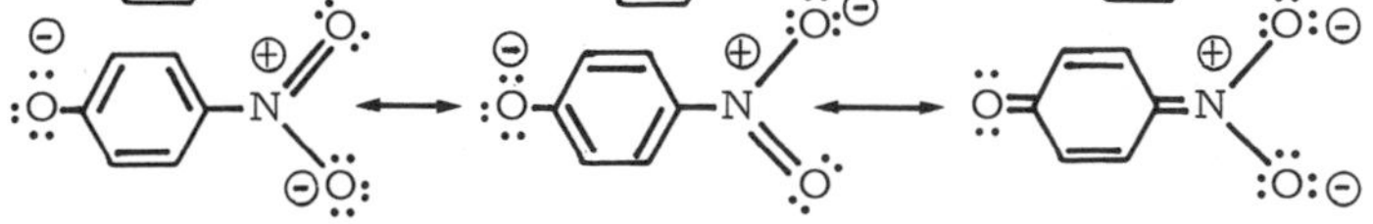

f.

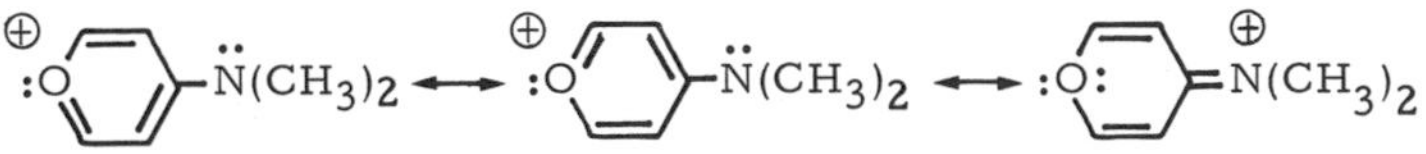

6A-2   a. 1,3-cyclopentadiene (See p. 997 of text for valence bond structures)
        b. hydrogen cyanide $\ominus$:C≡N: ⟷ :C=N̈:$\ominus$
        c. ethanoic (acetic) acid. (The valence bond structures of the conjugate
base are of equal energy.)

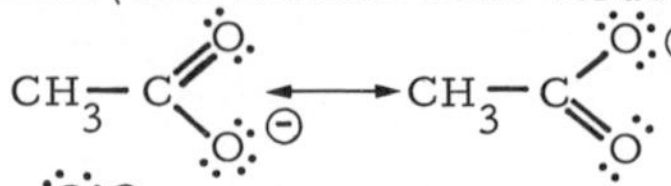

6A-3
a.

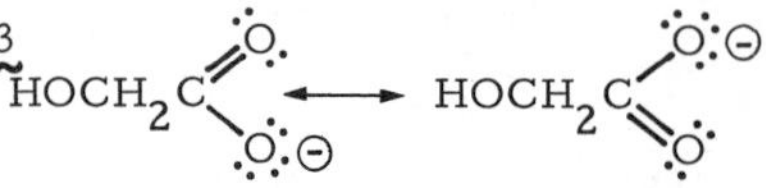

b.

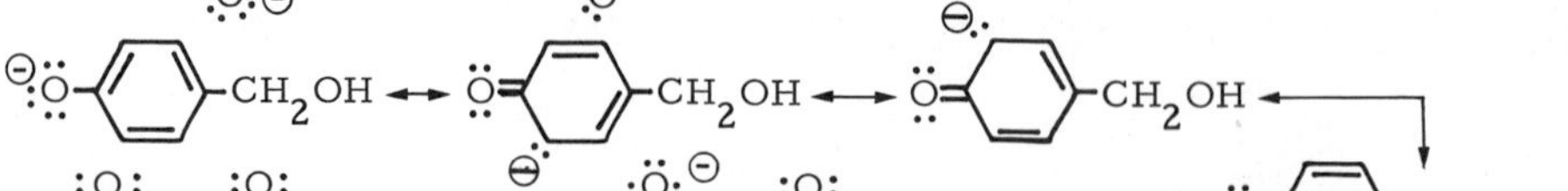

c.

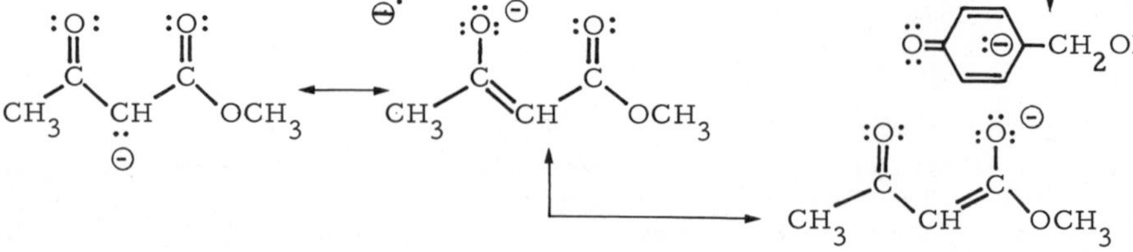

6A-4   a. $(CH_3)_2\dot{C}-C≡N:$ ⟷ $(CH_3)_2C=C=\dot{N}:$

b.

c.

6A-5   a. $CH_2^{\oplus}-CH=\overset{\ominus}{\ddot{N}}H_2$ terminal carbon has only six electrons
                                terminal nitrogen has ten electrons

        b. :Ö-CH=CH-CH$_2$ this structure could not contribute to the hybrid
                                of O=CH-CH=CH$_2$ because it has unpaired
                                electrons

        c.

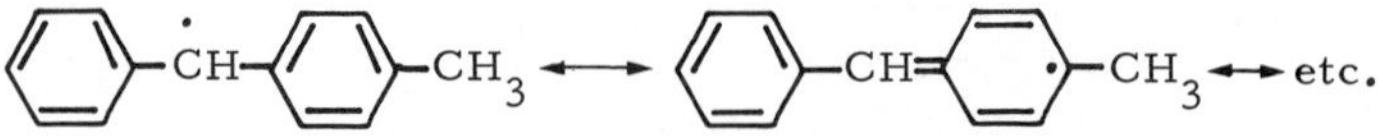

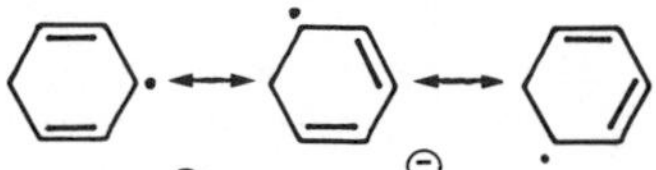

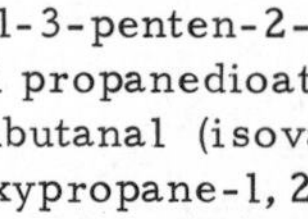

represents a grossly distorted bonding
                                arrangement

Chapter 7

a. 4-hydroxy-4-methyl-2-pentanone (diacetone alcohol)
b. 4-methyl-3-penten-2-one (mesityl oxide)
c. dimethyl propanedioate (methyl malonate)
d. dimethylbutanal (isovaleraldehyde)
e. 2-hydroxypropane-1,2,3-tricarboxylic acid (citric acid) or
    3-carboxy-3-hydroxypentanedioic acid
f. 1,2,3,4,5,6-hexanehexol (sorbitol)
g. 2-hydroxybutanedioic acid (malic acid)
h. 2-oxobutanedioic acid (oxalacetic acid)
i. 2-oxopentanedioic acid (α-ketoglutaric acid)
j. 2-amino-3-methylbutanoic acid (valine)
k. 2-amino-3-hydroxypropanoic acid (serine)
l. 2-amino-3-mercaptopropanoic acid (cysteine)

m. 2, 6-diaminohexanoic acid (lysine)

n. sodium 5-amino-5-carboxybutanoate (monosodium glutamate)

o. N, N, N-trimethyl-N-(2-methylcarbonyloxyethyl)-ammonium chloride (acetylchloline chloride)

p. 2-butynediamide (acetylenecarboxamide)

q. 2-(methylcarbonyloxy)benzenecarboxylic acid (acetylsalicylic acid)

r. 1,1-bis(4-chlorophenyl)-2,2,2-trichloroethane (DDT)

s. 2-methylpropenenitrile (methacrylonitrile)

t. 1-methylethenyl ethanoate (acetate) (isopropenyl acetate)

u. 2-amidobenzenesulfonic acid

v. 3,4,5-trihydroxy-1-cyclohexanecarboxylic acid (configuration unspecified) (shikimic acid)

w. 4-hydroxy-3-methoxybenzenecarbaldehyde (vanillin)

x. 2,4,6-trinitrobenzenol (2,4,6-trinitrophenol) (picric acid)

y. trans-2-methyl-2-butenoic acid (tiglic acid)

   cis-2-methyl-2-butenoic acid (angelic acid)

z. 3,7-dimethyl-2,6-octadienyl trans-2-methyl-2-butenoate (oil of geranium)

Chapter 8

8A-1 $CH_3CH_2CH_2C(CH_3)CH_2CH_3$ with Cl as one possibility; chloride must be tertiary and with the chiral center at the chlorine-bearing carbon

8A-2 $C_6H_5-\overset{\overset{CH_3}{|}}{\underset{\underset{CH_3}{|}}{C}}-Cl$ **8A-3** $CH_3CH_2CH_2CH_2Cl$

8A-4 $CH_3\overset{}{\underset{\underset{OH}{|}}{C}}HCH_2OH + (C_6H_5)_3Cl \xrightarrow{S_N1} CH_3CH(OH)CH_2OC(C_6H_5)_3 + HCl$

8A-5

a. $CH_3CH_2\overset{\overset{CH_3}{|}}{\underset{\underset{Cl}{|}}{C}}CH_2CH_3 \xrightarrow[S_N1-E1]{50\% \ H_2O, \ CH_3COCH_3}$

$CH_3CH_2\overset{\overset{CH_3}{|}}{\underset{\underset{OH}{|}}{C}}CH_2CH_3 + CH_3CH_2\overset{\overset{CH_3}{|}}{C}=CHCH_3 + (CH_3CH_2)_2C=CH_2$

b. $CH_3CH_2\overset{\overset{CH_3}{|}}{C}HCH_2CH_2Cl \xrightarrow[S_N2]{KI, \ CH_3COCH_3} CH_3CH_2\overset{\overset{CH_3}{|}}{C}HCH_2CH_2I$

c. $CH_3CH_2\overset{\overset{CH_3}{|}}{\underset{\underset{Cl}{|}}{C}}CH_2CH_3 \xrightarrow[E2]{NaOC_2H_5} CH_3CH_2\overset{\overset{}{}}{\underset{\underset{CH_3}{|}}{C}}=CHCH_3 + (CH_3CH_2)_2C=CH_2$

d. $(CH_3)_2CHCH_2CH_2CH_2Cl \xrightarrow[S_N2]{KI \ in \ CH_3COCH_3} (CH_3)_2CHCH_2CH_2CH_2I$

e. $CH_3CH=CHCH_2Br \xrightarrow[S_N2]{NaCN} CH_3CH=CHCH_2CN$

f. $BrCH_2C(CH_3)=CH_2$ $\xrightarrow[\text{-AgBr}]{AgNO_3}$ $CH_2\overset{\oplus}{-}\underset{\underset{CH_3}{|}}{C}=CH_2$ $\xrightarrow{C_2H_5OH}$ $C_2H_5OCH_2\underset{\underset{CH_3}{|}}{C}=CH_2$

g. $C_6H_5CH_2Cl$ $\xrightarrow[\text{-AgCl}]{AgNO_3}$ $C_6H_5\overset{\oplus}{CH_2}$ $\xrightarrow{C_2H_5OH}$ $C_6H_5CH_2OC_2H_5$

h. $C_6H_5CH_2I + CH_3\overset{\ominus}{S}\overset{\oplus}{Na}$ $\xrightarrow[S_N2]{\text{-NaI}}$ $C_6H_5CH_2SCH_3$

i. $C_6H_5CH_2OS(O_2)CH_3$ $\xrightarrow[(CH_3)_2S=O]{NaCN}$ $C_6H_5CH_2CN$

j. $C_6H_5OCH_3 + HI \longrightarrow C_6H_5\underset{\underset{H}{|}}{\overset{\oplus}{O}}CH_3 + I^{\ominus} \longrightarrow C_6H_5OH + CH_3I$

k. $C_6H_5\underset{\underset{Cl}{|}}{C}(CH_3)_2 + KOH$ $\xrightarrow[E2]{\text{-KCl, -}H_2O}$ $C_6H_5\underset{\underset{CH_3}{|}}{C}=CH_2$

l. $(CH_3)_3CCl + HONa$ $\xrightarrow{E2}$ $(CH_3)_2C=CH_2 + NaCl + H_2O$

m. $C_6H_5\underset{\underset{Cl}{|}}{C}HCH_3 + (CH_3)_3CO^{\ominus}K^{\oplus}$ $\xrightarrow{E2}$ $C_6H_5CH=CH_2 + (CH_3)_3COH + KCl$

n. $(CH_3)_3CCl + (CH_3)_2N^{\ominus}Li^{\oplus}$ $\xrightarrow{E2}$ $(CH_3)_2C=CH_2 + (CH_3)_2NH + LiCl$

o. $CH_3(CH_2)_4OH + LiBr + H_2SO_4$ $\xrightarrow{S_N2}$ $CH_3(CH_2)_4Br + H_2O + LiHSO_4$

p. $C_6H_5COCH_2Br + CH_3S^{\ominus}Na^{\oplus} \longrightarrow C_6H_5COCH_2SCH_3 + NaBr$

8A-6   a. 2-propanone   b. diethyl ether   c. dimethyl sulfide

8A-7   A possible but not exclusive structure for A is 3-penten-2-ol.

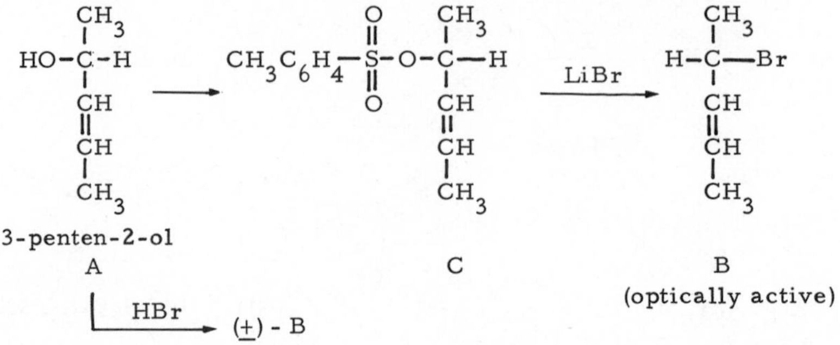

$\underset{\equiv}{L}$-3-penten-2-ol
A

C

B
(optically active)

HBr → (±) - B

Chapter 9

9A-1   a. $(CH_3)_2CHCH_2CH_3$

δ_1 δ_2 δ_3 δ_4

b.

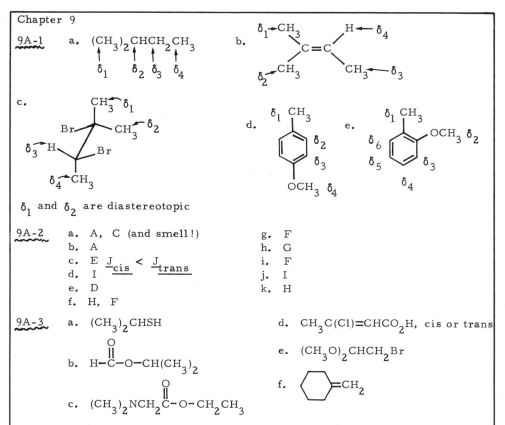

c.

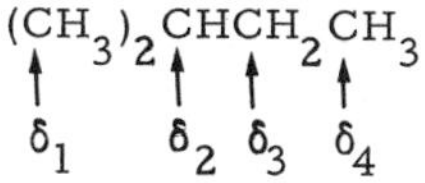

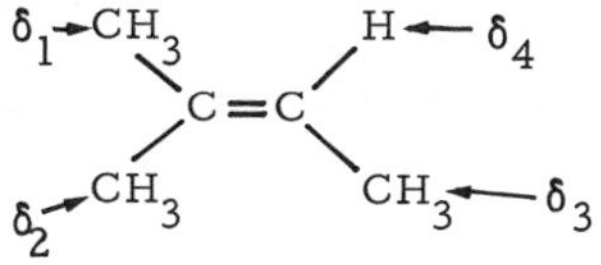

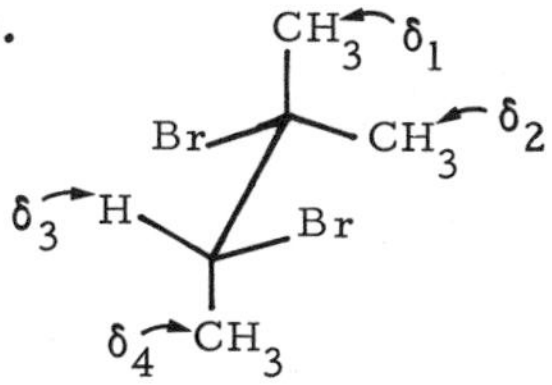

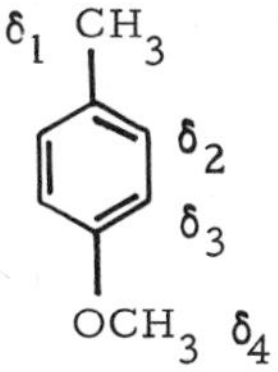

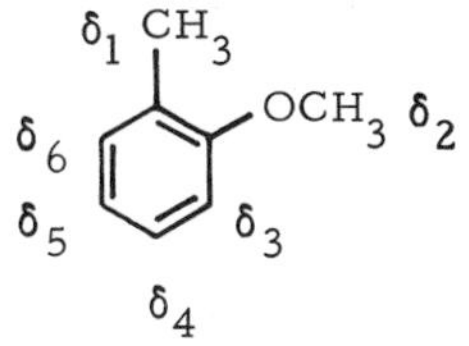

δ_1 and δ_2 are diastereotopic

9A-2   a.   A, C (and smell!)   g.   F
          b.   A   h.   G
          c.   E $J_{cis} < J_{trans}$ i.   F
          d.   I   j.   I
          e.   D   k.   H
          f.   H, F

9A-3   a. $(CH_3)_2CHSH$ d. $CH_3C(Cl)=CHCO_2H$, cis or trans

          b. $H-\overset{O}{\overset{\|}{C}}-O-CH(CH_3)_2$ e. $(CH_3O)_2CHCH_2Br$

          f.

          c. $(CH_3)_2NCH_2\overset{O}{\overset{\|}{C}}-O-CH_2CH_3$

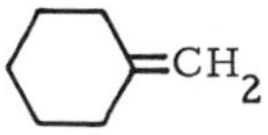

9A-4   Compare your answers with the spectra of these compounds given in the
reference spectra catalogues available to you in the library. The most useful
catalogues are the Sadtler nmr spectra, Aldrich nmr spectra, and Varian
Associates nmr spectra.

Chapter 11

11A-1   a.   1. BH_3,   2. H_2O_2, $HO^{\ominus}$
          b. H_2O, $H^{\oplus}$
          c.   HBr, hν or peroxides
          d.   1. Br_2,   2. KOH
          e.   HBr, in dark, no peroxides
          f. CH_3SH, peroxides
          g.   Product from a $\xrightarrow{\text{1. NaH}}$ ⬡$-CH_2O^{\ominus}Na^{\oplus}$ $\xrightarrow{\text{2. } CH_3I}$ ⬡$-CH_2OCH_3$

          h.   Product from c → ⬡$-CH_2Br$ $\xrightarrow{\text{NaCN}}$ ⬡$-CH_2CN$

          i.   Product from c → ⬡$-CH_2Br$ $\xrightarrow{\text{NaC}\equiv\text{CH}}$ ⬡$-CH_2C\equiv CH$

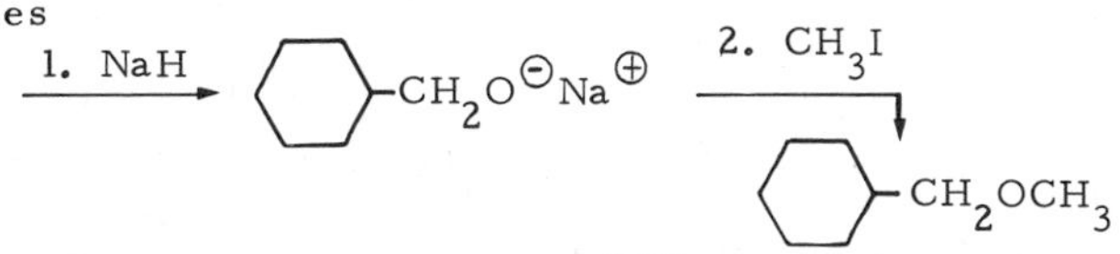

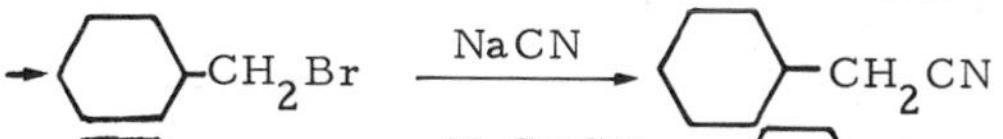

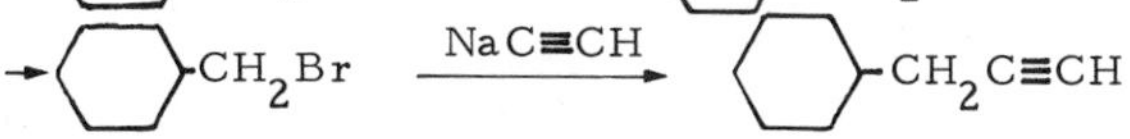

j. 1. O_3, 2. Zn, or H_2, Pt

k. methylenecyclohexane $\xrightleftharpoons{H^{\oplus}}$ 1-methylcyclohexene

l. Product from k $\xrightarrow{\text{1. BH}_3 \quad \text{2. H}_2O_2, \text{ HO}^{\ominus}}$

<u>11A-2</u> a. $CH_3CH=CHCH_3$ e. $H_2C=CH_2$

b. $CH_3CH=CHCH_3$ f. $CH_3C\equiv N$ reacts with HCl to add a proton to the nitrogen

c. $(CH_3)_2C=CH_2$

d. $(CH_3)_2C=CH_2$ g. $CH_2=CHCH_2NO_2$

h. $CH_2=C(CH_3)O\underset{\underset{O}{\|}}{C}CH_3$

<u>11A-3</u>

$CH_2=CH-\langle\text{ring}\rangle-C\equiv CH \xrightarrow{H_2, \text{ Pd-Pb}} CH_2=CH-\langle\text{ring}\rangle-CH=CH_2 \xrightarrow[\text{2. H}_2, \text{ Pt}]{\text{1. O}_3}$

$O=CH-\langle\text{ring}\rangle-CH=O + 2CH_2=O \longleftarrow$

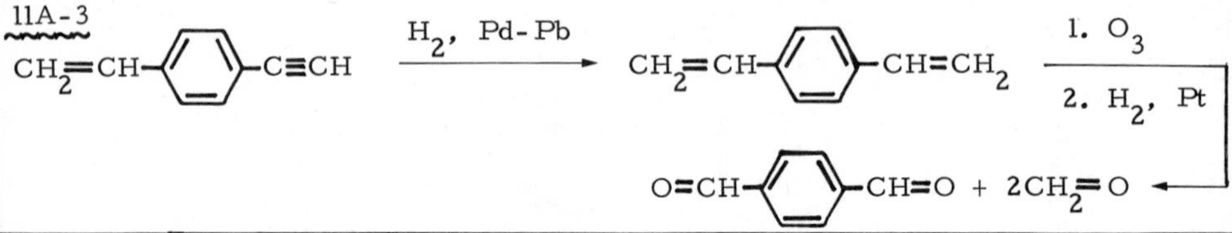

Chapter 12

<u>12A-1</u>

a.

b.

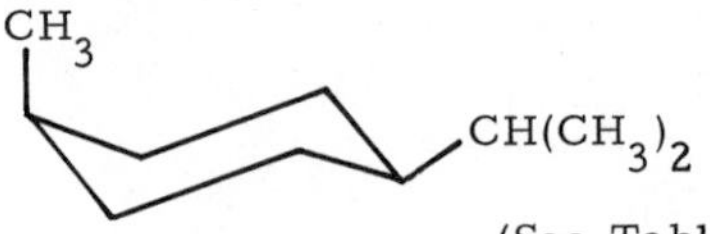

(See Table 12-2)

c.

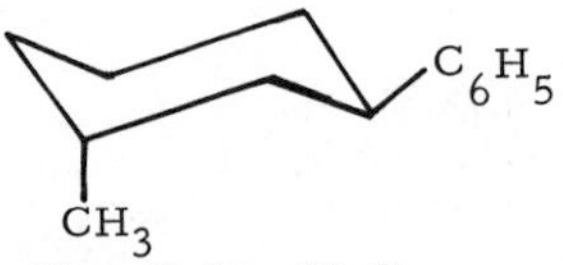

(See Table 12-2)

d.

trans 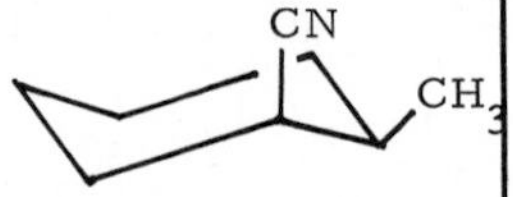   cis

e.

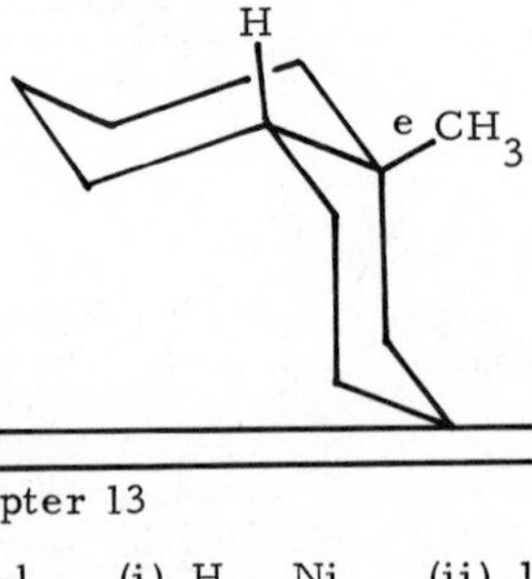

f.

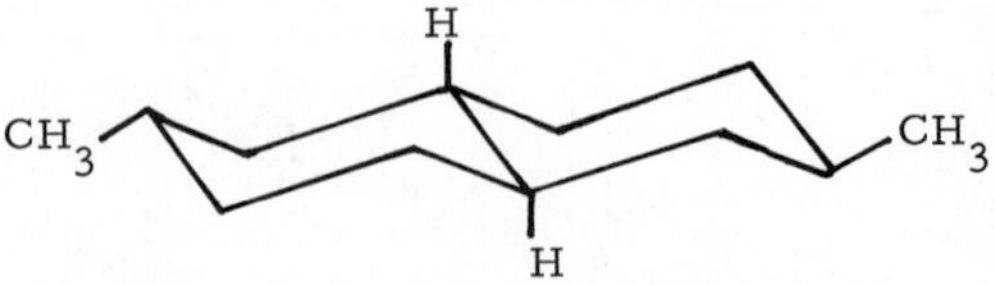

Chapter 13

<u>13A-1</u> (i) H_2, Ni   (ii) 1. Cl_2, hν 2. KOH   (iii) N-bromobutanimide (N-bromosuccinimide), peroxides 2. KOH

<u>13A-2</u> a. cyclohexene $\xrightarrow[\text{Cl}_2, \text{ h}\nu]{\text{1. O}_3 \quad \text{2. H}_2, \text{ Pt}}$

b. cyclohexane $\longrightarrow$ chlorocyclohexane $\xrightarrow{\text{NaCN}}$

c. cyclohexene $\xrightarrow{\text{H}_2\text{O, H}^{\oplus}}$ cyclohexanol $\xrightarrow{\text{1. NaH} \quad \text{2.}} \xrightarrow{\text{CH}_3\text{I}}$

d. cyclohexene $\xrightarrow{\text{CH}_3\text{SH, peroxides}}$

e. cyclohexadiene $\xrightarrow{\text{1. CH}_2\text{=CHCN} \quad \text{2. H}_2, \text{ Pt}}$

13A-3

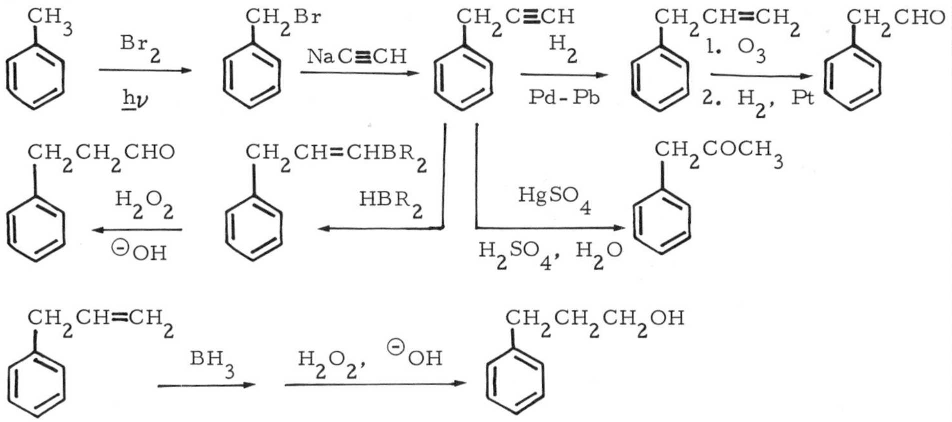

13A-4

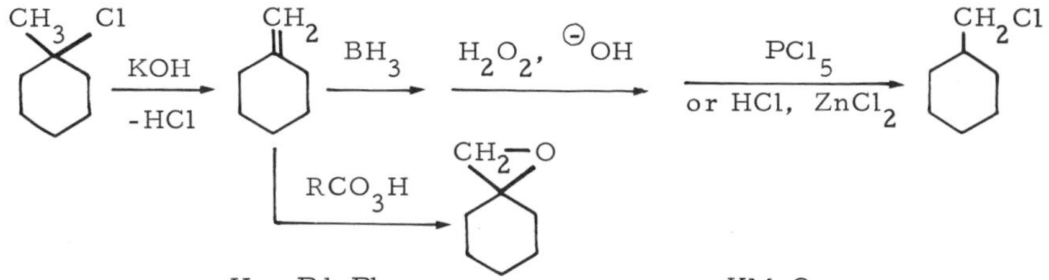

c, d. $\text{CH}_3\text{C}\equiv\text{CCH}_3$ $\xrightarrow[\text{suprafacial}]{\text{H}_2, \text{ Pd-Pb}}$ $\text{CH}_3\text{CH=CHCH}_3$ $\xrightarrow[\substack{\text{pH} > 7 \\ \text{suprafacial}}]{\text{KMnO}_4}$

cis

HOCl

antarafacial

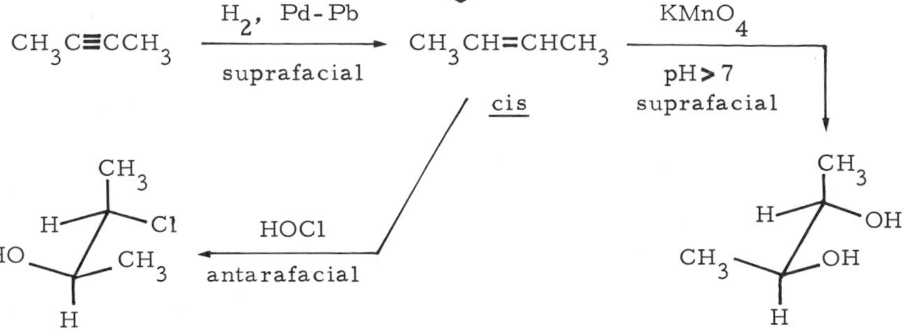

(and enantiomer)

USE OF THE CHEMICAL LITERATURE IN ORGANIC CHEMISTRY
(an Overview)

During your studies of chemistry you may have become curious as to the primary source of the many chemical principles, experimental observations and theories you have read about. Information in your textbooks is derived for the most part from the original chemical literature, which dates from about 1828 to the present. But retrieval of specific chemical information from this huge pool of literature can be a difficult and time-consuming task. It is the purpose of this concluding chapter to present an overview of the chemical literature and to indicate the ways in which its contents may be searched for specific facts.

You may be fortunate enough to have a course on the use of the chemical literature offered at your College or University; but most of us have to learn to use the literature on a self-taught, trial-and-error basis. While it is hardly a pleasurable exercise, it is one that few can deny as important because, without access to the literature, we lose the resource of chemical knowledge and experience gained over more than a century. In searching for information, it surely helps to minimize errors and time-wasted if you know beforehand how the literature is organized and how to set about the search. Several books and articles have been written on the subject, and a few selected examples follow; the three-part article by J. E. H. Hancock is particularly recommended.

1. J. E. H. Hancock, "An Introduction to the Literature of Organic Chemistry," Journal of Chemical Education, 45, 193, 260, 336 (1968). This is an exceptionally readable account of the literature of organic chemistry. Parts I and II categorize the literature into eighteen classes, with a description of each; Part III gives a brief account of the use of Science Citation Index, Beilstein, and Chemical Abstracts.

2. R. F. Gould, Ed., "Searching the Chemical Literature," Advances in Chemistry Series, No. 30, American Chemical Society, Washington, D. C., 1961. This interesting volume is in need of revision.

3. E. J. Crane, A. M. Patterson, and E. B. Marr, "A Guide to the Literature of Chemistry," 2nd ed., John Wiley and Sons, Inc., New York, 1957.

Classification and Use of the Chemical Literature

The literature can be divided into three main groups — the primary literature, the secondary literature, and the abstracting literature (see Chart I).

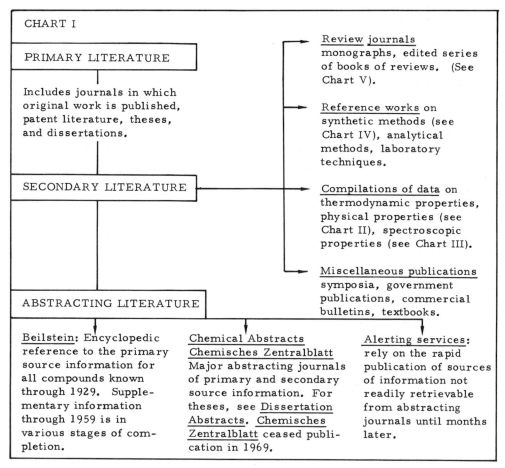

CHART I

PRIMARY LITERATURE

Includes journals in which original work is published, patent literature, theses, and dissertations.

SECONDARY LITERATURE

ABSTRACTING LITERATURE

Review journals monographs, edited series of books of reviews. (See Chart V).

Reference works on synthetic methods (see Chart IV), analytical methods, laboratory techniques.

Compilations of data on thermodynamic properties, physical properties (see Chart II), spectroscopic properties (see Chart III).

Miscellaneous publications symposia, government publications, commercial bulletins, textbooks.

Beilstein: Encyclopedic reference to the primary source information for all compounds known through 1929. Supplementary information through 1959 is in various stages of completion.

Chemical Abstracts Chemisches Zentralblatt Major abstracting journals of primary and secondary source information. For theses, see Dissertation Abstracts. Chemisches Zentralblatt ceased publication in 1969.

Alerting services: rely on the rapid publication of sources of information not readily retrievable from abstracting journals until months later.

The primary literature of chemistry is the documentation of original work and is found mainly in research journals, often with experimental details. There has been a great proliferation of research journals in recent years — to the point where it is impractical to provide a complete listing in this chapter. Rather we refer you to the listings of journals available to you in your campus library. (See also the listing of journals in J. E. H. Hancock, J. Chem. Educ., 45, 193, 1968).

Secondary sources of information include all manner of review publications, reference texts, data compilations, encyclopedias, handbooks, textbooks - in fact any publication that derives its information from the primary literature.

Systematic retrieval of information from both primary and secondary sources is achieved by way of the abstracting literature. Abstracting journals are catalogues of chemical information with references to the source of the original information. The most important abstracting journals are Chemical Abstracts, Chemisches Zentralblatt, and the encyclopedia Beilstein's Handbuch der Organischen Chemie. Their use is described in more detail later.

How best to use the literature depends on the type of information sought. Perhaps you need to know the documented melting point or boiling point of a particular compound. You could go to the primary literature, but it is usually simpler and quicker to go to secondary sources for such information. The most useful of these secondary sources are summarized in Chart II.

CHART II   NEED PHYSICAL CONSTANTS?

Fast

1. Handbook of Chemistry and Physics, R. C. Weast (Editor), 56th ed., C.R.C. Press, 1975-1976.

2. N. A. Lange, Handbook of Chemistry, latest ed., McGraw-Hill, New York.

3. The Merck Index, 9th ed., Merck and Co., New Jersey, 1976. This useful encyclopedia of chemicals and drugs describes nearly 10,000 compounds by name, composition, source, synthesis, physical properties, pharmacology, toxicology and therapeutic use. Literature references are given.

Little slower

4. I. Heilbron, A. H. Cook, H. M. Bunbury, D. H. Hey (Eds.), Dictionary of Organic Compounds, 4th ed., Vol. 1-5, Oxford University Press, 1965, with yearly supplements. Data include physical constants, solubility, and properties of important derivatives. Literature references are given. A Formula Index is published as a separate volume.

5. Beilstein's Handbuch der Organischen Chemie (See pp. 620-622 for a description of Beilstein's Handbuch).

Slow

6. Chemical Abstracts (See pp. 616-620).

Other

7. J. Timmermans, Physico-Chemical Constants of Pure Organic Compounds, Elsevier, New York, 1950.

8. W. Utermark and W. Schicke, Melting Point Tables of Organic Compounds, 2nd ed., Interscience Publishers, New York, 1963.

9. Selected Values of Properties of Chemical Compounds, Thermodynamics Research Center Data Project, Texas A and M University, Texas, 1975. This is a major compilation of physical and thermochemical properties. The introduction should be consulted for the layout of the tables.

CHART II contd.

10. Other useful multi-volume reference works include:

   (i) Rodd's Chemistry of Carbon Compounds; a Modern Comprehensive Treatise, 2nd ed., E. S. Coffey, Elsevier Publishing Co., New York, 1964 - (continuing series).

   (ii) Elsevier's Encyclopedia of Organic Chemistry, Ed. F. Radt, Series 1 and 2, Vols. 1-11 (incomplete), Series 3, Vols. 12-14, Springer Verlag, New York, 1969 - .

   (iii) Weissberger Series, Technique of Organic Chemistry, 3rd ed., Volumes I-XIV, Ed. A. Weissberger, Interscience Publishers, 1951-69; Techniques of Chemistry, Volumes I-XI (continuing series), Wiley-Interscience, New York, 1971 - .

   (iv) Organic Chemistry, Series Two (International Review of Science). Consultant Ed. D. H. Hey, Vols. 1-10, Butterworths (Publishers) Inc., Boston, 1976 - .

   (v) Comprehensive Organic Chemistry, Vols. 1- , Consultant Ed. D. H. Barton, Pergamon Press, Elmsford, New York, 1978 - .

Maybe you need to verify the spectral properties reported for a known compound. There are valuable reference compilations of spectral data available, and these are listed in Chart III.

CHART III   NEED SPECTROSCOPIC DATA?

Infrared

1. C. J. Pouchert, Ed., The Aldrich Library of Infrared Spectra, 2nd ed., Aldrich Chemical Company, Milwaukee, Wisconsin, 1975.

2. Standard Infrared Grating Spectra, Sadtler Research Laboratories, Philadelphia, 1966- to date by supplements. Consult the Total Spectra Chemical Classes Index or the Total Spectra Molecular Formula Index.

3. Standard Infrared Prism Spectra, Sadtler Research Laboratories, Philadelphia.

4. Coblentz Society, Coblentz Society Spectra, Sadtler Research Laboratories, Philadelphia, 1973 - .

5. H. A. Szymanski, Interpreted Infrared Spectra, Plenum Press, New York, 1964 - , Vols. 1-3.

6. American Petroleum Institute, Project 44, Infrared Spectral Data, Vols. I - IX, TRC, Texas A and M.

Nmr

1. Nuclear Magnetic Resonance Spectra, Sadtler Research Laboratories, Philadelphia, 1966 - to date with supplements. Consult the Total Spectra Indexes.

CHART III contd.

2. N. S. Bhacca, L. F. Johnson and J. N. Shoolery, High Resolution NMR Spectra Catalog, Vol. I and II, Varian Associates, Palo Alto, California, 1962.

3. E. Breitmaier, G. Haas, W. Voelter, Atlas of Carbon-13 NMR Data, Heyden and Son, Ltd., London, 1975.

4. F. A. Bovey, NMR Data Tables for Organic Compounds, Vol. I, Interscience Publishers, New York, 1967.

5. American Petroleum Institute, Project 44, NMR Spectral Data, Vols. I-TRC, Texas A and M. Includes 40, 60, and 100 proton nmr data and carbon-13 nmr data.

Ultraviolet-Visible Absorption Spectra

1. Standard Ultraviolet Spectra, Sadtler Research Laboratories, Philadelphia, 1966 - to date with supplements.

2. UV Atlas of Organic Compounds, Plenum Press, New York, 1966 - .

3. American Petroleum Institute, Project 44, Ultraviolet Spectral Data, Vols. I - III, TRC, Texas A and M.

General

1. J. W. Robinson, Ed., Handbook of Spectroscopy, Chemical Rubber Company Press, Vols. I - II, 1974 - .

2. J. G. Grasselli (Ed.), CRC Atlas of Spectral Data and Physical Constants for Organic Compounds, CRC Press, Cleveland, 1973.

Mass Spectra

1. A. Tatematsu, T. Tsuchiya, Structure-Indexed Literature of Organic Mass Spectra, Academic Press of Japan, 1968 - to date. This index lists the literature sources of mass spectra that were published in 1966, and in each succeeding year. Each volume covers one calendar year.

2. E. Stenhagen, S. Abrahamsson, F. W. McLafferty, Eds., Atlas of Mass Spectral Data, Interscience Publishers, New York, 1969 - .

3. A. Cornu and R. Massot, Eds., Compilation of Mass Spectral Data, 2nd ed., Heyden and Son, Ltd., London, 1966, 1975.

4. American Petroleum Institute, Project 44, Selected Mass Spectral Data, Vol. I - VII, TRC, Texas A and M.

A common problem is where to find a suitable method or technique for the synthesis and purification of a specific compound or type of compound. Usually one requires sufficient experimental detail to enable the procedure to be executed successfully in the laboratory. Useful secondary sources of synthetic information are summarized in Chart IV.

CHART IV   NEED SYNTHETIC METHODS AND PROCEDURES?

1. Organic Reactions, John Wiley and Sons, New York, 1943 - to date. New volumes appear approximately every two years.
2. Organic Syntheses, John Wiley and Sons, New York, 1932 - to date. This series comprises more than 50 volumes. Each volume presents several detailed and tested procedures for the synthesis of particular compounds not commercially available at the time of publication. Several collective volumes have now been published.
3. W. Theilheimer, Ed., Synthetic Methods of Organic Chemistry, Vols. I-IV in German, Vols. V - to date in English, S. Karger, Basle, John Wiley and Sons, New York, 1946 - .
4. W. Foerst, Ed., Newer Methods of Preparative Organic Chemistry, Vols. I-III, Academic Press, New York, 1948-1964.
5. Houben-Weyl, Die Methoden der Organischen Chemie, 4th ed., G. Thiene Stuttgart, 1952 - to date. Useful compendium of methodology in organic synthesis - in German.
6. A. Weissberger, Technique of Organic Chemistry, all volumes, Interscience Publishers, Inc., New York, 1949 - to date.
7. L. F. Fieser and M. Fieser, Reagents for Organic Synthesis, Vols. I - VI, John Wiley and Sons, Inc., New York, 1967- 1976.
8. Annual Reports in Organic Synthesis, Academic Press, New York. Published annually since 1970, each volume is an organized review of synthetically useful information published during the year.
9. H. O. House, Modern Synthetic Reactions, 2nd ed., W. A. Benjamin, Inc., Menlo Park, California, 1972.
10. Beilstein's Handbuch der Organischen Chemie (See pp. 620-621).

A rapid way of screening information on a particular topic is by reference to recent reviews and monographs on the subject. The most important review journals and review publications are listed in Chart V.

CHART V   NEED A REVIEW ARTICLE?

1. Accounts of Chemical Research (Acc. Chem. Res.), published monthly since January 1968. Each issue contains 4-6 review articles, 6-10 pages long, on a subject of current research interest. The articles are supposedly addressed to the nonspecialist.
2. Chemical Reviews (Chem. Rev.); published bimonthly since 1924. Each issue contains 1-3 comprehensive reviews on a specific topic. They are usually designed to give major and sometimes exhaustive coverage of the literature.
3. Quarterly Reviews (Quart. Rev. (London)); published quarterly since 1947. Called Chemical Society Reviews since 1972. Similar coverage to Chemical Reviews.
4. Annual Reports of the Chemical Society (London). A digest of the most significant reports in the literature for the year.

CHART V contd.

5. Angewandte Chemie (Angew. Chem.), International Edition, published monthly.

   In addition to the above-listed review journals, there are numerous journals that publish occasional reviews, monographs and edited volumes that appear at intervals as a continuing series. Examples include: the Advances in Chemistry series published by the American Chemical Society, Advances Series published by Academic Press, and the Progress Series published by Butterworths, London. Chemical Abstracts is the best source for tracing reviews. See also reference works in Chart II - 10.

Valuable though the secondary sources of chemical literature are, a complete search of the literature must inevitably involve reference to the primary sources through use of the abstracting services. We therefore turn to describe the use of Chemical Abstracts.

Chemical Abstracts

In the early publication history of Chemical Abstracts, namely the period 1907-1939, the coverage of the literature provided by this journal was not as complete as either Beilstein's Handbuch or the German-language abstracting service Chemisches Zentralblatt. For this reason, a thorough search of the literature prior to 1939 should utilize the German-language abstracts as well as Chemical Abstracts (See Chart VI). However, from 1939 to the present,

---

CHART VI   YEAR COVERAGE OF FORMULA INDEXES

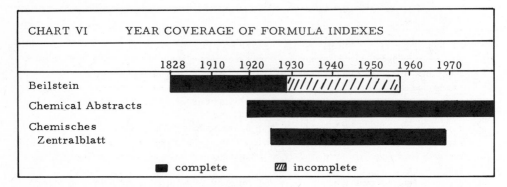

Chemical Abstracts is recognized as the most important abstracting journal in the field of chemistry. Chemisches Zentralblatt ceased publication in 1969.

One great advantage of Chemical Abstracts is the wide range of indexing it provides. At the present time, a new issue of Chemical Abstracts appears weekly; the issues for the year are compiled into two volumes (January - June, and July - December), each of which is indexed under the following headings:

Author Index

General Subject Index } Prior to 1972 these two indexes are combined in one index called the Subject Index

Chemical Substance Index }

Formula Index and Index of Ring Systems*

Numerical Patent Index and Patent Concordance

A second advantage of Chemical Abstracts is the convenience of its Collective indexes. Thus, abstracts are indexed in five ten-year increments from 1907 to 1956, and in five-year increments thereafter.

1907 - 1916	First Decennial Index
1917 - 1926	Second Decennial Index
1927 - 1936	Third Decennial Index
1937 - 1946	Fourth Decennial Index
1947 - 1956	Fifth Decennial Index
1957 - 1961	Sixth Collective Index
1962 - 1966	Seventh Collective Index
1967 - 1971	Eighth Collective Index
1972 - 1976	Ninth Collective Index
1977 -	

In using Chemical Abstracts, or any abstracting journal, you should be prepared to encounter frustration with indexes. Indexing is a science that has evolved and continues to evolve as the volume and complexity of information grow. The indexing system used by Chemical Abstracts has undergone change over the years, particularly with respect to the index headings of chemical substances. Major changes were introduced in 1972 with Volume 76, which necessitated publication of a comprehensive Index Guide to cover the Ninth Collective Index period (1972-76). The organization and nomenclature systems used are explained in detail in this Index Guide. To give you some idea of the complexity of the indexing problem, a Supplementary Index Guide appeared in

---

*See also the Parent Compound Handbook, which is a recent Chemical Abstracts Service publication that contains information about all known ring systems.

Volume 83, and it is a compilation of the numerous entries that have been added, deleted, or corrected since the publication of Volume 76. [*]

Index Guide to Chemical Abstracts for the period 1972-76, vols. 76-85, see Vol. 76.

Supplementary Index Guide, see Vol. 83.

With this introduction, you should be prepared to find numerous changes in indexing between the eighth and ninth collective volumes. For example, the Chemical Substance Index in the Ninth Collective Volume indexes the common substance CH_3COCH_3 under the heading 2-propanone rather than under the name acetone. Likewise, $CH_3CH_2NH_2$ is indexed under ethanamine - not as ethylamine. However, the eighth and earlier collective volumes index these compounds under the headings acetone and ethylamine.[**]

Examples of how representative entries appear in the Chemical Substance Index and the General Subject Index are shown below. The symbolism used is explained in the key.

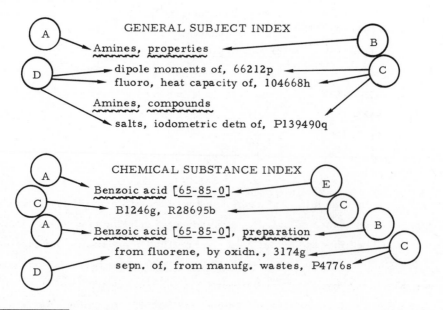

GENERAL SUBJECT INDEX

A

B

D

C

Amines, properties

dipole moments of, 66212p
fluoro, heat capacity of, 104668h

Amines, compounds

salts, iodometric detn of, P139490q

CHEMICAL SUBSTANCE INDEX

A

E

C

C

A

B

D

C

Benzoic acid [65-85-0]

B1246g, R28695b

Benzoic acid [65-85-0], preparation

from fluorene, by oxidn., 3174g
sepn. of, from manufg. wastes, P4776s

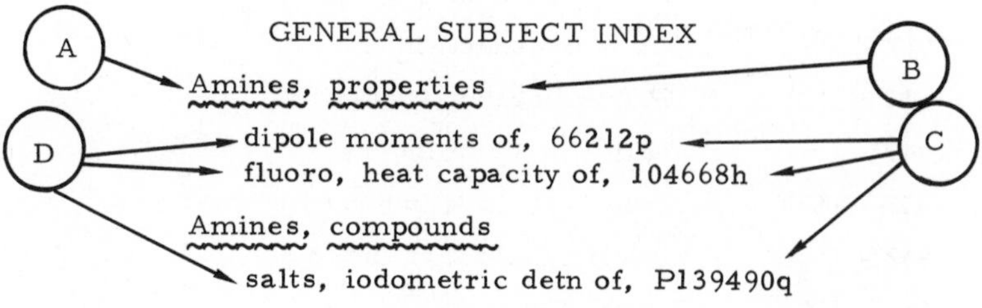

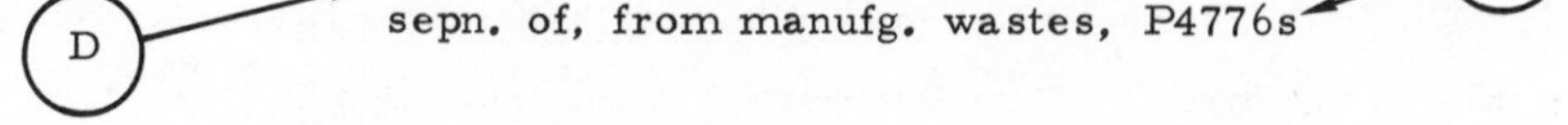

[*] Chemical Abstract Services have published Index Guides with regularity since 1968. The comprehensive CA Index Guide that appears with the first volume of a collective is supplemented annually during the collective period and then completely reissued at the end of the collective period.

[**] If you need help in finding the CA Index Name of a ring compound, look at the Parent Compound Handbook, a CAS publication.

KEY

A.   The index heading appears in boldface.

B.   The heading subdivision appears in boldface, following a comma.

C.   The abstract number locates the abstract within the semi-annual volume. Abstracts are numbered sequentially from the beginning of the volume to the end.   The capital letter preceding the number denotes that the source document is:

> B . . . . . a book
>
> P . . . . . a patent
>
> R . . . . . a review

The lower-case letter following each number is a computer-generated check-letter and need not concern you.   However, do not confuse this letter with similar letters used in abstract numbers prior to Volume 66; such letters in earlier volumes designate the approximate location of the index heading in the column of abstracts on the page.

In Collective Indexes, the abstract number is preceded by the Volume number in boldface.

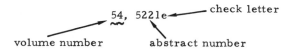

54, 5221e ←——— check letter

volume number   abstract number

Having located the numbered abstract in the correct volume, the next step is to read the abstract to determine whether the source document is likely to contain information of interest.   If you decide to refer to the source document, the reference needed to the original literature is given at the beginning of each abstract.   An example follows:

volume no. ——86: 139707e   Pericyclic synthesis and study of select π
abstract no. —— frames.   Anastassiou, Apostolos G. (dep. Chem., Syracuse
              Univ., Syracuse, N. Y.).   Acc. Chem. Res. 1976, 9(12),
              453-8 (Eng).   A review of pericyclic synthesis done by the
              author, with 56 refs.

D.   The Index modification appears in lightface, and adds further description to the index heading.

E.   The number appearing in italics within square brackets is called the Chemical Abstracts Service (CAS) Registry Number, and refers to the specific compound named in the index heading.   A system has been developed whereby numbers can be used to code for the two-dimensional structure of a chemical substance, plus means to record additional descriptive data, such as stereochemistry.   Thus, the number is a unique identifying number assigned to a specific chemical substance on the basis of a computer-language description of its molecular structure.   More than 4 million unique structures have been coded in this way.   While the utility of CAS registry numbers have not yet been exploited, it is clear that they will eventually be used to facilitate data storage and literature searching.

A rapid way to search for information regarding a specific compound is by way of the Formula Indexes. An illustrative key to Formula Index Entries follows:

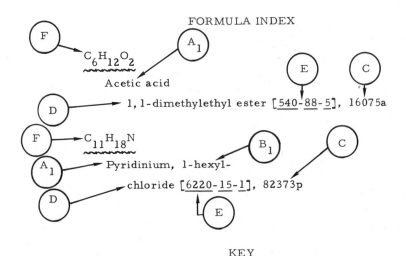

FORMULA INDEX

KEY

F. The molecular formula is the heading in boldface. Elements are listed from left to right in the order C, H, and alphabetically thereafter. Within a given sequence, entries are arranged in order of increasing complexity, CH ahead of CH_2, CH_2 ahead of C_2H, and so on.

A_1, B_1 and D have almost the same meaning as A, B and D in the chemical substance index. C and E have the same meaning as in the chemical substance index. However, notice that in the case of simple derivatives (esters) the molecular formula heading lists __all__ atoms in the derivative ($C_6H_{12}O_2$ for 1,1-dimethylethyl acetate) whereas salts (1-hexylpyridinium chloride) do not include the counter ion ($Cl^{\ominus}$).

More precisely:

A_1 The index heading parent.

B_1 The substituent prefix identifies chemical groups attached to the index heading parent.

D   The name modification completes the index name by citation of a function derivative (e.g., an ester) or an associated ion.

Beilstein

Organic Chemistry became a major science as the result of the pioneer work of early chemists, mostly German chemists. As the number of known compounds increased, the need to catalog their preparation and properties became increasingly important. This need led to the publication of the Handbuch der Organischen Chemie, which is better known as Beilstein after its first editor. This German-language work is the most complete secondary source for information about the properties, preparation and reactions of all organic compounds known through 1929.

Beilstein consists of three complete sets of volumes. The main series (Hauptwerk; H) covers the early literature from around 1828 through 1909. The first supplementary series (Erstes Ergänzungswerk; EI) covers the period 1910-1919 in an organization parallel to that of the Hauptwerk. A second supplementary series (Zweites Ergänzungswerk; EII) covers the period 1920-1929. There is a third supplementary series (Drittes Ergänzungswerk; EIII) that is only partially complete but which covers, or will cover, the literature from 1930-1949, and a fourth (yet incomplete) supplement through 1950-59.

The simplest way to find a compound in Beilstein is by reference to the cumulative formula indexes. Several compounds will usually be named under a particular molecular formula, and it may not be a trivial matter to decide which name refers to the compound of interest. The difficulty stems partly from language differences and partly from different nomenclature systems. Hopefully, you will be able to recognize the indexed name that corresponds to your compound even though you may not have thought of the name independently.

Suppose that you are searching Beilstein for information on 1,3-cyclohexadiene, which has the formula C_6H_8. The cumulative formula indexes give the entry as:

C_6H_8
> Cyclohexadien-(1,3) (1,2-dihydrobenzol)
>> 5, I 60, II 79, III 310

This means that reference is made to the compound in volume 5 in the first supplement on page 60, in the second supplement on page 79, and in the third supplement on page 310. The volume number (5) is the same for each supplement. A second example follows:

$C_6H_{10}O$
> Cyclohexanon
>> 7, 8, I 6, II 5, III 14

Thus cyclohexanone is listed in volume 7, on page 8 in the main series, page 6 of volume 7 in the first supplement, page 5 of volume 7 in the second supplement, and page 14 of volume 7 in the third supplement.

As a point of interest, the tables of physical constants of organic compounds listed in the CRC Handbook of Chemistry and Physics give the Beilstein reference in the far right column.

An alternative method of searching for information in Beilstein requires that you understand the organization of entries according to chemical structure. The system is logical but is complex to explain. The following references describe the use of Beilstein in more detail:

1. E. H. Huntress, A Brief Introduction to the Use of Beilstein's Handbuch der Organischen Chemie, John Wiley and Sons, Inc., New York, 1938.
2. O. A. Runquist, "Programmed Guide to Beilstein's Handbuch," Burgess, Minneapolis, 1966.
3. J. B. Hendrickson, D. J. Cram, G. S. Hammond, Organic Chemistry, McGraw-Hill, New York, 1970, Chapter 28, 3rd ed.
4. O. Weissbach, "The Beilstein Guide. A Manual for the Use of Beilstein's Handbuch der Organischen Chemie," Springer-Verlag, Berlin-Heidelberg, New York, 1976.

Searching the Current Literature

If you take a close look at the recent issues of Chemical Abstracts, you will find that they refer to literature originally published at least four months ago. If you then consider that the indexes to the current volume may not be completed for 7-12 months after completion of the volume, you will realize that there is at least a significant time lag between the publication of current chemical literature and its indexed citation in Chemical Abstracts. This time lag presents a serious problem to the research chemist who must keep up with the latest developments in the field. (Considering the enormous volume of information that accrues, it is amazing that the time lag is not years instead of months.)

There are ways to search the most current literature. One way is by way of the Keyword Index in each issue of Chemical Abstracts since 1963. The idea is to provide ready reference to the entries in the current issue through the use of key words. Prior to the inclusion of the key word index, issues of the current abstracts had no index whatever. The choice of key words is a bit arbitrary, and the index is unrelated to the permanent indexes of the chemical abstracts volume.

Another way to search the current literature is through specialized publications that are described as "alerting services" or "forward-looking services". These include:

Chemical Titles
Current Contents
Current Abstracts of Chemistry and Index Chemicus
Science Citation Index

To learn the objectives and use of these publications, you would be wise to consult with a professional librarian. Your librarian will also advise you of the services available for literature-searching by means of a computer. As the indexing of information becomes more amenable to machine storage, we can expect to see increasing emphasis on literature searches by computer methods.